Human Biology

Personal, Environmental, and Social Concerns

Human Biology

Personal, Environmental, and

Social Concerns

Judith Goodenough
University of Massachusetts

Robert A. Wallace

Betty McGuire
Smith College

HARCOURT COLLEGE PUBLISHERS

Fort Worth   Philadelphia   San Diego   New York   Orlando

San Antonio   Toronto   Montreal   London   Sydney   Tokyo

Vice President/Publisher: Emily Barrosse
Acquisitions Editor: Edith Beard Brady
Product Manager: Erik Fahlgren
Developmental Editor: Lee Marcott
Project Editor: Vanessa Ray
Production Manager/Director of EDP: Joanne Cassetti
Art Director, Text Designer, and Cover Designer: Ruth Hoover
Illustration Supervisor: Sue Kinney
Photo Researcher: Jane Sanders

Cover Image Credit: *The Dance II*, by Henri Matisse, © 1997 Succession H. Matisse, Paris/Artists Rights Society (ARS), New York; additional rights from the Hermitage Museum, St. Petersburg, Russia/Giraudon, Paris/SuperStock.

Essay Art Credits: (Social Concerns Essay) *Frenzied People*, by Diana Ong, © D. Ong/ SuperStock; (Personal Concerns Essay) *Femme Cousant*, by Henri LaBasques, Christie's Images/SuperStock; (Environmental Concerns Essay) *Summer*, by Eric Isenberger, Private Collection/SuperStock.

Printed in the United States of America

HUMAN BIOLOGY: PERSONAL, ENVIRONMENTAL, AND SOCIAL CONCERNS

0-03-001281-3

Library of Congress Catalog Card Number: 97-72125

0123456   032   10 9 8 7 6 5 4 3 2

Harcourt College Publishers

About the Authors

Judith Goodenough with her daughters, Heather (left) and Aimee (right), and her husband, Steve.

Judith Goodenough

Judith received her B.S. from Wagner College (Staten Island, NY) and her doctorate from New York University. She has more than 20 years of teaching experience at the University of Massachusetts, Amherst, specializing in introductory level courses. The insights into student concerns and problems gained from 17 years of teaching Human Biology and more than 10 years of team-teaching The Biology of Social Issues have helped shape this book. In 1986, Judith was honored with a "Distinguished Teaching Award." In addition to teaching, she coordinates the introductory biology laboratories at UMass. Judith has written articles in peer-reviewed journals, contributed chapters to several introductory biology texts, and written numerous laboratory manuals. With the author team of McGuire and Wallace, she wrote *Perspectives on Animal Behavior*.

Robert A. Wallace

The late Robert Wallace received a B.A. in Fine Art and Biology from Harding University, an M.A. in Muscle Histochemistry from Vanderbilt University, and a Ph.D. in Behavioral Ecology from the University of Texas at Austin. He subsequently taught at a number of colleges and universities in the United States and Europe, including the Richard Bland College of William and Mary, the University of Maryland—Overseas Division, Duke University, and the University of Florida. He is the author of seven previously published biology textbooks, including the well-known *Biology: The World of Life*, and two mass-market science books, *The Genesis Factor* and *How They Do It*, as well as numerous scientific articles on a variety of subjects. Robert was also a Fellow of the Explorers Club of New York and the Royal Geographical Society of London. He received the Orellana Medal from the government of Ecuador in recognition of his work with the medicinal plants of vanishing tribes in that country.

Robert Wallace with his wife, Jayne.

Betty McGuire

Betty received her B.S. in Biology from Pennsylvania State University, where she also played varsity basketball. She went on to receive an M.S. and Ph.D. in Zoology from the University of Massachusetts at Amherst, and then spent two happy years as a postdoctoral researcher at the University of Illinois, Champaign—Urbana. Her field and laboratory research emphasize the social behavior and reproduction of small mammals. She has published numerous research papers and co-authored the text *Perspectives on Animal Behavior*. Betty has taught Introductory Biology and upper division courses in Evolution, Anatomy, and Animal Behavior at the University of Massachusetts and Smith College. She lives in western Massachusetts in an old farmhouse with her husband, two children, and many pets.

Betty McGuire with her children Owen (left) and Kate (right), and her husband, Willy Bemis.

Humans are, by nature, curious, and this book is intended to stimulate the curiosity of students toward gaining an appreciation for the intricacy of human life and our place in the ecosphere. We have set out to provide information that will help students understand their everyday experiences with their bodies and with the world around them.

Our first goal is to present the basic principles of human anatomy, physiology, development, genetics, evolution, and ecology. Then, after explaining the basic concepts, we apply them in ways that will both interest and benefit the student. We discuss how the healthy system functions, how that system can malfunction, preventative measures, and what current medicine can offer when systems are compromised or fail.

We have tried to answer some very practical questions, including: What type of exercise benefits the heart? How do you cope with insomnia? How does one protect against unwanted pregnancy and prevent the spread of sexually transmitted diseases? Each of us enters this world with a most intricate machine—our body—but we do not come equipped with an owner's manual. In a sense, this book can be your owner's manual. Understanding the information in it and applying it to our own lifestyle and health choices can help us live longer, happier, and therefore more productive lives.

A second goal is to help students develop reasoning skills, so that information gained from this book can be used in actual situations that they face in life. Instructors will find throughout critical thinking questions that ask students to apply information to new situations. When a topic is controversial, we present both sides of the argument, together with the supporting evidence. In this way, we hope to foster the practice of thinking through issues and making decisions based on the best available information.

A third goal is to help students understand how the choices they make can affect the quality of life for themselves, society, and the planet. The information learned here, or in the classroom, often bears on issues that are important to us all. We want to prepare students to be responsible citizens and voters. Society is currently grappling with many pressing biological issues—the cloning of adult mammalian cells, gene therapy, organ transplants, defining death, and preventing and treating HIV infections, among others.

How the Book Is Organized

This text begins with a discussion of the chemistry of life and then moves to cells, tissues, organs, organ systems, and finally to ecosystems. Because different teachers may present topics in a different order, the chapters do not depend heavily on material covered in earlier chapters. Each part is described as follows:

Part 1: The Organization of the Body

Chapter 1 introduces the scientific method, the characteristics of life, and the process of natural selection. We build a bridge between science and society that will become a recurring theme throughout the text. Chapter 2 gives a brief overview of the chemical basis of life, followed by an introduction to life on the cellular level in Chapter 3. Here we encounter cell structure and function, and we are introduced to energetics of the cell and the cell cycle. Chapter 4 continues the trend toward increasing complexity with discussions of tissues, organs, and organ systems. The recurring theme of homeostasis first appears here.

Part 2: Control and Coordination of the Body

This unit presents the two great regulatory systems of the body—the nervous system and the endocrine system. The structure and function of neurons are covered in Chapter 7 and are applied to help in our understanding of the neurological changes that occur in Alzheimer's disease (Chapter 7A). In Chapter 8 we discuss the central nervous system, which is composed of the spinal cord and brain, and the peripheral nervous system. This knowledge is applied in Chapter 8A, where the effects of certain psychoactive drugs are explored. Chapter 9 focuses on sensation and perception and the special senses that engage us with the world.

Part 3: Maintenance of the Body

These are the organ systems most responsible for the maintenance of homeostasis. Chapters 10 and 11 deal with blood and the cardiovascular system. Chapter 12 introduces the array of body defense mechanisms—from the simplicity of the barrier created by skin to the complexity of the immune response. The digestive system coverage in Chapter 13 is augmented by Chapter 13A, on the thorny and controversial topics of nutrition and weight control. Chapter 14 covers the structure of the respiratory system and is followed in Chapter 14A by a discussion of the major threat to respiratory (and cardiovascular) health—smoking. Chapter 15, the Urinary System, rounds out the coverage of body maintenance.

Part 4: Reproduction and Development

Chapter 16 considers the reproductive system and the various means of contraception available today. Chapter 16A applies this knowledge to learning about sexually transmitted diseases and AIDS and how to protect against them. Chapter 17 discusses development, beginning at conception and continuing throughout the lifespan.

Part 5: Genetics

This unit explains the basic concepts of inheritance and genetic disease (Chapter 18) and the mechanisms by which genes work, as well as the growing field of biotechnology (Chapter 19). Chapter 19A takes up the mechanisms and ramifications of cancer—a topic that has an impact on many students and their families.

Part 6: Evolution and Ecology

Chapter 20 explains the principles of natural selection and evolution, together with the evidence of evolution. It details the current thinking on the probable pathways that human evolution has taken. Chapter 21 introduces basic concepts of ecology, noting ways that human activities are upsetting the ecological balance and giving some ideas for constructive action. Chapter 22 explains the principles of population growth, focusing on factors that affect human population growth. Consequences of uncontrolled population growth include world hunger, pollution, resource depletion, and loss of biodiversity.

Features of the Book

The features of this book include the following:

Chapter Outlines

Each chapter begins with an outline that provides a framework on which the student can organize the information presented. An outline identifies the important concepts and serves as a map of the relationships among these concepts.

Applications Chapters

The applications chapters (7A, Alzheimer's Disease; 8A, Drugs and the Mind; 13A, Nutrition and Weight Control;

14A, Smoking and Disease; 16A, Sexually Transmitted Diseases and AIDS; and 19A, Cancer) expand "pure biology" to cover topics that are likely to be of personal interest to groups of the reader population. The topics are developed with an interest in personal health and are more thoroughly developed than they could be in an essay. Educators who prefer to teach an applied human biology course can use these chapters as a "hook" to grab the attention of students. Even if these chapters are not assigned reading, however, it is hoped that the topics are so pertinent to issues facing the readers in their personal lives that they will read or at least refer to these chapters as guides to healthy lifestyles.

Critical Thinking Questions

These questions are scattered throughout each chapter and are intended to engage students in the learning process and to promote active learning. They ask the student to apply information that has been presented in the text to a new situation. They provide periodic checks for the student to determine whether he or she understands the basic concepts.

Social Issue Questions

These questions are also scattered throughout each chapter and raise ethical questions about issues that society faces today. These help the student see the relevance of information learned in a biology classroom to real life problems or decisions that society must make, including fluoridation of water, routine screening for prostate cancer, use of animals' organs to save human lives, the export of pesticides to developing countries, and the means of slowing the growth of human populations. There is no "right" answer to any of these questions. They simply point out to the student that there are broad implications to many of the topics discussed.

Essays

The essays fall into one of three categories. The Personal Concerns essays deal primarily with personal health issues. These essays provide current information on health topics to help students understand certain health problems that they, their family, or their friends might encounter. It is hoped that these essays will help students to better understand what their physicians may be telling them. The Social Concerns essays explore some of the ethical or social issues related to the topics under consideration. Finally, the Environmental Concerns essays deal with the ways in which human activities alter the environment, or the ways in which the environment influences human health or well-being.

Supplements

To further facilitate learning and teaching, a package of supplemental material has been carefully designed for the student and instructor. It includes:

1. An *Instructor's Manual with Test Bank* prepared by Patricia Matthews, Grand Valley State University. This manual

includes chapter outlines, chapter objectives, key terms, and lecture outlines. The Test Bank includes multiple-choice questions, true-false questions, matching exercises, fill-in-the-blank questions, and short-essay questions. All answers are provided. The **Computerized Test Bank** is available for both IBM PC and Apple Macintosh platforms.

2. A *Study Guide* has been prepared by Douglas Light, Ripon College. Each chapter begins with a description of the significance that each chapter has to the text and to the course. Each chapter includes chapter objectives, section-by-section summaries of key material, significant concepts, and key words. The student self-tests are composed of multiple-choice questions, short-answer questions, and essay and discussion questions.

3. The *Laboratory Activities for Human Biology*, by Craig Clifford, Northeastern State University, is a fully illustrated laboratory manual that follows the systemic organization of Goodenough/Wallace/McGuire: *Human Biology: Personal, Environmental, and Social Concerns*. More than 45 hands-on, non-dissection activities draw on students' everyday experiences. These are creative, observational activities that can be done by students without prior science background.

4. Other important components of the supplements package for *Human Biology* include 150 **Overhead Transparencies** taken from illustrations in the book and **Bio-Art**, a set of 100 black-and-white unlabeled line drawings from the text.

5. **Biology MediaActive, Version III B** (Version III 1998) The CD-ROM Biology Media Bank contains imagery from Goodenough/Wallace/McGuire: *Human Biology* and other Saunders biology texts. This CD-ROM is available as a presentation tool to be used in conjunction with commercial presentation packages, such as Powerpoint™ and Persuasion™, as well as the upgrade of Saunders LectureActive™ Presentation Software. Version III will be available on the Biology Mediactive CD-ROM. Available for both Macintosh and Windows platforms.

Saunders College Publishing may provide complimentary instructional aids and supplements or supplements packages to those adopters qualified under our adoption policy. Please contact your sales representative for more information. If, as an adopter or potential user, you receive supplements you do not need, please return them to your sales representative or send them to:

Attn: Returns Department
Troy Warehouse
465 South Lincoln Drive
Troy, MO 63379

Please visit our Human Biology Website in the fall at **http:///www.saunderscollege.com/lifesci/.**
Click on Goodenough/Wallace/McGuire: *Human Biology*.

Acknowledgments

It takes more than authors to get a book to the readers and many dedicated people have helped get this text to your hands.

The project was enthusiastically launched by Julie Levin Alexander, Executive Editor at the start of this project. Her vision helped shape the book. Edith Beard Brady took over the project after Julie's departure and never missed a beat.

Our Developmental Editor, Lee Marcott is without a doubt the best of the best. She helped us develop a plan for the book that incorporated our teaching experience, Bob's writing expertise, and her knowledge of the market. She kept that plan in mind through the entire project. Lee was a source of ideas on how to maintain quality and still keep to the schedule. She is a problem-solver par excellence with a can-do attitude and experience in many aspects of publishing. She put out many a fire before we even smelled smoke, and was always good company.

Vanessa Ray, Project Editor, kept track of everything with careful diligence and a keen eye for schedules. Throughout the project she was patient, calm, and helpful. She was responsive to our needs and rare complaints. Sue Kinney, Illustration Supervisor, and Joanne Cassetti, Director of Production, were always working behind the scenes to get the illustrations created and the final book produced.

The art program is essential not only to the appearance of the book but also to its usefulness as an instrument for learning and teaching. The Designer and Art Director, Ruth Hoover, developed an attractive design that incorporated all pedagogical elements, linked main chapters with application chapters, and emphasized the personal, social, and environmental themes of our text.

Carlyn Iverson, our Art Developmental Editor and Illustrator, created visual images that expressed our ideas and emphasized the important concepts. We must confess that she knows more about biology than we know about rendering art. Nonetheless, Carlyn was receptive to our suggestions and our interactions were educational and fun.

Jane Sanders, the photo researcher, was diligent in her pursuit of striking and pedagogically important photographs.

Two of us, Judy and Betty, owe a great deal to our late friend and colleague Bob Wallace. He brought an energy to this project that even his death could not extinguish. We learned so much, probably more than we realize, from his massaging of our manuscript and his gentle chiding to "just tell the story." He loved life and loved to tell its story—biology. Bob made writing fun. His insane editorial asides written in the margins of our manuscript never failed to bring smiles to our faces. We both had some personal tribulations during the development of this book, and Bob was always there to listen, to comfort, and make us laugh in spite of our problems. We hope that we've learned enough from him and that his spirit will be expressed in the way we write.

We would also like to thank Bob's wife, Jayne Wallace, for her help in completing this project.

For Judith Goodenough: Writing a text is often a family affair, at least the way I work. I could never have written this book without the support and help of each member of my family. My husband, Steve, was an enormous help. He kept things running smoothly at home, helped me balance the four entirely different schedules of our family members, and helped me choose esthetically pleasing photographs. Most importantly, he made me laugh, and that kept me sane. My daughters, Aimee and Heather, were patient and encouraging. They helped me remember that the truly important things in life are the people you love. They are the bright spots in my life and make each day worth living. I especially thank Aimee for her help in doing research for some of the chapters. My parents, Betty and Ray Levrat, instilled in me a love of learning that has taken (or driven) me to where I am today. They have always been willing to help me in any way, especially on this book. No task was too mundane, if it would allow me to focus on writing and still find time to participate in family events. They have always encouraged and supported me.

I would also like to thank my friends and colleagues in the Biology Department at UMass for lending their expertise to various aspects of this book. Gordon Wyse was extremely helpful in fine-tuning the chapters involving the nervous system. Arthur and Elaine Mange were an endless and willing source of information and examples concerning human genetics. Because they have more experience as writers than I have, I found myself seeking their counsel on many aspects of this book, ranging from design to the placement of commas. Chris Woodcock taught me about nucleosomes, patiently reviewed one figure multiple times, and supplied a much needed photograph of nucleosomes. Larry Schwartz provided photographs of cell death and many impromptu lectures on a variety of topics. Jack Palmer offered moral support, a multitude of current references, and many interesting discussions. I'd also like to thank the many students, especially Nirvana Filoramo, Alison Pitt, and Lori Bleumer, who helped me by proofreading and doing research. Finally, I'd like to thank Margaret Tillson for lifting my spirits while at work.

For Betty McGuire: I would like to thank Willy Bemis, my husband and best friend, for his support and help throughout this project. James, Dora, Kevin, and Cathy McGuire were always encouraging and supportive. My good friend and student Erika Henyey provided invaluable assistance at home with our children and pets. Jane Bemis and Carol Bigelow helped in innumerable ways with children and moral support. Carisa Zampieri and Sara Sullivan were my faithful proofreaders and day-to-day helpers. Richard Briggs provided histological expertise and Larry Schwartz taught me all about cell death, among other things. . . . Lowell Getz waited patiently as I undertook yet another project. Finally, I thank Robert Wallace for making me laugh, even when it seemed impossible.

Judith Goodenough
Betty McGuire
August, 1997

Reviewers

Amelia Ahern-Rindell	Weber State University
Hugo Leigh Auleb	San Francisco State University
Iona Baldridge	Lubbock Christian University
Chandra Basu	Erie Community College
Edmund Bedecarrax	City College of San Francisco
Michele Miller Bever	Wabash College
Ann Boyd	Angelo State University
Jim Brammer	North Dakota State University
Richard Connett	Monroe Community College
Lisa Danko	Mercyhurst College
Tom Denton	Auburn University—Montgomery
Linda Fagan Dubin	Loyola Marymount University
Rebecca Ferrell	Metropolitan State College of Denver
Sheldon Gordon	Oakland University
Kenneth Gregg	Winthrop University
Gail Hall	Trinity College
Madeline Hall	Cleveland State University
Richard Hall	Cypress College
Julie Harless	Long Island University—C.W. Post Campus
Clare Hays	Metropolitan State College of Denver
Robert Hollenbeck	Metropolitan State College of Denver
Mitrick Johns	Northern Illinois University
Dan Johnson	East Tennessee State University
R.W. Krohmer	St. Xavier University
Martin Levin	East Connecticut State University
Douglas Light	Ripon College
Mary Katherine Lockwood	University of New Hampshire
Patricia Matthews	Grand Valley State University

Tom McKinney	Mohave Community College	John Sherman	Erie Community College
Karen McMahon	University of Tulsa	Richard Shippee	Vincennes University
Eugene Mesco	Moorhead State University	Lori Smolin	University of Connecticut
Stacia Moffett	Washington State University	Greg Stewart	State University of West Georgia
David P. S. Mork	St. Cloud State University		
Donald Mykles	Colorado State University	Robert Sullivan	Marist College
John Natalini	Quincy University	Carol Summers	Rosary College
Emily Oaks	SUNY—Oswego	William Thieman	Ventura College
T. Lon Owen	Northern Arizona University	Karin Van Meter	Des Moines Area Community College
Beryl Packer	Des Moines Area Community College		
		Richard Walker	Des Moines Area Community College
David Polcyn	California State University— San Bernadino		
		Gary Wassmer	Earlham College
Barbara Roller	Florida International University	David Weisbrot	William Paterson University
Lynette Rushton	South Puget Sound Community College	Susan Whittemore	Keene State College
		Chester Wilson	University of St. Thomas
John Scheide	Central Michigan University	Leonard Yannielli	Naugatuck Valley Community— Technical College
Donna Schroeder	College of St. Scholastica		
Geri Seitchik	La Salle University	Debra Zehner	Wilkes University

Contents Overview

Contents

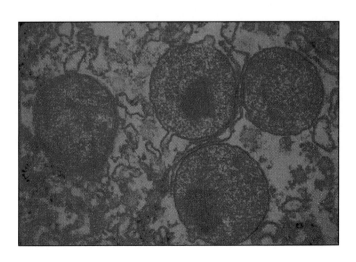

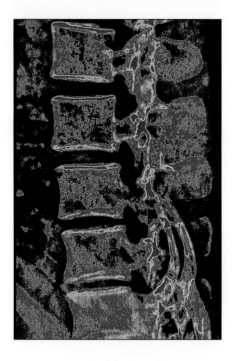

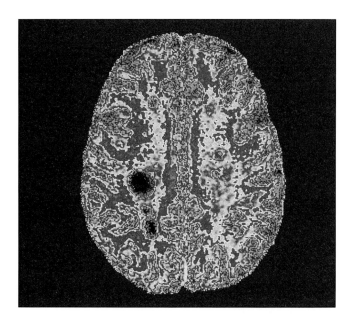

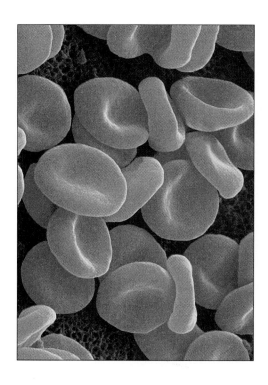

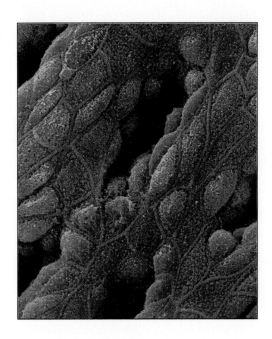

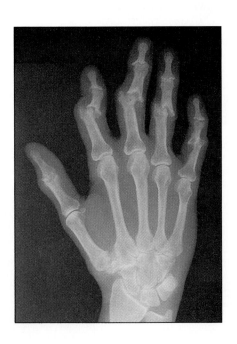

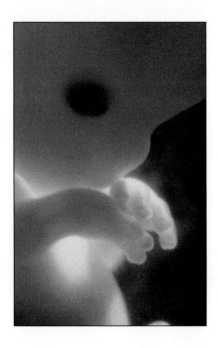

Special-Interest Essays

The Organization of the Body

The human body provides an example of nature's exquisite design and organization. As a result of this organization, the body's many parts and systems work together harmoniously, allowing this child to laugh and play while her body grows and maintains a healthy state. *(William E. Bemis)*

Chapter 1

Science and Society

Humans share many characteristics with other forms of life, but culture—social influences that produce an integrated pattern of knowledge and beliefs—distinguishes us from other species. Culture can be passed from generation to generation, as in the four generations of women in this Native American Family.   (©John Eastcott/YVA Momatiuk/Photo Researchers, Inc.)

The dugout canoe turns from the course of the muddy river and thrusts as far as possible onto the slippery bank. A tribesman cheerfully steps off into mud up to his knees and, laughing at his friend pushing with a pole from the back of the canoe, he pushes and heaves and moves the boat as far as possible onto the bank, then ties it to a tree with a braided chambira rope.

You carefully make your way forward, then stop and reach back for a plastic-wrapped bundle that your friend, in the rear of the canoe, hands you. You pass it along to the tribesman. He carries it to a grassy area, then comes back for another one, then another, until all the equipment is ashore. The tribesman helps you and your friend to the grassy area, knowing North Americans are not too good at this sort of thing.

Once the four of you are ashore, you shoulder your packs and with the tribesmen carrying the larger loads because of their greater strength, you peer through the tangled, viney vegetation into the dark interior of the forest. An insect whines. Already you're soaked with sweat. Still, this is what you came for, to explore this part of the Amazon rain forest, searching for medicinal plants (Figure 1–1).

You've trained, learned some Spanish and a little of the tribal dialect, talked to missionaries, and made your plans. And now here you are, finally entering the Amazon rain forest, an area boasting the densest accumulation of species on earth. Your guides are *curanderos,* or healers, and they know many of the medicinal plants of the area. But one of them has lost an eye and both are scarred, so you know that while this may be a promising and exciting endeavor, it also will have its risks. The question arises, since this obviously is a dangerous expedition, why are you here? The answer may take a variety of forms, but it comes down to one thing—because you are human.

Humans are an irrepressibly curious species. Our intelligence has been coupled with an inquisitiveness that leads us in innumerable directions, each according to our own tendencies, background, and abilities. Furthermore, as a scientist, your investigation will follow certain prescribed steps that have been developed over centuries of investigation. Let's take a look, then, at just how scientific investigation proceeds and then at how you came to apply these steps in your own investigation.

Figure 1–1 Explorers searching for medicinal plants in the Amazon rain forest. *(© Jane Thomas/Visuals Unlimited)*

The Scientific Method

Now that we have so boldly titled this section, we have to tell you that there is no such thing as *the* scientific method—not if it refers to a formalized ritual for performing experiments. Instead, the scientific method involves a certain logic in an approach to gathering information and reaching conclusions. It often begins with an observation that raises a question. Next, an educated guess is made about the answer to the question and, importantly, that guess must be testable. Generally, the tentative explanation will lead to a prediction, which will support the explanation if it holds true when it is tested.

Careful, quantified observation is part of the scientific method. If, for example, you were walking in a rain forest and came across a plant growing alone, with no other plants growing around it, that observation would be noted and described. Locally, this plant is known as the lemon ant plant. It grows quite close to the ground and it harbors small

Figure 1–2 Plant species in the Amazon rain forest are normally densely packed. Only rarely does one plant secrete products that interfere with the growth of others. *(© Richard Thom/Visuals Unlimited)*

ants considered delicious by the local tribesman. Considering the density of plants that you see in nearby areas of the forest floor, this observation of an isolated plant strikes you as curious and so you want to learn more about this plant (Figure 1–2). You make a few measurements to see whether your impression was correct. You may want to measure the size of the plant (its height, girth, and the combined area of its leaves) and the size of the cleared area around the plant. You might also want to look for other such plants, make the same measurements, and determine whether there is some correlation between the plant's size and the area of the clearing.

Hypotheses are part of the scientific method. Once you have your measurements and find that the larger the plant is, the larger the area cleared, you may assume that something about the plant inhibits the growth of other plants in the area. But what? Your first guess is that the plant is producing a chemical that inhibits the growth of other plants. This is your **hypothesis**, a testable explanation for your observation.

Controlled experiments are part of the scientific method. Now you must test your hypothesis with an experiment. If your hypothesis is correct, you might predict that the chemical alone, without the intact plant, would inhibit growth of other plants. Since there are a number of these lemon ant plants around, you pick one and homogenize its tissues in alcohol, using a blender. You now have an extract—a mixture of all the chemicals in the plant that are

soluble in alcohol. This extract can then be used to treat plant seeds to determine whether it affects growth.

You must design your experiment so that there can be only one explanation for the results. So you run a **controlled experiment**, an experiment in which the subjects (in this case, plant seeds) are divided into two groups, usually called "control" and "experimental" (Figure 1–3a). Both groups are treated in exactly the same way, except for the special factor, the variable, whose effect you want to determine. In this case, the variable is the plant's chemicals. You want to know if those chemicals inhibit the growth of other plants, so you prepare two identical plots of ground, seed them both with the same variety of plants, and then spread the plant extract over a specified area of the experimental plot and leave the control plot untreated. Sure enough, no new plants grow in the area that you treated with the alcohol extract, but they grow well in the untreated areas.

Is your hypothesis supported? Yes, but not completely. It is possible that the alcohol, not the plant's chemicals in the extract, was inhibiting plant growth. In the first experiment, you had two variables, the alcohol and the extract itself. So you experiment again. This time, however, you have three plots, as identical to one another as possible. You till the soil in each, plant the seeds in each, and water each. Next, you treat a specific area in one plot with the alcohol extract, you treat a second area the same size in another plot with alcohol only, and you leave the third plot untreated (Figure 1–3b). Then you wait, and sure enough, plant growth is retarded only in the area treated with the extract. Plants grow in the plot treated only with alcohol and in the untreated plot. It is evident that the chemicals in the plant tissue are inhibiting the growth of other plants.

Conclusions are part of the scientific method. Next, you draw a conclusion. Your conclusion, in this case, may be that the lemon ant plant produces a chemical that inhibits the growth of other plants.

Note, however, that although your results support your hypothesis, they do not *prove* your hypothesis. There may be other hypotheses that would make the same prediction, and there may be other explanations for the results. For instance, suppose the lemon ants that live on the plant deposit a chemical on the surface of the host plant that washes off with rain and inhibits plant growth in the surrounding area. The ant-produced chemical might also be in your extract along with the plant's chemicals. Thus, experiments often lead to revised hypotheses and new experiments (Figure 1–4).

Also, what if the plant extract had not inhibited the growth of other plants? This event would call for some additional thinking on your part. So, now think carefully and create another explanation. Remember, this explanation is a new hypothesis and will call for new experiments (with new controls) and new conclusions. Maybe the inhibitory chemical isn't soluble in alcohol. Maybe the chemical is a gas. Maybe the inhibition of growth isn't caused by a chemical. Essentially, you must now look for new variables. For example, you might want to go back to the forest and note

(a) **Experiment I**

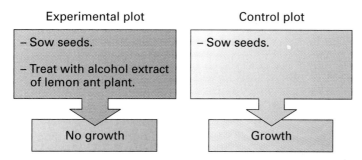

(b) **Experiment 2**

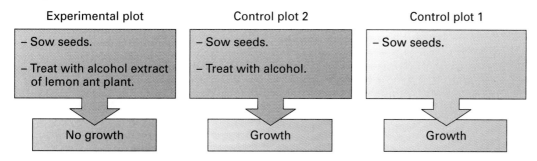

Figure 1-3 A controlled experiment is set up in duplicate. The two groups should be exactly alike except for the factor to be tested, which is called the variable. (a) In this case, the plant seeds are the subjects. The variable is the extract of the lemon ant plant. The experimental plot is treated with the extract. The control plot is not. The results are compared to determine whether the variable had an effect. In this case, the extract did inhibit plant growth. (b) A second control is added to the design to determine whether the alcohol used to prepare the extract was responsible for the inhibition of growth. Three plots are prepared: an experimental, an alcohol control, and a control.

Figure 1-4 Scientific process involves creating testable hypotheses, experimentation, conclusions, revised hypotheses, and new experiments.

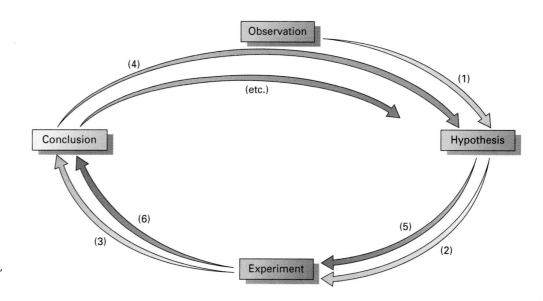

whether the plant was always found growing under another kind of plant. Were the two plants interacting? The herbicidal plant was host to lemon ants. Was the inhibitory effect due to the ants? How could you test the ideas?

Another important part of scientific inquiry is that experiments be repeated and yield similar results. Other scientists should be able to follow your procedure and obtain similar results, not just in that part of the rain forest but any place where the lemon ant plant is found.

As we have seen, the formation of a hypothesis represents one level of scientific process. In some cases, related hypotheses fit together and form a **theory**, a broad-ranging explanation for some aspect of the universe. Because of its breadth, a theory cannot be tested by a single experiment. Instead, a theory is the result of many observations, hypotheses, and experiments.

Inductive and Deductive Reasoning

As you looked at the bare patch around that plant on the jungle floor and wondered what was causing it, your mind was positioning itself to attack the problem, probably without your knowing it. It probably would have taken one of two approaches, depending on previous experience with problems or specific training in problem-solving.

The two approaches are formally referred to as inductive reasoning and deductive reasoning. **Inductive reasoning** involves the accumulation of facts through observation until the sheer weight of the evidence forces some general statement about nature. In the case of the cleared area around the plant, someone using this approach would collect as much information as possible about this plant, others like it, the soil, the leaves, and anything else that might have

Figure 1–5   The signs of life are abundant in a rain forest.   *(Paul Kratter)*

a bearing. When as much information as possible was gathered, it might be possible to draw a general conclusion, such as "some plants are able to interfere with the growth of other plants."

The type of reasoning you used in your research on the lemon ant plant, called deductive reasoning, is a powerful tool in scientific inquiry. **Deductive reasoning** involves making some general statement and then drawing, or deducing, conclusions from it. The statement is usually in the form of an "if . . . then" premise. *If* this plant is making something that interferes with the growth of other plants, *then* there should be a cleared area around this plant; *then* this plant should have a way of producing some chemical; *then* that chemical should be detectable; *then* the plant should have a delivery system to get it to the soil, and so on. Each of these should be a testable statement.

Good science can proceed by either technique, or perhaps a mixture of both. You can be sure that any conclusion will be tested by other scientists who, themselves, will take one (or both) of these approaches.

Signs of Life

As you continue to make your way deeper into the rain forest, you are keenly aware of the diversity of life around you. In fact, that diversity is equaled in very few places, such as a few tropical coastal areas and perhaps some coral reefs. The hothouse atmosphere emphasizes the existence of the life around you, much of it unfamiliar (Figure 1–5).

If something you encounter is unfamiliar to you, you may wonder if it is alive. How would you know? In some cases, the question is an easy one. You may not recognize the leaf but you know it's a leaf all the same. And a tree is a tree. That fuzzy thing on your neck is undoubtedly a caterpillar of some kind. Many forms of life easily fall within the realm of the living. But what about that gray thing on that tree trunk? It looks as if someone wedged a rock into a crack in the bark. Is it a rock? It feels soft. It's not a rock. So, is it alive? And more importantly, how can you tell?

Defining life would seem to be an easy thing, but it isn't. In fact, there is probably no single definition that would suit all life scientists. For example, if we say you can tell something is alive if it reproduces, someone is likely to note that oil droplets reproduce when they join together and grow so big they fall apart and form smaller droplets. If we say you can tell something is alive if it grows, how about crystals? They grow and they are not alive. Many living things use oxygen, but so does a rusting nail. And so it goes. It seems there is no single defining feature of life.

So how do we characterize life, this great pageant of which we are a part? We cannot do it in a phrase. No single definition covers all life. So we find that instead of defining life, we can only characterize it. That is, we can only list the traits associated with life, and even this list can vary from one biologist to the next. However, most biologists would agree that, in general, life can be characterized by the following statements:

1. **Living things reproduce.** Living things have ways of generating new individuals that carry some of the genetic material of the parent. Some organisms reproduce simply by making new and rather exact copies of themselves, as do bacteria (Figure 1–6a). Others reproduce by combining genetic material with another individual, as do humans (Figure 1–6b). DNA is the genetic material and it carries the instructions for the development of the offspring as well as the directions for maintaining life.

2. **Living things are composed of cells.** Some have but a single cell, such as the *Paramecium*, while others are composed of many cells, such as ourselves. It is interesting that the cells in most forms of life, no matter how divergent, share many characteristics. (The bacteria, however, differ markedly from other life forms.)

3. **Living things metabolize and grow.** *Metabolism* refers to the sum total of all chemical reactions that go on within the cells of living things. There are two basic aspects of metabolism: *anabolism* involves building processes, as when cells build complicated molecules from simple ones; *catabolism* involves the breakdown of complex molecules into simpler ones, releasing energy from chemical bonds.

   Metabolic activities allow organisms to grow. Some living things are composed only of a single cell and grow very little, as do paramecia (Figure 1–6c), whereas other organisms never stop growing, such as the giant sequoia tree (Figure 1–6d).

4. **Living things respond,** as when a fighter avoids the blow of an opponent or when a flower bends toward the light (Figure 1–6e). For a living thing to respond, it must first detect the stimulus and then have a way of responding. As we will see, our sensory systems detect stimuli, our nervous system processes sensory input, and our skeletal and muscular systems allow us to respond. This biological responsiveness is called irritability.

5. **Living things tend to maintain a relatively constant internal environment, an ability called homeostasis.** We will have a lot more to say about homeostasis as we discuss each physiological system of the body. We will generally find that life can only exist within certain limits and that living things tend to behave in ways that will keep their body systems functioning within those limits. For example, if we become too cold, we shiver (a metabolic response) and the activity produces heat that warms our bodies. Alternatively, if we become too hot, we may seek ways of cooling ourselves (a behavioral response) (Figure 1–6f).

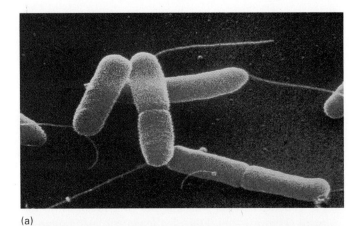

(a)

(b)

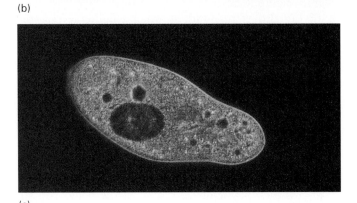

(c)

(d)

(e)

Figure 1-6   (a) Bacteria reproducing, (b) human couple with child, (c) *Paramecium*, (d) *Sequoia sempervirens*, (e) a field of sunflowers bending toward the sun shows responsiveness, and (f) a child playing in water to cool himself on a hot day. *(a, © David M. Phillips/Visuals Unlimited; b, Ken Fisher/© Tony Stone Images; c, © Dwight R. Kuhn, 1986; d, © 1995 Barbara Gerlach/Dembinsky Photo Associates; e, © 1993 Robert Pettit/Dembinsky Photo Associates; f, © 1992 Dan Dempster/Dembinsky Photo Associates)*

(f)

You see, then, we have not defined life, but instead have only given some of its characteristics—the *signs* of life. All life we are aware of shares at least these characteristics, and so we are linked to all the other life forms, from bacteria to whales, redwoods, and red-headed cousins.

Critical Thinking

If you were an explorer on a space ship who discovered an unidentified material, what experiments would you design to test the hypothesis that this material was alive?

The Unity and Diversity of Life

In these examples, we have stressed both *the unity and diversity of life*. In rain forests, we may find about 155,000 of the 250,000 known species of seed plants, yet every one of them will have the five traits we have just mentioned. Some of those plants will be tiny, living close to the earth. Others, though, form great trees and among their leaves we may find large sloths hanging by their claws from branches (Figure 1–7). The neck structure of sloths enables them to turn their heads almost entirely around and so they can stare at the people below them. Their coats may also appear green, making them hard to see among the leaves. The green comes from algae that grow in pits along the animal's hair shafts. Those algae, and the sloth they grow on, are very different from each other and both are different from the people below, yet they all must reproduce, grow, be cellular, metabolize, respond, and maintain homeostasis.

Evolution and Adaptation

Each of the amazing varieties of life forms you see around you has **adaptive traits**—that is, traits that help an organism survive and reproduce in its natural environment. Among the species of plant life in the rain forest, you are likely to see many adaptations that allow plants to compete for sunlight needed to convert carbon dioxide and water to sugars. For instance, the trees are incredibly tall. Vines twine around the trees and epiphytes, including orchids, are rooted on the surfaces of other plants. Many animals have adaptations that enable them to reach the plants for food—the ability to fly or climb, for example. Hummingbirds have long slender beaks that extend into exotic flowers to obtain pollen.

Adaptation is a product of evolution by natural selection, which is differential survival and reproduction. Consider how natural selection works. Within a population, in-

Figure 1–7 A sloth hanging from a branch. *(Barbara Magnuson/Photo/NATS)*

dividuals differ and some of the variations are due to genetic differences that can be passed on to offspring. Moreover, some of these heritable variations lead to increased survival and reproduction. Individuals with beneficial traits often leave more offspring than do individuals who lack such traits. Because of this differential reproduction, certain beneficial traits increase in frequency and other less helpful traits decrease in frequency. It is largely the particular set of environmental conditions to which an organism is exposed that determines whether a trait is beneficial. We say, then, that nature selects the individuals that will leave the most offspring. As a result, different lines of organisms living in different conditions acquire variations that help them survive and reproduce. And so we see how natural selection works on populations, leading to the evolution of individuals with adaptations to their environment.

Are Humans Different?

As you stand on the forest floor looking up at the sloth who stares back for a moment and then moves slowly away, you may be glad you can pick up your backpack and move on. You're not a sloth. You're human and you have things to do.

Figure 1–8 Social interactions are an important thread in the fabric of human life. *(E. A. Heiniger/Photo Researchers, Inc.)*

Yes, you're glad you're not a sloth. But what sets you apart? Why are you different? You are aware that you share the five signs of life with the sloth. Furthermore, you know that you both have hair (even if yours is not green) and can give milk, that you both have a four-chambered heart that pumps red blood, that you both have bones and muscles and reproduce sexually, and that you both eat plants and have eyes on the front of your face. Yet, he's a sloth and you're not. So what sets you apart?

Brain size, an opposable thumb, and a two-legged gait are human characteristics, but probably nothing distinguishes humans more than culture. *Culture* may be regarded as social influences that produce an integrated pattern of knowledge, belief, and behavior. Other animals, of course, have social interactions, as we see in everything from competition and cooperation to territoriality, hierarchies, and mating behavior. But in other species social influences are not so pronounced as to be called culture.

How important are social interactions? The famed student of chimpanzee behavior, Jane Goodall, once remarked that *one* chimpanzee is no chimpanzee at all. She was underscoring the futility of studying chimpanzee behavior by watching one caged animal. Her point was that chimpanzees need a social environment if they are to develop and behave normally.

We occasionally hear of some child being deprived of social interactions by being raised in a cellar or by wild animals or whatever. We can assume that these people, too, are not sharing the full human experience. Humans need social interaction, and because of our intelligence that interaction can produce the integrated pattern called culture (Figure 1–8).

Our intelligence allows us to analyze our behavior, to determine what works and what doesn't, to form rules, and to remember those rules, which may even become codified as laws or religious dogma. It is from such processes that true human culture is produced.

Keep in mind, however, that there is no single human culture. As you follow your guides through the rain forest you will eventually collect your plants and then return to their village. You may quickly learn that their culture is very tightly structured, that it's very functional, and that it differs from yours.

You may try to describe your culture to them and it may elicit anything from astonishment to howls of laughter. In the process of interacting with other cultures, you may learn just how malleable the human condition is, but that there are certain constants as well, such as rank, love of children, and political alliances. In every way, culture powerfully defines the human condition.

Ethics and the Human Condition

Of all the species, we are probably the most aware of our own condition. Thanks to culture, which can be perpetuated in various ways from tradition and storytelling to the written language, we are aware not only of our present situation but of our history as well. Also, because of our ability to communicate, we have some knowledge of the situation of other people in other places.

It can be argued that with this kind of information we have an increased responsibility to ourselves, to future generations, and to other cultures. For example, we know that each year we are cutting an area of tropical rain forest the size of Florida. We also know that, because species in these forests may exist in pockets, we are causing countless species to become extinct each year. The tropical rain forest has provided humankind with about 25% of our medicinal drugs. Only about 1% of the known species has been tested for their medicinal value. According to many researchers, we are not even aware of the existence of most species in the forest.

For a time, North Americans encouraged the destruction of the rain forests because the cleared area could be used to grow beef inexpensively. Other areas were cleared for farming. In each case, the land remained fertile for only a few years, until the torrential rains washed nutrients from the soil. Yet other clearing was done to extract hardwoods for good flooring and fine furniture. It is said that, using current methods, roughly 15 acres of forest must be destroyed to harvest a single hardwood tree. Also, as oil is found, roads must be built to bring in the people and equipment. There are, indeed, many pressures on the earth's rain forests.

We might ask ourselves, do we really need the hardwoods? Should we demand the furniture? How about the oil? A recent oil discovery in Ecuador brought a construction boom to the jungle and created great mucky swamps of crude oil. Rivers were polluted, fish were killed, and vast areas of land belonging to the Waorani and Cofan tribes were destroyed, even as the hunters abandoned their traditional ways to work for the "petroleros."

We have this information at our fingertips. We can see what is happening and what will happen, given our present course. So what should be our ethical response? Ethics deal with right and wrong. What is right for us? What should we, as human beings do? If we do nothing, is that a decision?

Social Issue

Powerful interests seek to missionize or westernize the Indians of the rain forest, getting them to give up their old ways. The Indians are often alarmingly willing to trade their ways for ours, and they now often visit infirmaries when they are sick, abandoning the healing properties of the forest. The cost of having them retain their knowledge is often to abandon them to superstition, some ineffective treatments, frequent hunger, parasites, and predators. If you could make the decision, would you allow them to be westernized, losing their old values and traditional knowledge (and losing it for all humankind), in order to bring them our medicines, our values, and our culture if those are the things they want?

As we journey through the pages of this book, we will revisit the unity and diversity of life as we find out just how much we share with the other species on earth. We will also be introduced to our distinctive place on the planet. Because we are intelligent, cultural, and informed, do we bear a special responsibility to help look after not only the other species but also our fellow humans?

SUMMARY

1. The "scientific method" is a sequence of activities that involves observation, hypothesis (a testable explanation for the observation), experimentation (performed with controls), and the conclusion. The conclusion may lead to further experimentation.
2. As evidence in support of a hypothesis mounts, it may become so impressive that the hypothesis becomes a theory, a well-supported explanation of nature.
3. Life cannot be defined, only characterized. Living things reproduce, grow, are made of cells, metabolize, respond, and maintain a steady state.
4. Humans share many traits with other species, but they also have many unique features, among the most important of which is culture. Our culture stems from our experience and intelligence defining for us those customs that should be retained.
5. Because of our awareness of our condition, as well as that of other species and the earth itself, humans tend to have an ethical standard, based on right and wrong, that helps us meet the challenges we find facing us.

REVIEW QUESTIONS

1. What is a hypothesis? How does it differ from a theory?
2. Define a controlled experiment.
3. Differentiate between inductive and deductive reasoning.
4. List five features that characterize life.
5. Define culture. Why is it so important to humans?

The Chemistry of Life

Our food contains biological molecules such as carbohydrates, proteins, and lipids, that provide fuel or become incorporated into our bodies. *(© Peter Cade/Tony Stone Images, Inc.)*

In this chapter we introduce some basic principles of chemistry and lay the foundation for what follows. We begin with atoms, small entities, and end our discussion with the major molecules of life.

Chemistry Fundamentals

When many of us hear the word chemistry, we think of very serious people in white lab coats writing numbers on a blackboard while something bubbles in a beaker behind them. In view of this image, it may seem farfetched to say that understanding basic chemistry is essential to understanding human biology. We only need a second to remember that we are a concoction of chemicals—substances that may have once been part of the stars. We share the basic makeup of everything else in the universe, and the chemicals that make up our very bodies are also found in rocks, water, and the air we breathe. By the end of this discussion you will see how some knowledge of chemistry aids not only in comprehending the structure and function of the human body but also can help make life easier, safer, and more understandable. For example, understanding basic chemistry makes food labels more comprehensible. (What is a saturated fat?) It can also help us avoid minor discomforts such as acid indigestion and more serious problems such as atherosclerosis (formation of fatty deposits in the arteries). Finally, background in biological chemistry is critical for all voters who must evaluate regional and global issues such as nuclear power, pollution, agricultural policies, and the changing rain.

Atoms

Atoms are units of matter that cannot be further broken down by chemical means. Atoms are composed of subatomic particles, which include protons (positively charged particles), neutrons (with no charge), and the much smaller electrons (with their negative charge). Subatomic particles can be characterized not only by charge but also by their mass and location within the atom.

Each atom consists of two regions, the nucleus and surrounding electrons. Protons and neutrons are found in the nucleus, and electrons orbit around the nucleus (Figure 2–1). Neutrons are uncharged, and because protons have a positive charge the nucleus of an atom is positively charged. That charge is balanced by the negatively charged electrons spinning around the nucleus. Finally, whereas both protons and neutrons have masses equal to 1 atomic mass unit, electrons have almost no mass. Table 2–1 summarizes the basic characteristics of protons, neutrons, and electrons.

Atoms are characterized by their numbers of protons, called the atomic number. When a substance is composed of atoms with the same number of protons (i.e., all with the same atomic number), then that substance is called an element. An **element** cannot be broken down into simpler substances by ordinary chemical means. Gold, for example, is an element. Each atom within a piece of gold contains the same number of protons. The same is true for iron or carbon, both of which are elements. There are 108 known elements, 92 of which are naturally occurring and the rest synthesized

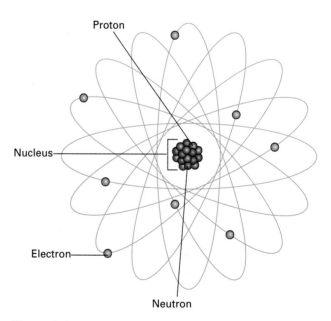

Figure 2–1 The atom. Electrons orbit around the nucleus, which contains protons and neutrons.

Table 2–1   Subatomic Particles

Particle	Location	Charge	Mass
Proton	Nucleus	One positive unit	One atomic mass unit
Neutron	Nucleus	None	One atomic mass unit
Electron	Shells	One negative unit	Negligible

in laboratories. Only about 20 elements are found in the human body, the most common being carbon, oxygen, hydrogen, and nitrogen.

The periodic table of elements lists the elements and their characteristics. Figure 2–2 depicts a simplified periodic table, including the elements most common to living things. Note that each element has a name and a one- or two-letter symbol. The symbol for the element carbon, for example, is C and that for chlorine, Cl. In addition to a name and symbol, the atomic number and the atomic weight are given for each element. Recall that the atomic number is simply the number of protons in the nucleus of an atom. The atomic weight of an element is the average mass of its constituent atoms, measured in atomic mass units (amu). An atomic mass unit is equal to one-twelfth the mass of a carbon atom. Protons and neutrons each have a mass of about one amu, and electrons have a mass of 0.0005 amu. The mass of an electron is so small that it is usually considered to be zero. Because electrons have negligible mass and protons and neutrons each have an atomic mass of one, the atomic weight for any element equals the number of protons plus the number of neutrons. Carbon (atomic number 6), for example, has 6 protons and usually 6 neutrons, and thus the atomic weight of carbon is about 12 amu.

Although the atoms of a particular element all contain the same number of protons, they can differ in the number of neutrons. Such differences result in atoms of the same element having slightly different weights. Atoms that have the same number of protons but differ in number of neutrons are called **isotopes**. The element carbon, for example, has three isotopes. Recall that all carbon atoms have six protons. Although most carbon atoms have six neutrons, some have seven or eight neutrons. The isotopes of carbon thus have atomic weights of 12, 13, and 14 (depending on the number of neutrons) and are designated ^{12}C, ^{13}C, and ^{14}C.

In some cases, isotopes are unstable and spontaneously disintegrate, or decay, emitting radiation in the form of gamma rays and alpha and beta particles. Isotopes that emit radiation are said to be radioactive and are termed **radioisotopes**. When radioactive atoms emit gamma rays or high-energy particles, their atomic numbers and masses often change. Because each element has a specific atomic number, if that number changes, the element itself changes to another element. For example, carbon-14 (^{14}C), the radioactive isotope of carbon, emits beta particles and in the process changes to another element, nitrogen. About 60 of the 320 naturally occurring isotopes emit radiation. Many more radioisotopes have been manufactured in laboratories.

Radiation can be very dangerous. Damage to the body from radiation can take two forms, direct damage to the person receiving the radiation or damage to that person's reproductive cells so that illness and defects are only apparent in

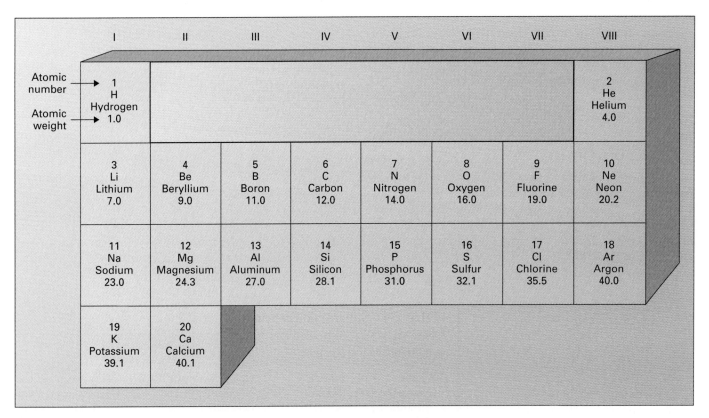

Figure 2–2   A simplified version of the periodic table showing elements commonly found in living things.

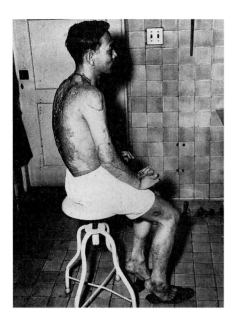

Figure 2-3 An example of direct damage caused by radiation. This individual survived the atomic explosion at Nagasaki, Japan, on August 9, 1945. Damage to the reproductive cells of people exposed to radiation has proven more difficult to observe than direct damage, and children of bomb survivors are still being monitored. *(UPI/Corbis Bettmann)*

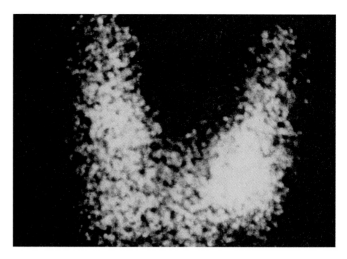

Figure 2-4 An image of the thyroid gland made possible through the use of radioactive iodine. Such images may be used to diagnose metabolic disorders. *(Science VU/© UCLBL/Visuals Unlimited)*

future offspring. The results of direct damage include low white blood cell counts, development of some cancers, and damage to organs and glands (Figure 2-3). In some cases, radiation may not produce any noticeable injury to the individual receiving the radiation, but it may alter the hereditary material carried in the cells of his or her reproductive system. This damage may cause physical or mental defects in the individual's offspring.

In stark contrast to the harmful effects of radiation are its diagnostic and therapeutic uses. Medical professionals use small doses of radiation to generate visual images of internal body parts such as organs, glands, and bones. Such images may then be used to diagnose irregularities in the body's structure or function. Radioactive iodine, for example, is often used to identify disorders of the thyroid gland. Located in the neck, the thyroid gland normally accumulates iodine and uses this element to regulate growth and metabolism. Small doses of iodine-131, a radioactive isotope of iodine that emits both beta particles and gamma rays, may be given to patients suspected of having metabolic problems (in adults, for example, undersecretion of thyroid hormone may cause decreased alertness, body temperature, and heart rate). When the radioactive iodine is taken up by the thyroid gland, medical instruments are then used to detect the radiation and translate its pattern into an image of the thyroid gland (Figure 2-4). This image can then be used in the diagnosis of disorders. Although the amount of radioactive iodine administered for diagnostic purposes is so small that it doesn't damage the thyroid gland or surrounding structures,

larger doses can be used to kill thyroid cells when the gland is enlarged and overactive, a condition known as hyperthyroidism. A more familiar use of small doses of radiation for diagnostic purposes is the x-ray.

Radiation can also be used to kill cancer cells. Because of their more rapid rates of division and higher levels of metabolic activity, cancer cells are more susceptible than normal cells to the destructive effects of radiation. For example, through precise focusing, beams of gamma rays directed at a tumor kill cancer cells and usually cause relatively minor damage to surrounding healthy tissue (Figure 2-5). Because radiation can be used either to kill or to heal, research into its effects and uses often raises ethical concerns, particularly when the subjects of such experiments are human (Social Concerns).

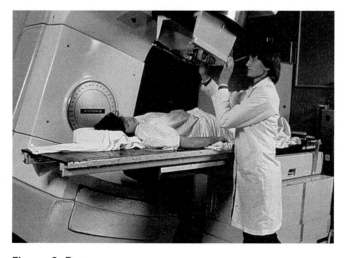

Figure 2-5 A patient receiving radiation therapy to combat cancer. *(Science VU-NIH/Visuals Unlimited)*

(Text continued on page 18)

The Ethics of Radiation Research on Humans

In the United States between 1945 and 1971, a number of people—some terminally ill hospital patients, some retarded school children, some prisoners, some scientists—were exposed to radiation as part of research into its effects and uses (Figure A). Did the research subjects or their families understand the risks of these experiments? Did anyone warn them? In other words, did the participants, or someone speaking for them, give their informed consent?

In at least one set of radiation experiments, the military's security concerns took precedence over fully informing study participants. Between 1945 and 1947, eighteen terminally ill volunteers received several injections of a substance called "product." The substance, not identified to the study participants, was plutonium, an element used to make one type of atomic bomb. The rationale for the study was that possibly many workers had already been exposed to plutonium during construction of the bomb and many more might be exposed at nuclear facilities in the years ahead, yet little was known about the effects of plutonium on the human body. The military needed to know about such effects on people. Plutonium is metabolized differently by different species, and so experiments on the effects of plutonium on rats, dogs, and rabbits could not be directly applied to humans. Also, although several thousand workers at atomic facilities were probably exposed to plutonium, the precise levels of exposure were unknown because the exposures had been accidental. Thus, in order to determine whether plutonium was rapidly excreted by humans or held in their tissues for years, the researchers believed it necessary to inject known quantities of the element into human subjects and monitor its movement through the body.

Regardless of the apparent need for such research and the potential value of the results (such as setting safety standards for plutonium exposure), failure to fully inform the volunteers in the plutonium injection experiments raised ethical questions. Is it ethical to use terminally ill volunteers as subjects in potentially harmful experiments? Were terminally ill patients chosen for study because the researchers could get around the issue of long-term harm? Were terminally ill patients convenient subjects because any plutonium remaining in the body could be measured at a not-too-distant autopsy? Could the need for an autopsy interfere with the doctor-patient relationship by making a quick death desirable? And what if, as apparently was the case in these experiments, some of the patients were misdiagnosed and were not close to death after all? Because of such questions, terminally ill people are generally no longer used in research, unless the research is an experimental treatment.

What about prisoners? Is it permissible to use them? Prisoners served as subjects in experiments on the effects of high doses of radiation on sperm production. The idea for this research grew out of a 1962 accident at a nuclear facility in which three male workers were exposed to high doses of gamma radiation. Officials at the plant found they could tell the men nothing about the possible impact of such exposure on their ability to father children. Prisoners were attractive research subjects because their confinement made follow-up easy. Although they volunteered to participate in the study, the prisoners sometimes received, as a result of their participation, letters of cooperation in their files and benefits such as improved food or housing. Such "enticements" raise the issue of whether the prisoners truly consented or were somehow coerced into participating in the research. Today many people believe that prisoners should not be used as research subjects in medical experiments because their situation makes them easily pressured to "volunteer."

The principle of informed consent was legally established in the United States in the 1930s and received substantial international attention during the mid to late 1940s when Nazi doctors were put on trial for torturing, in the name of science, inmates of concentration camps. However, for years afterward, various research groups in the United States violated that principle. In a number of other radiation studies with human subjects, the consent was questionable; but in others, informed consent was properly obtained. (Extreme examples of the latter include studies in which researchers conducted the experiments on themselves.)

Although today researchers working with human subjects must obtain the informed consent of all participants in their studies, new ethical questions continue to arise. Should physicians, for example, provide patients with AIDS with the experimental drugs they sometimes demand? Such patients may be more than willing to participate in research trials, but the issue of informed consent may be clouded by lack of adequate information on the risks of the experiment. Without the experiment, of course, researchers might never discover new treatments. So how can we get the data and know the risks without the experiments? Is informed consent always necessary? Should doctors be able to try out new methods to bring people out of a coma—people who cannot consent? Where should the lines be drawn?

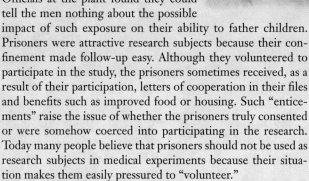

Figure A   A sample of radiation experiments performed on people.   *(Reprinted with permission from Mann, C.C. "Radiation: Balancing the Record," Science, 263: 470–473 1994.)*

ETHICS IN RADIOACTIVITY EXPERIMENTS ON HUMANS

 In some of the radiation studies cited in newspapers in late 1993 and in an earlier congressional report, known as the Markey Report, subjects did not freely consent to the experiments. In other studies it is doubtful whether informed consent was obtained. But in some of the studies informed consent was truly given. Here are examples from each category.

Date	Experiment

Possible Infliction of Harm or No Informed Consent

Date	Experiment
1945-47	18 supposedly terminal patients were injected with high doses of plutonium to learn whether the body absorbed it.
1946-47	6 hospital patients were injected with uranium salts to determine the dose that produced injury to the kidneys.
1963-70	64 prison inmates had testicles exposed to x-rays to relate radiation damage to sperm production.
1963-71	67 prison inmates had testicles exposed to x-rays to measure radiation damage to production of sperm.

Questionable Consent

Date	Experiment
1946	17 retarded teenagers at the Fernald School in Waltham, Massachusetts, ate meals with trace amounts of radioactive iron to learn about iron absorption in body.
1953-57	11 comatose brain cancer patients were injected with uranium to learn whether it is absorbed by brain tumors.
1954-56	32 retarded teenagers at the Fernald School drank milk with trace amounts of radioactive calcium to learn whether oatmeal impeded its absorption by the body.

Informed Consent

Date	Experiment
1945	10 researchers and workers at Clinton Laboratory, in Oak Ridge, Tennessee, voluntarily exposed patches of their skin to radioactive phosphorus.
1951	14 researchers at Hanford Nuclear Reservation voluntarily exposed patches of their skin to gaseous tritium.
1963	54 hospital patients volunteered to take trace amounts of radioactive lanthanum in effort to measure effects on large intestine.
1965	Trace doses of radioactive technetium were given to 8 healthy volunteers to determine its utility as medical diagnostic tool.

Molecules and Compounds

In nature, most atoms do not exist alone. Instead, they combine to form molecules. A **molecule** is a substance comprised of two or more atoms. Sometimes two atoms of the same element combine to form a molecule, but more typically atoms of different elements join together. A molecule that contains atoms of different elements is called a **compound**.

The characteristics of compounds are almost always very different from those of the elements that comprise them. Consider, for example, what happens when you join the element sodium (Na), a silvery metal that explodes when it comes in contact with water, with the element chlorine, a deadly yellow gas once used to kill troops in World War I. The result is a crystalline solid called NaCl, plain table salt.

A molecule is described by a formula that contains the symbols for all of the elements that are in that molecule. If more than one atom of a given element is present in the

(a)

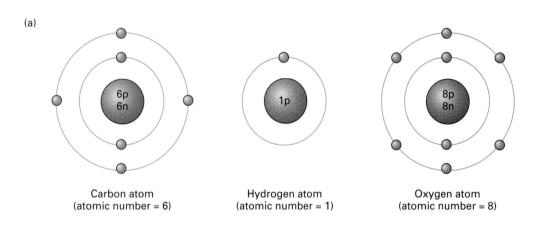

Carbon atom
(atomic number = 6)

Hydrogen atom
(atomic number = 1)

Oxygen atom
(atomic number = 8)

(b)

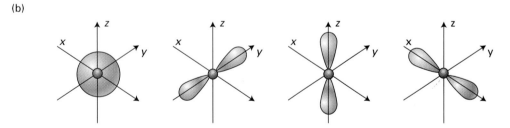

Figure 2–6   Depictions of atoms. (a) Atoms of carbon, hydrogen, and oxygen drawn according to the Bohr model, one way of representing the movement of electrons about the nucleus. Each of the concentric circles around the nucleus represents a shell occupied by electrons. The shell closest to the nucleus can hold up to two electrons. The next shell out can hold up to eight. Atoms with more than ten electrons have additional shells. (b) Although Bohr models help us to visualize the structure of atoms, the locations of electrons are more accurately represented by orbitals rather than by concentric circles. An orbital is the space in which an electron is likely to be found 90% of the time. Whereas the first shell consists of a single spherical orbital containing up to two electrons, the second shell has four orbitals each containing up to two electrons, for a maximum of eight electrons in that shell. Shown here are the four orbitals (one spherical and three in the shape of dumbbells) of the second shell.

molecule, subscripts are used to denote the precise number of atoms. For example, sucrose, table sugar, has the molecular formula $C_{12}H_{22}O_{11}$. A molecule of sucrose thus contains 12 atoms of carbon, 22 atoms of hydrogen, and 11 atoms of oxygen. Numbers are placed in front of the molecular formula when more than one molecule is present. For example, three molecules of sucrose are described by the formula $3C_{12}H_{22}O_{11}$.

Chemical Bonds

The atoms in a molecule are held together by chemical bonds. These bonds come in three types—ionic, covalent, and the weaker hydrogen bonds. In the first two kinds of bonds, the atoms behave as if they are trying to complete their "shells," which are the specific atomic areas in which electrons are most likely to be found. Shells are usually designated by circles in depictions of atoms (Figure 2–6). The innermost shell can hold up to two electrons and the next shell out can hold up to eight. Atoms with more than ten electrons have additional shells. The outermost shell is called the **valence shell**. It is the number of electrons in the valence shell that determines the type of chemical bond that forms between atoms.

Ionic Bonds

An **ionic bond** results from the mutual attraction of oppositely charged ions. An **ion** is an atom or group of atoms that carries an electric charge. Electric charges result from the transfer of electrons between atoms. An atom that loses an electron becomes positively charged and an atom that gains an electron becomes negatively charged. Oppositely charged ions are attracted to one another and an ionic bond results.

Atoms tend to react so as to fill their valence shells. Consider, again, the atoms of sodium and chlorine that join to form sodium chloride (Figure 2–7). Whereas an atom of sodium has one electron in its outer shell, an atom of chlorine has seven electrons in its outer shell. A molecule of sodium chloride is formed when the sodium atom transfers its single outer shell electron to the chlorine atom. The sodium atom, having lost an electron, becomes positively charged (Na^+) and the chlorine atom, having gained an electron, becomes negatively charged (Cl^-). These oppositely charged ions are attracted to one another and an ionic bond forms.

Covalent Bonds

Whereas ionic bonds form as a result of the transfer of outer shell electrons between atoms, **covalent bonds** form when outer shell electrons are shared between atoms. Consider the compound methane (CH_4), also known as "marsh gas," formed by the sharing of electrons between one atom of carbon and four atoms of hydrogen (Figure 2–8). Although the outermost shell of a carbon atom can hold eight electrons, it contains only four. Carbon can thus fill its valence shell by binding to four atoms of hydrogen, each with one electron in its outer shell. What does this sharing arrangement do for the hydrogen atoms? The outermost shell of a hydrogen atom can hold two electrons, but only has one. Thus, by covalently bonding to carbon with four outer shell electrons, each hydrogen atom can fill its outer shell. We see, then, that the covalent bonds between the carbon and hydrogens

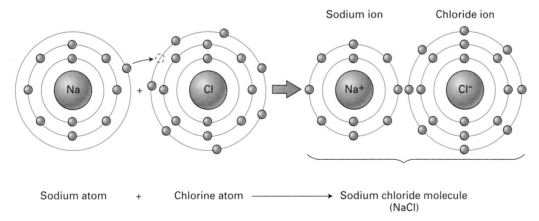

Sodium ion   Chloride ion

Sodium atom   +   Chlorine atom ⟶ Sodium chloride molecule
                                              (NaCl)

Figure 2–7   An ionic bond is formed when an atom of sodium transfers its outer shell electron to an atom of chlorine. Having given up an electron, sodium becomes positively charged, and having received an electron, chlorine becomes negatively charged. Atoms that have lost or gained electrons have electric charges and are called ions. The attraction between oppositely charged ions constitutes the ionic bond.

(a)

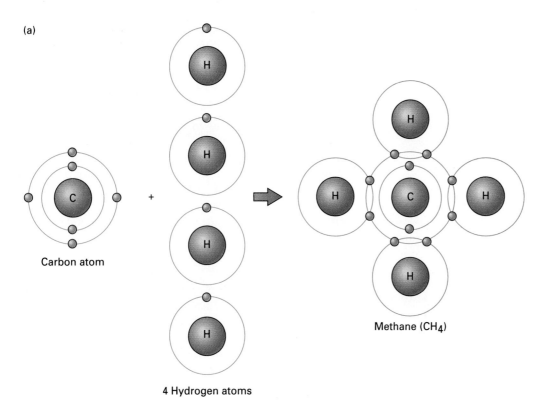

Carbon atom

4 Hydrogen atoms

Methane (CH₄)

Figure 2–8 **Figure 2–8** Covalent bonding in methane. (a) The molecule methane (CH₄) is formed by the sharing of electrons between one carbon atom and four hydrogen atoms. Note that this arrangement, referred to as covalent bonding, results in filled valence (outermost) shells for all five atoms (each of the four hydrogen atoms has two electrons in its shell and the carbon atom has eight electrons in its outer shell). (b) Two diagrams depicting the methane molecule. The diagram on the left, called the structural formula, shows that the bonds between the carbon atom and each hydrogen atom are single (one pair of electrons is shared in each case). The diagram on the right, referred to as the Lewis dot formula, shows the number and arrangement of electron pairs that are shared.

(b)

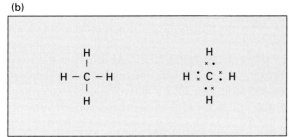

of methane result in filled valence shells for all atoms involved.

In methane, a single pair of electrons is shared between the carbon atom and each of the four hydrogen atoms (Figure 2–8). The methane molecule thus contains four single covalent bonds. Sometimes, however, atoms share two or three pairs of electrons and these bonds are called double or triple covalent bonds, respectively. A double covalent bond is found, for example, between the carbon atoms of ethene (C₂H₄; Figure 2–9). Ethene, also known as ethylene, is released by fruits and vegetables as they ripen and promotes further ripening.

The sharing of electrons between atoms may be equal (nonpolar covalent bond) or unequal (polar covalent bond).

Hydrogen Bonds

When the sharing of electrons during covalent bonding is unequal, different parts of the same molecule bear opposite charges. In water (H₂O), for example, the electrons shared by oxygen and hydrogen spend more time circling around the oxygen atom than they do around the hydrogen atom. As a result, the oxygen atom has a slightly negative charge and each hydrogen atom a slightly positive charge. Because of their slight positive charge, the hydrogen atoms of one wa-

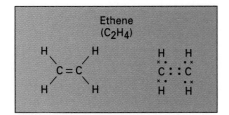

Figure 2–9 The molecule ethene contains a double covalent bond between its carbon atoms. In double bonds, two pairs of electrons are shared. Each carbon is also covalently bonded to two hydrogen atoms; these bonds are single because only one pair of electrons is shared between the atoms.

(a)

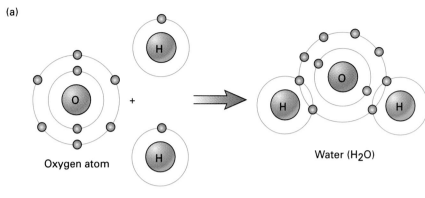

Oxygen atom

2 Hydrogen atoms

Water (H₂O)

Figure 2–10 Hydrogen bonds of water. (a) Water is formed when an oxygen atom covalently bonds (shares electrons) with two hydrogen atoms. (b) The sharing of electrons is unequal, however, and as a result, oxygen carries a slight negative charge and the hydrogens a slight positive charge. The hydrogen atoms from one water molecule are attracted to the oxygen atoms of other water molecules, and this relatively weak attraction (shown by dashed lines) is called a hydrogen bond.

(b)

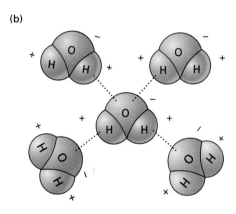

ter molecule are attracted to the negatively charged oxygen atoms of other water molecules. This attraction, called a **hydrogen bond**, is weaker than that which occurs in either ionic or covalent bonds, and for this reason it is illustrated by a dashed rather than solid line (Figure 2–10). Hydrogen bonds, formed between a positively charged hydrogen atom and a negatively charged atom nearby (oxygen in the case of water, but sometimes atoms of elements such as nitrogen, chlorine, or iodine), are important not only in water but also in maintaining the shape of proteins and our hereditary material, DNA. Table 2–2 summarizes the basic attributes of ionic, covalent, and hydrogen bonds.

Table 2–2   Chemical Bonds

Type	Basis for Attraction	Strength	Example
Ionic	Transfer of electrons between atoms creates oppositely charged ions that are attracted to one another.	Strong	NaCl (table salt)
Covalent	Sharing of electrons between atoms; the sharing between atoms may be equal (nonpolar covalent bonds) or unequal (polar covalent bonds).	Stronger than ionic	CH₄ (methane)
Hydrogen	The attraction is between a partially positively charged hydrogen atom in a molecule and some partially negatively charged atom in another molecule, or another region of the same molecule.	Weak	Between a hydrogen atom on one water molecule and an oxygen atom on another

Water, Acids, Bases, and the pH Scale

Water is such an integral part of our everyday lives that we often overlook its unusual qualities that include cohesiveness, a high heat of vaporization, and ability to act as a dissolving agent. As it turns out, it is precisely these unusual qualities, many of which can be traced to the presence of hydrogen bonds, that make water a critical component of the human body.

The Remarkable Water Molecule

Although weak in comparison to either ionic or covalent bonds, hydrogen bonds cause water molecules to stick together. When water is a liquid, the hydrogen bonds are very fragile, frequently forming, breaking, and re-forming. Also related to the polarity of the water molecule is the ease with which chemical substances such as the nutrients, gases, and wastes carried by the vessels of the circulatory and lymphatic systems dissolve in water. Water's positive and negative aspects enable it to interact with a variety of other charged molecules causing them to separate as the water moves between them.

Another important attribute of water is its high heat capacity, which simply means that a great deal of heat is required to raise its temperature. About 67% of the human body is water. (To put this figure in perspective, if a person weighs 150 pounds, then water makes up about 100 pounds of the body weight). Because humans, as well as other organisms, are made up largely of water, they are well-suited to resist changes in body temperature and to keep a relatively stable internal environment.

Another attribute of water that has important consequences for human health is its high heat of vaporization. Because a great deal of heat is required to make water evaporate (to change it from a liquid to a gas), water molecules that evaporate from a surface carry away a lot of heat when they leave, thus cooling the surface. We rely on the evaporation of water in sweat, for example, to cool the body surface and prevent overheating. Water, then, permits many of the processes of life because of its fragility and its stability.

Acids, Bases, Buffers, and the pH Scale

What happens when an individual water molecule breaks up? A molecule of water can dissociate (break up) into a positively charged hydrogen ion (H^+) and a negatively charged hydroxide ion (OH^-):

$$H-O-H \rightleftharpoons H^+ + OH^-$$

water   hydrogen   hydroxide
               ion   ion

This dissociation is not easily accomplished, and so we find that in the human body, water molecules are much more common than H^+ and OH^-. In fact, the amount of H^+ in the body must be precisely regulated because even slight changes in its concentration can be disastrous, disrupting chemical reactions within cells.

Acids and bases may be defined by what happens when each is added to water. To illustrate, an **acid** is anything that donates hydrogen ions (H^+) when placed in water and a **base** is anything that produces hydroxide ions (OH^-) when placed in water. Hydrochloric acid (HCl), for example, dissociates in water to produce hydrogen ions (H^+) and chloride ions (Cl^-). Because HCl has increased the concentration of H^+ in solution, it is classified as an acid. Sodium hydroxide (NaOH), on the other hand, dissociates in water to produce sodium ions (Na^+) and hydroxide ions (OH^-). Because NaOH has increased the concentration of OH^- in solution, it is classified as a base. The OH^- produced when NaOH dissociates reacts with H^+ to form water molecules and hence reduces the concentration of H^+ in solution. Thus, whereas acids increase the concentration of H^+ in solution, bases decrease the concentration of H^+ in solution.

Often we want to know more than simply whether a substance is an acid or a base. For example, how strong an acid is HCl? Asked in a different way, how acidic is the solution created when HCl is added to water? Questions such as these can be answered using a system of measurement called the **pH scale** (Figure 2–11). This scale ranges from 0 to 14, with a pH of 7 being neutral, a pH of less than 7 being

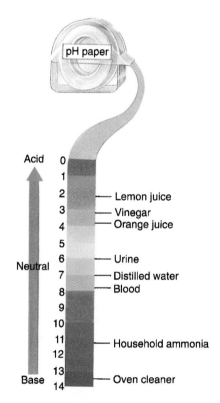

Figure 2–11   The pH scale and the pH of commonly encountered substances.

Table 2–3 Acids and Bases Compared

Characteristic	Acid	Base
Taste*	Sour	Bitter
Feel*	Not applicable	Slippery
Dissociation in water	Release H^+	Release OH^-
pH	Less than 7	Greater than 7
Example	HCl (hydrochloric acid)	NaOH (sodium hydroxide)

*Tasting or feeling an unknown substance to determine whether it is an acid or base is dangerous and should not be done.

acidic, and a pH of greater than 7, basic. According to the pH scale, then, the lower the pH, the greater the acidity or concentration of H^+ in a solution. Each lower pH unit has ten times the amount of H^+ than the preceding unit, so a solution with a pH of 4 is ten times more acidic than a solution with a pH of 5.

Hydrochloric acid is considered a strong acid and sodium hydroxide a strong base because they dissociate completely in water. Substances that dissociate reversibly in water are considered weak acids and bases. Ammonia (NH_3), for example, is a weak base because although it binds H^+ producing ammonium ions (NH_4^+), the ammonium ion also releases some H^+, forming NH_3 and H^+ again. Thus, once the reaction reaches equilibrium, there will be some NH_4^+ and some NH_3 and H^+ in solution, rather than all NH_4^+. Some of the characteristics of acids and bases, including their values on the pH scale, are summarized in Table 2–3.

Typically, biological systems function within a narrow range of pH values. Dramatic changes in pH are prevented by substances called **buffers** that remove excess H^+ from solution when concentrations increase and add H^+ when concentrations decrease. For example, an important buffering system that keeps the pH of blood at about 7.4 is the carbonic acid (H_2CO_3)/bicarbonate (HCO_3^-) buffering system. When added to water, carbonic acid dissociates into hydrogen ions and bicarbonate:

$$\underset{\text{carbonic acid}}{H_2CO_3} \quad \rightleftharpoons \quad \underset{\text{hydrogen ion}}{H^+} \quad + \quad \underset{\text{bicarbonate}}{HCO_3^-}$$

The buffering action of carbonic acid and bicarbonate results from the fact that when levels of H^+ decrease in the blood (causing an increase in pH), carbonic acid dissociates, adding H^+ to solution, but when levels of H^+ increase in blood (causing a decrease in pH), the H^+ combines with bicarbonate and are thus removed from solution. Such action is essential because even slight changes in the pH of blood—

say, a drop from 7.4 to 7.0 or an increase to 7.8—can cause death in a few minutes. The critical link between pH and life is also illustrated by the impact of acid rain on our environment and health (Environmental Concerns).

In the human body, almost all biochemical reactions occur around pH 7, kept there by powerful buffering systems. An important exception occurs in the stomach where pH values from about 1 to 3 are found. Hydrochloric acid (HCl), secreted by cells that line the stomach, promotes the breakdown of proteins by gastric juices. Although normal digestion requires an acid stomach, sometimes excess acid is produced and "acid indigestion" is the uncomfortable result. The discomfort of acid indigestion can be treated by taking an antacid such as Alka-Seltzer, Rolaids, or Tums. These products consist of weak bases that relieve the pain of excess stomach acid by neutralizing some of the hydrochloric acid.

Biological Molecules

Most of the molecules we have discussed so far have been rather small and simple. However, many of the molecules of life are enormous by comparison, with complex architecture. Some proteins, for example, are made up of thousands of atoms linked together in a chain that repeatedly coils and folds upon itself. The giant molecules of life are termed **macromolecules**.

Macromolecules that consist of many small molecules linked together in a chain-like fashion are called **polymers**. The small molecules or subunits that form the building blocks of the polymer are called **monomers**. We might think of a polymer as a pearl necklace, with each pearl being a monomer. A protein, as we shall see, is a polymer or chain of amino acids linked together. And glycogen, the storage form of carbohydrates in animals, is a polymer of glucose molecules linked together.

What's Happening to the Rain?

Statues that have withstood not only the critics but the ravages of time are among our vast heritage. Some have stood for centuries gazing over city streets and sidewalks at the generations of passersby. But now the statues have had to be moved, and many have been virtually destroyed (Figure A). Rain has eaten away the features—rain that has in recent years become a solution of acid.

Acid rain, usually defined as rain with a pH lower than 5.6—the pH of pure rainwater—is caused largely by the burning of fossil fuels in cars and factories. The sulfur and nitrogen oxides produced by these activities react with water in the atmosphere to form sulfuric and nitric acids (H_2SO_4 and HNO_3, respectively), which fall to earth as rain or snow with pH values sometimes as low as 1.5.

The effects of acid rain on the environment have been devastating (Figure B). On land, acid rain has been linked to damage to statues and the decline of

(a)

(b)

Figure A Acid rain, caused by the release into the atmosphere of nitrogen and sulfur oxides during the burning of fossil fuels, is slowly eating away our works of art (a and b). The damage to this statue on the Lincoln Cathedral in England took place over a period of about 70 years. *(Dean and Chapter of Lincoln)*

Polymers can form through **dehydration synthesis**. In this process, the monomers are linked together through the removal of a water molecule: One monomer donates OH and the other donates H. With the removal of water, the two monomers covalently bond to one another. By the same token, polymers can be broken apart by **hydrolysis**, the addition of water across the covalent bonds. The H from the water molecule attaches to one monomer and the OH to the adjoining monomer, thus breaking the covalent bond between the two. Hydrolysis plays a critical role, for example, in the process of digestion. Because most foods consist of polymers too large to pass from our digestive tract into the

bloodstream and on to our cells, the polymers are first hydrolyzed into their component monomers that can then be absorbed into the bloodstream for transport throughout the body. Dehydration synthesis and hydrolysis are summarized in Figure 2–12.

Figure 2–12 Formation and breaking apart of polymers. (a) Polymers are formed by dehydration synthesis, the removal of a water molecule and joining together of two monomers. (b) Polymers are broken apart by hydrolysis, the addition of a water molecule and breaking of bonds between monomers. ▶

forests and songbirds, and in aquatic environments, to the decrease and sometimes total elimination of populations of fish and amphibians. In most instances, the destruction appears to be caused by subtle changes in soil or water chemistry rather than by the direct "burning" of vegetation or animals. Trees, for example, become stressed and more susceptible to disease when nutrient uptake is disrupted by the increased acidity of soils. High soil acidity has also been linked to the increased incidence of reproductive failure among forest birds. Acid deposition apparently decreases the calcium content of soils causing decreases in the populations of snails upon which the birds rely for dietary calcium. With fewer snails in their diet, the birds are deficient in calcium and produce eggs that break or dry out because they are too thin and porous.

Like the case for forest birds, the harmful effects of acid rain on the human body appear linked to its impact on the element calcium. Apparently, the sulfur dioxide found in acid rain absorbs some of the rays of the sun that normally cause us to produce vitamin D. A deficiency in vitamin D seems to interfere with the metabolism of calcium, causing cells lining the colon to proliferate at an abnormal rate, producing cancer of the colon. A similar mechanism may link acid rain to breast cancer. Although researchers have suspected for some time that acid rain and other forms of air pollution contribute to poor respiratory health and lung cancer, these new data linking air pollution to cancer of the colon and breast demonstrate that the connection between harming our environment and harming our health may be somewhat less obvious, but no less deadly.

Because most acid rain is caused by human activity (a small percentage of the pollutants that cause acid rain are released into the atmosphere during volcanic eruptions and other natural processes), it is within our power to reduce, if not eliminate, the problem. Tighter controls on industrial emissions and the installment of antipollution devices in our cars could reverse many of the devastating effects of this new threat to life called acid rain.

(a)

(b)

Figure B Acid rain has also destroyed parts of our forests (a), reduced or eliminated populations of aquatic organisms in some of our lakes, and led to reproductive failure in forest birds such as the great tit (b).
(a, © 1996 Bill Lea/Dembinsky Photo Associates; b, Roger Wilmshurst/Dembinsky Photo Associates)

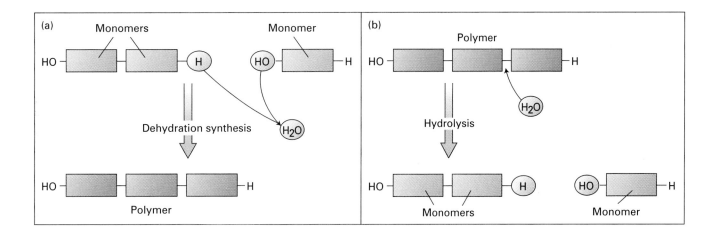

Table 2–4 Common Carbohydrates

Carbohydrate	Molecular Formula	Source	Component Monosaccharides
Monosaccharide			
Glucose	$C_6H_{12}O_6$	Blood, fruit, honey	—
Fructose	$C_6H_{12}O_6$	Fruit, honey	—
Galactose	$C_6H_{12}O_6$	From break-up of lactose (milk sugar)	—
Disaccharide			
Sucrose	$C_{12}H_{22}O_{11}$	Sugar cane, maple syrup	Glucose and fructose
Maltose	$C_{12}H_{22}O_{11}$	From break-up of starch; ingredient in beer	Glucose
Lactose	$C_{12}H_{22}O_{11}$	Component of milk	Glucose and galactose
Polysaccharide*			
Starch	—	Potatoes, corn, some grains	Glucose
Glycogen	—	Stored in muscle and liver cells	Glucose
Cellulose	—	Cell walls of plants	Glucose
Chitin	—	Exoskeletons of arthropods	Glucose

*These complex carbohydrates consist of chains containing hundreds of glucose molecules joined to each
other in long strings.

Carbohydrates

Carbohydrates are organic molecules (molecules that contain the element carbon) that provide fuel for the human body. From the standpoint of chemical composition, carbohydrates are made up entirely of carbon, hydrogen, and oxygen, having twice as many hydrogen atoms as oxygen atoms. These molecules, which we know as sugars and starches, can be more formally classified by size into the monosaccharides, disaccharides, and polysaccharides. Some common carbohydrates are described in Table 2–4.

Monosaccharides

Monosaccharides, also called simple sugars, are the smallest molecular units of carbohydrates. Containing from three to seven carbon atoms, these molecules readily dissolve in water where they form ring-like structures. Monosaccha-

rides can be classified according to the number of carbon atoms they contain; thus, five carbon sugars are pentoses, six carbon sugars are hexoses, and so on. Monosaccharides that contain the same number of carbons can, however, differ in structure. Glucose, galactose, and fructose, for example, are six carbon sugars with the molecular formula $C_6H_{12}O_6$ but slightly different structures (Figure 2–13). Glucose, formed by plants during photosynthesis, is a main source of fuel for cells.

Disaccharides

Disaccharides or double sugars are formed when two monosaccharides covalently bond to each other through dehydration synthesis. The disaccharide sucrose (table sugar) is formed by the joining together of the two monosaccharides glucose and fructose (Figure 2–14). Similarly, the di-

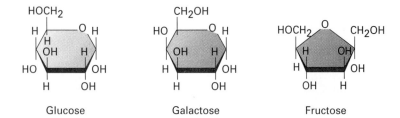

Figure 2–13 Three hexose sugars—glucose, galactose, and fructose—are examples of monosaccharides.

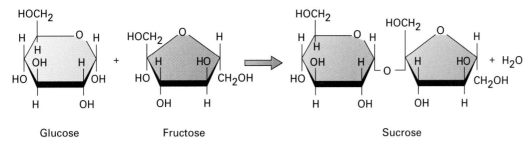

Glucose   Fructose   Sucrose

Figure 2-14 Disaccharides are formed by the joining together of two monosaccharides. Here, a molecule of glucose and one of fructose combine to form sucrose.

saccharide maltose, an important ingredient of beer, is formed by the joining together of two glucose molecules. A final example of a disaccharide is lactose, the principal carbohydrate of milk and milk products, formed by the joining together of glucose and galactose. Although infants and children usually produce adequate amounts of lactase, the substance needed to break lactose into its component parts for digestion and use by the body, many adults do not. For lactase-deficient adults, consumption of milk and milk products can lead to diarrhea caused by the passage of undigested lactose into the intestines where it undergoes chemical reactions to produce gases and lactic acid that irritate the bowels. The milk industry has responded to the problems associated with digesting the lactose in milk by producing Lactaid, a product with about 70% less of the sometimes problematic sugar.

Polysaccharides

Polysaccharides are complex carbohydrates formed when large numbers of monosaccharides (most commonly glucose) join together to form long chains through dehydration synthesis. Most polysaccharides serve to store energy or to give structure. In plants, the storage polysaccharide is starch, and in animals, glycogen. Humans store glycogen mainly in the cells of the liver and muscles where it can be broken down to release energy-laden glucose molecules.

Two structural polysaccharides are **cellulose**, found in the cell walls of plants, and **chitin**, found in the exoskeletons (hard outer coverings) of animals such as insects, spiders, and crustaceans (Figure 2–15). Humans lack the enzymes necessary to digest cellulose, and thus it passes unchanged through our digestive tract. Although this polysaccharide has no value as a nutrient, it is an important form of dietary fiber, known to facilitate the passage of feces through the large intestines, possibly reducing the incidence of colon cancer. Like humans, grazing animals such as cows and sheep do not produce the enzymes necessary to digest cellulose. Unlike humans, however, these animals harbor bacteria and other microorganisms in their digestive tracts that can digest cellulose, thereby freeing glucose molecules for use by the host animal.

In recent years, scientists have begun to exploit chitin, the second most common organic compound found in nature (cellulose being the most common), using it, or a chemically modified version of it called chitosan, to make bandages, burn dressings, cosmetics, and food additives. Also, because of its chemical properties, chitosan is an effective clumping agent with the potential for use in removal of suspended particles from waste water and drinking water. Much of the recent interest in chitin as a raw material comes from the fact that it is a "biologically friendly" substance; chitin is nontoxic, degrades naturally in the environment, and is a renewable resource that is generated as a waste product when animals such as lobsters, shrimp, and crabs are processed for eating. By recycling the chitin from crustaceans, we can reduce the amount of space taken up at landfill sites by the waste exoskeletons.

Lipids

Lipids have an important trait in common—they don't dissolve in water. This is because lipids are nonpolar (bearing no charges) and water is polar. (Recall that the electrons shared by the oxygen and hydrogen atoms of water molecules are shared unequally, resulting in the oxygens having a slightly negative charge and the hydrogens a slightly positive charge.) Because of this difference, water shows no attraction for lipids and vice versa and so they don't interact and push each other apart. Three types of lipids that are important to human health are the triglycerides, phospholipids, and steroids.

Triglycerides

Triglycerides are molecules comprised of one molecule of glycerol and three fatty acids. They are known to us as fats

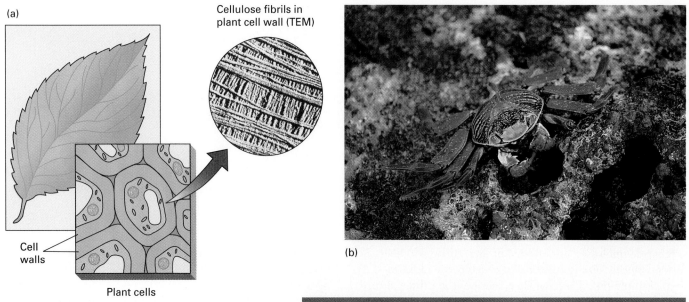

(a)

Cellulose fibrils in plant cell wall (TEM)

Cell walls

Plant cells

(b)

(c)

Figure 2–15 Sources of the two structural polysaccharides cellulose and chitin. (a) Cellulose is found in the cell walls of plants. Chitin is found in the exoskeletons of crustaceans such as crabs (b) and insects such as cicadas (c). The cicada shown here is in the process of shedding its old exoskeleton. *(a, photo, Biophoto Associates; b, Barbara Gerlach/Visuals Unlimited; c, © 1992 Skip Moody/Dembinsky Photo Associates)*

when solid and oils when liquid. The fatty acids bond to glycerol through dehydration synthesis (Figure 2–16). Triglycerides can be described as saturated, unsaturated, or polyunsaturated, depending on the number of double bonds that exist between the carbon atoms of their component fatty acids. Fatty acids are saturated when there are no double bonds linking the carbon atoms; such fatty acids are, in effect, "saturated with hydrogen" because the carbon atoms are bonded to as many hydrogen atoms as possible. Fatty acids that contain one double bond within their carbon skeleton are unsaturated and those containing two or more double bonds are polyunsaturated. Basically, these fatty acids are "not saturated with hydrogen" because they could bond to more hydrogen atoms if the double bonds between their carbon atoms were broken.

The distinction between saturated and unsaturated fatty acids helps to explain why, at room temperature, some triglycerides are solid (fats) and others liquid (oils). Triglycerides with saturated fatty acids are solid at room temperature because the absence of double bonds allows the fatty acids to straighten and the molecules to pack tightly together. In contrast, presence of one or more double bonds causes bending in the fatty acids that prevents the tight packing of molecules required for a solid, and so the triglyceride is a liquid (see Figure 2–16). Generally, the triglycerides of animals are saturated and therefore solid at room temperature (such as butter), while those of plants are unsaturated and liquid (such as corn oil and olive oil).

Fats and oils are rich sources of energy, providing about twice as much energy per gram as carbohydrates or proteins.

The high energy density of fat makes it an ideal way for the body to store energy in the long term. Our bulk would be much greater, indeed, if over the long run, we stored excess energy as carbohydrates or proteins, given their relatively low energy yield compared with fat. (The carbohydrate glycogen, described earlier, functions in short-term rather than long-term energy storage, basically getting us from one meal to the next rather than from one day to the next.)

Let's consider the role of fat in long-term energy storage. First, we find that excess triglycerides, carbohydrates, and proteins from the foods we consume are converted into small globules of fat that are deposited in the cells of adipose tissue. Here, the fat remains until broken down by cells to release the energy needed to keep vital processes going, even if we skip a meal or two. In addition to its role in long-term energy storage, fat also serves a protective function in the body. Thin layers of fat surround major organs like the kidneys, cushioning them against physical shock from falls or blows. Fat also serves as insulation. Despite the importance of fats and oils to human health, in excess they can be very dangerous, particularly to our circulatory system.

Phospholipids

Phospholipids are composed of a molecule of glycerol bonded to two fatty acids and a negatively charged phosphate group; other small molecules, usually polar and designated by the letter R, are linked to the phosphate. This chemical composition results in phospholipids having two regions with very different characteristics (Figure 2–17). One region, made up of the fatty acids, is nonpolar and forms the hydrophobic or "water-hating" tail. The other region, made up of the R group, glycerol, and phosphate, is polar (due to the R group) and forms the hydrophilic or "water-loving" head. The tails, being hydrophobic, are excluded from water whereas the heads, being hydrophilic, interact with water. The regional separation of phospholipids into hydrophilic heads and hydrophobic tails is important in

(a)

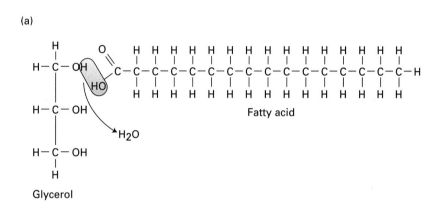

Glycerol

Figure 2–16 Triglycerides are composed of a molecule of glycerol joined to three fatty acids. (a) Each fatty acid bonds to glycerol through dehydration synthesis. (b) This triglyceride contains two saturated fatty acids (note the absence of any double bonds between the carbon atoms) and one unsaturated fatty acid (note the presence of one double bond between the carbon atoms).

(b)

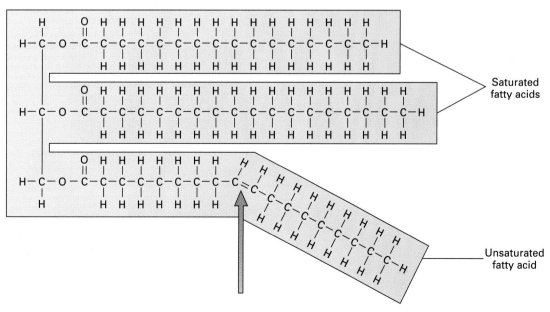

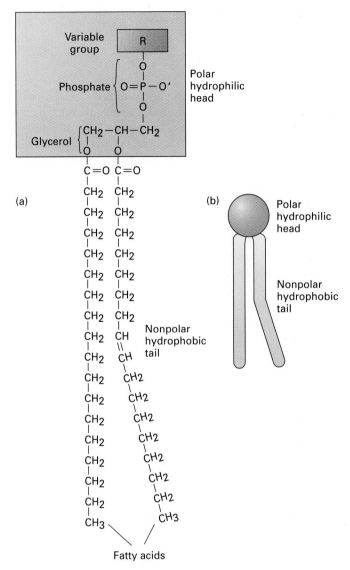

Figure 2-17 Structure of a phospholipid. (a) Phospholipids consist of glycerol, two fatty acids, a phosphate, and a variable group designated by the symbol R. (b) Because the R group is usually polar and the fatty acids nonpolar, phospholipids have a polar hydrophilic head and a nonpolar hydrophobic tail.

the structure of cell membranes. At the surface of cells, phospholipids are arranged in a double layer, called a bilayer (Figure 2–18). In this arrangement, the hydrophilic heads point outward, contacting extracellular fluid (the aqueous solution outside the cell) and cytosol (the aqueous solution inside the cell). The hydrophobic tails, on the other hand, point inward and help to hold the membrane together.

Steroids

In contrast to phospholipids and triglycerides that are made of either two or three fatty acids attached to glycerol, **steroids** consist of four carbon rings attached to functional

groups, molecules that vary from one steroid to the next. Cholesterol, one of the most familiar steroids, is shown in Figure 2–19. This steroid is a component of the plasma membrane and the foundation from which other steroid hormones, such as testosterone and estrogen, are made. High levels of cholesterol in the blood have been linked to elevated blood pressure and heart disease.

Cholesterol, Saturated Fats, and Heart Disease

The disease **atherosclerosis** is caused by the build-up of deposits of cholesterol in the walls of the arteries. These deposits, called **plaques**, cause the vessels to stiffen and thicken, eventually restricting blood flow. It is the restriction of blood flow that ultimately leads to high blood pressure, heart attacks, and strokes.

How can we reduce the levels of cholesterol in our blood? Although most of the cholesterol in our blood is synthesized by the liver, some comes from our diet. Thus, one step that we can take to reduce the blood levels is to lower our dietary intake of cholesterol-rich foods such as egg yolks, red meat, and butter. These foods can be replaced with low cholesterol animal products such as yogurt and fat-free milk.

Can we influence the amount of cholesterol produced by our own livers? Several factors, not all of which we can control, influence the level of cholesterol produced from

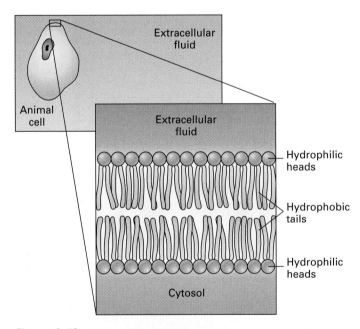

Figure 2-18 Phospholipids are important components of plasma membranes, the outermost boundary of cells that separates the internal aqueous environment (cytosol) from the external aqueous environment (extracellular fluid). Within the phospholipid bilayer, the hydrophobic tails point inward and help hold the membrane together. The outward-pointing hydrophilic heads mix with the aqueous environments inside and outside the cell.

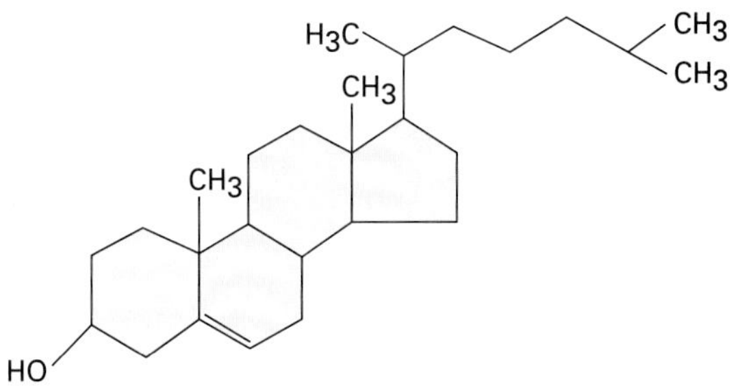

Figure 2-19 The steroid cholesterol is a component of cell membranes and the substance from which other steroids are made. All steroids have a structure consisting of four carbon rings, but they differ in the groups attached to these rings.

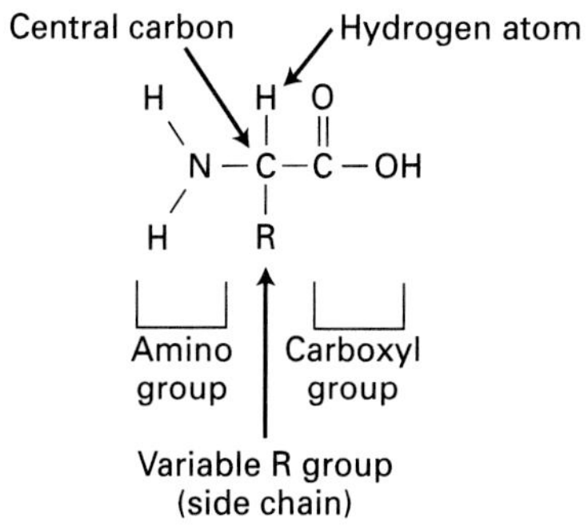

Figure 2-20 Structure of an amino acid. Amino acids differ from one another in the type of R group they contain.

saturated fats by our liver. Genetics, exercise, stress, and diet all seem to influence the body's production of cholesterol. Diet is particularly important because the liver makes cholesterol from the saturated fat that we eat. As described above, an often recommended step toward reducing levels of cholesterol in the blood is to reduce consumption of animal fat, particularly that associated with red meat. Such fat, while adding taste to the meat, poses the double risk of (1) being, itself, high in cholesterol and (2) providing the liver with the saturated fats needed for the manufacture of more cholesterol.

Critical Thinking

The phrase "partially hydrogenated" appears on the labels of foods such as margarine and vegetable oils. What does this phrase indicate has been done to the foods during processing and what impact might the change have on our health?

Proteins

Proteins are composed of one or more chains of amino acids. Within the chains, the amino acids are held together by peptide bonds. Thousands upon thousands of proteins are found in the human body. Protein structure, as we will see, can be quite elaborate, consisting of chains that twist, turn, and fold upon themselves. The functions of proteins

include structural support (as we see in collagen that comprises connective tissues and is a major component of bone), transport (as in hemoglobin molecules that transport oxygen in the blood), movement (such as in the actin and myosin filaments that move muscles), and regulation of chemical reactions (as when enzymes speed up reactions). Despite such diversity in structure and function, all proteins are polymers made from a set of only 20 monomers, the amino acids.

Amino Acids

Amino acids, the building blocks of proteins, consist of a central carbon atom (called the alpha carbon) bound to a hydrogen atom (H), an amino group (NH_2), a carboxyl group (COOH), and a side chain designated by the letter R (Figure 2-20). There are twenty different amino acids important to human life; these amino acids differ from each other in their side chains. Some amino acids can be synthesized by our bodies (nonessential amino acids), whereas others cannot be synthesized and must be obtained from the foods we eat (essential amino acids).

Peptides, Polypeptides, and Proteins

The monomers that form proteins, like the monomers that form carbohydrates and lipids, are linked together through dehydration synthesis. Basically, a bond, called a peptide bond, forms at the site of dehydration synthesis, linking the carboxyl group (COOH) of one amino acid to the amino group (NH_2) of the adjacent amino acid (Figure 2-21). Chains containing only a few amino acids are called

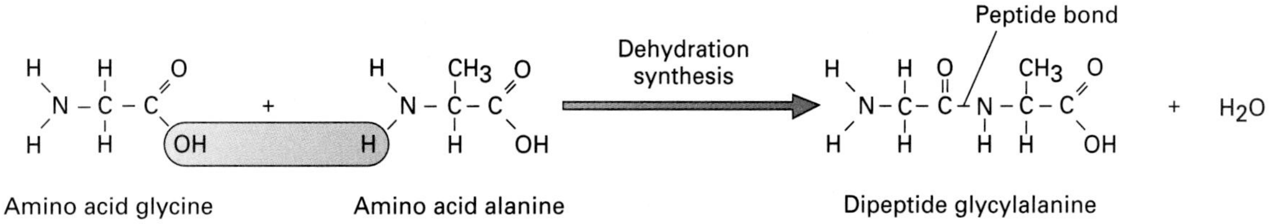

Figure 2-21 Formation of a peptide bond between two amino acids through dehydration synthesis. The carboxyl group (COOH) of one amino acid bonds to the amino group (NH_2) of the adjacent amino acid, releasing water.

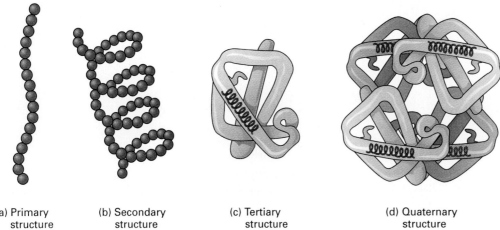

(a) Primary structure   (b) Secondary structure   (c) Tertiary structure   (d) Quaternary structure

Figure 2–22   Levels of protein structure. (a) Primary structure is the specific sequence of amino acids (each amino acid is depicted here as a bead within the polypeptide chain). (b) Secondary structure, such as the spiral shown here, results from the bending and coiling of the chain of amino acids. (c) This protein has bent back into itself and compacted further, forming a globular protein. Other proteins (not shown) are made of parallel strands of protein molecules entwined around one another and are called fibrous proteins. Tertiary structure is the three-dimensional shape of globular or fibrous proteins. (d) Some proteins have two or more polypeptide chains, each chain forming a subunit. Quaternary structure results from the attractive forces between subunits. The quaternary structure depicted here consists of four globular subunits held together by attractive forces.

peptides (dipeptides contain two amino acids, tripeptides contain three amino acids, and so on), and those containing ten or more amino acids are called **polypeptides**. By convention, the term **protein** is used to describe polypeptides with molecular weights greater than about 10,000 amu, which typically means a chain of fifty or more amino acids.

Levels of Protein Structure

Four levels of protein structure are recognized: primary, secondary, tertiary, and quaternary. The **primary structure** of a protein is defined by its precise sequence of amino acids. This sequence, determined by the genes, dictates a protein's structure and function. Even slight changes in the primary structure can alter the shape of a protein and its ability to function. The disease sickle-cell anemia, an inherited blood disorder that primarily affects populations in central Africa as well as about 1 in 400 African Americans, results from the substitution of one amino acid for another during synthesis of hemoglobin, the protein that carries oxygen in our red blood cells. This single mistake creates a misshapen hemoglobin molecule that then alters the shape of red blood cells, changing them from flattened disks to sickle-shaped cells. Death can result when these oddly shaped red blood cells clog the tiny vessels of the brain and heart.

The **secondary structure** of proteins is described through the bending and folding of the chain of amino acids. The bending and folding can produce coils, spirals, and pleated sheets. These shapes form as a result of hydrogen

bonding between different parts of the polypeptide chain. All three types of secondary structure may be found at different locations within a given protein. Those proteins whose secondary structures include spirals or coils, but not pleated sheets, may bend back into themselves and compact further into globular proteins. Fibrous proteins, on the other hand, are made of parallel strands of protein molecules that entwine like the fibers of a rope.

Tertiary structure is described as the three-dimensional shape of these globular and fibrous proteins. Hydrogen, ionic, and covalent bonds between different side chains may all contribute to tertiary structure. Changes in the environment of a protein, such as increased heat or changes in pH, can cause the protein to unravel and lose its three-dimensional shape. This process is called **denaturation**. Change in the shape of a protein results in loss of function. Finally, some proteins consist of two or more polypeptide chains, each chain forming a subunit. **Quaternary structure** results from the mutually attractive forces between the subunits. The forces that hold the subunits together are largely the result of oppositely charged side chains. The four levels of protein structure are summarized in Figure 2–22.

Although in many cases scientists have determined both the sequence of amino acids and the three-dimensional shapes of proteins, understanding the precise rules by which proteins fold—just how a certain amino acid sequence leads to a specific three-dimensional shape—has proven elusive. Much of the difficulty stems from the fact that most proteins pass through several intermediate stages on their way to a stable shape. Some of these intermediate steps last for a second, providing only a fleeting glimpse of the folding process. Recently developed techniques in which the intermediates are labeled and then visualized may help us to understand the rules of protein folding and lead to substantial benefits. Researchers in biotechnology industries, intent on making new proteins to carry out specific tasks, have designed genes to direct the synthesis of some proteins. Such efforts often fail, however, when the proteins fail to fold properly.

Enzymes

Most of the chemical reactions within our cells occur far too slowly to sustain life. Life is possible, however, because of enzymes. **Enzymes** are substances—almost always proteins, but sometimes RNA molecules—that speed up chemical reactions without being consumed in the process. Typically, reactions with enzymes proceed 10,000 to 1,000,000 times faster than the same reactions without enzymes. How do enzymes work?

The basic process by which an enzyme accelerates a chemical reaction can be summarized by the following equation:

$$\text{Substrate} \xrightarrow{\text{enzyme}} \text{Product}$$

The particular substance that an enzyme works on is called its substrate. For example, the enzyme sucrase speeds up the reaction in which sucrose is broken down into glucose and fructose. In this reaction, sucrose is the substrate and glucose and fructose are products. Similarly, the enzyme maltase speeds up the breakdown of maltose (the substrate) into molecules of glucose (the product). From these examples you can see that an enzyme's name may resemble the name of its substrate.

During reactions such as those described above, the substrate binds to the enzyme at a specific location on the enzyme known as the **active site**. While bound to one another the enzyme and substrate are known as an **enzyme-substrate complex**. At this time, the substrate is converted to products that then leave the active site. The entire process occurs very rapidly. In fact, one estimate suggests that within one second the typical enzyme converts about 1000 molecules of substrate into product. Figure 2–23 summarizes the steps involved in an enzymatic reaction, in this case the breakdown of maltose to glucose.

Enzymes are very specific, capable of binding to one or at most a few substrates. Maltase, for example, will act only on maltose and not on sucrose, a structurally similar compound (recall that both sucrose and maltose are disaccharides with the molecular formula $C_{12}H_{22}O_{11}$). This type of specificity is due to the unique shape of each enzyme's active site. Basically, the shape of an enzyme's active site corresponds to the shape of its substrate—the two fit together like pieces of a jigsaw puzzle.

The activity of enzymes may be regulated by the end products of chemical reactions. Sometimes the end products bind directly to the active site and prevent the substrate from entering. More typically, however, the end products bind to a site away from the active site, called an **allosteric**

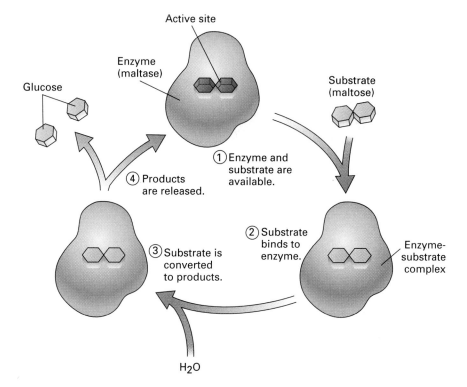

Figure 2–23 The working cycle of an enzyme. Here, the enzyme maltase speeds up the chemical reaction by which the substrate maltose is broken down into molecules of glucose (products) through the addition of water. The cycle begins when the active site of the enzyme is unoccupied and the substrate is present (step 1). The substrate binds to the active site of the enzyme, forming an enzyme-substrate complex (step 2). The substrate is converted to products (step 3). The products are released from the active site of the enzyme and the cycle can begin again (step 4).

site, and influence the activity of the enzyme by causing changes in its shape. Changes in the shape of an enzyme may, for example, make its active site nonfunctional. The activity of some enzymes can be selectively inhibited with certain chemicals. Some of these chemicals resemble the normal substrate and thus compete with it for access to the active site. Such is the case for the antibiotic penicillin. **Antibiotics** are chemicals used either to kill or to inhibit the growth of bacteria that have invaded the body. Bacteria have a somewhat rigid cell wall. Penicillin works by blocking the active site of an enzyme needed by the bacteria to make their cell walls.

Sometimes enzymes need **cofactors**, nonprotein substances that help them convert substrate to product. These cofactors may permanently reside at the active site of the enzyme or may bind to the active site at the same time as the substrate. Some cofactors are inorganic substances like zinc or iron, while others are organic substances such as vitamins. Organic cofactors are called **coenzymes**.

Nucleic Acids

In our discussion of protein structure we mentioned that genes determine primary structure, the sequence of amino acids. Genes, our units of inheritance, are parts of a long polymer called deoxyribonucleic acid (DNA), one of two members of the class of macromolecules known as the nucleic acids.

Nucleotides, DNA, and RNA

The nucleic acid deoxyribonucleic acid (DNA) forms the genes and directs protein synthesis. Ribonucleic acid (RNA), a similar molecule, also plays a major role in protein synthesis. Both DNA and RNA are polymers of smaller units called nucleotides. A **nucleotide** consists of a five-carbon (pentose) sugar bonded to one of five nitrogenous bases and at least one phosphate group (Figure 2–24). The five nitrogenous bases can be separated according to their structure into two families, the **pyrimidines** that contain a six-membered

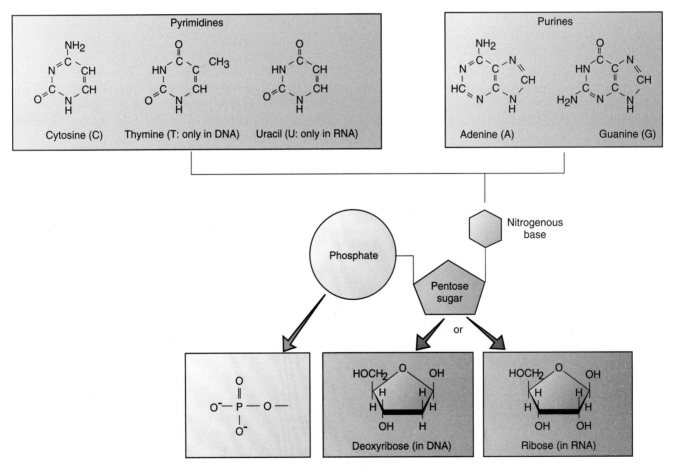

Figure 2–24   Structure of a nucleotide. Nucleotides, the building blocks of nucleic acids, consist of a five-carbon (pentose) sugar bonded to a phosphate molecule and one of five nitrogenous bases. Structurally, the nitrogenous bases fall into two families, the pyrimidines and the purines.

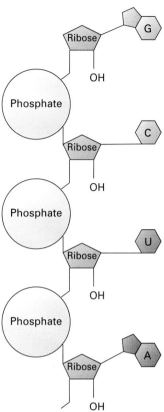

Figure 2–25   RNA is a single-stranded nucleic acid formed by the linking together of nucleotides comprised of the sugar ribose, a phosphate group, and the nitrogenous bases adenine (A), uracil (U), cytosine (C), and guanine (G).

Characteristic	RNA	DNA
Sugar	Ribose	Deoxyribose
Bases	Adenine, guanine, cytosine, uracil	Adenine, guanine, cytosine, thymine
Number of strands	One	Two; twisted to form double helix

ring comprised of carbons and nitrogens and **purines** that contain a five-membered ring attached to a six-membered ring. Nucleic acids, like the other polymers we have discussed, are formed when monomers, in this case nucleotides, join together through dehydration synthesis to form long chain-like molecules.

There are several key differences in the structures of RNA and DNA. RNA is a single strand of nucleotides. The five-carbon sugar in RNA is ribose and the nitrogenous bases are adenine, guanine, cytosine, and uracil (Figure 2–25). In contrast, DNA is a double-stranded chain. The two strands of DNA, held together by hydrogen bonds formed between the nitrogenous bases, twist around one another to form a double helix. The five-carbon sugar in DNA is deoxyribose and the nitrogenous bases include adenine, guanine, cytosine, and thymine (Figure 2–26). The structural differences between RNA and DNA are summarized in Table 2–5.

Adenosine Triphosphate (ATP)

The reason you're able to sit there so attentively, reading this book, is because within your cells, ATP is losing a phosphate. **Adenosine triphosphate** (ATP) is a nucleotide that

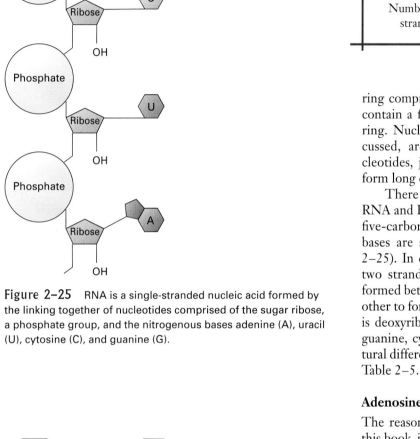

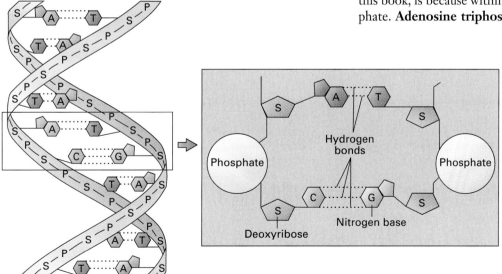

Figure 2–26   DNA is a nucleic acid containing two chains of nucleotides twisted around one another to form a double helix. The two chains are held together by hydrogen bonds between the nitrogenous bases. Note that each nucleotide of DNA contains the pentose sugar deoxyribose, a phosphate group, and one of the following four nitrogenous bases: adenine (A), guanine (G), cytosine (C), and thymine (T).

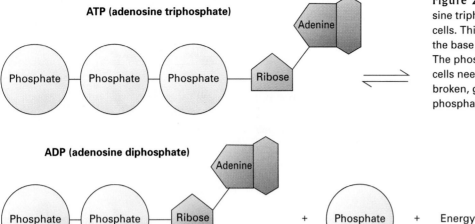

ATP (adenosine triphosphate)

Phosphate — Phosphate — Phosphate — Ribose — Adenine

Figure 2–27 Structure and function of adenosine triphosphate (ATP), the energy currency of cells. This nucleotide consists of the sugar ribose, the base adenine, and three phosphate groups. The phosphate bonds of ATP are unstable. When cells need energy, the last phosphate bond is broken, giving off energy and yielding ADP and a phosphate molecule.

ADP (adenosine diphosphate)

Phosphate — Phosphate — Ribose — Adenine   +   Phosphate   +   Energy

consists of the sugar ribose, the base adenine, and three phosphate groups. ATP is formed during an energy-requiring reaction in which an inorganic phosphate covalently bonds to adenosine diphosphate (ADP). The energy absorbed during the reaction is stored in the phosphate bond of the ATP molecule. The phosphate bonds of ATP molecules are very unstable because of repulsive forces among the three negatively charged phosphate groups. When cells require energy, the phosphate bond is broken, releasing energy and leaving ADP and inorganic phosphate (Figure 2–27). The energy released by the splitting of ATP is then available for other chemical reactions occurring at the same time in the cell.

ATP is often described as the energy currency of cells because all energy from the breakdown of molecules such as glucose must be channeled through ATP before it can be used by the body.

SUMMARY

1. Atoms consist of subatomic particles called protons, neutrons, and electrons. Whereas protons have a positive charge, neutrons are uncharged; both of these subatomic particles are found in the nucleus and have masses equal to one. Electrons, on the other hand, carry a negative charge, weigh almost nothing, and orbit around the nucleus.

2. An element, made of atoms all of which contain the same number of protons, cannot be broken down by ordinary chemical means. The atomic number of an element is the number of protons in one of its atoms, and the atomic weight equals the number of protons plus the number of neutrons. Elements are listed in the periodic table.

3. Isotopes are atoms that have the same number of protons but different numbers of neutrons. Some isotopes, called radioisotopes, decay and emit radiation in the form of energy and subatomic particles. Radiation can cause illness or death to the person receiving the radiation or damage his reproductive cells so that physical or mental defects are only apparent in future generations. Radiation can also be used to diagnose and treat certain illnesses.

4. Atoms combine to form molecules. A molecule that contains atoms of different elements is a compound. The characteristics of compounds are usually different from the characteristics of the elements that comprise them.

5. When atoms come together to form molecules, chemical bonds form between them. Ionic bonds are formed when electrons are transferred between atoms. Atoms that have lost or gained electrons have an electric charge and are called ions. The attraction between oppositely charged ions constitutes an ionic bond. Covalent bonds form when atoms share, rather than transfer, electrons. Sometimes the sharing of electrons during covalent bonding is unequal and polar covalent bonds result, forming polar molecules. Hydrogen bonds are weak attractive forces between the atoms of these polar molecules.

6. Water is an important component of the human body due, in part, to its unusual characteristics. As a result of its ability to act as a dissolving agent, water is a major component of body fluids and functions well in the transport of dissolved substances to cells and tissues. Whereas the high heat capacity of water makes possible the maintenance of a constant internal body temperature, its high heat of vaporization prevents overheating through perspiration and facilitates cooling of the body surface.

7. When added to water, an acid increases the number of H^+ and a base increases the number of OH^-. The OH^- produced by the dissociation of a base reacts with H^+ to form water molecules. Thus, whereas acids increase the concentration of H^+ in solution, bases decrease the concentration of H^+ in solution. Strong acids and bases dissociate completely in water.

8. The strengths of acids and bases can be measured on the pH scale. This scale ranges from 0 to 14, with a pH of 7 being neutral, a pH of less than 7 being acidic, and a pH greater than 7

being basic. Biological systems usually function within a narrow pH range. Dramatic changes in pH are prevented by buffers, substances that remove excess H^+ from solution when concentrations increase and add H^+ when concentrations decrease.

9. A polymer is a large molecule made of many smaller molecules, called monomers, linked together in a chain. Polymers form through dehydration synthesis (removal of a water molecule) and are broken apart by hydrolysis (addition of a water molecule).

10. Carbohydrates, the sugars and starches, provide fuel for the human body. Monosaccharides (simple sugars) such as glucose, galactose, and fructose are the smallest molecular units of carbohydrates. Disaccharides (double sugars) are formed when two monosaccharides join together through dehydration synthesis. Examples of disaccharides include sucrose, maltose, and lactose. Polysaccharides (many sugars) are complex carbohydrates formed when large numbers of monosaccharides (usually glucose) join together through dehydration synthesis. Examples of polysaccharides include starch and glycogen, two energy-storage carbohydrates, and cellulose and chitin, two structural carbohydrates.

11. Lipids, such as triglycerides, phospholipids, and steroids, are nonpolar molecules that do not dissolve in water. Triglycerides (fats and oils) are made of glycerol and three fatty acids. Fat functions in long-term energy storage and the protection of organs from physical shock. Phospholipids are made of glycerol bonded to two fatty acids and a phosphate group; other small molecules, usually polar and designated by the letter R, are attached to the phosphate. Phospholipids are important components of cell membranes. Steroids consist of four carbon rings attached to functional groups. Cholesterol, one of the most familiar steroids, is a component of the plasma membrane and the foundation from which steroid hormones are made. Elevated levels of cholesterol in the blood can cause atherosclerosis, a disease in which blood flow is restricted due to the build up of cholesterol deposits (plaque) in the walls of the arteries.

12. Although proteins have a diverse array of structures and functions, they are all polymers made from a set of 20 amino acids. Amino acids consist of a central carbon bound to a hydrogen atom, an amino group, a carboxyl group, and a side chain (R). Amino acids, linked together through dehydration synthesis, form chains that are called peptides when they contain only a few amino acids, polypeptides when they contain 10 or more amino acids, and proteins when they contain 50 or more amino acids.

13. There are four levels of protein structure. Primary structure is the specific sequence of amino acids in a protein. Secondary structure results from the bending and folding of the amino acid chain into coils, spirals, or pleated sheets. Proteins whose secondary structures include spirals or coils often bend back into themselves and compact into globular units. Such proteins are called globular proteins. In contrast, fibrous proteins are made of parallel strands of protein molecules that entwine with one another. Tertiary structure is the three-dimensional shape of these globular and fibrous proteins. Some proteins consist of two or more polypeptide chains, each chain forming a subunit. Attractive forces between the subunits of these proteins hold the subunits together and produce quaternary structure.

14. Enzymes are proteins, or sometimes RNA molecules, that speed up chemical reactions without being consumed in the process. During a chemical reaction, the substrate binds to the enzyme at a specific location on the enzyme called the active site. While bound to one another the enzyme and substrate are known as the enzyme-substrate complex. At this time the substrate is converted to products that then leave the active site.

15. Deoxyribonucleic acid (DNA) and ribonucleic acid (RNA) are polymers of nucleotides joined through dehydration synthesis. A nucleotide consists of a five-carbon sugar bonded to one of five nitrogenous bases and a phosphate group. DNA, a double-stranded molecule, forms the genes and directs protein synthesis. RNA is a single-stranded molecule that also plays a major role in protein synthesis. Whereas the sugar in DNA is deoxyribose, that in RNA is ribose. The nitrogenous bases adenine, guanine, and cytosine are found in both DNA and RNA. The base thymine, however, is only found in DNA, and the base uracil only in RNA.

16. Adenosine triphosphate (ATP), the energy currency of cells, is a nucleotide made of the sugar ribose, the base adenine, and three phosphate groups. When cells require energy, one of the unstable phosphate bonds is broken and energy is released.

REVIEW QUESTIONS

1. What is an atom?
2. Compare protons, neutrons, and electrons with respect to their charge, mass, and location within an atom.
3. What is an isotope?
4. Describe the damage to the body caused by radiation.
5. How is radiation used to diagnose or cure illness?
6. Describe ionic, covalent, and hydrogen bonds and give an example of each.
7. What characteristics of water make it a critical component of the body?
8. Define acids and bases according to what happens when they are added to water.
9. What is acid rain and how does it affect the environment and our health?
10. How are polymers formed and broken? Give three examples of polymers and their component monomers.
11. Name two important energy storage and two important structural polysaccharides.
12. Describe the structure of phospholipids.
13. What are saturated, unsaturated, and polyunsaturated triglycerides?
14. What is atherosclerosis and what steps can be taken to reduce its incidence?
15. Describe the four levels of protein structure.
16. Compare the structures of RNA and DNA.
17. Describe the structure and function of ATP.

Chapter 3

The Cell

A dividing cell. Cell division plays an important role in the growth and repair of body tissues. *(Francis Leroy, Biocosmos/Science Photo Library)*

Someone once calculated that we are made up of well over 10 trillion cells. These include nerve cells, striated muscle cells, cartilage cells, and more than 100 other types of cells. Simply learning all the cells and tissues of the human body can be an exercise in memorization that would tax even the most dedicated cell biologist. You might also be thinking, if I want to know about people, why should I study cells? The answer to this question is simple: Cells closely reflect the functions of the tissues and organs that they make up, and thus they provide essential clues about how the human body is assembled and coordinated and how it functions as a whole.

Rather than running through the types of cells and their specializations, we will look at some of the structures they have in common. Beginning with the plasma membrane and working inward, we will review the various organelles, small components within cells that have specialized functions. Then we will turn to how cells get the energy they need and how they reproduce. By the end of this chapter it should be apparent that cells are integrated units within the body and that the tiny roles each plays in our lives are coordinated by a constellation of chemical and neural controls.

Structure of the Cell

We should begin by noting that, within the world at large, cells come in two types—prokaryotic cells and eukaryotic cells. **Prokaryotic cells,** unique to bacteria and cyanobacteria (blue-green algae), are much simpler and typically much smaller than eukaryotic cells. Most are surrounded by a rigid cell wall. The DNA of prokaryotic cells is circular and not set apart from the rest of the cell in a membrane-bound or-

ganelle. Indeed, prokaryotic cells lack the membrane-enclosed organelles found in eukaryotic cells (prokaryotic cells do, however, have some specialized membranes). **Eukaryotic cells,** on the other hand, are found in plants, animals, and all other organisms except bacteria and cyanobacteria. The cells in your body are all eukaryotic. Eukaryotic cells have extensive internal membranes that divide the cell into many compartments. Some organelles within eukaryotic cells are enclosed by membranes, whereas others lack membranes. In addition, eukaryotic cells have DNA that is found within a well-defined membrane-bound region called the nucleus. Table 3–1 summarizes the major differences between prokaryotic and eukaryotic cells.

The Plasma Membrane

We begin our examination of the cell at the plasma membrane. This remarkably thin film is the outer boundary of the cell and is in position to control the movement of substances in and out of the cell. A cell's interior is a critically balanced area, and substances cannot be permitted to move into and out of it randomly. Eukaryotic cells also contain several internal membranes. In general, the principles of structure and function described below for the plasma membrane also apply to the membranes inside the cell.

Structure of the Plasma Membrane

The plasma membrane is composed of lipids, proteins, and carbohydrates in a carefully structured arrangement. Recall from Chapter 2 that phospholipids are major components of the plasma membrane and that these molecules have hydrophilic (water-loving) heads and hydrophobic (water-hating) tails. The phospholipid molecules form a double layer at the surface of the cell called the lipid bilayer (Figure 3–1). The hydrophilic heads of the outer layer point outward and contact the aqueous solution outside the cell, called the **extracellular fluid;** the hydrophilic heads of the inner layer point inward, contacting the aqueous solution inside the cell, the **cytosol.** Within the bilayer, the hydrophobic tails point toward each other and help to hold the plasma membrane together.

Table 3–1 Comparison of Prokaryotic and Eukaryotic Cells

Feature	Prokaryotic Cells	Eukaryotic Cells
Source	Bacteria, cyanobacteria	Plants, animals, fungi, protists
Size	$1–10 \ \mu m$ across	$10–100 \ \mu m$ across
Membrane-bound organelles	Absent	Present
DNA form	Circular	Coiled, linear strands
Location of DNA	Cytoplasm	Nucleus
Internal membranes	Some	Many
Cytoskeleton	Absent	Present

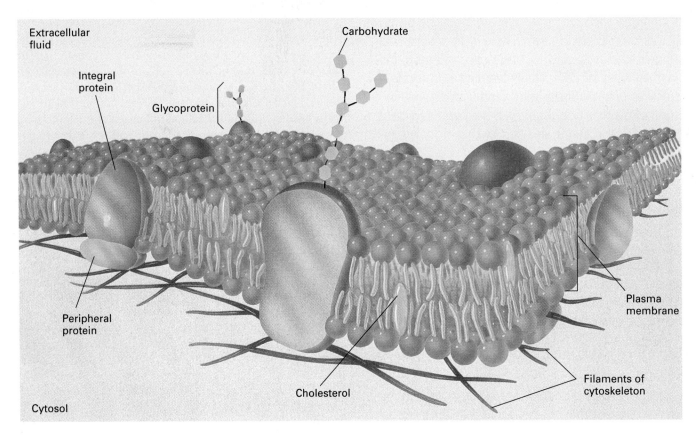

Figure 3–1   The structure of the plasma membrane of a cell according to the fluid mosaic model.

Interspersed in the phospholipid bilayer are a diverse array of proteins, as seen in Figure 3–1. Although molecules of phospholipids usually outnumber proteins about 50:1, the proteins are so large that they sometimes make up half the mass of a membrane. Some proteins, called **integral proteins,** are embedded in the membrane, either completely or incompletely spanning the bilayer. Other proteins, called **peripheral proteins,** are simply attached to the inner or outer surface of the membrane, often to an exposed portion of an integral protein. Molecules of cholesterol are also interspersed throughout the lipid bilayer.

The carbohydrates of the plasma membrane attach only to its outer surface, as shown in the figure. Some of these carbohydrates are covalently bonded to lipids, forming **glycolipids,** but most are covalently bonded to proteins, forming **glycoproteins.**

The structure of the plasma membrane is often described as a **fluid mosaic** because the proteins that are interspersed throughout the lipid molecules give the membrane its mosaic quality, and many protein and lipid molecules are able to move sideways, to some degree, giving the membrane its fluid quality.

Functions of the Plasma Membrane

The plasma membrane performs several functions for the cell. First, by forming the boundary between a cell's internal and external environment, the plasma membrane maintains the cell's structural integrity (basically, as someone has said, by keeping "cell stuff" in and "non-cell stuff" out). Second, the plasma membrane regulates the movement of ions and molecules into and out of the cell. Although traffic across the membrane is extensive, it is not indiscriminate. For this reason the membrane is often described as being **selectively permeable,** permitting the movement of some substances across while denying access to others. In addition, the rates at which substances move into and out of the cell are controlled by the plasma membrane.

The plasma membrane also functions in cell-cell recognition. Cells distinguish one type of cell from another by recognizing molecules on the surface of the plasma membrane, often the glycoproteins. Membrane carbohydrates differ from one species to another, among individuals of the same species, and even among different cell types within a given individual. Such variation allows the body to recognize foreign invaders such as bacteria, which also seek out particular surface carbohydrates when settling preferentially on certain cells in the body.

Cell-cell recognition, in fact, is the basis for rejection of tissue grafts and organ transplants. Because each individual has a characteristic glycoprotein composition on the surface of his cells, often referred to as a "cellular fingerprint," the transplanted cells are viewed as foreign (unless they come

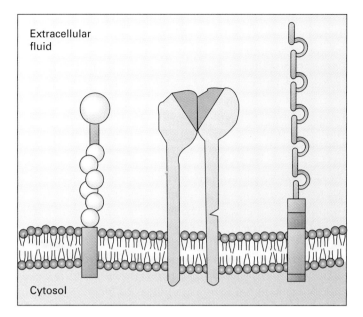

Figure 3-2 Examples of cell adhesion molecules (CAMs). These uniquely shaped molecules stick through the plasma membrane and help hold cells together to form tissues and organs. CAMs are also involved in many processes involving cellular interaction and movement.

Table 3-2 Functions of the Plasma Membrane

Maintains structural integrity of the cell
Regulates movement of ions and molecules into and out of the cell
Cell-cell recognition
Communication between cells
Sticks cells together to form tissues and organs

from an identical twin) and will tend to be rejected by the recipient's immune system. Although controversial, the transplantation of cells taken from aborted fetuses solves the problem of tissue rejection. Because the cellular fingerprints of fetal cells are incompletely developed, the transplanted cells are accepted by the recipient's immune system. For example, fetal cell transplants hold promise for people with diabetes. This disease involves an inability to control levels of glucose in the blood and often results from an insufficient output of the hormone insulin from the pancreas. Many diabetics rely on insulin injections to treat their illness. By transplanting healthy cells from a fetal pancreas, a patient may no longer need to rely on such injections, for once established in the recipient, the fetal pancreas cells start secreting appropriate levels of insulin.

The plasma membrane also functions in communication between cells. For example, hormones secreted by one group of cells may bind to specific proteins, called **receptors,** that span the plasma membranes of other cells. These proteins, in turn, relay the message to chemical intermediaries, some of which are known as G proteins, inside the cell. These proteins then transmit the message to other molecules inside the cell (often enzymes) that through a series of chemical reactions ultimately initiate a change in the behavior of the cell, perhaps causing it to release glucose or a hormone.

The plasma membrane plays an important role in sticking cells together. **Cell adhesion molecules** (CAMs, for short) poke through the plasma membranes of most cells and help hold cells together when forming tissues and organs (Figure 3–2). Such molecules are also involved in the forma-

tion of tissues and organs during embryonic development and in many processes in the adult that have as their basis cellular movement and interaction. For example, CAMs play a role in wound healing. Apparently the cells of blood vessels near a wound increase their production of CAMs, and these sticky molecules promote healing by snagging repair cells, such as white blood cells, as they move by. In fact, a gel currently being evaluated by the United States Food and Drug Administration is laced with the CAMs involved in healing. This gel is said to close wounds in 30% less time and to leave less of a scar. Although the wound healing gel encourages interactions among cells through the use of CAMs, many of the efforts of biotechnology companies are aimed at blocking harmful contact among cells by blocking CAMs. Some companies, for example, are trying to block the CAMs on bone cells because these molecules allow bone-destroying cells to attach and lead to the bone loss associated with osteoporosis, a disease primarily found in postmenopausal women. The functions of the plasma membrane are summarized in Table 3–2.

Critical Thinking

Cancer is a disease characterized by the uncontrolled proliferation of cells. Metastasis is a process in which cancerous cells separate from a primary tumor and move throughout the body creating new tumors. How might cell adhesion molecules be involved in metastasis?

A Closer Look at Membrane Transport

Recall that an important function of the plasma membrane is to control which substances move into and out of the cell and at what rates. There are several ways in which materials cross the plasma membrane. Perhaps the most straightforward method is **simple diffusion,** the spontaneous movement of a substance from a region of higher concentration to a region of lower concentration. Once the concentration of the substance is the same on both sides of the membrane, movement of the substance does not simply stop, however. Instead, there is continued random movement of the substance back and forth across the membrane, but the rate of movement in each direction is the same. Lipid-soluble

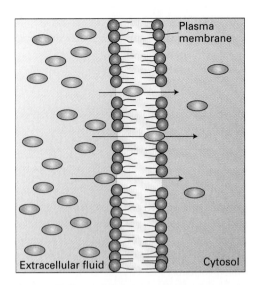

Figure 3-3 Diffusion is the spontaneous movement of a substance from a region of high concentration to a region of lower concentration. Here a lipid-soluble substance diffuses across the plasma membrane by moving directly through the lipid bilayer.

substances like carbon dioxide and oxygen can diffuse right through the lipid bilayer (Figure 3–3).

Water-soluble substances, on the other hand, cannot move through the bilayer by simple diffusion. Such molecules, and other polar molecules, may get across the plasma membrane by attaching to specific proteins in the plasma membrane that either transport them from one side of the membrane to the other or form channels through which they can move. **Facilitated diffusion** is the movement of a substance from a region of higher concentration to a region of lower concentration with the aid of such a membrane protein. Molecules of glucose, for example, enter fat cells by facilitated diffusion involving a carrier protein (Figure 3–4).

Osmosis is a special case of diffusion in which water moves across the plasma membrane or any other selectively permeable membrane. Consider what happens when a substance like table salt (the solute) is dissolved in water (the solvent) in a membranous bag through which water, but not salt, can move. If the bag is placed into a **hypotonic** solution, that is, one in which the concentration of salt is lower than that inside the bag, then there is net movement of water into the bag, causing the bag to swell and possibly burst (Figure 3–5a). If, on the other hand, the bag is placed into a **hypertonic** solution in which the concentration of salt is higher than inside the bag, then there is net movement of water out of the bag, causing the bag to shrivel (Figure 3–5b). When the bag is placed into an **isotonic** solution, one with the same concentration of salt as inside the bag, there is no net movement of water in either direction and the bag maintains its normal shape (Figure 3–5c). The bag in our example can be replaced by a red blood cell and the solution by the surrounding blood plasma. Note the changes in the shape of red blood cells in response to different levels of salt concentration in the plasma (Figure 3–5, a–c).

Another mechanism of membrane transport is active transport. **Active transport** is the movement of molecules across the plasma membrane with the aid of a carrier protein and energy supplied by the cell. Adenosine triphosphate (ATP) provides the energy for most instances of active transport (Figure 3–6). So far in our discussion we have described substances moving from regions of higher concentration to lower concentration, a pattern described as moving "down the concentration gradient." In active transport, however, substances move from regions of lower concentration to higher concentration, such as when cells need to concentrate certain substances, essentially moving "up the concentration gradient." For example, compared with their surroundings, the cells in our bodies have higher concentrations of potassium ions (K^+) and lower concentrations of sodium ions

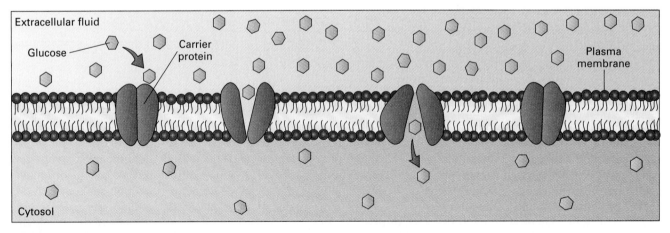

Figure 3-4 Facilitated diffusion is the movement of a substance from a region of higher concentration to a region of lower concentration with the aid of a membrane protein that acts as a channel or a carrier protein. Here, glucose enters a fat cell by means of a carrier protein.

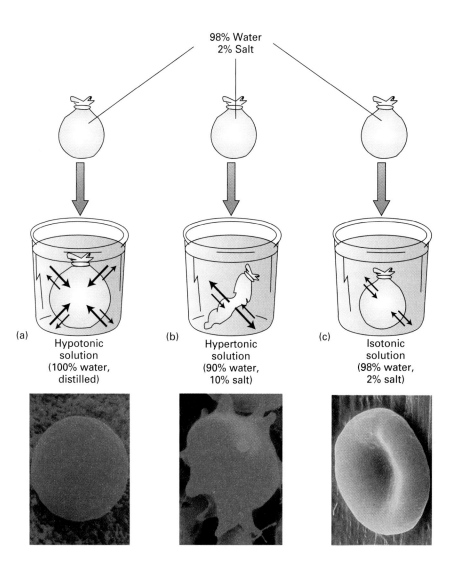

98% Water
2% Salt

(a)
Hypotonic
solution
(100% water,
distilled)

(b)
Hypertonic
solution
(90% water,
10% salt)

(c)
Isotonic
solution
(98% water,
2% salt)

Figure 3–5 Osmosis is a special case of diffusion in which water moves across a selectively permeable membrane. The drawings at the top of the figure show what happens when a membranous bag through which water but not salt can move is placed in solutions that are either (a) hypotonic (b) hypertonic or (c) isotonic. The width of the arrows in each case corresponds to the amount of water moving across the membrane. The photographs below show what happens to red blood cells when placed in similar solutions. Red blood cells are normally shaped like flattened disks as in (c).
(a, b, Visuals Unlimited/© David M. Phillips; c, Visuals Unlimited/© Stanley Flegler)

(Na^+). Through active transport, the plasma membrane helps maintain these conditions by pumping potassium ions into the cell and sodium ions out of the cell. In this example, both potassium and sodium are moving from regions of lower concentration to higher concentration, and thus carrier proteins and energy are needed to make the journey.

Most small molecules traverse the plasma membrane by way of diffusion, facilitated diffusion, or active transport. Large molecules and single-celled organisms like bacteria, however, enter cells through **endocytosis,** a process in which a region of the plasma membrane surrounds the substance to be ingested and then pinches off from the rest of

Figure 3–6 Active transport is the movement of molecules across the plasma membrane with the aid of a carrier protein and energy supplied by the cell, usually in the form of ATP. Active transport involves the movement of a substance from a region of lower concentration to higher concentration, such as when cells need to concentrate certain substances.

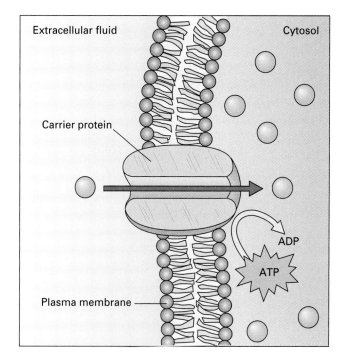

Extracellular fluid

Cytosol

Carrier protein

ADP

ATP

Plasma membrane

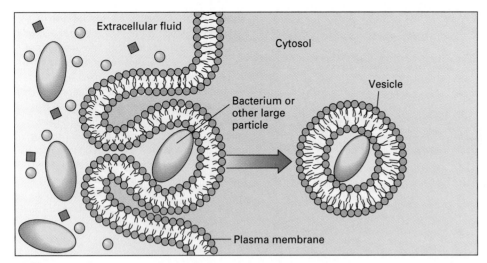

(a) Phagocytosis

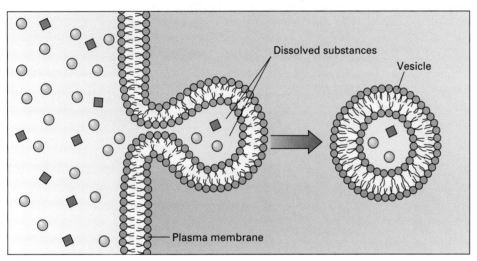

(b) Pinocytosis

Figure 3–7 Large molecules or bacteria move inside the cell by endocytosis. There are two types of endocytosis. (a) Phagocytosis ("cell eating") occurs when cells engulf large particles or bacteria. (b) Pinocytosis ("cell drinking") occurs when cells engulf droplets of extracellular fluid and the dissolved substances therein. Although the vesicles in (a) and (b) are shown as the same size, vesicles formed during pinocytosis are actually much smaller than those formed during phagocytosis.

the membrane, enclosing the substance in a sac-like structure called a **vesicle.** The vesicle is then released into the cell. Two types of endocytosis are phagocytosis ("cell eating") and pinocytosis ("cell drinking"). Whereas **phagocytosis** occurs when cells engulf large particles like bacteria (Figure 3–7a), **pinocytosis** occurs when cells engulf droplets of fluid, thus bringing into the cell all of the dissolved substances therein (Figure 3–7b).

The process by which large molecules leave cells is called **exocytosis.** Some cells that produce hormones, for example, package their products in membrane-bound vesicles that move within the cell toward the plasma membrane. Upon reaching the plasma membrane, the membrane of the vesicle fuses with it and spills its contents outside the cell (Figure 3–8). Table 3–3 summarizes the ways in which substances move across the plasma membrane. In the next section, we will look across the plasma membrane and examine the components inside the cell.

Critical Thinking

People with kidney disorders often undergo dialysis. During this process the patient's blood is passed through a long coiled tube that is submerged in a tank filled with a fluid called dialyzing fluid. The tubing is made of porous cellophane that allows small molecules to diffuse out of the blood into the surrounding fluid. Urea is a waste product that must be removed from the blood. Glucose molecules, on the other hand, should remain in the blood. How would you manipulate the concentrations of urea and glucose in the dialyzing fluid to achieve these goals?

Organelles

We have seen the many roles that the plasma membrane plays in the life of the cell. Inside the eukaryotic cell, how-

Table 3–3 Mechanisms of Transport Across the Plasma Membrane

Mechanism	Description
Simple diffusion	Spontaneous movement from high to low concentration
Facilitated diffusion	Movement from high to low concentration with the aid of a carrier or channel protein
Osmosis	Movement of water from high to low concentration
Active transport	Movement from low to high concentration with the aid of a carrier protein and energy usually from ATP
Endocytosis	Materials are engulfed by plasma membrane and drawn into cell in a vesicle.
Exocytosis	Membrane-bound vesicle from inside the cell fuses with the plasma membrane and spills contents outside the cell.

ever, membranes play a variety of different roles, primarily by delineating compartments. Specific chemical processes critical to the life of the cell occur within these compartments. This segregation or compartmentalization is thought to have evolved in eukaryotic cells because it enhances cellular efficiency as particular combinations of molecules carry out specific tasks. Also within the cell are components called

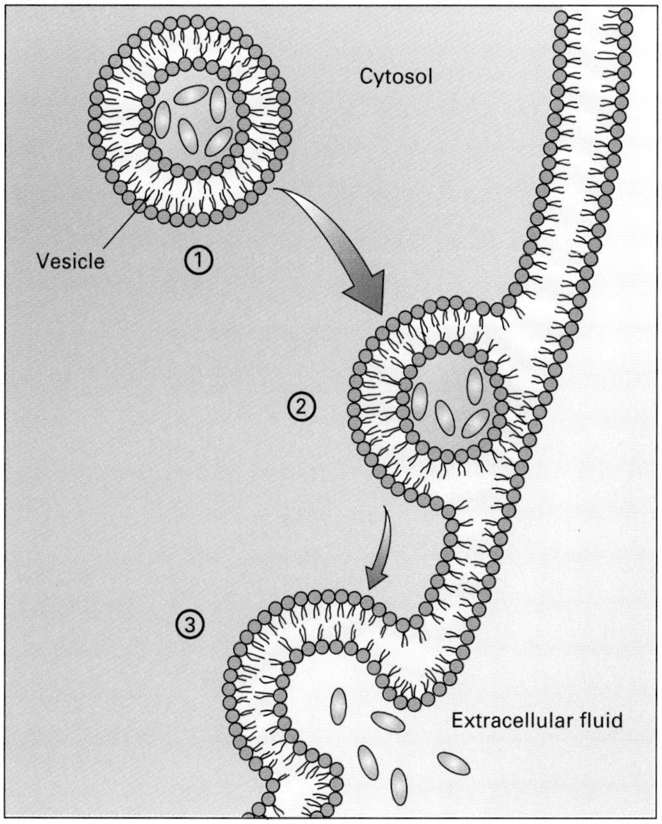

Figure 3–8 Large molecules leave cells by exocytosis. The cells package the molecules in membrane-bound vesicles that move through the cytosol toward the plasma membrane (step 1). Upon reaching the plasma membrane, the membrane of the vesicle fuses with it (step 2) and spills its contents outside the cell (step 3).

organelles ("little organs"). Some organelles are bounded by membranes, whereas others are not enclosed by membranes. Organelles are distributed in an aqueous environment called **cytosol.** The term **cytoplasm** is used to describe the cytosol plus all the organelles with the exception of the nucleus.

Nucleus

The nucleus has been likened to a command center and contains almost all the genetic information of the cell (Figure 3–9). It provides information for cellular structure and function through its DNA that directs which proteins should be made by the cell and when. Because all our cells contain the same genetic information, the character of a particular cell, whether it be a muscle cell or a liver cell, is determined in large part by the directions it receives from its nucleus.

The nucleus is surrounded by a double membrane called the **nuclear envelope** (see Figure 3–9). Although the envelope separates the nucleus from the cytosol, communication between the two areas occurs through **nuclear pores** that perforate the envelope. The traffic of selected materials back and forth across the nuclear envelope allows the cytosol to influence the nucleus and the nucleus to influence the cytosol. It seems probable, for example, that molecules made in the cytosol of a growing cell move through the pores into the nucleus where they prompt cell division by conveying information on cell size. As a cell grows, the nucleus might receive information from the cytosol indicating that it is time to divide.

The genetic information within the nucleus is in the form of **chromosomes,** thread-like structures comprised of DNA and associated proteins. Individual chromosomes are only visible with a light microscope during cell division, a time when they shorten and condense (Figure 3–10a). At all other times, the chromosomes are extended and not readily visible; in this dispersed state the genetic material is called **chromatin** (Figure 3–10b). The chromatin and the aqueous environment within the nucleus constitute the **nucleoplasm.** We will have more to say about chromosomes and cell division a little later in the chapter.

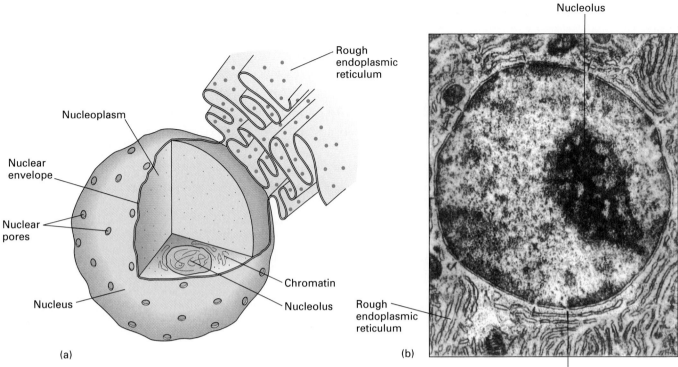

Figure 3-9 The nucleus is the command center of a cell. (a) Diagram of the nucleus. In some areas, the nuclear membrane is continuous with another organelle, rough endoplasmic reticulum. (b) Electron micrograph of the nucleus and surrounding cytosol. Arrows indicate nuclear pores. *(b, Don Fawcett/Photo Researchers, Inc.)*

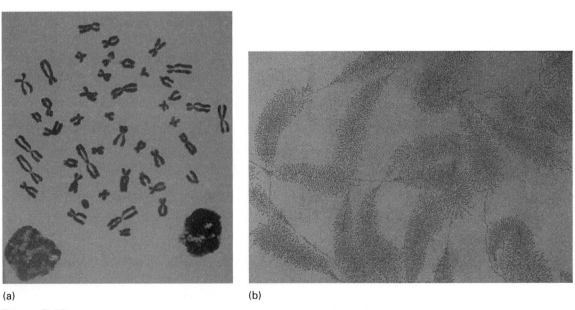

Figure 3-10 Chromosomes are composed of DNA and associated proteins. (a) Individual chromosomes are visible during cell division, when they shorten and condense. (b) At all other times, the chromosomes are extended and not readily visible, occurring in a dispersed state called chromatin. *(a, Visuals Unlimited/© L. Lisco, D. W. Fawcett; b, Visuals Unlimited/© O. L. Miller/B. R. Beatty/D. W. Fawcett)*

The **nucleolus** is a specialized region within the nucleus that forms and disassembles during the course of the cell cycle (see Figure 3–9). This structure plays a role in the generation of ribosomes, organelles involved in protein synthesis. In contrast to the cytoplasmic organelles, the nucleolus is not surrounded by a membrane. It is simply a region of DNA involved in the production of a type of RNA called ribosomal RNA (rRNA), one of the components of ribosomes. The protein components of ribosomes are made in the cytosol and then shipped into the nucleus. Inside the nucleus the rRNA and proteins join, then move into the cytosol where they become functional ribosomes.

Mitochondrion

Most cellular activities require energy. Energy is needed, for example, to transport certain substances across the plasma membrane and to fuel the many chemical reactions that take place in the cytosol. Specialized cells like muscle cells and nerve cells also require energy to carry out their particular activities. The energy needed by cells is provided by **mitochondria** (singular, *mitochondrion*), organelles within which most of cellular respiration occurs. Cellular respiration is the process by which oxygen and an organic fuel such as glucose are consumed and energy in the form of ATP is released. The first series of reactions takes place in the cytosol and the remaining three series occur in the mitochondria.

The number of mitochondria varies considerably from cell to cell, being roughly correlated with a cell's demand for energy. It is safe to say, however, that most cells have several hundred mitochondria. Like the nucleus, but unlike other organelles, mitochondria are bounded by a double membrane (Figure 3–11). The inner and outer membranes create two separate compartments within a mitochondrion, the internal matrix and the intermembrane space. The major working regions of the mitochondrion are the matrix that contains a concentrated mixture of enzymes needed for the citric acid cycle, which is a series of reactions in cellular respiration, and the highly folded inner membrane that is the site of the electron transport chain, another reaction sequence in cellular respiration. The infoldings of the inner membrane of a mitochondrion are called **cristae.**

Mitochondria contain ribosomes and a small percentage of a cell's total DNA (the rest being found in the nucleus). Because mitochondria have their own DNA and protein-synthesizing machinery, they can reproduce and they can make some of the proteins they need. Critical proteins that they cannot synthesize are encoded by genes in the nucleus, produced in cytosol, and then transported inside the mitochondrion. Nevertheless, the somewhat independent nature of mitochondria is thought to be linked to their past. Most scientists now believe that mitochondria arose by the symbiotic (mutually beneficial) association of a free-living bacterium (a prokaryote) and a primitive eukaryotic cell. This hypothesis, called the **endosymbiont hypothesis,** suggests that mitochondria were once free-living organisms that were somehow engulfed by eukaryotic cells. Over time, it is

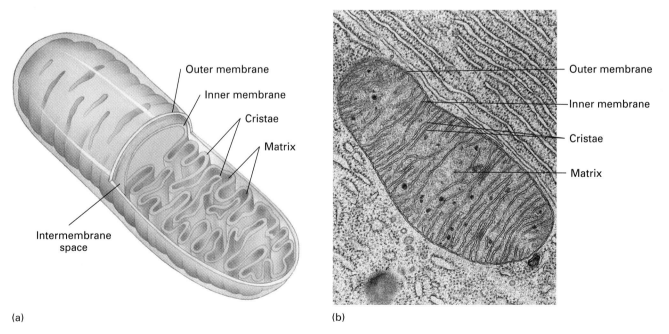

(a)

(b)

Figure 3–11 Mitochondria are sites of energy conversion in the cell. (a) A mitochondrion showing the double membrane that creates two compartments, the intermembrane space and the matrix. The folds of the inner membrane are called cristae. (b) Electron micrograph of a mitochondrion. *(b, © Keith Porter/Photo Researchers, Inc.)*

believed, these precursors to mitochondria transferred some of their genetic material to the host nucleus. The relationship may have been mutually beneficial because the mitochondria provided ATP for the host cell and the host cell provided its "guest" with food. What evidence is there to support the endosymbiont hypothesis? Most importantly, many aspects of mitochondrial structure resemble those of bacteria. For example, both mitochondria and bacteria have circular DNA and small ribosomes.

Ribosomes

Ribosomes are sites of protein synthesis. Not surrounded by a membrane and usually considered the smallest organelle of a cell, a ribosome consists of RNA and protein organized into two subunits within the nucleolus. These subunits join to become a functional ribosome only after moving from the nucleus into the cytosol where they come in contact with messenger RNA (Figure 3–12a). Because protein synthesis is covered in detail in Chapter 19, we include only the basics here.

In the cell's nucleus, genes containing the information needed to make proteins are transcribed into a type of RNA called messenger RNA (mRNA). Messenger RNA moves through the pores in the nuclear envelope and enters the cytosol where it contacts ribosomes. The ribosomes read the instructions from mRNA and translate the information into a string of amino acids, thus forming the appropriate protein, as shown in Figure 3–12a. The first several amino acids that the ribosomes link together signal the ribosomes to either remain suspended in the cytosol or attach to another organelle called the endoplasmic reticulum. Ribosomes that float in the cytosol are called **free ribosomes.** Free ribosomes usually make proteins to be used inside the cell. Ribosomes that are attached to the endoplasmic reticulum are called **bound ribosomes.** Bound ribosomes typically make proteins for use either outside the cell or in membranes. Figure 3–12b shows both free and bound ribosomes in a cell from the pancreas.

Endoplasmic Reticulum

The **endoplasmic reticulum** (ER) is an extensive network of membranous channels connected to the plasma membrane, nuclear envelope, and some organelles. There are two well-defined, though continuous, regions of endoplasmic reticulum known as rough ER and smooth ER, and each will be considered in turn (Figure 3–13).

Rough endoplasmic reticulum (RER) has ribosomes attached to its cytosol surface and is involved in the synthesis of proteins to be secreted by cells. The proteins made by the attached ribosomes are threaded through the RER's membrane (probably through pores) to its internal space. Once proteins are inside the lumen of the RER, enzymes attach carbohydrates to many of them, forming glycoproteins. The proteins or glycoproteins are then enclosed in vesicles

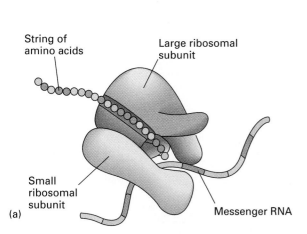

(a)

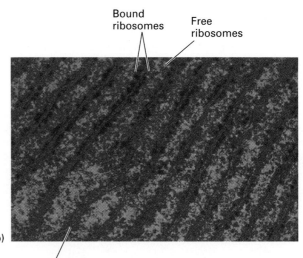

(b)

Figure 3–12 Ribosomes are sites of protein synthesis. (a) A ribosome consists of RNA and protein organized into two subunits, one large and one small. These subunits become a functional ribosome engaged in protein synthesis once they leave the nucleus and enter the cytosol where they come in contact with messenger RNA (mRNA) carrying information needed to make the proteins. The ribosomes read the instructions from mRNA and translate the information into a string of amino acids, thus forming the appropriate protein. (b) An electron micrograph showing ribosomes both suspended in the cytosol (free ribosomes) and attached to the endoplasmic reticulum (bound ribosomes).   (b, Visuals Unlimited/© Fred Hossler)

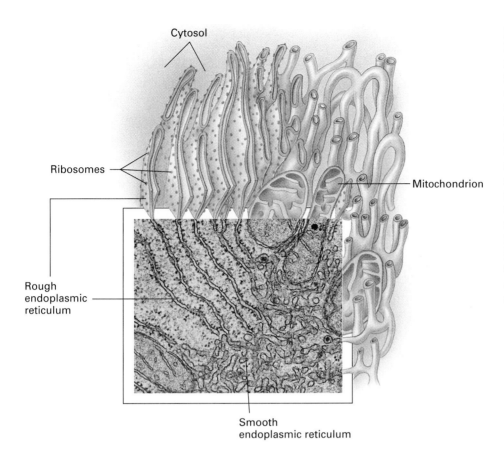

Cytosol

Ribosomes

Rough
endoplasmic
reticulum

Mitochondrion

Smooth
endoplasmic reticulum

Figure 3-13 The endoplasmic reticulum (ER) as depicted in the diagram and electron micrograph. A network of flattened membranous canals, this organelle consists of two continuous regions, rough ER and smooth ER, and is connected to other cell membranes, including the nuclear envelope. Rough endoplasmic reticulum (RER) has ribosomes attached on its cytosol surface and is involved in the synthesis of proteins to be secreted by cells or used in their membranes. Smooth endoplasmic reticulum (SER) lacks ribosomes and is involved in the production of phospholipids for incorporation into membranes and the detoxification of certain drugs. *(Photo, Visuals Unlimited/R. Bolender and D. W. Fawcett)*

formed from RER membrane and transferred to the Golgi complex for further processing and packaging. The precise molecular signals that direct the vesicles from the RER to the Golgi complex are not yet known.

RER is also involved in membrane production. Some of the proteins made by attached ribosomes remain in the RER membrane rather than continuing through to the inside of the organelle for eventual shipment to the Golgi complex. Because the RER membrane is continually used to form vesicles for shipping, it must constantly be replenished. Where does the new material come from? Recall that proteins and phospholipids are major components of cell membranes. The protein components of the RER membrane come from the RER itself and the lipid components come mostly from the smooth ER.

Smooth endoplasmic reticulum (SER) lacks ribosomes. One function of SER, mentioned above, is the production of phospholipids for incorporation into cell membranes. The SER also functions to detoxify alcohol and other drugs, such as the sedative phenobarbital, that may enter the system. Apparently, enzymes of SER help to chemically modify these substances, making them more water soluble and easier to flush from the body. Both alcohol and phenobarbital induce an increase in SER and its enzymes,

leading to greater tolerance. This is why, with repeated exposure to such drugs, higher and higher doses are needed to gain the same effect. The liver is the major detoxification center in the body and its cells are packed with SER.

Golgi Complex

Resembling a stack of dinner plates, the **Golgi complex** is the protein processing and packaging organelle of the cell (Figure 3–14). Protein-filled vesicles from the RER arrive at the "receiving side" of the Golgi complex, fuse with its membrane, and empty their contents to the inside. The Golgi complex then chemically modifies many of the proteins as they move, by way of vesicles, from one membranous disk in the stack to the next. After such processing is complete, the Golgi complex sorts the proteins, much like a postal worker would sort letters, and sends them to their specific destinations. Some of these proteins emerging from the "shipping side" are packaged in vesicles and sent to the plasma membrane either for export out of the cell or for incorporation into the plasma membrane. Other proteins are packaged in lysosomes. How is such sorting accomplished? Most cell biologists believe that each protein has a specific targeting signal containing the "address" of its final destination.

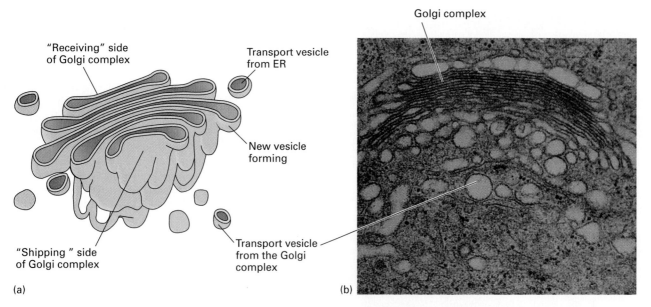

Figure 3-14 The Golgi complex. (a) This organelle consists of stacks of membranous disks and serves as the site for protein processing and packaging in the cell. Vesicles carrying proteins from the ER arrive at the receiving side of the Golgi complex and empty their contents to the inside where the proteins are modified. The Golgi complex then sorts the proteins and sends them in vesicles released from the "shipping side" to their specific destinations. (b) Electron micrograph showing the Golgi complex and its associated vesicles.   *(b, Visuals Unlimited/© Don Fawcett)*

Lysosomes

Lysosomes are the principal sites of digestion within the cell. These organelles are roughly spherical in shape, surrounded by a single membrane, and packed with about 40 different digestive enzymes. The enzymes and membranes of lysosomes are made by the rough endoplasmic reticulum and then sent to the Golgi complex for further treatment and packaging. Eventually, intact, enzyme-filled lysosomes bud off from the shipping side of the Golgi complex and begin their diverse roles in intracellular digestion.

Lysosomes are responsible for the breakdown of macromolecules or foreign invaders like bacteria that enter the cell by endocytosis. Consider, for example, what happens when a cell engulfs a particle of food (Figure 3–15). During the process of phagocytosis, a vesicle encircles the food, forming a food vacuole. Lysosomes, released from the Golgi complex, then fuse with the food vacuole, releasing their enzymes. The digestive enzymes break the food down into smaller molecules that diffuse out of the vesicle into the cytosol where they can be used by the cell. Indigestible residues may persist in the cell in residual bodies or may be expelled from the cell by exocytosis.

Although lysosomes are present in most cells, they are particularly numerous in cells that specialize in phagocyto-sis. Certain white blood cells, for example, ingest microorganisms like bacteria that have invaded the body. These microorganisms are usually destroyed once lysosomes fuse with the phagocytic vacuoles that contain them. Certain bacteria, however, have developed ways to evade destruction once engulfed by white blood cells. The bacteria that cause leprosy and tuberculosis have substances on their outer surfaces that prevent lysosomes from fusing with the vacuole that contains them. Thus, instead of being destroyed by lysosomal enzymes, these bacteria thrive and reproduce inside the cell, eventually killing it.

Lysosomes also destroy obsolete parts of the cell itself (see Figure 3–15). Old organelles and macromolecules are broken down and the resultant materials can be used by the cell to make new structures. Mitochondria, for example, last only about 10 days in a typical liver cell before being destroyed by lysosomes. Such "housecleaning" ensures proper functioning of the cell and promotes the recycling of essential materials.

Finally, lysosomes appear to play a role in the destruction of old or defective cells. Within such cells, lysosomes are thought to break open, releasing their enzymes and destroying the cells. There is some question, however, about the involvement of lysosomal enzymes in cell death. Because

Figure 3-15 Lysosome formation and functions in cellular digestion. Lysosomes, released from the Golgi complex, digest substances imported from outside the cell and obsolete parts of the cell itself. In the pathway on the left, the cell engulfs a food particle through phagocytosis (step 1). The food particle is then enclosed within a vesicle called a food vacuole that fuses with a lysosome (step 2). Lysosomal enzymes break the food down into smaller molecules that diffuse into the cytosol to be used by the cell (step 3). Indigestible substances may remain in the cell in residual bodies (step 4) or leave the cell by exocytosis (step 5). In the pathway on the right, lysosomes digest old or defective organelles and recycle the resulting materials.

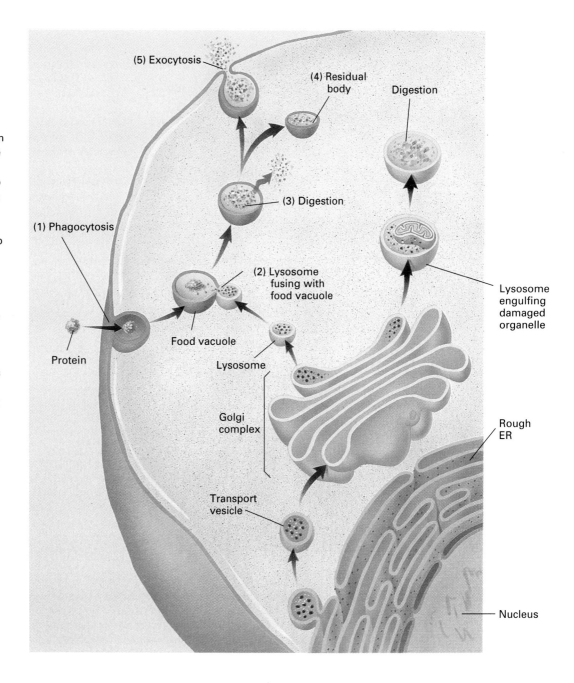

lysosomal enzymes require an acidic environment in order to function, there is speculation that they would become nonfunctional once released into the less acidic environment of the cytosol.

We have seen the important roles played by lysosomal enzymes in the digestion of macromolecules, organelles, and possibly even cells themselves. It should not be surprising to learn, then, that the absence of a single lysosomal enzyme can have devastating consequences. Called **lysosomal storage diseases,** these disorders result from the build-up of the molecules that would normally be degraded by the missing enzyme. Such molecules accumulate in the lysosomes (hence the name lysosomal storage diseases), causing them to swell, and ultimately interfere with cell functioning. More than 30 lysosomal storage disorders have been identified in humans and other animals. These diseases are inherited and progress with age. Focus, for a moment, on the lysosomal storage disorder called Tay-Sachs disease.

Tay-Sachs disease is more common in Jewish children of eastern European ancestry than non-Jewish children and is caused by the absence of the lysosomal enzyme hexosaminidase (Hex A). This enzyme is responsible for the

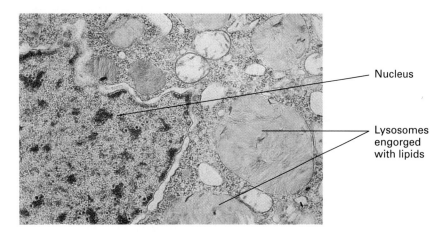

Nucleus

Lysosomes
engorged
with lipids

Figure 3-16   An electron micrograph of a nerve cell from the brain of a person with Tay-Sachs disease. This fatal disease is caused by the absence of the lysosomal enzyme hexosaminidase (Hex A), responsible for the breakdown of lipids in nerve cells. Typically, infants with Tay-Sachs show progressive deterioration in their motor skills and senses as abnormal amounts of fat accumulate in the cells of the nervous system. (Note the lysosomes engorged with undigested lipids in the nerve cell shown here.) Death usually occurs by about 3 or 4 years of age.   (© 1992 IMS Creative/ Custom Medical Stock Photo)

Environmental Concerns

Asbestos: The Deadly Miracle Material

Since the beginning of this century, over 30 million tons of asbestos have been used in the United States. By the end of this century, about half a million people will have died as a result. What is asbestos? Why do we use it? And why do we die from it?

Asbestos is a fibrous mineral containing silica (Figure A). It is strong, flexible, and resistant to heat and corrosion. Because of these properties, asbestos has proven extremely valuable in manufacturing and construction. It has been used, for example, in the manufacture of brake linings for automobiles and as an insulator on ceilings and pipes. Asbestos has been added to cement to increase resistance to weathering and sprayed on the walls of schools to make them more soundproof and fireproof.

The very same properties of asbestos that make it an ideal building material—its fibrous nature and durability—also make it deadly. Fibers of asbestos insulation are easily dislodged, producing small dust particles that can be inhaled into the lungs. Because these particles cannot be broken down, they remain in the lungs for life. Inhalation of asbestos particles over even a short period of time can cause lung cancer and mesothelioma, a form of cancer specific to the lining of the lungs that usually causes death within 1 year of diagnosis. Our focus, here, will be on a third disorder, asbestosis, which results from the dangerous interaction between asbestos and lysosomes.

Asbestosis, also known as pulmonary fibrosis, is the most common disease caused by exposure to asbestos. It usually occurs in people who have worked either directly with asbestos or in buildings with exposed insulation, and often takes 10 to 20 years to develop. Apparently, small particles of asbestos inhaled into lungs are engulfed by cells responsible for cleaning the respiratory passages. Lysosomes inside the cleaning cells then fuse with the vacuoles containing the asbestos particles. Rather than being broken down by lysosomal enzymes, however, the asbestos particles destabilize the membranes of the lysosomes causing massive release of enzymes. These lysosomal enzymes

Figure A   The deadly miracle material asbestos. The qualities of asbestos that make it a valuable material for manufacturing and construction—strength, durability, and flexibility—also make it hazardous to our health. Particles of asbestos released into the air and inhaled into the lungs cause lung cancer, mesothelioma (cancer of the lung lining), and asbestosis (scarring of lung tissue).   (Runk/Schoenberger, Grant Heilman Specimen from North Museum, Franklin and Marshall College)

breakdown of lipids in nerve cells. When Hex A is missing, the lysosomes become engorged with undigested lipids (Figure 3–16). Infants with Tay-Sachs disease appear normal at birth, but begin to show signs of deterioration by about 6 months of age due to the accumulation of abnormal amounts of lipid in the nervous system. Typically, motor skills such as crawling, grasping, and rolling over progressively deteriorate as well as senses such as sight and hearing. Eventually, by age 3 or 4, Tay-Sachs causes paralysis and death. Although at present there is no cure for this disease, identification of the missing enzyme has led to the development of a blood test to detect individuals that carry the gene for Tay-Sachs. Called **carriers,** these individuals do not have the disease but could pass the gene on to their offspring. It is also possible to diagnose Tay-Sachs in a fetus.

Although in theory it should be possible to treat people with lysosomal storage disorders such as Tay-Sachs disease by injecting the missing enzyme into their bloodstream, this kind of treatment, called **enzyme replacement therapy,** has not been very successful to date. A future possibility is **gene replacement therapy** in which the genes for the missing enzyme are inserted into the deficient cells. Once there, the genes could direct synthesis of the missing enzyme.

So far in our discussion of lysosomal diseases we have considered only those caused by genetic errors leading to missing enzymes. A number of environmental factors, however, also interfere with lysosomal function causing disease. In *Environmental Concerns: Asbestos—The Deadly Miracle Material,* we describe the impact of asbestos on lysosomes and health.

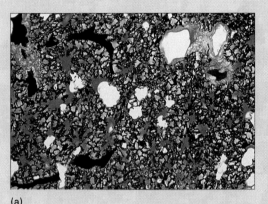
(a)

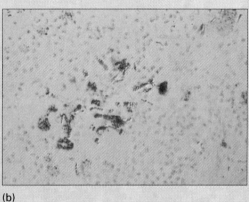
(b)

Figure B   A comparison of lung tissue from a healthy adult (a) and from an adult with asbestosis (b). The destruction of lung tissue characteristic of asbestosis results from the release of lysosomal enzymes inside the cells responsible for cleaning the respiratory tract. The release of enzymes is triggered by engulfed asbestos particles that destabilize the lysosomal membranes.   *(a, ©1993 SPL/Custom Medical Stock Photo; b, ©1992 J.L. Carson/Custom Medical Stock Photo)*

destroy the cells of the respiratory tract, causing irreversible scarring of lung tissue (Figure B). Such scarring eventually interferes with the exchange of gases in the lungs, making breathing difficult. People with asbestos-damaged lungs are more susceptible to lung infections such as pneumonia and bronchitis, and these infections can be life threatening to those whose respiratory health is already compromised.

At present, there is no effective treatment for asbestosis. The focus, therefore, has been on prevention. In the last 20 years or so, governments have begun to regulate the use and removal of asbestos. In the United States, for example, the use of asbestos for purposes such as insulation and fireproofing has been banned since 1974. Almost all other uses were banned in 1989, with complete bans scheduled to be in effect by 1997.

But what can be done about the asbestos already present in our schools and work places? It is generally recommended and often required that exposed and crumbling asbestos be either removed, enclosed by other building materials, or covered with an effective sealant. Experts should determine which method is best for dealing with asbestos in a particular setting and should be involved in the sealing or removal. This is critical because the greatest risk of asbestos exposure occurs when it is improperly removed causing large numbers of particles to be released into the air. Finally, people at high risk of being exposed to asbestos—plumbers, construction workers, firemen, and building custodians, to name a few—should be informed of the health risks associated with their jobs and should insist upon frequent testing of the air in their work place.

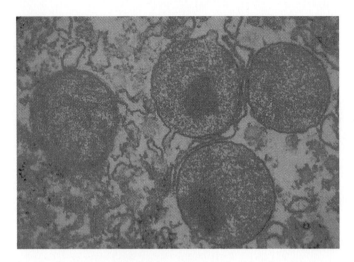

Figure 3-17 Cluster of peroxisomes in a liver cell. Also called microbodies, these organelles are sites of diverse chemical reactions ranging from the breakdown of fats to the detoxification of poisons and alcohol. *(Visuals Unlimited/© D. Friend/D. Fawcett)*

Peroxisomes

The **peroxisome** is a roughly spherical organelle surrounded by a single membrane (Figure 3–17). Within its matrix are enzymes that produce hydrogen peroxide (H_2O_2) as a by-product. Because H_2O_2 is itself toxic, it is converted to water by yet other enzymes within the peroxisome. The chemical reactions that take place in peroxisomes have diverse functions ranging from the breakdown of fats, particularly those made of very long-chain fatty acids, to the detoxification of poisons and alcohol. This latter function explains why peroxisomes are especially common in cells of the liver and kidneys, two locations in the body where molecules that enter the bloodstream are detoxified. In contrast to lysosomes, peroxisomes are not budded off from the Golgi complex but instead grow by taking in proteins and lipids produced in cytosol. Once peroxisomes reach a certain size, they split in two.

At present there are seven or eight genetic disorders known to disrupt the functioning of peroxisomes. In the disorder called Zellweger syndrome, for example, peroxisomal enzymes are synthesized normally in the cytosol but fail to move into the peroxisomes. Not protected within peroxisomes, the enzymes are soon destroyed by other enzymes in the cytosol. Zellweger syndrome is characterized by severe problems of the nervous system and malfunctioning of the liver and kidneys. The fact that this disorder is fatal attests to the importance of the chemical reactions that normally take place within peroxisomes. Table 3–4 summarizes the functions of organelles.

The Cytoskeleton

The organelles we have just described are suspended in cytosol, the aqueous environment within the cell. Within this fluid is a complex network of protein filaments called the **cytoskeleton.** One function of the cytoskeleton is to give the cell its shape. Some cells change their shape by dismantling the fibers of the cytoskeleton in one area and reassembling them in another. The cytoskeleton also functions in the movement of entire cells or certain organelles or vesicles within cells. In addition, some elements of the cytoskeleton serve to anchor organelles in place. We will have more to say about the functions of the cytoskeleton when we discuss its specific components. Apparently absent from bacteria, development of the cytoskeleton is thought to have been an important step in the evolution of eukaryotic cells.

Three types of filaments make up the cytoskeleton: microtubules, microfilaments, and intermediate filaments. **Microtubules** are the thickest fibers and are made from the globular protein tubulin. **Microfilaments** are the thinnest of the three fibers and are made from the globular protein actin. **Intermediate fibers** are a diverse group of fibers made from fibrous proteins. As their name suggests, intermediate fibers have diameters that fall in between those of microtubules and microfilaments.

Table 3-4 Overview of Organelle Functions

Organelle	Function
Nucleus	Contains almost all the genetic information and thus controls cellular structure and function
Nucleolus	Makes ribosomal RNA
Mitochondria	Provide cell with energy through the breakdown of glucose during cellular respiration
Ribosomes	Location of protein synthesis
Endoplasmic reticulum	
Rough (RER)	Site of synthesis of proteins to be secreted by cells; membrane production
Smooth (SER)	Detoxification of drugs; membrane production
Golgi complex	Sorts, modifies, and packages proteins produced by RER
Lysosomes	Digest substances imported from outside the cell; destroy old or defective cell parts or cells
Peroxisomes	Digest fats; detoxification of poisons and alcohol

Microtubules

Microtubules are straight, hollow rods made of the protein **tubulin.** Often, microtubules radiate out from a region near the nucleus called the **centrosome** (Figure 3–18a). Within the centrosome is a pair of **centrioles,** and each centriole is composed of nine sets of triplet microtubules arranged in a ring (Figure 3–18b). These microtubules near the nucleus are believed to provide support to the cell. Other microtubules near the plasma membrane function in maintenance of cell shape. Microtubules also serve as tracks along which organelles or vesicles may travel (Figure 3–19). For example, secretory vesicles budded off the Golgi complex make their way to the plasma membrane by attaching to a protein "motor" that propels them along a microtubule track to the membrane. Finally, microtubules play a role in cell division, the process by which the nucleus and cytoplasm of a cell split into two daughter cells. Cancer, a disease characterized by uncontrolled cell division, is sometimes treated by administering drugs that halt cell division. These highly toxic drugs, called chemotherapeutic agents, work either by dismantling

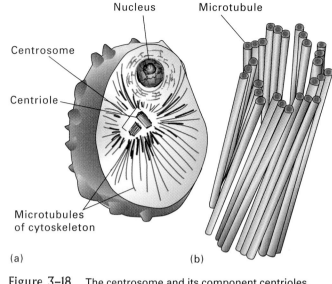

Figure 3–18 The centrosome and its component centrioles. (a) The centrosome, located near the cell nucleus, is an organizing center from which microtubules radiate outward. Within the centrosome are two centrioles arranged at right angles to one another. (b) Each centriole is composed of nine sets of triplet microtubules arranged in a ring.

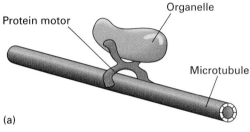

(a)

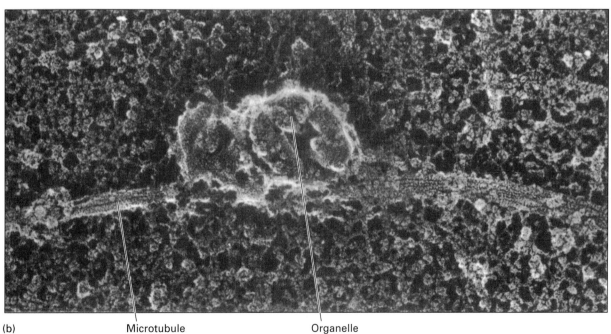

(b)   Microtubule   Organelle

Figure 3–19 Microtubules serve as tracks along which organelles or vesicles move. (a) An organelle attached to a protein motor travels along a microtubule. (b) An organelle moving along a microtubule in a nerve cell of a crayfish. (b, Courtesy of Dr. R. Vale)

microtubules or preventing their assembly and thereby stopping cell division. We will consider cancer and its treatments again in Chapter 19A.

Critical Thinking

Why do side effects, such as nausea and hair loss, often accompany the use of chemotherapeutic agents when treating cancer?

Microtubules are responsible for the movement of **cilia** and **flagella,** extensions of the plasma membrane found on some cells. Cilia are usually shorter and much more numerous than flagella. Resembling oars, cilia move in a rowing motion. Cilia are found, for example, on the surfaces of cells lining the respiratory tract (Figure 3–20). Here, cilia function to sweep debris trapped in mucus away from the lungs. Smoking destroys these cilia and thus hampers cleaning of respiratory structures. A flagellum, on the other hand, resembles a whip and performs an undulating motion. Most cells with flagella have one, although others may have a few. Each human sperm is propelled by a single flagellum (Figure 3–21).

Despite differences in length, number per cell, and pattern of movement, cilia and flagella have a similar arrangement of microtubules at their core. This arrangement, called a 9 + 2 pattern, consists of nine doublets of microtubules arranged in a ring with two single microtubules in the center of the ring (Figure 3–22). The microtubules of a cilium or flagellum are anchored to the cell by a structure called the

Figure 3-21 Sperm cells. Each sperm cell has a single whip-like flagellum. The undulating motion of the flagellum propels the sperm cell. Microtubules are responsible for movement of flagella. *(© Tony Stone Images/Douglas Struthers)*

basal body (Figure 3–23). Like the centrioles discussed above, basal bodies contain nine triplets of microtubules arranged in a ring.

Immotile cilia syndrome, a genetic disorder found in about 1 person out of every 20,000, is characterized by total or severe loss of function in cilia and flagella throughout the body. People with immotile cilia syndrome thus experience chronic respiratory problems such as bronchitis and sinusi-

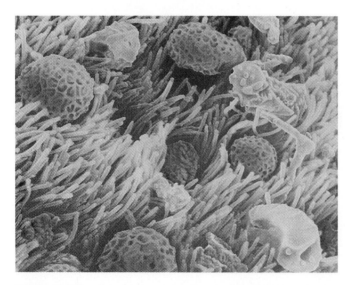

Figure 3-20 Cilia on cells lining the respiratory tract. The cilia move in a rowing motion and sweep debris trapped in mucus away from the lungs. Microtubules are responsible for the movement of cilia. *(Eddy Gray/Science Photo Library/Photo Researchers, Inc.)*

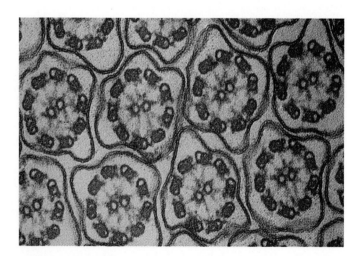

Figure 3-22 Structure of cilia. Several cilia in cross-section showing the 9 + 2 arrangement of microtubules. Flagella (not shown) have a similar arrangement of microtubules. *(© Omikron/Photo Researchers Science Source)*

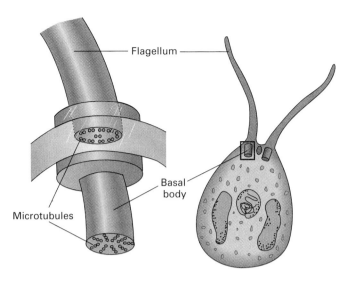

Figure 3-23 Basal bodies. Cilia and flagella are anchored to cells by basal bodies that contain an arrangement of microtubules identical to that found in centrioles.

tis, and males may be sterile because of immotile sperm. Loss of ciliary and flagellar function appears to be caused by the absence of arm-like structures that link the nine pairs of microtubules in the 9 + 2 pattern (Figure 3–24).

Microfilaments

Microfilaments are solid rods made of the globular protein **actin.** These fibers are best known for their role in muscle contraction when they slide past thicker filaments made of the protein **myosin,** causing muscle cells to shorten. Microfilaments also provide cells with support and are responsible for localized movements of the plasma membrane such as when extensions of the cell called **pseudopodia** (meaning "false foot") produce cell movement along a surface (Figure 3–25). This type of motion is called **amoeboid movement** and is characteristic of neutrophils, white blood cells that creep through infected tissue devouring bacteria and debris. Finally, microfilaments play a role in cell division when a band of them contracts and pinches the cell in two.

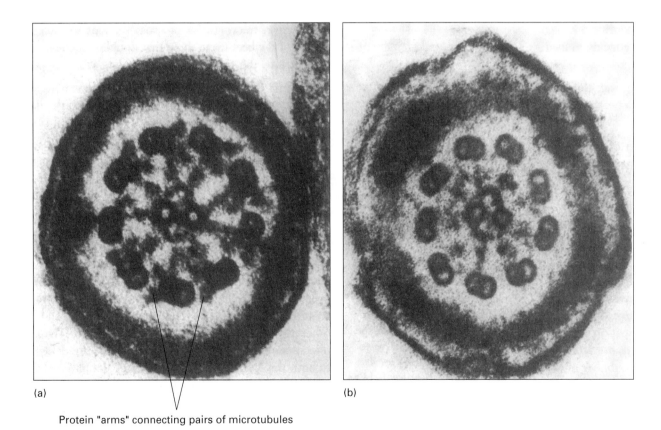

(a)

(b)

Protein "arms" connecting pairs of microtubules

Figure 3-24 Loss of function of cilia and flagella. (a) Cross-section of a flagellum of a normal sperm. (b) Cross-section of a sperm from a man with immotile-cilia syndrome, a disease characterized by total or severe loss of function in both cilia and flagella. Loss of ciliary and flagellar function is believed to result from the absence of protein "arms" connecting the nine pairs of microtubules. Men with immotile-cilia syndrome are usually sterile. *(Reprinted with permission from* Cell Biology *by David Sadava, Jones & Bartlett Publishers.)*

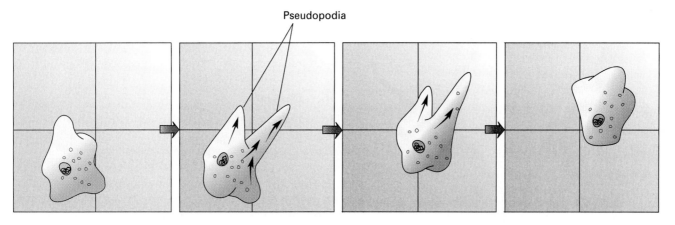

Pseudopodia

Figure 3–25 Amoeboid movement. Extensions of the cell called pseudopodia attach to a surface. When cytoplasm flows into the pseudopodia, the entire cell moves forward.

Intermediate Filaments

Intermediate filaments are a diverse group of fibers whose protein composition varies from one type of cell to another, although all are made of fibrous rather than globular proteins. Whereas microtubules and microfilaments are often disassembled and reassembled, intermediate filaments tend to be more permanent. These fibers help to maintain cell shape and to anchor certain organelles in place. The nucleus, for example, usually sits in a basket of intermediate filaments fixed in place by extensions to the outskirts of the cell. Table 3–5 summarizes the structures and functions of microtubules, microfilaments, and intermediate filaments.

Although we have discussed microtubules, microfilaments, and intermediate filaments as separate entities, these three components of the cytoskeleton are interconnected both physically and functionally. Cell movement and changes in shape undoubtedly require coordinated action from all three types of cytoskeletal elements.

Cell Structure Reflects Cell Function

The anatomy of a cell exquisitely reflects its functions (Figure 3–26). For example, few human cells are more specialized than sperm or egg, the cells that carry genetic and other information needed to make a new individual of the next generation. We have already discussed the highly specialized structure of the sperm's tail (Figure 3–26a). In addition, the head of a sperm cell contains a highly condensed nucleus and a greatly modified secretory vesicle known as the acrosome, which is essential for fertilization to occur. Eggs are much larger than typical cells because they are literally packed with nutrients in the form of yolk, as well as with mitochondria and billions of ribosomes. Thus, the embryo's protein-synthesizing machinery can start without delay (Figure 3–26b). Another example of cells highly specialized for specific functions are mature red blood cells, in which lack of a nucleus provides additional space for the hemoglobin needed for oxygen transport (Figure 3–26c).

Table 3–5 Structures and Functions of Cytoskeletal Elements

Element	Structure	Protein Monomer	Functions
Microtubules	Hollow rods	Tubulin	Cell movement (cilia and flagella) Support Tracks for organelles and vesicles Chromosome movements in cell division
Microfilaments	Solid rods	Actin	Muscle contraction Support Cell movement (as pseudopodia) Pinches cell in two during cell division
Intermediate filaments	Rope-like fibers	Several different proteins; depends on cell type	Maintains cell shape Support Anchors organelles

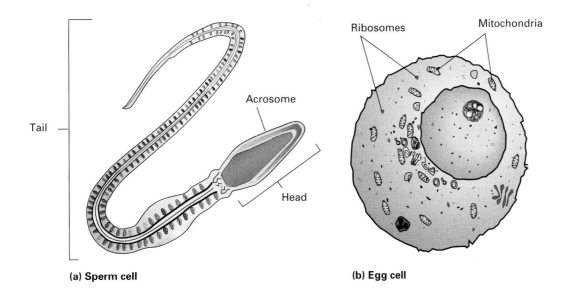

(a) Sperm cell

Tail

Acrosome

Head

Ribosomes

Mitochondria

(b) Egg cell

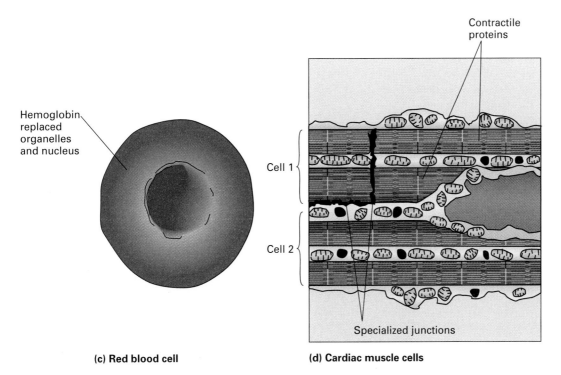

Hemoglobin replaced organelles and nucleus

Contractile proteins

Cell 1

Cell 2

Specialized junctions

(c) Red blood cell

(d) Cardiac muscle cells

Figure 3–26   Cell structure reflects cell function as illustrated by comparing the appearances of a sperm cell (a), egg cell (b), red blood cell (c), and cardiac muscle cells (d). See text for details.

Consider, also, cardiac muscle cells. These cells are specialized for contraction and propagation of the signal for contraction from one muscle cell to the next, and thus they are filled with contractile proteins and joined to adjacent cells by specialized junctions (Figure 3–26d). In each of these cases, careful study of the cell's structure provides excellent clues to its function and vice versa.

Critical Thinking

What specializations might you expect to see in cells that secrete steroid hormones (lipids) or connective tissue cells that secrete the protein collagen?

Energetics of the Cell

Living requires work and work requires energy. Logicians tell us, therefore, that living requires energy. So, where do we get our energy? The short answer is that we get our energy from the food we eat. Essentially, the carbohydrates and fats (and proteins under special circumstances such as starvation) in our food are broken down and the energy once stored in their chemical bonds is released. Some of this energy is used to form ATP, the form of energy that cells can use to do their work, and the rest is given off as heat. Our focus here will be on how ATP is generated during the breakdown of the carbohydrate glucose. We will trace the two pathways used by cells—cellular respiration and fermentation—to systematically break apart glucose molecules for energy. Whereas cellular respiration requires oxygen, fermentation does not.

Cellular Respiration

Cellular respiration is the oxygen-requiring pathway by which glucose is broken down by cells to yield carbon dioxide, water, and energy. The pathway is an elaborate series of chemical reactions that involves the passage of energy from one molecule to another. Why is cellular respiration so critical to cells? Basically, because the energy stored in the covalent bonds of glucose cannot be directly used by cells. Instead, this energy must be released during the breakdown of glucose, and then used by cells to make ATP, the form of energy that they can use. (Recall from Chapter 2 that ATP is the energy currency of cells and is formed from ADP and inorganic phosphate (P*i*), a process that requires energy.) The complete breakdown of glucose during cellular respiration is shown by the following chemical equation:

$$C_6H_{12}O_6 \ + \ 6O_2 \ \Rightarrow \ 6CO_2 \ + \ 6H_2O \ + \ \text{energy}$$

glucose   oxygen   carbon   water
                              dioxide

There are four phases or series of reactions in cellular respiration: (1) glycolysis (2) the transition reaction (3) the citric acid cycle and (4) the electron transport chain. Whereas glycolysis takes place in the cytosol of the cell, the remaining three phases take place in mitochondria.

Glycolysis

Glycolysis (literally "sugar-breaking") involves the splitting of glucose, a six-carbon sugar, into two three-carbon molecules called pyruvate, or pyruvic acid (Figure 3–27). This

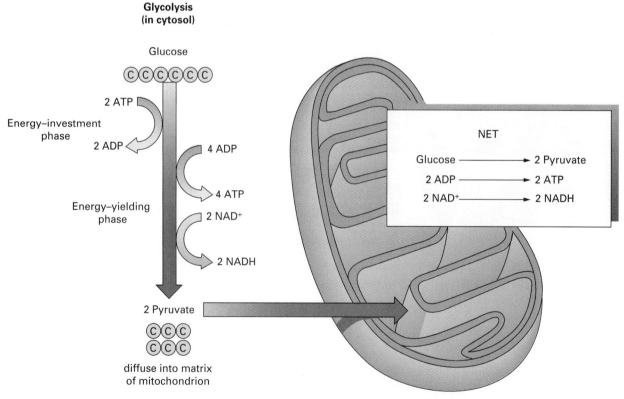

Figure 3–27 Glycolysis is a several-step sequence of reactions in the cytosol in which glucose, a six-carbon sugar, is split into two three-carbon molecules of pyruvate. During the first steps, two molecules of ATP are consumed. During the remaining steps, four molecules of ATP are produced, for a net yield of two ATP molecules. Two molecules of nicotine adenine dinucleotide (NADH) are also produced. The two molecules of pyruvate then diffuse from the cytosol into the inner compartment of the mitochondrion, where they enter the next phase of cellular respiration, the transition reaction.

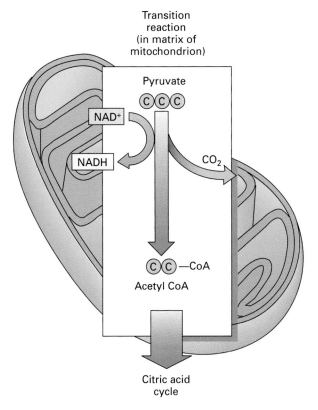

Transition
reaction
(in matrix of
mitochondrion)

Pyruvate

NAD^+

NADH

CO_2

—CoA

Acetyl CoA

Citric acid
cycle

Figure 3–28   The transition reaction takes place in the matrix of the mitochondrion. During this reaction, pyruvate reacts with a substance called coenzyme A (CoA), resulting in the removal of one carbon (in the form of CO_2) from each pyruvate. The resulting two-carbon molecule binds to CoA forming acetyl CoA. Acetyl CoA then enters the next phase of cellular respiration, the citric acid cycle. Two molecules of NADH are also produced during the transition reaction.

splitting occurs in several steps, and each step is catalyzed by a different enzyme. During the first steps, two molecules of ATP are consumed, but during the remaining steps four molecules of ATP are produced, for a net gain of two ATP. Glycolysis also produces two molecules of nicotine adenine dinucleotide (NADH), a molecule that participates in chemical reactions by picking up either hydrogens or electrons. Glycolysis does not require oxygen and releases only a small amount of the chemical energy stored in glucose. Most of the energy remains in the two molecules of pyruvate that move from the cytosol into the inner compartment of the mitochondrion, called the matrix, where they enter the next phase of cellular respiration, the transition reaction.

Transition Reaction

Once inside the matrix of the mitochondrion, pyruvate reacts with a substance called coenzyme A (CoA). This reaction, called the **transition reaction,** results in the removal of one carbon (in the form of CO_2) from each pyruvate (Figure 3–28). The resulting two-carbon molecule, called

an acetyl group, then binds to CoA forming acetyl CoA. A molecule of NADH is also produced from each pyruvate. Acetyl CoA then enters the next phase of cellular respiration, the citric acid cycle.

Citric Acid Cycle

Still in the matrix of the mitochondrion, acetyl CoA, the two-carbon compound formed during the transition reaction, then enters a cyclic series of chemical reactions known as the **citric acid cycle,** named after the first product (citric acid or citrate) formed along its route (Figure 3–29). This cycle is sometimes called the Kreb's cycle, after the scientist Hans Krebs who described many of the reactions. Rather than considering each of the chemical reactions in this eight-step series, let us simply say that the citric acid cycle yields two molecules of ATP (one from each acetyl CoA that enters the cycle) and several molecules of NADH and FADH₂ (flavin adenine dinucleotide), carriers of high-energy electrons that enter the electron transport chain, the final phase of cellular respiration.

Electron Transport Chain and Chemiosmosis

During the final phase of cellular respiration, the molecules of NADH and FADH₂ produced by glycolysis and the citric acid cycle pass their electrons to a series of carrier proteins embedded in the inner membrane of the mitochondrion. This series of protein molecules is known as the **electron transport chain** (Figure 3–30). (Recall that the inner membrane of the mitochondrion is highly folded, thus creating space for literally thousands of copies of the carrier proteins.) During the transfer of electrons from one molecule to the next, energy is released and this energy is then used to make ATP. Eventually, the electrons are passed to oxygen, the final electron acceptor. Now we see the critical role of oxygen in cellular respiration. In the absence of oxygen, electrons accumulate in the carrier molecules, halting the citric acid cycle and cellular respiration. When oxygen is present, however, electron transfer and the citric acid cycle continue. Upon accepting electrons, oxygen then combines with two hydrogens to form water.

How is the energy released when electrons move down the transport chain used to make ATP? The released energy is used to pump hydrogen ions formed during some of the reactions of the citric acid cycle across the inner membrane of the mitochondrion and into the intermembrane space. The tendency for hydrogen ions to move back across the inner membrane is strong, however, due in part to their lower concentration in the matrix. Substances tend to move from regions of higher to lower concentration; in this case, then, they move from the intermembrane space back into the matrix. Here is the catch: the structure of the inner mitochondrial membrane is such that hydrogen ions can be pumped out across its entire surface, but they can return to the matrix only at certain locations. These locations contain ATP synthetase, the enzyme responsible for synthesis of ATP. In

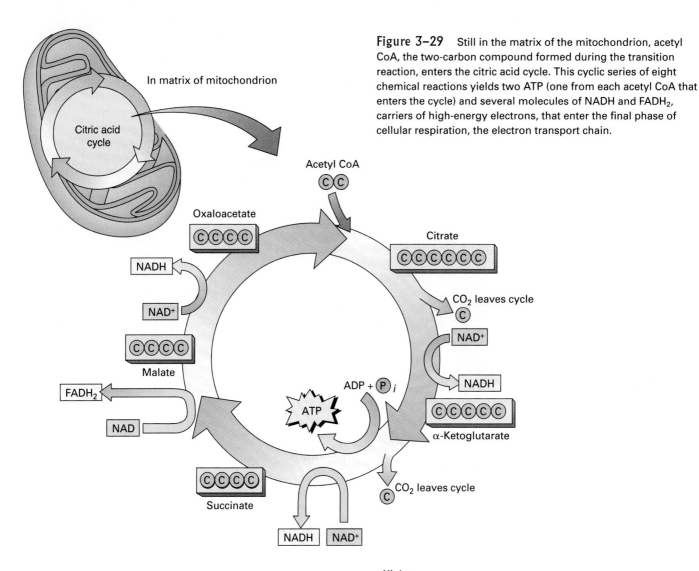

In matrix of mitochondrion

Citric acid cycle

Figure 3-29 Still in the matrix of the mitochondrion, acetyl CoA, the two-carbon compound formed during the transition reaction, enters the citric acid cycle. This cyclic series of eight chemical reactions yields two ATP (one from each acetyl CoA that enters the cycle) and several molecules of NADH and $FADH_2$, carriers of high-energy electrons, that enter the final phase of cellular respiration, the electron transport chain.

Acetyl CoA

Oxaloacetate

Citrate

CO_2 leaves cycle

NADH

NAD^+

NAD^+

NADH

Malate

α-Ketoglutarate

$FADH_2$

$ADP + P_i$

ATP

NAD

CO_2 leaves cycle

Succinate

NADH NAD^+

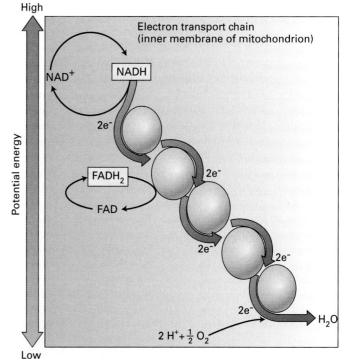

High

Potential energy

Low

Electron transport chain (inner membrane of mitochondrion)

NAD^+

NADH

$2e^-$

$FADH_2$

$2e^-$

FAD

$2e^-$

$2e^-$

$2e^-$

H_2O

$2 H^+ + \frac{1}{2} O_2$

Figure 3-30 The electron transport chain. During the final phase of cellular respiration, the molecules of NADH and $FADH_2$ produced by glycolysis, the transition reaction, and the citric acid cycle pass their electrons to a series of protein molecules (shown as colored circles) embedded in the inner membrane of the mitochondrion. As the electrons are transferred from one protein to the next, energy is released. Through chemiosmosis (see text and Figure 3–31) this energy is used to make ATP from ADP and inorganic phosphate (Pi). Eventually, the electrons are passed to oxygen, which combines with two hydrogens to form water.

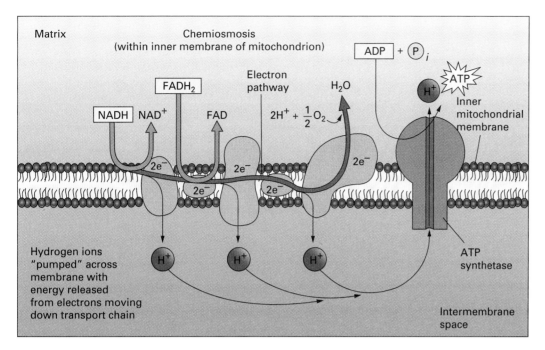

Figure 3-31 Chemiosmosis is the link between the electron transport chain and synthesis of ATP. The energy released as electrons move down the transport chain is used to pump hydrogen ions across the inner membrane of the mitochondrion and into its intermembrane space. There is a strong tendency for hydrogen ions to move back across the inner membrane caused in part by their lower concentration in the matrix. However, the structure of the inner mitochondrial membrane is such that although hydrogen ions can be pumped out across its entire surface, they can only return to the matrix at select points. These select points of entry house ATP synthetase, the enzyme responsible for making ATP. Thus, it is the inward flow of hydrogen ions that drives the synthesis of ATP from ADP and phosphate. The electron transport chain and chemiosmosis yield 32 ATP per molecule of glucose.

a process called **chemiosmosis,** the inward flow of hydrogen ions drives the synthesis of ATP from ADP and inorganic phosphate, Pi. Thus, chemiosmosis is the link between electron transport and ATP synthesis (Figure 3–31). Through chemiosmosis, the electron transport chain produces 32 molecules of ATP per molecule of glucose.

The entire process of cellular respiration is summarized in Figure 3–32, and basic descriptions of each of the four steps can be found in Table 3–6. All together, cellular respiration produces about 36 ATP per molecule of glucose (2 ATP from glycolysis, 2 ATP from the citric acid cycle, and 32 ATP from the electron transport chain). The value of 36 is an estimate, however, since each molecule of NADH yields somewhere between two and three molecules of ATP and each molecule of FADH$_2$ yields about two ATP.

Fermentation

We have seen that cellular respiration depends on oxygen as the final electron acceptor in the electron transport chain and that without it, the transport chain comes to a halt, stopping the citric acid cycle and cellular respiration. The question arises, then, is there a way for cells to harvest energy in the absence of oxygen? The answer is yes, and the pathway is called fermentation.

Fermentation begins with glycolysis. Glycolysis, you recall, occurs in the cytosol, does not require oxygen, and produces from one molecule of glucose two molecules each of ATP (although four ATP are produced, two are consumed for a net gain of two ATP), pyruvate, and the electron carrier NADH. The chemical reactions of fermentation that follow glycolysis also take place in the cytosol and involve a transfer of electrons from NADH to pyruvate or a derivative of pyruvate. This transfer of electrons is critical because it regenerates NAD$^+$, a molecule essential for the production of ATP through glycolysis. Thus, in contrast to cellular respiration in which oxygen is the final electron acceptor in the electron transport chain, in fermentation it is pyruvate or one of its derivatives that accepts the electrons. An important point about fermentation is that it is a very inefficient way for cells to harvest energy since it nets only 2 molecules of ATP as compared with the approximately 36 ATP molecules produced by cellular respiration.

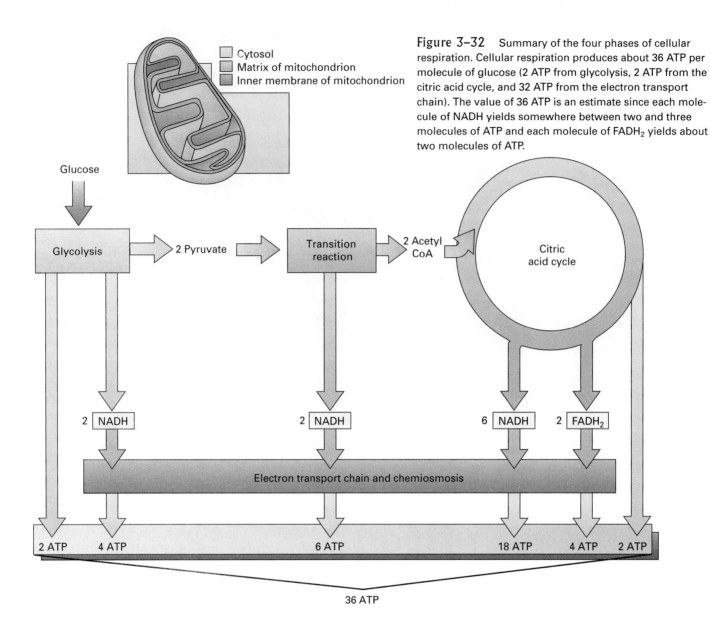

Figure 3–32 Summary of the four phases of cellular respiration. Cellular respiration produces about 36 ATP per molecule of glucose (2 ATP from glycolysis, 2 ATP from the citric acid cycle, and 32 ATP from the electron transport chain). The value of 36 ATP is an estimate since each molecule of NADH yields somewhere between two and three molecules of ATP and each molecule of $FADH_2$ yields about two molecules of ATP.

Table 3–6 The Four Phases of Cellular Respiration

Element	Location	Description	Main Products
Glycolysis	Cytosol	Several-step process by which glucose is split into 2 molecules of pyruvate	2 pyruvate 2 ATP 2 NADH
Transition reaction	Matrix of mitochondria	One CO_2 is removed from each pyruvate and the resulting molecules bind to CoA, forming 2 molecules of acetyl CoA	2 acetyl CoA 2 NADH
Citric acid cycle	Matrix of mitochondria	Cyclic series of 8 chemical reactions by which acetyl CoA is broken down	2 ATP 2 $FADH_2$ 6 NADH
Electron transport chain and chemiosmosis	Inner membrane of mitochondria	Electrons from NADH and $FADH_2$ are passed from one protein to the next, releasing energy for ATP synthesis through chemiosmosis	32 ATP H_2O

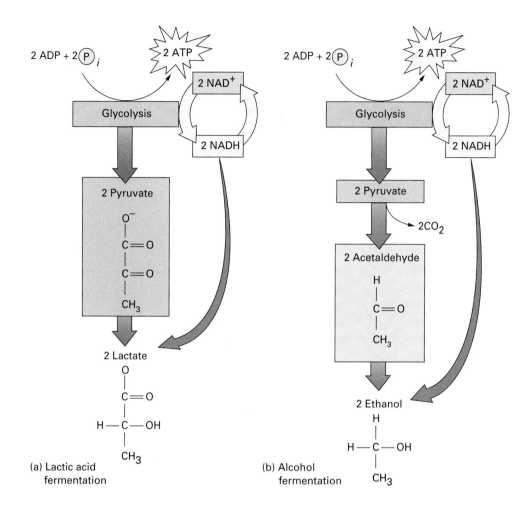

Figure 3–33 Two types of fermentation. Fermentation does not require oxygen and yields only two molecules of ATP per molecule of glucose. (a) Lactic acid fermentation occurs in the cytosol and consists of glycolysis plus chemical reactions in which NADH passes electrons directly to pyruvate, forming the waste product lactate or lactic acid. This type of fermentation occurs in our muscle cells during strenuous exercise. (b) Alcohol fermentation occurs in the cytosol of organisms such as yeast and consists of glycolysis plus a two-step chemical process. In the first step, CO_2 is released from pyruvate, leaving a two-carbon derivative of pyruvate called acetaldehyde. In the second step, NADH passes electrons to acetaldehyde, generating ethyl alcohol or ethanol. This process does not occur in our cells but is used by yeast, a microorganism used in the making of bread and beer.

There are several types of fermentation, distinguished from one another by the waste products formed from pyruvate. We will consider two types of fermentation that have some relevance to our daily lives, lactic acid fermentation and alcohol fermentation.

Lactic Acid Fermentation

During **lactic acid fermentation,** NADH passes electrons directly to pyruvate, forming the waste product lactate or lactic acid (Figure 3–33a). During strenuous exercise, the oxygen supply in our muscle cells is low and the cells switch from cellular respiration to lactic acid fermentation to ensure continued production of ATP. The muscle pain we often experience after intense exercise is caused, in part, by the accumulation of lactic acid produced by fermentation. In time the soreness disappears as the lactic acid diffuses out of the muscle cells and into the bloodstream where it is carried to the liver. Once in the liver, lactic acid is converted to pyruvate and eventually glucose.

Alcohol Fermentation

Unlike lactic acid fermentation, **alcohol fermentation** is a two-step process (Figure 3–33b). First, CO_2 is released from pyruvate, leaving a two-carbon derivative of pyruvate called acetaldehyde. In the second step, NADH passes electrons to acetaldehyde, generating ethyl alcohol or ethanol. Thus, in this type of fermentation, electrons are not passed directly to pyruvate, but to a derivative of pyruvate, acetaldehyde. Yeast, a single-celled fungus, switches to alcohol fermentation when oxygen becomes scarce in its environment. We often add yeast cells to bread dough and it is within this low-oxygen environment that yeast uses alcohol fermentation. The CO_2 produced during the first step of alcohol fermentation causes the bread to rise. The ethanol produced as a waste product dissipates during baking. Brewers don't see the ethanol as a waste, but as a product (beer) to be bottled and sold with warning labels.

The Cell Cycle

We begin life as a single cell, called a zygote, formed by the union of an egg and a sperm. By adulthood, our bodies consist of trillions of cells. What happened in the intervening years? How did we go from a single cell to the multitude of

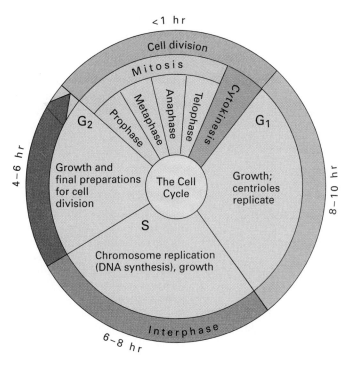

Figure 3–34 The cell cycle.

cells that make up the tissues of a fully functional adult? To put it simply, cell division happened, over and over and over again.

Mitosis, one type of cell division, is the process in which two identical cells, called daughter cells, are generated from a single cell. The original cell, as we will see, first replicates its genetic material and then distributes it equally among its daughter cells. Along with a full complement of genetic material, each daughter cell receives organelles, macromolecules, and cytosol from the original cell. Mitosis occurs in **somatic cells,** which are all the cells of the body with the exception of eggs and sperm. Eggs and sperm are considered **sex cells** or gametes and they are formed through **meiosis,** a second type of cell division in which daughter cells end up with half the amount of genetic material in the original cell. We will have more to say about meiosis in Chapter 18.

Mitosis is only one phase during the life of a dividing cell. The entire sequence of changes that a cell repeatedly goes through is called the **cell cycle** (Figure 3–34). The cell cycle consists of two major phases, interphase and cell division.

Interphase

Interphase is the period between cell divisions. It usually accounts for about 90% of the time that elapses during a cell cycle, but it is not a "resting period" as once thought. Indeed, it is now known that interphase is a time of intense preparation of cell division, a time when the DNA, cytosol, and organelles are duplicated. Such preparations ensure that

when the cell divides, each of its daughter cells will receive the essentials for survival.

Interphase consists of three parts: G_1, S, and G_2. Here the "S" stands for "synthesis" (DNA synthesis), and the "G" in G_1 and G_2 can stand either for "growth" or "gap" (see Figure 3–34). All three parts of interphase are times of cell growth, characterized by the production of organelles and the synthesis of proteins and other macromolecules. There are, however, some events specific to certain parts of interphase. For example, during G_1, the centriole is duplicated, and during the S phase, DNA is replicated. The details of DNA replication are described in Chapter 19. Our discussion here introduces some basic terminology regarding the stages of the cell cycle.

DNA, as we know, is the genetic material. **Chromosomes** are made of DNA and associated globular proteins called **histones** and are found in the cell nucleus. Depending on the phase of the cell cycle, chromosomes may be either single- or double-stranded. With few exceptions (sperm, eggs, and mature red blood cells in adults are some examples), human cells have 46 chromosomes. These 46 chromosomes carry an estimated 100,000 **genes,** segments of DNA that influence cell structure and function by specifying proteins to be synthesized.

During G_1, the chromosomes consist of a single strand of DNA and proteins. When DNA is replicated during the S phase, the single-stranded chromosomes change to double-stranded structures (Figure 3–35). Each strand of a particular chromosome, called a **chromatid,** has the same genetic information as the other strand. The two chromatids

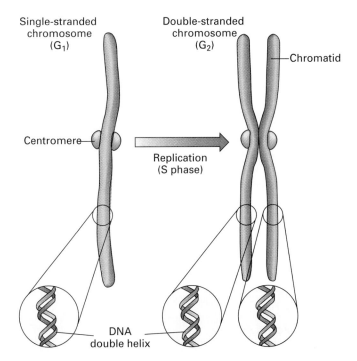

Figure 3–35 Changes in chromosome structure as a result of DNA replication during the S phase of interphase.

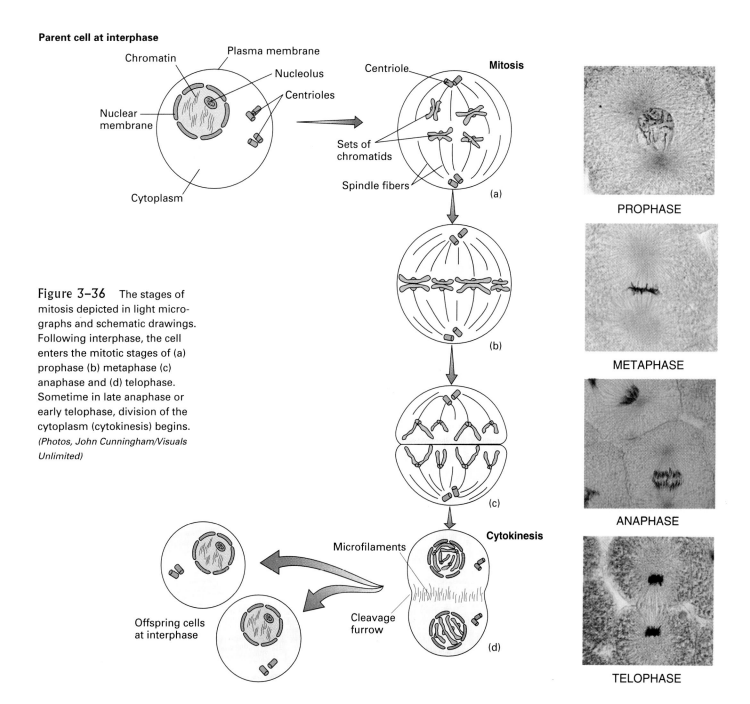

Parent cell at interphase

Chromatin · Plasma membrane · Nucleolus · Centrioles · Nuclear membrane · Cytoplasm

Centriole · **Mitosis** · Sets of chromatids · Spindle fibers · (a)

(b)

(c)

Microfilaments · **Cytokinesis** · Cleavage furrow · (d)

Offspring cells at interphase

PROPHASE

METAPHASE

ANAPHASE

TELOPHASE

Figure 3–36 The stages of mitosis depicted in light micrographs and schematic drawings. Following interphase, the cell enters the mitotic stages of (a) prophase (b) metaphase (c) anaphase and (d) telophase. Sometime in late anaphase or early telophase, division of the cytoplasm (cytokinesis) begins. *(Photos, John Cunningham/Visuals Unlimited)*

are held together at a region of the chromosome known as the **centromere.** Once all the preparations of interphase have been completed, the cell is ready for the next stage, cell division.

Somatic Cell Division

Somatic cell division occurs continually in the developing embryo and fetus and plays an important role in the growth and repair of body tissues in children. In the adult, some cells, such as nerve cells, have been arrested in interphase for some time, having completely lost their ability to divide. Other adult cells, such as liver cells, stop dividing, but they retain the ability to undergo cell division should the need for

tissue repair and replacement arise. Still other cells actively divide throughout life. Skin cells, for example, continue to divide in adults, the ongoing cell division serving to replace the enormous numbers of cells worn off each day.

Somatic cell division consists of two processes that overlap in their timing. The first process, division of the nucleus, is called **mitosis.** The second process, division of the cytoplasm, is called **cytokinesis,** and it occurs toward the end of mitosis.

Mitosis

For convenience, mitosis is usually divided into four stages: prophase, metaphase, anaphase, and telophase. The major events of each stage are depicted in Figure 3–36 and described next.

During **prophase,** changes occur both in the nucleus and in the cytoplasm (Figure 3–36a). In the nucleus, the chromosomes begin to thicken and shorten, becoming visible when stained and viewed under a light microscope. Before reaching this condensed state, the genetic material is often called **chromatin.** About this time, the nucleolus disappears, its absence reflecting the shutdown of rRNA production in the cell. The nuclear envelope also begins to break down. Outside the nucleus in the cytoplasm, a structure called the mitotic spindle is formed. The mitotic spindle is made of microtubules associated with the centrioles. Some of these microtubules run from a centriole to the centromere of a chromosome and are called **chromosome-to-pole fibers.** Others run from one centriole to the other and are called **pole-to-pole fibers.** The fibers of the mitotic spindle help draw the duplicated chromosomes apart, providing each daughter cell with a complete set of genetic material. The centrioles, duplicated during interphase, move away from each other toward opposite ends of the cell.

The next stage of the cell cycle is **metaphase,** a time when the chromosomes, guided by the fibers of the mitotic spindle, form a line at the center of the cell (Figure 3–36b). The actual location is called the **metaphase plate** and it is equidistant between the two poles of the mitotic spindle. As a result of this alignment, when the chromosomes split at the centromere, each daughter cell receives one chromatid from each chromosome and thus a complete set of the parent cell's chromosomes.

Anaphase begins when the chromatids of each chromosome begin to separate, splitting at the centromere. Now separate entities, the chromatids are considered chromosomes in their own right and they move toward opposite poles of the cell (Figure 3–36c). Although the fibers of the mitotic spindle are known to be involved in the drawing apart of the chromosomes, many details of the process are unclear. By the end of anaphase, equivalent collections of chromosomes are located at the two poles of the cell.

During **telophase,** nuclear envelopes form around the group of chromosomes at each pole, the mitotic spindle is disassembled, and nucleoli reappear (Figure 3–36d). The chromosomes also become less condensed and more thread-like in appearance.

As we have seen, one of the major features of cell division is shortening and thickening of the chromosomes. Chromosomes in this condensed state are visible with a light microscope and can be used for diagnostic purposes, such as when potential parents want to check their own chromosomal makeup for defects. One often-used method involves taking a blood sample, separating out white blood cells from the sample, and then growing the cells in containers with the nutrients and other substances required by cells. This culture is then treated with colchicine, a drug that destroys the mitotic spindle. Destruction of the spindle prevents separation of the chromosomes, halting cell division at metaphase. The cells in the culture are then fixed, stained, and photographed. The images of the chromosomes can then be arranged in groups based on physical characteristics such as length and location of the centromere. This arrangement of chromosomes is called a **karyotype** (Figure 3–37). Karyotypes can then be checked for defects in number or structure of chromosomes.

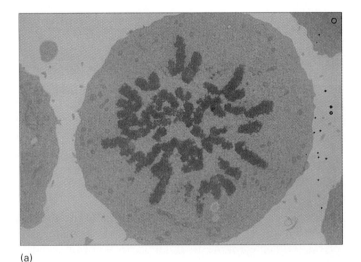

(a)

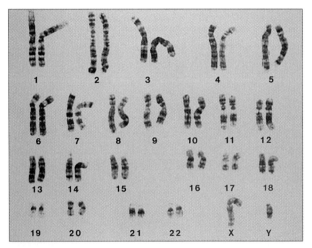

(b)

Figure 3–37 Chromosomes in dividing cells can be examined for defects in number or structure. (a) Photomicrograph of metaphase chromosomes from a human white blood cell. (b) A karyotype constructed by arranging the chromosomes from photographs like that in (a). The chromosomes have been arranged on the basis of size and location of centromere.

(a, Visuals Unlimited/© Robert Caughey; b, Department of Clinical Cytogenetics, Addenbrookes Hospital, Cambridge/Science Photo Library/Photo Researchers, Inc.)

Table 3–7 Phases of the Cell Cycle

Phase	Major Events
Interphase	
G₁	Production of macromolecules and organelles; duplication of centriole
S	Production of macromolecules and organelles; replication of DNA
G₂	Production of macromolecules and organelles
Cell division	
Mitosis	
Prophase	Chromosomes condense; nucleolus disappears; nuclear envelope breaks down; mitotic spindle formed
Metaphase	Chromosomes line up at metaphase plate
Anaphase	Centromere splits and chromatids begin to separate, each now becoming a chromosome
Telophase	Chromosomes reach opposite poles and nuclear envelopes form; mitotic spindle disassembled; nucleolus reappears; chromosomes uncoil
Cytokinesis	Cytoplasm divides

Cytokinesis

Cytokinesis—division of the cytoplasm—begins sometime during late anaphase or early telophase. At this time, a band of microfilaments located just below the plasma membrane in the area that was the metaphase plate contracts and forms a furrow. The furrow deepens, eventually pinching the cell in two (early stages of cytokinesis can be seen in Figure 3–36d). The phases of the cell cycle are summarized in Table 3–7.

Control of Cell Division

What controls the proliferation of cells in our bodies? How is it that damage to the liver prompts surrounding liver cells to divide? How do hormones stimulate the cells of the uterus to proliferate rapidly for a few days so that the tissue lost by menstruation is replaced? As it turns out, there are very specific controls that govern proliferation of the various cell types in the human body, controls that maintain a delicate balance of cell numbers according to need. Because it is difficult to study control of cell division in the complex environment of the human body, most studies have involved cells grown in culture. The environments of these cells can be precisely manipulated and the responses of the cells to various treatments can be continuously monitored.

Studies of cells in culture have shown that different cell types require different proteins, called growth factors, in order to divide. In fact, most cell types need a specific combination of growth factors rather than a single growth factor to stimulate cell division. Balanced against these growth factors are also factors that inhibit cell division. When cells are placed in a dish in the presence of their appropriate growth factors they divide until forming a monolayer, a layer in which neighboring cells are in contact with one another (Figure 3–38a). Once this monolayer is formed, the cells stop dividing. This phenomenon, known as **density-dependent contact inhibition** of cell division, is thought to result from competition among the cells for growth factors and nutrients. In addition, most cells in culture stop dividing when they lose their contact with a solid surface.

Many of the controls described above for normal cells appear to be lost in cancer cells. In culture, rather than stopping cell division once a monolayer is formed, cancer cells continue to divide, piling up on one another, until nutrients run out (Figure 3–38b). The resultant mass of cancer cells in the culture dish illustrates how tumors form when such frenzied cell division occurs in the body. Not only do cancer cells continue to divide despite having reached unusually high densities but they also seem to be less dependent on growth factors and attachment to a solid surface.

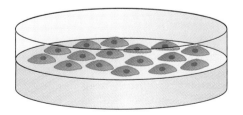
(a) Normal cells

(b) Cancer cells

Figure 3–38 Density-dependent contact inhibition of cell division in cells grown in culture. (a) Normal cells stop dividing once they form a monolayer. (b) Cancer cells do not display density-dependent contact inhibition and continue to divide, piling on top of one another.

SUMMARY

1. There are two main types of cells: prokaryotic cells, unique to bacteria and cyanobacteria, and eukaryotic cells, found in all other organisms. The structural differences between prokaryotic and eukaryotic cells are summarized in Table 3–1.

2. The plasma membrane is composed of phospholipids arranged in a lipid bilayer. Proteins and molecules of cholesterol are interspersed throughout the bilayer; carbohydrates attach only to the outer surface of the membrane. The structure of the plasma membrane is often described as a fluid mosaic. This is because the proteins are interspersed throughout the lipid molecules, giving the membrane its mosaic quality, and both protein and lipid molecules can display substantial sideways movement, giving the membrane its fluid quality. The functions of the plasma membrane are summarized in Table 3–2.

3. The plasma membrane controls which substances move into and out of the cell and at what rates they move. Table 3–3 summarizes the mechanisms by which substances move across the plasma membrane.

4. Within the eukaryotic cell, membranes delineate compartments within which specific chemical processes occur. Organelles, small components within cells, are suspended in the cytosol, the aqueous environment inside the cell. Cytoplasm is a term used to describe the cytosol plus all the organelles except the nucleus. The organelles and their functions are listed in Table 3–4.

5. Rather than being an amorphous fluid, the cytosol is actually a highly structured environment filled with a complex network of protein filaments called the cytoskeleton. Three types of filaments make up the cytoskeleton: microtubules, microfilaments, and intermediate filaments. The structures and functions of these filaments are summarized in Table 3–5.

6. Cells require energy to work. We get this energy from the food we eat. Carbohydrates, fats, and sometimes proteins that we consume are broken down and the energy stored in their chemical bonds is released. Some of this energy is used to make ATP and some is given off as heat.

7. Cells use two pathways to break down the carbohydrate glucose: cellular respiration and fermentation. Cellular respiration requires oxygen and yields about 36 molecules of ATP per molecule of glucose. The four series of reactions of cellular respiration are summarized in Table 3–6. Fermentation, on the other hand, does not require oxygen and nets only about two molecules of ATP per molecule of glucose. Lactic acid fermentation occurs in our muscle cells during periods of strenuous exercise when oxygen supplies are low. The lactic acid produced as a by-product of this pathway is responsible, in part, for the muscle pain we often experience after intense exercise.

8. The cell cycle consists of two major phases: interphase and cell division. Interphase consists of three parts (G_1, S, and G_2) and is the period between cell divisions when DNA, cytosol, and organelles are replicated in preparation for the cell to divide and produce two identical daughter cells. Somatic cell division consists of mitosis (division of the nucleus) and cytokinesis (division of the cytoplasm). The phases of the cell cycle are summarized in Table 3–7.

9. Mitosis occurs in somatic cells, which are all the cells of the body except eggs and sperm. It is a form of cell division in which two identical daughter cells are generated from a single cell. The original cell first replicates its genetic material and then distributes it equally among its daughter cells. There are four stages of mitosis: prophase, metaphase, anaphase, and telophase. The major events of each stage are summarized in Table 3–7. The sex cells—eggs and sperm—are formed through meiosis, a second type of cell division in which daughter cells end up with half the amount of genetic material in the original cell.

10. Cytokinesis, division of the cytoplasm, begins sometime during late anaphase or early telophase. A band of microfilaments, located just below the plasma membrane at the midline of the cell, contracts and forms a furrow that deepens and eventually pinches the cell in two.

11. Studies of normal cells grown in culture have illuminated factors that control cell division. Most cell types require a specific combination of proteins called growth factors in order to divide. When cells are placed on a dish in the presence of their appropriate growth factors, they divide until forming a monolayer. This is called density-dependent contact inhibition of cell division. Attachment to a substrate also seems to influence cell division; most cells stop dividing when they lose contact with a solid surface.

12. Many of the controls that regulate cell division in normal cells are lost in cancer cells. Cancer cells do not exhibit density-dependent contact inhibition and seem less dependent on growth factors and attachment to a substrate.

REVIEW QUESTIONS

1. How do prokaryotic and eukaryotic cells differ?
2. Describe the structure of the plasma membrane. Why is the plasma membrane often described as a fluid mosaic?
3. List five functions of the plasma membrane.
4. What is the difference between simple and facilitated diffusion? Give examples of each.
5. Describe endocytosis and exocytosis.
6. Why is the nucleus considered the command center of the cell?
7. What is the endosymbiont hypothesis? What evidence is there to support this hypothesis?
8. List three functions of lysosomes. What are lysosomal storage diseases?
9. Name the three types of filaments that make up the cytoskeleton. Describe their structures and functions.
10. What is the basic function of cellular respiration?
11. How do cellular respiration and fermentation differ with respect to the number of ATP molecules produced and the requirement for oxygen?
12. Define mitosis and cytokinesis.
13. What factors control cell division? What happens to these controls in cancer cells?

Chapter 4

Body Organization and Homeostasis

Cells are arranged in tissues, and tissues, in turn, form organs. The skin, our largest organ, protects underlying tissues and helps to regulate body temperature. *(Phil Jude/Science Photo Library/Photo Researchers, Inc.)*

A walk on a crowded beach will convince anyone that there is no such thing as "the human body." Not only do we come in two basic and quite distinct models, but within each sex or even within each age-group, ethnic origin, or dietary group, for example, we find incredible variation among humans. These differences, though, are not enough to mask fundamental similarities that, taken together, enable us to discuss the basic organization of the human body.

In Chapters 2 and 3 we introduced some fundamentals of body chemistry and cellular organization that characterize not only human cells but also those of other vertebrates, invertebrates, plants, and even simpler organisms. Yet these fundamentals cannot tell the entire story, for the many cell types found in humans are arranged in a bewildering array of tissues. These tissues, in turn, form organs, and organs function as part of organ systems. As usual, in biology, there is more to be examined.

In this chapter, we will take a closer look at ourselves, beginning with a broader discussion of cells and the tissues they form. After that we will take an extended look at a single organ system, the integumentary system. We will end with a discussion of how the body maintains relatively constant internal conditions at its many organizational levels.

Basic Organization of the Human Body

Cells and Tissues

Tissues are groups of cells that work together to serve a common function, for example, protection of the body from external environmental conditions. The study of tissues is called **histology**. Tissues come in four main types: epithelial tissue, connective tissue, muscle tissue, and nervous tissue. Epithelial tissue covers body surfaces, lines body cavities and organs, and forms glands. Connective tissue serves as a storage site for fat, plays an important role in immunity, and provides the body and its organs with protection and support. Muscle tissue is responsible for movement and nervous tissue for the coordination of body activities that results from initiation and transmission of nerve impulses. Each of the major types of tissues will be considered in more detail.

Epithelial Tissue

There are several key features of epithelial tissue. Adjacent cells of epithelial tissue are organized into a tight, coherent sheet with extensive cell-cell contact. The most familiar examples of epithelia are probably the skin and the lining of the digestive and respiratory tracts. The lining layer of the body cavity is also epithelium, as are lining layers of the urinary and reproductive tracts and the inner walls of blood vessels. The cells of epithelial tissue typically have **junctional complexes**, membrane specializations that attach adjacent cells to each other to form a contiguous sheet. There are three kinds of junctional complexes: tight junctions, desmosomes, and gap junctions (Figure 4–1). In **tight junctions**, the membranes of neighboring cells are actually attached, forming a seal. Tight junctions, such as those in the lining of the urinary tract or intestine, prevent fluid from flowing across the epithelium through the minute spaces between adjacent cells. **Desmosomes** are junctional complexes that resemble rivets holding adjacent cells together. There are often so many desmosomes along the membranes of adjacent epithelial cells that they are visible with the light microscope. **Gap junctions**, on the other hand, link the cytoplasm of adjacent cells through small holes, allowing physical and electrical continuity between cells.

Another key feature of epithelial tissue is that its cells have distinct **basal** and **apical** surfaces. Whereas the basal surfaces are anchored, the apical surfaces are free and typically face into a body cavity, the lining of an internal organ, or to the outside of the body. Cells with distinct basal and apical surfaces are said to be **polarized**.

The basal surfaces of epithelial cells are anchored to connective tissue by a basement membrane (Figure 4–2). This membrane contains two layers, the basal lamina and the reticular lamina. The **basal lamina**, secreted by the epithelium, is analogous to a concrete foundation strip upon which bricks are laid to build a wall. Like a foundation beneath a wall of bricks, the basal lamina is continuous beneath adjacent cells and serves to tie the overlying layer together. The **reticular lamina** is secreted by cells of the connective tissue to which the epithelial cells are anchored. Together, these two layers provide support and attachment for epithelial tissue.

Epithelial tissue comes in two types, covering and lining epithelium and glandular epithelium. Whereas **covering and lining epithelium** covers some internal organs, makes up the outer layer of the skin, and forms the inner lining of blood vessels and several body systems (digestive, urinary, reproductive, and respiratory), **glandular epithelium** is a component of endocrine and exocrine glands.

Covering and lining epithelium can be classified by the number of cell layers and shape of the cells. With respect to number of layers, epithelial tissues may be **simple** (cells form a single layer), **stratified** (cells stacked in several layers), or **pseudostratified** (cells form a single layer, but because some cells do not extend all the way to the surface, the tissue has a multilayered appearance). Cell shapes include **squamous** (flat and scale-like), **cuboidal** (cube shaped), **columnar** (cylindrical or column shaped), and **transitional**

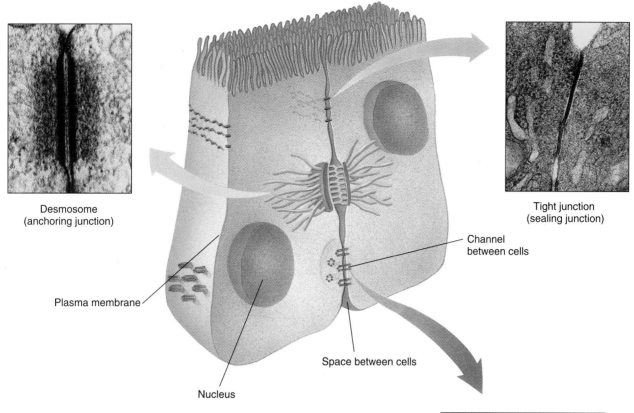

Figure 4-1 Junctional complexes in intestinal epithelium. Tight junctions are sites where the membranes of adjacent cells are attached to form tight seals. Desmosomes resemble rivets holding cells together. Gap junctions are channels between adjacent cells. *(Photos: left, © Don Fawcett/Visuals Unlimited; right top, © David M. Phillips/Visuals Unlimited; right bottom, © Allbertini/D. Fawcett/Visuals Unlimited)*

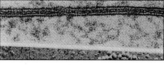

Gap junction
(communicating junction)

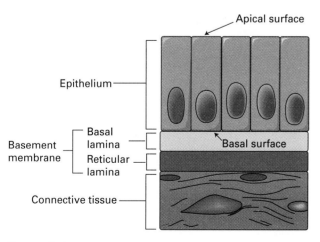

Figure 4-2 Structure and location of the basement membrane, the site of attachment of epithelial cells to underlying connective tissue.

(cells change shape as tissue stretches; such cells may exhibit any of the previous shapes possibly with some variation).

Considering, in combination, number of cell layers and shape of cells, there are about six different types of covering and lining epithelium in the human body (Figure 4–3). **Simple squamous epithelium**, consisting of a single layer of flattened scale-shaped cells, lines the air sacs of the lungs and forms the inside walls of blood vessels. **Simple cuboidal epithelium**, a single layer of cube-shaped cells, covers the ovaries and lines the ducts of the kidneys and many glands. **Simple columnar epithelium**, consisting of a single layer of cylindrical cells, lines the digestive tract and some portions of the upper respiratory tract. **Stratified squamous epithelium** contains several layers of cells. Although cells in the uppermost layers have a flattened scale-like appearance, cells in deeper layers may be cube-shaped or cylindrical. Stratified squamous epithelium lines the surfaces of the

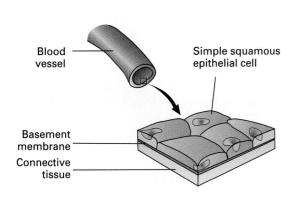

(a) Simple squamous epithelium

Blood vessel

Simple squamous epithelial cell

Basement membrane

Connective tissue

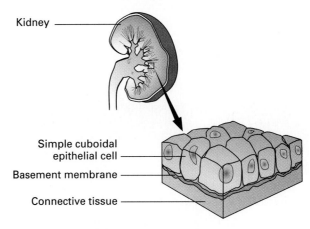

(b) Simple cuboidal epithelium

Kidney

Simple cuboidal epithelial cell

Basement membrane

Connective tissue

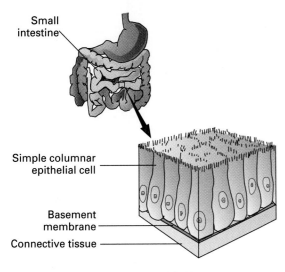

(c) Simple columnar epithelium

Small intestine

Simple columnar epithelial cell

Basement membrane

Connective tissue

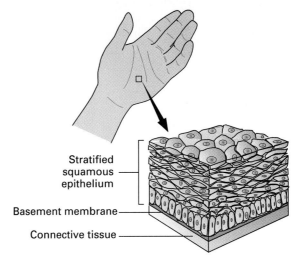

(d) Stratified squamous epithelium

Stratified squamous epithelium

Basement membrane

Connective tissue

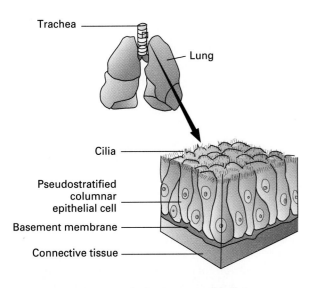

(e) Pseudostratified columnar epithelium

Trachea

Lung

Cilia

Pseudostratified columnar epithelial cell

Basement membrane

Connective tissue

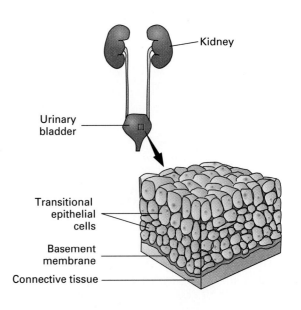

(f) Transitional epithelium

Kidney

Urinary bladder

Transitional epithelial cells

Basement membrane

Connective tissue

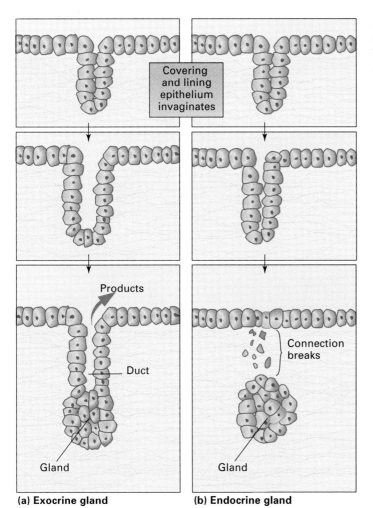

Figure 4–4 Formation of exocrine glands (a) and endocrine glands (b) by the invagination of covering and lining epithelia. Whereas exocrine glands maintain their connection to the surface epithelium via ducts, endocrine glands lose their connection.

Covering and lining epithelium invaginates

Products

Duct

Gland

(a) Exocrine gland

Connection breaks

Gland

(b) Endocrine gland

mouth, throat, and vagina and forms the outer layer of skin. **Pseudostratified columnar epithelium** consists of a single layer of cells and lines most of the upper respiratory tract. Although all cells of the single layer are anchored to the basement membrane, not all of them reach the surface, and thus this tissue has a multilayered appearance. Finally, **transitional epithelium** consists of several layers, the cells of which vary in appearance with the degree to which the tissue is stretched. This type of epithelium lines the bladder and parts of the uterus and urethra.

Glandular epithelium forms the secretory parts of glands. Such glands form during embryonic development by invagination of covering and lining epithelium. Glands that maintain their connection to the surface epithelium by way of ducts are called exocrine glands (Figure 4–4a). **Exocrine**

glands secrete their products via ducts onto body surfaces or into body cavities or organs and include the salivary glands of the mouth and the oil and sweat glands of the skin. Sometimes glandular epithelial cells lose their connection to the surface epithelium and form endocrine glands (Figure 4–4b). **Endocrine glands** lack ducts and secrete their products (hormones) into the spaces just outside the cells. Ultimately, hormones diffuse into the bloodstream and are carried throughout the body.

Connective Tissue

Sometimes described as the body's glue, **connective tissue** binds together and supports tissues of the body. Certain connective tissues are involved in transport (blood) and energy storage (adipose tissue). All connective tissues contain

◀ Figure 4–3 Types of covering and lining epithelial tissue. (a) Simple squamous epithelium. (b) Simple cuboidal epithelium. (c) Simple columnar epithelium. (d) Stratified squamous epithelium. (e) Pseudostratified columnar epithelium. (f) Transitional epithelium.

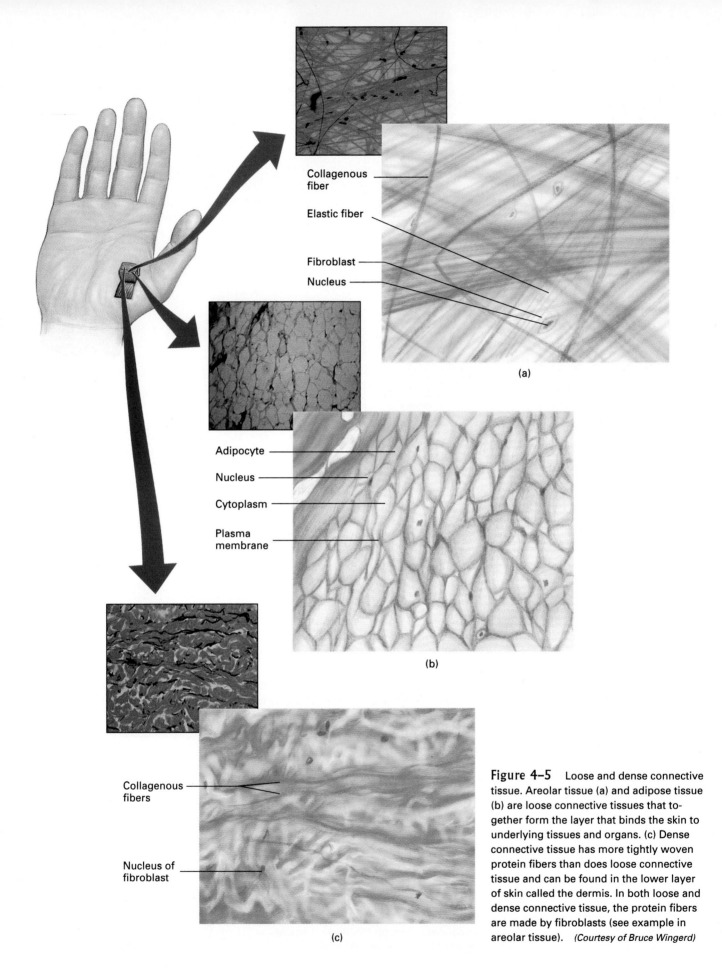

Collagenous fiber

Elastic fiber

Fibroblast

Nucleus

(a)

Adipocyte

Nucleus

Cytoplasm

Plasma membrane

(b)

Collagenous fibers

Nucleus of fibroblast

(c)

Figure 4–5 Loose and dense connective tissue. Areolar tissue (a) and adipose tissue (b) are loose connective tissues that together form the layer that binds the skin to underlying tissues and organs. (c) Dense connective tissue has more tightly woven protein fibers than does loose connective tissue and can be found in the lower layer of skin called the dermis. In both loose and dense connective tissue, the protein fibers are made by fibroblasts (see example in areolar tissue). *(Courtesy of Bruce Wingerd)*

cells and a matrix composed of protein fibers and ground substance. The ground substance is secreted by connective tissue cells and nearby cells, and it may be solid, fluid, or gelatinous. From this brief description we can predict that the organization of connective tissue differs dramatically from that of epithelial tissue. Typically, adjacent cells in connective tissue do not contact each other and there are few, if any, junctional complexes between them. Connective tissue cells do not have distinct basal and apical surfaces, and extracellular materials (the matrix), rather than cells, dominate. The many types of connective tissue can be grouped as connective tissue proper (loose and dense connective tissue) and specialized connective tissue (cartilage, bone, blood).

Loose and dense connective tissues differ in the ratio of cells to extracellular fibers. **Loose connective tissue** contains many cells, and the fibers of the matrix are fewer in number and more loosely woven than those in dense connective tissue (Figure 4–5). Examples of loose connective tissue include areolar connective tissue, a widely distributed tissue that contains many cells embedded in a gelatinous matrix, and adipose tissue, a tissue that contains cells called adipocytes that are specialized for fat storage. Together, these two tissues form the subcutaneous layer that anchors the skin to underlying tissues and organs. **Dense connective tissue** contains many tightly woven fibers and is found in ligaments (structures that join bone to bone), tendons (structures that join muscle to bone; Figure 4–6), and the dermis (layer of skin below the epidermis; Figure 4–5c). In both loose and dense connective tissue, the protein fibers are produced by fibroblasts, connective tissue cells that also repair tears in body tissues. For example, when the skin is cut, fibroblasts move to the area of the wound and produce collagen fibers, protein fibers that help close the wound and provide a surface upon which the outer layer of skin can grow and cover the damage.

Cartilage, one form of specialized connective tissue, consists of cartilage cells (chondrocytes) that sit within spaces called lacunae. The lacunae sit within a matrix that contains

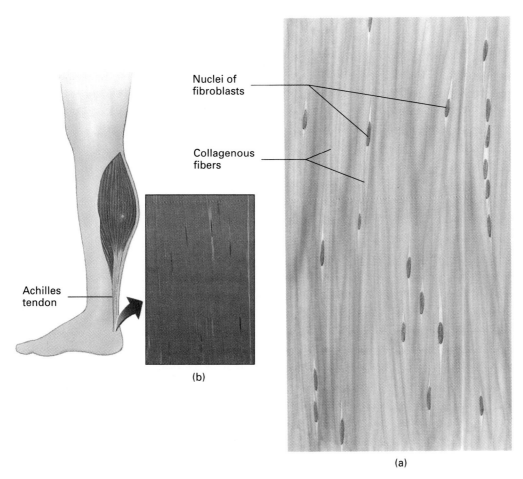

Nuclei of
fibroblasts

Collagenous
fibers

Achilles
tendon

(b)

(a)

Figure 4–6   Dense connective tissue. (a) This tissue contains many tightly packed collagen fibers (note also the fibroblasts responsible for making the collagen fibers). (b) Photomicrograph of a tendon showing components of dense connective tissue. The nuclei of fibroblasts appear red and the collagen fibers appear orange-pink.
(Courtesy of Bruce Wingerd)

protein fibers for strength and a somewhat gelatinous ground substance responsible for the tissue's resilience. All cartilage lacks blood vessels and nerves and is surrounded by a layer of dense connective tissue called the **perichondrium**. The perichondrium contains blood vessels, and thus nutrients reach cartilage cells by diffusion from the capillaries of the perichondrium inward to cartilage cells. However, this is a fairly slow process and may explain why cartilage heals more slowly than bone, a tissue with a rich blood supply.

We have three types of cartilage in our bodies: hyaline, fibrocartilage, and elastic cartilage. **Hyaline cartilage**, the most abundant of the three types, contains numerous cartilage cells in a matrix of collagen fibers and a bluish-white, gel-like ground substance. Known commonly as "gristle," hyaline cartilage is found at the ends of long bones (look carefully at your next drumstick) where it allows one bone to easily slide over another and as part of the nose, ribs, larynx, and trachea (Figure 4–7). Hyaline cartilage provides support and flexibility.

Elastic cartilage is more flexible than hyaline cartilage because of the large number of wavy elastic fibers in its matrix. Elastic cartilage is found in the external ear where it provides strength and elasticity (Figure 4–8).

Fibrocartilage contains fewer cells than either hyaline or elastic cartilage. Like hyaline cartilage, its cells sit in a matrix that contains collagen fibers. Fibrocartilage forms the outer part of the shock-absorbing disks between the vertebrae of the spine (Figure 4–9). If the protective fibrocartilage ring of an intervertebral disk tears, the inner spongy tissue bulges outward and presses on nerves, causing great pain and a condition known as a slipped or herniated disk.

Bone is another form of specialized connective tissue, which together with cartilage and joints makes up the skeletal system. To many people's surprise, bone is an active, ever-changing tissue with a good blood supply that promotes prompt healing. The functions of bone are many: protection and support for internal structures; movement by working with muscles; storage of lipids (in yellow marrow), calcium,

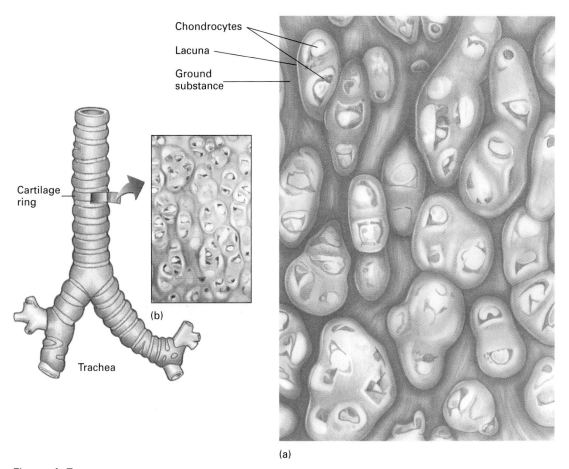

(a)

Figure 4–7 Hyaline cartilage. (a) Cartilage cells (chondrocytes) sitting within lacunae surrounded by a gel-like ground substance. (b) Photomicrograph of a section of the trachea revealing hyaline cartilage. *(Courtesy of Bruce Wingerd)*

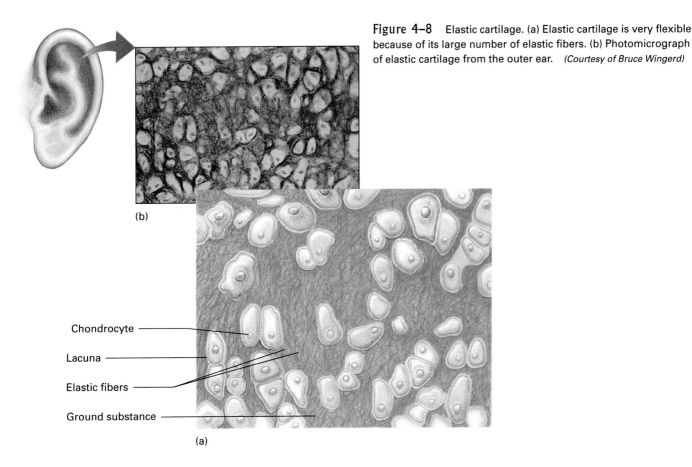

Figure 4–8 Elastic cartilage. (a) Elastic cartilage is very flexible because of its large number of elastic fibers. (b) Photomicrograph of elastic cartilage from the outer ear. *(Courtesy of Bruce Wingerd)*

(b)

Chondrocyte

Lacuna

Elastic fibers

Ground substance

(a)

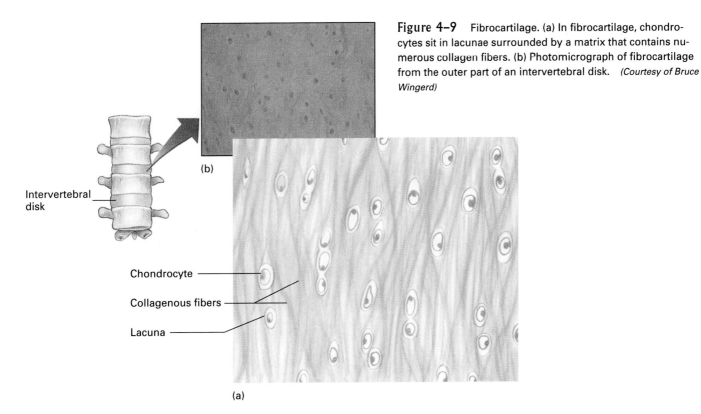

Figure 4–9 Fibrocartilage. (a) In fibrocartilage, chondrocytes sit in lacunae surrounded by a matrix that contains numerous collagen fibers. (b) Photomicrograph of fibrocartilage from the outer part of an intervertebral disk. *(Courtesy of Bruce Wingerd)*

(b)

Intervertebral disk

Chondrocyte

Collagenous fibers

Lacuna

(a)

and phosphorus; and production of blood cells (in red marrow).

Bone tissue is classified as compact or spongy (cancellous), depending upon its underlying organization (Figure 4–10). **Spongy bone** is less dense than compact bone and is made of an irregular network of collagen fibers surrounded by a calcium matrix. Cells called osteoblasts secrete the collagen, but once surrounded by the calcium matrix they are called osteocytes. In **compact bone**, mature bone cells (osteocytes) sit in tiny spaces called lacunae. Lacunae are located between lamellae, concentric rings of matrix comprised of collagen fibers and mineral salts. Large canals (central or Haversian canals) containing nerves and blood vessels run through compact bone. Tiny canals called canaliculi connect osteocytes to central canals, providing routes by which nutrients can reach the cells and wastes can leave.

Some of the spaces and cavities within bone contain yellow marrow, a storage site for lipids, and others contain red marrow, a production site for blood cells. Other connective tissue structures associated with bone include an outer covering called the **periosteum** and an inner lining of spaces that house yellow marrow called the **endosteum**, as seen in Figure 4–10.

Blood, the final example of a specialized connective tissue, consists of formed elements (cells and platelets) suspended in plasma, a liquid matrix (Figure 4–11). **Formed elements** include **erythrocytes** (red blood cells) that function in transporting oxygen to cells and carrying carbon dioxide away, **leukocytes** (white blood cells) responsible for fighting infections, and **platelets** that function in blood clotting. Plasma is mostly water plus dissolved substances such as oxygen, carbon dioxide, nutrients, and hormones.

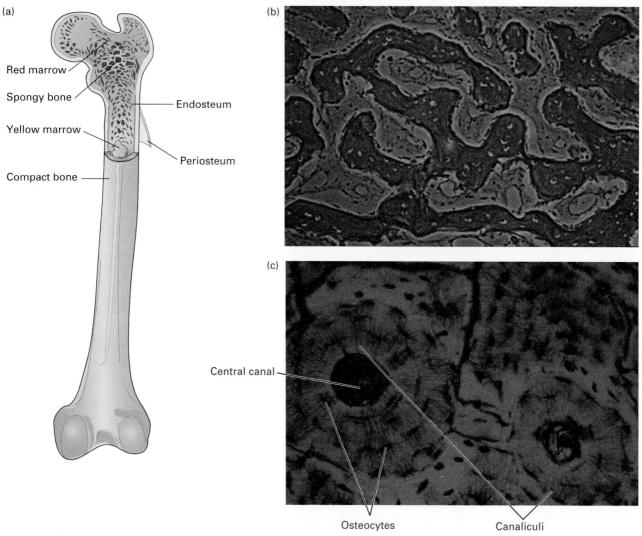

Figure 4-10 Bone. (a) Structure of a long bone. (b) Spongy bone. (c) Underlying structure of compact bone. *(b, © George J. Wilder/Visuals Unlimited: c, © John D. Cunningham/Visuals Unlimited)*

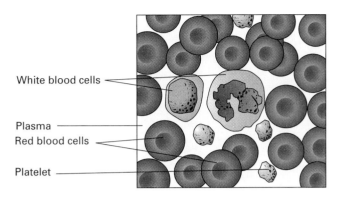

Figure 4–11 Like cartilage and bone, blood is a specialized connective tissue. Blood contains formed elements (red blood cells, white blood cells, and platelets) suspended in plasma, the liquid matrix.

Labels: White blood cells, Plasma, Red blood cells, Platelet

Muscle Tissue

Muscle tissue is composed of muscle cells (fibers) that contract when stimulated, generating force. There are three types of muscle tissue: skeletal, smooth, and cardiac. **Skeletal muscle tissue**, so-named because it is usually attached to bones, consists of cylinder-shaped cells, each of which contains several nuclei and exhibits striations, alternating light and dark bands that are visible under a light microscope (Figure 4–12). Because skeletal muscle is under conscious control (messages from the brain cause contraction), it is described as voluntary muscle. The functions of skeletal muscle include generating movement and heat and maintaining posture. **Smooth muscle tissue** is involuntary and is found in the walls of blood vessels and airways, where its contraction reduces the flow of blood or air. Smooth muscle is also found in the walls of organs such as the stomach, intestines, and bladder where it aids in propelling food through the digestive tract and eliminating wastes. The cells of smooth muscle tissue are spindle-shaped, contain a single nucleus, and lack striations (Figure 4–13a). **Cardiac muscle tissue** is found in the walls of the heart where its contractions are responsible for pumping blood to the rest of the body. Cardiac muscle cells resemble branching cylinders and have striations and typically only one nucleus (Figure 4–13b). The cells of cardiac muscle are attached to one another at **intercalated disks**, thickenings of the cells' plasma membranes that strengthen cardiac tissue and promote rapid conduction of impulses throughout the heart. Cardiac muscle tissue is not under conscious control and thus is considered involuntary.

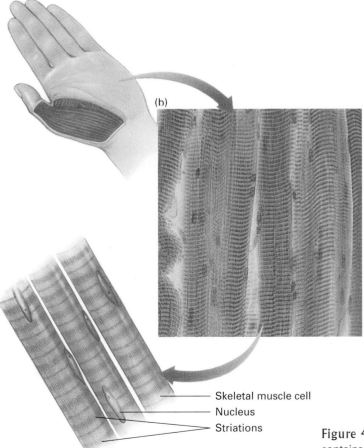

(b)

(a)

Labels: Skeletal muscle cell, Nucleus, Striations

Figure 4–12 Skeletal muscle. (a) Each skeletal muscle cell (fiber) contains several nuclei and exhibits striations. (b) Photomicrograph of skeletal muscle. *(Photo, © John D. Cunningham/Visuals Unlimited)*

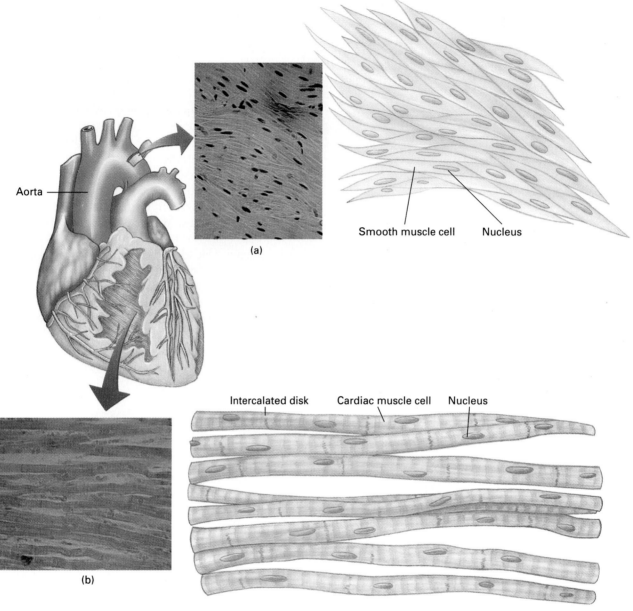

Figure 4–13 Smooth muscle and cardiac muscle. (a) Smooth muscle cells, such as these in the wall of the aorta, are spindle-shaped and contain a single nucleus. (b) Cardiac muscle cells resemble branching cylinders, usually have a single nucleus, and attach to one another by intercalated disks, thickenings of the plasma membrane. *(Courtesy of Bruce Wingerd)*

Nervous Tissue

The final major type of tissue is nervous tissue. **Nervous tissue** makes up the brain, spinal cord, and nerves and consists of two types of cells, neurons and neuroglia (Figure 4–14). **Neurons** convert stimuli (for example, pain, light, or thermal energy) into nerve impulses that they conduct to other neurons, cells of muscle tissue, or glands. Although neurons come in many shapes and sizes, most have a cell body (loca-

tion of the nucleus and most organelles), dendrites (highly branched processes of the cell body), and an axon (single, long process that usually conducts impulses away from the cell body). **Neuroglia** are found in the central nervous system (brain and spinal cord) and they come in four basic types (astrocytes, oligodendrocytes, microglia, and ependymal cells). Neuroglia do not generate or conduct nerve impulses, but they do increase the rate at which impulses are conducted by neurons and provide neurons with nutrients from nearby

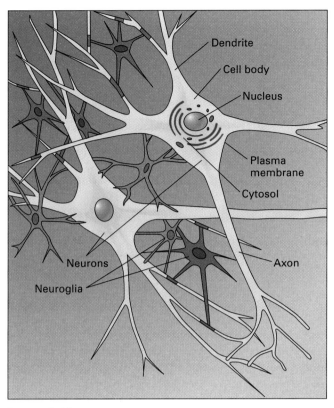

Figure 4–14   Neuroglia and neurons.

blood vessels. How the various tissues join together to form organs with particular functions will be considered next.

Organs

An **organ** is a structure composed of two or more different tissues with a specific function. All organs share four key features: (1) **parenchymal cells**, parenchyma being a general term to describe cells related to the organ's major function; (2) a **stroma** or connective tissue capsule that connects the tissues within the organ to each other and to the rest of the body; (3) a blood supply for transporting respiratory gases, hormones, nutrients, and wastes to and from the organ; and (4) a nerve supply to regulate the function of the organ.

Organ Systems

Organs usually do not function as independent units, but instead they work as part of a group of organs with a common function, called an **organ system**. For example, organs such as the trachea, bronchi, and lungs constitute the respiratory system. The common function of these organs is to bring oxygen into the body and remove carbon dioxide from the body. Table 4–1 lists the ten major organ systems and their functions.

In the next section, we will consider the structure and function of the integumentary system. Details about the

Table 4–1   The Ten Organ Systems

System	Basic Function	Major Components
Integumentary	Protect underlying tissues from abrasion and dehydration Provide cutaneous sensation Regulate body temperature Immune function Synthesize vitamin D Excretion	Epidermis, dermis, and derivatives
Skeletal	Attachment for muscles Enclose and protect organs Store calcium and phosphorous Produce blood cells Fat storage	Bones, cartilages, and ligaments
Muscular	Move body and maintain posture Internal transport of fluids Generation of heat	Striated muscle, smooth muscle, cardiac muscle
Digestive	Physical and chemical breakdown of food Absorb, process, store, and control release of digestive products	Teeth, tongue, salivary glands, stomach, small intestine, large intestine, liver, gallbladder, pancreas
Circulatory	Transport nutrients, respiratory gases, wastes, and heat Transport cells and antibodies for immune response Transport hormones Regulate pH through buffers	Heart, blood vessels, lymphatic vessels, spleen, tonsils, lymph nodes, bone marrow

(continued)

Table 4–1   The Ten Organ Systems (continued)

System	Basic Function	Major Components
Respiratory	Exchange respiratory gases with the environment	Nose, larynx, bronchial tree, lungs
Urinary	Maintain constant internal environment	Kidneys, ureters, bladder, urethra
Reproductive	Produce and secrete hormones Produce and release egg and sperm cells and accessory secretions Form placental attachment with fetus (females only)	Ovaries, oviduct, uterus, vagina, testes, prostate, penis
Nervous	Regulate and integrate body functions via neurons	Sense organs, brain, spinal cord, and peripheral nervous system
Endocrine	Regulate and integrate body functions via hormones	Cells and organs secreting hormones directly into the circulation

structure and functions of the other organ systems are presented in subsequent chapters.

An Example: The Integumentary System

We've all been told that our skin is as deep as our beauty goes, but it does more than make us attractive. It is the part of us that directly encounters the environment. In so doing it acts as a shield against all sorts of environmental dangers. We will see that this is only one of the many important functions of skin.

Functions of the Integumentary System

The skin and its derivatives—hair, nails, sweat glands, and oil glands—perform a variety of functions. One of the main functions is protection. Essentially a physical barrier between the contents of the body and the outside world, the skin protects the tissues it covers from dehydration, bacterial invasion, ultraviolet radiation, and physical insult. The skin also plays a role in the regulation of body temperature. Although we perspire almost constantly, our sweat glands dramatically increase their output of perspiration during times of strenuous exercise or high environmental temperatures. It is the evaporation of perspiration from the surface of the skin that helps rid the body of excess heat. Later in the chapter we will see how changes in the flow of the blood to the skin also help to maintain body temperature. Perspiration contains water, salts, and wastes such as urea, and thus the integumentary system also functions in excretion.

Some components of the integumentary system are part of the immune system. **Langerhans cells** of the skin, for example, recognize and ingest **antigens**, foreign substances that have found their way into the body. The Langerhans cells then present these foreign substances to specialized white blood cells, called lymphocytes, for destruction.

Also located in the epidermis are modified cholesterol molecules that are converted to vitamin D when exposed to ultraviolet radiation. Vitamin D then travels by way of the blood stream to the liver and kidneys where it is chemically modified to assume its role in stimulating the absorption of calcium and phosphorus from the food we eat.

The skin also contains components of the nervous system that detect temperature, touch, pressure, and pain stimuli. Some sensory receptors in the skin are free nerve endings and others are nerve endings encapsulated by one or more layers of cells (such as Meissner's and Pacinian corpuscles, which can be seen in Figure 4–15). By receiving such stimuli, receptors in our skin help keep us informed about the conditions around us. The functions of the integumentary system are summarized in Table 4–1. With these functions in mind, we will now study the structure of skin and its derivatives.

Structure of the Skin

The skin has two layers, the epidermis and the dermis (Figure 4–15). The **epidermis** is the outermost layer and is composed of epithelial cells. The **dermis,** a much thicker layer in most areas of the body, lies just below the epidermis and is composed of connective tissue. Blood vessels are present in the dermis but not in the epidermis. Nutrients reach the epidermis by moving out of dermal blood vessels and diffusing through tissue fluid into the layer above. Such tissue fluid is probably quite familiar to you. Trauma to the skin, such as that caused by a burn or vigorous snow shoveling, causes this fluid to accumulate in areas between the epidermis and dermis, forming blisters.

Just below the two skin layers is a layer of loose connective tissue known as the **hypodermis** (or superficial fascia; Figure 4–15). The hypodermis anchors the skin to underlying tissues and organs and contains about half of the body's fat stores. Interestingly (and often alarmingly), the distribu-

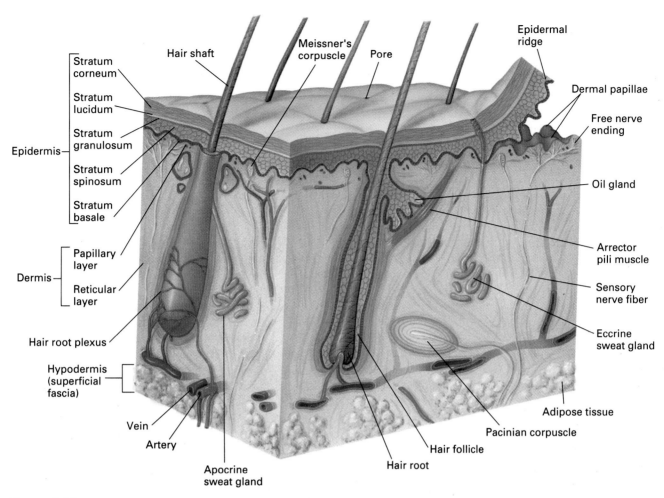

Figure 4-15 Structure of the skin and underlying hypodermis.

tion of these fat stores changes as we mature. As infants and toddlers, this layer of subcutaneous fat—often called "baby fat"—covers the entire body, but as we mature, some fat stores are redistributed in a gender-specific manner. In men, subcutaneous fat may accumulate, for example, in the neck or in the lower back where it produces "love handles." In women, fat often settles in the breasts, hips, and thighs. Both sexes, however, must contend with the tendency of fat cells to accumulate in the abdominal hypodermis where they help sculpt the all too familiar "pot belly." Although usually not considered part of the skin, the hypodermis shares some of the skin's functions, including cushioning blows and helping to prevent extreme changes in body temperature.

Critical Thinking

Burns—tissue damage caused by exposure to heat, radiation, electric shock, or chemicals—can be classified accord-ing to the depth to which the tissue damage penetrates. First-degree burns are those that are confined to the upper layers of the epidermis. In second-degree burns, damage extends through the epidermis into the upper regions of the dermis. Third-degree burns extend through the epidermis, dermis, and into underlying subcutaneous tissues. Severe burns, particularly those covering large portions of the body, are life threatening. Given your knowledge of skin functions, what would you predict the immediate medical concerns to be when third degree burns are suspected?

The Epidermis

The epidermis is composed of stratified squamous epithelium and is organized into the following layers, moving from deepest to most superficial: stratum basale, stratum spinosum, stratum granulosum, stratum lucidum, and stratum corneum (Figure 4–16). The **stratum basale** (basal layer) is the deepest layer of the epidermis. It consists of a single row of cuboidal to column-shaped cells characterized by rapid cell division. Certain viruses, called papilloma viruses, sometimes cause these cells to increase their already high rate of multiplication, and small, raised masses known as **warts** are the result. Warts may be removed by various means including cryosurgery with liquid nitrogen (essentially freezing the tissue to death), surgical excision, and chemical destruction with nonprescription drops. The best news, however, is that after a period of time, most warts go away without any treatment.

The next layer of epidermis, the **stratum spinosum** (spiny layer), contains several layers of cube-shaped cells. These cells have tiny "bridges" that connect them to adjacent cells. Interestingly, when such cells are prepared for study, they shrink, but the bridges between them hold tight, forming tiny projections that give the layer its name.

Continuing upward, we come to the **stratum granulosum** (granular layer) that consists of from three to five layers of flattened cells with granules in their cytoplasm. These granules are keratohyalin, a substance that contributes to the formation of **keratin**, the tough fibrous protein that gives the epidermis its protective properties.

The **stratum lucidum** (clear layer; not shown in Figure 4–16) contains three to four rows of flattened dead cells that contain droplets of a translucent substance called eleidin. Eleidin is formed from keratohyalin and eventually will become keratin. The stratum lucidum is found only in the palms of the hands and soles of the feet, areas of the epidermis known as "thick skin." The rest of the body is covered by "thin skin" containing the other four epidermal layers.

The outermost layer, the **stratum corneum** (horny layer), consists of from 25 to 30 rows of flat, dead cells. Completely filled with keratin, these cells prevent water loss and form an effective barrier against bacteria, chemicals, heat, and light. Eventually these cells are sloughed off. **Dandruff** occurs when dry, flattened patches of epidermal cells flake off from the scalp. Especially common in middle age, dandruff seems to be associated with stress, a high fat diet, and increased production of sebum, the oily substance produced by oil glands at the base of hair follicles (see the following discussion).

The cells of the epidermis include keratinocytes, Merkel cells, Langerhans cells, and melanocytes (see Figure 4–16). **Keratinocytes**, the most numerous cells, undergo keratinization, the process in which keratin gradually replaces the contents of maturing cells as they are pushed from the basal layer of the epidermis toward the skin surface.

The epidermis is in a constant state of renewal, faced with replacing the millions of cells sloughed off at the body's surface each day. As new cells are produced in the actively dividing basal layer, older cells are pushed toward the skin surface. On their way to the surface, the cells flatten and die as a result of their increasing distance from the dermis, their supplier of nutrients. Also along this death route, keratin gradually replaces the cytoplasmic contents of the cells. At the surface, the dead cells are eventually rubbed off. About 2 weeks to a month passes from the time a cell is formed in the basal layer to the time it is lost from the surface of the skin.

The other epidermal cells are much less numerous than keratinocytes (refer again to Figure 4–16). **Merkel cells** are found in association with sensory neurons in the area where the epidermis meets the dermis. The Merkel cell–neuron combination, known as a **Merkel disk**, functions as a sensory receptor, providing information about objects contacting the skin. **Langerhans cells**, mentioned before in our discussion of the skin's role in immunity, are macrophages that arise in bone marrow and migrate to the epidermis where they recognize and ingest foreign substances, only to hand them over to white blood cells for final destruction. Last, but not least, are the **melanocytes**, spider-shaped cells located at the base of the epidermis. These cells manufacture and store melanin, a pigment involved in skin color and absorption of the sun's ultraviolet radiation. Melanocytes eventually transfer melanin to other skin cells.

The Dermis

The dermis lies below the epidermis. Unlike the epidermis, the dermis does not wear away. This explains why tattoos—

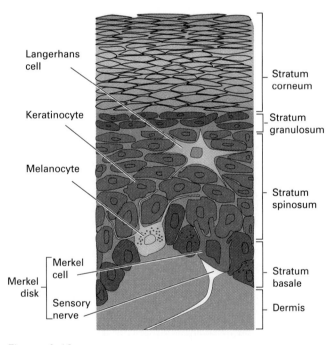

Figure 4–16 The layers and cell types of the epidermis. Keratinocytes are the most numerous cells. Also shown are the melanocytes, Langerhans cells, and Merkel cells. The Merkel cell is shown in association with a sensory nerve ending, forming a Merkel disk.

designs created when tiny droplets of ink are injected into the dermal layer—are meant to be permanent (Figure 4–17). (Tattoos, by the way, may be removed through surgical means including the use of lasers or "shaving" or abrading of the skin.) The dermis also differs from the epidermis in that it is laced with nerves and sensory receptors, and this is why getting a tattoo hurts. Largely connective tissue and fibers, the dermal layer also contains blood vessels, adipose tissue, hair follicles, oil glands, and the ducts of sweat glands.

The dermis contains two layers, the papillary layer and the deeper reticular layer (refer again to Figure 4–15). The **papillary layer** consists of loose connective tissue with elastic fibers. The upper surface of this layer is thrown into folds, called **dermal papillae**, that reach up into the epidermis, causing ridges on the skin surface. These **epidermal ridges** increase friction and help our fingers and feet obtain secure grips. Because the patterns of these ridges are genetically determined and unique to each individual, they have been used for many years in criminal investigations. Finger-

prints are the outlines of sweat left on surfaces by the epidermal ridges of our fingertips.

The **reticular layer** consists of dense connective tissue with collagen and elastic fibers. This combination of fibers allows the skin to stretch and then return to its original shape, but not always or forever, unfortunately. Pregnancy or a substantial weight gain may exceed the ability of skin to stretch and return to its earlier size, and stretch marks, tears in the dermis, result. The resilience of our skin also decreases as we age, the most pronounced effects beginning in the late forties. Around this time, collagen fibers begin to stiffen and decrease in number, and elastic fibers thicken and lose their elasticity. These changes, combined with reductions in moisture and the amount of fat in the subcutaneous layer, produce wrinkles and sagging skin.

Critical Thinking

Recently the drug tretinoin (Retin-A) has been used to treat wrinkles and sun-damaged skin. Originally developed to treat acne, Retin-A is a derivative of vitamin A that decreases the rate of wrinkle formation when applied to the skin as a cream or gel. What effect might Retin-A have on the reticular layer of the dermis?

The structure of the reticular region is of particular interest to surgeons as well as their patients. The collagen and elastic fibers of this region occur in bundles and the spaces between the bundles form **cleavage lines** in the skin (Figure 4–18). Whenever possible, surgeons make their incisions parallel to these lines because such cuts tend to heal more rapidly and leave less scarring than those made across lines of cleavage. Scars do not form when injury is confined to the epidermis. It is only when an injury or surgical incision penetrates down to the dermal layers that scar tissue may form. Such tissue forms when collagen-producing cells increase their activity and the newly produced material is pushed to the skin surface. Scar tissue lacks an epidermis, and when compared with normal skin, usually has denser collagen fibers, fewer blood vessels and sensory receptors, and no hair.

Skin Color

Two interacting factors produce skin color: (1) the quantity and distribution of pigments and (2) blood flow. The first factor concerns melanocytes. As described above, melanocytes are found at the base of the epidermis and are the cells responsible for production of the pigment melanin. **Melanin**, which actually comes in two forms, a yellow-to-red form called pheomelanin and the more common black-to-brown form called eumelanin, is formed within membrane-bound structures called **melanosomes**. These melanosomes move to the ends of the spider-like processes of melanocytes and re-

Figure 4–17   Tattoos—designs created when droplets of ink are injected into the dermis—are essentially permanent. This is because the dermis is not shed like the epidermis.   *(© Charles Sykes/Visuals Unlimited)*

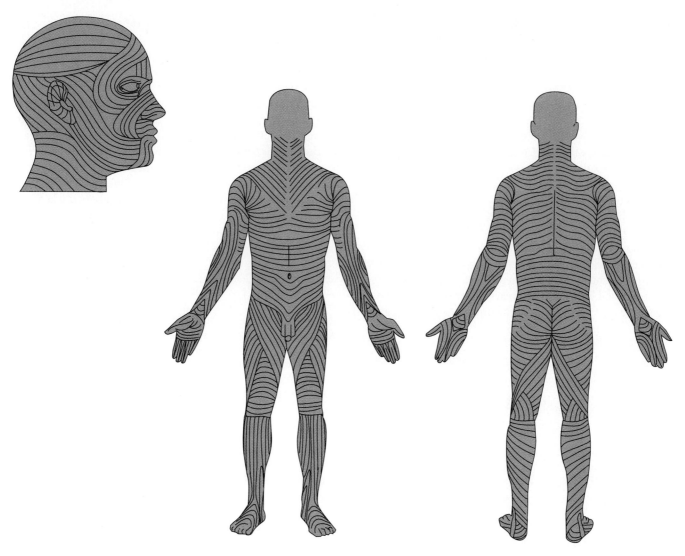

Figure 4–18 Cleavage lines on the head and body. These lines are actually spaces between the bundles of collagen and elastic fibers in the reticular region of the dermis. Whenever possible, surgeons make incisions parallel to these lines because such placement promotes rapid healing and less scarring.

lease melanin to the outside where it is ingested by surrounding epidermal cells. It is in this way that melanin, produced only in melanocytes, comes to color the entire epidermis.

All people have about the same number of melanocytes, so differences in skin color are due to differences in the form and amount of melanin produced and the way it is dispersed. A person's genetic makeup determines whether pheomelanin or eumelanin is produced. Those whose melanocytes produce pheomelanin have blond or red hair, blue eyes, and light complexions, often with freckles that are simply local concentrations of pigment. Those whose melanocytes pro-

duce eumelanin have skin that is yellow, brown, black, or white, depending on the amount of pigment produced and the size and dispersion of pigment granules. **Albinism**, an inherited condition in which an individual's melanocytes are incapable of producing melanin, occurs in about 1 person in 10,000 and is not specific to a particular race. Such individuals, known as albinos, lack pigment in their eyes, hair, and skin (Figure 4–19a). The absence of pigment in the eyes and skin leads to a pinkish color because the underlying blood supply shows through. **Vitiligo** is a condition in which melanocytes disappear either partially or completely from

(a)   (b)

Figure 4–19 Conditions involving melanocytes. (a) Albinism, a genetic condition in which an individual's melanocytes cannot produce melanin, occurs throughout the world. Here, an albino Hopi girl stands between her two sisters. (b) Vitiligo is a condition in which melanocytes disappear from areas of the skin, leaving white patches. *(a, © The Field Museum, Chicago, IL.; b, © Emil Muench, APSA/Photo Researchers, Inc.)*

certain areas of the skin, leaving white patches of skin in their wake (Figure 4–19b).

Another pigment that influences skin color is **carotene**. This orange-yellow pigment accumulates in the dermis and in cells of the outermost layer of the epidermis. Found in foods such as carrots, apricots, and oranges, carotene can lend an orange color to the skin of those who over indulge in carotene-rich fruits and vegetables. In addition, carotene is naturally found in the skin of people of Asian ancestry where, together with melanin, it lends a yellowish hue to the skin. Finally, **hemoglobin**, the pigment that carries oxygen in blood, is responsible for the pinkish color of Caucasian skin. (Recall that the epidermis does not have a blood supply, so the pinkish color comes from hemoglobin in the blood within capillaries of the dermis.)

Circulation can influence skin color. When well-oxygenated blood flows through vessels in the dermis, the skin has a pinkish or reddish tint that is most easily seen in light-skinned people. During times of intense embarrassment, this color may heighten as blood flow to the skin, particularly in the areas of the face and neck, increases. This re-

sponse, known as **blushing**, is impossible to stop. Intense emotions may also cause color to disappear from the skin. A sudden fright, for example, may cause a rapid drop in blood supply to the skin, causing the skin to blanch. Skin color may also change in response to changing levels of oxygen in the blood. In contrast to well-oxygenated blood that is bright red, poorly oxygenated blood is a much deeper red and gives the skin a bluish appearance. This condition, known as **cyanosis**, occurs in response to extremely cold temperatures (this is why your lips appear blue after emerging from a cold dip in the ocean) or as a result of disorders of the circulatory or respiratory systems.

Some people change their skin color by sitting in the sun. Melanocytes respond to the ultraviolet radiation of sunlight by increasing production of melanin that absorbs some of the radiation before it reaches the lower layers of the epidermis and dermis. **Tanning** is the build-up of melanin in the skin in response to UV exposure. Although some ultraviolet radiation is useful for the production of vitamin D, too much can be harmful. In Personal Concerns: Fun in the Sun? we discuss some of the dangers of sunbathing.

Fun in the Sun?

It is a beautiful sunny day at the shore and the beach is packed with people—some jump in the surf, some make sand castles along the water's edge, and others sit propped up in chairs or stretched out on blankets and towels across the sand. It is hard for us to equate this scene with disfigurement and death, but we should make the connection, and make it soon, for skin cancers are increasing at an alarming rate. At present, about one in six Americans will get skin cancer, and the incidence continues to rise annually. Here, then, we will first consider how the skin responds to sunlight and then describe what skin cancers are, how they form, and what can be done to treat and prevent them.

The ultraviolet (UV) radiation of sunlight causes the melanocytes of the skin to increase their production of the pigment melanin. Melanin is taken up by surrounding epidermal cells where it functions to absorb the UV before the radiation can travel any deeper into tissues. A tan is the build-up of melanin in the skin in response to sunlight. Unfortunately this somewhat protective build-up is not instantaneous, and sunburn—damage to skin accompanied by reddening and peeling—often follows the first day at the beach. Too much sun destroys the cells at the surface of the skin by inhibiting their DNA and RNA synthesis. The redness of sunburned skin results from the dilation of blood vessels in the dermis, a response prompted by the sun-damaged epidermal cells. Sunburned skin peels because the large scale destruction of skin cells caused by the overexposure to the sun's rays provokes an increased rate of production of new cells. These new cells push the burned cells off the skin surface, sometimes in small strands, and in severe cases, in large sheets.

Painful though a sunburn may be, its discomfort cannot compare to that of some skin cancers, conditions in which the DNA of skin cells is altered by UV radiation such that the cells grow and divide uncontrollably, forming a tumor. Three types of skin cancer are caused by overexposure to the sun: (a) basal cell carcinoma (b) squamous cell carcinoma and (c) melanoma (Figure A). Basal cell carcinoma arises in the rapidly dividing cells of the basal layer of the epidermis and is the most common type of skin cancer. Squamous cell carcinoma arises in the keratinocytes as they flatten and move toward the skin surface and is the second most common form of skin cancer. Melanoma, the least common and most dangerous type of skin cancer, arises in melanocytes, the pigment-producing cells of the skin. Unlike basal or squamous cell carcinomas, melanomas, when left untreated, often metastasize or spread rapidly throughout the body, first infiltrating the lymph nodes and later vital organs. If a melanoma is found before it has metastasized, the survival rate is about 90%. This rate drops to about 14%, however, if the cancerous cells have spread throughout the body. Melanomas can be caught at an early stage by carefully examining your skin while keeping the ABCD mnemonic of The American Cancer Society in mind:

A stands for asymmetry: Most melanomas are irregular in shape.

B stands for border: Melanomas often have diffuse, unclear borders.

C stands for color: Melanomas usually have a mottled appearance and contain colors such as brown, black, red, white, and blue.

D stands for diameter: Growths with diameters of more than 5 mm (about 0.2 inches) are threatening.

All three types of skin cancer are more common in fair-skinned than in dark-skinned people, particularly those who are middle-aged or elderly, live in the tropics or at high altitude, and who spend large amounts of time outdoors. People whose immune systems are suppressed also appear more susceptible to skin cancer. In addition, episodes of severe sunburn during childhood or adolescence seem to predispose people to developing melanomas years, even decades, later. Whatever your skin color, lifestyle, or past history of exposure to the sun, an unusual growth on your skin warrants a visit to a dermatologist.

Most skin cancers are treated either by cutting out the tumor and nearby tissue (surgical excision) or by radiation therapy. In some cases cancerous cells may be destroyed through laser surgery, cryosurgery (freezing), or electrodesiccation (exposing the tumor to an electric current and thereby causing the cells to dry out and die). Many skin cancers occur on the face, and the disfigurement caused by the removal of large chunks of the nose or ears may necessitate reconstructive surgery. The unattractive results of too much exposure to the sun do not, however, stop here. Excessive exposure to UV radiation also causes the skin to wrinkle, sag, and take on a leathery appearance.

What can be done to prevent skin cancer? The best way to avoid getting skin cancer is to avoid prolonged exposure to the sun. However, if you must be out in the sun, then apply a sunscreen and avoid the midday hours when the rays are the strongest. Sunscreens that have a sun protection factor (SPF) of at least 15 and that contain PABA (para-aminobenzoic acid), a chemical that binds to the outer layer of the epidermis and helps to absorb ultraviolet light, are best. Apply sunscreen about 45 minutes before going out into the sun and reapply after swimming or perspiring. Sunscreens, it is important to realize, are not foolproof. Most sunscreens block the higher energy portion of the sun's ultraviolet radiation known as UV-B, while only providing limited protection against the lower energy portion called UV-A. Whereas UV-B causes skin to burn, recent research suggests that exposure to UV-A weakens the body's immune system, possibly impairing its ability to fight melanoma. Ironically, then, by providing protection from sunburn, sunscreens have had the potentially devastating effect of

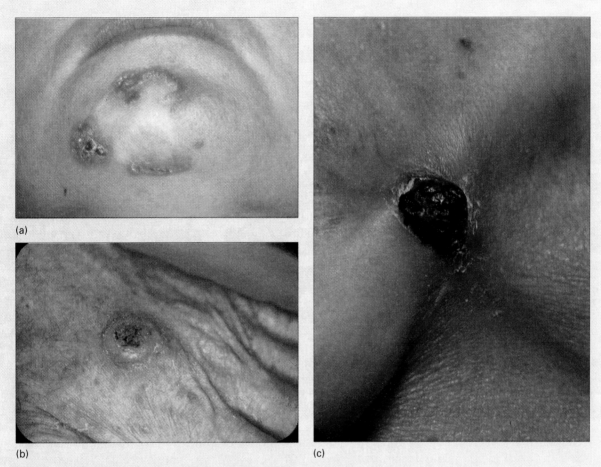

Figure A Three skin cancers. (a) Basal cell carcinoma. (b) Squamous cell carcinoma. (c) Melanoma.
(a, c, Custom Medical Stock Photo; b, Biophoto Associates/Science Source)

enabling people to spend more time in the sun, possibly increasing their risk of developing melanoma. Sunblocks such as zinc oxide, the white ointment used by lifeguards and others who spend long hours in the sun, differ from sunscreens in that they totally deflect ultraviolet rays. For this reason sunblocks are usually applied to particularly sensitive or already burned areas of the face such as the nose and lips.

In addition to using chemical protection, it is helpful to wear a wide-brim hat while out in the sun—this will reduce the rays that hit you by 70%. Tightly woven clothing also offers protection from UV. In fact, without such protective clothing it is hard to avoid the sun's rays. You should know that ultraviolet rays can pass through clouds, can penetrate water up to about 3 feet, and reflect off surfaces such as the sand and patio deck to reach you under your umbrella. A final preventative measure is to avoid tanning salons. For many years tanning salons have claimed to use "safe" wavelengths of ultraviolet radiation because they did not use skin reddening UV-B. But these "safe" wavelengths are actually UV-A. Given the apparent link between UV-A and increased risk of melanoma, the danger of these claims is now obvious.

Sadly, even if everyone follows these suggestions, the incidence of skin cancer will likely continue to increase. This dire prediction is based on our progressive destruction of the ozone layer, the protective layer of the earth's atmosphere that absorbs ultraviolet light. We are destroying this layer through our use of chlorofluorocarbons (CFCs), chemicals used as refrigerants and spray-can propellants. Scientists estimate that for every 1% decrease in the ozone layer, we will see at least a 2% increase in the incidence of skin cancer. The connection between skin cancer and destruction of the ozone layer illustrates the intimate relationship between our health and the health of our environment.

Derivatives of the Skin

Many seemingly diverse structures are derived from the epidermis—hair, nails, oil glands, sweat glands, and teeth. Next we consider the first four of these epidermal derivatives in terms of their structure, functions, and roles in everyday life. Teeth are discussed in Chapter 13.

Hair

Hair is simply an outgrowth of skin. It covers almost all of our body, the few exceptions being areas such as the lips, palms of the hands, and soles of the feet. Hair comes in three basic types—lanugo, vellus, and terminal. **Lanugo** is the soft, fine hair that covers the fetus beginning around the third or fourth month after conception. This first covering of lanugo falls off about a month before birth, only to be replaced by a second coat that lasts until a few months after birth. **Vellus hair** is also soft and fine, but it grows and persists throughout life, covering most of our body surface. **Terminal hair** is thick and strong and is found on the scalp, eyebrows, and eyelashes. During adolescence many of the vellus hairs of the armpits and pubic area are replaced by terminal hairs. In males the same is true for areas such as the face, chest, legs, forearms, back, and shoulders.

A major function of hair is protection. Whereas hairs on the scalp protect the head from UV radiation, those in the nostrils and external ear canals keep particles and bugs from entering these structures. Our eyes are protected by eyebrows and eyelashes. Hair also has a sensory role. Receptors associated with hair follicles are sensitive to touch.

A hair consists of a shaft and a root (Figure 4–20a). The shaft projects above the surface of the skin and the root extends below the surface into the dermis or subcutaneous layer where it is embedded in a structure called the hair follicle (discussed later in this section). In the dermis a tiny smooth muscle called the **arrector pili** is attached to the hair follicle. This muscle can pull on the follicle, causing the hair to stand up. Contraction of the muscle is usually associated not only with fear but with cold. The tiny mound of flesh that forms at the base of the erect hair is sometimes called a "goose bump." Each hair is also supplied with an oil gland that opens onto the follicle and supplies the hair with an oily secretion that makes it soft and pliant.

A coarse hair contains three layers of keratinized cells, moving from inside to out: (1) medulla (2) cortex and (3) cuticle (Figure 4–20b). The medulla, absent from fine hairs, contains eleiden granules and air spaces. The cortex forms the bulk of the shaft and contains melanin and air spaces. Melanin, produced by melanocytes at the base of the hair follicle and then transferred to the cortex, is in large part responsible for hair color. The outermost layer of a hair, the cuticle, consists of a single layer of thin, flat cells that are scale-like in appearance and packed with keratin. Exposed to the elements and abrasion, the cuticle sometimes wears off at the tip of hair shafts, causing the underlying cortex to frizz, forming the dreaded "split ends." The cuticle may also be damaged or worn away by exposure to chlorine in swimming pools, and this may explain why many swimmers wear bathing caps or wash their hair soon after emerging from the water.

Hair color is genetically determined and is based largely on the amount of melanin in the cortex. A large amount of melanin makes hair dark brown or black, while lesser amounts yield light brown or blond hair. On the other hand, red hair is not produced by melanin but by a reddish pigment. True redheads, in fact, have only the red pigment. People with auburn hair, however, have the red pigment as well as large amounts of melanin, and the melanin tends to drown out most of the red. Strawberry blondes have the red pigment and small amounts of melanin. Hair color often changes as we age, turning gray or white. This is caused by decreases in the production of melanin and corresponding increases in the number and size of air pockets in the hair shaft. Some hair tonics restore or darken the color of hair by filling in these air spaces. Hair dyes may be temporary, permanent, or semipermanent. Temporary dyes contain large molecules that cannot get through the cuticle to the inside of the hair, and thus these dyes only coat the surface of the hair and wash off with shampooing. Permanent dyes, on the other hand, are made up of smaller molecules that move through the cuticle to the cortex of the hair where they fill in air spaces. Once inside the hair they cannot be washed away. Continued growth of hair, however, will eventually reveal undyed roots. Semipermanent dyes penetrate hair to a certain extent and usually last a few weeks. In contrast to the few minutes to an hour that dyeing usually takes, natural changes in hair color occur quite gradually. Contrary to what some horror stories would have us believe, it is impossible for hair to "turn white overnight," unless of course some bleaching agent is used.

The hair follicle surrounds the root of a hair (see Figure 4–20a). Nerve endings called the **hair root plexus** surround the follicle and are sensitive to touch. This is why we are aware of even slight movements of the hair shaft. (Try to move just one hair without feeling it.) We have already mentioned that each follicle also has an oil gland and arrector pili muscle associated with it. The hair follicle itself has an outer connective tissue sheath, and moving inward, an external root sheath and an internal root sheath (Figure 4–20c). Whereas the **external root sheath** is simply a continuation of some layers of the epidermis, the **internal root sheath** is formed by actively dividing cells of the **matrix**, the region of the hair deep within the follicle that is responsible for producing new cells. These new cells are nourished by blood vessels in the **papilla**, an indentation of connective tissue. As new cells are produced in the matrix they push older cells upward toward the skin surface—this is how the hair on

Figure 4-20 Basic structure of a hair. (a) Section of skin showing hairs (whole and in longitudinal section) in follicles. (b) Enlarged longitudinal section of a hair showing the three cell layers. (c) Enlarged longitudinal section of a hair follicle.

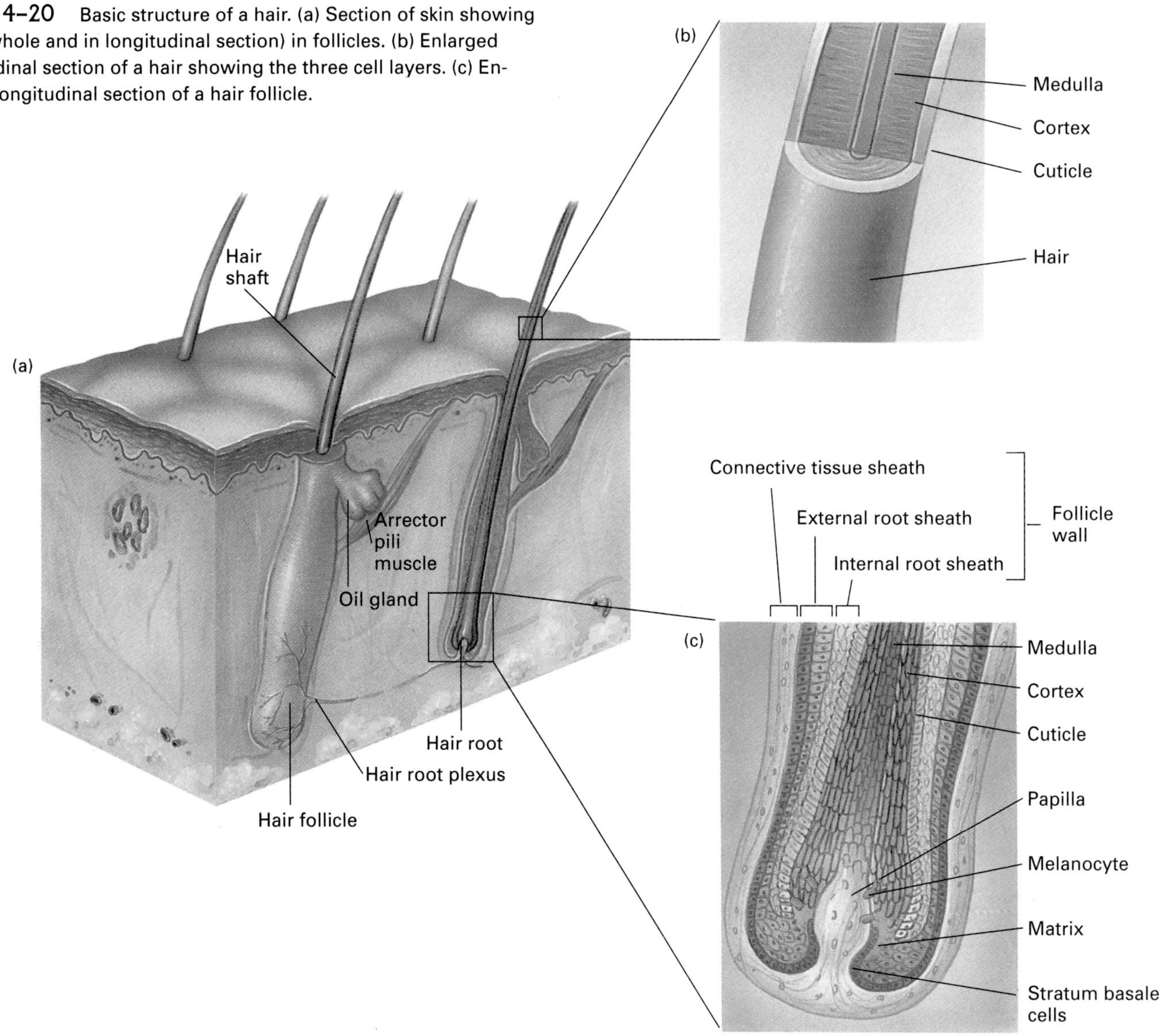

your scalp gets longer, growing about 5 or 6 inches a year. Along their upward route these cells die, and thus the part of a hair extending above the skin surface is simply a column of dead cells. A hair on the scalp typically produces new cells for 2 to 6 years and then stops, entering a resting phase that will last a few months. At the end of this time, the root of the hair detaches from the base of the follicle and the hair falls out, often aided by brushing or combing. A few months later this same follicle will start to grow another hair.

Hair growth and hair loss are two processes that we often try to stop or at least modify. Indeed some people want to get rid of hair while others wish they had more. Individuals in the first group often use depilatories—products like Nair—that work by dissolving the protein in the hair shaft. A few minutes after application of a depilatory, the remains of hair shafts can be wiped away to leave a hairless surface.

Because such products do not reach the root and follicle of the hair—the areas responsible for growth—the hair, in time, will once again protrude above the skin surface. A longer lasting option for hair removal is electrolysis, permanently destroying the bases of hair follicles with an electric current and thereby preventing future growth.

Consider the predicament of those who want more hair. Many of these people have **common baldness**, also known as **male pattern baldness**. Medical professionals call this condition androgenetic alopecia, a term that highlights the two factors thought to be involved in this type of hair loss—male hormones (androgens) and heredity. During balding, the hair follicles shrink and each hair produced is smaller than the one that preceded it. Basically, the thick terminal hairs of the scalp are replaced over time by soft, fine vellus hairs. This replacement occurs in a characteristic pattern, beginning at the forehead and temple and eventually reaching the crown (Figure 4–21). Such hair loss is more common in men than in women because women produce lower levels of androgens.

People who are unhappy with their hair loss have several options. The option chosen by many people today is the wig, traditionally a symbol of status. A more time-consuming alternative to wearing a "piece" is to undergo a hair transplant. This procedure, also known as a "punch graft," is usually done in a doctor's office under local anesthesia. The first step is to remove and discard small plugs of bald scalp (each plug is about half the diameter of a drinking straw). Plugs punched from an area of the scalp with active hair follicles are then inserted into the holes. Because the area of the scalp that donated the plugs will never again grow hair, it must be carefully selected to ensure that surrounding hair will cover it. Several sessions are usually needed to complete the transplanting process, the final results of which will be visible in about 9 to 12 months. Typically, by this time, only tiny scars remain in the donor areas. Another treatment for baldness involves the lotion minoxidil (Rogaine). Originally a medication to treat high blood pressure, minoxidil may restimulate hair growth when applied twice a day for several months to a scalp with minimal hair loss. The lotion appears to work by increasing blood flow to the scalp and stimulating activity in existing hair follicles.

Nails

Like hair, nails are modified skin tissue hardened by the protein keratin. Nails differ from hair, however, in that they grow continuously (recall that an individual hair will produce new cells for several years, enter a resting phase that lasts a few months, and then fall out). Compared with the rate at which hair grows, the pace of nail growth is quite slow. Whereas a hair may grow 5 or 6 inches a year, fingernails and toenails typically grow only about 1.5 and 0.5 inches per year, respectively.

There are three basic parts to a nail: the free edge, the body, and the root (Figure 4–22). The **free edge** extends over the tip of the finger or toe and is the part that we cut or clip. At the opposite end of the nail, embedded in the skin, is the **root**. The **nail body** lies between the free edge and the

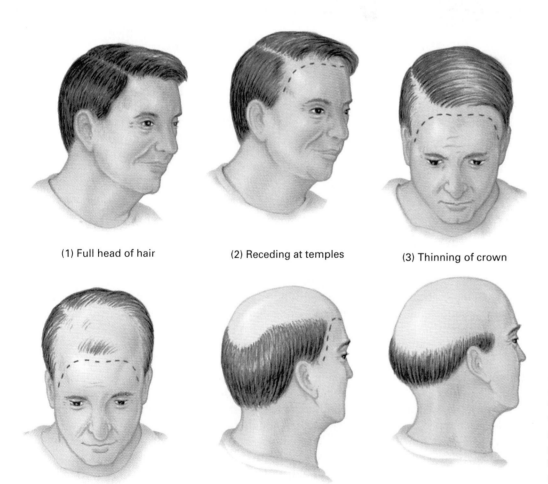

(1) Full head of hair

(2) Receding at temples

(3) Thinning of crown

(4) Pronounced bald spot

(5) Balding at back of head

(6) Fringe at sides and back

Figure 4–21 Male pattern baldness. The replacement of thick terminal hairs by soft vellus hairs occurs in a characteristic pattern. Hair recedes first at the forehead, then the temple, and then the crown. Bald areas tend to join until only a fringe of hair around the ears and back of the head remains.

root. The stratum basale layer of the epidermis (recall that this is the deepest layer of the epidermis and is characterized by rapid cell division) extends beneath the nail forming the **nail bed** (Figure 4–22b). New nail cells are produced here in the thickened proximal portion known as the **matrix**. Cells produced by the matrix acquire keratin, then harden, and the addition of new cells pushes the nail body forward over the nail bed. Most of a nail appears pink because of blood vessels in the dermis below. At the base of the nail, however, the underlying vessels are obscured by the thickened matrix and the area appears as a white crescent called the **lunula** (see Figure 4–22a). Skin folds, called **nail folds,** overlap the base and sides of the nail. The fold overlapping the base extends over the nail body and is called the **cuticle** (or eponychium). Nails provide protection for the tips of our fingers and toes and help us grasp and manipulate small objects.

Glands

Three types of glands—oil, sweat, and wax—are found in the skin. Although all three types are epidermal derivatives, they differ in their locations, structures, and functions.

Oil (sebaceous) glands, found virtually all over the body except the palms of the hands and soles of the feet, secrete **sebum**, an oily substance made of fats, cholesterol, proteins, and salts. The secretory part of these glands is located in the dermis. In some instances oil glands open directly onto the surface of the skin, but in most cases they open onto hair

follicles (see Figure 4–15). Sebum lubricates hair and skin, protects skin against desiccation, and contains substances such as fatty acids that inhibit growth of certain bacteria. Sometimes, however, the duct of an oil gland becomes blocked, sebum accumulates, and bacteria invade the gland and hair follicle. The resulting condition is called **acne**. In Personal Concerns: Acne: The Misery, the Myths, and the Medications, we discuss the causes and treatments of acne.

Sweat (sudoriferous) glands come in two basic types, eccrine and apocrine. **Eccrine sweat glands** are numerous, simple coiled tubules that discharge their secretions, made in portions of the glands located in the dermis or subcutaneous layer, directly onto the skin surface. These glands function throughout life and are most common in the skin of the forehead, palms of the hands, and soles of the feet. The few areas of the body devoid of eccrine glands include the margins of the lips, eardrums, nail beds, and portions of the genitalia. Eccrine glands secrete the substance we call sweat. Sweat is largely water plus some salts, lactic acid, vitamin C, and metabolic wastes such as urea. Although some wastes are eliminated through sweating, the principal function is to help in the regulation of body temperature through the evaporation of sweat from the skin surface (see the section Regulation of Body Temperature). The structure of an eccrine sweat gland can be seen in Figure 4–15.

Apocrine sweat glands, found mostly in the armpits, pubic area, and the pigmented region around the nipples, begin to function at puberty. These glands are larger and less numerous than eccrine glands and produce a thicker secretion that they discharge onto hair follicles (see Figure 4–15). The secretions of apocrine glands contain the same basic components of true sweat plus some fatty substances and proteins. Although the secretions are initially odorless, they may take on a musky odor as a result of the bacteria that thrive on them on the skin surface. Antiperspirants and hygiene sprays are designed either to inhibit such secretions or to mask their odor.

Wax (ceruminous) glands are modified apocrine sweat glands found in the tissue of the external ear canal. These glands consist of coiled tubules that open either directly onto the surface of the ear canal or into a nearby oil gland associated with a hair follicle. The secretion of wax glands is called ear wax or **cerumen;** it is a mixture of substances secreted by wax and oil glands. In combination with tiny hairs in the external ear canal, cerumen functions to prevent foreign material from reaching the eardrum. Some people overproduce cerumen and it becomes impacted. Because impacted cerumen may block sound waves from reaching the eardrum, it should be removed either by a medical professional using a blunt instrument or by periodic irrigation with a syringe.

Mammary glands are also modified sweat glands. Although technically part of the integumentary system, we consider mammary glands in our discussion of the female reproductive system in Chapter 16.

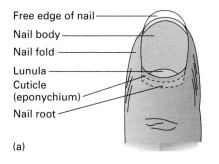

(a)

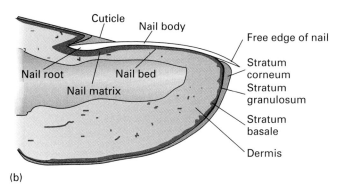

(b)

Figure 4–22 Structure of a nail. (a) Fingernail seen from above. (b) Sagittal section of the end of a finger and a fingernail.

Acne: The Misery, the Myths, and the Medications

Acne and adolescence go hand in hand. In fact, about four out of five teenagers have acne, a skin condition that will probably dog them and distress them well into their twenties and possibly beyond. What is acne? What causes this most common of skin disorders? Is it really the chocolate, potato chips, and sodas that often make up the teenage diet? Why do males usually experience more severe cases of acne than do females? Can acne be treated?

Simple acne goes by the telling technical name, acne vulgaris. It is a condition that affects hair follicles associated with oil glands. When adolescence begins, oil glands increase in size and step up their production of the oil, sebum. These changes are prompted by increasing levels in the blood of both males and females of "male" hormones called androgens. These hormones are secreted by the testes, ovaries, and adrenal glands, and the changes they induce in the activity and structure of oil glands set the stage for development of acne. It should come as no surprise, then, that acne occurs on areas of the body where oil glands are largest and most numerous—the face, chest, upper back, and shoulders are all hot spots for acne. Because most females have lower levels of androgens than do males, their acne is typically less severe. Flare-ups in acne around the time of menstruation are common, however, and may be linked to progesterone, a hormone secreted after ovulation. Unfortunately, the precise details of the link between progesterone and acne are unknown.

Acne is basically the inflammation that results when sebum and dead cells clog the duct where the oil gland opens onto the hair follicle (Figure A). A follicle obstructed by sebum and cells is called a whitehead. Sometimes the sebum in plugged follicles oxidizes and mixes with the skin pigment melanin, forming a blackhead. It is melanin and not dirt or bacteria that lends the dark color to these blemishes. The next stage of acne is pimple formation. Red raised bumps, often with a white dot of pus at the center, appear when obstructed follicles rupture and spew their contents into the surrounding epidermis. Rupture of such a follicle may be induced either by the general build-up of sebum and cells or by squeezing the area (this is why squeezing pimples is a bad idea). The sebum, dead cells, and bacteria that thrive on such things cause a small infection, called a pimple or pustule, that will usually heal within a week or two without leaving a scar. In severe cases of acne, the rupture of plugged follicles produces large cysts called boils. Boils extend into the dermis and thus leave scars when they heal.

Acne, then, is not caused by eating nuts, chocolate, pizza, potato chips, or any of the other "staples" of the teenage diet. Some of the afflicted, however, find that they have to avoid certain foods. Acne is also not caused by poor hygiene. Follicles plug from below, so dirt or oil on the skin surface is not responsible for causing acne. (Most doctors do, however, recommend washing the face two or three times a day with hot water to help open plugged follicles.) Two factors that do appear to contribute to acne are heredity and stress. Individuals are more likely to have acne if their parents had acne, and acne often flares up during times of stress, presumably as a result of stress-induced changes in the levels of hormones.

Treatments for acne fall into two main categories: topical (those applied to the skin) and oral (those taken by mouth). Medicines applied to the skin usually require several weeks of treatment before improvements are noticeable. Benzoyl peroxide 5% is a useful topical preparation that is marketed under several names, some of which are sold over the counter (such as Oxy-5 or OxyClear). Benzoyl peroxide, when applied once or twice a day as a cream after washing the area affected by acne, is a powerful antibacterial agent that kills bacteria living in the follicles. Benzoyl peroxide also induces peeling and stimulates blood flow—in fact a 2.5% cream is available if the 5% is too irritating. Another option is Tretinoin (retinoic acid, Retin-A), available only by prescription. This topical medication is a derivative of vitamin A that helps thin the epidermis, reduce the stickiness of dead cells, and increase the rate at which such cells are sloughed off. These actions help follicles remain clear and push out existing whiteheads and blackheads. Retin-A is usually

Integration with Other Organ Systems

Many of the skin's functions link it to other organ systems. For example, vitamin D synthesized in the skin is essential for the digestive system's absorption of calcium and phosphorus from the food we eat. Calcium and phosphorus, in turn, are important to bone growth and maintenance, and calcium plays a role in muscle contraction. Here, then, is an example of links among four organ systems—integumentary, digestive, skeletal, and muscular.

The integumentary system also works closely with the nervous system. Recall, for example, that there are receptors

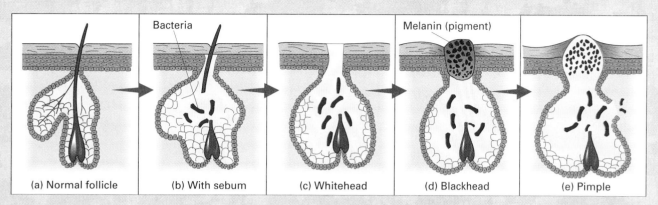

Figure A The stages of acne. (a) Normal follicle showing the tiny hair and cells of the oil gland. (b) A follicle whose canal has become clogged with sebum, dead cells, and bacteria. (c) A whitehead showing the accumulated sebum, cells, and bacteria. (d) A blackhead forms when the sebum in the clogged follicle oxidizes and mixes with melanin. (e) An inflamed pimple forms when the follicle wall ruptures releasing the contents of a whitehead or blackhead into the surrounding epidermis.

applied nightly as a cream or gel. It too causes inflammation of the skin and those who find it too irritating may apply it less frequently. Use of sunscreens is essential for those using Retin-A because this medication makes the skin hyper-sensitive to sunlight. Finally, Retin-A is a strong medication and should not be used during pregnancy.

Severe cases of acne often call for oral, or systemic medicines, such as tetracycline or minocycline that work by inhibiting follicle-inhabiting bacteria. After an initial large dose, a low dose is given over a long period of time. Few problems are caused by long-term use of tetracycline or minocycline, although some women experience a yeast infection of the vagina. Neither antibiotic, however, should be taken during pregnancy because they cause discoloration of the teeth in the developing fetus (both antibiotics are deposited along with new bone). Isotretinoin (Accutane), another derivative of vitamin A available only with a prescription, is probably the most effective treatment for severe cystic acne. This drug works by poisoning the oil glands, causing them to shrink and reduce their output of sebum. Some of this shrinkage is permanent, so Accutane differs from all other treatments in that it suppresses acne even after the treatment has stopped. Most patients take Accutane for 4 months, at a cost of

about $1000; about 30% may need a second 4-month course of the drug. Although highly effective, Accutane should be used with extreme caution, particularly by women of child-bearing age; increased birth defects have been found in children born to women using Accutane. More general side effects of Accutane result from its overall drying of the body, leading to such things as chapped lips, flaking skin, nosebleeds, and hair loss. Finally, estrogen in the form of birth control pills is sometimes prescribed for women with severe acne. Like Accutane, estrogen seems to work by shrinking oil glands (not permanently though as Accutane does) and reducing sebum production. The side effects associated with birth control pills are discussed in Chapter 16.

There are also ways to reduce the scarring that sometimes results from severe acne. Once the acne has subsided, doctors may recommend dermabrasion (skin planing) in which the scarred area is frozen and the top layer is sanded off with a rapidly rotating brush. Dermabrasion is usually restricted to the face where it helps to remove shallow pitted scars. Other alternatives include grafting onto the scarred area of the face a small amount of unscarred skin from behind the ear or using chemical peeling agents to make a depressed scar more level with the surrounding skin surface.

in the skin sensitive to pressure, touch, temperature, and pain, and that many aspects of the skin's role in the regulation of body temperature are influenced by the nervous system (e.g., activity of sweat glands and regulation of blood flow). Consider one more example of the close ties between organ systems. The activity of oil glands and apocrine sweat glands, as well as the distribution and growth of hair, are all influenced by sex hormones, thus linking the integumentary, endocrine, and reproductive systems. The close working relationships among organ systems will become especially apparent in the next section on homeostasis.

Homeostasis

Despite the sometimes dramatic changes in the environment around us, our bodies maintain a relatively constant internal environment. The pH of blood, for example, is about 7.4, the concentration of glucose in the blood about 0.1%, and body temperature usually hovers slightly below 37°C (98.6°F). This internal constancy is called **homeostasis**, and it occurs at all levels of body organization, from cells to organ systems.

Homeostasis is maintained primarily through **negative feedback mechanisms**—mechanisms in which the substance produced feeds back on the system, shutting down the production process. Homeostatic mechanisms do not maintain absolute constancy, but rather dampen fluctuations around a set point. Thus, homeostasis is a dynamic rather than a static state. Homeostatic mechanisms have three components: a **receptor** (to detect change in the internal or external environment), a **control center** such as the brain (to integrate information coming in from receptors and select a response), and an **effector** such as a muscle or gland (to carry out the response). The regulation of body temperature illustrates these features of homeostatic mechanisms.

Regulation of Body Temperature

Core temperature is the temperature in body structures below the skin and subcutaneous layers. The body's core temperature differs ever so slightly from one person to the next, but it averages just below 37°C (98.6°F). This temperature, then, is the set point. It is essential that the core temperature not vary too far from this mark because even slight changes in the core temperature have dramatic effects on the body's metabolism. A change in core temperature, for example, of 1.5°F alters metabolism by 20%. So how do our bodies maintain a relatively constant internal temperature?

Let's begin by considering what happens when we find ourselves in an environment where the temperature is above our set point, say 38°C (100.4°F). Thermoreceptors in the skin detect heat and activate nerve cells that send a message to an area of the brain called the hypothalamus. The hypothalamus then sends nerve impulses to the sweat glands to produce more perspiration. As perspiration evaporates, the surface of the skin cools, lowering body temperature. When body temperature drops below 37°C, signals are no longer sent from the hypothalamus to the sweat glands. In this homeostatic system, thermoreceptors in the skin are the receptors, the hypothalamus is the control center, and the sweat glands are the effectors (Figure 4–23). It is a negative feedback mechanism because the effect (cooling of the skin) feeds back on the system and inhibits further lowering of body temperature.

In addition to hot weather, strenuous exercise can increase body temperature. In this situation, heat sensors in the hypothalamus respond to increases in the temperature of the blood and initiate responses such as increased sweat gland activity and dilation of blood vessels in the dermis. The latter response releases more heat to the surrounding air and gives our skin the flushed appearance we get during strenuous exercise. In addition, the arrector pili muscles relax so that damp cooling hair lies close to the skin. Other mechanisms by which body temperature may be lowered include lowering metabolic rate and decreasing skeletal muscle tone, both of which reduce heat production, and behav-

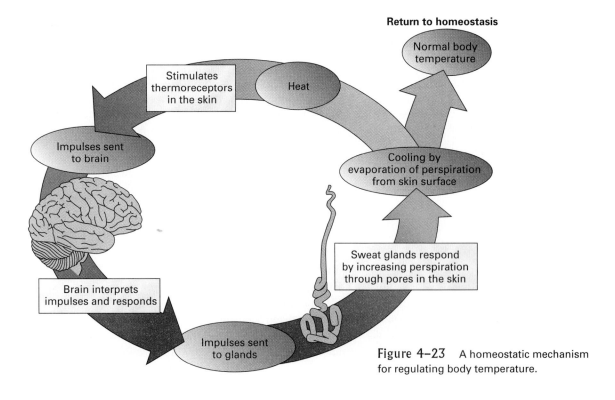

Figure 4–23 A homeostatic mechanism for regulating body temperature.

ioral responses such as seeking shade or removing your sweatshirt.

Sometimes the body, despite its many means for lowering higher than normal core temperatures, fails, and potentially deadly **hyperthermia**—abnormally elevated body temperature—results. For example, some marathon runners and people immersed in hot tubs with water of too high temperature (say up around 114°F—the recommended setting is about 39–40°C, or 102–104°F) have died as a result of elevated core temperatures. In both situations, efficient sweating and evaporation of perspiration from the body surface was prevented by high humidity (in the case of the runners) or the surrounding water (in the case of the hot tubbers). The condition, commonly called heat stroke, is marked by confusion and dizziness. If the core temperature reaches about 42°C (107°F), heart beat becomes irregular, oxygen levels in the blood drop, the liver ceases to function, and unconsciousness and death soon follow. Few people can survive core temperatures of 43°C (110°F).

Our bodies are capable of resetting the thermostat to a higher level, causing **fever**. Changes in the set point are brought on by pyrogens, chemicals released by infection-fighting white blood cells. Pyrogens travel in the blood to the hypothalamus where they raise the set point, apparently by influencing other chemicals in the brain, namely the fatty acids called prostaglandins. Prostaglandins are a class of chemical messengers that exert their effects at, or close to, their site of release. Although severe fevers (those above 40°C, 105°F) can be dangerous, mild fevers appear to be beneficial because they speed healing by increasing metabolism and the activity of white blood cells, and in some cases by creating more heat than the infectious organisms can stand. In addition, fever seems to prompt the liver and spleen to step up their removal of iron from the blood. Iron is an element required by many pathogenic bacteria to reproduce. As our body struggles to meet the new set point, we shiver to generate more heat and experience the chills typical of early fever. Often at this time we take aspirin, acetaminophen, or ibuprofen. These medications help the set point return to normal by interfering with the production or circulation of prostaglandins. As our set point drops back to normal, we suddenly feel hot and begin sweating to get rid of the excess heat.

What happens when the core body temperature drops? Subtle drops in body temperature are detected largely by thermoreceptors in the skin, which send their message up the spinal cord to the hypothalamus. The hypothalamus then sends nerve impulses to sweat glands, ordering a decrease in their activity. Impulses are also sent to vessels in the dermis, telling them to constrict. Constriction of these vessels reduces blood flow to the extremities, conserving heat for the internal organs and giving credence to the saying "cold hands, warm heart." Another response to decreasing body temperature is contraction of arrector pili muscles. As described previously, contraction of these tiny muscles causes hairs to stand on end, thereby trapping an insulating layer of air near the body. This response, known as **piloerection**, is less effective in humans than in more heavily furred animals. The body also responds to cooling by increasing metabolic activity to generate heat. This response is most noticeable in the increased contractions of skeletal muscles and is known as shivering. Finally, behavioral responses such as folding arms across the chest may also help combat a drop in core body temperature.

What happens if the body's heat-preserving and heat-generating mechanisms do not prevail and the core temperature continues to drop? **Hypothermia**—decreases in internal body temperature to 35°C (95°F) or below—results. Continued drops in body temperature disrupt normal functioning of the nervous system and temperature-regulating mechanisms. People suffering from hypothermia usually become giddy and confused, and they lose consciousness when their core temperature drops to 33°C (91.4°F); when the core temperature drops to 30°C (86°F), blood vessels are completely constricted and temperature-regulating mechanisms are fully shut down. Death follows shortly thereafter as a result of impaired nerve transmission and failure of the circulatory system. If detected soon enough, however, hypothermia can be treated, the most severe cases by using dialysis machines to artificially warm the blood and pump it back into the body.

Critical Thinking

Frostbite is damage to tissues exposed to cold temperatures. Given what you know about the body's response to cold temperature, why are fingers and toes particularly susceptible to frostbite?

We have seen the important role the skin plays in maintaining one feature of homeostasis, a relatively constant internal body temperature (summarized in Figure 4–24). We have also seen that many organ systems (the integumentary, circulatory, nervous, and muscular systems) cooperate to keep the body temperature around its set point. Our discussion, however, may have given the impression that body temperature and other physiological set points remain the same throughout the day. This impression is incorrect. Indeed, in the next section on rhythmic homeostasis, we will see that many physiological processes—body temperature included—change rhythmically and predictably throughout the day.

Rhythmic Homeostasis: The Biological Clock

You are not the same person you were this morning, and by bedtime you will have changed again. But, rest assured, the chances are very high that at this time tomorrow, you will be very much the same as you are right now. Most of your physiological processes, as well as many of your physical abilities and even your mental sharpness, vary predictably throughout the day. And, day after day, these variations repeat

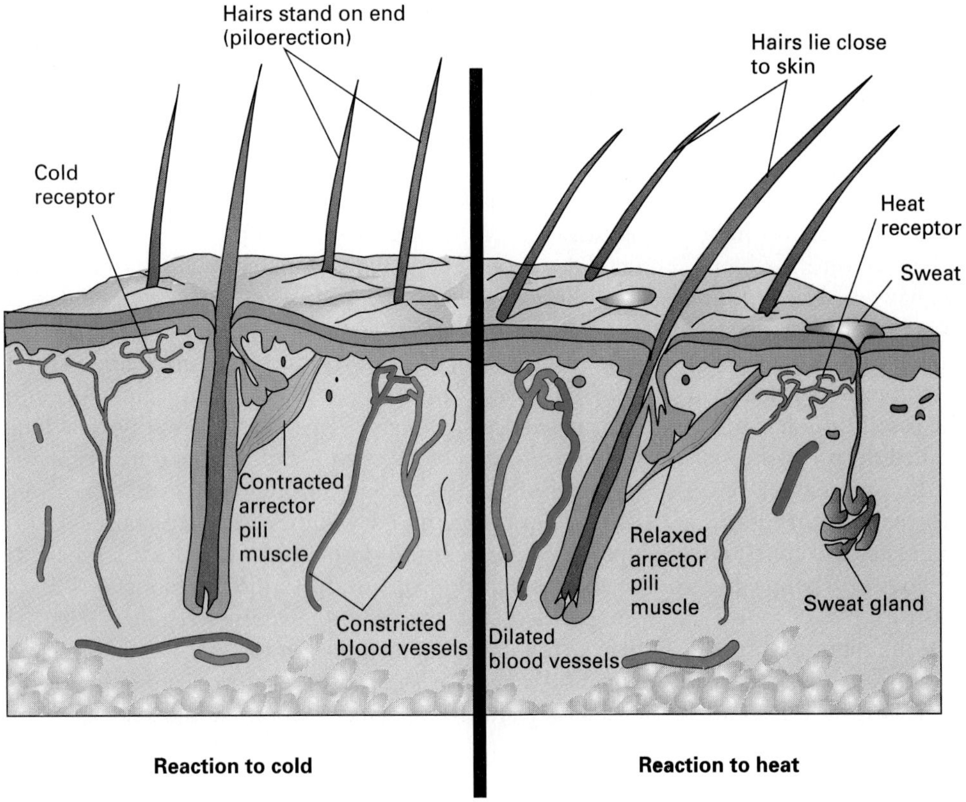

Figure 4-24 Summary of changes in the skin in response to cold and heat.

Hairs stand on end (piloerection)

Hairs lie close to skin

Cold receptor

Heat receptor

Sweat

Contracted arrector pili muscle

Relaxed arrector pili muscle

Constricted blood vessels

Dilated blood vessels

Sweat gland

Reaction to cold

Reaction to heat

themselves with such regularity that they are called **biological rhythms**. Homeostasis, then, does not describe a constant internal state and is more than the regulation of body functioning so that the body systems are regulated within certain limits. Instead, homeostasis is often rhythmic.

When you awaken in the morning, your body temperature is probably at the lowest point it will be all day. It will continue to climb until it peaks in the early evening, when it begins to slowly decline again. Thus, a temperature of 38°C (100°F) at 8 AM is more likely to indicate illness than it would at 8 PM. You might at first suspect that the rhythm in body temperature is caused by changing activity levels throughout the day or by eating. Activity and eating do, indeed, raise body temperature; however, body temperature continues to fluctuate in this predictable manner even when a person is confined to bed and fasts. Conversely, the rhythm in body temperature is not caused by the nighttime interval of sleep; it continues with beat-like regularity even when a person is kept awake continuously for days (Figure 4–25). Your heart beats faster as body temperature increases, resulting in a rhythm in heart rate. The chemical composition of your blood and urine also varies throughout the day, as do the levels of many hormones. In fact, nearly all physiological processes are rhythmic.

The most obvious of our rhythms is the ebb and flow of consciousness that occurs as we awaken each morning and drift off to sleep each night. You might at first guess that fatigue continues to increase the longer you are awake, but this is not entirely correct. Typically, we become increasingly drowsy after midnight until some time around dawn, when we begin to feel slightly more alert. Thus, if you stay up all night cramming for an exam, you may perk up enough to make it through a morning exam. After the exam, you may find it difficult to fall asleep, even though you are extremely tired, because it is the wrong biological time for sleep.

Your ability to perform fine motor skills and mental activities also varies throughout the day. Among the activities

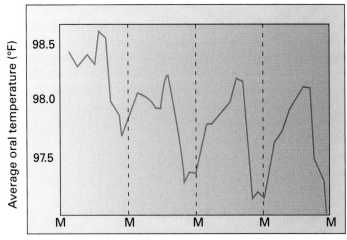

Figure 4–25 Body temperature varies predictably throughout the day. Although body temperature is influenced by activity levels, the daily cycle in body temperature continues in people who are confined to bed or who are prevented from sleeping. The curve shown here is the average body temperature (taken orally) of 15 men who were deprived of sleep for 98 consecutive hours.

that show daily fluctuations are calculation speed and accuracy in addition and multiplication, response time to a light stimulus, card-dealing speed, and vigilance while monitoring a radar screen. It is interesting to note that one's ability to perform these tasks and others that involve little memory is best when body temperature peaks. In contrast, tasks that require memorization are performed best when body temperature is at its lowest point. As a student, you might make the best use of study time by scheduling it with knowledge of your own body temperature rhythm.

Because the cycle length in each of the rhythms we have discussed is 24 hours, it seems reasonable to assume that biological rhythms are instilled by fluctuating environmental factors, such as light-dark cycles. But this is not the case. We know this because the rhythms continue without environmental time cues. Scientists have studied many individuals who have volunteered to live alone, without clocks, in places such as caves or windowless, soundproof laboratory rooms for weeks or months at a time. While the person lives in isolation, scientists monitor physiological functions and sleep-wakefulness without providing any time cues. Almost all of these rhythms continue in the absence of external time cues (Figure 4–26). We must conclude, therefore, that the rhythms are driven by a mechanism within the body, called the **biological clock**.

Without time cues, rhythms persist, but with an important difference: They typically become slightly longer (or shorter) than 24 hours. The rhythms are, therefore, described as circadian (*circa*, about; *diem*, day). It is assumed that the cycle length reflects the rate at which the biological clock is running.

Under more natural conditions, time cues are available. The sun lights up the sky each morning and as it sets each evening, darkness falls. When time cues are present, biological rhythms become "locked onto" a 24-hour cycle, because it is reset each day by the light-dark cycle.

Rhythms exist at every level of biological organization. Within a cell, there may be cycles in enzyme production or in biochemical reaction rates. As cells are grouped to form tissues and organs, their timing becomes synchronized. Thus, the processes they govern may also be rhythmic. We see that a single cell—perhaps every cell in the body—may have its own "wristwatch."

The body's many clocks are set to the same time by a master clock. In mammals, including humans, a master clock seems to be in a region of the brain called the **suprachiasmatic nucleus (SCN)** of the hypothalamus. The master clock is synchronized with the day-night cycle via light information from the eyes. (Interestingly, the light receptors involved in setting the master clock seem to be different from those involved in vision.) The SCN, in turn, communicates with the **pineal gland**, a gland within the brain that produces the hormone melatonin. In this way, the master clock in the SCN instills a rhythm in melatonin production.

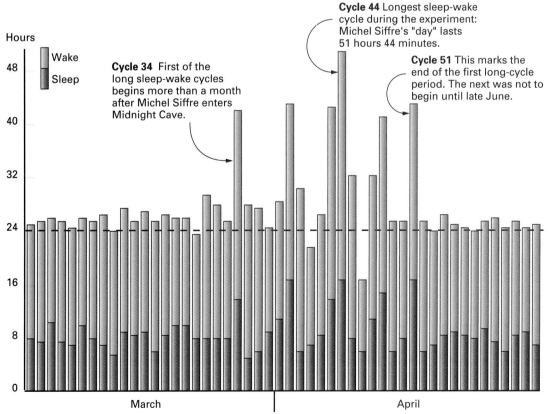

Cycle 34 First of the long sleep-wake cycles begins more than a month after Michel Siffre enters Midnight Cave.

Cycle 44 Longest sleep-wake cycle during the experiment: Michel Siffre's "day" lasts 51 hours 44 minutes.

Cycle 51 This marks the end of the first long-cycle period. The next was not to begin until late June.

Figure 4–26 Studying human biological rhythms. We look to a study carried out by Michel Siffre in 1975 ("Six months alone in a cave," *National Geographic* 147 [March,1975]: 426–435). An underground cave served as a live-in laboratory in which Siffre lived alone for months without any time cues. His activities and physiological functions were monitored from the surface. Most of his rhythms, including sleep and wakefulness as shown here, persisted in spite of the lack of time cues. Under these conditions, each cycle was typically slightly longer than 24 hours. The timing mechanism that instills such rhythms is thought to be within the body and is called the biological clock. *(Graph, © National Geographic)*

Melatonin is secreted into the bloodstream and can serve as a time signal for other physiological processes. For instance, melatonin induces sleep. As a result, we become drowsy as melatonin levels increase at night.

Although the same time signal governs many physiological processes, those processes may peak at different times of day, as we have seen. Indeed, each rhythmic process peaks at the most appropriate time of day and at the best time relative to the others.

Proper coordination among rhythmic life processes is essential to health and homeostasis. Consider the following analogy. A symphony is pleasing to the ear when the horn, string, wind, and percussion sections play their different parts in phase, that is, at the right time relative to one another. But when the sections are out of phase with one another, cacophony results, even if each section plays its part perfectly. The same is true in the body. Like a section in the orchestra, each body system must play its part with the proper timing relative to the others or internal chaos results. We see the importance of the synchronization of internal timing most clearly when it is disrupted, as might occur following rapid travel across time zones or in shift workers.

When we travel east or west across time zones, say between California and New York, the first thing we usually do when we step off the airplane is to reset our watch to the new local time. This adjustment helps us shift our eating and sleeping habits so that they are appropriate at our destination. Unfortunately, the body's many biological clocks cannot be reset as quickly as a watch. On average, a biological clock can be reset by about 2 to 3 hours a day. (Adjustment following eastward travel is generally more rapid than it is after westward travel.) Consequently, for several days after the trip, there is a mismatch between the body's clocks and the one on the wall. As the body's clocks slowly readjust to the new local time, they do so at slightly different rates, throwing the body's rhythms temporarily off beat. Until body time and real time are once again "in synch," one experiences symptoms commonly referred to as jet lag. One of the initial symptoms of jet lag is an all-consuming sense of fatigue that affects one's concentration, memory, and performance. Soon afterward, other symptoms begin. These include gastrointestinal problems, such as diarrhea or constipation, sleep disturbances, loss of appetite, impaired night vision, and limited peripheral vision. Therefore, if it is important that you perform well—mentally or physically—at the new location, you should allow plenty of time to adjust before you perform.

When peak performance at the new location is important and it is not possible to arrive a few days early to allow for adjustment, plan for jet lag. If possible, stay on home time and schedule events at times that would be comfortable in the home time zone. After traveling across three time zones from California to New York on business, for instance, a "power lunch" might be scheduled for noon New York time, which would be 9 AM California time. Alterna-tively, try to preset your body clock to the new time zone before you leave home. For several days before the trip, wake up and go to bed at times that would be appropriate at your destination.

Most of us, however, would find these suggestions unrealistic. So, here are some simple hints for minimizing jet lag. Begin your trip rested. While traveling, drink juice or water instead of an alcoholic or caffeinated beverage. Pace yourself when you reach your destination. Schedule activities lightly. Avoid overeating and drinking, because your digestive system is not prepared for action at the new times. And, since extroverts seem to get over jet lag faster than introverts, socializing may help you adjust to the new time zone. Spend as much time as possible outdoors. Exposure to sunlight helps reset one's biological clock. Finally, some researchers suggest that special diets or drugs will ease the symptoms of jet lag. Some studies have shown, for instance, that travelers who take pills containing the hormone melatonin, which can be purchased at health food stores, suffer less severely from jet lag.

Shift work, especially if it involves rotating shifts, has the same effect on the biological timing system as would endless globe-trotting. Many jobs, including health care, air-traffic control, and emergency and security services, must be performed 24 hours a day to ensure public safety. In industry, round-the-clock staffing is a way to maximize the use of expensive buildings or equipment. The industrialized nations of the world are, indeed, 24-hour societies.

Typically, the work day is divided into three 8-hour shifts—a day shift (e.g., 7 AM to 3 PM), an evening shift (e.g., 3 PM to 11 PM), and a night shift (e.g., 11 PM to 7 AM). An employee will often continually rotate from one shift to another with a few days off in between. Over 27% of American male workers and 16% of female workers have jobs in which they must rotate between day and night shifts.

Shift work wreaks havoc on the body's internal timing system. When employees switch to a new shift, their biological clocks must suddenly make an 8-hour adjustment from their old working schedule. This is the equivalent of traveling across eight time zones. A complete adjustment of all body rhythms to an 8-hour schedule change is likely to require 8 to 12 days. Thus, the rhythms of an employee who changes shifts more often than this may never be in their proper phase relations.

Such chronic disruption of the biological timing system can have serious health consequences. Night-shift workers and employees who rotate through shifts get more colds and have more heart attacks than do those who always work during the day. And, they are more likely to be plagued with gastrointestinal problems, such as peptic ulcers or constipation.

In addition, shift work disrupts sleep cycles. People who work the night shift or rotating shifts often complain of difficulty falling asleep or staying asleep. At work, then, they are mentally sluggish and more likely to make mistakes. Worse yet, they may fall asleep on the job. In a confidential survey of rotating shift workers in industrial plants, one-third to two-thirds of both the day and night employees reported that they fell asleep at work at least once a week.

There is a higher incidence of error among shift workers. The most important factor leading to these errors is disrupted sleep patterns. Obviously, on-the-job accidents and employees dozing off can have serious consequences for all of us, whether the employee is an airline pilot, air-traffic controller, physician, or operator in the control rooms of a nuclear plant.

SUMMARY

1. Tissues are groups of cells that work together to perform a common function. There are four main types of tissues in the human body: epithelial (covers body surfaces, forms glands, and lines internal cavities and organs), connective (storage site for fat, plays role in immunity, and provides protection and support), muscle (movement), and nervous tissue (coordinates body activities through initiation and transmission of nerve impulses).

2. Cells in epithelial tissues are usually polarized, having distinct basal and apical surfaces. The basal surfaces are anchored to underlying connective tissue by a basement membrane. One layer of the basement membrane, called the basal lamina, is secreted by the epithelial cells; the other layer, the reticular lamina, is secreted by cells of the connective tissue. The apical or free surfaces of epithelial cells face into body cavities, the linings of internal organs, or to the outside of the body. Neighboring cells in epithelial tissue are often connected by junctional complexes.

3. Epithelial tissue comes in two types, covering and lining epithelium (covers organs and forms outer layer of the skin and inner lining of blood vessels and several body systems) and glandular epithelium (forms secretory portions of exocrine and endocrine glands). Covering and lining epithelium is classified by the number of cell layers and shape of the cells. In terms of number of cell layers, epithelial tissue may be simple (cells form single layer), stratified (cells stacked in several layers), or pseudostratified (cells form a single layer, but because some cells do not reach the surface, the tissue has a multilayered appearance). Categories of cell shape include squamous (flat and scale-like), cuboidal (cube shaped), columnar (column shaped), and transitional (shape changes as tissue stretches).

4. All connective tissues contain cells and an extracellular matrix composed of protein fibers and ground substance. Cells in connective tissues typically are not polarized and have few, if any, junctional complexes between them. The various types of connective tissue can be grouped as connective tissue proper (loose and dense connective tissue) and specialized connective tissue (cartilage, bone, and blood). Many connective tissues function in binding and supporting structures within the body, while others function in transport (blood) and fat storage (adipose tissue).

5. Loose connective tissues, such as areolar and adipose tissue, have more cells and fewer, more loosely woven protein fibers in their matrix than do dense connective tissues. Dense connective tissues are found in ligaments (structures that join bone to bone), tendons (structures that join muscle to bone), and the dermis (the layer of skin below the epidermis).

6. Cartilage consists of cartilage cells (chondrocytes) embedded in a matrix that contains protein fibers for strength and a gelatinous ground substance for resilience. We have three types of cartilage in our bodies: hyaline cartilage (found at the ends of long bones), elastic cartilage (found in the external ear), and fibrocartilage (found in intervertebral disks).

7. Bone can be classified as compact or spongy. Compact bone has an underlying organization in which bone cells (osteocytes) sit in spaces called lacunae amid concentric rings of matrix (lamellae) comprised of collagen fibers and mineral salts. Large central canals contain blood vessels, and nutrients reach the osteocytes by moving from these canals through smaller canals called canaliculi. Spongy bone is less dense than compact bone and is made of an irregular network of collagen fibers covered by a calcium matrix.

8. Blood consists of formed elements (red blood cells, white blood cells, and platelets) suspended in a liquid matrix (plasma). Red blood cells transport oxygen and carbon dioxide, white blood cells aid in fighting infections, and platelets function in blood clotting.

9. Muscle tissue is composed of muscle fibers that contract when stimulated, generating force. There are three types of muscle tissue: skeletal, cardiac, and smooth. Skeletal muscle tissue is found attached to bone, is voluntary, and has cross striations and several nuclei per cell. Cardiac muscle tissue is found in the walls of the heart, is involuntary, and has cross striations and usually only one nucleus per cell. Smooth muscle tissue is found in the walls of blood vessels, airways, and organs, is involuntary, lacks striations, and each spindle-shaped cell has a single nucleus.

10. Nervous tissue contains two basic cell types, neurons and neuroglia. Whereas neurons convert stimuli into nerve impulses that they conduct to glands, muscles, or other neurons, neuroglia increase the rate at which impulses are conducted by neurons and provide neurons with nutrients from nearby blood vessels.

11. An organ is a structure composed of two or more different tissues with a specialized function. All organs may be characterized by having parenchymal cells, a connective tissue capsule, and a blood and nerve supply.

12. Organs that participate in a common function are collectively called an organ system. The ten major organ systems of the human body include: integumentary, skeletal, muscular, digestive, circulatory, respiratory, excretory, reproductive, nervous, and endocrine.

13. The integumentary system includes the skin and its derivatives such as hair, nails, and sweat and oil glands. The functions of these structures include protection of underlying tissues from abrasion and dehydration; excretion in the form of sweat; regulation of body temperature; synthesis of vitamin D; detection of stimuli associated with touch, temperature, and pain; and immune function.

14. The outermost layer of the skin is composed of epithelial cells and is called the epidermis. Just below the epidermis is the

dermis, a much thicker layer composed of connective tissue and containing nerves and blood vessels. Below the dermis is the hypodermis, a layer of loose connective tissue that anchors the skin to underlying tissues.

15. The epidermis has five layers (moving from deepest to most superficial): stratum basale, stratum spinosum, stratum granulosum, stratum lucidum, and stratum corneum. Cells produced in the deepest layer are pushed toward the skin surface. During this journey the cells flatten and die as a result of their increasing distance from the blood supply of the dermis and the replacement of their cytoplasmic contents with keratin. These cells, called keratinocytes, are the most numerous cells of the epidermis and are eventually sloughed off at the skin surface. Also found in the epidermis are Merkel cells (part of sensory receptors), Langerhans cells (macrophages), and melanocytes (manufacture and store melanin).

16. The dermis contains two layers, the papillary layer consisting of loose connective tissue and elastic fibers and the deeper reticular layer consisting of dense connective tissue with collagen and elastic fibers.

17. Skin color is determined, in part, by the quantity and distribution of pigments, primarily melanin. Melanin, released by melanocytes at the base of the epidermis, is ingested by neighboring keratinocytes and thus colors the entire epidermis. Albinism is an inherited condition in which an individual's melanocytes cannot produce melanin.

18. Circulation influences skin color through changes in either the amount of blood flow to certain regions of the body or oxygen content of the blood.

19. Hair, a derivative of skin, comes in three basic types—lanugo (found on fetuses and newborns), vellus, and terminal. The primary function of hair is protection. Male pattern baldness occurs when the thick terminal hairs of the scalp are replaced over time by soft fine vellus hairs.

20. A hair consists of a shaft and a root. The shaft projects above the skin surface and the root extends below the surface into the dermis or subcutaneous layer where it is embedded in a follicle. A terminal hair has three layers of keratinized cells, moving from inside to out: medulla, cortex, and cuticle. New hair cells, produced deep within the follicle, push older cells upward toward the skin surface. On their way to the surface these cells die. The part of a hair extending above the skin surface is thus a column of dead cells.

21. Nails are modified skin tissue hardened by keratin. The three basic parts of a nail include the free edge, the body, and the root. Nails provide protection for the tips of our fingers and toes and help us grasp and manipulate small objects.

22. Oil, sweat, and wax glands are derivatives of the skin. Oil glands open either directly onto the surface of the skin or more typically onto a hair follicle. Sebum, the oily substance secreted by these glands lubricates the skin and hair, prevents desiccation, and inhibits certain bacteria. Sweat glands may be either eccrine or apocrine glands. Eccrine sweat glands function throughout life and aid in regulation of body temperature through their secretion of sweat. Apocrine sweat glands begin functioning at puberty and discharge a thicker secretion composed of sweat, fatty substances, and proteins. Wax glands are modified sweat glands that secrete wax in the external ear canal.

23. The integumentary system does not function alone but instead works closely with other organ systems.

24. Homeostasis is the relative internal constancy that occurs at all levels of body organization. It is a dynamic state (small fluctuations occur around a set point) maintained primarily through negative feedback mechanisms. For example, thermoreceptors in the skin detect heat and activate nerve cells that send a message to the hypothalamus. In response, the hypothalamus sends impulses to the sweat glands to produce more perspiration. As perspiration evaporates, the body surface cools and body temperature decreases. This is an example of a negative feedback mechanism because the "product" of the system (cooling of the skin) feeds back on the system and inhibits further lowering of body temperature. Note also that homeostatic mechanisms consist of receptors (here, thermoreceptors in the skin), a control center (here, the hypothalamus), and effectors (here, the sweat glands).

25. The set points of many physiological processes change rhythmically throughout the day. Thus, homeostasis is often rhythmic. The body's many clocks are set to the same time by a master clock probably in the suprachiasmatic nucleus (SCN) of the hypothalamus.

REVIEW QUESTIONS

1. What are the four basic levels of organization found in the human body?
2. Compare the organization of epithelial and connective tissues. How do differences in the underlying organization of these tissues relate to their functions?
3. Why does bone heal more rapidly than cartilage?
4. Describe the structure of blood.
5. Compare and contrast skeletal, cardiac, and smooth muscle with respect to structure and function.
6. What types of cells are found in nervous tissue? What are their functions?
7. What four key features do all organs share?
8. List the functions of the integumentary system.
9. Compare the basic structures of the epidermis, dermis, and hypodermis.
10. What are the four types of cells found in the epidermis? What are their basic functions in the skin?
11. Describe keratinization.
12. What are cleavage lines?
13. Describe the roles of pigments and blood flow in determining skin color.
14. What are the three basic types of hair?
15. Describe how hair grows.
16. What is male pattern baldness?
17. What are oil glands? What role do these glands play in acne?
18. Compare eccrine and apocrine sweat glands.
19. Give an example of interaction between the integumentary system and another organ system.
20. What is homeostasis?
21. Describe the body's homeostatic mechanisms for raising and lowering core temperature.
22. Where is the master clock located in humans? How does it function?

Control and Coordination of the Body

(© Gerard Vandystadt/Photo Researchers, Inc.)

Chapter | 5

Skeletal and Muscular Systems

SOCIAL CONCERNS *Building Muscle, Fair and Square? Anabolic Steroid Abuse*

(© Art Stein/Photo Researchers, Inc.)

The skeletal and muscular systems largely determine what you look like and how you are able to move. Muscles put flesh on the bony framework of your body. Together they give your body shape and allow you to ski down a mountain slope, stroll along a beach, or write a letter home. In this chapter, we will examine the structure of bone and see why it is able to support our bodies against gravity, holding us away from the earth so that we don't just lie there, flattened against the ground. We'll see that bone is a dynamic, living tissue and consider how bones grow and a few reasons why some of us are taller than others (Figure 5–1). Following this we will see how the skeleton provides attachments for muscles, enabling us not only to move the skeleton but also to move from one place to another. Then, we will study the muscles themselves and how they function.

Figure 5–1 The skeletal and muscular systems are largely responsible for physical appearance. *(© Focus on Sports.)*

The Human Skeleton

The human skeleton has a special place in the thoughts of people, and historically, that place has been in the shadows. For one thing, if we see a skeleton, or even a bone, that means bad news for someone. If all is going well, bones stay on the inside, nicely covered by other tissues. In any case, let's consider the importance of the skeleton as part of the human body.

Functions of the Skeleton

The **skeleton** can be defined as a framework of bones and cartilage that functions in movement and in the protection of internal organs. Specifically, the skeleton functions in:

1. **Support.** It provides a rigid framework that supports soft tissues.
2. **Movement.** It provides places of attachment for muscles.
3. **Protection.** It encloses internal organs, such as the heart and lungs, which are within the chest cavity, and the brain, which lies within the skull.
4. **Storage of minerals.** It stores minerals in its bones, particularly calcium and phosphorus, which can be released to the rest of the body when needed.
5. **Storage of fat.** It stores energy-rich fat in yellow bone marrow (the soft tissue within some bones). The fat can be metabolized and the energy released when needed.
6. **Blood cell production.** It produces blood cells in the red marrow of certain bones.

Structure of the Skeleton

The 206 bones that form the internal scaffolding and girders of the human body have a variety of shapes and sizes. They are generally classified on the basis of their shape—short, flat, or irregular. Bones generally contain some degree of both compact and spongy bone, the exact proportion depending on the shape and size of the bone.

Compact bone is very dense bone, containing few internal spaces. Compact bone is what you see when you look at the outside of any bone (Figure 5–2). For instance, it forms most of the shaft of long bones, such as those of the arms and legs. Compact bone is covered by a **periosteum**, the membrane covering that nourishes bone. The periosteum is composed of two layers—an outer, dense layer through which run blood vessels, and nerves, and an inner layer containing elastic fibers, blood vessels, and various kinds of bone cells. When a bone is bruised or fractured, most of the pain results from injury to the periosteum.

Spongy bone is formed from a latticework of thin layers of bone with open areas between. These layers form internal struts that brace the bone from within and are called **trabeculae** (singular, *trabeculum*). Spongy bone is largely found in small, flat bones and in both the head (enlarged end) and near the ends of the shaft of long bones. In adults, the small spaces between the trabeculae of some spongy bones (including the ribs, backbone, skull, and long-bone ends) are filled with the blood cell–forming connective

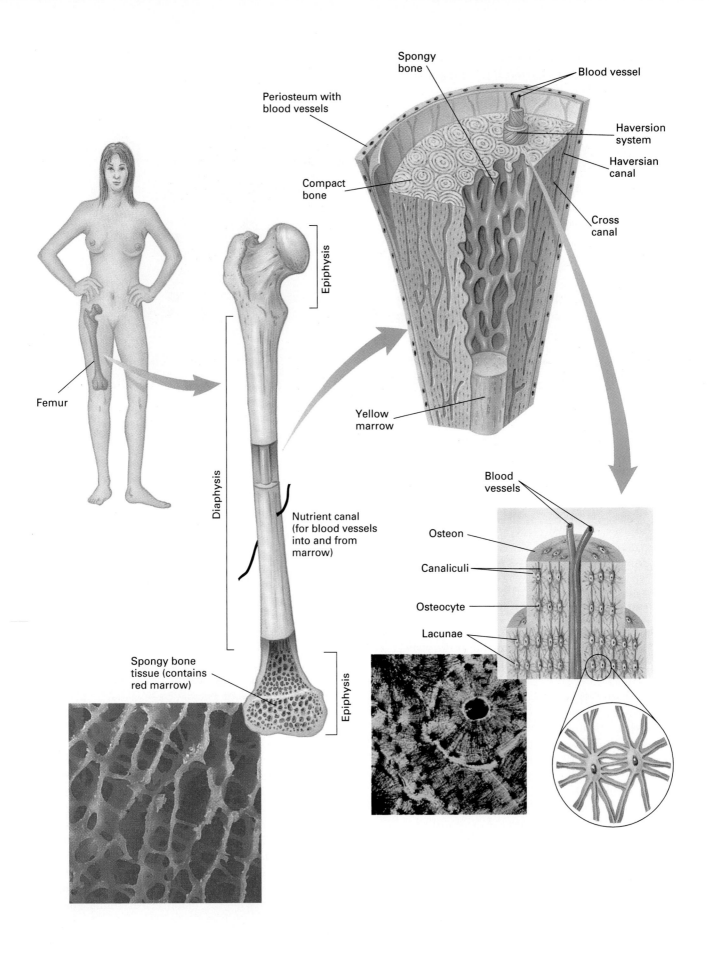

Spongy bone

Blood vessel

Periosteum with blood vessels

Haversion system

Haversian canal

Compact bone

Cross canal

Femur

Epiphysis

Diaphysis

Yellow marrow

Nutrient canal (for blood vessels into and from marrow)

Blood vessels

Osteon

Canaliculi

Osteocyte

Lacunae

Spongy bone tissue (contains red marrow)

Epiphysis

tissue called **red marrow**. In the shaft of long bones is a cavity, the medullary cavity, that is filled with **yellow marrow**, a fatty tissue for energy storage.

Interestingly, spongy bone, though light, is remarkably strong. One reason is that as the bone develops, the trabeculae are laid down along lines of stress. Bone tends to build in response to stress. For instance, exercise may cause bones to grow new tissue as areas of muscle attachment respond to the stress caused by muscle contractions.

Bone Tissue

Compact bone is highly organized living tissue. The structural unit of compact bone, called an **osteon** or **Haversian system**, appears as a series of concentric circles around a central canal (Figure 5–2). The central canal, also called an **Haversian canal**, runs longitudinally through the bone, one in the center of each osteon. Within tiny cavities (lacunae) in the hardened matrix are the mature bone cells, called **osteocytes**. The osteocytes are arranged in concentric rings around the Haversian canal. Tiny canals, called **canaliculi**, project outward from the lacunae. Within the canaliculi minute extensions of the osteocytes touch one another, allowing materials to be transferred from one bone cell to the next, as in a bucket brigade. The canaliculi connect with nearby lacunae and eventually with the Haversian canal. In this way, oxygen and nutrients can pass from the blood vessels of the Haversian canal to the embedded osteocytes and wastes can be carried away.

Bone is, indeed, a living tissue, but its characteristics come from its nonliving component—the matrix. Secreted by the bone cells, the matrix makes bone both hard and resilient. Nearly two-thirds of the matrix consists of mineral salts, mostly of calcium and phosphorous. The mineral salts are arranged in rod-shaped crystals that make bone hard and rigid. Woven through the matrix are strands of the elastic protein collagen. Collagen is extremely strong. A 20-pound weight could dangle from a 1-millimeter-thick strand of collagen. Without the calcium and phosphorus salts, bone would be rubbery and flexible like a garden hose. Indeed, bending of bones does occur, causing bowlegs, in disorders such as rickets, in which the amount of calcium salts in the bones is greatly reduced (Figure 5–3). Without collagen, however, bone would be brittle and crumble like chalk. In

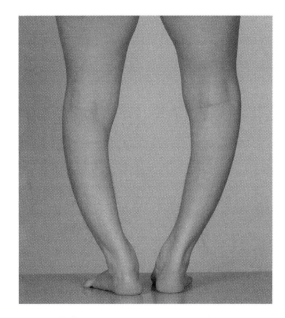

Figure 5–3 The legs of this child are bent because of rickets, a condition caused by insufficient vitamin D, which is needed for proper absorption of calcium from the intestines. Calcium salts make the matrix of bone very hard. Collagen fibers in the matrix add strength. In rickets, the bones become soft and somewhat pliant. *(© Biophoto Associates/Photo Researchers, Inc.)*

bone, calcium salts are organized around collagen fibers. Together, they create a substance that can withstand 25,000 lb/in.^{2} of compression and 15,000 lb/in.^{2} of tension.

Critical Thinking

Strontium-90 is a radioactive material that enters the atmosphere after atomic explosions and can end up in human bodies via milk from cows that grazed on contaminated grass. Strontium-90 can replace calcium in bone and then blast nearby tissues with beta particles, destroying cells and altering the genetic material of nearby cells. Explain why exposure to strontium-90 can lead not just to bone cancer but also to disruption of blood cell formation.

◀ **Figure 5–2** The structure of bone. A long bone, here the femur, consists of a shaft (diaphysis) and two heads or enlarged ends (epiphyses). The part of the bone seen from the outside is compact bone. Spongy bone is found in the heads. The structural unit of compact bone is called an osteon. Living bone cells (osteocytes) are found in small spaces within the hard matrix. Projections of the bone cells extend outward through tiny canaliculi and touch one another. In this way, materials can be exchanged with the blood supply in the central canal. The trabeculae of spongy bone form struts that brace a bone from within. Spongy bone is lighter than compact bone, but the trabeculae are laid down following lines of stress, making it well designed to resist loads. *(© left photo, Courtesy of Bruce Wingerd; right photo, © Astrid and Hanns Frieder Michler/Science Photo Library Custom Medical Stock Photo)*

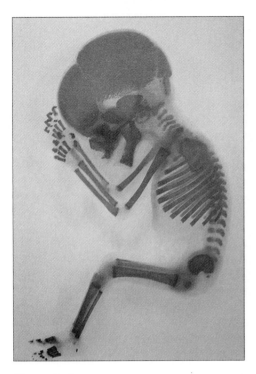

Bone Development

During embryonic development, the skeleton is first formed of cartilage, a firm yet flexible connective tissue (Figure 5–4). Unlike mature bone cells, cartilage cells are able to divide and form new cartilage cells. Thus, the cartilage model can grow and expand as the fetus grows rapidly. Beginning in the third month of development, the cartilage of the long and short bones is gradually replaced by bone in a process called **endochondral ossification**.

The formation of a long bone, such as those of the arms or legs, begins with the formation of a collar of bone around the shaft of the cartilaginous model. The collar is formed by bone-forming cells, called **osteoblasts** (*osteo*, bone; *blast*, beginning or bud). The osteoblasts form from cartilage cells that are transformed as blood vessels grow around the cartilage. The bony collar supports the shaft as the cartilage within it breaks down, forming the marrow cavity. Once a cavity is produced within the shaft, osteoblasts migrate into the space and form spongy bone. Osteoblasts secrete both the salts and the collagen (as well as other organic materials)

Figure 5–4 During fetal development, the skeleton is first laid down in cartilage. The cartilaginous model is gradually replaced by bone. *(© Photo Researchers, Inc.)*

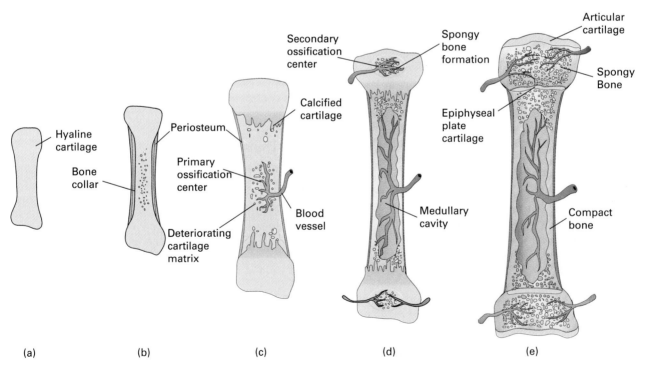

(a) (b) (c) (d) (e)

Figure 5–5 Endochondral bone formation. (a) In the embryo, a cartilaginous model of the future bone forms. (b) Osteoblasts form a collar of bone around the shaft of the model. (c) The shaft of the cartilage model begins to hollow out and spongy bone fills the space. Blood vessels continue to penetrate the area and the region of bone formation expands. (d) Secondary centers of bone formation develop in the ends of the bone. (e) Cartilage remains only on the articular surfaces that rub against other bones and in the epiphyseal plates.

of the matrix. These cells form bone, but they cannot undergo cell division. After they form the matrix around themselves, they are isolated. Once they entomb themselves in bone, they are called **osteocytes** (*cyte*, cell), which are mature bone cells and the principal cells in bone tissue.

At about the time of birth, bone growth centers form in the heads (epiphyses) of the bone and spongy bone begins to fill these regions (Figure 5–5). Two regions of cartilage will remain. One is a cap of cartilage that covers the surfaces that rub against other bones in a joint. The second is a plate of cartilage, called the **epiphyseal plate**, that separates the head of the bone from the shaft. Cartilage cells within the epiphyseal plate divide, forcing the end of the bone farther away from the shaft. As bone replaces the newly formed cartilage in the region closest to the shaft, the bone becomes longer.

Regulation of Bone Growth

Parents proudly measure the growth of a child, inch by inch, on a growth chart. During childhood, bone growth is powerfully stimulated by **growth hormone** (**GH**) released by the anterior pituitary gland. GH prompts the liver to release **growth factors** that produce a surge of growth in the epiphyseal plate. The activity of growth hormone is, in turn, modified by thyroid hormones. Thyroid hormones ensure that the skeleton grows with the proper proportions.

At puberty, a growth spurt often occurs. At this time, the legs of pants can seem to shrink almost weekly. These dramatic changes are orchestrated by the increasing levels of male or female sex hormones (testosterone and estrogen, respectively) that accompany puberty. Initially the sex hormones stimulate the cartilage cells of the epiphyseal plates into a frenzy of cell division. But, we generally stop getting taller toward the end of our teenage years (age 18 in females and 21 in males), because of later changes initiated by the sex hormones. The cartilage cells in the epiphyseal plates gradually divide less frequently. The plates get thinner then, as cartilage is replaced by bone, and finally the bone of the ends fuses with that of the shaft.

Critical Thinking

Keeping in mind the way the long bones grow and the mechanism by which growth hormone causes the growth of long bones, why would it be ineffective for a short, middle-aged person to be treated with growth hormone to stimulate growth?

Bone Fractures and Healing

In spite of the great strength of bones, they break. Fortunately, bone tissue can heal. The first thing that happens with a break is bleeding (even with a simple fracture), and a clot forms at the break (Figure 5–6). Within a few days, fibroblasts invade the clot. These are connective tissue cells

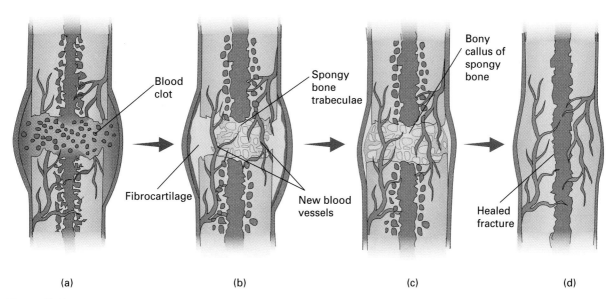

(a)   (b)   (c)   (d)

Figure 5–6 The progress of healing in a bone. (a) A blood clot forms (b) a cartilaginous callus begins to appear (c) osteoblasts begin to form new bone (d) the fracture is healed, although for a time it may be larger than normal.

that grow inward from the periosteum. The fibroblasts secrete collagen fibers that form a mass called a **callus**, which links the ends of the broken bone. Some of the fibroblasts then transform into cartilage-producing cells and these secrete cartilage in the callus.

Next, osteoblasts from the periosteum invade the callus and begin to transform the cartilage into new bone material. As this happens, the callus becomes larger than the bone itself and protrudes from it. In time, however, the extra material will be broken down and the bone will return to normal size.

Bone Remodeling

After the initial growth process is completed, the ongoing process of bone deposition and absorption is called **remodeling**. As we have seen, bone is deposited by osteoblasts. Another kind of bone cell, called an **osteoclast**, breaks down bone and the minerals are reabsorbed by the body. The process of remodeling is continuous and occurs at different rates in different parts of bones. For example, the end of the femur (thighbone) closest to the knee is completely replaced every 5 to 6 months, but the shaft of the same bone is replaced more slowly.

An important factor in determining the rate and extent of bone remodeling is the degree of stress to which a bone region is exposed. Bone forms in response to stress and is destroyed when it is not stressed. Weight-bearing exercise, such as walking or jogging, thickens the layer of compact bone, leading to stronger bones. Frequently used bones may actually become misshapen. Due to continual practice, the knuckles of pianists and the big toes of ballet dancers can enlarge. On the other hand, bones that are not stressed lose mass. For instance, after a few weeks without stress, as might occur in an astronaut in the weightless environment of space flight or right here on earth while using crutches with a leg in a cast, a bone could lose nearly a third of its mass.

Two hormones, calcitonin and parathyroid hormone, play an important part in both the control of bone remodeling and the regulation of blood levels of calcium. When blood levels of calcium are high, as might occur after a meal, the hormone **calcitonin**, which is released from the thyroid gland, removes calcium from the blood and causes it to be stored in bone. Calcitonin brings about these effects by stimulating the activities of osteoblasts while inhibiting those of osteoclasts. In contrast, **parathyroid hormone (PTH)**, which is released by the parathyroid glands found embedded in the tissues of the thyroid gland, raises blood calcium levels by stimulating the activities of osteoclasts. The interplay of these two hormones keeps blood calcium levels fairly steady.

Osteoporosis—Fragility and Aging

Osteoporosis is a decrease in bone density that occurs when the destruction of bone outpaces the formation of new bone, causing bones to become thin, brittle, and susceptible to fracture. One of the greatest problems with osteoporosis is the constant threat of broken bones. We frequently hear of old people falling and breaking a hip, and indeed, such accidents are all too frequent, but sometimes the bones break simply from the stress of daily routine. A fall, blow, or lifting action that would not bruise a person with healthy bones could easily cause bone fracture or crushing in a person with severe osteoporosis. Osteoporosis affects about 20 million to 25 million Americans, most of them elderly white women.

As we have seen, bone remodeling occurs throughout life. Until we reach about age 35, osteoblasts lay down new bone faster than osteoclasts remove it. Thus, bones are strongest and most dense in our mid-thirties. Peak bone density is influenced by a number of factors, including sex, race, nutrition, exercise, and overall health. In men, bone mass is generally 30% higher than in women. The bones of African Americans are generally 10% more dense than those of Caucasian and Asian Americans. Within each group, bone density will be influenced by diet, especially the dietary levels of calcium and vitamin D, which are essential for forming hard bones, and by the history of weight-bearing exercise, such as walking, which stresses bones and promotes bone formation.

Nearly everyone begins to lose bone density after their mid-thirties, but the degree to which bones will become dangerously weakened depends largely on how dense the bones were at their peak. Osteoporosis is sometimes very apparent. The afflicted person becomes hunched and stooped and, as vertebrae lose mass, may become even shorter (Figure 5–7). It is sometimes startling to come across someone, in middle age or older, whom you haven't seen in a while and see that they're actually shorter!

Women are at greater risk for developing osteoporosis than men, not just because they have less bone mass at peak but also because the rate of bone loss is accelerated for several years after menopause. Menopause, the time in a woman's life when she no longer produces mature eggs or menstruates, marks a severe reduction in the amount of the female hormone, estrogen, being produced. Estrogen stimulates osteoblasts to form new bone. Without the high levels of estrogen, the osteoblasts reduce their rate of bone formation. The osteoclasts continue their destruction of bone and the imbalance results in the reduction of bone mass. Estrogen is also important in the absorption of calcium from the intestines, and calcium is needed to make bones hard.

It is noteworthy that young women who undergo athletic training extensive enough to stop menstruation are also placing their bones at risk. It is not uncommon for intense physical exercise to disrupt menstrual cycles because of alterations in hormone levels. Researchers have shown that, in such cases, bone density can also be affected. In one study, 20% of the young female athletes who failed to ovulate during even one cycle out of the whole year suffered as much as a 4% loss in bone density.

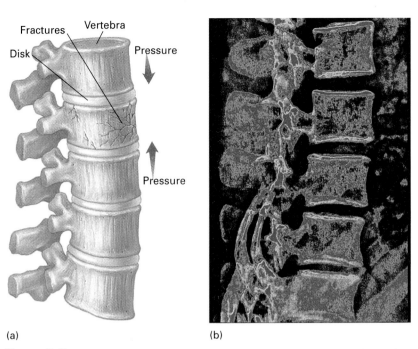

(a)

(b)

(c)

Figure 5–7 Osteoporosis is a loss of bone density that occurs when, during the continuous process of bone remodeling, bone destruction outpaces bone deposition. Bones become brittle and are easily fractured. (a) Normal bending can place pressure on vertebrae that can cause small fractures. (b) Osteoporosis can be seen in this colored x-ray image of vertebrae. (c) A loss of height and a stooped posture result as the weakened bones of the vertebrae become compressed. *(b, Alfred Pasieka/Science Photo Library/Custom Medical Stock Photo; c, © Jim Pickerell/Tony Stone Images, Inc.)*

A number of factors other than estrogen levels have been implicated in the onset of osteoporosis. For example, size is important. Short people are at greater risk, perhaps because they generally start with less bone mass. Those people with a good supply of body fat are less at risk because fat can be converted to estrogen. Heavy drinkers are at higher risk because alcohol interferes with estrogen function. Smoking is also bad for bones, because it can reduce estrogen levels. People who do not take in enough calcium, or who cannot absorb it, have thinner bones because calcium is necessary for bone growth. Calcium-rich foods include not only milk products but also spinach, shrimp, and soybean products. Certain drugs, such as caffeine (a diuretic), tetracycline, and cortisone can promote osteoporosis. Finally, sedentary people have thinner bones. Exercise places stress on the bones and the bones respond by becoming thicker, so weight-bearing exercise is recommended.

Women who have passed menopause are advised to take calcium supplements, to avoid smoking and drinking, to avoid certain drugs and diuretics, to enter a program of weight-bearing exercise, and to get plenty of vitamin D, which helps in the absorption of calcium. But these measures alone are generally not enough to halt bone loss after menopause. Hormone replacement therapy, which involves administering estrogen, usually along with progesterone, is often recommended. A newer treatment involves high doses of calcium fluoride to stimulate bone development. Unfortunately, fluoride in high doses can cause abnormal bone development and stomach bleeding. In the fall of 1995, a new drug, alendronate sodium (trade name Fosamax), gained FDA approval for the treatment of osteoporosis. Although possible long-term side effects are not yet known, the drug has been shown not just to slow bone loss but to actually increase bone density.

We can expect osteoporosis and other afflictions associated with aging to become more common in the population as people tend to live longer. Thus, it would seem prudent for each of us to start an early program of prevention.

Parts of the Human Skeleton

The earliest solid information about the human skeleton was gained under some rather remarkable, even macabre, circumstances. In the second century, the great physician Galen dissected animals to learn about the skeleton, but one day, he found the skeleton of an executed robber, picked clean by vultures. He summoned a group of medical men to observe the bones and note their shape and texture. This began his serious study of the human skeleton. As time passed, Greek surgeons began to examine other corpses of criminals, left beside the road as an encouragement to walk the straight and narrow. By the Renaissance, many great European universities were encouraging the dissection of corpses. Thus, by the eighteenth century, the skeleton had been thoroughly described.

As can be seen in Figure 5–8, the bones of the human body are arranged into the axial skeleton and the appendicular skeleton. The **axial skeleton** includes the skull, the vertebral column (backbone), and the bones of the chest region (sternum and rib cage). The **appendicular skeleton** includes the pectoral girdle (shoulders), the pelvic girdle (pelvis), and the limbs (arms and legs).

The Axial Skeleton

We will focus on only the major bones of the 80 that comprise the axial skeleton, starting from the top—that is, be-

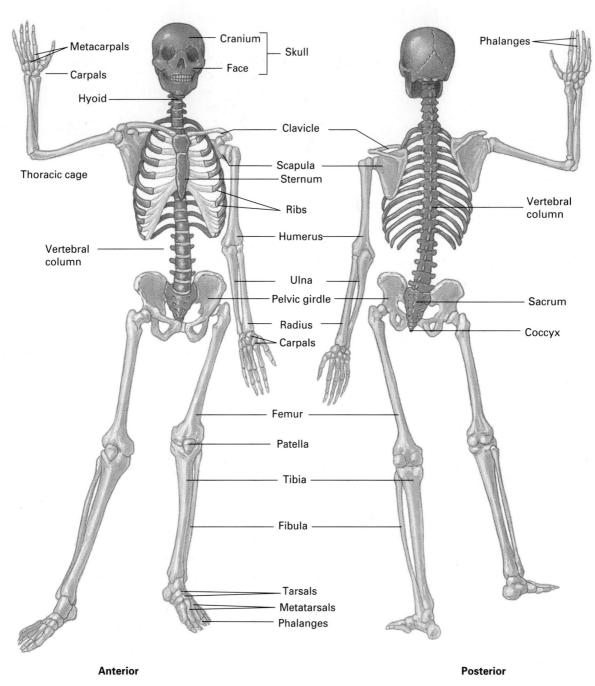

Anterior

Posterior

Figure 5–8 The major bones of the human body. The darker bones comprise the axial skeleton, the lighter bones, the appendicular skeleton.

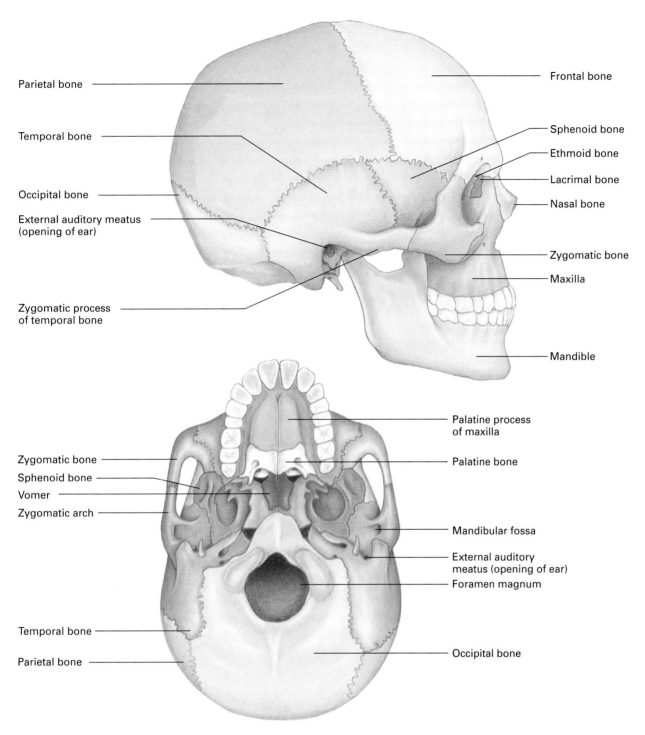

Parietal bone

Frontal bone

Temporal bone

Sphenoid bone

Ethmoid bone

Lacrimal bone

Occipital bone

Nasal bone

External auditory meatus (opening of ear)

Zygomatic bone

Maxilla

Zygomatic process of temporal bone

Mandible

Palatine process of maxilla

Zygomatic bone

Palatine bone

Sphenoid bone

Vomer

Zygomatic arch

Mandibular fossa

External auditory meatus (opening of ear)

Foramen magnum

Temporal bone

Parietal bone

Occipital bone

Figure 5-9 The major bones of the skull.

ginning with the skull. The skull is divided into two areas— the bones of the cranium and the facial bones (Figure 5–9). The cranium houses the brain and the structures of hearing. The facial bones give us our facial features and house certain sensory systems of the head, including vision, smell, and taste.

The Cranium. The cranium is formed from eight (or sometimes more) flattened bones. The single **frontal bone** forms the forehead and the anterior part of the brain case. Posteriorly, the frontal bone joins the two **parietal bones** that join at the midline. Posterior to these, at the back of the head lies the **occipital bone**, which surrounds the **foramen**

magnum, the opening through which the spinal cord passes.

Before and shortly after birth, these bones are held together by membranous areas called the **fontanels**, often referred to as the "soft spots" (Figure 5–10). These accommodate the rapid growth of the brain during fetal growth and infancy. During birth, the fontanels allow the skull to be compressed, easing the passage of the head through the birth canal. The fontanels will be replaced by bone by the age of 2.

On either side of the cranium are the **temporal bones**, part of which form what we think of as our temples. The **sphenoid**, with its bow-tie shape, forms the anterior floor of the cranium. Deep within the head, an indentation in the sphenoid forms a pocket called the **sella turcica** (Turk's saddle), in which the pituitary gland (often called the "master gland") nestles. The **ethmoid**, the smallest bone in the cranium, separates the cranial cavity from the nasal cavity. The nerves from the nasal cavity communicate with the brain through tiny holes in the ethmoid.

The Facial Bones. The face is composed of 14 bones. They support several sensory structures and serve as attachments for most muscles of the face.

Nasal bones form the bridge of the nose. They are paired and fused at the midline. Boxers, by the way, often have had the nasal bones removed after being broken frequently, suggesting the name "pug" for an older, broken-down pugilist. Inside the nose, a partition called the **nasal septum** (or vomer) divides the left and right chambers.

Cheekbones are formed largely from the paired **zygomatic bones**. Flattened areas of these bones form part of the bottom of the eye sockets. The zygomatic bone also gives rise to an extension that joins with an extension from the temporal bone to form the zygomatic arch (the "cheekbone" itself).

The smallest facial bones are the **lacrimal bones**, located at the corners of the eyes near the nose. A groove passes through each bone and drains tears from the eyes into the nasal chambers, which explains why we sniff when we cry.

The jaw is formed by two pairs of bones. The upper jaw is composed of two **maxillae**, fused at the midline, directly below the nasal septum. The importance of the maxillae to facial structure is revealed in the fact that most other facial bones are joined to them. Horizontal plates, projecting posteriorly, form the anterior part of the hard palate, the roof of the mouth. Behind the palate lie two **palatine bones** that form the rest of the roof. When the maxillae fail to join as the face develops, a cleft palate results. Here, the mouth cavity and the nasal cavity are not fully separated. This condition is easily corrected surgically today.

The lower jaw is called the **mandible**. It, too, is formed from two bones connected at the midline. The mandible is connected to the skull at the temporal bone, forming a hinge called the temporomandibular joint. (Joints are places where bones meet and will be discussed later in this chapter.) This joint allows the mouth to open and close. Emotional stress causes some people to clench or grind their teeth, sometimes unconsciously. This can cause physical stress on the temporomandibular joint, causing headaches, toothaches, or even earaches. The condition is known as temporomandibular joint (TMJ) syndrome.

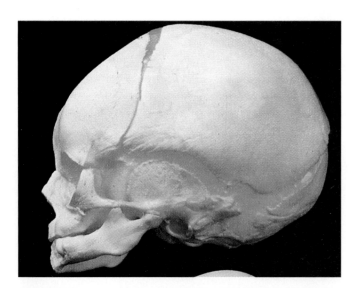

Figure 5–10 The bones of the skull of a human newborn are not fused, but are instead held together by fibrous connective tissue. These "soft spots" allow the skull bones to move during the birth process, easing the passage of the skull through the birth canal. By the age of 2, the soft spots will be replaced by bone.
(From Color Atlas of Human Anatomy, Second Edition by R. M. H. McMinn and R. T. Hutchings, © 1988 Year Book Medical Publishers, Inc.)

The Vertebral Column. The **vertebral column**, more familiarly known as the backbone, is a series of bones descending from the cranium, through which the spinal cord passes (Figure 5–11). Each of these bones, 26 in all, is called a **vertebra**. The vertebrae are divided according to where they lie along the length of the vertebral column. There are 7 **cervical** (neck) **vertebrae**, 12 **thoracic** (chest) **vertebrae**, 5 **lumbar** (lower back) **vertebrae**, 1 **sacrum** (formed by the fusion of five vertebrae), and 1 **coccyx** (or tailbone), formed by the fusion of four vertebrae.

The added strength of the fused sacral vertebrae is necessary because the sacrum joins the pelvic girdle and great stress is placed on it due to the weight of the vertebral column and the powerful movements of the leg. The coccyx, on the other hand, may be fused for precisely the opposite reason. It serves no function, being regarded as a vestigial tail—an evolutionary relic—whose bones may have joined as a side effect of growth without movement. (The upper part of

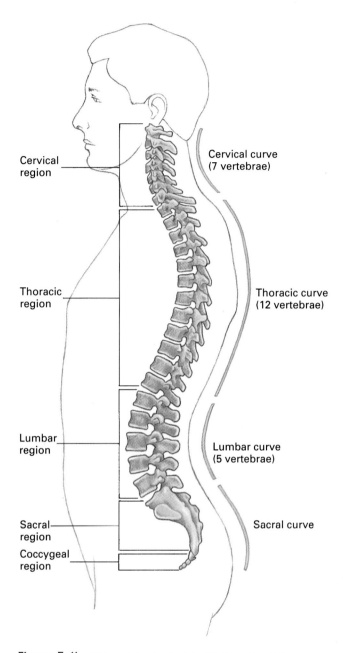

Figure 5-11 The vertebral column with its areas and curves.

Cervical region

Cervical curve
(7 vertebrae)

Thoracic region

Thoracic curve
(12 vertebrae)

Lumbar region

Lumbar curve
(5 vertebrae)

Sacral region

Sacral curve

Coccygeal region

the coccyx, however, is supplied with nerves and can cause extreme pain if somehow broken.)

Between the vertebrae of the cervical, thoracic, and lumbar regions lie **intervertebral disks**, pads of cartilage that help cushion the bones of the vertebral column and, because of their smooth, lubricated surfaces, help in the movement of the column. As the intervertebral disks become compressed over the years, the person may become shorter. This, alone or coupled with the effects of osteoporosis, can have a pronounced effect on height.

Other problems can also arise from the abnormalities of these disks. For example, if too much pressure is applied to the vertebral column, the disks can bulge from between the vertebrae. If the disks bulge inward, pressing against the spinal cord itself, they can cause problems with perceiving incoming stimuli or interfere with muscle control. If the bulging disk presses against a spinal nerve that branches from the spinal cord, great pain can result. The sciatic nerve, a large nerve that extends down the back of the leg, is one of the more frequently affected nerves. Sciatica, the resulting inflammation, can be painfully crippling.

A particularly mysterious human ailment is lower back pain. It accounts for more missed workdays than any medical problem besides colds. It is estimated that the annual costs for medical expenses and lost time are as much as $30 billion a year. The source of the pain is notoriously difficult to pinpoint. People with slipped disks may or may not have pain, and others whose backs look perfectly normal may complain of terrible pain. Part of the problem seems to be that the pain may originate from muscle rather than bone and so may not show up on diagnostic tests, such as a CAT scan or x-ray. One source of problems, however, may be weak abdominal muscles that cannot counteract the pull of the powerful back muscles, a situation that can cause the vertebrae to misalign.

The Rib Cage. Twelve pairs of ribs attach at the back to the thoracic vertebrae. At the front, the upper ten pairs of ribs attach directly or indirectly to the breastbone (**sternum**) by cartilage. Their flexibility permits the ribs to take some blows without breaking and permits the ribs to move during breathing. The last two ribs do not attach to the sternum at all and are called **floating ribs**.

The Appendicular Skeleton

The **appendicular skeleton** is composed of the pelvic and pectoral girdles along with the attached limbs. A **girdle** is a skeletal structure that supports the arms (the **pectoral girdle**) or the legs (the **pelvic girdle**). The pectoral girdle, then, connects the arms to the rib cage and the pelvic girdle connects the legs to the vertebral column.

The Pectoral Girdle. The pectoral girdle is composed of the **scapulae** (shoulder blades) and the **clavicles** (collar bones). Each clavicle makes a rigid connection between a scapula and the sternum. The clavicles are more curved in males than females (one way to tell the sex of a skeleton). One corner of the roughly triangular scapula bears a socket into which fits one end of the **humerus**, the upper arm bone. The humerus joins at the elbow with the **radius** and **ulna**, the bones of the lower arm. As the wrist turns, as when making a "so-so" motion, the distal (far) end of the radius forms a circle, thus the name. The ulna extends past the junction with the humerus to form the elbow. The radius and ulna meet the eight **carpals** at the wrist, and these join with the five **metacarpals**, which are the bones of the hand. The hand gives rise to three **phalanges** in each finger and two in each thumb.

There is a narrow opening, or tunnel, through the carpal bones that form the wrist, and through this so-called carpal tunnel passes a nerve that controls sensations in the fingers and some muscles in the hand (Figure 5–12). Tendons, bands of connective tissue that attach muscles to bones, also pass through the carpal tunnel. Repeated motion

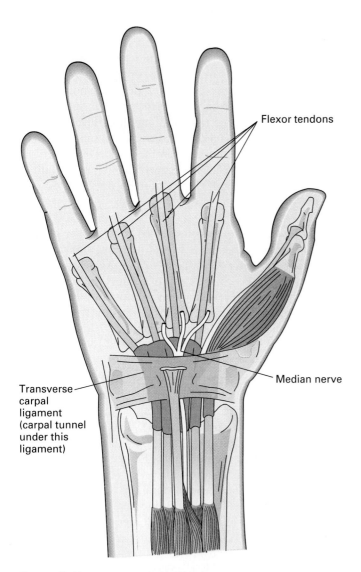

Figure 5-12 The small tunnel through the carpal bones forming the wrist is called the carpal tunnel. The median nerve that controls sensation in the fingers and some of the muscles of the hand passes through the carpal tunnel, along with many tendons. Repetitive wrist movements can inflame the tendons, causing them to swell and press the nerve against the bone. The result can be numbness of the hand or wrist pain, known as carpal tunnel syndrome.

Labels in figure: Flexor tendons; Median nerve; Transverse carpal ligament (carpal tunnel under this ligament)

in the hand or wrist can cause the tendons to become inflamed and press against the nerve. As a result, there may be numbness or tingling in the affected hand and pain that may affect the wrist, hand, and fingers. This condition, known as **carpal tunnel syndrome**, is becoming increasingly common as increasing numbers of people operate the keyboard of a computer or play video games. The same problem has been noticed by barbers, cab drivers, needlepointers, and pianists. The increased incidence of carpal tunnel syndrome has caused the computer industry to redesign their keyboards and games and employers to provide adjustable work stations and have mandated rest and exercise periods to alleviate the problem. The cost in terms of lost workdays and medical expenses is estimated at $20 billion a year.

Social Issue

In recent years, there has been a rash of lawsuits against computer manufacturers and video-game designers by people who have developed carpal tunnel syndrome as a result of endless hours spent using keyboards and video-game controllers. Do you think manufacturers should be held liable for injury due to overuse of their product?

The Pelvic Girdle. The pelvic girdle is much more rigid than is the pectoral girdle, as you know if you have ever tried to shrug your hips. The pelvic girdle consists of two **pelvic bones**, which are attached to the sacrum. These arch around to the front where they join in the **pubic symphysis**. The male and female hips are easily distinguishable with the opening in the female being wider to facilitate childbirth (Figure 5–13). During fetal development, the pelvis is formed from three bones, the **ilium**, **ischium**, and **pubis**.

The **femur**, or thigh bone, rotates within a socket in the pelvis. At the knee, the femur joins the **tibia** (the shinbone) and the **fibula**, a smaller bone that runs down the side of the lower leg. The junction where they join the femur is covered by a **patella** (kneecap). At the ankle, the lower leg bones meet the **tarsals**, or ankle bones. These are connected to the **metatarsals**, or foot bones, and these are, in turn, connected to the phalanges, the toe bones.

Joints

Joints are the places where bones meet. They can be classified according to the degree of movement they permit. Some allow no movement, others permit slight movement, and still others are freely moveable.

An example of immovable joints is provided by the joints between the skull bones of an adult. In this case, the joints, called sutures, are immovable because the bones are interlocked and held together tightly by fibrous connective tissue (Figure 5–14).

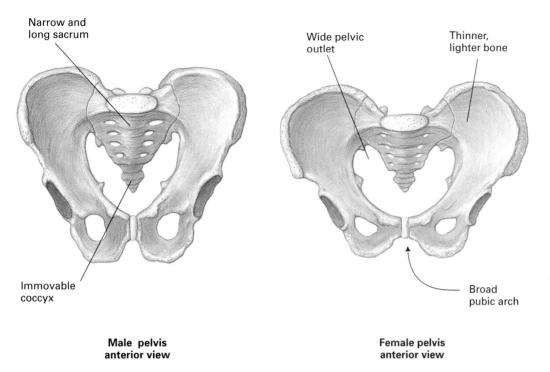

Narrow and
long sacrum

Wide pelvic
outlet

Thinner,
lighter bone

Immovable
coccyx

Broad
pubic arch

**Male pelvis
anterior view**

**Female pelvis
anterior view**

Figure 5-13   The wider opening of a female pelvis is an adaptation that facilitates childbirth.

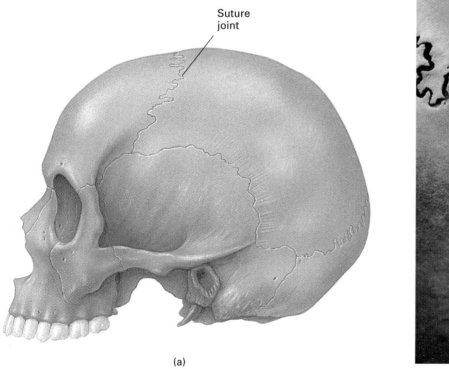

Suture
joint

(a)

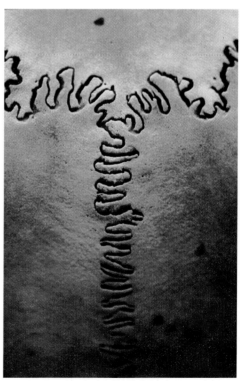

(b)

Figure 5-14   The bones of the skull are joined by suture joints. (a) Lateral view of the skull. (b) Photograph of suture joints in an adult skull.   *(b, © Michael Gabridge/Visuals Unlimited.)*

Other joints allow slight movement. In many of these joints, adjacent bones are held together with cartilage, which is rather rigid. We find such joints between vertebrae, in the attachment of ribs to the sternum, and in the pubic symphysis. In a pregnant woman, hormones loosen the cartilage of the pubic symphysis, allowing the pelvis to widen to ease childbirth.

Synovial Joints

Synovial joints are freely moveable. Most of the joints in the body are synovial joints. All synovial joints share certain common features (Figure 5–15):

1. The surfaces of the joints that move past one another have a thin layer of hyaline cartilage. The cartilage reduces friction, allowing the bones to slide over one another without grating and grinding.
2. Synovial joints are surrounded by a fluid-filled space, called a **synovial cavity**. The cavity is filled with a viscous, clear fluid (synovial fluid), which acts both as a shock absorber and as a lubricant between the bones.
3. Synovial joints have an articular capsule, a two-layered structure that connects the bones of the joint. The outer layer of the articular capsule is continuous with the covering membranes (periostea) of the bones forming the joint. The synovial membrane forms the inner layer and secretes the synovial fluid.

4. The entire synovial joint is reinforced with **ligaments**, which are strong straps of connective tissue that serve to hold the bones together, support the joint, and direct the movement of the bones.

A **sprain** is a tear in a ligament. Slight sprains may be caused by a ligament simply being stretched to some degree, but a torn ligament results in swelling and enough pain to inhibit movement. Like tendons, ligaments have few blood vessels and heal slowly. They are covered with a concentration of pain receptors that are very sensitive to stretching and swelling. Thus, the injury may not be as severe as the pain would suggest. As with most musculoskeletal swelling, first treatment may involve reducing the swelling with ice.

In regions of the body where movements might cause friction between moving parts, such as might occur around synovial joints, the body has its own form of "ball bearings"—fluid-filled sacs called **bursae** (sing., *bursa*; meaning pouch or purse). Bursae surround and cushion certain joints. They resemble synovial sacs in that they are lined with synovial membranes, but they are found in places where skin rubs over bone as the joint moves and between tendons and bones, muscles and bones, and ligaments and bones.

Repeated pressure on a bursa or injury to a nearby joint can cause the bursa to become inflamed and swell with excess fluid, a condition called bursitis ("itis" refers to an in-

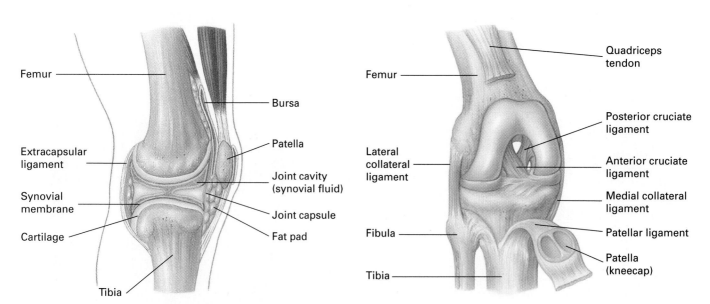

Figure 5–15 Synovial joints, such as the knee shown here, permit a great range of movement. (Left) All synovial joints have an articular capsule that connects the bones of the joint. The outer layer of the capsule is formed of fibrous connective tissue. The inner layer is the synovial membrane, which secretes synovial fluid. (Right) Ligaments hold the bones together, support the joint, and direct the movement of the bones.

flammation). Bursitis is characterized by intense pain that becomes worse when the joint is moved and cannot be relieved by resting in any position. Nonetheless, bursitis is not serious and usually subsides on its own within a week or two. In severe cases, a physician may drain some of the excess fluid to remove the pressure. The condition known as dart-player's elbow can be caused by bursitis of the elbow joint, and housemaid's knee is bursitis of the bursa that covers the kneecap.

Arthritis

Arthritis is an inflammation of a joint. However, there are more than 100 kinds of arthritis, some far more serious than others. **Osteoarthritis** is a degeneration of the surfaces of a joint, caused by wear and tear. Over time, any articular (joint) surface that causes friction is bound to wear down. Finally, the slippery cartilage at the ends of the bones forming a joint begins to disintegrate until the bones themselves come into contact and grind against each other, causing intense pain and stiffness. Osteoarthritis is most likely to occur in the weight-bearing joints, such as the hip, knee, and spine, and occasionally in the finger joints or the wrist.

Another far more threatening form of arthritis (and, unfortunately, one of the most common) is **rheumatoid arthritis**. Rheumatoid arthritis is marked by an inflammation of the synovial membrane. As the synovial membrane becomes inflamed, excess synovial fluid accumulates in the joint, causing swelling, pain, and stiffness. Eventually, the constant irritation can lead to the destruction of the cartilage, which may then be replaced by fibrous connective tissue that further impedes the movement of the joint.

Rheumatoid arthritis differs from the others in that it is apparently an autoimmune disease. That is, it is caused by an immune response of the body toward its own synovial membranes. The body attempts to reject its own tissue just as it would some invasive foreign matter. Rheumatoid arthritis can vary in its severity, but it is a permanent condition, normally affecting the joints of the fingers, wrist, knees, neck, ankles, and hips. Sometimes the only effective treatment is an artificial replacement of the joint (Figure 5–16).

The Muscular System

There are generally regarded to be three kinds of muscles—skeletal, cardiac, and smooth. They have distinct qualities and functions, but they all have four traits in common.

1. Muscles are excitable. They respond to stimuli.
2. Muscles are contractile. They have the ability to shorten.
3. Muscles are extensible. They have the ability to stretch.
4. Muscles are elastic. They can return to their original length after being shortened or lengthened.

Smooth and cardiac muscles are considered in Chapters 4 and 11, respectively, so here we will concentrate on skeletal muscles, the ones we usually think of when we hear the word muscles. Unlike cardiac and smooth muscle, skeletal muscle is under voluntary control.

Skeletal Muscles and Movement

There are more than 600 skeletal muscles in the body and most are arranged in pairs or groups that cooperate in movement (Figure 5–17). Those that work together to cause movement in the same direction are called synergistic muscles. Most muscles, however, are arranged in pairs so that the actions of the members of the pair are opposite to one another; that is, muscles are usually arranged in **antagonistic pairs**. As any muscle contracts and pulls on a bone, a muscle with the opposite action must relax. With any movement, the muscle moving the bone is called an **agonist**. The muscle that must relax to permit the movement is called the **antagonist**. An example of such cooperation is provided by the **biceps brachii**, located on the top of the forearm (the ones people like to show off), contraction of which causes the arm to flex, and the **triceps brachii** (on the back of the

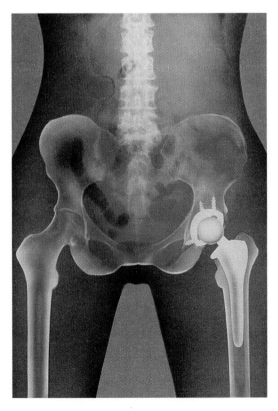

Figure 5–16 Replacement prosthetic of the hip. *(© Chris Bjornberg/Photo Researchers, Inc.)*

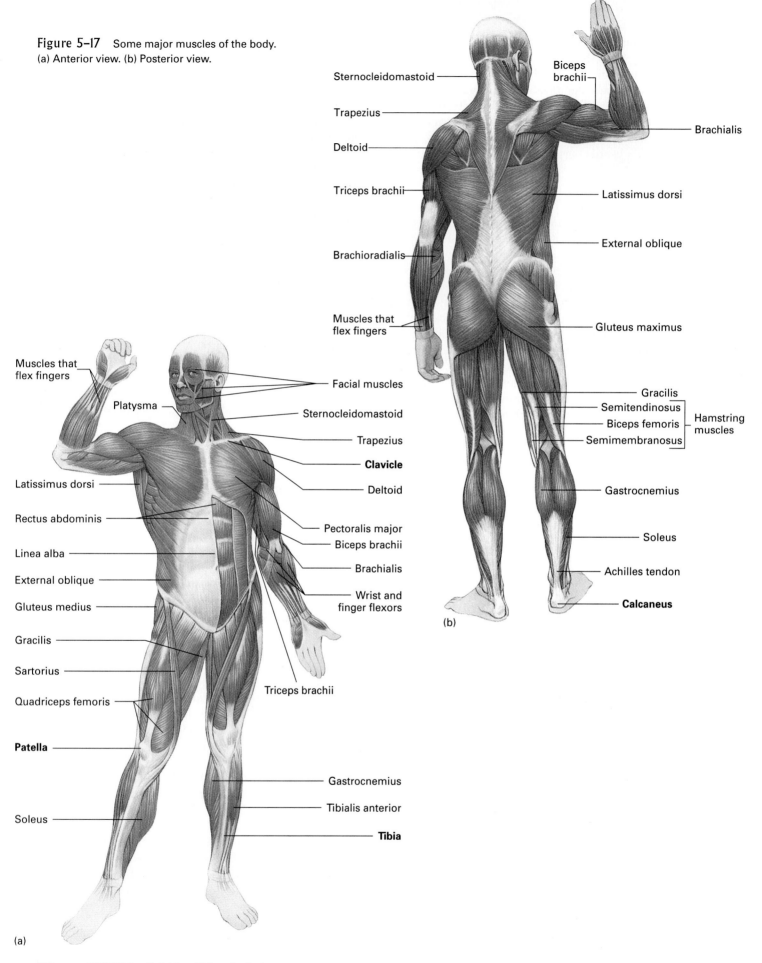

Figure 5-17 Some major muscles of the body. (a) Anterior view. (b) Posterior view.

Sternocleidomastoid

Trapezius

Deltoid

Triceps brachii

Brachioradialis

Muscles that flex fingers

Biceps brachii

Brachialis

Latissimus dorsi

External oblique

Gluteus maximus

Gracilis

Semitendinosus

Biceps femoris

Semimembranosus

Hamstring muscles

Gastrocnemius

Soleus

Achilles tendon

Calcaneus

(b)

Muscles that flex fingers

Platysma

Latissimus dorsi

Rectus abdominis

Linea alba

External oblique

Gluteus medius

Gracilis

Sartorius

Quadriceps femoris

Patella

Soleus

Facial muscles

Sternocleidomastoid

Trapezius

Clavicle

Deltoid

Pectoralis major

Biceps brachii

Brachialis

Wrist and finger flexors

Triceps brachii

Gastrocnemius

Tibialis anterior

Tibia

(a)

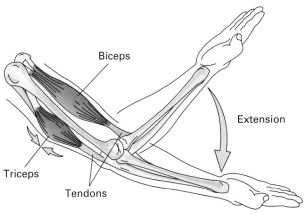

Origin of muscle:
less moveable attachment
of muscle to skeleton

Flexion

Contracting biceps
while relaxing triceps:
flexing of forearm

Insertion of muscle:
more moveable
attachment of muscle
to skeleton

Biceps

Extension

Triceps

Tendons

Figure 5–18 The antagonistic action of the triceps and biceps muscles during flexion and extension, showing origins and insertions.

Critical Thinking

Keeping in mind that muscles work in antagonistic pairs, explain why the excessive development of the biceps from lifting weights unwisely, for example, without balanced development of the triceps can make a person "muscle-bound," making it difficult to quickly straighten the arm.

Most muscles involved in the general movement of the body are attached to bones. Indeed, each end of the muscle is attached to a bone by a tendon, which is a band of connective tissue. One end of the muscle, the **origin**, is attached to the bone that remains relatively stationary during a movement. The other end, the **insertion**, is attached to the bone

that moves. Thus, as the muscle contracts, one part of the skeleton moves more than the other. In the case of the forearm, the biceps brachii flexes the arm and the triceps brachii extends the arm, but in both cases, the muscle origin is in the shoulder region and the insertion is on the lower arm.

Contraction of a muscle pulls on the tendon that connects that muscle to a bone. Excessive stress on a tendon can cause it to become inflamed, a condition called **tendinitis**. In some cases, calcium deposits can appear in tendons in the form of gritty hydroxyapatite crystals that can form quite large accumulations. The painful condition is called calcified tendinitis. The crystals are sometimes squeezed out of the tendon by movement into the lining of the bursa, causing sharp, intense episodes of acute bursitis.

Unfortunately, the causes of tendinitis are not well understood, but it is probably largely due to overuse, to misuse (as when lifting improperly), and age. Overuse, by the way, is relative and paradoxical. People who use the muscles and joints most often suffer less from overuse. Sedentary people who move relatively little are more likely to overuse a joint that is not accustomed to use. Because tendons are poorly supplied with blood vessels, they heal slowly and one of the most effective treatments is rest. The general rule is, if it hurts, don't use it.

The Structure of Muscle

Now we'll look at the brawny bulk that gives shape to the body, gradually delving deep into the fine structure forming the mechanism responsible for muscle contraction. An entire, intact muscle is formed from individual muscle fibers grouped in increasingly larger bundles, each wrapped in a connective tissue sheath (Figure 5–19). Each individual muscle, the biceps brachii for instance, is covered by a membrane made of fibrous connective tissue. The muscles themselves are formed of smaller bundles called **fascicles**, each wrapped in its own connective tissue sheath. The connective tissue sheaths merge and condense at the ends of the muscles to form the tendons that attach the muscle to bone.

Inside the fascicles are the **muscle fibers** themselves, the actual muscle cells. Skeletal muscle fibers can be enormously long as cells go, up to several centimeters. Indeed, muscle fibers in a thigh muscle may be a foot in length. Each muscle fiber is encased in its own fine connective tissue sheath.

When viewed under the microscope, skeletal muscle fibers have pronounced bands that look like stripes, so this type of muscle is also called **striated** (striped) **muscle**. The striations are formed from the orderly arrangement of many elongated **myofibrils** within the muscle fiber. Each myofibril contains groups of long **myofilaments**, which are composed of two protein filaments—the thicker **myosin filaments** and a greater number of the thinner **actin filaments**. Myofilaments make up about 80% of the cell volume.

If we view the myofibrils from their surface, we find rather well-defined regions and bands. These have been

arm—the ones involved in push-ups—Figure 5–18), the contraction of which causes the arm to extend.

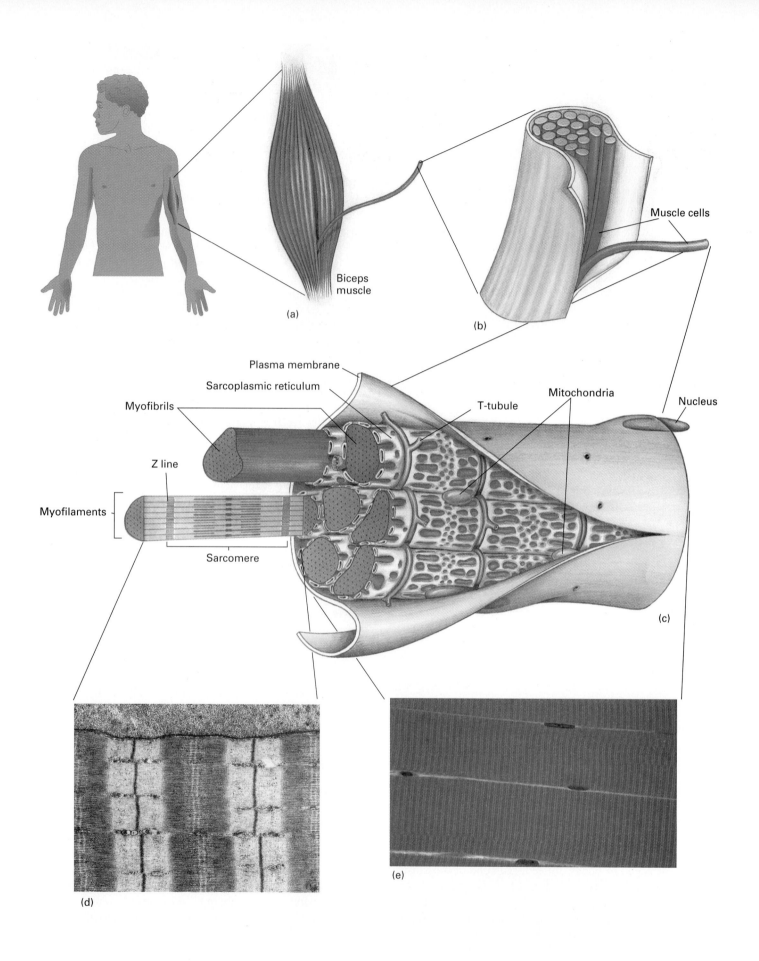

Biceps muscle

(a)

Muscle cells

(b)

Plasma membrane
Sarcoplasmic reticulum
Myofibrils
Mitochondria
T-tubule
Nucleus
Z line
Myofilaments
Sarcomere

(c)

(d)

(e)

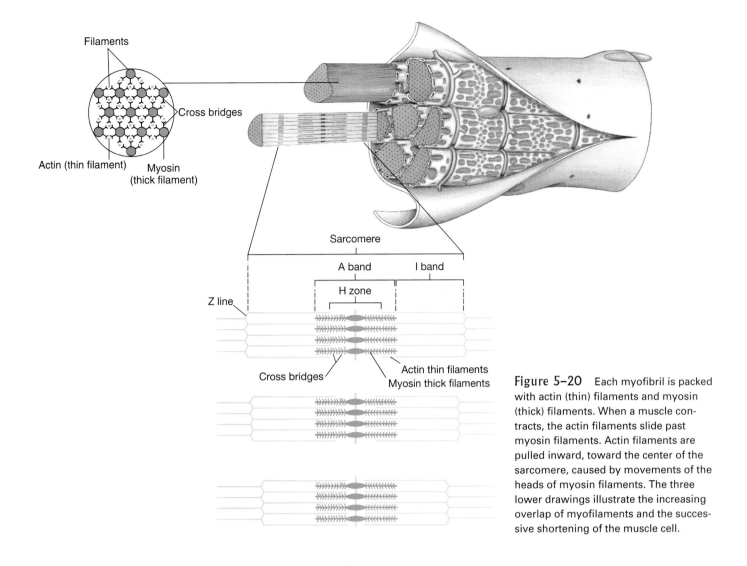

Filaments

Cross bridges

Actin (thin filament)

Myosin (thick filament)

Sarcomere

A band   I band

H zone

Z line

Cross bridges

Actin thin filaments
Myosin thick filaments

Figure 5-20 Each myofibril is packed with actin (thin) filaments and myosin (thick) filaments. When a muscle contracts, the actin filaments slide past myosin filaments. Actin filaments are pulled inward, toward the center of the sarcomere, caused by movements of the heads of myosin filaments. The three lower drawings illustrate the increasing overlap of myofilaments and the successive shortening of the muscle cell.

designated by letter, as we see in Figure 5–20. Bands called Z lines run perpendicularly to the length of the myofibrils. A Z line is composed of strands of protein. One end of each actin filament is attached to a Z line. The area between the Z lines is the smallest contractile unit, called the **sarcomere**. The actin and myosin filaments within a sarcomere partially overlap. Near the ends of each sarcomere, we find the light-colored I band, formed entirely from the thin actin filaments. Farther toward the center of the sarcomere, where the actin and myosin filaments overlap, we see the wide,

dark A band. In the center of the A band is the lighter H zone, composed of only the myosin filaments. The early physiologists were particularly interested in the H zone, because it became smaller when the muscle contracted.

How Skeletal Muscle Contracts

Muscle is able to contract because the actin filaments slide past the myosin filaments, increasing the degree of overlap between actin and myosin filaments and shortening the sar-

◀ **Figure 5-19** In a muscle, such as (a) the biceps in the upper arm, there are many bundles of muscle cells. (b) Each bundle is wrapped in connective tissue. (c) Each muscle cell is packed with myofibrils. The myofibrils contain the contractile myofilaments, actin and myosin. It is the orderly arrangement of myofilaments that gives skeletal muscle its striped, or striated, appearance. The contractile unit of a muscle cell is called a sarcomere. An electronmicrograph of striated muscle is shown in (d) and a light photomicrograph is shown in (e).   *(d, © D. W. Fawcett; e, Ed Reschke.)*

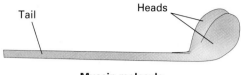

Myosin molecule

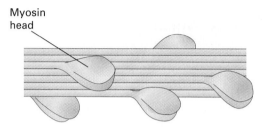

Section of a thick filament

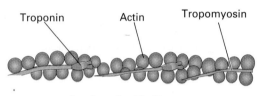

Section of a thin filament

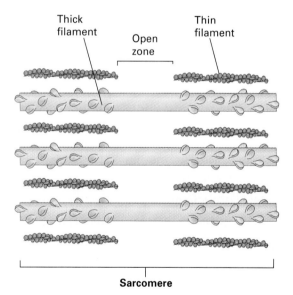

Sarcomere

Figure 5–21   The structure of the thin (actin) filaments and the thick (myosin) filaments.

comere. (As they slide toward each other, the Z lines draw closer together and the H zone becomes smaller.) This is known as the **sliding filament model**. When enough sarcomeres shorten, the entire muscle contracts.

The question arises, then, how do they do it? How do the actin filaments move? To answer this question, we must consider the structure of the thin and thick myofilaments in more detail. Spherical actin molecules make up almost all of a thin myofilament. Many actin molecules join together to form a chain that resembles a string of beads. A thin myofilament consists of two strands of actin twisted together to form a helix. In the grooves of the helix sit two proteins, troponin and tropomyosin, that are important in regulating muscle contraction. The thick filaments are composed of the protein myosin. Each myosin molecule is shaped like a golf club, but with two heads side by side on the same shaft (Figure 5–21). The "shafts" of several hundred myosin molecules adhere to one another, forming a bundle with "heads" protruding from each end of the bundle in a spiral pattern. Actin filaments are arranged around the thick filaments. Generally, each myosin filament is surrounded by six actin filaments.

The club-shaped ends of the myosin molecule, the so-called **myosin heads**, are essential to the movement of actin filaments and, therefore, to muscle contraction. A myosin head is attached to the shaft with a hinge-like connection that allows the head to rock back and forth relative to the shaft. Myosin heads have other features important to muscle contraction. They can bind both actin and the energy-laden molecule ATP. In addition, myosin heads have an enzyme that can split ATP and release its stored energy to power contraction.

Muscle contraction involves cyclic interactions between myosin and actin (Figure 5–22). First, myosin heads bind to ATP and an enzyme in the myosin heads splits the ATP into ADP and inorganic phosphate, releasing energy that activates myosin. The myosin heads swivel in a way that causes them to stretch toward the borders of the sarcomere. This step is analogous to cocking a pistol. Second, the myosin heads attach themselves to the nearest actin filament. When the myosin head is bound to an actin molecule, it forms a bridge between the thick and thin filaments. For this reason, myosin heads are also called **cross bridges**. Binding to actin causes the energy stored in the activated myosin heads to be released, causing the heads to snap forcefully back to their original positions. Because actin filaments are bound to the myosin heads, they are pulled toward the midline of the sarcomere. This so-called power stroke is analogous to pulling the trigger on a pistol. Then the ADP and inorganic phosphate pop off the myosin head. New ATP molecules can then bind to the myosin heads, causing the heads to disengage from the actin. Myosin heads then flip back to their original position, attach farther along the actin filament, pull, and release. This cycle of events is repeated hundreds of times each second.

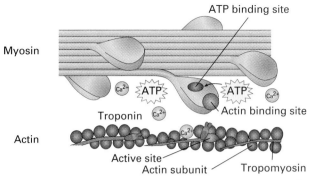

(a) ATP binds with myosin head. Actin sites activated by Ca^{2+}.

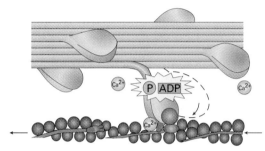

(b) Cross bridge forms. ATP splits into ADP and phosphate.

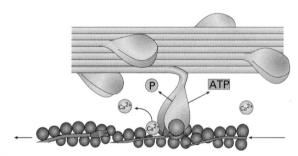

(c) Myosin head pulls actin filament toward center of sarcomere.

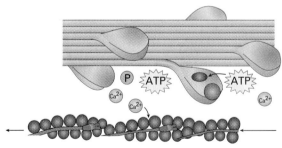

(d) Myosin head picks up a new ATP and detaches from the first active site.

Figure 5-22 The sliding filament model of muscle contraction. (a) At rest, ATP is bound to the myosin thick filament. Contraction is initiated when calcium ions bind to troponin, changing its shape and displacing tropomyosin, which in turn, exposes the binding site on actin. (b) The myosin heads then attach to actin. (c) The ATP provides the energy for the myosin heads to swivel, which slides the thin filament past the thick filament. (d) As another ATP molecule binds to the myosin thick filament, the cross bridge releases from actin and returns to its original position.

Critical Thinking

It takes one ATP molecule for the myosin heads to pull against the actin and another ATP molecule to break the cross bridges so that a new one may be formed. Thus, without ATP, cross bridges cannot be broken and the muscle becomes stiff. When a person dies, ATP is no longer formed. How does this explain the stiffening of muscles, known as rigor mortis, that begins about 3 to 4 hours after death?

How the Contraction Is Controlled

It seems that in biology, every answer leads to the next question. In this case, the next question is, what causes the myosin heads to form bridges? We might also ask, what controls the timing? After all, muscles cannot contract constantly or randomly. Actually, both questions can be answered with the same information.

The actin filament is composed of two proteins besides actin itself. These are called **tropomyosin** and **troponin**, and together they form the **tropomyosin-troponin complex.** Tropomyosin is rod shaped and spirals along the actin filament. Troponin molecules bind tropomyosin to actin at regular intervals along its length, like thumbtacks. The troponin molecule is able to bind to calcium ions (Ca^{2+}). The tropomyosin-troponin complex functions by blocking the actin-myosin binding sites—where the cross bridges of myosin filaments attach to the actin filaments—thereby inhibiting contraction.

But what causes them to release the binding sites, allowing the bridges to form and contraction to occur? It turns out that they release the binding site when calcium ions are present. When calcium appears in the sarcomere, it binds to troponin, causing it to change shape. This results in tropomyosin shifting position, thereby exposing the myosin binding sites on actin. Actin is then able to form new cross bridges with myosin.

So, what controls calcium–ion availability? Where do calcium ions come from? Calcium ions are stored in the **sarcoplasmic reticulum**, an elaborate form of smooth endoplasmic reticulum found in muscle fibers. Like the sleeve of a loose lace shirt, the sarcoplasmic reticulum surrounds each myofibril in a muscle fiber. The calcium ions are moved into the sarcoplasmic reticulum by membrane pumps powered by the energy of ATP. The pumps keep the calcium isolated there until it is needed for contraction, when it is released. Also scattered through the fiber are a number of **transverse tubules** (T-tubules), which are tiny, cylindrical inpocketings of the muscle fiber's plasma membrane. The T-tubules carry signals from motor neurons deep into the muscle fiber to virtually every sarcomere.

Let's try to summarize all this and put it in some kind of perspective. Muscles are signaled to contract by impulses

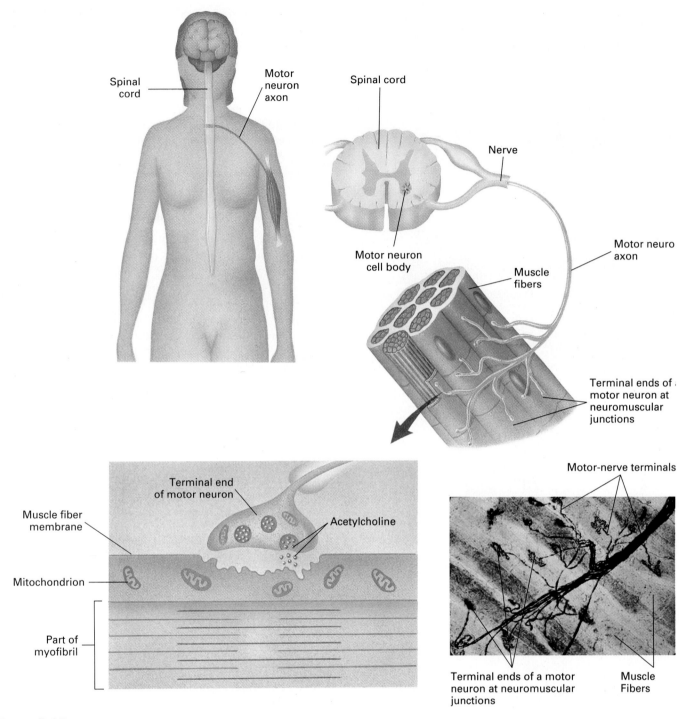

Figure 5–23   A motor unit includes a motor neuron and the muscle fibers it stimulates. The connection between a motor neuron and a muscle fiber is called a neuromuscular junction. When a nerve impulse reaches a neuromuscular junction, it causes the release of the chemical acetylcholine. The acetylcholine binds to receptors on the muscle fiber membrane, causing changes in membrane permeability similar to those that occur during a nerve impulse. These changes result in calcium ions being released from the sarcoplasmic reticulum and, therefore, muscle contraction results.   *(Photo © Courtesy of Bruce Wingerd.)*

carried by motor nerve cells. A motor neuron branches when it is close to the muscle. The tip of each branch forms a **neuromuscular junction** on a muscle fiber (Figure 5–23). When a nerve impulse reaches a neuromuscular junction, it causes the release of the chemical acetylcholine from small packets onto the surface of the muscle fiber. The acetylcholine binds to special receptors on the muscle fiber membrane. Acetylcholine causes changes in the permeability of the muscle fiber membrane, creating an electrochemical message similar to a nerve impulse. This message travels along the plasma membrane, into the T-tubules and then to the sarcoplasmic reticulum where calcium ions are stored. The message causes channels in the sarcoplasmic reticulum to open, releasing calcium ions. The calcium ions then combine with troponin, the myosin binding sites are exposed, and the muscle contracts. All this happens every time you absent-mindedly scratch your head, and it happens a lot faster than it takes to tell about it (Figure 5–24).

When the impulse stops, the events are reversed. Membrane pumps quickly clear the sarcomere of calcium ions and the tropomyosin-troponin complexes block the binding sites, resulting in the muscle relaxing.

Neural Control of Muscle Contraction

Skeletal muscles are stimulated to contract by motor nerve cells. A nerve cell that brings an impulse from the brain telling a muscle fiber to contract does, in fact, stimulate a number of muscle fibers in that area. The single axon[1] from a motor neuron and all the muscle fibers it stimulates is called a **motor unit**. All the muscle fibers in a given motor unit contract together. On average, there are 150 muscle fibers in a motor unit, but this number is quite variable. Muscles responsible for finely controlled movements, such as those of the fingers or eyes, have a small number of muscle fibers in a motor unit. In contrast, muscles with less precise movements, such as the hip or calf, have many muscle fibers in a motor unit. Whereas a motor unit in tiny eye muscles may have only three muscle fibers, a motor unit in a calf muscle may have thousands of muscle fibers.

The nervous system increases the strength of muscle contraction by increasing the number of motor units involved. The muscle fibers of a given motor unit are generally spread throughout the muscle. Thus, if a single motor unit is stimulated, the entire muscle contracts weakly. Although the same muscles are involved in lifting a table and lifting a fork, the number of motor units summoned is greater when lifting the table.

Our movements are generally smooth and graceful, rather than jerky, because the nervous system carefully choreographs the involvement of motor units. Movement is smooth because not all motor units are active simultaneously. In addition, different motor units are active for different amounts of time.

Muscle Contraction

If a neuron is artificially stimulated briefly in the laboratory, all the muscle fibers it innervates will contract, causing a

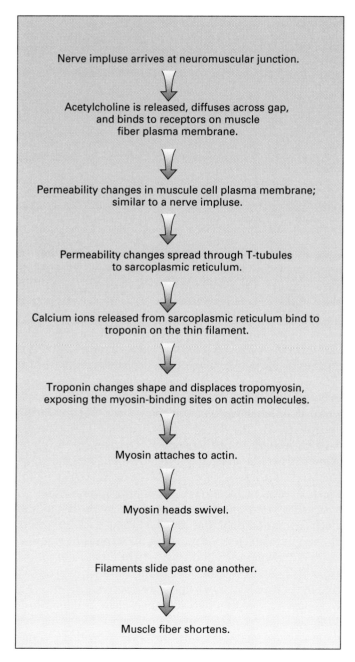

Nerve impulse arrives at neuromuscular junction.

Acetylcholine is released, diffuses across gap, and binds to receptors on muscle fiber plasma membrane.

Permeability changes in muscle cell plasma membrane; similar to a nerve impulse.

Permeability changes spread through T-tubules to sarcoplasmic reticulum.

Calcium ions released from sarcoplasmic reticulum bind to troponin on the thin filament.

Troponin changes shape and displaces tropomyosin, exposing the myosin-binding sites on actin molecules.

Myosin attaches to actin.

Myosin heads swivel.

Filaments slide past one another.

Muscle fiber shortens.

Figure 5–24 A summary of the events involved in muscle contraction.

[1] An axon of a motor neuron is a long extension that carries the nerve impulse away from the cell body.

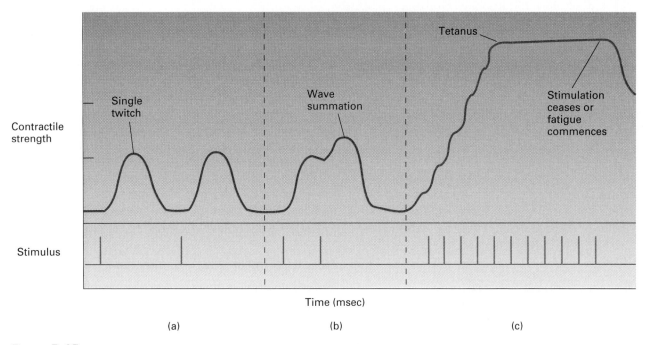

Figure 5-25 Muscle contraction shown graphically: (a) muscle twitch (b) wave summation and (c) tetanus.

muscle twitch (Figure 5–25). The interval between the reception of the stimulus and the time when contraction begins is called the latent period. The contraction phase is quite rapid and is followed by a slower relaxation phase as the muscle returns to its resting state.

If a second stimulus is given before the muscle is fully relaxed, the second twitch will be stronger than the first. Because the second contraction is added to the first, the phenomenon is described as **wave summation**.

When stimuli arrive even more frequently, the muscle twitches are added one onto the next and the contraction becomes increasingly strong. If the stimuli occur so frequently that there is no time for relaxation before the next stimulus arrives, the muscle goes into a constant, powerful contraction called **tetanus**. Tetanus cannot continue indefinitely. Eventually, the muscle will become unable to produce sufficient ATP to fuel contraction and lactic acid accumulates. As a result, the muscle becomes unable to contract in spite of continued stimulation, a condition called **fatigue**.

Energy Sources for Muscle Contraction

When it is forcefully contracting, a single muscle fiber can require as much as 600 trillion ATP molecules per second simply to form and break the cross bridges responsible for contraction. Even small muscles contain thousands of muscle fibers.

A fundamental question presents itself: Since ATP is the only source of energy for muscle contraction, where

does all the ATP come from? There are a number of sources typically employed in a particular sequence depending on the duration and intensity of exercise: (1) ATP in muscle cells (2) creatine phosphate (3) anaerobic respiration and (4) aerobic respiration.

A resting muscle stores some ATP, but this reserve is used quickly. When you start to exercise vigorously, the ATP reserves in the active muscles are depleted within about 6 seconds. You are able to exercise for another 25 seconds or so because when the muscles were resting energy was transferred to another high-energy compound, called **creatine phosphate**, which is stored in muscle tissues. This molecule has a high-energy bond between the creatine and the phosphate parts of the molecule. When needed, creatine phosphate releases its stored energy to convert ADP to ATP, which, in turn, can be used to power muscle contraction.

Although a resting muscle contains about six times as much creatine phosphate as stored ATP, this back-up reserve of energy is depleted within about 30 seconds of vigorous activity. Activities such as diving, weight lifting, and sprinting, which require a short burst of intense activity, are powered by ATP and creatine phosphate reserves.

Once the supply of creatine phosphate is diminished, ATP must be generated from either anaerobic or aerobic respiratory pathways and the primary fuel for either respiratory pathway is glucose. The glucose that fuels muscle contraction comes mainly from glycogen, a large polysaccharide chain of glucose molecules. Indeed, a muscle fiber contains a great deal of glycogen, about 1.5% of its total weight. Muscles depend on these glycogen stores to fuel contraction. Endurance sports may deplete glycogen reserves. The accompanying feeling of overwhelming fatigue is known to runners as "hitting the wall," to cyclists as "bonking," and to boxers as becoming "arm weary."

When an active muscle fiber runs short of ATP and creatine phosphate, enzymes convert glycogen to glucose. The circulatory system can supply enough oxygen for aerobic respiratory pathways to produce ATP to power *low* levels of activity, even if continued for a prolonged time. However, anaerobic pathways produce ATP two and one-half times faster than aerobic pathways. Therefore, during strenuous activity lasting 30 or 40 seconds, anaerobic pathways supply the ATP to fuel muscle contraction. Indeed, we find that activities such as tennis or soccer rely nearly completely on anaerobic pathways of ATP production. In anaerobic respiratory pathways, the pyruvic acid produced in glycolysis is converted to lactic acid.

During more prolonged muscular activity, the body gradually switches back to aerobic pathways for producing ATP. Aerobic pathways require oxygen, which can come from either of two sources. One oxygen source is the blood supply. As activity continues, heart rate increases, pumping blood more quickly. At the same time, blood is shunted to the most needy tissues. Another source is myoglobin, an oxygen-binding pigment in muscle fibers. Aerobic pathways produce more than 90% of the ATP required for intense activity lasting more than 10 minutes and nearly 100% of the ATP that powers a truly prolonged intense activity, such as running a marathon.

After prolonged exercise, a person continues to breathe heavily for several minutes. The extra oxygen, called an **oxygen debt**, is used to return the body to pre-exercise conditions. Lactic acid is converted back to pyruvic acid and then oxidized to carbon dioxide and water through aerobic respiratory pathways. In addition, the oxygen that was released by myoglobin is replaced and glycogen and creatine phosphate reserves are restored.

Fast and Slow Muscle Fibers—Twitch Do You Have?

There are three types of muscle fibers: slow-twitch, fast-twitch, and intermediate fibers with properties between the others. We will consider only the slow- and fast-twitch fibers. **Slow-twitch fibers** contract slowly when stimulated, but with enormous endurance. These fibers are dark and reddish in color because they are packed with the oxygen-binding pigment myoglobin and because they are richly supplied with capillaries. Slow-twitch fibers predominate in the dark meat of certain fowl. Long-distance migrators, such as ducks and geese, have dark breast meat (flight muscles), while chickens, which get around by walking, have dark leg meat. Slow-twitch fibers also contain abundant mitochondria, the cellular structures essential for aerobic production of ATP. As a result, slow-twitch fibers are specialized to deliver prolonged, strong contractions.

In contrast, **fast-twitch fibers** contract rapidly and powerfully, with far less endurance. Fast-twitch fibers have a form of an enzyme that can split ATP bound to myosin more quickly than the same enzyme in slow-twitch fibers. Because they can make and break cross bridges more quickly, they can contract more rapidly than slow-twitch fibers. In addition, compared with their slow-twitch cousins, fast-twitch fibers have a wider diameter because they are packed with more actin and myosin, allowing them to contract powerfully. However, fast-twitch fibers, rich in glycogen deposits, depend more heavily on anaerobic means of producing ATP. As a result, fast-twitch fibers tire more quickly than slow-twitch fibers.

The two kinds of fibers are distributed unequally throughout any human body. The abdominal muscles do not need to contract rapidly, but they need to be able to contract steadily to hold our paunch in at the beach and to balance the powerful, slow-twitch back muscles. Fast-twitch fibers, on the other hand, are more common in the legs and arms, structures that must sometimes move quickly.

Human athletes show some of these biochemical cellular differences as well. Whereas the muscles of endurance athletes, such as marathoners, are made up of about 80% slow-twitch fibers, those of sprinters are about 60% fast-twitch fibers. By the way, if you are a "fast-twitch person," you can build a certain level of endurance, but you will be limited to a degree because such differences are genetic. Endurance runners, on the other hand, dread those times when, at the end of a long race, some well-trained competitor, loaded with fast-twitch fibers, sprints past them at the finish line.

Exercise and the Development of Muscles

In adolescents and adults, exercise can have great influence on the further development of muscle. Different kinds of exercise, though, can produce different results. For example, **aerobic exercise** such as walking, jogging, or swimming (exercises in which enough oxygen is delivered to the muscles to keep them going for long periods) will foster the development of new capillaries that service the muscles and more mitochondria to facilitate energy usage. It also increases muscle coordination, improves digestive tract movement, and increases the strength of the skeleton by placing force on the bones. Cardiovascular and respiratory system improvements help muscles to function more efficiently and the heart is enlarged so that each stroke pumps more blood.

Aerobic exercises, however, generally do not increase muscle size. Muscular development, the kind that helps you look good on the beach, comes mostly from **resistance exercise**, such as you get from lifting heavy weights. To build muscle mass, one must force muscles to exert more than 75% of their maximum force. These exercises can be very brief because only three sets of six contractions each are needed to build bulk and increase strength. The added size apparently comes from an increase in diameter of existing

Building Muscle, Fair and Square?
Anabolic Steroid Abuse

We have always appreciated strength, probably because it's associated with a "can-do" ability. Of course, such appreciation has led us not only to strength itself, but to the appearance of strength, that is, body building. Unfortunately, some people are always looking for shortcuts.

One of the biggest dangers to young people is the use of anabolic (building) steroids. Anabolic steroids are synthetic hormones that mimic the male sex hormone, testosterone. Anabolic steroids, taken in large amounts, stimulate the body to build muscle and to increase strength dramatically (Figure A). They do this by stimulating protein formation in muscle fibers and by reducing the amount of rest needed between workouts.

Although anabolic steroids are available only by prescription, recent studies have shown that 6.5% of adolescent boys and 1.9% of adolescent girls obtain and regularly use the drugs without a prescription. Commonly called roids, juice, or slop, steroids are swallowed as a pill or injected. Steroid use in the United States is a $500 million business and growing. Since most steroids are obtained illegally through the black market or from foreign countries, the purity and quality of the drugs is questionable at best.

We might ask, then, why steroid use is so attractive. One reason seems to be that steroids provide an easy route to the body image so prized by society. Nearly one third of high school males who use steroids admit that they use the drugs to acquire a muscular, well-built look. Steroid use seems to be particularly widespread among high school senior boys. Perhaps they are unsure of themselves as they prepare to leave the nest and enter the "real world."

Among athletes, however, the reason for steroid use is clear—it gives them a distinct competitive edge and everyone loves a winner. Steroid use is prohibited by almost all international athletic committees, and more than a few medals have had to be returned when the cheating was discovered. Cheating is the right word because it gives the user an unfair advantage.

Unfortunately, steroid use has more than 70 possible unwanted side effects that can range in severity from liver cancer to acne. Among the risks are injuries that result from the intended effect of the drug—increased muscle strength—without the accompanying increase in strength of tendons or ligaments. Such injuries may take a long time to heal. But most commonly and most seriously affected by steroid abuse are the liver, the cardiovascular system, and the reproductive system. Effects on the cardiovascular system, including heart attacks and strokes, may not show up for years. However, the effects on the reproductive system are more immediate. The testicles of male steroid users often become smaller. Males may also be-come sterile and impotent (unable to achieve an erection). Female steroid abusers develop irreversible masculine traits, such as a deeper voice, growth of body hair, loss of scalp hair, smaller breasts, and an enlarged clitoris (during embryological development, the clitoris develops from the same structure that develops into the penis in a male).

Steroid use can also have psychological effects. For instance, it may promote aggression and a feeling of invincibility, often referred to as "roid rage." Users may also develop severe depression. Also, steroids are addictive.

The person who abuses steroids is the one immediately responsible for the behavior, but one cannot help but wonder how great an influence the ideals of a society that idolizes winners may have been in shaping the behavior.

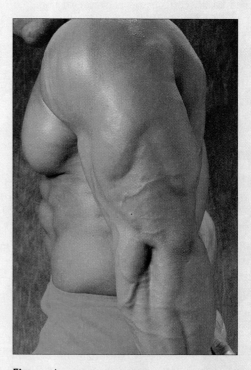

Figure A Anabolic steroids are often abused as an easy way to build muscle and strength. (*© Tom Raymond/The Stock Shop.*)

muscle fibers. Some researchers, however, suggest that heavy exercise splits or tears the muscle fibers and that each of these parts then regrows to its previous size.

To build and firm muscles, people head to the gyms to pump "snails" (as using Nautilus equipment may be called) or lift free weights (generally more dangerous and more effective) (Figure 5–26). Nautilus machines automatically adjust the resistance that muscles encounter during an exercise. Muscles are weaker at some parts of their range of motion. A cam (built on a "snail shape") in the Nautilus machine makes the lifting easier where the muscle is weaker and more difficult in the range the muscle can handle. The hard-core body builders concentrate on "free weights," the barbells and dumbbells that are much harder to control.

One problem with building muscle is that you have to keep at it. If you just train in the spring to look good at the beach, you have an uphill battle, because all the mass of muscle you built last year began to disappear only 2 weeks after the training sessions stopped.

A frequent question is, are men naturally stronger than women? Generally, yes. Women have about 35% muscle mass and men have 42%. Also, men are generally larger than women and so have more muscle mass to begin with. However, the individual muscle fibers of men and women have the same strength.

In this chapter, then, we have learned something about the support, attachment, and contractile elements of the human body. Although we have discussed them more or less independently of other systems, such categorizing is too simplistic because the human body is a symphony of systems,

each part dependent on the rest. In the following chapters we will look at some other systems of the human body, keeping in mind their interdependence in keeping us alive and functioning.

Figure 5–26 Muscles get larger when they are repeatedly forced to exert more than 75% of their maximum force. *(© Ron Chapple, 1991/FPG International.)*

SUMMARY

1. The skeleton is a framework of bone and cartilage that supports and protects the internal organs and allows for movement. In addition, it serves as a storage site for minerals (calcium and phosphorus) and fat. Blood cells are produced in the red marrow of certain bones.

2. Compact bone is dense bone found on the outside of all bones. It is covered by a membrane, the periosteum, that nourishes the bone cells. Spongy bone is a latticework of bony struts found in flat bones and near the ends of long bones.

3. Each enlarged end of a long bone is called an epiphysis and the shaft is called the diaphysis. The medullary cavity in the shaft of a long bone is filled with fatty yellow marrow. The spongy bone of certain bones in adults is filled with red marrow.

4. The structural unit of bone, called an osteon or Haversian system, consists of a central canal with bone cells (osteocytes) arranged around it in concentric circles. Cellular extensions of osteocytes touch one another through tiny canals called canaliculi, allowing for the exchange of materials between cells and the central canal. Bone matrix is hardened by calcium salts and strengthened by strands of collagen.

5. In an embryo, the skeleton first forms in cartilage. The cartilage is gradually replaced by bone in a process called endochondrial ossification. The process begins when osteoblasts form a bony collar around the shaft of the bone. Cartilage within the shaft then begins to break down and is replaced by bone. The cartilage in the heads (epiphyses) is then replaced by bone. Epiphyseal plates of cartilage remain and allow bones to grow in length until the person reaches ages 18 to 21.

6. Bone growth is stimulated by growth hormone from the anterior pituitary gland. Sex hormones (estrogen and testosterone) initially stimulate growth, but later they cause the epiphyseal plates to disappear, ceasing growth.

7. Repair of a bone fracture involves the formation of a blood clot and then a cartilaginous callus that links the broken ends of the bone. The cartilage is gradually replaced by bone.

8. Bone is a living tissue that is constantly being remodeled. Osteoblasts deposit new bone and osteoclasts break down bone. Stress is an important factor in bone remodeling because bone forms when it is stressed and is destroyed when it is not stressed. The hormones calcitonin (produced by the thyroid

gland) and parathyroid hormone (produced by the parathyroid glands) regulate both bone remodeling and blood calcium levels.

9. Osteoporosis is a condition in which bones become thin, brittle, and easily fractured because more bone is destroyed than is rebuilt. The rates of bone building and destruction generally remain about equal until about age 35, when bone destruction begins to occur at a faster rate than deposition. The danger this poses to bones depends on their peak density and the rate of loss. Bone density is greater in men than women. It is also influenced by race and diet. In women, bone loss is accelerated after menopause because of a decline in estrogen levels.

10. The skeleton can be divided into the axial skeleton (the skull, the vertebral column, and the bones of the chest region) and the appendicular skeleton (the pectoral and pelvic girdles with their associated limbs.)

11. Joints, the places where bones meet, can be classified according to the degree of movement they permit. Some joints, such as the sutures between the skull bones, allow no movement. Others allow slight movement. Synovial joints are freely moveable. Synovial joints have cartilage on the adjoining bone surfaces. They are surrounded by a synovial cavity filled with synovial fluid and are held together by ligaments.

12. Arthritis is inflammation of a joint. Osteoarthritis occurs when the surface of a joint degenerates because of use. Rheumatoid arthritis is an autoimmune disease.

13. Many of the skeletal muscles of the body are arranged in antagonistic pairs so that the actions of the members of the pair are opposite to one another. For instance, one member of a pair usually causes flexion and the other extension.

14. Muscles are attached to bones by tendons. Whereas the muscle origin is the end attached to the bone that is more stationary during a movement, the insertion is the end attached to the bone that moves.

15. An entire muscle is wrapped in a connective tissue sheath. The muscle is composed of smaller bundles, called fascicles. A fascicle is a bundle of muscle fibers, which are muscle cells. A muscle cell is packed with myofibrils, which are composed of myofilaments (the contractile proteins actin and myosin).

16. A muscle contracts when a myosin head binds to actin, swivels, and pulls actin toward the midline of the cell, causing the filaments to slide past one another and increasing their degree of overlap.

17. Contraction is controlled by the availability of calcium ions. Calcium ions interact with two proteins on the actin filament—troponin and tropomyosin—that in turn determine whether myosin can bind to actin. Calcium ions are stored in the sarcoplasmic reticulum and released when a motor nerve sends an impulse.

18. Motor nerves contact muscle cells at neuromuscular junctions. When the nerve impulse reaches a neuromuscular junction, acetylcholine is released and causes a change in membrane permeability of the muscle fiber that, in turn, causes calcium ions to be released from the sarcoplasmic reticulum.

19. A motor neuron and all the muscle fibers it stimulates are collectively called a motor unit.

20. The response of a muscle fiber to a single brief stimulus is called a twitch. If a second stimulus arrives before the muscle has relaxed, the second contraction builds upon the first. This phenomenon is known as wave summation. Frequent stimuli cause a sustained contraction, called tetanus.

21. It requires two ATP molecules each time a cross bridge between myosin and actin forms and is broken. The sources of ATP are (1) stored ATP (2) creatine phosphate (3) anaerobic respiration and (4) aerobic respiration.

22. Slow-twitch muscle fibers contract slowly, but with enormous endurance. Fast-twitch muscle fibers contract rapidly and powerfully, but with less endurance.

23. Muscle development is a result of resistance exercise. When muscles are forced to exert more than 75% of their maximal force, myofilaments are added to existing muscle fibers and so their diameter is increased.

REVIEW QUESTIONS

1. List six functions of the skeleton.
2. Compare compact and spongy bone.
3. Describe the structure of a long bone. Where are the yellow and red marrow found?
4. Diagram an Haversian system (osteon), including the osteocytes, central canal, and canaliculi.
5. Describe the formation of bone in a fetus. Explain how bone growth continues after birth.
6. Explain how a bone heals after it has been fractured.
7. What is bone remodeling? Explain the role of osteoblasts and osteoclasts. How does stress affect remodeling? What roles do calcitonin and parathyroid hormone play in the process?
8. Define osteoporosis. What factors increase a person's risk of developing osteoporosis?
9. Describe the axial and appendicular parts of the skeleton.
10. What are the five common features of all synovial joints?
11. Why are most skeletal muscles arranged in antagonistic pairs? Give an example that illustrates the roles of each member of an antagonistic pair of muscles.

12. Describe a skeletal muscle including definitions of fascicles, muscle fibers, myofibrils, and myofilaments.
13. What is the sliding filament model of muscle contraction?
14. Describe the structure of actin and myosin and in doing so explain how actin moves during muscle contraction.
15. Explain the roles of troponin, tropomyosin, and calcium ions in regulating muscle contraction.
16. Explain how the events that occur when a motor nerve impulse reaches a neuromuscular junction and the release of calcium ions from the sarcoplasmic reticulum lead to muscle contraction.
17. Define a motor unit. Explain how motor units vary depending on the degree of control one has over a particular muscle.
18. Define muscle twitch, wave summation, and tetanus.
19. List the sources of ATP for muscle contraction and explain when each source is typically called upon.
20. Differentiate between slow- and fast-twitch muscle fibers.
21. What type of exercise can build muscles? What accounts for the increase in muscle size?

Chapter 6

The Endocrine System

Basics of the Endocrine System

Endocrine Glands

Stress

Milk production and ejection from the mammary glands are stimulated by hormones, chemical messengers of the endocrine system.   *(© Erika Stone/Photo Researchers, Inc.)*

135

Long ago humans believed that certain organs were endowed with special attributes. In fact, it was not unusual for victors in war to eat organs such as the heart, brain, or sexual organs of those conquered, believing consumption of such parts would confer those attributes on the victor. They might have believed that hearts, for example, could confer bravery and that testicles could increase strength and vigor. Although such actions may have fallen from popularity in most modern societies, the practice could be viewed as a crude and ineffective forerunner to modern hormone replacement therapy, treating glandular conditions by using extracts from healthy glands. (Advances in technology, however, have made synthetic forms of many of those extracts available; see the Personal Concerns essay on hormone replacement.) Indeed, we are still using the bodies of others to improve ourselves.

In this chapter we will consider the endocrine glands and their secretions. We will also focus on what those secretions do. We will see that organs communicate with each other and with other tissues by such secretions. We will learn that once these secretions are produced, they enter the bloodstream and are carried along to their destinations where they initiate activities that produce a vast range of changes in our bodies, such as growth, development, and sexual behavior. In the end we will see that proper functioning of the body depends on this rather leisurely system of internal communication. In Chapters 7 and 8 we will discuss our more rapid system of internal communication, the nervous system.

Basics of the Endocrine System

Recall from Chapter 4 that our bodies contain two types of glands, exocrine glands and endocrine glands (Figure 6–1). **Exocrine glands** secrete their products into ducts that empty onto the body surface, into the spaces within organs, or into a body cavity. Oil (sebaceous) glands, for example, are exocrine glands that secrete oil into ducts that open onto the surface of the skin. Salivary glands, another familiar example of exocrine glands, secrete saliva into ducts that open into the mouth. **Endocrine glands** are made up of secretory cells that release their products, called hormones, into the fluid just outside the cells. Unlike exocrine glands, they do not secrete their products into ducts. Instead, the hormones diffuse from the extracellular fluids directly into the bloodstream where they contact virtually all the cells of the body.

Glands and Organs of the Endocrine System

The endocrine system is made up of endocrine glands and organs that contain some endocrine tissue, but whose primary function is not hormone secretion (Figure 6–2). The major endocrine glands include the pituitary gland, thyroid gland, parathyroid glands, adrenal glands, and pineal gland.

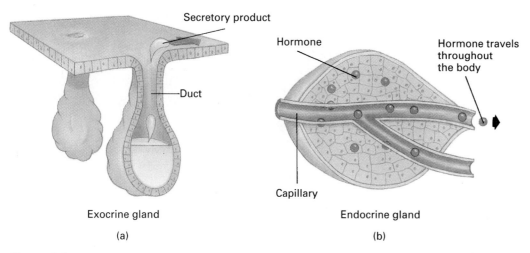

Secretory product

Hormone

Hormone travels throughout the body

Duct

Capillary

Exocrine gland

Endocrine gland

(a)

(b)

Figure 6–1 Exocrine and endocrine glands. (a) Exocrine glands secrete their products into ducts that open onto the surface of the body, into the spaces within organs, or into cavities within the body. (b) Endocrine glands release their products, called hormones, into surrounding tissues. The hormones then diffuse into the bloodstream to be transported throughout the body.

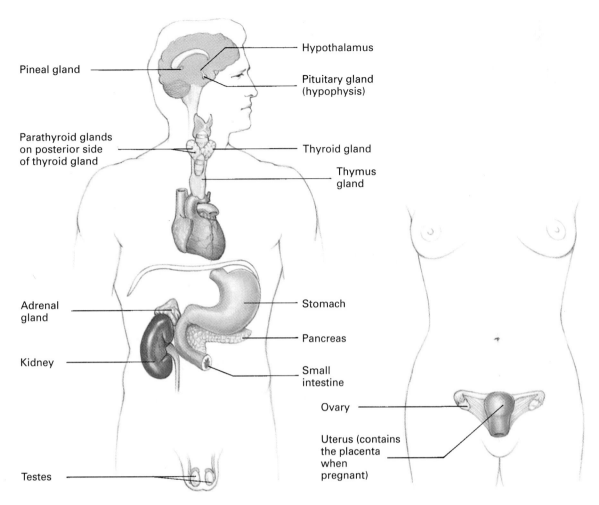

Figure 6-2 The endocrine system. The endocrine system is made up of endocrine glands and organs that contain some endocrine tissue.

Labels in figure:
Pineal gland — Hypothalamus — Pituitary gland (hypophysis) — Parathyroid glands on posterior side of thyroid gland — Thyroid gland — Thymus gland — Adrenal gland — Stomach — Pancreas — Kidney — Small intestine — Ovary — Uterus (contains the placenta when pregnant) — Testes

Organs that contain some endocrine tissue include the hypothalamus, thymus, pancreas, ovaries, testes, and placenta, as well as other organs associated with the digestive and excretory systems, such as the stomach and kidneys. Our discussion here will be limited to the major endocrine glands and the first three organs with endocrine tissue listed above. The remaining organs with endocrine function will be discussed in subsequent chapters.

Types of Hormones and Their Modes of Action

Hormones are the chemical messengers of the endocrine system. Released in very small amounts by cells of endocrine glands and some organs, hormones enter the bloodstream and travel throughout the body. Although hormones contact virtually all cells, most affect only a particular type of cell, called a **target cell**. Target cells have **receptors**, protein molecules located either in the cell's cytosol or embedded in its plasma membrane. The receptors recognize and bind to specific hormones. Once a hormone binds to and activates its specific receptor, it begins to exert its effects on the cell, perhaps altering protein synthesis, secretory activity, or properties of the plasma membrane. Cells other than target cells lack the correct receptors and are unaffected by the hormone.

The mechanisms by which hormones influence target cells depend in large part on the chemical makeup of the hormone. Generally, hormones are classified as being either lipid soluble or water soluble. Lipid-soluble hormones include steroid and thyroid hormones. **Steroid hormones** are a group of closely related hormones chemically derived from cholesterol and secreted primarily by the ovaries (such as estrogen), testes (including testosterone), and adrenal glands (for example, corticosterone). Lipid-soluble hormones move easily through the plasma membrane because it is a lipid bilayer. Once inside the target cell, a steroid hormone combines with receptor molecules in the cytosol. The hormone-receptor complex then moves into the nucleus of the cell where it attaches to DNA and activates genes causing transcription of messenger RNA (in this process, information is transferred from the DNA molecule to an RNA molecule). The coded sequence of information in mRNA is then converted, by means of a process known as translation, into a particular sequence of amino acids, thus forming polypeptide chains or proteins. In short, steroid hormones direct the synthesis of specific proteins by the target cell. These pro-

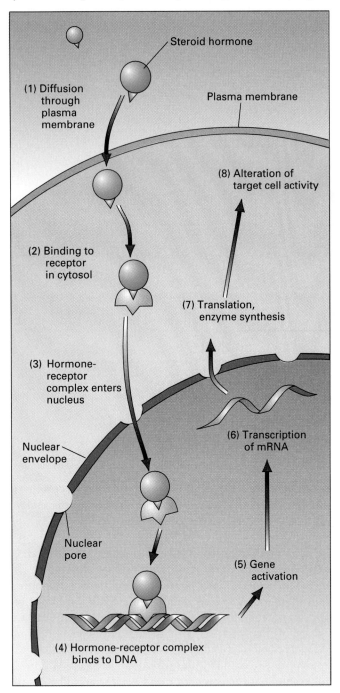

Figure 6–3 Mode of action of steroid hormones. Steroid hormones move through the plasma membrane, bind to receptors in the cytosol of target cells, and enter the nucleus where they attach to DNA and activate transcription of messenger RNA (mRNA). During transcription, information from a molecule of DNA is transferred into a molecule of RNA. The coded sequence of information on mRNA is then translated into a particular sequence of amino acids to form proteins, including enzymes, which alter the activities of the cell.

teins may include enzymes that stimulate or inhibit particular metabolic pathways. In the case of thyroid hormones (such as thyroxine), the hormones cross the cell membrane and enter the nucleus without binding to a receptor in the cytosol. Once in the nucleus, thyroid hormones bind to a receptor that attaches to DNA and alters protein synthesis as described for steroid hormones. Figure 6–3 summarizes the mode of action of steroid hormones. The roles of transcription and translation in protein synthesis are discussed in detail in Chapter 19.

Water-soluble hormones include molecules made of amino acids, the building blocks of proteins. Some water-soluble hormones are relatively small molecules, consisting of only slightly modified amino acids. For example, norepinephrine and epinephrine, two hormones secreted by the inner region of the adrenal glands, result from small modifications to the amino acid tyrosine. These structurally simple hormones are often called amines or amino acid derivatives. Other water-soluble hormones consist of chains of amino acids that form larger molecules called peptides or proteins. Examples of peptide or protein hormones include insulin, secreted by the pancreas, and oxytocin, manufactured by the hypothalamus and secreted by the pituitary gland.

Water-soluble hormones cannot pass through the lipid bilayer of the plasma membrane, and thus they exert their effects indirectly by binding to receptors on the surface of the cell and activating **second messenger systems**. Basically, the hormone, considered the first messenger, binds to a receptor on the plasma membrane, and this in turn activates a second messenger in the cytosol. This second messenger carries the hormone's message within the cell where it influences the activity of enzymes and chemical reactions. Intermediaries, called G proteins, are enzymes attached to the receptor in the plasma membrane and often link first and second messengers.

Let us consider the second messenger system involving cyclic adenosine monophosphate (cAMP; Figure 6–4). When a water-soluble hormone binds to its receptor on the outer surface of the plasma membrane, specific G proteins link the hormone-receptor complex to adenylate cyclase molecules on the inner surface of the plasma membrane. Adenylate cyclase is the enzyme responsible for converting ATP into cAMP, the second messenger that will alter cellular activity. Depending on the specific hormone, synthesis of cAMP in the cytosol may be increased or decreased. Changes in cAMP levels alter the activity of a group of enzymes called protein kinases that change metabolic pathways, protein synthesis, membrane characteristics, or rates of secretion by the cell.

Control of Hormone Release

Now that we know how hormones work at the cellular level, we turn our attention to which factors stimulate and regulate the release of hormones from endocrine glands.

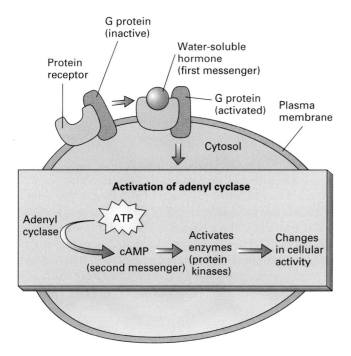

Figure 6-4 Mode of action of water-soluble hormones: the second messenger system of cAMP. Water-soluble hormones cannot cross the plasma membrane, and thus they indirectly affect the activities of target cells. Typically, the hormone, or first messenger, binds to a receptor on the outer surface of the plasma membrane. This process then activates a particular G protein associated with the receptor that then activates adenyl cyclase, the enzyme responsible for producing cAMP from ATP. The second messenger, cAMP, influences enzymes called protein kinases, which alter cellular activity.

Stimuli That Prompt Hormone Release

Stimuli that prompt endocrine glands to manufacture and release hormones include chemical changes in the blood, other hormones, and signals from the nervous system. First, chemical changes in the blood include changes in the levels of certain ions (such as when decreases in calcium ions in the blood trigger the release of parathyroid hormone from the parathyroid glands) or nutrients (as when increases in blood glucose stimulate the pancreas to secrete insulin). Secondly, hormones released by one endocrine gland may stimulate the release of hormones from another gland. Later in the chapter, for example, we will see how adrenocorticotropic hormone (ACTH) released from the anterior pituitary gland stimulates the release of corticosteroid hormones by the adrenal glands, thereby helping our bodies resist stress. Thirdly, the release of some hormones is stimulated by messages from the nervous system. For example, signals traveling along neurons in the hypothalamus stimulate the release of the hormone oxytocin from the pituitary gland. Although we have described separate examples for the three types of stimuli that trigger hormone release, some endocrine glands respond to multiple stimuli.

Feedback Control

Recall from Chapter 4 that body homeostasis, a relatively constant internal environment, is often accomplished through **negative feedback mechanisms**. The same is true for homeostasis with specific regard to hormonal secretions. Typically, one or more of the stimuli described in the previous section triggers a gland to release a hormone, and then rising blood levels of that hormone inhibit its further release. The feedback is described as negative because it acts to inhibit further production and release of the hormone. When negative feedback mechanisms are functioning properly, blood levels of most hormones fluctuate within a fairly narrow range. Hormonal disorders result when negative feedback mechanisms fail and hormone production and release becomes excessive or deficient.

Occasionally, secretion of hormones is regulated by **positive feedback mechanisms**. During childbirth, for example, the pituitary gland releases oxytocin that stimulates the uterus to contract. Uterine contractions then stimulate further release of oxytocin, which stimulates even more contractions. In this example, the feedback is described as positive because it acts to stimulate, rather than inhibit, production and release of oxytocin.

Modulation by the Nervous System

At times it is necessary for the nervous system to override the controls of the endocrine system. For example, during times of severe stress the nervous system overrides the mechanism controlling release of the hormone insulin. Under normal conditions, insulin is released from the pancreas in response to high levels of glucose in the blood (such as might occur after a meal). As glucose is taken up by cells, the level of blood glucose begins to fall, and this, in turn, inhibits the rate of insulin secretion. As a result of this negative feedback system, levels of glucose in the blood are normally maintained within a relatively narrow range. Under conditions of severe stress, such as those associated with strong emotional reactions, heavy bleeding, or starvation, activation of certain areas of the nervous system permits levels of blood glucose to rise much higher than normal, with the result that large amounts of energy are made available to cells.

Local Signaling Molecules

Before we go on to describe the endocrine glands and their functions, let's consider a class of chemical messengers that unlike hormones do not travel to distant sites within the body, but instead they exert their effects locally, acting either on the secreting cells themselves or nearby tissues. Prostaglandins are one example of these so-called local signaling molecules.

Prostaglandins are lipid molecules found in and continually released by the plasma membranes of most cells. Different types of cells secrete different prostaglandins, and at least 16 different prostaglandin molecules are known to function within the human body. While not considered "true hormones" because they produce their effects locally rather than at distant sites, prostaglandins have remarkably diverse effects, including prevention of blood clotting, regulation of body temperature, opening of airways to the lungs, and contributing to the body's inflammatory response. Prostaglandins also affect the reproductive system. Menstrual cramps, for example, are thought to be caused by prostaglandins released by cells of the uterine lining. These prostaglandins act on the smooth muscle of the uterus, causing muscle contractions and cramping. Drugs such as aspirin inhibit the initial synthesis of prostaglandins and thus may lessen the discomfort of menstrual cramps. Prostaglandins are also found in semen, the fluid discharged from the penis at ejaculation. Again, prostaglandins cause the smooth muscles of the uterus to contract, perhaps helping sperm continue their journey further into the female reproductive tract. In fact, the name prostaglandin, given to the substances in semen by one of the scientists who first identified them, came from the mistaken belief that these compounds were secreted by the prostate gland. We now know that the source of prostaglandins in semen is actually the seminal vesicles, pouch-like structures near the urinary bladder that secrete components of semen, but the name still stands.

Endocrine Glands

We now study the endocrine glands. For each gland we will consider location and general structure, hormones secreted and their effects, and any disorders associated with the gland and hormone secretion. We begin with the pituitary gland, a pea-sized gland that despite its diminutive size is often described as "the conductor of the endocrine orchestra." We will see that this title comes from the pituitary's role in controlling the secretions of several endocrine glands.

Pituitary Gland

The pituitary gland (hypophysis) is suspended from the base of the brain, just above the roof of the mouth, by a short stalk called the **infundibulum** (Figure 6–5). The gland consists of two lobes, the anterior lobe and the posterior lobe. These lobes differ in size, development, relationship with the hypothalamus (the area at the base of the brain to which the pituitary is connected by its stalk), and the hormones secreted.

The Anterior and Posterior Lobes of the Pituitary Gland

The anterior lobe is the larger of the two lobes, originating from the embryonic growth of epithelial tissue upward from the roof of the mouth. A network of tiny blood vessels called capillaries runs from the base of the hypothalamus through the stalk of the pituitary where it connects to veins that, in turn, lead into a capillary bed in the anterior lobe of the pituitary gland (Figure 6–6). This system whereby a capillary bed drains to veins that drain to another capillary bed is called a **portal system**, in this case the **pituitary portal sys-**

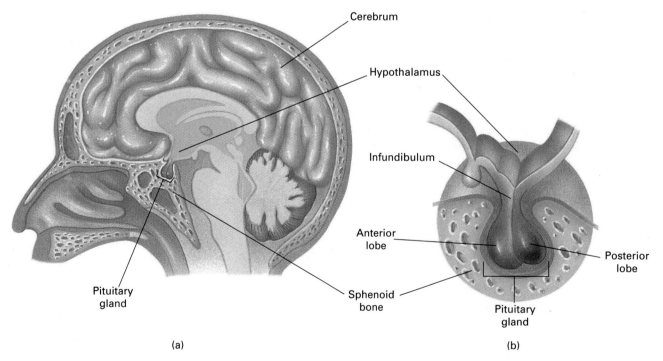

(a)   (b)

Figure 6–5   The pituitary gland. (a) Lateral view. (b) Close-up showing how the pituitary gland is attached to the hypothalamus by means of the infundibulum.

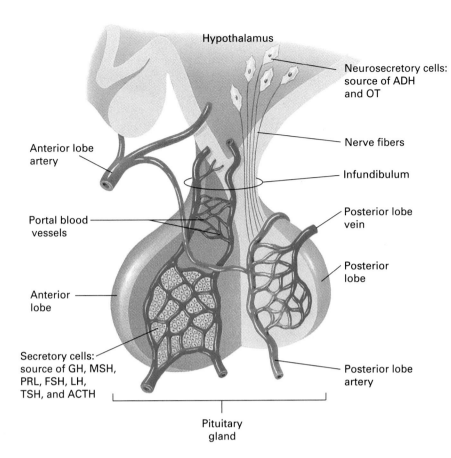

Hypothalamus

Neurosecretory cells: source of ADH and OT

Nerve fibers

Anterior lobe artery

Infundibulum

Portal blood vessels

Posterior lobe vein

Posterior lobe

Anterior lobe

Secretory cells: source of GH, MSH, PRL, FSH, LH, TSH, and ACTH

Posterior lobe artery

Pituitary gland

Figure 6–6 The two lobes of the pituitary gland and their connections with the hypothalamus. The anterior lobe receives releasing hormones and inhibiting hormones from the hypothalamus by way of portal blood vessels. In response to releasing and inhibiting hormones, the anterior pituitary modifies secretion of its seven hormones—growth hormone (GH), melanocyte-stimulating hormone (MSH), prolactin (PRL), follicle-stimulating hormone (FSH), luteinizing hormone (LH), thyroid-stimulating hormone (TSH), and adrenocorticotropic hormone (ACTH). The posterior lobe receives neurosecretory cells that extend from the hypothalamus. These cells manufacture oxytocin (OT) and antidiuretic hormone (ADH), which are transported to the posterior lobe for release.

tem. It is this portal system that allows the hypothalamus to control the secretion of hormones from the anterior pituitary. Basically, nerve cells in the hypothalamus release substances into the pituitary portal system and these substances travel to the anterior lobe where they stimulate or inhibit hormone secretion. Those substances that stimulate hormone secretion are called **releasing hormones** and those that inhibit hormone secretion are called **inhibiting hormones.** In response to releasing and inhibiting hormones from the hypothalamus, the anterior pituitary modifies its synthesis and secretion of the following seven hormones: growth hormone (GH), thyroid-stimulating hormone (TSH), adrenocorticotropic hormone (ACTH), follicle-stimulating hormone (FSH), luteinizing hormone (LH), prolactin (PRL), and melanocyte-stimulating hormone (MSH).

The posterior lobe of the pituitary is only slightly larger than the head of a pin, and it is formed when nerves from the brain grow downward. At some point in fetal development, the anterior and posterior lobes join together to form the pituitary gland. In contrast to the circulatory connection between the hypothalamus and the anterior lobe, nerve cells of the hypothalamus project directly into the posterior lobe (see Figure 6–6). Thus, neurosecretory cells (nerve cells that synthesize and secrete hormones) that originate in the hypothalamus extend down into the posterior pituitary where

they release the two posterior lobe hormones, oxytocin (OT) and antidiuretic hormone (ADH).

Hormones of the Anterior Pituitary

The anterior pituitary produces and secretes seven major hormones. We will begin with **growth hormone (GH)**, the primary function of which is to stimulate growth through the enlargement of cells and increases in rates of cell division. Cells of bone and muscle tissue are most susceptible to GH, but cells of other tissues are affected as well. (The effects of GH on body growth are thought to occur indirectly, through intermediary hormones called somatomedins. Basically, the presence of GH stimulates the cells of the liver to release somatomedins and these hormones, in turn, increase protein synthesis, thereby enhancing cell growth and division.) GH also plays a role in glucose conservation by making fats more readily available as a source of fuel.

The synthesis and release of GH is regulated by two hormones of the hypothalamus. Growth hormone–releasing hormone (GHRH) stimulates release of GH and growth hormone–inhibiting hormone (GHIH) inhibits release of GH. Although levels of GH are normally maintained within an appropriate range, excesses or deficiencies of the hormone can produce dramatic effects on growth. Increased production of GH in childhood, when the bones are still capable of growing in length, results in **giantism,** a condition characterized by rapid growth and eventual attainment of heights up to 8 or 9 feet (Figure 6–7). Increased production of GH in adulthood, when the bones can thicken but not lengthen, causes **acromegaly.** Literally meaning "enlarged extremities," acromegaly is characterized by a gradual thickening of the bones of the hands, feet, and face and enlargement of the tongue (Figure 6–8). Giantism and acromegaly are usually caused by a tumor of the pituitary. Because both conditions are associated with decreased life expectancy, the tumor should be treated with radiation, drugs, or surgery. Finally, insufficient production of GH in childhood results in **pituitary dwarfism.** Typically, pituitary

Figure 6–7 Robert Wadlow, a pituitary giant, was born at a normal size but developed a pituitary tumor as a young child. The tumor caused increased production of growth hormone and Robert never stopped growing until his death at 22 years of age. *(© UPI/Corbis-Bettman)*

(a)

(b)

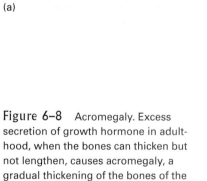

(c)

(d)

Figure 6–8 Acromegaly. Excess secretion of growth hormone in adulthood, when the bones can thicken but not lengthen, causes acromegaly, a gradual thickening of the bones of the hands, feet and face. The disorder was not apparent in this female at ages (a) 9 or (b) 16, but became apparent by ages (c) 33 and (d) 52. *(© Dr. William H. Daughaday, University of California, Irvine/American Journal of Medicine (20) 1956.)*

Figure 6-9 Pituitary dwarfism is caused by decreased production of growth hormone in childhood. *(© Daniel Margulies)*

dwarfs are sterile and attain a maximum height of about 4 feet (Figure 6–9). Pituitary dwarfism, but not other forms of dwarfism, can be treated by administering GH in childhood.

Social Issue

In the past, GH for treatment of medical conditions was extremely scarce, the small amounts available having been extracted from the pituitary glands of cadavers. As a result of its scarcity, use of GH was limited to the treatment of conditions such as pituitary dwarfism. Beginning in the late 1970s, however, GH could be synthesized in the laboratory and with its new abundance came research on its uses in the treatment of a variety of conditions including aging and below average height in children. Many aspects of normal aging, including reduced lean body mass, thinning of the skin, and increased body fat, appear to be reversed by administration of GH. For children within the normal height range for their age, but of below average height, GH holds the promise of increased stature. Opponents of using GH for the treatment of aging and for below average height in children believe it is wrong to give a powerful and potentially harmful hormone to people who are basically healthy. Rather than administering GH to such people, opponents suggest working instead to increase societal acceptance of those that are elderly or short. What do you think?

As its name indicates, **thyroid-stimulating hormone (TSH)** acts on the thyroid gland in the neck to stimulate synthesis and release of the thyroid hormones thyroxine and triiodothyronine. TSH itself is controlled by thyrotropin-releasing hormone from the hypothalamus. If TSH causes hypersecretion (oversecretion) of thyroid hormones, then Graves' disease may result (see section on the thyroid gland).

Adrenocorticotropic hormone (ACTH) controls the synthesis and secretion of glucocorticoid hormones from the outer portion (cortex) of the adrenal glands (see Figure 6–2). Addison's disease results when the actions of ACTH cause hyposecretion (undersecretion) of glucocorticoids. Most often, however, this disease is caused by a defect in the adrenal gland itself (see section on adrenal glands). Release of ACTH is controlled by ACTH-RH, a releasing hormone from the hypothalamus.

Follicle-stimulating hormone (FSH) promotes, in females, development of egg cells and secretion of ovarian estrogen. In males, FSH promotes production of sperm. Because of its impact on the gonads (ovaries and testes), FSH is considered a gonadotropin. Gonadotropin-releasing hormone (GnRH) from the hypothalamus controls release of FSH.

Luteinizing hormone (LH) causes ovulation, the release of a future egg cell by the ovary, and stimulates the ovaries to secrete estrogen and progesterone. These two hormones prepare the uterus for implantation of a fertilized ovum and the breasts for production of milk. In males, LH stimulates interstitial cells within the testes to develop and secrete testosterone. Like FSH, LH is a gonadotropin controlled by GnRH. The discovery of the gonadotropins LH and FSH by two German gynecologists in the late 1920s led to development of the first pregnancy test. Selmar Aschheim and Bernhardt Zondek discovered that the urine of pregnant women contained substances similar to LH and FSH and that when such urine was injected into a female frog, it caused the frog's ovaries to produce eggs. Urine of nonpregnant women did not contain these gonadotropin-like substances and did not cause ovulation in frogs. Thankfully, from the standpoints of frogs and prospective parents, pregnancy tests have come a long way in recent years.

Prolactin (PRL) is produced mainly during lactation, the period in which milk is produced and ejected by the mammary glands. Prolactin stimulates a mother's mammary glands to produce milk. Oxytocin, a hormone secreted from the posterior pituitary, causes the ducts of the mammary glands to eject milk. Prolactin interferes with female sex hormones and this is why most mothers fail to have regular menstrual cycles and to conceive while nursing their newborns. (Lactation should not, however, be relied upon as a method for birth control because it cannot be depended upon.) Growth of a pituitary tumor may cause excess secretion of prolactin resulting in infertility and the production of milk when birth has not occurred. Such a tumor in men can cause impotence. Hormones from the hypothalamus stimulate and inhibit production and secretion of prolactin.

Table 6–1　Hormones Secreted by the Anterior Lobe of the Pituitary Gland

Hormone	Function
Growth hormone (GH)	Stimulates growth, particularly of muscle and bone, through increases in protein synthesis, cell size, and rates of cell division Stimulates breakdown of fat
Thyroid-stimulating hormone (TSH)	Stimulates synthesis and release of thyroxine and triiodothyronine from thyroid gland
Adrenocorticotropic hormone (ACTH)	Stimulates synthesis and release of glucocorticoid hormones from adrenal glands
Follicle-stimulating hormone (FSH)	Stimulates gamete production in males and females Stimulates secretion of estrogen by ovaries
Luteinizing hormone (LH)	Causes ovulation and stimulates ovaries to secrete estrogen and progesterone Stimulates interstitial cells of testes to develop and secrete testosterone
Prolactin (PRL)	Stimulates breasts to produce milk
Melanocyte-stimulating hormone (MSH)	Precise function in humans unknown; repeated administration causes darkening of skin

Although the exact function of **melanocyte-stimulating hormone (MSH)** in humans is unknown, repeated administration causes a darkening of the skin, perhaps stimulating pigment-producing cells of the skin called melanocytes to increase their production of the pigment melanin. (See Chapter 4 for a discussion of skin pigments and color.) The hormones produced by the anterior pituitary are summarized in Table 6–1.

Hormones of the Posterior Pituitary

The posterior pituitary gland does not synthesize any hormones. However, neurons of the hypothalamus manufacture **antidiuretic hormone (ADH)** and **oxytocin (OT)** that travel down the nerve cells into the posterior pituitary where they are stored and subsequently released. The main function of ADH is to conserve body water by decreasing urine output. ADH accomplishes this task by causing the kidneys to remove water from the fluid destined to become urine and returning it to the blood. Alcohol inhibits secretion of ADH and this is why urination increases following alcohol consumption. The increased output of urine causes dehydration and the resultant headache and dry mouth typical of many hangovers. ADH is also called **vasopressin** for its role in constricting arterioles, the blood vessels between arteries and capillaries, and raising blood pressure, particularly during times of severe blood loss. A deficiency of ADH, often caused by damage to either the posterior pituitary or the

area of the hypothalamus responsible for the hormone's manufacture, results in **diabetes insipidus**. This condition is characterized by excessive urine production and resultant dehydration and thirst. Many children with this condition wet their beds. Although mild cases may not require treatment, severe cases may cause fluid losses of up to 10 liters a day (normal urine output is 1 to 2 liters per day) and death through dehydration can result. Treatment usually involves administration of synthetic ADH in a nasal spray. **Diabetes insipidus** (*diabetes* = overflow, *insipidus* = tasteless) should not be confused with **diabetes mellitus** (*mel* = honey), a condition in which large amounts of glucose are lost in the urine as a result of an insulin deficiency. Both conditions, however, are characterized by increased production of urine. (We will look at diabetes mellitus again when we discuss the hormones of the pancreas.) Look, again, at what their names mean. Can you guess how physicians in the past distinguished between these two forms of diabetes?

Oxytocin (OT) is the second hormone produced in the hypothalamus and released by the posterior pituitary. The name oxytocin (*oxy* = quick, *tokos* = childbirth) reveals one of its two main functions, stimulating the uterine contractions of childbirth (Figure 6–10). During pregnancy, the cells of the uterus become increasingly sensitive to oxytocin as the number of OT receptors increases. Eventually the uterus begins to contract in response to OT. During labor, when the baby's head begins to stretch the narrow portion of

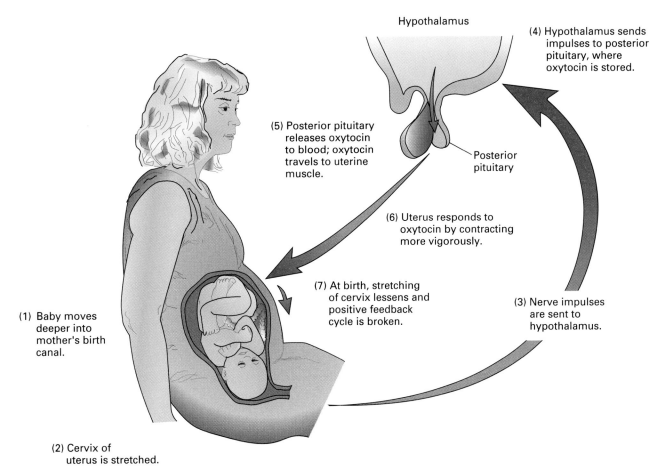

(4) Hypothalamus sends impulses to posterior pituitary, where oxytocin is stored.

Hypothalamus

(5) Posterior pituitary releases oxytocin to blood; oxytocin travels to uterine muscle.

Posterior pituitary

(6) Uterus responds to oxytocin by contracting more vigorously.

(3) Nerve impulses are sent to hypothalamus.

(7) At birth, stretching of cervix lessens and positive feedback cycle is broken.

(1) Baby moves deeper into mother's birth canal.

(2) Cervix of uterus is stretched.

Figure 6-10   The neuroendocrine reflex by which oxytocin stimulates uterine contractions during childbirth.

the uterus called the cervix, nerve impulses are sent to the hypothalamus which, in turn, signals the posterior pituitary to release additional OT. The released OT travels in the blood to the uterus where it stimulates ever stronger contractions of the uterus until finally the baby is pushed down the birth canal and birth occurs. As described earlier, the control of OT during the induction of labor is an example of a positive feedback mechanism—OT stimulates uterine contractions that, in turn, stimulate further release of OT. Once the baby is born, stretching of the cervix lessens and the positive feedback cycle is broken. Because this cycle has input from the nervous system (stretch receptors in the cervix send nerve impulses to the hypothalamus) and output from the endocrine system (releases of OT), it is called a **neuroendocrine reflex**. Pitocin, a synthetic form of OT, is sometimes used to induce labor.

The second major function of OT is to stimulate milk ejection ("letdown") from the mammary glands in response to a suckling infant (Figure 6–11). Recall that prolactin se-creted by the anterior pituitary stimulates the mammary glands to produce, but not eject, milk. Once a baby begins to suck at the nipple, touch receptors in the nipple send nerve impulses to the hypothalamus, which prompts the posterior pituitary to release OT (this is another example of a neuroendocrine reflex). Once in the bloodstream, OT travels to the mammary glands where it stimulates the smooth muscle cells surrounding milk-producing cells to contract and eject milk. Milk ejection occurs about one minute after initial stimulation of the nipple. The hormones released by the posterior pituitary are summarized in Table 6–2.

Critical Thinking

Women who have just given birth are often encouraged to nurse their babies as soon as possible thereafter. How might an infant's suckling promote completion of, and recovery from, the birth process?

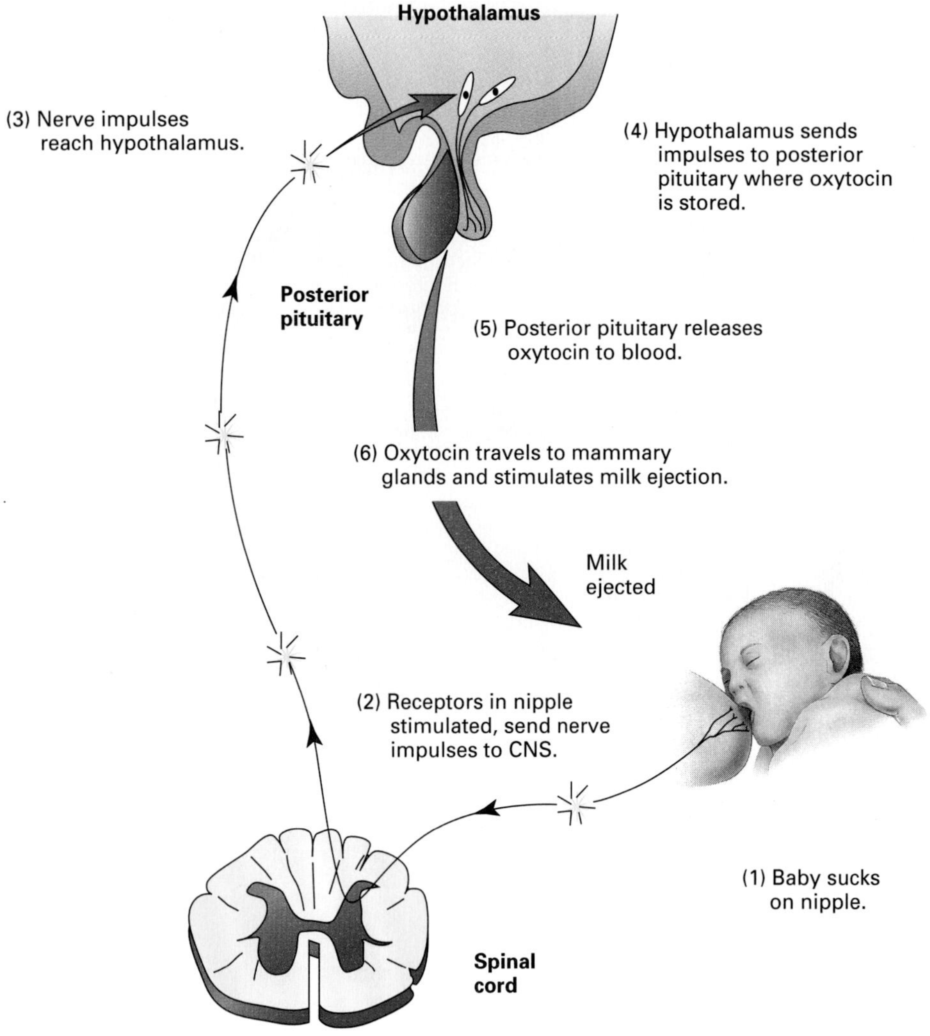
Hypothalamus
(3) Nerve impulses reach hypothalamus.
(4) Hypothalamus sends impulses to posterior pituitary where oxytocin is stored.
Posterior pituitary
(5) Posterior pituitary releases oxytocin to blood.
(6) Oxytocin travels to mammary glands and stimulates milk ejection.
Milk ejected
(2) Receptors in nipple stimulated, send nerve impulses to CNS.
(1) Baby sucks on nipple.
Spinal cord

Figure 6–11 The neuroendocrine reflex by which oxytocin stimulates milk ejection (letdown).

Table 6–2 Hormones Secreted by the Posterior Lobe of the Pituitary Gland

Hormone	Function
Antidiuretic hormone (ADH)	Promotes water reabsorption by kidneys
Oxytocin (OT)	Stimulates milk ejection from the breasts and uterine contractions during childbirth

Thyroid Gland

The thyroid gland is a shield-shaped structure at the front of the neck, overlying the trachea and positioned just below the larynx (Figure 6–12a). It has two lateral lobes connected by a medial band of tissue called the **isthmus**. The characteristic deep red color of the thyroid gland stems from its prodigious blood supply. Within the thyroid are small spherical chambers called follicles (Figure 6–12b and c). Follicular cells line the walls of the follicles and produce thyroglobulin, a glycoprotein from which **thyroid hormone (TH)** is made. (Thyroid hormone is the name given to two very similar hormones, thyroxine—also called T4 because it has four iodine atoms—and triiodothyronine—also called T3 because it has three iodine atoms.) Other endocrine cells in the thyroid, called parafollicular cells, secrete the hormone **calcitonin** (Figure 6–12c).

Thyroid hormone affects almost all cells of the body. Not surprisingly, then, the hormone has broad effects, ranging from regulating the body's metabolic rate (a measure of the pace of chemical reactions within the body) and production of heat to maintaining blood pressure and promoting normal development and functioning of the nervous, muscular, skeletal, and reproductive systems. Thyroid hormone

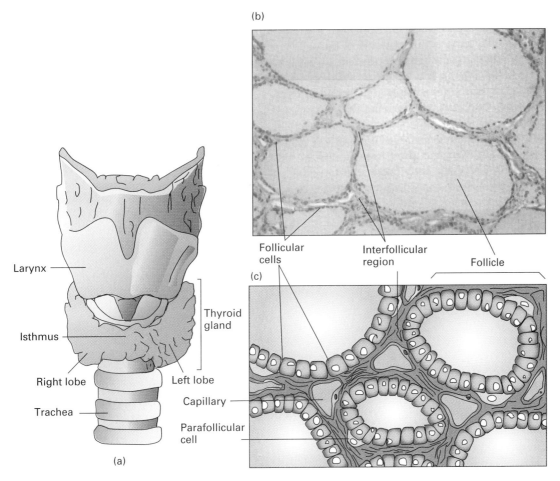

Larynx

Isthmus

Right lobe

Trachea

Thyroid gland

Left lobe

(a)

(b)

Follicular cells

Interfollicular region

Follicle

(c)

Capillary

Parafollicular cell

Figure 6–12 Structure of the thyroid gland. (a) The thyroid gland lies over the trachea, just below the larynx. (b) and (c) Histology of the thyroid gland showing the follicular cells that produce the precursor to thyroid hormone and the parafollicular cells that produce calcitonin. *(b, © Martin Rotker/Phototake, NYC)*

affects cellular metabolism by stimulating protein synthesis, the use of glucose for production of ATP, and the breakdown of lipids. The pituitary gland and hypothalamus control release of thyroid hormone. Falling levels of thyroid hormone in the blood prompt the hypothalamus to secrete thyrotropin-releasing hormone (TRH) that, in turn, stimulates the anterior pituitary to release thyroid-stimulating hormone (TSH). In response to TSH, the thyroid releases larger amounts of thyroid hormone.

Hyposecretion of thyroid hormone during development of the fetus or infant causes **cretinism**, a condition characterized by dwarfism, mental retardation, and slowed sexual development (Figure 6–13). Providing that a pregnant woman produces sufficient thyroid hormone, many of the symptoms associated with cretinism do not appear until after birth when the infant relies solely on its own malfunctioning thyroid to supply the needed hormones (maternal thyroid hormones cross the placenta during pregnancy and reach the fetus). Most infants are tested shortly after birth for proper thyroid function, and cretinism can be prevented

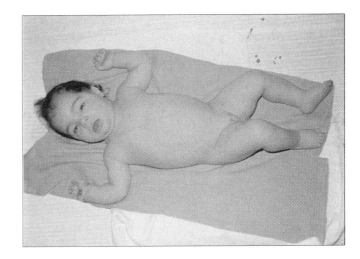

Figure 6–13 Cretinism, a disorder of the thyroid gland. This condition is characterized by mental retardation, dwarfism, and delayed sexual development and is caused by undersecretion of thyroid hormone during fetal life or infancy. *(© Lester V. Bergman and Associates)*

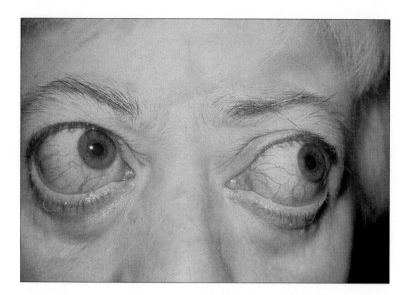

Figure 6-14 Exophthalmos, a disorder of the thyroid gland. Oversecretion of thyroid hormone leads to an accumulation of interstitial fluid behind the eyes, causing the eyes to bulge out. This condition is often associated with Graves' disease. *(© Ralph Eagle/Science Source/Photo Researchers, Inc.)*

by oral administration of thyroid hormone. In adulthood, hyposecretion of thyroid hormone causes **myxedema,** a condition characterized by swelling of the facial tissues due to **edema,** the accumulation of interstitial fluid. Other symptoms associated with undersecretion of thyroid hormone include decreased alertness, body temperature, and heart rate, and all can be alleviated by oral administration of thyroid hormone.

Hypersecretion of thyroid hormone causes **Graves' disease,** an autoimmune disorder made famous when the irregular heart rhythms experienced by former President George Bush were attributed to it. Basically, a person's own immune system produces antibodies that mimic the action of TSH causing the thyroid gland to enlarge and overproduce its hormones. The symptoms of Graves' disease include increased metabolic rate and heart rate, sweating, nervousness, and weight loss. Many patients with Graves' disease also have **exophthalmos,** protruding eyes caused by the accumulation of interstitial fluid (Figure 6-14). Treatment of Graves' disease entails either the use of certain drugs to block synthesis of thyroid hormones or reduction of thyroid tissue through surgery or administration of radioactive iodine. Because the thyroid gland accumulates iodine, ingestion of radioactive iodine selectively destroys thyroid tissue.

Iodine is needed for production of thyroid hormone. A diet deficient in iodine can produce **simple goiter,** an enlarged thyroid gland (Figure 6-15). When intake of iodine is inadequate, levels of thyroid hormone are low and this triggers secretion of TSH that stimulates the thyroid to increase production of thyroglobulin. The lack of iodine, however, prevents formation of thyroxine and triiodothyronine from

the accumulating thyroglobulin. Responding to the continued low levels of thyroid hormones, the pituitary continues to release increasing amounts of TSH, causing the thyroid to enlarge in a futile effort to filter more iodine from the

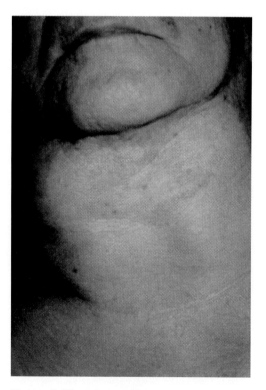

Figure 6-15 Simple goiter. In response to an iodine-deficient diet, the thyroid gland enlarges causing a goiter. *(© Ken Greer/ Visuals Unlimited)*

blood. Before iodine was added to most table salt, goiters were quite common, particularly in parts of the midwestern United States (deemed the "goiter belt") where iodine-poor soil and little access to iodine-rich shellfish led to diets deficient in iodine. Simple goiter can be treated by iodine supplements or administration of thyroid hormone.

Calcitonin (CT), the hormone produced by the parafollicular cells of the thyroid gland, maintains low blood levels of calcium and phosphates by stimulating their absorption by bone and inhibiting the breakdown of bone. Too little calcitonin can cause kidney stones and formation of calcium deposits in soft tissue. The pituitary gland does not control release of calcitonin. Instead, levels of calcium in the blood directly influence secretion of calcitonin in a feedback system involving the parathyroid glands, which we discuss next.

Parathyroid Glands

The parathyroid glands are four small, round masses of tissue attached to the back of the thyroid gland (Figure 6–16). These glands secrete **parathyroid hormone (PTH)**, also called **parathormone**. As mentioned above, calcitonin from the thyroid gland lowers levels of calcium in the blood. There is, however, another side to the balancing act of maintaining proper blood calcium levels. This other side involves PTH, the antagonist to calcitonin (that is, PTH and calcitonin have opposite effects). Low levels of calcium in the

blood stimulate the parathyroid glands to secrete PTH that causes calcium to move from bone and urine to the blood. PTH causes the breakdown of bone and consequent release of calcium and phosphates to the blood by stimulating bone-destroying cells called osteoclasts. In the kidneys, PTH stimulates (1) the removal of calcium and magnesium from the urine and the return of these substances to the blood (2) the excretion of phosphates and (3) the secretion of **calcitriol**, a hormone that increases the rates at which calcium, magnesium, and phosphate are absorbed into the blood from the gastrointestinal tract. In summary, the overall effect of PTH is to enhance blood levels of calcium and magnesium. With respect to phosphates, however, there is a net loss because more phosphate is lost in the urine than is gained by the breakdown of bone. The feedback system by which PTH and calcitonin regulate levels of calcium in the blood is summarized in Figure 6–17.

Surgery on the neck or thyroid gland sometimes results in accidental removal of, or damage to, the parathyroid glands. The resultant decrease in parathyroid hormone causes decreased blood calcium that produces nervousness and muscle spasms. In severe cases, death may result from spasms of the larynx and paralysis of the respiratory system. Because parathyroid hormone is difficult to purify and therefore expensive, deficiencies are usually treated not by administering the hormone but by giving calcium, either in tablet form or through increased dietary intake of calcium-rich foods. A tumor of the parathyroid gland can cause

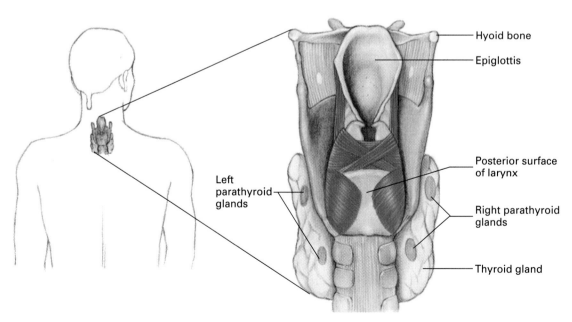

Hyoid bone

Epiglottis

Posterior surface of larynx

Right parathyroid glands

Thyroid gland

Left parathyroid glands

Figure 6–16 The parathyroid glands. Attached to the back of the thyroid gland, the four small parathyroid glands secrete parathyroid hormone.

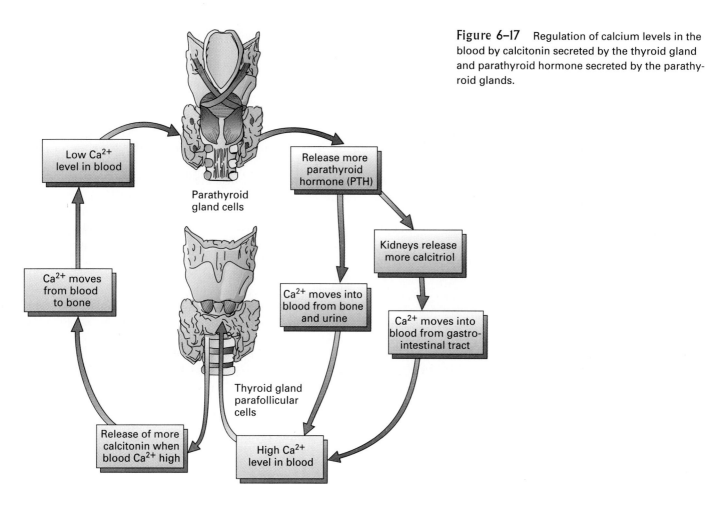

Figure 6–17 Regulation of calcium levels in the blood by calcitonin secreted by the thyroid gland and parathyroid hormone secreted by the parathyroid glands.

Low Ca²⁺ level in blood

Release more parathyroid hormone (PTH)

Parathyroid gland cells

Kidneys release more calcitriol

Ca²⁺ moves from blood to bone

Ca²⁺ moves into blood from bone and urine

Ca²⁺ moves into blood from gastro-intestinal tract

Thyroid gland parafollicular cells

Release of more calcitonin when blood Ca²⁺ high

High Ca²⁺ level in blood

excess secretion of parathyroid hormone. Hypersecretion of parathyroid hormone pulls calcium from bone tissue, producing increased blood calcium. High levels of calcium in the blood may lead to decreased activity of the nervous system, kidney stones (excess calcium salts precipitate out in the tubules of the kidneys), and the formation of calcium deposits in other soft tissue. Hormones released by the thyroid and parathyroid glands are summarized in Table 6–3.

Adrenal Glands

Each of us has two adrenal glands. Each gland is about the size of an almond, sits perched on top of a kidney (*ad*, upon; *renal*, kidney), and has two distinct regions (Figure 6–18). The outer region, the **adrenal cortex**, secretes more than 20 different steroid hormones that fall into three groups, the glucocorticoids, mineralocorticoids, and gonadocorticoids. The inner region, called the **adrenal medulla**, secretes epinephrine and norepinephrine.

The **glucocorticoids** secreted by the adrenal cortex affect glucose homeostasis and thereby influence metabolism and resistance to stress. In terms of their influence on me-

tabolism, glucocorticoids act on the liver to promote **gluco-neogenesis**, the conversion of noncarbohydrate molecules, such as amino acids, to glucose. The glucocorticoids also act on adipose tissue to prompt the breakdown of fats to fatty acids. As a result of these activities, glucose and fatty acids are released into the bloodstream where they are available for uptake and use by the body's cells. The extra glucose may also be used to combat stresses related to temperature, fasting, infection, or bleeding. Another function of glucocorticoids is to inhibit the inflammatory response. These hormones slow the movement of white blood cells to the site of injury and inhibit the activity of those cells already present, thus preventing further swelling and inflammation. The downside of glucocorticoids is that they inhibit wound healing. This is why steroid creams containing glucocorticoids are applied only to the surface of the skin to treat superficial rashes, such as those caused by allergies or poison ivy, and should not be applied to open wounds. Three examples of glucocorticoids are **cortisol**, **corticosterone**, and **cortisone**.

Hypersecretion of glucocorticoids, particularly cortisol and cortisone, causes **Cushing's syndrome**. Individuals

Table 6–3 Hormones Secreted by the Thyroid Gland and Parathyroid Glands

Gland and Hormone	Function
Thyroid gland	
Thyroid hormone (TH) (includes thyroxine, T4, and triiodothyronine, T3)	Regulates metabolism and heat production Promotes normal development and functioning of nervous, muscular, skeletal, and reproductive systems
Calcitonin	Lowers blood levels of calcium and phosphates by stimulating their absorption by bone and inhibiting breakdown of bone
Parathyroid glands	
Parathyroid hormone (PTH)	Increases blood levels of calcium and magnesium and decreases blood levels of phosphates by stimulating breakdown of bone and causing kidneys to remove calcium and magnesium from urine and to excrete phosphates

with this condition display an accumulation of fluid in the face and a redistribution of body fat, having a pendulous abdomen often accompanied by stretch marks and relatively thin arms and legs (Figure 6–19). Additional symptoms include high blood glucose levels, poor wound healing, and decreased resistance to stress. Although sometimes caused by a tumor on either the adrenal cortex or anterior pituitary

(recall that the anterior pituitary secretes adrenocorticotropic hormone that stimulates release of hormones from the adrenal cortex), Cushing's syndrome usually results from administration of steroids for disorders such as asthma and rheumatoid arthritis (inflammation of joint membranes). Treatment in medically induced cases of Cushing's syndrome typically entails a gradual reduction of the glucocor-

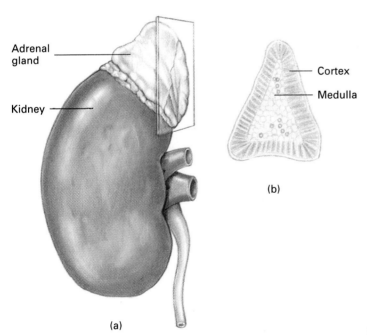

(a)

(b)

Figure 6–18 Location and structure of an adrenal gland. (a) Each adrenal gland sits on top of a kidney. (b) A section through the adrenal gland reveals two regions, the outer adrenal cortex and inner adrenal medulla.

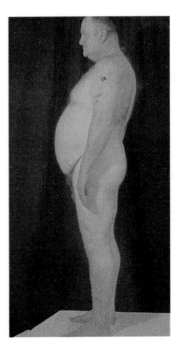

Figure 6–19 Cushing's syndrome. Hypersecretion of glucocorticoids by the adrenal cortex causes fluid to accumulate in the face and fat to accumulate in the abdomen. Most often Cushing's syndrome is caused by the administration of glucocorticoids for allergies or inflammation. *(© Science vu/Visuals Unlimited)*

ticoid dose. Tumors are treated with radiation, drugs, or surgery.

The **mineralocorticoids** secreted by the adrenal cortex affect mineral homeostasis and water balance. The primary mineralocorticoid is **aldosterone**. This hormone acts on cells of the kidneys to increase reabsorption of Na^+, preventing depletion of Na^+ and increasing water retention. Aldosterone also acts on kidney cells to promote excretion of K^+ in urine. **Addison's disease** is caused by hyposecretion of glucocorticoids and aldosterone. This disease appears to be an autoimmune disorder in which the body's own immune system perceives cells of the adrenal cortex as foreign and destroys them. The resultant loss of hormones from the adrenal cortex causes weight loss, fatigue, loss of appetite, poor resistance to stress, and electrolyte imbalance. Also associated with this disorder is a peculiar bronzing of the skin. Addison's disease can be treated by oral administration of the missing hormones in tablet form.

Critical Thinking

High blood pressure can signal abnormal aldosterone secretion. Would high blood pressure be associated with hyposecretion or hypersecretion of aldosterone?

The adrenal cortex also secretes **gonadocorticoids**, male and female sex hormones known as **androgens** and **estrogens**. These hormones appear to be important in the growth spurt that occurs before puberty and in development of secondary sex characteristics such as growth of pubic hair. In normal adult males, androgen secretion by the testes far surpasses that by the adrenal cortex, and thus the effects of adrenal androgens are probably insignificant. Similarly, in females, estrogens are also produced by the ovaries and placenta. During menopause, however, when the ovaries decrease their secretion of estrogens, gonadocorticoids from the adrenals may be important. At this time, female androgens from the adrenals are converted to estrogens and this may somewhat alleviate the effects of decreased secretion of hormones by the ovaries. (Even though androgens are often described as "male" hormones and estrogens as "female" hormones, hormonal output is not so rigidly determined by sex as these labels suggest. Normal females generally produce small amounts of androgens and normal males produce small amounts of estrogens.) Female androgens are thought to play a role in sex drive. One option for menopausal women is estrogen replacement therapy. The advantages and disadvantages of hormone replacement therapy, with particular regard to estrogen replacement therapy in menopause, are considered in the Personal Concerns essay.

The adrenal medulla produces **epinephrine** (adrenaline) and **norepinephrine** (noradrenaline). These hormones help us respond to stress and are critical in the **fight-or-flight response**, our body's physiological reaction to threatening situations. Imagine, for example, that you are walking home alone late at night and a stranger suddenly steps out in front of you from the bushes. Impulses received by your hypothalamus are sent by neurons to your adrenal medulla causing cells in your medulla to increase their output of epinephrine and norepinephrine. In response to these hormones, your heart rate, respiratory rate, and blood glucose levels increase. Blood vessels associated with the digestive tract constrict because digestion is not of prime importance, while vessels associated with skeletal muscles dilate, allowing more blood, glucose, and oxygen to reach them. These substances also reach your brain in greater amounts, leading to the increased mental alertness essential to fleeing or fighting. The hormones secreted by the adrenal glands are summarized in Table 6–4.

Table 6–4   Hormones Secreted by the Adrenal Glands

Region and Hormone	Examples	Function
Adrenal cortex		
Glucocorticoids	Cortisol, corticosterone, cortisone	Regulate metabolism and resistance to stress and inhibit the inflammatory response
Mineralocorticoids	Aldosterone	Increase sodium reabsorption and potassium excretion by kidneys
Gonadocorticoids	Androgens, estrogens	Amounts secreted by adults are so low that effects are probably insignificant
Adrenal medulla		
Epinephrine Norepinephrine		Fight-or-flight and stress responses

Is It Hot in Here or Is It Me? Hormone Replacement Therapy and Menopause

Many endocrine disorders are characterized by little or no secretion of certain hormones. Cretinism, for example, a condition characterized by dwarfism and mental retardation, is caused by undersecretion of thyroid hormone during fetal life or infancy. In other cases, hormone secretion declines as part of the normal aging process. Estrogen, a hormone produced primarily by the ovaries, is secreted in decreasing amounts beginning when women are about 30, eventually leading into menopause, the termination of menstrual cycles, about 15 to 20 years later. Hyposecretion of hormones, whether caused by malfunctioning glands or normal aging, is often treated with hormone replacement therapy (HRT), a medical technique in which the deficient hormone is replaced through injections, pills, patches, or creams. HRT is most controversial when used to combat normal declines of hormones. Here we consider HRT, its curious history and its controversial use in combating the miseries of menopause.

Charles Edward Brown-Sequard, perhaps the earliest practitioner of HRT, attempted in 1889 at the age of 72 to achieve self-revitalization through injections of extracts from the testes of dogs and guinea pigs. He reported, after several weeks of self-treatment, that he experienced increased muscle strength, more energy (assessed by greater ability to work long hours in the laboratory), and enhanced bladder tone (measured by how far he could urinate). These astonishing achievements, which we now know to have resulted from the powers of autosuggestion, led some physicians to administer testicular extracts to male patients in need of rejuvenation. On other fronts, Brown-Sequard's work continued to inspire physicians and researchers in endocrinology such as George Murray, an English doctor who in 1891 injected thyroid extract into patients suffering from hypothyroidism. Murray's patients showed dramatic improvement, leading him to conclude that when a condition is caused by the absence of a particular substance in the body, then the only "rational treatment" is to compensate for that deficiency by injecting extracts from a normal gland. And so HRT began. About 80 years after Murray's revelations laid the theoretical groundwork for HRT, technical advances such as the ability to synthesize and mass-produce artificial hormones sent HRT to the forefront of treating endocrine disorders.

Despite HRT's many successes, it is not without controversy, particularly when used to treat normal declines in hormone levels. Menopause, as mentioned above, is caused by declining levels of ovarian estrogen as women age. In conjunction with the eventual cessation of menstrual cycles, women face a host of symptoms including hot flashes, night sweats, and vaginal dryness and atrophy. About 85% of women will experience one or more of these symptoms, some lasting for as long as 5 years. Declining estrogen levels have also been linked to osteoporosis (increased bone loss leading to increased risk of fracture), heart disease caused by reduced pliability of blood vessels and a low ratio of good cholesterol to bad cholesterol, heightened mood swings and memory loss, increased risk of colon cancer, and declines in skin elasticity. Given all this, it is not surprising that doctors in recent years have been administering estrogen in the form of pills, patches, or creams to millions of women at or past menopause. Many women are thrilled with the results of estrogen replacement therapy (ERT), claiming a greater sense of well-being, better sex lives, and absence of those dreadful hot flashes and cold sweats.

There is, however, a dark side to ERT. Long-term use of estrogen (as is necessary for this hormone to work its wonders) has been linked to increased risk of gallstones and cancer of the breast and uterus, abnormal blood clotting, weight gain, and headaches. Some doctors believe that such risks, particularly those related to cancer, have been diminished by reducing dosages and adding the hormone progesterone to the estrogen prescription. However, recent data suggest that while the combined estrogen-progesterone treatment appears to reduce the risk of uterine cancer, it may actually put women at a higher risk for breast cancer than did the estrogen-only treatment. In view of these risks, some physicians urge women to consider the many other ways available to stave off heart disease and osteoporosis, including avoiding smoking, eating a diet rich in calcium and low in fat, and exercising regularly. Some alternative practitioners advise getting estrogen from dietary sources such as yams and soybeans. These foods are certainly weaker sources than prescription estrogen and may reduce the risks.

Every woman will face the choice of whether or not to take estrogen in middle age. The final decision should be based on personal philosophy (is it OK to take a drug to fight off a natural process?), careful consideration of current health, and a long hard look at family history of cancer, heart disease, and osteoporosis. Unfortunately, even when fully aware of the risks and benefits of ERT, we must all acknowledge that current data for both are woefully inadequate.

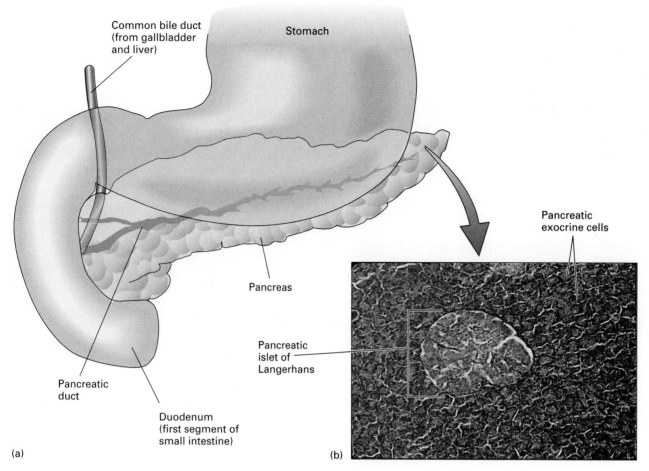

Figure 6-20   The pancreas. (a) Gross anatomy of the pancreas. Exocrine cells of the pancreas secrete digestive enzymes into the pancreatic duct which unites with the common bile duct from the gallbladder and liver before entering the small intestine. (b) Histology of the pancreas. Endocrine cells of the pancreas are found in clusters called islets of Langerhans. Surrounding the islets are exocrine cells.   *(Astrid and Hanns Frieder Michler/Science Photo Library/Custom Medical Stock Photo)*

Pancreas

The pancreas, located in the abdomen just behind the stomach, contains endocrine cells and exocrine cells (Figure 6–20). The exocrine cells, called **acinar cells**, secrete digestive enzymes into ducts that empty into the small intestine. The role of the pancreas in digestion will be discussed in Chapter 13. The endocrine cells occur in small clusters called **islets of Langerhans**. These clusters contain two major types of hormone-producing cells, **alpha cells** that produce the hormone **glucagon** and **beta cells** that produce the hormone **insulin**.

Glucagon increases glucose in the blood. Glucagon accomplishes this task by prompting cells of the liver to increase conversion of glycogen to glucose and formation of glucose from lactic acid and amino acids. The resultant glucose molecules are released by the liver into the bloodstream, causing a rise in blood sugar level. Because amino acids in the blood are taken up by the liver to form new glucose molecules, a secondary effect of glucagon is a lowering of blood levels of amino acids. Glucagon secretion is mainly stimulated by declining levels of blood glucose, but it is also stimulated by increasing levels of amino acids in the blood as might occur following consumption of a large steak.

In contrast to glucagon, **insulin** decreases glucose in the blood. Insulin produces a lowering of blood sugar by (1) stimulating transport of glucose into muscle cells, white blood cells, and connective tissue cells (2) inhibiting the breakdown of glycogen to glucose and (3) preventing conversion of amino and fatty acids to glucose. As a result of these actions, insulin promotes protein synthesis, fat storage, and the use of glucose for energy. Insulin release is triggered by rising blood levels of glucose, amino acids, or fatty acids. Once cells begin to take up these substances, and their levels in the blood fall, insulin secretion is inhibited. Insulin secretion is also indirectly regulated by other hormones that influence blood glucose level. Growth hormone, for example, increases the amount of glucose in the blood and this, in turn, stimulates the pancreas to secrete insulin.

Pancreatic hormones, particularly insulin, have dramatic effects on our health. More than 100 million people worldwide, about 12 million of them Americans, suffer from **diabetes mellitus**, a disorder that causes **hyperglycemia**, an elevated level of glucose in the blood. Excessive urine production, as well as excessive thirst and eating, characterize this disorder. There are two major types of diabetes mellitus. **Type I diabetes**, representing about 10% of all diabetes cases, is an autoimmune disorder in which a person's own immune system attacks the beta cells of the pancreas interrupting insulin production and causing insulin deficiency. Type I diabetes is often called either juvenile-onset diabetes or insulin-dependent diabetes mellitus because it usually develops in people younger than 20 years of age and is treated by daily injections of insulin. (Ingestion of insulin is not yet useful because insulin is a protein hormone and would be broken down in the digestive tract. Work is underway, however, to package insulin in pill form in microcapsules that would be resistant to stomach acids and digestive enzymes.) In the absence of insulin, glucose does not readily move into body cells, and thus cells turn instead to fatty acids for energy. The increased breakdown of fats causes many problems, including (1) **ketoacidosis**, a lowering of blood pH resulting from the accumulation of breakdown by-products of fats (2) weight loss and (3) formation of fatty deposits on the walls of blood vessels as a result of increased transport of lipids by blood to starving cells. Problems with vision also commonly occur with Type I diabetes. **Cataracts**, for example, result when excessive glucose attaches to proteins of either the lens or lens capsule of the eye leading to loss of lens transparency and clouded vision. Recent research has focused on how to prevent Type I diabetes through transplanting either the entire pancreas or clusters of islets of Langerhans cells. Because such transplants carry with them the risk of rejection by the recipient's immune system, attempts have been made to use either fetal cells or cells encapsulated by a substance that allows insulin out but prevents the elements responsible for rejection from getting in.

Type II diabetes mellitus, representing about 90% of diabetes cases, is characterized by a decreased sensitivity to insulin possibly caused by decreased numbers of insulin receptors on target cells. Most people with Type II diabetes are overweight and the decrease in insulin receptors is believed to result from cells, constantly overloaded with food, adjusting downward their number of insulin receptors. Of-

ten the pancreas reacts by releasing even more insulin and thus some afflicted individuals have a surplus of insulin in their blood. Type II diabetes is also called mature-onset diabetes or non–insulin-dependent diabetes because it typically develops after age 40 and usually does not require insulin injections. Treatment of Type II diabetes involves diet, exercise, weight loss and sometimes antidiabetic drugs. Recent research with animals suggests the link between obesity and diabetes has to do with a hormone-like substance called tumor necrosis factor-alpha (TNF-alpha). Cells of obese animals secrete high levels of TNF-alpha that decrease the expression of genes responsible for producing a protein that enables glucose to cross cell membranes. If it should be shown that obese human diabetics also overproduce TNF-alpha, then blood sugar level may eventually be controlled by administering drugs that block TNF-alpha.

Too much insulin (hyperinsulinism) sometimes results from a tumor of the pancreas but more typically occurs when a diabetic person injects too much insulin. The result is **hypoglycemia**, depressed levels of glucose in the blood. Low levels of blood glucose cause symptoms such as anxiety, sweating, hunger, weakness, and disorientation and prompt the secretion of glucagon and other hormones such as growth hormone. In severe cases **insulin shock** results when brain cells, starved of glucose, fail to function properly causing convulsions and unconsciousness. Such shock may prove fatal unless blood sugar levels are raised. The hormones secreted by the pancreas are summarized in Table 6–5.

Thymus Gland

The thymus gland lies just behind the breastbone, on top of the heart (see Figure 6–2). It is more prominent in infants and children and decreases in size as we age. The thymus

Table 6–5 Hormones Secreted by the Pancreas

Hormone	Function
Glucagon	Increases blood glucose level by prompting the liver to increase conversion of glycogen to glucose and formation of glucose from fatty and amino acids
Insulin	Decreases blood glucose level by stimulating transport of glucose into cells, inhibiting breakdown of glycogen to glucose, and preventing conversion of fatty and amino acids to glucose

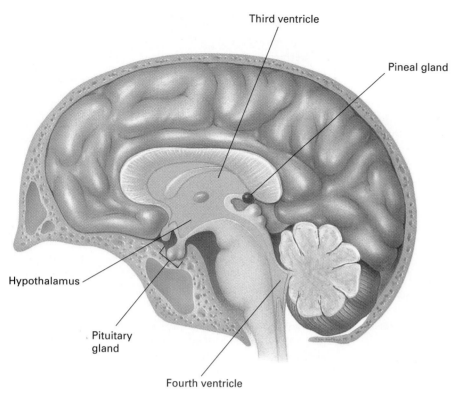

Third ventricle

Pineal gland

Hypothalamus

Pituitary gland

Fourth ventricle

Figure 6–21 The pineal gland. Located at the center of the brain, the pineal gland secretes the hormone melatonin.

secretes **thymopoietin, thymosin, thymic humoral factor (THF),** and **thymic factor (TF),** hormones involved in the production and maturation of white blood cells called T lymphocytes. T lymphocytes, also known as T cells, are involved in the body's defense mechanisms. The thymus and T lymphocytes play an important role in immunity and will be discussed in greater detail in Chapter 12.

Pineal Gland

The pineal gland, named for its resemblance to a pine cone, lies at the center of the brain, attached to the roof of the cavity within the brain known as the third ventricle (Figure 6–21). About the size of a pea, this tiny gland contains secretory cells called **pinealocytes** that produce the hormone **melatonin**. Levels of circulating melatonin are greater at night than during daylight hours. This pattern can be explained by the observation that the pineal gland receives input from visual pathways. Basically, when neurons of the retina are stimulated by light entering the eye, impulses are sent to the hypothalamus and these messages eventually reach the pineal gland where they inhibit secretion of melatonin.

Although the functions of melatonin in humans still remain somewhat of a mystery, research in the past few decades has suggested wide-ranging roles for melatonin including (1) inhibiting production of the pigment melanin by melanocytes of the skin (2) establishing or maintaining daily rhythms of physiological processes (3) triggering sleep and seasonal changes in depression and lethargy (4) slowing the aging process and (5) inhibiting hormones involved in fertility. One disorder associated with melatonin is **seasonal affective disorder (SAD),** a form of depression associated with winter months when overproduction of melatonin is triggered by short daylengths. Overproduction of melatonin causes symptoms such as an inability to concentrate, lethargy, long periods of sleep, and low spirits. Treatment of SAD patients often involves repeated exposure to sunlamps that produce the full-spectrum light needed to inhibit melatonin production (standard interior lighting apparently does not depress melatonin production). The links between melatonin and sleep, aging, and fertility have led in the 1990s to what some have described as "melatonin mania," wild, over-the-counter purchasing of melatonin and self-medication. In a Personal Concerns essay, we examine the promise and possible risks of taking melatonin.

Melatonin: Miracle Supplement or Potent Drug Misused by Millions?

Melatonin, a hormone secreted by the pineal gland, is being billed as the wonder drug of the nineties, capable of incredible medical feats including curing insomnia, preventing pregnancies, boosting immunity, and slowing the aging process. All this and melatonin is apparently nontoxic, even at extremely high doses. No wonder health food stores can barely keep their shelves stocked. What are the bases for these claims regarding the miracles of melatonin, and should this potent hormone be sold as an over-the-counter dietary supplement to self-prescribing consumers?

For those who toss and turn and watch the clock on the nightstand hoping for a few hours of shut-eye, melatonin can bring on sleep. Research, beginning in the 1980s and continuing today, has shown that remarkably low doses of melatonin can induce sleep without producing the "hangover-like feeling" that sometimes occurs the morning after having taken a sedative. An added advantage of melatonin is that the sleep it induces has dream phases of normal timing and duration. While most agree that melatonin is an effective sleep aid, there is debate among scientists concerning the precise mechanism by which melatonin produces its soporific effect. Some believe melatonin induces sleep by acting on the body's biological clock while others speculate that the mechanism is independent of the biological clock. A few complaints have also surfaced—it seems nightmares, disrupted sleep, and next-day grogginess have occurred in a small number of instances. It is not yet known whether melatonin was fully or partially responsible for these conditions.

In the Syrian hamster, melatonin functions in birth control, so why not give it a try in humans? Syrian hamsters are seasonal breeders, being infertile during winter months and sexually active in the spring (Figure A). This cycle appears to be regulated by the pineal gland through seasonal changes in photoperiod (day-night cycle). Beginning in late September when daylight hours start to wane, the reduced light reaching the pineal gland stimulates greater production of melatonin that in turn causes ovarian function to go on hold and the testes to shrink in size and testosterone output. Winter is a sexually quiet time for Syrian hamsters, and it is not until early spring that their reproductive systems spontaneously awaken and breeding begins.

(continued)

(a)

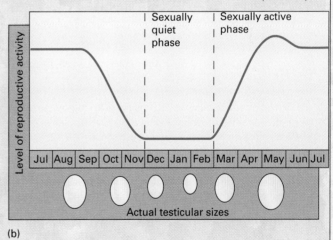

(b)

Figure A Melatonin influences reproduction in hamsters. (a) A Syrian hamster. (b) The reproductive cycle of the Syrian hamster. In September, level of exposure to daylight drops below a certain daily level and the pineal increases its production of melatonin that in turn causes the reproductive systems of female and male hamsters to shut down for the winter (note, for example, the decrease in testis size shown at the bottom of the figure). In early spring the reproductive systems spontaneously activate and breeding begins. The effects of melatonin on the reproductive biology of Syrian hamsters have spurred interest in the hormone's possible use as a human contraceptive. *(a, Photo Researchers, Inc.)*

Melatonin's prospects for birth control in humans stem from its observed role in regulating fertility in other mammals and its apparent lack of toxicity. Some hormones currently being used for contraception appear, over the long term, to have carcinogenic effects. Synthetic forms of the hormones estrogen and progesterone, for example, are used in the "combination" birth control pill. Recall that estrogen stimulates development of breast tissue during each reproductive cycle. Recent data suggest that risk of breast cancer is linked to a woman's cumulative lifetime exposure to estrogen. These data, together with melatonin's documented effects on reducing fertility in some animals, have sparked research in Holland devoted to replacing the estrogen in combination birth control pills with fairly high doses of melatonin. The contraceptive success rate of melatonin-progesterone pills appears similar to that of estrogen-progesterone pills, and clinical trials in the United States are scheduled to begin soon. Research is also underway on the possibility of combining melatonin with testosterone to produce a male contraceptive that stops sperm production and at the same time reduces the risk of prostate cancer.

Melatonin also seems to boost immunity and to slow aging. Recall that the thymus gland, responsible for the production of infection-fighting white blood cells called T lymphocytes, is largest in infants and young children and shrinks as we age. Old mice also show shrinkage of the thymus. Such shrinkage can be reversed in geriatric mice by injections of melatonin and the immune systems of injected animals show signs of revitalization. Similarly dramatic results were also obtained in aging research when the pineal glands of ten old mice were switched with the pineal glands of ten young mice. The young mice with the "old" pineals aged rapidly and died in what would normally be middle age. The old mice with the "young" pineals lived about 30% longer than untreated mice. These results led researchers to describe the pineal as the "aging clock," speculating that melatonin is the pineal's way of translating its timekeeping into changes in the body and raising the possibility of melatonin supplements to compensate for aging pineal glands (the amount of melatonin secreted by the pineal declines with age).

Melatonin may also slow aging through its effective scavenging of free radicals, molecular fragments that contain an unpaired electron. Free radicals have been implicated in many chronic diseases of old age such as cancer and heart disease. Substances like melatonin that destroy free radicals are called antioxidants. Free radicals are normally generated by some cells of the body, and their role is to destroy bacteria and old cells. Environmental agents such as drugs, toxins, and radiation also generate free radicals. What evidence is there to support melatonin as an effective antioxidant? Well, to begin with, white blood cells incubated with melatonin sustained 70% less damage from radiation than did untreated cells. Also, in rats, injections of melatonin before and after exposure to a toxic herbicide protected these animals from the horrendous liver and lung damage found in animals that did not receive melatonin treatment. At the very least, these studies raise the possibility that melatonin could be used to protect people from cellular damage caused by the free radicals generated by radiation and environmental toxins.

Melatonin certainly seems to be the answer for many of life's ills. But should consumers be gobbling down melatonin pills and lozenges purchased at health food stores? At present melatonin is being sold as a dietary supplement and thus is not subject to intense scrutiny by the Federal Drug Administration. No one, for example, knows the possible long-term effects of high doses of melatonin. Another concern is that production of the hormone by companies not subjected to regulation raises questions about the strength and purity of the final marketed product. And, questions remain about the benefits of melatonin. Health claims of the miracles of melatonin have been piling up at such a fast rate that the FDA has barely had time to evaluate whether such claims are real. As a result, some people in the medical profession caution that all consumers should wait until melatonin is subjected to further scrutiny and tighter regulation. Others argue, however, that while melatonin is probably safe for most people, certain groups should avoid it until further information is available. Specifically, avoidance of melatonin is urged for pregnant or nursing women (the effects of high doses on fetuses or infants are unknown), children (who normally produce melatonin in large amounts), and those with autoimmune disorders or cancers of the immune system (melatonin stimulates the immune system and thus could worsen these conditions). Finally, given melatonin's apparent effectiveness as a contraceptive agent, women trying to conceive should also avoid it.

Stress

Rarely does a day go by that we are not subjected to some form of **stress**, broadly defined as mental or physical tension. For example, awaiting the start of an exam, some personal performance, or an interview can be stressful. Homeostatic mechanisms can usually deal with everyday stresses and successfully maintain the relative constancy of the body's internal environment. Sometimes, however, stress is extreme in its intensity and duration and homeostatic mechanisms prove inadequate. At such times, stress triggers the hypothalamus to initiate the **general adaptation syndrome**

(GAS), a series of physiological adjustments made by our bodies in response to extreme stress. Rather than maintaining homeostasis, the general adaptation syndrome resets certain internal conditions, such as blood glucose level, to allow the body to respond to the crisis at hand. Interestingly, although **stressors**, the stimuli that produce stress, come in a variety of forms including strong emotional reactions, extreme heat or cold, heavy bleeding, and starvation, our bodies respond to each in the same manner. The basic physiological adjustments to stress constitute the general adaptation syndrome.

The General Adaptation Syndrome

The general adaptation syndrome consists of the alarm reaction and the resistance reaction. Whereas the alarm reaction consists of a series of immediate physiological responses, the resistance reaction kicks in if the stress continues for more than a few hours. If the resistance phase fails to alleviate stress, then exhaustion and possibly death result.

The **alarm reaction**, also known as the **fight-or-flight response**, was discussed earlier in our consideration of the hormones of the adrenal medulla. Recall that this response, initiated primarily by epinephrine from the adrenal medulla, immediately funnels huge amounts of glucose and oxygen to the organs most critical in responding to danger—the brain, the heart, and skeletal muscles—and away from nonessential sites such as digestive and reproductive organs. The increased glucose needed by cells to combat the stressor comes from glycogen transformed by the liver and pumped into the bloodstream. The additional oxygen needed by cells for chemical reactions comes from increases in breathing rate and widening of respiratory passages.

Sometimes the adjustments of the alarm reaction are sufficient to overcome stress. At other times, however, stress caused by extreme anxiety, severe illness, or starvation is so intense and long-lasting that the individual enters the **resistance phase**. In contrast to the short-lived nature of the physiological responses of the alarm reaction, changes wrought by the resistance reaction are more long-term. Also, rather than being stimulated by nerve impulses from the hypothalamus, the resistance reaction is largely initiated by releasing hormones from the hypothalamus. These releasing hormones stimulate the anterior pituitary to secrete (1) adrenocorticotropic hormone (ACTH) that in turn stimulates the adrenal cortex to secrete more cortisol (2) thyroid-stimulating hormone (TSH) that stimulates the thyroid to secrete more thyroid hormone and (3) growth hormone (GH). The overall effects of these hormonal changes include increased availability of ATP from the breakdown of fats and conversion of glycogen to glucose, conservation of body fluids, and the maintenance of normal blood pH despite increased catabolism (Na^+ is conserved and H^+ and K^+ are excreted in urine).

The resistance phase, sustained by the body's fat reserves, may last for weeks or months, but it cannot go on indefinitely. Sooner or later lipid reserves are exhausted and structural proteins must be broken down to meet energy demands. Also, the enhanced excretion of K^+ during the resistance phase eventually disrupts the ability of cells to control water concentration in their cytosol and cells begin to weaken and die. Finally, the heart, adrenal glands, and blood vessels, unable to meet the heavy metabolic and endocrine demands of the resistance phase, begin to fail. This is the **exhaustion phase**, and without immediate attention, death may result from collapse of one or more organ systems. The stages of the general adaptation syndrome are summarized in Figure 6–22.

Stress and Health

Stress, particularly when prolonged and uncontrollable, can have dramatic effects on our health. It is now known that stress depresses wound healing, increases our susceptibility

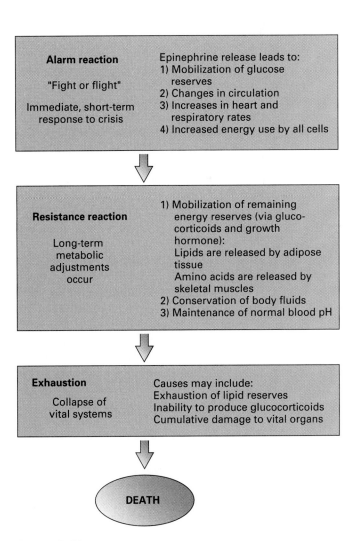

Figure 6–22 The general adaptation syndrome.

to infections by inhibiting components of our immune system, and leads to disorders such as hypertension, gastritis, irritable bowel syndrome, and asthma. Some studies have shown that stress puts people at greater risk for developing chronic diseases and leads to shortened life span. Risk of cardiovascular disease, for example, appears to be enhanced by the circulatory changes that occur as part of the stress response. These changes accelerate the accumulation of cholesterol on the walls of the arteries, increasing the likelihood of heart attack or stroke.

Coping with Stress

Given the connection between stress and health, it is important, when possible, to reduce the stress in our lives and to develop ways to cope with unavoidable stress. A first step in reducing stress is to be realistic in assessing the levels of stress associated with various life events. Several years ago two psychiatrists asked a broad cross-section of people to rank common events, such as marriage, change in residence, and vacation, in terms of stress. On the basis of the responses received, they developed a scale from 1 (low stress) to 100 (high stress), arbitrarily assigning marriage a score of 50 (Figure 6–23). These data were helpful not only in pointing out the many life events that can cause stress but also in allowing people to anticipate stressful situations and develop coping strategies beforehand.

Stress and response to stress, of course, are more complicated than can be summarized in a single chart. Everyone is not affected in the same way by the same life event. When we choose to minimize stress by selecting low stress environments and lifestyles, we are making very personal choices—what is stressful for one individual, say living in a big city, may be enjoyable and challenging for another. Consider what you find stressful and take positive steps to reduce your exposure to stress.

There are several mental and physical strategies for coping with stress. One mental strategy is to develop things in your life that make you feel good and optimistic and help you to function more effectively. Many people achieve such support from developing close relationships with family or friends and from jobs or hobbies that they enjoy. With such support systems in place, people are often better able to cope with stress, feeling that they are part of something larger than themselves. Another option is **biofeedback**, a procedure used to help a person recognize the symptoms of stress and learn how to control them. During a biofeedback session, a health care professional hooks a patient up to a machine that monitors physiological indicators of stress such as heart rate, respiratory rate, or muscle tension (Figure 6–24). The health care worker then discusses a stressful situation with the patient and when the patient begins to show signs of stress, signals are given off by the machine. Increased tension in muscles, for example, might prompt a clicking sound. The patient may be able to decrease muscle tension,

Stress scale	
Life event	**Points**
Death of spouse	100
Divorce	73
Marital separation	65
Jail term	63
Death of close family member	63
Personal injury or illness	53
Marriage	50
Fired at work	47
Marital reconciliation	45
Retirement	45
Change in family member's health	44
Pregnancy	40
Sex difficulties	39
Addition to family	39
Business readjustment	39
Change in financial state	38
Death of close friend	37
Change to different line of work	36
Change in number of marital arguments	35
Mortgage or loan for major purchase (home etc.)	31
Foreclosure of mortgage or loan	30
Change in work responsibilities	29
Son or daughter leaving home	29
Trouble with in-laws	29
Outstanding personal achievement	28
Spouse begins or stops work	26
Starting or finishing school	26
Change in living conditions	25
Revision of personal habits	24
Trouble with boss	23
Change in work hours, conditions	20
Change in residence	20
Change in schools	20
Change in recreational habits	19
Change in church activities	19
Change in social activities	18
Mortgage or loan for lesser purchase (car, TV, etc.)	17
Change in sleeping habits	16
Change in number of family gatherings	15
Change in eating habits	15
Vacation	13
Christmas season	12
Minor violations of the law	11
Total score	_____

Figure 6–23   A stress scale developed by psychiatrists Thomas Holmes and Richard Rahe.   *(Reprinted with permission from the Journal of Psychosomatic Research, 11(2), 1967:213–218, Elsevier Science, Inc.)*

through deep breathing and relaxation, and this decrease can be monitored by decreases in the frequency of clicks. Eventually patients are able to recognize and cope with signs of stress without the help of the machine. Other ways by which coping skills may be improved include obtaining help through counseling or self-help groups. Finally, regular exercise and relaxation training (periodic tightening and relaxation of muscles) can dramatically alleviate the effects of stress.

Figure 6-24 Biofeedback, monitoring the body's physiological response to stress, is one way in which people can learn to recognize the symptoms of stress and how to cope with them. *(© Will and Deni McIntyre/Photo Researchers, Inc.)*

SUMMARY

1. Exocrine glands secrete their products into ducts that open onto the body surface, into spaces within organs, or into a body cavity. In contrast, endocrine glands lack ducts and release their products (hormones) into the spaces just outside cells where they diffuse into the bloodstream. Endocrine glands and organs that contain some endocrine tissue constitute the endocrine system.

2. Hormones, the chemical messengers of the endocrine system, contact virtually all cells within the body but affect only target cells, those cells with receptors that recognize and bind specific hormones.

3. Steroid hormones are lipid soluble (as are thyroid hormones), derived from cholesterol, and secreted primarily by the ovaries, testes, and adrenal glands. Steroids move easily through the plasma membrane of target cells into the cytosol where they combine with a receptor molecule, forming a hormone-receptor complex. This complex moves into the nucleus of the cell where it directs synthesis of specific proteins, including enzymes that may stimulate or inhibit particular metabolic pathways.

4. Water-soluble hormones include amino acid derivatives, peptides, and proteins. These hormones cannot pass through the lipid bilayer of the plasma membrane and thus exert their effects indirectly by activating second messenger systems. The hormone, considered the first messenger, binds to a receptor on the plasma membrane, and this in turn activates a second messenger that carries the hormone's message inside the cell where it changes the activity of enzymes and chemical reactions.

5. Endocrine glands are stimulated to manufacture and release hormones by chemical changes in the blood, hormones released by other endocrine glands, and messages from the nervous system.

6. Hormone secretion is usually regulated by negative feedback mechanisms. Typically, a gland is stimulated to release a hormone and then rising blood levels of that hormone inhibit its further release. Sometimes, however, secretion of hormones is regulated by positive feedback mechanisms whereby release of a hormone causes a change that acts to stimulate further production and release of that hormone.

7. Prostaglandins are lipid molecules released by the plasma membranes of many cells. Rather than traveling to distant sites within the body as hormones do, prostaglandins exert their effects locally by affecting nearby cells or tissues. Prostaglandins are known to influence physiological processes including regulation of body temperature, blood clotting, and contraction of smooth muscle.

8. The pituitary gland is suspended from the hypothalamus at the base of the brain by a short stalk called the infundibulum. The gland has two parts, an anterior lobe and a posterior lobe. The anterior lobe is influenced by the hypothalamus through a circulatory connection known as the pituitary portal system. Nerve cells in the hypothalamus release factors into the portal system, and these factors travel to the anterior lobe where they stimulate or inhibit hormone release.

9. In contrast to the circulatory connection between the hypothalamus and the anterior lobe of the pituitary gland, the connection between the hypothalamus and the posterior lobe is neural. Neurosecretory cells from the hypothalamus extend down into the posterior lobe where they release oxytocin and antidiuretic hormone.

10. The thyroid gland lies over the trachea at the front of the neck. This gland produces thyroid hormone and calcitonin. Thyroid hormone has broad effects, including regulating metabolic rate, heat production and blood pressure, and promoting normal development and functioning of nervous, skeletal, and reproductive systems. Calcitonin maintains low levels of calcium and phosphates in the bloodstream by promoting the absorption of these substances by bone and inhibiting the breakdown of bone.

11. The parathyroid glands are four small masses of tissue attached at the back of the thyroid gland. The parathyroids secrete parathyroid hormone (PTH or parathormone), an antagonist to calcitonin. As such, PTH is responsible for raising blood levels of calcium by stimulating the movement of calcium from bone and urine to the blood.

12. Each of two adrenal glands sits perched on top of a kidney and contains two regions, an outer region known as the adrenal cortex and an inner region called the adrenal medulla. The adrenal cortex secretes a large number of hormones that fall into three groups, the glucocorticoids, mineralocorticoids, and gonadocorticoids. The adrenal medulla, on the other hand, produces epinephrine (adrenaline) and norepinephrine (noradrenaline), two hormones that initiate the fight-or-flight response.

13. In addition to exocrine cells that secrete digestive enzymes into ducts that empty into the small intestine, the pancreas has endocrine cells that secrete the hormones glucagon (increases glucose in the blood) and insulin (decreases glucose in the blood). Type I diabetes mellitus is an autoimmune disorder in which a person's own immune system attacks the insulin-producing cells of the pancreas, causing insulin deficiency. This disorder usually develops in people younger than 20 years old and is treated by daily insulin injections. Type II diabetes mellitus is characterized by a decreased sensitivity to insulin that develops after 40 years of age, is much more common than Type I, and is usually treated through diet, exercise, and weight loss.

14. The thymus gland lies on top of the heart and plays an important role in immunity through its secretion of thymopoietin, thymic humoral factor (THF), and thymic factor (TF), hormones that influence the production and maturation of certain white blood cells.

15. The pineal gland lies at the center of the brain and secretes melatonin, a hormone that appears responsible for establishing biological rhythms, triggering sleep, inhibiting fertility, and slowing aging.

16. Intense long-term stress triggers the general adaptation syndrome (GAS), a series of physiological adjustments that reset internal conditions. Initially the body responds to stress by means of the alarm reaction (also known as the fight-or-flight response). This reaction is initiated primarily by epinephrine from the adrenal medulla. If the stress is severe and persists for more than a few hours, then the individual may enter the next phase, the resistance reaction. Releasing hormones from the hypothalamus stimulate the anterior pituitary to release ACTH, TSH, and GH. The heavy metabolic and endocrine demands of the resistance reaction cannot be sustained forever, and death may result from failure of one or more organ systems. This final stage of the GAS is called the exhaustion phase.

17. Stress has dramatic effects on health that include inhibiting wound healing, depressing the immune system, increasing risks for developing chronic diseases, and shortening life span. In view of the connection between stress and health, we should take steps to reduce the stress in our lives and to develop mental and physical strategies to cope with unavoidable stress.

REVIEW QUESTIONS

1. How do endocrine glands differ from exocrine glands? Give examples of each.
2. Given that hormones contact virtually all cells in the body, why are only certain cells affected by a particular hormone?
3. How do lipid-soluble and water-soluble hormones differ in their mechanism of action?
4. What stimuli prompt release of hormones?
5. Compare negative and positive feedback mechanisms with regard to regulation of hormone secretion. Provide an example of each.
6. What are prostaglandins?
7. How do the anterior and posterior lobes of the pituitary gland differ in size, development, and relationship with the hypothalamus?
8. List the hormones secreted by the anterior pituitary and their functions.
9. List the hormones secreted by the posterior pituitary and their functions.
10. What are the effects of thyroid hormone?
11. Describe the feedback system by which calcitonin and parathyroid hormone regulate levels of calcium in the blood.
12. What are the major functions of the glucocorticoids, mineralocorticoids, and gonadocorticoids secreted by the adrenal cortex?
13. What is the fight-or-flight response? Which hormones are critical in initiating this response?
14. What hormones are secreted by the pancreas? What are their functions?
15. Explain the differences between Type I and Type II diabetes mellitus.
16. What is the basic function of hormones secreted by the thymus gland?
17. What roles might melatonin play in the body?
18. Describe the stages of the general adaptation syndrome.
19. How does stress affect health?
20. What steps can we take to reduce the stress in our lives?

Neurons: The Matter of the Mind

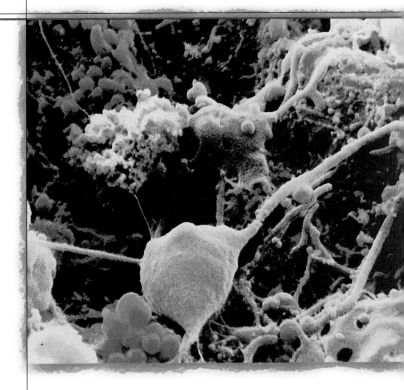

This colored scanning electron micrograph shows the cells of the brain. The large gray cells are neurons. The reddish orange cells are glial cells, which form a special connective tissue that nourishes and supports the neurons.   (*Prof. P. Motta/Dept. of Anatomy/University "La Sapienza," Rome/Science Photo Library/Custom Medical Stock Photo*)

Who are you? You may have wondered about that yourself. As you struggle to come up with an answer, you are relying on your nervous system, particularly the brain. And a wonderful structure it is. Within the human brain there are over 100 billion nerve cells that interact in ways that make you uniquely you. The intricate connections among neurons give rise to your thoughts and desires, consciousness and creativity, and reason and intelligence. The interactions among these special cells permit such things as learning and memory—both central to your sense of who you are. The nervous system orchestrates all your intricate movements, such as chewing on a pencil while you think about the question, and it keeps all body systems operating smoothly while you do so, without your giving it a thought. Furthermore, the sensations handled by the nervous system provide a window to the outside world. Additionally, the communication among the neurons makes possible a rich rainbow of emotions, such as your puzzlement at not being able to come up with an answer.

We will explore the universe of the mind by looking more closely at the cells from which it is built. We will next consider how those cells work and how they communicate with one another. Finally, we will consider how they are organized into networks. In the next chapter, we will see how the neural networks are organized into specific, coordinated pathways.

An Overview of Nervous System Functioning

The nervous system allows you to do many complex actions. Most actions have only three basic components: sensory input, integration, and motor output (Figure 7–1). Sensory input occurs when sensory receptors, for instance the light-detecting cells in the eye, send neural messages over sensory nerve cells to the brain or spinal cord. Nerve cells within the brain and spinal cord then "interpret" or "weigh" the sensory signals, thereby "deciding" on the appropriate motor response. Motor output consists of signals from the brain or spinal cord to the appropriate effectors, muscles or glands, that will carry out the instructions from the command center.

The Cells of the Nervous System

The nervous system has two types of specialized cells: There are **neurons** (nerve cells), excitable cells that generate and transmit messages, and **glial cells**, supporting cells that outnumber the neurons by about 10 to 1.

Glial Cells

There are several types of glial cells within the nervous system, each with different jobs to do.

1. **Structural Functions.** Some glial cells help hold neurons together. Indeed, glial cells get their name from the Greek word for glue. In addition, glial cells create a three-dimensional scaffolding that provides structural support for the neurons of the brain and spinal cord.
2. **Protection.** In the brain, one type of glial cell has numerous processes that wrap around the capillaries so completely that they help form a living barrier, called the **blood-brain barrier,** that slows or prevents the passage of unwanted material from the blood to the neurons. The blood-brain barrier often frustrates physicians by keeping potentially life-saving, infection-fighting, or tumor-suppressing drugs away from brain tissue. Another type of glial cell protects the central nervous system from disease by engulfing disease-causing organisms and clearing away dead cells.
3. **Nurture of Neurons.** Glial cells also provide a steady supply of chemicals called nerve growth factors that stimulate nerve growth. Without nerve growth factors, neurons die. It has been suggested that glial cells may be important in preventing certain degenerative nerve diseases, such as Alzheimer's disease and Parkinson's disease (discussed in the next two chapters), by promoting the growth and well-being of neurons.
4. **Housekeeping functions.** The activity of neurons alters the environment around them. Glial cells "tidy up" after the neurons, maintaining the proper environment for neuronal functioning. One way they do this is by controlling the types of ions surrounding neurons. As we will see, the proper ionic environment is essential to nerve cell functioning. Glial cells also "mop up" the chemicals the neurons use to communicate with one another, called neurotransmitters. If one of these neurotransmitters, glutamate, were to accumulate, the neurons would be killed.
5. **Insulation of nerves.** A particularly interesting role of glial cells is the formation of insulating sheaths around the long projections extending from certain nerve cells. This sheath, called the myelin sheath, has several important roles and will be discussed in more detail later in this chapter.
6. **Development of the nervous system.** As the nervous system develops, strands of glial cells guide neurons along their proper routes so that the correct connections are made.

Neural messages are sent from the eyes to the brain informing the brain of the position of the cup.

Motor messages are sent from the brain to the arm that cause the arm to move toward the cup.

Sensory messages from touch receptors in the hand inform the brain that contact has been made with the cup.

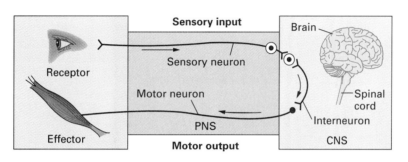

The brain sends motor impulses that cause the hand to grasp the cup and the arm to raise it to the mouth.

Sensory impulses from the eyes and arm keep the brain informed of the cup's exact position as it is raised.

Sensory impulses from the lips tell the brain the cup has reached the mouth.

Figure 7-1 Drinking a cup of coffee involves sensory input from the receptors, integration of that input by the brain, and motor output from the brain that directs the activity.

7. **Communication.** Glial cells may also have a part in the communication that goes on within the nervous system. The molecules on the surface of glial cells suggest that they are sensitive to the messages of neurons. Although glial cells cannot generate their own signals equivalent to those of neurons, some observations suggest that glial cells may be able to pass certain impulses from one to the next and may even be able to influence the messages of neurons in some small way.

Unlike neurons, glial cells are able to reproduce. In fact, when neurons die, glial cells multiply, filling up the vacant space. A consequence of this trait is that most brain tumors are formed by the uncontrolled division of glial cells.

Neurons

Now we will take a closer look at the neurons themselves. As we have seen, the basic unit of the nervous system is the nerve cell, or neuron. Neurons are responsible for an amazing variety of functions, but we can group these into three general categories: (1) **Sensory** (or **afferent**) **neurons** conduct information from the receptors *toward* the central nervous system. (2) **Motor** (or **efferent**) **neurons** carry information *away from* the central nervous system to an effector, either a muscle or gland. (3) **Interneurons**, also called **association neurons**, are located between sensory and motor neurons. They are found only within the central nervous system, where their job is to integrate information. Accounting for over 99% of the body's neurons, interneurons are by far the most numerous nerve cells in the body.

Structure of Neurons

Nerve cells are specialized for communicating with other cells and their shape reflects their functions (Figure 7–2).

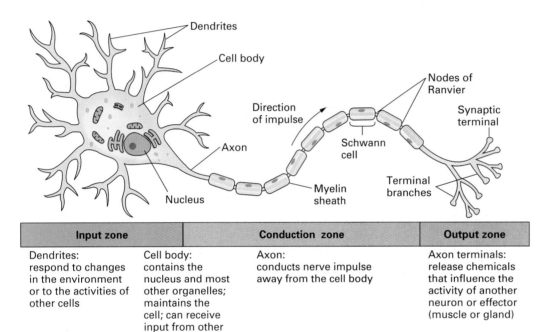

Input zone		Conduction zone	Output zone
Dendrites: respond to changes in the environment or to the activities of other cells	Cell body: contains the nucleus and most other organelles; maintains the cell; can receive input from other neurons	Axon: conducts nerve impulse away from the cell body	Axon terminals: release chemicals that influence the activity of another neuron or effector (muscle or gland)

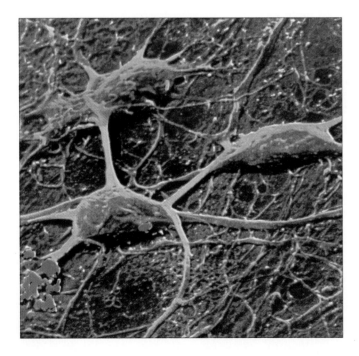

Figure 7–2   Generalized neurons. The highly branched dendrites form a vast receptive surface. Dendrites transmit information toward the cell body. The cell body contains the organelles needed to keep the cell alive and functioning. The cell body may also receive input from other neurons. If a neuron transmits a message to another cell, the information is conducted away from the cell body along the axon. Thus, whereas the dendrites and cell body are the receiving portions of the cell, the axon is the sending portion.
(Photo, © Secchi-Lecaque/Roussel-UCLAF/CNRI/Science Photo Library/Photo Researchers)

Numerous short branching projections from the nerve cell, called **dendrites**, create a huge surface for receiving signals from other cells. The information then travels toward an enlarged central region called a **cell body**, which has all the normal machinery, including a nucleus, for maintaining a cell. When a neuron responds to input, it transmits its message along the **axon**, a single long extension from the cell body, toward either another neuron or an effector, such as a muscle or a gland. In some cases, the axon allows the neuron to communicate over long distances. The end of the axon branches and the tip of each branch forms an **axon terminal** that releases a chemical (neurotransmitter) that alters the activity of the target cell.

In order to appreciate the dimensions of a neuron, we will pretend for a moment that there is such a thing as a "typical neuron" and choose as our example a motor neuron, one that carries a message from the spinal cord to a muscle. Imagine an enlarged cell body of a typical neuron to be about the size of a tennis ball. The axon of this neuron could be about a mile long and about one half-inch wide. The dendrites, the shorter but more numerous projections of the neuron, would fill an average-sized living room.

There is tremendous variation in the sizes and shapes of neurons. In spite of this anatomical diversity, neurons have been classified according to their structure (Figure 7–3). The "typical" neuron we have just described is an example of a **multipolar neuron**, one that has at least three processes—an axon and a minimum of two dendrites. Most motor neurons and association neurons are multipolar. Most sensory neurons, however, are **unipolar**, which means that there is a single, continuous process and the cell body lies off to one side. The long process is the axon and the dendrites are branches at one of the tips of the process. The two processes of a **bipolar** neuron, the axon and the dendrite, extend from opposite sides of the cell body. Bipolar neurons are receptor cells found only in some of the special sensory organs, such as in the retina of the eye and in the olfactory membrane of the nose.

Parallel axons from many neurons are often bundled together, forming a tract called a **nerve**. Nerves are covered with tough connective tissue.

The Myelin Sheath

Most of the axons outside the central nervous system and some of those within are enclosed in an insulating layer called the **myelin sheath**. The myelin sheath is composed of multiple wrappings of the plasma membrane of glial cells. Within the brain and spinal cord, the myelin sheath is formed by a type of glial cell called an **oligodendrocyte**. In the peripheral nervous system (the part outside the brain and spinal cord), the type of glial cell that forms the myelin sheath is the **Schwann cell**. A Schwann cell wraps around the axon many times, forming a spiral of membrane that looks somewhat like a jelly roll (Figure 7–4). Thus, the myelin sheath serves as a kind of living electrical tape that insulates individual axons, preventing messages from short-circuiting between neurons. The myelin sheath is kept alive by the nucleus and cytoplasm of the Schwann cell, which are squeezed to the periphery as the sheath forms.

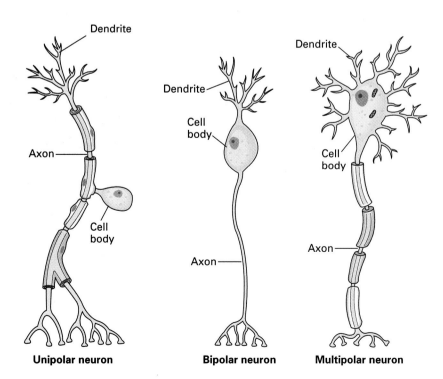

Unipolar neuron **Bipolar neuron** **Multipolar neuron**

Figure 7–3 Structural variations among neurons.

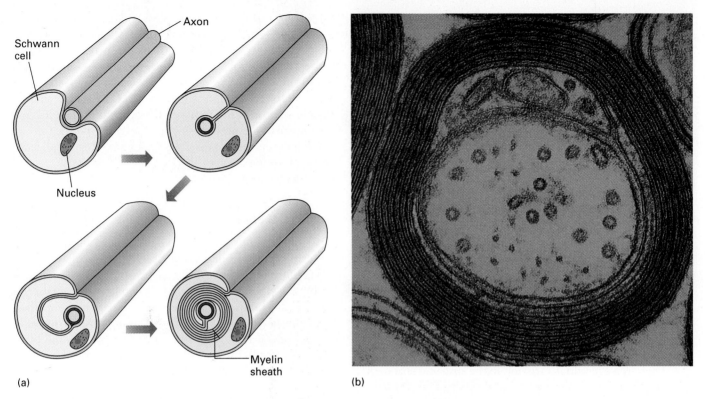

Figure 7–4 The myelin sheath forms from multiple wrappings of Schwann cells. (a) Diagrams depicting the formation of the myelin sheath. (b) An electron micrograph of the cut end of a myelinated axon. The myelin sheath greatly increases the rate at which nerve impulses are conducted, insulates neurons from one another, and plays a role in the repair of damaged nerves. *(b, © C. Raines/Visuals Unlimited)*

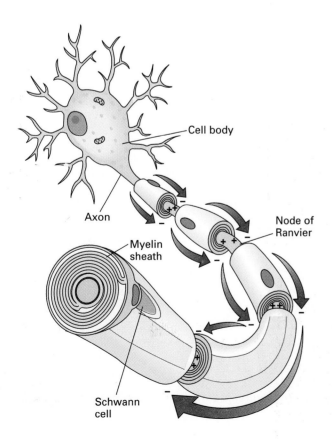

A single Schwann cell encloses only a small region, about 1 mm, of an axon. Along the longest axons in your body, those that conduct impulses from the spinal cord to your toes for instance, there may be as many as 500 Schwann cells. The regions along an axon between adjacent Schwann cells are exposed to the extracellular environment. Such an unsheathed region is called a **node of Ranvier**. This arrangement is very important to the speed at which a neuron transmits messages because, with the myelin sheath in place, a nerve impulse "jumps" successively from one node of Ranvier to the next. This type of transmission, called **saltatory conduction** (*saltare*, jump), is up to 100 times faster than signal conduction would be on an unmyelinated axon of the same diameter (Figure 7–5). A nerve impulse races along certain myelinated axons at 100 meters per second (over 200 miles per hour). For this reason, the axons

◀ Figure 7–5 A space between adjacent Schwann cells where the axon's membrane is exposed is called a node of Ranvier. In myelinated axons, the nerve impulse "jumps" from one node of Ranvier to the next. This mode of nerve impulse transmission, called saltatory conduction, is many times faster than impulse transmission along unmyelinated axons.

that are involved in conduction of signals over long distances are typically myelinated.

It is easy to see how this "jumping" mode of transmission increases the speed at which the message travels. We frequently see a similar effect during a basketball game. With seconds left in the half, dribbling the ball the length of the court would take too much time. Passing through a series of players is faster. Likewise, passing the impulse from one node to the next, as occurs in myelinated nerves, is faster than traveling the length of the axon, as occurs in unmyelinated nerves.

The myelin sheath also plays a role in helping the repair of a cut or crushed nerve (Figure 7–6). When a nerve in the peripheral nervous system is cut, the part that has been separated from the cell body begins to degenerate within a few minutes of the injury because it can no longer receive life-sustaining materials from the cell body. Several days after the injury, the separated region of axon is completely fragmented and the pieces are engulfed by Schwann cells. Once the Schwann cells have cleared all the axon fragments from the myelin sheath, which usually occurs within a week, the axon stump begins "sprouting" from the cell body, stimulated by chemicals released from the Schwann cells. The Schwann cells divide and bridge the gap, aligning themselves to form a tube that will guide the regenerating axon to the previous target cell.

If normal functioning is to be restored after a nerve is severed, the cut ends must be properly aligned within a millimeter or two of one another and there can't be any obstacles, such as scar tissue, between them. Unfortunately, these conditions are not always met. Neurosurgeons can assist the repair process by trimming the damaged nerve ends and sewing them together, but the procedure is not always successful. Furthermore, in addition to severing nerves, accidents often cause them to be crushed or destroyed. If too long a section is missing, it may not be possible to bring the intact ends close enough together to allow them to regenerate, because axons do not stretch easily.

There is hope on the horizon, however. Researchers have developed a synthetic sleeve that can guide axons across

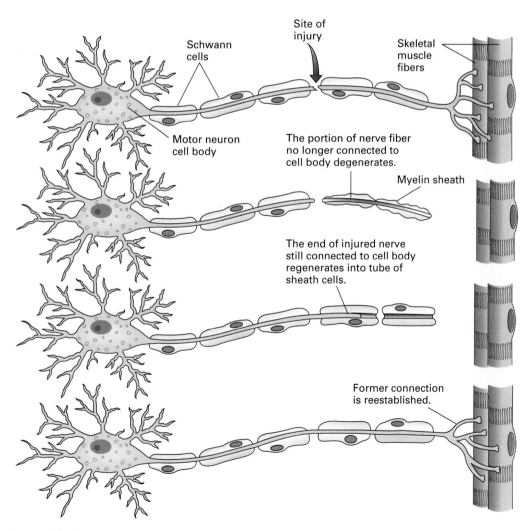

Figure 7–6 The myelin sheath helps repair a damaged nerve.

gaps ten times greater (20 mm) than is possible without assistance. Although this technique has yet to be tried on humans, it has been successful in rats and it is now being tried on nonhuman primates.

Axons within the brain and spinal cord do not regenerate. The glial cells that form the myelin sheath in these areas do not form tunnels to guide regenerating axons to their proper connections. Instead, the glial cells die and are replaced by scar tissue that blocks regeneration. As a result, injuries within the central nervous system (the brain and spinal cord) cannot be repaired. Researchers are investigating ways of stimulating nerve regeneration within the spinal cord by using chemicals that stimulate nerve growth, called nerve growth factors, and ways of suppressing the release of chemicals that inhibit nerve growth by the cells within the spinal cord. Perhaps someday these techniques will help people with spinal cord injury to walk again.

The importance of the myelin sheath becomes dramatically clear in people with **multiple sclerosis**. This disease, which affects 250,000 Americans, involves the progressive destruction of myelin sheaths within the brain and spinal cord (Figure 7–7). The damaged regions of myelin then become hardened scars called scleroses (hence the name of the disease). The scleroses interfere with the transmission of

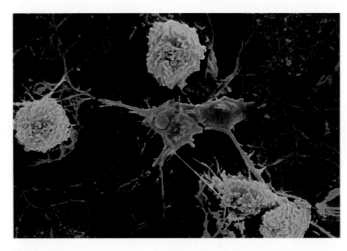

Figure 7–8   Myelin-forming cells (oligodendrocytes), shown in purple, are being attacked by the brain's immune cells, shown in yellow-orange. Observations such as this support the idea that multiple sclerosis is an autoimmune disease.   *(Dr. John Zajicek/ Science Photo Library/Photo Researchers, Inc.)*

nerve impulses and allow short-circuiting between normally unconnected conduction paths, which delays or completely blocks the signals from one brain region to another. The result can be paralysis or the loss of sensation, depending on the part of the brain affected. The first symptoms of multiple sclerosis generally begin in young adulthood, around 20 to 30 years of age. Depending on the region of the brain or spinal cord affected, the symptoms may include partial or complete paralysis of parts of the body, a loss of coordination, dizziness, or sensory disturbances such as blurred vision or numbness. The variety of possible symptoms makes multiple sclerosis difficult to diagnose. The symptoms occur in waves, with periods of remission between new attacks on myelin. Each time lesions form, some of the neurons are damaged by the hardening of their myelin sheaths. Others, however, escape without permanent injury and as these recover the symptoms may sporadically disappear. Unfortunately, as more neurons are damaged, the symptoms get progressively worse.

Multiple sclerosis is caused by an autoimmune response in which the body's own defense mechanisms attack myelin (Figure 7–8). It is not clear what triggers the attack, although an idea that is gaining support is that a previous infection with the measles virus may be the culprit.

Although there is currently no way to halt the slow progression of the disease, researchers are actively searching for a cure. Some researchers are looking for the chemical that triggered myelination when the brain originally developed. Because nerve cells are not irreparably damaged as soon as the myelin is stripped away, it is hoped that normal functioning could be restored if new myelin could be introduced early in the disease. Other researchers are trying to coax the body's defense system into tolerating proteins that it consid-

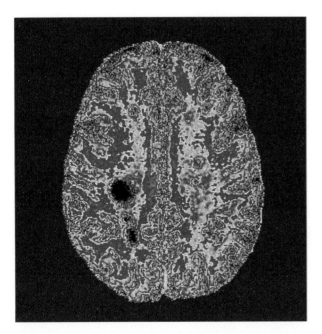

Figure 7–7   Multiple sclerosis is a disease in which the body's immune system attacks the myelin sheaths within the brain or spinal cord. The affected area can be seen as dark regions on the lower left side in this magnetic resonance image (MRI) of the brain of a person with multiple sclerosis. The symptoms vary depending on the region of the central nerv-ous system that is affected. Paralysis results when motor areas of the brain are affected. Loss of sensation results when sensory brain regions are affected.   *(Department of Clinical Radiology, Salisbury District Hospital/Science Photo Library/Photo Researchers, Inc.)*

ers to be foreign. Trial studies in which half the patients swallowed a capsule containing myelin from cows and the other half swallowed capsules containing a protein from cow's milk were promising. There were significantly fewer major attacks of multiple sclerosis in the myelin-treated group than in the control group.

Critical Thinking

A low-fat diet is generally recommended for adults to reduce the risk of heart attack and stroke. Infants and toddlers have rapidly developing nervous systems. Why do you think that a low-fat diet is not recommended during the first few years of life?

The Nerve Impulse

We said earlier that neurons are specialized for communication, so next we will consider the nature of the message, which is called a nerve impulse or action potential. The **action potential** is an electrical signal generated when positively charged ions (atoms that have lost an electron) rush across the neuron's plasma membrane.

The key to the informational signals of a neuron is its membrane. Like most living membranes, the nerve cell plasma membrane is selectively permeable, meaning that it allows some substances through but not others. Furthermore, the membrane controls the passage of substances that are permeable. The membrane contains many pores, called channels, that provide the only means for ions to cross it without using cellular energy (Figure 7–9). Each channel is

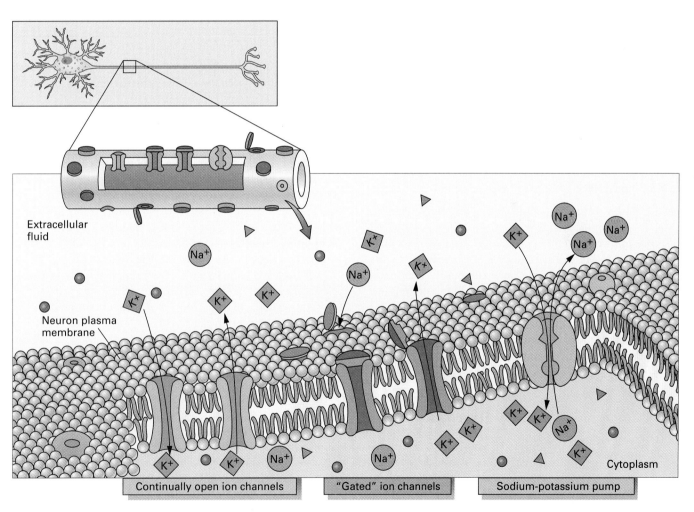

Extracellular fluid

Neuron plasma membrane

Cytoplasm

| Continually open ion channels | "Gated" ion channels | Sodium-potassium pump |

Figure 7–9 The plasma membrane of a neuron contains two types of pathways through which ions can pass. One type of pathway is an ion channel. Ions pass through ion channels without using cellular energy. Ion channels are protein-lined pores through the membrane that allow one or very few types of ions to cross the membrane. Some ion channels are normally open. Others have a molecular "gate" that is opened to allow ions to pass through or closed to prevent them from crossing the membrane. The second pathway, the sodium-potassium pump, uses cellular energy to pump three sodium ions out of the cell while pumping two potassium ions into the cell.

specific for one or very few types of ions. Thus, ion channels function as molecular sieves, selecting which ions can cross the membrane. Some of the channels are normally open, while the opening of other channels is regulated by a gate. A change in the shape of the protein comprising the gate either opens the channel, allowing ions to pass through, or closes it, preventing ions from crossing the membrane.

When permitted to cross the membrane through ion channels, ions move passively in response to two forces. The first force, the ion's **concentration gradient**, will tend to push the ions away from a region where that type of ion is more concentrated to an area where it is less concentrated. The second force is an **electrical gradient** that pushes ions away from others with a similar charge and attracts them to ions with an opposite charge. Positive charges repel positive charges, but attract negative charges. During certain times, the concentration and electrical gradients act in the same direction and their effects are additive. At others, the forces are opposing and the effect they have on the movement of ions is a compromise.

In addition to moving passively through specific ion channels, sodium and potassium ions are also moved actively across the membrane by transport proteins embedded in the membrane called **sodium-potassium pumps**. These pumps use cellular energy in the form of ATP to pump ions against their concentration gradients. Typically, each pump ejects three sodium ions from the cell while bringing in two potassium ions.

The Resting State

It will be easier to understand how and why the ions move during an action potential if we first consider a neuron that is not transmitting an action potential, that is, a neuron in its **resting state**. As we will see, however, resting is hardly the word to describe what is going on at this stage. The membrane of a resting neuron maintains a charge difference across its surface such that the inside surface is more negative than the outside. This charge difference results from the unequal distribution of ions across the membrane. We describe the membrane of a neuron in this state as being **po-larized**. The separation of positive and negative charges constitutes an **electrical potential difference** that can be used to do work. A potential difference can cause the ions to flow across the membrane, thereby generating an electric current. This separation of charge is called the **resting potential** of the neuron.

The three types of ions most important in establishing the resting potential are sodium ions (Na^+), potassium ions (K^+), and certain negatively charged ions to which the membrane is impermeable (Figure 7–10a). These negatively charged ions are inside the cell and because the membrane is not permeable to them, they are trapped inside. The sodium and potassium ions are also unequally distributed across the membrane of the resting neuron. There are roughly ten times more sodium ions outside the membrane than inside and there are about 30 times more potassium ions inside than outside. The result of these ion distributions is that the inner surface of a resting neuron's membrane is typically about 70 millivolts (mv) more negative than the outer surface. This is about 5% of the voltage in a size AA flashlight battery.

The membrane is the key to establishing and maintaining the unequal distribution of ions. When a neuron is at rest, most of its gated ion channels are closed. Furthermore, the great majority of the normally open ion channels only let through potassium ions. As a result, the membrane lets potassium ions through fairly easily but is not very permeable to sodium ions. Recall that ions move in response to the balance of the electrical gradient and the concentration gradient. The negative charge within the cell pulls K^+ inward, but the K^+ concentration gradient slightly exceeds the electrical force. When the neuron is resting, then, there is a small outward driving force on K^+. Sodium ions that leak in are pumped out by sodium-potassium pumps, while K^+ ions are moved back in. At this point, the interior of the neuron is still negative because as K^+ ions leak out, the negatively charged ions are held inside and cannot follow.

So, we will call it "resting," but neurons at this stage are quite active. Although neurons consume a lot of energy to maintain the resting potential, the energy is not wasted. The resting potential allows the neuron to be ready to respond

Figure 7–10 The resting state and the propagation of an action potential along an axon. (a) ▶ During the resting state, the membrane is polarized; that is, the inner surface of the membrane is more negative than is the outer surface. (b) Depolarization is caused by sodium ions entering the cell. A local disturbance may open gates to some sodium channels in a certain region of the membrane. When this occurs, Na^+ ions enter the cell and begin to reduce the negative charge within. When this disturbance causes the voltage difference across the membrane to reach a critical value, called threshold, many voltage-sensitive sodium gates open. Sodium ions then flood to the inside, reversing the voltage difference across the membrane. Within 0.5 msec, the gates automatically snap shut. (c) Repolarization is caused by potassium ions leaving the cell. The voltage changes caused by sodium entering the cell trigger the opening of voltage-sensitive potassium gates. Potassium ions then exit the cell, restoring a negative charge to the cell's interior. (d) The sodium-potassium pump then restores the original distributions of sodium and potassium ions.

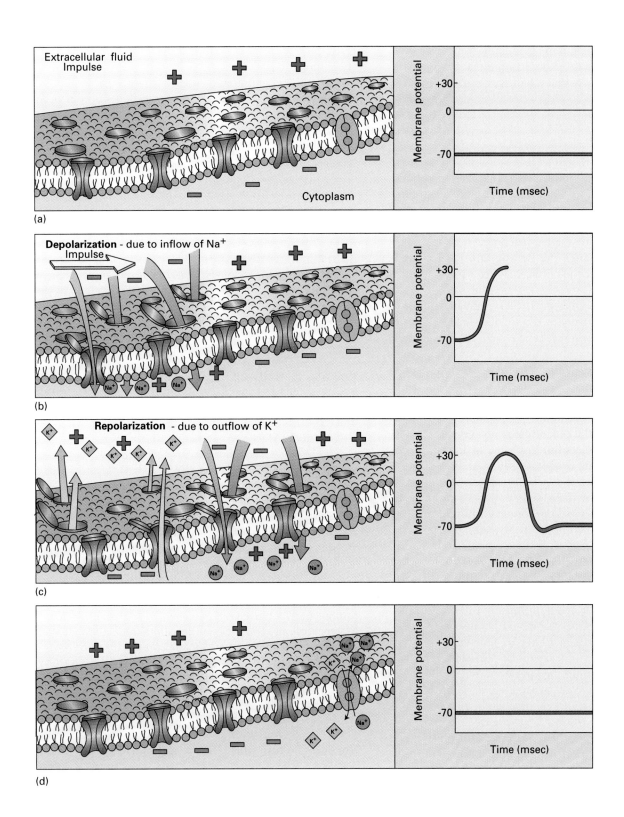

(a)

Depolarization - due to inflow of Na⁺

(b)

Repolarization - due to outflow of K⁺

(c)

(d)

more quickly than it could if the membrane were electrically neutral in its resting state. This is somewhat analogous to charging a car's battery so that it will start as soon as the key is turned.

The Action Potential

What happens, then, when a neuron is stimulated, as might occur when another neuron releases a chemical called a neurotransmitter? Shortly, we will consider in more detail some of the ways in which a neuron may be stimulated to fire. Here we will gloss over that part of the story and consider the effect of an excitatory signal reaching a neuron's membrane. Sodium ions enter the neuron and their positive charge begins to reduce the negative charge within, an event called **depolarization**.

The action potential begins when a great number of sodium gates in the vicinity of the stimulus are opened and membrane depolarization reaches a certain value, called the **threshold**. The opening and closing of these sodium gates depends on the voltage, that is, the charge difference across the membrane. For this reason, these channels are called **voltage-sensitive channels**. Sodium ions flood inward through the open gates in numbers sufficient to set up a net positive charge inside the cell in that region (Figure 7–10b). Despite the reversed electrical gradient, sodium ions continue to enter for a brief instant, driven by their concentration gradient. However, the inward rush of sodium ions is quickly halted because these sodium gates automatically snap shut (or inactivate) within about half a millisecond.

Repolarization, the return of the membrane potential to close to its resting value, follows immediately due to potassium ions leaving the cell. As depolarization nears its peak, membrane permeability to potassium ions increases as voltage-sensitive gates to potassium channels open. Potassium ions now leave the cell both because they are repelled by the net positive charge within and because they follow their concentration gradient. (They are much more concentrated within the cell.) The exodus of potassium ions with their positive charge causes the interior of the neuron to become negative relative to the outside again. Thus, with the outward flow of potassium ions the membrane potential is returned to close to its resting value (Figure 7–10c).

The action potential, then, is a wave of depolarization caused by the inward flow of sodium ions, followed immediately by a wave of repolarization caused by the outward flow of potassium ions. These changes actually involve a small number of ions. The concentrations of specific ions inside the cell usually change by less than 0.1% and outside the cell the change is too small to measure. Notice that at the end of an action potential, the charge distribution across the membrane is returned to the resting potential, but inside the cell there are slightly more sodium ions and slightly fewer potassium ions than before. This is corrected as the sodium-potassium pump restores the original ion distribution (Figure 7–10d). The action of the sodium-potassium pump

is slow and does not contribute directly to the events of the action potential.

The action potential is described as a wave of depolarization and repolarization because the events do not occur simultaneously along the entire length of the cell. Instead, depolarization in one region of the cell triggers the opening of sodium gates in an adjacent region. Then, as sodium enters in this region, the gates in the next adjacent region are opened. As a result, depolarization and repolarization spread along the axon. Therefore, once started, neural impulses do not diminish, just as the last domino in a falling row falls with the same energy as the first. And, the intensity of the nerve impulse doesn't vary with the strength of the triggering stimulus. If a nerve cell fires at all, the action potential is always of the same intensity. The "all-or-none" principle of nerve cell conduction is similar to the firing of a gun in that the force of the bullet is not changed by how hard you pull the trigger.

Refractory Period

Immediately following an action potential, the neuron cannot be stimulated again for a brief instant, called the **refractory period**. During this period the sodium channels are said to be inactivated (closed and unable to be opened by depolarization). Moreover, potassium channels are still open. Consequently, a second action potential cannot be generated until a millisecond after the first ends. Up to a point, the frequency of impulses increases with increasing strength of the stimulus.

Critical Thinking

"Red tides" are caused by a bloom of dinoflagellates, single-celled marine protozoans. These protozoans contain a chemical called saxitoxin (STX). Extremely small concentrations of STX block voltage-sensitive sodium channels. Clams, scallops, and mussels consume the dinoflagellates. Because the shellfish are insensitive to the toxin, STX accumulates in their tissues. What effect would you expect STX to have on nerve transmission in humans who accidentally consume tainted shellfish?

Encoding of Stimulus Quality and Intensity

All action potentials in a particular neuron are basically identical. This fact raises at least two questions about the way in which neurons function. One of these questions is how are we able to distinguish among environmental stimuli that are qualitatively different from one another. In other words, how do we tell the difference between light and sound? How do we know whether we are touching or tasting the stimulus? The answer is that the information about the presence of a particular stimulus, say light, is picked up by receptors specialized to detect only that type of stimulus and the messages sent by those receptors are routed along their own pathways to separate systems within the brain. For example, signals from the eyes travel over separate pathways and end in a different region of the brain than do signals from the ears. Whenever signals are generated within the visual system, the brain interprets them as light. Thus, if you press gently on your closed eyelid, the pressure usually stimulates visual receptors, as well as pressure receptors, and you see a spot of colored light where your finger pressed your eyelid.

A second question arising from the fact that all action potentials generated in a neuron are the same intensity is how we are able to distinguish varying intensities of stimuli. How do we determine whether our companion is whispering or shouting? Actually, the brain often has two cues regarding stimulus intensity—the frequency of action potentials and the number of neurons stimulated. Up to a point, the number of action potentials moving along the axon in a given amount of time (the frequency) increases with the intensity of the stimulus (in this case, the loudness of the sound). Also, different neurons have different thresholds. Because some neurons are easily stimulated and others harder to fire, the number of neurons stimulated increases with the strength of the stimulus.

Synaptic Transmission

When a nerve impulse reaches the end of an axon, in almost all cases the message must be relayed to the adjacent cell across a small gap. Although the gap is just billionths of a meter wide, the impulse cannot spark across it. Communication with the adjacent cell requires a change in the nature of the message. The action potential in the first cell causes the release of a chemical from the tips of its axon. That chemical, called a **neurotransmitter**, travels across the gap, usually in less than half a millisecond, and conveys a message to the adjacent cell.

The junction between a neuron and another cell is called a **synapse**. The structure of a synapse between two neurons is shown in Figure 7–11. The gap between the cells is called the **synaptic cleft**. Recall that the axon branches near the end of its length. Each branch ends with a small bulb-like swelling called an axon terminal or **synaptic knob**. Within the synaptic knobs of the neuron sending the message, the **presynaptic neuron** (before the synapse), are tiny sacs containing between 10,000 and 100,000 molecules of a neurotransmitter. These sacs are called **synaptic vesicles**. When the nerve impulse reaches a synaptic knob, it causes a temporary change in the knob's permeability. (It does this by opening voltage-sensitive calcium gates.) Driven by their concentration gradient, calcium ions rush into the knob. There they activate enzymes that cause the membranes of the synaptic vesicles to fuse with the plasma membrane and dump the enclosed neurotransmitter into the synaptic cleft. The neurotransmitter then diffuses across the gap and binds with receptors on the membrane of the **postsynaptic neu-**

ron (after the synapse), the neuron receiving the message. A receptor is a protein that is specific for a particular neurotransmitter, recognizing the neurotransmitter much as a lock "recognizes" a key. A neurotransmitter can only affect a cell that has receptors specific for it. Thus, only certain neurons can be affected by a specific neurotransmitter. Furthermore, the more receptors on the postsynaptic cell, the more vigorous its response. When the neurotransmitter binds with the receptor, the interaction triggers a response. The ways that neurotransmitters affect the postsynaptic cell fall into two main categories: ionic and metabolic. In some cases, a particular neurotransmitter acts through both mechanisms by binding to different receptors that are specific for that neurotransmitter.

Direct Ionic Synaptic Interactions

When a neurotransmitter with direct ionic effects binds to its receptor, an ion channel is opened. The effect this will have on the cell receiving the message depends on the type of ion channel opened. Note that it is really the receptor that determines which ion channels will open and, therefore, the effect a given neurotransmitter will have. If it is an **excitatory synapse**, the likelihood of an action potential beginning in the postsynaptic cell is increased, because binding of the neurotransmitter to its receptor opens channels that allow sodium ions to enter. The result is a temporary local depolarization of the postsynaptic membrane. If enough receptor sites bind with neurotransmitter to cause depolarization to threshold value, an action potential is generated in the postsynaptic cell. Thus the localized graded depolarization of the postsynaptic membrane acts like pressure on the trigger of a gun. Slight pressure on the trigger won't cause the gun to fire, but when the pressure on the trigger reaches a critical threshold level, the bullet is always discharged from the gun with the same force.

In contrast, activity at an **inhibitory synapse** decreases the likelihood that an action potential will be generated in the postsynaptic neuron. The postsynaptic cell is inhibited because its resting potential becomes more negative than usual. Consequently, it will require more neurotransmitter with an excitatory effect than usual to reach threshold. The inhibitory effect is generally caused either by the opening of chloride gates, which permit chloride ions (Cl^-) to enter the cell, or by the opening of potassium gates, which allow potassium ions (K^+) to leave the cell. Either way, the inner surface of the membrane is left more negative than it was previously.

Indirect Synaptic Interactions

Certain neurotransmitters can affect membrane potential indirectly by affecting the level of other chemicals, known as second messengers. In many cases, the second messenger is cyclic AMP. The role of the second messenger can be to open ion channels, which results in changes in membrane potential. However, this is not the only possible action.

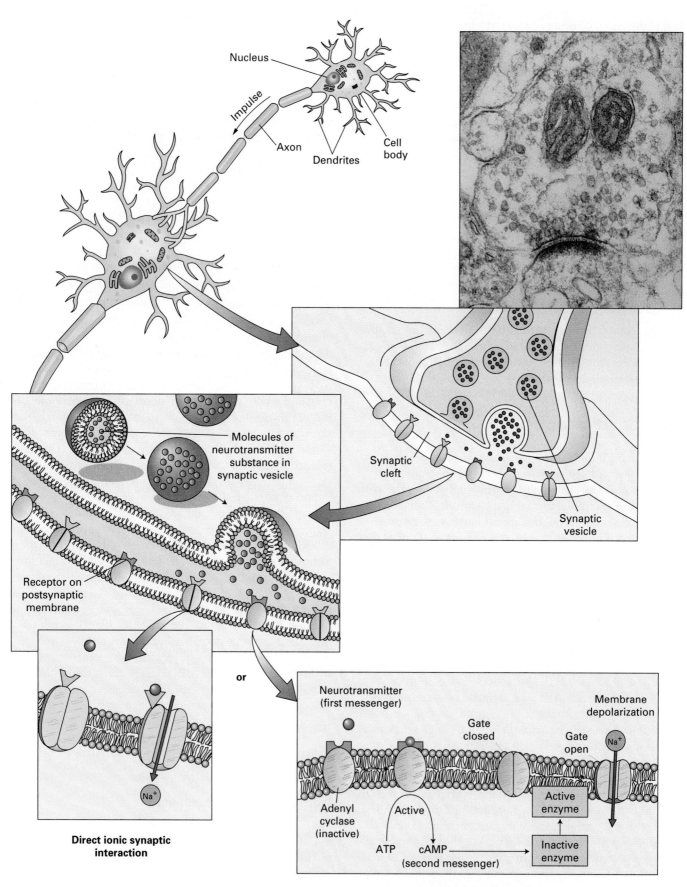

Nucleus

Impulse

Axon

Dendrites

Cell body

Molecules of neurotransmitter substance in synaptic vesicle

Synaptic cleft

Synaptic vesicle

Receptor on postsynaptic membrane

or

Na⁺

Direct ionic synaptic interaction

Neurotransmitter (first messenger)

Gate closed

Membrane depolarization

Gate open

Na⁺

Adenyl cyclase (inactive)

Active enzyme

Active

Inactive enzyme

ATP → cAMP (second messenger)

Indirect synaptic interaction

Indeed, second messengers can affect many metabolic functions. For instance, the second messenger may activate certain enzymes within the cell. These enzymes may then cause changes in the postsynaptic neuron, including changes that increase or decrease its responsiveness to other synaptic stimuli. Because their effects may last minutes or even hours, these are sometimes referred to as indirect synaptic interactions. When a synaptic chemical has long-lasting effects on the cell's responsiveness, it is usually called a **neuromodulator**. Among the neuromodulators are endorphins and enkephalins, which inhibit the perception of pain.

Synaptic Integration

Synapses provide interaction points in the nervous system. Postsynaptic cells integrate synaptic input, "deciding" whether to generate an action potential according to the balance of input. A single nerve impulse in a single synaptic knob will not generate an impulse in the postsynaptic cell. Generally, it requires at least 50 nerve impulses arriving almost simultaneously before the postsynaptic cell will fire. Furthermore, the dendrites and cell body of a neuron may have as many as 10,000 synapses with other neurons (Figure 7–12). Some of these will have excitatory effects on the postsynaptic membrane. Others will have inhibitory effects. Thus, there is often a "tug-of-war" on the postsynaptic membrane and the outcome will determine whether the cell fires. In this way, the integration of synaptic input provides an opportunity to modify the response of the receiving cell. An important result of this integration is finer control over responses, just as having both an accelerator and a brake gives you finer control over the movement of a car.

Removal of Transmitter at the Synapse

The effects of neurotransmitters are temporary because they are removed from the synapse. If they were not, they would continue to excite or inhibit the postsynaptic membrane indefinitely. Depending on the neurotransmitter, disposal may be accomplished in one of two ways. First, a neurotransmitter can be deactivated by enzymes. For example, the neurotransmitter **acetylcholine** is broken into its inactive component parts, acetate and choline, by the enzyme

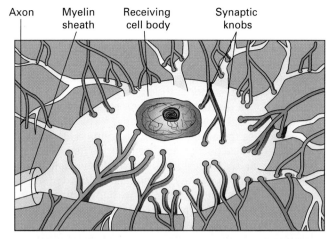

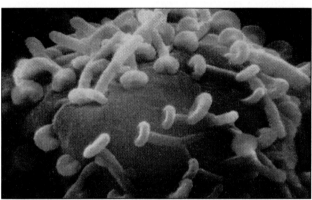

Figure 7–12 Synapses are interaction points within the nervous system. A neuron may have as many as 10,000 synapses. Some synapses have an excitatory effect on the postsynaptic membrane and make it more likely that the neuron will fire. Other synapses have an inhibitory effect and reduce the likelihood that the neuron will fire. The net effect of all synapses will determine whether an action potential is generated in the postsynaptic cell. (The shape of the synaptic knobs shown in this electron micrograph is distorted as a result of the preparation process.)
(b, © Omikron/ Science Source/Photo Researchers, Inc.)

acetyl-cholinesterase. A second means of removing neurotransmitter from the synapse is by using cellular energy to transport it back into the presynaptic cell. This is the primary method of disposal for the neurotransmitters serotonin, dopamine, norepinephrine, and epinephrine.

◄ **Figure 7–11** Structure of a synapse. The axon ending of one neuron (the presynaptic neuron) is separated from the dendrite or cell body of the next neuron (the postsynaptic neuron) by a small gap called a synaptic cleft. Within the axon ending (synaptic knob) are small sacs, called synaptic vesicles, filled with neurotransmitter molecules. When the action potential reaches the synaptic knob, synaptic vesicles fuse with the presynaptic membrane and dump their load of neurotransmitter into the gap. Molecules of neurotransmitter diffuse across the gap and join with receptors on the postsynaptic membrane. This causes ion channels to open and makes it more (or less) likely that an action potential will be generated in the postsynaptic cell. Alternatively, neurotransmitter may bind to receptors that alter the activity of intracellular enzymes. *(Photo, © Biophoto Associates/Science Source/Photo Researchers, Inc.)*

Critical Thinking

The neurotransmitter acetylcholine triggers contraction of voluntary muscles. Myasthenia gravis is an autoimmune disease in which the body's defense mechanisms attack the acetylcholine receptor at the junction between nerve and muscle. As a result, a person with myasthenia gravis is weak and repeated movements become feeble quite rapidly, because the amount of acetylcholine released with each nerve impulse decreases after neurons have fired a few times in rapid succession. Although acetylcholine levels also decline in healthy people, the reduced number of acetylcholine receptors in people with myasthenia gravis makes them extremely sensitive to even the slightest decline in the availability of acetylcholine. Why would you expect drugs that inhibit acetylcholinesterase to be helpful in treating myasthenia gravis?

Roles for Various Neurotransmitters

There are three principal neurotransmitters used by neurons outside the central nervous system: acetylcholine, epinephrine, and norepinephrine. However, there are about 50 different substances involved in communication between neurons within the brain. Why so many? One reason seems to be that different neurotransmitters are involved with different behavioral systems (Figure 7–13). Norepinephrine, for instance, is important in the regulation of mood, in the pleasure system of the brain, arousal, and dreaming sleep. Norepinephrine is thought to produce an energizing "good" feeling. It is also thought to be essential in hunger, thirst, and sex drive. Serotonin, another neurotransmitter, is thought to promote a generalized feeling of well being. Some neurons that communicate with the neurotransmitter dopamine are thought to be involved in regulating emo-

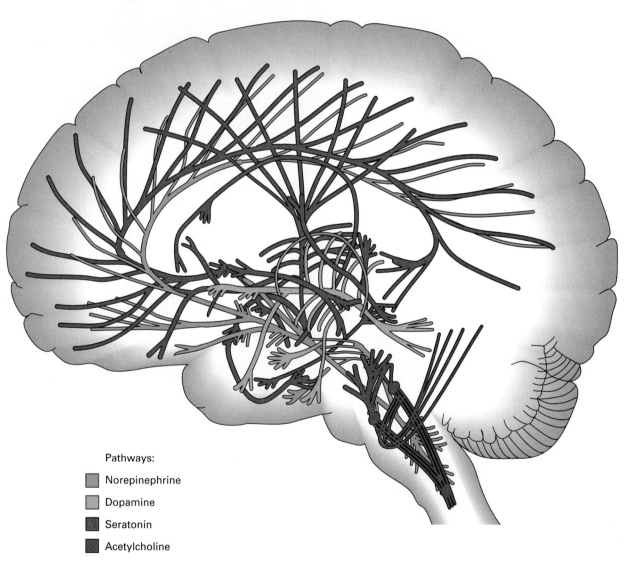

Pathways:

- Norepinephrine
- Dopamine
- Seratonin
- Acetylcholine

Figure 7–13 Circuits in the central region of the brain in which neurons that communicate with a particular neurotransmitter are in high concentration.

tions, and others are important in the brain pathways that control complex movements.

Moving on, we will consider neurotransmitters that are involved with the sensation of pain. When an injury occurs, the bad news is relayed to the brain by neurons that use a transmitter called substance P. Fortunately, the body has its own painkillers that modify the perception of pain. These are the endorphins (short for *endo*genous mor*phine*-like substances) and enkephalins ("in the brain"). The enkephalins are 200 times better than morphine at relieving pain. When a pain message reaches the brain, it stimulates neurons that release endorphins or enkephalins onto the pain-transmitting neurons. The endorphins or enkephalins then bind to the so-called opiate receptors on the pain-transmitting neurons, reducing the release of substance P and quelling the pain.

When the level of a neurotransmitter changes, as may occur in certain diseases or as a result of taking certain drugs (see Chapter 8A), it affects the behaviors controlled by neurons that communicate using that neurotransmitter. An insufficient amount of dopamine, norepinephrine, and serotonin is thought to play a role in depression and suicidal behavior. The brains of people with Alzheimer's disease have too little acetylcholine. Alzheimer's disease is progressive and results in loss of memory, particularly for recent events, followed by sometimes severe personality changes. A formerly good-natured person may become moody and irritable. We will focus our attention on diseases that result from abnormal levels of the neurotransmitter dopamine, specifically schizophrenia and Parkinson's disease, to see how this affects behavior.

Schizophrenia, a mental illness characterized by hallucinations and disordered thoughts and emotions, is caused by excessive activity at dopamine synapses in one part of the brain (the midbrain). As a result, dopamine is no longer in the proper balance with glutamate, a neurotransmitter released by neurons in another brain region (the cerebral cortex). A schizophrenic's world is not the real one. Many hear voices that urge them to act in irrational ways. Many are extremely paranoid, as in one case in which a victim believed that the telephone was robbing his brain. Delusions are also common, as in the case in which another schizophrenic believed that Czar Nicholas II lived in his home, so he always dressed in either a suit and tie or a historical costume. This person also avoided the gaze of people pictured on magazine covers. Others report they are tormented by demons.

Schizophrenia strikes in the prime of one's life. The first signs usually occur between the ages of 17 and 25. It rarely begins as early as age 15 or after age 30.

Schizophrenia cannot be cured but it can be treated with drugs. In the 1950s, Thorazine became the first drug used to treat schizophrenia. Thorazine acts by blocking dopamine release. Unfortunately, less than half of the schizophrenics treated with Thorazine respond to it. And, most of those whose symptoms are eased by Thorazine suffer serious side effects, such as dulled emotions, twitching or jerking movements, and a clumsy way of walking, called the Thorazine shuffle. A more recent drug, clozapine, which lowers the levels of both dopamine and serotonin, has helped rescue some schizophrenic people from their illness. Clozapine is very expensive, however, and patients who take it must be monitored because it can trigger a serious blood disease, agranulocytosis, in which the number of the body's infection-fighting cells drops suddenly and dramatically.

Social Issue

To verify that a drug is useful in treating schizophrenia, controlled experiments must be performed. In some experiments this means giving patients powerful antipsychotic drugs with unknown side effects. In other experiments this means taking patients off medication to see whether they would suffer a psychotic relapse. Participation in such experiments now requires written, informed consent. Every detail of the experiment and its potential risks must be presented in writing. Do you think that someone with schizophrenia will understand the consent form? Should researchers be held accountable if they did not know about certain risks? Should the research be published if the participants did not give informed consent?

Whereas too much dopamine is at least partially responsible for schizophrenia, too little dopamine is primarily to blame for Parkinson's disease. Parkinson's disease, which affects half a million Americans and about 50 million people worldwide, is a progressive disorder that results from the death of dopamine-producing neurons that lie in the heart of the brain's movement control center, the substantia nigra (Figure 7–14). A person with Parkinson's disease generally moves slowly, usually with a shuffling gait and a hunched posture. In addition, involuntary muscle contractions may interfere with intended movements, causing tremors (involuntary rhythmic shaking) of the hands or head when the muscles alternately contract and relax. Involuntary muscle contractions may also cause muscle rigidity when some muscles contract continuously. This muscle rigidity may cause someone with Parkinson's disease to suddenly "freeze" in the middle of a movement.

As the dopamine-producing neurons in the substantia nigra die, dopamine levels begin to fall. Initially, the symptoms are subtle and are written off as part of the aging process. By the time that symptoms of Parkinson's disease are clear, 80% of the neurons in this small area of the brain have already died.

Attempts to treat Parkinson's disease have focused on replacing dopamine or helping the brain get by with the remaining dopamine. Unfortunately, treatment isn't as easy as swallowing a few dopamine pills, because dopamine is prevented from reaching the brain by the blood-brain barrier

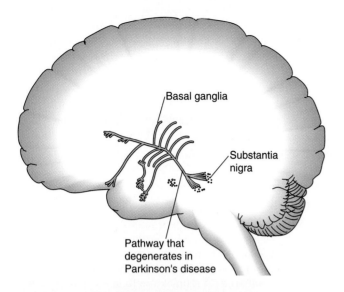

Basal ganglia

Substantia nigra

Pathway that degenerates in Parkinson's disease

Figure 7-14 Parkinson's disease is a progressive debilitating disease, characterized by slowed movements, tremors, and rigidity. The symptoms are caused by an insufficient amount of the neurotransmitter dopamine, which is brought on by the death of dopamine-making nerve cells in a region of the brain called the substantia nigra. *(Photo, Catherine Pouedras/Eurelios/Science Photo Library/Photo Researchers, Inc.)*

that shields the brain from many substances. Therefore, patients are instead given other substances that can reach the brain. The most common and effective treatment combines L-dopa, an amino acid that the brain converts to dopamine, and carbidopa, which prevents dopamine formation outside of the brain where it causes undesirable side effects. Unfortunately, because this drug cannot stop the steady loss of dopamine-producing neurons, it loses effectiveness as the

disease progresses. Some patients are treated with drugs that inhibit the enzyme that breaks down dopamine, thereby enhancing the level of dopamine.

A controversial experimental treatment for Parkinson's disease—the transplantation of dopamine-producing tissue from aborted fetuses into adult brains of patients with Parkinson's disease—shows some promise. (See Chapter 12 for further discussion of fetal cell transplantation studies.) The number of studies using this technique is still small. However, the symptoms of the treated patients seem to decrease gradually and the patients are able to use less L-dopa than before. Some transplant patients have been able to take over some of the normal tasks of daily living, such as dressing and feeding themselves. In some cases, the transplanted tissue is still pumping out dopamine 3 years after surgery. The next step would be to find a substitute for fetal cells. The ethical considerations of using tissue from aborted fetuses aside, there simply are not enough fetal cells to help a large number of patients. Each transplant must include neurons from the brains of several aborted fetuses. Skin or muscle cells that have been genetically altered so that they are able to produce dopamine have been suggested.

Social Issue

Drugs are available to treat severe psychological problems such as schizophrenia and depression. These drugs work by altering the level of the neurotransmitter(s) responsible for causing the disturbances. For instance, Prozac, a drug used to treat severe depression, works by increasing the amount of serotonin in the synapse. There is also some evidence that Prozac changes aspects of the treated person's personality. For example, it helps the person feel more comfortable in social situations. Few would question the wisdom of helping someone who is depressed or who has schizophrenia to live a normal life. There is, however, a spectrum of severity for each personality disorder—from a nail-biting habit to obsessive-compulsive behavior, for example. Researchers are even looking for the biological roots of personality traits, such as shyness or impulsiveness, and believe that these traits are also affected by the levels of key neurotransmitters. If this is true, it means that we may someday be able to design our own personalities.

Should minor personality problems be treated with drugs? Should a personality "flaw" be treated? What do you think?

Neuronal Pools

Disney World is a place where many fantasies come true for millions of visitors. Thousands of employees work day and night to make those fantasies come true. The park runs like

Environmental Toxins and the Nervous System

Every day many of us are exposed to factors that can harm our nervous systems. The exposure may occur because of our jobs, our lifestyles, the foods we eat, or the drugs we take. Since we are born with all the neurons we will ever have, the death of neurons can lead to irreversible loss of function. Some damage may take years to become apparent, but other effects are immediate. Some of the neural and behavioral effects caused in animals and humans by environmental toxicants are listed in the table.

Pesticides are chemicals that are sprayed on plants to kill organisms, such as insects, that might damage the plant. Organophosphate pesticides, such as malathion, parathion, and diazinon, are poisons that kill insects by excessively boosting activity at certain synapses in the insect's body. These pesticides work by inhibiting acetylcholinesterase, the enzyme that breaks down the neurotransmitter acetylcholine. As a result, acetylcholine accumulates in the synapses and has a continuous effect.

These pesticides work the same way in the human body. Approximately half a million people around the world, primarily farm workers, are accidentally poisoned by pesticides each year and 5,000 to 14,000 die as a result. One symptom of pesticide poisoning is muscle spasms. These continuous involuntary contractions occur because acetylcholine is the transmitter that triggers contraction of voluntary muscles, such as those in the arms or legs.

Neurotoxic pesticides may also have less immediate and more subtle effects on the functioning of the nervous system. In 1992, the National Research Council summarized evidence that environmental chemicals such as pesticides may be at least partly responsible for the increasing incidence of nervous disorders such as amyotrophic lateral sclerosis (more commonly called Lou Gehrig's disease), Alzheimer's disease, and Parkinson's disease.

The idea that certain chemicals could cause symptoms similar to those of Parkinson's disease was hurled to the forefront in 1982 because of a mistake made by an underground chemist attempting to synthesize a cheap heroin substitute for the illegal drug trade. The chemist was attempting to produce a compound called MPPP, but accidentally synthesized a similar compound called MPTP. Within a few days of injecting MPPP tainted with MPTP, several drug users developed symptoms indistinguishable from those of advanced Parkinson's disease—the drug users were virtually paralyzed and speechless.

It turns out that MPTP and the compounds formed when it is broken down in the body are chemically very similar to certain industrial chemicals, the most noteworthy of which is the herbicide paraquat. For this reason, some people have suggested that Parkinson's disease may have an environmental cause. André Barbeau in Montreal, Canada, has gathered data linking pesticides and other industrial chemicals to Parkinson's disease. For instance, in a study of 5000 cases of Parkinson's disease in Quebec Province, there was a very strong correlation between the incidence of Parkinson's disease and the level of pesticide use. In a rural area southwest of Montreal, for instance, the example of Parkinson's disease in a farming area where a great deal of pesticide was used was nearly five times higher than in regions where pesticide use was low. However, pesticides may not be the only problem. The industrialized sections of Montreal also had a higher incidence of Parkinson's disease than did the residential sections.

Effects on the Nervous System or Behavior of Animals and Humans Caused by Environmental Toxins

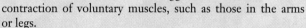

Motor
Activity Changes
Convulsions
Unsteadiness
Clumsiness
Paralysis
Reflex abnormalities
Tremor, twitching
Weakness

Sensory
Auditory disorders
Equilibrium changes
Olfactory disorders
Pain
Visual disorders
Tactile disorders

Cognitive
Confusion
Memory problems
Speech impairment

Personality
Lethargy
Delirium
Depression
Excitability
Hallucinations
Irritability
Nervousness, tension
Restlessness
Sleep disturbances

From Anger, W. K. 1986. Workplace exposures. In *Neurobehavioral Toxicology*. Z. Annau, ed. Baltimore Md.: Johns Hopkins University Press. pp. 331–347. As cited in: *Environmental Neurotoxicology*. National Research Council, 1992. Washington D.C.: National Academy Press. p. 11.

a well-oiled machine because the employees' activities are organized into units, or pools, such as entertainment, maintenance, and food service. And so it is with neurons. The 100 billion neurons in your nervous system are organized into **neuronal pools**, each dedicated to a specific task. The neurons within each pool are arranged in pathways called **circuits**. Some of the synapses in the circuits will be excitatory and others inhibitory. The pattern of neuronal connections within each circuit determines how the pool will function. The greater the number of neurons in the pathway, the more flexible the response can be.

Most neuronal pools contain thousands of neurons, but we can gain an appreciation of how neurons interact by simplifying the story and identifying three of the basic types of circuits (Figure 7–15). In diverging circuits, one neuron synapses with many, perhaps as many as 25,000, other neurons. In this way, for instance, a single neuron firing in the region of the brain that controls voluntary movement may synapse with hundreds of interneurons in the spinal cord, each of which may synapse with hundreds of motor neurons so that thousands of muscle fibers then contract. We see, then, that diverging circuits usually have an amplifying effect. In contrast, a converging circuit, one in which a neuron receives multiple inputs from one or many other neurons, have a concentrating effect. Converging systems integrate input from several sources. Thus, the same emotional re-

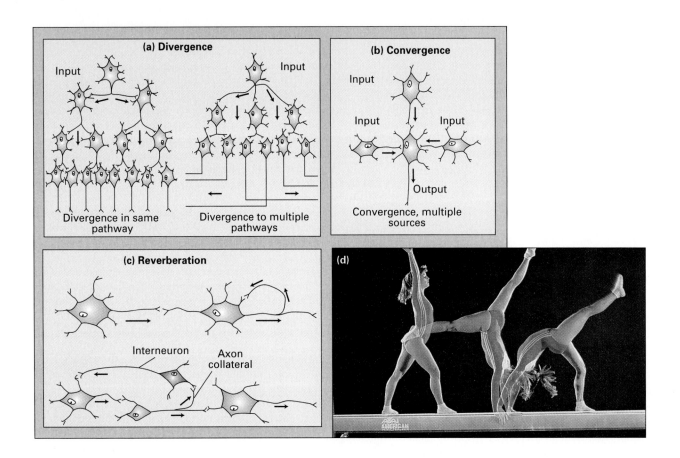

Figure 7–15   Three basic types of circuits within the nervous system. (a) A divergent circuit is one in which one neuron's input is distributed to many others. (b) A convergent circuit is one in which one neuron receives multiple inputs from one or many other neurons. (c) A reverberating circuit is one in which neurons are arranged in a series. A branch from one of the later neurons of the circuit stimulates one of the earlier neurons. In this way, the circuit remains continuously active. (d) Performing on a balance beam depends on the coordinated activities of many neuronal pools that contain convergent and divergent circuits. Convergent circuits will integrate sensory information from the eyes, the organs of balance within the ears, and receptors in the joints that tell the brain the position of the arms and legs.   *(b, © Jerry Wachter/ Photo Researchers, Inc.)*

sponse may be generated in a beach lover by the sound of breaking waves, sand moving between the toes, or the smell of suntan lotion. Reverberating circuits are those in which neurons are arranged in a series so that branches from a later neuron synapse with an earlier one, restimulating it and sending the impulse through the circuit again. This arrange- ment causes an impulse to cycle through the series continuously until one of the neurons in the circuit is inhibited or fails to fire because of fatigue. Reverberating circuits are thought to be important in maintaining wakefulness and in short-term memory.

SUMMARY

1. Most actions involve only three components. Sensory neurons bring information toward the central nervous system from receptors. That information is integrated in the brain and spinal cord where directives for the appropriate actions are generated. Motor neurons then conduct information from the brain or spinal cord to the effectors, the muscles or glands that will respond.

2. The nervous system has two types of specialized cells: neurons (nerve cells) and glial cells. Glial cells outnumber neurons and have several important jobs: They help form the protective blood-brain barrier, they support nerve cells, they protect the central nervous system from disease-causing organisms, and they form myelin sheaths around certain axons.

3. Neurons are the excitable cells of the nervous system that conduct information. Sensory (or afferent) neurons conduct information from the receptors toward the central nervous system. Motor (or efferent) neurons conduct information away from the central nervous system to an effector. Interneurons (association neurons) are positioned between sensory and motor neurons and are located in the central nervous system.

4. A "typical" motor neuron has a cell body that houses the organelles that maintain the cell. Two types of processes extend from the cell body: numerous branching fibers called dendrites, which conduct messages toward the cell body, and a long axon, which conducts impulses away from the cell body.

5. Many axons are enclosed in an insulating layer called the myelin sheath. The myelin sheath forms from multiple wrappings of the plasma membrane of glial cells. The myelin sheath greatly increases the rate at which impulses are conducted along axons and it plays a role in the regeneration of cut axons in the peripheral nervous system.

6. The message conducted by a neuron is called a nerve impulse or action potential. In the resting state, that is, when a neuron is not conducting an impulse, there is an electrical potential difference across the membrane, called the resting potential. It is generated by the unequal distribution of ions across the membrane that results from their different permeabilities. There is a greater concentration of sodium ions outside the neuron than inside and a greater concentration of potassium ions inside than outside. There are also many negatively charged ions held inside the cell by the membrane. The resting potential makes the neuron about 70 mv more negative inside than outside. The resting neuron's membrane is not very permeable to sodium ions. If some leak in, they are pumped out by the sodium-potassium pump.

7. The action potential begins when a region of the membrane suddenly becomes permeable to sodium ions. If enough sodium ions enter to reach threshold, an action potential begins. The gates on sodium channels open and many of these ions then en- ter the cell, making the interior of the cell in that region of the membrane temporarily positive (depolarization). Potassium ions then leave the interior, making the inside once again more negative than the outside (repolarization). The change in the distribution of ions sweeps along the axon as a wave of depolarization and repolarization called an action potential. Once initiated, an action potential sweeps to the end of the axon without diminishing in strength. Immediately following an action potential, there is a brief period, called the refractory period, during which the neuron cannot be stimulated again.

8. Action potentials are essentially identical. We experience different sensations because the impulses are sent to different regions of the brain. The brain interprets the intensity of the stimulus by either the frequency of impulses or by the number of neurons firing.

9. The point where a neuron meets another one is called a synapse. At the axon tip, an impulse causes calcium ions to enter the cell. These ions cause small packets (called synaptic vesicles) storing neurotransmitters to fuse with the plasma membrane and release their contents into the gap between the cells. The neurotransmitter then diffuses across the gap and binds to receptors on the postsynaptic membrane. When a neurotransmitter with ionic effects binds to its receptors, an ion channel is opened. If it is an excitatory synapse, sodium ions enter the cell and increase the likelihood that the postsynaptic neuron will generate a nerve impulse. If it is an inhibitory synapse, the threshold for firing in the receiving neuron is raised. Some neurotransmitters work through a second messenger, usually cyclic AMP. In addition to opening ion channels, the second messenger may cause other, longer-lasting effects. The neurotransmitter is quickly removed from the synapse either by enzymatic breakdown or by transporting it back into the presynaptic cell.

10. Many neurotransmitters are used in the brain. Different neurotransmitters are involved with different behavioral systems. Disturbances in brain chemistry affect mood and behavior. High levels of dopamine in one brain region cause schizophrenia, a mental illness characterized by hallucinations and disordered thought and emotions. Low levels of dopamine in another brain region cause Parkinson's disease, which is characterized by slow movements, tremors, and muscle rigidity.

11. Neurons are organized into groups or pools, each with specific functions. The neurons in neuronal pools are organized into circuits. A diverging circuit is one in which a single neuron synapses with many others. In converging circuits, a single neuron receives input from many neurons. In a reverberating circuit, the neurons are arranged in a series such that branches from a later neuron synapse with an earlier one, restimulating it and sending the impulse through the circuit again.

REVIEW QUESTIONS

1. List four functions of glial cells.
2. List the three types of neurons and give their general functions.
3. Draw a motor neuron and label the following: cell body, nucleus, dendrites, and axon.
4. Explain how a myelin sheath is formed. What are the functions of the myelin sheath?
5. Describe the distribution of sodium ions, potassium ions, and large negatively charged proteins during a neuron's resting state. What factors account for this distribution?
6. Why is the resting potential important?
7. What happens to sodium ions at the beginning of an action potential? How does this affect the membrane potential?
8. Describe the events that bring about repolarization.
9. How is the original ion distribution restored?
10. What is the refractory period?
11. How do we distinguish among stimuli of different senses, such as touch and taste? How do we distinguish stimuli of different intensities?
12. Draw a chemical synapse between two neurons. Label the following: presynaptic cell, postsynaptic cell, synaptic cleft, synaptic vesicles, neurotransmitter molecules, and receptors.
13. Briefly describe the events that occur at a synapse, beginning with the arrival of the action potential at a synaptic knob and ending with the binding of the neurotransmitter to the receptor. Compare ionic synaptic interactions with metabolic synaptic interactions.
14. How do the events at an excitatory synapse differ from those at an inhibitory synapse?
15. How is the action of a neurotransmitter terminated?
16. Describe three types of neuronal circuits and explain the general effects of each.

Chapter **7A**

Alzheimer's Disease

Structural Abnormalities of the Brain
Acetylcholine Deficiency
Amyloid Plaques
Neurofibrillary Tangles
The *ApoE* Gene
Microglia and Inflammation

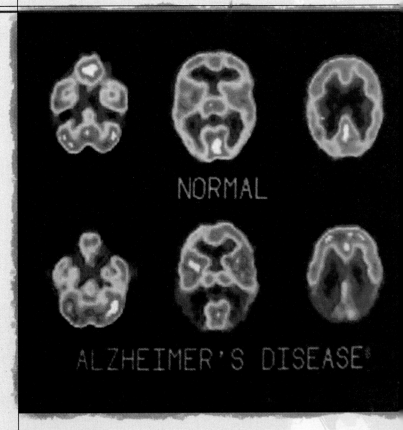

PET scans comparing a normal brain with the brain of a
person with Alzheimer's disease. There is cell loss in the
basal forebrain in the brain of the person with Alzheimer's
disease. (*Science VU/Visuals Unlimited*)

Alzheimer's disease slowly robs a person of a sense of "self" by wiping out memories. Recent memories are the first to go. At first the memory lapses may be only an inconvenience—forgetting where the car is parked, for instance. Gradually, however, the memory lapses worsen and the person may need extensive lists just to get through the day. Then there may be memory losses with more severe consequences. The person may go out for a stroll and forget how to get home. A forgotten pot on a gas stove may cause a fire. Later, older and more deep-seated memories are erased. Loved ones and family members are no longer recognized. Sadly, even those closest to the person with Alzheimer's disease cannot prevent the "person" from fading away (Figure 7A–1). There is currently no cure or effective treatment for Alzheimer's disease.

Memory loss is just part of the story. The forgetfulness is generally accompanied by changes in personality. The person often becomes moody, irritable, and belligerent. Control of bodily functions, such as bladder control, is also gradually lost. The person requires help dressing, eating, and in other daily activities. Often a person with Alzheimer's disease must be waited on and watched by family members or placed in a nursing home. Eventually, he or she lapses into a vegetative state and usually dies of some complication that affects bedridden people, such as pneumonia.

Named after the German physician, Alois Alzheimer, who first described the disorder in 1907, Alzheimer's disease affects about 1% of the population between ages 65 and 74, 7% of the population between the ages of 75 and 84, and 25% of those older than 85. It affects roughly 4 million Americans and kills 100,000 of them each year. Some researchers predict that Alzheimer's disease will soon affect half of all people who live past 85 years of age.

Structural Abnormalities of the Brain

Alzheimer's disease is diagnosed through a process of elimination. Doctors assume that if the symptoms are not caused by a stroke, brain tumor, or a bad reaction to a drug, then it is probably Alzheimer's disease. However, it is impossible to be sure until the person dies and an autopsy is performed. The diagnosis is confirmed if the brain tissue shows three distinct structural abnormalities: (1) There is a tremendous loss of neurons in the hippocampus and the cerebral cortex, particularly the frontal and temporal lobes. These are parts of the brain important in memory and intellectual functioning. It is also known that other neurons (in the nucleus basalis of the cerebrum) don't produce enough of the neurotransmitter acetylcholine. The neurons in the nucleus basalis send axons throughout the cerebral cortex. It is thought that the declining levels of acetylcholine are linked to the loss of cerebral function. (2) The brains are pocked with amyloid plaques. These plaques consist of dying or abnormal neurons surrounding a core of fibrils of the protein β-amyloid. (3) Abnormal nerve cells are clogged with neurofibrillary tangles, which are twisted fibers of a protein called tau (Figure 7A–2).

A new diagnostic test for Alzheimer's disease may be available soon. This test, called AD7C, measures the concentration of neural thread protein (NTP) in the fluid surrounding the brain and spinal cord (cerebrospinal fluid). Large amounts of NTP are associated with the neurofibrillary tangles of Alzheimer's disease. NTP is thought to be important in the repair of nerve cells in the brain and so its presence may reflect the brain's defense against the cellular damage of Alzheimer's disease. Early studies on the accuracy of AD7C in diagnosing Alzheimer's disease, released during the summer of 1996, suggested that it matched the accuracy of postmortem examination. Since many patients who show symptoms of dementia do not have Alzheimer's disease, this test could eliminate prolonged anxiety in patients and their families.

The symptoms of Alzheimer's disease are associated with the loss of neurons, but what kills them? Is the inadequate supply of acetylcholine to blame? Are either (or both)

Figure 7A–1 A man with Alzheimer's disease and his wife. Alzheimer's disease is characterized by a progressive loss of memory that cannot be stopped, even by those closest to the person. (*Will and Deni McIntyre/Photo Researchers, Inc.*)

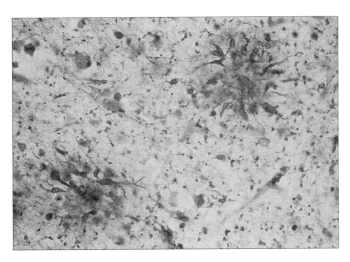

Figure 7A–2 Amyloid plaques and neurofibrillary tangles are the hallmarks of Alzheimer's disease. An amyloid plaque is a region of an area of dark-staining amyloid protein containing tangled masses of filaments and glial cells. Neurofibrillary tangles are masses of thickened filaments within the cytoplasm of neurons. (© *Martin M. Rotker/Photo Researchers, Inc.*)

the amyloid plaques or the neurofibrillary tangles the murder weapons or are they simply tombstones of dying cells? Scientists are finding an increasing number of pieces to the Alzheimer's disease puzzle, but they still are not sure how the pieces fit together. Nonetheless, each new piece offers the hope of finding a successful treatment for Alzheimer's disease. We will take a look at some of these pieces and consider how they were discovered so that we may glimpse at the process of science and so we can understand some of the approaches to developing treatments.

Acetylcholine Deficiency

We know that acetylcholine levels fall in certain areas of the brain during Alzheimer's disease, so it seems reasonable to wonder whether this is related to nerve health. There are at least two ways that acetylcholine is beneficial to neuronal health and functioning: It seems to direct the growth of axons and it keeps certain supporting cells within the brain, called microglia, from secreting substances that are harmful to neurons.

If the brain is to function properly, the correct connections among neurons must be formed and maintained. As it turns out, axons grow toward a source of acetylcholine. When a nerve cell grows, it sprouts axons and dendrites. On the leading edge of the growing nerve fiber are cellular fingers that sense the chemicals around them. Some of the chemicals may mark the trail to the proper connection. Researchers hypothesize that when acetylcholine is released, it causes calcium ions to enter the growing cell. These, in turn,

cause many cellular fingers to form on the side of the growth cone closest to the source of acetylcholine and eventually cause the growing fiber to turn.

Acetylcholine can also stop microglia, cells forming the brain's private immune system, from destroying neurons. As we will see, chemicals secreted by microglia can either hurt or help a neuron. Acetylcholine shuts down the production of the harmful ones.

Some researchers have tried to treat Alzheimer's disease with drugs that will raise acetylcholine levels. One such drug, tacrine, raises acetylcholine levels by inhibiting acetylcholinesterase, the enzyme that breaks down the neurotransmitter. The results of studies on the effectiveness of tacrine in treating Alzheimer's disease has been a mixed bag. Tacrine did improve the memory and intellectual ability of some people with Alzheimer's disease. For instance, some of them could once again recognize family members after taking the drug. Unfortunately, there were no dramatic recoveries. Furthermore, tacrine has side effects, including nausea, vomiting, and diarrhea. There is also concern that tacrine may cause liver damage.

A chemical called huperzine A has been isolated from the club moss (*Huperzia serrata*). Like tacrine, huperzine A is a potent inhibitor of acetylcholinesterase, the enzyme that stops the action of acetylcholine in a synapse. However, the effects of huperzine A are more specific than tacrine, so researchers are hoping that studies will show that huperzine A is also a more effective treatment. Aricept, another drug that inhibits acetylcholinesterase, was approved by the Food and Drug Administration late in 1996 for the treatment of mild to moderate Alzheimer's disease. Normally, the mental functions of patients with Alzheimer's disease would decline considerably over a six-month period. However, in early clinical tests of Aricept, 84% of those treated had no decline in mental functions during the six months of treatment and half of the patients experienced some improvement of memory and alertness.

Amyloid Plaques

Everyone's nerve cells produce some β-amyloid, but the β-amyloid in the brains of people with Alzheimer's disease is abnormal—they produce too much, don't break down enough, or produce an abnormal form. Outside the nerve cell, β-amyloid often forms deposits or plaques.

Although the amyloid plaques are always found in the brains of people with Alzheimer's disease, there is considerable controversy about whether they are a cause or a result of the disease. Evidence that β-amyloid plaques are involved in causing cell death includes the following: (1) One form of Alzheimer's disease strikes people at an early age. People with this form of Alzheimer's disease have a mutant and defective gene that codes for the protein APP (amyloid precursor protein), from which β-amyloid is made. Cells containing the mutant form of this gene grown in tissue culture produce excessive amounts of β-amyloid. (2) Insoluble fibrils of β-amyloid kill nerve cells in test tubes. (3) Injecting β-amyloid into brains of mice caused them to forget tasks they had recently learned.

Evidence can also be presented against a role of amyloid plaques in causing Alzheimer's disease: (1) Many older people develop these plaques, but they do not get Alzheimer's disease. Thus, plaques may simply accompany aging. (2) Everyone's brain produces a great deal of β-amyloid. However, most β-amyloid exists in a soluble form rather than as plaques. (3) Sometimes there are healthy nerve cells in the middle of plaques. (4) The results of experiments on animals are variable. Injection of β-amyloid into animal brains doesn't always kill nerve cells. Even when it does kill cells, the lesions vary depending on the solvent and the source of β-amyloid. Therefore, the injection itself could cause the damage.

If β-amyloid does kill nerve cells, *how* does it do this? One idea is that β-amyloid may kill nerve cells by disrupting their ability to regulate their internal calcium levels. Another idea is that β-amyloid kills nerve cells indirectly by initiating an immune response. We will consider the possible role of the immune system in Alzheimer's disease shortly.

Neurofibrillary Tangles

Another structural abnormality of the brains of people with Alzheimer's disease is neurofibrillary tangles. Some scientists now believe that these tangles are in some way involved in the death of neurons. The number and distribution of tangles correlates with the degree of brain function. Furthermore, the density of synapses is lowest in nerve cells where tangles have developed.

How could the neurofibrillary tangles impair the health of nerve cells? The tangles form from a protein called tau. Tau normally helps build the microtubules that transport nutrients within the nerve cell. Thus, the microtubules form a freight system within the cell that is essential to the cell's health. First, tau encourages copies of another protein (tubulin) to link together and form microtubules (the metro system of neurons). Tau then steadies the microtubules, in a manner similar to the way ties anchor the rails of train tracks. Tau's function depends on its shape, which is altered by the number of attached phosphate groups. When there are too many phosphates attached, the tau molecules cannot bind to the microtubules and they are released from tubulin. The results may be that not enough microtubules form. The loss of too many taus, therefore, triggers the breakdown of the cell's transportation system, causing the axon or the entire cell to die. The excess tau molecules may then intertwine and form a paired helical filament (because tau is hydrophobic). These filaments congregate to form neurofibrillary tangles. The tangles clog the cell.

The *ApoE* Gene

Another piece in the Alzheimer's disease puzzle is the gene *ApoE*, which influences a person's risk of developing Alzheimer's disease. Everyone has two copies of this gene, which has at least three variants: (1) a rare form that seems to have a protective effect, *ApoE-2* (2) the most common and a healthy form, *ApoE-3* and (3) an unhealthy form, *ApoE-4*. A person who has two copies in the form of *ApoE-4* is eight times more likely to develop Alzheimer's disease. Two copies of *ApoE-3* may cause dementia to set in by age 85. However, with two copies of *ApoE-4*, dementia may occur by 69. People with even a single copy of *ApoE-2*, however, are much less likely to develop Alzheimer's disease. If Alzheimer's disease does affect a person with *ApoE-2*, it is usually at an advanced age.

The *ApoE* gene codes for a protein that combines with a lipid to form lipoprotein ApoE, which ferries cholesterol around in the bloodstream. In addition, ApoE is important to the health of neurons. When peripheral nerves are damaged, white blood cells in the area make more ApoE, which, in turn, takes up lipids released by deteriorating nerve cells. The lipids can be used later to rebuild nerve tissue. ApoE-3 also helps nerve cells construct microtubules that are needed to transport materials within cells and to ends of long axons.

We are not yet sure how the ApoE piece of the puzzle fits with the others, but it is beginning to look as though it might be linked to several other pieces. ApoE is found in both plaques and tangles, and it may be involved in their formation (Figure 7A–3).

ApoE-4 leads to denser β-amyloid deposits, and receptors for ApoE are densest in areas where most plaques form.

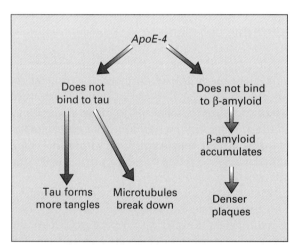

Figure 7A–3 The gene *ApoE* codes for a lipoprotein that is one factor in a person's risk of developing Alzheimer's disease. There are at least three variants of this gene. *ApoE-2* is rare and protective. *ApoE-3* is the healthy form. *ApoE-4* increases one's risk of developing Alzheimer's disease. There are at least two hypothetical ways that *ApoE-4* may bring about that increased risk.

β-amyloid is hydrophobic and, therefore, doesn't dissolve easily in the watery environment of nerve cells. ApoE may transport β-amyloid back into the cell for processing. If the transport mechanism doesn't work correctly, as with ApoE-4, β-amyloid might accumulate outside the cell and condense into plaques. Alternatively, ApoE may sequester β-amyloid, keeping it in solution.

Another beneficial role of ApoE-3 is that it binds to the protein tau, which forms neurofibrillary tangles and stops them from forming. ApoE-4 doesn't bind to tau. Consequently, the neurofibrillary tangles form.

Some researchers think *ApoE* has nothing to do with Alzheimer's disease. Instead, the increased risk of Alzheimer's disease may be due to a gene near *ApoE* on the chromosome.

Social Issue

It is technically possible to test people to determine the type of ApoE alleles they have. The forms of ApoE that a person has are correlated with that person's predisposition to develop Alzheimer's disease. Do you think that ApoE testing should become routine? Some insurance companies have already expressed interest in ApoE testing. Under what circumstances, if any, do you think a person should be tested?

Microglia and Inflammation

It has also been suggested that Alzheimer's disease is an inflammatory response, perhaps to the plaques or neurofibril-

lary tangles. Small glial cells called microglia make up the brain's private immune system (Figure 7A–4). Most of the time, these cells with their radiating appendages form a lacy network throughout the white and gray regions of the central nervous system. However, when part of the brain becomes injured or infected, the microglia seem to know which cells are doomed. They withdraw their appendages and move toward the site of injury, arriving on the spot within 20 minutes. Once at the site of injury, microglia engulf unwanted substances.

Microglia also secrete several substances that affect the health of nerve cells. Some of these chemicals, for instance highly reactive charged molecules, called free radicals, accelerate or cause cell death. Other chemicals help heal injured nerve cells. Studies on nerve cells grown in tissue culture suggest that environmental substances can determine whether harmful or healing substances are released by microglia. For example, acetylcholine can shut down the production of harmful chemicals by microglia.

Microglia surround the β-amyloid plaques in the brains of people with Alzheimer's disease. This observation has led to at least two hypotheses about the role of microglia in Alzheimer's disease. One hypothesis is that microglia do the dirty work themselves. The amyloid plaques may indicate a chronic inflammatory response. Microglia may be attracted to the region of inflammation, and the nerve cells may be innocent victims of the cross fire. However, a second hypothesis is that microglia may be involved in producing or processing β-amyloid. The amyloid plaques then kill the nerve cells. Exactly how the amyloid plaques kill the nerve cells is unclear, but there is some evidence that another aspect of the immune response, the complement system, may be triggered by β-amyloid (Figure 7A–5). The complement system is a group of about 25 proteins in the immune system that help destroy disease-causing microorganisms (see Chapter

Figure 7A–4 Microglia small cells with numerous radiating appendages that function as the brain's own immune system. Some scientists are now wondering whether an over zealous immune response directed by the microglia is a cause of Alzheimer's disease. (© *John D. Cunningham/Visuals Unlimited.*)

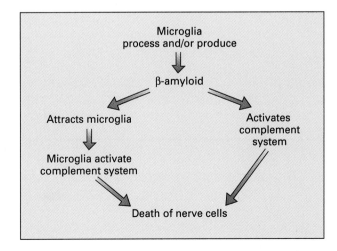

Figure 7A–5 Two hypothetical pathways involving the immune system that could lead to Alzheimer's disease.

12). Some researchers suspect that while complement proteins in a healthy brain help clean up debris when neurons die of old age, the response simply gets out of hand in the brains of people with Alzheimer's disease.

If it is true that Alzheimer's disease is caused by an over zealous immune response, then it should be possible to treat or prevent the disease with anti-inflammatory drugs. Some studies do indicate that anti-inflammatory drugs may reduce the incidence of Alzheimer's disease. Rheumatoid arthritis is a disease known to be caused by the body's immune system turning against itself. Therefore, this form of arthritis is usually treated with anti-inflammatory drugs. When hospital records of 12,000 patients with either Alzheimer's disease or rheumatoid arthritis were examined, a surprisingly small number of them had both conditions. This would be expected if the anti-inflammatory drugs taken to treat arthritis also prevented Alzheimer's disease. In addition, some members of a group of 4000 leprosy patients in Japan were treated with dapsone, an antibiotic that is also an effective anti-inflammatory drug. Dapsone significantly reduced the incidence of Alzheimer's disease. Whereas 6.25% of the patients that had not taken dapsone suffered from dementia, only 2.9% of those taking the drug had dementia. The habits of 50 sets of elderly twins were also studied to see whether anti-inflammatory drugs reduced the incidence of Alzheimer's disease. At least one twin in each set had Alzheimer's disease. In some pairs, the other sibling developed Alzheimer's disease at a later date and in others the twin remained healthy. All the twins, and sometimes their close family members, were questioned about the use of anti-inflammatory drugs, ranging from arthritis medication to aspirin. Statistical analysis revealed that a person who took an anti-inflammatory drug on a regular basis for at least 1 year was four times more likely than his or her twin who had not taken these drugs to remain healthy and not to develop Alzheimer's disease later in life. Clinical trials using another drug, indomethacin, are underway. Initial studies on this drug are promising.

REVIEW QUESTIONS

1. What are the symptoms of people with Alzheimer's disease?
2. What are the three structural abnormalities found in the brains of persons with Alzheimer's disease that have died?
3. How is acetylcholine involved in making connections among neurons?
4. In what beneficial way (for neurons) does acetylcholine interact with microglial cells?
5. What is the evidence for and against the role of amyloid plaques in causing Alzheimer's disease?

6. Excess tau proteins that are lost from microtubules form neurofibrillary tangles. How does their loss from the microtubules affect the microtubule system and, thus, the health of the nerve cells of the brain?
7. Which form of *ApoE* is thought to make a person more susceptible to Alzheimer's disease by possibly being involved in the production of plaques and tangles?
8. What are the two aspects of an immune response that are thought to possibly lead to Alzheimer's disease?

Chapter | 8

The Nervous System

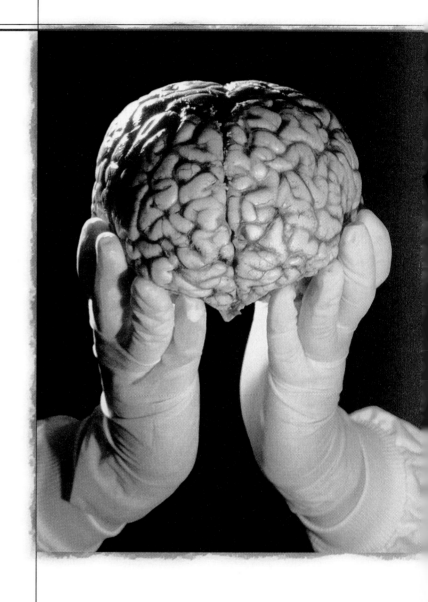

(Geoff Tompkinson/Science Photo Library/Photo Researchers, Inc.)

N ow that we have a basic understanding of how nerve cells function individually and in pools, we can consider the component parts of the nervous system. We will begin with the central nervous system, which is composed of the spinal cord and the brain. Then we will discuss the peripheral nervous system, the link between the body and the central nervous system.

Divisions of the Nervous System

If we were to view the nervous system apart from the rest of the body we would see dense areas of neural tissue, in the head and forming a cord extending down the middle of the back (Figure 8–1). These are the brain and spinal cord and they comprise the **central nervous system** (CNS), which is concerned with integrating and coordinating all voluntary and involuntary nervous functions. Extending from the brain and spinal cord we see many communication "cables," the nerves that carry messages to and from the central nervous system. The nerves branch extensively, forming a vast network. We also see small clusters of nerve cell bodies, called **ganglia**, at certain locations outside the CNS. The nerves and ganglia make up the **peripheral nervous system**, which keeps the central nervous system in continuous contact with almost every part of the body.

The peripheral nervous system can be further subdivided on the basis of function into the somatic nervous system and the autonomic nervous system. We are often aware of the activities of the **somatic nervous system**. Messages

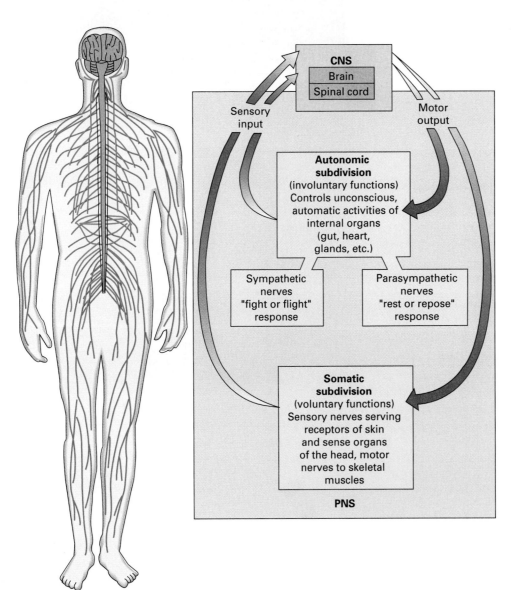

Figure 8–1 Although the human nervous system functions as an integrated whole, it is often convenient to talk about its functional divisions. The brain and spinal cord, which together form the central nervous system (CNS) (shown in blue), integrate incoming information and direct appropriate responses. Communication cables called nerves carry information to and from the central nervous system. These nerves, together with collections of nerve cell bodies called ganglia, comprise the peripheral nervous system (PNS). The PNS may be divided into the somatic and the autonomic nervous systems. The somatic nervous system (shown in green) carries sensory information from the skin, skeletal muscles, and joints to the CNS and also directs our voluntary movements, such as those of the arms or legs. The autonomic nervous system (shown in red) serves as an "automatic pilot" that regulates our involuntary bodily activities, such as heart beat and breathing rate. The autonomic nervous system is composed of the parasympathetic nervous system, which governs activities during restful conditions, and the sympathetic nervous system, which prepares the body to face stressful or emergency conditions.

carried on somatic sensory nerves result in sensations, including light, sound, or touch. The somatic nervous system also controls our voluntary movements, allowing us to smile, stomp a foot, sing a lullaby, or frown as we sign a check. The **autonomic nervous system**, on the other hand, governs the involuntary, unconscious activities that maintain a relatively stable internal environment. Its activities alter digestive activity, open or close blood vessels to shunt blood to areas that need it most, and alter heart rate and breathing rate. The autonomic nervous system has two divisions that generally cause opposite effects on the muscles or glands they control. The sympathetic division is in charge during stressful or emergency conditions, such as fear, rage, or vigorous exercise. During calm or restful times, however, the parasympathetic division reverses the actions of the sympathetic system.

Although we have referred to various divisions of the nervous system, we should keep in mind that these all function as a single unit, as can be seen in the following scenario. Imagine for a moment that you are sunbathing on the beach. While you are relaxing, the parasympathetic branch of the autonomic nervous system is ensuring that your life-sustaining bodily activities continue. Then you feel someone grasp your hand. Sensory receptors in the skin (part of the somatic nervous system) respond to the pressure and warmth of the hand and send messages over sensory nerves to the spinal cord. Neurons within the spinal cord relay the messages to the brain. The brain integrates incoming sensory information, "deciding" on an appropriate response. The brain may then generate messages that result in your eyes opening. If the sight of the person holding your hand generates strong emotion, the sympathetic nervous system may cause your heart to beat faster and perhaps even your breathing rate to increase.

The Central Nervous System

The Brain

In a sense, your brain is more "you" than any other part of your body, because it holds your memories and the keys to your personality. Yet, if you were to see it, you would probably deny any relationship to it (Figure 8–2). It is a grayish mass of tissue with the consistency of raw calf's liver and weighs less than 1600 grams (3 pounds), which is probably less than 3% of your body weight. But it is the essential "you." It is the origin of your secret thoughts and desires, it remembers your most embarrassing moment, and it regulates your other body systems so that they function harmoniously while you concentrate on other activities. So, let's look at the brain to appreciate better how its many circuits are organized to accomplish these amazing feats (Figure 8–3).

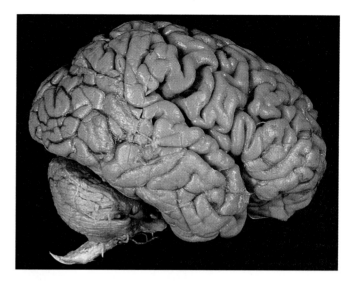

Figure 8–2   As you study this photograph of a human brain, you are experiencing a most unusual phenomenon—the human brain pondering the human brain. Notice that most of the surface is highly folded. This is the cerebrum, the area of the brain that makes you a unique person. The two smaller lobes at the bottom form the cerebellum, which functions in sensory motor coordination.   (*Visuals Unlimited/© Fred Hossler*)

Cerebrum

The **cerebrum** is the largest and most prominent part of the brain. It is, quite literally, your "thinking cap." Accounting for 83% of the total brain weight, it is the "conscious" part of the brain, the part that gives us most of our human characteristics.

The many ridges and grooves on the surface of the cerebrum make it appear wrinkled. Some furrows are deeper than others. The deepest indentation is in the center and runs from front to back. This groove, called the longitudinal fissure, separates the cerebrum into two hemispheres. Each hemisphere receives sensory information from and directs the movements of the opposite side of the body. Furthermore, as we will see later, the hemispheres process information in slightly different ways and are, therefore, specialized for slightly different mental functions. There are other grooves that also serve as anatomical landmarks, indicating the boundaries of four lobes on each hemisphere: the frontal, parietal, temporal, and occipital. A fifth lobe, the insula, is hidden in the walls of the groove separating the temporal and parietal lobes. Each of these lobes has its own specializations (Figure 8–4).

Each hemisphere has three regions. The outer layer is the **cerebral cortex**. (Cortex means bark or rind.) In fresh brain tissue the cortex appears gray and is, therefore, called the gray matter. It consists of billions of glial cells, nerve cell

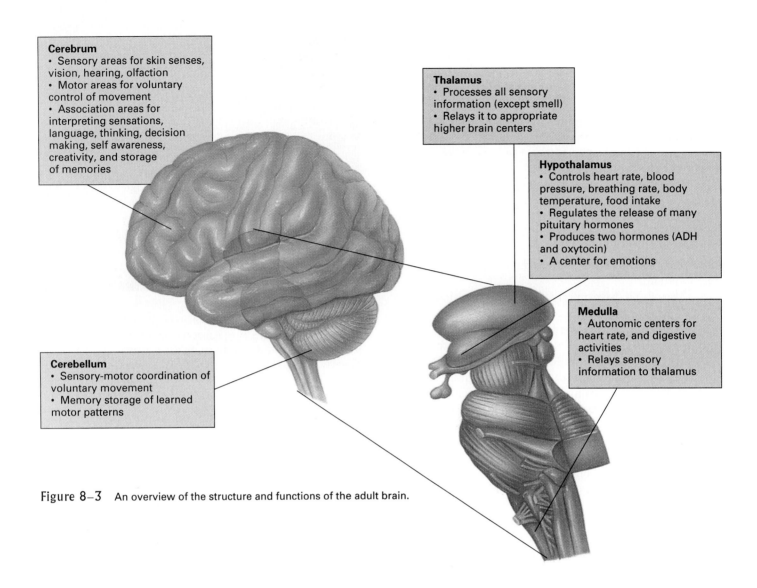

Cerebrum
• Sensory areas for skin senses, vision, hearing, olfaction
• Motor areas for voluntary control of movement
• Association areas for interpreting sensations, language, thinking, decision making, self awareness, creativity, and storage of memories

Thalamus
• Processes all sensory information (except smell)
• Relays it to appropriate higher brain centers

Hypothalamus
• Controls heart rate, blood pressure, breathing rate, body temperature, food intake
• Regulates the release of many pituitary hormones
• Produces two hormones (ADH and oxytocin)
• A center for emotions

Medulla
• Autonomic centers for heart rate, and digestive activities
• Relays sensory information to thalamus

Cerebellum
• Sensory-motor coordination of voluntary movement
• Memory storage of learned motor patterns

Figure 8–3   An overview of the structure and functions of the adult brain.

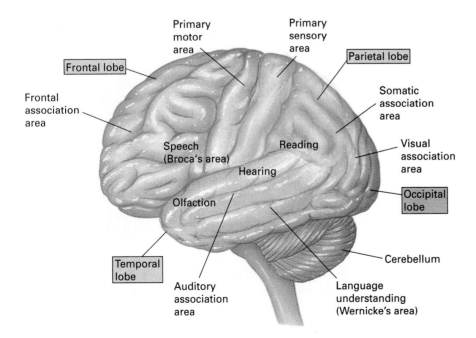

Primary motor area

Primary sensory area

Frontal lobe

Parietal lobe

Frontal association area

Somatic association area

Speech (Broca's area)

Reading

Visual association area

Hearing

Olfaction

Occipital lobe

Temporal lobe

Cerebellum

Auditory association area

Language understanding (Wernicke's area)

Figure 8–4   Four lobes of the cerebral cortex and some functions associated with each.

bodies, and unmyelinated axons. Although the cerebral cortex is only 1- to 2-mm (1/8-in.) thick, it is highly folded. These folds, or convolutions, increase the surface area of the cortex threefold. Beneath the cortex is the **cerebral white matter** (Figure 8–5). It appears white because it consists primarily of myelinated axons. Recall that myelin sheaths increase the rate of conduction along axons and are, therefore,

found on axons that conduct information over distances. The axons of the cerebral white matter are grouped into tracts that allow various regions of the brain to communicate with one another. Some tracts permit communication within a hemisphere and others allow communication between the cortex and lower brain centers or the spinal cord. A very important band of white matter, called the **corpus callosum**,

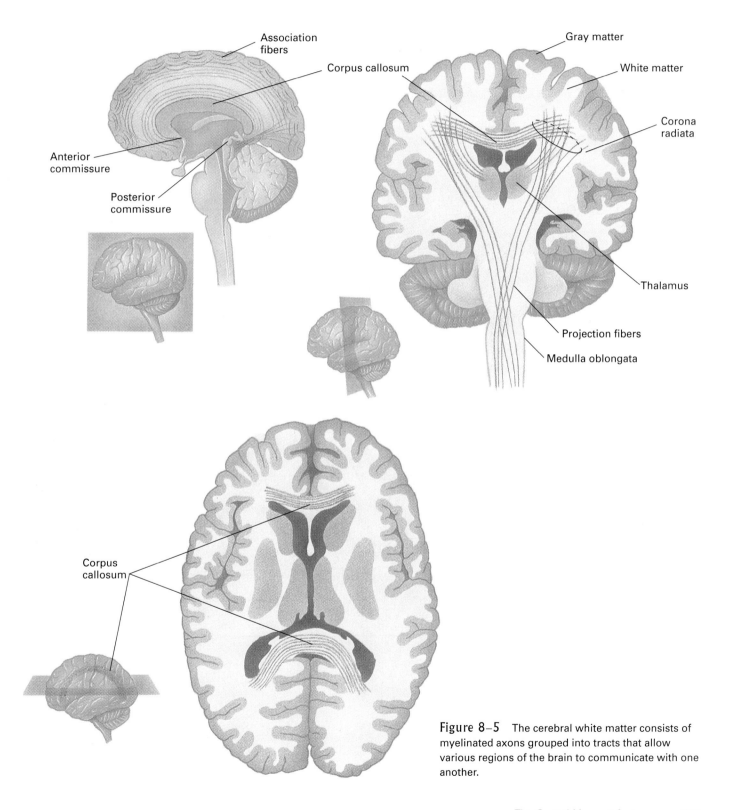

Figure 8–5 The cerebral white matter consists of myelinated axons grouped into tracts that allow various regions of the brain to communicate with one another.

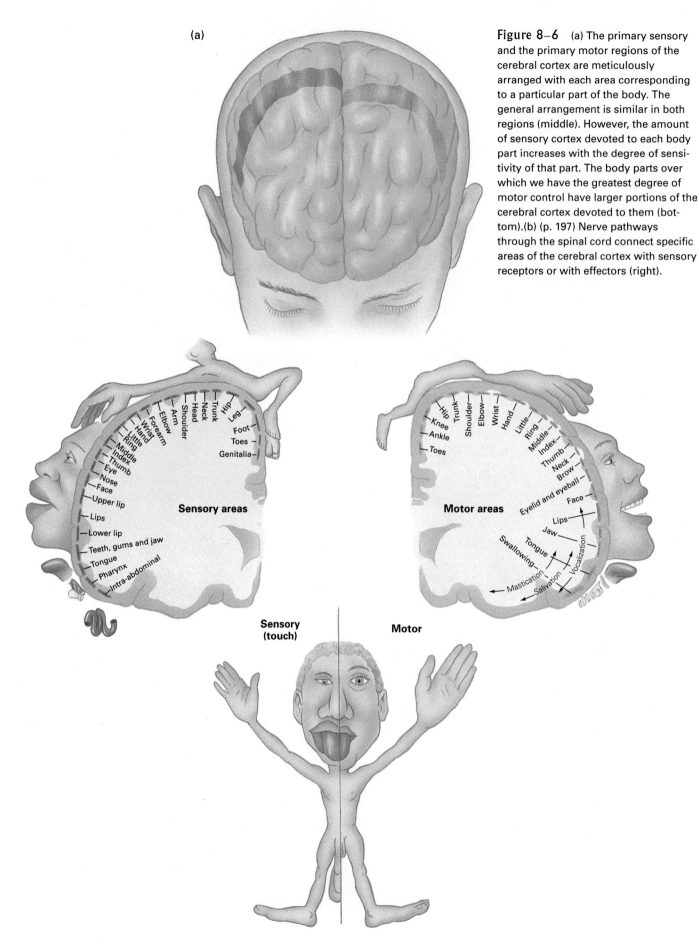

(a)

Figure 8–6 (a) The primary sensory and the primary motor regions of the cerebral cortex are meticulously arranged with each area corresponding to a particular part of the body. The general arrangement is similar in both regions (middle). However, the amount of sensory cortex devoted to each body part increases with the degree of sensitivity of that part. The body parts over which we have the greatest degree of motor control have larger portions of the cerebral cortex devoted to them (bottom).(b) (p. 197) Nerve pathways through the spinal cord connect specific areas of the cerebral cortex with sensory receptors or with effectors (right).

Sensory areas

Trunk · Hip · Leg · Foot · Toes · Genitalia · Neck · Head · Shoulder · Arm · Elbow · Forearm · Wrist · Hand · Little · Ring · Middle · Index · Thumb · Eye · Nose · Face · Upper lip · Lips · Lower lip · Teeth, gums and jaw · Tongue · Pharynx · Intra-abdominal

Motor areas

Hip · Trunk · Knee · Ankle · Toes · Shoulder · Elbow · Wrist · Hand · Little · Ring · Middle · Index · Thumb · Neck · Brow · Eyelid and eyeball · Face · Lips · Jaw · Tongue · Swallowing · Vocalization · Salivation · Mastication

Sensory
(touch)

Motor

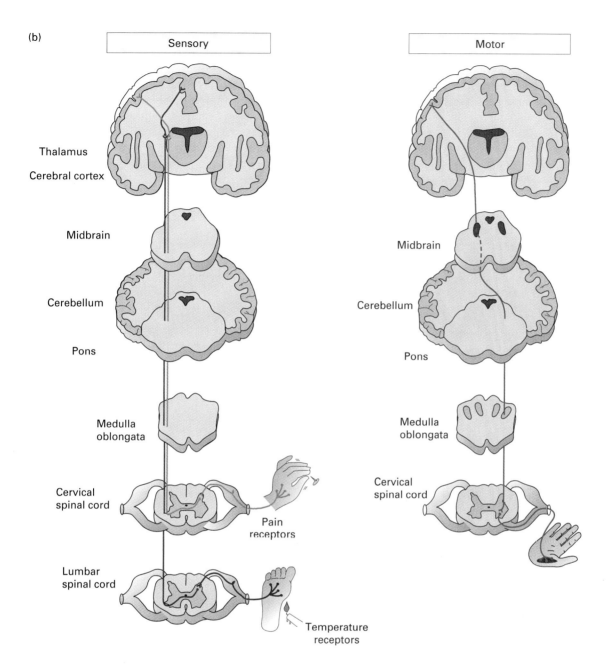

(b)

Sensory

Motor

Thalamus
Cerebral cortex

Midbrain

Cerebellum

Pons

Medulla oblongata

Cervical spinal cord

Pain receptors

Lumbar spinal cord

Temperature receptors

Midbrain

Cerebellum

Pons

Medulla oblongata

Cervical spinal cord

connects the two cerebral hemispheres so they can communicate with one another. Deep within the white matter of the cortex are collections of cell bodies called the **basal ganglia**. (Collections of nerve cell bodies within the central nervous system are more correctly called nuclei, but for some reason, the term basal ganglia remains more common.) The basal ganglia are important in coordinating movement and may also play a role in cognition and memory of learned skills.

Although the assignment of a specific function to a particular region of the cerebral cortex is imprecise, it is generally agreed that there are three types of functional areas: sensory, motor, and association. Sensory areas of the cerebral cortex are responsible for our awareness of sensations. The various sensory receptors send information to sensory areas of the cortex, each sense to a different region (refer back to Figure 8–4). If you stand on the street corner watching a pa-

rade go by, you would hear the band play because information from your ears is sent to the auditory area in the temporal lobe. You would see the flag wave because information from your eyes is sent to the visual area in the occipital lobe. When you catch a whiff of popcorn it is because information is sent from the olfactory receptors in the nose to the olfactory area located in the temporal lobe of the cortex. As you eat that popcorn, you know it is too salty because information from the taste receptors is sent to gustatory areas in the parietal lobe. You know that you are standing and that your belt is too tight because information from receptors in the skin regarding touch, temperature, and pain and in the joints and skeletal muscles is sent to the **primary sensory area**. This region forms a band in the parietal lobes that stretches over the cortex from the region of one ear to the other (Figure 8–6). Sensations from different parts of the body are represented in different regions of the primary sensory area (of the hemisphere on the opposite side of the body). The greater the degree of sensitivity, the greater the area of cortex devoted to that body part. Thus, if areas of equal size are considered, our most sensitive body parts, such as the tongue, hands, face, and genitals, have more of the cortex devoted to them than do less sensitive areas, such as the forearm.

Should you decide to join the parade, the **primary motor area** of the cerebral cortex will initiate messages that allow you to direct your skeletal muscles. This motor area forms a band in the frontal lobe, just anterior to the primary sensory area. Like the sensory area, the motor area is meticulously arranged. Each point on the surface corresponds to the movement of a different part of the body. The parts of the body over which we have finer control, such as the

tongue and fingers, have greater representations on the motor cortex than do regions with more restricted movements, such as the trunk of the body.

The rest of the cortex consists of association areas, which communicate with the sensory and motor areas to analyze and act on sensory input. Neighboring each primary sensory area is an association area that analyzes input. As the sensory association areas communicate with one another and with other parts of the brain, you begin to recognize what you are sensing. Each sensory association area communicates with the general interpretation area, also called the **gnostic area**. The gnostic area blends the input from sensory association areas with stored sensory memories and assigns meaning to the experience (Figure 8–7). For example, on a dark night, your eyes may detect a small moving object. If the object then rubs against your legs and purrs, you may

Figure 8–7 As the squares get smaller in successive figures of this series, association areas are given more clues. The activity of association areas results in your eventual recognition of the figure as a face. You may even recognize the face as Leonardo da Vinci's painting of the Mona Lisa. (© *Blocpix Image®* by Ed Manning, Stratford, CT.)

recognize it as the neighbor's friendly cat. However, if it turns away from you and raises its tail, you will recognize it as a skunk. Information is then sent to the most complicated of all association areas, the **prefrontal cortex** (the most anterior part of the frontal lobe), which will decide how you should respond. In making the decision, the prefrontal cortex will predict the consequences of possible responses and judge which response will be best for you in that situation. The prefrontal cortex is also important in reasoning, long-term planning, producing abstract ideas, judgment, complex learning, intellect, and personality.

In our everyday lives, both cerebral hemispheres gather pretty much the same sensory information about the world around us. However, that input is often processed in slightly different ways by each hemisphere. The left hemisphere processes information analytically; it dissects the situation to understand the component parts and considers one aspect at a time. It is best in verbal and mathematical skills. It is interesting that language abilities are generally restricted to one hemisphere. About 95% to 99% of right-handed people and 60% to 70% of left-handed people have their language centers in the left hemisphere. The right hemisphere, on the other hand, puts the parts together to understand the whole. The right hemisphere excels at generating visual images and mental images of sound, touch, and smell to compare relationships. As a result, it is better at space and pattern perception, identifying objects on the basis of shape, and recognizing faces. Perhaps this is why the right hemisphere is generally better at music and art.

You get the benefit of both ways of thinking, however, since the two hemispheres communicate with one another through the corpus callosum. Therefore, when you find your new puppy with one of your favorite shoes in his mouth, the right cerebral hemisphere may be the first to recognize him, and then it informs the left hemisphere. It is the left hemisphere that allows you to respond, "Bad dog!"

Thalamus

The cerebral hemispheres sit comfortably on the thalamus. The **thalamus** is the gateway to the cerebral cortex. All messages to the cortex must pass through the thalamus first. It is important in sensory experience, motor activity, stimulation of the cerebral cortex, and memory. It is composed of many nuclei (clusters of neurons) and nerve fibers, each one specializing in a different job. There is two-way communication between each nucleus of the thalamus and a specific region of the cerebral cortex. Sensory input from every sense (except smell) and all parts of the body is delivered to at least one of the thalamic nuclei. The thalamus sorts the information, groups it according to function, and relays it to the appropriate brain regions, such as the sensory cortex and sensory association areas, for processing. Some regions of the thalamus do more than just relay information to the cortex: They integrate information. At the thalamic level of processing, we have a general idea whether the sensation is

pleasant or unpleasant. If you step on a tack, for instance, you may experience pain by the time the messages reach the thalamus, but you won't know where it hurts until the message is directed to the cortex.

Hypothalamus

Below the thalamus is the **hypothalamus** (*hypo* means under), a small region of the brain that is essential to maintaining a stable environment within the body. The hypothalamus influences blood pressure, heart rate, digestive activity, breathing rate, and many other vital physiological processes. Because the hypothalamus receives input from the cerebral cortex, it can make your heart beat faster when you just see or think of something exciting or dangerous—a rattlesnake about to strike, for instance. In addition, it acts as the body's "thermostat." When the hypothalamus senses that the body is too cool, it initiates physiological responses such as shivering to raise the body temperature. The hypothalamus initiates responses such as sweating when it senses an elevated body temperature. By regulating food intake, hunger, and thirst, the hypothalamus helps maintain the body.

Another important function of the hypothalamus is to coordinate the activities of the nervous system and the endocrine (hormonal) system (see Chapter 6). The pituitary gland is an endocrine structure connected to the hypothalamus. Because the pituitary produces hormones that regulate the release of hormones by so many other endocrine structures throughout the body, it is sometimes called the "master gland." However, the hypothalamus is the real master because it regulates the release of many of the pituitary's hormones by producing its own releasing hormones. In addition, the hypothalamus produces two of its own hormones, ADH (antidiuretic hormone) and oxytocin, which are released from the pituitary.

As part of the limbic system, the hypothalamus is also part of the circuitry for emotions. Certain nuclei within the hypothalamus play a role in the sex drive and are involved with the perception of pain, pleasure, fear, and anger. When electrodes were implanted in a certain region of a rat's hypothalamus and wired so that the animal could electrically stimulate itself by pressing a lever, it quickly learned to press the lever. After having learned how to "turn itself on," the rat would press the lever thousands of times an hour for hours without stopping. One rat pressed the lever more than 2000 times an hour for 26 hours and then collapsed from fatigue. Another one pressed the lever steadily for the entire 48-hour period that the lever was made available! Because the rats worked so hard to stimulate themselves, researchers assumed that the rats "liked" the experience. Consequently, this is often called the "pleasure center." It appears to be part of the brain's reward system. A few humans have had electrodes implanted in this area of their brains as medical treatment for severe pain or depression. Each one has said that stimulation felt good. They often described it as a sexual kind of pleasure.

Another area of the hypothalamus (the suprachiasmatic nuclei) serves as a "master biological clock" that synchronizes all the other biological clocks within the body. In this way, the hypothalamus sets the timing for all our biological rhythms (discussed in Chapter 4).

Cerebellum

The **cerebellum** looks somewhat like a small cerebrum because it too has two hemispheres and a surface with many folds. Indeed, its name literally means "small brain." The primary function of the cerebellum is sensory-motor coordination. It acts as an automatic pilot that produces smooth, well-timed voluntary movements and controls both equilibrium and posture. It receives sensory information regarding the position of joints and the degree of tension in both muscles and tendons throughout the body. It integrates this information with input from the eyes and the equilibrium receptors in the ears. As a result, the cerebellum is able to determine the body's position at any given instant, as well as where it is going. When you decide to stand up from a sitting position, the activity is planned in the frontal lobe of the cerebral cortex, which informs the motor areas of the cortex. The cerebellum is also notified of the intended movements. Using the sensory information about body position and movement, the cerebellum calculates the best way to coordinate muscle contractions to keep you from falling and then it modifies the directions from the motor cortex.

The coordination of sensory input and motor output by the cerebellum involves two important processes—comparison and prediction. During every move you make, the cerebellum continuously compares the actual position of each part of the body with where it *ought* to be at that moment (with regard to the intended movement) and makes the necessary corrections. If you try to touch the tips of your two index fingers together above your head, you will probably miss on the first attempt. However, the cerebellum makes the necessary corrections and you will likely succeed on the next attempt. At the same time, the cerebellum calculates future positions of a body part during a movement. Then, just before that part reaches the intended position, the cerebellum sends messages to stop the movement at a specific point. Therefore, when you scratch an itch on your cheek, the hand stops before slapping your face.

The cerebellum stores learned motor patterns. Therefore, when we practice movements, they become automatic. This is the purpose of training or rehearsing. "Muscle memory" is important to many activities, ranging from football to dancing. A well-rehearsed skating routine, for instance, may be performed flawlessly in spite of the nervousness resulting from competition. After an action has become automatic, thinking about its execution usually disrupts the pattern. We say an athlete "choked" when a moment of hesitation in a response that should have been automatic costs the victory.

Some researchers have suggested that besides its role in sensory-motor coordination the cerebellum may play a role in higher mental processes, including language. Studies using PET scans (see page 204), which show the parts of the brain active during an activity, have supported this suggestion. One region of the cerebellum (the neodendate) is active during language tasks that require higher thinking, but not during simple speech. In another study, the same area of the cerebellum was active when the subjects had to move pegs around a board to solve a difficult problem. The area was not involved, however, when the subjects merely moved the pegs across the board randomly. Thus, the cerebellum may be involved in mental dexterity as well as motor dexterity.

Medulla

The **medulla oblongata** is often simply called the medulla. This marvelous inch of nervous tissue contains reflex centers for some of life's most vital physiological functions: the pace of the basic breathing rhythm, the force and rate of heart contraction, and blood pressure.

The medulla connects the spinal cord to the rest of the brain. Therefore, all sensory information going to the upper regions of the brain and all motor messages leaving the brain are carried by nerve tracts running through the medulla.

Critical Thinking

Why would a brain tumor that destroyed the functioning of nerve cells in the medulla lead to death more quickly than a tumor of the same size on the cerebral cortex?

Functional Systems of the Brain

The limbic system is our emotional brain. The **limbic system** is a collective term for several structures, including parts of the anterior thalamus, parts of the cerebral cortex, the hypothalamus, the olfactory bulb, the amygdala, and the hippocampus (Figure 8–8). The limbic system allows us to experience a rainbow of emotions including rage, pain, fear, sorrow, joy, and sexual pleasure. Emotions are important because they motivate behavior that will increase the chance of survival. Fear, for example, may have evolved to focus the mind on the threatening things in the environment. Connections between the higher brain centers and the limbic system allow us to have *feelings* about *thoughts*. These connections also keep us from responding to emotions, such as rage, in ways that would be unwise. The limbic system includes the hypothalamus and is also connected to lower brain centers, which control the activity of our internal organs. Therefore, we also have gut responses to emotions.

The **reticular activating system** (RAS) is a very extensive network of neurons that runs through the medulla and projects to the cerebral cortex. Its name describes its functions. The term reticular comes from the Latin word *rete*, which means "net." The RAS functions as a net, or filter, for sensory input. Our brain is constantly flooded with a tremendous amount of sensory information, about 100 million impulses each second, most of them trivial. The RAS

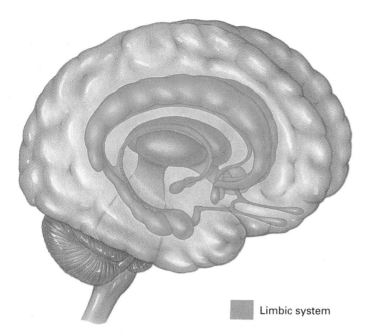

Limbic system

Figure 8–8 The limbic system is our emotional brain. It allows us to feel fear, sorrow, joy, and sexual arousal. These emotions lead to behaviors that increase the chance of survival.

filters out the repetitive familiar stimuli—the sound of street traffic, paper rustling, the person next to you coughing; the pressure of clothing, a breeze blowing, or the air temperature. However, infrequent or important stimuli pass through the RAS to the cortex and, therefore, reach our consciousness. Consequently, if you are very tired, you might fall sleep with the television on but wake up when someone whispers your name.

In addition, the RAS is an activating center. Unless inhibited by other brain regions, the RAS activates the cerebral cortex, keeping it alert and "awake." Thus, although consciousness comes from activity in the cerebral cortex, this only occurs while the RAS stimulates it. When sleep centers in other regions of the brain inhibit activity in the RAS, sleep results. In essence, then, the cerebrum "sleeps" whenever it is not stimulated by the RAS. Sensory input to the RAS stimulates the cortex and raises consciousness levels. This is why it is usually easier to sleep in a dark quiet room than in an airport terminal. Conscious activity in the cerebral cortex can also stimulate the RAS, which will, in turn, stimulate the cerebral cortex. Therefore, thinking about a problem may keep you awake all night.

Critical Thinking

When a boxer is hit in the jaw very hard, his head and, therefore, his medulla and RAS are twisted sharply. Why might this result in a knockout in which the boxer loses consciousness?

Visualizing the Brain

It is a difficult thing to know what goes on inside someone's head. However, medical technology has made great strides forward and now noninvasive techniques allow us to view the structure of body tissues, including those of the brain, as well as the activity of the brain.

Computed tomography, or CT scanning, creates pictures of brain tissues based on their relative densities. The brain is imaged using an x-ray source that moves in an arc around the head, thereby providing different views of the brain (Figure 8–9). The amount of radiation absorbed by

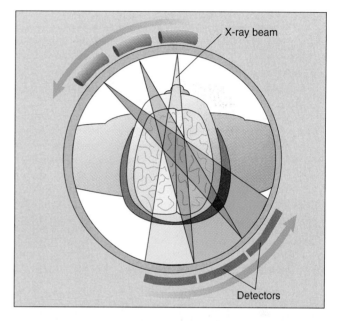

X-ray beam

Detectors

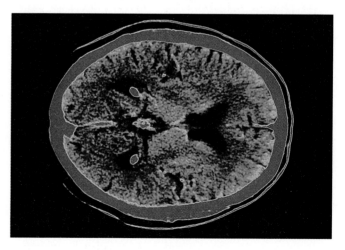

Figure 8–9 In a CT scan, x-ray images are taken through a section of the patient's head from different angles within an arc. The x-rays are absorbed to different degrees, depending on the density of the tissue. The computer then uses the information to construct an image of a cross-section through the brain. Serial cross-sections can be used to create a three-dimensional image of the brain. (© Scott Camazine/Photo Researchers, Inc.)

To Sleep, Perchance to Dream

The activities of the brain's billions of neurons produce a "hum" of electrical impulses that can be measured by placing electrodes at various locations on the scalp. The record of this activity over time is called an **electroencephalogram**, or EEG. The patterns of brain activity that are recorded in an EEG, the so-called **brain waves**, are correlated with the person's state of alertness (Figure A).

During each day, you experience a variety of states of consciousness. If your brain waves were recorded now, as you read this text, the EEG would show beta waves, which occur during a mentally alert state when the eyes are open. However, when you begin to get drowsy and close your eyes, alpha waves appear, indicating that you are still awake but are relaxed.

During a typical night's sleep, each hour and a half you will cycle through five stages of sleep that are identified by certain kinds of brain waves. *Stage 1* sleep occurs while you "drift off to sleep." During this period of transition from wakefulness to sleep, you become less aware of your surroundings and alpha brain waves give way to slower theta waves. Within the next few minutes, stage 2 sleep usually occurs. *Stage 2* sleep is characterized by two distinctive types of brain waves: bursts of rapid waves, called sleep spindles, interrupted occasionally by large slow waves called K-complexes. Muscle tension is lower than it is during wakefulness and breathing and heart rate also decrease. More than half the night's sleeping time is spent in stage 2 sleep. Soon afterward, delta waves appear. You are in *stage 3* sleep when the delta waves take up 20% to 50% of the recording time and in *stage 4* sleep if they represent more than 50%. These two stages are considered deep sleep. After about 20 minutes of stage 4 sleep, you usually switch back to stage 3 followed by stage 2. Then, you will begin an interval of a fifth stage of sleep, called **paradoxical sleep**, so named because the EEG shows erratic patterns of beta waves, the type of brain waves also seen during alert mental activity. Therefore, it seems that the brain is quite active at this time, but it is more difficult to awaken a person in paradoxical sleep than it is at any other stage. Paradoxical sleep is also called **rapid-eye-movement (REM) sleep** because, dur-

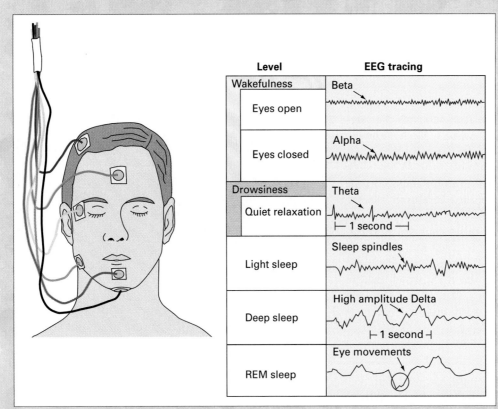

Level		EEG tracing
Wakefulness		Beta
	Eyes open	
	Eyes closed	Alpha
Drowsiness		Theta
	Quiet relaxation	⊢ 1 second ⊣
	Light sleep	Sleep spindles
	Deep sleep	High amplitude Delta ⊢ 1 second ⊣
	REM sleep	Eye movements

Figure A Brain waves are correlated with different levels of consciousness.

ing this interval, your eyes move rapidly behind closed eyelids. At this time, your heart and breathing rate can be quite variable. During REM sleep, almost every skeletal muscle in your body except for those of eyes and ears, are virtually paralyzed. This paralysis may keep us from hurting ourselves or others.

Adults repeatedly cycle through these stages all night (Figure B). The first bout of REM sleep usually occurs about 90 minutes after falling asleep and lasts about 10 minutes. The REM period of each successive cycle is a little longer. By morning, the interval of REM sleep may be about an hour long. So, the second half of a night's sleep is not identical to the first half. There is much more REM in the second half. For this reason, several short "cat naps" don't provide the same quality of sleep as a single, longer period of sleep, even if the total number of hours of sleep is identical.

REM sleep is significant because most dreams that have a story-like progression of events occur at this time. About 80% of the people who are awakened during REM sleep and asked what was going on say they were dreaming. Many fewer, about 20%, will report that they were dreaming if awakened during other stages of sleep. However, the dreams that occur outside of REM sleep are more likely to be a thought, an image, or an emotion than a story. Indeed, most nightmares occur during stage 3 and 4 sleep.

Some people may think that they do not dream, but this is not true. Everyone dreams. Generally, dreams occur at least once in every REM period. So, during a good night's sleep, a person may have a dozen or more dreams. Furthermore, dreams last about as long as they *seem* to last. That is, they proceed in real time. Nonetheless, dreams will be forgotten unless the person awakens during or immediately after a dream. Sometimes, the memory of a dream will fade away even in the middle of telling its story to someone.

How much sleep does one need to remain healthy? There's no set answer to this question. You need less sleep as you get older. A newborn sleeps 18 hours a day, young adults generally sleep 7 to 8 hours, and elderly people need only 4.5 to 6.5 hours. Sleep requirements also vary between individuals. A few people need as little as 3 hours of sleep a night. Others don't feel their best unless they get 10 hours or more. So, you need as much sleep as it takes to feel well and function efficiently the next day—and no more.

Millions of Americans cheat on sleep, however. Students pull "all-nighters" studying for exams. Some students must also fit jobs into their study schedules. Parents try to juggle jobs and family, and sometimes school as well. Workers cope with long shifts and long commutes. And, when does *fun* fit into the schedule? With such busy schedules, there often isn't enough time for a good night's sleep.

Without enough sleep, people can usually manage to get through the day doing simple things—walking, seeing, hearing—but they can't think clearly. They can't make appropriate judgments and their attention spans are shortened. Thus, sleepiness interferes with the ability to learn. Sleepy students often sit through class in a daze, experiencing it like a dream. Some nod off.

Sleep deprivation can also be hazardous to health. Drowsiness is a major cause of industrial accidents and traffic fatalities. Driving on Friday night is a greater risk than it is on Monday night, because so many drivers have been sleep deprived all week.

Ironically, because of the pressures of life, sleep does not come easily. Tens of millions of Americans will lie awake tonight, some of them for hours, suffering from insomnia. Insomnia occurs in different patterns—difficulty falling asleep, waking up during the night, or waking up earlier than desired. If it happens to you, don't worry. Most people have insomnia at some time. Occasional bouts of poor sleep will not harm your health. Furthermore, scientific studies have shown that people who complain about an inability to sleep are poor judges of the amount of sleep they actually get. About one-third of those who think they are lying awake are actually asleep. Some even dream that they are awake and trying to fall asleep. Most insomniacs overestimate the amount of time it takes to fall asleep and underestimate the duration of sleep. Most elderly people report problems with sleep. For some of them, there is a physical cause, such as chronic pain. However, many elderly people have not adjusted to their reduced need for sleep and worry when their customary sleep schedule changes, even though they are getting all the sleep they need.

What should you do if you occasionally have trouble sleeping? Here are some suggestions:

1. Establish a regular bedtime and a regular waking time. Don't sleep late, even on weekends. Don't nap.
2. Don't lie in bed if you can't sleep. The harder you try to fall asleep, the more aroused you become and the less likely you'll be able to fall asleep. Learn to associate the bedroom with sleep. To do this, get out of bed if you haven't fallen

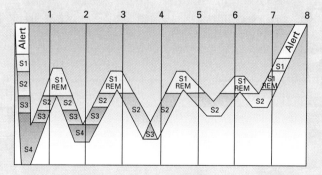

Figure B During a good night's sleep a person generally cycles through the five stages of sleep every 90 minutes.

asleep in 10 to 15 minutes and do something else until you are sleepy again.

3. Relax before bedtime—read a book, watch television, take a warm shower—whatever helps you unwind.

4. Avoid caffeinated beverages, such as coffee, tea, and cola, for 6 hours before bedtime. Chocolate also contains caffeine, so a cup of hot cocoa at bedtime is not a good idea.

5. Establish a regular pattern of exercise. Mild exercise promotes sleep and reduces stress. However, exercise in the late afternoon, not too close to bedtime.

6. Avoid drinking alcohol for 2 hours before bedtime. A small amount of alcohol makes you drowsy, but alcohol also disrupts sleep patterns. It doesn't increase the duration of sleep and it decreases REM sleep.

7. Don't take sleeping pills, even those sold over-the-counter in the drug store. At best, they only decrease the amount of time it takes to fall asleep by 10 to 20 minutes and lengthen the night's sleep by only 20 to 40 minutes. Like alcohol, sleeping pills decrease REM sleep and the REM sleep that occurs is not normal. Sleeping pills generally lose effectiveness within 2 weeks, so there is a tendency to increase the dosage. The effects of sleeping pills may linger into the next day, causing difficulty with coordination or memory. Furthermore, they have side effects, some of which are dangerous. The antihistamines in over-the-counter sleeping pills can provoke or worsen episodes of asthma, urinary retention, or glaucoma (increased pressure within the eye that can cause blindness).

the tissues depends on their densities. The x-rays that pass entirely through the head are converted to electrical signals by detectors. The electrical signals are then passed to the scanner's computer, which converts them to an image. A CT scan allows a physician to view successive sections of the patient's brain. The procedure takes only a few minutes and is painless. It is very useful in detecting brain tumors and areas in which there is hemorrhaging within the brain. Because a CT scan shows the skull bones, it is particularly useful for revealing the relationship between a part of the skull and the area of injury.

Another technique, magnetic resonance imaging (MRI), provides even clearer images of brain tissue because the picture has greater contrast and more gradations of contrast for soft tissue than does a CT scan. There is greater contrast in an MRI because the view of brain tissue is not obscured by the skull bones, which show up in CT scans. The picture of an MRI results from differences in the way the hydrogen nuclei in the water molecules within the tissues of the brain vibrate in response to a magnetic field around the head.

A new type of MRI, called a functional MRI, provides a way to see what the brain is doing. It shows the regions of the brain that are active while the person performs different activities. The blood flow to active regions of the brain is greater than it is to inactive regions. To produce a functional MRI, a computer compares the MRI scans of a brain at rest with an MRI scan while the person is performing some activity. When the image of the active brain is subtracted from the image of the resting brain, the active area stands out. It is even possible to take a series of images of the brain and create a "movie" of brain activity. Presently, there are only a dozen or so medical centers with the technology to produce

functional MRIs. However, it is thought that this will be *the* tool of the future.

Positron emission tomography (PET) is another method that can be used to measure the activity of various brain regions. The person being scanned is injected with a radioactive-labeled nutrient, usually glucose, which is tracked as it flows through the brain. The radioisotope emits positively charged particles, called positrons. When the positrons collide with electrons in the body, gamma rays are released. The gamma rays can be detected and recorded by PET receptors. Computers then use the information to construct a PET scan that shows where the radioisotope is being used in the brain. The more active regions of the brain use more glucose and they receive greater blood flow. In this way, it is possible to determine which areas of the brain are most active during specific tasks (Figure 8–10). PET scans can detect weaker signals than can a functional MRI. However, PET scans have at least two drawbacks: a person cannot be given frequent PET scans because the radioactivity could become a health hazard and the PET scanners cost about five times as much as an MRI machine.

Social Issue

Electrodes implanted in the pleasure centers of the brain can alleviate pain and depression. Under what conditions, if any, would it be ethical to use this as medical treatment?

The Spinal Cord

The other major component of the central nervous system besides the brain is the spinal cord. The **spinal cord** is a tube of neural tissue that is continuous with the medulla at the base of the brain and extends about 45 cm (17 in.) to just below the last rib. For most of its length, the spinal cord is about the diameter of your little finger. The spinal cord is slightly thicker in two regions, just below the neck and at the end of the cord, because of the large group of nerves connecting these regions of the cord with the arms and legs.

The spinal cord is encased in and protected by the stacked bones of the vertebral column (Figure 8–11). Spinal nerves (part of the peripheral nervous system) arise from the spinal cord and exit through the openings between each of the vertebrae to serve a specific part of the body. The vertebrae are separated by a cushion of cartilage. When one of these disks slips out of place (slipped disk), the vertebrae may no longer be properly aligned and may press on the nerve between them, causing back pain.

The cord itself does not extend the entire length of the vertebral column. The nerves that leave between the vertebrae of the lower back are part of a group of nerves at the base of the spinal column. Because this collection of nerves resembles a horse's tail, it has been named the cauda equina. The vertebral column of the lower back contains only these nerves and some cerebrospinal fluid, which bathes the entire central nervous system. A spinal tap, in which a needle is inserted between the vertebrae of the lower back, is always done in the region of the cauda equina because the rubbery nerves slide away from the needle. Physicians may perform a spinal tap in order to administer spinal anesthesia, which deadens the nerves serving the lower regions of the body.

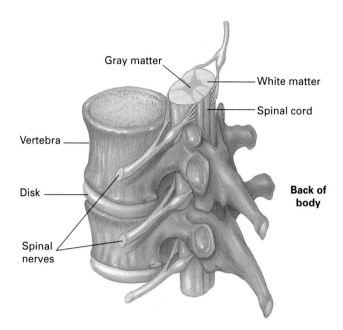

Figure 8–11 A cross section of the spinal cord showing the spinal nerves leaving through the gap between two vertebrae.

Spinal anesthesia is often administered to ease the pain of childbirth. A spinal tap may also be done to withdraw a sample of cerebrospinal fluid. Analysis of the fluid helps in diagnosing certain disorders of the central nervous system, including infection, multiple sclerosis, or stroke.

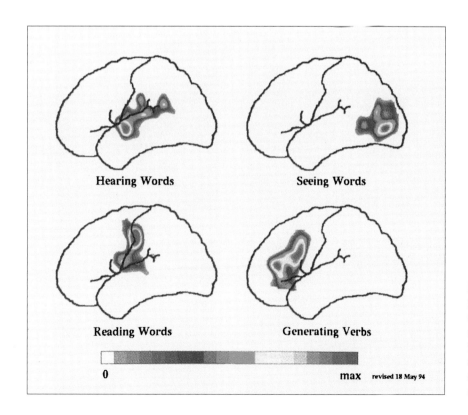

Figure 8–10 These PET scans of the brain indicate the regions that are active during different activities. PET scans are, therefore, a way to determine the functions of certain regions of the brain, in health and in disorders, as well as for diagnosing disorders such as brain tumors and epilepsy. (© *Marcus E. Raichle, M.D., Washington University School of Medicine, St. Louis, MO*)

The spinal cord has two functions: to transmit messages to and from the brain and to serve as a reflex center. The first function is performed primarily by the white matter, which is found in the outer regions of the spinal cord. Within the white matter are myelinated nerves grouped into tracts (Figure 8–12). Ascending tracts carry sensory information up to the brain. Descending tracts carry motor information from the brain to each spinal nerve.

The second main function of the spinal cord is to serve as a reflex center. Spinal reflexes are essentially "decisions" made by the spinal cord. A reflex action is an automatic response to a stimulus. Reflexes are pre-wired in a circuit of neurons, called a **reflex arc**, which consists of a receptor, a sensory neuron (that brings information from the receptors toward the central nervous system), usually at least one interneuron, a motor neuron (that brings information from the central nervous system toward an effector), and an effector (Figure 8–13). Shaped somewhat like a butterfly in the central region of the spinal cord, the gray matter houses the interneurons and the cell bodies of motor neurons involved in reflexes.

Spinal reflexes are beneficial when a speedy reaction is important to a person's safety. Consider, for example, the withdrawal reflex. When you step on a piece of broken glass, impulses speed toward the spinal cord over sensory nerves. Within the gray matter of the spinal cord, the sensory neuron synapses with an interneuron. The interneuron, in turn, synapses with a motor neuron that sends a message to the appropriate muscle to contract and lift your foot off the glass.

Note that as you lift one foot, you must maintain your balance while shifting your weight to the other leg. A reflex called the cross-extensor reflex accomplishes this. Within a fraction of a second of the initiation of the withdrawal reflex in one leg, the opposite leg begins to stiffen. Thus, the reflexive withdrawal of a foot may, in the end, involve hundreds of muscle fibers.

While the spinal reflexes were removing the foot from the glass, pain messages from the cut foot were sent to the brain through ascending tracts in the spinal cord. However, it takes longer to get a message to the brain than it does to the spinal cord because the distance and number of synapses involved is greater. Therefore, by the time pain messages reach the brain, you have already withdrawn your foot. Nonetheless, once the sensory information reaches the conscious brain, decisions can be made about how to care for the wound.

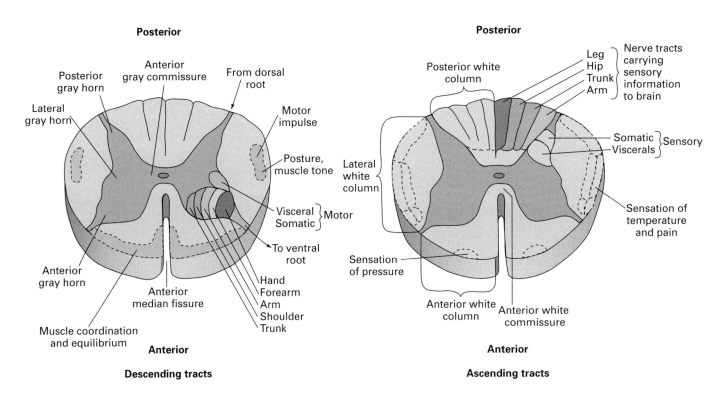

Figure 8–12 The white matter of the spinal cord consists of bundles, or tracts, of myelinated axons carrying sensory information up to the brain (ascending tracts) or motor messages down from the brain (descending tracts).

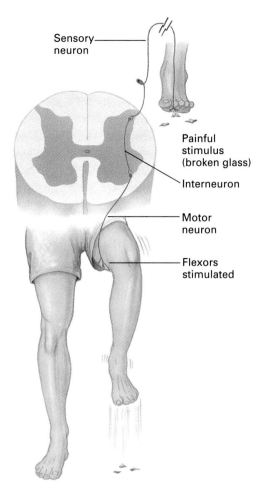

Sensory neuron

Painful stimulus (broken glass)

Interneuron

Motor neuron

Flexors stimulated

Figure 8–13 A reflex arc consists of a sensory receptor, sensory neuron, usually at least one interneuron, a motor neuron, and an effector. When you step on a piece of broken glass, sensory messages are carried to the spinal cord by a sensory neuron, which synapses with an interneuron in the gray matter of the spinal cord. The interneuron synapses with a motor neuron that sends a message to the muscles of the leg, causing them to contract and lift your foot off the glass.

Protection and Support of the Central Nervous System

Neurons cannot divide and produce new cells. Therefore, if a neuron is damaged or dies, it cannot be replaced. Neurons are very fragile. The brain and spinal cord are protected by bony cases (the skull and vertebral column), membranes (the meninges), and a fluid cushion (cerebrospinal fluid).

Meninges

The meninges consist of three connective tissue membranes. The outermost layer, the **dura mater** (Latin meaning "hard mother"), is tough and leathery. Around the brain, the dura mater has two layers that are separated by a fluid-filled space containing blood vessels. Additional blood ves-

sels are found in the space beneath the dura mater. Severe head injury can damage either of these sets of blood vessels and allow blood to accumulate in the spaces. The blood presses on the soft tissue of the brain and, depending on the quantity, can damage or kill the neurons. The most dangerous head injuries cause an epidural hemorrhage, in which an artery in the space between the two layers of the dura mater breaks. An untreated epidural hemorrhage is always fatal. If the blood is removed and the damaged vessel repaired, the person has a 50% chance of survival. Beneath the dura mater is the **arachnoid** (Latin, meaning "like a cobweb"). The arachnoid is anchored to the next lower layer of meninges by thin thread-like extensions that resemble a spider's web (hence the name of the layer). The innermost layer is the **pia mater** (Latin meaning "tender mother"), and it is molded around the brain. Fitting like a leotard, the pia mater dips into every irregularity on the brain's surface.

Certain bacteria and viruses can cause inflammation of the meninges, a condition called **meningitis** when it affects the meninges around the spinal cord and **encephalitis** when it affects the meninges of the brain. These are very serious conditions because the infection can spread to the underlying nervous tissue. Diagnosis is usually done by studying the cerebrospinal fluid (discussed in the next section). If bacteria are the cause, the person is treated with antibiotics. The treatment for viral meningitis includes medicines to alleviate pain and fever and to keep the patient comfortable while the body's immune system fights the virus.

Cerebrospinal Fluid

The **cerebrospinal fluid**, which fills the space between the arachnoid and the pia mater as well as the internal cavities (ventricles) of the brain and spinal cord, has several important functions: (1) It serves as a shock absorber. Just as an air bag protects the driver of a car by preventing impact with the steering wheel, the cerebrospinal fluid protects the brain by cushioning its impact with the skull during blows or other head trauma. (2) The cerebrospinal fluid supports the brain. Because the brain floats in the cerebrospinal fluid, it is not crushed under its own weight. (3) The cerebrospinal fluid also nourishes the brain, delivers chemical messengers, and removes waste products.

An average adult has between 120 and 150 ml of cerebrospinal fluid, slightly more than half a cup. It is continuously produced by clusters of capillaries in certain chambers (ventricles) within the brain. It circulates from there through the central canal of the spinal cord to the space beneath the arachnoid and back to the ventricles, where it is reabsorbed into the blood. The rates of production and drainage of cerebrospinal fluid are normally equal. If the rates are not matched, the accumulation of fluid causes increased pressure on the brain that can damage the brain. The bones of an infant's skull are not yet fused. Therefore, if fluid accumulates in an infant's brain, the head enlarges. This condition, known as hydrocephalus, can be treated by

Figure 8–14 The enlarged heads of these 18-month-old twins are caused by hydrocephalus. Hydrocephalus results when cerebrospinal fluid is produced at a greater rate than it is drained. The head of an infant or young child enlarges with the increased fluid pressure because the skull bones have not yet fused. *(Dr. P. Marazzi/Science Photo Library/Custom Medical Stock Photo)*

installing a drainage system that channels the fluid into the abdominal cavity where it can be absorbed (Figure 8–14).

Social Issue

Head injuries are the leading cause of bicycle-related deaths. In spite of this, fewer than 10% of bicycle riders in America wear helmets. The failure to protect the head from injury while riding a bicycle results in one death every day and one head injury every 4 minutes. Most of those who die are children. Do you think that there should be laws requiring bicyclists to wear helmets? If so, what should the penalty be for not wearing a helmet? Who should pay the penalty if the cyclist without a helmet is a child?

The Peripheral Nervous System

The peripheral nervous system links the brain and spinal cord to the real world. It includes (1) the sensory receptors (2) the peripheral nerves and ganglia and (3) the specialized motor endings that stimulate the effectors, that is, the muscles or glands that respond to stimulation. We will discuss the sensory receptors in Chapter 9. The motor endings were described in Chapter 5. Here we will concentrate primarily on the nerves and the ganglia, which are collections of neurons and their associated nerve fibers.

The vast network of nerves of the peripheral nervous system originates as 12 pairs of cranial nerves, which arise from the brain and service the structures of the head and certain body parts such as the heart and diaphragm, and 31 pairs of spinal nerves, each of which services a specific region of the body. Some cranial nerves carry only sensory fibers, others carry only motor fibers, and others carry both types of fibers (Figure 8–15). All spinal nerves carry both sensory and motor fibers. The fibers from sensory neurons enter the spinal cord from the dorsal, or posterior, side. They are grouped and the bundle is called the dorsal root of the spinal nerve. The cell bodies of the sensory neurons are located in a ganglion in the dorsal root. Motor neurons have their cell bodies in the gray matter of the spinal cord and their axons leave the ventral (or belly side) of the spinal cord in a bundle called the ventral nerve root (Figure 8–16). The dorsal and ventral roots join to form a single spinal nerve, which passes through the opening between the vertebrae.

The Somatic Nervous System

The peripheral nervous system is subdivided into the somatic nervous system and the autonomic nervous system. The somatic nervous system carries sensory messages that tell us about the world around us. We are generally aware of these sensations. In addition, the somatic nervous system carries the commands from the brain to our voluntary muscles.

The Autonomic Nervous System

The autonomic nervous system automatically adjusts the functioning of our body organs so that an internal stability is maintained and the body is able to meet the demands of its interactions with the world around it.

The autonomic nervous system consists of two branches—the sympathetic and the parasympathetic nervous systems. The **sympathetic nervous system** gears the body to face an emergency or stressful situation. In contrast, the **parasympathetic nervous system** adjusts bodily function so that energy is conserved during nonstressful times.

Both the parasympathetic and the sympathetic nervous systems send nerve fibers to innervate most, but not all, internal organs (Figure 8–17). When both systems innervate an organ, they have opposite effects on its functioning. If one stimulates, the other inhibits. The antagonistic effects are brought about by different neurotransmitters. Whereas sympathetic neurons secrete norepinephrine at their target organs, parasympathetic neurons secrete acetylcholine at their target organs.

The sympathetic nervous system acts as a whole, bringing about all its effects at once, because its neurons are connected through a chain of ganglia. A unified response is exactly what is needed in an emergency. To face the threat, certain cells of the body will have an increased need for energy. For example, the skeletal muscles will need energy to carry out the appropriate response. That energy will come

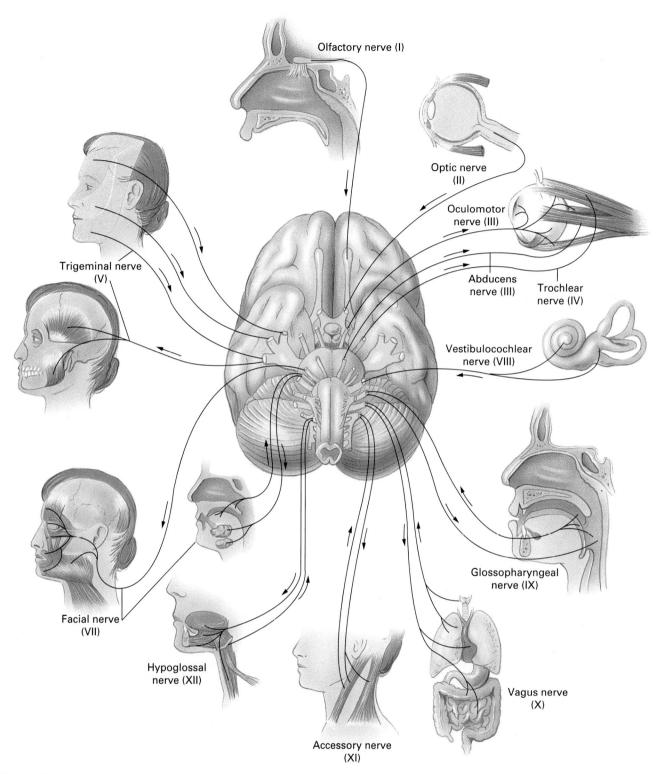

Olfactory nerve (I)

Optic nerve (II)

Oculomotor nerve (III)

Abducens nerve (III)

Trochlear nerve (IV)

Vestibulocochlear nerve (VIII)

Trigeminal nerve (V)

Facial nerve (VII)

Hypoglossal nerve (XII)

Accessory nerve (XI)

Vagus nerve (X)

Glossopharyngeal nerve (IX)

Figure 8–15   The 12 pairs of cranial nerves can be seen in this view of the underside of the brain. Most cranial nerves service structures within the head, but some, such as the vagus, service the lower body organs.

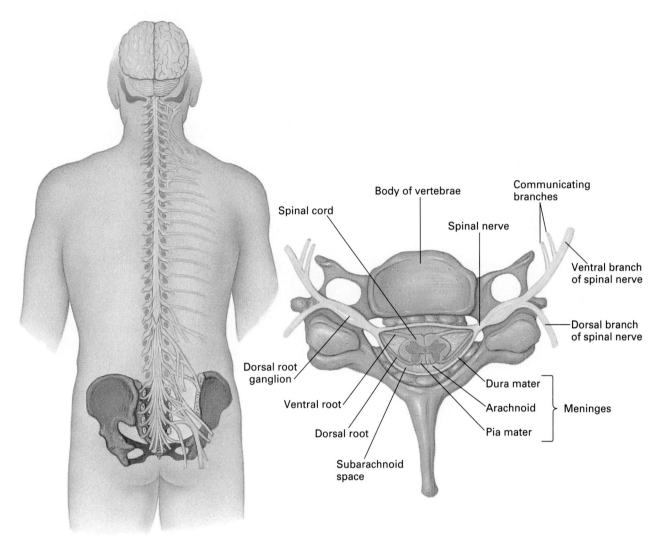

Figure 8-16 Each spinal nerve is formed from a dorsal root that carries sensory information to the spinal cord and a ventral motor root that carries motor messages away from the spinal cord.

from the glucose released from the liver. In the presence of oxygen, the energy in glucose is converted to a form that cells can use. The increased oxygen demand is met by faster, deeper breathing. The oxygen is delivered to the cells more rapidly because the heart beats faster and pressure increases. Blood is directed to the brain and skeletal muscles. The adrenal medulla is stimulated to release two hormones, epinephrine and some norepinephrine, into the bloodstream. These hormones back up and prolong the effects of sympathetic stimulation. To be most effective, all these responses must occur together. In a crisis, digesting the previous meal is hardly a priority. The digestive activity is, therefore, inhibited. The pupils of the eyes dilate and let more light reach the receptors so that vision is improved.

The effects of the parasympathetic nervous system occur more independently of one another. After the emergency, organ systems return to a relaxed state at their own pace. This is also based on the anatomy of the system. The parasympathetic neurons that stimulate the target organs originate in ganglia that are located near the organs.

Possible Problems with the Nervous System

Headaches

Excessive exercise may make your muscles hurt. However, thinking too much can't cause a headache. The brain has no pain receptors, so a headache is not a brain ache. But, this fact is a small consolation to someone who has a headache and has to function normally. Headaches often present a no-

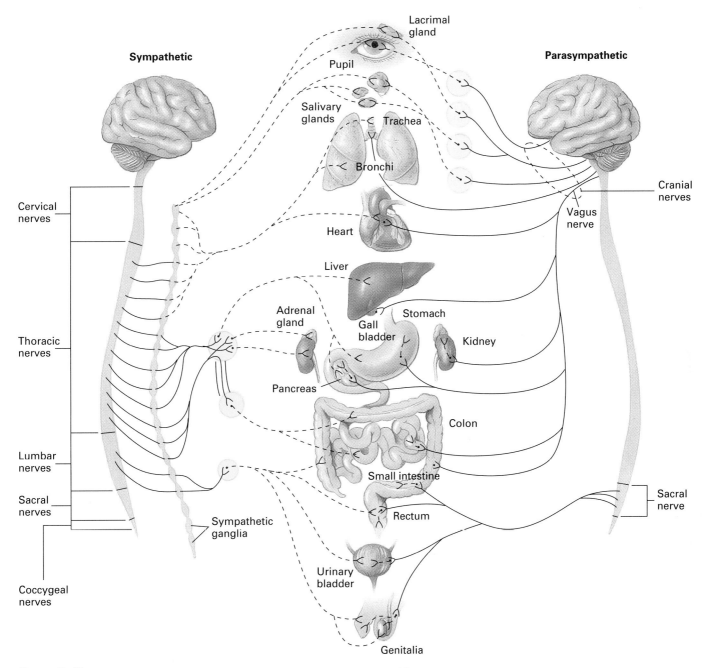

Figure 8–17 The structure and function of the autonomic nervous system. Most organs are innervated by fibers from both the sympathetic and the parasympathetic nervous system. When this occurs, the two branches of the autonomic nervous system have opposite effects on the level of activity of that organ. The sympathetic nervous system is linked by a chain of ganglia. It usually acts as a unit, with all its effects occurring together. The ganglia of the parasympathetic nervous system are near the organ serviced. Its effects are more localized.

win situation—they can be caused by stress or by relaxation, by hunger or by eating the wrong thing, or by too much or too little sleep.

The most common type of headache is a tension headache, affecting some 60% to 80% of people who suffer frequent headaches. In response to stress, most of us uncon-sciously contract our head, face, and neck muscles. There-fore, the pain of a tension headache is usually a dull, steady ache that is often felt as a tight band around the head, as if your head were in a vise.

Migraine headaches are so named because about 70% of the time they are confined to one side of the head, often centered behind one eye. (The term migraine comes from the Greek word meaning "half a head.") A migraine headache typically causes a throbbing pain that increases with each beat of the heart. It is sometimes called a "sick headache" because it often causes nausea and vomiting. Some migraine sufferers experience an aura, a group of sensory symptoms, just before an attack. The aura may be visual disturbances (a blind spot, zigzag lines, flashing lights); auditory hallucinations, or numbness. The characters described by Lewis Carroll in *Alice in Wonderland* were visual hallucinations he experienced before a migraine attack.

The disturbances associated with a migraine headache are centered in cranial blood vessels. The aura results when the blood vessels constrict, reducing the oxygen supply to the brain. The pain is caused by the stretching of blood vessels, which sets off a cascade of events including an inflammatory response that results in fluid accumulation in the brain tissue and a drop in endorphins, the body's natural painkillers.

Migraines are set off by an imbalance in the brain's chemistry. The level of one of the brain's chemical neurotransmitters, serotonin, is low. With too little serotonin, pain messages flood the brain.

Most migraine sufferers have a family history of migraine and most are women. In women, migraines can be triggered by hormones. They are much more common around the time of the menstrual period. Migraines can also be caused by foods that contain tyramine, a protein that causes the widening of blood vessels and the release of norepinephrine, the stress-related hormone. Tyramine-containing foods include aged cheese, pickled herring, and lima beans. Other vasodilators, such as alcohol, the food preservative sodium nitrate, or the flavor enhancer MSG (monosodium glutamate) are other common causes of migraines.

Cluster headaches are also vascular headaches, but unlike migraines, they tend to occur in groups. They occur two or three times a day for days or weeks. The headache causes severe pain that lasts for a few minutes to a few hours and often awakens the person from a sound sleep. Although different attacks may affect opposite sides of the head, each bout affects only one side. Besides the pain, there is tearing of the eye on that side of the head and a runny nose.

Stroke

Neurons have a high demand for both oxygen and glucose. Therefore, when the blood supply to a portion of the brain is shut off, the affected neurons begin to die within minutes. This is called a stroke or a cerebrovascular accident. The extent and location of the mental or physical impairment caused by a stroke depend on the region of the brain involved. If the left side of the brain is affected, the person may lose sensations in or the ability to move parts of the right side of his or her body. Since the language centers are usually on the left hemisphere, the person may have difficulty speaking. When the stroke damages the right rear of the brain, some people show what is called the neglect syndrome and behave as if the left side of everything, even their own bodies, does not exist. The person may comb only the hair on the right side of the head or eat only the food on the right side of the plate.

Common causes of strokes include hemorrhage from the rupture of a blood vessel in one of the protective layers surrounding the brain (the pia mater), blood clots that block a vessel, or atherosclerosis, the formation of fatty deposits that block a vessel. The risk of stroke is increased by high blood pressure, heart disease, diabetes, smoking, obesity, and excessive alcohol intake.

Coma

Although a person in a coma seems to be asleep—eyes closed and no recognizable speech—a coma is not deep sleep. A comatose person is totally unresponsive to all sensory input and cannot be awakened. Although the cerebral cortex is most directly responsible for consciousness, damage to the cerebrum is rarely the cause of coma. Instead, coma is caused by trauma to neurons in regions of the brain responsible for stimulating the cerebrum, particularly those in the reticular activating system or thalamus. Coma can be caused by mechanical shock, such as might be caused by a blow to the head, tumors, infections, drug overdose (from barbiturates, alcohol, opiates, or aspirin), or failure of the liver or kidney.

Spinal Cord Injuries

Spinal cord injuries afflict about 10,000 Americans each year. Most of the injuries occur during sporting events such as diving or skiing, in automobile accidents, or from gunshot wounds. Immediately after a severe injury there is a period of spinal shock, during which the person is paralyzed, has no sensations of touch, pain, heat or cold, and has no reflex activity. Spinal shock is often seen in whiplash injuries of the neck. The extent and location of the injury will determine how long these symptoms persist, as well as the degree of permanent damage. Injury to the spinal cord affects the body below the region of injury. Depending on which tracts are damaged, injury results in the loss of sensation, paralysis, or both. If the cord is completely severed, there is a complete loss of sensation and voluntary movement below the level of the cut.

Damage to the spinal cord continues for about 2 days after injury. Dying nerve cells release toxins that attack neighboring neurons that managed to survive. Additional neurons are killed because their vital blood supply is shut down as the injured area becomes inflamed. However, if a new anti-inflammatory steroid drug, methylprednisolone, is administered within 8 hours of the injury, the amount of secondary damage might be cut in half. Sometimes, this makes the difference of whether the person will be able to walk again.

SUMMARY

1. The nervous system is divided into the central nervous system (the brain and spinal cord) and the peripheral nervous system (all the neural tissue outside the central nervous system). Ganglia are clusters of nerve cell bodies located outside the central nervous system. The peripheral nervous system can be further subdivided into the somatic nervous system and the autonomic nervous system.

2. The cerebrum is the thinking, conscious part of the brain. It consists of two hemispheres. Each hemisphere receives sensory impressions from and directs the movements of the opposite side of the body. The cerebrum consists of (1) an outer layer of gray matter called the cortex (2) an underlying layer of white matter consisting of myelinated nerve tracts that allow communication between various regions of the brain and (3) basal ganglia, which are important in the initiation of skeletal movements. The cerebral cortex has three types of functional areas: sensory, motor, and association.

3. The thalamus is an important relay station for all sensory experience except smell. It also plays a role in motor activity, cortical arousal, and memory.

4. The hypothalamus is essential in maintaining a stable environment within the body. It regulates many vital physiological functions, such as blood pressure, heart rate, breathing rate, digestion, and body temperature. The hypothalamus coordinates the activities of the nervous and endocrine systems through its connection to the pituitary gland. As part of the limbic system, the hypothalamus is a center for emotions. It also serves as a "master biological clock."

5. The primary function of the cerebellum is sensory-motor coordination. It integrates information from the motor cortex and sensory pathways to produce smooth movements. The cerebellum also stores memories of learned motor skills.

6. The medulla regulates breathing, heart rate, and blood pressure. It also serves as a pathway for all sensory messages to higher brain centers and motor messages leaving the brain.

7. The limbic system, which includes several brain structures, is largely responsible for emotions.

8. The reticular activating system filters sensory input and keeps the cerebral cortex in an alert state.

9. Computed tomography (CT scanning) and magnetic resonance imaging (MRI) are noninvasive ways to see the structures of the brain. They are helpful in detecting brain tumors and cerebral hemorrhages. Positron emission tomography (PET) reveals the areas of the brain that are most active during different intellectual or physical tasks.

10. The spinal cord is a tube of nerve tissue extending from the medulla to approximately the bottom of the rib cage. Spinal nerves arise from the cord and exit through the openings between the stacked vertebrae of the vertebral column. The spinal cord has two functions: to conduct messages between the brain and the body and to serve as a reflex center.

11. The brain and spinal cord are protected by the bones of the vertebral column and skull, three connective tissue membranes that form the meninges, and cerebrospinal fluid.

12. The peripheral nervous system includes the sensory receptors, the peripheral nerves and ganglia (collections of nerve cell bodies and their associated nerve fibers), and specialized motor endings that stimulate the effectors.

13. The peripheral nervous system is divided into the somatic nervous system, which governs conscious sensations and voluntary movements, and the autonomic nervous system, which is concerned with our unconscious, involuntary internal activities.

14. The autonomic nervous system can be divided into the sympathetic and parasympathetic nervous systems, two branches with antagonistic actions. The sympathetic nervous system gears the body to face stressful or emergency situations. The parasympathetic nervous system adjusts body functioning so that energy is conserved during restful times.

REVIEW QUESTIONS

1. Distinguish between the central nervous system and the peripheral nervous system. List the components of each.

2. List the three regions of the cerebrum and give the general function of each.

3. What are the three types of functional areas of the cerebral cortex?

4. In what way is the organization of the primary sensory area and the primary motor area of the cerebral cortex similar? In what way do these areas differ?

5. Describe the specializations of the left and right cerebral hemispheres.

6. List five functions of the hypothalamus.

7. What is the function of the cerebellum?

8. Which functional system of the brain is responsible for emotions?

9. Describe the two functions of the reticular activating system.

10. What can a PET scan reveal that an MRI cannot?

11. List the two functions of the spinal cord and relate each function to the structure of the spinal cord.

12. You are cooking dinner and carelessly touch the hot burner on the stove. You remove your hand before you are even aware of the pain. Explain how this occurs using the anatomy of a spinal reflex arc.

13. What are the functions of cerebrospinal fluid?

14. What are the two divisions of the peripheral nervous system? What type of response does each control?

15. Compare and contrast the functions of the sympathetic and parasympathetic nervous systems.

16. List some effects of sympathetic stimulation and explain how these prepare the body for an emergency.

C h a p t e r

8A

A law enforcement officer sleeping on a pile of confiscated marijuana. *(© Frank Oberle/Photo Resources)*

Drugs and the Mind

As soon as humans realize they have a mind, they set about to alter it—or so it sometimes seems: Young children will twirl around until they fall to the ground in dizziness; the pain of a simple headache will drive most of us to reach for a bottle of aspirin; we often reach for a cup of coffee in the morning to clear the cobwebs from our heads; and, we often raise champagne glasses in celebration. Indeed, the use of mind-altering drugs has been common throughout history and among most cultures. So, let's take a closer look at some of the more common mind-altering drugs.

Drugs that alter one's mood or emotional state are often described as psychoactive drugs. The mind-altering effects of these drugs result from their ability to alter communication between nerve cells. As we saw in Chapter 7, nerve cells (neurons) communicate with one another using chemicals called neurotransmitters. Neurotransmitters are released by one neuron, diffuse across a small gap, and bind to specific receptors on another neuron, triggering changes that alter the activity of the second neuron. The action of the neurotransmitter is stopped almost immediately, either because it is broken down by enzymes or because it is taken back into the cell that released it.

A psychoactive drug may alter this communication in any of several ways. It may stimulate the release of neurotransmitter, thereby enhancing the response. Alternatively, it may inhibit the release of neurotransmitter and dampen the response. A drug may also augment and prolong the effect of a neurotransmitter by interfering with its removal from the synapse. If the drug is chemically similar to the normal neurotransmitter, it may bind to the receptor and would affect the activity of the receiving neuron in the same manner as the neurotransmitter. Finally, a drug may bind to the receptor and prevent the neurotransmitter from acting at all.

One of the problems with using psychoactive drugs is that the user may develop some level of dependence on the drug, so before we consider the drugs themselves, let's consider the matter of dependence. A summary of the health risks and likelihood of developing dependence to the mind-altering drugs discussed in this chapter is found in Table 8A–1.

Drug Dependence

It is difficult to precisely define drug dependence. A loose interpretation might be, "It is what causes a person to continue using a drug."

Tolerance is a progressive decrease in the effectiveness of a drug. As tolerance develops, larger or more frequent doses of the drug must be taken to produce the same effect. Tolerance develops partly because the body steps up its production of enzymes that break down the drug and partly because of changes in the nerve cells that make them less responsive to the drug. When a drug elevates the level of neurotransmitter in the synapse for a prolonged period of time not only can the supply of neurotransmitter be depleted but the nerve cells can begin to make fewer receptors for that neurotransmitter. With fewer receptors, the nerve cell becomes less sensitive to stimulation. The normal level of neurotransmitter is no longer sufficient to adequately stimulate the nerve cells. The drug is then needed to maintain normal functioning. At this point, we describe the person as being **physically dependent** on the drug. **Cross tolerance** is the development of tolerance for a second drug, not taken, by taking another, usually similar drug. If one abuses codeine, for instance, tolerance develops not just for codeine but also for the other opioids, such as morphine and heroin.

Another reason that certain drugs cause users to continue using them is that the drugs stimulate the "pleasure" centers in the limbic system of the brain. An animal with electrodes implanted in the pleasure center will quickly learn to press a lever to stimulate this brain region. If permitted, it will self-stimulate repeatedly, sometimes hundreds of times an hour, until it is exhausted. When the experiment is changed so that a small dose of a drug is released into the blood of an animal when it presses a lever, the animal will learn to press the lever to self-administer the drug—if the drug stimulates the pleasure center. Among the drugs that stimulate the pleasure center are cocaine, amphetamine, morphine, and nicotine. (Nicotine is discussed further in Chapter14A, Smoking.) The stimulation of the pleasure center, then, is one reason that people continue to use these drugs.

When deciding whether to use a drug, there are several issues that should be considered. One is safety—short-term and long-term. Safety issues include not only health risks associated with the use of the drug but also the degree to which it leads to tolerance and dependence. Another issue to be considered is whether the drug interferes with the ability to meet goals. Is the drug legal? If not, its use could have serious consequences on the rest of a person's life because a felony conviction is serious and a misdemeanor offense looks bad on one's record. If a drug is legal, it can still be abused, sometimes seriously abused. Therefore, as we consider some of the psychoactive drugs, we will discuss their

Table 8A–1   The Mind-Altering Effects of Drugs

Drug	How Used	Effects/Risks	Dependence
Marijuana (*Cannabis*)	Smoked or eaten	Mild euphoria; lung damage (if smoked); accumulation in fatty tissues	Might be slightly psychological and physical
Depressants Alcohol	Swallowed	Euphoria; damage to liver, gastrointestinal system, nerve cells, heart, and skeletal system; slow reflexes; high dose leads to loss of motor coordination	Physical and psychological
Barbiturates	Taken orally	Reduced anxiety, decreased alertness; drowsiness; loss of motor coordination; death	Great potential for dependence
Quaaludes	Taken orally	Reduced anxiety; relaxation; sleep	Physical
Stimulants Cocaine	Inhaled through nasal passages; injected; smoked	Euphoria; increased energy; increased blood pressure	Physical and psychological
Amphetamines	Taken orally; injected; inhaled through nasal passages	Increased energy; alertness; loss of appetite	Physical and psychological
Caffeine	Swallowed in tablets or beverage	Heightened alertness; increased blood pressure; high doses lead to irritability	Some psychological
Psychedelics LSD	Taken orally; licked off paper	Distortion of time, space, and sensory input; "bad trips"	Might be some psychological
Narcotics Heroin	Injected	Euphoria; muscle relaxation	Strong, physical dependence

habit-forming potential as well as possible health risks associated with their use. We will begin with what seems to be society's old standby as the drug of choice.

Alcohol

Beer, wine, scotch, or gin, whatever the form of the "booze" chosen, the alcohol is ethanol and it is produced as a by-product of fermentation when yeast cells break down sugar to release energy for their own use. The taste of the potion is determined by where the sugar comes from, such as which fruit or vegetable is fermented: grapes in wine, juniper berries in gin, barley in beer, and malt in scotch.

Ethanol is ethanol, no matter where it comes from, and its effects depend on how much of it there is in the blood. The blood level in turn depends on several factors, including how much is consumed, how rapidly it is consumed (over what period of time), and its rate of absorption, distribution, and metabolism.

A "drink" can mean different things to different people. For some it is a can of beer, for others a glass of wine, and for others scotch on the rocks. Whatever the drink, the effect on one's body will be determined by the *amount of ethanol* in the beverage and this can vary tremendously. The strength of each brew depends on how it has been treated. Natural fermentation cannot yield more than 15% alcohol, because the alcohol kills the yeast cells producing it. The alcohol content of beers generally varies from 4.5% in light beers to 6% or 7% in the dark brews, such as stout, porter, or bock. Most American wines contain 12% to 14% alcohol. French and German wines usually contain less alcohol, between 8.5% and 10%. Sherry and port are more potent, averaging between 18% and 21% alcohol. Liquor is produced by distillation, a process that concentrates the alcohol. The alcohol content of distilled spirits is measured as "proof." One degree of proof equals 0.5% alcohol. Most distilled spirits (vodka, gin, scotch, whiskey, rum, brandy, and cognacs) are 80 proof, which means they contain 40% alcohol, but some brands are 90 proof (45% alcohol) or even 100 proof (50%). Clearly, a 3-oz Manhattan, which contains only distilled spirits and therefore about 1.2 oz of alcohol, is more intoxicating than an 8-oz mug of beer, which would contain about 0.4 oz of alcohol.

We see, then, that there is a need to standardize the meaning of a "drink" when discussing the effects of alcohol. In most studies, the standard "drink" contains 0.5 oz of alcohol. So, any of the following would be considered to be equivalent to one drink:

1 jigger (1.5 oz)	80 proof distilled spirits
1 oz	100 proof distilled spirits
12 oz	lager beer
8 oz	stout
5 oz	French wine
4 oz	American wine
3 oz	sherry

Absorption

The intoxicating effects of alcohol begin when it is absorbed from the digestive system into the blood and is delivered to the brain. As a rule, the rate of absorption of alcohol depends on the concentration of alcohol. The higher the concentration, the faster is the absorption. So, wine or beer or distilled spirits diluted with a mixer will be absorbed more slowly than pure liquor. The choice of mixer will also influence the rate of absorption. Carbonated beverages speed the rate of absorption because of the pressure of the gas bubbles.

Although very few substances are absorbed across the walls of the stomach, about 20% of the alcohol consumed is absorbed here. The remaining alcohol is absorbed through the intestines. Because alcohol can be absorbed from the

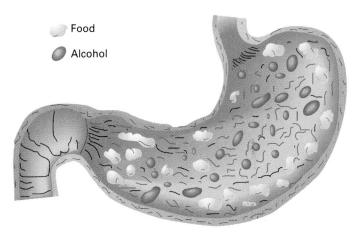

Food

Alcohol

Figure 8A–1 Food in the stomach slows the absorption of alcohol. Food dilutes the alcohol and covers part of the surface through which alcohol is absorbed. It also keeps alcohol in the stomach longer, slowing the rate of absorption from the intestines. Therefore, if you do drink, it is wise to have food in your stomach.

stomach, one begins to feel the effects of a drink fairly quickly, usually within about 15 minutes. The presence of food in the stomach slows alcohol absorption because it dilutes the alcohol, covers some of the stomach membranes through which alcohol would be absorbed, and slows the rate at which the alcohol is released to the intestines (Figure 8A–1). Thus, it is never a good idea to drink on an empty stomach.

Distribution

Ethanol is a small molecule that is soluble in both fat and water, so it is distributed to all body tissues. Consequently, after consuming equal amounts of alcohol, a large person, whether obese or muscular, would have a lower blood alcohol level and, therefore, be less intoxicated than would a small, slender person. (Figure 8A–2).

Figure 8A–2 Alcohol consumption can impair driving. Blood level of alcohol depends on both the number of drinks consumed and body size. A small person has a higher blood alcohol level than a large person after consuming the same amount of alcohol. It's the blood level of alcohol that determines the effect on the nervous system.

Drinks (2 hr. period)
1½ oz liquor or 12 oz beer

Weight												
100	1	2	3	4	5	6	7	8	9	10	11	12
120	1	2	3	4	5	6	7	8	9	10	11	12
140	1	2	3	4	5	6	7	8	9	10	11	12
160	1	2	3	4	5	6	7	8	9	10	11	12
180	1	2	3	4	5	6	7	8	9	10	11	12
200	1	2	3	4	5	6	7	8	9	10	11	12
220	1	2	3	4	5	6	7	8	9	10	11	12
240	1	2	3	4	5	6	7	8	9	10	11	12

Be careful driving
BAC to .05%

Driving will be impaired
.05–.09%

Do not drive
.10% & up

Effects of drinking on driving

Elimination

Ninety-five percent of the alcohol that enters the body is metabolized (broken down) before it is eliminated. Most of that metabolism occurs in the liver, which converts alcohol to carbon dioxide and water. The rate of metabolism is slow, about one-third of an ounce of pure ethanol per hour, and unlike most other drugs, the rate does not increase with concentration. In practical terms, this means that it takes slightly more than an hour for the liver to break down the alcohol contained in one standard "drink"—a can of beer or a glass of wine. Since alcohol cannot be stored in the body, it continues to circulate in the bloodstream until it is metabolized. Therefore, if more alcohol is consumed in an hour than is metabolized, both the blood alcohol level and the degree of intoxication increase.

There is no way to increase the rate of alcohol metabolism by the liver and, therefore, no way to sober up quickly. A cup of coffee may slightly counter the drowsiness caused by alcohol, but it does not reduce the level of intoxication. Giving coffee to a person who has had too much to drink merely produces a wide-awake drunk. Furthermore, since alcohol is not metabolized by muscles, exercise doesn't help. So, walking around the block won't make a person sober either. Nor will a cold shower. If you or a friend have had too much to drink, the best thing to do is to sleep it off.

Men can usually outdrink women. Part of the explanation for this is based on anatomy. Women are usually smaller than men. In addition, a woman's body contains a higher percentage of fat than does a man's. As a result, the alcohol is diluted more slowly in a woman's body and the effects are prolonged. But there's more to the story—it also has to do with differences in metabolism. Although the liver is the primary site for alcohol metabolism, some is broken down by an enzyme found in the stomach lining. Alcohol that is metabolized in the stomach never enters the bloodstream and cannot cause intoxication. It turns out that women have less of this enzyme in their stomach linings than do men. Consequently, a woman absorbs about 30% more of the alcohol in a drink than a man does. When weight differences between an average man and woman are also taken into account, it can mean that 2 oz of liquor can have approximately the same effect on a woman as 4 oz would have on a man.

A small amount of alcohol, about 5%, is eliminated from the body unchanged through the lungs or in the urine. It is the alcohol eliminated from the lungs that forms the basis of the breathalyzer test that may be administered by law enforcement officers who want an on-the-spot sobriety check.

Health-Related Effects

Roughly 70% of the Americans over the legal drinking age consume alcohol. Nearly 10% of those who do drink do so in excess or have lost some degree of control over their habit. One-third of the teenagers in America consume alcohol at least once a week. In 1994, 35% of college women reported that they sometimes drink alcohol expressly to get drunk, a percentage that almost matches their male classmates. Neither group seems to realize that alcohol abuse is a serious problem, both to society and to personal health. Alcohol has a negative effect on virtually every organ of the body. People don't have to be alcoholics for alcohol to impair their health.

The Nervous System

There seems to be the general notion that alcohol is a stimulant. However, alcohol is a depressant. It inhibits the activity of all the neurons of the brain, beginning with the higher cortical, or "thinking," centers (Table 8A–2). Alcohol is of-

Table 8A–2 Behavioral Effects of Alcohol

Number of Drinks*	Ounces of Alcohol	Blood Alcohol Content (g/100 ml)†	Approximate Time for Removal	Effects
1	½	0.02	1 hr	Relaxation; begin to lose inhibitions
2½	1¼	0.05	2½ hr	Feeling "high"; impaired judgment; increased confidence
5	2½	0.10	5 hr	Impaired muscular coordination; slurred speech; impaired memory; extreme emotions (happy or sad)
10	5	0.20	10 hr	Greatly slowed reflexes; erratic swings in emotion
15	7½	0.30	15 to 16 hr	Loss of consciousness; little sensation; complete loss of coordination
20	10	0.4	20 hr	Coma possible; may stop breathing
25–30	15–20	0.5	26 hr	Usually fatal

*1 drink = 1 beer, glass of wine, or mixed drink.

†For a person weighing approximately 150–160 pounds.

ten mistakenly thought to be a stimulant because it depresses the inhibitory neurons first, allowing the excitatory ones to take over. As alcohol removes the "brakes" from the brain, normal restraints on behavior may be lost. Release from inhibitory controls also tends to reduce anxiety and this often creates a sense of well-being. However, discrimination, control of fine movements, memory, and concentration are gradually lost as well.

The lower brain centers that are involved in balance and coordination are affected next, causing a staggering gait. Numbed nerve cells send slower messages, resulting in slower reflexes. The brain regions responsible for consciousness are eventually inhibited, causing a person to black out. Yet higher concentrations can cause coma and death from respiratory failure.

Finally, alcohol kills nerve cells and nerve cells cannot be replaced. As they die, the brain actually gets smaller. The frontal lobes of the cerebral cortex, where judgment, thought, and reasoning are centered, are the first to shrink. In a Danish study of 37 alcoholic men, all less than 35 years old, 22 of them (59%) had impaired intellectual functioning when tested on tasks involving memory and comprehension. Eleven of the men (about 30%) had enough brain damage to be considered "occupationally disabled." The researchers concluded that the loss of intellectual ability may be the first complication of chronic alcohol use.

Nutrition

Alcohol is considered a nutrient because it supplies energy (calories). In fact, it is high in calories and downright fattening (Table 8A–3). (The term "beer belly" is well deserved.) But it is not just the calories in booze that makes one fat. Part of the problem is that the body "prefers" to utilize alcohol over other nutrients. When alcohol is present in the body, it is metabolized for energy before fats are used. Indeed, when alcohol is added to the diet, fat metabolism slows down by about 36% and substituting alcohol for other foods calorie-for-calorie slows fat breakdown by 31%. Unused fat is then stored in unsightly places such as thighs, hips, and bellies.

Alcohol supplies little else than fuel and even robs the body of other nutrients. It contains too few vitamins or minerals to be of any nutritional value. In addition, alcohol deprives the body of certain other nutrients in the food eaten. For instance, alcohol decreases the absorption of certain vitamins, including folate, thiamine, B_{12}, and B_6. In addition, it causes the kidneys to pump out of the body important substances such as potassium ions, zinc, calcium, magnesium, and folate.

The Liver

Alcohol consumption damages the liver, an organ that performs over 200 different functions in the body. Severe damage to the liver is, therefore, a serious threat to life. Liver damage occurs because alcohol metabolism preempts fat metabolism in the liver, causing fats to accumulate in the liver.

Table 8A–3 The Caloric Content of Alcoholic Drinks

Beverage	Amount	Approximate Calories
Beer		
Regular beer and ale	12 oz	140–150
Light beer	12 oz	95
Wine		
Sweet	5 oz	200
Dry table, red	5 oz	110
Dry table, white	5 oz	115
Liquor		
Gin, rum, vodka, whiskey		
80 proof	1½ oz (jigger)	97
86 proof	1½ oz (jigger)	105
90 proof	1½ oz (jigger)	110
94 proof	1½ oz (jigger)	116
100 proof	1½ oz (jigger)	124
Vermouth, sweet	1½ oz (jigger)	70
Vermouth, dry	1½ oz (jigger)	55
Cordials and Liqueurs	1 oz	70–115
Brandy and Cognac	1 oz	65
Mixers		
Ginger ale	8 oz	72
Club soda	8 oz	0
Cola	8 oz	96
Quinine water (tonic)	8 oz	72
Tom Collins mixer	8 oz	112
Fruit Juice		
Orange	6 oz	90
Tomato	6 oz	35
Grapefruit	6 oz	75
Pineapple	6 oz	105

Four or five drinks daily for several weeks is enough to begin fat accumulation in liver cells. At this early stage, however, the liver cells are not yet harmed and, with abstinence, they can be restored to normal. With continued drinking, accumulating fat causes liver cells to enlarge, sometimes so much so that the cells rupture or grow into cysts that replace normal cells (Figure 8A–3). The fat reduces blood flow through the liver, causing inflammation known as alcoholic hepatitis. Signs of hepatitis include fever and tenderness in the upper abdominal region. As the liver finds it increasingly difficult to perform its many functions, jaundice and changes in blood chemistry occur, including abnormal numbers of white blood cells in the circulation. Gradually, fibrous scar tissue may form, a condition known as cirrhosis, which further impedes

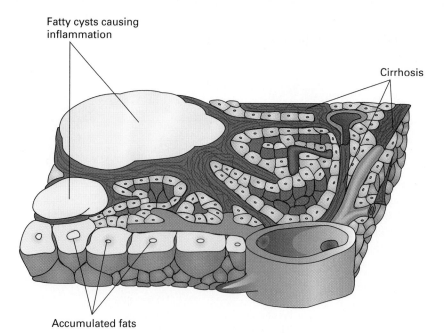

Fatty cysts causing inflammation

Cirrhosis

Accumulated fats

Figure 8A–3 Alcohol damage to the liver, an organ that performs over 200 functions within the body, can threaten life. Since alcohol is burned for energy before fats, the fats accumulate in liver cells, impairing their ability to function. Blood flow through the liver is hampered and this causes inflammation of the liver (hepatitis) and eventually causes scar tissue to form (cirrhosis). The consequences can be deadly.

blood flow and impairs liver functioning. Cirrhosis can lead to intestinal bleeding, kidney failure, fluid accumulation, and eventually death, if drinking continues. Indeed, cirrhosis of the liver, the ninth leading cause of death in the United States, is most often caused by alcohol abuse.

Cancer

A person who drinks heavily is at least twice as likely to develop cancer of the mouth, tongue, or esophagus than is a nondrinker. There is also evidence that a person who both drinks and smokes cigarettes is at greater risk of cancer than the sum of the risks caused by either habit alone (Figure 8A–4).

Heart and Blood Vessels

Here's the good news. *Moderate* amounts of alcohol can be good for the heart. Teetotalers are more likely to suffer heart attacks than are those who drink moderately, say a drink a day. One reason for this may be that alcohol's relaxing effect helps relieve stress. But alcohol also seems to raise the levels of the "good" form of a cholesterol-carrying particle—HDL—in the blood. This form of cholesterol reduces the likelihood that fats in the blood will be deposited in the walls of blood vessels, clog the vessels, and reduce the blood supply to vital organs such as the heart or brain. In addition, moderate amounts of alcohol reduce the likelihood that blood clots will form when they shouldn't. Such clots can block a blood vessel nourishing the heart muscle, thereby causing a heart attack. Thus, those who imbibe moderately generally live longer.

When alcohol is consumed in more than moderate quantities, it damages the heart and blood vessels. It weakens the heart muscle itself, reducing the heart's ability to pump blood. It also promotes the deposit of fat in the blood vessels, making the heart work harder to pump blood

through them. And, it raises blood pressure. Although a drink a day has little impact on blood pressure, consuming larger quantities may have a substantial effect. In one study, people with high blood pressure were asked to drink six drinks a day for 5 days. Their blood pressures rose steadily during this period. For the next 5 days, they abstained and their blood pressures fell again. Even people who regularly consume three drinks a day can measurably lower their

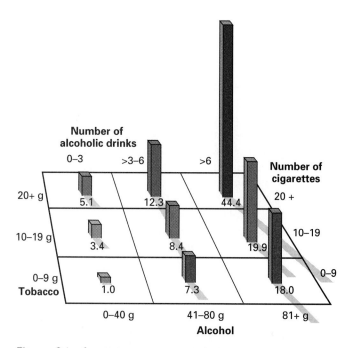

Figure 8A–4 Consumption of alcohol increases one's risk of developing cancer of the mouth, throat, and esophagus. If one also smokes cigarettes, the risks are even greater.

Figure 8A–5 The alcoholic's heart on the right is nearly twice the normal size. *(© George Steinmetz)*

blood pressure by going on the wagon for a week. Together these effects—damage to heart muscle, blood vessels clogged with fatty materials, and high blood pressure—can enlarge the heart to twice its normal size (Figure 8A–5).

Accidents

Accidents—at home or on the road—are more likely to happen when people have been drinking. Indeed, drunk driving causes more than 20,000 deaths a year in the United States, nearly half of the nation's traffic fatalities. It is a serious risk, not just to the drunk driver but also to innocent people who are injured or killed by that driver. A blood alcohol level of only 0.04% to 0.05%, which can result from only two or three drinks, decreases peripheral vision, decreases the light sensitivity of the eye by 30% (equivalent to wearing sunglasses at night), slows recovery from headlight glare, and reduces reaction time by as much as 25%. At the same time, it impairs the ability to concentrate and the judgment of the distance and speed of objects. If the driver is lucky, these impairments may not have dire consequences. However, considering that it has been estimated that in a single mile of city driving a driver must make roughly 300 split-second decisions, such impairments can make the difference between life and death.

Reproduction and Sexual Performance

Alcohol consumption can put a damper on sex life, too. Although the first drink may reduce inhibitions and stimulate one's interest in sex, continued drinking can impair a man's performance. About three-quarters of all male alcoholics report that they suffer from a reduced sex drive and/or impotence. At least part of the reason for this may be that chronic alcohol use causes the liver to produce excessive amounts of an enzyme that breaks down the male sex hormone, testosterone. Furthermore, prolonged testosterone deficiency causes the testes to shrink in size.

Effects on Fetal Development

If a pregnant woman consumes even one or two drinks a day, she increases the chance that her baby will be of lower than normal birth weight. Low birth weight, in turn, is associated with many complications soon after birth. Heavy drinking during pregnancy increases the risk of miscarriage and stillbirth. After Down syndrome and spina bifida, alcohol is the third leading cause of birth defects associated with mental retardation. Of the three leading causes, alcohol use is the only cause that is preventable.

About 20 years ago, researchers began to notice a pattern of growth abnormalities and birth defects common among children of women who drank heavily during pregnancy. This collection of effects has since been named the fetal alcohol syndrome (FAS), although all characteristics are not always present in any one infant. FAS is associated with mental retardation, growth deficiency, and characteristic facial features (Figure 8A–6).

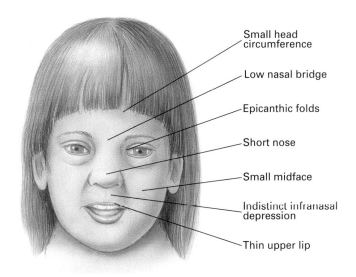

Small head circumference

Low nasal bridge

Epicanthic folds

Short nose

Small midface

Indistinct infranasal depression

Thin upper lip

Figure 8A–6 Characteristic facial features of children with fetal alcohol syndrome. *(Photo, © Dennis Drenner)*

These effects may be caused by chronic alcohol abuse throughout pregnancy or by binge drinking at critical times of development. During the first 3 months of pregnancy, most of the fetus's organ systems are formed. The first trimester, then, is a time when the fetus is particularly sensitive to chemicals that alter development, such as alcohol. The best advice for women who are pregnant is to avoid alcohol consumption entirely, since safe levels are not known.

Alcoholism

Although alcohol is a legal drug, alcoholism is America's number one drug problem. The problem with identifying alcoholism is that there is no such thing as a typical alcoholic. Alcoholics can be young or old, rich or poor, or of any race, economic status, or profession.

Different alcoholics have different drinking patterns, some binge and some chronically overindulge—but what they all share is a loss of control over their drinking. When an alcoholic takes the first sip of booze, he or she cannot predict how much or how long drinking will continue.

Marijuana

Marijuana (grass, pot, hashish, ganja, dagga) is the most widely used illegal drug in the United States today. It consists of the leaves, flowers, and stems of the Indian hemp plant, *Cannabis sativa*. The principal psychoactive ingredient (the component that brings about marijuana's mind-altering effects) is delta-9-tetrahydrocannibinol (THC).

The effects of marijuana depend on the concentration of THC in the preparation and the amount consumed. The average THC content of marijuana smoked in the United States has increased over the last 20 years from about 0.4% to as much as 10%. In small to moderate doses, THC produces feelings of well-being and euphoria. In large doses, it can cause hallucinations and paranoia. Anxiety may even reach panic proportions at very high doses. People who have previously suffered from mental problems, such as depression or schizophrenia, would be well advised to avoid marijuana.

Marijuana is not addicting in the sense that it produces severe, unpleasant withdrawal symptoms. Nonetheless, 100,000 people in the United States seek treatment for marijuana dependence each year. The dependence may result from a feeling of anxiety caused by the release of a brain chemical called corticotropin-releasing factor (CRF) when marijuana is withdrawn.

The Amotivational Syndrome

Long-term, heavy use of marijuana is sometimes associated with certain personality changes, including a loss of motivation, work ethic, and goal direction, that have come to be called the "amotivational syndrome." The amotivational syndrome is an example of the classic chicken-and-the-egg question: Which comes first? Does heavy marijuana use *cause* a loss of motivation or is it a symptom of a problem the user had to begin with? Since THC generally has sedating effects, it is not unreasonable to think that chronic use throughout the day could cause people to lose motivation. On the other hand, people who are depressed, anxious, or feel inadequate may be more likely to try to dull their pain by turning to drugs. In any case, a person who smokes marijuana during the schoolday or workday has a problem and needs help. It doesn't matter whether drug use is a cause or an effect of the problem.

Determining cause and effect is important, however, in forming public policy and in personal decisions about marijuana use. But, it is not easy to determine. On one hand, there are clinical reports of people who smoke marijuana heavily and who become listless and apathetic. On the other hand, studies in Jamaica, Costa Rica, and Greece, countries where marijuana use is much more prevalent than in the United States, failed to show any differences between users and nonusers of marijuana in the willingness to work and participate in society. Studies of American college students at universities such as the University of California at Berkeley found that users and nonusers of marijuana were equally successful academically.

The Stepping Stone Hypothesis

Through the years, it has been suggested that marijuana use leads to the use of harder drugs, such as heroin or cocaine. Such a relationship could be expected since a person who uses one illegal drug is more likely to be interested in using others than is someone who has never used any drug. It is argued that buying marijuana might also introduce a user to dealers who could supply other drugs. Thus, it should not be surprising to learn that marijuana use often precedes the use of other, more dangerous drugs. However, this connection does not show that the use of one drug *causes* the use of another. For example, whereas most people who smoke marijuana have previously consumed alcohol, most people who drink alcohol never smoke marijuana. Likewise, most people who smoke marijuana never develop an interest in other drugs.

Driving

Marijuana has a detrimental effect on driving skills and performance. In particular, it impairs the ability to track a moving object, the ability to judge time, speed, and distance, and it slows reaction time and reduces coordination. The decline in driving ability has been clearly demonstrated in experiments on restricted driving courses. Also, marijuana use has been shown to be a factor contributing to death and injury in car accidents.

Learning

Marijuana intoxication impairs short-term memory and slows learning. It interferes with the ability to pay attention, as well as with acquiring and storing information. It also impairs problem-solving skills. While under the influence of marijuana, users have trouble organizing their thoughts and conversing. It would be a waste of time, therefore, to attend class or to study while under the influence of marijuana.

Mode of Action

We are just beginning to understand how marijuana brings about its effects. When THC binds to certain receptors on nerve cells, it triggers a series of chemical reactions within the cell by increasing the levels of a chemical messenger within the cell called cyclic AMP (Figure 8A–7). The cascade of events that follow ultimately lead to the "high" that the user experiences. Once the receptors were discovered, researchers quickly set about to find the natural substance that binds to the receptors, reasoning that the receptors didn't evolve millions of years ago just in case someone decided to smoke marijuana. Researchers then sorted through thousands of chemicals in pulverized pig brains to find one that would bind to THC receptors. The natural substance that binds the THC receptors is a newly discovered chemical messenger, named *anandamide* after the Sanskrit word meaning internal bliss (ananda). Anandamide functions as the brain's own THC. The normal jobs of anandamine probably include the regulation of mood, memory, pain, appetite, movement, and other activities. In addition, THC seems to stimulate the release of dopamine in the reward pathways of the brain.

Effects of Long-Term Use of Marijuana

The media has popularized many claims that long-term use of marijuana carries with it a plethora of health hazards. However, the evidence for most of these claims is still inconclusive. Let's consider what is known about the risks of marijuana use.

Respiratory System

Marijuana and hashish are usually smoked, either in a pipe or in a hand-rolled cigarette called a joint or a blunt. It should not be surprising, then, that the most clearly harmful effects of marijuana are on the respiratory system. However, the damage, it seems, is done by the residual materials in the smoke not by the THC itself. It is true that people who smoke marijuana on a daily basis smoke fewer joints than cigarette smokers smoke cigarettes. However, compared to a regular cigarette, a joint has 50% more tar, which contains cancer-causing chemicals. Also, marijuana smoke is usually inhaled deeply and held within the lungs. As a result, three times as much tar is deposited in the airways and five times as much carbon monox-

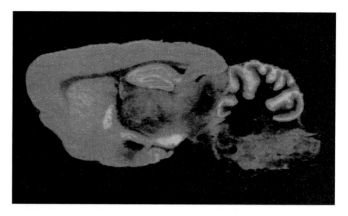

Figure 8A–7 The map of marijuana receptors shows that they are most highly concentrated in brain regions where THC is thought to be active. The areas with receptors appear yellow in this photograph showing a section of a rat's brain. There are few receptors in the brain stem. The receptors are highly concentrated in brain regions involved in thinking and memory (the cerebral cortex and hippocampus) and in the motor control areas such as the cerebellum. *(© Miles Herkenham/National Institute of Mental Health)*

ide is inhaled. Carbon monoxide prevents red blood cells from carrying needed oxygen to the cells of the body.

The end result is marked trauma to the respiratory system. In one study, people who smoked marijuana, but not tobacco, were twice as likely to report symptoms such as coughing, excess phlegm, and wheezing as were people who did not smoke at all. Like cigarette smoke, marijuana smoke inflames air passages and reduces breathing capacity.

Several experimental studies have sounded the alert that marijuana smoke can cause lung cancer. However, because lung cancer usually takes decades to develop, it is still too soon to say. We do know, though, that the tissues of the lung and airways of a marijuana or hashish smoker show the same progression of changes that are observed in cigarette smokers who develop lung cancer. Furthermore, some researchers have suggested that a combination of tobacco and marijuana smoke may be more likely to lead to cancer than either one alone.

The Heart

Smoking marijuana makes the heart beat faster, sometimes to double its normal rate. In some people, it also increases blood pressure. Either change increases the heart's workload. These changes could pose a threat to people with pre-existing cardiovascular problems, such as high blood pressure or atherosclerosis (fatty deposits in the arteries).

Reproduction

Some studies have shown that THC can interfere with reproductive functions in both males and females. At least some of these disturbances may result from the structural similarity between THC and the female hormone estrogen. Males who smoke marijuana tend to have lower levels of the male sex hormone, testosterone, and to produce fewer sperm. In one study of men who smoked marijuana, testosterone levels decreased 44% and sperm numbers were down by 35%. There is, however, no evidence that either change has any effect on a male's fertility. So, whereas marijuana should not be relied on as a means of male birth control, if a man is having difficulty becoming a father and smokes marijuana, quitting might boost fertility. Testosterone and sperm levels return to normal when THC is cleared from the body.

The effects of marijuana on the female reproductive system are not clear. Although we know that THC interferes with ovulation in female monkeys, we know less about what it does to human females. There are, however, some doctors' reports of menstrual problems and reproductive irregularities in women who smoke marijuana.

Effects on Development

Research on the effects of marijuana on reproduction has centered on the effects of THC on the developing embryo. We know, for example, that THC does cross the placenta, the organ through which materials are exchanged between the blood supply of the mother and that of the developing fetus, and it enters breast milk. Although we know that THC does reach the fetus, it has been difficult to identify the effects, if any, of THC on development.

At this point, in fact, there is no clear evidence that marijuana use adversely affects pregnancy. It does not appear to increase the incidence of miscarriage, stillbirth, or complications of either pregnancy or delivery.

Does marijuana use during pregnancy cause birth defects? In animals such as rabbits and hamsters, high doses of a crude extract of marijuana administered to a pregnant female can cause birth defects, including malformations of the central nervous system, the forelimbs, and the liver. The majority of studies on humans, however, have failed to find an increase in birth defects in infants born to mothers who smoked marijuana during pregnancy. But, one large study did find a link between minor birth defects and marijuana use during pregnancy. In this study, 2% of the marijuana-exposed infants had some facial features characteristic of fetal alcohol syndrome (FAS), representing a fivefold increase in the risk of birth defects. However, more than half of the women in this study who smoked marijuana also drank alcohol during pregnancy. Nonetheless, we cannot be sure there are no risks, so it would be wise to abstain from smoking marijuana during pregnancy.

Critical Thinking

One study performed in Jamaica on children born to mothers who smoked marijuana during pregnancy showed that these children scored higher on certain developmental tests, at 1 month and 5 years of age, than did children whose mothers did not smoke marijuana. However, the study went on to show that there were significant lifestyle differences between the mothers who smoked marijuana during pregnancy and those who did not. For instance, most mothers who smoked marijuana during pregnancy also received extra income from selling the drug. How could the extra income earned by these mothers account for the superior performance of their children compared with the children of nonsmokers even if marijuana does have a harmful effect on a child's development?

Stimulants

Cocaine

When the drug problem in the United States is mentioned, cocaine is usually the first drug that comes to mind. Indeed, cocaine is the second most popular illegal drug (after marijuana) and it is currently the most abused drug. Cocaine use has mushroomed with the advent of crack, a more potent, more addicting, and less expensive form of cocaine.

Cocaine is extracted from the leaves of the coca plant (*Erythroxylon cocca*), which grows naturally in the mountainous regions of South America. Before the early 1980s, cocaine was used in the form in which it is extracted, the hydrochloride salt. The salt is soluble in water, so it can be injected or absorbed through mucous membranes of the nose. When the cocaine powder is inhaled into the nasal cavity ("snorted"), it reaches the brain within a few seconds and produces an effect almost as intense as when it is injected. Smoking the drug is an even more effective delivery route. However, cocaine cannot be smoked as the hydrochloride salt because it breaks down with extreme heat. Forms of cocaine that can be smoked—specifically, freebase and crack—are obtained by further extraction and purification of the cocaine. Crack is an extremely potent form of cocaine used by two-thirds of the cocaine addicts (several hundred thousand) in the United States.

Cocaine brings about a rush of intense pleasure (euphoria), a sense of self-confidence and power, clarity of thought, and increased physical vigor. It does this by increasing the amount of certain chemicals (neurotransmitters) released by nerve cells to communicate with other nerve cells. Specifically, cocaine affects the levels of two "feel good" neurotransmitters, dopamine and norepinephrine. The euphoria is primarily caused by cocaine's effect on dopamine, a neurotransmitter used by nerve cells in the pleasure centers of the brain. Normally, dopamine is almost immediately reabsorbed into the nerve cell that released it and its effect on the next nerve cell ceases. Cocaine, however, interferes with the re-uptake of dopamine, thus increasing and prolonging dopamine's effect. As a result, cocaine loads the brain's pleasure centers and stimulates them. Cocaine also augments the effects of another neurotransmitter, norepinephrine. Norepinephrine brings about the effects of the sympathetic nervous system, the part of the nervous system that prepares the body to face an emergency (see Chapter 8). Thus, cocaine also triggers the bodily responses that would equip the body to face stress—increased heart rate and blood pressure, narrowing of certain blood vessels, dilation of pupils, a rise in body temperature, and a reduction of appetite. Cocaine, then, makes the user feel alert, energetic, and confident.

The effects of cocaine are short-lived and are followed by a "crash." A cocaine high may last from 2 to 90 minutes, depending on how it enters the body. When the cocaine high wears off, it is generally followed by a period of deep depression, anxiety, and extreme fatigue. To relieve these uncomfortable feelings, the user often craves more cocaine. The higher the high, the lower the crash and, therefore, the more intense the craving for cocaine. For this reason, crack, which is estimated to be about 75% pure as compared with the 10% to 35% purity of street cocaine, causes higher highs and harder crashes. It is extremely addicting.

It is not uncommon for a person to consume alcoholic beverages while using cocaine and the combination can be deadly. In the presence of alcohol, the liver metabolizes cocaine to form cocaethylene, which produces the same feeling of euphoria as cocaine, but the feeling is more intense and lasts longer. Cocaethylene remains in the body four times as long as cocaine alone. It can cause heart attacks or strokes hours after cocaine use. Indeed, cocaethylene is now thought to be primarily responsible for deaths that occur among cocaine users. People who combine alcohol and cocaine also have a higher risk of addiction to the combination of substances and have more severe withdrawal symptoms than those who use either substance alone.

Cardiovascular Risks

Cocaine may also cause heart attack or stroke, conditions that occur when an interruption of blood flow deprives heart cells or brain neurons of oxygen and nutrients (Figure 8A–8). One way that cocaine blocks blood flow is by constricting arteries. It can cause spasms in the arteries supplying the heart, for instance, bringing on a heart attack. Blood flow to the heart can also be disturbed by irregularities in heartbeat. By disabling the nerves that regulate heartbeat, cocaine can cause disturbed heartbeat rhythms that result in chest pain and heart palpitations (that uncomfortable feeling of being aware of your heart beating). It may even cause the heart to stop beating completely. In addition, cocaine increases blood pressure and this may cause a blood vessel to burst.

Respiratory Risks

It is not well known, but cocaine can cause respiratory failure that can lead to death. Stimulation of the central nervous

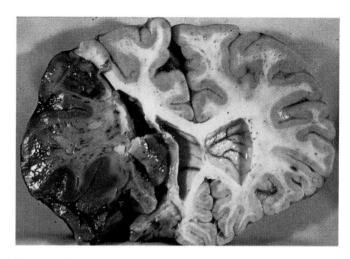

Figure 8A–8 A stroke triggered by cocaine use. *(© WRS Group, Inc., Waco, TX/Health Edco.)*

system is always followed by rebound depression. Thus, as the stimulatory effects of cocaine wear off, the respiratory centers in the brain that are responsible for breathing become depressed or inhibited. Breathing may become shallow and slow and may stop completely.

Damage to the respiratory system itself caused by cocaine use depends largely on how the drug is taken. If it is snorted, it causes damage to the nerves, lining, and blood vessels of the nose. It can dry out the nose's delicate mucous membranes until they crack and bleed almost constantly. Symptoms of a sinus infection—a perpetually runny nose and a dull headache spanning the bridge of the nose—are common. On a more startling level, the partition between the two nasal cavities may even disintegrate. Because cocaine is a painkiller, considerable damage to the nose may occur before it is discovered.

Smoking crack, on the other hand, damages the lungs and airways. The chronic irritation leads to bronchitis and may also cause lung damage from reduced oxygen flow into the lungs or blood flow through the lungs.

Seizures

The brain may respond to cocaine with epileptic-like seizures. Repeated use increases susceptibility to seizures.

Effects on Fetal Development

The number of pregnant women who use cocaine during pregnancy has been increasing since 1970. In this way, some 92,000 to 240,000 fetuses are exposed to cocaine each year in the United States.

When a pregnant woman uses cocaine, the drug does reach her developing fetus. We know this because the fetal urine contains breakdown products of cocaine. Breakdown products of cocaine can be detected in the infant's hair for several months after birth.

The Controversy over the Legalization of Marijuana for Medicinal Use

Many claims have been made that marijuana has medicinal uses. In fact, physicians now have FDA approval to prescribe synthetic THC as an antinausea drug for cancer patients and to combat the weight loss (wasting syndrome) that afflicts some people with AIDS. THC is also effective in the treatment of glaucoma, a condition in which the pressure within the eyeball increases to the point where it can cause blindness. Some patients with multiple sclerosis or phantom limb pain (pain that seems to originate in an amputated limb) report that marijuana eases their suffering.

Should marijuana be legalized for medicinal purposes? The Drug Enforcement Agency, the Public Health Service, and the American Medical Association all answer emphatically, "No!" On the other hand, many cancer patients and AIDS patients say it eases their suffering. One survey of over 1000 oncologists (doctors who treat cancer patients) revealed that 44% have already suggested marijuana use to their cancer patients. The debate is often heated. So, let's step back from the fray and consider the arguments of both the opponents and the proponents for the legalization of marijuana for medical use.

Opponents of legalization often argue that there is no reason to smoke marijuana to gain its effects, because synthetic THC is available and it already has FDA approval for use in certain circumstances. However, many chemotherapy patients claim that smoking marijuana is more effective in combating nausea than the latest antinausea drug, Zofran, that is usually given. Furthermore, they point out that it can be difficult to swallow a capsule during a bout with nausea.

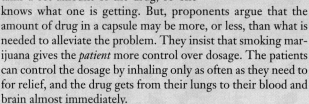

Opponents of legalization also contend that it is impossible to control the dose of THC when marijuana is smoked, because batches of marijuana differ in potency and because some people inhale more deeply than others. On the other hand, a capsule always contains a set amount of the drug, so one knows what one is getting. But, proponents argue that the amount of drug in a capsule may be more, or less, than what is needed to alleviate the problem. They insist that smoking marijuana gives the *patient* more control over dosage. The patients can control the dosage by inhaling only as often as they need to for relief, and the drug gets from their lungs to their blood and brain almost immediately.

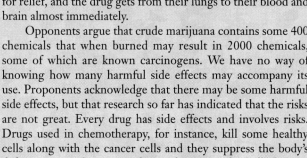

Opponents argue that crude marijuana contains some 400 chemicals that when burned may result in 2000 chemicals, some of which are known carcinogens. We have no way of knowing how many harmful side effects may accompany its use. Proponents acknowledge that there may be some harmful side effects, but that research so far has indicated that the risks are not great. Every drug has side effects and involves risks. Drugs used in chemotherapy, for instance, kill some healthy cells along with the cancer cells and they suppress the body's defense system. Some even destroy heart muscle. Still, they are

---

Direct exposure to cocaine is not the only way that the developing fetus can be harmed when its mother uses cocaine. The blood vessels in the umbilical cord constrict, shutting down the lifeline through which oxygen and nutrients reach the fetus.

It isn't always easy to determine the effects of prenatal exposure to cocaine because pregnant women who use cocaine often also indulge in other activities that are known to be harmful to fetal development, such as smoking cigarettes or drinking alcohol. Early studies on the effect of cocaine on fetal development did not take into account other substances that the pregnant women used, such as tobacco and alcohol, and both of these may have a greater effect on fetal development than does cocaine.

It is safe to say, however, that cocaine use by a pregnant woman does increase the risk of certain problems: miscarriage, premature birth, and low birth weight. Because cocaine produces uterine contractions by interfering with the metabolism of stress hormones, a quarter of the mothers who use cocaine during pregnancy give birth prematurely, increasing the risks of complications shortly before or after birth. Crack babies are also twice as likely to be of lower than average birth weight, which also increases the risk of complications.

There are also neurological changes caused specifically by cocaine. For example, cocaine retards the growth of the brain, causing the heads of cocaine-exposed newborns to be smaller than normal. In addition, there is an increased likelihood that the nerve cells won't make the proper connections in the brain. The blood vessels of the brain of a crack baby may also have been blocked, either because they have constricted or because a vessel has ruptured.

used because the benefits outweigh the risks. Proponents argue that it is the same with marijuana.

Opponents argue further that marijuana has never been *proven* safe and effective for anything. The evidence that it works is anecdotal—the testimony of people who have tried it.

Since marijuana has been around for thousands of years, it seems reasonable to wonder why scientific data on its effectiveness in medicine is so scarce. One thing that discourages research on the medical uses of marijuana is that it is an illegal drug. Marijuana is classified as one of the most stringently controlled substances. Other drugs, such as cocaine and morphine, have a much higher potential for abuse than marijuana, but they are still considered to be useful in medicine.

Another factor hampering research on the medicinal uses of marijuana is the lack of interest of drug companies. Marijuana is a plant. It cannot be patented, so it is not attractive to drug companies. Clinical trials cost money. Who will foot the bill?

Even when these two problems can be overcome, it is not easy to design and conduct a good scientific study on the efficacy of marijuana. Pain relief and antinausea are more difficult to evaluate than effects such as heart rate. Many doctors question the ethics of performing a well designed controlled experiment on critically ill patients to properly test the efficacy of the drug. For example, a study has been proposed to determine the relative effectiveness of three doses of smoked marijuana compared with dronabinol, the drug currently used as an appetite stimulant to combat HIV wasting syndrome. The experiment should have a control, that is, a group treated with a placebo.

Would it be ethical to give patients with HIV wasting syndrome an inert substance for 12 weeks so that they could serve as a control? Their bodies are already severely compromised. Since the blood level of THC may vary with the batch of marijuana and the manner in which it is inhaled, it would be best to measure the blood level. Would it be ethical to draw extra blood from these already critically ill patients?

Even if there were scientific evidence that it worked, it is unlikely that the FDA would approve marijuana because it is an herbal medicine. Dosage cannot be controlled. Also, we don't know the other ingredients in the plant.

The controversy over the medical use of marijuana has recently intensified. In November 1996, voters in California and Arizona approved propositions that would allow physicians to prescribe marijuana for certain medical conditions. Soon afterward, the Clinton administration announced that doctors in these states who do prescribe or recommend marijuana may face criminal prosecution under federal law. Public opposition to the federal government's response to these state laws then prompted the Office of National Drug Control Policy to commit nearly a million dollars to fund a National Academy of Sciences review of the literature regarding the therapeutic use of marijuana. Clearly, the issue has yet to be resolved.

How much research is enough and how much credence should be given to anecdotal evidence? When people say they feel better, is that enough evidence that they do? What other factors should be considered? So, should marijuana be legalized for medical purposes? For testing purposes? What do you think?

Social Issue

Cocaine use by pregnant women is estimated to add $504 million annually to the cost of health care in America. It has been suggested that much of this money could be saved by mandatory treatment programs for maternal cocaine abusers. Should pregnant women who use cocaine be forced to undergo drug treatment? If they refuse, should they go to prison?

There is also a possibility that the father's use of cocaine may influence the development of his child. Cocaine binds to human and animal sperm in test tubes without destroying them or affecting their motion. This raises the possibility that cocaine might do the same within the body. If so, it could damage the sperm without preventing fertilization, which could lead to birth defects. The idea is consistent with the observation that men who use cocaine and other drugs produce an increased number of offspring with abnormalities.

Amphetamines

There are many forms of amphetamine: Benzedrine, Dexedrine, Methedrine, Desoxyn, and now "ice." All are stimulants of the central nervous system. Like cocaine, amphetamines make the user feel good—exhilarated, energetic, talkative, and confident. They suppress appetite and the need for sleep. Amphetamines are active for a longer period of time than is cocaine, hours as opposed to minutes.

Amphetamines bring about their physical and psychological effects by causing the release of certain newly synthesized neurotransmitters, such as norepinephrine and especially dopamine. Norepinephrine is used for communication between certain neurons within the brain and it is the neurotransmitter that brings about the effects of the sympathetic nervous system, which prepares the systems of the body to face an emergency. Thus, amphetamine elevates blood pressure, speeds heart rate, and opens up airways in the lungs. High intravenous doses administered during a "spree" of several days may cause a paranoid mania, similar to that associated with schizophrenia. The personality changes are caused by several factors including excessive stimulation of norepinephrine and dopamine neurotransmission as well as the chronic lack of sleep.

Amphetamines can be swallowed in a pill or injected intravenously. A crystalline form of methamphetamine, called "ice," is smoked to produce effects similar to crack cocaine. Methamphetamine can produce hazardous effects, including blood vessel spasm, blood clot formation, insufficient blood flow to the heart, and accumulation of fluid in the lungs.

Tolerance to amphetamine develops (accompanied by cross tolerance to cocaine) along with both physical and psychological dependence. Withdrawal symptoms include extreme fatigue, depression, and increased appetite. Also, the pleasurable feelings caused by amphetamine use can lead to a compulsion to overuse the drug.

Amphetamines cross the placenta and can affect the developing fetus. Women who use amphetamines during pregnancy are more likely to give birth prematurely and to have low-birth-weight babies.

Caffeine

Coffee, tea, and cola drinks are known for their arousing effects. The sense of wakefulness and increased energy come primarily from caffeine, although other substances may also contribute. Caffeine occurs naturally in more than 60 different plants, but Americans overwhelmingly prefer theirs from coffee. Indeed, Americans drink more than half of all the coffee produced in the world.

Unlike the other drugs we have discussed, caffeine does not act by altering the communication between neurons. Instead, caffeine acts by speeding up the metabolic rate of all cells, including neurons. It increases cellular metabolism by increasing the production of glucose from more complex molecules within cells. The glucose, in turn, serves as the fuel to support higher rates of cellular activity. When metabolism within neurons is increased, they are stimulated. Thus, caffeine stimulates the central nervous system. Because the neurons of the cerebral cortex (the thinking part of the brain) are so sensitive to caffeine, it serves as a "pick-me-up." With caffeine, as with many other stimulants, dosage is important. A low dose, such as is obtained from one or two cups of coffee, increases wakefulness and mental alertness,

resulting in faster and clearer thought. Studies have shown, for instance, that two cups of coffee improve both driving and typing skills. But, higher doses can cause "coffee nerves" that make one jumpy and restless. Very high doses (12 or more cups a day) can cause agitation, anxiety, tremors, rapid breathing, and irregular heart beats.

The heart is affected by caffeine in several ways. First, it increases heart rate. (Take your pulse before and after a cup or two of coffee.) Second, it dilates (widens) the arteries supplying blood to the heart, resulting in increased oxygen and nutrients reaching the heart muscle, making it a more efficient pump.

Caffeine affects other muscles too. It relaxes the muscles in the digestive tract and in the walls of the airways in the respiratory system. The increased glucose supply spurred by caffeine's effect on metabolism makes voluntary muscles, such as those in the arms and legs, tire less easily.

Caffeine is addicting, even in low to moderate doses such as a cup or two of coffee a day. Withdrawal symptoms include headache (sometimes severe), depression, anxiety, and occasionally, flu-like symptoms, all of which lead to irritability. People who haven't had their cup of coffee can be crabby, indeed.

The idea that caffeine consumption endangers health is still highly debated. Most human studies looking for effects of caffeine compare coffee drinkers with nondrinkers, because coffee is a primary source of caffeine. But, this comparison may not be valid. For one thing, coffee contains many ingredients other than caffeine. The exact ingredients in a cup of coffee vary among brands and with the method of preparation. Therefore, even when an effect is demonstrated between coffee drinkers and abstainers, we can't say for sure that the caffeine is responsible. Also, among cigarette smokers, lighting up goes so naturally with a cup of coffee that separating the effects of smoking and drinking coffee can be difficult.

Cancer

Studies investigating a link between cancer and caffeine have produced conflicting results. The Ames test, a standard laboratory test screen for cancer-causing potential, has indicated that caffeine increases mutation rate in bacteria. But, increasing coffee consumption among laboratory animals does not seem to make them more prone to cancer. Nonetheless, some (but not all) studies comparing people who drink coffee with those who don't have suggested that coffee drinkers may be at higher risk of developing cancer of the colon, the bladder, or the ovary. (These results are highly debated.)

Reproduction

Although the studies are not yet conclusive, there is enough evidence linking caffeine to reproductive difficulties, such as reduced fertility, miscarriage, and birth defects, that it would be wise for pregnant women and those who would like to become pregnant to avoid caffeine.

Heart

Caffeine's biggest threat to the health of the heart may be through its effect on blood cholesterol level. High blood cholesterol is linked to atherosclerosis, a condition in which fatty deposits clog the arteries. Atherosclerosis is a common cause for heart attack and stroke. In one study, men who drank two or more cups of coffee a day had higher blood cholesterol levels than men who didn't drink coffee. In another much larger study in Norway, higher blood cholesterol levels were also reported among men who drank coffee. But in this study, the men were *big* coffee drinkers. Two-thirds of them drank five or more cups of coffee a day. Not only was blood cholesterol elevated, so was the associated risk of heart attack. The risk of heart attack among men who drank five or more cups of coffee a day was 250 times that of non–coffee drinkers.

Psychedelic Drugs

The psychedelic drugs are grouped together because they have similar effects, in spite of their diverse chemical structures. These effects include visual, auditory, or other distortions of sensation, as well as vivid, unusual changes in thought and emotions.

The psychedelic drugs include some natural and some synthetic forms. There are about six natural psychedelic drugs, the best known of which are mescaline, which comes from peyote cactus, and psilocybin, which is found in certain mushrooms. The most famous synthetic form is undoubtedly LSD, lysergic acid diethylamide.

These drugs are thought to act by augmenting the action of serotonin, norepinephrine, or acetylcholine. Some of the more well-known psychedelic drugs, including LSD, psilocybin, DMT (dimethyltryptamine), and bufotenin, bind to the serotonin receptors in the brain, thereby mimicking the natural effects of serotonin. Mescaline, on the other hand, is structurally similar to norepinephrine.

The normal physiological reactions to psychedelic drugs are not especially harmful, but the distortions of reality that are created by the drug experience may lead to some actions that are quite dangerous. A "bad trip" are the words most often used to describe an unpleasant reaction to a psychedelic drug. The symptoms may include paranoia, panic, depression, and confusion. The trip usually ends within 24 hours, when drug levels decrease. It is nearly impossible to determine who will have a bad trip, but the likelihood is greater if the person using the drug has pre-existing emotional problems or is uncomfortable with the company or setting.

Brief recurrences of sensory or emotional alterations produced by the drug are called "flashbacks." Almost any aspect of the drug experience may be relived during a flashback. They are usually brief and merely annoying, but they can be frightening.

Tolerance for psychedelic drugs develops quickly. Furthermore, cross tolerance is the rule; that is, a person who has become tolerant of one psychedelic drug will be tolerant to others as well. Craving and withdrawal reactions are unknown and laboratory animals offered LSD will not take it voluntarily.

Sedatives

Sedatives are drugs that depress the central nervous system. As with alcohol, the depressant drugs affect inhibitory neurons first. As a result, low doses first produce a relief from anxiety and a mild euphoria. As in so many cases involving drugs, however, the effects are dose-dependent. In the case of sedatives, higher doses also inhibit excitatory neurons and sleep follows.

The effects of different depressants are additive. Thus, if you drink alcohol and also take a sedative, the depressant effects will be greatly intensified. Combining depressant drugs is always a bad idea, because it can easily lead to overdose. The respiratory system can, in fact, become so depressed that breathing stops. An overdose due to depressants cannot be reversed by administering a central nervous system stimulant, amphetamine, for example. Although the drugs may have antagonistic effects, each brings about its effects by binding to different receptors or by affecting different regions of the brain. So, taking a stimulant and a depressant at the same time is like simultaneously stepping on the brake and the accelerator of a car. Slamming the accelerator doesn't remove the brake. To stop an overdose, then, it would be necessary to remove the drug from its receptors. A stimulant may temporarily arouse the person, but when it wears off, the resulting rebound depression added to the residual depression caused by the sedative could be fatal.

Continued use of sedatives leads to tolerance as well as both psychological and physical dependence. Cross tolerance also develops, so tolerance to one sedative also reduces the response to other sedatives. Tolerance develops because the liver produces more of the enzymes that break down the drug and, to some extent, because the nerve cells become less sensitive to its effects. As tolerance develops, overdose becomes a greater risk. Increased doses are required for the drug's sedating effect, but the neurons in the breathing center are still sensitive to a lethal dose.

Another problem is breaking the habit of taking sedatives. When a sedative has been used with regularity over a period of time, withdrawal results in hyperexcitability of the nervous system, which could lead to convulsions and death. Therefore, withdrawal should be done under a physician's care.

The barbiturates were the sedative of choice about 20 years ago, but their popularity is waning. Barbiturates such as amobarbital (Amytal, Tuinal), seconbarbital (Seconal), pentobarbital (Nembutol), and phenobarbital were often prescribed as sleeping pills. However, the sleep induced by barbiturates is not normal sleep. REM sleep, the type of sleep during which most dreaming occurs, is suppressed. Although we don't know the function of REM sleep, it is thought to be important. When REM sleep is reduced on one night, the amount of REM increases during the following night's sleep.

The risk of overdose from barbiturates is high. Part of the reason is that the effect depends not just on the dose and the mode of administration but also on the person's state of mind when it is taken. A barbiturate will have a more profound effect on a person who is depressed or physically tired. Also, a lethal dose is only three times higher than a therapeutic dose. For this reason, barbiturates are often used by those wishing to commit suicide.

Today, physicians are more likely to prescribe one of the benzodiazepines (Valium, Librium, or Dalmane, to name just a few) to treat anxiety and insomnia. Because benzodiazepines lack the reinforcing effects of barbiturates and opiates, they are less likely to be abused. However, the withdrawal symptoms can be uncomfortable, which may prompt some people to continue using the drug. Overdose is much less likely to occur with benzodiazepines than with barbiturates, partly because they don't depress the heart and respiratory system to as great an extent.

The Opiates

The opiates and related drugs have two different faces. Until the 1970s, when cocaine use became widespread, the opiates were the most commonly abused illegal drugs. Although opiates hold a potential for abuse, they are medically important because they alleviate severe pain.

Among the first opiates used were morphine and codeine, which come from the opium poppy. Heroin is a synthetic derivative of morphine that is more than twice as effective. When heroin is injected intravenously, it reaches the brain very quickly, producing a feeling that can only be described in ecstatic or sexual terms. The rush of euphoria produced by heroin has made it the drug preferred by opiate abusers and invokes its continued use.

The opiates, including heroin, exert their effects by binding the receptors for the body's endogenous (natural, internally produced) opiates—endorphins, enkephalins, and dynomorphins. These are the neurotransmitters involved in the perception of pain and fear, among other things.

Heroin and other commonly abused opiates have effects similar to those of morphine: euphoria, pain suppression, and reduction of anxiety. They also slow the breathing rate. An overdose may cause the user to fall into a coma and stop breathing. Overdose is always a potential problem because heroin is bought on the street for illegal use. The buyer has no idea of its strength. Street supplies are often diluted with sugar. If a heroin addict who is accustomed to a diluted drug injects heroin that is much more potent than he or she thinks, death due to overdose often occurs. Extremely constricted pupils in an unconscious person is a sign of heroin overdose.

Surprisingly, habitual use of heroin has few directly toxic effects. Many of the problems arise because heroin addicts suffer from a general disregard of good health practices. Other problems are associated with the injections themselves. Frequent intravenous injection of any drug is associated with certain ailments. Because of the constant puncturing, veins can become inflamed. In addition, shared needles may spread infections, including HIV, viral hepatitis, and syphilis. Because heroin is often cut with other substances, the batch may be contaminated.

Tolerance, and cross tolerance, to opiates develops. However, the rate at which tolerance develops varies with the manner in which the drug is used. It develops more slowly if the drug is used in sprees, say on the weekends only, with drug-free periods between. However, the duration of each spree tends to increase over time. So, what began as a weekend habit begins to carry over into weekdays. As the use becomes more frequent, tolerance develops quickly.

The symptoms of heroin withdrawal have been so publicized that they are often the first thing people think of when the word addiction is mentioned, but misconceptions abound. There is no doubt that the withdrawal symptoms are extremely uncomfortable: craving for the drug, sweating, anxiety, fever, chills, violent retching (dry heaves) and vomiting, panting, cramping, insomnia, explosive diarrhea, and extreme aches and pains. However, they are rarely fatal, unlike withdrawal from barbiturates or sometimes from alcohol.

REVIEW QUESTIONS

1. What are the ways in which a psychoactive drug can alter the communication between neurons?
2. What does it mean to be physically dependent on a drug?
3. Why does tolerance to a drug develop?
4. Why is it a good idea to have eaten a meal before consuming alcohol?
5. What is the difference between a stimulant and a depressant?
6. Is alcohol a stimulant or a depressant?
7. How does alcohol consumption cause cirrhosis of the liver?
8. What is the psychoactive ingredient of marijuana?
9. What are some negative effects of marijuana on the respiratory system?
10. How does marijuana affect learning while under the influence?
11. Why does cocaine bring about a feeling of pleasure?
12. Why can cocaine cause heart attacks and strokes?
13. What are the physiological effects of amphetamines?
14. Why does caffeine serve as a "pick-me-up"?
15. How do sedatives affect the central nervous system?
16. How do opiates such as heroin affect the central nervous system?

Sensory Systems

The sensory information provided by this virtual reality mask creates an artificial environment that is perceived as real.
(© Geoff Tompkinson/Science Photo Library/Photo Researchers, Inc.)

At the end of your life, if you were asked what the world is like, you would have some answers. You might describe green forests and cool streams and mention how the sky turns gray over cities. You might say something about traffic sounds, hugs, and the tinkling of ice in a glass. Your neighbors may paint a different picture of the world. However, most of us will agree that our knowledge of the world comes to us through our senses.

Sometimes, though, those senses can mislead us (Figure 9–1). For example, the stately Parthenon appears to be rectangular. Our senses tell us that. But it is not. If it were, we would perceive it as sagging in the middle. It was intentionally constructed to correct for the illusion. Also, look at the spiral. Actually, there is no spiral in this drawing. Look again, they are concentric circles. You may need to trace one with your finger to convince yourself of this. The four small colored disks are the same color in each column. Only the color of the background differs. There are no gray dots at the corners of the black lines. So what is reality? And what is our world really like?

Perhaps sensations are more than just a window to the outside world. Perhaps they are the basis of our inner world as well. Philosophers have argued that sensations are the basis of consciousness itself. A person cannot be conscious without *feeling* something. In his novel, *Immortality*, Milan Kundera expressed it in the words, "'I think, therefore I am,' is the statement of an intellectual who underrates toothaches."

General Principles of Sensory Reception

Information about the external and internal world comes to us in different energy forms, including light, sound, chemicals, and pressure. **Sensory receptors** are structures that are specialized to respond to changes in their environment (stimuli) by generating electrochemical messages that are eventually converted to nerve impulses if the stimulus is strong enough. The nerve impulses are then conducted to the brain, where they are interpreted to build our perceptions of the world.

Sensory receptors share several characteristics. For one thing, they are selective. That is, each type responds best to one form of energy. Photoreceptors, for instance, respond to light. The response of a sensory receptor, called a **receptor potential**, is an electrochemical message (a change in the degree of polarization of the membrane) that varies in magnitude with the strength of the stimulus. Thus, sensory receptors can produce graded signals, that is, signals that vary in intensity. For instance, the louder the sound, the larger the receptor potential—up to a point. When the receptor potential reaches threshold level, an action potential (nerve impulse) is generated. Most types of sensory receptors gradually stop responding when they are continuously stimulated. This phenomenon is called **sensory adaptation**. As receptors adapt, we become less aware of the stimulus. For example, the musty smell of an antique store may be obvious to a person who just walked in, but it goes unnoticed by the salesclerk working there. Some receptors, such as those for pressure and touch, adapt quickly. For this reason, we quickly become unaware of the feeling of our clothing against the skin. Other receptors adapt slowly or not at all. The receptors in muscles and joints that report on the position of body parts, for instance, never adapt. Their continuous input is essential for coordinated movement and balance.

Classes of Receptors

Receptors can be classified according to the stimulus to which they respond. Several classes of receptors are traditionally recognized:

1. **Mechanoreceptors** respond to distortions in the receptor itself or in nearby cells. Mechanoreceptors are responsible for the sensations we describe as touch, pressure, hearing, and equilibrium. In addition, mechanoreceptors detect changes in blood pressure and tell us the position of our body.
2. **Thermoreceptors** detect changes in temperature.
3. **Photoreceptors** detect changes in light intensity.
4. **Chemoreceptors** respond to chemicals. We describe the input from the chemoreceptors of the mouth as taste (gustation) and from those of the nose as smell (olfaction). Other chemoreceptors monitor levels of chemicals, such as carbon dioxide, oxygen, and glucose, in body fluids.
5. **Pain receptors** respond to noxious stimuli, which usually result from physical or chemical damage to tissues. Instead of being classified in a category of their own, pain receptors are sometimes classified as chemoreceptors, because they often respond to chemicals liberated by damaged tissue, and occasionally as mechanoreceptors, because they are stimulated by physical changes, such as swelling, in the damaged tissue.

Receptors may also be classified by their location. **Exteroceptors** are located near the surface of the body and respond to changes in the environment. **Interoceptors** are in-

(a)

How the Parthenon appears to us.

How the Parthenon would appear if it were actually built as a rectangle.

The compensations made in the architecture to make it appear rectangular.

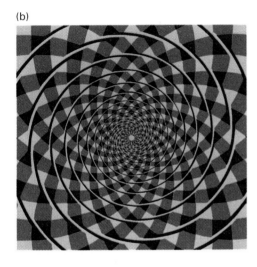
(b)

What appears to be a spiral is actually a series of concentric circles.

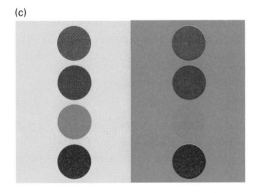
(c)

The colors of the circles in both columns are actually identical. They appear to be different because the background colors differ.

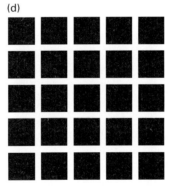
(d)

There are no gray dots at the corners of the black squares.

Figure 9–1 Although we know the world only through our senses, we cannot always trust them. (*b, Fraser's Spiral, reprinted from* Seeing: Illusion, Brain, and Mind *by John Frisby (1979) by permission of Oxford University Press*)

side the body and monitor conditions there. Although we are often unaware of the activity of interoceptors, they play a vital role in maintaining homeostasis. They are an important part of the feedback loops that regulate blood pressure, blood chemistry, and breathing rate. Interoceptors may also cause us to feel pain, hunger, or thirst, thereby prompting us to take appropriate action.

The **general senses**—touch, pressure, vibration, temperature, a sense of body and limb position, and pain—arise from receptors in the skin, muscles, joints, bones, and inter-

nal organs. The **special senses** include smell, taste, vision, and hearing, as well as the sense of balance or equilibrium.

The General Senses

The receptors for the general senses are widespread throughout the body. As you can see, some monitor conditions within the body and others provide information about the world around us.

Touch

During the seventh week of pregnancy, long before the eyes and ears have formed, a sense of touch is already functioning in the fetus. Right from the start, infants explore their world using the sense of touch. Throughout life, touch is a way of learning about the world and of communicating with one another. In fact, humans seem to have a need to be touched by others. Social bonds are established through touch.

Light touch, such as might occur when the cat brushes past your legs, depends on several types of receptors. Wrapped around the base of the fine hairs of the skin are free nerve cell endings, which detect any bending of the hairs. **Merkel disks**, flattened endings of dendrites of sensory neurons that contact the outer layer of skin, also sense light touch. They are found on both the hairy and the hairless parts of the skin. Free nerve cell endings and Merkel disks tell us that something has touched the skin. **Meissner's corpuscles** tell us exactly where we have been touched. These encapsulated nerve cell endings are common on the hairless, very sensitive areas of skin, such as the lips, nipples, and fingertips.

Pressure

The sensation of pressure generally lasts longer than does touch and it is felt over a larger area. **Pacinian corpuscles**, which consist of layers of tissue surrounding a nerve ending, are scattered in the deeper layers of skin and the underlying tissue. They respond only when the pressure is first applied and, therefore, are important in sensing vibration.

Temperature

Thermoreceptors respond to changes in temperature. They are widely distributed throughout the body and are especially common around the lips and mouth. In humans, thermoreceptors may simply be two types of specialized free nerve cell endings. The so-called cold receptors are most active between 10°C (50°F) and 20°C (68°F). The "heat" receptors are active as the temperature climbs to 25°C (77°F), but become inactive at temperatures above 45°C (113°F). You may have noticed that the sensation of hot or cold fades rapidly. For example, because the thermoreceptors adapt rapidly, the water in a hot tub may feel scalding at first, but it very quickly feels comfortably warm.

Sense of Body and Limb Position

Whether you are at rest or in motion, the brain "knows" the location of all the body parts. It continuously scans the signals from muscles and joints to check body alignment and ensure coordination. **Muscle spindles** are specialized muscle fibers with sensory nerve cell endings wrapped around them that report to the brain whenever a muscle is stretched. **Golgi tendon organs** measure the degree of muscle tension. They are highly branched nerve fibers located in the tendons, the connective tissue bands that connect muscles to bones. The information from muscle spindles and Golgi tendon organs is used along with information from the inner ear (as we will see later in this chapter) to coordinate our movements.

Pain

The receptors for pain are free nerve cell endings that are found in almost every tissue of the body. Any stimulus strong enough to damage tissues, including heat, cold, touch, and pressure, will cause pain. Tissue damage activates enzymes that form a chemical called bradykinin. Bradykinin then triggers the release of inflammatory chemicals, such as histamine and prostaglandins, that begin the healing process. Together, bradykinin, histamine, and prostaglandins alert free nerve cell endings of the injury. Those neurons then carry the message to the brain where it is interpreted as pain. Aspirin reduces pain because it interferes with the production of prostaglandins.

Our perceptions of pain involve both the sensation of pain and our emotional response to that pain. Pain is an important mechanism that warns and protects the body from injury. For example, it usually prevents a person with a broken leg from causing additional damage by moving the limb. Nonetheless, few of us appreciate the value of pain while we are experiencing it. Also, pain that persists long after the warning is needed can be debilitating.

Critical Thinking

The venom of many wasps contains bradykinin. How does this explain why wasp stings are particularly painful?

The Special Senses

Now let's consider the special senses, those that usually come to mind when we think of the senses, largely because we are so dependent on them as we try to perceive the nature of our world.

Vision

We, as humans, are very visual creatures, and whereas we may not see detail as well as an eagle does or movement as well as an insect, we see far better than most other mammals.

The Structure of the Eye

Each of your eyeballs is an irregular sphere about 25 mm (1 in.) in diameter (Figure 9–2; Table 9–1). The wall of the eyeball consists of three layers. The outermost layer is a tough, fibrous covering with two obviously different regions: the sclera and the cornea. The **sclera**, often called the "white of the eye," protects and shapes the eyeball and serves as an attachment site for the muscles that move the eye. In the front and center of the eye, the clear, transparent **cornea** bulges slightly outward and provides the window through which light enters the eye. As light passes through the curved surface of the cornea, it is bent toward the light-sensitive surface at the rear of the eye. Unlike most tissues in the body, the cornea lacks blood vessels. Therefore, the cells of the immune system that can cause the rejection of transplanted tissue cannot reach the cornea. Although the cornea

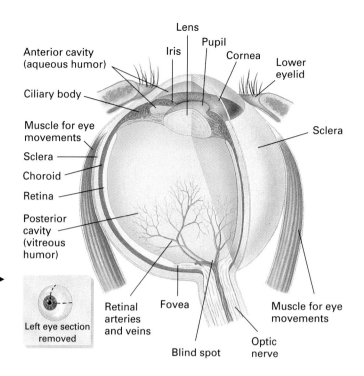

Figure 9–2 The structure of the human eye. Light enters through the transparent cornea and then passes through the pupil, the opening through the center of the pigmented iris. Changes in the shape of the lens focus light on the light-sensitive layer of the eye, the retina. The photoreceptors, the rods and cones, are located in the retina.

Table 9–1   Structures of the Eye and Their Functions

Structure	Description	Function
Sclera	Outer layer of the eye	Protection of the eyeball
Cornea	Transparent dome of tissue forming the outer layer at the front of the eye	Refracts light, focusing it on the retina
Pupil	Opening at the center of the iris	Opening for incoming light
Iris	Colored part of the eye	Regulates the amount of light entering the eye through the pupil
Lens	Transparent, semispherical body of tissue behind the iris and pupil	Fine focusing of light onto retina
Ciliary body	Encircles lens; contains the ciliary muscles	Controls shape of lens; secretes aqueous humor
Aqueous humor	Clear fluid found between the cornea and the lens	Refracts light and helps maintain shape of the eyeball
Vitreous humor	Gelatinous substance found within the chamber behind the lens	Refracts light and helps maintain shape of the eyeball
Retina	Layer of tissue that contains the photoreceptors rods and cones; also contains bipolar and ganglion cells involved in retinal processing	Rods and cones receive light and generate neural messages
Fovea	Small pit in the retina that has a high concentration of cones	Detailed color vision
Optic nerve	Group of axons from the eye to brain	Transmits impulses from retina to brain

has no blood supply, it does have pain receptors. Indeed, a tear in the cornea is extremely painful.

Three distinct regions—the choroid, the ciliary body, and the iris—make up the middle layer of the eye. The **choroid** is a layer that contains many blood vessels that supply nutrients and oxygen to the tissues of the eye. The choroid layer contains a brown pigment, melanin, that acts like the black paint on the inside of a camera. That is, the pigment in the choroid absorbs light after it strikes the light-sensitive layer, helping to prevent the reflection of light within the eye, which would cause visual confusion.

The **ciliary body** is a ring of tissue that encircles the lens, holding it in place and controlling its shape. As we will see, the shape of the lens is important in focusing light on the light-sensitive layer of the eye.

The **iris**, the colored portion of the eye that can be seen through the cornea, regulates the amount of light that enters the eye. Eye color is determined by the thickness of the iris and the amount of pigment it contains. The only pigment in the iris is brown in color. When this pigment is highly concentrated, brown eyes result. However, when the iris has less pigment, some light passes through it and is reflected off the back surface of the eyeball, giving the iris a bluish tint. Thus, decreasing pigment concentrations result in hazel, green, blue, or gray eyes. Whatever its concentration, the pigment prevents light from passing through it. The iris is shaped like a flat doughnut; the doughnut hole is the opening through the center of the iris, called the **pupil**, through which light enters the eye. The pupil looks black because light within the eye is bent and it does not leave the eye again through the pupil.

The iris contains smooth muscle fibers that automatically adjust the size of the pupil to admit the appropriate amount of light to the eye. The pupil gets larger (dilates) in dim light and gets smaller (constricts) in bright light. In fact, its size can vary 16-fold. Pupil size is also affected by emotions. The pupils dilate when you are frightened or when you are very interested in something. They constrict when you are bored. Candlelight creates a romantic setting partly because its dimness dilates the pupils, making the lovers appear more attentive and interested. On the other hand, your eyes may betray feelings that you would rather hide. Clever salesclerks trying to guess how high a price you'll pay and magicians guessing the card you've picked need only look into your eyes to see the dilation of your pupils.

The innermost layer of the eye is the **retina**. It contains almost a quarter-billion photoreceptors, the structures that generate electrical signals in response to light. There are two types of photoreceptors, rods and cones. The **rods** are more numerous and are responsible for our black and white vision. Being exceedingly sensitive to light, they are capable of responding to the light produced by one 10-billionth of a watt, the equivalent of a match burning 50 miles away on a clear, pitch-dark night! (Of course, the ideal conditions necessary to actually *see* a burning match at such a distance are impossible to attain.) Nonetheless, the rods do allow us to see in dimly lit rooms and in pale moonlight. By detecting changes in light intensity across the visual field, rods contribute to the perception of movements. The **cones** are responsible for color vision. Unlike the rods, the cones produce sharp images. The cones are most concentrated in a small region in the center of the retina called the **fovea**. Only the fovea contains a sufficient number of cones, about 150,000 per mm^2 to produce detailed color vision. Therefore, when we want to see the fine details of something, the image must be focused on the fovea. However, the fovea is only the size of the head of a pin. At any given moment, therefore, only about a thousandth of our visual field is in sharp focus. Eye movements bring different parts of the visual field to the fovea.

Critical Thinking

There are no rods in the fovea. Rods are more concentrated at the periphery of the retina. How does this explain why you can see better at night if you do not look directly at the object?

The processing of electrical signals from the rods and cones begins before they leave the retina. The messages are first sent to bipolar cells and then on to ganglion cells (Figure 9–3). Between these two cell layers are horizontal cells and amacrine cells, which send signals laterally across the retina. Together, these cells convert the dot-like input from the retina to patterns, such as edges and spots. The axons of the ganglion cells bundle together, forming the **optic nerve** that will bring processed messages from the eye to the brain, where it is interpreted. The region where the optic nerve leaves the retina has no photoreceptors. As a result, we cannot see an image that strikes this area. It is, therefore, called the **blind spot**. You are not usually aware of your blind spot, because the missing parts are automatically "filled in" as the visual information is processed by the brain.

Critical Thinking

A single bipolar cell usually receives input from several rods. However, a single bipolar cell receives input from only one cone in the fovea. How might this explain why rods permit us to see at lower light intensities than do cones? How might it explain why the sharpest images are formed when the object is focused on the fovea?

The eye is filled with fluid. The posterior cavity of the eye, found between the lens and the retina, is filled with a jelly-like fluid called **vitreous humor**. It helps to keep the eyeball from collapsing and holds the thin retina against the wall of the eye. Between the cornea and the lens is the anterior cavity, which is filled with a fluid called **aqueous humor**. It supplies nutrients and oxygen to the cornea and lens

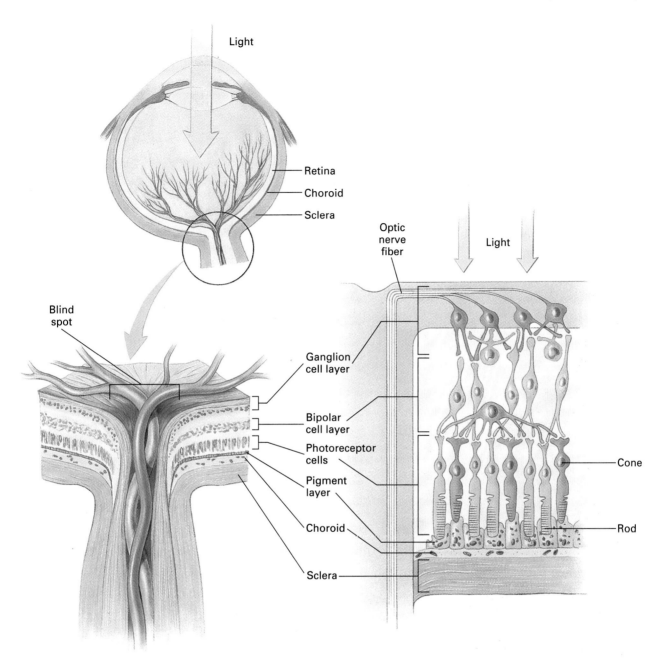

Figure 9–3 The complex neural pathways of the retina convert light to nerve signals. When light is focused on the retina, it must first pass through several layers of retinal cells before reaching the photoreceptors, the rods and cones. The rods are very sensitive to dim light, but provide only black and white vision. Although the cones require brighter light, they provide sharp images in color. In response to light, the rods and cones generate electrical signals that are sent to bipolar cells and then to ganglion cells. These cells begin the processing of visual information. Thus, before leaving the eye, simple patterns and contours are extracted from the dot-like input from the photoreceptors. The axons of the ganglion cells leave the eye at the blind spot, carrying nerve signals to the brain by means of the optic nerve.

and carries away their metabolic wastes. In addition, the pressure it creates within the eye helps maintain the shape of the eyeball. Unlike the vitreous humor, which is produced during embryonic life and is never replaced, aqueous humor is replaced about every 90 minutes. It is continuously produced from the capillaries of the ciliary body, circulates

through the anterior cavity, and is drained into the blood through a network of channels that surround the eye.

If the drainage of aqueous humor is blocked, the pressure within the eye may increase to dangerous levels. This condition, called **glaucoma**, is the second most common cause of blindness after cataracts. The accumulating aqueous humor pushes the lens partially into the posterior cavity of the eye, increasing the pressure there and compressing the retina and optic nerve. This, in turn, collapses the tiny blood vessels that nourish the rods and cones and the fibers of the optic nerve. Deprived of nutrients and oxygen, the photoreceptors and nerve fibers begin to die and vision fades. Unfortunately, glaucoma is painless and progressive. Late signs include blurred vision, headaches, and seeing halos around objects. Because many people don't realize they have a problem until some vision has been lost, it is recommended that everyone over 40 years of age be tested for glaucoma every year. Early glaucoma is treated with eye drops and pills that either reduce the rate of formation of aqueous humor or enhance its drainage. If these fail to reduce the pressure in the eye, an artificial drainage channel can be created surgically.

Focusing the Image

Just as a clear photograph requires that the light be focused on the film, clear vision requires that light rays be focused on the retina. This is accomplished by bending the light rays. Light rays are bent, or refracted, when the speed at which they are traveling changes, as occurs when they enter a medium of a different density (Figure 9–4). It is this bending of light rays that focuses the image on the retina.

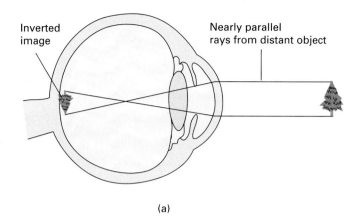

(a)

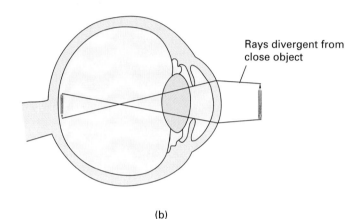

(b)

Figure 9–5 The lens changes shape to view objects at varying distances. (a) Light from a distant object approaches the eyes in nearly parallel rays. (b) In contrast, the rays of light from an object within 6 meters diverge. To view closer objects the lens becomes rounder and thicker. This increases the degree to which light rays are bent and keeps the image focused on the retina.

Figure 9–4 When light passes from a medium, such as air, to one of a different density, in this case water, the light rays are bent. This bending of light makes the pencil appear broken. Similar bending of light rays occurs when light travels from the air to the cornea and as it enters and leaves the lens. The bending of light rays in the eye focuses the image on the retina. *(Visuals Unlimited/© SIU)*

The most important structures in focusing the image are the cornea and the lens. Because the cornea is the only transparent structure of the eye exposed directly to air, most of the light bending, better than 75% of it, occurs here. But the cornea has a fixed shape, so it always bends light to the same degree. It cannot make the adjustments needed to focus on objects at varying distances.

However, the lens is elastic and its shape can be changed to focus on both near and distant objects, a process called **accommodation**. When focusing on a nearby object, the lens becomes rounder and thicker, thereby increasing the degree of bending in the light ray (Figure 9–5).

These changes in lens shape are caused by contractions of the circular ciliary muscle (Figure 9-6). The lens is attached to the ciliary muscle by ligaments. When the muscle contracts, its diameter gets smaller, much like pursed lips. This relaxes the tension on the ligaments and the lens is free

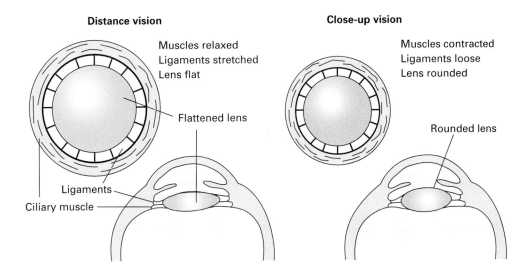

Distance vision

Muscles relaxed
Ligaments stretched
Lens flat

Flattened lens

Ligaments

Ciliary muscle

Close-up vision

Muscles contracted
Ligaments loose
Lens rounded

Rounded lens

Figure 9–6 The lens is attached to the ciliary muscle by ligaments. Because the ciliary muscle is circular, its diameter gets smaller when it contracts. This loosens tension on the ligaments and allows the lens to become rounder to focus on a nearby object. When the ciliary muscle relaxes, its diameter increases and so does the tension on the ligaments. Thus, relaxation of the ciliary muscle flattens the lens so that distant objects come into focus.

to assume the spherical shape needed to focus on nearby objects. Relaxation of the ciliary muscle increases the tension on the ligaments and the lens. Consequently, the lens flattens and focuses light from more distant objects on the retina. As we age, the lens becomes less elastic and doesn't "round up" to focus on nearby objects as easily. This explains why we hold the newspaper farther away as we get older.

A cataract is a lens that has become cloudy or opaque. Cataracts are the most common eye problem that affects men and women who are older than 50. Typically, the lens takes on a yellowish hue that blocks light on its way to the retina. In the beginning, the cataracts may cause a person to see the world through a haze that can limit activities and cause automobile accidents, early retirement, and falls that can result in hip fractures. As the lens becomes increasingly opaque, the fog thickens. Indeed, cataracts are the leading cause of blindness, not just in the United States but also worldwide. When cataracts make it impossible for a person to perform everyday tasks, the clouded lens can be surgically removed and replaced with an artificial lens. There is even a bifocal lens that has allowed 70% of cataract patients to avoid using glasses after surgery. Cataract surgery has a success rate of more than 90%.

Most cataracts form as a natural part of the aging process, but they can also be a consequence of diabetes mellitus, drug reactions, or injuries. People who smoke 20 or more cigarettes a day are twice as likely as a nonsmoker to develop cataracts. Exposure to the ultraviolet B portion of sunlight, the same part of sunlight that causes sunburn and an increased risk of skin cancer, is thought to be a major factor in the formation of cataracts. It is generally recommended that adults and children wear sunglasses with specially treated lenses that absorb nearly all of the ultraviolet rays. Regular tinted glasses reduce the amount of visible light reaching the eyes, but not UV light. In fact, untreated sunglasses may do more harm than good, because they re-

duce the intensity of light entering the eye enough that people don't squint in bright sunlight, thus allowing more UV light in. So, it is worth the extra money for sunglasses that will filter out UV. Unfortunately, the current labeling code for sunglasses is not as informative as it could be. Someday, sunglasses may be rated with a numerical system similar to the one used for sunscreens.

Critical Thinking

Water is nearly as dense as the tissue of the cornea. Why would this cause vision to be blurry when the eyes are opened under water? Goggles place a pocket of air in front of the cornea. Why would this improve underwater vision?

In addition to changes in lens shape, sharp vision requires coordinated eye movements to keep the image focused on the retina's fovea. Both eyes focus on the same group of objects. When looking at a distant object, both eyes are directed at the same angle. However, as we move closer to the object, our eyes must be directed toward the midline if they are to remain fixed on the object, movement called **convergence**. Convergence is important because it keeps the image focused on the fovea. The nearer the object, the greater the degree of convergence required.

When you have been reading or doing other close work for a long time, the constant contraction of both the ciliary muscles needed for focusing over short distances and of the eye muscles causing convergence can cause eye strain. This can be relieved by gazing into the distance for a while, allowing these muscles to relax.

Focusing Problems. The three most common visual problems—farsightedness, nearsightedness, and astigmatism—are all focusing problems and therefore normal vision can be

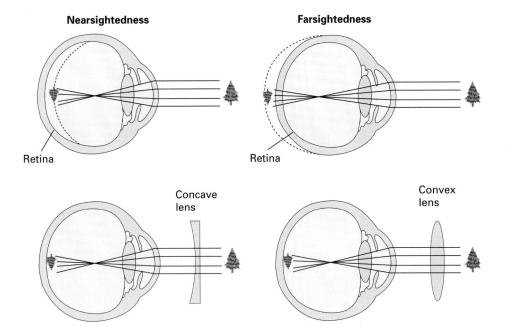

Nearsightedness

Retina

Farsightedness

Retina

Concave lens

Convex lens

Figure 9–7 Focusing problems. In nearsightedness, the image is focused in front of the retina. A concave lens will cause the light rays from a viewed object to diverge slightly before entering the eye. As a result, the image will be focused on the retina. In farsightedness, the image is focused behind the retina. The problem can be corrected with a convex lens, which will cause the divergent light rays reflected from a nearby object to converge slightly and bring the object into focus on the retina.

restored with corrective lenses. In **farsightedness**, the eyeball is too short or the lens is too thin, causing the image to be focused behind the retina (Figure 9–7). Although distant objects can be seen clearly, the lens cannot round up enough to bend the light sufficiently to focus on nearby objects. Corrective lenses that are thicker in the middle than at the edges (convex) cause the light rays to converge a bit before they enter the eye. The lens can then focus the image on the retina.

About 25% of the American population are nearsighted. **Nearsightedness** (myopia) occurs when the eyeball is elongated or when the lens is too thick. This causes the image to focus in front of the retina. Nearsighted people see nearby objects clearly, because the lens rounds up enough to focus the image on the retina. However, the lens simply cannot flatten enough to bring the focused image to the retina, and so distant objects appear blurred. Nearsightedness can be corrected by lenses that are thinner in the middle than at the edges (concave). These lenses will cause the light rays to diverge slightly before entering the eye.

Although genetics undoubtedly plays a role in the development of nearsightedness, frequent close work such as reading or working at a computer terminal is also a cause. (Could it be that your mother was right to nag you to use a good light and to occasionally "unglue" yourself from the television?) When you do a lot of close work, the frequent contraction of the ciliary muscles that change the shape of the lens increases the pressure within the eye. This pressure can cause the eye to stretch and elongate, causing nearsightedness. Indeed, studies have shown a relationship between the amount of time spent at close work and the incidence of nearsightedness. For example, the incidence of nearsightedness among school children in Taiwan has increased with the additional educational demands placed on them. Today, about 70% of Taiwanese school children are nearsighted.

When Eskimo children first started going to school, the incidence of nearsightedness skyrocketed. It is therefore recommended you look up from the page or away from the computer screen at frequent intervals. This is particularly important if nearsightedness runs in your family.

Irregularities in the curvature of the cornea or lens will cause distortion of the image because it causes the light rays to converge unevenly. This condition is called **astigmatism**. A simple test for astigmatism is shown in Figure 9–8. Vision can be restored to normal by corrective lenses that compensate for the asymmetrical bending of light rays.

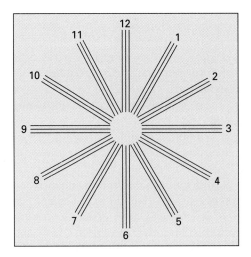

Figure 9–8 Astigmatism, caused by irregularities in the curvature of the cornea, will cause asymmetrical blurring or darkening of the lines in this figure.

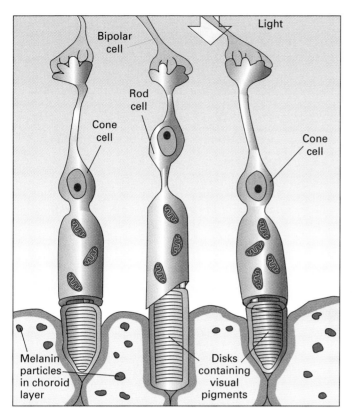
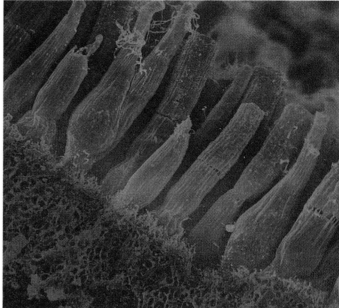

Figure 9–9   The photoreceptors, the rods and cones, are named for their shapes. Their function is to generate neural messages in response to light. The outer portion of each photoreceptor is packed with membrane-bound disks containing pigment molecules. The light absorbing part of a pigment molecule, retinal, is bound to one of four types of proteins called opsins. When light is absorbed, chemical changes in the pigment molecules trigger permeability changes in the membranes that cause the electrochemical changes constituting neural messages.   *(photo, © Omikron/Photo Researchers, Inc.)*

Light and Neural Messages

We said earlier that humans are very visual creatures, but it may surprise you to know that the eye contains 70% of all the sensory receptors in the human body. The function of all photoreceptors, the rods and cones, is to respond to light with neural messages that are sent to the brain where they are translated into images of our surroundings (Figure 9–9).

Our world is, in fact, so visual that few of us ever stop to wonder how vision works (Figure 9–10). So, let's take a moment to consider the process. First, light waves in the visible spectrum, that is, the wavelengths that our eyes can detect (380–750 nanometers), strike an object and at least some wavelengths are reflected off the object to the retina of an eye. The light is absorbed by pigment molecules in the photoreceptors, causing a chemical change in the pigment molecules. This, in turn, changes the membrane permeability of the photoreceptor, resulting in electrochemical changes in the receptor (the receptor potential mentioned earlier) that constitute a neural message. The optic nerve carries nerve signals from the retina to the back of the brain, where the

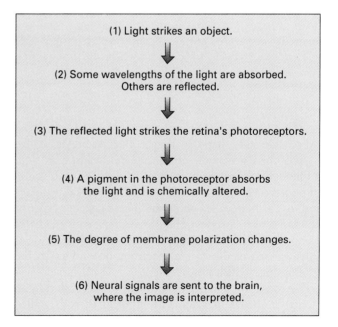

Figure 9–10   An overview of the visual process.

Correcting Vision Problems

Although most people wear glasses to correct vision problems, almost 20 million Americans use contact lenses instead. The contact lens sits on the cornea over a layer of tears. The outer surface is the corrective surface and the inner surface fits snugly on the cornea. There are two types of contact lenses. Hard lenses are made of hard plastic. These are tough and hard-wearing, but some people find that they irritate the eyes. Soft lenses are made of a more malleable plastic. Although soft lenses are gentler on the eyes, they become scratched and damaged more easily.

It is very important that contact lenses be cleaned and disinfected regularly. Infections, caused by bacteria, fungi, or *Acanthamoeba* (a type of amoeba), can develop behind the lens. Tears leave deposits of protein, fats, and calcium that help infectious organisms to adhere to the contact lens. At best, these deposits are difficult to remove. The more infrequently the lenses are cleaned, the more difficult it becomes to remove the deposits, and the more likely infection becomes.

Extended-wear contact lenses have been developed, some of which can be left in place for a month or more. There are even disposable varieties that can be left in place for up to a week. Initially it was thought that less frequent placing and removal of lenses would reduce trauma to the cornea. However, it is now known that leaving lenses in overnight increases the risk of ulcerative keratitis, a condition in which the cells of the cornea may be rubbed away by the contact lens. This sometimes leads to infection and scarring. If not promptly treated, ulcerative keratitis can lead to blindness. The problem with leaving contact lenses in overnight seems to be a reduced oxygen supply to the cornea, which causes it to swell. Although hard lenses let in enough air, they are usually too uncomfortable to sleep with. No soft lens, including the extended-wear and disposable varieties, will allow enough oxygen through to the eye when the lids are closed during sleep. Almost everyone who wears disposable lenses leaves them in overnight, but only half of those with extended-wear lenses do. As a result, the risk of developing ulcerative keratitis is about 7 times greater for those with extended-wear lenses than for those with daily-wear lenses and 14 times greater for those with disposable lenses than for those with daily wear lenses. Nonetheless, the absolute risk of keratitis is small, less than 1%, even if lenses are left in overnight.

A new, increasingly controversial, corrective trend for curing nearsightedness is a type of surgery called radial keratotomy. Commercials promise nearsighted people a life without glasses or contact lenses—no more foggy lenses and the chance to play sports without needing elastic to hold your glasses in place. What is this procedure and why is it controversial?

Radial keratotomy is a procedure in which an ophthalmologist makes four to eight incisions that radiate outward, like the spokes of a wheel, from the central area of the cornea that remains untouched (see the figure to the right). (Keratotomy means "cutting the cornea.") As the incisions heal, the cornea flattens and, therefore, does not bend light rays as much as it did previously. Consequently, the image becomes focused on the retina, instead of in front of it. Radial keratotomy does not use special surgical devices and, therefore, does not require the approval of the Food and Drug Administration.

Does it work? There is a reasonably high percentage of satisfied customers. In one study, 6 years after surgery, 74% of

---

world outside takes shape. In a sense, then, seeing is an illusion, since there are no pictures inside our heads.

The light-absorbing portion of the pigment molecules in all photoreceptors is a compound called **retinal**, which is bound to a protein called an **opsin**. There are four types of opsins and thus four types of photoreceptors: rods and three types of cones. Retinal preferentially absorbs different wavelengths (colors) of the visible spectrum, depending on the type of opsin to which it is bound.

The rods, you may recall, allow us to see in dim light, but only in shades of gray. The pigment in rods, called **rhodopsin**, is packaged in membrane-bound disks, which are stacked like coins in the outer segment of the rod. When light strikes a rod, the retinal portion of rhodopsin changes shape and splits from the protein portion, which is called scotopsin. The splitting of rhodopsin triggers events that reduce membrane permeability to sodium ions and eventually lead to changes in the activity of bipolar cells and ganglion cells. In the dark, rhodopsin is resynthesized.

Dark Adaptation

When you first enter a dark movie theater from a well-lit area, it is very difficult to see and you tend to stumble over people's feet. You have the problem because the cones cannot function in dim light and the rhodopsin in the rods has been split by the bright light. After a few moments, however, you can see the shadowy outlines of people. Within 20 to 30 minutes, you can even see the features on the face of the per-the patients said that their preoperative goals had been met. However, that does not mean that they all had perfect vision, just better vision than before surgery. About 60% of the patients in the largest study to date had "normal" vision 5 years after surgery. Ten percent of the patients still required corrective lenses. Often, contact lenses cannot be tolerated after surgery and glasses must be worn. In about one-third of all radial keratotomies the cuts are too deep or too shallow and the patient ends up with undercorrection or overcorrection. Nearly 20% of radial keratotomies leave the patient overcorrected, and they actually become *farsighted*. About 10% of the patients have eyesight that gets worse as the day wears on. Furthermore, 3% of those having radial keratotomies had worse vision after surgery, even with glasses.

There are also some less serious side effects. The patient may see starburst patterns at night, when the pupil dilates enough to see the edges of the incision catch the light. These effects can impair night driving. If the two eyes heal differently, the vision in each may be different, and headaches can result.

Also, such surgery weakens the cornea. As a result, the cornea is more likely to rupture if it is struck. This means that protective goggles should be worn when there is any danger of being hit in the eye, which might be while playing the sport you wanted to play without glasses. Also, if the job carries the risk of eye injury, some employers, the FBI for instance, won't hire anyone who has had radial keratotomy.

A newer procedure, photorefractive keratectomy, is believed to be less risky. Short bursts of a laser beam are used to shave a microscopic layer of cells off the corneal surface and flatten it. A computer calculates and controls the laser exposure. People who are too nearsighted to benefit from radial keratotomy can benefit from photorefractive keratectomy. During the Food and Drug Administration trials leading to approval of the procedure, 90% of those treated regained "functional" vision and two-thirds of them regained perfect vision.

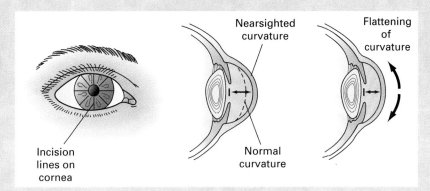

Incision lines on cornea

Nearsighted curvature

Normal curvature

Flattening of curvature

Radial keratotomy is a procedure in which spoke-like slits are made in the cornea. As they heal, the cornea flattens. Light rays passing through the cornea are then bent less so they focus on, or closer to, the retina.

son sitting in the seat next to you. The adjustment to dim light, known as **dark adaptation**, occurs as rhodopsin is resynthesized, thereby making the rods responsive to light again. The rods are particularly insensitive to red light. Therefore, a photographer will often make last-minute adjustments in the darkroom under red light. The pigment in the red cones will be split, but the rods will remain dark-adapted and the photographer will be able to see after the red light has been turned off.

Color Vision

What makes you see red? Actually, red, or any other color, is determined by wavelengths of light. White light, such as that emitted by an ordinary light bulb or the sun, consists of all wavelengths (Figure 9–11). (Rainbows are created when the various wavelengths of white light are separated as they pass through tiny droplets of water in the air.) When light strikes an object, some of the wavelengths may be absorbed and others reflected. We see the ones that are reflected. Thus, an apple looks red because it reflects mostly red light and absorbs most of the other wavelengths (Figure 9–12).

We see color because we have three types of cones, which are called "blue," "green," and "red." The cones are named for the wavelengths they absorb best, not for their color. When light is absorbed, the cone is stimulated. As you can see in Figure 9–13, each type of cone absorbs a range of wavelengths and the ranges overlap quite a bit. As a result, colored light stimulates each type of cone to a different

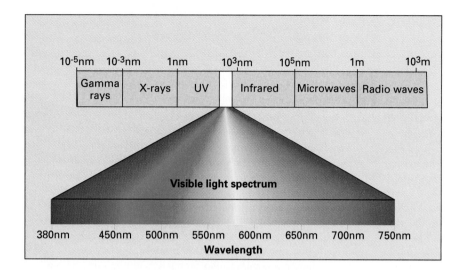

Figure 9-11 The electromagnetic spectrum. We see wavelengths of light between 380 and 750 nanometers. The color of light is determined by its wavelength.

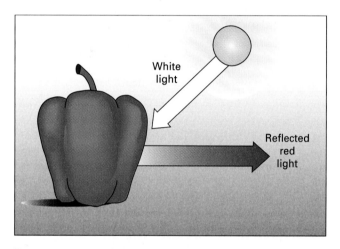

Figure 9-12 When light strikes an object, some wavelengths may be absorbed and others reflected. We see the reflected light. The apple looks red because it reflects mostly red light and absorbs most of the other wavelengths.

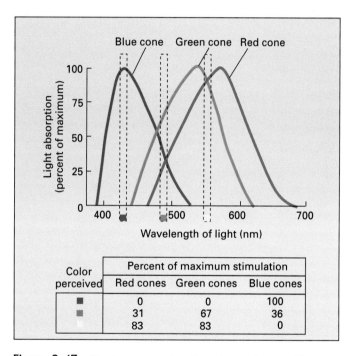

Color perceived	Percent of maximum stimulation		
	Red cones	Green cones	Blue cones
■	0	0	100
■	31	67	36
	83	83	0

Figure 9-13 The human eye has three types of cones. They are called blue, green, and red cones to describe the wavelengths of light they absorb best. Each type of cone absorbs a range of wavelengths and these ranges overlap. Thus, light of a specific wavelength will activate each cone type to a different degree. The brain compares the input from all three types of cones and interprets this as color.

extent. For instance, when we look at a bowl of fruit, light reflected by a red apple stimulates red cones; light reflected by a ripe yellow banana stimulates both red and green cones; blueberries stimulate both blue and green cones; and purple concord grapes stimulate blue cones. The brain then interprets color according to how strongly each type of cone is stimulated.

Most people who are color-blind see some colors, but they tend to confuse certain colors with others. The confusion is caused by a lack, or a reduced number, of one of the types of cones. A person who lacks red cones sees deep reds as black. In contrast, a person who lacks green cones, sees deep red, but cannot distinguish among reds, oranges, and yellows. Absence of blue cones is extremely rare.

People who are color-blind generally function normally in everyday life. They compensate for the inability to distinguish certain colors by using other cues, such as intensity, shape, or position. Red and green, the universal traffic colors for stop and go, are the most commonly confused colors. However, it is usually possible to pick out the brightest of the lights in the traffic signal and all traffic signals are arranged the same way—red on top and green on the bottom. It is a good idea to test school children for color blindness, however, so that both students and teachers are aware of the condition. It may cause difficulty with classwork, such as reading maps or graphs that are presented in color. If the teacher is aware of the condition, steps can be taken to avoid frustration and confusion for the student faced with such tasks.

Critical Thinking

Retinal is a derivative of vitamin A. How does this explain why eating foods that are rich in vitamin A, such as carrots, can improve night vision?

Depth Perception

In addition to seeing objects, it is important to know where we are in relation to them and their relative positions in our visual field. The brain's ability to determine how far away something is, is called **depth perception** and it involves the conversion of the two-dimensional image on each retina to a three-dimensional scene. The brain uses a number of cues to make such calculations, including: (1) overlap—the object in front usually covers an object behind it (2) relative size and height—nearby objects appear larger and taller than distant ones (Figure 9–14) (3) linear perspective—parallel lines in a scene appear to converge as they get farther away (4) movement parallax—as you move, nearby objects appear to move past more quickly than do distant ones and (5) the degree of accommodation and convergence needed to focus on the object.

One of the most important cues in depth perception, however, is the difference in the view of the world we get from each eye (Figure 9–15). Different views occur because the pupils (and foveae) are about 7 cm apart, causing them to

Figure 9–14 The brain uses relative size as a cue in judging distance. The person in the front who appears to be tiny is the same size as the one in back. However, because we know the one in the back is farther away, he doesn't look small. *(George Semple)*

see an object from different angles. You will notice these different views if you focus on a nearby object and close first one eye and then the other. The object will appear to move. However, when you focus on an object with both eyes open, the brain uses the difference in the views to figure out how far away the object is. At the same time, the brain fuses the

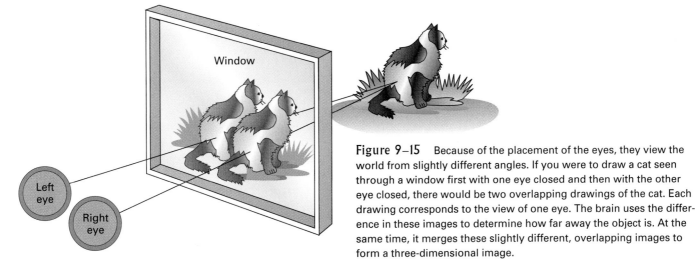

Figure 9–15 Because of the placement of the eyes, they view the world from slightly different angles. If you were to draw a cat seen through a window first with one eye closed and then with the other eye closed, there would be two overlapping drawings of the cat. Each drawing corresponds to the view of one eye. The brain uses the difference in these images to determine how far away the object is. At the same time, it merges these slightly different, overlapping images to form a three-dimensional image.

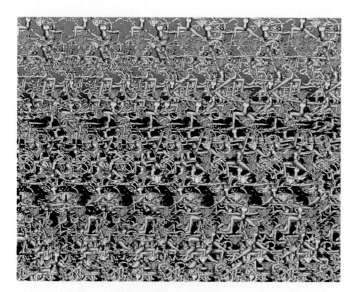

Figure 9–16 If you gaze at this picture long enough, a hidden 3-D image of aerobic dancers emerges. To find the hidden image, keep your gaze *unfocused* as you move the page away from your eyes. It may require many attempts. Seeing the hidden image requires that you "uncross" your eyes, thereby eliminating depth cues associated with eye convergence. Therefore, it may help to put a sheet of glass in front of the image so that it catches your reflection. This computer-generated art form is called an autostereogram. When we see any object as three-dimensional, it is because the brain has merged the slightly different, but overlapping, images from each eye that result from the separation of the eyes. To create an autostereogram, a computer first creates a 3-D model of the object and then uses this to determine the depth of the image at various points. Next, the computer produces a pattern of dots in which the dots are close together for the parts of the image that will appear close to the viewer and farther apart for the parts that will appear more distant. That pattern is then shifted slightly to form the image for the other eye. *(©1994 Digi-Rule, Inc.)*

Figure 9–17 In his painting *The Gossips* Norman Rockwell visually reveals the importance of human speech. Speech does indeed bind together friends, families, and society. *(©1948 The Norman Rockwell Family Trust)*

two slightly different images from the eyes and creates a single image that appears solid, that is, three-dimensional (Figure 9–16).

Hearing

From a social point of view, hearing is perhaps the most important of our senses because speech plays an important role in the communication that binds society together (Figure 9–17). But hearing can also make us aware of things in the external world that we cannot see—the approach of friend or foe on a pitch-black night, for instance. Hearing can also help us better understand events that we can see, as when a mother determines by the sound of the cry whether her baby's cry is caused by pain or hunger. And, hearing can enrich the quality of our lives, such as when we listen to music.

In every instance, what we hear are sound waves produced by vibration. Vibrating objects, such as guitar strings, the surface of the stereo speaker, or vocal cords, move

rapidly back and forth. As they do, each pushes repeatedly against the surrounding air, creating sound waves. Whereas the loudness of sound is determined by the amplitude of the sound wave, its pitch is determined by its frequency (the number of cycles per second) (Figure 9–18). The more cycles, the higher is the pitch. The sound waves travel through the air (or water) to reach our ears.

The Ear

The ear has three parts—the outer, middle, and inner ear (Figure 9–19; Table 9–2). The **outer ear** functions as a receiver. The visible part of the outer ear consists of a fleshy flap visible on the side of the head, called the **pinna**. Shaped like a funnel, the pinna gathers the sound and channels it into the **external auditory canal**. The pinna also accentuates the frequencies of the most important speech sounds, making speech easier to pick out from background noise. Because of their shape and placement at the sides of the

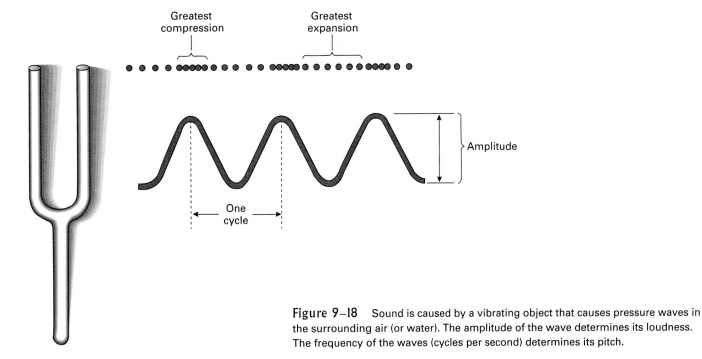

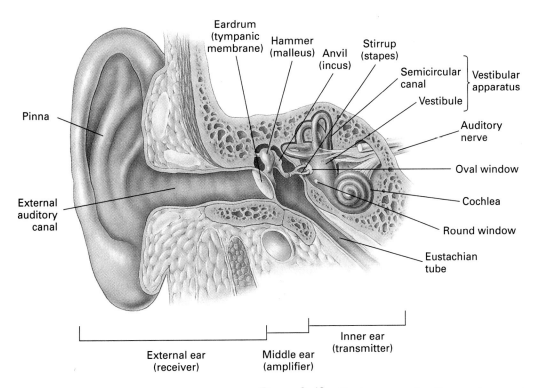

Figure 9–18 Sound is caused by a vibrating object that causes pressure waves in the surrounding air (or water). The amplitude of the wave determines its loudness. The frequency of the waves (cycles per second) determines its pitch.

Figure 9–19 The human ear has three parts: a receiver (the outer ear), an amplifier (the middle ear), and a transmitter (the inner ear). The outer ear consists of the pinna, which gathers sound and funnels it into the external auditory canal to the eardrum. The eardrum is a thin sheet of tissue between the outer and middle ear that vibrates synchronously with sound waves. The three bones of the middle ear convey the vibrations of the eardrum to the inner ear. The pressure waves caused by sound waves are amplified more than 20-fold as they are transmitted through the middle ear. In the inner ear, the pressure waves are converted to neural messages that are sent to the brain for interpretation.

Table 9–2 Structures of the Ear and Their Functions

Structure	Description	Function
Outer Ear		
Pinna	Part of the ear protruding from side of head	Collects and directs sound waves
Auditory canal	Canal between pinna and tympanic membrane	Directs sound to the middle ear
Middle Ear		
Tympanic membrane (eardrum)	Membrane spanning the end of the auditory canal	Vibrates in response to sound waves
Malleus Incus Stapes	Three tiny bones of the middle ear	Amplify the vibrations of the tympanic membrane and transmit them to inner ear
Inner Ear		
Utricle Saccule	Fluid-filled chambers	Equilibrium
Semicircular canals	3 fluid-filled chambers arising from utricle at right angles to one another	Equilibrium
Cochlea	Fluid-filled, bony, snail-shaped chamber	Houses organ of Corti
Organ of Corti	Contains hair cells	The organ of hearing

head, the pinnas also help us to determine the direction from which a sound comes.

The sheet of tissue that separates the outer and middle ears is the eardrum, or the **tympanic membrane**. It is as thin as a sheet of paper and as taut as the head of a tambourine. When sound waves strike the eardrum, it vibrates at the same frequency and transfers the vibrations to the middle ear.

Critical Thinking

As we age, the tissues of the eardrum thicken and become less flexible. How does this partially explain why sensitivity to high-frequency sounds is usually lost as we get older?

The **middle ear** serves as an amplifier. It consists of an air-filled cavity within the temporal bone of the skull that is spanned by the three smallest bones of the body—the **malleus** (hammer), **incus** (anvil), and **stapes** (stirrup). Together these bones function as a system of levers to convey the airborne sound waves from the eardrum to the **oval window**, a sheet of tissue that forms the threshold of the inner ear. The malleus is attached to the inner surface of the eardrum. Therefore, the vibrations of the eardrum in response to sound cause the malleus to rock back and forth. The rocking of the malleus, in turn, causes the incus and then the stapes to move. The base of the stapes fits into the oval window. When the stapes moves, it pushes against the oval window, conveying the vibrations of the eardrum to the inner ear. When very loud sounds strike the ear, two small muscles in the middle ear contract and restrain the movements of these bones to prevent irreparable damage to the delicate receptor cells of the inner ear.

The force of vibrations of the eardrum are amplified 22 times in the middle ear. The magnification of force is necessary to transfer the vibrations to the fluid of the inner ear. The amplification occurs both because of the lever system arrangement of the bones of the middle ear and because of the size difference between the eardrum and the oval window. The eardrum is much larger than the oval window and this concentrates the pressure against the oval window.

If the air pressure is not about equal on both sides of the eardrum, it cannot vibrate freely. The reason is simple: Unequal pressure makes the eardrum bulge inward or outward, holding it in place and causing discomfort, pain, and difficulty hearing. For instance, when a person ascends quickly to high altitudes, the atmospheric pressure is lower than the pressure in the middle ear and the eardrum bulges outward. Unequal pressure is usually alleviated by the **eustachian tube**, which connects the middle ear cavity with the upper region of the throat. Most of the time, the eustachian tube is closed and flattened. However, swallowing or yawning opens it briefly, allowing the air pressure to equalize with the pressure in the middle ear cavity. The sensation of pressure suddenly equalizing is often described as ear popping. It is likely to occur at times when the pressure on the external eardrum changes, such as when riding up or down in an elevator, taking off or landing in an airplane, or scuba diving.

The **inner ear** is a transmitter that generates neural messages in response to pressure waves caused by sound waves, which are sent to the brain for interpretation. The inner ear contains two sensory organs only one of which, the **cochlea**, is concerned with hearing. The other sensory organ, the **vestibular apparatus**, is concerned with sensations of body position and movement and will be discussed in the next section.

The cochlea is the true seat of hearing, so we should consider some of the details of its structure. Keep in mind that the cochlea is actually a cavity, so the representation in Figure 9–20 can be compared with a plaster cast of the hollow space inside the temporal bone.

About the size of a pea, the cochlea is a bony tube about 35 mm long and is coiled about two and a half times, somewhat like the shell of a snail (Figure 9–20). (Cochlea is from Latin for snail.) The wider end of the tube, where the snail's head would be, has two membrane-covered openings. We have already mentioned the upper opening. It is the oval window into which the stapes fits. The lower opening, called the **round window**, serves to relieve the pressure created by the movements of the oval window.

The internal structure is easier to picture if we imagine the cochlea uncoiled so that it forms a long straight tube (Figure 9–21). We would then see that two membranes divide the interior of the cochlea into three longitudinal chambers, each filled with fluid. Like the finger on a glove, the central chamber ends blindly and does not extend completely to the end of the cochlea. As a result, the upper and lower chambers are connected at the end of the tube nearest to the tip of the coil. The floor of the central chamber is formed by the basilar membrane, which supports the **organ of Corti**, the portion of the cochlea most directly responsible for the

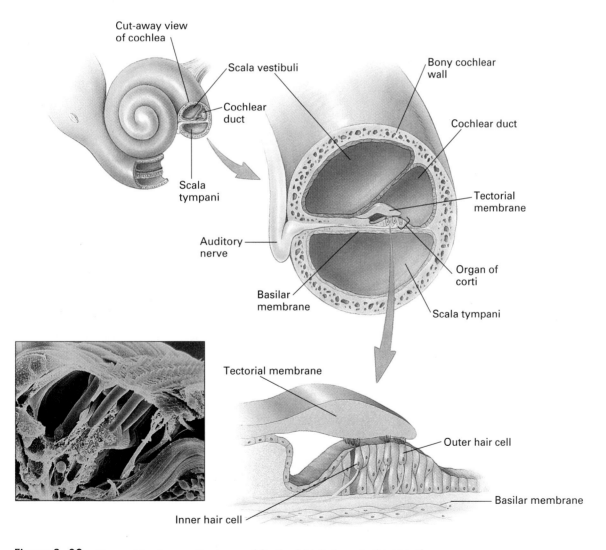

Figure 9–20 The cochlea houses the organ of Corti, which is the actual organ of hearing. The organ of Corti rests on the basilar membrane. The hair cells on the basilar membrane are the receptors for hearing. (photo, Dr. Oran Bredberg/Science Photo Library/Photo Researchers, Inc.)

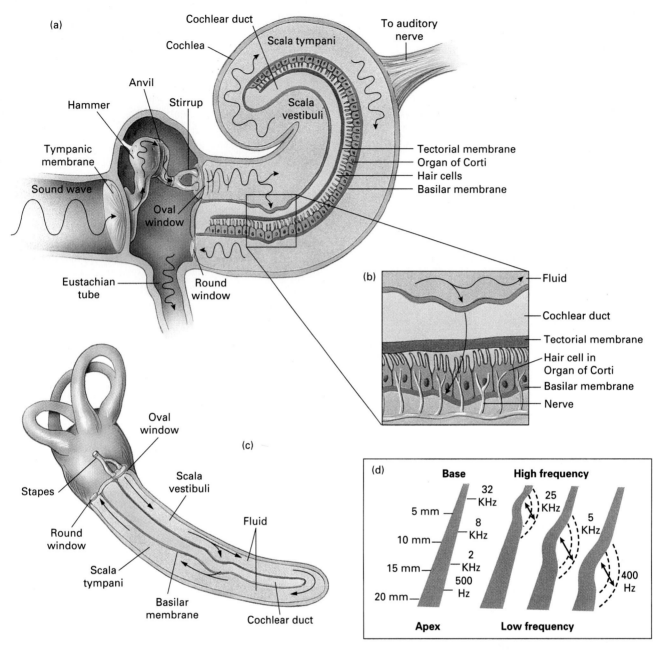

Figure 9-21 The sequence of events from sound vibration to a nerve impulse begins when sound enters the external auditory canal. (a) The sound waves cause the eardrum to vibrate and these vibrations are transmitted through the bones of the middle ear to the oval window. The movements of the oval window cause pressure changes in the fluid of the upper chamber of the cochlea and these are, in turn, transmitted to the lower chamber, where they cause movements of the basilar membrane. (b) When the basilar membrane vibrates, the hairs on the receptors are pushed against the overhanging tectorial membrane. This ultimately results in nerve impulses that are carried to the brain. (c) The width and stiffness of the basilar membrane vary along its length. (d) As a result, different regions of the basilar membrane will vibrate maximally in response to tones of different pitches. The brain interprets input from different regions of the basilar membrane as different pitches.

sense of hearing. The receptors for hearing, called **hair cells** because each has about 100 "hairs" protruding from its upper surface, are positioned on the basilar membrane. The 25,000 hair cells are arranged in rows that resemble miniature picket fences. The hair cells stimulate the nerve cells that carry nerve impulses to the brain. The roof of the organ of Corti is

formed by the **tectorial membrane**, which projects over and is in contact with the hair cells.

When the stapes moves to and fro against the oval window, it sets up corresponding movements in the fluid of the inner ear. The pressure waves are transmitted to the fluid of the topmost canal and then to the lower canal, because they are continuous. The movements of the fluid cause the basilar membrane to swing up and down, which in turn causes the processes on the hair cells to be pressed against the tectorial membrane. The bending of the hairs ultimately results in nerve impulses in the auditory nerve.

Loudness and Pitch Determination

The louder the sound, the greater the pressure changes in the fluid of the inner ear and, therefore, the stronger the bending of the basilar membrane. Because hair cells have different thresholds of stimulation, more vigorous vibrations in the basilar membrane stimulate more hair cells. The brain interprets the increased number of impulses as louder sound.

How do we determine the pitch of a sound? Very low pitched sound, below 150 Hz (Hz, Hertz = cycles per second), is interpreted by the brain by the frequency of impulses in the auditory nerve. The brain's interpretation of the pitch of sounds above 150 Hz is based on the region of the basilar membrane that is stimulated by the sound. In a sense, the cochlea is like a spiral piano keyboard. The basilar membrane varies in width and flexibility along its length (see Figure 9–21c, d). Near the oval window the basilar membrane is narrow and stiff. Like the shorter strings on a piano, this region vibrates maximally in response to high-frequency sound. At the tip of the cochlea, the basilar membrane is wider and floppier. Low-frequency sounds cause this region of the basilar membrane to vibrate maximally. Thus, sounds of different pitches activate hair cells at different places along the basilar membrane. The brain then interprets input from different regions of the basilar membrane as sounds of different pitch.

Hearing Loss

It is estimated that 28 million Americans have some degree of hearing loss and 2 million of them are completely deaf. Hearing loss that is severe enough to interfere with social and job-related communication is among the most common chronic neural impairments in the United States.

There are two types of hearing loss: conductive loss and sensorineural loss. Conductive loss results when airborne sounds are not conducted through the auditory canal to the eardrum or over the bones of the middle ear to the inner ear. An obstruction anywhere along this route can prevent sound messages from stimulating the nerves in the inner ear. The auditory canal can become clogged with wax or other foreign matter. In this case, the insertion of cotton-tip swabs usually aggravates the situation by further clogging and should, therefore, be avoided. Other possible causes for conductive hearing loss are thickening of the eardrum, which might occur with chronic infection, or perforation of the eardrum, which might occur as a result of trauma. Excess fluid in the middle ear, which often occurs with middle ear infections, can also cause conductive hearing loss. A common age-related cause of conduction deafness is otosclerosis, a condition in which the bones of the middle ear become fused and can

no longer transmit sound. In this case, the only way that sound can be transmitted to the inner ear is through the bones of the skull, which is much less efficient. Ludwig von Beethoven suffered from otosclerosis. He would prop a piece of wood between the piano and his skull, just behind his ear. This arrangement helped him hear himself play because the vibrations from the piano traveled through the wood to his inner ear, bypassing the bones of the middle ear. Today otosclerosis is routinely corrected by surgery.

Sensorineural deafness is caused by damage to the hair cells in the inner ear or to the nerve supply of the inner ear. Sensorineural deafness is commonly caused by the gradual loss of hair cells throughout life. Some of the loss is simply due to aging. After age 20, we lose about 1 Hz of our total perceptive range of 20,000 Hz every day. However, much of the damage to hair cells is caused by exposure to loud noise (see Environmental Concerns: Noise Pollution). Studies have shown that people in societies where loud noises are rare, such as in the Mabaan tribe in Africa, have far more acute hearing in their fifties than does the average young American. But, neural damage can also be caused by certain infections, including mumps, rubella (German measles), syphilis, and meningitis. Drugs such as those used for treating tuberculosis and certain cancers can also cause neural damage, as can certain antibiotics.

One of the first signs of sensorineural damage is the inability to hear high-frequency sounds. Because consonants, which are needed to decipher most words, include higher frequencies than do vowels, speech becomes difficult to understand. So, beware if you find yourself asking, "Could you repeat that?"

Critical Thinking

A tuning fork is often useful for determining whether a hearing loss is conductive or sensorineural. A tuning fork is struck and placed on the top of the head. Thus, sound reaches the ears in two ways: through the bones of the skull and through the air. The sound is equally loud in both ears of a person with normal hearing. If the person has a conductive hearing loss, would the sound be louder in the good ear or in the deaf ear? In which ear would the sound seem louder if the person has a sensorineural loss?

The most effective way of combating hearing loss is the use of a hearing aid. The basic job of a hearing aid is to amplify sound. One type of hearing aid presents amplified sound to the eardrum. Another type uses a vibrator or stimulator that is placed in the bone of the skull behind the ear. Sound is then conducted through the bones of the skull to the inner ear.

Hearing aids are often unable to help profoundly deaf people, but cochlear implants can pierce the silence for some of them. A cochlear implant is an array of electrodes that transforms sound into electrical signals that are delivered to the nerve cells near the cochlea. It is surgically inserted into

Noise Pollution

I t is difficult to escape the din of modern life—noise from airports, city streets, loud appliances, stereos. Noise pollution threatens your hearing and your health. Exposure to excessive noise is to blame for the hearing loss of one-third of all hearing-impaired people.

Loud noise damages the hairs on the hair cells of the inner ear. When they are exposed to too much noise, the hairs become worn down, lose the ability to move, and can become fused together (see the figure below). Unfortunately, there is no way to undo the damage. You can't get spare parts for your ears.

The loudness of noise is measured in decibels (dB). The decibel scale is logarithmic. An increase of 10 dB generally makes a given sound twice as loud. The decibel ratings of some familiar sounds are given in the table, *Effects of Noise Pollution*. Most people judge sounds over 60 dB to be intrusive, over 80 dB to be annoying, and over 100 dB to be extremely bother-

some. The Federal Occupational Safety and Health Administration has set 85 dB as the sound safety limit for 8 hours of exposure. The threshold for physical pain is 140 dB.

The louder a sound, the shorter the exposure time necessary to damage the ear. Even a single, explosively loud sound is capable of damaging hair cells. More commonly, however, hearing loss results from prolonged exposure to volumes over 55 dB. Hearing can be damaged by exposure to noise loud enough to make it difficult to converse with someone. Some damage probably occurred if sounds seem muffled after you leave the noisy area. Your ears can endure sound at 90 dB for about 8 hours. For every 5 dB above that, it

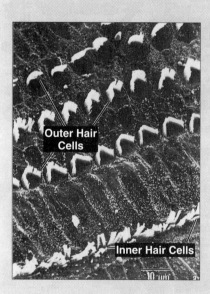

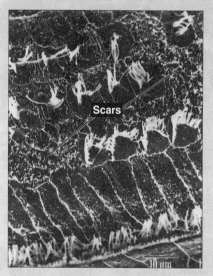

The hair cells of the inner ear can be permanently damaged by loud noise. The micrograph on the left shows a portion of the organ of Corti from a normal guinea pig, showing the characteristic three rows of outer hair cells and a single inner row. The micrograph on the right shows the organ of Corti from a guinea pig who had been exposed to loud (120 dB) sound. Sound this loud is common at most rock concerts. Note that some injured cells have been replaced by scars. On most of the hair cells that have survived, the hairs are no longer in an orderly pattern. Hearing is permanently damaged because the lost hair cells cannot be replaced, and the damaged cells may not recover. *(Robert Preston and Joseph E. Hawkins, Kresge Hearing Institute, University of Michigan)*

the cochlea through the round window. A cochlear implant is not a miracle cure. Only about 20% of implant recipients can hear well enough to understand most spoken sentences and to use the telephone easily. The use of a telephone is especially difficult for deaf people, because they cannot get any visual cues, such as reading lips to help them understand the spoken words. Another 20% of those with cochlear implants are not helped at all. For most recipients, the benefit is somewhere in between these two extremes. They are able to hear enough sound to make lip reading easier.

Although it is still years from reality, there is a new glimmer of hope on the horizon for people whose hearing

Social Issue

Many deaf people consider themselves to be a cultural group, not a medical anomaly. It has been argued, for instance, that communication is the real problem faced by the deaf and that the solution to the problem should, therefore, be a social one, not a medical one. If medical technology is available, should we use it to "fix" conditions such as deafness? If you had a child that was born deaf, would you want a cochlear implant for that child? Why? (Assume that the child is not yet able to make that decision for himself or herself.)

Effects of Noise Pollution

Example of Sound Source	dbA	Effect from Prolonged Exposure
Jet plane at takeoff	150	Eardrum rupture
Deck of aircraft carrier	140	Very painful; traumatic injury
Rock-and-roll band (at max)	130	Irreversible damage
Jet plane at 500 ft	110	Loss of hearing
Subway, lawn mower	100	
Electric blender	90	Annoying
Washing machine, freight train at 50 ft	80	
Traffic noise	70	Intrusive
Normal conversation	65	
Chirping bird	60	
Quiet neighborhood (daytime)	50	
Soft background music	40	Quiet
Library	30	
Whisper	20	Very quiet
Breathing, rustling leaves	10	
	0	Threshold of hearing

takes half as long for damage to begin. Thus, sound at 95 dB will damage your ears in only 4 hours. At 110 dB, the average rock concert or stereo headset at full blast, can damage your ears in as little as half an hour.

We have come to expect hearing loss in the elderly, but a surprising number of young people also have impaired hearing. A study of entering freshmen at the University of Tennessee revealed that 60% of them had hearing loss. In fact, the hearing of 14% of the young men tested was about equal to that expected in the average 65 year old! Since fewer than 4% of the sixth graders had hearing loss, it suggests that the damage was done during the teen years. The culprit is most likely noise—probably in the form of music.

If you have ever walked away from a noisy area, such as a concert or a construction site, and your ears were ringing or everything sounded as if you were under water, you've experienced what is called a temporary threshold shift. It's a sign that some of the hair cells in the inner ear were damaged.

Noise has other harmful effects. The stress responses to daytime noise can carry over into the night, causing sleep disturbances that can make you groggy, tense, and forgetful the next day. Noise triggers stress responses that may increase heart rate and blood pressure. Even common kitchen appliances produce sound levels that cause pupils to dilate, the mouth to become dry, muscles to tense, and the digestive system to slow down. People become irritable when exposed to noise for prolonged periods and this may aggravate tension within a family. Many an argument has flared up when a person who has spent the day in a noisy work place comes home to find the television or stereo blaring or children crying instead of the peace and quiet that was craved.

So, how can you protect yourself from bad vibes? Don't listen to loud music. Keep the tunes low enough that you can still hear other sounds. No one else should be able to hear the music you are listening to with earphones. When you can't avoid loud noise, such as when you are mowing the lawn, vacuuming, or attending a rock concert, wear ear plugs to protect your hearing. You'll find ear plugs at most drug stores, sporting good stores, and music stores.

Social Issue

For many years the guideline for workplace noise was set at 90 dB, a value known to cause hearing loss after years of exposure. Studies showed that it would cost American industry billions of dollars to reduce the noise level to 85 dB. That cost would inevitably be passed on to the consumer. What price should be paid to protect the hearing of workers? How much more are you willing to pay for goods?

loss is due to the loss of hair cells. It was once thought that hair cells were formed only once during life and that they could not regenerate, but it now seems that this is wrong—at least for birds, fish, and rodents. Some studies on rats suggest that retinoic acid, a chemical that helps the brain take shape during embryonic development, stimulates the regeneration of hair cells. (It is also the effective component in certain wrinkle creams, such as the highly advertised Retin-A.) This discovery opens the possibility that it may someday be possible to design a drug that will stimulate ear cells to form new hair cells and, in this way, to cure some kinds of deafness. However, even if human hair cells can be made to

regenerate, they would still have to make functional connections to the brain. Furthermore, hearing loss often involves additional damage that would have to be repaired before hearing was restored. Nonetheless, there is now hope where none existed before.

Ear Infections

An external ear infection is one in the ear canal that leads to the eardrum. Swimmer's ear, the most common type, is precipitated by water trapped in the ear canal that results in events that are favorable for the growth of bacteria. The first symptom of swimmer's ear is itching, which is usually followed by pain that can become intense and constant. Chewing food or touching the earlobe sharpens the pain. The usual treatment consists of antibiotic ear drops, heat, and pain medication. To prevent swimmer's ear, always be sure that water drains from the ear canal after swimming or bathing, by hopping or shaking the head if need be. Also, several drops of 97% alcohol can be added to the ear canal to help dry it.

Middle ear infections usually result from infections of the nose and throat that work their way through the eustachian tubes connecting the throat and middle ear. At least half of all children get an ear infection at some time and some children get four or five ear infections a year. Middle ear infections are more common in children than in adults because the eustachian tubes become tilted during growth. In children, the tubes are nearly horizontal, making it easier for infectious organisms to travel through them than through the tilted, curved tubes of adults. The signs of a middle ear infection are a stabbing earache, impaired hearing, and a feeling of fullness in the ear, often accompanied by a fever. A middle ear infection is usually treated with an antibiotic—amoxicillin being the drug of choice.

Chronic ear infections can lead to a build-up of fluid in the middle ear that impairs hearing. The loss of hearing may be gradual and go unnoticed, but it may interfere with language development, lead to problems in school, and cause irritability. In some cases, the fluid is reduced by treatment with antibiotics and/or decongestants. However, if noninvasive treatments don't work, the fluid can be drained by a simple operation in which tubes are placed through a tiny opening in the eardrum. The tubes are kept in for 6 weeks to several months, usually until they come out on their own. This is the most common surgery performed on children. Although the results are immediate and sometimes dramatic, the decision to have tubes put into a child's ear should be considered carefully. Recently, a study of 6600 such oper-

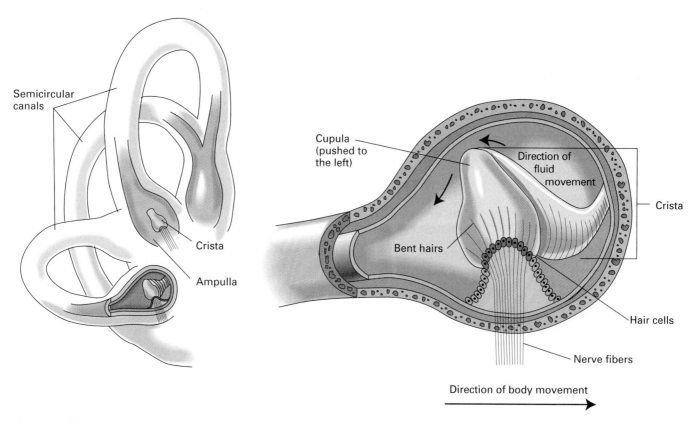

Figure 9–22 Our sense of dynamic equilibrium is due to the semicircular canals and the ampullae. Together they report any sudden changes in the head, including those caused by acceleration or deceleration. The three semicircular canals are oriented perpendicularly to one another. At the base of each canal is a chamber, called an ampulla. Within the ampulla is a gelatinous mass called the cupula in which hair cells are embedded. The canals and chambers are filled with a fluid that moves when the head does. The movement of fluid pushes the cupula and stimulates hair cells that send messages to the brain regarding body position and movement. The brain interprets the signals and maintains our balance.

ations was performed to assess whether this operation is appropriate. The conclusions were that 23%, nearly a quarter of them, were inappropriate and 35% were equivocal. Surgery is more likely to be appropriate when the child has had fluid build-up lasting 90 days or more and when the fluid has caused a hearing loss that resulted in delayed language development.

Balance

The **vestibular apparatus**, a fluid-filled maze of chambers and canals within the inner ear, is responsible for monitoring the position and movement of the head. The receptors in the vestibular apparatus are hair cells similar to those in the cochlea. When the hairs on these cells are bent, as will occur during head movements or changes in velocity, the hair cells send messages to the brain. The brain uses this input to maintain balance.

The two components of the vestibular apparatus are the semicircular canals and the vestibule. The **semicircular canals** monitor precisely any sudden movements of the head, including those caused by acceleration and deceleration (Figure 9–22). At the base of each canal is a tuft of hair cells. The hair-like projections from these cells are embedded in a pointed cap of stiff, pliable gelatinous material, called the **cupula**. When the head is moved, the fluid in the canal lags, causing the cupula to bend the hair cells and stimulate them.

In each ear there are three semicircular canals, oriented at right angles to one another (Figure 9–22a). One is parallel to the sides of the head, another to the plane of the face, and the third to the horizon. Head movements will set the fluid in the canals into motion. The fluid moves in the opposite direction of head movement in the same way that sudden acceleration slams passengers against the car seat. Whereas nodding your head to say yes will cause the fluid in the canals parallel to the sides of the head (the anterior semicircular canals) to swirl, shaking it from side to side to say no will cause fluid in the canals parallel to the horizon (the lateral semicircular canals) to move. Tilting your head to look under the bed will cause fluid in the canals that are parallel to your face (the posterior semicircular canal) to move back and forth. Everyday movements are rarely in a single plane, but even the most complex movement can be analyzed in terms of motion in three planes. Indeed, every movement of your head causes fluid movement in at least one of the semicircular canals. This stimulates the hair cells and they send messages to the brain that are interpreted as head movement in the opposite direction as hair-cell bending. Within a few seconds after the rotation of the body stops, the movement of fluid in the semicircular canals also ceases. Therefore, the hair cells are no longer stimulated when we travel at a constant velocity and, without visual cues, we cannot tell that we are moving.

Critical Thinking

When you stop moving, there is a time lag before the fluid in the semicircular canals stops swirling. How does this account for the sensation of dizziness after you have been twirling around for a while?

The **vestibule**, the other part of the vestibular apparatus, consists of the utricle and the saccule, two fluid-filled cavities. Thanks to the utricle and the saccule, you know which end is up, literally (Figure 9–23). They tell the brain the position of the head with respect to gravity when the body is not moving. They also respond to acceleration and deceleration, but not to rotational changes as the semicircular canals do. The utricle and the saccule are two cavities that contain hair cells overlaid with a gelatinous material in which small granules of calcium carbonate, a chalk-like substance, are embedded. These granules, called **otoliths**, make the gelatin heavier than the surrounding material and, therefore, make it slide over the hair cells whenever the head is moved. The movement of the gelatin stimulates the hair cells, which send messages to the brain about the position of the head relative to gravity.

The utricle and saccule sense different types of movement. The utricle senses the forward tilting of the head, as well as forward motion, because its hair cells are on the floor of the chamber, oriented vertically when the head is upright. In contrast, the hair cells of the saccule are on the wall of the chamber, oriented horizontally when the head is upright. They respond when we move vertically, as when we jump up and down.

Motion sickness—that dreadful feeling of dizziness and nausea that sometimes causes vomiting—is thought to be caused by a mismatch of sensory input from the vestibular apparatus and the eyes. For example, many people get carsick from reading in the car. When looking down at a book, your eyes tell your brain that your body is stationary. However, as the car changes speed, turns, and hits bumps in the road, the vestibular system detects motion. Seasickness results when the vestibular apparatus tells the brain your head is rocking back and forth, but the deck under your feet looks level. The brain is somehow "confused" by the conflicting information and the result is motion sickness. The feeling can sometimes be relieved by staring at the horizon, so you can see that you are moving. Over-the-counter drugs to prevent motion sickness, such as Dramamine, work by inhibiting the messages from the vestibular apparatus.

Finally, high-impact aerobics and other jarring activities, such as long-distance running and volleyball, may damage the inner ear, disturbing equilibrium and causing dizziness, imbalance, and motion sickness. Although most people who regularly engage in high impact activities have no ill effects, 20% to 25% of them may experience some symptoms of inner ear damage. It seems that repeated jarring may

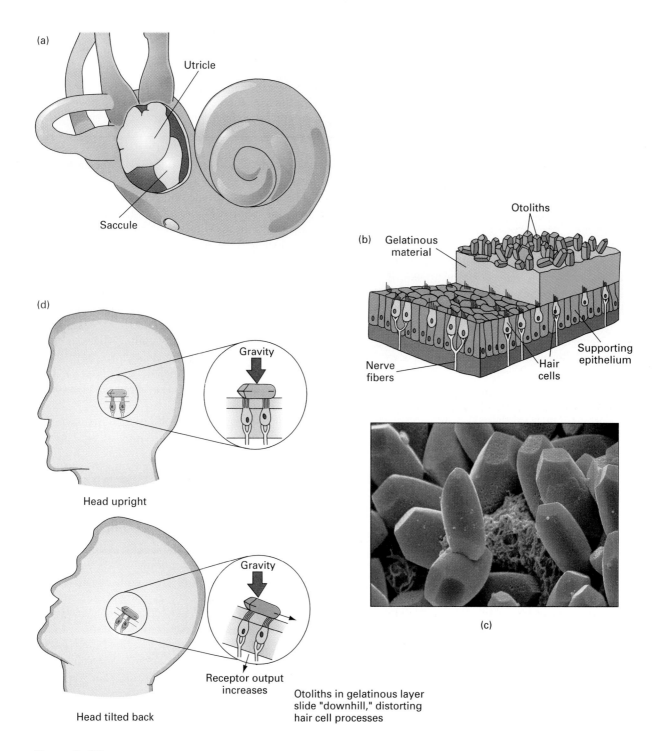

(a) Utricle

Saccule

(b) Gelatinous material

Otoliths

Nerve fibers

Hair cells

Supporting epithelium

(d) Gravity

Head upright

Gravity

Receptor output increases

Head tilted back

Otoliths in gelatinous layer slide "downhill," distorting hair cell processes

(c)

Figure 9–23 Our sense of static equilibrium is due to receptors in the vestibule. (a) The vestibule is comprised of two chambers, the utricle and the saccule. The utricle and saccule sense not only the position of the head with respect to gravity but also changes in linear acceleration. (b) Within the utricle and saccule, hair-like projections of hair cells are embedded in a jelly-like cap. (c) Small crystals of calcium carbonate, called otoliths, make the cap heavier than the surrounding fluid. (d) When the head is tilted, the otoliths shift position, causing the jelly-like cap to move and to stimulate the hair cells embedded within it.   *(c, © Lennart Nilsson)*

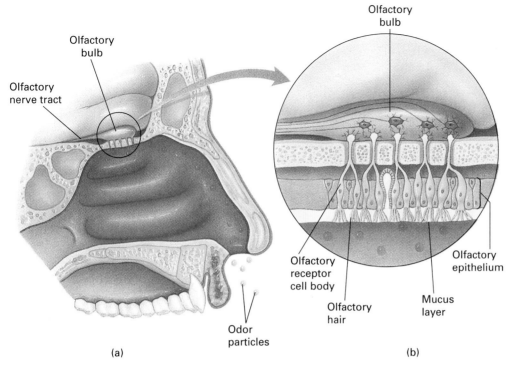

Figure 9–24 Our sense of smell resides in a small patch of tissue in the roof of each nasal cavity within which are some 5 million olfactory receptors.

loosen the otoliths and jam them down among the hair cells, which then send incorrect signals to the brain.

Critical Thinking

Many astronauts report extreme motion sickness during their first few days in orbit. How might the absence of gravitational pull on the vestibular apparatus explain this?

Smell and Taste

Smell is perhaps the least appreciated of our senses. You could easily get along without it, but life wouldn't be as interesting. For one thing, about 80% of what we usually think of as the flavor of a food is really due to our sense of smell. This is why food often seems bland when your nose is congested.

You have about 5 million olfactory (smell) receptors, located, not in the nostrils, but in a small patch of tissue the size of a postage stamp in the roof of each nasal cavity (Figure 9–24). The receptors are neurons with long cilia (or cell hairs) that project outward from the cells that line the nasal cavity and are covered by a coat of mucus, which keeps them moist and is a solvent for odorous molecules. The hairs move, gently swirling the mucus. Olfactory receptors,

by the way, are the only neurons known to be replaced during life—about every 60 days.

An odor is detected when it is carried into the nasal cavity in air, dissolves in the mucus, and binds to receptors on the hairs of the olfactory receptors, thereby stimulating the receptor. If threshold is reached, the message is carried to the two olfactory bulbs of the brain. There the information is processed and passed on to the limbic system and to the cerebral cortex, where it is interpreted. Interestingly, the limbic system is a center for emotions and memory. Thus, we rarely have neutral responses to odors. Generally, the smell of fresh baked bread is pleasing, but the smell of a skunk is repulsive. The perfume industry makes a fortune because of the association between scents and sexuality. Also, odors can trigger a flood of long-forgotten memories. If your first kiss was near a blooming lilac tree, for instance, a simple sniff of lilac may take you back in time and place.

We have about 1000 different types of olfactory receptors, with which we can distinguish about 10,000 different odors. Each receptor responds to several different odors. Thus, the brain relies on input from more than one type of receptor to identify an odor.

There are four primary tastes: sweet, salty, sour, and bitter. However, four tastes are enough to answer the important question about food or drink—should we swallow or spit it out. We are prompted to swallow if it tastes sweet, because sweetness implies a rich source of calories and thus

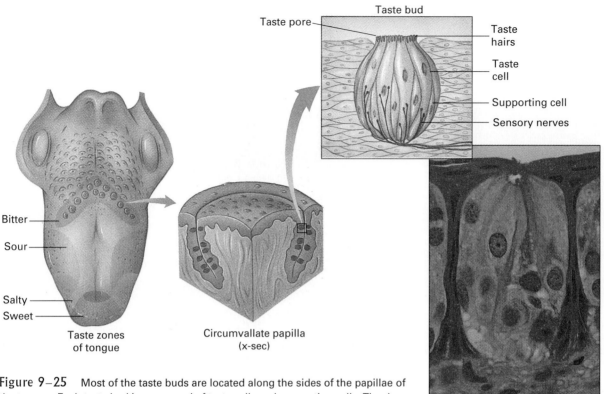

Figure 9–25 Most of the taste buds are located along the sides of the papillae of the tongue. Each taste bud is composed of taste cells and supporting cells. The dendrites of sensory nerve cells are coiled around the base of the taste cells. The tip of each taste cell bears many taste hairs that project through a pore at the tip of the taste bud. The taste hairs respond to certain food molecules dissolved in water and send messages to the sensory neurons, which then deliver the message to the brain. (photo, © Cabisco/Visuals Unlimited)

energy. Salty tastes also prompt swallowing because salts will replace those lost in perspiration. Sourness is tough to call. Sour is the taste of unripe fruit that would have more food value later on. As fruit ripens, starches break down into sugars that create a sweet taste and mask the sourness. So it is often better to reject sour fruits and wait for them to ripen. However, some of them, such as oranges, lemons, and tomatoes, are rich sources of vitamin C, an essential vitamin. Bitter is easy. We reject foods that taste bitter. Bitterness usually indicates that food is poisonous or spoiled.

We have about 10,000 **taste buds**, the structures responsible for our sense of taste. Most of them are on the tongue, but some are scattered on the inner surface of the cheeks, on the roof of the mouth, and in the throat. The taste buds on the tongue are located along the sides of the papillae, those small bumps that give the tongue a slightly rough feeling. Each papilla contains 100 to 200 taste buds. The cells of a taste bud are completely replaced about every 10 days, so you need not worry about losing your sense of taste for life when you burn your mouth eating hot pizza, as almost always happens.

A taste bud is the interface between chemicals dissolved in saliva and the sensory neurons that will convey information to the brain. Each taste bud is a lemon-shaped structure containing about 40 modified epithelial cells (Figure 9-25). Some of these cells are the taste cells that respond to chemicals and others are supporting cells. The taste cells have long microvilli, called **taste hairs**, that project into a pore at the tip of the taste bud. These taste hairs bear the receptors for certain chemicals found in food. Dissolved in water, the food molecules enter the pore and stimulate the taste hairs. Although taste cells are not neurons, they do generate electrical signals, which are then sent to the dendrites of sensory nerve cells wrapped around the taste cell.

Each taste bud responds to all four basic tastes, but it is usually more sensitive to one or two of them. Because of the way the taste buds are distributed, different regions of the tongue have *slightly* different sensitivity to these tastes. In general, the tip of the tongue is most sensitive to sweet tastes, the back to bitter, and the sides to sour. Sensitivity to salty tastes is fairly evenly distributed.

SUMMARY

1. Changes in external and internal environments stimulate sensory receptors, which in turn generate electrochemical messages that are eventually converted to nerve impulses and are conducted to the brain.

2. There are five classes of sensory receptors:
   a. Mechanoreceptors are responsible for touch, pressure, hearing, equilibrium, blood pressure, and body position.
   b. Thermoreceptors detect changes in temperature.
   c. Photoreceptors detect light.
   d. Chemoreceptors are responsible for taste and smell and levels of carbon dioxide, oxygen, and glucose in body fluids.
   e. Pain receptors respond to noxious stimuli that result from tissue damage.

3. Receptors can also be classified by their location. Exteroceptors are located near the body surface and respond to environmental changes. Interoceptors are located inside the body and monitor internal conditions.

4. The general senses include touch, pressure, temperature, sense of body and limb position, and pain. Merkel disks, flattened dendrite endings that contact the outer layer of skin, detect when the skin has been touched. Meissner's corpuscles, encapsulated nerve cell endings, tell us exactly where we have been touched. Pacinian corpuscles, layers of tissue surrounding a nerve cell ending, are scattered in the deeper layers of the skin and respond to pressure. Thermoreceptors respond to changes in temperature. Cold receptors are most active between 10°C and 20°C, and heat receptors are active at temperatures between 25°C and 45°C. Specialized muscle fibers with sensory nerve cell endings wrapped around them, called muscle spindles, alert the brain whenever a muscle is stretched. Golgi tendon organs, highly branched nerve fibers in tendons, measure the degree of muscle tension. This information coupled with the information from the inner ear coordinates our movements. A stimulus that results in damaged tissue causes pain receptors to report to the brain. Chemicals involved in healing damaged tissue such as prostaglandins and their precursor bradykinin alert free nerve cell endings of injury. This results in the sensation of pain.

5. There are three layers to the eye. The sclera and the cornea make up the outer layer. The choroid, ciliary body, and iris make up the middle layer. The retina, which contains the photoreceptors, the rods and cones, is the innermost layer.

6. The cornea and the lens work together to focus images on the retina. The lens can accommodate (change its shape) in order to focus on near or far away objects.

7. The three most common visual problems are farsightedness, nearsightedness, and astigmatism. They are all focusing problems and can be rectified with corrective lenses. In farsightedness, the eyeball is too short or the lens is too thin, causing the image to be focused behind the retina, making it hard to see nearby objects although distant ones can be seen clearly. In nearsightedness, the eyeball is elongated or the lens is too thick, causing the image to focus in front of the retina. Distant objects are blurred, although close objects can be seen clearly. Astigmatism comes from irregularities in the curvature of the cornea, which causes a distortion of the image because the light rays converge unevenly.

8. When light is absorbed by pigment molecules in photoreceptors, a chemical change occurs in the pigment molecule. This results in a change in membrane permeability of the photoreceptor that sets up a neural message. The optic nerve then carries nerve signals from the retina to the back of the brain.

9. The three types of cones are named for the color of light they absorb—green, red, and blue—and allow us to see colors. Color blindness is caused by the lack of a sufficient number of one of these types of cones.

10. Dark adaptation is the adjustment of eyes to dim light. It occurs when rhodopsin is resynthesized in the rods.

11. The brain converts a two-dimensional image on each retina to a three-dimensional image to give us depth perception. The separate images received from each eye are the most important cue the brain uses to do this.

12. The ear is made up of three regions: the outer ear, the middle ear, and the inner ear. The outer ear consists of the pinna and the external auditory canal. The middle ear consists of the eardrum and three bones: the malleus, incus, and stapes. The inner ear consists of the cochlea and vestibular apparatus.

13. Hearing is the sensing of sound waves caused by vibration. Sound enters the outer ear and vibrates the eardrum, or tympanic membrane. These vibrations move the malleus, which moves the incus and the stapes. The stapes conveys these vibrations to the inner ear by means of the oval window.

14. Sound is amplified 22 times in the middle ear due to the arrangement of the middle ear bones and the larger size of the eardrum relative to the oval window.

15. Pressure on either side of the eardrum is regulated by the eustachian tube.

16. The cochlea is a bony, coiled tube about 35 mm long. It is separated into three longitudinal tubes. The middle tube contains the organ of Corti. It is lined with hair cells and is the portion of the cochlea most responsible for hearing.

17. The ear detects loudness because loud sounds stimulate a higher percentage of the hair cells on each section of the basilar membrane than do soft sounds. Pitch is detected because sounds of different pitches (over 150 Hz) stimulate different areas of the basilar membrane.

18. The two types of hearing loss are conductive and sensorineural. Conductive loss occurs when sounds are prevented from reaching the inner ear. Sensorineural hearing loss occurs when there is damage to the hair cells or nerve supply of the inner ear. Cochlear implants, amplifiers, vibrators, and possibly some methods for regenerating hair cells are means of combating hearing loss.

19. Balance is controlled by the vestibular apparatus, which consists of the semicircular canals and the vestibule. The semicircular canals monitor sudden movements of the head. The vestibule is made of two components, the saccule and the utricle, which tell the brain the position of the head with respect to gravity.

20. Smell receptors are located in the nasal cavity. They are lined with cilia and are coated by mucus. Odorous molecules dissolve in the mucus and bind to the receptors' hairs, stimulating the receptor. Information is passed to the olfactory bulbs and then to the limbic system and the cerebral cortex.

21. Taste buds are responsible for our sense of taste. They are on the tongue, inside the cheeks, on the roof of the mouth, and in the throat. They sense the four basic tastes of sweet, salty, sour, and bitter.

REVIEW QUESTIONS

1. What is a receptor potential?
2. Define sensory adaptation and give two examples.
3. What are the five classes of receptors? Which would give rise to the general senses? Which would give rise to special senses?
4. When stimulated, what do Merkel disks tell us?
5. What are the two distinct regions of the outer layer of the eye?
6. How is light focused on the retina?
7. How is light converted to a neural message?
8. How are sound waves produced? How are loudness and pitch determined?
9. What is the visible part of the outer ear and how does its shape and placement help in the reception of sound? Give two examples.
10. What is the function of the eardrum?
11. Why is it necessary for the force of the vibrations to be amplified in the middle ear? How is it accomplished?
12. What are the effects of unequal air pressure on either side of the eardrum? How does the eustachian tube help to regulate air pressure?
13. What are the two sensory organs of the inner ear? Explain the structure of each and how it allows them to sense body position and motion.
14. How does the basilar membrane respond to pitch?
15. What are the two types of hearing loss? Explain how they differ.
16. What causes motion sickness? How can it be "cured"?
17. Where are the olfactory receptors located? Explain their structure as it relates to their function.
19. What are the four primary tastes?
20. What are the structures responsible for taste? Where are they found?
21. Describe the structure of a taste bud.

Maintenance of the Body

(© Jeff Greenberg/Visuals Unlimited)

Blood

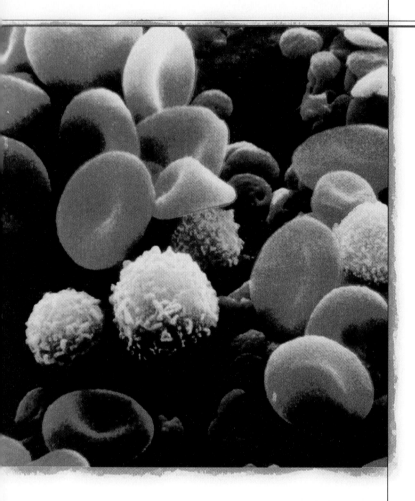

(© Ken Eward/Biografx/Photo Researchers, Inc.)

Blood is often used as a symbol of life. Perhaps the association between life and blood developed when it was noticed that life drains from the body when enough blood is lost. Indeed, blood is truly a life-sustaining fluid. In this chapter we will consider the functions and composition of blood.

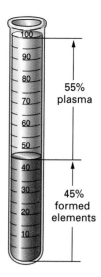

Figure 10–1   Whole blood consists of a straw-colored liquid, called plasma, in which cellular elements, called formed elements, are suspended. Blood can be separated into its major components if it is placed in a test tube with an anticoagulant and then spun in a centrifuge. The formed elements, primarily red blood cells (erythrocytes), are packed at the bottom of the test tube. Just above the red blood cells is a thin layer consisting of white blood cells (leukocytes) and platelets. The uppermost layer consists of plasma.

The Functions of Blood

Blood is sometimes referred to as the "river of life." The title is apt because, like a river, it serves as the body's transportation system. It carries vital materials to the cells and carries away the wastes that cells produce. But blood does more than passively move its precious cargoes around the body. Its white blood cells help to protect us against disease-causing organisms and its clotting mechanisms help protect us from blood loss when a vessel is damaged. In addition, buffers in the blood help to regulate the acid-base balance of body fluids. Blood also helps regulate body temperature by absorbing heat produced in metabolically active regions and distributing it to cooler regions and to the skin, where heat can be dissipated. We see, then, that the diverse functions of blood can be grouped into three categories: transportation, protection, and regulation.

The Composition of Blood

Blood *is* thicker than water. Not only is it thicker, it's also denser. An important reason for the difference in properties between blood and water is that blood contains cells suspended in its watery fluid. In fact, a single drop of blood contains more than 250 million blood cells. You may recall from Chapter 4 that blood is classified as a connective tissue because it contains cellular elements suspended in a matrix. The liquid matrix is called *plasma* and the cellular elements are collectively called the *formed elements* (Figure 10–1).

Plasma

Plasma, a straw-colored liquid that makes up about 55% of whole blood, serves as the medium for transporting materials within the blood. Plasma consists of water with substances dissolved in it (7%–8%). Almost every substance that is transported within the blood is dissolved in the plasma. Thus, the plasma contains nutrients (e.g., simple sugars, amino acids, lipids, vitamins), ions (e.g., Na^+, K^+, Cl^-), dissolved gases (e.g., carbon dioxide, nitrogen, and a small amount of oxygen), and every hormone. In addition to

transporting materials to the cells, the plasma also carries away cellular wastes, such as urea from protein breakdown and uric acid from nucleic acid breakdown, to the sites, such as the kidneys, where they can be removed from the body.

Most of the dissolved substances (solutes) in the blood are **plasma proteins**. The plasma proteins help balance water flow between the blood and the cells. You may recall from Chapter 3 that water moves across biological membranes from an area of lesser solute concentration to an area of greater solute concentration. Without the plasma proteins, water would be drawn out of the blood by the proteins in cells. As a result, fluid would accumulate in the tissues, causing swelling (Figure 10–2).

Most of the 50 or so types of plasma proteins fall into one of three general categories. First, are the albumins. Because the albumins make up more than half of the plasma proteins, they are most important in the blood's water-balancing ability. Second, are the globulins, a group of proteins with a variety of functions. Some globulins transport lipids, including fats and cholesterol as well as fat-soluble vitamins. Other globulins are antibodies, which provide protection against many diseases. The third category of plasma protein is clotting proteins, such as fibrinogen.

Formed Elements

The plasma is the vehicle for transporting many substances around the body. However, the **formed elements**—

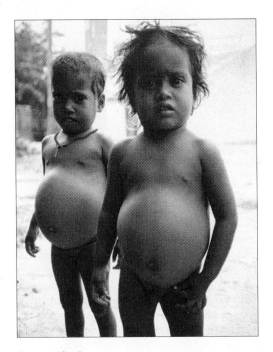

platelets, white blood cells, and red blood cells—perform some of the key functions of the blood.

Red bone marrow, a connective tissue framework that fills the cavities within many bones, is the birth place and nursery for the formed elements. On this framework sit fat cells along with undifferentiated cells, called **stem cells**, which divide and give rise to all blood cells (Figure 10–3). In infants and children, red marrow fills most bone cavities. With age, however, the marrow of most long bones fills with fat. In this way, red marrow becomes yellow marrow and serves as a reserve of energy that can be used if body fat becomes depleted. In adults, then, red bone marrow is found primarily in the flat bones of the skull and pelvis, the sternum, the ribs, and in the heads of the humerus (bone of the upper arm) and of the femur (thigh bone). However, following a severe and sustained blood loss, yellow marrow can be converted back to red marrow, increasing the rate of blood cell production.

Platelets

Platelets, sometimes called thrombocytes, are essential to blood clotting. Platelets are actually fragments of larger precursor cells, called megakaryocytes. Platelets are formed in the red bone marrow when these precursor cells break apart. The fragments are released into the blood at the astounding rate of about 200 billion a day. Platelets mature during the course of a week and then circulate in the blood for about 10

Figure 10–2   These children are suffering from kwashiorkor, a condition that results when the diet is extremely deficient in proteins. Insufficient dietary protein causes a deficit in plasma proteins. Notice the swelling in the abdomens of these children. The swelling occurs because there are not enough plasma proteins to draw water back into the blood from the tissues as it would in a well-nourished person.   *(© Food and Agriculture Organization of the United Nations)*

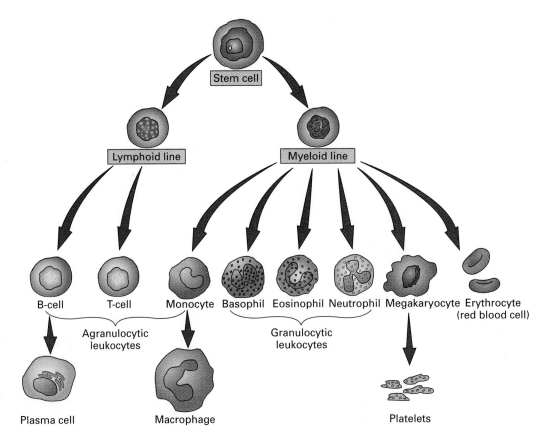

Figure 10–3   All blood cells originate in the red bone marrow from an undifferentiated precursor, called a stem cell. Stem cells divide and give rise to two lines of cells, which through additional cell division and differentiation give rise to all types of blood cells.

Stem Cells

If blood is considered to be the river of life, then stem cells are the springs that feed that river. Stem cells give rise to red cells, white cells, and platelets. Indeed, they produce about 260 billion new cells (about an ounce of blood) every day.

Because stem cells are able to produce an endless supply of blood cells, they hold a promise for treating a host of blood conditions. For instance, inherited disorders such as sickle-cell anemia and thalassemia, in which abnormal forms of hemoglobin are produced, could become curable. Stem cells of the same tissue type but that produce normal hemoglobin could be transplanted into a person with sickle-cell anemia. Then, voilà, normal red cells could begin replacing abnormal ones. Stem cells have already been used to successfully treat some cases of beta-thalassemia, a form of anemia caused by a defective gene.

Stem cells could also be used to regenerate important immune system cells that form the foundation of the body's defense system. This would open the door for new cancer therapies, especially for leukemia. The use of stem cells even provides a glimmer of hope for new approaches to the treatment of HIV infections that cause AIDS. Transplants of stem cells might be able to restore some of the lymphocytes that are killed by HIV, thereby slowing the course of the infection or perhaps even bringing the patient back to an earlier stage in the condition. Furthermore, if a gene were identified that could make a lymphocyte resistant to HIV, it could then be inserted into stem cells. The genetically altered stem cells would then generate lymphocytes that would not be killed by an HIV infection and the patient's dwindling lymphocyte supply could be replenished.

Although researchers had no difficulty dreaming up life-saving uses for stem cells, the trick was to find them. Fewer than one in every thousand marrow cells is a stem cell. Nonetheless, techniques have been developed to identify and isolate the stem cells from the other marrow cells. Umbilical cord blood has also been identified as a good source of stem cells. It turns out that stem cells don't settle into the bone marrow until a few days after birth. During fetal development, stem cells circulate in the bloodstream and, therefore, travel through vessels in the umbilical cord when circulating to and from the placenta. Because the umbilical cord blood is rich in stem cells, it is possible to collect whole cord blood and know that it contains some stem cells.

The transplantation of stem cells to replace bone marrow cells destroyed by cancer chemotherapy or radiation may someday be routine. Stem cells are already being used successfully to treat a variety of inherited blood disorders. Transplanted stem cells must be of the same tissue type (have the similar molecules on their cell surfaces) as that of the recipient. Umbilical cord blood of siblings is most likely to provide a match of tissue types, but not everyone in need of stem cells has a sibling who is still a fetus. Should we begin now to prepare for the possibility of wide-ranging uses of stem cells by extracting umbilical cord blood at the time of birth and freezing it for possible future use? Most of the patients who have been successfully treated with cord blood are children. We are not yet certain that the number of stem cells in cord blood is sufficient to reconstitute the bone marrow of an adult. Should we wait until more adults have been successfully treated before routinely freezing the cord blood of all newborns? If cord blood banks are established, should that blood be reserved only for the possible future need of the donor or should it be made available for anyone in need whose tissue type matches?

to 12 days before dying. Platelets contain several substances important in stopping the loss of blood through damaged blood vessels. This vital function of platelets will be considered later in this chapter.

White Blood Cells

White blood cells, or **leukocytes** (*leuk-*, white; *-cyte*, cell), perform certain mundane housekeeping duties, such as removing wastes, toxins, or damaged, abnormal cells, as well as serving as warriors in the body's fight against disease. Although leukocytes represent less than 1% of whole blood, we simply could not live without them. We would succumb to the microbes that surround us. Since their numbers increase when the body responds to microbes, white blood cell counts are often used as an index of infection.

White blood cells are also produced in the red bone marrow. (One type, the lymphocytes, may also be produced in the lymphoid tissues, such as lymph nodes.) They are nucleated cells that circulate in the bloodstream. However, white blood cells are not confined to the bloodstream. By squeezing between the cells that form the walls of blood vessels, they can leave the circulatory system and move to the site of infection,

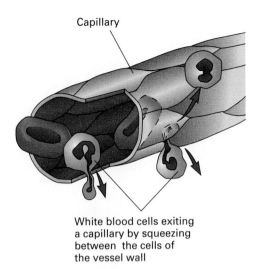

Capillary

White blood cells exiting
a capillary by squeezing
between the cells of
the vessel wall

Figure 10–4 White blood cells squeeze between the cells forming the wall of a capillary, in a process called diapedesis. They then enter the fluid surrounding body cells and, attracted by chemicals released by microbes or damaged cells, gather at the site of infection or injury.

tissue damage, or inflammation[1] (Figure 10–4). Having slipped out of the capillary into the fluid bathing the cells, white blood cells roam through the tissue spaces by amoeboid movement. That is, they form cytoplasmic extensions in the direction they are traveling and then flow into these extensions. White blood cells gather in areas of tissue damage or infection because they are attracted by chemicals released by invading microbes or by damaged cells. Certain types of white blood cells may then engulf the "offender" in a process called **phagocytosis** (*phago-*, to eat; *-cyt-*, cell; *-sis*, process of). Here we will only briefly consider the white blood cells, but we will discuss them and their many tactics for defending our bodies in more detail in Chapter 12.

Types of White Blood Cells. The five types of white blood cells can be classified into one of two groups based on cytoplasmic differences (Table 10–1). Members of one group, the granulocytes, have granules in their cytoplasm. The granules are sacs containing chemicals that are used as weapons to destroy invading pathogens, especially bacteria. The other group, the agranulocytes, is comprised of white blood cells that lack cytoplasmic granules or have very small granules.

There are three kinds of **granulocytes**: neutrophils, eosinophils, and basophils. **Neutrophils**, the most abundant of all white blood cells, are the blood-cell soldiers on the front lines. Arriving at the site of infection before the other

[1]The process by which white blood cells leave the circulatory system is called diapedesis (*dia-*, through; *pedesis*, leaping).

types of white blood cells, they immediately begin to phagocytize the microbes, thus curbing the spread of the infection. After engulfing a dozen or so bacteria, a neutrophil dies. But, even in death, it helps the body's defense by releasing chemicals that attract more neutrophils to the scene. Dead neutrophils, along with bacteria and cellular debris, make up pus, the yellowish liquid we usually associate with infection. **Eosinophils** are important in the body's defense against parasitic worms and they play a role in allergic reactions. **Basophils** release histamine, a chemical that both attracts other white blood cells to the site and causes blood vessels to widen, thereby increasing blood flow to the affected area. They play a role in some allergic reactions.

The **agranulocytes** include the monocytes and the lymphocytes. The largest of all blood cells, **monocytes**, are active in fighting chronic infections, viruses, and intracellular bacterial infections. They mature in various tissues and become macrophages, which are phagocytic cells. There are two major categories of lymphocytes. The **B lymphocytes** give rise to the plasma cells, which, in turn, produce antibodies. Antibodies are proteins that recognize specific molecules (called antigens) on the surface of microbes that have invaded the body. Antibodies bind with the antigens and help prevent the microbe or other foreign cell from harming the body. There are several types of **T lymphocytes**. Some attack and destroy cells that are not recognized as belonging in the body. Typically, the attacked cells are infected with a virus or have become cancerous. Other T lymphocytes help activate the immune system and yet others suppress the immune system.

Disorders of White Blood Cells. Infectious mononucleosis, or simply mono, is a viral disease caused by the Epstein-Barr virus and is highly contagious. Because it is often spread from person to person by oral contact, it is sometimes called "the kissing disease." However, mono can be spread just as easily by sharing eating utensils or drinking glasses. Mono is most common among teenagers and young adults, particularly those away from home at school. It often strikes at stressful times, such as during final exams, when resistance is low.

The initial symptoms of mono are similar to those of influenza—fever, chills, headache, sore throat, and an overwhelming awareness of illness. Within a few days, the glands in the neck, armpits, and groin become painfully swollen. Mono must simply run its course. The major symptoms generally subside within a few weeks, but fatigue may linger much longer.

Leukemia is a cancer of the white blood cells that causes white blood cell numbers to increase. The cancerous cells, all descendants of a single abnormal cell, remain unspecialized and are, therefore, unable to perform their normal duties. However, they divide more rapidly and live longer than do normal cells. Thus, although there is an abundance of white blood cells, they do not effectively defend the body against infectious agents. The abnormal cells "take over" the bone

Table 10–1 The Formed Elements of Blood

Type of Cell	Diagram	Description	No. of Cells/mm³	Life Span	Cell Function
Erythrocyte Red blood cells (RBC)		Biconcave disk; no nucleus	4–6 million	About 120 days	Transport oxygen and carbon dioxide
Leukocyte White blood cell (WBC)					
Granulocytes Neutrophil		Multilobed nucleus; clear-staining cytoplasm; inconspicuous granules	3000–7000	6–72 hours	Phagocytizes bacteria
Eosinophil		Large granules in cytoplasm; pink-staining cytoplasm; bilobed nucleus	100–400	8–12 days	Phagocytizes antibody-antigen complex; attacks parasites
Basophil		Large purple cytoplasmic granules; bilobed nucleus	20–50	3–72 hours	Releases histamine
Agranulocytes Monocyte		Gray-blue cytoplasm with no granules; U-shaped nucleus	100–700	Several months	Gives rise to macrophages that phagocytize bacteria, dead cells, and cell parts
Lymphocyte		Round nucleus that almost fills the cell	1500–3000	Many years	Attacks damaged or diseased cells or produces antibodies
Platelet		Fragments of megakaryocyte; small; purple-stained granules in cytoplasm	250,000	5–10 days	Plays a role in blood clotting

marrow, preventing the development of normal blood cells, including red cells, white cells, and platelets.

Although it is still not clear what causes leukemia, several environmental factors seemed to be linked to it. For instance, chemicals, such as benzene (a chemical used as motor fuel and as a solvent in dry cleaning), have caused a form of anemia that, in some persons, developed into leukemia. In addition, the rate of certain kinds of leukemia is higher among persons exposed to heavy doses of x-rays or atomic radiation. However, there are still some unanswered questions about the connection between radiation and leukemia. In 1986, the world's worst nuclear accident occurred in Chernobyl, Ukraine. However, the increase in leukemia among children exposed to the radiation was not as high as expected.

Symptoms of leukemia are generally caused either by the insufficient number of normal blood components or by the abnormal white cells invading organs. Insufficient numbers of platelets cause gum bleeding and frequent bruising. Anemia, which causes chronic fatigue, breathlessness, and pallor, results from reduced red cell levels. Because their white cells do not function properly, leukemia patients may suffer from repeated chest or throat infections, herpes, or skin infections. As the immature white cells fill the lymph nodes, the nodes in the armpits, groin, and neck may become swollen. Also, bone tenderness may be experienced because the immature white cells pack the red marrow. Headaches, another symptom, may be caused by anemia or by the effects of abnormal white cells in the brain.

The treatment of leukemia usually involves radiation therapy and chemotherapy to kill the rapidly dividing cells. In addition, transfusions of red cells and platelets may be given to alleviate anemia and prevent excessive bleeding.

Today, children with acute leukemia are often cured with bone marrow transplants. A person, most often a family member, whose tissue type closely matches that of the patient must be located and agree to serve as a donor. After a donor has been found, the leukemic bone marrow in the leukemia patient must be destroyed by irradiation and drugs. This drastic treatment leaves the patient completely vulnerable to any type of infection. The donor is then anesthetized and some bone marrow is removed from his or her pelvic bones. Next, the donor's bone marrow is given to the patient intravenously, just like a blood transfusion. The marrow cells find their way to the patient's marrow and begin to grow. Soon, the donated bone marrow cells begin to produce new healthy white cells. As white cell numbers reach normal levels, usually within 2 months, the patient regains resistance to infection.

Red Blood Cells

The **red blood cells**, also called **erythrocytes**, serve to pick up oxygen in the lungs and ferry it to all the cells of the body (Figure 10–5). Red cells also carry about 23% of the total carbon dioxide, a metabolic waste product. They are by far the most numerous cells in the blood. Indeed, they number 4–6 million/mm^3 of blood and constitute approximately 45% of the total blood volume, a percentage known as the **hematocrit**.

The form of red blood cells is marvelously suited to the function of picking up and transporting oxygen. They are quite small, only about 7 μm in diameter. Each is shaped like a biconcave disk. That is, it is a flattened cell indented on each side. These two features maximize the surface area of

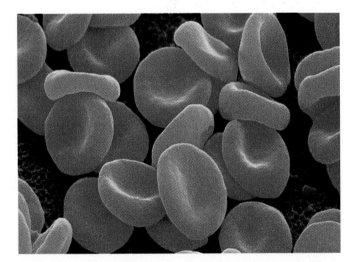

Figure 10–5 Red blood cells serve to ferry oxygen from the lungs to the needy tissues. Each is a small biconcave (indented on both sides) disk. This design maximizes surface area for gas exchange. Lacking a nucleus and other organelles, a red blood cell is essentially a bag packed with the oxygen-binding pigment, hemoglobin. *(Andrew Syred/© Tony Stone Images)*

the cell. With a greater surface area, oxygen can enter the red blood cell more rapidly, which is important because a red blood cell spends only a second or two in the capillaries of the lungs. A red blood cell is also unusually flexible. Thus, it can bend and twist to fit through small capillaries. Each cell is packed with **hemoglobin,** the oxygen-binding pigment. As a red blood cell matures in the red bone marrow, it loses its nucleus and most organelles. Thus, it is scarcely more than a sac of hemoglobin molecules. With roughly a third of its weight contributed by hemoglobin, each red blood cell can transport about a billion molecules of oxygen. That precious load of oxygen is not used by the red blood cells because they lack mitochondria and, therefore, can produce ATP by only anaerobic mechanisms.

Each red blood cell is packed with approximately 280 million molecules of hemoglobin. Once inside a red blood cell, oxygen binds to those molecules. It is important that the hemoglobin be packaged in blood cells because more of it can be transported this way than if it were free in the plasma. If all our hemoglobin were free-floating, our blood would be thick and hemoglobin molecules would clog the filtering system of the kidneys.

As can be seen in Figure 10–6, each hemoglobin molecule is made up of four subunits. Each of the subunits has a protein chain (the globin) and a **heme** group. The heme group includes an iron ion, which actually binds to the oxygen. Therefore, each hemoglobin molecule can carry four molecules of oxygen. The compound formed when hemoglobin binds with oxygen is called, logically enough, **oxyhemoglobin**.

As wonderfully adapted as the hemoglobin molecule is for carrying oxygen, it has more than 200 times the affinity for carbon monoxide as it does for oxygen. In other words, if concentrations of carbon monoxide and oxygen in inhaled air were identical, for every one molecule of hemoglobin that picked up an oxygen molecule, 200 molecules of hemoglobin would bind to carbon monoxide, which is a product of the incomplete combustion of any carbon-containing fuel. The primary source of carbon monoxide is automobile exhaust. The reason that carbon monoxide can be deadly is that it binds to the oxygen-binding site on hemoglobin and prevents blood from carrying the life-giving oxygen molecule to the cells. As a result, the cells cannot carry out cellular respiration and the person can die. Carbon monoxide is a particularly insidious poison because it is odorless and tasteless.

Life Cycle of a Red Blood Cell

The birth process of a red cell, which takes about 6 days to complete, involves many changes in the cell's activities and structure. First, the very immature red cell becomes a factory for hemoglobin molecules. After the cell is packed with hemoglobin, its nucleus is pushed out. A structural metamorphosis then occurs, culminating in a cell with the biconcave shape typical of a red blood cell. These mature red cells then leave the bone marrow and enter circulation. In this way, red

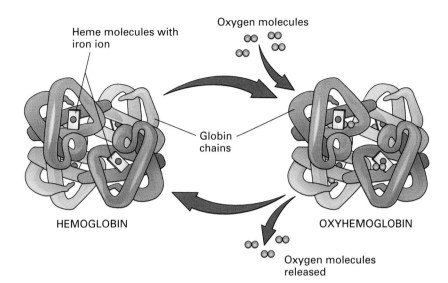

Heme molecules with iron ion

Oxygen molecules

Globin chains

HEMOGLOBIN

OXYHEMOGLOBIN

Oxygen molecules released

Figure 10–6 The structure of hemoglobin and of heme. Each hemoglobin molecule consists of four subunits. Each of these subunits is composed of a protein chain and a heme group. The heme group contains an iron ion that binds to oxygen. Since each heme group can bind oxygen, each hemoglobin molecule can carry four molecules of oxygen.

marrow produces roughly 2 million red cells a second, for a cumulative total of more than half a ton in your lifetime.

A red blood cell lives for only about 120 days. During that time it travels through approximately 100 kilometers (62 miles) of blood vessels, being bent, bumped, and squeezed repeatedly. Its life span is probably limited by the lack of a nucleus to maintain the cell and direct needed repairs. Without a nucleus, for instance, protein synthesis needed to replace key enzymes cannot take place and so the cell becomes increasingly rigid and fragile.

Worn red blood cells are removed from circulation in the liver and spleen, the "graveyards" for red blood cells. The inflexible red cells tend to become stuck in the tiny circulatory channels of these organs. Macrophages (large phagocytic cells) then engulf and destroy the dying red cells. The liver degrades the hemoglobin released from destroyed red cells to its protein (globin) component and heme. The protein is digested to amino acids, which can be used to make other proteins. The iron from the heme is salvaged and sent to the red marrow for recycling.

The remainder of heme is degraded to a yellow pigment, called bilirubin, which is excreted by the liver in bile. Bile is released into the small intestine, where it assists in the digestion of fats. It is carried along the digestive system to the large intestines with the undigested food and becomes a component of feces. The color of feces is partly due to bilirubin that has been broken down by intestinal bacteria.

Products formed by the chemical breakdown of heme also create the yellowish tinge in a bruise that is healing. A bruise, or "black and blue" mark, results when tiny blood vessels or capillaries are ruptured and blood leaks into the surrounding tissue. As the tissues use up the oxygen, the blood becomes darker in color and, viewed through the overlying tissue, looks black or blue. Gradually, the red cells degenerate, releasing hemoglobin. The breakdown products of hemoglobin then make the bruise appear yellowish.

Jaundice is a condition in which the skin develops a yellow tone. It signifies that the liver is not handling bilirubin adequately. The excess bilirubin is absorbed into the bloodstream, carried throughout the body, and deposited in certain tissues, such as the skin. One form of jaundice sometimes develops in newborns, particularly if they were born prematurely. Because fetal red cells are broken down rapidly during the first few days after birth, the infant's liver may not be able to process the bilirubin fast enough to prevent its accumulation in the blood (Figure 10–7). Bilirubin may be

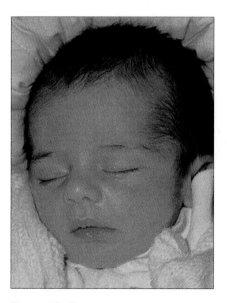

Figure 10–7 Newborns, especially if they are born prematurely, may develop jaundice. Newborn jaundice occurs because the infant's liver is unable to handle all the bilirubin that results from the breakdown of fetal hemoglobin. The bilirubin then causes a yellowish tint in the skin. Brain damage can result if the bilirubin levels become too high. *(© 1996 SPL/Custom Medical Stock Photo)*

deposited in the brain, where it can damage nerve cells. To prevent brain damage, the jaundiced infant may be placed under a bank of fluorescent lights, which helps break down the bilirubin. In adults, a diseased or damaged liver can also cause jaundice.

Critical Thinking

Between 25 ml and 65 ml of blood is lost in each menstrual cycle. Why would this make women in their reproductive years more likely than men of the same age to suffer from iron-deficiency anemia (a condition in which an insufficiency of iron causes reduced hemoglobin synthesis)?

Red cell production is regulated according to the needs of the body, especially the need for oxygen (Figure 10–8). Most of the time, red cell production matches red cell destruction. There are circumstances, blood loss for instance, that trigger a homeostatic mechanism that speeds up the rate of red cell production. This mechanism is initiated when the oxygen supply to the body's cells drops. The reduced oxygen is sensed by certain cells in the kidney and they respond by producing an enzyme that converts a protein in the blood to the hormone **erythropoietin**. Erythropoietin then travels to the red marrow where it steps up both the division rate of stem cells and the maturation rate of im-

mature red cells. When maximally stimulated by erythropoietin, the red marrow can increase red cell production tenfold—to 20 million cells per second! The resulting increase in red cell numbers should soon be adequate to meet the oxygen needs of body cells.

Anemia, a condition in which the blood's ability to carry oxygen is reduced, can result from too little hemoglobin, too few red blood cells, or both. The symptoms of anemia include fatigue, headaches, dizziness, paleness, and breathlessness. In addition, an anemic person's heart often beats faster to help compensate for the blood's decreased ability to carry oxygen. The accelerated pumping can cause heart palpitations—the uncomfortable awareness of one's own heart beat. Although anemia is not usually life threatening, it does increase susceptibility to illness and it stresses the heart and lungs. It can also affect the quality of life because the lack of energy and low levels of productive activity often go hand-in-hand.

Worldwide, the most common cause of anemia is an insufficiency of iron in the body, which leads to an inadequate hemoglobin production. Iron-deficiency anemia can be caused by a diet that contains too little iron, an inability to absorb iron from the digestive system, or from blood loss, such as might occur due to menstrual flow or peptic ulcers. Treatment for iron-deficiency anemia involves dealing with the cause of the depletion of iron in the body and restoring iron levels to normal by taking pills that contain iron.

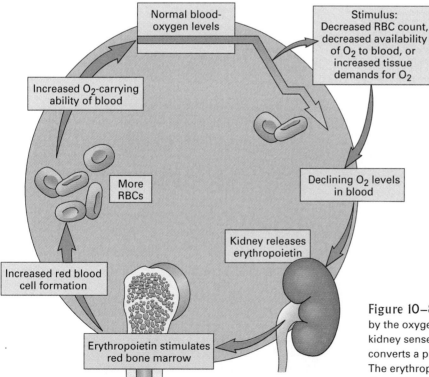

Figure 10–8 The production of red blood cells is regulated by the oxygen levels in the blood. When certain cells in the kidney sense a drop in oxygen, they produce an enzyme that converts a protein in the blood to the hormone erythropoietin. The erythropoietin then stimulates red blood cell production by the red bone marrow.

Blood loss will obviously lower red cell counts, but so will any condition that results in red cell destruction exceeding red cell production. For example, in hemolytic anemias, red cells are ruptured because of infections, defects in the red cell membrane, transfusion of mismatched blood, or hemoglobin abnormalities. Sickle-cell anemia is an example of a hemolytic anemia caused by abnormal hemoglobins. In sickle-cell anemia the abnormal hemoglobin (hemoglobin S) causes the red cells to become deformed to a crescent (or sickle) shape when the oxygen content is low. The distorted cells are fragile and rupture easily. They also clog small blood vessels and promote clot formation. Because these events prevent oxygen-laden blood from reaching the tissues, they can cause episodes of extreme pain. Interestingly, although sickle-cell anemia is an inherited condition, the *fetal* hemoglobin[2] of those who later develop sickle-cell anemia does not cause sickling. Early trials of a drug called hydroxyurea have shown that it reduces both the frequency of these painful episodes and the need for blood transfusions in people with sickle-cell anemia. It is thought that the new drug works by increasing the production of fetal hemoglobin in red blood cells.

Red cell numbers also drop when the production of red blood cells is halted or impaired as occurs in pernicious anemia. The production of red cells depends on a supply of vitamin B_{12}. Absorption of this vitamin from the small intestines depends on a chemical called intrinsic factor, which is produced by the stomach lining. People with pernicious anemia do not produce intrinsic factor and are, therefore, unable to absorb vitamin B_{12}. They are treated with injections of B_{12}.

Social Issue

Blood doping is a practice that boosts red cell numbers and, therefore, the blood's ability to carry oxygen. It involves withdrawing and storing some blood, which triggers stepped-up production of red cells by the body, and then reinfusing the stored blood. Some athletes who compete in aerobic events, like swimming or cycling, have used blood doping to increase their endurance and speed. Although blood doping is considered unethical and is prohibited in Olympic competition, it is very difficult to detect. It could be detected if the athlete had an exceptionally high number of red blood cells or if the presence of a hormone that stimulates red cell production (erythropoietin) were found in the urine. Tests can detect only about 50% of the athletes who practice blood doping. Considering that roughly half of the athletes who practice blood doping will escape detection even if they are tested, do you think that those who are caught should be punished?

[2]During embryonic growth in the womb, red blood cells contain a slightly different form of hemoglobin than is found in adults.

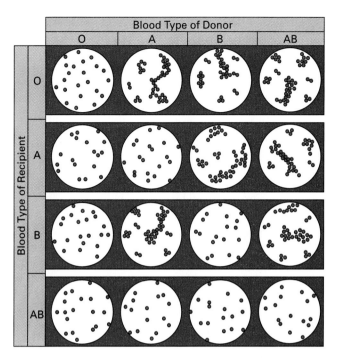

Figure 10-9   Blood is typed by mixing it with serum known to contain antibodies specific for a certain antigen. If blood containing that antigen is mixed with the serum, the blood will clump. Thus, a drop of blood is mixed with serum containing anti-A and another drop with serum containing anti-B. Clumping reveals the presence of the antigen.

Blood Types

In the 1600s, Europeans began experimenting with blood transfusions. The results were disastrous. Indeed, so many patients died that transfusions were outlawed in many European countries. It was not until 1900 that people knew that blood types must be compatible for a transfusion to be successful.

Blood types are determined by the presence of certain glycoproteins (proteins attached to carbohydrates) on the surface of red blood cells. These molecules act as antigens and will be recognized as foreign if they are transfused into a person whose red cells lack this protein. Blood types are determined by the specific antigens found on the surface of the red blood cells.

Although there are at least 30 common varieties of antigens, those of the ABO and Rh blood groups cause serious, often fatal, complications if improperly transfused. We will, therefore, consider these blood types more carefully.

When asked about your blood type, you will probably respond by indicating one of the types in the ABO series—A, B, AB, or O. Red blood cells with only the antigen A on their surface are type A. When only the B antigen is on the red cell surface, the blood is type B. Blood with both A and B antigens on the red cell surface is designated type AB. When neither A nor B antigens is present, the blood is type O (Figure 10-9).

Normally, a person has antibodies in the plasma against those antigens that are not on his or her own red blood cells. Thus, individuals with type A blood have antibodies against the B antigen and those with type B blood have antibodies against A. Since individuals with type AB blood have both antigens on their red blood cells, they have neither antibody. Those with type O blood will have both A and B antibodies in their blood. It is not certain why these antibodies form without exposure to red blood cells bearing the foreign antigen. It may be that either bacteria that invade our bodies or the food we eat contains a small amount of A and B antigens, enough to stimulate antibody production.

When a person is given a blood transfusion with donor blood containing foreign antigens, the antibodies in the recipient's blood will cause the donor's cells to **agglutinate** (clump) (Figure 10–10). This clumping of the donor's cells is damaging, perhaps even fatal. The clumped cells get stuck in small blood vessels and block blood flow to body cells. Or, they may break open, releasing their cargo of hemoglobin, which then clogs the filtering system in the kidneys and death follows.

It is important, therefore, to be sure that the blood types of the donor and recipient are compatible, which means that the recipient's blood does not contain antibodies to antigens on the red cells of the donor. The plasma of the donor's blood may contain antibodies against antigens on the recipient's red blood cells, but these will be diluted as they enter the recipient's circulation. Therefore, the donor's antibodies are not a major problem. To determine whether a blood transfusion will be safe, then, we must consider the antigens on the donor's cells and the antibodies in the recipient's blood. For example, if a person with blood type A is given a transfusion of blood type B or of type AB, the naturally occurring anti-B antibodies in the recipient's blood will cause the red blood cells of the donor to clump because they have the B antigen. The transfusion relationships among blood types in the ABO series are shown in Table 10–2.

The A and B antigens are not the only important antigens found on the red blood cell surface. The Rh factor, which is actually a series of antigens, is also important. The name Rh comes from the beginning of the species name, *Rhesus*, of the monkey in which the antigen was first discovered. People who have any of the Rh antigens on their red blood cells are considered Rh-positive. In the case that Rh antigens are missing from the red blood cell surface, the individual is considered Rh-negative.

An Rh-negative person will not have the antibodies to the Rh antigens unless he or she has been exposed to the Rh antigen. This is unlike the case with the ABO series, where no prior exposure to foreign red blood cell antigens is required to stimulate the production of antibodies. For this reason, an Rh-negative individual should be given only Rh-negative blood in a transfusion. If he or she is mistakenly given Rh-positive blood, it will stimulate the production of antibodies to the Rh antigen. Following a second transfusion of Rh-positive blood, the antibodies in the recipient's plasma will react with the antigens on the red blood cells of the donated blood. This reaction may lead to the death of the patient.

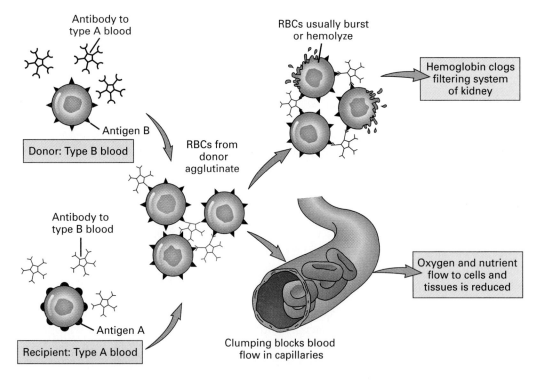

Figure 10–10 When a transfusion with mismatched blood is given, the antibodies in the recipient's blood react with antigens on the donor's red blood cells, causing the donor's blood to clump. The clumped cells may block nutrient and oxygen delivery to cells by blocking capillaries. The cells may lyse (burst). Each red blood cell contains roughly 280 million molecules of hemoglobin that, when released, can block the filtering system in the kidneys. Kidney failure and death can follow.

Antibody to type A blood

Antigen B

Donor: Type B blood

Antibody to type B blood

Antigen A

Recipient: Type A blood

RBCs from donor agglutinate

RBCs usually burst or hemolyze

Hemoglobin clogs filtering system of kidney

Oxygen and nutrient flow to cells and tissues is reduced

Clumping blocks blood flow in capillaries

Table 10-2 Transfusion Relationships Among Blood Types

Blood Type	Antigens on Red Blood Cells	Antibodies in Plasma	Blood Types That Can Be Received in Transfusions
A	A	Anti-B	A, O
B	B	Anti-A	B, O
AB	A and B	None	A, B, AB, O (Universal recipient)
O	None	Anti-A, anti-B	O (Universal donor)

The Rh factor may also be of medical importance in pregnancies when the mother is Rh-negative and the father and fetus are both Rh-positive (Figure 10–11). Ordinarily the maternal and fetal blood supplies do not mix during pregnancy. However, some mixing may occur during delivery when some blood vessels are damaged. If the baby's red blood cells, which bear Rh antigens, accidentally pass into the bloodstream of the mother, she will produce Rh antibodies. There are usually no ill effects associated with the first introduction of the Rh antigen. However, if antibodies are present in the maternal blood from a previous pregnancy with an Rh-positive child or from a transfusion of Rh-positive blood, the anti-Rh antibodies may pass into the blood of the fetus. This can occur because antibodies, unlike red blood cells, can cross the placenta. These antibodies may destroy the fetus' red blood cells. As a result, the child may be stillborn or else it may be very anemic at birth. This condition is called **hemolytic disease of the newborn**.

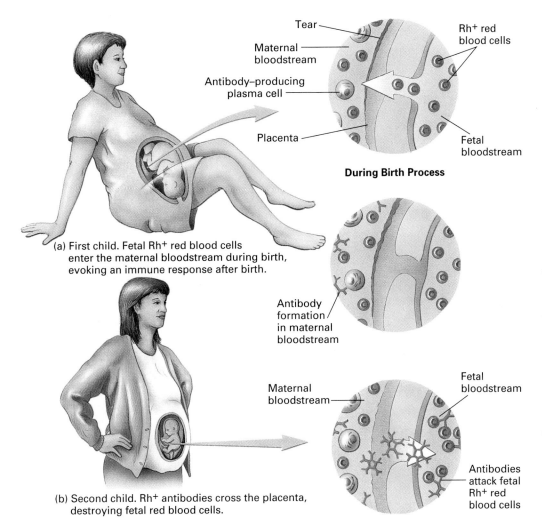

Tear

Maternal bloodstream

Antibody–producing plasma cell

Placenta

Rh⁺ red blood cells

Fetal bloodstream

During Birth Process

(a) First child. Fetal Rh⁺ red blood cells enter the maternal bloodstream during birth, evoking an immune response after birth.

Antibody formation in maternal bloodstream

Maternal bloodstream

Fetal bloodstream

Antibodies attack fetal Rh⁺ red blood cells

(b) Second child. Rh⁺ antibodies cross the placenta, destroying fetal red blood cells.

Figure 10–11 Rh-incompatibility can result when an Rh-negative woman is pregnant with an Rh-positive baby. If the mother has never been exposed to Rh-positive blood, she will not have antibodies against it and there will be no incompatibility. However, during the birth process, some Rh-positive fetal cells may enter the mother's circulatory system and stimulate the formation of antibodies. If this occurs and the woman subsequently becomes pregnant with another Rh-positive child, the mother's antibodies may cross the placenta and destroy the fetal red blood cells.

Transfusions and Artificial Blood

The need for blood for transfusion often exceeds the amount supplied by volunteer donations. The need is indeed great. In the United States alone, 12 million pints of blood are transfused each year. This need, coupled with concerns about the safety of the blood supply, particularly the possibility of the transmission of hepatitis or HIV, the virus that causes AIDS through tainted blood, have led to some changes in thinking about transfusions.

Today before elective surgery, patients are likely to have blood withdrawn and stored so that it can be reinfused, if necessary, during surgery. Although they have become quite common because of the patient's concern about disease transmission, these so-called autologous transfusions can be very expensive for hospitals and health insurance companies. Why? The most important reason is that the patient does not always need all the blood that was donated ahead of time. The unneeded blood is usually discarded, but the cost of collecting and storing the blood remains. Furthermore, the safety of blood collected from volunteers has increased greatly in recent years. Thus, autologous transfusions may only pay off when the anticipated surgery has a high probability of requiring a blood transfusion, such as in a total hip replacement or coronary artery bypass.

If a person is given a transfusion of an incompatible blood type, it almost always causes death. Although the risk of receiving the wrong blood type is low, mistakes are sometimes made, especially in situations in which the medical personnel are hurried, such as in emergency rooms or intensive care units. For this reason, some emergency rooms stock only type O blood, the universal donor. The problem with this plan is that it can lead to a shortage of type O blood and wasted units of types A and B.

It is expected that before the turn of the century, however, the risk of receiving a transfusion of the wrong blood type will plummet to nearly zero thanks to a new technology. Using a genetically engineered enzyme, it will be possible to convert type A and type B blood to type O. The enzymes simply snip off part of the glycoprotein on the red blood cell surface, leaving only the part of the molecule found on type O cells. The enzyme needed for the B to O conversion can already be produced in bulk. In early trials, the converted cells seem to be accepted just like normal type O cells when transfused into a recipient of another blood type. This potentially lifesaving enzyme is awaiting approval by the Food and Drug Administration and may reach the hospitals by the end of 1998. The A to O conversion is slightly more difficult and will probably take an additional year before it becomes available.

Whole blood—your own or someone else's—has some disadvantages when used for transfusions. It must be typed and matched. It must be screened for a host of diseases, among them syphilis, HIV-1, HIV-2, human T-cell lymphotrophic virus type 1, and both hepatitis B and C. At best, its shelf life is slightly over a month. These drawbacks to the use of whole blood have spurred attempts to produce synthetic blood components.

When a person has lost blood, it is often important to boost the oxygen-carrying ability of the remaining blood. For this purpose, blood substitutes that contain cell-free hemoglobin have been developed. There are several types. One type is a genetically engineered version of one of the subunits of normal human hemoglobin. Another type is human hemoglobin released from red cells and bound to inert carrier molecules to prevent loss in the kidneys. Purified cattle hemoglobin has also gained approval of the Food and Drug Administration as a blood substitute. However, a problem with infusing independent molecules of hemoglobin in any form is that they tend to break down to a form that clogs the kidney's filtering system. Scientists have recently managed to link more than a million hemoglobin molecules to form an oxygen-filled microbubble. Although it is still too soon for clinical trials to have proven the safety of this blood substitute, it is expected that the clumping will prevent the breakdown of individual hemoglobin molecules and thus prevent kidney damage. Unfortunately, there is yet another possible problem associated with blood substitutes containing cell-free hemoglobin—they may increase a patient's chances of developing bacterial infections. It is thought that infection may develop because the cell-free hemoglobin provides an easy iron source for reproducing bacteria.

Whole blood substitutes are also being sought actively. The liquid portion of this artificial blood contains large carbohydrate molecules (instead of proteins) to maintain water balance. The liquid serves as a "plasma expander" that can help replace lost blood volume. In whole blood substitutes, oxygen is not carried by hemoglobin in any form, but rather by oxygenating agents such as perfluorochemical (PFC) emulsions. These compounds can carry roughly 70% of the oxygen that could be transported by whole blood.

The incidence of the disease has been decreased in recent years by the development of a means of destroying any Rh-positive fetal cells in the maternal blood supply, before they can stimulate the mother's cells to produce her own anti-Rh antibodies. The Rh-positive cells are killed by injecting RhoGAM, a serum containing antibodies against the Rh antigens before she gives birth or shortly afterward. Rh antigens are thus prevented from being "set" in the memory of the mother's immune system. The injected antibodies disappear after a few months. Therefore, there are no lingering antibodies to affect the fetus in a subsequent pregnancy.

Blood Clotting

When a blood vessel is cut, a series of reactions stop blood flow (Figure 10–12). These reactions are not unlike those you might initiate if the garden hose you were using sprang a leak. Your initial response might be to squeeze the hose, in hopes of stopping water flow. Likewise, the body's immediate response to blood vessel injury is for the vessel to squeeze shut. This constriction of the injured vessel is caused by chemicals released from the injured tissue.

The next response is to plug the hole. While your thumb might do the job on your garden hose, platelets form the plug that seals the leak in a vessel. The **platelet plug** is formed when platelets cling to cables of collagen, a protein fiber, in the exposed blood vessel surface. When they attach to collagen, platelets change in several ways: They swell, form many cellular extensions, and stick together. Platelets produce a chemical called thromboxane that, besides making platelets stick to one another, also attracts other platelets to the wound. Aspirin prevents the formation of thromboxane and, therefore, inhibits clot formation. This is how a daily dose of aspirin may prevent the formation of blood clots that can block

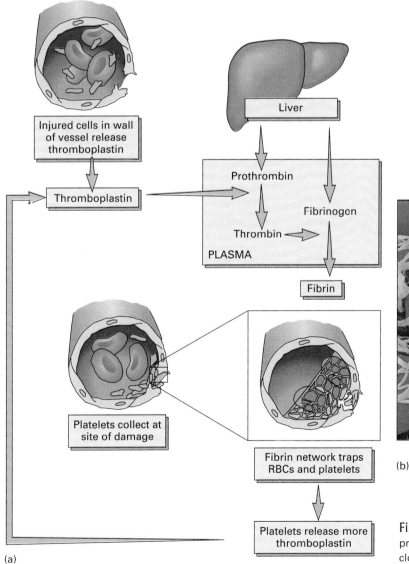

(a)

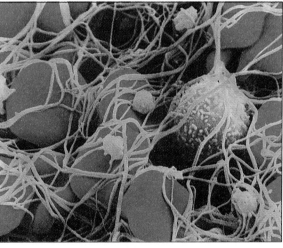

(b)

Figure 10-12   (a) Selected steps in the blood clotting process. (b) A scanning electron micrograph of a blood clot.   *(b, CNRI/Science Photo Library/Photo Researchers, Inc.)*

blood vessels nourishing heart tissue, resulting in a heart attack. It also explains how aspirin can cause excessive bleeding.

The next step in stopping blood loss through a damaged blood vessel is the formation of the clot itself. There are over 30 steps in the process of clot formation, but here we will simplify the process. Clot formation begins when a chemical called **thromboplastin** is released from injured tissue and from platelets. Thromboplastin converts an inactive blood protein, **prothrombin**, to an active form, **thrombin**. Thrombin then causes a remarkable change in another plasma protein produced by the liver, **fibrinogen**. The altered fibrinogen forms long strands of **fibrin**, which form a web that traps blood cells and forms the clot. The clot is a barrier that prevents further blood loss through the wounded vessel.

Critical Thinking

Thromboplastin, a chemical important in the initiation of clot formation, is released from both damaged tissue and platelets. How does this explain why a scrape, which causes a great deal of tissue damage, generally stops bleeding more quickly than a clean cut, such as a paper cut or one that might occur with a razor blade?

If even one of the many factors needed for clotting is lacking, the process can be slowed or completely blocked. Vitamin K is needed for the liver to synthesize prothrombin and three other clotting factors. As we've seen, prothrombin is one of the first links in the complicated chain of events leading to clotting. Thus, without vitamin K, clotting doesn't occur. We have two sources of vitamin K. One is the diet. Vitamin K is found in leafy green vegetables, tomatoes, and vegetable oils. A second source is from bacteria living in our intestines. These bacteria manufacture vitamin K, some of which we then absorb for our own use. Antibiotic treatment for serious bacterial infections can kill gastrointestinal bacteria and lead to a vitamin K deficiency in as few as 2 days. Vitamin K is used rapidly by body tissues, and so both sources are needed for proper blood clotting.

Hemophilia is an inherited condition in which the affected person bleeds excessively due to a faulty gene needed to produce a clotting factor. Because of the way hemophilia is inherited, the condition usually occurs in males. Symptoms appear when the affected child first becomes active. Crawling, for instance, causes bruises on the elbows and knees and cuts tend to bleed longer than usual. Excessive internal bleeding can damage nerves or, when it occurs in joints, permanently cripple the hemophiliac.

Treatment for hemophilia involves restoring the missing clotting factor. Bleeding episodes can be controlled with repeated transfusions of concentrated clotting factor. Each transfusion is prepared from 2000 to 5000 individual donations. In the first few years of the AIDS epidemic, before blood supplies were screened, many hemophiliacs became infected with HIV through their clotting factor transfusions. The clotting factor VIII, which is missing in patients with the most common form of hemophilia (hemophilia A), has been produced using genetically engineered hamster cells.[3] Although the product has the potential to provide a limitless supply of uncontaminated clotting factor, its cost is *twice* that of the human blood product. The high cost makes it difficult for some clinics to take advantage of genetically engineered clotting factor. In the future, gene therapy may be able to provide hemophiliacs with the genes needed to produce their own clotting factors.

Blood clots can be life-saving or life-threatening, depending on when and why they form. Clots are beneficial when they stop the loss of blood from an injured vessel. When they linger too long or form when not needed, however, they can be harmful because they can disrupt blood flow. A blood clot in an unbroken blood vessel that stays in place is called a **thrombus**. Clots can also drift through the circulatory system until they become lodged in a narrow vessel. When the tiny vessels that nourish the heart or brain become clogged with a clot, the consequences can be severe indeed—disability or even death. A blood clot that drifts through the circulatory system is called an **embolus**.

After a wound has healed, clots are normally dissolved by an enzyme called **plasmin**, which is formed from an inactive protein, **plasminogen**. Plasmin dissolves clots by digesting the fibrin strands that form the framework of the clot.

Critical Thinking

Heparin is a drug that inactivates thrombin. It is sometimes administered to patients when it is desirable to inhibit the clotting response. How would heparin act to achieve these ends?

[3] Genetic engineering and gene therapy are discussed in Chapter 19.

SUMMARY

1. Blood is a type of connective tissue containing formed elements suspended in a plasma matrix.
2. Plasma is the liquid portion of the blood. It contains the substances transported by the blood, including nutrients, ions, dissolved gases, hormones, and waste products. Plasma proteins, which constitute most of the substances dissolved in the plasma, aid in the balance of water flow between blood and cells. Plasma proteins are grouped into one of three categories: albumins, globulins, or clotting proteins (fibrinogen).
3. The formed elements of the blood are platelets, leukocytes (white blood cells), and erythrocytes (red blood cells). Platelets play an important role in blood clotting. Leukocytes help the body fight off disease and help to remove wastes, toxins, and damaged cells. Erythrocytes carry oxygen and some carbon dioxide.
4. There are five types of leukocytes: neutrophils, eosinophils, basophils, monocytes, and lymphocytes. Neutrophils are the most abundant type and immediately phagocytize foreign microbes. Eosinophils help the body defend against parasitic worms and play a role in allergic reactions. Basophils can increase the flow of blood by releasing histamines. Monocytes are the largest leukocytes and they actively fight chronic infections. Lymphocytes give rise to antibodies or attack foreign cells.
5. Erythrocytes are flexible cells packed with hemoglobin, an oxygen-binding pigment. Originating in the red bone marrow, erythrocytes have a life span of about 120 days. Red cell production is controlled by erythropoietin, a hormone formed by the action of enzymes secreted by certain kidney cells in response to low oxygen. Worn and dead red cells are removed from circulation and broken down in the liver and the spleen.
6. Anemia is a reduction in the blood's ability to carry oxygen. There are several forms of anemia, including iron-deficiency anemia, hemolytic anemias, such as sickle-cell anemia, and pernicious anemia, caused by a lack of the intrinsic factor produced by the stomach lining that is needed to absorb the vitamin B_{12} necessary for red cell production.
7. Blood types are determined by the presence of certain antigens (glycoproteins) on the surface of erythrocytes. There are at least 30 common varieties of antigens, the ones with which we are most familiar with are those of the ABO and Rh groups. The plasma contains antibodies against A and B antigens if the antigens are not present on the red blood cells. Antibodies to Rh antigens are formed only after exposure to Rh positive blood. If blood containing foreign antigens is introduced during a transfusion, a reaction between the antibodies and antigens can cause agglutination and clogging in the recipient's bloodstream, which can lead to death. The Rh incompatibility becomes important to an Rh-negative woman who has been previously exposed to Rh-positive blood and who is pregnant with an Rh-positive fetus. Antibodies to the Rh antigens can cross the placenta and destroy the red blood cells of the fetus, a condition called hemolytic disease of the newborn.
8. The prevention of blood loss involves three mechanisms: blood vessel spasm, platelet plug formation, and clotting. Clotting (coagulation) is initiated when platelets and damaged tissue release a chemical called thromboplastin, which converts the blood protein prothrombin to thrombin. Thrombin then converts the blood protein fibrinogen to fibrin. Strands of fibrin form a mesh that traps red blood cells and forms the clot. Hemophiliacs are missing an important clotting factor and, therefore, bleed excessively.

REVIEW QUESTIONS

1. What is plasma? What are its functions?
2. What are the three categories of plasma proteins?
3. List the three types of formed elements and describe the function of each.
4. Compare the size, structure, and relative numbers of leukocytes with that of erythrocytes.
5. List the five types of white blood cells and describe the function of each.
6. What is leukemia and what are its effects on the body?
7. List the characteristics of red blood cells that make it specialized for delivering oxygen to the tissues.
8. Describe the structure of hemoglobin.
9. Where are red blood cells produced? How is the production of red blood cells controlled?
10. Describe what happens to worn or damaged red blood cells and their hemoglobin.
11. Why would pernicious anemia be treated with regular injections of vitamin B_{12} instead of by dietary supplements of this vitamin?
12. How are blood types determined? What happens if a person is given a blood transfusion with blood of an incompatible type?
13. What is hemolytic disease of the newborn? What causes it?
14. After a blood vessel is cut, what mechanisms prevent blood loss?
15. Describe the steps involved in blood clotting.

Chapter | **11**

The Circulatory System

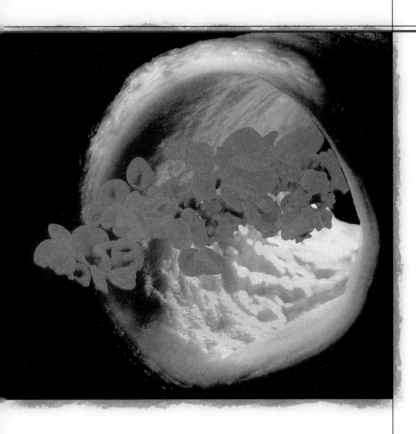

(© Goivaux Communication/Phototake, Inc.)

The importance of the heart to life has long been recognized. Four hundred and fifty years ago the Chinese Emperor physician Hung Ti proclaimed, "The heart is the root of life." Through the ages, even the character of people has been described with reference to the heart: An enemy is cold-hearted, while a friend is warmhearted, young people are often lighthearted, and at a funeral the heart is heavy.

Although our understanding of the importance of the heart to life has grown considerably over the years, diseases of the heart and blood vessels (cardiovascular diseases) are the leading cause of death and disability after the age of 35 in industrialized countries. Indeed, in the United States, heart disease kills more people than all forms of cancer combined. Why is the cardiovascular system so critical to survival? It provides the pump and the pipelines so that blood can deliver a continuous supply of oxygen and nutrients to the cells of the body and wash away metabolic waste products before they poison the cells.

The good news is that there is a great deal that each of us can do to reduce the risk of cardiovascular disease and, at the same time, improve our quality of life. In this chapter, then, we will consider the circulatory system to enhance our appreciation of its vital functions and marvelous design. An understanding of how the system functions might help us to properly care for it and maintain its health.

The Blood Vessels

Once every minute, or about 1440 times each day, the blood circulates through a life-sustaining loop of vessels (Figure 11–1). This loop is extremely long, indeed. To illustrate, if

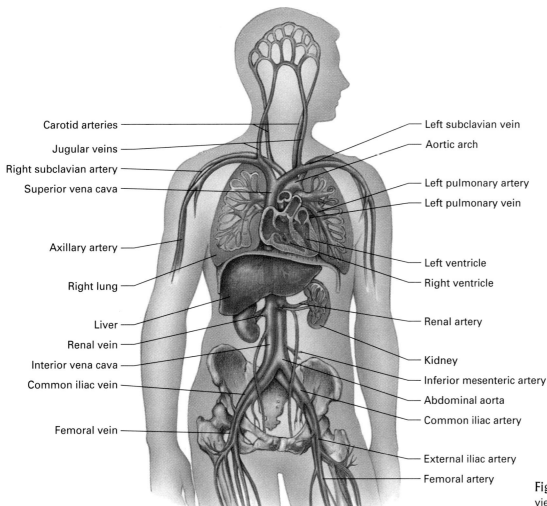

Carotid arteries
Jugular veins
Right subclavian artery
Superior vena cava
Axillary artery
Right lung
Liver
Renal vein
Interior vena cava
Common iliac vein
Femoral vein

Left subclavian vein
Aortic arch
Left pulmonary artery
Left pulmonary vein
Left ventricle
Right ventricle
Renal artery
Kidney
Inferior mesenteric artery
Abdominal aorta
Common iliac artery
External iliac artery
Femoral artery

Figure 11–1 A diagrammatic view of the principal arteries and veins.

all the vessels in an adult's body were placed end-to-end they would stretch about 100,000 km (60,000 miles), long enough to circle the Earth's equator more than twice! The vessels don't form a single long tube, however. Instead, they are arranged in branching networks. With each circuit through the body, blood leaves the heart in an *artery*, which branches and becomes narrower, giving rise to *arterioles*. In turn, an arteriole leads to a network of microscopic vessels in the tissue, called *capillaries*. The capillaries eventually merge to form *venules* and these eventually join to form larger tubes

called *veins*. The veins return the blood to the heart. We see, then, that the path of blood is

heart → artery → arteriole → capillaries → venule → vein → heart

These blood vessels share some common features, but each type has its own traits and each is marvelously adapted for its function (Figure 11–2). The hollow interior of a blood vessel, through which the blood flows, is called the lumen. The vessel layer in contact with the blood flowing through it is composed of flattened, tight-fitting cells. This

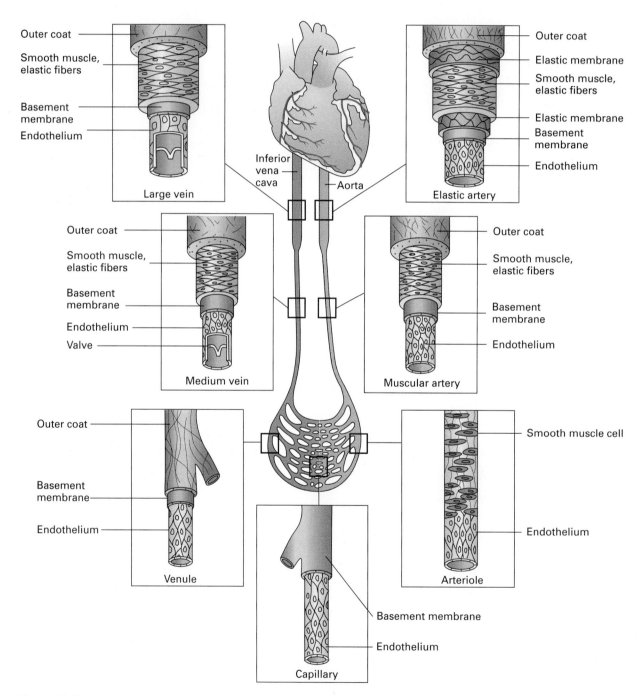

Figure 11–2 The structure of blood vessels.

lining, called the endothelium, forms a smooth surface that minimizes friction and allows the blood to flow over it easily.

Arteries

Arteries are muscular tubes that transport blood away from the heart, delivering it rapidly to the body tissues. The innermost layer of an arterial wall is formed by the endothelium. Immediately outside the endothelium is the middle layer of the arterial wall, which contains elastic fibers and circular layers of smooth muscle. The elastic fibers allow an artery to stretch and then return to its original shape. Thus, elasticity both enables the artery to tolerate the pressure shock caused by blood surging into the artery when the heart contracts and helps maintain a relatively even pressure within the artery, in spite of large changes in arterial blood volume. Consider, for instance, what happens when the heart contracts and sends blood into the **aorta**, the body's main artery. Each beat of the heart causes 70 ml (about ¼ cup) of blood to pound against the wall of the aorta like a tidal wave. A rigid pipe could not withstand the repeated pressure surges, but the elastic arterial walls stretch with each wave of blood and recoil when the surge has moved along, resulting in the intermittent waves being dampened into a continuous stream.

The alternate expansion and recoiling of arteries create a pressure wave, called a **pulse**, that moves along the arterial system with each heartbeat. The pulse can be felt by slightly compressing with the fingers any artery that lies near the body's surface, such as the one at the wrist or the one under the angle of the jaw. With each beat of the heart, the wave of expansion begins, moving along the artery at the rate of 6 to 9 meters per second.

In addition to elastic fibers, the middle layer of an artery wall contains smooth muscle, which provides the artery with another important property—the ability to contract. Upon contraction, called **vasoconstriction**, the diameter of the lumen becomes narrower. On the other hand, when the smooth muscle relaxes and the arterial lumen increases in diameter, the process is called **vasodilation**. Vasoconstriction reduces blood flow through the artery. Conversely, vasodilation increases blood flow through the artery. Thus, the small to medium size arteries, those in which the muscle is most well developed, serve to regulate the distribution of blood, adjusting flow to suit the needs of the body. Vasoconstriction is also important when an artery is cut. Because arterial blood is under great pressure, much blood can be lost in a short time through the wound. In this situation, the artery may contract, helping to minimize blood loss.

The outer layer of an arterial wall is a connective tissue sheath that contains elastic fibers and collagen. This layer adds strength to the arterial wall and anchors the artery to surrounding tissue.

When the wall of an artery becomes weakened, as may be caused by disease, inflammation, injury, or a congenital defect, the pressure of the blood flowing through the weakened area may cause it to swell outward like a balloon, forming an aneurysm (Figure 11–3). Commonly, an aneurysm does not cause symptoms, but it can be threatening just the same. The primary risk is that the aneurysm will burst, caus-

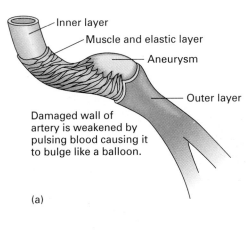

Damaged wall of artery is weakened by pulsing blood causing it to bulge like a balloon.

(a)

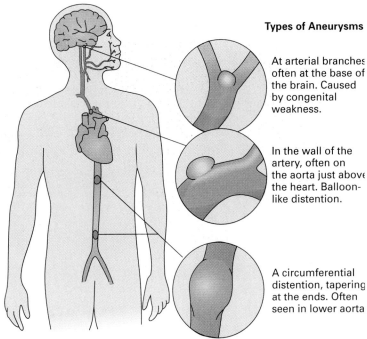

Types of Aneurysms

At arterial branches often at the base of the brain. Caused by congenital weakness.

In the wall of the artery, often on the aorta just above the heart. Balloon-like distention.

A circumferential distention, tapering at the ends. Often seen in lower aorta

(b)

Figure 11–3 (a) An aneurysm is a permanent distention of a weakened spot in an artery wall caused by the pressure of the blood flowing past the weakened area. (b) Common sites of aneurysms include the aorta (the body's main artery) and the arteries that deliver blood to the brain. Rupture of an aneurysm can cause fatal blood loss. In some cases, aneurysms can be surgically repaired.

ing blood loss and depriving tissues of oxygen and nutrients, a situation that can be fatal. Even if it does not rupture, an aneurysm can cause life-threatening blood clots to form. A clot can break free of the site of formation and float through the circulatory system until it lodges in a small vessel, where it can block blood flow and cause tissue death beyond that point. In some cases, an aneurysm can be surgically repaired.

The smallest arteries, called **arterioles**, are barely visible to the naked eye. Their walls have the same three layers found in arteries, but the middle layer is primarily smooth muscle with only a few elastic fibers. Indeed, the tiny arterioles that lead to capillary beds are little more than endothelial lining encircled by a single layer of smooth muscle.

Arterioles have two extremely important regulatory roles. First, they are the prime controllers of blood pressure, which will be discussed later in this chapter. Second, they serve as gatekeepers to capillary beds. That is, a capillary bed can be open or closed, depending on whether the smooth muscle of the arteriole leading to it allows blood through. In this way, metabolically active cells can receive more blood than metabolically inactive ones. Minute by minute, arterioles respond to input from hormones, the nervous system, and local conditions, constantly modifying blood pressure and flow to meet the body's changing needs.

Capillaries

Capillaries are microscopic vessels that connect arterioles and venules. They are well suited to their primary function—the exchange of materials between the blood and the body cells. For instance, because they are gossamer thin, substances need to cross only a single layer of endothelial cells to move between the blood and the fluid surrounding the cells. In spite of their delicate construction, the plasma membrane of the endothelial cells is an effective selective barrier that determines whether various substances can cross by diffusion, active transport, or pinocytosis. Most substances, however, cross the capillary walls, not by passing *through* the endothelial cells, but by filtering through small slits *between* adjacent cells. The slits are just large enough for some fluids and small dissolved molecules to pass through.

In addition, taken together, capillaries provide a tremendous surface area for the rapid exchange of materials between body and blood. In fact, in an adult's body, there are about 10 billion capillaries with a combined length of over 5000 miles. As you can see in Figure 11–4, they are arranged in highly branched networks. The networks bring capillaries very close to nearly every cell. There are so many capillaries that few cells are more than 125 μm (0.005 in.) from a capillary. Your fingernails provide windows that can allow you to appreciate the vast capillary network in the body. You may have noticed that the tissue beneath a fingernail normally has a pink tinge. This is because of the blood flowing through a profusion of capillaries there. Gentle pressure on the nail causes the tissue to turn white as the blood is pushed from those capillaries.

Although a single capillary is scarcely wide enough for a single red blood cell to squeeze through it, there are so many capillaries that their *combined* cross-sectional area is enormous, much greater than that of the arteries or veins. As a result, the blood flows much more slowly through the capil-

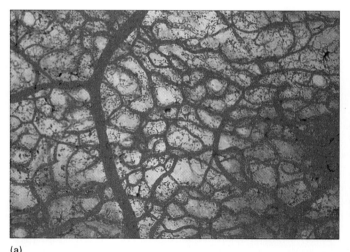

(a)

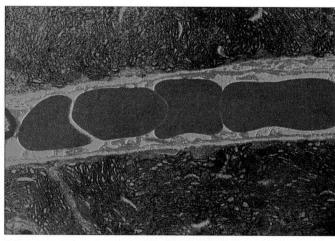

(b)

Figure 11–4   (a) A capillary bed is a network of capillaries servicing a particular region. (b) Capillaries are so narrow that red blood cells must travel through them in single file.   *(a, © Biophoto Associates/Science Source/Photo Researchers, Inc.; b, © David Phillips/Science Source/Photo Researchers, Inc.)*

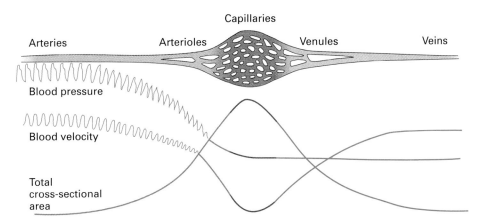

Capillaries

Arteries   Arterioles   Venules   Veins

Blood pressure

Blood velocity

Total
cross-sectional
area

Figure 11–5 The capillaries are so numerous that their total cross-sectional area is much greater than that of arteries or veins. Thus, blood pressure drops and velocity of flow slows as blood passes through a capillary bed. The slower rate of flow allows time for the exchange of materials between the blood and the tissues.

laries than through the arteries or veins. The slower rate of flow in the capillaries provides more time for the exchange of materials (Figure 11–5).

Furthermore, blood flow through capillaries can be adjusted to the body's metabolic needs. The adjustments are made possible by the design of capillary networks. The network of capillaries servicing a particular area is called a **capillary bed** (Figure 11–6). A capillary bed generally has a short vessel, called a **thoroughfare channel**, that can shunt blood directly between the arteriole and the venule at opposite ends of the bed when the capillaries are not being used. Most capillaries in the bed emerge from the thoroughfare channel, but some may spring directly from the arteriole. The number of capillaries in a bed generally ranges between 10 and 100, depending on the type of tissue. At the point

where a capillary branches from the thoroughfare channel or arteriole, a ring of smooth muscle called a **precapillary sphincter** surrounds the capillary and regulates blood flow into it. Contraction of the precapillary sphincter squeezes the capillary shut and directs blood through the thoroughfare channel to the venule.

The precapillary sphincters act as valves that open and close capillary beds. For instance, while you are resting on the beach after finishing a picnic lunch, the capillary beds servicing the digestive organs will be open and nutrients from the food will be absorbed. If you then decide to go for a swim, the capillary beds of the digestive organs will close down and those in the skeletal muscles will open. This redirection of blood flow can lead to abdominal cramps, which could have unfortunate consequences, especially if you are in deep water.

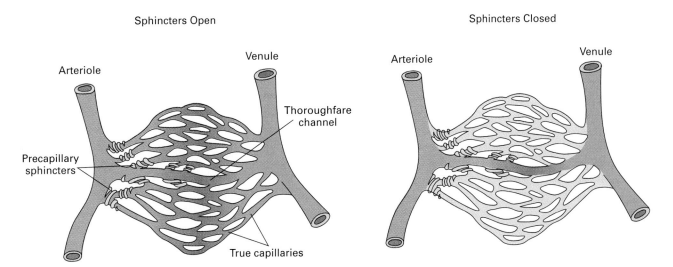

Sphincters Open   Sphincters Closed

Arteriole   Venule

Thoroughfare channel

Precapillary sphincters

True capillaries

Arteriole   Venule

Figure 11–6 The entrance to each capillary is guarded by a ring of muscle called a precapillary sphincter, and blood flow through a capillary bed is regulated according to the body's metabolic needs. When the sphincter is relaxed, blood flows into the capillary bed. Contraction of precapillary sphincters closes the capillaries and shunts blood directly from the arteriole to the venule through a thoroughfare channel.

Capillary Exchange

Materials can move between the capillaries and the body cells in a number of ways. Simple diffusion is the most important way. Gases, such as oxygen and carbon dioxide, hormones, glucose, amino acids, and other small molecules simply move down their concentration gradient directly through a capillary cell or the small gaps between the capillary cells.

The movement of many materials gets a boost from the flow of water in and out of the capillary (Figure 11–7). Two competing forces—blood pressure and osmotic pressure—determine the direction of water movement. Blood pressure is the force of blood against the vessel that contains it. You may recall from Chapter 3 that osmosis is the movement of water across a semipermeable membrane in response to differing concentrations of solutes (dissolved particles). Osmotic pressure, then, indicates the force of that water movement. These competing pressures affect the flow of fluid across capillaries. First, the fluid is pushed out of the capillary at its beginning by the high blood pressure there. As blood flows through the capillary bed, blood pressure falls. At the same time, osmotic pressure increases because the blood becomes more concentrated as proteins are left behind when water is pushed out. At the end of the bed, water is drawn back into the capillary because the osmotic pressure there exceeds blood pressure. Substances that are small enough to fit through the gaps between capillary cells can simply "go with the flow" of water in or out of the capillary.

Finally, a few materials are actively ferried across the capillary wall in tiny vesicles. In this process, a pocket forms on the membrane of a capillary cell. Fluid and suspended materials are drawn into that pocket. The pocket then breaks off, forming a vesicle that crosses the cytoplasm and unloads its contents on the other side.

Veins

Capillaries merge, forming the smallest of veins, called **venules.** Venules then join and form veins. **Veins** are blood vessels that return the blood to the heart.

Although veins share some structural features with arteries, there are some important differences. The walls of veins have the same three layers found in arterial walls, but the middle layer in veins has little smooth muscle. As a result, the walls of veins are thinner and more collapsible than those of comparable arteries (Figure 11–8). Because the blood pressure in veins is generally low, there is no danger that the thin walls will burst. The lumens of veins are also larger than those of arteries of equal size. Together, the thin walls and large lumens allow veins to hold a large volume of blood. Thus, veins serve as blood reservoirs, holding up to 65% of the body's total blood supply.

The same amount of blood that is pumped out of the heart must be conducted back to the heart, but it must be moved along without assistance from the high pressure generated by the heart's contractions. How is this possible? Two design features of veins help. First, the relatively large lumens of veins reduce resistance to blood flow. But this feature alone would not be sufficient to move blood against the force of gravity—from the foot back to the heart, for instance (unless, perhaps, you had your foot in your mouth). Thus, a second design feature often found in veins is the

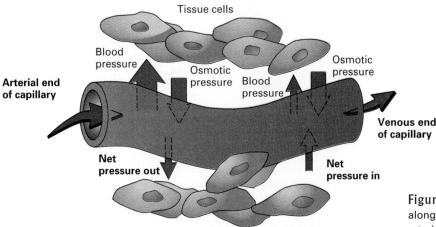

Tissue cells

Blood pressure

Osmotic pressure

Osmotic pressure

Blood pressure

Arterial end of capillary

Venous end of capillary

Net pressure out

Net pressure in

Tissue cells

Figure 11–7 Many substances are passively swept along with fluid that is pushed out of the capillary at the arteriole end by blood pressure and drawn back into the capillary at the venous end by osmotic pressure.

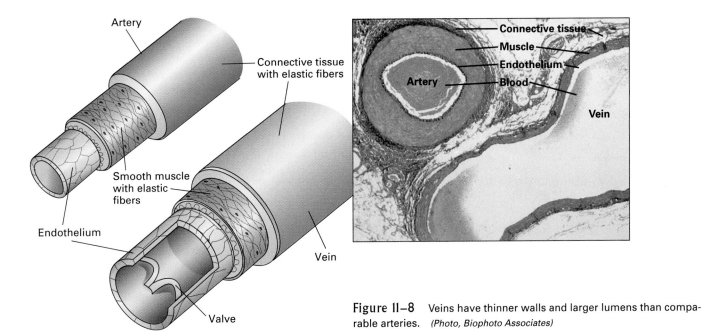

Figure 11-8 Veins have thinner walls and larger lumens than comparable arteries. *(Photo, Biophoto Associates)*

presence of valves that act as one-way turnstiles, allowing blood to move toward the heart but preventing it from flowing backwards. Venous valves are pockets of connective tissue that emerge from the lining of the vein (Figure 11–9). Still, we haven't identified a force that could move blood along back to the heart. In the head and neck, of course, that force is gravity. But what happens in the lower parts of the body? It turns out that the force is produced by simple muscle contraction. Virtually every time a skeletal muscle contracts, it squeezes nearby veins. This pressure pushes blood past the valves toward the heart. The mechanism propelling the blood is not unlike the one that causes toothpaste to squirt out the uncapped end of the tube regardless of where the tube is squeezed. When skeletal muscles relax, any blood that moves backward fills the valves. As the valves fill with blood, they extend into the lumen of the vein, closing the vein and preventing blood from reversing direction (Figure 11–10). Thus, the skeletal muscles are always "milking" the veins and driving blood toward the heart.

A simple experiment can demonstrate the effectiveness of venous valves. Allow your hand to hang by your side until the veins on the back of your hand become distended. Place two fingertips from the other hand at the end of one of the distended veins nearest to the knuckles. Then, leaving one fingertip pressed on the end of the vein, move the other toward the wrist, pressing firmly and squeezing the blood from the vein. Lift the fingertip near the knuckle and notice that blood immediately fills the vein. Repeat the procedure, but this time lift the fingertip near the wrist. You will see the vein remain flattened, because the valves prevent the backwards flow of blood.

The return of blood to the heart from the lower torso is also aided by pressure differences generated by breathing.

During inhalation, the diaphragm, a broad sheet of muscle between the thoracic (chest) and abdominal cavities, contracts and flattens. This motion simultaneously decreases the pressure in the thoracic cavity and increases pressure in the abdominal cavity. These pressure differences create a

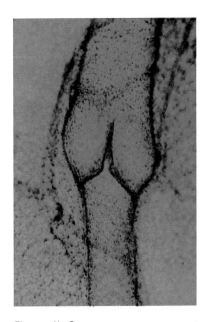

Figure 11-9 Pocket-like valves on the lining of veins assist the return of blood to the heart against gravity by preventing the backflow of blood. The contraction of skeletal muscle compresses nearby veins, pushing blood past the open valves toward the heart. When skeletal muscles relax, blood fills the valves and closes them. The valves thus prevent blood from flowing away from the heart. *(John D. Cunningham/VU)*

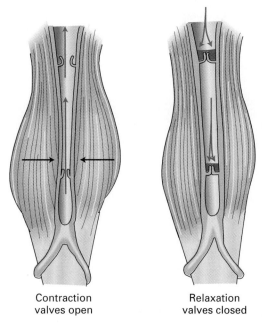

Contraction
valves open

Relaxation
valves closed

Figure 11–10 Blood is "milked" back toward the heart from the lower torso. When skeletal muscles near the vein contract, the bulging muscles squeeze the veins, propelling blood in the direction permitted by the valves. When the skeletal muscles relax, the valves in the veins prevent the backflow of blood.

pump that sucks blood upward toward the heart. During exhalation, backflow of blood is prevented by valves.

Critical Thinking

Coronary bypass surgery is often performed when a region of a coronary artery (an artery that nourishes heart tissue) is blocked. A section of vein from the patient's leg is then removed and sutured so that it makes a connection between the aorta and the coronary artery, past the region that is blocked. The vein provides a new channel of blood flow to reach the heart tissue. Why is it important for the surgeon to suture the vein in place in the correct orientation? What would happen if the piece of vein were sutured in backwards?

About 15% of adults have **varicose veins**, which are veins that have become distended because blood is prevented from flowing freely and so it accumulates, or "pools" in the vein. As you may recall, the walls of veins are relatively thin. Thus, they may be easily stretched, especially if the vein is near the body surface where there is little surrounding tissue to support it. Stretching is also more likely to occur in veins, such as those in the legs, that must support a great column of blood. Once a section of a vein has been stretched, extra blood can accumulate there and place stress on the valve be-

low. Eventually, that valve can become stretched and ineffective and so the blood it once held falls back to the next valve. As a result, varicose veins may appear as bulging veins or as blue-purple, spider-like lines (Figure 11–11).

Many factors can contribute to the formation of varicose veins. Some people inherit veins with weak walls or valves, for instance. In some cases, however, varicose veins are an occupational hazard. Jobs that require standing or sitting for long time periods favor the formation of varicose veins because there is little muscle contraction to squeeze blood back toward the heart. People in such occupations can reduce their chances of developing varicose veins by contracting their leg muscles regularly. Obesity is another predisposing factor, as is pregnancy. In both these cases, the enlarged abdomen creates a downward pressure on the vessels in the groin that impedes the return of blood to the heart.

In most cases, the varicose veins don't cause serious problems. They're unattractive, that's all. When appearance is the only concern, the only treatment needed is often wearing elastic support hose, which provides outside pressure to help move blood against gravity. Regular walking is advised. And, when sitting is the only option, elevate the legs whenever possible.

Critical Thinking

Why would sitting with your legs crossed favor the formation of varicose veins?

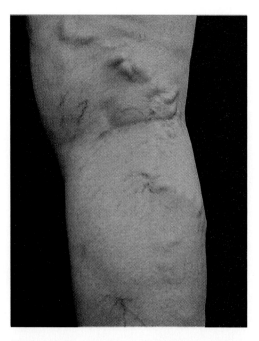

Figure 11–11 Varicose veins are veins that have become stretched and distended by pooled blood. Weak venous valves can be a contributing factor. (© 1993 Rick Brady/Medichrome/The Stock Shop, Inc.)

When we think of varicose veins, we generally picture twisted blue lines in legs. However, varicose veins also form in the anal canal, where they are called hemorrhoids. Here, too, varicose veins form because of increased pressure. In this case, the pressure is often associated with straining during bowel movements. Hemorrhoids can be painful if they protrude through the anal opening. In many cases, they bleed and this can mask more serious problems, such as colon cancer. Mild cases can be controlled by adhering to the motto, "no strain, no pain." In other words, reduce constipation and, therefore, straining during bowel movements by eating a high fiber diet and drinking plenty of fluids. Treatments for more severe cases of hemorrhoids include injecting them with caustic substances that cause scar tissue to form and block the vein, tying them off with tiny rubber bands, or simply snipping them off.

The Heart

The heart is an incredible pump. It beats about 72 times a minute every hour of every day. To appreciate this, form a fist and alternately clench and relax it 70 times a minute. How many minutes does it take before the muscles of your hand are too tired to continue? In contrast, the heart does not fatigue. It beats over 100,000 times each day. That adds up to about 2 billion beats over a lifetime. The pumping done by the heart is equally remarkable. It pumps 10 pints of blood a minute through its chambers, which adds up to 2500 to 3000 gallons per day.

Structure of the Heart

The heart is enclosed within a double-layered sac, called the **pericardium** (*peri-* around; *cardia*, from the Greek for heart), that holds the heart in the center of the thoracic (chest) cavity without hampering its movements, even when the heart is contracting vigorously. The outer layer, which is composed of tough fibrous connective tissue, is tethered to nearby structures, such as the diaphragm, and so it provides a highly flexible and protective case that helps to hold the heart in place. The inner layer of the pericardium is actually an integral part of the wall of the heart. Between the two layers of the pericardium is a thin film of slippery, watery liquid. This liquid serves as a lubricant that allows the heart to move freely within the pericardium as the heart's contractions cause it to move back and forth.

The wall of the heart itself is composed of three layers. The outer layer is part of the pericardium, as was just described. The middle layer, called the **myocardium,** is composed of cardiac muscle tissue and makes up the bulk of the heart. The contractility of the myocardium is responsible for the heart's incredible pumping action. The innermost layer, the **endocardium,** is a thin layer that lines the cavities of the heart. The endocardium includes a layer of endothelial cells that is continuous with the endothelial lining of all blood vessels. The endothelium reduces friction and, therefore, the resistance to blood flow through the heart.

Critical Thinking

An inflammation of the pericardium (pericarditis) can cause a build-up of fluid between the layers of the pericardium or bleeding into the pericardium. Although the pericardium is flexible, it is not elastic and cannot stretch. Explain why pericarditis places a strain on the heart and can even cause the heart to stop beating.

The heart appears to be a single structure, but the right and left halves function as two separate pumps (Figure 11–12). As we will see shortly, the right side of the heart pumps blood to the lungs where it picks up oxygen and the left side drives the blood to the body cells. The two pumps are physically separated by a partition called a septum. Each pump, or side of the heart, consists of two chambers: an upper chamber, called an **atrium** (plural, *atria*), and a lower chamber, called a **ventricle.** The two atria function as receiving chambers for the blood returning to the heart. The two ventricles function as the main pumps of the heart. It is the contraction of the ventricles that forces blood out of the heart under great pressure.

The atria and ventricles differ in size and in the thickness of their walls. The atria are smaller chambers in both respects. Most of the blood flows directly through the atria to the ventricles. The role of the atria is to pump the blood only a relatively short distance to the ventricles. In contrast, the ventricles are the main pumps of the heart. When we think about the work of the heart, we are, in fact, thinking about the work of the ventricles. It should not be surprising, then, that the ventricles are much larger chambers than the atria and have thicker, more muscular walls.

Two pairs of valves ensure that the blood flows in one direction through the heart. The first pair, known as the **atrioventricular (AV) valves,** is located between each atrium and ventricle (Figure 11–13). The AV valves are connective tissue flaps, called cusps, anchored to the wall of the ventricle by strings of connective tissue known as the **chordae tendinae**—the heart strings. These strings prevent AV valves from flapping back into the atria under the pressure developed when the ventricles contract. The AV valve on the right side of the heart has three flaps and is called the tricuspid valve. The AV valve on the left side of the heart has two flaps and is called the bicuspid valve or the mitral valve.

The second pair of valves, known as the **semilunar valves,** is located between each ventricle and its connecting artery. Whereas the cusps of the AV valves are flaps of connective tissue, those of the semilunar valves are small pockets of tissue attached to the inner wall of their respective arteries. Like the valves in veins, these valves fill with blood, preventing the backflow of blood into the relating ventricles from the aorta or pulmonary artery.

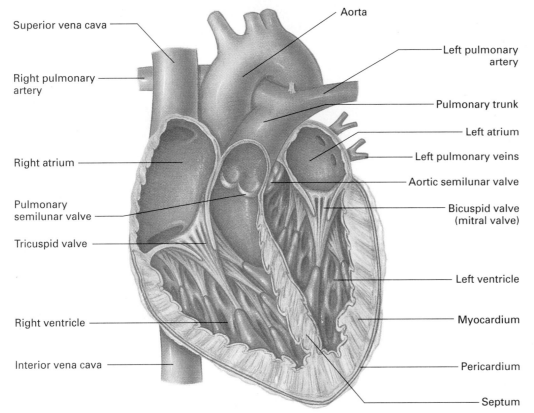

Superior vena cava

Right pulmonary artery

Right atrium

Pulmonary semilunar valve

Tricuspid valve

Right ventricle

Interior vena cava

Aorta

Left pulmonary artery

Pulmonary trunk

Left atrium

Left pulmonary veins

Aortic semilunar valve

Bicuspid valve (mitral valve)

Left ventricle

Myocardium

Pericardium

Septum

Figure II–I2 The human heart is an incredible, tireless pump. Beating about 72 times each minute, it pumps several thousand gallons of blood during an average lifetime. This diagram of the cross-section of a human heart shows the four chambers, the major vessels connecting to the heart, and the heart valves.

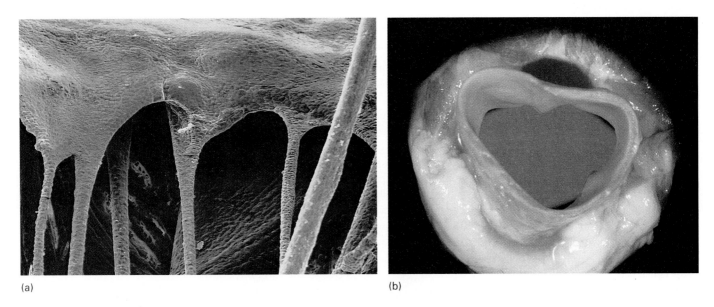

(a)   (b)

Figure II–I3 Two pairs of heart valves ensure that blood flows through the heart in one direction. (a) The atrioventricular valves are located between each atrium and ventricle. An atrioventricular (AV) valve consists of flaps of connective tissue anchored to the wall of the ventricles by strings of connective tissue called chordae tendinae, or heart strings. (b) The semilunar valves are located between each ventricle and its artery. The valve shown here is between the right ventricle and the pulmonary artery. *(a, Prof. P. Motta/Dept. of Anatomy/University "La Sapienza," Rome/Science Photo Library/Custom Medical Stock Photo; b, Science Photo Library/Photo Researchers, Inc.)*

(a)

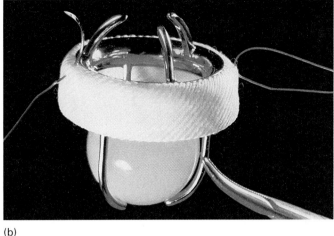

(b)

Figure 11–14 Defective heart valves can be replaced by artificial valves made either from animal tissues or from synthetic materials. Here you see (a) a pig's aortic semilunar valve sewn onto a frame and (b) the ball-and-cage model. *(a, © SIU/Visuals Unlimited; b, VU/SIU © Visuals Unlimited)*

The familiar sounds of the heart, which are often described as "lubb dup," are associated with the closing of the valves. The first heart sound (lubb) is produced when the AV valves snap shut as the ventricles begin to contract. The higher pitched second heart sound (dup) represents the closure of the semilunar valves and the beginning of ventricular relaxation.

Heart murmurs, which are heart sounds other than "lubb dup," are created by turbulent blood flow. Although heart murmurs are sometimes heard in normal, healthy people, they can indicate a heart problem. For instance, malfunctioning valves often disturb blood flow through the heart, causing the swishing or gurgling sounds of heart murmurs. Several conditions can cause valves to malfunction. In some cases, valves become thickened, which narrows the opening and impedes blood flow. In other cases, the valves don't close properly and, therefore, allow the backflow of blood. In either case, the heart is strained because it must work harder to move the blood.

Defective valves can be replaced with artificial valves. Replacement heart valves can be made of animal tissues, such as a pig's aortic valve, or from synthetic materials (Figure 11–14). One popular model is designed as a ball-and-cage. Blood can flow in one direction by forcing the ball away from the ring into the cage. Blood flow in the opposite direction is prevented because blood forces the ball firmly into the ring.

Critical Thinking

Abnormally short (or long) chordae tendinae of the mitral (bicuspid) valve can cause a condition known as mitral valve prolapse. Why would mitral valve prolapse cause heart murmurs?

The Pathway of Blood Through the Heart

You may recall that the heart functions as two separate pumps (Figure 11–15). Each pump circulates the blood through different routes. The right side of the heart drives blood through the **pulmonary circuit**, which transports blood to and from the lungs. The left side of the heart pumps blood through the **systemic circuit**, which transports blood to and from body tissues.

The pulmonary circuit begins in the right atrium, as three veins return blood that is low in oxygen from the systemic circuit. The blood then moves from the right atrium to the right ventricle. Contraction of the right ventricle pumps poorly oxygenated blood to the lungs through the pulmonary trunk, which divides to form two **pulmonary arteries**. In the lungs, oxygen diffuses into the blood and carbon dioxide diffuses out. The oxygen-rich blood is delivered to the left atrium through four **pulmonary veins**, two from each lung. (Note that the pulmonary circulation is an *exception* to the general rule that arteries carry oxygenated blood and veins carry blood low in oxygen. Exactly the opposite is true of vessels in the pulmonary circulation.)

The pathway of blood pumped through the pulmonary circuit by the right side of the heart is

right atrium → AV valve (tricuspid) → right ventricle →
semilunar valve → pulmonary arteries → lungs →
pulmonary veins → left atrium

The systemic circuit begins when oxygen-rich blood enters the left atrium. Blood then moves to the left ventricle. When the left ventricle contracts, oxygenated blood is pushed through the largest artery in the body, the **aorta**. The aorta arches over the top of the heart and gives rise to the smaller arteries that will eventually feed the capillary beds of the body tissues. The venous system collects the

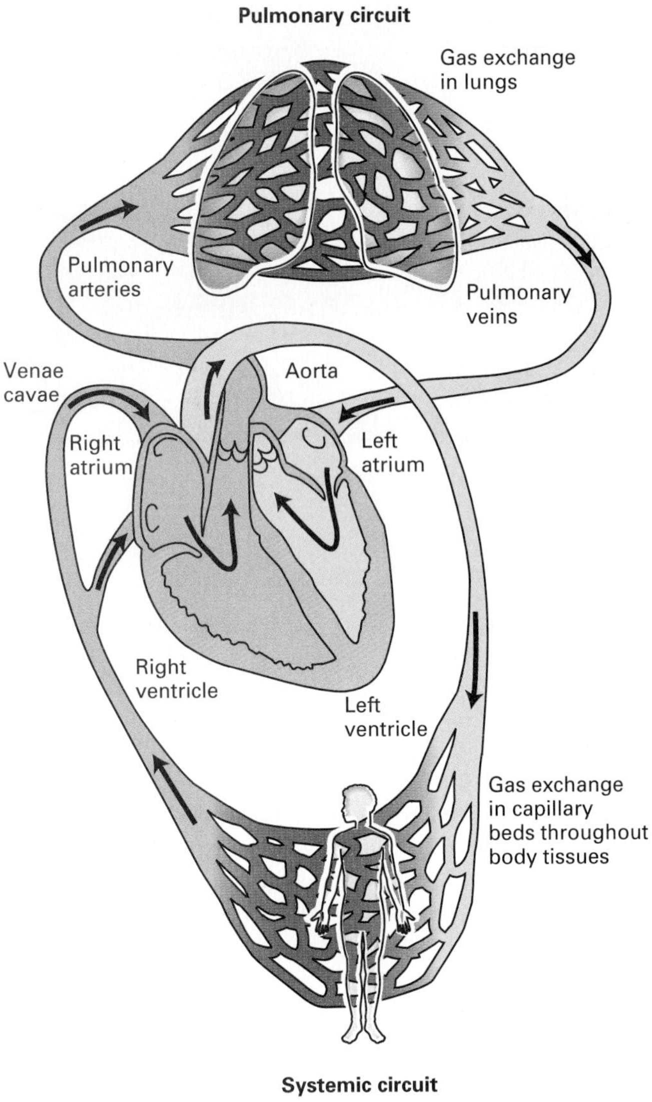

Pulmonary circuit

Gas exchange
in lungs

Pulmonary
arteries

Pulmonary
veins

Venae
cavae

Aorta

Right
atrium

Left
atrium

Right
ventricle

Left
ventricle

Gas exchange
in capillary
beds throughout
body tissues

Systemic circuit

Figure 11–15   The right side of the heart pumps blood through the pulmonary circuit, which carries blood to and from the lungs. The left side of the heart pumps blood through the systemic circuit, which conducts blood to and from the body tissues.

oxygen-depleted blood, and eventually culmimates in three veins returning the blood to the right atrium. These are the **superior vena cava**, which delivers blood from regions above the heart, the **inferior vena cava**, which returns blood from regions below the heart, and the **coronary sinus**, which returns blood collected from the heart muscle itself.

Thus, the pathway of blood through the systemic circuit pumped by the left side of the heart is

left atrium → AV (bicuspid or mitral) valve → left ventricle → semilunar valve → aorta → body tissues → inferior vena cava, superior vena cava, or coronary sinus → right atrium

Coronary Circulation

Heart muscle cells are not nourished by blood flowing through the chambers. Instead, an extensive network of vessels, known as the **coronary circulation**, services the tissues of the heart itself. The first two branches off the aorta are the coronary arteries. These then give rise to numerous branches, ensuring that the heart receives a rich supply of oxygen and nutrients (Figure 11–16). After passing through the capillary beds that nourish the heart tissue, blood enters cardiac veins, which join to form the coronary sinus. Blood flows from the coronary sinus to the right atrium.

Cardiac Cycle

Although the sides of the heart pump blood through different circuits, they work in tandem. First, the two atria contract. Next, the ventricles contract, squeezing blood into their respective arteries.

We see, then, that a "heartbeat" is not a single event. Each beat involves contraction, which is called **systole**, as well as relaxation, which is called **diastole**. All the events associated with the flow of blood through the heart chambers during a single heartbeat are collectively called the **cardiac cycle**. First, all chambers relax and blood passes through the atria and enters the ventricles. When the ventricles are about 70% filled, the atria contract and push their contents into the ventricles. Then the atria relax and the ventricles begin their contraction phase. Upon completion of this con-

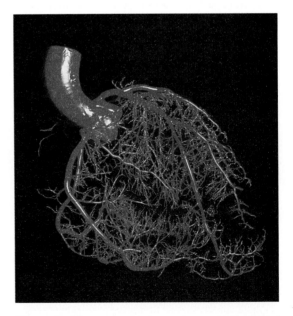

Figure 11–16   This cast of the coronary blood vessels reveals the complexity of the coronary circuit. The coronary vessels deliver a rich supply of oxygen and nutrients to the heart muscle cells and remove the metabolic wastes.   *(Martin Dohrn/Royal College of Surgeons/Science Photo Library/Photo Researchers, Inc.)*

traction, the whole heart again relaxes. If we were to add the contraction time of the heart during a day and compare it with the relaxation time during a day, the heart's workday might turn out to be equivalent to yours. In 24 hours, the heart spends a total of about 8 hours working (contracting) and 16 hours relaxing. However, unlike your workday, the heart's day is divided into repeating cycles of work and relaxation.

The Heart's Conduction System

If a human heart is removed, as in a transplant operation, and placed in a dish it will continue to beat, keeping a lonely and useless rhythm until its tissues die. In fact, if a few cardiac muscle cells are grown in the laboratory, they too will beat on their own, each twitch a reminder of the critical role the intact organ plays. Clearly, then, the heart muscle does not require outside stimulation in order to beat. Instead, the tendency is intrinsic, within the heart muscle itself.

Another remarkable observation has been made of heart muscle cells grown in a laboratory dish. Although isolated heart cells twitch independently of others, if two cells should touch, they will begin beating in unison. This, too, is part of their basic nature and it has a lot to do with the type of connections between cells. The cell membranes of adjacent cardiac muscle cells interdigitate extensively, forming specialized boundaries called **intercalated disks** (Figure 11–17). Cell junctions in the intercalated disks mechanically and electrically couple the connected cardiac muscle cells. Adjacent cells are held together so tightly that they don't rip

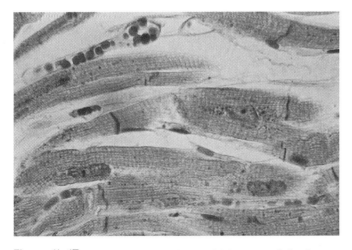

Figure 11–17 The intercalated disks, which are specialized junctions between cardiac muscle cells, can be seen in this photograph of cardiac muscle as dark purple vertical bars across the cells. The intercalated disks lock adjoining cells together tightly, allowing the cells to work together during a contraction and not be ripped apart by the force. At the same time, intercalated disks allow electrical communication between the cells, permitting the signals for contraction to spread rapidly from cell to cell. *(© Biophoto Associates/Photo Researchers, Inc.)*

apart during contraction and the pull of contraction is transmitted from one cell to the next. At the same time, the junctions permit electrical communication between adjacent cells, allowing the electrical events responsible for contraction to spread rapidly over the heart by passing from cell to cell. So, because of the heart's construction, the cells contract in a coordinated manner. However, this intrinsic tendency to contract still needs some outside control in order to contract at the proper rate.

The tempo of the heartbeat is set by a cluster of specialized cardiac muscle cells, called the **sinoatrial (SA) node**, located in the right atrium near the junction of the superior vena cava (Figure 11–18). Because the SA node sends out impulses that initiate each heartbeat, it is often referred to as the **pacemaker**. About 70 to 80 times a minute, the SA node sends out an electrical signal that spreads through the muscle cells of the atria, causing them to contract. The signal reaches another cluster of specialized muscle cells known as the **atrioventricular (AV) node**, located in the partition between the two atria, and stimulates it. The AV node then relays the stimulus by means of a bundle of specialized muscle fibers, called the **atrioventricular bundle**, that runs along the wall between the ventricles. The bundle forks into right and left branches and then divides into many other specialized cardiac muscle cells, called **Purkinje fibers**, which penetrate the walls of the ventricles. Thus, the electrical signal is conducted to the ventricles through the atrioventricular bundle and quickly fans out through the ventricle walls, carried by the Purkinje fibers. The rapid spread of the impulse through the ventricles ensures that they contract smoothly.

When the heart's conduction system is faulty, cells can begin to contract independently. Such cellular independence can result in rapid, irregular contractions of the ventricles, called **ventricular fibrillation**, which renders the ventricles useless as pumps and stops circulation. The brain no longer receives the blood it needs to function. Death occurs unless an effective heartbeat is restored quickly. To stop ventricular fibrillation, a strong electric shock is often delivered to the heart, causing the entire heart to contract at once, in the hopes that the SA node will then begin to function normally.

Problems with the conduction system of the heart can sometimes be treated with an artificial pacemaker (Figure 11–19). A pacemaker is a small device that monitors the heart rate and rhythm and responds to abnormalities if they occur. For instance, if the heart rate becomes too slow, the pacemaker will send electrical signals to the heart through an electrode. Today, many types of pacemakers can even vary the heart rate to suit the body's needs. The pacemaker can be worn on a belt, if it will only be needed for a short time, such as during the temporary disturbance of the heart's rhythm that can follow a heart attack. However, if the need for a pacemaker will be permanent, it can be inserted under the skin on the chest wall.

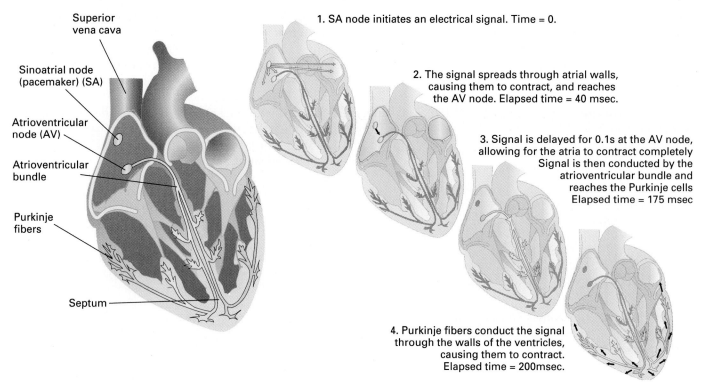

1. SA node initiates an electrical signal. Time = 0.

2. The signal spreads through atrial walls, causing them to contract, and reaches the AV node. Elapsed time = 40 msec.

3. Signal is delayed for 0.1s at the AV node, allowing for the atria to contract completely Signal is then conducted by the atrioventricular bundle and reaches the Purkinje cells Elapsed time = 175 msec

4. Purkinje fibers conduct the signal through the walls of the ventricles, causing them to contract. Elapsed time = 200msec.

Superior vena cava

Sinoatrial node (pacemaker) (SA)

Atrioventricular node (AV)

Atrioventricular bundle

Purkinje fibers

Septum

Figure 11–18 The conduction system of the heart consists of specialized cardiac muscle cells that speed electrical signals through the heart. The sinoatrial (SA) node serves as the heart's internal pacemaker that determines the heart rate. Electrical signals from the SA node spread through the walls of the atria causing them to contract. The signals then stimulate the atrioventricular (AV) node, which in turn sends the signals along the atrioventricular branch to its forks and finally to the many Purkinje fibers that penetrate the ventricular walls. The Purkinje fibers distribute the signals to the heart muscle cells, causing them to contract.

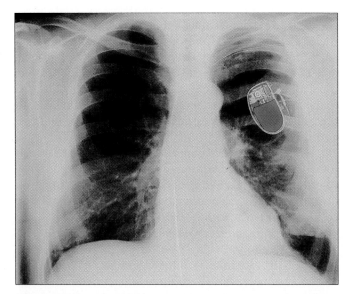

Figure 11–19 An artificial pacemaker is a small device that monitors the heart's rate and rhythm and sends corrective electrical signals to the heart when necessary. *(Dept. of Clinical Radiology; Salisbury District Hospital/Science Photo Library/Photo Researchers, Inc.)*

Regulation of the Heart Rate

The pace or rhythm of the heartbeat changes constantly in response to activity or excitement. The autonomic nervous system and certain hormones make the necessary adjustments so that heart rate suits the body's needs. During times of stress, the sympathetic nervous system increases the rate and force of heart contractions. At this time, the adrenal medulla produces epinephrine. This hormone can then prolong the effects of the sympathetic nervous system. In contrast, when restful conditions prevail, the parasympathetic nervous system dampens heart activity, bringing it in line with the body's more modest metabolic needs.

The Electrocardiogram

The electrical events that spread through the heart with each heartbeat actually travel throughout the body because body fluids are good conductors. These electrical events can be detected by electrodes on the body surface. The electrical changes associated with each heartbeat cause deflections (movements) in the tracing made by the recording device.

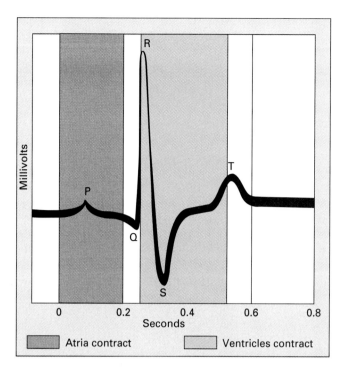

Figure 11–20 The electrical activity that accompanies each heartbeat can be visualized in an electrocardiogram (ECG). As the electrical signals from the SA node spread across the atria and cause them to contract, the P wave is generated. The QRS wave represents the spread of the signal through the ventricles and ventricular contraction. The T wave occurs as the ventricles recover and return to the electrical state that preceded contraction.

An **electrocardiogram** (an **ECG** or **EKG**) is a recording of the electrical activities of the heart.

A typical ECG consists of three distinguishable deflection waves (Figure 11–20). The first wave, called the *P wave*, accompanies the spread of the electrical signal over the atria and the atrial contraction that follows. The next, the *QRS wave*, reflects the spread of the electrical signal over the ventricles and ventricular contraction. Because the muscle mass of the ventricles is much greater than that of the atria, the QRS wave is larger than the P wave. The third, the *T wave*, represents ventricular repolarization, that is, the return of the ventricles to the electrical state that preceded contraction. The recovery (repolarization) of the atria is not seen in an ECG because it occurs during ventricular contraction and is, therefore, masked by the QRS wave. Because the pattern and timing of these waves is remarkably consistent in a healthy heart, abnormal patterns can indicate heart problems (Figure 11–21).

Blood Pressure

We hear a lot about blood pressure, usually when someone frets about how high it is or brags about how low it is. We have already referred to blood pressure, but what exactly is

it? **Blood pressure** is the force exerted by the blood against the walls of the blood vessels. When the ventricles contract, they push blood into the arteries under great pressure. This pressure is the driving force that moves blood through the body and the same pressure pushes outward against vessel walls. Thus, it is important that blood pressure be great enough to circulate the blood, but not so great that it stresses the heart and blood vessels.

Blood pressure in the arteries varies predictably during each heartbeat. It is highest during the contraction of the ventricles (systole), when blood is being forced into the arteries. The **systolic pressure**, the highest pressure in the artery during each heartbeat, of a typical, healthy adult is about 120 mm Hg.[1] Blood pressure is lowest when the

[1] Pressure is measured as the height to which that pressure could push a column of mercury (Hg).

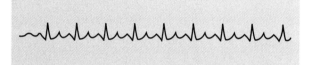

Tachycardia- the heart rate is over 100 beats per minute, but the pattern of the waves is normal.

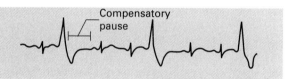

Premature contraction of the ventricles. This ECG shows the QRS wave (caused by ventricular contraction) occurring too soon after the P wave (associated with atrial contraction). There is also a longer than usual pause before the next heartbeat.

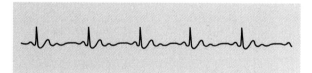

A heart block. In this ECG, several P waves (atrial contractions) occur before a QRS wave (ventricular contraction) because damage to the conduction system, such as might have been caused by a heart attack, is partially obstructing the spread of the electrical signals to the AV node.

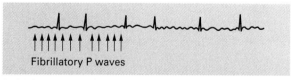

Atrial fibrillation. The rapid, irregular contractions of the atria that occur in atrial fibrillation are seen as a series of small P waves in this ECG.

Figure 11–21 An ECG can help a physician detect and diagnose various heart problems.

ventricles are relaxing (diastole). In a healthy adult, the lowest or **diastolic pressure** is about 80 mm Hg. Thus, a person's blood pressure is usually expressed as two values—the systolic then the diastolic. For instance, normal adult blood pressure is said to be 120/80. (Do you know what yours is?)

Blood pressure is measured with a device called a **sphygmomanometer**. Basically, a sphygmomanometer consists of an inflatable cuff that wraps around the upper arm attached to a device that can measure the pressure within the cuff. The idea is to measure blood pressure by applying a counter pressure within the cuff by pumping air into it. Using a stethoscope, the person measuring blood pressure listens to blood flow through the brachial artery, which runs along the inner surface of the arm (Figure 11–22). The pressure in the cuff is increased until it exceeds the pressure in the artery and blood can no longer force its way through the artery, even when the heart is contracting and exerting maximal pressure against the cuff. Blood flow through the artery is completely shut off and no

sound can be heard through the stethoscope. Then the cuff pressure is slowly released. A point is reached when the pressure in the cuff is slightly lower than that in the artery, allowing a small amount of blood to push past the cuff, but only when the heart is contracting and exerting maximal force. At this point, tapping sounds are heard through the stethoscope, each "tap" representing a spurt of blood past the cuff. The pressure in the cuff when these tapping sounds are first heard is the systolic pressure, and it represents the highest pressure in the artery during each heartbeat. As cuff pressure continues to drop, it eventually becomes lower than arterial pressure, even when the heart is relaxing. At this point, the artery is no longer constricted and blood flows freely through it. Blood still exerts pressure on the arterial walls when the ventricles are relaxing because of the elastic recoil of the arteries. The sounds heard through the stethoscope disappear. The pressure at which the sounds are no longer heard is the diastolic pressure.

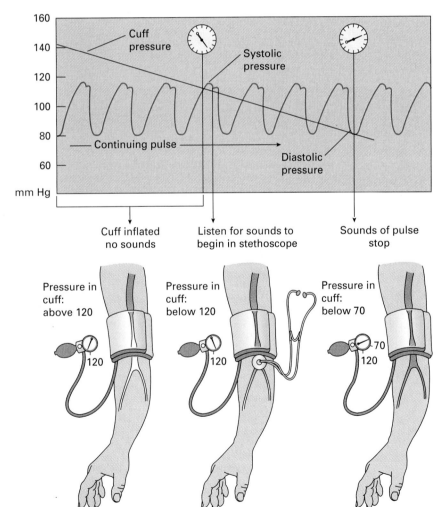

Figure 11–22 Blood pressure is measured with a sphygmomanometer, which consists of an inflatable cuff and a means of measuring the pressure within the cuff. The cuff is placed around the upper arm and, when inflated, compresses the brachial artery. The pressure in the cuff is slowly released. As the cuff pressure is released, blood is able to spurt through the constricted artery only at the points of highest blood pressure. The pressure when thudding sounds are first heard is the systolic pressure, the blood pressure when the heart is contracting. As the pressure in the cuff drops, a point is reached when the sounds disappear. The blood is now flowing continuously through the brachial artery. The pressure in the cuff at this point is the diastolic pressure, the blood pressure when the heart is relaxing.

Cardiovascular Disease

Cardiovascular disease is the single biggest killer of men and women in the United States. It affects slightly more men than women because, until menopause, women gain some natural protection from the female hormone estrogen. Ironically, although slightly fewer women than men have heart attacks, those who do are twice as likely to die within the following weeks than their male counterparts. Because heart attacks are commonly thought of as a male problem, women and their physicians often fail to recognize the symptoms, which delays treatments that could be lifesaving. So, let's examine common problems with the heart and blood vessels with an eye to how we can reduce our risks of developing them.

Hypertension

High blood pressure, or **hypertension**, plagues over 58 million Americans. It accounts for more outpatient prescriptions than any other disease in America.

Hypertension is often called the "silent killer." It is *silent* because it doesn't produce any telltale symptoms. It's a *killer* because it can cause fatal problems, usually involving the heart, brain, blood vessels, or kidneys. Hypertension damages the heart in a number of ways, but primarily by causing the heart to work harder to keep the blood moving. In response, the heart muscle thickens and the heart enlarges and works less efficiently, making it difficult to keep up with the body's needs. At the same time, the increased workload increases the heart's need for oxygen and nutrients. If these cannot be delivered rapidly enough, a heart attack can result.

In addition to the harm high blood pressure does to the heart, it can also damage the blood vessels—by leading to clogging or by causing rupture. High blood pressure promotes the development of atherosclerosis, which, as we will see shortly, is a condition in which fatty deposits form in the arteries and obstruct blood flow. Atherosclerosis is a leading cause of heart attacks and strokes (death of brain cells). Furthermore, the added stress placed on the arteries when the blood pressure is high can cause them to rupture leading to blood loss, which can be fatal. The blood vessels of the brain generally have less supporting tissue than other blood vessels do. Thus, these are particularly likely to burst with high blood pressure. The kidneys are also vulnerable to the effects of hypertension. The damage begins as the walls of the arterioles that supply the filtering system of the kidney thicken and the lumens become narrowed, thus reducing blood flow. This can cause kidney damage. The kidneys make matters worse by responding to the reduced blood flow by secreting renin, a chemical that leads to further increases in blood pressure in an ever-escalating cycle.

Although about 90% of the cases of hypertension have no *known* cause, many contributing factors have been identified. Between 10% and 40% of them have a genetic basis. In other cases, the kidneys have an impaired ability to handle sodium and this results in fluid retention. This increases blood pressure by increasing blood volume. Yet, in other people it seems that the sympathetic nervous system reacts too strongly to stressful conditions by narrowing blood vessels and increasing heart rate. Thus, more blood per minute is pumped through vessels that provide a greater resistance to flow.

At what point should blood pressure be considered "high"? This is one of the medical profession's most vexing problems. Most physicians would agree that a blood pressure of 160/105 is high and should be treated. But, uncertainty clouds the treatment issue when the person's diastolic pressure is between 90 and 99. Although drug treatment may help in borderline cases, it usually must be continued for life and a diagnosis of high blood pressure may influence other aspects of a person's life, such as life insurance premiums. So, sometimes only lifestyle changes are recommended for borderline cases of hypertension.

A high upper (systolic) value usually suggests that the person's arteries have become hardened, a condition known as arteriosclerosis, and are no longer able to dampen the high pressure of each heartbeat. The lower (diastolic) value is generally considered more important because it indicates the pressure when the heart is relaxing.

When the diagnosis of hypertension is clear, there are many drugs that can be prescribed, each of which combats a specific mechanism that contributes to high blood pressure. Some drugs, the diuretics for instance, decrease blood volume by increasing the excretion of sodium and fluids. Yet others cause the blood vessels to dilate (become wider), which decreases peripheral resistance by making blood flow easier.

There are a number of changes in lifestyle that are recommended to treat or prevent hypertension.

1. **Control weight.** There is a close relationship between being overweight and high blood pressure. Thus, maintaining normal body weight can help control blood pressure. Many overweight people with high blood pressure benefit from shedding just a few extra pounds. The best way to lose weight is to eat a moderate balanced diet, reduce fat intake, and increase physical activity.
2. **Exercise regularly.** Aerobic exercise, such as brisk walking, jogging, swimming, or cycling, performed for at least 20 minutes three times a week, helps lower blood pressure and keeps it low.
3. **Don't smoke.** Cigarette smoke contains nicotine, a drug that increases heart rate and constricts blood vessels, both of which increase blood pressure.
4. **Limit dietary salt.** Some people with hypertension are able to lower their blood pressure by lowering the amount of salt in their diet.
5. **Don't drink alcohol.** Alcohol consumption can elevate blood pressure. It can also interact with medications prescribed to treat hypertension.

The Cardiovascular Benefits of Exercise

What would you say if you were offered something that would reduce your risk of heart attack, stroke, diabetes, and cancer, while controlling your weight, strengthening your bones, and relieving both anxiety and tension. "Impossible!" you might say. "What's the catch?" There is none. This seemingly magical elixir of life is regular aerobic exercise.

Although exercise has many beneficial effects throughout the body, here we will consider only the benefits to the cardiovascular system. Exercise benefits the heart in several ways: It makes the heart a more efficient pump, thus reducing its workload. A well-exercised heart beats more slowly than that of a sedentary person—during both exercise and rest. For instance, the heart rate of a well-trained athlete is about 40 to 60 beats per minute, as compared with the average 72 beats per minute. This gives the heart more time to rest between beats. At the same time, however, the well-exercised heart enlarges and so it pumps more blood with each beat.

Exercise also increases the oxygen supply to the heart muscle, by widening the coronary arteries, thus increasing blood flow to the heart. Also, because the capillary beds within the heart muscle become more extensive following regular exercise, oxygen and nutrients can be delivered to the heart cells and wastes can be removed more quickly.

Furthermore, exercise helps to ensure continuous blood flow to the heart. One way it accomplishes this is by increasing the body's ability to dissolve blood clots that can lead to heart attacks or strokes. It stimulates the release of a natural enzyme, called tissue plasminogen activator (tPA), that prevents blood clotting. The effect of tPA is immediate and lasts for as long as 1½ hours after you stop exercising. Besides this, exercise stimulates the development of collateral circulation, that is, additional blood vessels that provide alternative pathways for blood flow. As a result, blood flows continuously through the heart, even if one vessel becomes blocked.

At the same time, there are blood changes that allow more oxygen to be delivered to the cells. The amount of hemoglobin, the oxygen-binding protein in red blood cells, increases. In addition, the blood volume and the numbers of red blood cells increase.

The blood vessels also profit from regular exercise because it lowers the risk of coronary artery disease (CAD) by lowering blood pressure. Exercise lowers blood pressure by keeping arteries elastic and relieving nervous tension.

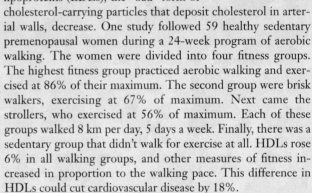

Besides lowering blood pressure, exercise lowers the risk of CAD by shifting the balance of lipids in the blood. High density lipoproteins (HDLs), which are the "good" form of cholesterol-carrying particles that remove cholesterol from the arterial walls, increase with exercise. Low density lipoproteins (LDLs), the "bad" form of cholesterol-carrying particles that deposit cholesterol in arterial walls, decrease. One study followed 59 healthy sedentary premenopausal women during a 24-week program of aerobic walking. The women were divided into four fitness groups. The highest fitness group practiced aerobic walking and exercised at 86% of their maximum. The second group were brisk walkers, exercising at 67% of maximum. Next came the strollers, who exercised at 56% of maximum. Each of these groups walked 8 km per day, 5 days a week. Finally, there was a sedentary group that didn't walk for exercise at all. HDLs rose 6% in all walking groups, and other measures of fitness increased in proportion to the walking pace. This difference in HDLs could cut cardiovascular disease by 18%.

Not all exercise promotes cardiovascular fitness. Cardiovascular conditioning requires that the exercise be aerobic, which means that it promotes the use of oxygen and can be sustained for at least 2 minutes without causing shortness of breath. Thus, to reap cardiovascular benefits, you must exercise hard enough and long enough. How hard? The exercise must be vigorous enough to elevate your heart rate to the target zone. The heart rate target zone is between 70% and 85% of your maximal attainable heart rate, which can be determined by subtracting your age in years from 222 beats per minute. How long? The exercise must continue for at least 20 minutes and be performed at least 3 days a week, with no more than 2 days between bouts.

Doing *something* active on a regular basis can improve the quality of life during your lifetime. Moderate activity helps you feel better—emotionally and physically. It is a difference that has been likened to traveling first class instead of coach.

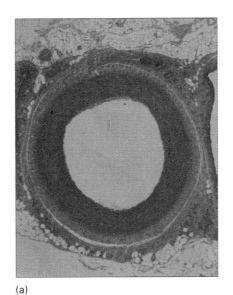

(a)

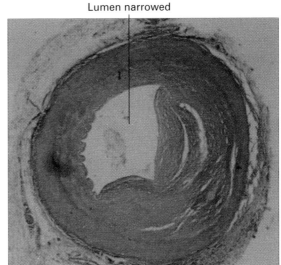

Lumen narrowed

(b)

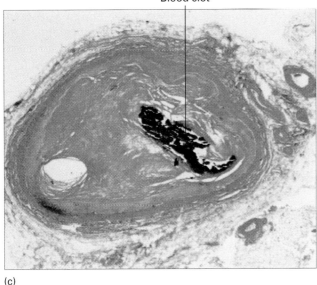

Blood clot

(c)

Figure 11–23 In atherosclerosis, the arterial wall thickens due to the proliferation of smooth muscle and the accumulation of lipids, primarily cholesterol. Together the smooth muscle and cholesterol form plaque, which is typically covered over with a protein cap. Plaque obstructs blood flow through the artery, thus depriving the cells that would be fed by the artery of life-sustaining blood. Plaque can also snag platelets and cause a blood clot to form. The clot may then completely clog the vessel and cause the death of tissue downstream. *(a, © SIU/Peter Arnold, Inc.; b, © W. Ober/Visuals Unlimited; c, © Sloop-Ober/Visuals Unlimited)*

Atherosclerosis

Atherosclerosis is a narrowing of the arteries caused by thickening of the arterial walls and a build-up of fatty deposits. It causes problems because it reduces blood flow through the vessel, choking off the vital supply of oxygen and nutrients to the tissues served by that vessel. The starved cells may even die, an event that can be fatal, especially if the affected cells are those of the heart (a heart attack) or of the brain (a stroke).

The process begins with minor damage to the lining of the artery, as might occur because of excessively rapid or turbulent blood flow. The injury attracts white blood cells called monocytes, which then transform into phagocytic cells called macrophages and these enlarge by engulfing lipids. Platelets also gather in the region and, along with the macrophages,

secrete a growth factor that attracts smooth muscle cells from the middle layer of the artery and causes them to proliferate. Together the macrophages and smooth muscle cells produce the first telltale signs of atherosclerosis—a *fatty streak* on the innermost lining of arteries. As macrophages continue to scavenge lipids, the fatty streak enlarges and forms *plaque*, a bumpy layer that bulges into the channel of the artery. Much of the plaque consists of smooth muscle cells engorged with fatty material, especially cholesterol. The plaque is normally covered with a protein cap that keeps pieces from breaking away (Figure 11–23). But a small break in that protective cap can allow the plaque to rupture. Besides obstructing blood flow through an artery, plaque makes the artery less elastic and its rough surface can snag platelets, leading to clot formation. Then, white blood cells and platelets will rush to the scene in an attempt to heal the damaged arterial wall. The resulting jelly-like clot can further block blood flow. Calcium salts may then be added to the plaque. These harden the artery wall, making it less elastic and weakening it.

Atherosclerosis is a self-perpetuating problem. Plaque narrows the channel of the artery and makes the blood flow more turbulent. The turbulence further damages the artery wall, thus leading to additional plaque formation. Eventually blood flow through the artery can become completely clogged.

Any artery can be affected by atherosclerosis. In some people, it occurs mainly in the vessels of the legs. Although the leg arteries are partially clogged in about 12% of those between the ages of 65 and 70 and 20% of those older than 75, few of them experience symptoms. When there is a symptom, it's usually leg pain—pain that can be severe enough to prevent a person from working or even crossing the street. The carotid arteries that deliver blood to the brain may also be clogged with plaque. In this case the risk is that the reduced flow will cause brain cells to die, thus resulting in a stroke. For some reason, however, the most vulnerable arteries are those of the heart. Therefore, we will consider this problem of the heart in more detail.

Coronary Artery Disease

Coronary artery disease (**CAD**), the underlying cause of the vast majority of heart attacks, is a condition in which fatty deposits associated with atherosclerosis form within coronary arteries, obstructing the flow of blood.

Basically, the problem of supplying blood to the heart muscle is one of supply and demand. The harder the heart works, the more oxygen it needs to continue pumping. If the blood flow through the coronary arteries can't meet the heart's demand, warning symptoms, such as chest pain or irregularities in heartbeat, may occur. A temporary reduction in blood supply due to obstructed blood flow, called *ischemia*, causes reversible damage to heart muscle.

A temporary shortage of oxygen to the heart is accompanied by *angina pectoris*—chest pain, usually experienced in the center of the chest or slightly to the left. The name angina comes from the Latin word *angere*, meaning "to strangle." The name is apt, since the pain of angina is often described as suffocating, vise-like, or choking. Typically, the pain comes on during physical exertion or emotional stress, when the demands on the heart are increased and the blood flow to the heart muscle can no longer meet the needs. The pain stops after a period of rest. This type of angina is called stable angina. However, blood flow can also be shut down by spasms (intense contractions) of the smooth muscle in the walls of the coronary arteries. The spasms also produce chest pain, but in this case it is called variant angina, because the pain occurs during rest rather than during exertion.

Angina, then, serves as a warning that part of the heart is receiving insufficient blood through the coronary arteries, but it doesn't damage the heart permanently. The warning should be taken seriously, however, since each year up to 15% of those people who have angina die suddenly from a heart attack. It seems that the lack of oxygen stimulates a type of white blood cell called neutrophils to release a flood of harmful substances. These can then trigger ventricular fibrillation, a condition in which the ventricles beat extremely rapidly but ineffectively. As a result, the brain no longer gets oxygen and the victim slips into unconsciousness. Death follows if a normal heartbeat is not quickly restored by cardiopulmonary resuscitation (CPR) or by a defibrillator (a device that electrically shocks the heart so that all its cells contract at once, allowing the natural pacemaker to restart the heart).

Although coronary artery disease is usually diagnosed from the symptoms of angina and a physical examination, another procedure allows a physician to spot areas in the coronary arteries that have become narrowed by atherosclerosis. In this procedure, which is called coronary angiography, a contrast dye that is made visible by x-rays is released in the heart so that the coronary vessels can be seen on film. A catheter is inserted into an artery in the arm or leg and then threaded through the blood vessels until it reaches the heart. The dye is then squirted into the openings of the coronary arteries. The movement of the dye through the arteries is then recorded in a series of high-speed x-rays.

Social Issue

According to the guidelines of the American College of Cardiology and the American Heart Association, coronary angiography should be performed only on people thought to have coronary artery disease whose occupations make them responsible for the safety of others. Nonetheless, most coronary angiography is performed on people with stable angina, the kind associated with exertion. Some experts suggest that this is an inappropriate use of the procedure since there are simpler, less risky, and less costly ways of diagnosing this type of angina. One study concluded that as many as 80% of those who underwent coronary angiography did not need it. Other experts argue that patients should be offered all the options and allowed to make their own decisions. How do you think the decision should be made?

Coronary artery disease can be treated with medicines or with surgery. Among the medicines commonly used are those that dilate (widen) blood vessels, nitroglycerin for instance. Certain other drugs (e.g., Verapamil) specifically dilate the coronary arteries. Wider blood vessels make it easier for the heart to pump blood through the circuit, and when the coronary arteries dilate, more blood is delivered to the heart muscle. Also used are drugs such as Inderal or Lopressor, which dampen the heart's response to stimulation from the sympathetic nervous system, thus decreasing its need for oxygen.

Three surgical operations used to treat CAD are balloon angioplasty, atherectomy, and coronary artery bypass. **Bal-**

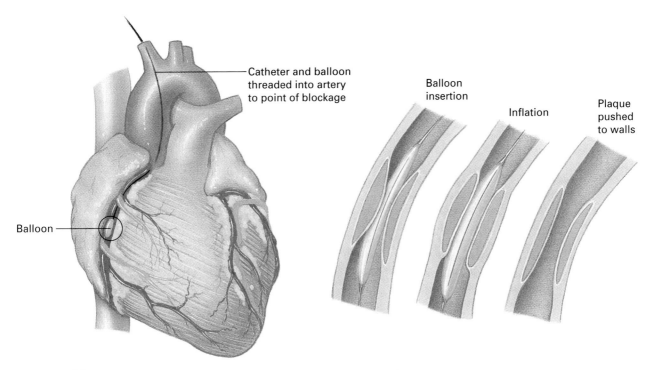

Figure 11–24   In balloon angioplasty, a deflated balloon is threaded through arteries to the region of a coronary artery that is blocked with plaque. The balloon is then inflated, thus stretching the artery and pushing plaque against the wall.

loon angioplasty involves widening the channel of an artery that is narrowed by plaque from the inside by inflating a tough, plastic balloon (Figure 11–24). The tiny, deflated balloon is attached to the end of a long tube, called a catheter. It is inserted through an artery in the arm or leg and then, using x-rays to follow its progress, it is pushed to the blocked spot in a coronary artery. Then the balloon is inflated under pressure, stretching the artery and pressing the soft plaque against the wall. The artery is immediately widened.

Balloon angioplasty is a mixed bag of costs and benefits. The success rate is over 90% and death rate during the procedure is low—only about 1%. Unfortunately, in about 20% of the patients, the artery narrows again within a few months and additional treatment is required.

A more direct approach to widening the blocked artery is removing the deposits responsible for blockage, a procedure called **atherectomy**. Again, a catheter is guided to the site of blockage and then one of several surgical instruments can be slid through the catheter. With one type of instrument, plaque can be scraped away from the artery wall. Another type attacks the plaque with laser beams. The plaque fragments are then sucked into the catheter so that they don't block smaller vessels downstream.

An alternative surgical procedure is a **coronary artery bypass**, in which a segment of a leg vein is removed and grafted so that it provides a shunt between the aorta and a coronary artery—past the point of obstruction (Figure 11–25). Blood flow then "bypasses" the clogged region of

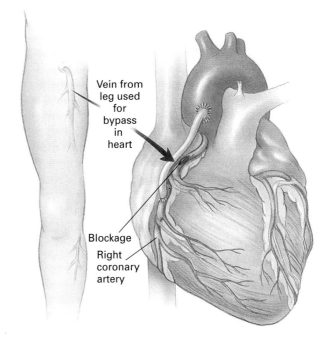

Figure 11–25   In coronary bypass surgery, a section of a leg vein is removed and one end is attached to the heart's main artery, the aorta, and the other to a coronary artery, but past the obstructed region. The grafted vein then provides a pathway through which blood can reach the previously deprived region of heart muscle.

the artery. It is possible to reroute as many as four coronary arteries in a single operation. A coronary bypass is usually performed if the blockage in a coronary artery is located in a critical spot or if there are simply too many bottlenecks to correct using balloon angioplasty. Although coronary bypass surgery provides great relief from angina, it does not seem to lengthen the person's lifetime.

Heart Attack

In a heart attack, technically known as a **myocardial infarction**, a part of the heart muscle dies because of an insufficient blood supply. (*Myocardial* refers to heart muscle and *infarct* refers to dead tissue.) Choked off from their essential blood supply for more than 2 hours, heart muscle cells begin to die. Depending on the extent of damage, the effects can spread quickly throughout the body—the brain receives insufficient oxygen, the lungs fill with fluid, and the kidneys fail. Within a short time, white blood cells swarm in to remove the damaged tissue. Then, over the next 8 weeks or so, scar tissue replaces the dead cardiac muscle (Figure 11–26). Since scar tissue can't contract, part of the heart loses its pumping ability.

What causes a heart attack? The most common type of heart attack is a coronary thrombosis, which means that it is caused by a blood clot blocking a coronary artery. This generally doesn't happen unless the artery is already obstructed with lipid deposits, as occurs in coronary artery disease. In some instances, the blood clot is formed elsewhere in the body but is swept along in the bloodstream and lodges in a coronary artery. In still other instances, the blockage may be temporary, caused by a constriction of a coronary artery, called a coronary artery spasm.

The most common symptom of a heart attack is chest pain. In some cases, the pain is a severe, crushing pain beginning in the center of the chest and often spreading down the inside of one or both arms (most commonly the left one), as well as up to the neck and shoulders. Although the pain is severe enough to cause the victim to stop whatever he or she is doing, it isn't always so overwhelming that it is recognized as a heart attack. A heart attack may also cause nausea and dizziness, which can prompt the victim to write off the symptoms as an upset stomach. (Interestingly, many of the founding fathers of our country were declared as having died of indigestion. Either the quality of American food has greatly improved over the years, or physicians in the late 1700s, unaware of the symptoms of a heart attack, misdiagnosed the problem.)

Oddly, those who experience severe pain may be the lucky ones, because they are more likely to realize the cause is a heart attack and seek immediate help. Doubt about the cause of the symptoms and denial of a heart attack often go hand-in-hand. Denial is unfortunate because it often delays treatment, and treatment within the first few hours after a heart attack can make the difference between life and death. Unfortunately, the average time period between the start of a

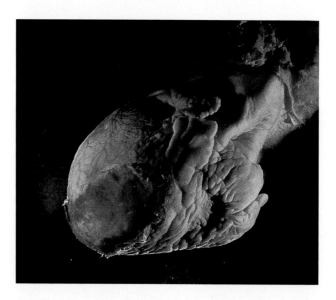

Figure 11–26 The ravages of a prior heart attack are visible as scar tissue at the bottom of this lifeless heart. Scar tissue replaced cardiac muscle when the blood supply to the heart muscle was shut down. Scar tissue cannot contract. Therefore, that part of the life-sustaining pump becomes ineffective. The most common cause of a heart attack is a coronary thrombosis, in which a blood clot blocks blood flow through a coronary artery.   *(© Lennart Nilsson)*

heart attack and seeking medical help is 2 hours. Roughly one quarter of heart attack victims die within that time period, usually because the heart develops an abnormal rhythm.

If administered in time, clot-busting drugs can restore circulation to the needy heart muscle before it dies. There are two important drugs used to dissolve blood clots. The older, less expensive drug is streptokinase. The newer, genetically engineered drug is tissue plasminogen activator (tPA). Studies have indicated that tPA, administered with another clot-busting drug, heparin, saves about one extra life for each 100 patients treated than does treatment with streptokinase. However, the cost of tPA is about ten times greater than streptokinase. Although some physicians continue to debate whether tPA is worth the extra money, others are quick to point out that the timing of treatment is much more important than the choice of drug. If the drug is given to the patient within 2 hours, it has a chance of clearing the clot from the coronary artery before permanent damage is done to the heart muscle. But, if treatment is delayed for 2 to 4 hours from the start of a heart attack, the heart will almost certainly suffer some permanent damage. Thus, treatment should be given as soon as possible.

If a large enough section of heart muscle is damaged by a heart attack, the heart may no longer be able to continue pumping blood to the lungs and the rest of the body at an adequate rate. When the heart becomes an inefficient pump, the condition is known as **heart failure**. The symptoms of heart failure include shortness of breath, fatigue, weakness, and fluid accumulation in the lungs or limbs. Although the heart can never be restored to its former health, the symptoms of heart failure can be treated with drugs. For instance, digitalis can increase the strength of heart contractions, diuretics reduce fluid accumulation, thus lessening the heart's workload, and vasodilators relax constricted arteries, thereby reducing resistance to blood flow and blood pressure. Together these drugs help a weakened heart pump more efficiently.

Sometimes, drugs and surgery cannot halt progressive heart failure. In this case, heart transplant surgery may provide hope for some patients, but generally only for those younger than 60 years of age. First, a donor heart that provides an acceptable tissue match must be found. During transplant surgery, a heart-lung machine takes over the circulation while the weakened heart is removed and the new heart sewn in place. The patient is then treated with drugs that suppress the immune system, lessening the chances of rejection. Nonetheless, although 80% of heart transplant recipients survive at least a year after surgery, few live as long as 10 years.

A newer surgical technique for treating heart failure holds a promise of using the person's own skeletal muscle to boost the heart's pumping ability. The damaged region of the heart wall is surgically removed and the cut edges are sewn together. A section of a skeletal muscle from the back is then cut free at one end and stretched through the body and around the heart to reinforce the weakened area. The other end of the back muscle is left in place so that its nerve and blood supply are intact. A pacemaker is attached to the skeletal muscle now strapped around the heart, causing it to contract intermittently. Since the skeletal muscle is from the patient's own body, rejection is not a problem. However, skeletal muscle can't always be coaxed to contract at the rate of heart muscle, nonstop. Researchers are working on ways to condition the muscle so that it can perform as needed.

The Lymphatic System

The lymphatic system consists of a fluid, called **lymph**, that is essentially identical to interstitial fluid (the fluid that bathes all the cells of the body), a system of **lymphatic vessels** through which the lymph flows, and various **lymphoid tissues** and organs scattered throughout the body (Figure 11–27).

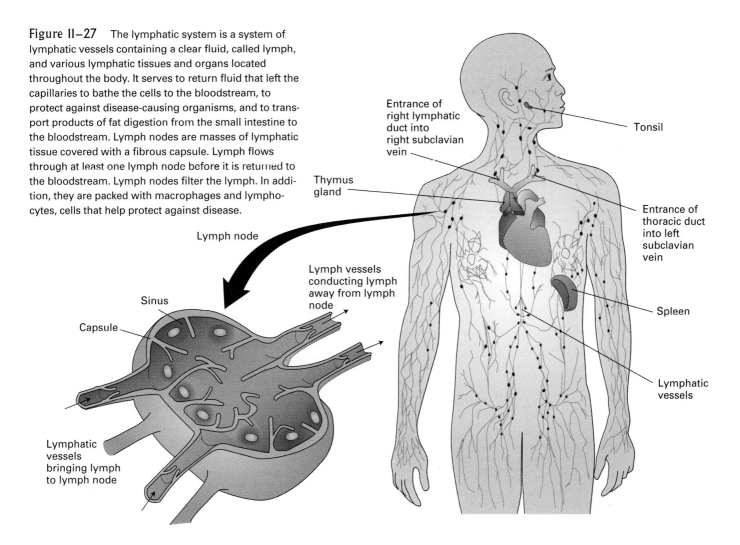

Figure 11–27 The lymphatic system is a system of lymphatic vessels containing a clear fluid, called lymph, and various lymphatic tissues and organs located throughout the body. It serves to return fluid that left the capillaries to bathe the cells to the bloodstream, to protect against disease-causing organisms, and to transport products of fat digestion from the small intestine to the bloodstream. Lymph nodes are masses of lymphatic tissue covered with a fibrous capsule. Lymph flows through at least one lymph node before it is returned to the bloodstream. Lymph nodes filter the lymph. In addition, they are packed with macrophages and lymphocytes, cells that help protect against disease.

Lymph node

Sinus

Capsule

Lymphatic vessels bringing lymph to lymph node

Lymph vessels conducting lymph away from lymph node

Entrance of right lymphatic duct into right subclavian vein

Thymus gland

Tonsil

Entrance of thoracic duct into left subclavian vein

Spleen

Lymphatic vessels

The functions of the lymphatic system are as diverse as they are essential to life:

1. **Return excess interstitial fluid to the bloodstream.** The lymphatic system maintains blood volume by returning excess interstitial fluid to the bloodstream. It turns out that only 85% to 90% of the fluid that leaves the blood capillaries and bathes the body tissues as interstitial fluid is reabsorbed by the capillaries. The rest of that fluid, which amounts to 2 or 3 liters a day, is absorbed by the lymphatic system and returned to the circulatory system. This is an important job. If the surplus interstitial fluid were not drained, it would cause the tissue to swell, the volume of blood would drop to potentially fatal levels, and the blood would become too viscous (thick) for the heart to pump.

A dramatic example of the importance of the lymphatic system returning fluid to the blood is provided by elephantiasis, a condition in which a parasitic worm blocks lymphatic vessels (Figure 11–28). The blockage can cause a substantial build-up of fluid in the affected body region followed by the growth of connective tissue. Elephantiasis is so named because it results in massive swelling and the darkening and thickening of the skin in the affected region, making the region resemble the skin of an elephant. The swelling can, indeed, be substantial. In one instance, the scrotum of a man grew to a weight of 18 kg (nearly 40 pounds) before it was surgically removed! Elephantiasis is a tropical disease, transmitted by mosquitoes, that affects over 400 million people.

2. **Transport products of fat digestion from the small intestine to the bloodstream.**
3. **Help defend against disease-causing organisms.** The lymphatic system produces lymphocytes, white blood cells that defend the body against specific disease-causing organisms or abnormal cells. In this way, it helps protect against disease and cancer.

Let's now consider the structures of the lymphatic system in more detail to shed some light on how it carries out such varied jobs. The structure of the lymphatic vessels is crucial to their ability to absorb the excess interstitial fluid. The extra fluid enters microscopic tubules, called **lymphatic capillaries**, which form a branching network that penetrates between the cells and the capillaries in almost every tissue of the body (except teeth, bones, bone marrow, and the central nervous system) (Figure 11–29). The lymphatic capillaries differ from the blood capillaries in two ways. First, they end blindly, like the fingers of a glove. In essence, they serve as drainage tubes, since fluid enters the "fingertips" easily and lymph moves through the system in one direction only. Second, they are much more permeable than blood capillaries, a feature that is key to their ability to absorb the digestive products of fats as well as excess interstitial fluid. The permeability is largely due to the unique structure of lymphatic capillaries. Whereas a layer called the basement membrane generally prevents proteins from crossing blood capillary walls, the membrane is either incomplete or lacking in the walls of lymphatic capillaries. Also, the endothelial cells that form the wall of lymph capillaries loosely overlap one another and each is anchored to surrounding tissue by fine filaments. As a result, the endothelial cells function as one-way valves that respond to the pressure of accumulating interstitial fluid. When fluid builds up in the tissue surrounding a capillary, the anchoring filaments pull on the edges of the endothelial cells, causing the flaps to separate and forming gaps in the wall of the lymphatic capillary. Interstitial fluid then enters the capillary. (After the interstitial fluid enters a

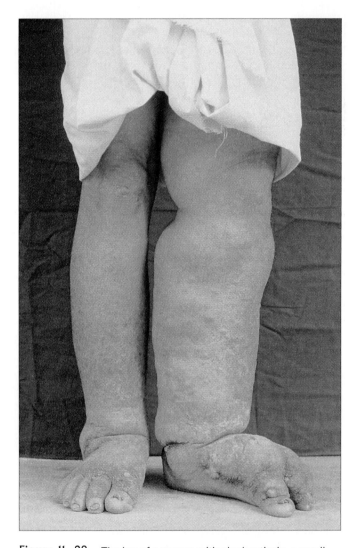

Figure 11–28 The leg of a person with elephantiasis, a condition in which a parasitic worm plugs lymphatic vessels and prevents the return of fluid from the tissues to the circulatory system. *(© R. Umesh Chandran, TDR, WHO/Science Photo Library/Photo Researchers, Inc.)*

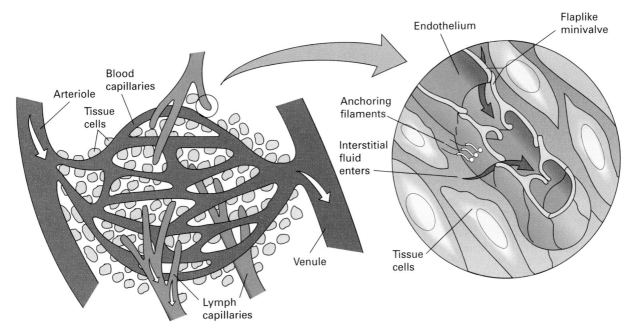

Figure 11-29   The lymphatic capillaries are microscopic, blind-ended tubules through which surplus interstitial fluid enters the lymphatic system to be returned to the bloodstream. They weave between the cells and the blood capillaries. The endothelial cells that form the walls of lymphatic capillaries function as one-way valves that open in response to pressure from accumulating tissue fluid.

lymphatic vessel, it is called lymph.) When the pressure of interstitial fluid drops below that of the lymphatic capillary, the flap is forced shut and the lymph cannot leave the capillary.

The lymphatic capillaries drain into larger lymphatic vessels. These become progressively larger tubes with thicker walls. Lymph is eventually returned to the circulatory system through one of two large ducts that join with the great veins at the base of the neck.

With no pump to drive it, lymph flows slowly through the lymphatic vessels, driven by the same forces that move blood through the veins. That is, the contractions of nearby skeletal muscles compress the lymphatic vessels, pushing the lymph along. Backflow is prevented by one-way valves similar to those in veins. Pressure changes in the thorax that accompany breathing also help pull the lymph upward from the lower body. Gravity assists the flow from the upper body.

The lymphatic vessels are studded with **lymph nodes**, which cleanse the lymph as it slowly filters through. The lymph nodes contain macrophages and lymphocytes, cells that play an essential role in the body's defense system. Macrophages engulf bacteria, cancer cells, and other debris, thus clearing them from the lymph. Lymphocytes serve as the surveillance squad of the immune system. They are continuously on the lookout for specific disease-causing invaders, as we will see in Chapter 12. Swollen lymph nodes indicate infection.

Besides the lymph nodes, there are several other **lymphoid organs**. Among these are the **tonsils**, which form a ring around the entrance to the throat where they help protect against disease organisms that are inhaled or swallowed. The **thymus gland**, located in the chest region, is another lymphoid organ. It plays its part during early childhood by processing T lymphocytes, thus enabling them to function against specific germs. On the left side of the abdominal region is the largest lymphoid organ—the **spleen**. In addition to serving as a birthplace for lymphocytes, it clears the blood of old and damaged red blood cells and platelets. Finally, isolated clusters of lymph nodules along the small intestine, known as **Peyer's patches**, keep bacteria from breaching the intestinal wall.

Critical Thinking

Cancer cells often break loose from their original site, a process called metastasis. The cancer cells have easy access to the highly permeable lymphatic capillaries. The lymphatic vessels then provide a route for the cancer cells to spread to nearly every part of the body. Explain why the lymph nodes are often studied to determine whether the cancer has spread. Why are the lymph nodes near the original cancer site often removed?

SUMMARY

1. Blood circulates through a branching network of blood vessels that forms a long loop. The path of blood through this loop is from the heart to arteries to arterioles to capillaries to venules to veins and then back to the heart.

2. Arteries are elastic, muscular tubes that withstand the high pressure of blood as it is pumped from the heart by stretching and then returning to their original shape. These changes help maintain a relatively even blood pressure within the arteries even with large changes in the volume of blood within them. The pressure change along an artery as it expands and recoils is called a pulse. Contraction of the smooth muscle in arterial walls, which is called vasoconstriction, narrows the space through which blood flows and reduces blood flow through the artery. Vasodilation occurs when the smooth muscle in arterial walls relaxes, increasing blood flow through the vessel.

3. Arteries branch to form narrower tubules called arterioles. Arterioles are important in the regulation of blood pressure and they regulate blood flow through capillary beds.

4. The exchange of materials between the blood and tissues takes place across the thin walls of capillaries. Capillaries are arranged in highly branched networks that provide a tremendous surface area for the exchange of materials. Each network of capillaries is called a capillary bed. A ring of muscle called a precapillary sphincter determines whether blood flows through a capillary bed or is shunted past it through a short vessel called a thoroughfare channel.

5. The most important way that materials are exchanged across capillary walls is by simple diffusion. The flow of water in or out of the capillary, depending on the balance of the force of blood pressure and the force of osmotic pressure, assists the movement of many substances. Other materials are actively moved across capillary cells in tiny vesicles.

6. Capillaries merge to form venules and these, in turn, merge to form veins, which conduct blood back to the heart. Because veins have thin walls and large lumens, they function as blood reservoirs. Blood is returned to the heart against gravity as nearby skeletal muscles contract and push blood along within the veins. Valves within the veins prevent blood from flowing backwards while the skeletal muscles are relaxing. Pressure differences generated by breathing also help to draw blood toward the heart from the lower torso.

7. Every minute, the heart beats about 72 times and moves about 10 pints of blood through its chambers.

8. The heart is enclosed in a double-layered sac called the pericardium, which allows the heart to beat while still confining it near the midline of the thoracic cavity. The heart wall has three layers: the pericardium, the myocardium (composed of cardiac muscle), and the endocardium (a thin inner lining).

9. The right and the left halves of the heart function as two separate pumps. Each side consists of two chambers—an upper chamber called the atrium and a lower, thick-walled chamber called the ventricle. The smaller, thin-walled atria function primarily as receiving chambers that accept the blood returning to the heart and pump it a short distance to the ventricles. When the ventricles contract, they push the blood through the arteries, toward the body cells.

10. Blood circulates in one direction through the body because of the action of two pairs of valves. The atrioventricular (AV) valves are located between each atrium and ventricle. The semilunar valves are located between each ventricle and its connecting artery. The valves function by responding passively to pressure differences on either side. The heart sounds, lubb dup, are caused by blood turbulence associated with the closing of the heart valves. Defective valves can produce abnormal heart sounds, called heart murmurs.

11. The right side of the heart pumps blood to the lungs through a loop of vessels called the pulmonary circuit. The left side of the heart pumps blood through a loop of vessels called the systemic circuit.

12. The heart has its own network of vessels, called the coronary circuit, that services the heart tissue itself.

13. Each heartbeat consists of contraction (systole) and relaxation (diastole). The atria contract together and then the ventricles contract in unison. The events associated with each heartbeat are collectively called the cardiac cycle.

14. Heart muscle cells contract without outside stimulation. Contraction spreads throughout the heart because cardiac muscle cells are connected by junctions called intercalated disks. A cluster of specialized cardiac muscle cells, called the sinoatrial (SA) node sets the tempo of the heartbeat and is, therefore, called the pacemaker. The electrical signal to contract begins in the SA node, which is located in the right atrium near the superior vena cava, and spreads over the atria, causing them to contract. When the electrical signal reaches another cluster of specialized muscle cells, called the atrioventricular (AV) node, the stimulus is quickly relayed along the atrioventricular bundle that runs through the wall between the two ventricles and then fans out into the ventricular walls through the Purkinje fibers.

15. The rate of heartbeat is regulated by the autonomic nervous system to suit the body's level of activity. Whereas the sympathetic branch of the autonomic nervous system speeds up heart rate during stressful times, the parasympathetic nervous system slows heart rate during restful times.

16. An electrocardiogram (ECG or EKG) is a recording of the electrical events associated with each heartbeat. Each heartbeat produces three distinguishable deflections in the ECG. The P wave is produced by the spread of the electrical signal over the atria and atrial contraction. Next is the QRS wave, which is produced by the spread of the electrical signal through the ventricles and by ventricular contraction. Finally, the T wave represents ventricular repolarization. An ECG can be used to detect many heart abnormalities.

17. Blood pressure, the force created by the heart to drive the blood around the body, is measured as the force of blood against the walls of blood vessels. Blood pressure in an artery peaks when the ventricles contract. This is called the systolic pressure. In contrast, the lowest blood pressure of each cardiac cycle, called the diastolic pressure, occurs due to the elastic recoil of arteries while the heart is relaxing between contractions. Blood pressure is measured with a device called a sphygmomanometer.

18. Hypertension (high blood pressure) is called a silent killer because it does not always produce symptoms and can have fatal consequences, involving the heart, brain, blood vessels, or kidneys. The heart enlarges, but it works less efficiently. High blood pressure damages blood vessels by promoting the forma-

tion of fatty deposits in arteries that block blood flow. Vessels may rupture due to high blood pressure. The kidneys can be damaged when the arteries leading to their filtering system are narrowed. Most cases of hypertension have no known cause.

19. Atherosclerosis is a narrowing of blood vessels due to thickening of the walls and a build-up of fatty deposits. The resulting decrease in blood supply to cells downstream can cause either heart attack or stroke.

20. When the vessel affected by atherosclerosis is a coronary artery that supplies blood to heart tissues, the condition is known as coronary artery disease (CAD). A temporary reduction in blood supplied to the heart (ischemia) is accompanied by chest pain, called angina pectoris. Angina is a warning of a heart problem, but it does not cause permanent damage to the heart. When the heart receives inadequate oxygen, neutrophils in the area release chemicals that can trigger ventricular fibrillation, which can quickly cause death.

21. In coronary angiography a dye that can be seen on x-ray films is released into coronary arteries, allowing physicians to see where a coronary artery is blocked. After the blockage is located, CAD can be treated with balloon angioplasty, atherectomy, or coronary bypass surgery.

22. A region of heart muscle dies during a heart attack (myocardial infarction). In a surviving patient, the dead cells are gradually replaced by scar tissue. The most common cause of a heart attack is a coronary thrombosis, in which a blood clot blocks a coronary artery. CAD makes a coronary thrombosis more likely. Another cause of a heart attack is a coronary artery spasm. Symptoms of a heart attack include severe chest pain and, often, nausea. Drugs that break up blood clots can often restore circulation to the heart muscle and save the person's life.

23. In heart failure the heart becomes an inefficient pump. Symptoms include shortness of breath, fatigue, and fluid accumulation in lungs and limbs. Heart failure is commonly treated with drugs that strengthen the heartbeat, reduce fluid retention, and widen blood vessels.

24. The lymphatic system consists of lymph, lymphatic vessels, and lymphoid tissue.

25. Three vital functions of the lymphatic system are to return interstitial fluid to the bloodstream, to transport products of fat digestion from the digestive system to the bloodstream, and to defend the body against disease-causing organisms or abnormal cells.

26. Tissue fluid enters lymphatic capillaries, microscopic tubules that end blindly and are more permeable than blood capillaries. The fluid, then called lymph, is moved along larger lymphatic vessels by contraction of nearby skeletal muscles. The lymphatic vessels have valves to prevent the backflow of lymph.

27. Lymph nodes filter lymph and contain cells that actively defend against disease-causing organisms.

28. Lymphoid organs include the lymph nodes, tonsils, thymus gland, spleen, and Peyer's patches.

REVIEW QUESTIONS

1. Trace the flow of blood from the heart and back by naming, in order, the general types of vessels through which the blood flows.

2. What is a pulse?

3. Explain how vasoconstriction and vasodilation regulate blood flow through arteries.

4. What are two important functions of arterioles?

5. What determines whether blood flows through any particular capillary bed?

6. List three mechanisms responsible for the exchange of materials across capillary walls.

7. Compare the structure of arteries, capillaries, and veins and explain how the structure is suited to the function of each type of vessel.

8. Explain how blood is returned to the heart from the lower torso against the force of gravity.

9. Describe the structure of the heart and explain how it functions as two separate pumps.

10. Describe the structure of the heart valves and explain how they function.

11. Trace the path of blood from the left ventricle to the left atrium, naming each major vessel associated with the heart and the heart chambers in the correct sequence.

12. Describe the cardiac cycle.

13. Explain how clusters or bundles of specialized cardiac muscle cells coordinate the contraction associated with each heartbeat.

14. List the three important factors that determine blood pressure.

15. What is hypertension? In what ways does it damage the body, sometimes causing death?

16. What is atherosclerosis? Why is it harmful? How does it develop?

17. Explain the relationships among coronary artery disease, angina pectoris, and heart attack.

18. What is a heart attack? What are the common causes?

19. List three important functions of the lymphatic system.

20. Compare the structure of lymphatic capillaries with those of the circulatory system. How do these structural differences allow lymphatic capillaries to absorb tissue fluid?

21. What is the function of lymph nodes?

Chapter | 12

Body Defense Mechanisms

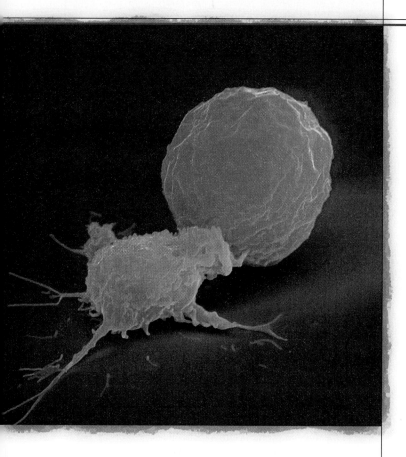

This natural killer cell has made contact with a cancer cell (shown in orange) and is beginning to destroy it. *(Meckes/Ottawa/Photo Researchers, Inc.)*

On a good day, the world can seem a magnificent place, indeed. You look around and see beauty and gentleness at every turn. On such a day, you might find a place and lie in the grass as a cool breeze wafts across your face. At these times, you might not like to think of the fact that you are lying there in a sea of bacteria, viruses, protozoa, fungi, and parasitic worms. Still, they are in the air, water, soil, and on surfaces all around us. Some even live in us or on us (Figure 12–1). Most of the microbes are harmless. Some are beneficial. But, others can cause infections or diseases that can even kill us. Generally, though, we can withstand the threats of disease-causing organisms, because over the course of time we have developed defenses against them. Before considering those defense mechanisms, however, we will think about some of the possible threats to our well-being.

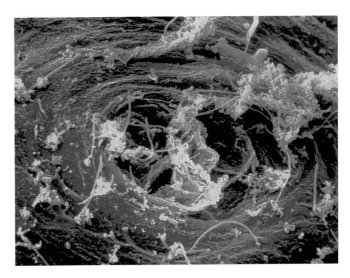

Figure 12–1   Millions of bacteria live on your skin surface. Even after a thorough shower, there may be as many as 20 million bacteria per square inch of skin. Most of these bacteria are harmless. Because they are so well-entrenched, they can often outcompete dangerous microbes and, in this way, help defend you from infection.   *(© David Phillips/Science Source/Photo Researchers, Inc.)*

Targets of the Body's Defense System

Your body generally defends you against anything that is not recognized as belonging in your body. Common targets of the defense system include organisms that cause disease or infection, as well as body cells that have turned cancerous.

Pathogens

There are a variety of **pathogens** (disease-causing organisms), each with specific effects, and some are more familiar than others.

Bacteria

Bacterial cells differ from those that compose our bodies. Bacteria are prokaryotes, which means they lack a nucleus and other membrane-bound organelles. Nearly all bacteria have a cell wall composed of a strong mesh of peptidoglycan, which consists of sugars and amino acids, that makes the cell wall semirigid. The cell wall maintains most types of bacteria in one of three common shapes: a sphere (coccus), which can occur singly, in pairs, or in chains; a rod (bacillus), which usually occurs singly; or a helix or spiral (spirillum).

Some bacteria have long, whip-like structures called flagella that allow them to move. Bacteria may also have filaments called pili that help them attach to the cells they are attacking, including those of the human body. Outside the bacterial cell, there is often a capsule that provides protection as well as a means of adhering to a surface.

Bacteria reproduce asexually in a type of cell division called **binary fission**, in which the bacterial genetic material (DNA) is copied and the cell pinches in half with each new cell containing a complete copy of the original genetic material. Under ideal conditions, this fission can occur every 20 minutes. Thus, if every descendent lived, a single bacterium could result in a massive infection of trillions of bacteria within 24 hours.

Bacteria can cause disease in a number of ways, but most do their damage by releasing toxins (poisons) into the bloodstream or the surrounding tissues. If the toxins enter the bloodstream, they can be carried throughout the body and disturb body functions. For instance, the toxin produced by the bacterium that causes botulism, a type of food poisoning caused by eating improperly canned food, interferes with nerve functioning.

Fortunately, bacteria can be killed. The human body has its own array of defenses, and when the body needs help we can call on antibiotics. Some antibiotics work by killing bacteria directly by causing them to burst, and others work by slowing their growth. Bacterial growth can be slowed by preventing the synthesis of their cell walls or by blocking protein synthesis. Antibiotics are able to prevent bacterial protein synthesis without interfering with host cell protein synthesis by taking advantage of certain differences in the process between bacterial and host cells.

Viruses

Viruses, which are much smaller than bacteria, are responsible for many human illnesses. Some of these, such as the

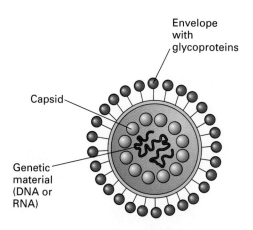

Figure 12–2 The structure of a typical virus—a protein coat, called a capsid, surrounds a core of genetic information, with DNA or RNA. Some viruses have an outer membranous layer, called the envelope, from which glycoproteins project.

Labels: Envelope with glycoproteins; Capsid; Genetic material (DNA or RNA)

common cold, the flu, chickenpox, measles, and mumps, are usually not very serious. Others, such as yellow fever, can be deadly.

A virus is not considered to be a living organism because, on its own, it cannot perform any life processes. A virus requires another organism, called the host, in order to copy itself. It exploits the host's nutrients and metabolic machinery to make copies of itself that then go on to infect other host cells.

A virus consists of a strand or strands of genetic material, either DNA or RNA, surrounded by a coat of protein, called a capsid (Figure 12–2). The genetic material carries the instructions for making new viral proteins. These proteins may actually become part of the new viruses, they may serve as enzymes that help carry out biochemical functions important to the virus, or they may be regulatory. Some regulatory proteins trigger the specific viral genes that will be active under a certain set of conditions, whereas other regulatory proteins may convert the host cell into a virus-producing factory.

Some viruses have an envelope, an outer membranous layer that is studded with glycoproteins. In some viruses, the envelope is actually a bit of plasma membrane from the previous host cell that became wrapped around the virus as it left the host cell. The envelope of certain other viruses, those in the herpes family for instance, is from a previous host cell's nuclear membrane. In any case, the glycoproteins on the envelope are produced by the virus.

A virus can replicate (make copies of itself) only when its genetic material is inside a host cell. It gains entry by binding to a receptor (a protein, or other molecule of a certain configuration) on the host cell surface. Such binding is possible because the viral surface has molecules of a specific shape that fit into the host's receptors.

Viruses generally attack only certain kinds of cells in certain species because a particular virus can only infect cells that bear a certain kind of receptor. Whereas the virus that causes the common cold infects only cells in the respiratory system, the virus that causes hepatitis infects only liver cells.

After the virus has bound to a receptor, the entire virus may enter the host cell where it is stripped of its capsid, leaving only its genetic material, or the virus may simply inject its genetic material into the host cell, leaving the capsid outside. The virus may then follow one of two courses. In one, new viruses are formed and released very quickly. As viruses are released, the host cell bursts (lyses) and dies. In the other course, the virus inserts its genetic material into a chromosome of the host cell. There it can stay for years, either in a completely inactive state or slowly producing new viruses. In the latter case, the viral DNA can leave the host chromosome at any time and begin actively replicating.

Viruses can cause disease in several ways. Some cause disease when they kill the host's cells or cause the cells to malfunction. In such cases, the nature of the disease will depend on which cells are infected. Other viruses cause cancer when they insert themselves into the host chromosome near a cancer-causing gene and, in doing so, activate that gene. Still other viruses bring cancer-causing genes with them into the host cell.

Unfortunately, viruses are not as easy to destroy as bacteria. Nonetheless, some antiviral drugs are now available to slow viral growth and others are being developed.

Protozoa

Protozoa are single-celled organisms with a well-defined nucleus similar to ours. The protozoa can cause disease by producing toxins or by releasing enzymes that prevent host cells from functioning normally. Protozoa are responsible for many diseases, including malaria, sleeping sickness, amebic dysentery, and giardiasis. Giardiasis is a diarrheal disease that can last for weeks. There are frequent outbreaks of giardiasis in the United States, most of them resulting from contaminated water supplies. Fortunately, drugs are available to treat protozoan infections.

Fungi

Like the protozoans, fungi are organisms with a well-defined nucleus in their cells. Some fungi exist as single cells and others are organized into simple multicellular forms, with not much difference among the cells comprising them. There are over 100,000 species of fungi, but less than 0.1% of them cause human ailments. Some fungi cause serious lung infections, such as histoplasmosis and coccidioidomycosis. Others, though, are less threatening, such as the one that causes athlete's foot, which grows on the skin and secretes enzymes that digest cells. Most fungal infections can be cured.

Parasitic Worms

The parasitic worms are multicellular animals. They include flukes, tapeworms, and roundworms, such as hookworm and pinworm. They can cause illness by releasing toxins into the bloodstream, feeding off blood, or competing for food with the host. Parasitic worms cause many human diseases, including schistosomiasis and trichinosis.

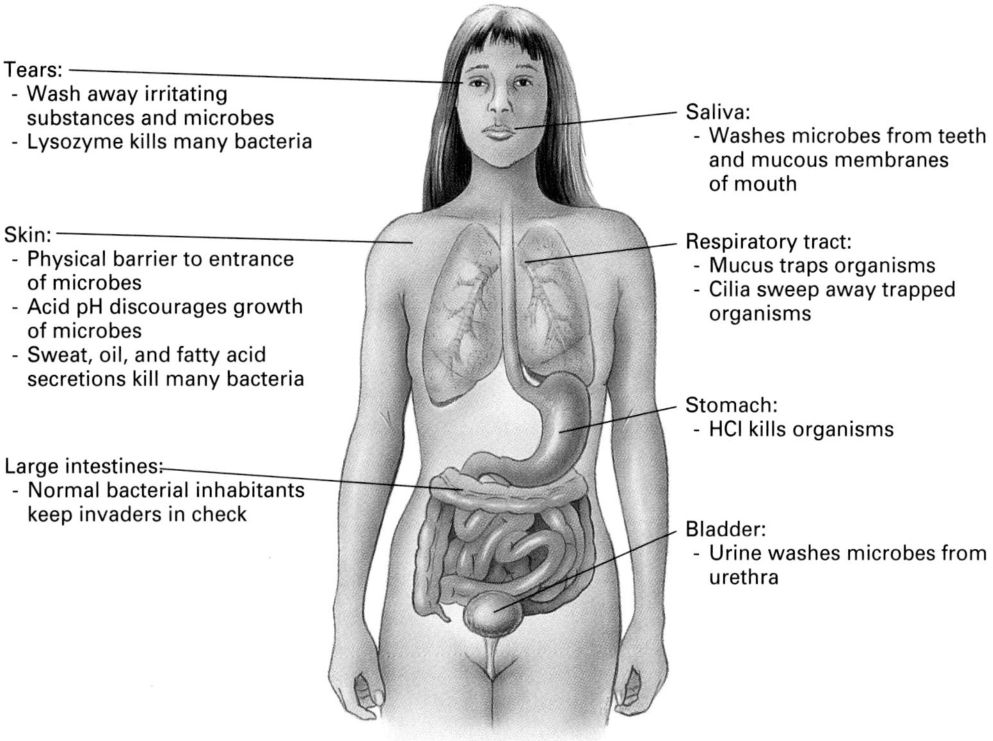

Figure 12-3 Non-specific defenses protect against *any* threats to our well-being. Collectively they prevent the threat from entering the body, confine it to a local region, kill it, remove it, or slow its growth.

Tears:
- Wash away irritating substances and microbes
- Lysozyme kills many bacteria

Saliva:
- Washes microbes from teeth and mucous membranes of mouth

Skin:
- Physical barrier to entrance of microbes
- Acid pH discourages growth of microbes
- Sweat, oil, and fatty acid secretions kill many bacteria

Respiratory tract:
- Mucus traps organisms
- Cilia sweep away trapped organisms

Stomach:
- HCl kills organisms

Large intestines:
- Normal bacterial inhabitants keep invaders in check

Bladder:
- Urine washes microbes from urethra

Body Covering	Chemical	Cellular
Skin	Histamine (increases permeability of capillaries and attracts phagocytes to area of injury)	Phagocytic cells such as neutrophils, macrophages and eosinophils (engulf bacteria)
Mucous membranes	Kinins (increase permeability of capillaries and attract phagocytes to area of injury)	Natural killer cells (kill many microbes and certain cancer cells)
Toxic secretions	Complement (stimulates histamine release; promotes phagocytosis; kills bacteria)	
Normal bacteria inhabitants	Interferon (protects uninfected cells from viral infection)	

Although parasitic worms are not likely to be the first cause of disease to come to mind, they have plagued humans for millennia. The Bible tells of "fiery serpents" that afflicted the Egyptians and later the Israelites as they approached the Red Sea. Most likely, these serpents were roundworms called *Dracunculus medinensis*. Indeed, at that time, the main undertaking of surgeons in the Middle East was to remove these parasites, which was accomplished by winding them on a stick.

Cancerous Cells

Cancerous cells also threaten our well-being. A cancer cell was once a normal body cell, but because of changes in its genes it can no longer regulate its cell division. These renegade cells can then multiply until they take over the body, upsetting its balance, choking its pathways, and ultimately causing great pain.

The Body's Defense Mechanisms

The body has two types of defense mechanisms. Those that are nonspecific are effective against *any* foreign organisms or substances and those that are specific act on a certain target.

Nonspecific Defenses

Nonspecific defenses provide barriers that help keep foreign substances from entering the body and limit the spread of the organisms that manage to breach the barriers (Figure 12-3).

Surface Barriers

Like a suit of armor, unbroken skin shields us from the hostile world by providing an effective barrier to foreign substances. A layer of dead cells forms the tough, horny outer layer of the skin. These cells are filled with the fibrous protein, keratin, that makes the skin nearly impenetrable, waterproof, and resistant to the disruptive toxins and enzymes of would-be invaders. Some of the strength of this barrier results from the tight connections binding the cells together. Furthermore, the dead cells are continuously shed and are replaced at the rate of about a million every 40 minutes. As dead cells flake off, they take with them any microbes that have somehow managed to latch on.

The skin also provides some chemical protection against invaders. The secretions of the sweat and sebaceous glands wash away microbes and their acidity slows bacterial growth. In addition, the sebaceous glands produce chemicals that kill bacteria.

The inner surfaces of the body are guarded by mucous membranes, which provide both physical and chemical barriers to entry. The linings of the digestive and respiratory passages produce sticky mucus that traps many microbes. The lining of the stomach produces hydrochloric acid and protein-digesting enzymes that destroy pathogens. The secretions of the vaginal lining are also acidic. Furthermore, the acidity of urine slows bacterial growth and urine periodically flushes microbes from the lower urinary tract. Saliva and tears contain an enzyme called **lysozyme** that kills bacteria by disrupting their cell walls.

Phagocytic Cells

Phagocytes (*phage*, to eat; *cyte*, cell) are scavenger cells specialized to engulf and destroy particulate matter, such as pathogens, damaged tissue, or dead cells. In this way, phagocytes serve as the front-line soldiers in the body's defense system and also as janitors that clean up debris on the battlefield. During the process, cytoplasmic extensions flow from the phagocytic cell, binding to the particle and pulling it inside, enclosed within a membrane-bound vesicle (Figure 12–4). The vesicle fuses with a lysosome, an organelle that contains digestive enzymes. The pathogen is then quickly killed by digestive enzymes and lethal free-radical oxidants.

There are several types of phagocytes. The type of white blood cell called **neutrophils** attacks primarily bacteria. Other white blood cells, **eosinophils**, attack parasitic worms and foreign proteins. **Macrophages** (*macro*, big; *phage*, to eat) have hearty and less discriminating appetites and will attack virtually anything that is not recognized as belonging in the body—including pathogens and damaged tissue.

Natural Killer Cells

Natural killer (NK) cells are lymphocytes that roam the body in search of abnormal cells, which are then quickly killed. In a sense, then, NK cells function as the body's po-

lice walking a beat. They are not seeking a specific villain. Instead, they respond to *any* suspicious character, which, in this case, is a cell whose cell membrane has been altered by the addition of proteins that are unfamiliar to the NK cell. The prime targets of NK cells are cancerous cells and cells infected with viruses. It is thought that cancerous cells routinely form, but their quick destruction by NK cells prevents them from spreading (Figure 12–5).

When an NK cell touches a cell with an abnormal surface, it immediately attaches to its target and delivers a "kiss of death," which comes in the form of proteins called **perforins.** The perforins create numerous pores in the target cell, making it leaky. Fluid is then drawn into the leaky cell due to the high salt concentration within and the cell bursts.

Defensive Proteins

A cell that has been infected with a virus can do little to help itself, but infected cells can do something to help as yet uninfected cells. Before a virally infected lymphocyte, macrophage, or one of certain other types of body cells dies, it secretes small proteins called **interferons**, which act to slow the spread of viruses already in the body. There are actually several types of interferons, each secreted by a differ-

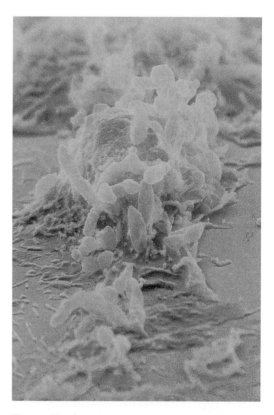

Figure 12–4 A macrophage ingesting bacteria (the rod-shaped structures). The bacteria will be pulled inside the cell within a membrane-bound vesicle and quickly killed. *(© David M. Phillips/Photo Researchers, Inc.)*

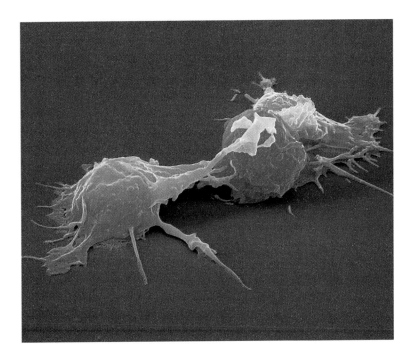

Figure 12–5 Two natural killer (NK) cells (shown in yellow-orange) attacking a leukemia cell (shown in red). NK cells patrol the body, bumping and touching other cells as they do. When they contact a cell with an altered cell surface, such as a cancer cell or a virus-infected cell, a series of events is immediately initiated. The NK cell attaches to the target cell and releases proteins called perforins that create pores in the target cell, making the membrane leaky and causing the cell to burst.
(© Meckes/Ottawa/Photo Researchers, Inc.)

ent type of cell. Interferons mount a two-pronged attack. By attracting macrophages and natural killer cells, interferons attack the source of the problem—virus-infected cells, which serve as reservoirs for the virus or the source of cancerous cells. Macrophages and natural killer cells then destroy the infected or cancerous cells immediately. In addition, interferons protect cells that are not yet infected with the virus. As the name implies, interferons interfere with viral activity. When released, an interferon diffuses to neighboring cells and stimulates them to produce proteins that prevent viruses from making copies of themselves in those cells. Since viruses cause disease by replicating inside body cells, preventing replication curbs the disease.

Interferon helps protect uninfected cells from *all* strains of virus, not just the one responsible for the initial infection. Consider the possible medical uses of a chemical with this ability. Indeed, in 1957, when interferons and their broad-spectrum antiviral activity were first discovered, researchers hoped that they could be used to treat viral diseases and even to cure cancer. During the 1980s, genetic-engineering techniques made it possible to produce quantities of interferon large enough to test the idea. Although it hasn't turned out to be the wonder drug people had hoped for, it has been shown to be effective against certain cancers and viral infections. For instance, interferon is usually successful in combating a rare form of leukemia (hairy cell leukemia) and Kaposi's sarcoma, a form of cancer that often occurs in people living with AIDS. Interferon has also been approved for treating the hepatitis C virus, which invades and destroys liver cells and can lead to cirrhosis and liver cancer; the human papilloma virus that causes genital warts; and the herpes virus that causes genital herpes. One form of interferon

slows the progression of multiple sclerosis, a debilitating disease of the nervous system. Also, tests are now underway to determine whether interferon can slow the progression of an HIV infection before AIDS develops.

The **complement system**, or simply **complement**, is a group of at least 20 proteins whose activities complement, or enhance, the other body defense mechanisms. Complement can act *directly*, by punching holes in the target cell's membrane (Figure 12–6). Fluid from the target's surroundings then floods the cell, causing it to swell and burst, just as we saw when the NK cells secreted proteins that made the target cell's membrane leaky. In addition, complement can act *indirectly*, by working in concert with other defense mechanisms, including phagocytosis and inflammation. For instance, complement enhances phagocytosis in two ways. First, complement proteins attract macrophages and neutrophils to the site to remove the foreign cells. Second, one of the complement proteins binds to the surface of the microbe, making it easier for macrophages and neutrophils to "get a grip" on the intruder and devour it. Activated complement proteins also promote inflammation by causing blood vessels to become wider and more permeable.

Inflammation

Inflammation is an important aspect of the body's nonspecific responses to injury or invasion by foreign organisms. We are all familiar with the four cardinal signs of inflammation that occur at the site of a wound—redness, swelling, heat, and pain. These signs announce that certain cells and chemicals have combined efforts to contain infection, clean up the battlefield, and heal the wound. Let's consider the

Figure 12–6 The proteins of the complement system can form membrane attack complexes that punch holes in the membrane of the target cell. The resulting inflow of fluid from the surrounding environment causes the cell to swell and burst (lyse). *(Photo, Podack, E.R. and Dennert, G. 1983, Nature 307: 442. Used with permission.)*

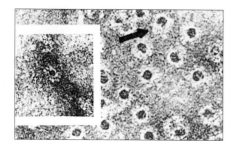

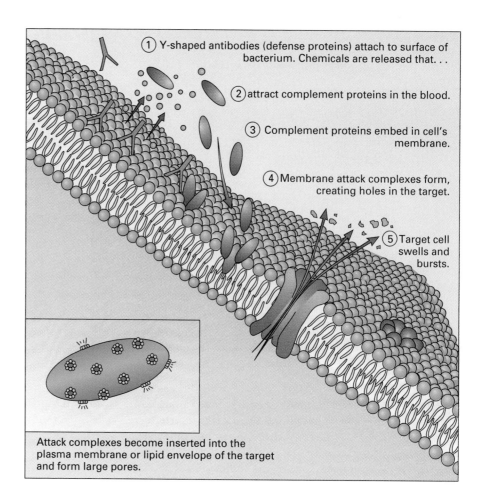

① Y-shaped antibodies (defense proteins) attach to surface of bacterium. Chemicals are released that. . .

② attract complement proteins in the blood.

③ Complement proteins embed in cell's membrane.

④ Membrane attack complexes form, creating holes in the target.

⑤ Target cell swells and bursts.

Attack complexes become inserted into the plasma membrane or lipid envelope of the target and form large pores.

causes of the cardinal signs and how they relate to the benefits of inflammation (Figure 12–7).

The redness and heat of an inflamed area are due to an increased blood flow to the region that occurs because the blood vessels dilate (become wider). This occurs because, immediately after the injury, **mast cells**, which are small, mobile connective tissue cells, and **basophils**, which are a type of white blood cell, release histamine. The increase in blood flow to the site of injury delivers phagocytes, blood-clotting proteins, and defensive proteins, including complement and antibodies. At the same time, the increased blood flow washes away dead cells and toxins produced by the invading microbes. The elevated temperature in the area of injury is beneficial because it increases the metabolic rate of the body cells in the region, thus speeding up healing, and because it increases the activities of phagocytic cells and other defensive cells.

The injured area swells because histamine also makes capillaries more permeable (leaky) than usual. Fluid seeps into the tissues from the bloodstream, bringing with it many beneficial substances. For instance, blood-clotting factors enter the injured area and begin to wall off the region, which helps to protect surrounding areas from injury and prevent

the loss of blood. The seepage also increases the oxygen and nutrient supply to the cells. In addition, swelling can help healing directly. If the injured area is a joint, swelling can hamper movement, which might seem to be an inconvenience, but it permits the injured joint to rest and recover.

There are several causes for the pain in an inflamed area. For example, the excessive fluid that has leaked into the tissue presses on cells and contributes to the sensation; some soreness can be caused by bacterial toxins; and injured cells also release pain-causing chemicals, such as prostaglandins.

Soon after the inflammatory response begins, phagocytes swarm to the injured site, attracted by chemicals released when tissue is damaged. First on the scene are the neutrophils. Other chemicals released in the injured area cause the neutrophils to stick to the walls of the blood vessels. Within minutes, the neutrophils squeeze through the walls of the vessels and begin engulfing pathogens, toxins, and dead body cells. A neutrophil can engulf about 25 bacteria before being killed by the poisons produced by the engulfed bacteria and by its own powerful chemicals produced to destroy those bacteria.

Not long after the neutrophils begin the battle, reinforcements arrive—additional neutrophils and monocytes,

Figure 12–7 The inflammatory response is a general response to tissue injury or the invasion of foreign microbes. It serves to defend against pathogens and to clear the injured area of pathogens and dead body cells, allowing repair and healing to occur. The four cardinal signs of inflammation are redness, warmth, swelling, and pain.

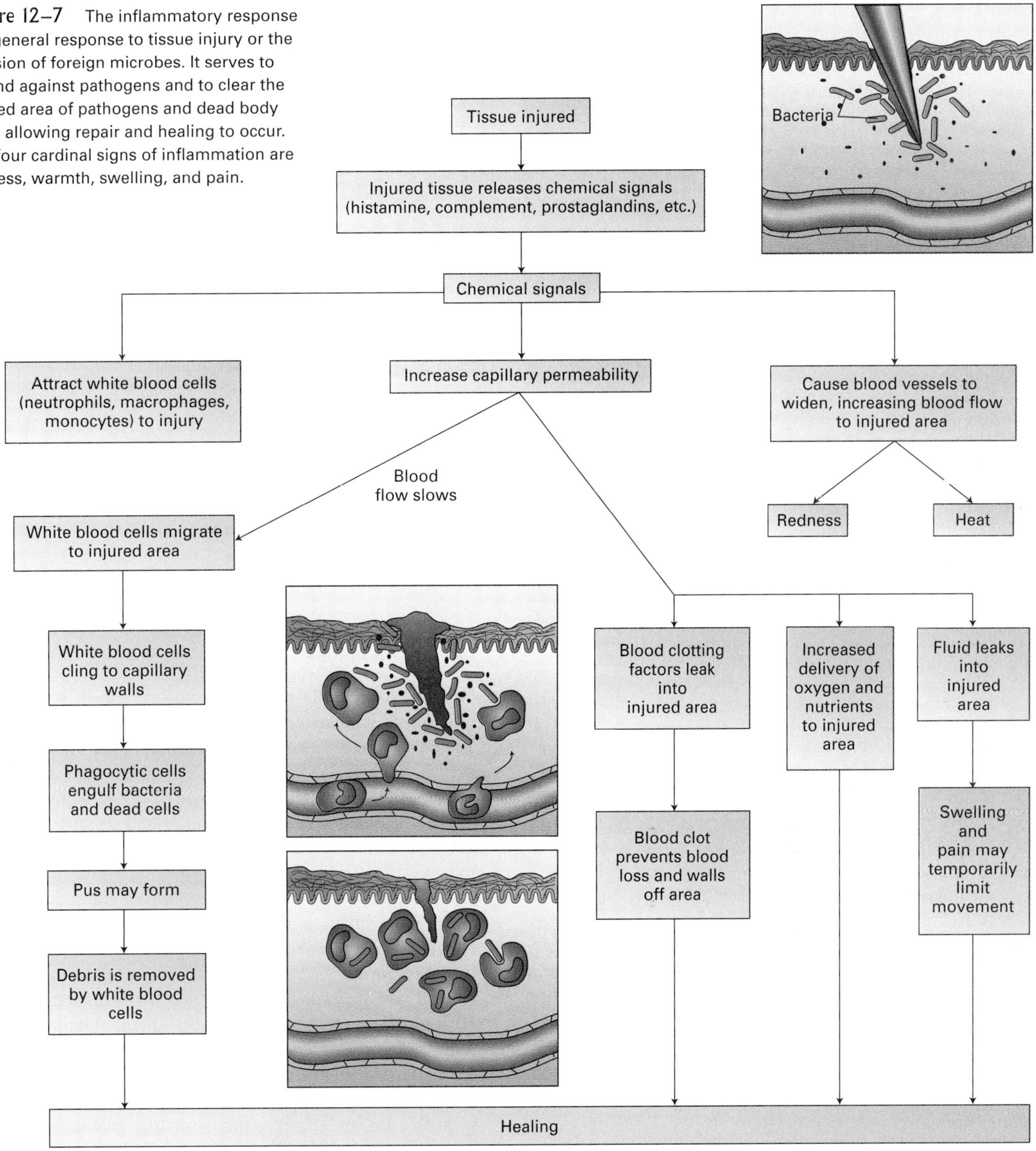

```
                    Tissue injured
                         │
                         ▼
        Injured tissue releases chemical signals
     (histamine, complement, prostaglandins, etc.)
                         │
                         ▼
                  Chemical signals
```

- Attract white blood cells (neutrophils, macrophages, monocytes) to injury
- Increase capillary permeability
- Cause blood vessels to widen, increasing blood flow to injured area
  - Redness
  - Heat

Blood flow slows

- White blood cells migrate to injured area
- White blood cells cling to capillary walls
- Phagocytic cells engulf bacteria and dead cells
- Pus may form
- Debris is removed by white blood cells

- Blood clotting factors leak into injured area
  - Blood clot prevents blood loss and walls off area
- Increased delivery of oxygen and nutrients to injured area
- Fluid leaks into injured area
  - Swelling and pain may temporarily limit movement

Healing

---

another type of white blood cell. Within 8 to 12 hours, the monocytes transform into macrophages and these continue the body's counterattack over the long haul. Macrophages are also important in cleaning debris, such as dead body cells, from the battlefield. There are many casualties in this battle. As the infection continues, dead cells, including mi-crobes, body tissue cells, and phagocytes, may begin to ooze from the wound as **pus**.

Fever

A fever is an elevated body temperature. A high fever is dan-gerous because it can inactivate enzymes. However, a mild

Figure 12–8 Although it might make us feel uncomfortable, a fever can help the body fight disease-causing organisms. *(Stephen Goodenough)*

or moderate fever helps the body fight disease-causing invaders in a number of ways (Figure 12–8). The benefits of a higher body temperature include:

1. Phagocytes attack pathogens with increased speed and vigor.
2. Interferons work more efficiently.
3. The production of T cells, white blood cells involved in defense against specific intruders (discussed shortly), is stepped up.
4. Pathogens don't survive or reproduce as well. Each type of organism has evolved to survive best within a certain temperature range, and fever apparently raises body temperature above the optimal point for most pathogens.
5. The liver and spleen remove iron from the bloodstream, triggered by the same chemicals that cause fever. Many disease-causing bacteria require iron to reproduce.

Body temperature is regulated by a brain region known as the hypothalamus. Within the hypothalamus is a group of cells that act as the body's thermostat; that is, they determine the "set point," (the body temperature that will be maintained) and initiate responses that will help return body temperature to that set point if it should deviate.

Pyrogens are chemicals that set the thermostat to a higher point. Toxins released by pathogens sometimes act as pyrogens. It is interesting to note, however, that the body produces its own pyrogens, as part of its defensive strategy. The main source of these pyrogens is macrophages that have been exposed to bacteria or other foreign substances. The pyrogens are carried by the bloodstream to the brain, where they stimulate certain neurons in the hypothalamus to secrete prostaglandins. The prostaglandins, in turn, raise the set point of the hypothalamus. Physiological responses, such as shivering, are then initiated to raise body temperature. Thus, we have the "chills" while the fever is rising. When the fever breaks, the set point is lowered and physiological responses, such as perspiring, lower the body temperature until it reaches the new set point.

Critical Thinking

Aspirin inhibits the synthesis and release of prostaglandins. Explain how this contributes to aspirin's pain-killing effect. Why might it be wise to avoid taking aspirin to reduce a fever?

Specific Defenses—The Immune System

You may have noticed that after you recover from a certain disease, say chickenpox, you are not likely to get it again. You might get measles or the flu afterward, but you won't get chickenpox. We say that you are *immune* to chickenpox.

From this simple observation, you can deduce several important characteristics of the immune system. First, it is *specific* for one particular invader, in this case for the virus that causes chickenpox. The job of the immune system is to recognize an intruder as not belonging in the body and then act to immobilize, neutralize, or destroy it. You know how effective the immune system is because you recovered from the chickenpox. Second, the immune system has *memory*. Even if you are exposed to the virus that causes chickenpox 20 years after you succumbed to the illness, the immune system will remember the virus and attack it so quickly and vigorously that you won't get ill again.

Certain white blood cells, called lymphocytes, are responsible for both the specificity and the memory of the immune response. There are two principal types of lymphocytes. They are the **B lymphocytes**, or more simply **B cells**, and the **T lymphocytes**, or **T cells**. Both types form in the bone marrow, but they mature in different organs of the body. It is thought that B cells mature in the bone marrow. (The B stands for bursa of Fabricius, a pouch of lymphatic tissue in a chicken's digestive system where B cells were first identified.) The T cells, on the other hand, mature in the thymus gland, a butterfly-shaped gland that overlies the heart. (The T stands for thymus gland.)

Specificity

Substances that trigger an immune response are called **antigens**. Antigens provoke an immune response because they are not recognized as belonging in the body. The response is then directed against the specific antigen that provoked it. Typically, antigens are large molecules, such as proteins, polysaccharides, or nucleic acids. Antigens are often found on the surface of an invader, embedded in the plasma membrane of an invading bacterial cell or part of the protein coat of a virus, for instance. However, pieces of invaders and chemicals secreted by invaders, such as bacterial toxins, can also serve as antigens. Each antigen is recognized by its characteristic shape.

As lymphocytes mature, they are programmed to recognize only one particular antigen and this recognition is the basis of the specificity of the immune response. Each lymphocyte develops receptor molecules with a unique shape. Indeed, thousands of *identical* receptor molecules pepper the surface of each lymphocyte. When an antigen fits into a lymphocyte's receptors, the body's defenses target that particular antigen. Because of the tremendous diversity of receptor molecules, each type on a different lymphocyte, a few of the billions of lymphocytes in your body are programmed to respond to each of the thousands of different antigens that you will be exposed to in your lifetime.

Besides being tailored to attack a specific antigen during their maturation in the thymus gland, T cells also learn to distinguish between cells that belong in the body and those that do not. Each cell in your body has special molecules embedded in the plasma membrane that label the cell as "self." These molecules then serve as flags declaring the cell as a "friend." Any substance or organism that lacks this label is considered to be "nonself" or "foe." The immune system uses these labels to distinguish between what is part of your body and what is not (Figure 12–9).

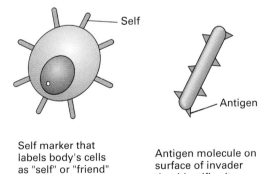

Self marker that labels body's cells as "self" or "friend"

Antigen molecule on surface of invader that identifies it as "nonself" or "foe"

Figure 12–9   All the cells in your body have molecular markers on their surface that label them as "self." Foreign substances, including potential disease-causing organisms, have molecules on their surface that are not recognized as belonging in the body. Foreign molecules that are capable of triggering an immune response are called antigens.

Memory

When a foreign substance is detected, B cells and T cells bearing receptors specific for that particular invader are stimulated to divide repeatedly, forming two lines of cells. One line of descendent cells is made up of **effector cells** that are responsible for the attack on the enemy. Effector cells generally live for only a few days. Thus, after the invader has been eliminated from the body, the number of effector cells declines. The other line of descendent cells is composed of **memory cells**, long-lived cells that "remember" that particular invader and mount a rapid, intense response to it if it should ever appear again. The quick response of memory cells is the mechanism that prevents you from getting the same illness twice.

Overview of the Immune Response

There are some striking similarities between the body's immune defenses and a nation's military defense system. The body has scouts, called macrophages, that roam the body looking for an invader. If an invader is found, the scout will alert the commander-in-chief of the military forces and provide an exact description of the villain. The scout must, of course, provide the appropriate password so that the commander knows it is not a spy providing misinformation. The immune system's commander-in-chief is a subset of T cells, called helper T cells. When properly alerted, helper T cells call out the body's military forces.

A nation's military often has various branches, the Army and the Navy, for instance. Primed to respond to slightly different forms of enemy invasion, each branch is armed with certain types of weapons. When activated, the military is usually looking for a specific threat, say little green people with purple fur. The Navy may be called into action if the enemy is encountered at sea, whereas the Army will come to the defense if the enemy is on land.

The body also has two types of specific defense. **Antibody-mediated immune responses** defend primarily against enemies that are free in body fluids, including toxins or extracellular pathogens such as bacteria or free viruses. The warriors of this branch of immune defense are the B cells and their weapons are antibodies. When called into action a B cell transforms into a type of effector cell called a plasma cell, whose job it is to secrete into the body fluids many copies of a specific antibody tailored to recognize and bind to the antigen posing the threat. As we will see shortly, antibodies help eliminate the antigen from the body. **Cell-mediated immune responses** involve living cells and protect against cellular threats, including body cells that have become infected with viruses or other pathogens and cancer cells. The soldiers responsible for cell-mediated immune responses are T cells. Once activated, T cells quickly destroy the infected or cancerous cells.

Fetal Tissue Transplants

A major problem with organ transplants is rejection of the transplanted tissue. Rejection occurs when the immune system recognizes the molecular markers on the surface of the transplanted cells as foreign. T cells then attack and kill the transplanted cells.

Cells from a 6- to 8-week-old fetus have not yet formed their "self" labels nor have they become fully differentiated to carry out the jobs they will do when they mature. Because they lack self markers, fetal cells can be injected into other humans without fear of rejection. In their new home, the cells can take on their predestined duties and so benefit the recipient.

In experiments using mice, rats, and monkeys, researchers have discovered that fetal cells can cure a wide array of conditions. Transplanted insulin-producing cells have cured diabetes. Transplanted neurons have allowed rats with certain spinal cord injuries to run again. Learning and memory were improved when fetal nerve cells were transplanted to the brain, offering hope that someday fetal cells may be used to cure Alzheimer's disease. Sight has been restored by transplanted retinal cells.

Fetal tissue transplants may also cure certain human ailments. For instance, fetal cell transplants have helped some people with Parkinson's disease. In Parkinson's disease a small region of the brain called the substantia nigra is slowly destroyed. The cells of the substantia nigra normally produce a crucial chemical, called dopamine. The dopamine is supplied to the striatum, a brain region important in controlling movements. Without dopamine, the movements of people with Parkinson's disease become jerky and may freeze completely.

Several hundred people with Parkinson's disease have received transplants of dopamine-producing fetal brain cells. The brain cells were collected from about six aborted fetuses, between 6 and 8 weeks old. Because the nerve cells of the striatum are alive, but in need of dopamine, unlike the cells of the substantia nigra which are dying, the fetal brain cells are inserted into the striatum of the patient with Parkinson's disease. With luck, the fetal cells will begin to grow and form a thick mat of dopamine-producing cells. Roughly one-third of the patients who have had transplants have returned to nearly normal life—to walk without falling, drive a car, or go cross-country skiing. Another third show more modest gains. The rest benefit for only a short time. However, there is a remote chance that something can go drastically wrong and the patient may die.

Transplants of another type of fetal cell, in this case insulin-producing cells from the pancreas, offer the 12 million Americans with Type I diabetes the hope of a cure. People with this type of diabetes do not produce the hormone insulin and, therefore, cannot regulate the blood level of sugar. The condition can be controlled with insulin injections, but blindness and kidney failure are two common complications.

The best fetal cells come from elective abortions. Fetuses from spontaneous miscarriages and ectopic pregnancies (in which the embryo implants in the fallopian tube and must be aborted to save the mother's life) are often abnormal.

There are approximately 1.5 million elective abortions in the United States each year. Few of them yield tissue that would be useable in fetal cell transplants. If only 10% of the abortions provided tissue that could be used, these would help about 18,000 to 25,000 people with Parkinson's disease and 7500 to 15,000 people with Type I diabetes.

The idea of fetal cell transplants has sparked one of the most heated ethical and political controversies of the decade. Surprisingly, one's position on abortion does not necessarily dictate one's position on fetal cell transplants. Some pro-life advocates argue that since, in their opinion, abortion is immoral, any use of tissue from aborted fetuses must also be immoral. Other pro-life advocates maintain that, although abortion is immoral, it is legal. Fetal tissue that is now discarded could, and should, be used to save lives. At the same time, people who support a woman's right to abortion, worry that some women would become pregnant with the intention of aborting the fetus so that the tissue might be used to alleviate the pain or suffering of a relative. More mercenary women, they fear, might be willing to simply sell the aborted fetus.

Next, we'll see how these pieces fit together to form your body's highly effective immune response (Table 12–1).

Sounding an Alarm

You may recall that macrophages are phagocytic cells that rove the body, engulfing any foreign material or organisms they may encounter. Within the macrophage, the engulfed material is digested into smaller pieces. The macrophage then moves some of these pieces to its own surface where they are bound to the same molecules that identify the macrophage as "self." The "self markers" function like a secret password that identifies the macrophage as a "friend." However, the pieces of antigen that the macrophage places

Table 12–1 Summary of Cells Involved in Immune Responses

Cell	Functions
Macrophage	Engulfs and digests antigens; places a piece of antigen on its plasma membrane; identifies the exact identity of the "invader" by presenting antigen to appropriate helper T cells; secretes chemicals (e.g., interleukin 1) to activate helper T cells; primes cytotoxic T cells by presenting antigen to them
Helper T cell	After activation by an antigen-presenting macrophage, it secretes chemicals (e.g., interleukin 2) to activate B cells and cytotoxic T cells.
Cytotoxic T cell, effector	Destroys cellular targets, such as virus-infected body cells, bacteria, fungi, parasites, and cancer cells
B cell	When stimulated by antigen and helper T cell, it divides to produce clones of plasma cells (effector B cells) and memory B cells.
Plasma cell	Produces huge amounts of antibodies specific for one particular antigen
Memory cell	Produced during initial exposure to antigen from B cell or any class of T cell; provides quick, efficient response to subsequent exposure to an antigen
Suppressor T cell	Turns off the immune response; dampens activity of B cells and T cells as antigen level declines and threat to body lessens

on its surface function as a kind of "wanted poster" that tells lymphocytes not only that there is an invader but also reveals what the invader looks like. Thus, the displayed pieces are antigens that trigger the immune response. Although a few other kinds of cells can display antigens, the macrophage is the most important type of **antigen-presenting cell**.

The macrophage then presents the antigen to a **helper T cell** (also known as a *T4 cell* or a *CD4 cell*, after the receptors on its surface), the kind of T cell that serves as the main switch for the entire immune response. But, the macrophage can't alert just *any* helper T cell. It must alert a helper T cell bearing receptors that recognize the antigen being presented and these constitute only a tiny fraction of the entire T cell population. Finding the right one is like looking for a needle in haystack. The macrophage wanders through the body until it literally bumps into an appropriate helper T cell. The encounter is most likely to take place in a lymph node, a bean-shaped structure that contains huge numbers of lymphocytes of all kinds (Chapter 11).

Drafting the Soldiers: B Cells and T Cells

When the antigen-presenting macrophage meets the appropriate helper T cell and binds to it, the macrophage secretes a chemical, called **interleukin 1**, which activates the helper T cell (Figure 12–10). Within hours an activated helper T cell begins secreting its own chemical messages in the form of **interleukin 2**. The helper T cell's message calls into active duty the soldiers of both lines of immune defense, the appropriate B cells and T cells. The appropriate cells are, of

course, those with the ability to bind to the particular antigen that triggered the whole response. Before they can be activated to mount a full-fledged defense by interleukin 2, the appropriate B cells and T cells must be primed to respond. In the case of T cells, priming is accomplished by another cell, usually a macrophage, but they can be primed by any infected body cell that presents the antigen bound to the cell's self marker. Priming of a T cell requires simultaneous recognition of self and nonself. B cells, on the other hand, can bind to antigens that are free in body fluids. This binding sensitizes a B cell and makes it more responsive to activation from helper T cells. (Priming can be likened to starting the engine of a car. Activation by interleukin 2 would then be analogous to putting the car in gear.)

When these "virgin"[1] B cells or T cells are activated, they begin to divide repeatedly. The result is a clone (a population of genetically identical cells) that is specialized to protect against that particular antigen.

We see, then, that the body produces highly specialized armies of B cells and T cells designed to eliminate a specific antigen from the body. The process by which this occurs, called **clonal selection**, underlies the entire immune response. Each lymphocyte is equipped to recognize a specific antigen and there are only a few lymphocytes able to recognize each of thousands of different antigens. By binding to the receptors on the lymphocyte surface, an antigen *selects* a lymphocyte that was preprogrammed during its maturation to recognize that particular antigen. That lymphocyte is

---

[1]A "virgin" cell is one that has been preprogrammed to respond to a particular antigen, but it has not been previously activated to respond.

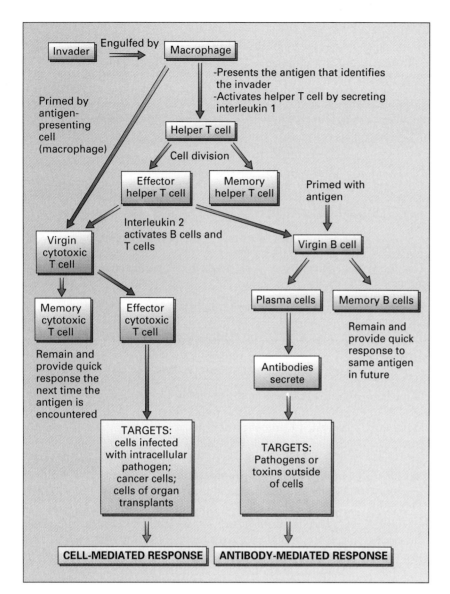

Figure 12–10 When an antigen-presenting macrophage encounters the appropriate helper T cell, it binds to it. The binding causes the macrophage to secrete a chemical called interleukin 1, which activates a helper T cell. The activated helper T cell then locates an appropriate B cell or T cell and secretes interleukin 2, a chemical that activates the B cells and T cells to begin a full-fledged defense against the invader.

Figure 12–11 Clonal selection is the process by which the specificity of the immune response develops. As a B cell or T cell matures, it develops receptors capable of recognizing and binding to one specific antigen. When an antigen enters the body it binds to a receptor on a B cell or T cell, and the reaction causes the cell to proliferate. In this way, millions of exact copies (a clone) of cells specialized to recognize that antigen are created.

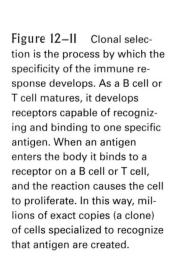

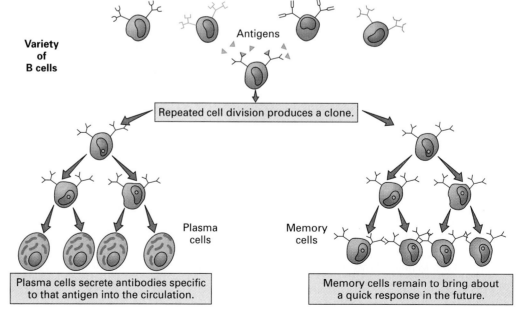

then stimulated to divide and produces a clone of millions of cells tailored to recognize that antigen (Figure 12–11).

We have already mentioned that there are two types of cells in such a clone: memory cells and effector cells. Before discussing the role of memory cells, let's look more closely at exactly *how* the effector cells protect us.

Antibody-Mediated Immune Response

The effector cells produced when activated B cells divide are called **plasma cells**, and they secrete antibodies into the bloodstream (Figure 12–12). Antibodies are Y-shaped proteins that recognize a specific antigen by the shape of the molecule. Each antibody is specific for one particular antigen. The specificity results from the shape of the proteins forming the tips of the Y (Figure 12–13). Because of their shapes, the antibody and antigen fit together like a lock and a key.

Antibodies can only bind to antigens that are free in body fluids or on the surface of a cell. Thus, their main targets are extracellular microbes, including bacteria, fungi, and protozoans. Antibodies can also bind to free virus particles, but they can't touch viruses that have already entered a host cell. Likewise, antibodies can bind to free toxins, but not toxins within a cell.

There are five classes of antibodies, each with a special role to play in protecting against invaders. Antibodies are

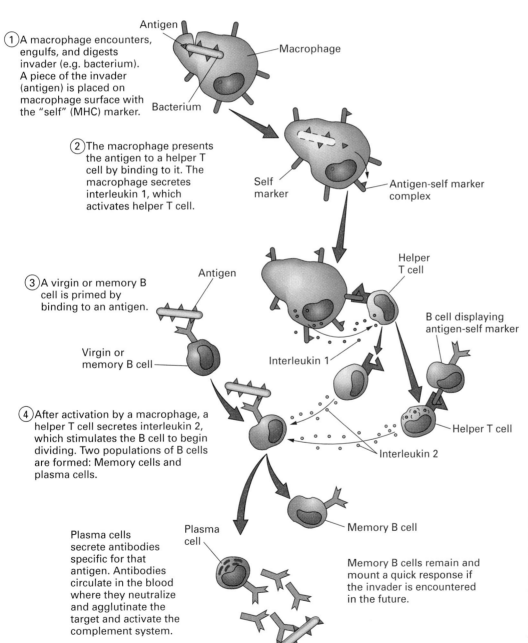

① A macrophage encounters, engulfs, and digests invader (e.g. bacterium). A piece of the invader (antigen) is placed on macrophage surface with the "self" (MHC) marker.

② The macrophage presents the antigen to a helper T cell by binding to it. The macrophage secretes interleukin 1, which activates helper T cell.

③ A virgin or memory B cell is primed by binding to an antigen.

④ After activation by a macrophage, a helper T cell secretes interleukin 2, which stimulates the B cell to begin dividing. Two populations of B cells are formed: Memory cells and plasma cells.

Plasma cells secrete antibodies specific for that antigen. Antibodies circulate in the blood where they neutralize and agglutinate the target and activate the complement system.

Memory B cells remain and mount a quick response if the invader is encountered in the future.

Antigen
Macrophage
Bacterium
Self marker
Antigen-self marker complex
Helper T cell
B cell displaying antigen-self marker
Antigen
Virgin or memory B cell
Interleukin 1
Helper T cell
Interleukin 2
Plasma cell
Memory B cell

Figure 12–12 B lymphocytes are responsible for the antibody-mediated immune response. When properly activated, B cells divide to form memory cells and effector cells. The effector cells, called plasma cells, secrete antibodies that bind to and inactivate the specific invader that triggered the response.

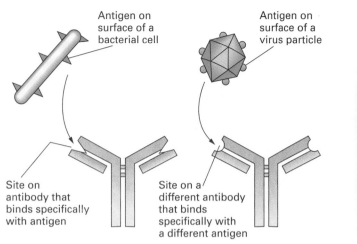

Antigen on surface of a bacterial cell

Antigen on surface of a virus particle

Site on antibody that binds specifically with antigen

Site on a different antibody that binds specifically with a different antigen

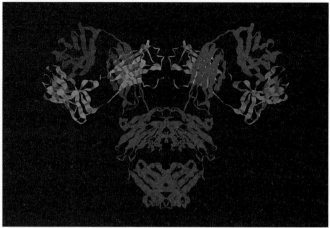

Figure 12–13   An antibody is a Y-shaped protein designed to recognize a specific antigen. The recognition occurs because of the shape of the tips of the Y.   *(Photo, © Eward/Biografx/S/Photo Researchers, Inc.)*

Table 12-2   The Classes of Antibodies (Immunoglobulins) (In Order of Abundance)

Class	Structure	Location	Characteristics	Protective Functions
IgG	Monomer	Blood, lymph, and the intestines	Most abundant of all antibodies in body; involved in primary and secondary immune responses; can pass through placenta from mother to fetus and provides passive immune protection to fetus and newborn	Enhances phagocytosis; neutralizes toxins; triggers complement system
IgA	Dimer or monomer	Present in tears, saliva, and mucus as well as in secretions of gastrointestinal system and excretory systems; present in breast milk	Levels decrease during stress, raising susceptibility to infection.	Prevents pathogens from attaching to epithelial cells of surface linings
IgM	Pentamer	Attached to B cell where it acts as a receptor for antigens; free in blood and lymph	First Ig class released by plasma cell during primary response	Powerful agglutinating agent (10 antigen binding sites); fixes complement
IgD	Monomer	Surface of many B cells; blood and lymph	Life span of about 3 days	Thought to be involved in recognition of antigen and in activating B cells
IgE	Monomer	Secreted by plasma cells in skin, mucous membranes of gastrointestinal and respiratory	Become bound to surface of mast cells and basophils	Involved in allergic reactions by triggering release of histamine and other chemicals from mast cells or basophils

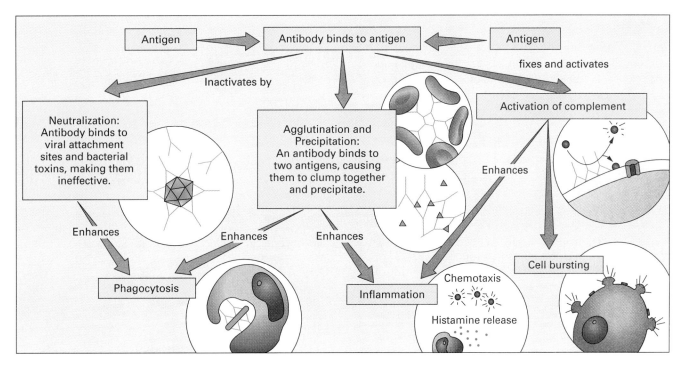

Figure 12–14 Antibodies bind to the antigen and help render it harmless by neutralization, agglutination and precipitation, or by activating the complement system. Each Y-shaped antibody can bind to two antigens. Cross-linking can cause large antigen-antibody complexes to form. These complexes may help remove small molecules such as toxins from the body by causing them to precipitate out of solution, where they can be engulfed by phagocytes.

also called **immunoglobulins** (Ig) and each class is designated with a letter: IgG, IgM, IgE, IgA, and IgD. As you can see in Table 12–2, in some classes, the antibodies exist as single Y-shaped molecules (monomers), in one class they exist as two attached molecules (dimers), and in one class they exist as five attached molecules (pentamers) radiating outward like the spokes of a wheel.

There are three general ways that antibodies help defend against invaders (Figure 12–14):

1. **Neutralization.** You may recall that viruses enter a cell by binding to a receptor on the cell's surface. Once inside, the virus damages the cell by taking over the host's cellular machinery to make copies of the virus. There are specific molecules on the surface of the virus that allow it to bind to the cell's receptor. Some antibodies bind to these viral molecules and mask them. As a result, the virus is neutralized and cannot enter the host cell. Antibodies can also bind to toxins produced by pathogens, forming a coat around them and preventing them from damaging host tissues. The antibody coating attracts phagocytic cells to the area and makes the viruses or toxins more eas-

ily engulfed and removed from the body by those phagocytic cells.

2. **Agglutination and precipitation.** Each antibody can bind two or more antigens, one at the end of each of the tips of the Y. The interactions of a large number of antibodies and antigens can form a huge three-dimensional structure. When the antigens are on the surface of a cell or a virus, the antigen-antibody interactions cause the cells or viruses to clump together, a process called agglutination. It is the agglutination of foreign blood cells that causes problems in a mismatched blood transfusion. Small molecules, such as toxins, are often soluble in body fluids. However, when antibodies bind to them, large complexes that are no longer soluble may form. These may precipitate out of the blood. The insoluble complexes can then be removed by phagocytic cells.

3. **Activation of complement.** When antibodies bind to cellular targets, the shape of the antibody changes. As a result, a region of the antibody molecule that can bind complement is exposed. Complement proteins then bind to this site and the complement system is activated. As we saw earlier, the complement system pokes holes through the membrane of the target cell, causing it to burst.

Cell-Mediated Immune Response

Cytotoxic T cells[2] are the basis of the cell-mediated immune response. Each is programmed to recognize the same antigen, in this case, an antigen that is bound to self markers on the surface of a body cell that has become infected or cancerous or on cells of a tissue transplant. A cytotoxic T cell becomes activated to destroy a target cell when two events occur simultaneously (Figure 12–15). First, the cytotoxic T cell must recognize and bind to the combination of a self marker plus an antigen on the surface of an antigen-presenting cell (either a macrophage or an infected body cell). Second, it must receive additional stimulation from interleukin 2 released from a helper T cell. When sufficiently activated, the cytotoxic T cell divides, producing memory cells and effector cytotoxic T cells. An effector cytoplasmic T cell releases chemicals, called **perforins**, that cause holes to form in the target cell membrane. As water rushes into the target cell, it bursts. The cytotoxic T cell then detaches from the target cell and seeks another cell having the same antigen.

Cytotoxic T cells and the natural killer cells we described earlier use the same methods to destroy their targets. The difference between the attacks of these two types of cells is specificity—cytotoxic T cells destroy cells bearing a specific antigen and natural killer cells attack any cell that isn't recognized as belonging in the body.

Critical Thinking

Rejection of an organ transplant occurs when the recipient's immune system attacks and destroys the cells of the transplanted organ. Why would this attack occur? Which branch of the immune system would be most involved?

Turning Off the Immune Response

As the immune system begins to conquer the invading organism and the level of antigens declines, another type of T cell, called **suppressor T cells**, releases chemicals that dampen the activity of both B cells and T cells. Thus, suppressor T cells turn off the immune response when the antigen no longer poses a threat.

Immunological Memory

As we have seen, the first time an antigen enters the body there are only a few lymphocytes that can recognize it. Those lymphocytes must be located and stimulated to divide to produce an army of lymphocytes ready to eliminate that particular antigen. As a result, the **primary response**, that is, the one that occurs during the body's first encounter with a particular antigen, is slow. It is possible to monitor the im-

[2]Other sources may refer to these cells as cytolytic cells, T8 cells, or CD8 cells.

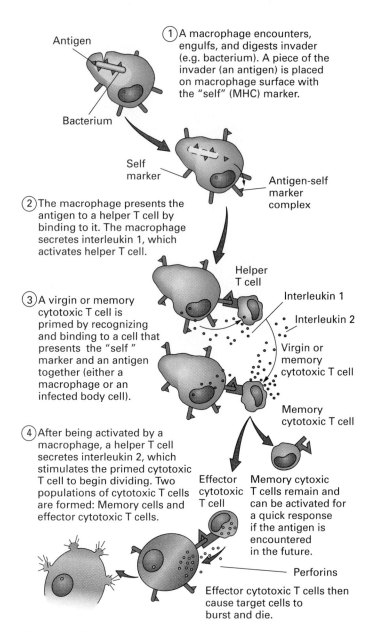

① A macrophage encounters, engulfs, and digests invader (e.g. bacterium). A piece of the invader (an antigen) is placed on macrophage surface with the "self" (MHC) marker.

② The macrophage presents the antigen to a helper T cell by binding to it. The macrophage secretes interleukin 1, which activates helper T cell.

③ A virgin or memory cytotoxic T cell is primed by recognizing and binding to a cell that presents the "self" marker and an antigen together (either a macrophage or an infected body cell).

④ After being activated by a macrophage, a helper T cell secretes interleukin 2, which stimulates the primed cytotoxic T cell to begin dividing. Two populations of cytotoxic T cells are formed: Memory cells and effector cytotoxic T cells.

Antigen

Bacterium

Self marker

Antigen-self marker complex

Helper T cell

Interleukin 1

Interleukin 2

Virgin or memory cytotoxic T cell

Memory cytotoxic T cell

Effector cytotoxic T cell

Memory cytoxic T cells remain and can be activated for a quick response if the antigen is encountered in the future.

Perforins

Effector cytotoxic T cells then cause target cells to burst and die.

Figure 12–15   Cell-mediated immune responses are directed against cellular targets such as infected body cells or cancerous cells. When properly activated, virgin cytotoxic T cells divide forming two lines of descendent cells—effector cytotoxic T cells and memory cells. The effector cytotoxic T cells release perforins, which cause the target cell to burst and die.

mune response by measuring the concentration of circulating antibodies. During the primary response, there is a lapse of several days before the antibody concentration begins to rise and it doesn't peak until 1 to 2 weeks after the initial exposure to the antigen (Figure 12–16).

Following subsequent exposure to the antigen, the so-called **secondary response** is strong and swift. Recall, that when virgin B cells and T cells were stimulated to divide, they produced both effector cells that actively defended

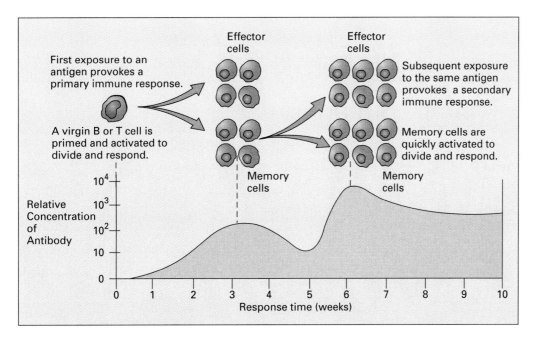

Figure 12–16 The primary and secondary immune responses. During the primary response, which occurs following the first exposure to an antigen, there is a delay of several days before the concentration of circulating antibodies begins to increase and it takes a week or two for the antibody concentration to peak. (The T cells show a similar pattern of response.) This is because the few lymphocytes programmed to recognize that particular antigen must be located and activated. The secondary response following a subsequent exposure to an antigen is swifter and stronger than the primary response. This difference is caused by the long-lived memory cells produced during the primary response. The memory cells create a larger pool of lymphocytes programmed to respond to that particular antigen.

against the invader and memory cells. These memory B cells and T cells live for years or even decades. Thus, the number of lymphocytes programmed to respond to that particular antigen is greater than it was before the first exposure. When the antigen is encountered again, each of those memory cells can divide and produce effector cells and memory cells specific for that antigen. As a result, the number of effector cells rises quickly and reaches a higher peak during the secondary response.

Active and Passive Immunity

In **active immunity** the body actively participates by producing memory B cells and T cells following exposure to an antigen. This happens naturally whenever a person gets an infection. Fortunately, active immunity can also develop through **vaccination**, a procedure that introduces a harmless form of the disease-causing organism into the body to stimulate immune responses against that antigen. In some vaccines, those for whooping cough and typhoid fever for instance, the microbe is inactivated before the vaccine is prepared. These inactivated vaccines usually consist of the cell

walls of bacteria or the protein coats of viruses. Other vaccines must be made from live organisms to be effective. In these cases, the microbe is first weakened by some treatment, such as exposure to a chemical. (This is the way the Sabin oral polio vaccine is prepared.) Some vaccines, including the one against smallpox, are prepared from living microbes that cause related but milder diseases. Today, genetic engineering techniques are used to produce certain vaccines by putting antigens of dangerous pathogens into the cell walls of harmless bacteria.

Because memory cells are produced, active immunity is relatively long-lived. The first dose of a vaccine causes the primary immune response, and antibodies and some memory cells are generated. In certain cases, especially when inactivated antigens are used in the vaccine, the immune system may "forget" its encounter with the antigen with time. To prevent this, a booster is administered periodically. The booster results in a secondary immune response and enough memory cells to provide for a quick response should a potent form of that pathogen ever be encountered.

Passive immunity results when a person receives antibodies that were produced by another person or animal. For instance, antibodies produced by a pregnant woman can

Pollutants and Immune Suppression

Widespread populations of dolphins began dying during the late 1980s and 1990s. In 1987, white-sided dolphins began dying in the waters off Lubec, Maine. Then, during 1987 and 1988, bottle-nosed dolphin corpses were found on beaches ranging from New Jersey to Florida. Many of the dolphins died of pneumonia. Others succumbed to skin lesions similar to acid burns. Next, in 1989, 274 bottle-nosed dolphins were found on U.S. beaches along the Gulf of Mexico. In this instance, the problem appeared to be a fungal growth covering their skin. Between 1992 and 1993, a virus similar to the canine distemper virus, so deadly to dogs, killed more than 1000 Mediterranean striped dolphins.

Dolphins aren't the only animals mysteriously dying. Beginning in 1986, green sea turtles began to develop tumors, called fibroadenomas, that are associated with a herpes virus infection. The next year, many seals in Siberia's Lake Baikal died because of a distemper virus. Others succumbed to skin lesions similar to acid burns. In 1988, the distemper virus struck again, wiping out 60% to 70% of the population of harbor seals living in the North and Baltic Seas. Fifteen members of a population of 500 beluga whales living in the St. Lawrence River wash ashore each year, 40% of them with tumors.

Is there a common thread linking the demise of so many animals? The answer appears to be, "Yes." And that thread appears to be that the immune systems of animals are suppressed by certain pollutants—toxic organochlorines such as polychlorinated biphenyls (PCBs) and dioxins. Organochlorines include long-lived fat-soluble pesticides such as DDT. Also highly persistent, PCBs were used as insulating material, dyes, lubricants and in the production of plastics until their dangers were recognized. These substances accumulate in fatty tissues and, therefore, build up as one organism eats another in the food chain.

At first the link between organochlorines and immune suppression was merely circumstantial. In each die-off, high levels of these pollutants were found in the corpses and the populations lived in areas known to contain these pollutants.

A team of Dutch scientists led by Albert D.M.E. Osterhaus demonstrated that pollutants such as PCBs do, indeed, directly impair immune functioning. They collected harbor seal pups from an unpolluted area along the northeast coast of Scotland. Half of the pups were fed herring from the clean North Atlantic waters and the other half were fed herring from PCB-contaminated waters of the Baltic. The herring from the Baltic contained ten times the amount of organochlorines as did the North Atlantic herring.

At the beginning of the study, all the pups' immune systems were healthy. Soon, however, the responses of natural killer cells and T cells began to slow in the seals eating contaminated fish. The numbers of certain other white blood cells, granulocytes that fight bacterial infections, began to rise in the group eating contaminated fish, suggesting that these seals were fighting chronic bacterial infections.

As is often the case, what we humans do to the creatures with whom we share the Earth, we do to ourselves. Confirmation of the harm we are doing to ourselves comes from many corners of the world. In the mid-1980s the Environmental Protection Agency estimated that the body of nearly everyone in the United States had at least a trace of PCBs. The breast milk of Inuit women living in the Canadian Arctic contains the highest levels of PCBs ever measured in humans. During their first year of life, breast-fed infants of these Inuit women have 20 times more infectious diseases, such as meningitis, measles, and severe ear infections, as do breast-fed infants of women living in Southern Quebec who have no PCBs in their breast milk. Furthermore, the breast-fed Inuit infants had smaller ratios of helper T cells to suppressor T cells than did the breast-fed infants of other women. Some of the Inuit children could not even be effectively vaccinated, because they don't produce antibodies.

Social Issue

Although many laws have been passed that ban the use of certain pesticides in the United States, many harmful pesticides are still used in Third World countries. Farmworkers in these countries are often not taught and, therefore, do not use, proper safety precautions to avoid exposure to pesticides. Infectious diseases are the leading cause of death in many Third World countries. Can you think of measures that would help to protect people in these countries from exposure to pesticides?

cross the placenta and give the growing fetus some immunity. These maternal antibodies remain in the infant's body for as long 3 months, after which the infant can produce its own antibodies. Antibodies in breast milk also provide passive immunity to nursing infants, especially against pathogens that might enter through the intestinal lining. The mother's antibodies can be a blanket of protection because most of the pathogens that can threaten the health of a newborn have been seen previously by the mother's immune system.

Passive immunity is also possible in adulthood when antibodies produced in another person or animal are injected into another person. In this case, passive immunity is a good news/bad news situation. The good news is that the effects are immediate. For this reason, gamma globulin (a preparation of antibodies) is used to help people who have been exposed to diseases such as hepatitis B or who are already infected with the microbes that cause tetanus, measles, or diphtheria. Gamma globulin is often given to travelers before they visit a country where viral hepatitis is common. The bad news is that the protection is short-lived. The borrowed antibodies circulate for 3 to 5 weeks before being destroyed in the recipient's body. Since the recipient's immune system was not stimulated to produce memory cells, protection disappears with the antibodies.

Critical Thinking

The viruses that cause influenza (the flu) mutate rapidly, and so the antigens in the protein coat continually change. Why does this make it difficult to develop a vaccine against the flu that will be effective for several consecutive years?

Social Issue

The cost of childhood immunizations has been skyrocketing. It rose 1000% in a single decade. Drug companies say that most of this rise is because of lawsuits against the companies resulting from a small number of people who suffered harmful side effects from immunizations. For example, in the United States during 1976 and 1977, the large-scale immunization program against swine flu resulted in the paralysis of a number of people. Still, many lives have been saved by the program. Do you think such efforts are wise? If they help more people than they hurt, should they be continued?

Monoclonal Antibodies

Suppose you wanted to determine whether a particular antigen were present in a solution, tissue, or even somewhere in the body. An antibody specific for that antigen would be just the tool you would need. Because of such specificity, any such antibody would go directly to that target. If a label were attached to the antibody (a radioactive tag or a molecule that fluoresced for instance), the antibody could then reveal the location of the antigen. We see, then, that it is sometimes desirable to have a supply of identical antibodies that will react with a specific antigen. Such groups of identical antibodies are called monoclonal antibodies.

Monoclonal antibodies are produced by the descendants of a single B cell and, therefore, are identical antibodies that will react with one specific antigen. Lymphocytes and plasma cells don't ordinarily grow well in tissue culture, but they can be "coaxed" into dividing by fusing them with cells that divide without restraint—cancer cells. A normal, antibody-producing lymphocyte can be fused with a myeloma cell (a special type of lymphocyte cancer cell). The resulting cell, called a **hybridoma**, is an antibody-producing cell with the capacity to divide and reproduce virtually forever.

Monoclonal antibodies have many uses in diagnosis. Home pregnancy tests contain monoclonal antibodies produced to react with a hormone (human chorionic gonadotropin) produced by membranes associated with the developing embryo. When the antibodies bind to this hormone in the woman's urine, the material changes color.

Monoclonal antibodies have also proven useful in screening for certain diseases, including Legionnaire's disease, hepatitis, certain sexually transmitted diseases, and certain cancers, including those of the lung and prostate. Monoclonal antibodies can not only reveal whether cancer has spread from the initial tumor, but they can also detect cancers that are too small to be seen using other methods.

Monoclonal antibodies are often used not just to find the problem but also to solve it. They can be used to deliver drugs to exactly the cells that need them. This is an especially useful trick when it comes to killing cancer cells. When administered to the whole body, cancer treatments such as irradiation or chemotherapy affect all cells. Healthy dividing cells are killed along with the cancerous ones. However, if the radioactive material or chemical treatment is attached to a monoclonal antibody, it acts as a "smart bomb" that homes in on the tumor cells but has little effect on other cells. In a similar manner, toxic drugs can be delivered to a specific strain of bacteria.

Autoimmunity

Autoimmune disorders occur when the immune system fails to distinguish between self and nonself and attacks the tissues or organs of the body. Thus, if the immune system can be called the body's military defense, then autoimmune disease is the equivalent of "friendly fire."

Table 12–3   Selected Autoimmune Disorders

Autoimmune Disorder	Target of Immune System Attack	Effect
Organ-Specific		
Hashimoto's thyroiditis	Thyroid gland	Increased production of thyroid hormone
Pernicious anemia	Cells in stomach lining that produce intrinsic factor	Decreased production of red blood cells
Addison's disease	Adrenal glands	Adrenal failure
Diabetes mellitus Type I	Insulin-producing cells in the pancreas	Elevated blood sugar
Grave's disease	Thyroid gland	Increased rate of chemical reactions in body
Multiple sclerosis	Myelin sheath of nerve cells	Short circuiting of neural impulses resulting in sensory and/or motor deficits
Ankylosing spondylitis	Joints between vertebrae	Spine bent and fused; inflammation
Myasthenia gravis	Connections between nerve and muscle	Muscle weakness
Glomerulonephritis	Kidney	Kidney failure
Encephalitis	Brain	Impaired brain function; headache; irritability; double vision; impaired speech
Non–Organ-Specific		
Lupus erythematosus	Connective tissue	Butterfly-shaped rash on face; skin lesions; joint pain
Rheumatoid arthritis	Collagen fibers of joints	Joint pain
Dermatomyositis	Skin inflammation	Body rash
Rheumatic fever	Heart valves, joints	Joint pain; kidney failure

As we have seen, during their development, lymphocytes are programmed to attack a specific foreign antigen while still tolerating self antigens. Lymphocytes that do not learn to make this distinction are usually destroyed. Unfortunately, some lymphocytes that are primed to attack self antigens escape destruction. These cells are like time bombs ready to attack the body's own cells at the first provocation. For example, if these renegade lymphocytes are activated by a virus or bacterium, they may direct their attack against healthy body cells as well as the invading organism.

Autoimmune disorders are often classified as organ-specific or non–organ-specific (Table 12–3). As the name implies, organ-specific autoimmune disorders are directed against a single organ. The thyroid gland, for example, is attacked in Hashimoto's thyroiditis. Organ-specific autoimmune disorders are usually caused by T cells that have gone awry. In contrast, non–organ-specific autoimmune disorders tend to have effects throughout the body. In systemic lupus erythematosis, for instance, connective tissue is attacked. Since connective tissue can be found throughout the body, almost any organ can be affected. There may be skin lesions or rashes, especially a butterfly-shaped rash centered on the nose and spreading to both cheeks. It may affect the heart (pericarditis), joints (arthritis), kidneys (nephritis), or nervous system (seizures). Non–organ-specific autoimmune disorders are generally caused by antibodies (Figure 12–17).

A number of autoimmune disorders occur because portions of disease-causing organisms resemble antigens found on normal body cells. If the immune system mistakes the body's antigens for the foreign antigens, it may attack them. For instance, the body's attack on certain streptococcal bacteria causing a sore throat may result in the production of antibodies that target not only the streptococcal bacteria but also similar molecules that are found in the valves of the heart and joints. The result is an autoimmune disorder known as rheumatic fever.

Treatment of autoimmune disorders is usually two-pronged. First, deficiencies caused by the disorder are corrected. In diabetes mellitus Type 1, insulin-producing cells in the pancreas are destroyed by the immune system. Treatment for diabetes would, therefore, include replacement of insulin. Second, immune system activity is suppressed with drugs. A more recent treatment for certain autoimmune disorders involves coaxing the immune system to be more accepting.

Allergy

An **allergy** is a strong immune response to an antigen, in this case called an **allergen**, that is not usually harmful to the body (Table 12–4). Roughly 20% of Americans and

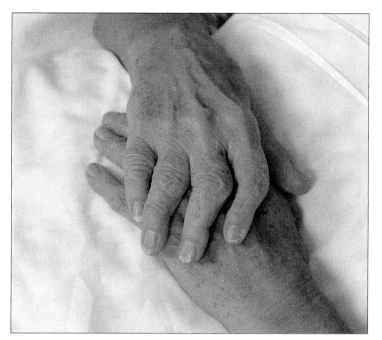
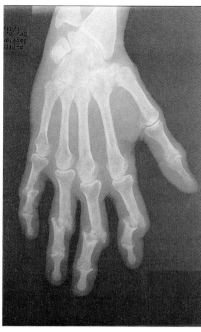

Figure 12–17   Rheumatoid arthritis is an autoimmune disease—a condition caused when the body's immune cells turn the attack against the body's own cells.   *(© David York/Medichrome/The Stock Shop, Inc.)*

Table 12–4   Common Allergies

Type of Allergic Response	Common Causes	Location of Reactive Mast Cells	Symptoms
Hay fever (allergic rhinitis)	Pollen, mold spores, animal dander (bits of skin and hair), feces of dust mites	Lining of nasal cavity	Sneezing, nasal congestion
Asthma	Pollen, mold spores, animal dander	Airways of lower respiratory tract	Difficulty breathing
Food allergy	Chicken, eggs, fish, milk, nuts, (especially peanuts), shellfish, soybeans, and wheat	Lining of digestive system	Nausea, vomiting, abdominal cramps, and diarrhea
Hives	Foods (especially shellfish, strawberries, chocolate, nuts, and tomatoes); insect bites; certain drugs (especially penicillin and aspirin); and chemicals, such as food additives, dyes, and cosmetics	Skin	Patches of skin become red and swollen
Anaphylactic shock	Insect stings (especially from bees, wasps, hornets, yellow jackets, and fire ants); medicines (especially penicillin and tetracycline); and certain foods (especially eggs, seafood, nuts, and grains)	Throughout the body	Widening of blood vessels, causing blood to pool in capillaries and resulting in dizziness, nausea, diarrhea, and unconsciousness; death

Canadians suffer from some form of allergy. The most common allergy is hay fever, which, by the way, isn't caused by hay and doesn't cause a fever. Hay fever is more correctly known as allergic rhinitis (*rhino*, a nose; *-itis*, inflammation of). The symptoms of hay fever—sneezing and nasal congestion—occur when an allergen is inhaled, triggering an immune response in the respiratory system. Mucous membranes of the eyes may also respond, causing red, watery eyes. Common causes of hay fever include pollen, mold spores, animal dandruff, and the feces of dust mites, creatures that are found all over your house (Figure 12–18). The same allergens, however, can cause asthma. During an asthma attack, the small airways in the lung (bronchioles) constrict and make breathing difficult. In food allergies the immune response occurs in the digestive system and may cause nausea, vomiting, abdominal cramps, and diarrhea. Foods can also cause hives, a skin condition in which patches of skin temporarily become red and swollen.

Anaphylactic shock is an extreme allergic reaction that occurs within minutes after exposure to the substance that a person is allergic to and that can causes pooling of blood in capillaries, which causes dizziness, nausea, and sometimes unconsciousness as well as extreme difficulty in breathing. Anaphylactic shock can cause death, but fortunately this does not happen often. Common triggers of anaphylactic shock include certain foods, medicines, including antibiotics such as penicillin and tetracycline, and insect stings, especially stings from bees, wasps, yellow jackets, and hornets.

Although there are many substances that can cause allergies, the mechanisms by which they affect the body are similar. An allergy begins when a person becomes sensitized to a particular allergen (Figure 12–19). It develops in the same way any immune response is initiated: The allergen is engulfed by a macrophage, which presents it to the appropriate lymphocytes. Soon, plasma cells churn out IgE antibodies. These IgE antibodies bind to either the type of white blood cell called basophils or to mast cells, which are cells found in connective tissue along blood vessels that are thought to be transformed basophils that have left the bloodstream. A person is said to be sensitized to an allergen when basophils and mast cells have IgE antibodies specific for a particular allergen fixed to their surfaces.

After a person has been sensitized to an allergen, subsequent exposures to that substance cause an allergic response. This time the allergen binds to IgE antibodies on the surface of basophils or mast cells. When binding occurs, granules containing histamine release their contents. Histamine then causes the swelling, redness, itching, and other symptoms of an allergic response.

When histamine is released from mast cells or basophils, it triggers a dramatic series of events. The blood vessels widen and so the blood flows through them more slowly. At the same time, the blood vessels become leaky, allowing fluid to flow from the vessel into spaces between tissue cells. The fluid causes edema (swelling) of the surrounding tissues. The result in hay fever is a stuffy nose and puffy, red eyelids. Histamine also causes the release of large amounts of mucus. Hence, the nose begins to run. In addition, histamine can cause smooth muscles of internal organs to contract. In an asthma attack, contraction of smooth muscle in the respiratory airways causes these passageways to constrict and makes breathing difficult. When the smooth

(a)

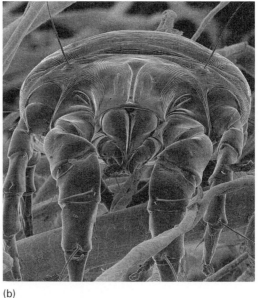

(b)

Figure 12–18   Pollen and the feces of dust mites are common causes of hay fever.   *(a, © Dr. Jeremy Burgess/Science Photo Library/Photo Researchers, Inc.; b, David Scharf/Peter Arnold, Inc.)*

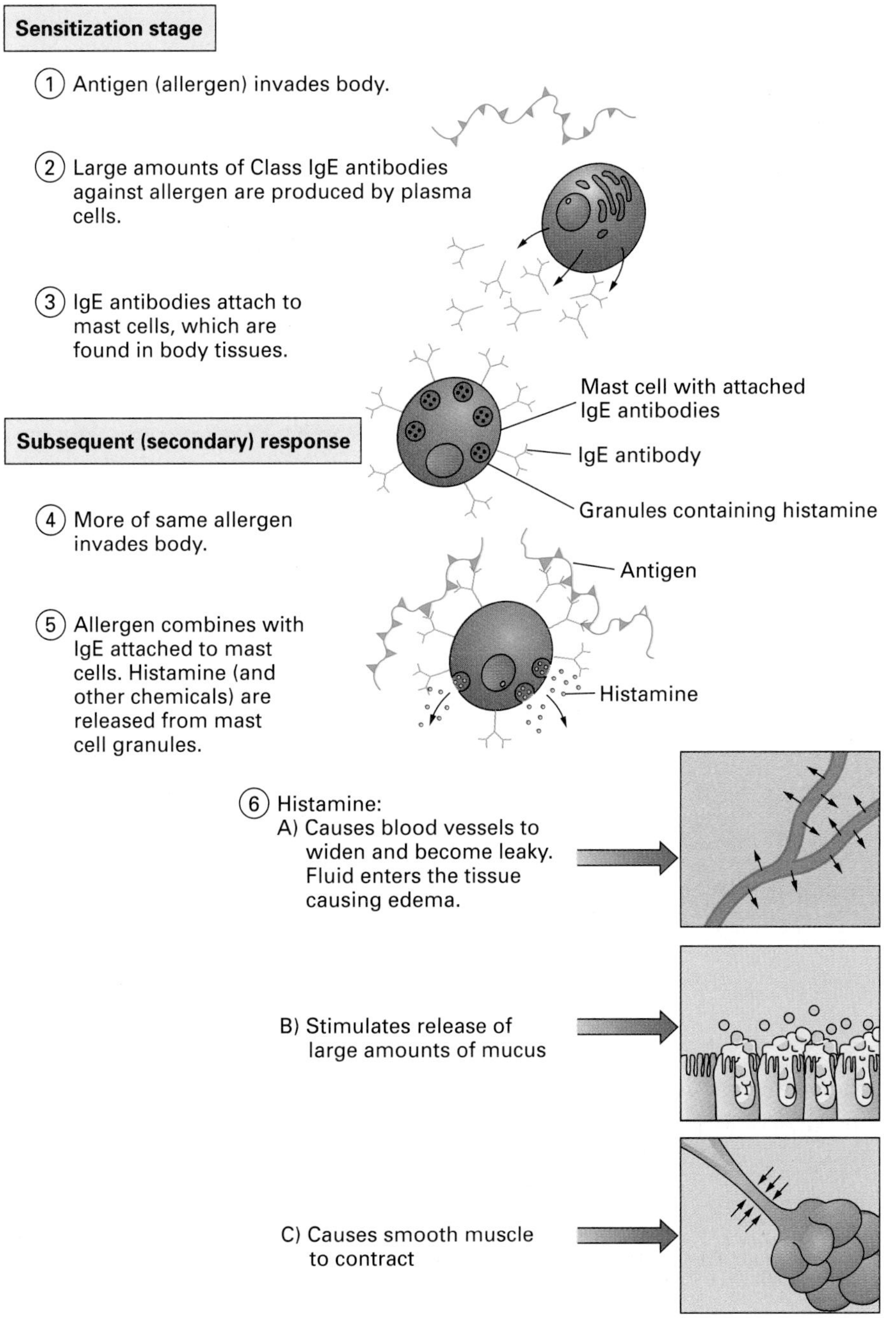

Sensitization stage

1. Antigen (allergen) invades body.

2. Large amounts of Class IgE antibodies against allergen are produced by plasma cells.

3. IgE antibodies attach to mast cells, which are found in body tissues.

Subsequent (secondary) response

4. More of same allergen invades body.

5. Allergen combines with IgE attached to mast cells. Histamine (and other chemicals) are released from mast cell granules.

6. Histamine:
   A) Causes blood vessels to widen and become leaky. Fluid enters the tissue causing edema.

   B) Stimulates release of large amounts of mucus

   C) Causes smooth muscle to contract

Mast cell with attached IgE antibodies

IgE antibody

Granules containing histamine

Antigen

Histamine

Figure 12–19 A person becomes sensitized to an allergen when exposure to the allergen causes plasma cells to produce large quantities of IgE antibodies. The IgE antibodies then bind to mast cells located in connective tissue near blood vessels and to a type of white blood cell called basophils. When the antigen enters the body again, it binds to the IgE antibodies on mast cells, causing the granules in the cells to release histamine and other chemicals. Histamine causes blood vessels to widen and become leaky. Fluid leaving the vessels enters the surrounding tissue, causing it to swell. It also causes the release of mucus from cells that produce it. In addition, it causes the contraction of smooth muscle. If the smooth muscle is in the small airways of the lung, breathing becomes difficult. Contraction of smooth muscle in the digestive system can cause cramps, nausea, and diarrhea.

muscle of the digestive system contracts, the result can be abdominal cramps and stomach pain. If the allergen spreads from the area where it entered the body, its effects can be widespread and the result can be anaphylactic shock.

People with allergies often know which substances cause their problems. When the culprits are not known, doctors can identify them using a crude but effective technique. Small amounts of suspected allergens are injected into the skin. If the person is allergic to one of the suspected allergens, a red welt will form at the site of injection (Figure 12–20).

The simplest way to avoid the miseries of allergies is to avoid exposure to the substances that cause problems.

During pollen season, spend as much time as possible indoors, using an air-conditioner to filter pollen out of the incoming air. Unfortunately, spores from molds growing in air-conditioners and humidifiers are also common triggers of allergies. Some foods, for instance strawberries, may be easy to avoid. Others, such as peanut oil, can show up in some unlikely foods, including stew, chili, or meat patties.

Certain drugs may also reduce allergy symptoms. As their name implies, antihistamines block the effects of histamine. They do this by binding to receptors for histamine on the surface of responsive cells, thereby preventing histamine from binding. Antihistamines are most effective if they are

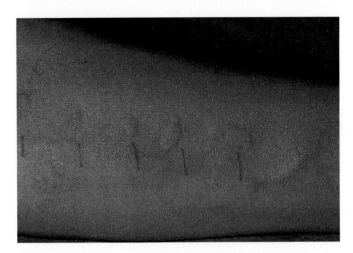

Figure 12-20 In an effort to identify substances to which a person is allergic, doctors can inject small amounts of the suspected materials into the skin. Swelling and redness at the site of injection indicates that the person is allergic to that substance. *(© Dr. P. Marazzi/Science Photo Library/Photo Researchers, Inc.)*

taken before the allergic reaction begins. Unfortunately, antihistamines tend to become less effective over time and most of them cause drowsiness, which can impair performance on the job or in school and can be extremely hazardous when driving a car.

Finally, allergies can be treated by gradually desensitizing the person to the offending allergens. Allergy shots inject gradually increasing amounts of a known allergen into the person's bloodstream. The allergen then causes the production of another class of antibodies—IgG. After this, when the person is exposed to the naturally occurring allergen, IgG antibodies bind to it. In this way, the allergen is prevented from binding to IgE antibodies on mast cells and triggering an allergic reaction.

Organ Transplantation and Tissue Rejection

Each year, tens of thousands of people receive a gift of life in the form of a kidney, heart, lungs, liver, or pancreas. Although these transplants seem commonplace today, they have only been performed for about 30 years.

As we have seen, the effector T cells of the immune system attack and kill cells that lack self markers. When transplanted tissue is killed by the host's immune system, we say that the transplant has been rejected.

We see, then, that the success of the transplant depends on the similarity between the host and transplanted tissues. Thus, it is clear that the most successful transplants would involve tissue taken from one part of a person's body and trans-

planted to another part. In cases of severe burns, for example, healthy skin can replace badly burned areas of skin. Since identical twins are genetically identical, their cells have the same self markers and organs can be transplanted from one twin to another with little fear of rejection. But, few of us have identical twins. The next best source for tissue for a transplant, and the most common source, would be from a person whose cell surface markers closely match those of the host. Usually the transplanted tissue comes from a person who has recently died. However, in some cases, living people can donate organs—one of two healthy kidneys can be donated to a needy recipient, as can sections of liver.

Social Issue

Forty thousand Americans are on waiting lists for organ transplants. Eight of these people die every day because suitable organs cannot be located in time. The best chance of finding an organ with compatible tissue type is in a family member. Should the law require that family members be tested for tissue compatibility when another member needs an organ? Fewer than one in a thousand living liver donors die as a result of their generosity. Is there any degree of risk that living donors should be forced to accept? Is there a level of risk for a living donor that would make the procedure unethical? Who should decide?

Regardless of the improving odds for successful transplants, the waiting list of patients in need of an organ from a suitable donor has outpaced the supply. Some researchers believe that, in the future, organs from nonhuman animals may fill the gap between supply and demand for organs. So far, attempts to transplant animal organs into people have failed. The biggest obstacle is hyperacute rejection. Within minutes to hours after transplant, the animal organ dies because its blood supply is choked off by the human immune system. The more genetically similar the animal species is to humans, the less likely is rejection. Chimpanzees, the animal most genetically similar to humans, are endangered and, therefore, cannot be used for transplantation experiments. Transplants using organs from baboons, another close relative, have not been successful. In 1984, the heart of a young baboon was used to replace the heart of an infant who was born with a fatal heart defect. The infant died within 3 weeks. Eight years later, baboon livers were transplanted into two dying patients. These transplants were also unsuccessful.

In 1995, researchers began trying a different approach: Instead of using a close primate relative as an organ donor, they genetically altered the genes of pigs to make the pig cell membrane proteins more similar to those of humans. Genetically altered pig organs have been transplanted into baboons. When the baboons were treated with drugs to suppress their immune systems, some genetically altered transplanted hearts survived as long as 2 months. Although no pig organs have been permanently placed in human recipients yet, some researchers are hopeful that this procedure will someday be possible.

Social Issue

Medical researchers who transplant animal organs into humans claim that these experiments are done out of commitment to human life. Critics of animal-human transplants charge that transplanting baboon livers into dying individuals was unethical because the people may have (incorrectly) felt that experimental surgery was their only hope for survival. Furthermore, some animal rights activists oppose animal-human transplants, regardless of human need, because it unjustly exploits animals. Do you believe that human need for organs justifies the use of animals? Who should make these decisions? Researchers? Patients? Government? Animal rights activists?

SUMMARY

1. The targets of the body's defense system include anything that is not recognized as belonging in the body, such as disease-causing organisms and cancerous cells.

2. Pathogens (disease-causing organisms) include bacteria, viruses, protozoans, fungi, and parasitic worms.

3. Bacteria are prokaryotic cells that can replicate at high rates. Bacteria cause diseases by releasing toxins into the bloodstream. Antibiotics can kill bacteria by causing them to burst or by slowing their growth.

4. A virus is a parasite that requires a host to reproduce and can infect any susceptible cell. A virus consists of a protein coat surrounding some genetic information in the form of either DNA or RNA. Once inside a host cell, a virus can use the host's metabolic machinery to make new copies of itself. If new viruses are made quickly and released from the cell, they often cause the cell to burst and die. However, a virus can also insert its genetic material into the host cell chromosome where it can remain dormant. Viruses can cause disease by impairing or killing the host cell. Some cause cancer by turning on host cell genes that regulate cell division.

5. Protozoa are single-celled eukaryotic organisms that cause disease by secreting toxins (poisons) or enzymes.

6. Fungi are eukaryotic organisms that exist as single cells or multicellular organisms. They cause illness by secreting enzymes that digest cells.

7. Parasitic worms include flukes, round worms, and tapeworms. They can cause illness by releasing toxins into the bloodstream, feeding off blood, competing for food with the host, or blocking passageways.

8. Whereas nonspecific defense mechanisms are effective against any foreign organisms or substances, specific defense mechanisms target one specific threat.

9. Nonspecific defenses include surface barriers, phagocytic cells, natural killer (NK) cells, defensive proteins, inflammation, and fever.

10. The first line of defense against pathogens is provided by the surface barriers—the skin and mucous membranes. These barriers are nearly impenetrable and provide some chemical protection.

11. Phagocytes are scavenger cells specialized to engulf and destroy particulate matter, such as pathogens, damaged tissue, or dead cells. Phagocytes include neutrophils, which attack bacteria, eosinophils, which attack parasitic worms and foreign proteins, and macrophages, which engulf almost any pathogen and/or damaged cells.

12. Natural killer (NK) cells are lymphocytes that find and kill abnormal cells in the body, primarily cancerous cells and virus-infected cells. NK cells kill by secreting chemicals called perforins that poke holes in the target cell, causing it to burst.

13. Two types of defensive proteins are interferons and complement. Interferons are proteins released by certain cells that are infected by viruses to help curb viral diseases. Complement is a group of at least 20 proteins whose activities enhance other body defense mechanisms.

14. The inflammatory response occurs in response to tissue injury or invasion by foreign microbes. It begins when cells in the injured area release histamine, which increases blood flow by dilating blood vessels to the region and by increasing the permeability of capillaries there. Increased blood flow results in redness and warmth in the region. Fluid leaking from the capillaries causes swelling. Phagocytes and other protective cells are attracted to the area.

15. Fever, a body temperature that is elevated above normal, helps the body fight invading microbes by enhancing several body defense mechanisms and slowing the growth of many pathogens.

16. Immune responses are specific for a particular invader and they have a memory for that invader. Substances that trigger immune responses are called antigens. Lymphocytes are white blood cells that are responsible for immune responses. Both B lymphocytes (B cells) and T lymphocytes (T cells) develop in the bone marrow. The B cells mature in the bone marrow, but the T cells mature in the thymus gland. During maturation B cells and T cells develop receptors on their surfaces that allow them to recognize one specific antigen. Maturing B cells and T cells also learn to recognize the cells that belong in the body.

17. Macrophages are phagocytic cells that engulf any foreign material or organism they encounter. After engulfing the material, the macrophage places a part of it on its own surface, where it serves as an antigen to alert lymphocytes to the presence of an invader and also reveals what the invader looks like. Macrophages also have molecular markers on their membranes that identify them as belonging in the body, that is, as "self."

18. A macrophage then presents the antigen to a helper T cell, a cell that serves as the main switch to the entire immune response. The helper T cell must be one that recognizes the

antigen on the surface of the macrophage. When this encounter occurs, the macrophage secretes interleukin 1, which activates the helper T cell. The helper T cell, in turn, secretes interleukin 2, which activates the appropriate B cells and T cells (those specific for the antigen that the macrophage engulfed).

19. B cells are responsible for antibody-mediated immune responses, which defend against enemies that are free in body fluids, including bacteria, free virus particles, and toxins. When called into action by a helper T cell, a B cell divides repeatedly, forming two lines of descendent cells—effector cells that transform into plasma cells and memory B cells. Plasma cells secrete Y-shaped proteins called antibodies into the bloodstream. Antibodies bind to the particular antigen and inactivate it or help remove it from the body. Antibodies work by neutralizing the antigen, agglutinating or precipitating antigens so that they are more easily engulfed by phagocytes, or by activating complement.

20. T cells are responsible for cell-mediated immune responses, which are effective against cellular threats, including infected body cells or cancer cells. When a T cell is properly activated, it divides forming two lines of descendent cells—effector cells, called cytotoxic T cells, and memory T cells. Cytotoxic T cells secrete perforins that poke holes in the foreign or infected cell, causing it to burst and die. Suppressor T cells dampen the activity of B cells and T cells when antigen levels begin to fall.

21. After the first encounter with a particular antigen, the primary response is initiated, which may take several weeks to be effective against the antigen. However, because of memory cells, a subsequent exposure to the same antigen triggers a quicker response, called a secondary response.

22. In active immunity the body actively participates in forming memory cells against a particular antigen. Active immunity may occur when the antigen infects the body or it may occur through vaccination, a procedure that introduces a harmless form of the antigen into the body. Passive immunity results when a person receives antibodies that were produced by another person or animal. Passive immunity is short-lived.

23. Monoclonal antibodies are identical antibodies derived from a clone of genetically identical cells descending from a genetically engineered cell called a hybridoma. A hybridoma is a cell resulting from the fusion of an antibody-producing lymphocyte with a myeloma cell (a special type of cancerous white blood cell). Monoclonal antibodies are useful in diagnosis and treatment of diseases.

24. Autoimmune disorders occur when the immune system mistakenly attacks the body's own cells. Organ-specific autoimmune disorders are directed against a single organ and are caused by T cells. Non–organ-specific autoimmune disorders are caused by antibodies and have effects that are widespread throughout the body. Treatments of autoimmune disorder involve correcting the deficiency caused by the disorder and suppressing the immune system.

25. An allergy is a strong immune response against an antigen (called an allergen). An allergy occurs when the allergen causes plasma cells to release large numbers of IgE antibodies. These IgE antibodies bind to mast cells or basophils, causing them to release histamine. Histamine, in turn, causes the redness, swelling, itching, and other symptoms of an allergic response.

26. Organ transplants are rejected when T cells identify the transplanted cells as foreign and attack them. The chances of rejection are minimized by finding a donor whose membrane markers are similar to the self markers on the recipient's cells. Success of the transplant is increased by treating the recipient with drugs that suppress the immune system.

REVIEW QUESTIONS

1. Describe the characteristics of the five types of common pathogens. Explain how each type causes disease.
2. Explain the difference between non-specific and specific defense mechanisms.
3. List six non-specific defense mechanisms and explain how each helps to protect us against disease.
4. How does a natural killer cell kill its target cell?
5. What are interferons? What type of cell produces them? How do they help protect the body?
6. What are the complement proteins? Explain how they act directly and indirectly to protect the body against disease.
7. Signs of inflammation include redness, swelling, and warmth. What causes these symptoms? How does inflammation help defend against infection?
8. What causes a fever? In what ways is fever beneficial?
9. What does an antigen-presenting cell do? What is the most common type of antigen-presenting cell? How do other cells recognize the antigen-presenting cell as a "friend"?
10. What cells are responsible for antibody-mediated immune responses? What are the targets of antibody-mediated immune responses?
11. Describe an antibody. How do antibodies inactivate or eliminate antigens from the body?
12. What is responsible for cell-mediated immune responses? What are the targets of cell-mediated immune responses?
13. How does a natural killer cell differ from a cytotoxic killer cell?
14. Why does a secondary response occur more quickly than the primary response?
15. Differentiate between active and passive immunity.
16. What are monoclonal antibodies? How are they produced? What are some medical uses for them?
17. What is an autoimmune disorder? Give some examples. How are autoimmune disorders treated?
18. What is an allergy? What causes the symptoms?
19. What causes a transplanted organ to be rejected? How can the chances of acceptance be increased?

Chapter | 13

The Digestive System

(© Timothy Shonnard/Tony Stone Images)

There is some truth to the saying, "You are what you eat." Rest assured, however, that no matter how many hamburgers you eat, you will never become one. Instead, the hamburger becomes *you*. The transformation is possible largely because of the activities of the digestive system. Like an assembly line in reverse, the digestive system takes the food we eat and breaks the complex organic molecules into their chemical subunits. The subunits are molecules small enough to be absorbed into the bloodstream and delivered to body cells. Once in the cells, food molecules ultimately meet one of two fates—they may be used to provide energy for daily activities or they may be put back together to form new molecules that become body components. Thus, the starch in the hamburger bun may fuel a jump for joy, the protein in the beef may be used to build muscle, and the fat may become sheaths insulating nerve fibers. Food, then, is needed to provide energy and to provide raw material for maintenance, repair, and growth. Without the digestive system, however, food would be useless to us because it could never reach the cells.

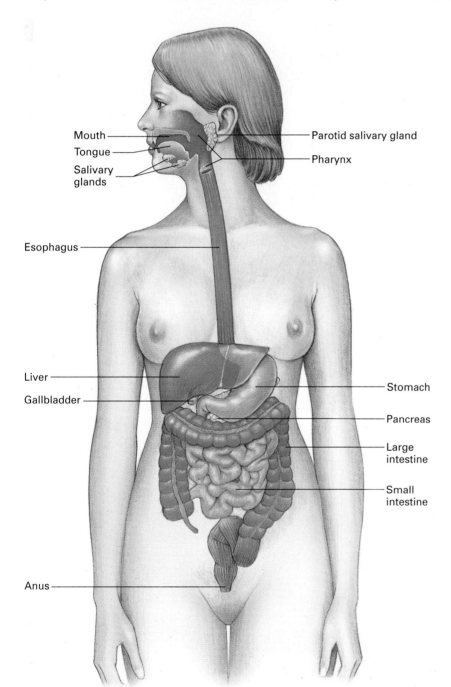

Mouth

Tongue

Salivary glands

Parotid salivary gland

Pharynx

Esophagus

Liver

Gallbladder

Stomach

Pancreas

Large intestine

Small intestine

Anus

Figure 13–1 The digestive system consists of a long tube, called the gastrointestinal tract, into which accessory glands release their secretions. As food moves through the gastrointestinal tract, it passes through the mouth, pharynx, esophagus, stomach, small intestine, and large intestine. Accessory structures include the salivary glands, liver, gallbladder, and pancreas. The function of the digestive system is to break complex food molecules into simpler component subunits that are small enough to be absorbed into the bloodstream and delivered to the cells.

The digestive system consists of a long tube, called the **gastrointestinal (GI) tract,** into which various accessory glands release their secretions (Figure 13–1). Beginning with the esophagus and continuing throughout, the walls of the digestive tube have four basic layers (Figure 13–2). The innermost layer is the moist, mucus-secreting layer called the **mucosa.** The mucus helps lubricate the tube, allowing food to slide through easily. Mucus also helps protect the lining cells from rough materials in food and from digestive enzymes. In some regions of the digestive system, cells in the mucosa also secrete digestive enzymes. The mucosa in some organs is highly folded, which increases the surface area for absorption. Next is the **submucosa,** a layer of connective tissue containing a blood supply and nerves. The blood supply maintains the cells of the digestive system and, in some regions, picks up and transports the products of digestion. The nerves are important in coordinating the contractions of the next layers, which are muscular and are called the **muscularis.** In most regions there is a double layer of muscle. The muscles of the inner layer circle the tube, causing a constriction when they contract, whereas the muscles in the outer layer run lengthwise, causing shortening when they contract. The muscle layers churn the food until it is liquefied, mix it with enzymes, and propel the food along the digestive system. The digestive tract is wrapped in a thin layer of connective tissue called the **serosa.** It secretes a fluid that reduces friction with contacting surfaces.

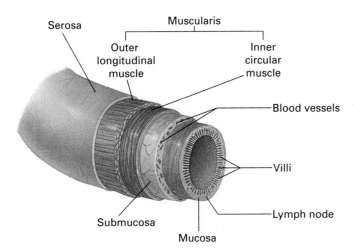

Figure 13–2 Along most of its length, the wall of the digestive system has four basic layers: the mucosa, the submucosa, the muscularis, and the serosa.

Regions of the GI tract are specialized to process food in particular ways. Part of that processing involves **mechanical digestion,** which is physically breaking food into smaller pieces, and part involves **chemical digestion,** which is breaking chemical bonds so that complex molecules are broken into their component subunits. Chemical digestion produces molecules that can be absorbed into the bloodstream and used by the cells. We will trace the path food travels along the GI tract to see how it is processed and absorbed.

Organs of the Digestive System

As food moves along the gastrointestinal tract, it passes through the mouth, pharynx, esophagus, stomach, small intestine, and large intestine. The salivary glands, liver, and pancreas add secretions along the way. Most nutrients are absorbed from the small intestine. Undigested and indigestible materials pass out the anus (Table 13–1).

The Mouth

The first stop on the journey through the digestive tract is the mouth. The mouth serves several functions: (1) mechanical, and to some extent, chemical digestion begins, (2) food quality is monitored, and (3) food is moistened and manipulated so that it can be swallowed.

Teeth and Mechanical Digestion

As we chew, our teeth prepare food for swallowing (Figure 13–3). The sharp, chisel-like incisors in the front of the mouth slice the food as we bite into it. At the same time, the pointed canines to the sides of the incisors tear the food. Foods, such as fruits and vegetables, are then ground, crushed, and pulverized by the premolars and molars that lie along the sides of the mouth. Thus, teeth mechanically break food into smaller fragments, making it easier to swallow.

Teeth are alive. In the center of each tooth is the **pulp,** which contains the tooth's life support systems—blood vessels that nourish the tooth and nerves that sense heat, cold, pressure, and pain (Figure 13–4). Surrounding the pulp is a hard, bone-like substance, called **dentin.** The **crown** of the tooth (the part visible above the gum line) is covered with **enamel,** a nonliving material that is hardened with calcium salts. The **root** of the tooth (the part below the gum line) is covered with a calcified, yet living and sensitive connective tissue called **cementum.** The gums of young people adhere tightly to the enamel, but the gums recede with age, exposing

Table 13–1 Structures of the Digestive System

Structure	Description/Functions	Mechanical Digestion	Chemical Digestion
Gastrointestinal Tract			
Mouth	Receives food; contains teeth and tongue; tongue manipulates food and monitors quality	Teeth tear and crush food into smaller pieces	Digestion of carbohydrates begins
Esophagus	Tube that transports food from mouth to stomach	None	None
Stomach	J-shaped muscular sac for food storage	Churning of stomach mixes food with gastric juice creating liquid chyme	Protein digestion begins
Small intestine	Long tube where digestion is completed and nutrients absorbed	Segmental contractions mix food with intestinal enzymes, pancreatic enzymes, and bile	Carbohydrate, protein, and fat digestion completed
Large intestine	Final tubular region of GI tract; absorbs water and ions; houses bacteria; forms and expels feces	None	None
Anus	Terminal outlet of digestive tract	None	None
Accessory Structures			
Salivary glands	Secrete saliva, a liquid that moistens food and contains an enzyme for digesting carbohydrates	None	Saliva contains enzymes that begin carbohydrate digestion
Pancreas	Digestive secretions include bicarbonate ions that neutralize acidic chyme and enzymes that digest carbohydrates, proteins, fats, and nucleic acids	None	Pancreatic enzymes assist in digestion of carbohydrates, proteins, fats, and nucleic acids
Liver	Digestive function is to produce bile, a liquid that emulsifies fats, making chemical digestion easier and facilitating absorption.	Bile emulsifies fats	Bile facilitates digestion and absorption of fats
Gallbladder	Stores bile and releases it into small intestine	None	None

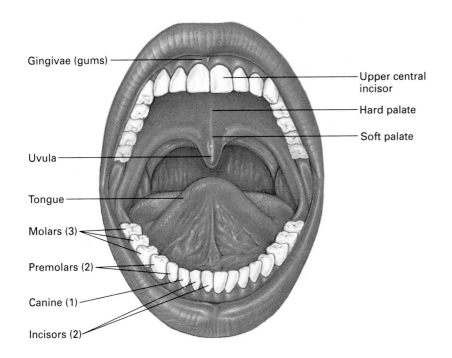

Gingivae (gums) —

Upper central incisor

Hard palate

Soft palate

Uvula —

Tongue —

Molars (3) —

Premolars (2) —

Canine (1) —

Incisors (2) —

Figure 13-3 The structures of the mouth help prepare the food for swallowing and monitor food quality. The teeth slice, tear, and grind food until it can be swallowed. Taste buds on the tongue and palate, combined with sensations from olfactory receptors in the nose, can enhance the pleasure received from eating or warn that the food is spoiled or possibly poisonous and should be rejected.

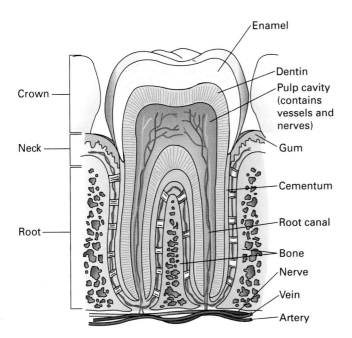

Enamel

Dentin

Pulp cavity (contains vessels and nerves)

Crown —

Gum

Neck —

Cementum

Root canal

Root —

Bone

Nerve

Vein

Artery

Figure 13-4 The structure of the human tooth is suited for its function of breaking food into smaller pieces. In the center of a tooth is the pulp, which contains blood vessels and nerves. A hard substance called dentin surrounds the pulp. On the crown of the tooth, the part visible above the gum line, the dentin is covered with an extremely hard substance called enamel. Cementum, a hard, bone-like material, covers the dentin on the part of the tooth below the gum line, which is called the root. Each tooth fits into a socket in the jawbone.

a region of the cementum. A receding gum line, therefore, increases the sensitivity of the tooth to temperature. The roots of the teeth fit into sockets in the jawbone. Blood vessels and nerves reach the pulp through a tiny tunnel through the root called the **root canal**. Attached to the cementum of the root, a periodontal ligament anchors the tooth to the jawbone.

Critical Thinking

Elderly people are sometimes described as being "long in the tooth." How might the changes in gum line that accompany aging be responsible for this expression?

Tooth decay is caused by acid produced by bacteria living in the mouth. When you eat, food particles become trapped between the teeth and in the regions where the teeth meet the gums. Bacteria in the mouth are nourished by the sugar in these food particles. As the sugar is broken down by bacteria, acid is produced. Although enamel is the hardest substance in the body, it can be eroded by acid produced by bacteria living on the tooth surface. The acid dissolves away the calcium and phosphate that make the enamel hard, causing a cavity to form. Plaque, an invisible film of bacteria, mucus, and food particles, promotes tooth decay

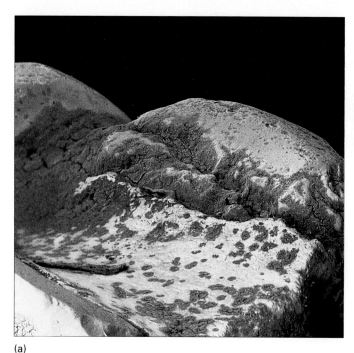

(a)

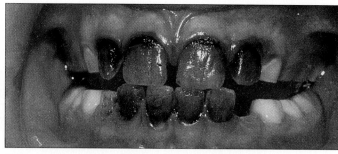

(b)

Figure 13–5 (a) Plaque is a layer of bacteria, mucus, and food debris that forms on the surface of teeth, especially along the gum line and between teeth. The plaque in this photo is rusty orange in color. (b) Although plaque is usually invisible, it is revealed when a person chews a disclosing tablet containing a vegetable dye that stains plaque. *(a, © Meckes/Ottawa/Photo Researchers, Inc. © Science Photo Library/Photo Researchers, Inc.)*

because it holds acid produced by bacteria against the enamel (Figure 13–5). After the enamel has been penetrated, bacteria can invade the softer dentin beneath. If the cavity is not filled by a dentist, the bacteria can infect the pulp (Figure 13–6). The body then responds by sending white blood cells to the pulp to fight the infection. Blood vessels in the pulp widen, increasing blood flow and pressing on nerves in the pulp. It's the pressure on the nerves in the pulp of an infected tooth that causes a toothache. Although it may be a great relief if a toothache goes away on its own, this is generally not a good sign—it may mean that the nerve fibers in the pulp have died. However, the infection originally responsible for the toothache persists and the bacteria responsible can form an abscess, a pus-filled sac in the tissue around the tip of the tooth's root. The bacteria from the abscess can affect the jawbone or enter the bloodstream and cause generalized blood poisoning.

Gum disease, which affects two of three middle-aged persons in the United States, is a major cause of tooth loss in adults. Gingivitis, (*gingiv*, the gums; *itis*, inflammation of) an early stage of gum disease, occurs when plaque that has formed along the gum line causes the gums to become inflamed and swollen. Although it isn't painful at this point, the swollen gums can bleed and they don't fit as tightly around the teeth. The resulting pocket that forms between the tooth and the gum traps additional plaque. The bacteria in the plaque can then destroy the bone and soft tissues around the tooth, causing periodontitis (*peri*, around; *dont*, teeth; *itis*, inflammation of). As the tooth's bony socket and the tissues that hold the tooth in place are eroded, the tooth becomes

loose. Once this occurs, nothing can be done to save the tooth, even one in perfect health without a single cavity.

There are several ways to keep your teeth and gums healthy. Choose a diet that is low in sugar and avoid foods that tend to stick to teeth, such as potato chips. This diet will reduce the food supply for the acid-producing bacteria in plaque. Brush your teeth at least twice a day and floss regularly. Flossing cleans the sides of teeth and removes food particles from crevices that a toothbrush can't reach, helping to reduce plaque formation. Fluoride hardens tooth enamel. So, use fluoridated toothpaste and, if your public drinking water does not contain sufficient fluoride, get advice from your dentist on fluoride treatments. It is also important to visit your dentist regularly to have your teeth professionally cleaned and cavities filled while they are still small.

Social Issue

For the last 50 years, some communities in regions where the water is not naturally fluoridated have added low doses of fluoride to the public water supply to reduce tooth decay. It has been estimated that 98% of Americans have, or have had, tooth decay. Over 6 million teeth are removed each year. Proponents of fluoridation point out that the incidence of tooth decay has declined following the addition of fluoride to drinking water. Opponents of the practice argue that (1) fluoridation of the water supply is a form of forced medication, (2) the dosage cannot be adequately controlled because it will vary with a person's body weight and the amount of water consumed, and that (3) excessive fluoride intake causes teeth to become brown and mottled.

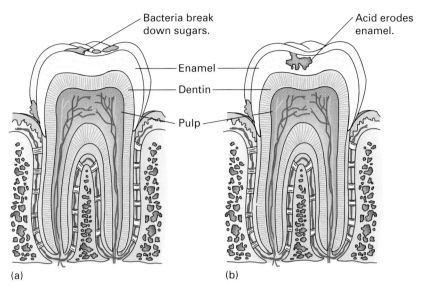

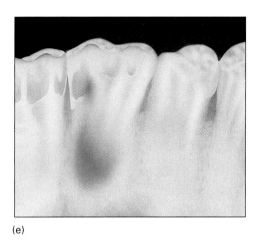

(e)

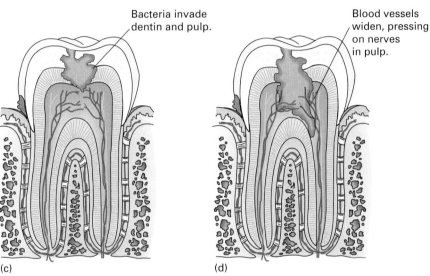

(a)

(b)

Bacteria break down sugars.

Enamel

Dentin

Pulp

Acid erodes enamel.

Bacteria invade dentin and pulp.

Blood vessels widen, pressing on nerves in pulp.

(c)

(d)

Figure 13–6 These drawings illustrate the development of a dental cavity. (a) Tooth decay is caused by acid produced as bacteria living on the tooth surface break down sugars in food particles adhering to the teeth. (b) The acid erodes the tooth's enamel, which causes a cavity to form. (c) Bacteria can then infect the softer dentin beneath the enamel and later the pulp at the heart of the tooth. (d) The body responds by widening blood vessels in the pulp, which increases delivery of white blood cells to fight the infection. When widened blood vessels press on nerves within the pulp, a toothache results. (e) A dental cavity can be seen as a red area in this x-ray of teeth. *(e, © Chris Bjornberg/Photo Researchers, Inc.)*

If you had to vote on fluoridation of your public water supply, what additional information would you want to vote intelligently? How would you vote? Explain your reasons.

Salivary Glands

Three pairs of salivary glands—the sublingual, submaxillary, and parotid—release their secretions, which are collectively called **saliva**, into the mouth (Figure 13–7). As we chew, food is mixed with saliva. Water in saliva moistens food and the mucus binds food particles together. The water in saliva also dissolves chemical components of food. Once dissolved, the chemicals can stimulate receptors in taste buds. Information from the taste buds, along with input from the olfactory receptors in the nose, helps us monitor the quality of food. For instance, spoiled or poisonous food usually tastes

bad and so we can spit it out before swallowing. Saliva contains an enzyme, called **salivary amylase**, that begins to chemically digest starches into shorter chains of sugars. You will notice the result of salivary amylase activity if you chew a piece of bread for several minutes: The bread will begin to taste sweet.

Tongue

The tongue is basically a large skeletal muscle studded with taste buds. Our ability to control the position and movement of the tongue is critical to both speech and the manipulation of food within the mouth. The tongue manipulates food so that it is crushed and ground by teeth, mixed with saliva, and then shaped into a small, soft mass, called a **bolus**, that is easily swallowed. The tongue also initiates swallowing by pushing the bolus to the back of the mouth.

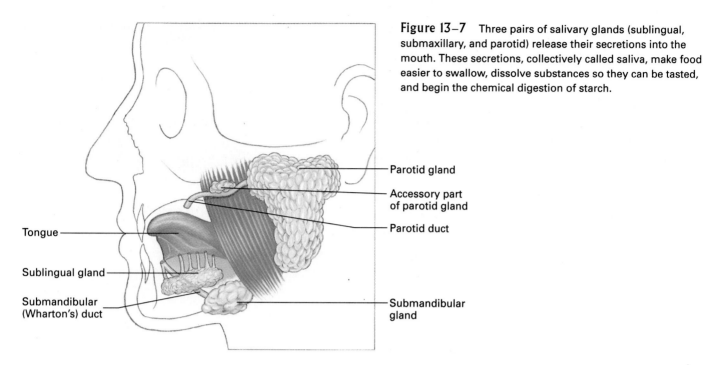

Figure 13-7 Three pairs of salivary glands (sublingual, submaxillary, and parotid) release their secretions into the mouth. These secretions, collectively called saliva, make food easier to swallow, dissolve substances so they can be tasted, and begin the chemical digestion of starch.

Parotid gland

Accessory part of parotid gland

Parotid duct

Tongue

Sublingual gland

Submandibular (Wharton's) duct

Submandibular gland

The Pharynx

When we swallow, food is pushed from the mouth, through the pharynx, and along the esophagus, the muscular tube that leads to the stomach. Swallowing consists of a voluntary component followed by an involuntary one (Figure 13–8).

When a person begins to swallow, the tongue pushes the food into the **pharynx**, which is the space shared by the respiratory and digestive systems that is commonly called the throat. Once food is in the pharynx, it's too late to change one's mind about swallowing. Sensory receptors in the wall of the pharynx detect the presence of food and stimulate the

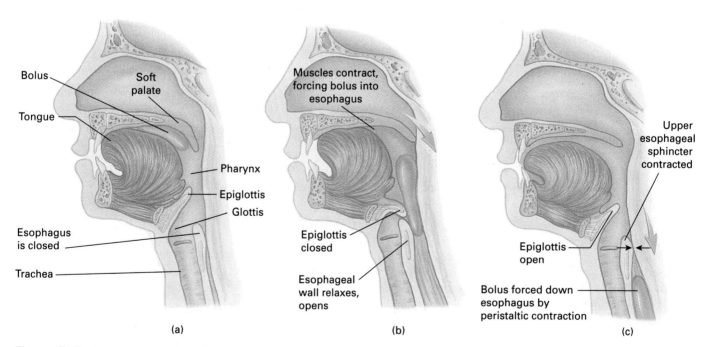

Bolus

Soft palate

Tongue

Pharynx

Epiglottis

Glottis

Esophagus is closed

Trachea

Muscles contract, forcing bolus into esophagus

Epiglottis closed

Esophageal wall relaxes, opens

Upper esophageal sphincter contracted

Epiglottis open

Bolus forced down esophagus by peristaltic contraction

(a)   (b)   (c)

Figure 13-8 Swallowing consists of voluntary and involuntary components. (a) The tongue begins by pushing the food bolus to the back of the mouth. (b) The soft palate rises and prevents food from moving toward the nose in the pharynx. At the same time, the epiglottis moves, covering the glottis (the opening of the respiratory system) and preventing food from entering the trachea (windpipe). (c) The food bolus is then pushed along the esophagus by waves of peristalsis.

involuntary swallowing reflex. The soft palate rises, preventing food from entering the nasal cavity. Immediately afterward, muscles in the wall of the pharynx contract and push the larynx (the voice box, commonly called the Adam's apple) upward. The movement of the larynx causes a cartilaginous flap called the **epiglottis** to move, covering the **glottis** (the opening to the airways of the respiratory system) and preventing food from entering the airways. Instead, as the muscles of the lower part of the pharynx relax, food is pushed into the esophagus.

The Esophagus

The **esophagus** is a muscular tube about 25 cm (10 in.) long that conducts food from the pharynx to the stomach. Food is moved along the esophagus and along the entire digestive tract by a wave of muscle contraction called **peristalsis**. The presence of food stretches the walls in one region of the tube and this triggers the contraction of circular muscles in the region of the tube immediately behind the food mass. Circular muscles are arranged in rings around the tube. So, when they contract, that region of the tube pinches inward, pushing food forward. The food then stretches the next adjacent region of the tube, again stimulating contraction of circular muscles behind it. At the same time, longitudinal muscles in front of the food contract, shortening this region and widening its walls to receive the food (Figure 13–9). We see, then, that gravity is not important in moving food along the digestive tract. It is possible, therefore, to swallow while standing on your head or in the weightless conditions of outer space.

The Stomach

The **stomach** is a muscular sac that is well designed to carry out its three important functions—storage of food, liquefaction of food, and the initial chemical digestion of proteins (Figure 13–10).

Storage of Food

Like any good storage compartment, the stomach is expandable and has adjustable openings that can close to seal the contents within or can open to fill or empty the compartment. When empty, the stomach is a small J-shaped sac, about the size of a sausage, that can hold only about 50 ml (a quarter of a cup). The empty stomach's wall has folds, called **rugae**, that can unfold, allowing the stomach to expand as it fills. The smooth muscle of the stomach wall is quite elastic and can stretch to accommodate large meals. When fully expanded, the stomach can hold several liters of food. Stretch receptors in the stomach wall signal the appetite center in the hypothalamus of the brain when the stomach is full. The appetite center then helps quell the urge to eat.

The opening at each end of the stomach is guarded by a band of circular muscle called a sphincter. Whereas contraction of a sphincter closes an opening, relaxation of a sphinc-

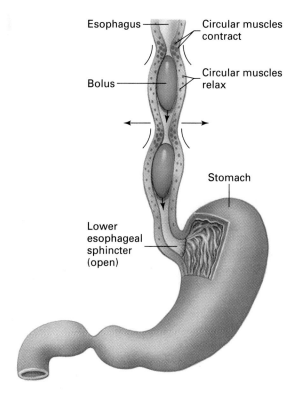

Figure 13–9 Peristalsis is a wave of muscle contraction that pushes food along the esophagus and the entire digestive tube. Peristalsis occurs when food stretches one region of the tube, triggering the contraction of circular muscles in the region immediately behind the food mass. When circular muscles contract, the tube is narrowed and food is pushed forward.

ter allows material to pass through. The **lower esophageal sphincter** is located at the juncture of the esophagus and the stomach. When it relaxes, swallowed food can enter the stomach. Contraction of the lower esophageal sphincter normally prevents food from re-entering the esophagus while it is being stored and churned within the stomach. The **pyloric sphincter** between the stomach and small intestine regulates the emptying of the stomach.

Liquefaction of Food

Food is generally stored and processed within the stomach for 3 to 5 hours. Mechanical digestion occurs during this time as the food is churned and mixed with gastric juices produced by the glands of the stomach until it is a soupy mixture called **chyme**. The stomach wall has three layers of smooth muscle, each oriented in a different direction. The coordinated contractions of these layers twist, knead, and compress the stomach contents, physically breaking food into smaller pieces.

Chemical Digestion

In the adult, chemical digestion in the stomach is limited to the initial breakdown of proteins. The lining of the stomach

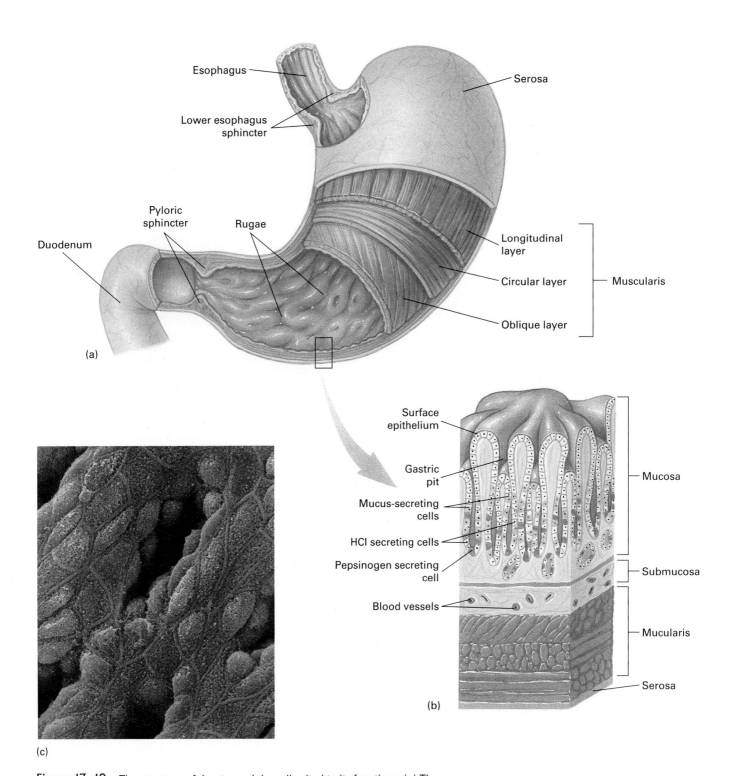

Esophagus

Lower esophagus
sphincter

Serosa

Pyloric
sphincter

Rugae

Duodenum

Longitudinal
layer

Circular layer

Oblique layer

Muscularis

(a)

Surface
epithelium

Gastric
pit

Mucus-secreting
cells

HCl secreting cells

Pepsinogen secreting
cell

Blood vessels

Mucosa

Submucosa

Mucularis

Serosa

(b)

(c)

Figure 13–10 The structure of the stomach is well suited to its functions. (a) The stomach wall has three layers of smooth muscle, each oriented in a different direction. The contractions of these muscles cause churning movements that physically break food into smaller pieces and mix it with gastric juice to form a soupy liquid called chyme. The stomach wall has many folds, called rugae, that can unfold and allow the stomach to expand as it fills with food. (b) Gastric glands in the wall of the stomach produce gastric juice, a mixture of hydrochloric acid and pepsin (a protein-splitting enzyme). (c) The gastric glands release their secretions through tiny openings in the stomach wall called gastric pits. The holes seen in this electron micrograph are the gastric pits. (© Fred Hossler/Visuals Unlimited)

has millions of gastric pits, within which are **gastric glands** that contain several types of secretory cells. Certain secretory cells produce **hydrochloric acid** (**HCl**), which kills most of the bacteria swallowed with food or drink, denatures proteins in food, breaks down the connective tissue of meat, and activates pepsinogen (the inactive form of **pepsin**, a protein-digesting enzyme). Pepsinogen is secreted by other cells in the gastric glands. While still within the gastric gland, HCl breaks off a portion of the pepsinogen molecule, forming pepsin. The mixture of pepsin and HCl, which is called **gastric juice**, is released into the stomach where the pepsin begins the chemical digestion of protein in food. Yet other cells within the gastric glands secrete **mucus**, which helps protect the stomach from the action of gastric juice. Although not related to the digestive function of the stomach, a very important material secreted by the gastric glands is **intrinsic factor**, a protein necessary for the absorption of vitamin B_{12} from the small intestine.

The stomachs of newborns contain two additional enzymes, rennin and gastric lipase, that are important in the digestion of milk. **Rennin** coagulates milk protein, making it curdy, like soured milk. **Gastric lipase** begins the digestion of milk fats.

The stomach wall is composed of the same materials that gastric juice attacks, and so you might wonder why the stomach does not digest itself. One reason is mucus, as mentioned. Mucus forms a thick protective coat that prevents gastric juice from reaching the cells of the stomach wall, and its alkalinity helps neutralize the HCl. In addition, pepsin is produced in an inactive form that cannot digest the cells that produce it. The stomach is also protected by neural and hormonal reflexes that regulate the production of gastric juice so that little is released unless food is present. Food absorbs and dilutes gastric juice. Finally, if the stomach lining is damaged, it is quickly repaired. Indeed, the high rate of cell division of the cells of the stomach lining replaces a half million cells every minute, creating a new stomach lining every 3 days!

Very little absorption of food materials occurs in the stomach. This is largely because food simply hasn't been broken down into molecules small enough to be absorbed. Notable exceptions are alcohol and aspirin. The absorption of alcohol from the stomach is the reason its effects can be felt so quickly, especially if there is no food present to dilute the alcohol. The absorption of aspirin can cause bleeding of the wall of the stomach, which is the reason aspirin should be avoided by people who have stomach ulcers (discussed in Personal Concerns: Heartburn and Peptic Ulcers).

When the food has been adequately prepared by the stomach and the small intestine is ready to receive it, the pyloric sphincter relaxes and allows peristaltic contractions of the stomach to squirt chyme into the small intestine. Through neural and hormonal reflexes, several factors work together to ensure that the small intestine is not overloaded with more chyme than it can process. On average, the stomach is emptied within 4 hours of finishing a meal, but there can be considerable variation. In general, the larger the meal, the greater the distention of the stomach and the more quickly it is emptied from the stomach. However, the degree to which food is liquefied also plays a role. Solid food generally stays until it has been mixed with gastric juice and converted to chyme. Thus, a bowl of soup will leave the stomach more quickly than an equal volume of hamburger. In addition, the nutritive content of food is important. When food entering the small intestine is high in fats, a hormonal message is sent to the stomach that slows emptying. As a result, a high fat meal can remain in the stomach for more than 6 hours. Foods rich in carbohydrates pass through the stomach the quickest, and protein-rich foods pass in an intermediate amount of time.

Critical Thinking

Many oriental foods contain mainly rice and vegetables (carbohydrates). How does this explain why you might feel hungry again an hour or two after eating Chinese food, such as chicken chow mein?

The Small Intestine

The major site of digestion and absorption of nutrients is the **small intestine**. The small intestine is a narrow tube, only about 2.5 cm (1 in.) in diameter. In a living person, it's about 2 m (6 ft) long, but since it has been more commonly measured in cadavers, it's usually described as 6 to 7 m (about 20 ft) long. The difference in length is due to muscle tone (the sustained, partial contraction of muscle) of the muscles of the intestinal wall in a living person. As food moves along this twisted tube, it passes through three specialized regions—the duodenum, the jejunum, and the ileum.

The duodenum receives acidic chyme from the stomach. There are several ways in which the walls of the duodenum are protected from the acid. The acid is neutralized by bicarbonate ions in the pancreatic juice that empties into the region. In addition, the duodenum secretes mucus, which forms a protective blanket preventing the acid from reaching the cells. Cells that are killed by the acid in spite of these protective mechanisms are rapidly replaced.

Critical Thinking

Radiation and chemotherapy are effective treatments for cancer because they kill rapidly dividing cells. These treatments also kill other cells that divide rapidly. How might this explain the nausea and diarrhea that often accompany radiation treatment or chemotherapy?

Heartburn and Peptic Ulcers—Those Burning Sensations

Heartburn is a burning sensation behind the breastbone that occurs when the acidic gastric juice backs up into the esophagus. At least 30% of all Americans suffer from heartburn on a monthly basis and 10% on a daily basis. Heartburn occurs when the pressure of the stomach contents overwhelms the lower esophageal sphincter. In some people, the lower esophageal sphincter is weak and is ineffective in keeping stomach contents out of the esophagus. A person with a normal lower esophageal sphincter will experience heartburn when the stomach contents exert a greater pressure than usual against the sphincter, as might occur after a large meal, during pregnancy, when lying down, or when constipated.

Acidic foods, such as tomatoes and citrus fruits, and spicy, fatty, or caffeine-containing products can aggravate the problem. Tight clothing also worsens heartburn.

There is now a variety of treatments for heartburn. Antacids will provide temporary relief. Unfortunately, long-term uses of certain antacids can have undesirable side effects. The fizzing bicarbonate types are high in sodium and should be avoided by persons with high blood pressure. Others are high

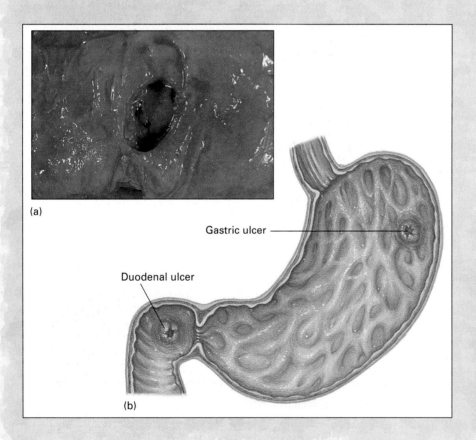

(a)

Gastric ulcer

Duodenal ulcer

(b)

Figure A A peptic ulcer is a raw area that forms when gastric juice erodes the lining of the esophagus, stomach, or most commonly the duodenum. The ulcer shown in this stomach wall is bleeding. The most common symptom is abdominal pain that occurs when the stomach is empty. *(a, © Dr. E. Walker/Science Photo Library/Photo Researchers, Inc.)*

in calcium, which might promote the formation of kidney stones and also stimulate the production of stomach acid. Newer medications, such as cimetidine (Tagamet), ranitidine (Zantac), famotidine (Pepcid), nizatidine (Axid), and omeprazole (Prilosec) shut down the production of stomach acid. Another medication, metoclopramide (Reglan) works by temporarily tightening a weakened lower esophageal sphincter. In addition to tightening the lower esophageal valve, cisapride (Propulsid) is thought to stimulate muscle contractions in the digestive system, causing the stomach to empty its contents into the small intestine more effectively.

Critical Thinking

Why would it be wise for a person who experiences heartburn frequently to avoid eating a large meal within 3 hours of bedtime?

In nearly 13% of Americans, the mechanisms that protect the stomach and duodenum from their acidic contents fail at some point in their lives, allowing the acidic gastric juice to erode the lining of some region in the gastrointestinal tract. The resulting sore resembles a canker sore of the mouth and is called a **peptic ulcer** (Figure A). Although a peptic ulcer may form in the esophagus or the stomach, the most common site is the duodenum. An ulcer is usually between 10 mm and 25 mm (0.33 in. and 1 in.) in diameter and may occur singly or in several places.

The symptoms of an ulcer are variable. A common symptom is abdominal pain, which can be quite severe. Vomiting, loss of appetite, bloating, indigestion, and heartburn are other common symptoms. However, some people with ulcers, especially those who are taking nonsteroidal anti-inflammatory agents (NSAIDS), have no pain. Unfortunately, the degree of pain is a poor indicator of the severity of ulceration. Often people with symptomless ulcers are unaware of the problem until more serious complications develop. Gastric juice can erode the gut lining until it bleeds and in some cases it can eat a hole completely through the gut wall (a perforated ulcer). Recurrent ulcers can cause scar tissue to form and this may narrow or block the lower end of the stomach or duodenum.

Although acidic gastric juice is the direct cause of peptic ulcers, factors that interfere with the mechanisms that normally protect the lining of the gastrointestinal tract from the acid are considered to be the real causes. NSAIDS, which include aspirin, ibuprofen, and naproxen, can cause ulcers because they slow the production of chemicals called prostaglandins, which normally help protect the gut lining from damage by acid.

Critical Thinking

Peptic ulcers often cause a burning pain in the upper abdomen or lower chest. If the ulcer is in the stomach, the pain worsens while eating. On the other hand, the pain from an ulcer in the duodenum typically begins 30 to 60 minutes after a meal. Explain this difference in the timing of the onset of pain between ulcers in the stomach and those in the duodenum.

The leading cause of peptic ulcers, however, is thought to be infection with the bacterium *Helicobacter pylori*. More than 80% of those people with ulcers in the stomach or duodenum are infected with *H. pylori*. These corkscrew-shaped bacteria live in the layer of mucus that protects the gut lining. Here, partially protected from gastric juice, the bacteria attract body defense cells, called macrophages and neutrophils, that cause inflammation leading to ulcer formation. Toxic chemicals produced by the bacteria also cause ulcers. *H. pylori* causes an infection that may last for years. It affects more than a billion people throughout the world and approximately 50% of the people in the United States who are over 60 years of age. For some reason, however, only about 10% to 15% of those who are infected actually develop peptic ulcers.

The initial treatment for ulcers is usually antacids. Stronger medicines to reduce acid production, such as those mentioned as heartburn treatment, may be necessary if antacids are not effective in relieving symptoms. Another drug, called Carafate, works by creating a protective coating over the ulcer that protects it from gastric juice and allows it to heal. Since the discovery that an infection with *H. pylori* causes most peptic ulcers, antibiotics have been used to treat ulcers. Antibiotic treatment has been shown to heal ulcers and prevent their recurrence. Besides ulcers, an *H. pylori* infection causes stomach cancer. It is hoped that it will soon be possible to vaccinate children against this bacterium, thus preventing both peptic ulcers and stomach cancer.

Table 13-2 Major Digestive Enzymes

Site of Production	Enzyme	Site of Action	Substrate	Main Digestive Products
Salivary Glands	Amylase	Mouth	Polysaccharides	Shorter polysaccharides
Stomach	Pepsin	Stomach	Proteins	Polypeptides (protein fragments)
Pancreas	Trypsin	Small intestine	Proteins and polypeptides	Peptides (smaller protein fragments)
	Chymotrypsin	Small intestine	Proteins and polypeptides	Peptides (smaller protein fragments)
	Amylase	Small intestine	Polysaccharides	Disaccharides
	Carboxypeptidase	Small intestine	Polypeptides	Amino Acids
	Lipase	Small intestine	Triglycerides (fats)	Fatty acids, monoglycerides, and glycerol
	Nucleases (deoxyribonuclease and ribonuclease)	Small intestine	DNA, RNA	Nucleotides
Small Intestine	Maltase	Small intestine	Maltose	Monosaccharides (glucose)
	Sucrase	Small intestine	Sucrose	Monosaccharides (glucose and fructose)
	Lactase	Small intestine	Lactose	Monosaccharides (glucose and galactose)
	Aminopeptidase	Small intestine	Peptides	Amino acids

Digestive Activities

Within the small intestine, a battery of enzymes completes the chemical digestion of virtually all the carbohydrates, proteins, fats, and nucleic acids in the food. Although both the small intestine and the pancreas contribute enzymes, most of the digestion that occurs in the small intestine is due to pancreatic enzymes (Table 13–2). The **pancreas** is an accessory organ that lies behind the stomach, extending toward the left from the duodenum. In addition to enzymes, pancreatic juice contains water and ions, including bicarbonate ions that are important in neutralizing the acid in chyme. Pancreatic juice drains into the pancreatic duct, which fuses with the common bile duct from the liver just before entering the duodenum (Figure 13–11).

The enzymes produced by the small intestine remain attached to the epithelial cells lining the small intestine. Together the pancreatic enzymes and intestinal enzymes break nutrients into their component building blocks: proteins to amino acids, carbohydrates to monosaccharides, and triglycerides (a type of fat) to fatty acids and glycerol.

Fats present a special digestive problem. Fats are insoluble in water, as you have observed when the oil quickly separates from the vinegar in your salad dressing. Thus, tiny fat droplets tend to coalesce into large globules. A problem arises because lipase, the enzyme that chemically breaks down fats, is soluble in water and not in fats. As a result, lipase can work only at the surface of a fat globule. Large fat globules have less combined surface area than do smaller droplets.

Bile, a mixture of water, ions, cholesterol, bile pigments, and bile salts produced by the liver, plays an important role in the digestion of fats. Bile salts emulsify fats, that is, they keep fats separated into small droplets. Tiny fat droplets become coated with bile salts and the negative

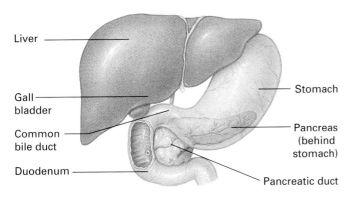

Figure 13–11 The pancreas, liver, and gallbladder are accessory organs of the digestive system. The pancreas produces several digestive enzymes that act in the small intestine. The liver produces bile, which is stored in the gallbladder before being released into the small intestine where bile eases fat digestion by emulsifying fats.

charges on the bile salts keep the droplets separated. The coated droplets expose a large combined surface area to lipase, facilitating fat digestion and absorption.

After its production by the liver, bile is stored in a muscular pear-sized sac, called the **gallbladder**. During storage, bile is modified and concentrated. When chyme is present in the small intestine, a hormone causes the gallbladder to contract, squirting its bile through the common bile duct into the duodenum.

Bile is rich in cholesterol and sometimes, if the balance of dissolved substances in bile becomes upset, a tiny particle precipitates out of solution. As cholesterol and other substances build up around the particle, a **gallstone** forms (Figure 13–12). Many people form several gallstones.

About 20 million people in the United States have gallstones and roughly a million new cases develop annually. One-third to one-half of these people do not experience symptoms. A problem can develop, however, if the gallstones block the flow of bile from the gallbladder. If a gallstone prevents the gallbladder from emptying after a meal as it normally would, the pressure within the gallbladder builds, causing intense pain in the right side or center of the upper abdomen. If a gallstone becomes lodged in the common bile duct, bile may build up behind the stone and cause jaundice, a condition in which the skin develops a yellowish tone caused by the build-up of bilirubin (a bile pigment) in the bloodstream and then in body tissues.

Although there are some nonsurgical treatments for gallstones, these are not always successful. For instance, there are medicines that can dissolve cholesterol gallstones. The stones do not always dissolve completely, however. Even when the stones do completely dissolve, they form again within 5 years in half of the patients. Lithotripsy, which is treatment with shock waves delivered through water, successfully breaks up a gallstone in 90% of the patients who have only one stone. Unfortunately, it is unusual to form only one gallstone. For these reasons, surgical removal of the gallbladder remains the most common treatment. If

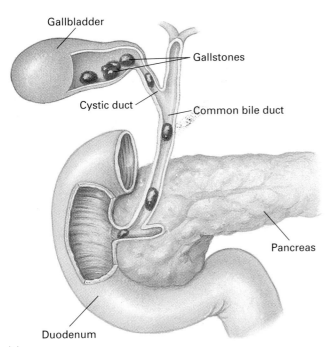

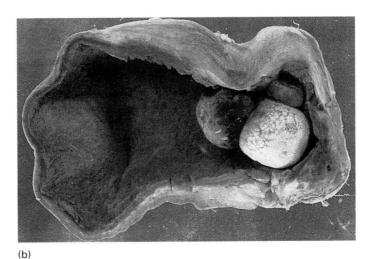

(b)

Figure 13–12 Gallstones are composed primarily of cholesterol that has precipitated out of bile during storage in the gallbladder. (a) A gallstone can intermittently or continuously block the ducts that drain bile into the small intestine. When a blocked gallbladder contracts after a meal, it is unable to empty its contents and the pressure within it increases, causing intense pain. (b) This photograph shows the gallbladder and several gallstones. *(b, © Biophoto Associates/Science Source/Photo Researchers, Inc.)*

the gallbladder is removed, bile drains directly from the liver into the small intestine. Thus, people who have had their gallbladders removed can still digest their food.

Absorption of Nutrients

Each day as much as 10 liters (approximately 2.5 gallons) of food and liquid enters the small intestine, but only 0.5 to 1 liter (approximately 2 to 4 cups) reaches the large intestine. We see, then, that the small intestine is the primary site of absorption in the digestive system.

The small intestine is able to absorb materials so effectively because it has a vast surface area due to several structural specializations, in addition to its length (Figure 13–13). First, the entire lining of the small intestine is

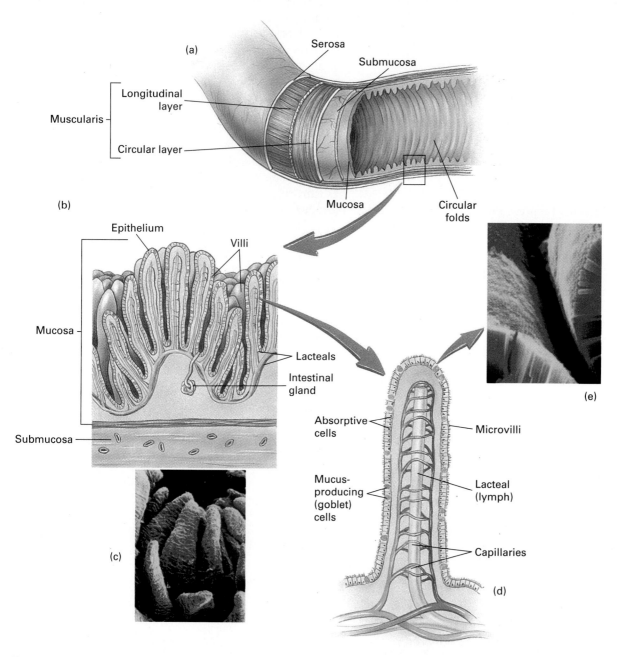

Figure 13–13 The small intestine is specialized for the absorption of nutrients by structural modifications that increase its surface area. (a) Its wall contains accordion-like pleats called circular folds. (b) The lining has numerous finger-like projections called villi. (c) This electron micrograph shows the intestinal villi. (d) The surface of each villus is bristled with thousands of microscopic projections of cell membranes called microvilli. In the center of each villus, a network of blood capillaries, which carries away absorbed products of protein and carbohydrate digestion as well as ions and water, surrounds a lacteal (a small vessel of the lymphatic system) that carries away the absorbed products of fat digestion. (e) An electron micrograph of the microvilli that cover each villus. *(c, © G. Shih-R. Kessel/Visuals Unlimited; e, © VU/SIU/Visuals Unlimited)*

pleated, like an accordion. These permanent folds not only increase the surface area for absorption, they also cause chyme to flow through the small intestine in a spiral pattern. The spiral flow helps mix the chyme with digestive enzymes and increases contact with the absorptive surfaces. Covering the entire lining surface are tiny projections, called **villi** (singular *villus*, tuft of hair). These give the lining a velvety appearance. Like the pile on a bath towel, the villi increase the absorptive surface. Indeed, these 1-mm projections increase the surface area of the small intestine tenfold. Also, thousands of microscopic projections, called **microvilli**, cover the surface of each villus, increasing the surface area of the small intestine another 600 times. It is estimated that every epithelial cell that functions in absorption within the small intestine is bristled with 3000 microvilli. The microvilli form a fuzzy border, known as the **brush border**, on the surface of absorptive epithelial cells. The circular folds, villi,

and microvilli create a surface area of 300 to 600 square meters—greater than the size of a tennis court!

The core of each villus is penetrated by a network of capillaries and a lacteal, which is a lymphatic vessel. Thus, as substances are absorbed from the small intestine, they must cross only two cell layers—the epithelial cells of the villi and the wall of either the capillary or the lacteal. Most materials enter the epithelial cells by active transport, facilitated diffusion, or diffusion (Figure 13–14). Monosaccharides, amino acids, water, ions, vitamins, and minerals then diffuse across the capillary wall into the bloodstream and are delivered to body cells. After forming, the products of fat digestion quickly combine with bile salts, creating particles called micelles. When a micelle contacts an epithelial cell of a villus, the products of fat digestion easily diffuse into the cell. Within an epithelial cell, glycerol and fatty acids are reassembled into triglycerides, mixed with cholesterol and phospholipids, and coated with special proteins, creating a complex known as a **chylomicron**. The protein coating makes the fat soluble in water, allowing it to be transported throughout the body. The chylomicrons leave the epithelial cell by exocytosis. Chylomicrons are too large to pass through capillary walls. However, they easily enter the more porous lacteal and enter the lymphatic system, which carries them to the bloodstream.

The Liver

The nutrient-laden blood from the capillaries in the villi goes first to the liver, the largest internal organ in the body, which has a variety of metabolic and regulatory roles (Figure 13–15). We have already seen that its primary role in digestion is the production of bile. One of the liver's other roles is to monitor the glucose level of the blood, removing excess glucose and storing it as glycogen or breaking down glycogen to raise blood glucose levels. Thus, the liver keeps the glucose levels of the blood at a steady level. The liver also packages lipids with protein-carrier molecules to form lipoproteins that transport lipids in the blood. In addition, the liver removes poisonous substances, including lead, mercury, and pesticides, from the blood and, in some cases, breaks them down into less harmful chemicals. Also, the liver converts the breakdown products of amino acids into urea, which can then be excreted by the kidney. These are but a few of the approximately 500 functions of the liver.

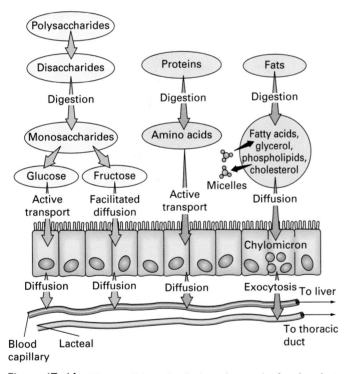

Figure 13–14 The small intestine is the primary site for chemical digestion and absorption. Disaccharides are digested to monosaccharides and proteins to amino acids. Triglycerides (the most common form of fat in the diet) are digested to fatty acids and glycerol. The digestive products enter absorptive epithelial cells of villi by active transport, facilitated diffusion, or diffusion. Monosaccharides, along with water, ions, and vitamins, then enter the capillaries within the villus and are carried to body cells by the bloodstream. Triglycerides are resynthesized from fatty acids and glycerol within the epithelial cells. The triglycerides are then mixed with cholesterol and phospholipids and coated in protein, forming tiny droplets called chylomicrons. Chylomicrons enter the lacteal, a lymphatic channel, in the core of the villus and are delivered to the bloodstream by the lymphatic system.

The Large Intestine

Materials that have not been absorbed in the small intestine move into the final segment of the gastrointestinal tract, the large intestine. The principal functions of the large intestine are to absorb most of the water remaining in the indigestible food residue (thereby adjusting the consistency of feces), to store the feces, and to eliminate them from the body. The large intestine is home to many types of bacteria, some of which produce vitamins that may be absorbed for use by the body.

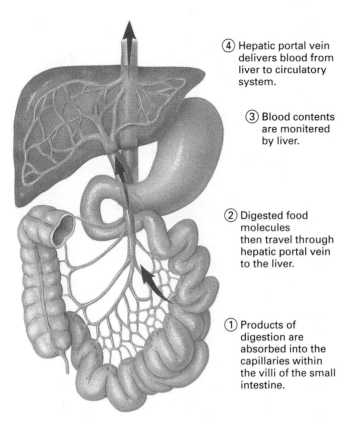

④ Hepatic portal vein delivers blood from liver to circulatory system.

③ Blood contents are monitered by liver.

② Digested food molecules then travel through hepatic portal vein to the liver.

① Products of digestion are absorbed into the capillaries within the villi of the small intestine.

Figure 13–15   Blood flows from the capillaries of the villi to the liver, which monitors blood content and processes nutrients before they are delivered to the bloodstream.

Structure of the Large Intestine

At 7 cm (nearly 3 inches), the diameter of the large intestine is more than two and a half times as great as that of the small intestine. It is slightly shorter than the small intestine, however. It measures 1.5 m (about 5 ft) compared with the small intestine's 2 m (6 ft) in a living adult.

The large intestine has four regions: the cecum, colon, rectum, and anal canal (Figure 13–16). The cecum is a pouch that hangs below the junction of the small and large intestines. Extending from the cecum is another slender, worm-like pouch, called the **appendix**. The appendix has no digestive function. (It is thought that the appendix plays a role in the immune system, which protects the body against disease.)

Each year about 1 of 500 people develops appendicitis, inflammation of the appendix. Appendicitis is usually caused by an infection that develops in the appendix after it becomes blocked by a piece of hardened stool, food, or a tumor. At first, appendicitis is usually experienced as vague bloating, indigestion, and a mild pain in the region of the navel (bellybutton). As the condition worsens, the pain becomes more severe and is localized in the region of the appendix, the lower right abdomen. The pain is typically accompanied by fever, nausea, and vomiting. Diagnosis of appendicitis is often difficult because symptoms can be variable, and other conditions, such as kidney stones, urinary infections and, in women, ovarian cysts, have similar symptoms. This is unfortunate because an untreated infection in the appendix usually causes the appendix to rupture, allowing its contents to spill into the abdominal cavity. This usu-

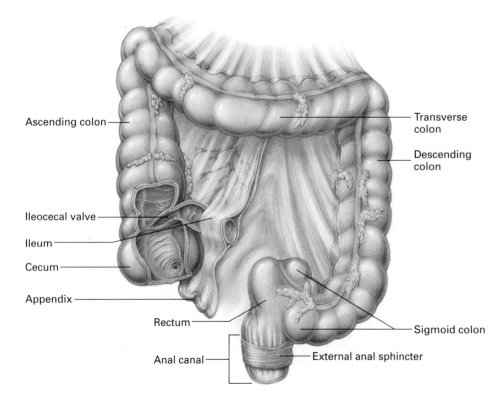

Ascending colon

Ileocecal valve

Ileum

Cecum

Appendix

Rectum

Anal canal

Transverse colon

Descending colon

Sigmoid colon

External anal sphincter

Figure 13–16   The large intestine consists of the cecum, colon, rectum, and anal canal. It absorbs water from undigested material, forming the feces, and houses bacteria.

ally leads to peritonitis, an infection and inflammation throughout the abdomen, which is potentially fatal.

The largest region of the large intestine, the **colon**, is composed of the ascending colon on the right side of the abdomen, the transverse colon across the top of the abdominal cavity, and the descending colon on the left side. A watery mixture of material that was not digested or absorbed in the small intestine enters the colon through the ileocolic valve. Although much of the water that was originally in chyme was absorbed in the small intestine, the material entering the colon is still quite liquid. As it passes through the colon, 90% of the remaining water and sodium and potassium ions are absorbed. The remaining material, known as **feces**, consists primarily of undigested food, sloughed-off epithelial cells, and millions of bacteria. The brown color of feces comes from bile pigments.

The bacteria, which account for nearly a third of the dry weight of feces, are not normally disease-causing and, in fact, are beneficial. Intestinal bacteria produce several vitamins that we are unable to produce on our own, including vitamin K and some of the B vitamins. Some of the vitamins are then absorbed from the colon for our own use. Indeed, there are roughly 50 species of bacteria, including the well-known *Escherichia coli*, living in the healthy colon. The bacteria are nourished by undigested food and material that we are unable to digest, including certain components of plant cells. When the intestinal bacteria use the undigested food for their own nutrition, their metabolic processes liberate gas that sometimes has a foul odor. Although most of the gas produced is absorbed through the intestinal walls, the remaining gas can produce some embarrassing moments when it is released as flatus.

People who are lactose intolerant lack the enzyme lactase, which is normally produced by and acts in the small intestine. Lactase breaks down lactose, the primary sugar in milk, into its component monosaccharides. Without lactase, then, lactose moves into the colon where it provides a nutritional bonanza for the bacteria living there. As a result, when people who are lactose intolerant consume milk products, the colonic bacteria ferment the lactose and produce carbon dioxide that, in turn, produces bloating, gas, and abdominal discomfort. Although lactose intolerance is common in adults, it is not dangerous. Problems can usually be avoided by swallowing capsules or tablets of lactase or by modifying the diet to avoid dairy products.

Critical Thinking

Certain foods, beans for instance, are notorious for producing intestinal gas. Beans contain large amounts of certain short-chain carbohydrates that our bodies are unable to digest. Explain how the nutritional content of beans is related to flatulence.

Periodic peristaltic contractions move material through the large intestine. These contractions are slower than those in the small intestine. Consequently, it generally takes about 18 to 24 hours for material to pass through the colon, but only 3 to 10 hours to pass through the small intestine. Eventually, the feces are pushed into the **rectum**, stretching the rectal wall and initiating the **defecation reflex**. Nerve impulses from the stretch receptors in the rectal wall travel to the spinal cord, which sends back motor impulses that stimulate muscles in the rectal wall to contract and propel the feces into the **anal canal**. Two rings of muscles, called sphincters, must relax to allow defecation, the expulsion of feces. The internal sphincter relaxes as part of the defecation reflex. The external sphincter is under voluntary control, allowing us to decide whether the situation is appropriate for defecation. If it is, conscious contraction of abdominal muscles can increase abdominal pressure and help expel the feces.

The water absorption that occurs in the colon adjusts the consistency of feces. When material passes through the colon too rapidly, as might occur when colon contractions are stimulated by toxins produced by microorganisms or by excess food or alcoholic drink, too little water is absorbed. As a result, the feces are very liquid. This condition, which results in frequent loose stools, is called diarrhea. Diarrhea can be dangerous, especially in an infant or young child, because it can lead to dehydration. It is one of the major causes of death worldwide.

On the other hand, if material passes through the colon slowly, too much water is absorbed, resulting in infrequent, hard stools, a condition called constipation. People who are constipated may find it difficult and painful to have a bowel movement.

Control of Digestive Activities

As we have seen, food material moves along the gastrointestinal tract, stopping for specific treatment at stations along the way. Both nerves and hormones play a role in orchestrating the release of digestive secretions, timing the release of each to prepare for or to correspond to the presence of food at each stop (Figure 13–17).

Food spends little time in the mouth and so, to be effective, saliva must be released quickly. Since nervous stimulation is faster than hormonal, it isn't surprising that the nervous system is in complete control of salivation. Some saliva is released before food is even in the mouth. Indeed, your mouth may begin to water simply at the *thought* of food and certainly begins at the sight or smell of it. Although some saliva is continuously produced to keep the mouth moist, this anticipatory salivation more adequately prepares the mouth to receive food. (At the same time that the brain initiates the secretion of saliva, the digestive juices begin to be secreted in the stomach and small intestines.) The major trigger for salivation, however, is the presence of food in the

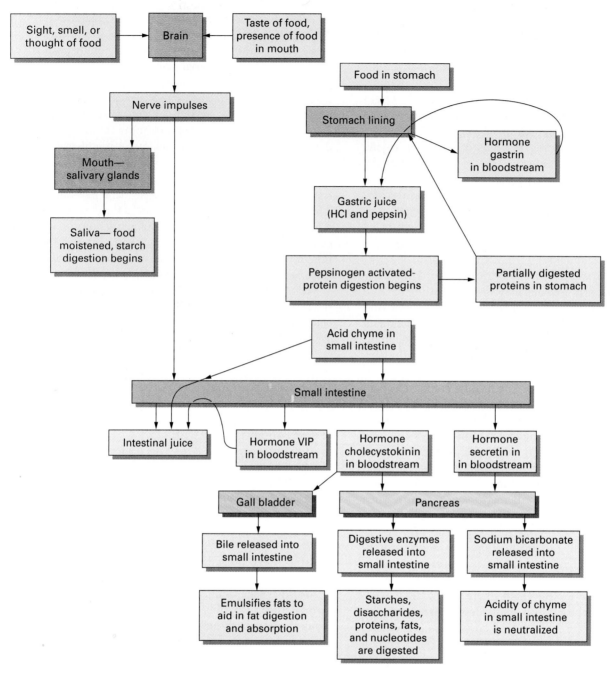

Figure 13–17   Nerves and hormones regulate the release of digestive secretions so that secretions important to digestion are present when and where they are needed.

mouth—its taste and pressure. Salivary juices continue to flow for some time after the food is swallowed, helping to rinse out the mouth.

While food is still being chewed, neural reflexes stimulate the stomach lining to begin secreting gastric juice and mucus. After swallowed food enters the stomach, its presence is detected by chemoreceptors and distention of the stomach stimulates stretch receptors. These receptors then initiate reflexes that cause the stomach to churn and to step up the production of stomach secretions. As gastric juice is mixed with food, proteins begin to be digested. The distention of the stomach, along with the presence of partially digested proteins, stimulate cells in the stomach lining to release the hormone **gastrin**. Gastrin enters the bloodstream and circulates throughout the body and back to the stomach where it increases the production of gastric juice.

The presence of acidic chyme in the small intestine is the most important stimulus for the release of enzymes from both the small intestine and the pancreas, as well as bile from the gallbladder. Local nerve reflexes that are triggered by acid chyme are the most important factor regulating motility and the release of intestinal secretions. Acid chyme also causes the small intestine to release several hormones that, in turn, are responsible for the release of digestive enzymes and bile. For instance, one hormone, **vasoactive intestinal peptide (VIP)**, is released from the small intestine into the bloodstream and is carried back to the small intestine where it causes the release of intestinal juices. At the same time, the small intestine releases a second hormone, **secretin**, which stimulates the release of sodium bicarbonate from the pancreas into the small intestine where sodium bicarbonate helps neutralize the acidity of chyme. A third hormone from the small intestine is **cholecystokinin**, which causes the pancreas to release its digestive enzymes and the gallbladder to contract and release bile.

SUMMARY

1. The digestive system consists of the gastrointestinal tract (mouth, esophagus, stomach, small intestine, and large intestine) and several accessory organs (salivary glands, pancreas, and liver).

2. The mouth serves several functions. Teeth tear and grind food, making it easier to swallow. The salivary glands produce salivary amylase, which is released into the mouth where it begins the chemical breakdown of starches. Taste buds help monitor the quality of food. Finally, the tongue manipulates food so that it can be swallowed.

3. When we swallow, food is pushed from the mouth and waves of muscle contraction called peristalsis push the food along the esophagus, a tube that leads to the stomach.

4. The stomach stores food, liquefies it by mixing it with gastric juice, begins the chemical digestion of proteins, and regulates the release of material into the small intestine. Gastric juice consists of hydrochloric acid (HCl) and pepsin, a protein-splitting enzyme. Pepsin is produced in an inactive form, called pepsinogen, that is activated by HCl.

5. The small intestine is the primary site of digestion and absorption. Enzymes produced by the small intestine and by the pancreas work here to chemically digest carbohydrates, proteins, and fats into their component subunits. Bile, which is produced by the liver and stored in the gallbladder, emulsifies fat (breaks it into tiny droplets) in the small intestine, thereby increasing the combined surface area of fat droplets. Bile's action makes fat digestion by water-soluble lipase faster and more complete.

6. The small intestine's surface area for absorption is increased by circular folds in its lining, finger-like projections called villi, and microscopic projections covering the villi, called microvilli. Products of digestion are absorbed into the epithelial cells of villi by active transport, facilitated diffusion, or simple diffusion.

Most materials, including monosaccharides, amino acids, water, and ions, then enter the capillary blood network in the center of each villus. However, fatty acids and glycerol are resynthesized into triglycerides, combined with cholesterol, and covered in protein, forming droplets called chylomicrons. The chylomicrons enter a lymphatic vessel called a lacteal in the core of each villus. They are then delivered to the bloodstream in lymphatic vessels.

7. The large intestine consists of the cecum, colon, rectum, and anal canal. The large intestine absorbs water, ions, and vitamins. It is home to millions of beneficial bacteria that live on undigested material that has passed on from the small intestine. The bacteria produce several vitamins, some of which we then absorb for our own use.

8. Material left in the large intestine after passing through the colon is called feces. Feces consist of undigested or indigestible material, bacteria, sloughed off cells, and water.

9. Neural and hormonal mechanisms regulate the release of digestive secretions. Neural reflexes trigger the release of saliva, initiate the secretion of some gastric juice, and are the most important factors regulating the release of intestinal secretions. When stimulated by the presence of food in the stomach, especially food containing partially digested proteins, cells of the stomach lining release the hormone gastrin into the bloodstream. Gastrin is the most important factor controlling the secretion of gastric juice. The small intestine releases several hormones: (1) vasoactive intestinal peptide (VIP), which triggers the release of intestinal juices; (2) secretin, which causes the release of sodium bicarbonate from the pancreas; and (3) cholecystokinin, which causes the release of pancreatic digestive enzymes and the contraction of the gallbladder, leading to the release of bile.

REVIEW QUESTIONS

1. List the structures of the gastrointestinal tract in the order in which food passes through each.
2. Describe how food is processed in the mouth. What are the functions of the teeth and tongue?
3. Describe the structure of a tooth. What causes tooth decay?
4. What are the functions of the stomach? What are the digestive functions of gastric juice?
5. Why are so few substances absorbed from the stomach?
6. Describe the structural features that increase the surface area for absorption in the small intestine.

7. Which structures produce the digestive enzymes that act in the small intestine?
8. How does bile assist the digestion and absorption of fats?
9. What are the functions of the large intestine?
10. Where are carbohydrates digested? proteins? fats?
11. Describe the neural and/or hormonal mechanisms that regulate the release of digestive juices from each structure involved in digestion.

*Apply
the
Concepts*

Chapter

13A

Nutrition and Weight Control

(Tony Craddock/Tony Stone Images)

We make decisions about what to eat all the time. Sometimes we simply settle for what is available. When we make conscious choices, a number of factors are usually involved, including our ethnic and religious background, family and peer pressure, personal preferences, media advertising, and how hungry we are at the time.

The choices we make can influence how healthy we will be in the future. A high-fat diet, for instance, promotes certain cancers as well as diseases of the heart and blood vessels. On the other hand, a high-fiber diet is good for the heart and the intestines.

What does the body do with the food you eat? Food provides fuel, building blocks, metabolic regulatory molecules, and water. Fuel is needed for all cellular activities, for instance keeping your heart beating. Building blocks are needed for cell division, maintenance, and repair and for metabolic regulators, such as enzymes, which cells need to coordinate life's processes. Water is necessary for the proper cellular environment and for certain cellular reactions.

As we saw in the previous chapter, the digestive system breaks complex molecules of carbohydrates, proteins, fats, and nucleic acids into their component subunits. This provides a pool of building blocks that can be used to build molecules in the body. Most cells of the body, especially those of the liver, are able to convert one type of molecule into certain others. There are some amino acids and fatty acids that your body cannot synthesize, at least not in quantities sufficient to meet body needs. These are called essential amino acids and essential fatty acids. "Essential" means that it must be included in the diet.

The Food Guide Pyramid

The Food Guide Pyramid, released in 1992 by the U.S. Department of Agriculture, helps us plan a well-balanced diet even if we don't understand the reasons for choosing certain foods (Figure 13A–1). It translates nutritional knowledge

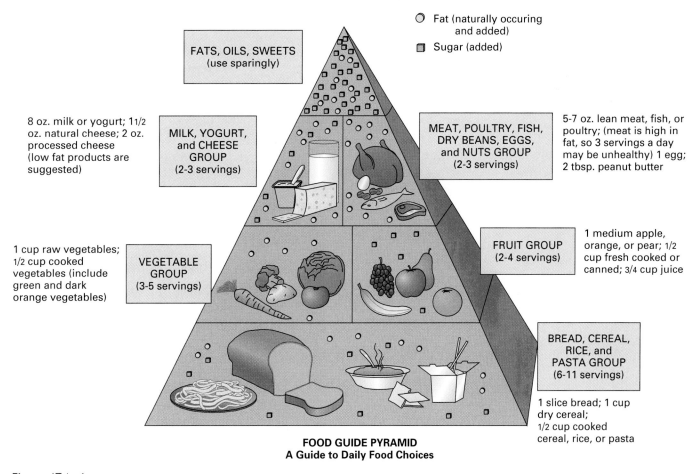

○ Fat (naturally occuring and added)

▨ Sugar (added)

FATS, OILS, SWEETS (use sparingly)

8 oz. milk or yogurt; 1 1/2 oz. natural cheese; 2 oz. processed cheese (low fat products are suggested)

MILK, YOGURT, and CHEESE GROUP (2-3 servings)

MEAT, POULTRY, FISH, DRY BEANS, EGGS, and NUTS GROUP (2-3 servings)

5-7 oz. lean meat, fish, or poultry; (meat is high in fat, so 3 servings a day may be unhealthy) 1 egg; 2 tbsp. peanut butter

1 cup raw vegetables; 1/2 cup cooked vegetables (include green and dark orange vegetables)

VEGETABLE GROUP (3-5 servings)

FRUIT GROUP (2-4 servings)

1 medium apple, orange, or pear; 1/2 cup fresh cooked or canned; 3/4 cup juice

BREAD, CEREAL, RICE, and PASTA GROUP (6-11 servings)

1 slice bread; 1 cup dry cereal; 1/2 cup cooked cereal, rice, or pasta

FOOD GUIDE PYRAMID
A Guide to Daily Food Choices

Figure 13A–1   The food guide pyramid helps us plan a well-balanced diet without being an expert nutritionist.

into recommended eating patterns. Take a moment now to consider how the food you ate today fits with the eating pattern recommended in this pyramid. Then, as you read this chapter, reflect back on the food guide pyramid to see how the recommendations for consumption of specific nutrients fit into this pyramid.

The Nutrients

A nutrient is a substance in food that provides energy or plays a structural or functional role to promote normal growth, maintenance, or repair (Table 13A–1). Three nutrients—fats, carbohydrates, and proteins—can provide energy. Although proteins can provide energy, they are usually used to build structures in the cell or for regulatory molecules, such as enzymes or certain hormones. Vitamins, minerals, and water do not provide energy, but they are essential to cellular functioning, as we will see.

Fat and Other Lipids

To weight-conscious Americans, fat is a dirty word. It is especially nasty because you don't have to eat fat to "get fat." Your body makes its own fat from proteins and carbohydrates whenever more calories are consumed than are used for energy on a regular basis. We'll now consider some of the types of molecules we've been referring to using the blanket term of "fat." The more technical name for what we've been calling fat is lipids. There are several types of lipids, as we will see.

Uses for Lipids

Although cultural attitudes regarding body fat have varied throughout history, the biological need for some fat in the diet has never changed. One reason is that fat is an excellent storage form of energy. It holds 9 calories of energy per gram, instead of the 4 calories per gram stored in carbohydrate or protein. (Energy is measured in a unit called a calorie, which is the amount of energy needed to raise one gram of water 1° Celsius. When discussing biochemical reactions,

Table 13A–1 Components of Food

Nutrient	Good Sources	Functions
Energy-Containing Nutrients		
Fats (lipids)	Milk, cheese, meat, vegetable oils, nuts	Provides 9 calories per gram; component of cell membranes; component of nerve sheaths, insulates body; forms protective cushions around vital body organs
Carbohydrate	Cereal, bread, pasta, vegetables, fruits, sweets	Provides 4 calories per gram; primary fuel for all cells
Protein and amino acids	Meat, poultry, fish, legumes, nuts	Provides 4 calories per gram; important component of all cells; structural proteins, including muscle fibers; regulatory proteins, including enzymes and certain hormones
Other Nutrients		
Vitamins	Many vegetables, fruits, whole grain, meats, dairy products	Most function as regulatory molecules that allow the cellular reactions of the body to take place fast enough to support life.
Minerals	Many vegetables, fruits, whole grain, meats, seeds, nuts	Structural roles, including hardness of bones and teeth; functional roles, including oxygen transport in blood; electrolyte balance; proper nerve and muscle function
Water	Nearly every food and all beverages	Solvent, transport of materials, medium for all and participant in some chemical reactions; lubricant; protective cushion; regulation of body temperature
Fiber (not absorbed from intestines)	Whole grains, vegetables, fruits	Soluble fiber good for health of heart and blood vessels; insoluble fiber promotes intestinal health

the energy exchanges are usually reported in kilocalories. A kilocalorie is 1000 calories of energy. In popular usage, the "kilo" is dropped. We will follow this tradition and refer to kilocalories as simply calories.[1]) Thus, fat is usually found between muscle cells, including those of the heart, to provide a ready source of energy. In addition, fat is a poor conductor of heat and so the layer of fat stored just beneath the skin provides some insulation against extreme heat or cold. Fat deposits also cushion and protect many vital organs, including the kidneys and the eyeballs. Certain lipids are essential components of all cell membranes and some are used in the construction of myelin sheaths that insulate nerve fibers. Lipids are also needed for the absorption of the fat-soluble vitamins A, D, E, and K. These vitamins commonly are absorbed from the intestines along with the products of fat digestion. Lipids can also carry these vitamins in the bloodstream to the cells that use them. Oils keep skin soft and prevent dryness. Certain other lipids provide the raw materials for the construction of chemicals used in communication among cells. For example, prostaglandins, which are regulatory molecules important in smooth muscle contraction and the control of blood pressure among other things, are derived from a particular fatty acid (linoleic acid). Another lipid, cholesterol, is the structural basis of the steroid hormones, including the sex hormones.

Critical Thinking

Olestra is a fat substitute that was recently approved by the FDA for limited use. Basically, Olestra is made from table sugar and vegetable oil, but its molecules are much larger than those of ordinary fats and cannot be broken down into small enough units to be absorbed by the body. Olestra, therefore, passes through the intestines and is eliminated with the feces. An important drawback to the use of Olestra is that it diminishes the body's absorption of fat-soluble vitamins (A, D, E, and K). Explain why this would occur.

Types of Lipids

You may recall from Chapter 2 that there are several types of lipids. One important lipid is a waxy substance called cholesterol. You have probably heard a great deal about cholesterol and may be surprised to learn that it makes up only a small percentage of the typical diet.

In fact, 95% of the lipids found in food are triglycerides—the neutral fats that we commonly think of when we hear the term fat. A triglyceride is a molecule made from three fatty acids (hence, *tri*) attached to a molecule of glycerol (hence, *glyceride*). When only one fatty acid is attached to the glycerol backbone, the molecule is called a

monoglyceride and when two fatty acids are attached, it is a diglyceride. The fatty acids in the triglyceride give the molecule its characteristics. An important way in which fatty acids can differ is in their degree of saturation, which refers to the number of places in the fatty acid where a carbon could hold additional hydrogen atoms (Figure 13A–2). A saturated fatty acid contains all the hydrogen it can hold. A polyunsaturated fatty acid could hold two or more additional hydrogens and a monounsaturated fatty acid could hold one more hydrogen. In general, saturated fats are solid at room temperature and they usually come from animal sources. In contrast, unsaturated fats are liquid at room temperature. They are the oils and usually come from plant sources.

Health Hazards Associated with Fats

Unfortunately for most of us, dietary fat quickly becomes too much of a good thing. Although the daily need for dietary fat is a mere tablespoon, the average American adult consumes 6 to 8 tablespoonsful of fat a day! Because fat contains more than twice the calories of an equal weight of carbohydrate or protein, it can quickly add inches to the waistline. By itself, obesity is associated with health problems such as high blood pressure and increased risk of diabetes.

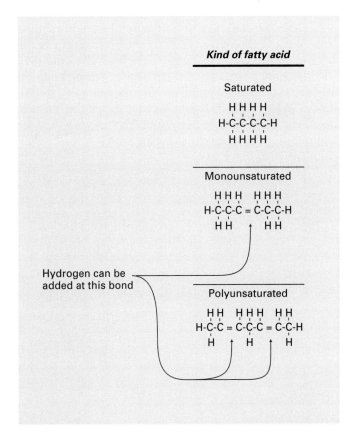

Figure 13A–2 A saturated fatty acid has all the hydrogen it can hold, but an unsaturated fatty acid could hold additional hydrogens.

[1]In technical writing kilocalories are written as Calories.

The high-fat diet that causes obesity is also related to certain cancers, including cancer of the colon, prostate, lungs, and perhaps the breast.

The clearest health risk of a high-fat diet is that of developing atherosclerosis, a condition in which fatty deposits form in the walls of blood vessels, thereby increasing the risk of heart attack and stroke. The risk of atherosclerosis increases with blood level of cholesterol (Table 13A–2). In general, blood cholesterol levels under 200 mg/dL are recommended for adults. The bad news is that the average blood cholesterol level for a middle-aged adult in the United States is 215 mg/dL. The good news is that people who lower their blood cholesterol levels can slow, or even reverse, atherosclerosis and, therefore, their risk of heart attack. It has been estimated that individuals who lower their blood cholesterol levels by 25% lower their risk of heart damage caused by atherosclerosis by 49%.

It turns out that total blood cholesterol doesn't paint a complete picture of a person's risk of atherosclerosis. Before cholesterol can be transported in the blood or lymph, it is combined with protein to form a lipoprotein, which makes it soluble in water (Figure 13A–3). There are different types of lipoproteins that differ in the relative proportions of cholesterol and protein. Low-density lipoproteins (LDLs) have a high proportion of cholesterol. LDLs are considered to be a "bad" form of cholesterol because, in addition to bringing cholesterol to the cells that need it to sustain life, they deposit cholesterol in artery walls (Figure 13A–4). In contrast, high-density lipoproteins (HDLs) contain more protein relative to cholesterol. HDLs carry cholesterol to the liver, where it is used to make bile. Bile is then released into the small intestine where it aids in the digestion of fats. Although some bile is reabsorbed, about half of the bile produced leaves the body in the feces. This is the only way our bodies lose cholesterol. In this way, HDLs help the body eliminate cholesterol.

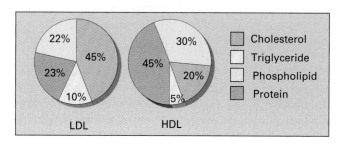

Figure 13A–3   (a) A lipoprotein is a cluster of lipids associated with protein that functions as a transport particle that allows lipids to be transported throughout the body in blood and lymph. (b) Low-density lipoproteins (LDLs) contain a higher proportion of cholesterol than do high-density lipoproteins (HDLs).

Because of the roles they play in transporting cholesterol, the ratio of HDLs to LDLs is considered more important than the total blood cholesterol level. The ratio of total cholesterol to HDLs should not be greater than 4:1. HDLs higher than 60 mg/dL are considered to protect against heart disease.

Blood cholesterol comes from one of two sources—the diet or the liver. Of the two, the liver is most significant. Indeed, *most* of the cholesterol in blood comes from the liver and not from the food we eat. Nonetheless, diet is still important and, surprisingly, saturated fat in the diet raises blood levels of cholesterol more than dietary cholesterol does. Why? It turns out that the body handles the cholesterol differently depending on a person's genetic makeup and the amount of cholesterol consumed in a meal. Both the amount of cholesterol absorbed from the intestines and the amount excreted from the body can vary. Furthermore, the liver can adjust the amount of cholesterol it produces according to the blood cholesterol level. As a result, there isn't a clear relationship between dietary and blood cholesterol. Saturated fat, however, consistently boosts blood levels of

Table 13A–2   Blood Cholesterol Levels and the Risk of Cardiovascular Disease

	Desirable	Borderline-High	High
Total Cholesterol/ 100 ml blood	<200 mg	200–239 mg	≧240 mg
LDL cholesterol/ 100 ml blood	<130 mg	130–159 mg	≧160 mg
Ratio total cholesterol/HDL cholesterol	<4:1		

L.A. Smolin & M.B.Grosvenor. *Nutrition Science and Applications, 2 ed.*, p 133.

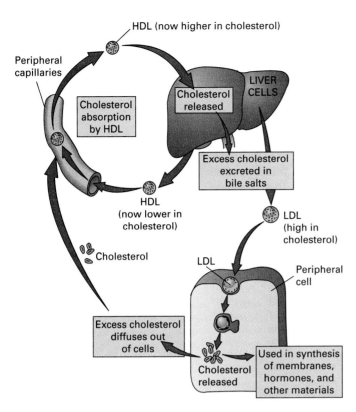

Figure 13A-4 LDLs are important in transporting cholesterol from the liver to the cells. They are often called the "bad" cholesterol because they play a role in atherosclerosis, the formation of fatty deposits in arteries, which can lead to heart attack and stroke. High-density lipoproteins (HDLs) have less cholesterol and more protein than do LDLs. HDLs transport cholesterol to the liver for disposal. HDL is considered the "good" cholesterol because it reduces the risk of atherosclerosis.

harmful LDL cholesterol both by directly stimulating the liver to step up its production of LDLs and by slowing the rate at which LDLs are cleared from the bloodstream.

Other types of fats also have been shown to alter blood cholesterol levels. Some studies have suggested that the so-called "trans" fatty acids may behave like saturated fatty acids and raise LDL levels. *Trans* fatty acids are formed when hydrogens are added to unsaturated fats (oils) to stabilize them or to solidify them, as when margarine is formed from vegetable oil. In contrast, monounsaturated fats and omega-3 fatty acids lower total blood cholesterol and LDLs. Monounsaturated fats are found in olive, canola, and peanut oils and in nuts. Omega-3 fatty acids are found in the oils of certain fish, such as Atlantic mackerel, lake trout, herring, tuna, and salmon. Thus, it would be a good idea to eat more fish. It is not wise, however, to take capsules containing omega-3 fatty acids. Such capsules would boost the total dietary fat intake. Furthermore, excessive quantities of fish oils can slow blood clotting and may lead to overdoses of vitamins A and D, which are stored in fish oil.

Dietary Recommendations

Not only does the typical American eat far too much fat, but the balance among the types of lipids is unhealthy. It is recommended that no more than 30% of calories in the diet come from fat, but fats make up about 34% of the calories in the average American diet.

It is further recommended that saturated, monounsaturated, and polyunsaturated fats be equally represented in the fat calories. For most of us, achieving this balance of fats would require a drastic reduction in the amount of saturated fat in the diet. This can be done by reducing animal fat in the diet. Foods high in saturated fat include meat (especially red meat), butter, cheese, whole milk, and other dairy products (Figure 13A–5).

Carbohydrates

Carbohydrates have gotten a lot of bad press over the years. As a result, many people erroneously think that carbohydrates are particularly fattening. In fact, carbohydrates provide the same number of calories per gram as pure protein and most animal sources of protein bring with them a good amount of fat, which boosts their caloric value tremendously. For instance, 5 ounces of steak contains 500 calories, four and a half times the number of calories of a medium-sized baked potato. Furthermore, unlike any other food cat-

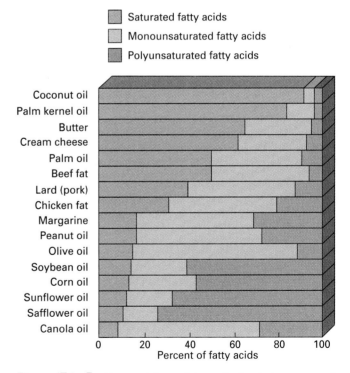

Figure 13A-5 Fats contain a mixture of saturated, monounsaturated, and polyunsaturated fats. Most Americans would be well advised to reduce their total fat intake, especially the saturated fats. Saturated fats increase blood levels of cholesterol.

egory, carbohydrates, such as whole grains, fruits, and vegetables, aren't associated with any killer diseases. Before we go on extolling the value of carbohydrates, however, we should become familiar with what they are.

Types of Carbohydrates

Carbohydrates in our diets include sweets (sugars), starches, and roughage (dietary fiber). The basic unit of a carbohydrate is a **monosaccharide**. Sugars are generally monosaccharides or disaccharides (two monosaccharides linked together). Sugars taste sweet and come in a variety of sweeteners (e.g., table sugar, honey, corn syrup, and molasses) as well as in certain foods, including candies, cookies, cakes, and pies. Starches are polysaccharides—long, sometimes branched chains of hundreds or thousands of linked molecules of the monosaccharide glucose. Plants store energy in starches, and common sources of starches include wheat, rice, oats, corn, potatoes, and legumes. Dietary fiber is a mixture of substances, mostly polysaccharides, that humans cannot digest into component monosaccharides. Since only monosaccharides can be absorbed from the small intestine, dietary fiber is passed along to the large intestine. Some fiber is digested by the bacteria living in the large intestine and the remaining fiber gives bulk to feces.

Dietary fibers are classified by their ability to dissolve in water. **Insoluble fiber** includes cellulose, hemicellulose, and lignin. Cellulose forms plant cell walls and together with hemicellulose it forms tough plant structures, including the strings of celery, the membranes dividing sections of citrus fruit, and the skins of corn kernels. Lignin, which is the only one of the seven types of dietary fiber that is not a polysaccharide, comes from the extremely woody parts of plants, the skins of fruits and vegetables, and whole grains. **Soluble fibers** have a gummy consistency and include the pectins, gums, mucilages, and some hemicelluloses. They are found in and between plant cells. Good food sources include oat bran, apples, and beans.

Uses and Health Benefits of Carbohydrates

The most important function of carbohydrates is to provide fuel for the body. Each gram of carbohydrate yields 4 calories. Of the carbohydrates, glucose is the fuel that cells use most of the time. Indeed, even a temporary shortage of blood glucose can impair brain function and kill nerve cells. Although the body can also use protein for energy, protein is better used to build and maintain tissues. Gram for gram, fat contains more than twice as much energy as carbohydrates. Without sufficient carbohydrates, however, fats can't be used completely, and so ketone bodies are formed during fat metabolism. The accumulation of ketone bodies in the bloodstream, a condition called ketosis, can cause nausea, loss of appetite, and dehydration. Ketosis can even cause death.

Although fiber cannot be digested or absorbed, it is still an important part of a healthful diet. Water-soluble fiber is good for the heart and blood vessels. It lowers LDLs, the "bad"

form of cholesterol-carrying particles and total cholesterol but does not lower the beneficial HDL cholesterol levels.

Several intestinal disorders—constipation, hemorrhoids, and diverticulosis—are related to the formation of hard, dry stools. You may recall from the previous chapter that the consistency of feces changes as it passes through the large intestine and water is absorbed. If material moves through the large intestine slowly, too much water may be absorbed. The resulting hard, dry stools are difficult to pass. This condition, commonly known as constipation, makes it necessary to strain during bowel movements. Straining increases pressure within veins in the rectum and anus, causing them to stretch and enlarge. The resulting varicose veins are hemorrhoids. The wall of the large intestine also experiences great pressure during a strained bowel movement. Then, like the inner tube of an old tire, the weaker spots in the intestinal wall can begin to bulge outward, forming small outpouchings called diverticula (singular, *diverticulum*)(Figure 13A–6). Diverticula are very common in people older

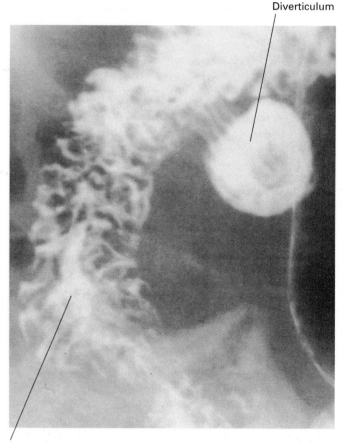

Diverticulum

Large intestine

Figure 13A–6 A diverticulum is a small pouch that forms in the wall of the large intestine, usually caused by repeated straining during bowel movements. A high fiber diet results in softer, more bulky stools that are easier to pass. Thus, fiber makes it less likely that diverticula will form. *(© 1995 Lester Bergman, Cold Springs, NY)*

than age 50. When diverticula don't cause problems or symptoms, the condition is called diverticulosis. But, if the diverticula become infected with bacteria and inflamed, the condition is then called diverticulitis and can cause abrupt, cramping abdominal pain, a change in bowel habits, fever, and rectal bleeding.

Constipation, hemorrhoids, and diverticulosis improve when the amount of fiber in the diet is increased. Soluble fiber can absorb an amazing amount of water, which softens stools and makes them easier to pass. The fiber in a carrot, for instance, can hold 20 to 30 times its weight in water. Insoluble fiber, on the other hand, gives bulk to the stools. The increased water and bulk of stools stretch the intestinal walls, thereby stimulating peristalsis and speeding up the movement of feces through the intestine. Thus, fiber reduces constipation and straining during bowel movements, which improves intestinal health.

Dietary fiber also reduces the risk of colon cancer, which is the second leading cause of cancer deaths in the United States, behind only lung cancer. Certain foods we eat may contain carcinogens (cancer-causing chemicals). In addition, bacteria living in the large intestine metabolize certain of the undigested material in feces, producing carcinogens in the process. As we have seen, fiber absorbs water in the colon and speeds transit time, that is, the time it takes material to move through the digestive system and exit as feces. As a result, potential carcinogens are diluted and the time they are in contact with colon walls is reduced.

Dietary Recommendations

It is recommended that 55% to 60% of the calories in your diet come from carbohydrates. It is further recommended that the bulk of those carbohydrate calories come from complex carbohydrates, such as starches, and that refined and processed sugars be minimized. Starches and sugars both provide the same amount of calories—4 per gram. However, starches are generally "packaged" with other nutrients, including vitamins and minerals, and with dietary fiber. Therefore, high-starch foods, including beans, peas, potatoes, whole grains, pasta, fruits, and vegetables, provide much more than the "empty" calories of sugar.

The typical American diet falls short of the recommended carbohydrate intake. On average, only 48% of the calories in our diets come from carbohydrates and far too many of those calories are from sugars instead of complex carbohydrates.

Complex carbohydrates vary in their nutritional value, and so we might wonder which ones are best. Nutritionists generally agree that the emphasis should be on grains, vegetables, and fruits. It is often suggested that the dinner plate should be one-half steamed vegetables, one-quarter a starch such as potato, rice, or pasta, and one-quarter a good protein source such as meat or legumes. Potatoes, rice, and pasta are not fattening by themselves, when eaten in moderation. It's

the butter, sour cream, and sauces we add to them that pile on the calories. Portion size still matters, however. Some people gain weight because they think that it's acceptable to eat an extra bagel or two, as long as they use low-fat cream cheese on them. Products made from whole-grain flours are more nutritious than those from refined flour. During refinement, two parts of the wheat grain, called the bran and the germ, are removed and with them go some of the vitamins, minerals, and fiber.

Nutritionists generally recommend that a healthy adult consume between 20 and 35 grams of dietary fiber each day. The average American diet falls short of this goal. Indeed, most of us consume only 17 grams to 19 grams of fiber a day. Figure 13A–7 shows some good sources of dietary fiber.

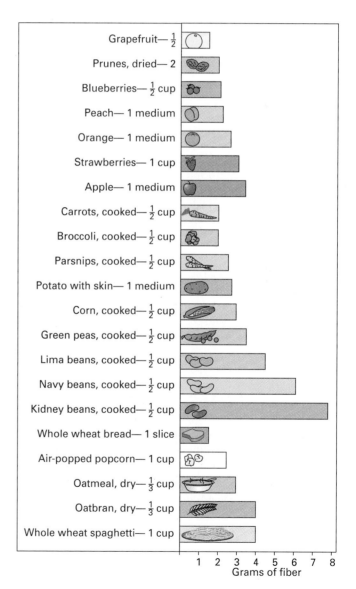

Figure 13A–7 Good sources of dietary fiber include fruits, vegetables, dried beans, and whole grains. It is recommended that we eat between 20 and 35 grams of fiber a day.

Sugar consumption in the United States averages 125 pounds per person each year. Although it seems unbelievable, it is possible to consume this much sugar because so much is hidden in processed foods, even ones that don't taste sweet. Who would expect sugar in ketchup or mayonnaise, for instance? Even people who faithfully read food labels may not be aware of the sugar content in processed food because simple sugars have so many different names, including honey, brown sugar, dextrose, maltose, molasses, fructose, levulose, and corn syrup. Foods that do taste sweet often contain whopping amounts of sugar. A single glazed donut, for example, contains 6 teaspoons of sugar.

The primary nutritional benefit of any form of sugar is energy. Although all sugars contain the same number of calories per gram, some taste sweeter than others. The fruit sugar fructose, for example, tastes about 70% sweeter than table sugar. Honey and table sugar contain the same two monosaccharides—glucose and fructose. Because honey contains slightly more fructose than sucrose does, it tastes sweeter. Thus, honey provides more sweetness for fewer calories.

Protein

Protein seems to have a privileged status in our culture. When we think of protein, images of vitality, strength, muscle mass, and energy often come to mind. Proteins were named from the Greek word *proteios*, which means "of prime importance." But what makes them so important?

Uses of Proteins

Protein is an essential structural component of every cell in your body. Indeed, if the water were removed, half of the remaining body weight would be protein. Protein forms the framework of bones and teeth. It forms the contractile part of muscle and helps red blood cells carry oxygen. Dietary protein provides the raw materials to replace or repair cells that are stressed by everyday wear and tear and to form new tissues as you grow.

Proteins also regulate body processes. Certain proteins are enzymes that help the chemical reactions of the body to take place at body temperature and at rates fast enough to support life. Other proteins are hormones, which are chemical messengers that enable cells in distant parts of the body to communicate with one another.

Proteins known as antibodies help the body defend itself against foreign invaders, such as bacteria. Also, proteins carried in the blood help maintain the water balance in the body.

Proteins also can be used for energy. When one gram of protein is oxidized, it yields 4 calories. However, protein is not the preferred fuel for the body. It is only used for energy if the supply of carbohydrates and fat is insufficient or if dietary protein exceeds the body's needs.

Chemical Nature of Protein

A protein consists of one or more chains of amino acids. Human proteins contain 20 different kinds of amino acids. The protein in the food you eat is digested into its component amino acids and then absorbed into the bloodstream and delivered to the cells, creating a pool of amino acids. Your cells then draw the amino acids needed to build the proteins in your body from those available in the pool. The body is able to synthesize 11 of the amino acids from nitrogen and molecules derived from carbohydrates or fats. There are others, though, called the **essential amino acids**, that the body cannot synthesize. The essential amino acids, then, must be supplied in the diet.

When your body makes a protein, the amino acids are put together in a specific order depending on which protein is being made. If a particular amino acid is needed for a certain protein, but is not available, then that protein cannot be synthesized. Using an analogy, a sign maker with a bag containing only one copy of the letter R and 100 copies of the other 25 letters in the alphabet could make only one "NO PARKING" sign. There is simply no way to substitute for the limiting letter, R. The same principle applies to protein synthesis—if a needed amino acid is lacking, protein production stops or the body breaks down existing proteins to get it.

Amino acids cannot be stored for use at a much later time. Those that are not used to produce new proteins are burned for energy or are converted to either fats or carbohydrates. Therefore, the pool of amino acids available for protein synthesis must always contain sufficient amounts of all the essential amino acids. The **complete proteins** that you eat are those that contain ample amounts of all of the essential amino acids. Animal proteins are generally complete proteins. **Incomplete proteins**, which are typically found in plants, are low in one or more of the essential amino acids. The essential amino acids most likely to be present in limited quantities in a particular plant are lysine, methionine, or tryptophan.

Incomplete proteins from two or more different plant sources can be combined so that, after digestion, the pool of amino acids available for protein synthesis will contain ample amounts of all the essential amino acids. Combinations such as these are called **complementary proteins**. For example, corn is low in lysine and tryptophan whereas beans are low in methionine, but together they supply enough of all the essential amino acids (Table 13A–3). Thus, a vegetarian must make careful food choices to be sure that complementary proteins are consumed.

Dietary Recommendations

The Dietary Guidelines for Americans suggest that protein calories represent 10% to 15% of the calories in the diet. Most North Americans have no trouble meeting this goal. Your specific protein requirement can be determined by

Table 13A-3 Complementary Vegetable Proteins

Grains	+	Legumes
Rice		Blackeyed peas; pinto beans; or tofu
Corn		Lima beans
Cornbread		Split-pea soup
Corn or flour tortillas		Beans
Wheat bread		Peanut butter
Brown bread		Baked beans

multiplying your body weight in kilograms[2] by 0.8. Thus, a person weighing 68 kg (about 150 lb) would require 55 grams of protein a day. In North America, however, the typical person consumes about 100 grams of protein a day, nearly twice the daily requirement.

Although the choice of protein foods must ensure an adequate supply of all the essential amino acids, the protein need not come from animal sources. Indeed, the tendency to make meat the focal point of every meal is the main reason that the typical diet in the United States contains excess fat. Animal protein usually comes with a lot of fat, whereas plant protein usually brings along carbohydrates (Figure 13A-8). Thus, eating a variety of plant proteins will not only supply all the essential amino acids but also will help reduce the percentage of fat calories in the diet and boost the percentage of calories from complex carbohydrates. When choosing among animal sources of protein, it is wise to cut back on the amount of red meat in the diet, in favor of chicken or fish. Beef has fat interspersed among the muscle fibers, which we call marbling. Much of the fat in chicken, on the other hand, tends to be just under the skin and is easily removed. Many types of fish, such as sole, are naturally low in fat.

Vitamins

A vitamin (*vita*, life giving) is an organic (carbon-containing) compound that, while essential for health and growth, is only needed in minute quantities—milligrams or micrograms. All the vitamins you need in a day would fill only an eighth of a teaspoon. The strict definition of a vitamin is a bit odd because it defines a vitamin by what happens when the substance is *not* present. To fit the strict definition of a vitamin, the lack of the substance for a prolonged time period must produce a set of symptoms, referred to as a deficiency disease, which is quickly cured when the substance is resupplied. Examples of deficiency diseases include: (1) rickets, a deficiency of vitamin D that causes bone deformation; (2) scurvy, a deficiency of vitamin C that results in poor

[2]Weight in kilograms is determined by dividing weight in pounds by 2.2.

wound healing and impaired immunity; and (3) pellagra, a deficiency of niacin (vitamin B_3) that causes damage to the skin, gastrointestinal tract, and nervous system.

Vitamins are needed in tiny amounts because they are not destroyed during use. Most function as coenzymes, which are nonprotein molecules that are necessary for certain enzymes to function. Enzymes and coenzymes are continuously recycled and, therefore, can be used repeatedly by the body.

The lack of one vitamin can cause a wide spectrum of effects. Because they function as coenzymes, vitamins function as regulatory nutrients that help direct the flow of chemical reactions in the body. In this role a single vitamin may be necessary for chemical reactions that occur in nearly every cell of the body, and different types of cells have different functions. Furthermore, vitamins can interact with other nutrients, especially minerals. Vitamin C, for instance, increases iron absorption from the intestines. Thus, a deficiency of vitamin C can also affect all the cellular functions that require iron. However, excesses of vitamin C can lessen the ability of cells to use copper ions, interfering with all cellular functions that depend on copper ions. Table 13A-4 lists the vitamins, their functions, good sources of them, and the problems associated with deficiencies or excesses.

There are two categories of vitamins: the **water-soluble vitamins**, which dissolve in water, and the **fat-soluble vitamins**, which are stored in fat. Of the 13 vitamins

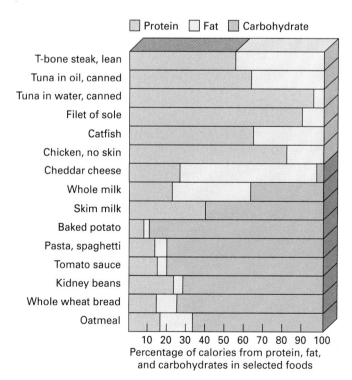

Figure 13A-8 The percentage of calories from protein, fat, and carbohydrate in certain plant and animal foods. Animal foods contain a greater percentage of fats than do plant foods.

Table 13A–4   Vitamins

Vitamin	Good Sources	Function	Effects of Deficiency	Effects of Excess
Fat-Soluble Vitamins				
A	Liver; egg yolk; fat-containing and fortified dairy products; formed from carotene, which is found in deep-yellow and deep-green leafy vegetables	Component of rhodopsin, the eye pigment responsible for black and white vision, maintains epithelia; cell differentiation	Night-blindness; dry, scaly skin; dry hair; skin sores; increased respiratory, urogenital, and digestive infections; xerothalmia, the leading cause of preventable blindness worldwide; most common vitamin deficiency in world	Drowsiness, headache, vomiting; dry, coarse, scaly skin, hair loss; itching, brittle nails; abdominal and bone pain
D	Fortified milk, fish liver oil, egg yolk; formed in skin when exposed to ultraviolet light	Increases absorption of calcium; enhances bone growth and calcification	Bone deformities in children; rickets; bone softening in adults	Calcium deposits in soft tissues; kidney damage; vomiting, diarrhea, and weight loss
E	Whole grains, dark green vegetables, vegetable oils, nuts, seeds	May inhibit effects of free radicals; helps maintain cell membranes; prevents oxidation of vitamins A and C in gut	Rare; possible anemia and nerve damage	Muscle weakness, fatigue, nausea
K	Primary source is from bacteria in large intestine; leafy green vegetables, cabbage, cauliflower	Important in forming proteins involved in blood clotting	Easy bruising; abnormal blood clotting; severe bleeding	Liver damage and anemia
Water-Soluble Vitamins				
C(ascorbic acid)	Citrus fruits, cantaloupe, strawberries, tomatoes, broccoli, cabbage, green pepper	Collagen synthesis; may inhibit free radicals; improves iron absorption	Scurvy; poor wound healing; impaired immunity	Diarrhea; kidney stones; may alter results of certain diagnostic lab tests

known to be needed by humans, 9 are water-soluble (C and the various B-vitamins) and 4 are fat-soluble (A, D, E, and K).

Except for vitamin D, our cells cannot make vitamins, and so we must obtain them in our food. A varied, balanced diet is the best way to ensure an adequate supply of all vitamins. No one food contains every vitamin, but most contain some. Vitamins are often more easily available for absorption when the foods containing them are cooked. Cooked carrots, for instance, are a better source of vitamin A than are raw carrots. However, water-soluble vitamins are just that and so these vitamins are likely to be lost if the vegetables containing them are cooked by boiling them in water. Steaming vegetables, then, is a better way to preserve their vitamin content.

The results of many recent studies have emphasized the importance of assuring that you obtain adequate daily supplies of certain vitamins. Folic acid, one of the B vitamins that is particularly abundant in leafy greens, plays a role in pre-

Vitamin	Good Sources	Function	Effects of Deficiency	Effects of Excess
Water-Soluble Vitamins (cont.)				
Thiamin (B_1)	Pork, legumes, whole grains, leafy green vegetables	Coenzyme in energy metabolism; nerve function	Water retention in tissues; nerve changes leading to poor coordination; heart failure; beriberi	None known
Riboflavinin (B_2)	Dairy products such as milk; whole grains; meat; liver; egg white; leafy green vegetables	Coenzyme used in energy metabolism	Skin lesions	None known
Niacin (B_3)	Nuts; can be formed from tryptophan found in meats; green leafy vegetables; potatoes	Coenzymes used in energy metabolism	Contributes to pellagra (damage to skin, gut, nervous system, etc.)	Flushing of skin on face, neck, and hands; possible liver damage
B_6	Meat, poultry, fish, spinach, potatoes, tomatoes	Coenzyme used in amino acid metabolism	Nervous, skin, and muscular disorders; anemia	Numbness in feet and poor coordination
Pantothenic acid	Widely distributed in foods, animal products, and whole grains	Coenzymes in energy metabolism	Fatigue; numbness and tingling of hands and feet; headaches; nausea	None known
Folic acid (folate)	Dark green vegetables, orange juice, nuts, legumes, grain products	Coenzyme in nucleic acid and amino acid metabolism	A type of anemia (megaloblastic anemia); gastrointestinal disturbances; nervous system damage; inflamed tongue; neural tube defects	None known
B_{12}	Poultry, fish, red meat, and dairy products except butter	Coenzyme in nucleic acid metabolism	Anemia (megaloblastic and pernicious); impaired nerve function	None known
Biotin	Legumes; egg yolk; widely distributed in foods; bacteria of large	Coenzyme used in energy metabolism	Scaly skin (dermatitis); sore tongue; anemia;	None known

venting birth defects that involve the brain and spinal cord, such as spina bifida. It now seems that folic acid, along with vitamins B_6 and B_{12}, may also help prevent heart disease. It is thought that five daily servings of fruits and vegetables would provide enough of these vitamins to protect the heart. Furthermore, it has been suggested that vitamins that act as antioxidants, including vitamin C, vitamin E, and beta-carotene, may slow the aging process, protect against cancer, atherosclerosis, and macular degeneration (the leading cause of irreversible blindness in people over age 65). Spinach, collard greens, and carrots are good sources of antioxidant vitamins.

Are vitamin supplements too much of a good thing? It depends. For one thing, the components of your everyday diet will influence how beneficial vitamin supplements will be. Even though you may know that eating a well-balanced diet, including vegetables, is the best way to obtain all the necessary nutrients, the pressures of time or force of habit may keep that from happening. In that case, a daily multi-

Table 13A–5   Minerals

Mineral	Source	Functions	Effects of Deficiency	Effects of Excess
Major Minerals				
Calcium	Milk, cheese, dark green vegetables, legumes	Hardness of bones; tooth formation; blood clotting; nerve and muscle action	Stunted growth; loss of bone mass; osteoporosis; convulsions	Impaired absorption of other minerals; kidney stones
Phosphorus	Milk, cheese, red meat, poultry, whole grains	Bone and tooth formation; component of nucleic acids, ATP, and phospholipids; acid-base balance	Weakness; demineralized bone	Impaired absorption of some minerals
Magnesium	Whole grains, green leafy vegetables, milk, dairy products, nuts, legumes	Component of enzymes	Muscle cramps; neurologic disturbances	Neurologic disturbances
Potassium	Available in many foods; meats, fruits, vegetables, and whole grains	Body water balance; nerve function; muscle function; role in protein synthesis	Muscle weakness	Muscle weakness; paralysis; heart failure
Sulfur	Protein-containing foods including meat, legumes, milk and eggs	Component of body proteins	None known	None known
Sodium	Table salt	Body water balance; nerve function	Muscle cramps; reduced appetite	High blood pressure in susceptible people
Chloride	Table salt, processed foods	Formation of HCl in stomach; role in acid-base balance	Muscle cramps; reduced appetite; poor growth	High blood pressure in susceptible people

vitamin supplement won't hurt. However, most of the 40% of adults in the United States who take vitamin supplements are not at risk of deficiency. Other factors influencing whether vitamin supplements will be beneficial or harmful is the amount and type of vitamin consumed. Excess water-soluble vitamins are usually excreted in the urine. In contrast, excess fat-soluble vitamins are stored in fat and can accumulate in the body, causing serious problems.

Critical Thinking

Why might a person who takes megadoses of vitamin B_2 (doses more than ten times that recommended as necessary for healthy persons) be less likely to suffer harmful consequences than a person who takes excessive supplements of vitamin A?

Minerals

A mineral is an inorganic substance that is essential to a wide range of life processes. We need fairly large, although not megadose amounts, of seven minerals: calcium, phosphorus, potassium, sulfur, sodium, chloride, and magnesium. In addition, we need trace amounts of about a dozen others (Table 13A–5). The average adult body contains about 5 pounds of minerals.

We obtain minerals from the foods we eat. As with vitamins, we must be careful how we cook foods to avoid losing minerals. Many minerals are water-soluble and can be lost during food preparation.

Water

Water is perhaps the most essential nutrient. We can live without food for about 8 weeks but without water for only

Mineral	Source	Functions	Effects of Deficiency	Effects of Excess
Trace Minerals				
Iron	Meat, liver, shellfish, egg yolk, whole grains, green leafy vegetables, nuts, dried fruit	Component of hemoglobin, myoglobin, and cytochrome (transport chain enzyme)	Iron-deficiency anemia; weakness; impaired immune function	Liver damage; heart failure; shock
Iodine	Marine fish and shellfish; iodized salt; dairy products	Thyroid hormone function	Enlarged thyroid	Enlarged thyroid
Fluoride	Drinking water, tea, seafood	Bone and tooth maintenance	Tooth decay	Digestive upsets; mottling of teeth; deformed skeleton
Copper	Nuts, legumes, seafood, drinking water	Synthesis of melanin, hemoglobin, and transport chain components; collagen synthesis; immune function	Rare; anemia; changes in blood vessels	Nausea; liver damage
Zinc	Seafood, whole grains, legumes, nuts, meats	Component of digestive enzymes; required for normal growth; wound healing, and sperm production	Difficulty in walking; slurred speech; scaly skin; impaired immune function	Nausea; vomiting; diarrhea; impaired immune functioning
Manganese	Nuts, legumes, whole grains, leafy green vegetables	Role in synthesis of fatty acids, cholesterol, urea, and hemoglobin; neural	None known	Nerve damage

about 3 days. Although you look solid, within your body is an internal sea. Even bone is about one-fourth water! A newborn is, in fact, 85% water. The water content gradually decreases with age to between 55% and 65% in an adult female and between 65% and 75% in an adult male. Females contain less water than males because their bodies contain a greater proportion of fat, which doesn't hold water as well as muscle.

Water is so common—found in nearly every food and beverage we consume—that we rarely think about its importance. It is an excellent solvent. Thus, in the blood and lymph, it transports materials, including nutrients, metabolic wastes, and hormones. Not only does water provide a medium in which chemical reactions can take place, it participates in many of those reactions. One example is hydrolysis (*hydro*, water; *lysis*, splitting) reactions, in which chemical bonds are broken by the addition of water. For instance, the food molecules we eat are split into their component subunits during digestion by hydrolysis. Water is also a lubricant. It keeps your joints from creaking and the soft tissues of your body from sticking together. Water forms a protective cushion within the eyes and around the brain and spinal cord. During pregnancy, water in the amniotic fluid protects the fetus. In addition, water plays an important role in regulation of body temperature. Perspiration, for instance, cools the body.

Body Energy Balance

As we have seen, our bodies obtain energy from the carbohydrates, proteins, and fats that we eat. Carbohydrates and proteins provide 4 calories per gram, and fats provide 9.

Although energy can be neither created nor destroyed, it can be changed from one form to another. These energy conversions take place in the biochemical reactions that occur in the body. They are the reason that the energy in a potato you eat can fuel muscle contraction, allowing you to run to catch the bus. There is a dynamic balance between the energy the body takes in (in food) and its expenditure of energy. Food energy that isn't used for the body's various activities is stored as fat or glycogen. This is why you gain weight when you consume more calories than you use.

$$\text{Energy intake} = \text{total energy output} \\ (\text{heat} + \text{work} + \text{energy storage})$$

Regulation of Body Weight

The principle is simple: Body weight remains stable as long as the calories consumed equal the calories expended. When intake and output are not in balance, weight is gained or lost.

But, *how* is body weight regulated? Or, rephrased to get at the *real* question of interest, how much control over body weight do we have? The answers are not as simple as the principle, because many factors are involved.

An important regulatory mechanism determining body weight is a feedback loop that turns appetite and metabolism up and down in much the same way that a thermostat controls temperature (Figure 13A–9). This "fatstat" involves a hormone called leptin, which is released into the bloodstream by fat cells. Since leptin is produced by fat cells, the amount of leptin indicates the amount of fat in the body. The brain's hypothalamus has receptors that bind leptin, thereby measuring the amount present in the blood. The hypothalamus then adjusts appetite and metabolic rate to keep the level of leptin constant and at an optimum level. The hypothalamus would maintain the amount of body fat needed to keep leptin at optimal levels. This would be the fat set point, analogous to the temperature setting on a thermostat. When leptin levels rise with fat gain, the hypothalamus causes appetite to decrease and metabolic rate to increase. Then, when fat is shed and leptin is low, appetite is stimulated and metabolism slows.

Leptin's role as a fat-signaling hormone may help explain why some people apparently are born to be fatter than others. Studies using mice have revealed genetic problems that can disturb the normal functioning of the leptin feedback loop (Figure 13A–10). One gene, the so-called *obese* (*ob*) gene, carries the instructions to make leptin. If the *ob* gene is defective, meager amounts of leptin are produced re-

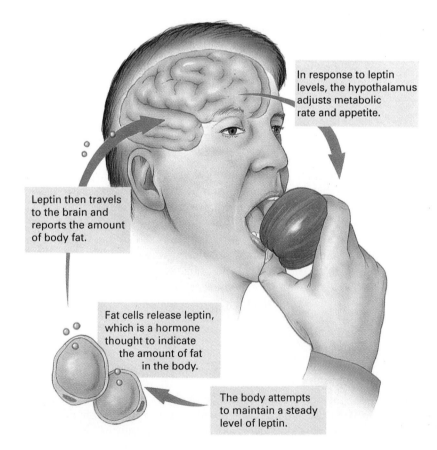

In response to leptin levels, the hypothalamus adjusts metabolic rate and appetite.

Leptin then travels to the brain and reports the amount of body fat.

Fat cells release leptin, which is a hormone thought to indicate the amount of fat in the body.

The body attempts to maintain a steady level of leptin.

Figure 13A–9 A feedback loop involving a hormone called leptin, which tells the brain's hypothalamus how much fat is present in the body, plays an important role in regulating body weight. In response to leptin levels, the hypothalamus adjusts metabolism and appetite to maintain body weight at a predetermined set point.

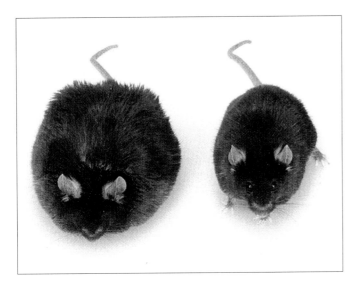

Figure 13A–10 Mice with a defect in the *obese (ob)* gene are remarkably overweight. Because of the defect, fat cells are unable to produce adequate amounts of the fat-signaling hormone, leptin. The brain equates low leptin levels with low fat stores and sets into motion physiological responses, such as increased appetite and slowed metabolism, that will result in weight gain. Although both of these mice have defective *ob* genes, the one on the right was given leptin injections. After 4½ weeks, the untreated mouse weighed 67 grams and the mouse given leptin weighed 35 grams. A normal adult mouse weighs about 24 grams. *(John Sholtis, Rockefeller University, New York/ © 1995 Amgen, Inc.)*

gardless of the amount of body fat. The brain interprets low leptin levels as an indication of low fat stores and stimulates appetite and slows metabolism. Defects in a second gene called the *diabetes (db)* gene result in defective leptin receptors, which predispose the brain to respond weakly to leptin. Mice and humans with defective *db* genes have very high leptin levels. However, their brains can't respond to the signal because of the defective receptors. The weak signal is interpreted as an indication of low fat stores and initiates responses that will result in weight gain. Thus, genes may have a lot to do with whether you fit in your jeans.

There are also signals from other nutrients that influence appetite. Blood glucose and amino acid levels rise following a meal. Whereas we generally stop eating when blood glucose or amino acids are high, low blood glucose or amino acids prompt us to look for food.

As those of us who are fighting the battle of the bulge know all too well, hunger often has nothing to do with why we eat. Emotional states, including depression, boredom, anxiety, and stress, stimulate eating. There's also the power of suggestion that can be caused by advertisements, companions, or simply the social situation. Sometimes, food just looks too good to resist.

When it comes to weight control, the expenditure of energy is just as important as the input of energy. The body requires energy for maintenance of basic body functions, physical activity, and processing the food that is eaten. The part of the need for body energy that goes to maintenance is called the **basal metabolic rate** (**BMR**); it's the minimum energy needed to keep an awake, resting body alive and it generally represents between 60% and 75% of the body's energy needs. A male usually has a higher metabolic rate than does a female. This is because a male's body has more muscle and less fat than a female's. Muscles use more energy than fat does. So, while a man and a woman of equal size sit on the couch and watch television together, he burns 10% to 20% more calories than she does. As you age, muscle mass and metabolic rate both decline. Together these factors reduce caloric needs. If other adjustments in lifestyle aren't made to compensate, these metabolic changes can add an extra pounds each year after age 35.

The second most important use of energy is physical activity (Table 13A–6). Exercise is an excellent way to burn calories. Not only does it boost your needs during the activity but it also speeds up metabolic rate for a while afterward.

Obesity

Although overweight and obese are both terms used to describe people who have excess body weight, they don't have exactly the same meaning. An obese person is overweight because of excess fat. An athletic person who weighs more than the desirable weight listed on height-weight tables because of well-developed muscle is not obese.

Insurance companies have used analyses of body weight and death rates of policyholders to develop desirable weight tables for height, build, and gender (Table 13A–7). Obesity is usually defined as being 20% or more above the desirable weight indicated in these tables. The tables were revised upward by about 10% in 1983, when it became apparent that even obese policyholders were living longer than they used to. However, in 1990, researchers began suggesting that the revised tables might give people at the upper end of their "ideal weight" range a false sense of security. They point out that individuals who are just slightly overweight still have an increased risk of having a heart attack.

As we have seen, a person can weigh more than the ideal according to a height and weight table and not be obese. The pinch test or the jump test can provide indications of the amount of body fat. To perform the pinch test, grab a fold of skin at the abdomen, waist, or back of the arm and measure the fold. If it is more than an inch, you could stand to lose some fat. For the jump test, stand in front of a mirror and jump. Anything that shakes is fat.

Although most dieters are motivated to slim down for "cosmetic" reasons, more important reasons are the health risks associated with obesity. For instance, obesity leads to disease of the heart and blood vessels. An extra mile of blood

Table 13A–6 Approximate Caloric Expenditure for Various Activities

Cal/min/lb of Body Weight	Activity	Cal/hr/lb. of Body Weight	Cal/min/lb of Body Weight	Activity	Cal/hr/lb of Body Weight
0.0234	House painting	1.40	0.023	Volleyball	1.38
.026	Carpentry	1.56	.026	Playing pingpong	1.56
.031	Farming, planting, hoeing, raking	1.86	.033	Calisthenics	1.98
.039	Gardening, weeding	2.34	.033	Bicycling on level roads	1.98
.045	Pick-and-shovel work	2.70	.036	Golfing	2.16
.050	Chopping wood	3.00	.046	Playing tennis	2.76
.062	Gardening, digging	3.72	.047	Playing basketball	2.82
.0078	Sleeping	0.47	.069	Playing squash	4.14
.0079	Resting in bed	0.47	.100	Running long distance	6.00
.0080	Sitting, normally	0.48	.156	Sprinting	9.36
.0080	Sitting, reading	0.48		Swimming	
.0089	Lying quietly	0.53	.032	Breast stroke 20 yd/min	1.92
.0093	Sitting, eating	0.56	.064	Breast stroke 40 yd/min	3.84
.0096	Sitting, playing cards	0.58	.026	Back stroke 25 yd/min	1.56
.0094	Standing, normally	0.56	.056	Back stroke 40 yd/min	3.36
.011	Classwork, lecture	0.66	.058	Crawl 45 yd/min	3.48
.012	Conversing	0.72	.071	Crawl 55 yd/min	4.76
.012	Sitting, writing	0.72	.033	Walking on level	1.98
.016	Standing, light activity	0.96	.093	Running on level (jogging)	5.58
.020	Driving a car	1.20			
.028	Cleaning windows	1.68			
.024	Sweeping floors	1.44			
.044	Walking downstairs	2.64			
.116	Walking upstairs	6.96			
.014	Lecturing	0.84			

EXAMPLE: 150-pound man sitting and reading for 60 min = 150 × 0.0080 × 60 = 72 calories expended, or
1 hour = 150 × 0.48 × 1 = 72 calories

Table 13A–7 Height and Weight Tables

Man's Height	Size of Frame			Women's Height	Size of Frame		
	Small	Medium	Large		Small	Medium	Large
5'2"	128–134	131–141	138–150	4'10"	102–111	109–121	118–131
5'3"	130–136	133–143	140–153	4'11"	103–113	111–123	120–134
5'4"	132–138	135–145	142–150	5'0"	104–115	113–126	122–137
5'5"	134–140	137–148	144–160	5'1"	106–118	115–129	125–140
5'6"	136–142	139–151	146–164	5'2"	108–121	118–132	128–143
5'7"	138–145	142–154	149–168	5'3"	111–124	121–135	131–147
5'8"	140–148	145–157	152–172	5'4"	114–127	124–138	134–151
5'9"	142–151	148–160	155–176	5'5"	117–130	127–141	137–155
5'10"	144–154	151–163	158–180	5'6"	120–133	130–144	140–159
5'11"	146–157	154–166	161–184	5'7"	123–136	133–147	143–163
6'0"	149–160	157–170	164–188	5'8"	126–139	136–150	146–167
6'1"	152–164	160–174	168–192	5'9"	129–142	139–153	149–170
6'2"	155–168	164–178	172–197	5'10"	132–145	142–158	152–173
6'3"	158–172	167–182	176–202	5'11"	135–148	145–159	155–176
6'4"	162–176	171–187	181–207	6'0"	138–151	148–162	158–179

vessels forms to support each extra pound of fat, increasing the work of the heart. Obesity also raises total cholesterol levels in the blood while lowering levels of the beneficial HDL cholesterol. In addition, it increases the risk of high blood pressure, which can lead to death from heart attack, stroke, or kidney disease. Obesity has harmful effects besides those on the heart and blood vessels. It can induce diabetes, which results in elevated blood glucose levels, is a major cause of gallstones, and can worsen degenerative joint diseases.

Weight Management

A successful weight loss program has three components: (1) a reduction in the number of calories consumed, while keeping the balance of calories in line with the recommended guidelines (58% of the calories from carbohydrates, mostly complex carbohydrates, 12% from protein, and not more than 30% from fat); (2) an increase in energy expenditure; and (3) behavior modification. Gradual changes in eating habits are most likely to lead to permanent lifestyle changes.

A pound of fat contains approximately 3500 calories, so to lose 1 pound a week, a person should reduce calorie consumption by 500 calories a day (500 calories times 7 days = 3500 calories) or increase calorie use by 500 calories a day or any equivalent combination (Figure 13A–11). The lowest daily caloric intake recommended for a female is 1200 calories and for a male, 1500 calories, unless they are in a medically supervised program.

The easiest way to reduce calorie intake is to cut back on fatty foods. Recall that fat contains more than twice as many calories as an equivalent weight of carbohydrate or protein.

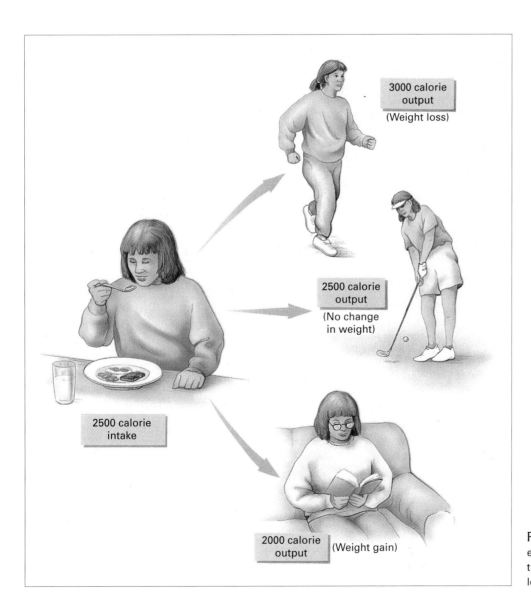

Figure 13A–11   The balance of energy intake and energy expenditure determines whether we gain, lose, or maintain body weight.

(a)   (b)   (c)

Figure 13A–12   Typical dieters change their energy balance temporarily and when old habits are resumed the weight is regained. The cycle is then repeated. The "yo-yo effect" makes weight loss more difficult with successive diets. When there is a severe reduction in calories, as occurs during a crash diet, the metabolic rate decreases so that calories are conserved. In addition, when weight is lost, it comes from both lean and fat tissue. However, when it is regained, it is in the form of fat. Fat is less metabolically active than protein. Together, these factors make it difficult to lose weight and pounds are regained even if fewer calories are consumed. The popular talk show host, Oprah Winfrey, has been vocal about her struggle with weight regain.   *(a, Douglas C. Pizac, 1986/AP/Wide World Photos; b, © Wide World Photos, Inc.; c, AP/Wide World Photos, Inc.)*

That is why fat calories add up quickly. Not only does fat contain more calories, it is also easier for the body to store as body fat than is protein or carbohydrate. Indeed, most of us do end up wearing the fat we eat. There's some truth to the quip, "Save time and spread that ice cream sundae on my thighs now. That's where it will end up after I eat it."

Another healthful diet tip is to increase the amount of fiber in the diet. High fiber foods, such as fruits and vegetables, tend to be low in calories and fat but also high in vitamins and minerals. Because the fiber is bulky, these foods are also filling.

Approximately 60% to 90% of dieters who lose weight will later regain all the weight they've lost (Figure 13A–12). Often this is because the weight loss was achieved by drastically cutting back calories, which can be unhealthy and is very difficult to continue for long periods of time. As old eating habits return, so do the pounds. When the determination to shed some pounds returns, the diet begins again. Dr. Jean Mayer, an obesity and nutrition expert, has called this "the rhythm method of girth control." It's more commonly known as "the yo-yo effect."

What causes the yo-yo effect? When there is a severe restriction in calories, as occurs in the typical crash diet, the body adopts a calorie-sparing defense by reducing the resting metabolic rate by as much as 45%. This response conserves energy and evolved long ago to help our ancestors survive during times of food scarcity. Today, it makes weight loss during successive diets progressively more difficult. Furthermore, with each diet, a person usually loses both lean and fat tissue, especially if increased exercise isn't part of the weight-loss program. When the weight is regained, most of it is fat. Fat is less metabolically active than lean muscle tissue, so the person's metabolic rate drops even lower. Thus, repeated crash dieting may be a "no-lose situation."

Eating Disorders

Dieting can, indeed, go too far. Most people with the eating disorders anorexia nervosa and bulimia begin by dieting. It isn't always clear *why* they begin dieting, because only 25% of anorexics and 40% of bulimics were overweight at the onset of the eating disorder. Whereas **anorexia nervosa** is a deliberate self-starvation, **bulimia** involves binge eating followed by purging by self-induced vomiting, enemas, laxatives, or diuretics. The change in eating habits associated with these eating disorders is thought to be the result of psychological, social, and physiological factors. Cholecystokinin (CCK), a hormone involved in regulating eating be-

Diet Fads

Permanent weight loss requires permanent changes in eating and exercise habits and many fad diets are too extreme to be maintained for a long time. As we have seen, all *rapid* weight-loss diets are self-defeating because they cause metabolic changes that will lead to weight gain, the so-called yo-yo effect. We will now consider how fad diets can be hazardous to health.

Low-Carbohydrate Diets

Low carbohydrate diets go by many names—the Atkins diet, the Air Force diet, the Scarsdale diet, the Stillman diet, and the "drinking man's" diet—to name a few. But the rationale is always similar. The idea is that, with a diet with minimal carbohydrate available for energy, the body will be forced to metabolize its own fat stores. Many of the low-carbohydrate diets are combined with a high-protein component in order to supply protein to meet the body's needs and prevent the excessive breakdown of body protein that accompanies a low-carbohydrate diet. In the absence of carbohydrate, fat is indeed burned, but it cannot be completely metabolized, and so ketone bodies are produced. The ketone bodies generated are thought to suppress appetite, so you don't feel hungry while on the diet. The kidneys increase urine volume as they excrete the ketone bodies and water loss looks like weight loss when you step on the scale. However, water is quickly regained when a normal diet is resumed. Risks associated with low-carbohydrate diets include high blood cholesterol, hypoglycemia (low blood sugar), mineral imbalances, and other metabolic disturbances. Excessive protein in the diet brings additional risks, ranging from kidney damage to osteoporosis.

Liquid Diets

First introduced in the 1970s, over-the-counter liquid formula diets have recently become popular again. Many people find it easier to control their intake of liquid than of food. These formulas are usually high in protein and low in fat. They are not nutritionally balanced by themselves, but they are not intended to replace all food. If the supplemental food is not too high in calories and *if* it is nutritionally balanced, a liquid formula may lead to weight loss. However, these diets are typically low in fiber, which although not a nutrient, is extremely beneficial in the diet.

Diuretics

Diuretics increase the amount of water excreted by the kidneys. However, water loss is not fat loss. The weight will be quickly regained.

Diet Pills

Many diet pills are intended to blunt the appetite. Those sold over-the-counter often contain phenylpropanolamine, which might slightly curb appetite for a while but soon loses effectiveness and normal appetite returns. Its side effects include increased blood pressure, nervousness, rapid heart beat, and difficulty sleeping. Many of these diet pills also contain caffeine because it is a stimulant and a diuretic.

A number of prescription diet pills are available, but they are not without risk and gradually lose effectiveness. Amphetamines, for instance, are strong appetite suppressants—at first. The effect begins to wane after about 2 weeks and the dosage must be increased to maintain the same effects on appetite. For this reason, amphetamines are considered to have a high potential for abuse. As dosage is increased, so are the side effects of nervousness and inability to sleep. Dieters who take amphetamines eventually regain their normal appetites and all the weight they lost.

Two newer drugs, fenfluramine and phentermine in combination (commonly called fen-phen), have been approved in some states for the treatment of genuine obesity, not just for shedding 10 or 20 pounds, and when other weight-loss treatments have failed. The drugs work by curbing appetite. There have been some remarkable success stories of people losing 50, 100, or even 200 pounds while on these medications. Unfortunately, the pounds begin to pile on again as soon as they stop taking the medication, which means that some people will need to take them for years. These drugs are ripe for abuse, not just by those who are only slightly overweight and want to lose a few pounds to look good on the beach but also by physicians who may prescribe them haphazardly. Users should beware. Although the risk is small (estimated at 1 in 45,000), the drugs can cause pulmonary hypertension, a condition of elevated blood pressure in the vessels that carry blood from the heart to the lungs. Pulmonary hypertension often produces no symptoms until it is quite advanced and it continues to worsen after the drugs are discontinued. Treatment may require a heart and lung transplant. This form of pulmonary hypertension usually causes death within 3 years of diagnosis. Thus, the benefits of using fen-phen may outweigh the risks for those who are severely overweight, but the use is probably not worth the risk for those who are only slightly overweight.

During the spring of 1996, the FDA approved a new drug, dexfenfluramine (Redux), for the treatment of obesity. People who take this drug are able to lose up to 5% to 10% of the body weight and keep it off *as long as they continue taking it*. Weight is regained when the medication is stopped. Questions about the risks associated with dexfenfluramine remain. Initial studies suggest that its use may be associated with depression and pulmonary hypertension.

havior among other things, may play a role in bulimia. In both disorders there is a preoccupation with body weight and shape.

Eating disorders are more common than might be imagined. Although they are not exclusively a female problem, they do affect about ten times more females than males. Approximately 1 in 200 white females between the ages of 12 and 18 in the United States has an eating disorder. The problems are most prevalent during the college years, when it is estimated that between 5% and 20% of college females are afflicted. Although their problem may not be severe enough to classify them as bulimic, 78% of college females report that they have had binge eating sprees and 8.2% have used self-induced vomiting to control weight. Anorexia is, in fact, the biggest health problem among women involved in sports.

The behavior patterns involved in anorexia and bulimia differ, but both result in a severe deficit in calories. An anorexic eats very little food and, therefore, consumes few calories. But, anorexia, which means lack of appetite, is misnamed. Although anorexics deny hunger and often refuse to eat, this is due to an intense fear of becoming fat, *not* from lack of appetite. Indeed, anorexics are often preoccupied with food and may develop strange rituals about eating. For instance, food may be measured into extremely small amounts or cut into tiny pieces. Excessive exercise is also typical in anorexia. No matter how much weight is lost, however, it can never be enough, because an anorexic has a distorted body image. The person perceives the body as fat even when emaciated (Figure 13A–13).

In contrast, a bulimic eats a huge amount of food but then eliminates it from the body. During a bulimic binge, which may last as long as 8 hours, as many as 20,000 calories may be consumed. Several shorter binges may occur in a single day. Each binge is followed by attempts to purge the body of the calories, usually by self-induced vomiting or by laxatives. Some bulimics take as many as 200 laxatives a week to rid the body of unwanted calories. It is estimated, however, that laxatives can eliminate only 10% of the calories consumed.

The consequences of these eating disorders can be disastrous. Those of anorexia, for instance, are not so different from those of starvation. In the early phase of the illness, an anorexic typically chooses a diet that is low in energy-dense foods but rather high in proteins and other essential nutrients. Dietary protein, combined with the high activity levels characteristic of a person with anorexia nervosa, has a nitrogen-sparing effect. As a result, the initial weight loss is almost entirely due to loss of fat tissue. However, when fat reserves are exhausted and refusal of food becomes more severe, the body begins to break down its own proteins to use as an energy source. The primary source of these proteins is skeletal and heart muscle. At the same time, water loss is accelerated, especially from the fluids *within* the body cells. The water loss then leads to disturbances in metabolism and electrolytes.

Figure 13A–13 Anorexia is a form of self-starvation. No matter how emaciated the person becomes, the individual still perceives the body as being fat. *(© 1993 B. Bodine/Custom Medical Stock Photo)*

Amenorrhea is the cessation of menstruation. It is caused by excessive exercise, low percentage of body fat, poor nutrition, and emotional stress—all characteristic of anorexics. A woman with amenorrhea does not ovulate and, therefore, has very low levels of the hormone estrogen. Estrogen is important in more than just the reproductive functions of the body. It is also associated strongly with the growth and maintenance of bones in women. Estrogen also has a protective effect on the heart by reducing the amount of fatty deposits in arteries.

A major side effect of anorexia is a severe decrease in bone health. Although the excessive exercise practiced by most anorexics may have a slight strengthening effect on bones, many other factors associated with anorexia work to weaken bones. Amenorrhea, malnutrition, and low body weight, particularly low body fat, are just several potential factors contributing to poor bone health. There is a loss of bone mass, particularly from vertebrae and the long bones of the arms and legs. This makes the bones fragile and increases the risk of fractures, particularly hip fractures.

Heart problems are the most common cause of death in anorexics. Starvation, dehydration, and electrolyte disturbances cause the heartbeat to slow (bradycardia) and blood pressure to fall (hypotension). Heart murmurs occur and are thought to be caused by the shrinkage of the heart muscle, which occurs as the patients lose muscle mass as well as body fat. The decrease in heart size can lead to abnormal blood flow through the heart. In addition, the person can develop congestive heart failure, if fluids are replaced too quickly during treatment.

A bulimic's practice of purging by either self-induced vomiting or overuse of laxatives brings with it special problems. Purging can result in dehydration and electrolyte disturbances that can lead to muscle cramping, neurological problems, and heart failure. When vomiting is self-induced frequently, it can cause serious problems. Glands in the face and neck may become swollen. Repeated bathing in acidic stomach contents can cause tooth decay, chronic heartburn, and sores in the mouth or lips. In addition, the acts involved in inducing vomiting can tear the throat or esophagus.

A potential cause of death associated with both eating disorders is hypoglycemia, an abnormally low blood glucose level. Since the brain depends entirely on glucose for its metabolism, hypoglycemia can cause unconsciousness and death.

Help is available for people with eating disorders. If you have an eating disorder, seek help. If you have a friend with an eating disorder, speak up and save a life. Be supportive, but don't try to deal with it alone.

There are four general elements of treatment. The dominant goals are: (1) to gain and maintain weight, (2) to resume normal eating patterns, (3) to evaluate and treat the necessary psychological issues in the patient, and (4) to educate the families about the disorder. Families are informed of their importance in the treatment and are assisted in developing methods to promote normal functioning of the patient.

REVIEW QUESTIONS

1. List the six nutrients and their functions. Which nutrients supply energy?
2. What are some health risks associated with a high-fat diet?
3. Why is the proper balance among the types of fat in the diet so important?
4. Explain the difference between a complete and an incomplete protein.
5. Explain what is meant by complementary proteins.
6. How can a deficiency in one vitamin have widespread effects in the body?

7. Explain the role that leptin is thought to play in regulating body weight.
8. Explain the difference in meaning of the terms overweight and obesity.
9. Why are rapid weight-loss diets doomed to failure? Explain the yo-yo effect.
10. Differentiate between anorexia and bulimia. What are the health problems associated with these eating disorders?

The Respiratory System

(© 1990 Kent Wood/Photo Researchers, Inc.)

"In goes the good air. Out goes the bad air." This old expression describes what we usually call breathing. But what makes air "good" or "bad"? The answer you may hear is that good air contains more oxygen than bad air and bad air has a higher level of carbon dioxide.

The next question is, "Why is oxygen good?" The simple answer is that we need it to extract energy from food molecules, energy that we store in a molecule called ATP (adenosine triphosphate). Our cells can make ATP without oxygen, but they can make 18 times as much with oxygen. The stored energy can later be released to do the work of the cell.

We can see, then, that oxygen is "good," but why is carbon dioxide "bad"? Essentially, carbon dioxide earned this label because it is a by-product of cell activity and no longer useful. The same chemical reactions that require oxygen for the production of ATP produce carbon dioxide. Furthermore, in solution it forms carbonic acid, which can be harmful to cells.

The role of the respiratory system is to exchange oxygen and carbon dioxide between the air and the blood, a process that provides essential oxygen, rids us of carbon dioxide, and regulates the acidity of body fluids. This exchange takes place across a moist membrane with an enormous surface area. This membrane is arranged within two structures called the lungs. Since the lungs lie within the body, the respiratory system must also provide a means of moving oxygen-rich air into them and carrying carbon dioxide-laden air away. This is the role of breathing, as we will see.

The Structures of the Respiratory System

One way to see how humans handle the processes of gaining oxygen and losing carbon dioxide is to follow the path of air from the nose to the lungs. We will do that and consider the structures the air passes along the way (Table 14–1).

Table 14–1   The Respiratory System

Structure	Description	Function
Nasal cavity	Cavity within nose, divided into right and left halves by nasal septum; has three shelf-like bones called nasal conchae	Warms, moistens, and filters incoming air; serves as resonance chamber for sound of voice
Pharynx (throat)	Muscular tube leading from nasal cavities to esophagus and larynx	Common passageway for air, food, and drink
Larynx	A cartilaginous box between the pharynx and trachea that contains the vocal cords and the glottis	Produces sound
Epiglottis	A flap of tissue reinforced with cartilage	Covers the glottis during swallowing; conducts air to the trachea
Trachea	A tube reinforced with C-shaped rings of cartilage that leads from the larynx to the bronchi	The main airway; conducts air from larynx to bronchi
Bronchi (primary)	Two large branches of the trachea reinforced with cartilage	Conduct air from trachea to the each lung
Bronchioles	Narrow passageways leading from bronchi to alveoli	Conduct air to alveoli; control airflow in lungs
Lungs	Two, lobed, elastic structures within the thoracic cavity containing surfaces for gas exchange	Exchange of oxygen and carbon dioxide between blood and air
Alveoli	Microscopic sacs within lungs, bordered by extensive capillary network	Provide immense, internal surface area for gas exchange
External intercostal muscles	Muscles between the ribs	Contract to lift the rib cage upward, increasing the size of the thoracic cavity to allow inhalation
Internal intercostal muscles	Muscles between the ribs	Contract to lower rib cage during forced exhalation
Diaphragm	Sheet of muscle between the thoracic and abdominal cavities	Contracts and flattens, increasing the size of the thoracic cavity during inhalation

The Nose and Its Passages

However large someone's nose might seem from the outside, the inside is not as roomy as you might imagine. One reason is that a thin partition of cartilage and bone called the **nasal septum** divides the inside of the nose into two **nasal cavities**. In addition, much of the space within the nasal cavities is taken up by three convoluted and shelf-like bones, the **nasal conchae** (Figure 14–1). These bones increase the surface area of the nasal cavities and divide each cavity into three narrow passageways through which the air flows. Mucous membrane covers the entire inner surface of the nasal cavities.

We all know what a nose looks like, but what does a nose do? A nose has three important functions:

1. **Hygiene.** The nose helps clear particles from the air that moves through its passages. Considering that each of us inhales about 150,000 bacteria along with a great deal of pollen and dust each day, this is an important job. Most

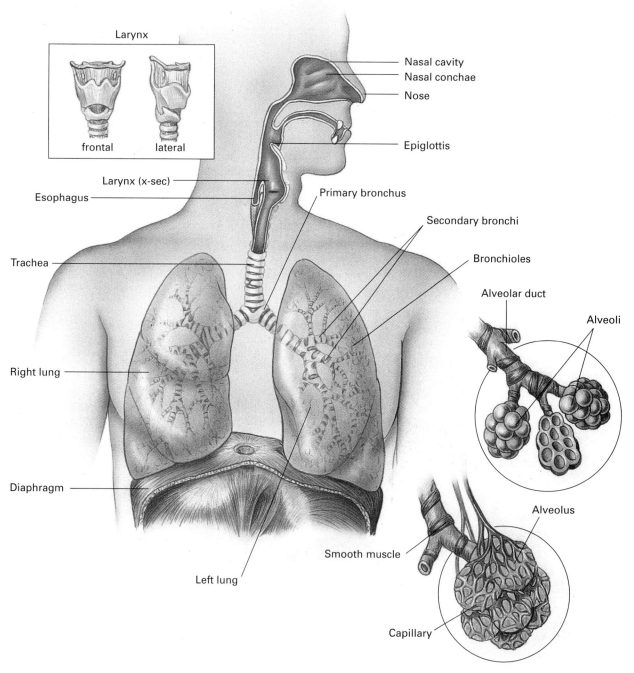

Figure 14–1 The respiratory system.

pollen and dust is removed from the air before it reaches the lungs; rarely do any particles greater than 3 to 5 μm, which is half the size of a red blood cell, escape entrapment and reach the lungs.

The nose and air tubules clean inhaled air in a variety of ways. The largest particles are filtered out by hairs inside the nose. In addition, certain cells in the membrane lining the surface of the nasal cavities and air tubules produce mucus, a sticky substance that catches dust particles. The effectiveness of the trapping process is enhanced by the curves of the nasal conchae, which cause turbulence in the incoming air so that it strikes many surfaces as it passes through. Then, cilia, tiny projections extending from the membranous lining that function like little brooms, sweep the mucus, along with the trapped dirt particles and bacteria, toward the throat where they can be swallowed and destroyed by digestive enzymes (Figure 14–2).

Those particles that do not become trapped in the nasal cavities or the air tubules are deposited in the lungs. Some of them may be engulfed by macrophages, large irregularly shaped cells that wander the surfaces of the lungs. However, if too many particles are inhaled or if the mechanisms for removing them fail, the particles may accumulate in the lungs and cover some of the surfaces for gas exchange, reducing their efficiency and setting the stage for infection.

2. **Conditioning the air.** The nose also warms and moistens the inhaled air before it reaches the delicate lung tissues. The blood in the extensive capillary system of the mucous membrane lining the nasal cavity warms the incoming air. The profuse bleeding that follows a blow to the nose is evidence of the vascularity of these mem-

branes. Warming the air before it reaches the lungs can be extremely important in cold climates. Frigid arctic air, which may be as cold as −70°F, would freeze lung cells if it were not warmed close to body temperature.

Moistening the inhaled air is also essential because oxygen cannot cross dry membranes. Mucus helps moisten the incoming air so that lung surfaces don't dry out.

3. **Olfaction.** Our sense of smell is due to the olfactory receptors located on the mucous membranes high in the nasal cavities behind the nose. We detect an odor when molecules of a substance in a gaseous state are delivered to the olfactory receptors where they are dissolved in fluids covering the membranes. Since olfactory receptors are on the roof of the nasal cavity, we can increase our exposure to an odor by sniffing, which brings more air into the upper regions of the nose. The sense of smell is discussed in more detail in Chapter 9.

The Sinuses

Connected to the nasal cavities are some large air-filled spaces in the bones of the face called the sinuses. Bone is heavy. Thus, one advantage of the sinuses is to make the head lighter. The sinuses also help warm and moisten the air we breathe because they, too, are lined with mucous membranes and some incoming air does pass through them. In addition, the sinuses are part of the resonating chamber that affects the quality of the voice. This is why a cold that causes the mucous membranes of the sinuses to swell and produce excess fluid results in a change in the sound of your voice.

Because the air spaces of the sinuses are continuous with those of the nasal cavities, the sinuses can normally drain. However, when their mucous membranes become inflamed, as they do in **sinusitis** (-*itis*, inflammation of), the swelling can block the connection to the sinuses, making it difficult for the sinuses to drain the mucous fluid they produce. The pressure caused by the accumulation of fluids in the sinuses causes pain over one or both of eyes or in the cheeks or jaws. Sinusitis may be caused by the virus responsible for a cold or by a subsequent bacterial infection.

The Pharynx

The **pharynx**, commonly called the throat, is space behind the nose and mouth. It is a passageway for both air and food.

Small passageways, called the **Eustachian tubes**, join the upper region of the pharynx with the middle ear. These passages help equalize the air pressure in the middle ear with that of the atmospheric air reaching the pharynx. (This is why your ears may pop on airplanes as the pressure in the middle ear suddenly equalizes with that of the pharynx.) Unfortunately, the Eustachian tubes may also allow the organisms causing a sore throat to spread to the ears, resulting in an ear infection.

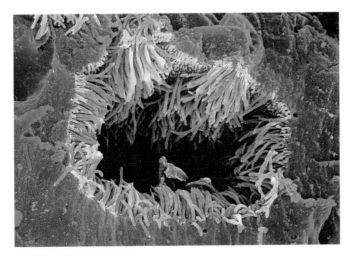

Figure 14–2   The respiratory passageways are lined with clumps of hair-like structures called cilia interspersed between mucus-secreting cells. The cilia are green in this color-enhanced electron micrograph.   *(Proff. Motta, Correr, and Nottola/University "La Sapienza," Rome/Science Photo Library/Photo Researchers, Inc.)*

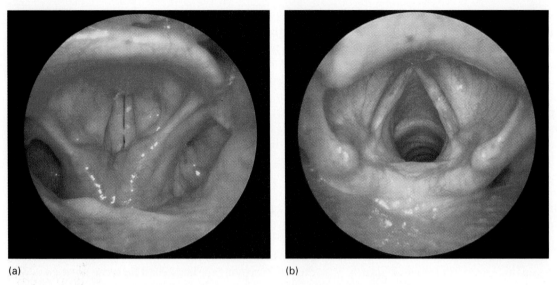

(a)　　　　　　　　　　　　　(b)

Figure 14–3 The vocal cords are folds of connective tissue that are stretched over the glottis seen here from above. (a) During speaking, the vocal cords cover the glottis and vibrate as air passes between them. The vibrations generate sounds that result in the speaking voice. (b) At other times, the cords separate and the glottis is open. *(© CNRI/Phototake, NYC)*

The Larynx

After moving through the pharynx, the air next passes through the **larynx**, which is commonly called the voice box or Adam's apple. The larynx is a box-like structure composed of muscles and cartilage held together by elastic tissue.

The larynx has two main functions. It is a traffic director for materials passing through the region, allowing air, but not other materials, to enter the lower respiratory system. Also, it is the source of the voice. Let's consider these two functions in more detail.

1. **An adjustable entrance to the respiratory system.** The larynx provides an opening to the respiratory system that can be adjusted—that is, it can be opened to allow air to pass to the lungs and closed to prevent other matter, such as food, from entering the lungs. Since the esophagus (the tube leading to the stomach) is behind the larynx, material must pass over the opening to the respiratory system to reach the digestive system. If foreign material enters the respiratory system, it could lodge in one of the tubes conducting air to the lungs and prevent air flow. Fluid entering the lungs is equally dangerous because it can cover the respiratory surfaces, decreasing the area of gas exchange. Normally, foreign material is prevented from entering the lower respiratory system because swallowing causes the larynx to move up and under a piece of cartilage called the **epiglottis**. In this way, the epiglottis covers the opening in the larynx that air passes through, the **glottis**. You can feel the movement of the larynx if you put your fingers on your Adam's apple while swallowing. Because of this movement, you cannot breathe and swallow at the same time. (Try it!)

2. **Production of the voice.** You should not be surprised to learn that the larynx, being the "voice box," is the source of the voice. The voice is generated by the vibration of the **vocal cords**, two thick strands of tissue stretched over the opening of the glottis (Figure 14–3). When you speak, muscles stretch the vocal cords across the air passageway, narrowing the opening of the glottis. Air passing between the stretched vocal cords causes them to vibrate and produce a sound, just as the edges of a balloon vibrate and make noise if you stretch its neck while allowing the air to escape. The vibrations of the vocal cords set up sound waves in the air spaces of the nose, mouth, and pharynx, and this resonation is largely responsible for the tonal quality of your voice.

The loudness of the sound is determined by the pressure of the air forced over the vocal cords—the greater the pressure, the louder the sound. So when you shout, you force a great deal of air under intense pressure over the vocal cords, but when you talk quietly, you gently pass less air between these cords. When you whisper, the vocal cords may not vibrate at all.

The pitch of the voice is altered by changing the tension on the vocal cords. The more they are stretched, making them thinner and more taut, the more rapidly they vibrate and the higher the pitch of the sound. You can demonstrate the relationship between thickness and pitch for yourself by plucking a rubber band stretched

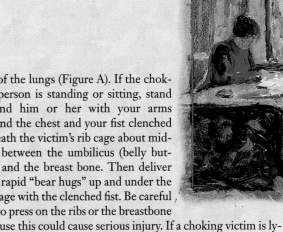

Food Inhalation and the Heimlich Maneuver

You are sitting in a restaurant, laughing, eating, and drinking, when suddenly someone in your party leans forward in obvious distress. You quickly see that the person is choking. What do you do?

First, assess the severity of the problem. If the person can speak, breathe, or cough, try to calm the victim but do not interfere with the person's efforts to cough out the object. Most choking incidents do not involve complete blockage of the air passageways, so even if coughing doesn't dislodge the object, it may be possible to move enough air in and out to keep a calm victim alive until a hospital can be reached.

Immediate action is needed, however, if the person cannot speak or breathe. First, support the victim's chest with one hand and hit the back between the shoulder blades with the heel of your other hand four times.

If the person is still unable to breathe, use the **Heimlich maneuver**, a procedure intended to force a large burst of air out of the lungs (Figure A). If the choking person is standing or sitting, stand behind him or her with your arms around the chest and your fist clenched beneath the victim's rib cage about midway between the umbilicus (belly button) and the breast bone. Then deliver four rapid "bear hugs" up and under the rib cage with the clenched fist. Be careful not to press on the ribs or the breastbone because this could cause serious injury. If a choking victim is lying on the ground, you can generate the same pressure changes by pushing inward and upward in the anterior part of the victim's abdomen. If you begin to choke yourself and there is no one to perform the Heimlich maneuver on you, it may be possible to dislodge the food yourself by throwing the upper abdominal region against a table, chair, or other stationary object.

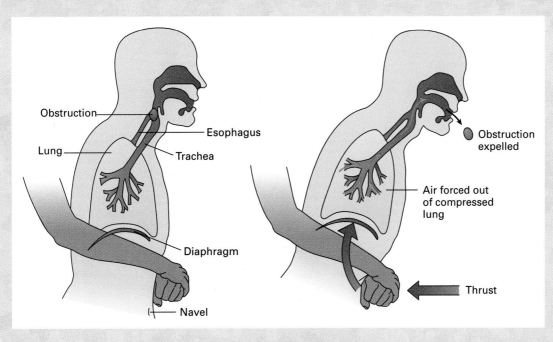

Figure A Heimlich maneuver. To assist a choking victim, push inward and upward in the center of the abdomen below the rib cage about midway between the "belly button" and the breast bone. This action pushes against the diaphragm and forces air out of the lungs. Air forced through the respiratory tubules expels the object that was blocking air flow.

between your thumb and forefinger. The more the rubber band is stretched, the higher the pitch of the twang.

The vocal cords of young boys and girls are quite thin, causing them to have high-pitched voices. When boys reach puberty, the vocal cords begin to thicken, resulting in that adolescent vocal quality somewhere between a croak and a shriek that later usually deepens into the voice of an adult male. The vocal cords of females normally do not thicken, so a woman's voice usually has a higher pitch than a man's.

When you suffer from **laryngitis**, an inflammation of the larynx, the vocal cords become swollen and thick. As a result, they cannot vibrate freely and the voice becomes deeper and more husky. Sometimes, when the vocal cords are particularly inflamed, it is difficult to speak at all because the cords cannot vibrate in that condition.

The Trachea

The trachea, or windpipe, is a tube roughly 12 cm (4.5 inches) long and 2.5 cm (1 inch) in diameter that conducts air between the environment and the lungs. It is held open by horseshoe-shaped rings of cartilage, giving it the general appearance of a vacuum cleaner hose (Figure 14–4). You can feel these rings of cartilage in your neck, just below the larynx.

These support rings are necessary in the trachea and its branches to prevent the air tubules from collapsing with the drop in pressure created by the rapid flow of air that accompanies each breath. Air (or fluid) passing rapidly over a surface causes a partial vacuum or "pull" on that surface. Maybe you have noticed that when you jump in the shower and turn on the water, the shower curtain is drawn in toward you. The shower curtain attacks you because the moving water creates a partial vacuum just as the rapid movement of air through the respiratory tubules does. If the trachea were not

supported open, the rapid flow of air would cause it to collapse or flatten.

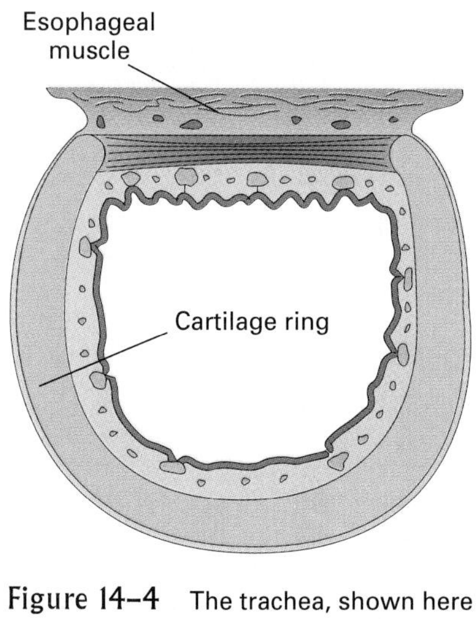

The Bronchial Tree

The trachea divides into two air tubes called primary **bronchi**, one bronchus (singular) going to each lung. These branch repeatedly, forming progressively smaller air tubes. The smallest bronchi divide to form yet smaller tubules called **bronchioles**, which finally end with the gas exchange surfaces, the alveoli (discussed next).

The repeated branching of air tubules in the lung is reminiscent of a maple tree in winter. In fact, the resemblance is so close that the system of air tubules is often called the **bronchial tree** (Figure 14–5).

All the bronchi are held open by cartilage, just as we saw in the trachea. However, the amount of cartilage decreases with the diameter of the tube.

The tiny bronchioles have no cartilage, but their walls do contain smooth muscle, which is controlled by the autonomic nervous system so that air flow can be adjusted to suit metabolic needs. During an emergency, certain cells such as muscle cells, may require more oxygen. Under these conditions, the sympathetic branch of the autonomic nervous system causes the bronchioles to dilate making air flow easier.

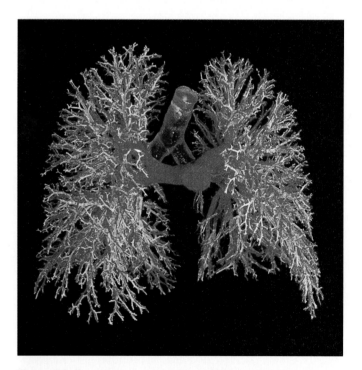

Figure 14–5 A resin cast of the bronchial or respiratory tree (shown in white) and the arteries of the lungs (shown in red). In the body, this branching system of air tubules is hollow and serves as a passageway for the movement of air between the atmosphere and the alveoli, where gas exchange takes place. (*© Martin Dohrn/Royal College of Surgeons/Science Photo Library/Photo Researchers, Inc.*)

Esophageal muscle

Cartilage ring

Figure 14–4 The trachea, shown here in cross section, contains C-shaped rings of cartilage that keep it open during breathing.

Although the contraction of the muscle in bronchial walls is usually marvelously adjusted to the body's needs, sometimes, such as with **asthma**, the bronchial muscles go into spasm, making air flow exceedingly difficult. The difficult breathing is worsened by persistent inflammation of the airways. Asthma is a chronic condition characterized by recurring attacks of wheezing and difficulty in breathing. Many asthma attacks are triggered by an allergy to things such as pollen, dog or cat dander (skin particles), or tiny mites in house dust. However, an attack can also be caused by a cold or respiratory infection, certain drugs, inhaling irritating substances, vigorous exercise, and psychological stress. (Some attacks, though, start for no apparent reason.) Some inhalants prescribed to treat asthma attacks work by relaxing the bronchial muscles. Other inhalants contain steroids that reduce the inflammation of the air tubules that occurs in asthma.

The Alveoli

Each bronchiole ends with an enlargement called an alveolus (plural, *alveoli*) or, more commonly, with a grape-like cluster of alveoli. Each **alveolus** is a thin-walled, rounded chamber surrounded by a vast network of capillaries (Figure 14–6). It is at this interface that oxygen diffuses from the alveoli into the blood to be delivered to cells and that carbon dioxide produced by the cells diffuses from the blood into the alveolar air (Figure 14–7).

Most of the lung tissue is composed of alveoli, making the structure of the lung much more similar to foam rubber than to a balloon, the way the lung is sometimes imagined. The surface area of simple, hollow (balloon-like) lungs the same size as our lungs would be roughly 0.01 square meter. However, each of our lungs contains approximately 300 million alveoli, and so the combined alveolar surface area is about 70 to 80 square meters, roughly the surface area of a tennis court. In other words, the alveoli increase the surface area of the lung about 8500 times.

For the alveoli to function properly as a surface for gas exchange, they must be kept open. Moist membranes, such as those of the alveolar walls, are attracted to one another because of the complex physical properties of water, which is the major constituent of the fluid moistening the alveoli. One consequence of this force of attraction between water molecules is surface tension. The high surface tension of water pulls water molecules together, causing small quantities of water to "round up" into droplets. You may have noticed that a water droplet falling from a faucet forms a sphere. This attraction between water molecules would also exist in the thin watery fluid covering the surface of the alveoli and

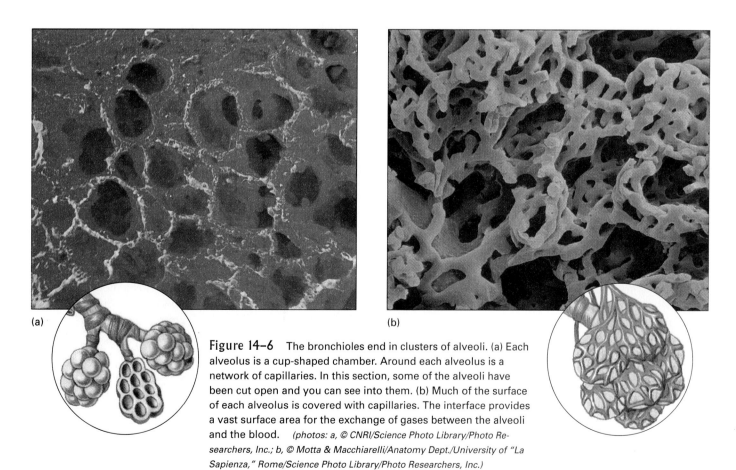

(a)

(b)

Figure 14–6 The bronchioles end in clusters of alveoli. (a) Each alveolus is a cup-shaped chamber. Around each alveolus is a network of capillaries. In this section, some of the alveoli have been cut open and you can see into them. (b) Much of the surface of each alveolus is covered with capillaries. The interface provides a vast surface area for the exchange of gases between the alveoli and the blood.   *(photos: a, © CNRI/Science Photo Library/Photo Researchers, Inc.; b, © Motta & Macchiarelli/Anatomy Dept./University of "La Sapienza," Rome/Science Photo Library/Photo Researchers, Inc.)*

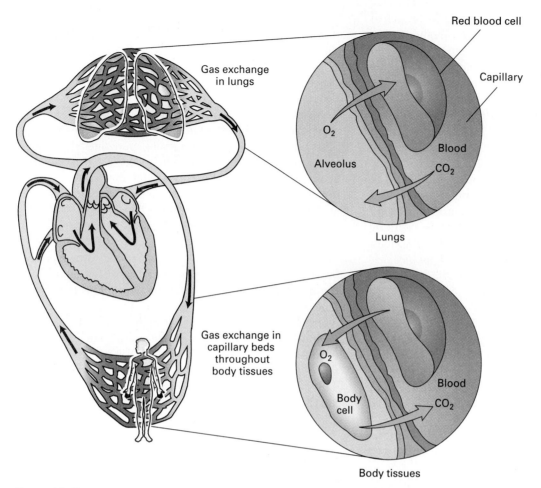

Figure 14–7   In the lungs, oxygen diffuses from the alveolus into the blood. Oxygen is carried to the cells in red blood cells. At the cells, oxygen diffuses from the blood to the body cells, which use the oxygen and produce carbon dioxide in the process. Carbon dioxide diffuses into the blood and is carried back to the lungs, where it diffuses from the blood into an alveolus and is exhaled.

would pull the alveolar walls together, collapsing the air chambers, if it were not prevented by phospholipid molecules called **surfactant**. Surfactant reduces the surface tension, just as a detergent would. With the attractive force between water molecules coating the alveoli reduced, the alveoli remain open.

Usually surfactant production begins during the eighth month of fetal life and enough is present to keep the alveoli open when the newborn takes its first breath. Unfortunately, some premature babies have not yet produced a sufficient amount of surfactant to reduce the high surface tension. As a result, their alveoli collapse after each breath. This condition, called **respiratory distress syndrome** (RDS), makes each breath difficult for the newborn. In order to fill collapsed alveoli, an infant must overcome the force of surface

tension, which is 15 to 20 times more difficult than ventilating expanded alveoli. Although a normal newborn must work this hard for its first breath, a newborn with RDS must repeat this effort for every breath. Some newborns with RDS die because the task of breathing is too difficult. However, many are saved by using mechanical respirators to keep them alive until their lungs mature and produce enough surfactant.

Critical Thinking

Pneumonia is a lung infection that results in an accumulation of fluid and dead white blood cells in the alveoli. Why might this result in lower blood levels of oxygen?

How We Breathe

The next question is, how do we get the "good" air in and the "bad" air out? The first thing we find is that the lungs themselves do not expend the energy and do the work. Instead, they respond passively to pressure changes in the thoracic (chest) cavity. We create the pressure changes by altering the size of the thoracic cavity. Thus, the mechanism of breathing is similar to the way a bellows works. As the handles of the bellows are spread apart, its internal volume is increased. This reduces the pressure inside the bellows and air moves in. When the process is reversed, pressure is increased and air is pushed out. As with a bellows, air moves between the atmosphere and the lungs in response to pressure gradients. Air moves into the lungs when the pressure in the lungs is lower than the air pressure outside the body, and it moves out when the pressure in the lungs is greater than the pressure in the atmosphere. Let's consider how the necessary changes in the size of the thoracic cavity are brought about.

Inspiration

Air moves into the lungs when the size of the thoracic cavity increases. The increase is due to the action of both the **diaphragm**, a broad sheet of muscle that separates the abdominal and thoracic cavities, and the muscles of the rib cage, called the **external intercostals** (*costa*, rib). This process is called **inspiration** (or inhalation). The external intercostals lie between the ribs so that when the muscles contract they pull the rib cage upward and forward. By placing your hands on your rib cage while you inhale, you can feel the rib cage move up and out. Raising the rib cage increases the size of the thoracic cavity from the front to the back. Meanwhile, the contraction of the diaphragm lengthens the thoracic cavity from top to bottom (Figure 14–8). The volume of the lungs is increased as the volume of the thoracic cavity increases and the pressure within the lungs falls below atmospheric pressure, allowing air to move into them.

Expiration

The muscles of the rib cage and the diaphragm relax. The elastic tissues of the lung then recoil, and the rib cage falls back to its former lower position and the diaphragm bulges into the thoracic cavity again. The pressure within the lungs increases as their volume decreases. When the pressure within the lungs exceeds atmospheric pressure, air moves out. The process is called **expiration** (or exhalation).

If it is necessary to exhale more air than usual, as in heavy breathing or coughing, other muscles assist the process. For example, there is another layer of muscles between the ribs. These lie behind the external intercostals and are called the **internal intercostals**. When the internal intercostals contract, they pull the sternum and rib cage even further down, increasing the pressure on the lungs. In addition, the muscles of the abdomen can be contracted. This pushes the organs in the abdomen against the diaphragm, thereby causing it to bulge even further into the thorax.

Critical Thinking

Victorian women often wore corsets containing whale bone that formed a band around their waists and lower chests. The corsets would be laced tightly to create a wasp-like waistline. These women frequently fainted. What is the most likely cause of their fainting spells?

Lung Volumes

You are very likely breathing quietly at this moment. This means you are moving about one-half liter, roughly one pint, of air in and out with each breath. The amount of air inhaled or exhaled during a normal breath is called the **tidal volume** (Figure 14–9).

If, after a normal inhalation, you were to inhale until you could not take in any more air, you would probably bring another 3 liters of air into your lungs. The additional volume of air that can be brought into the lungs after normal inhalation is called the **inspiratory reserve volume**.

After you have exhaled normally, you can still force slightly more than a liter (1100 ml) of additional air from the lungs. This additional volume of air that can be expelled from the lungs after the tidal volume is called the **expiratory reserve volume**.

The lungs can never be completely emptied, even with the most forceful expiration. The amount of air that remains in the lungs after a maximal exhalation, called the **residual volume**, is roughly 1200 ml of air.

If you were to take the deepest breath possible and exhale until you could not force any more air from your lungs, you would measure your **vital capacity**, the maximal amount of air that can be moved in and out of the lungs during forceful breathing. The vital capacity, therefore, equals the sum of the tidal volume, the inspiratory reserve, and the expiratory reserve. Although average values for college-age people are about 4800 ml in males and 3150 ml in women, the values can vary tremendously. They may be as high as 6500 ml in a well-trained athlete or as low as 3000 ml in a frail person.

Since some air is always left in the lungs, the vital capacity is not a measure of the total amount of air that the lungs can hold. The **total lung capacity**, that is, the total volume of air contained in the lungs after deepest possible breath, is calculated by adding the residual volume to the vital capacity. This volume is approximately 6000 ml in men and 4200 ml in women.

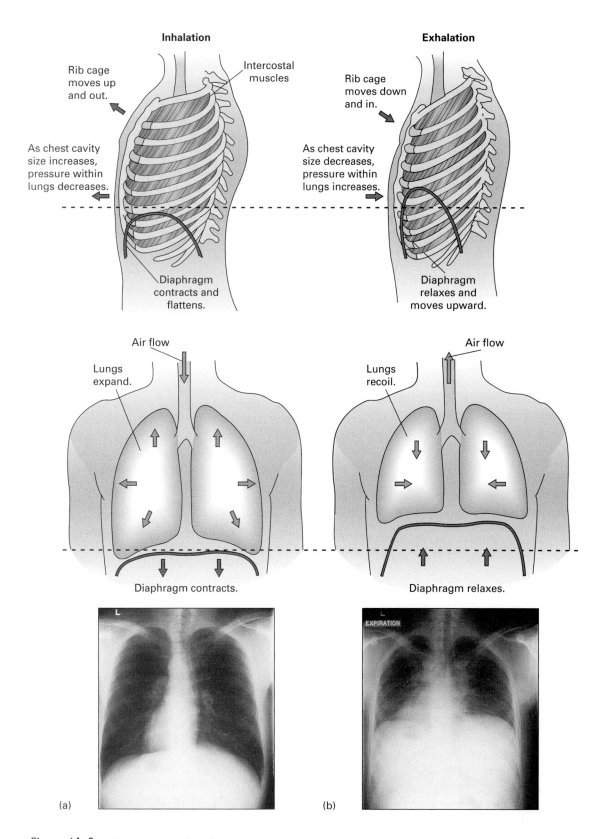

Figure 14–8 Changes in the size of the thoracic cavity bring about inhalation (*left*) and exhalation (*right*). The contraction of the external intercostal muscles pulls the rib cage upward and outward, thereby increasing the size of the thoracic cavity from front to back. The contraction of the diaphragm increases the size of the thoracic cavity from top to bottom. As the size of the thoracic cavity increases, air is pulled into the lungs. The relaxation of these muscles allows a decrease in the size of the thoracic cavity, and so exhalation results. Below, radiographs show actual changes in lung volume during inhalation (a) and exhalation (b). *(a, b, © SIU/Visuals Unlimited)*

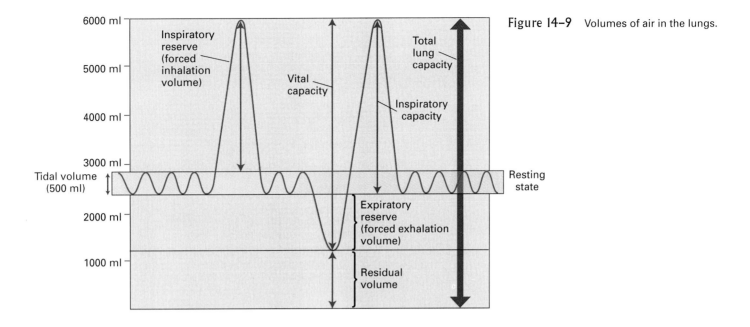

Figure 14–9 Volumes of air in the lungs.

Transport of Gases

Transport of Oxygen

Oxygen is carried from the alveoli throughout the body by the blood. Almost all, about 98.5%, of the oxygen that reaches the cells is carried there bound to hemoglobin in the red blood cells. The remaining 1.5% of the oxygen delivered to the cells is dissolved in the plasma. In fact, whole blood, which contains cells as well as plasma, carries 70 times more oxygen than an equal amount of plasma.

We know that hemoglobin picks up oxygen at the lungs and releases it at the cells, but what determines whether hemoglobin will bind or release oxygen? The concentration (partial pressure[1]) of oxygen is the most important factor determining whether hemoglobin will bind to oxygen. In the lungs, where the concentration is high, hemoglobin picks up oxygen. The oxygen is then released when the concentration is low, as it is near the cells.

Interestingly, oxygen delivery is responsive to the needs of the cells. Oxygenated blood passing metabolically inactive cells gives up less oxygen than it does when passing metabolically active ones. When there is serious need, the amount of oxygen delivered to the cells can be increased more than threefold even if the rate of blood flow remains constant.

What causes such increased delivery? It is changes in the conditions around the active cells, changes intimately associated with metabolism, that alter the behavior of hemoglobin so that oxygen is released when and where it is

needed. Carbon dioxide and heat are produced as by-products of cellular metabolism. When carbon dioxide dissolves in the water of tissue fluid and blood, carbonic acid is formed and the acidity of the blood rises. Under these conditions of increased temperature and acidity, hemoglobin releases its oxygen load more readily.

Transport of Carbon Dioxide

The carbon dioxide that is produced as the cells use oxygen is removed by the blood. As shown in Figure 14–10, carbon dioxide transport occurs in three fundamental ways:

1. **Dissolved in blood.** Some carbon dioxide, about 7%, is transported dissolved in the blood.
2. **Carried by hemoglobin.** A slightly higher percentage, roughly 23%, of the transported carbon dioxide is carried by hemoglobin molecules. When carbon dioxide combines with hemoglobin it forms a compound called **carbaminohemoglobin**.
3. **As a bicarbonate ion.** By far the most important means of transporting carbon dioxide is as a bicarbonate ion dissolved in the plasma. About 70% of the carbon dioxide is transported this way.

The process begins when the carbon dioxide produced by cells diffuses into the blood and into the red blood cells. Carbon dioxide reacts with water in the red blood cells and plasma and forms carbonic acid. Carbonic acid quickly dissociates to form hydrogen ions, H^+, and bicarbonate ions, HCO_3^-. Although these reactions occur in the plasma as well as in red blood cells, they occur hundreds of times faster in red blood cells. The increased rate of reaction is caused by an enzyme called **carbonic anhydrase** that is found within

[1]The pressure exerted by one of the gases in the mixture, called its partial pressure, is directly related to the concentration of that gas in the mixture.

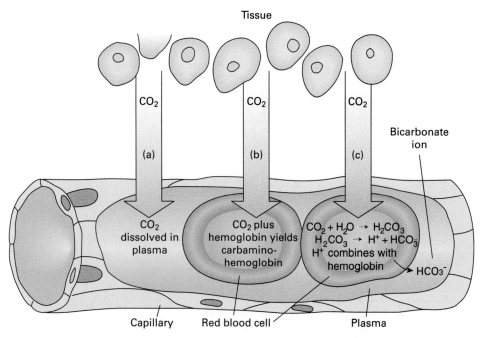

Figure 14–10 Three means of transporting carbon dioxide. (a) Dissolved in plasma. About 7% is transported this way. (b) In the red blood cells bound to hemoglobin. About 23% is transported this way. (c) In the plasma as a bicarbonate ion. Approximately 70% is transported this way.

red blood cells but not in the plasma. The hydrogen ions produced by the reaction then combine with hemoglobin. In this way, hemoglobin acts as a buffer and there is only a slight change in the acidity of the blood as it passes through the tissues. The bicarbonate ions diffuse out of the red blood cells into the plasma and are transported to the lungs. (As the negatively charged bicarbonate ions leave the red blood cells, chloride ions (Cl^-) diffuse in from the plasma, counterbalancing the negative charge that would have been left within the red blood cells.)

At the lungs, the process is reversed. When the blood reaches the capillaries of the lungs, carbon dioxide diffuses from the blood into the alveoli because the concentration (partial pressure) of carbon dioxide is comparatively low in the alveoli. This drives the chemical reactions we have just described in the reverse direction. The bicarbonate ions rejoin the hydrogen ions to form carbonic acid. In the presence of carbonic anhydrase within the red blood cells, carbonic acid is converted to carbon dioxide and water. The carbon dioxide then leaves the red blood cells, diffuses into the alveolar air, and is exhaled.

Besides providing a means of transporting carbon dioxide, bicarbonate ions are an important part of the body's acid-base buffering system. They help neutralize acids in the blood. If the blood becomes too acidic, the excess hydrogen ions are removed by combination with bicarbonate ions to form carbonic acid.

Control of Breathing

Neural Control

As you sit there, your breathing is probably rather rhythmic. The basic rhythm is controlled by the **medullary rhythmicity center** in the brain stem, within which are an inspiratory center and an expiratory center.

During quiet breathing, that is, when you are calm and breathing normally, the inspiratory center shows rhythmic bouts of neural activity (Figure 14–11). The activity of these neurons increases for about 2 seconds and then ceases for about 3 seconds. While the inspiratory neurons are active, impulses that stimulate contraction are sent to the muscles involved in inhalation, the diaphragm and the external intercostals. As we have seen, contraction of the diaphragm and the external intercostals causes the size of the thoracic cavity to increase, moving air into the lungs. When the inspiratory signals cease, the diaphragm and intercostals relax, and you passively exhale. Then, after the 3-second rest, the inspiratory neurons spontaneously begin to fire again, repeating the cycle.

During heavy breathing, breaths become deeper and faster, increasing the rate of oxygen delivery and carbon dioxide removal. At such times, for example during strenuous activity, exhalation is no longer a passive event. Instead,

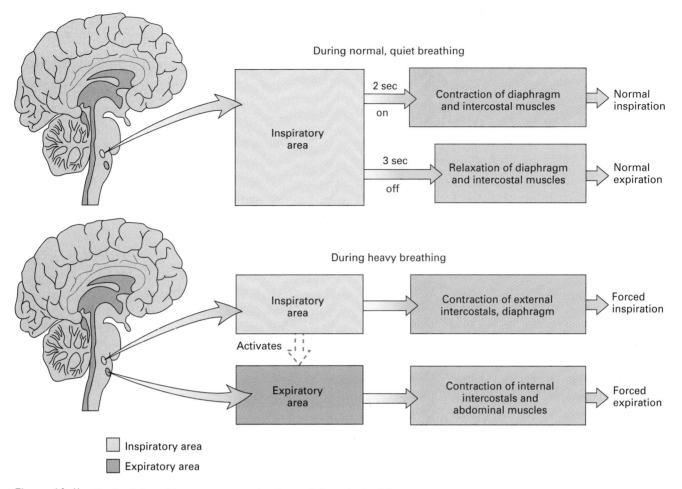

Figure 14-11 The basic breathing pattern is set by the medullary rhythmicity center. During quiet breathing, only the inspiratory area plays a role. It is on for 2 seconds, causing inhalation, and then off for 3 seconds, causing exhalation. Both the inspiratory and expiratory areas are active during heavy breathing.

it is an active process. As in quiet breathing, the inspiratory center is active and inhalation is caused by its impulses contracting both the diaphragm and intercostal muscles. During heavy breathing, however, the inspiratory center also sends impulses to the expiratory center. The expiratory center causes the contraction of both the internal intercostals and the abdominal muscles. In this way, air is quickly pushed out of the lungs.

Most of the time we breathe without giving it a thought. However, we can voluntarily alter our pattern of breathing through the impulses originating in the **cerebral cortex** (the "conscious" part of the brain). We control breathing when we speak or sigh, and we can voluntarily pant like a dog. Holding our breath while swimming underwater is obviously a good idea, and breath holding can also protect us from inhaling smoke or irritating gases.

Stretch receptors in the walls of the bronchi and bronchioles throughout the lung prevent the over inflation of adult lungs. When a deep breath greatly expands the lungs and stretches these receptors, they send impulses over the vagus nerve that inhibit the medullary inspiratory center, permitting exhalation. As the lungs deflate, the stretch receptors are no longer stimulated.

Although the stretch reflex was once thought to be an important mechanism for the control of lung ventilation in all people, it is now believed that its importance decreases as we mature so that, in adults, it may only function during exercise. Although this reflex does appear to be important in the regulation of a newborn's breathing pattern, normal breathing in adults (with an average tidal volume of 500 ml) probably does not activate the stretch receptors. When the tidal volume exceeds a liter, as it might during exercise, the stretch reflex protects the lungs from over inflation.

Chemical Control

The purpose of breathing is to control the blood levels of carbon dioxide and oxygen. So, we will now consider how the levels of these gases control the breathing rate, which, in turn, influences the levels of the gases.

Carbon Dioxide

The most important chemical influencing breathing rate is carbon dioxide. Most of carbon dioxide's effect on breathing is caused by hydrogen ions formed when the carbon dioxide goes into solution and forms carbonic acid ($CO_2 + H_2O \longrightarrow H_2CO_3 \longrightarrow H^+ + HCO_3^-$). Carbon dioxide is produced as a by-product of the energy-releasing reactions that require oxygen, so monitoring blood carbon dioxide level is a good way to see how quickly the cells are using oxygen.

Chemical control of the breathing rate is based on input from chemoreceptors in a region of the brain called the medulla and to some extent those in the aortic bodies (in the main blood vessel delivering blood to the body) and carotid bodies (in the blood vessels that deliver blood to the brain) (Figure 14–12). All these chemoreceptors respond to changes in hydrogen concentration, and those in the aortic arch and the carotid bodies also respond to the blood levels of carbon dioxide. The key areas in the chemical control of breathing are in the medulla. The medulla's receptors are near its surface where they are bathed in cerebrospinal fluid. Although the brain is protected from many changes in blood chemistry because most substances cannot pass through the specialized capillary walls that form a "blood-brain barrier," carbon dioxide is very soluble and can diffuse into the cere-

brospinal fluid, where it raises the hydrogen ion concentration by forming carbonic acid. Because the cerebrospinal fluid lacks the blood's ability to buffer changes in acidity, dissolved carbon dioxide changes the acidity of cerebrospinal fluid more than it does that of blood. When the chemosensitive areas are stimulated by rising carbon dioxide levels (or hydrogen ion concentration), breathing rate is increased, causing a decrease in the blood level of carbon dioxide.

Oxygen

Since oxygen, not carbon dioxide, is essential to survival, it may be somewhat surprising to learn that oxygen does not influence the breathing rate unless its blood level falls dangerously low. The firing of these oxygen-sensitive chemoreceptors serves as a warning that the blood oxygen level is at a critical point and initiates a last-minute call to the medulla to increase the breathing rate and raise oxygen levels. If the oxygen level falls much more, the neurons in the inspiratory center die from a lack of oxygen and do not respond well to impulses from the chemoreceptors. As a result, the inspiratory center begins to send fewer impulses to the muscles of inspiration and the breathing rate decreases and may even cease completely.

Critical Thinking

Divers sometimes take several deep and rapid breaths before diving. Why would this practice allow the diver to remain underwater longer? Occasionally, this diver loses consciousness while underwater. Why would this occur?

Respiratory Diseases

The Common Cold

The **common cold** is indeed common. Some 30 million Americans have a cold at this moment. Most people get their first cold before they are 1-year-old and continue to get several colds each year through adulthood. Therefore, if you are not now one of those suffering from the miseries of a cold, chances are good, about 75%, that you will catch one within a year.

Typically a cold begins with a runny nose, possibly a sore throat, and sneezing. In the beginning, the nasal discharge is thin and watery, but it becomes thicker as congestion increases. Although adults don't usually have a fever with a cold and a child's fever is usually low, there may be a feeling of chilliness at the onset of a cold. Almost any part of the respiratory system can be affected. Sneezing and a stuffy nose indicate that the infection is in the upper respiratory system. When the pharynx is affected, a sore throat results. The infection may spread to the bronchi, causing a cough, or to the larynx, making your voice hoarse.

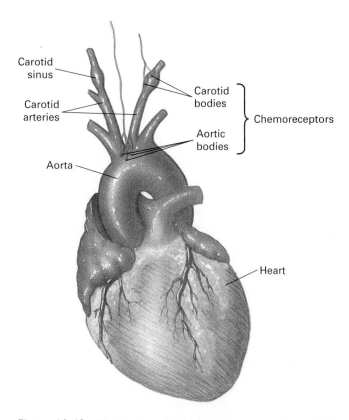

Figure 14–12 The location of peripheral chemoreceptors, which monitor the chemical composition of blood and bring about changes in breathing rate.

Carotid sinus

Carotid arteries

Carotid bodies

Aortic bodies

Chemoreceptors

Aorta

Heart

As miserable as you feel with a cold, you can take comfort from the fact that your suffering won't last forever. A cold is self-limiting, lasting only 1 to 2 weeks. Furthermore, colds are seldom fatal, except occasionally among the very young or very old or in those people already seriously ill with another malady.

A cold can be caused by any one of over 150 different viruses, a small fact that explains several things about colds. Because there are so many different cold-causing viruses, you can get several colds a year, each from a different virus. The variety of viruses that can cause a cold is also the reason that no effective cold vaccine can be developed. Interestingly, not all these viruses are prevalent in a single geographical area. Instead, a group of viruses may be common in one locale, so as you become immune to the viruses in your hometown, you get fewer colds. However, if you travel or meet new people, you are exposed to new viruses and are more likely to catch a cold. Many a long-awaited vacation trip has been spent sniffling and sneezing.

Colds are spread when the causative virus is transmitted from an infected person. The viruses, plentiful in nasal secretions, are picked up on the hands and then passed on with a handshake or other contact. The viruses may remain alive on the skin or an object for several hours, waiting for an unsuspecting person to touch them, thereby contaminating his or her own fingers. Then, when the virus-laden fingers are touched to the mucous membranes of the nose, the transfer is completed. Therefore, the best way to prevent a cold is to avoid being around those who have one. Furthermore, you are more likely to get a cold at the times when you feel you can cope with it the least. Studies have shown that both the risk of infection and the likelihood of developing cold symptoms are greatest at times of psychological stress.

There are many myths about colds. For example, we've all been told to stay warm and dry if we want to avoid catching a cold. However, there is absolutely no evidence that wet feet or sitting in a draft will bring on a cold. Colds *are* more common in the winter, but *not* because of the chill. During the cold winter months, we spend more time inside and are, therefore, more likely to be in closer contact with infected people. Also, the heated indoor air may be dry, causing nasal membranes to dry and crack, thereby allowing viruses easy access to the body.

The Flu

Flu is an abbreviation of **influenza**, so named because early Italian astrologers thought that the disease came from the influence of heavenly bodies (the name means "influence" in Italian). Modern science has ferreted out yet another inaccuracy, and so today we know that the stars don't cause the flu. Viruses do.

The flu is caused by viruses, but unlike the cold, there are relatively few viruses responsible. The viruses that cause the flu in humans are all variants of two major types—A and B. Influenza A is often more serious than B in that it is often

accompanied by severe complications and more frequently results in death. Although there are only two basic types of influenza viruses, there are hundreds of variants.

In some ways, getting the flu is similar to being hit by a speeding train. Both calamities seem to happen suddenly, but a train doesn't really come out of the blue and neither does the flu. Although the flu symptoms appear to start abruptly, the disease has actually been incubating for several days before it strikes. Typically the flu begins with chills and a high fever, about 103°F in adults and perhaps higher in children. Many flu victims experience aches and pains in the muscles, especially in the back. Other common ailments include a headache, sore throat, dry cough, weakness, pain and burning in the eyes, and sensitivity to light. Nausea and vomiting may also accompany the flu. When the flu hits, one usually feels sick enough to go to bed. The flu generally lasts for a week or 10 days, but it may take an additional week or more before you are completely back on your feet.

Unfortunately, the flu is often complicated by secondary infections, which follow the initial disease as other disease-causing organisms take advantage of the body's weakened state. The most common complication is pneumonia, an inflammation of the lungs. Bacteria taking advantage of the weakened state of the body can also cause bronchitis, sinusitis, and ear infections.

There are two ways to prevent the flu: flu shots and a drug called amantadine. The flu shot is a vaccine made from the strains of viruses that are causing the current outbreaks of the illness. Flu shots are only about 60% to 70% effective because the viruses they target change slightly with each outbreak and become unrecognizable to the body defenses that were programmed by the vaccine. Each flu season brings new strains of flu viruses. Thus, new vaccines must be developed to protect us. By mid-February, scientists must guess which viruses will be prevalent during the next winter's outbreak so the vaccine can be prepared in time. The effectiveness of the vaccine lasts only as long as that season's most prevalent strains. As a result, flu shots must be repeated each year.

In addition, the shots often cause an adverse reaction, such as a sore arm or a fever lasting several days. Nonetheless, people in the high risk group, such as the chronically ill, the elderly, and health-care professionals who are likely to be exposed to the virus, benefit from the protection shots provide.

A second way to ward off the flu is with the drug amantadine, which prevents the virus from infecting new cells. Some studies report that amantadine is 60% to 70% effective in preventing a severe case of the flu, about as effective as the flu shot. An added advantage is that amantadine relieves flu symptoms, allowing quicker return to normal activities. If the drug is taken within 24 hours of the onset of flu symptoms, most people will be back on their feet again within a day or two.

Amantadine may seem like a miracle drug, and it is if you have the flu or have been exposed to the virus, but it has

Surviving a Common Cold

The only thing more common than the cold is advice on how to treat it. Here we'll examine the validity of some frequently suggested treatments for a cold.

1. **Take large doses of vitamin C?** There is no evidence that vitamin C reduces the frequency of colds. Nonetheless, some studies suggest that some people who take vitamin C may experience less severe or shorter colds. However, this effect is so small that it may not warrant the use of a substance with possible side effects.
2. **Take an antibiotic?** Generally a physician will not prescribe an antibiotic for a cold because it is not effective against viruses. Since an antibiotic cannot cure a cold, it should be prescribed only to prevent or control bacterial infections such as a middle ear infection, bronchitis, or sinusitis that may accompany a cold. Unnecessary use of an antibiotic may cause side effects such as diarrhea and can lead to the development of bacterial resistance to the drug.
3. **Go to bed?** Bed rest enables the body to muster its resources and to fight secondary infections. Staying at home with a cold is also socially responsible, since it helps prevent the spread of the virus. But if you are too busy to spend a few days in bed because of a cold, you probably aren't hurting yourself; bed rest won't cure your cold or shorten its duration.
4. **Have some chicken soup?** Grandmothers have suggested this for years and doctors finally agree that the advice has some merit. One reason is that it is always good to consume plenty of fluids when you have a cold. They help loosen secretions in the respiratory tract. Hot fluids, such as chicken

soup, are more effective than cold drinks in increasing the flow of nasal mucus. By increasing the flow of nasal secretions, you reduce congestion. As a result, you can breathe more freely and the amount of time the cold viruses are in contact with the cells lining the respiratory system is reduced.

Although a cold can't be cured, there are ways to make it more bearable. Some ways to relieve the misery of a cold are as follows:

For Nasal Congestion. The congestion is caused when the mucous membranes of the nasal cavities become swollen and produce increased amounts of mucus caused by viral infection. The best ways to relieve the congestion are to drink plenty of extra fluids and to inhale moist air from a hot bath, shower, or vaporizer.

By constricting small blood vessels in the nose, decongestants may reduce the accumulation of fluid causing nasal congestion. Unfortunately, if taken orally, decongestants constrict small blood vessels throughout the body, which can raise blood pressure. Other possible side effects of oral decongestants include nervousness, sleeplessness, and dryness of the mouth. Therefore, if you must use a decongestant, it is wise to choose one in nose sprays or drops. Use these at the recommended dosage and only for a few days. Be aware that, if overused, nasal

some limitations. First, it is effective only against the influenza A viruses. Second, it only protects against the flu if it is taken at the time of exposure and only reduces symptoms if it is taken shortly after the flu strikes. Furthermore, the drug must be taken daily, which makes it less convenient and more expensive than a flu shot. Finally, like all drugs, amantadine causes side effects in some people. About 7% to 13% of those taking the drug suffer from insomnia, nervousness, difficulty in concentrating, or depression and anxiety.

Strep Throat

Strep throat, a sore throat that is caused by *Streptococcus* bacteria, is a problem mainly in children 5 to 15 years old. The soreness in the throat is usually accompanied by

swollen glands and a fever. The pain may be so mild that a doctor is never consulted.

Ignoring a strep infection can have serious consequences because, if untreated, the *Streptococcus* bacteria can spread to other parts of the body and cause rheumatic fever or kidney problems. The main symptoms of rheumatic fever are swollen, painful joints and a characteristic rash. About 60% of the rheumatic fever sufferers develop disease of the heart valves. Another possible consequence of a streptococcal infection is kidney disease (glomerulonephritis). The kidney damage is due to a reaction from the body's own protective mechanisms. The body fights the bacteria by producing antibodies that destroy the bacteria. If these antibodies persist after the bacteria have been killed, they can cause the kidneys to become inflamed. When this happens, the kid-

sprays may produce a "rebound" effect making the congestion even worse.

Although antihistamines are present in many cold remedies, there is no evidence that they are effective in reducing nasal congestion caused by colds. Furthermore, antihistamines may cause blurred vision, retention of urine, and dizziness. Drowsiness is a major side effect, so if you do take an antihistamine, avoid driving or other hazardous activities.

For Coughs. Coughing is a protective reflex controlled by a cough center in the brain. Besides clearing the respiratory tubules of foreign material, coughing loosens and removes phlegm and mucus. Therefore, it is not always a good idea to suppress a cough. If a cough persists for more than a week, it is advisable to see a physician.

The safest and cheapest cough remedies are substances such as hard candy or honey, which coat and soothe the throat, and drinking extra fluids, which loosens the mucus and helps relieve some irritation.

Cough medications fall into two general classes: suppressants and expectorants. Cough suppressants work by reducing the activity of the brain's cough center. These are helpful in easing dry hacking coughs. The most effective nonprescription cough suppressant is dextromethorphan. Codeine also suppresses coughs, but is available only by prescription. A cough that brings up sputum is performing a useful function and should not be suppressed. Instead, removal of the irritating substances should be assisted with an expectorant, which increases the flow of respiratory tract secretions.

For General Aches and Pain. The drug "most doctors recommend" is aspirin. It relieves pain and reduces fever. Aspirin is thought to reduce pain by blocking the production of prostaglandins, substances produced by tissues during inflammation. By sensitizing pain receptors, prostaglandins increase the feeling of pain.

Aspirin is aspirin. When you choose a brand of aspirin, look at the number of grains of aspirin per tablet and the cost. The cheapest brand is just as effective as the most expensive one.

Aspirin can cause stomach upset in some people. A National Academy of Sciences drug panel found no evidence that buffered aspirin reduces stomach irritation. If aspirin upsets your stomach, try taking it on a full stomach and with a full glass of water or milk.

Acetaminophen is just as effective as aspirin in reducing pain and fever and is a suitable alternative for those who can't use aspirin. However, some studies suggest that aspirin and acetaminophen may hinder the body's defense mechanisms against the cold virus and may even make cold symptoms worse.

There is also evidence that it is not a good idea to reduce fever. The additional body heat slows the growth and reproduction (or replication) of many disease-causing organisms, including cold viruses.

It is interesting that although no medication can prevent or shorten a cold, Americans spend $500 million each year on over-the-counter cold remedies to treat the symptoms of a cold.

neys may be unable to produce a normal amount of urine and may allow blood to leak into the urine.

Since many viruses can cause sore throats that look like strep infections, the only way to identify the disease is to test for the causative organism. If *Streptococcus* bacteria are found, an antibiotic, usually penicillin, is prescribed to prevent rheumatic fever and kidney disease. So, don't ignore a sore throat.

Tuberculosis

Tuberculosis, or TB, is a highly contagious disease caused by a rod-shaped bacterium, *Mycobacterium tuberculosis*. It is spread when the cough of an infected person sends bacteria-laden droplets into the air and the bacteria are inhaled into the lungs of an uninfected person. Because the bacteria are inhaled, the lungs are usually the first site attacked, but the bacteria can spread to any part of the body, especially to the brain, kidneys, or bone.

As a defense against the bacteria, the body forms fibrous connective tissue casings, called tubercles, that encapsulate the bacteria (hence, the name of the disease). Although the formation of tubercles slows the spread of the disease, it does not actually kill the bacteria. The immune system does destroy at least some of the walled-off bacteria and may, in fact, kill them all. When the body's defenses fail, however, the pockets of bacteria may persist, undetected, for many years. Later, the disease may progress to the secondary stage as pockets of bacteria become activated again. Furthermore, bacteria may escape from the tubercles and be carried by the

Air Pollution and Human Health

Caution: The air you breathe may be hazardous to your health. It may even kill you—especially if you have heart or respiratory problems. In most major cities, the poor air quality may be obvious at times as a brown haze. However, even in remote national parks, air pollution is often significant enough to reduce visibility. Although the effects of air pollution may be subtle and take a long time to occur, they damage the environment as well as human health.

What are the sources of the pollutants in our air? Two major human sources of air pollution are motor vehicle exhaust and industrial emissions. The exhaust pipes on our cars and the smokestacks on factories spew into the air oxides of sulfur, nitrogen, and carbon, a variety of hydrocarbons, and many particulates (such as soot and smoke). In the atmosphere, sulfur dioxide and nitrogen dioxide dissolve in the moisture and form an aerosol of strong acids—sulfuric acid and nitric acid, respectively. Some may drift upward, forming acidic clouds that may be blown hundreds of miles away by the prevailing winds and then fall as acid raindrops. On the other hand, the acidic aerosol may remain close to its source as a component of the haze created by air pollution. In addition, sunlight can cause hydrocarbons and nitrogen dioxide to react with one another and form a mix of hundreds of substances called photochemical smog. The most harmful component of photochemical smog is ozone, which attacks cells, destroys tissues, irritates the respiratory system, damages plants, and even erodes rubber. Paradoxically, the ozone that is so damaging when it is found at ground level in photochemical smog is the same chemical that is beneficial in the upper layers of the atmosphere, where it prevents much of the sun's ultraviolet radiation from reaching the Earth's surface. Unfortunately, the ozone found in smog does not make its way to the ozone layer of the upper atmosphere because it is converted to oxygen within a few days.

When we breathe polluted air, the respiratory system is, not surprisingly, the first to be affected. As air containing toxic substances fills the lungs, cells lining the airways and within the lungs are injured. Damaged cells release histamine, which causes nearby capillaries to widen and become more permeable to fluid. As a result, fluid leaks from the capillaries and accumulates within the tissues. Even a brief exposure to oxides of sulfur (5 ppm[2] for a few minutes), the oxides of nitrogen (2 ppm for 10 minutes), or ozone leads to fluid accumulation, increased mucus production, and spasms (intense involuntary contractions) of the bronchioles. These effects make air flow more difficult and reduce gas exchange.

People with asthma are usually among those who suffer the most from air pollution. A person with asthma experiences breathing difficulty after inhaling 0.1 ppm nitrogen dioxide for 1 hour. In contrast, a healthy person is not likely to experience the same degree of respiratory distress unless the air inhaled contains 25 times as much nitrogen dioxide (2.5 ppm) for several hours. Ground-level ozone is also a problem for asthmatics. One study reported that hospital admissions for asthma

[2]ppm = parts per million.

bloodstream to other parts of the body. As a result, whenever the victim becomes weak, ill, or poorly nourished, the disease may flare-up.

The initial symptoms of tuberculosis, if there are any, are similar to those of the flu. In the secondary stage, the patient usually develops a fever, loses weight, and feels tired. If the infection is in the lungs, as it usually is, it causes a dry cough that eventually produces pus-filled and blood-streaked phlegm.

If untreated, TB can be fatal. Indeed, more people in the United States have been killed by tuberculosis than died in both World Wars.

The development of effective treatment for tuberculosis, including effective antibiotics and widespread testing for the disease, caused a gradual 30-year decline in its prevalence in the United States beginning in the 1950s. However, beginning in 1985, the incidence of tuberculosis began to rise. A major cause of this increase is the spread of the HIV, the virus that causes AIDS. HIV attacks the body's defense system, leaving the person susceptible to opportunistic infections such as tuberculosis. However, HIV is not the only reason for the rising incidence of tuberculosis. Immigration of people from countries where tuberculosis is prevalent has also increased. Because the disease is so contagious, increased crowding in housing among the poor has contributed further to the increase in tuberculosis.

Even more alarming than the dramatic increase in the occurrence of tuberculosis is the appearance and spread of tuberculosis strains that are resistant to many antibiotics. Today, one in seven cases of tuberculosis in the United

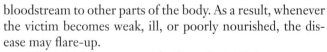

increased 23% in New York City and 29% in Buffalo on the days when the ozone levels were highest.

Long-term irritation of bronchi by pollutants is a cause of chronic bronchitis and emphysema. The process begins as irritation and leads to increased fluid accumulation that, in turn, stimulates mucus production and coughing. The cough and mucus, signs of chronic bronchitis, irritate lungs even more. As bronchitis continues, the air passageways become narrower, trapping air in the lungs. When the increased pressure accompanying a cough causes the over-inflated alveoli to rupture, emphysema begins.

Some pollutants can cause cancer. As the normal mechanisms that cleanse the respiratory system are damaged or overloaded by pollutants, cancer-causing chemicals in the inhaled air are no longer effectively removed. Within the airways these chemicals may then bring about changes in genetic material that can lead to cancer.

Critical Thinking

Studies have indicated that the rate of lung cancer deaths has increased with the degree of air pollution. In both the United States and England, the rate of death due to lung cancer has steadily increased since the industrialization of these countries. Why is this correlation insufficient evidence to link air pollution and lung cancer? What other factor(s) might be responsible for the increase in lung cancer deaths?

Air pollution can also decrease resistance to infectious disease. In one experiment, a group of mice was exposed to auto exhaust in concentration lower than that found in major cities. The exhaust fumes had been irradiated to create photochemical smog. Another group of mice was not exposed to smog. Both groups were then exposed to the type of streptococcal bacteria that causes a form of pneumonia. The mice that had been exposed to smog had a much higher rate of pneumonia than did the mice that were not previously exposed to air pollution.

With so many documented health consequences of air pollution, you may wonder why the world's great minds have not developed the technology to solve the problem. Indeed, they have. But, unfortunately it turns out that air pollution is not only a scientific problem. Instead, it is primarily a social, political, and economic problem. We have the technology to prevent it, but it costs money. Industry often claims that the cost of installing pollution control devices will have to be covered by increased price of goods or the loss of jobs. Efforts to improve air quality will require some increased personal cost and inconvenience. How much more are you willing to pay for low-sulfur fuels, for instance? The automobile is a wonderful convenience. Are you willing to take a bus, or better yet, to walk instead of drive in order to improve your own air quality? Since the pollutants in air are often blown hundreds of miles from their source, are you willing to make the same sacrifices to improve a stranger's air quality? Also, when the pollutants cross state or national borders, who makes the laws that would control the pollution? Who enforces those laws?

States is resistant to antibiotics that were effective previously. Multi-drug resistance occurs because many patients stop taking their antibiotic medicines too soon, before the recommended 6 to 12 months is up. As a result, only the susceptible bacteria are killed and the resistant ones remain. The resistant bacteria then grow unchecked and the patient suffers a relapse. When another course of antibiotics is prescribed, the patient may stop taking it too soon as well. In this way, resistance to several drugs has developed in tuberculosis bacteria.

Social Issue

New York City health officials are detaining certain tuberculosis patients in a modern-day tuberculosis sanitarium,

even though the patients are no longer contagious. The patients are being confined against their will because they have repeatedly failed to take their medicine. This failure can lead to the spread of strains of the tuberculosis bacterium that are resistant to antibiotics when the patient suffers a relapse and becomes contagious once again. Is this action a fair balance of protection of the public with personal rights? What do you think?

Bronchitis

Viruses, bacteria, or chemical irritation may cause the mucous membrane of the bronchi to become inflamed, a condition called **bronchitis**. The inflammation results in the production of excess mucus, which triggers a deep cough that produces greenish-yellow phlegm.

There are two types of bronchitis—acute and chronic. Acute bronchitis, which often follows a cold, is usually caused by the cold virus itself, but it may be caused by bacteria that take advantage of the body's lowered resistance and invade the trachea and bronchi. An antibiotic will hasten recovery, if the cause is bacterial.

When a cough that brings up phlegm is present for at least 3 months during 2 consecutive years, the condition is called chronic bronchitis. This is a more serious condition that is usually associated with cigarette smoking or air pollution. Some people with chronic bronchitis may lack an enzyme that normally protects the air passageways from such irritants. As the disease progresses, it becomes increasingly difficult to breathe. One reason for the labored breathing is that the linings of the air tubules thicken, narrowing the passageway for air. The contraction of muscles in walls of the bronchioles and the excessive secretion of mucus further obstructs the air tubules.

There can be serious consequences to chronic bronchitis. The degenerative changes in the lining of the air tubules make removal of mucus more difficult. As a result, the patient is more likely to develop lung infections such as pneumonia, which can be fatal, and degenerative changes in the lungs, such as emphysema.

Emphysema

Emphysema is one of the more common results of smoking, although it can have other causes as well. In **emphysema**, the alveoli become over inflated and break down, causing them to merge, thereby creating fewer and larger alveoli (Figure 14–13). This has two major effects—a reduction in the surface area available for gas exchange and an increase in the volume of residual or "dead" air in the lungs. Exhalation, you may recall, is a passive process that depends on the elasticity of lung tissue. In emphysema, the lungs become inelastic and air becomes trapped in the lungs. As the dead air space increases, adequate ventilation of the lungs requires more forceful inhalation. Forcing the air causes more alveolar walls to rupture, further increasing the dead air space. Lung size gradually increases as the residual volume of air becomes greater, giving a person with emphysema the characteristic barrel chest. However, gas exchange becomes more difficult. To get an idea of what poor lung ventilation caused by increased dead air space feels like, take a deep breath, then exhale only slightly, and repeat this several

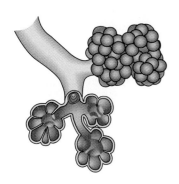

Normal alveoli

Emphysema causes
breakdown of alveolar walls

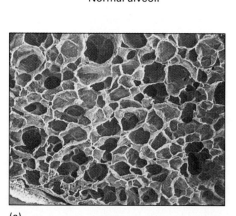

(a)

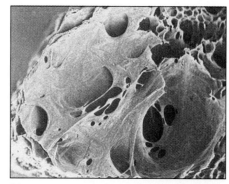

(b)

Figure 14–13 A comparison of (a) normal alveoli and (b) alveoli in an individual with emphysema. In emphysema, the alveolar walls rupture, increasing the dead air space in the lung. Micrographs of sections of normal lung tissue and lung tissue of an individual with emphysema. Notice that in emphysema there is a decrease in the surface area for gas exchange, an increase in the dead air space, and a thickening of the alveolar walls.   *(a, b, © Dr. Andrew Evan, Indiana University School of Medicine).*

times. Notice how quickly you feel an oxygen shortage if you continue taking very shallow breaths that leave the lungs almost completely filled with air.

Shortness of breath, the main symptom of emphysema, has several causes. We've mentioned two causes, the decreased surface area for gas exchange and increased dead air space. As the disease progresses, gas exchange becomes even more difficult because the alveolar walls thicken with fibrous connective tissue. The oxygen that does make it to the alveoli has difficulty crossing the connective tissue to enter the blood. Thus, a person with emphysema constantly gasps for air.

SUMMARY

1. The oxygen that we breathe is needed to maximize the number of energy-storing ATP molecules formed from food molecules. Exhaling carbon dioxide, a waste product formed by the same reactions, helps to regulate the acid-base balance of body fluids. The role of the respiratory system is to exchange oxygen and carbon dioxide between the air and the blood.

2. During inhalation, the first structure that air usually passes through is the nose, which serves to clean, warm, and moisten the incoming air. Olfactory receptors, located in the nasal cavities, are responsible for our sense of smell. The sinuses are air-filled spaces in the facial bones that also help warm and moisten the air. After leaving the nose, the inhaled air next passes through the pharynx, or throat, and then the larynx, or voice box. Reflex movements of the larynx prevent food from entering the airways and lungs. The larynx is also the source of the voice. The air passageways include the trachea, which branches to form bronchi and these branch extensively within the lung to form progressively smaller tubules called bronchioles. The bronchioles terminate at the gas exchange surfaces, the alveoli. The alveoli are thin-walled air sacs surrounded by a capillary network.

3. Pressure changes within the lungs caused by changes in the size of the thoracic cavity move air into and out of the lungs.

4. Oxygen and carbon dioxide are exchanged between the alveolar air and the capillary blood by diffusion along their concentration (partial pressure) gradients. Oxygen diffuses from the alveoli into the blood where it binds to hemoglobin within the red blood cells and is delivered to the body cells. A small amount of carbon dioxide is carried to the lungs dissolved in the blood plasma or bound to hemoglobin. Most, however, is transported to the lungs as bicarbonate ions.

5. The basic rhythm of breathing is controlled by the inspiratory center within the medullary rhythmicity center in the brain. The neurons within this center undergo spontaneous bouts of activity. When they are active, messages are sent causing contraction of the diaphragm and the muscles of the rib cage. As a result, the thoracic cavity increases in size and air is drawn into the lungs. When the inspiratory neurons are inactive, the diaphragm and rib cage muscles relax, and exhalation occurs passively. Also in the medullary activity center is an expiratory center that causes forceful exhalation during heavy breathing.

6. The most powerful stimulant to breathing is an increasing level of carbon dioxide (acting primarily because of the increased number of hydrogen ions formed from carbonic acid). Extremely low levels of oxygen also increase breathing rate.

7. Two fairly common respiratory diseases, the common cold and the flu, are caused by viruses. Strep throat, a sore throat caused by *Streptococcus* bacteria, can lead to rheumatic fever (and consequently to disease of the heart valves) or to kidney disease. Tuberculosis is caused by the *Mycobacterium tuberculosis*. The body resists the bacteria by walling them off in connective tissue capsules called tubercles. Acute bronchitis is caused either by bacteria or a virus. Chronic bronchitis is a persistent irritation of the bronchi. Emphysema is a breakdown of the alveolar walls and thus a reduction in the gas exchange surfaces. Chronic bronchitis and emphysema are usually caused by smoking or by air pollution.

REVIEW QUESTIONS

1. Why must we breathe oxygen?
2. Trace the path of air from the nose to the cells that use the oxygen.
3. How are most particles and disease-causing organisms removed from the inhaled air before it reaches the lungs?
4. Describe the reflex that normally prevents food from entering the lower respiratory system.
5. How is human speech produced?
6. What is the function of the cartilage rings in the trachea?
7. What is the bronchial tree?
8. How are the pressure changes in the thoracic cavity that are responsible for breathing created?
9. Is tidal volume or the vital capacity a larger volume of air? Explain.
10. How is most oxygen transported to the body cells?
11. How is most carbon dioxide transported from the cells to the lungs?
12. What region of the brain causes the basic breathing rhythm? How does the activity of the brain region differ during quiet breathing and heavy breathing?
13. Explain how blood carbon dioxide levels regulate the breathing rate.
14. Why are colds more common during the winter?
15. What are the causes of the shortness of breath experienced by people with emphysema?

Chapter

14A

(© *George Semple*)

Smoking and Disease

What's in the Smoke?
Health Risks of Smoking
   Lung Disease
   Cancer
   Heart Disease
   A Potpourri of Additional Hazards
Women and Smoking
Passive Smoking
Safe Cigarettes?
Benefits of Quitting

S moking is the greatest single preventable cause of disease, disability, and death in our society. Its dangers are so obvious that even the government knows about them. In fact, every cigarette pack and cigarette advertisement in the United States must bear a warning from the Surgeon General (Figure 14A–1). Yet, tobacco, which causes bodily harm when used exactly as intended, is still deemed a legal product to sell.

Let's put it this way. Each cigarette shortens the smoker's life by about 7 minutes, a little less than the time it takes to smoke it. So, on average, a person who smokes a pack a day, will lose roughly 7 years of life. Although some smokers may lose less, one-third of them will cut as much as 21 years off their lives. Consequently, each year more than 3 million lives around the world, over 400,000 of them American, are snuffed out prematurely because of smoking. Nonetheless, in 1996 there were still more than 45 million adults in the United States who smoked regularly. Whereas the number of adult smokers is decreasing or at least holding steady, the number of teenage smokers is increasing.

The economic loss because of lost work days, premature death, and health care costs caused by cigarette smoke is also staggering. This is one reason nonsmokers want smokers to quit. (Another reason is that sometimes they love them.) The Worldwatch Institute estimates that each pack of cigarettes sold in the United States costs Americans $1.25 to $3.17 in medical expenses alone. When medical costs are added to lost productivity, the cost of tobacco addictions soars to $65 billion a year in the United States.

Figure 14A–1 Cigarette packs are now required to display warnings about the dangers of smoking. (© *George Semple*)

What's in the Smoke?

The damage caused by cigarettes begins the instant the smoke touches the lips. Then, it harms every living tissue it touches—mouth, tongue, throat, esophagus, air passageways, lungs, and stomach. When autopsied, even light smokers show lung damage. The substances in the smoke are metabolized (broken down) in the liver, but even the breakdown products injure the bladder, pancreas, and kidneys. Smoking a pipe or cigar is somewhat safer than cigarettes because the smoke is generally not inhaled. However, the risk of cancer of the lips, mouth, and tongue is still greater than a nonsmoker's. A pipe or cigar smoker's risk of lung cancer or heart disease, although not as great as that of a cigarette smoker, is still above that of a nonsmoker.

The average American smoker consumes a pack and a half of cigarettes each day. That's about 300 puffs a day and over 109,000 puffs a year—year after year. With such exposure, then, we might wonder, what's in the smoke?

The answer is not simple because smoke contains at least 4000 different substances (Table 14A–1). Even after all these years, not all have been tested, but so far at least 50 have been shown to cause cancer. Also present are some well-known poisons—hydrogen cyanide (the poisonous gas used in gas chambers), carbon monoxide (common in auto exhaust), and cresols (chemicals similar to those used to preserve telephone poles). In addition, tobacco smoke contains significant amounts of radioactive substances (thorium-228, radium-226, and polonium-210). This exposure to radiation may account for as much as 40% of maximum permissible annual exposure to radiation.

The three most dangerous substances in smoke are nicotine, carbon monoxide, and tar. Nicotine is perhaps the most insidious of the three because it creates pleasurable feelings of relaxation, but it is actually a poison that has many harmful effects on the body. More important, it causes smokers to become "hooked" on cigarettes, thus ensuring continued exposure to the other injurious substances in the smoke. Depending on the brand, a cigarette contains between 0.5 and 2.0 mg of nicotine. Each puff delivers about 0.2 mg of the drug to the bloodstream and this reaches the brain within 6 to 7 seconds, twice as fast as injected heroin.

Nicotine causes a relaxed state of mind. Indeed, it increases the brain waves characteristic of a relaxed awake state, the alpha waves. In addition, it causes the release of endorphins, chemical messengers within the nervous system that cause a feeling of well-being. Some of the feeling of

Table 14A-1 Primary Toxic and Carcinogenic Components of Cigarette Smoke (Including Vapor-Phase and Particulate Phase–Components)

Agent	Toxic	Kills Cilia	Carcinogenic	Co-Carcinogenic/ Promoter
Carbon monoxide	x			
Nitrogen oxides (NO$_x$)	x			
Hydrogen cyanide	x	x		
Formaldehyde		x	x	
Acrolein		x		
Acetaldehyde		x		
Ammonia	x			
Hydrazine			x	
Vinyl chloride			x	
Urethane			x	
2-Nitropropane			x	
Quinoline			x	
Benzo[a]pyrene			x	x
Dibenz[a,h]anthracene			x	x
I-Methylindoles				x
Dichlorostilbene				x
Catechol				x
Aromatic Amines			x	
Aromatic nitrohydrocarbons			x	
Polonium-210			x	
Nickel			x	
Arsenic			x	
Cadmium			x	

Adapted from U.S. Surgeon General Reports on the Health Consequences of Smoking, summarizing the toxic components of cigarettes.

relaxation brought on by nicotine may come from the marked reduction in muscle tone it causes.

Although it brings on a feeling of relaxation, nicotine is actually a stimulant that affects the brain at all levels. Many of the responses are those associated with the fight/flight response. When a cigarette delivers nicotine, the heart beats as many as 33 more beats per minute. Yet, the blood is forced through a less receptive circulatory system because the smoke also causes the blood vessels to constrict. The result is an increase in blood pressure. In addition, the platelets, structures in the blood that contain chemicals that initiate clotting, become sticky. This increases the likelihood of abnormal clots forming that may lead to heart attacks or strokes.

Nicotine is roughly as poisonous as cyanide—a lethal dose of either is 60 mg. However, nicotine in tobacco does not kill because it is taken in small doses, which are meted out in several hundred puffs throughout the day and are metabolized and excreted before a lethal level is reached. Thus, a three-pack-a-day smoker does not die at bedtime.

Nonetheless, it is the nicotine that causes nausea in new smokers and, in high doses, will do the same to experienced smokers.

Nicotine is a powerfully addicting drug and 95% of smokers are physiologically dependent on it. Indeed, some opium addicts have reported that it was easier to do without opiates than nicotine. Why is nicotine so addicting? Nerve cells become hyperactive when nicotine is removed. When a smoker tries to quit, this hyperactivity causes withdrawal symptoms: irritability, anxiety, headache, nausea, constipation or diarrhea, craving for tobacco, and insomnia. Avoiding the withdrawal symptoms is a powerful incentive to continue smoking. Most withdrawal symptoms begin to lessen after a week without nicotine, but some may continue for weeks or months. Certain symptoms, such as drowsiness, difficulty concentrating, and craving, seem to get worse about 2 weeks after quitting. Consequently, many smokers return to their habit and continue exposing themselves to the harmful materials in smoke, including carbon monoxide and tar.

Carbon monoxide makes up about 4% of the smoke of an average cigarette made in the United States. The amount of carbon monoxide in cigarette smoke is 1600 ppm (parts per million), which greatly exceeds the 10 ppm considered dangerous in industry. Furthermore, carbon monoxide lingers in the bloodstream up to 6 hours after smoking a cigarette. You may recall from Chapter 10 that carbon monoxide is a poison because it prevents oxygen transport by the red blood cells. In fact, carbon monoxide reduces the oxygen-carrying capacity of the blood by an average of 12%, reducing oxygen delivery to every part of the body including the brain and heart. The diminished oxygen supply to the brain can impair judgment, vision, and attentiveness to sounds. For these reasons, smoking can be hazardous for drivers. Furthermore, whereas nicotine makes the heart beat faster, carbon monoxide makes it more difficult to deliver the oxygen needed for increased heart rate, straining the heart. Obviously, the reduced oxygen supply to muscles can hinder athletic performance.

Tar is a collection of thousands of substances in the smoke that settles out as a brown sticky substance when the smoke cools within the body. A pack-a-day smoker coats the respiratory system with about 50 mg of tar each day. Besides the cancer-causing chemicals in tar, there are others that destroy the elasticity of the lung.

Health Risks of Smoking

Most of the health risks of cigarette smoking are caused by increases in the risk of three diseases, all of which can kill: lung disease, cancer, and heart disease. We'll examine these now, keeping in mind that there are also many other dangers.

Lung Disease

Since the idea of smoking is to bring smoke to the lungs, obviously we can look for some of its most damaging effects there. In fact, even teenage smokers show some damage to airways and lungs, but because they are young and strong, this may only be apparent in breathing tests or in athletic ac-

tivities. Young smokers are more likely to be short-winded than their nonsmoking friends. If they continue to smoke, the effects become much more apparent. By age 60, most smokers have significant changes in airways and lungs.

The damage to the respiratory system of smokers is gradual and progressive. The injury begins as the smoke hampers the actions of two of the lung's cleansing mechanisms—cilia and macrophages. Even the first few puffs slow the movement of cilia, the hair-like structures on the membranes lining the airways, making them less effective in sweeping debris from the air passageways. Smoking an entire cigarette prevents the cilia from moving at all for an hour or more. With continued smoking the nicotine and

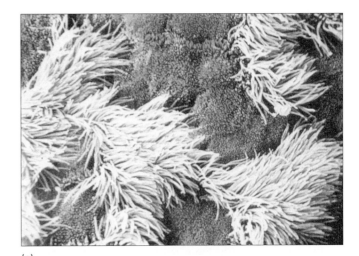

(a)

(b)

Figure 14A–2 Changes in the ciliated linings of the air passageways accompanying smoking. (a) The cilia on the cells lining the airways of a healthy nonsmoker cleanse the airways of debris. The cilia are seen as clumps of hairlike structures on the epithelial cells. (b) Cigarette smoke first paralyzes and then destroys the cilia. As a result, hazardous materials can accumulate on the surfaces of the air passageways. *(a, © CNRI/Science Photo Library/Photo Researchers, Inc.; b, © Dr. Andrew Evan, Indiana University, School of Medicine)*

sulfur dioxide in the smoke paralyze the cilia and the cyanide destroys them (Figure 14A–2).

Because a smoker's lungs are chronically inflamed, many macrophages, wandering cells that engulf foreign debris, enter in a vain attempt to clean up the lung surfaces. However, the smoke paralyzes the macrophages, further hampering the cleansing efforts.

As the cleansing mechanisms of the airways become hampered, greater quantities of tar and disease-causing organisms remain within the respiratory system. This is one reason that cigarette smokers lose 40% more work days per year than nonsmokers. As a group, in fact, they are sick in bed 88 million more days each year than nonsmokers.

At the same time that the cilia and macrophages are being slowed, the smoke stimulates the mucus-secreting cells in the linings of the respiratory passageways. As a result, the smaller airways become plugged with mucus, making breathing more difficult.

At this point, if not before, "smoker's cough" begins. Coughing is a protective reflex, and initially the smoker coughs simply because smoke irritates air passageways. However, as smoking continues and the cilia become increasingly less able to remove mucus and debris, the only way to remove the material from the passageways is to cough. The cough is generally worse in the morning as the body attempts to clear away the mucus that accumulated during the night.

Gradually the inflammation and congestion within the lungs, along with the constant irritation from smoke, lead to chronic bronchitis. The main symptom of chronic bronchitis is a persistent deep cough that brings up mucus. The air passageways become narrow because of the thickening of their linings caused by repeated infection, the accumulation of mucus, and the contraction of the smooth muscle in their walls. Airflow becomes difficult, resulting in breathlessness and wheeziness. Bronchial infections now become more common because the air passageways are not cleared of disease-causing organisms. Bacterial infections may be treated with antibiotics, bringing slight temporary relief, but chronic bronchitis will continue as long as smoke irritates the airways.

Emphysema is often the next stage in the progressive damage to the lungs. Emphysema is a lung disease in which the walls of alveoli, which are the surfaces for gas exchange, become destroyed. The lung damage begins as the elasticity of the airways and alveolar walls is lost because accumulating tar destroys elastin in the lung tissues. The increase in pressure that accompanies a cough is no longer absorbed by flexible walls of airways and alveoli. The pressure is, therefore, directed at the delicate alveolar walls and, like soap bubbles, they break.

With more and more alveoli destroyed, the surface area for gas exchange is reduced, so less oxygen is delivered to the body. Furthermore, the dead air space within the lungs increases and it becomes gradually more difficult to exhale. As a result, cyanide, formaldehyde, and carcinogens from the smoke remain in the lungs, killing even more cells.

As the alveoli are damaged, the small blood vessels to and from the alveoli rupture. If the blood vessels sustain enough damage, stress on the right ventricle of the heart increases because it must push the same amount of blood to lungs through fewer vessels.

Cancer

Here we consider the most familiar link of all—smoking and cancer. And, do not be misled, the link is proven. Cancers can develop because the smoke contains about 50 carcinogens, cancer-causing chemicals that stimulate cell division. Some change the structure of the genetic material, DNA. Others, such as formaldehyde, cause enzyme changes that allow cells to become malignant. Still other components of the smoke work as co-carcinogens by enhancing the action of other carcinogens or by promoting tumor growth once the cancer has begun.

Smoking is the major single cause of lung cancer, but it causes other cancers as well (Table 14A–2). In fact, it is responsible for 30% of all cancer deaths. Sadly, cancer-causing chemicals in tobacco damage every tissue they touch. Even the breakdown products of the components in tobacco may cause cancer.

It is sobering to realize that 85% of all cases of lung cancer are caused by smoking and are, therefore, preventable. Unfortunately, 90% of those persons diagnosed as having lung cancer die within 5 years. There are usually no symptoms of lung cancer until it is quite advanced. Therefore, it is not usually detected in time for cure (Figure 14A–3).

Admittedly, not every smoker gets lung cancer, but smokers are 15 to 25 times more likely to get it than lifetime nonsmokers. Indeed, one out of six men who smoke two or more packs a day will develop lung cancer. The likelihood of a smoker getting lung cancer depends on several factors, such as the number of cigarettes smoked, the number of years with the habit, the age at which smoking began, how

Table 14A–2   Types of Increased Cancer Risk Due to Smoking

Type	Increased Risk
Mouth and lips	4 times
Larynx	5 times in light smokers
	20–30 times in heavy smokers
Esophagus	2–9 times
Kidney and bladder	2–10 times
Pancreas	2–5 times (especially if alcohol is also consumed)

Figure 14A–3 Lung cancer. The tumor is the light-colored solid mass shown in the upper region of the lung. (© *James Stevenson/Photo Researchers, Inc.*)

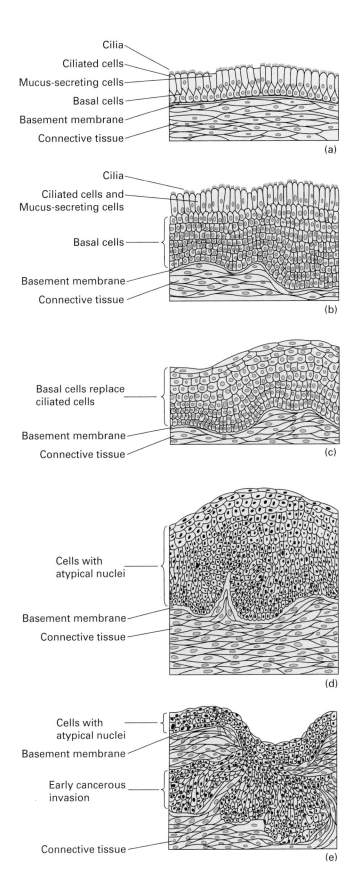

deeply the smoke is inhaled, and the amount of tar and nicotine in the brand of cigarette smoked. There are also individual differences in genetic and biological make-up that influence cancer risk.

The progression to lung cancer is marked by changes in the cells of the airway linings of smokers (Figure 14A–4). In a nonsmoker, the lining of air passageways has a basement membrane underlying basal cells and a single layer of ciliated columnar cells. In a smoker, however, one of the first signs of damage is an increase in the number of layers of basal cells. Next, the ciliated columnar cells die and disappear. The nuclei of the basal cells then begin to change and the cells become disorganized. This is the beginning of cancer. When cancer cells break through the basement membrane, they can spread to other parts of the lung and on to the rest of the body, a process called metastasis.

Figure 14A–4 Lung cancer developing in the lining of a bronchus, an air passageway in the lung. (a) The bronchial lining of a healthy nonsmoker has a single layer of columnar epithelial cells overlying a layer of basal cells and a basement membrane. (b) An early sign of damage to the lining is an increase in the number of layers of basal cells, (c) which gradually replace the columnar epithelial cells. (d) Cancer begins when the basal cells develop atypical nuclei and the arrangement of basal cells becomes disorganized. (e) The cancer spreads when these cells break through the basement membrane.

Heart Disease

When we think about the hazards of tobacco smoke, lung cancer generally leaps to mind, but the increased risk of cardiovascular disease is even more significant. Each year cardiovascular disease kills many more people than does lung cancer, and smokers have a twofold to threefold increase in the risk of heart disease (Figure 14A–5). The American Heart Association estimates that about 25% of all fatal heart attacks are caused by cigarette smoke. This translates to roughly 200,000 heart attacks a year in the United States that could have been prevented by not smoking.

Smoking stresses both the heart and blood vessels in many ways. Some of the effects are immediate and direct. For instance, nicotine makes the heart beat faster at the same time that carbon monoxide reduces oxygen delivery to it. Nicotine also constricts blood vessels, raising blood pressure.

A less immediate but no less important way that smoking leads to cardiovascular disease is by increasing atherosclerosis, a condition in which lipid deposits, primarily composed of cholesterol, form in the walls of blood vessels, restricting the flow of blood. Smoking influences atherosclerosis in two ways. One is by decreasing the levels of protective cholesterol-transport particles, called HDLs, that carry cholesterol to the liver, perhaps even removing it from cells,

so that cholesterol can be eliminated from the body. With fewer HDLs, more cholesterol begins to clog the arteries. A second way that smoking promotes cholesterol deposits is by raising blood pressure. The elevated blood pressure stresses the linings of the arteries, making them more susceptible to cholesterol deposit. This narrowing of blood vessels not only results in starved tissue downstream but in increased blood pressure as well. When these deposits form in the arteries that supply blood to the heart, as they often do, the blood supply to the heart may be reduced or shut down completely, causing heart cells to die. The death of heart cells is called a heart attack. When a vessel to the head is blocked so that brain cells are damaged, it is called a stroke.

Smoking also increases the chances of forming blood clots in the vessels. The clots form for several reasons. You may recall, for instance, that nicotine stimulates an increase in the number of platelets and makes them stickier. The levels of fibrinogen, a blood protein also important in clotting, increase as well. Either condition may promote clotting. These clots may break loose from the site where they form and travel through the bloodstream until they lodge in a small vessel where they block the blood flow. Thus, these clots may also result in a heart attack or stroke.

Smoking can also harm the heart in particular. It can initiate coronary artery spasms, sudden violent contractions of the smooth muscle in the arteries supplying the heart, that may lead to heart attack.

A Potpourri of Additional Hazards

Smoking causes wrinkles. The reduction in the amount of oxygen reaching tissues impairs the body's ability to make collagen, a main supportive protein in certain connective tissues, including skin and bones. One result is wrinkles. The extent of wrinkling increases with cigarette consumption and the duration of the smoking habit. Heavy smokers are 3.5 times more likely than nonsmokers to show crow's feet, wrinkles around the eyes. Impaired collagen production is also responsible for slower bone and skin wound healing among smokers.

Smoking can dim your vision. In the United States, smoking is to blame for 20% of all cases of cataracts, a clouding of the lens of the eye that leads to blindness if it is not corrected surgically. Unfortunately, ex-smokers continue to have a higher risk of cataracts, so unlike many of the other effects of smoking, quitting may not solve this problem.

Smokers have more car and industrial accidents. A traffic safety study in Massachusetts showed that, even when use of alcohol, age, driving experience, and education are taken into account, smokers have 50% more traffic accidents and are issued 46% more tickets than nonsmokers.

Urinary incontinence (lack of urinary control) is more common among smokers. This is partly because the hacking smoker's cough weakens the muscle that normally holds urine in the bladder and partly because nicotine causes the muscles of the bladder to contract, allowing urine to leak.

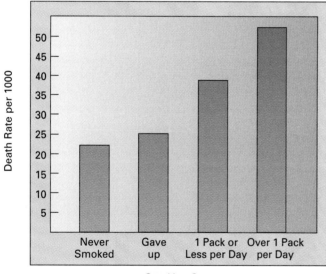

Figure 14A–5 Death rate due to heart disease among nonsmokers and smokers. Notice that the death rate from heart disease increases with the number of cigarettes smoked per day. Those who smoke over a pack per day have more than twice the risk of death due to heart disease than do people who have never smoked. In any case, a smoker who successfully quits is much less likely to die of heart disease than if the smoking habit continues.

(Reprinted with the permission of Simon & Schuster from The Well Adult *by Mike Samuels, MD and Nancy Samuels, © 1988 by Mike Samuels, MD and Nancy Samuels)*

Also, where there's smoke, there's fire. Smoking causes more fatal fires each year than any other source of combustion. As a result of these fires, 2500 people die and another 25,000 are injured each year. These fires cost more than $313 million in property damage.

Women and Smoking

More than a quarter of women of childbearing age smoke, although almost half of them have tried to quit. It is true that the number of female smokers seems to be decreasing, but it is not falling as rapidly as the number of male smokers. Between 1965 and 1987, smoking declined by 20% among men but only 6% among women.

Unfortunately, the hazards of smoking may be greater for women than for men. For instance, some studies suggest that women smokers have a greater chance of getting lung cancer than do male smokers.

Furthermore, some risks are unique to women. Women who smoke are three times more likely than nonsmokers to develop cervical cancer. Most cases of cervical cancer are directly caused by the papilloma virus that also causes genital warts. A substance in tobacco smoke activates the virus. In addition, women who smoke reach menopause 2 to 3 years earlier than nonsmokers. They also have higher rates of osteoporosis, a condition more common in women than men in which bones become less dense and, therefore, weaker.

Women of reproductive age may be particularly at risk because the combination of smoking and oral contraceptives can be deadly, since both elevate blood pressure and increase the likelihood of abnormal clot formation. The death rate of women smokers who are also taking the birth control pill is three times as high as that of nonsmokers who are on the pill. The increased death rate is due to a much higher incidence of strokes, heart attacks, and blood clots in the legs.

There are many adverse effects of smoking on the reproductive ability of a woman. A woman smoker who would like to become pregnant will have much better luck doing so if she quits smoking. Specifically, women who smoke more than a pack a day are half as fertile as women who don't smoke. Furthermore, if a woman smoker does become pregnant and continues to smoke, she is twice as likely to miscarry than a nonsmoker. Smoking during pregnancy increases risk of complications such as premature separation of the placenta from the uterus, which shuts off the oxygen supply to the fetus, and a poorly positioned placenta that may partially or completely block the birth canal. (Incidentally, exposure to someone else's cigarette smoke can also affect a fetus if its nonsmoking mother is exposed to it while she is pregnant.)

In a very real sense, tobacco smoke suffocates the growing fetus because nicotine constricts its lifeline, the blood vessels of the umbilical cord, and because carbon monoxide reduces the amount of oxygen carried by the blood. Smoking two packs a day blocks approximately 40% of the oxygen supply to the fetus. As a result, the fetus's heart rate and blood pressure increase and the acid-base balance of its blood is shifted.

It is estimated that smoking leads to 5000 preventable fetal deaths annually that occur during the 8 weeks before or 7 days after birth. The incidence of stillbirth among pregnant women who smoke is 7.8% as compared with 4.1% among nonsmoking women. Furthermore, a newborn of a woman who smoked during pregnancy has a one-third higher risk of dying soon after birth. Among the reasons for the increased infant mortality are premature birth and low birth weight. Newborns of women who smoke weigh an average of 200 grams (about 1/2 pound) less than those born to nonsmoking women.

Passive Smoking

It is not true that those who smoke are only hurting themselves. After all, their smoke pollutes the air of others who don't smoke, and a dangerous pollution it is. The pollution takes the form of exhaled smoke and the side-stream smoke that comes from the burning end of the cigarette. As a result, two-thirds of a cigarette's smoke actually enters the environment. And, because it is not filtered through tobacco, this smoke contains higher levels of many hazardous materials than the smoke that is inhaled through a cigarette. For example, the side-stream smoke, contains more cadmium, which is related to high blood pressure, chronic bronchitis, and emphysema. It also has twice the tar and nicotine of inhaled smoke and five times as much carbon monoxide. Because the concentration of cancer-causing nitrosamines in side-stream smoke is 50 times greater than inhaled smoke, after an hour in a smoke-filled room, a nonsmoker may inhale an amount of nitrosamines equal to smoking 15 filter cigarettes.

During a typical campus party the number of particulates in the air of a room is 40 times greater than the U.S. standard for air quality. After 30 minutes in a smoky room, a nonsmoker's heartbeat and blood pressure begin to increase. The carbon monoxide level in the blood begins to rise and is enough to hamper one's ability to distinguish time intervals or to distinguish the relative brightness of two lights. (This impairment could be of critical significance if the lights to be distinguished are the headlights of oncoming cars.) After a single hour in a very smoky room, a nonsmoker's blood level of carbon monoxide and nicotine match those of someone who smoked a cigarette. After leaving the room, a person's blood contains carbon monoxide for 3 to 4 hours.

Ironically, because we generally spend the most time with those we care about the most, the loved ones of smokers are often hurt the most and exposure to smoke most often occurs at home. In other cases, secondary smoke is encountered in the work place.

We have seen that the major health risks of smokers are increased risk of lung disease, cancer, and heart disease.

Long-term exposure to secondary tobacco smoke jeopardizes the health of nonsmokers in the same ways. In one study, 2100 healthy nonsmoking workers who were chronically exposed to smoke of others were found to have changes in two measures of lung function. The extent of these changes were comparable with those of people who smoke 1 to 10 cigarettes per day. Another study revealed that children whose mothers smoke have a 7% decrease in lung function and children in families where both parents smoke have twice the number of upper respiratory infections as children from nonsmoking homes.

Secondhand smoke is a carcinogen. The Environmental Protection Agency estimates that 500 to 5000 cases of lung cancer each year are caused by passive smoking. A Japanese study of nonsmoking wives of men who smoke revealed the women developed lung cancer at the same rate as people who smoke half a pack a day. These wives were four times more likely than wives of nonsmokers to get lung cancer. Another study showed that nonsmoking women who inhale sidestream smoke 3 or more hours a day show the same elevated risk of cervical cancer as women who smoke themselves.

Passive smoking also affects the cardiovascular system of nonsmokers. For instance, in those susceptible, passive smoking can cause angina, chest pain brought on when the oxygen supply to the heart is insufficient.

Safe Cigarettes?

There is no such thing as a safe cigarette. True, filters do reduce risk by trapping some tar, nicotine, carbon monoxide, and other poisonous gases. As a result, smokers who smoke cigarettes with filters have a slightly lower risk of lung cancer than those who smoke cigarettes without filters. However, the filters do not trap everything, and these smokers still have a risk of developing lung cancer that is 6.5 times that of a nonsmoker.

Since many of the harmful effects of cigarette smoke are caused by the tars and nicotine, the risks associated with low tar and nicotine cigarettes might be expected to be lower than those associated with other brands. However, this is not always the case. Because of the craving for nicotine, smoking habits often change so that the blood level of nicotine remains constant regardless of the type of cigarette smoked. For example, there is evidence that smokers who switch to low tar and nicotine cigarettes inhale more deeply or puff more often, and they may even smoke more cigarettes.

It is easier to smoke a cigarette with low tar and nicotine. Such cigarettes have made it easier for a whole new group—young girls and women—to become hooked. This is because females are more sensitive than males to unpleasant side effects of nicotine, and they tolerate the low levels of nicotine more easily than the levels in regular cigarettes.

Benefits of Quitting

Three of four smokers say they want to quit, and 60% say they have tried. On any one serious try, chances are only 1 in 5 of successfully quitting. The chances of success go up to 3 in 5 with repeated attempts. But, more than 30 million Americans have quit.

The health benefits of quitting are enormous and the major payoff is a longer life for you, your friends, your colleagues, and your loved ones. Much of the damage caused by smoke is reversible once you quit. Blood pressure and heart rate begin to decrease toward normal levels. The risk of heart attack begins to drop a year after quitting. After 5 years, the risk of heart attack for someone who smoked a pack a day is the same as someone who never smoked. The risk of lung cancer also drops (Figure 14A–6). In a study of 34,000 British physicians, the death rate due to lung cancer for those who continued to smoke was 16 times greater than that of a lifetime nonsmoker. The risk for those who had quit the habit for 5 to 9 years was only six times that of a nonsmoker. Fifteen years after quitting, the lung cancer risk was only twice that of someone who had never smoked.

Then there are the financial gains. Smokers not only spend money supporting their habit, they also spend more than nonsmokers on cold remedies because they are sick more often than nonsmokers. Smokers also pay higher premiums for life insurance.

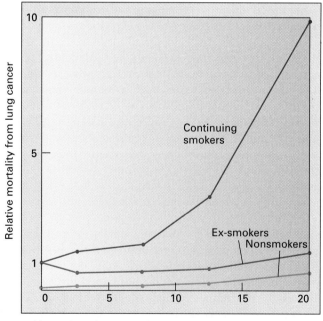

Years since the ex-smokers stopped smoking cigarettes

Figure 14A–6 A comparison of lung cancer deaths among lifetime nonsmokers, ex-smokers, and continuing smokers. Quitting dramatically reduces a smoker's risk of death due to lung cancer. *(Reprinted from Prescott and Flexer:* Cancer: The Misguided Cell, *Second Edition, 1986 with permission from Sinauer Associates, Inc.)*

Chapter 15

The Urinary System

(© W. Ormeron/Visuals Unlimited)

Our cells, like tiny chemistry laboratories, serve as sites for a multitude of chemical reactions that constitute metabolism. Just like the day-to-day activities of laboratories, the ongoing cellular synthesis and break down of molecules produces wastes that must be diligently disposed of. Examples of metabolic wastes include nitrogen-containing wastes (ammonia, urea, and uric acid), carbon dioxide, water, and heat. These wastes, along with excess essential ions such as hydrogen (H^+), sodium (Na^+), and chloride (Cl^-), are eliminated from our bodies through the actions of several organs. For example, lungs and skin eliminate carbon dioxide, heat, and water, and in the case of skin, some salts and urea are also excreted. Organs of our gastrointestinal tract eliminate solid wastes, in addition to many of the previously mentioned substances. Our focus in this chapter will be on the kidneys and their role in excreting nitrogen-containing wastes, water, carbon dioxide, inorganic salts, and hydrogen ions.

The two kidneys, along with two ureters, one urinary bladder, and one urethra, make up the **urinary system** (Figure 15–1). The main function of this system is to regulate the volume, pressure, and composition of the blood. The urinary system accomplishes this task by regulating the amounts of water and dissolved substances that are added to or removed from the blood. Wastes and excess materials removed from the blood form **urine**, the yellowish fluid produced by each kidney that travels down the ureters to the urinary bladder where it is stored until being excreted from the body through the urethra. The color of urine comes from urochrome, a yellow pigment produced as a waste product during the break down by the liver of hemoglobin in red blood cells. The urochrome travels in the bloodstream from the liver to the kidneys where it is filtered from the blood and excreted with urine. Some drugs and foods, however, can cause a change in urine color. Red beets, for example, change the color of urine from its characteristic yellow to a deep red.

We begin our study of the urinary system with an examination of kidney structure and then follow with a consideration of the critical role kidneys play in maintaining homeostasis. We discuss disorders of the kidneys (see, for example, the Personal Concerns essay on kidney stones) and ways to replace kidney function. The importance of properly functioning kidneys is

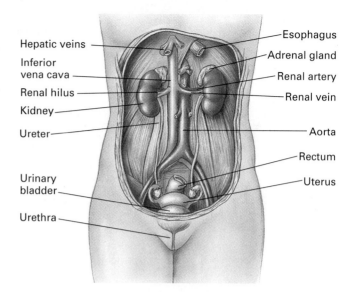

Figure 15–1   Organs of the urinary system and major blood vessels. Digestive organs are not shown.

evidenced by the number of times we are handed a small cup by medical personnel and asked to provide a urine sample. Indeed, as we will see in the Personal Concerns essay on urinalysis, the characteristics of our urine say much about us. Following our discussion of the kidneys, we describe the other organs of the urinary system and finish the chapter with a discussion of urination.

Components of the Urinary System

As mentioned previously, the urinary system includes the kidneys, the ureters, the urinary bladder, and the urethra.

Kidneys

Our kidneys are reddish in color, each about the size of a fist, and shaped like kidney beans (see Figure 15–1). They are located just above the waist between the back wall of the abdominal cavity and the parietal peritoneum, the membrane in front that lines the abdominal cavity. The slightly indented or concave border of each kidney faces the midline of the body. Perched on top of each kidney is an adrenal gland. The kidneys are covered and supported by several layers of connective tissue (Figure 15–2). The outermost layer, the **renal fascia**, anchors each kidney and its adrenal gland to the abdominal wall and surrounding tissues. Beneath the renal fascia is the **adipose capsule**, a protective cushion of fat.

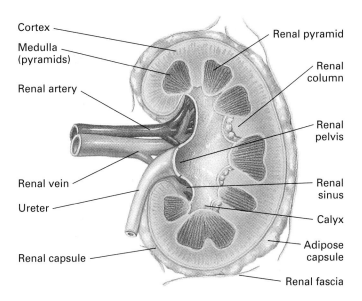

Cortex

Medulla (pyramids)

Renal artery

Renal vein

Ureter

Renal capsule

Renal pyramid

Renal column

Renal pelvis

Renal sinus

Calyx

Adipose capsule

Renal fascia

Figure 15–2 Structure of the kidney. This frontal section through the left kidney shows the basic internal structure and protective outer layers of connective tissue.

Sometimes, in very thin people, the renal fascia or adipose capsule is not substantial enough and a kidney will slip from its normal position. This painful condition, known as floating kidney or **nephrotosis**, is dangerous because the ureter of the displaced kidney may kink, preventing normal flow of urine from the kidney down the ureter to the bladder. The back-up of urine and increased pressure can damage the kidney. The innermost layer, the **renal capsule**, is a transparent layer of fibrous connective tissue that protects the kidneys from trauma and infection. The numerous protective barriers and cushions surrounding our kidneys highlight the importance of these organs to our daily existence.

Structure of the Kidneys

The **hilus** is a notch in the concave border of each kidney where the ureter leaves the kidney (see Figure 15–1). The hilus is also the area where nerves, lymphatic vessels, and blood vessels enter and exit the kidney. With respect to blood supply, the renal arteries branch off the aorta and bring blood to the kidneys, and the renal veins carry filtered blood away from the kidneys to the inferior vena cava that brings the blood to the heart.

Each kidney has an inner region, the **renal medulla**, and an outer region, the **renal cortex** (see Figure 15–2). The medulla contains cone-shaped structures called renal pyramids. The cortex begins at the outer border of the kidney and portions of it, called renal columns, extend between the renal pyramids of the medulla. The innermost region of the kidney is the **renal pelvis**, a cavity within an even larger space called the **renal sinus**. The renal sinus also contains

fat and connective tissue. The apex of each renal pyramid joins a cup-like extension of the renal pelvis called a **calyx**. As we will soon see, urine produced by the kidneys eventually drains to a calyx and then into the renal pelvis and out the ureter to the urinary bladder.

Nephrons are the functional units of the kidneys and are responsible for formation of urine (Figure 15–3). These microscopic tubules number 1 to 2 million per kidney and perform the following three functions: (1) filtration—only certain substances are allowed to pass out of the blood and into the nephron, (2) reabsorption—some useful substances are returned from the nephron to the blood, and (3) secretion—the nephron directly removes wastes and excess materials in the blood and adds them to the filtered fluid that becomes urine. We will have more to say about these processes when we discuss kidney function.

There are two basic parts to a nephron, the renal corpuscle and the renal tubule (see Figure 15–3). The **renal corpuscle**, the portion of the nephron where fluid is filtered, consists of a tuft of capillaries, the **glomerulus**, and a surrounding cup-like structure, **Bowman's (glomerular) capsule**. Bowman's capsule has an inner layer close to the capillaries of the glomerulus and an outer layer. The space between the two layers is known as **Bowman's space**. Blood enters a glomerulus by way of an afferent arteriole. Once within the glomerular capillaries, water and many solutes move from the plasma into Bowman's space and then into the renal tubule. Blood leaves the glomerulus by means of an efferent arteriole. The **renal tubule**, the site of reabsorption and secretion by the nephron, has three sections: the **proximal convoluted tubule**, the **loop of Henle**, and the **distal convoluted tubule**. The loop of Henle resembles a hairpin turn, having a descending limb and an ascending limb. The distal convoluted tubules of several nephrons empty into a single collecting duct (in a renal pyramid) that eventually drains into a calyx and then the renal pelvis where it exits the kidney by way of the ureter and moves on to the urinary bladder.

In our kidneys, about 80% of the nephrons have small loops of Henle and are confined almost entirely to the renal cortex. These nephrons are called **cortical nephrons**. The remaining 20% of our nephrons, called **juxtamedullary nephrons**, have large loops of Henle that extend from the cortex into the renal medulla. Once in the medulla, the loops of these nephrons then turn abruptly upward, back into the cortex, where they lead into distal convoluted tubules. As we will see, juxtamedullary nephrons play an important role in water conservation.

Function of the Kidneys

The kidneys are key to maintaining homeostasis. They are responsible for (1) filtering wastes and excess materials from the blood, (2) assisting the respiratory system in the regulation of blood pH, and (3) maintaining fluid balance by regulating the volume and composition of blood and urine.

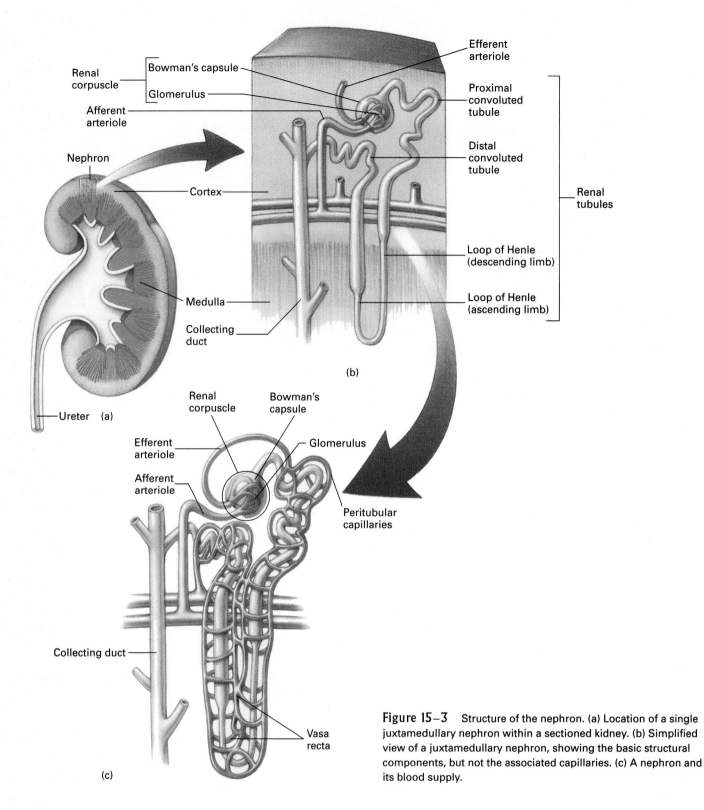

Figure 15–3 Structure of the nephron. (a) Location of a single juxtamedullary nephron within a sectioned kidney. (b) Simplified view of a juxtamedullary nephron, showing the basic structural components, but not the associated capillaries. (c) A nephron and its blood supply.

To understand the functions of the urinary system we must examine the work of nephrons. Recall that nephrons perform three functions—filtration, reabsorption, and secretion. As a general process, filtration involves pressure forcing fluid and dissolved substances through a membrane. In the kidneys, **filtration** occurs in the renal corpuscle of the nephron and involves blood pressure forcing water and dis-

solved substances across a filter that consists of two cell layers, the endothelium of the glomerular capillaries and the inner lining of Bowman's capsule, between which is sandwiched a basement membrane of extracellular material (Figure 15–4). Blood moves from an afferent arteriole into glomerular capillaries where pressure is high because of the smaller size of the exiting efferent arteriole relative to the

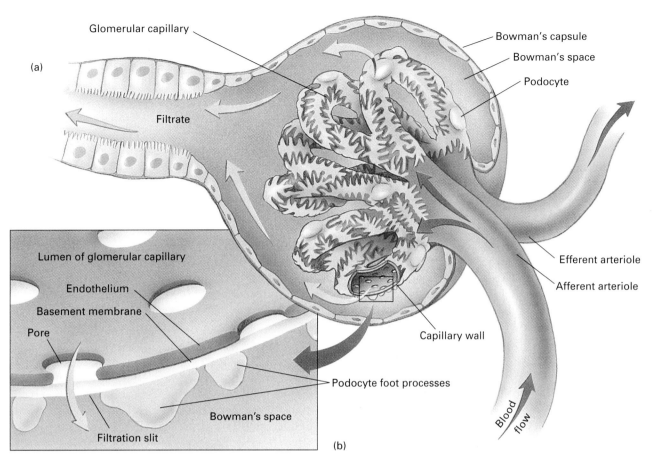

(a)

Glomerular capillary

Filtrate

Bowman's capsule

Bowman's space

Podocyte

Lumen of glomerular capillary

Endothelium

Basement membrane

Pore

Capillary wall

Efferent arteriole

Afferent arteriole

Podocyte foot processes

Bowman's space

Filtration slit

Blood flow

(b)

Figure 15–4 The renal corpuscle, site where filtration occurs. (a) The renal corpuscle consists of Bowman's capsule and a ball of capillaries called the glomerulus. Blood flow is indicated by red arrows and path of the filtrate by yellow arrows. (b) Cross-section through the wall of a glomerular capillary showing how water and dissolved substances in the blood move first through the pores in the endothelial lining of the capillary, then through the basement membrane, and finally through filtration slits in the inner lining of Bowman's capsule.

entering afferent arteriole. This high pressure forces water and dissolved substances in the blood through pores in the endothelial lining of the glomerular capillaries. Although water, ions, glucose, and some proteins pass through the pores, formed elements of the blood such as red blood cells, white blood cells, and platelets cannot pass through. The basement membrane of the glomerulus restricts the passage of large proteins. The second cell layer in the glomerular filter is the inner layer of Bowman's capsule. This layer contains cells, called **podocytes**, with highly branched extensions that wrap around the glomerular capillaries. Water and dissolved substances that have passed through the endothelial lining of a capillary and the basement membrane must then pass through the small openings between the branching extensions of the podocytes. These small openings are known as **filtration slits** and they prevent medium-sized proteins from entering the space within Bowman's capsule. Water, ions, and smaller molecules, now known collectively as glomerular filtrate, move into Bowman's space.

The rate of filtration by the glomerulus can be changed by altering the diameters of the afferent or efferent arterioles. An increase in the diameter of afferent arterioles or a reduction in the diameter of efferent arterioles produces higher pressure in the glomerular capillaries and results in higher filtration rates. Higher filtration rates can also be achieved through increases in systemic blood pressure that cause higher rates of blood flow into glomerular capillaries.

Reabsorption, the process by which water and some dissolved substances removed from the blood during glomerular filtration are then returned to the blood, occurs in the renal tubule. Remarkably, as the glomerular filtrate passes through the renal tubule, about 99% of it is returned to the blood, specifically to capillaries. These capillaries are called peritubular capillaries when they surround the proximal and distal convoluted tubules and are referred to as the vasa recta when they surround the loop of Henle (see Figure 15–3c). Thus, only about 1% of the glomerular filtrate is eventually excreted as urine. Put another way, of the approximately 180 liters (48 gallons) of filtrate that enters Bowman's space, almost 179 liters are returned to the blood by reabsorption, the remaining 1 liter being excreted as urine. Reabsorption results in the return to the blood of water, essential ions, and nutrients. For example, about 99% of water molecules and sodium ions and almost 100% of glucose molecules and amino acids are reabsorbed. Imagine how much water and food we would have to consume if we did not have reabsorption to offset the losses from glomerular filtration! In contrast to water, essential ions, and nutrients, some wastes are not reabsorbed at all and others, such as urea, are partially reabsorbed.

Reabsorption is carried out by epithelial cells lining the renal tubule and involves both active and passive processes. Whereas water reabsorption occurs by osmosis, the reabsorption of solutes occurs through active transport, passive diffusion, and pinocytosis. Although all parts of the tubule participate, the bulk of reabsorption occurs in the proximal convoluted tubule. The epithelial cells of the proximal convoluted tubule have specialized projections called microvilli that dramatically increase the surface area available for reabsorption, thereby allowing this region to make the largest contribution.

Secretion, the third process involved in the formation of urine by nephrons, results in the removal from the blood of drugs, such as penicillin, and wastes and excess ions that escaped glomerular filtration. These wastes are added to the filtered fluid that will become urine. For example, hydrogen ions (H^+), potassium ions (K^+), and ammonium ions (NH_4^+) present in the blood of peritubular capillaries are actively transported into the renal tubule where they become part of the filtrate to be excreted. Tubular secretion occurs along the proximal convoluted tubule (H^+ and NH_4^+), loop of Henle (urea), and collecting ducts (H^+ and K^+). Table 15–1 summarizes the regions of the nephron and their roles in filtration, reabsorption, and secretion.

In addition to removing wastes, secretion helps to regulate the pH of blood. Recall from Chapters 2 and 14 that blood pH must be precisely regulated for proper functioning of the body. This precise regulation is achieved through (1) the actions of the kidneys, (2) respiration, and (3) buffer systems that pick up or release hydrogen ions depending upon whether such ions are in excess or short supply. In previous discussions we described the importance of carbonic acid as a buffer in the blood. When added to water, carbonic acid (H_2CO_3) dissociates into hydrogen ions (H^+) and bicarbonate (HCO_3^-). The buffering action of carbonic acid results from the fact that when levels of H^+ increase in the blood, the H^+ combine with bicarbonate and are thus removed from solution, preventing dramatic changes in pH. Alternatively, when levels of H^+ decrease, carbonic acid dissociates, adding H^+ to solution, again preventing substantial changes in pH. The role of the kidneys in helping to maintain pH is twofold. First, through reabsorption of HCO_3^-, the kidneys aid in restoring the carbonic acid buffer system by resupplying bicarbonate to the blood. Second, through secretion of H^+ the kidneys remove excess H^+ from the blood.

Two fluids, blood and urine, leave the kidneys. By the end of filtration, reabsorption, and secretion, blood leaving the kidneys contains most of the water, nutrients, and essential ions that it contained upon entering the kidneys. Wastes and excess materials, however, have been removed, and thus the blood has been filtered. This purified blood within capillaries moves into small veins that eventually form the renal vein that exits each kidney and joins the inferior vena cava. From the inferior vena cava, the purified blood travels to the heart. Whereas purified blood leaves the kidneys and travels in an upward direction toward the heart, urine leaving the kidneys travels in a downward direction. Urine within the distal convoluted tubules of nephrons empties into collecting ducts (within renal pyramids) where more water may be reabsorbed, further concentrating the urine. From the collecting ducts, urine moves into a calyx and then into the renal pelvis. From here, urine leaves the kidney through the ureter, moving down to the urinary bladder for storage until elimination from the body through the urethra.

Factors That Influence Kidney Function

Several factors influence kidney function. Among the most important are antidiuretic hormone, the hormones aldosterone and atrial natriuretic peptide, and the proteins renin and angiotensin. **Antidiuretic hormone (ADH)** is manufactured by the hypothalamus and then travels to the posterior pituitary for storage and eventual release. Once released by the posterior pituitary gland, ADH regulates the amount of water reabsorbed by the distal convoluted tubules and collecting ducts of nephrons. Basically, the hypothalamus responds to changes in the concentration of water in the blood by increasing or decreasing production and secretion

Table 15–1   Regions of the Nephron and Their Roles in Filtration, Reabsorption and Secretion

Region of Nephron	Function
Renal corpuscle (Bowman's capsule and glomerulus)	Filters the blood, removing water, glucose, excess essential ions, nitrogenous wastes, and other small molecules
Proximal convoluted tubule	Reabsorbs water, glucose, amino acids, urea, Na^+, Cl^-, and HCO_3^- Secretes H^+, NH_4^+
Loop of Henle	Reabsorbs water, Na^+, Cl^-, and K^+ Secretes urea
Distal convoluted tubule	Reabsorbs water, glucose, Na^+, Cl^-, and HCO_3^-

of ADH. Decreases in the concentration of water in the blood stimulate increased secretion of ADH. Higher levels of ADH in the bloodstream increase the permeability to water of the final portions of the distal convoluted tubules and collecting ducts of nephrons with the result that more water is reabsorbed from the filtrate (in the absence of ADH, cells of the distal convoluted tubule and collecting duct have exceptionally low permeability to water). The movement of increased amounts of water from the filtrate back into the blood results in increased blood volume and pressure and production of small amounts of concentrated urine. Just the opposite occurs when the concentration of water in the blood increases. Increases in water concentration prompt decreased release of ADH, and this reduces water reabsorption from the filtrate, resulting in reduced blood volume and pressure and production of large amounts of dilute urine. Alcohol inhibits the secretion of ADH, causing reduced water reabsorption by the kidneys and production of large amounts of dilute urine. Thus, it makes little sense after exercising to try to quench your thirst and restore body fluids by drinking an alcoholic beverage. Substances like alcohol that promote urine production are called **diuretics**. **Diabetes insipidus** is a disease characterized by excretion of large volumes of dilute urine. This condition, caused by a deficiency of ADH, has been linked to damage to either the site of manufacture of ADH, the hypothalamus, or the site of release of the hormone, the posterior pituitary.

Aldosterone, a steroid hormone released by the adrenal glands (specifically, the adrenal cortex), increases reabsorption of sodium by the distal convoluted tubules and collecting ducts. This is important because water follows sodium. As more sodium is transported out of the nephron into the peritubular capillaries, increased amounts of water follow, resulting in increased blood volume and pressure and production of small amounts of more concentrated urine. Caffeine, a familiar chemical found in coffee, tea, and many soft drinks, has just the opposite effect. Caffeine decreases reabsorption of sodium, thus decreasing the amount of water moving out of tubules back into the bloodstream, resulting in decreased blood volume and pressure and production of large amounts of dilute urine. Because caffeine, like alcohol, increases urine output, it too is considered a diuretic.

What stimulates release of aldosterone? Unfortunately, as is sometimes the case in physiology, the answer to this straightforward question is somewhat complicated. Look at Figure 15–5 as we describe each step along the pathway that

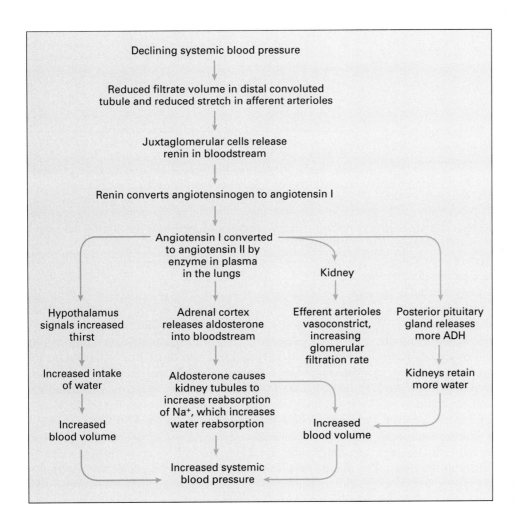

Figure 15–5 The renin-angiotensin system and the many routes by which it influences blood pressure.

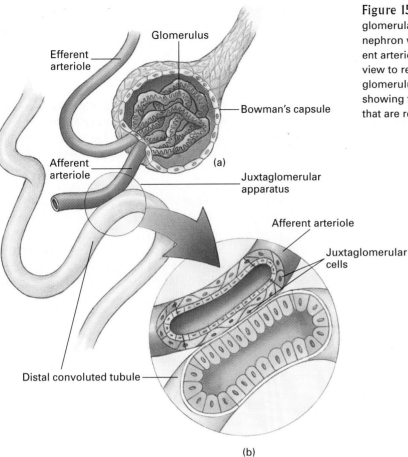

Glomerulus

Efferent arteriole

Bowman's capsule

Afferent arteriole

(a)

Juxtaglomerular apparatus

Afferent arteriole

Juxtaglomerular cells

Distal convoluted tubule

(b)

Figure 15–6 The juxtaglomerular apparatus. (a) The juxtaglomerular apparatus (within the circle) is a section of the nephron where the distal convoluted tubule contacts the afferent arteriole. The nearby renal corpuscle is shown in cutaway view to reveal its components, Bowman's capsule and the glomerulus. (b) Close-up view of the juxtaglomerular apparatus showing the juxtaglomerular cells within the afferent arteriole that are responsible for the secretion of renin.

leads to aldosterone release. The focus will be on the **juxtaglomerular apparatus**, a region of the nephron where the distal convoluted tubule contacts the afferent arteriole bringing blood into the glomerulus (Figure 15–6). Basically, declines in systemic blood pressure cause declines in the glomerular filtration rate resulting in reduced volume of filtrate within nephrons. Declines in filtrate volume cause cells within the juxtaglomerular apparatus to release renin. **Renin**, an enzyme, converts angiotensinogen, a protein produced by the liver and found in the plasma, into another protein, **angiotensin I**. Angiotensin I is then converted by a plasma enzyme in the lungs into **angiotensin II**, an active form that stimulates the adrenal gland to release aldosterone. Aldosterone increases reabsorption of sodium and water by the distal convoluted tubules and collecting ducts of nephrons, which increases blood volume and pressure, ultimately increasing the filtration rate within the glomerulus and volume of filtrate within the nephron. In addition to stimulating the release of aldosterone from the adrenal gland, angiotensin II also (1) constricts efferent arterioles causing increases in glomerular blood pressure and, hence, filtration rate, (2) stimulates the thirst center in the hypothalamus to increase intake of water, and (3) stimulates the posterior pituitary to release ADH.

Critical Thinking

Estrogens, female sex hormones, are chemically similar to aldosterone and thus have similar effects on the distal convoluted tubules and collecting ducts of the kidneys. How might this explain the water retention experienced by many women as their estrogen levels rise during the menstrual cycle?

A final hormone that influences kidney function is **atrial natriuretic peptide (ANP)**. This hormone, identified in 1983, is released by cells in the right atrium of the heart in response to stretching of the heart caused by increased blood volume and pressure. ANP decreases water and solute reabsorption by the kidneys, either by increasing permeability of the glomerular filter or by dilating afferent arterioles. Regardless of its precise site of action, ANP causes declines in blood volume and pressure and production of large amounts of urine. There are high hopes in the medical community that ANP can be used to treat edema, the accumulation of fluid in tissues, and high blood pressure. Finally, in addition to its direct effect on the kidneys, ANP also appears to influence kidney function indirectly by inhibiting secretion of ADH, aldosterone, and renin.

Before leaving the topic of kidney function, we should briefly review two functions of the kidneys that are important to homeostasis but not directly related to the urinary system. First, cells of the kidneys produce and release into the bloodstream an enzyme that results in the formation of **erythropoietin**, a hormone that travels to the red bone marrow where it stimulates production of red blood cells. Second, the kidneys transform vitamin D, a substance provided by certain foods in our diet or produced by the skin in response to sunlight, into an active form, called calcitriol, that promotes absorption and use of calcium and phosphorus by the body.

Water Conservation by the Kidneys

Our kidneys, specifically the juxtamedullary nephrons within, enable us to conserve water through the production of concentrated urine. The mechanism of urine concentration is based on an increasing concentration of solutes, particularly NaCl and urea, in the interstitial fluid from the cortex to the medulla of the kidneys. The loop of Henle is responsible for maintaining the gradient of NaCl in the interstitial fluid. The collecting ducts are sites where urea diffuses out of the filtrate into the interstitial fluid of the medulla. To understand the mechanism of urine concentration we will retrace the path of filtrate as it flows through the renal tubule, focusing on what happens in the loop of Henle and collecting ducts (Figure 15–7).

The solute concentration of filtrate passing from Bowman's capsule to the proximal tubule is about the same as that of blood. As the filtrate moves through the proximal convoluted tubule, large amounts of water *and* salt are reabsorbed, producing dramatic reductions in the volume of filtrate but little change in its solute concentration. Next the filtrate enters the descending limb of the loop of Henle, and this path takes it from the cortex to the medulla. Along the descending limb, water leaves the filtrate by osmosis, creating an increase in the concentration of solutes, including NaCl, within the filtrate. The concentration of salt in the filtrate peaks at the curve of the loop of Henle, setting the stage for the next step in the process of urine concentration. As the filtrate moves up the ascending limb of the loop of Henle, large amounts of NaCl move out of the filtrate into the interstitial fluid of the medulla (in the initial segment of the ascending limb, NaCl diffuses out, but in the thicker segment further along the limb, it is actively transported out). Water, however, remains in the filtrate because the ascending limb is impermeable to it. When the filtrate reaches the distal convoluted tubule in the cortex, it is quite dilute— in fact, the filtrate is hypotonic to body fluids, because it has a lesser solute concentration than body fluids. The filtrate then moves into a collecting duct and begins its descent, once again, toward the medulla, a pathway of increasing interstitial salt concentration. Collecting ducts are permeable

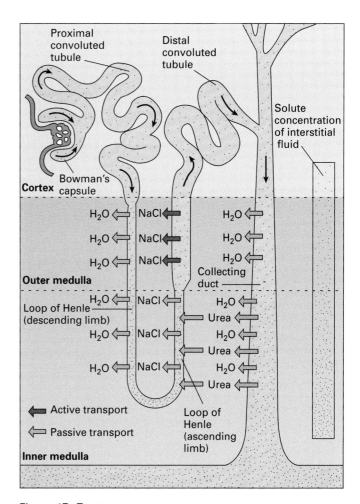

Figure 15–7 The mechanisms within juxtamedullary nephrons that enable our kidneys to produce concentrated urine and to thereby conserve water. See text for details. Stippling within the nephron indicates solute concentration.

to water but not to salt. Thus, as the filtrate encounters increasing concentrations of salt in the fluid of the inner medulla, water leaves the filtrate by osmosis. With the departure of large amounts of water, urea is now concentrated in the filtrate and some of it moves into the interstitial fluid of the medulla from lower portions of the collecting duct (the remaining urea is excreted). This leakage of urea contributes to the high solute concentration of the inner medulla and thus aids in concentrating the filtrate. At its most concentrated, urine is hypertonic to blood and interstitial fluid in all parts of the body except the inner medulla where it is isotonic. Together, then, the loop of Henle and collecting ducts maintain solute gradients in the interstitial fluid of the kidneys and thus make possible the concentration of urine and conservation of water.

Renal Failure and Replacement

Renal failure, a decrease or complete cessation of glomerular filtration, can be acute or chronic. **Acute renal failure** refers to an abrupt, complete or near complete, cessation of

Kidney Stones: An Ancient Problem with New (and Shocking) Solutions

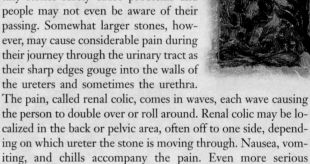

Ask anyone who has experienced passing a good-sized kidney stone—one with a diameter of at least 5 mm (0.2 inches)—and the words most often used to describe the event are "excruciatingly painful." Kidney stones, by the way, are not only a modern ailment. They have been found, for example, in Egyptian mummies, and we can only assume that their passing caused as much distress in ancient times as today. What are these pebble-sized "hell raisers" commonly found within the urinary tract, and what makes their passing often so memorable? How can they be treated and how can they be prevented?

Kidney stones, also called renal calculi, are small, hard crystals formed when substances like calcium (the most common constituent), uric acid, or magnesium ammonium phosphate precipitate out of urine as a result of higher than normal concentrations. For example, increased levels of calcium in the urine may result from increased levels of calcium circulating in the blood, caused perhaps by overactive parathyroid glands (glands in the neck that produce parathormone, a hormone that increases calcium in the blood), excessive dietary intake of calcium, or the destruction of bone by disease. More generally, higher than normal concentrations of stone-forming substances in the urine may result from dehydration and the production of small amounts of concentrated urine.

Often beginning as just a speck in the renal pelvis, kidney stones grow over a period of years as more and more material is deposited on the outside of the original particle (Figure A). Some stones may attain diameters of 25 mm (1 inch). Before reaching anything near that size, however, many stones are flushed out of the kidneys and down the ureters to the bladder where they are expelled with urine out the urethra. These tiny stones rarely cause problems and people may not even be aware of their passing. Somewhat larger stones, however, may cause considerable pain during their journey through the urinary tract as their sharp edges gouge into the walls of the ureters and sometimes the urethra. The pain, called renal colic, comes in waves, each wave causing the person to double over or roll around. Renal colic may be localized in the back or pelvic area, often off to one side, depending on which ureter the stone is moving through. Nausea, vomiting, and chills accompany the pain. Even more serious problems arise when large stones become lodged in the kidneys or ureters, blocking the flow of urine. The high internal pressure that results from such an obstruction can seriously damage the nephrons and impair kidney function.

In the past, stones lodged in the kidneys or ureters were removed through major abdominal surgery, with recovery taking several weeks. Today, however, doctors use ultrasound (or shock wave) lithotripsy, a relatively painless technique in which ultrasound waves directed at the stones pulverize them into tiny pieces while the patient lies in a water bath. These tiny particles, about the size of sand grains, are painlessly passed in the urine. Because this new technique does not involve making an incision, recovery is more rapid and the procedure less costly than surgery. Occasionally, however, particularly large

kidney function. This condition typically develops over a few hours or days and is characterized by little output of urine. Causes of acute renal failure include nephrons damaged by severe inflammation or ingestion of a poison, low blood volume resulting from profuse bleeding, and obstruction of urine flow caused by kidney stones (see the Personal Concerns essay on kidney stones). **Chronic renal failure**, on the other hand, is a progressive and often irreversible decline in rate of glomerular filtration that occurs over a period of months or years. Declines in filtration rate during chronic renal failure result from the destruction of nephrons associated with kidney diseases such as polycystic disease, an inherited and progressive condition in which fluid-filled cysts and tiny holes form throughout kidney tissue. Few symptoms are apparent at the beginning of chronic renal failure. This is because, although nephrons lost to kidney disease cannot be replaced, the remaining nephrons enlarge and take over for those that have been destroyed. With time, however, the loss of nephrons becomes so severe that symptoms of decreased glomerular filtration rate appear. For example, increased levels of nitrogenous wastes are found in the blood. By end-stage renal failure, about 90% of nephrons have been lost, making necessary a kidney transplant or the use of artificial kidney machines (see the following discussion).

Renal failure, whether acute or chronic, has many consequences, among them (1) acidosis—a decrease in blood pH caused by the inability of kidneys to excrete hydrogen

Kidney stone

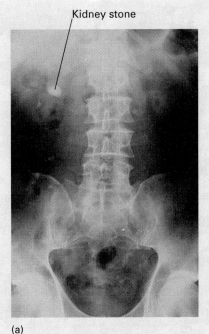

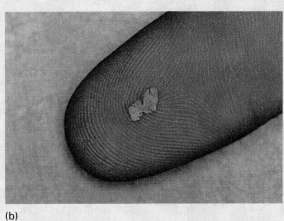

Figure A Kidney stones. (a) An x-ray image showing the location of a kidney stone. (b) Examples of kidney stones that have been removed through surgery. *(a, © 1990 Custom Medical Stock Photo; b, © Stephen J. Krasemann/Photo Researchers, Inc.)*

(a)   (b)

stones cannot be broken apart by lithotripsy and surgical removal is needed. Finally, certain stones, particularly those formed of uric acid, can be dissolved with drugs.

Who gets kidney stones? Kidney stones seem to run in families and are more common among people living in hot climates. In the tropics, people often do not take in enough fluid to counteract the large amounts of water lost through breathing and sweating, resulting in the production of small amounts of urine with high concentrations of stone-forming substances. In addition, men seem more susceptible to kidney stones than do women, and people more than 30 years of age are more likely to develop the disorder than are those less than 30.

What can be done to prevent formation of kidney stones? A simple solution is to drink large quantities of water, about 5 or 6 pints a day. This helps to keep the urine dilute and flushes any existing stones from the urinary tract. Although some people have suggested limiting dietary intake of calcium, at present there is no hard evidence to support the notion that reduced consumption of calcium-containing foods substantially reduces the incidence of kidney stones. Finally, people with a history of developing kidney stones can usually take specific medications to prevent formation of new stones once the old ones have been removed.

ions, (2) anemia—low numbers of red blood cells caused by failure of damaged kidneys to produce erythropoietin, (3) edema—build-up of fluid in the tissues because of water and salt retention, (4) hypertension—an increase in blood pressure caused by failure of the renin-angiotensin system and salt and water retention, and (5) accumulation of nitrogenous wastes in the blood. In short, failure of the kidneys severely disrupts homeostasis. In fact, when left untreated, kidney failure can lead to death within a few days. The cause of death is often cardiac arrest, brought on by disrupted rhythmic contractions of the heart, a result of high levels of potassium ions in the blood and tissue fluids.

A common way of coping with failure or severe impairment of the kidneys is **hemodialysis**, the use of artificial de-

vices to cleanse the blood. This process frequently involves use of the artificial kidney machine (Figure 15–8). Basically, a tube is inserted into an artery of the patient's arm and blood flows into the tube and on into the kidney machine where it is filtered and then returned to the body. (Sometimes a small amount of heparin, a drug that inhibits blood clotting, is added to the blood once the patient is connected to the machine.) Within the machine, the blood flows through tubing made of a selectively permeable membrane surrounded by a dialysis solution, the dialysate. The selectively permeable membrane permits wastes and excess small molecules to move from the blood into the dialysate, but it prevents the passage of blood cells and most proteins, which are too large. Nutrients are sometimes provided in the

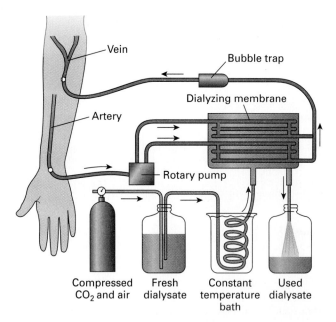

Figure 15–8 An artificial kidney machine used to cleanse the blood when kidneys fail. The path of the blood is shown in red and the path of the dialysate is shown in gold.

tients requiring hemodialysis typically undergo the procedure about three times a week, each session lasting several hours. Although in past years people went to hospitals for hemodialysis, today some patients dialyze themselves at home. For home dialysis to work, the patient must be highly motivated to learn and follow the precise routine, have a small room that can be devoted to the kidney machine and its associated paraphernalia, and have a friend or relative close by to help with setting up and coming off the machine. Another advance in hemodialysis is the miniature portable kidney. This small artificial kidney is available for those who wish to be away from home or the hospital for 2 to 3 weeks, thus making vacations possible.

Another option for removing wastes from the blood is **continuous ambulatory peritoneal dialysis (CAPD)**. In this procedure, the peritoneum, one of the body's own selectively permeable membranes, is used as the dialyzing membrane (Figure 15–9). The peritoneum lines the abdominal cavity and covers the internal organs. Dialyzing fluid held in a plastic container suspended over the patient flows down a tube inserted into the abdomen. As the fluid bathes the peritoneum, wastes move from the blood vessels that line the abdomen across the peritoneum and into the solution that is then returned to the plastic container and discarded. Typically, fluid is passed into the abdomen, left for a few hours, and is then removed and replaced with new fluid. Although CAPD involves about three to four fluid changes each day, the patient is free to move around between fluid changes while dialysis proceeds internally. CAPD can be done in either the hospital or at home.

dialysate and these move into the blood. The composition of the dialysate is precisely controlled to maintain proper concentration gradients between the solution and the blood. Once the blood has completed the circuit through the tubing, it is returned, free of wastes, to a vein in the arm. Pa-

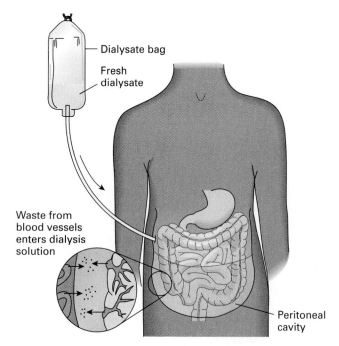
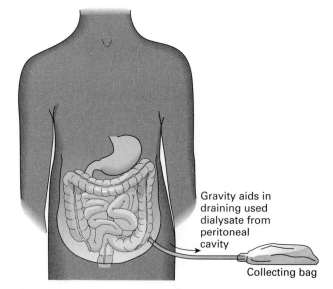

Figure 15–9 Continuous ambulatory peritoneal dialysis uses the peritoneum (the membrane that lines the abdominal cavity and covers the organs within) as the dialyzing membrane.

Compared with hemodialysis, CAPD is simpler to learn, less costly, and gentler in its waste removal. However, because CAPD requires several daily changes of dialysis fluid (each change requiring hookup of a new container of dialysate), there is ample opportunity for bacteria to move down the tube and into the abdomen where they may cause peritonitis, inflammation of the peritoneum. Even the most meticulous patients suffer at least one episode of peritonitis a year. Each episode requires hospitalization and treatment with an antibiotic, and as bouts with peritonitis accumulate, the peritoneum can become so scarred as to make CAPD impossible.

The ultimate hope for many people with end-stage renal failure is to receive a healthy kidney donated by another person. The kidney was the first organ to be successfully transplanted from one individual to another, and availability of dialysis was critical to the development and success of kidney transplantation techniques. In short, dialysis keeps people alive until a suitable donor organ can be found. Indeed, progress in the techniques of liver and heart transplantation has been much slower because of the difficulty in providing temporary mechanical replacements for these organs.

What makes a suitable donor organ? Basically, the main obstacle to successful acceptance of a transplanted organ comes from the patient's own immune system that rejects foreign tissue. Thus, the most suitable kidney would come from a patient's identical twin, with one from a close relative being the next best choice (removal of a healthy kidney is a safe operation, and parents or siblings of the patient who give

up one of their kidneys can get along fine with their remaining one). About 90% of kidneys donated by close relatives are still functioning 2 years after transplantation. Most donated kidneys, however, come from individuals unrelated to the patient who died suddenly in accidents and who agreed before their death to donate the organs (such decisions can also be made by surviving family members). About 75% of kidneys donated from unrelated individuals matched as closely as possible to the patient's tissue and blood type are still functioning 2 years after transplantation. The high success rate of kidney transplants is linked to the use of cyclosporin, a drug that suppresses the body's immune system and thus inhibits rejection of transplanted organs. Discovered in the 1970s, cyclosporin has revolutionized organ transplantation. In most situations, the kidneys of the patient needing a transplant are not removed, and the donor kidney is simply transplanted to a protected area within the pelvis (Figure 15–10). Finally, even if a transplant fails after a few years, successful second and third transplants may extend a patient's life.

Despite major advances in kidney transplantation, there is still room for improvement. For example, donor kidneys can only be kept alive and healthy for about 24 hours, necessitating rapid location of a recipient, shipment of the donor kidney, and transplantation (Figure 15–11). Perhaps most importantly, however, kidneys available for transplantation are always in short supply, causing many patients to wait for

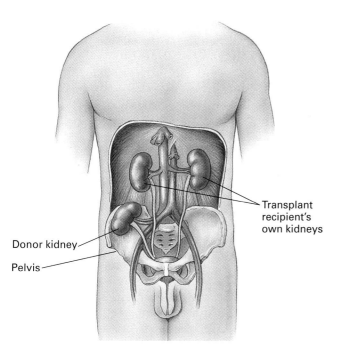

Figure 15–10   In most transplant situations, the donor organ is located in a safe region within the pelvis and the recipient's own kidneys are left in place.

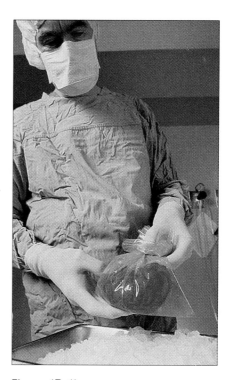

Figure 15–11   Before transplantation, donor kidneys must be kept in a cool salt solution under sterile conditions. Even under such conditions, kidneys will deteriorate after about 1 day.   *(© Will and Deni McIntyre/Science Source/Photo Researchers, Inc.)*

months or years for a new kidney. In fact, only about 10% of all suitable organs are made available for transplantation. We can all help lessen the shortage of organs available for transplantation by carrying an organ donor card that authorizes the use of our organs after death. At the very least, we can make our wishes regarding organ donation known to family members.

Social Issue

The constant shortage of donor organs has led some people to suggest a change in the laws regarding organ donation. At present, a person must give permission to use his organs after death, or if a potential donor is incapable of giving such permission, then medical personnel may consult close relatives. Recently, however, individuals concerned about severe shortages of organs have suggested changing the laws so that a person would have to provide written proof that they *did not* want their organs made available for transplantation after death. In other words, without anything in writing, the law would presume that the person had no objection to organ donation. Because many people who probably have no objections to organ donation are healthy and simply do not think about death and future use of their organs, it is likely that such a change in law would go a long way to solving organ shortages. There are, however, some concerns. Would such a system promote an image of organ transplantation as body snatching? Would relatives claim that doctors and nurses let their loved one die because his organs were needed by someone else? Would you recommend changing the laws regarding organ donation?

Ureters

Urine formed by nephrons moves into collecting ducts and on into calyces and the renal pelvis. Urine exits the kidneys at the renal pelvis, moving through tubular organs called ureters that carry it along their 30 cm length (12 inches) to the urinary bladder. We have two ureters, one for each kidney, and both enter the bladder from behind (see Figure 15-1). A small flap of membrane, not a true valve, borders the opening between each ureter and the bladder. As urine fills the bladder, increasing pressure causes the flaps to close, preventing back flow of urine from the bladder up into the ureters. Failure of this flap to close may allow bacteria associated with bladder infections to travel up the ureters and into the kidneys where they may cause serious harm.

The walls of the ureters have three layers. The innermost layer is a mucous membrane. Mucus secreted by this layer protects underlying cells from the urine that flows by. (Because the fluid within cells that line the ureters differs from urine in its pH and concentration of dissolved substances, such protection is needed.) The middle layer, called the muscularis, contains smooth muscle whose contractions

help propel urine through the ureters to the bladder (gravity also plays a role). The outermost layer is fibrous connective tissue that anchors each ureter in place and protects underlying layers. We see that the structure of the ureters is well-suited to their function, transport of urine from the renal pelvis of the kidneys to the urinary bladder.

Urinary Bladder

The bladder is a muscular sac-like organ that receives urine from the two ureters and temporarily stores it until release into the urethra (Figure 15-12). Located in the pelvic cavity, just behind the pubic bone, the bladder changes shape as it fills with urine. Empty, it is a collapsed, shriveled sac. As it fills, however, the bladder becomes spherical and eventually pear-shaped. A somewhat full bladder holds about 500 ml (1 pint), although up to 1000 ml can be stored if necessary.

The wall of the urinary bladder has four layers (seen in Figure 15-12). Like the ureters, the bladder's innermost layer is a protective mucous membrane. This membrane contains transitional epithelium, a tissue that is able to stretch, thus enabling the bladder to expand when filling with urine and to contract when urine is released to the urethra. The next layer consists of connective tissue that supports the inner membrane. External to the connective tissue layer is a muscular layer called the **detrusor muscle**. This layer contains smooth muscle whose contraction plays a role in the process of urination. Finally, with the exception of its top surface that is covered by the peritoneum, the rest of the bladder is covered by a layer of fibrous connective tissue that is continuous with the same layer on the ureters.

Inside the bladder are three openings, one for each ureter and one for the urethra (see Figure 15-12). These three openings outline a triangular region called the **trigone**. Compared with other areas within the bladder, this region has a smooth appearance and is a common site for urinary tract infections.

Urethra

The urethra, the end-of-the-line for the urinary system, is a muscular tube that transports urine from the floor of the urinary bladder to the outside of the body (as seen in Figure 15-12). Two muscular sphincters surround the urethra and play a role in urination. The first, the **internal urethral sphincter**, is a thickening of smooth muscle at the junction of the bladder and the urethra. This sphincter is involuntary and keeps urine from flowing into the urethra while the bladder is filling. The second sphincter, the **external urethral sphincter**, is a little further along the urethra. This sphincter is made of skeletal muscle and is therefore voluntary, helping to stop the flow of urine down the urethra should we wish to wait for a more opportune time to urinate.

The urethras of males and females are quite different in length (as seen in Figure 15-12). In females, the urethra is 3 to 4 cm long (1.5 inches), lies in front of the vagina, and

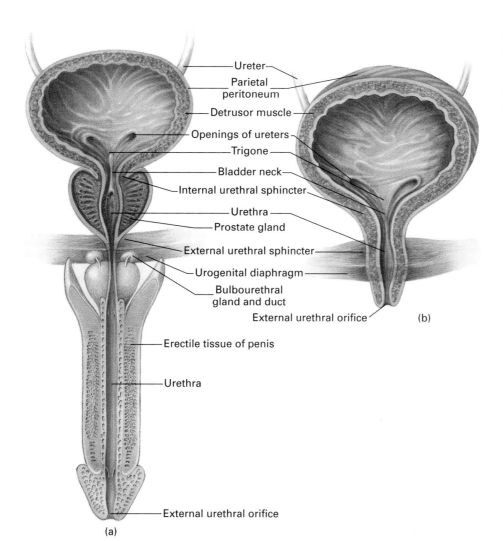

Ureter
Parietal peritoneum
Detrusor muscle
Openings of ureters
Trigone
Bladder neck
Internal urethral sphincter
Urethra
Prostate gland
External urethral sphincter
Urogenital diaphragm
Bulbourethral gland and duct
External urethral orifice
(b)

Erectile tissue of penis

Urethra

External urethral orifice
(a)

opens to the outside at the external urethral orifice located between the vaginal opening and the clitoris. In males, the urethra is about 20 cm long (8 inches), running from the bladder to the external urethral orifice at the tip of the penis. When the male urethra first emerges from the bladder, it is surrounded by the prostate gland. This gland contributes substances to semen and may become enlarged due to infection, causing painful urination and possibly obstructing the flow of urine. Finally, unlike the female urethra that only transports urine, the male urethra carries urine and reproductive fluids as well. The reproductive function of the male urethra is discussed in Chapter 16.

Healthy urinary tracts contain no microorganisms, and normal urine is sterile upon leaving the body (in fact, in the absence of conventional disinfectants, urine can be used as an emergency antiseptic). However, infectious microorganisms, particularly bacteria, can thrive in the urinary tract once introduced there. Although some bacteria may arrive at the kidneys by way of the bloodstream, most gain access to the urinary tract by moving from outside the body and up the urethra. Interestingly, the difference in the lengths of male and female urethras has a profound impact on suscep-

tibility of the sexes to urinary tract infections. The shorter urethras of females makes them more susceptible than males to such infections, because the bacteria need only travel a short distance from the external urethral orifice through the urethra to the bladder. In addition, because the external urethral orifice is closer to the anus in females than in males, improper (from back to front) wiping after defecation can easily carry fecal bacteria to the urethra. Bacteria may also enter the urethra during sexual intercourse, and this is why women are often advised to urinate soon after sex, essentially flushing any bacteria from the lower urinary tract.

Bacteria, whether introduced following defecation or sexual intercourse, may infect only the urethra, causing inflammation and a condition known as **urethritis**. Sometimes, however, the bacteria travel further up the urinary tract, causing **cystitis**, inflammation of the urinary bladder. Bacteria may also be introduced into the bladder by a catheter, a rubber or plastic tube used by medical personnel for a variety of purposes, including removal of an obstruction or draining urine from the bladder. Although the process of inserting a catheter does pose a risk of infection,

(*Text continued on page 424*)

Urinalysis: What Your Urine Says about You

It is often said that much can be learned about people by examining their handwriting, meeting their parents, or looking into their eyes. However, probably none of these things can reveal as much about someone as can urinalysis, an analysis of the volume, microorganism content, and physical and chemical properties of the person's urine. Ancient Greek and Roman physicians knew this to be true, and they routinely studied the urine of their patients when diagnosing ailments. Today, urine is still under intense scrutiny. Indeed, most of us, at one time or another, have been handed a small cup by medical personnel and asked to provide a urine sample. In this essay we will consider what a urine sample reveals about us.

Although daily urine volume varies considerably, most people expel between 1000 and 2000 ml (about 1 to 2 quarts) of urine each day. Factors such as diet, temperature, and blood volume and pressure influence urine volume. For example, high blood volume and pressure cause decreased reabsorption of water in the renal tubules of nephrons, resulting in increased urine output. Thus, when used in conjunction with other tests, an analysis of urine volume may indicate high blood pressure. Typically, volume is not measured at the time of providing a routine urine sample. It is only when a problem with the urinary tract is suspected that people are asked to monitor their urine output over a 24-hour period. Such monitoring may reveal polyuria, excessive urine production (well over 2 liters per day), or oliguria, inadequate urine production (less than 500 ml per day). Failure to produce any urine, an extreme health crisis, is called anuria.

Upon exiting the body, healthy urine is sterile, containing no microorganisms. The presence of bacteria in a properly collected urine sample usually signals infection of the urinary tract, although some bacteria may enter a urine sample during improper collection. For example, if the perineal area (in males, the area between the anus and the scrotum, and in females, the area between the anus and the vulva) has not been cleaned and the cup or urine is allowed to contact the area around the anus or vagina, bacteria may be introduced. Bacteria found in a sample may be cultured to determine their identity, and this information can then help in diagnosing the particular infection. Urine may also be screened for fungi or protozoans that cause inflammation within the urinary or reproductive tracts.

The following physical characteristics are routinely checked in urine samples: color, turbidity, pH, and specific gravity. The color of urine ranges from straw to yellow to amber, but it varies somewhat according to diet (beets lend urine a red color whereas asparagus cause a green tinge) and concentration (more concentrated urine, such as that collected first

thing in the morning, is darker in color). An abnormal color of urine—particularly red, when followed by microscopic confirmation of the presence of red blood cells—can indicate kidney stones and inflammation of or trauma to urinary organs. Because menstrual blood can contaminate urine samples, women should always inform their doctors if they are menstruating at the time of urine collection. Freshly voided urine is usually transparent. Cloudy or turbid urine may indicate a urinary tract infection. Healthy urine has a pH of about 6, although considerable variation may occur in response to diet (values ranging from 4.6 to 8.0 are not cause for alarm). Vegetarian diets, for example, produce alkaline urine and high-protein diets acidic urine. An alkaline pH is also associated with some bacterial infections. Urine color and turbidity are usually determined without specialized equipment,

Table 15A Some Abnormal Constituents of Urine

Abnormal Constituent	Clinical Implications
Albumin (plasma protein)	Increased permeability of filtration membranes of glomerulus and Bowman's capsule caused by high blood pressure, injury, or kidney disease
Bile pigments	Liver disease (cirrhosis, hepatitis) Obstruction of bile ducts
Glucose	Diabetes mellitus
Red blood cells	Inflammation of urinary tract from disease Kidney stones Tumors Trauma
White blood cells	Infection of urinary tract

and pH can be checked by recording color changes in test strips dipped in the urine sample. The specific gravity of a substance is the ratio of its weight to the weight of an equal volume of distilled water. Commonly known as density, specific gravity is used as a measure of the concentration of solutes in urine— basically, the higher the specific gravity, the higher the concentration of solutes. Thus, specific gravity provides a measure of the concentrating ability of the nephrons. When urine becomes highly concentrated for long periods of time, substances like calcium and uric acid may precipitate out, forming kidney stones (see the Personal Concerns essay on kidney stones). Specific gravity is measured with an instrument called a urinometer or densitometer, and samples are sometimes spun in a centrifuge to separate residue from fluid. The residue can then be examined under a microscope.

In the past, it was not unusual for physicians to taste urine to assess its sweetness or saltiness. Today, however, quantitative chemical tests are run to assess the levels of specific chemical constituents of urine. We know, for example, that water is the major constituent of urine, making up about 95% of its total volume. Solutes, from outside sources (such as drugs) or the metabolic activities of our cells, make up the remaining 5%. Tests run on urine specimens highlight abnormal constituents of urine (Table 15A) and normal constituents present in abnormal amounts (Table 15B).

The kidneys are our body's filtering system, and as such, their product, urine, contains substances that originate from the work of almost all of our organs. This is why a detailed analysis of the constituents of urine tells us not only how the organs of the urinary tract are functioning but also about the general health of our other organ systems. Taken together, then, a complete urinalysis involving measures of volume, microorganisms, and the physical and chemical properties of urine provides a basic check on our health.

Table 15B Some Normal Constituents of Urine and Their Clinical Implications when Present in Abnormal Amounts

Constituent	Clinical Implications
Ca^{++}	Values depend on dietary intake Increases signal overactive parathyroid glands, cancer of breasts or lungs Decreases signal vitamin D deficiency, underactive parathyroid glands
Cl^-	Values depend on dietary salt intake Increases signal Addison's disease (undersecretion of glucocorticoids by adrenal cortex), dehydration, starvation Decreases signal diarrhea, emphysema
Creatinine	Increases signal infection Decreases signal muscular atrophy, anemia, kidney disease
K^+	Increases signal chronic renal failure, dehydration, starvation, Cushing's syndrome (oversecretion of glucocorticoids by adrenal cortex) Decreases signal diarrhea, underactive adrenal cortex
Na^+	Values depend on dietary salt intake Increases signal dehydration, starvation, low blood pH from diabetes Decreases signal diarrhea, acute renal failure, emphysema, Cushing's syndrome
Urea	Increases signal high protein intake Decreases signal impaired kidney function
Uric acid	Increases signal gout (excessive uric acid in blood causes crystals to form in joints, soft tissues, and kidneys), leukemia (cancer of white blood cells), liver disease Decreases signal kidney disease

use of these devices can be lifesaving. Sometimes bacteria do not stop at the bladder and continue to move up the urinary tract, through the ureters, and into the kidneys. Once in the kidneys, bacteria may cause **pyelitis**, inflammation of the renal pelvis and calyces, or **pyelonephritis**, inflammation of the nephrons and renal pelvis.

Symptoms of urinary tract infections include fever, blood in the urine, painful and frequent urination, bedwetting in young children, lower abdominal pain (if the bladder is involved), and back pain (if the kidneys are involved). Such infections, usually diagnosed by checking the urine for bacteria and blood cells, can be treated with antibiotics. Inflammations of the lower urinary tract should be treated immediately and the prescribed antibiotics taken for their full term, to prevent the spread of infection to the kidneys where very serious damage can occur. Many disorders of the urinary tract, including infection, can be diagnosed by a detailed examination of the urine as described in the Personal Concerns essay on urinalysis.

Table 15–2 summarizes the components of the urinary system and their functions.

Urination

Urination, also called voiding or micturition, is the process by which the urinary bladder is emptied. This process involves both involuntary and voluntary actions. Recall that

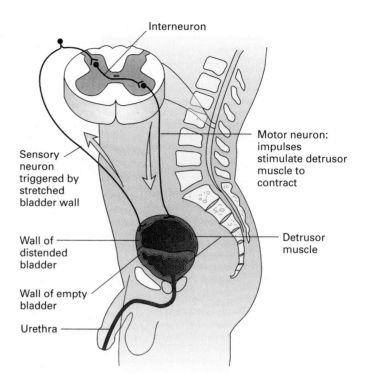

Figure 15–13   Pathway of the urination reflex. The male bladder is shown when empty and full.

Table 15–2 Components of the Urinary System and Their Functions

Component	Functions
Kidneys	Filter wastes and excess materials from the blood
	Help regulate blood pressure and pH
	Maintain fluid balance by regulating the volume and composition of blood and urine
	Secrete erythropoietin to stimulate production of red blood cells
	Participate in synthesis of calcitriol, the active form of vitamin D
Ureters	Transport urine from kidneys to urinary bladder
Urinary bladder	Stores urine
	Contracts and expels urine into urethra
Urethra	Transports urine from urinary bladder to outside the body
	In males, transports semen to outside the body

urine, produced around the clock by the kidneys, trickles down the ureters to the bladder where it is temporarily stored. When at least 200 ml of urine have accumulated, stretch receptors in the wall of the bladder send impulses along sensory nerves to the lower part of the spinal cord (Figure 15–13). From the spinal cord, impulses are sent along motor nerves back to the bladder where they cause the detrusor muscle in the wall of the bladder to contract and the internal urethral sphincter, located at the junction of the bladder and urethra, to relax. These combined actions push stored urine into the urethra. At about this time we feel the need to urinate (more specifically, upon arrival of sensory impulses in the lower spinal cord, information travels up to the brain and a few moments later initiates the desire to void). If the time for urination is appropriate, the brain permits voluntary relaxation of the external sphincter muscle and urine exits the body. (The external sphincter is usually kept in a constant state of contraction by impulses from the brain and spinal cord, and thus it is only when such impulses are inhibited that the muscle relaxes, letting urine flow). If, however, we deem the time to be inappropriate for urination, the brain does not permit the external sphincter to relax and we can, within reason, store the urine until a better time. Later, when we allow the external sphincter to relax, urine exits the body. Any urine left in the bladder is pushed out by further contractions of the detrusor muscle at the time of voiding. Then the detrusor muscle relaxes, allowing the bladder to once again fill with urine. Although urination is a reflex (a relatively rapid response to a stimulus that is mediated by the nervous system), it can be started and stopped voluntarily because of the control exerted by the brain over the external urethral sphincter.

Not everyone can control their external urethral sphincter. A lack of voluntary control over urination is called **urinary incontinence**. Incontinence is the norm for infants and children less than 2 or 3 years old, because nervous connections to the external urethral sphincter are incompletely developed. The very young, then, do not have conscious control over urination, and voiding occurs whenever the bladder fills with enough urine to activate its stretch receptors. Toilet training, a phase often dreaded by parents, occurs when toddlers learn to bring urination under conscious control, a step made possible by the development of complete neural connections to the external sphincter (Figure 15–14). In adults, incontinence may occur as a result of many conditions, including damage to the external sphincter (in men, often caused by surgery on the prostate gland), disease of the urinary bladder, and spinal cord injuries that disrupt the pathways along which travel impulses related to conscious control of urination. In any age-group, urinary tract infection can result in incontinence.

Mild incontinence, especially of the form called stress incontinence, is fairly common in adults. **Stress incontinence** is characterized by the escape of small amounts of urine when sudden increases in abdominal pressure, perhaps caused by laughing, sneezing, or coughing, force urine past the external sphincter. This condition is common in women, particularly following childbirth, an event that may stretch or damage the external sphincter making it less effective in controlling the flow of urine. Such damage can be minimized or repaired by doing exercises designed to strengthen the external sphincter and muscles along the floor of the pelvic cavity.

Urinary retention is the failure to completely or normally expel urine from the bladder. This condition may result from lack of sensation to urinate, such as might occur temporarily after general anesthesia, or from contraction or obstruction of the urethra, a condition caused in men by enlargement of the prostate gland. Immediate treatment for retention usually involves use of a catheter to drain urine from the bladder.

Figure 15–14 Conscious control over urination is usually acquired by the age of 3 when neural connections to the external sphincter muscle are completely developed. *(© Margaret Miller/Photo Researchers, Inc.)*

SUMMARY

1. Two kidneys, two ureters, one urinary bladder, and one urethra make up the urinary system. The main function of this system is to regulate the volume, pressure, and composition of the blood.

2. The kidneys are located just above the waist against the back wall of the abdominal cavity and are covered by three layers of connective tissue: the renal fascia, the adipose capsule, and the renal capsule, moving from outside to the inside. These layers support and protect the kidneys.

3. Each kidney has an inner region, the renal medulla, where cone-shaped structures called renal pyramids are found. The outer portion of each kidney is the renal cortex. Parts of the renal cortex, called renal columns, extend between the renal pyramids.

4. Properly functioning kidneys are critical to maintaining homeostasis. These organs filter wastes and excess materials from the blood, help regulate blood pH, maintain fluid balance by regulating the volume and composition of blood and urine, produce the hormone erythropoietin necessary for red blood cell production, and convert vitamin D into an active form that promotes the body's absorption and use of calcium and phosphorus.

5. The functional units of the kidneys are nephrons, tiny tubules responsible for formation of urine. There are two basic parts to a nephron. The first is the renal corpuscle that consists of a tuft of capillaries, the glomerulus, and a surrounding cup-like structure called Bowman's capsule. The second part of a nephron is the renal tubule, which contains three sections: the proximal convoluted tubule, the loop of Henle, and the distal convoluted tubule. The distal convoluted tubules of several nephrons empty into a single collecting duct. Collecting ducts are located within renal pyramids, and from there the urine moves into a cup-like cavity called a calyx and on into a much larger cavity, the renal pelvis. From the renal pelvis, urine leaves the kidneys, traveling down the ureters to the urinary bladder where it will be stored until expelled out the urethra.

6. Most nephrons have small loops of Henle and are restricted almost entirely to the renal cortex. About 20% of our nephrons, however, have long loops of Henle that extend into the renal medulla. Nephrons with well-developed loops of Henle are called juxtamedullary nephrons and are responsible for the water-conserving ability of the kidneys.

7. The renal arteries bring blood to the kidneys. After repeated branching of arterial vessels within the kidney, blood eventually moves through afferent arterioles into glomerular capillaries. High pressure within glomerular capillaries forces water, ions, and small molecules in the blood across the filter (the endothelium of the capillaries, the basement membrane, and the inner cell layer of Bowman's capsule) into Bowman's space within the nephron. Large molecules and formed elements of the blood normally cannot cross the filter. Filtration is the process by which only certain substances are allowed to pass out of the blood and into the nephron.

8. Once in Bowman's space, the filtrate—consisting of water, ions, and small molecules filtered from the blood—moves into the renal tubule of the nephron. As the filtrate passes through the renal tubule, about 99% of it is returned to the blood in a process known as reabsorption. Basically, almost all the water, ions, and nutrients in the filtrate are returned to the blood. Wastes are either partially reabsorbed or not reabsorbed at all. Most reabsorption occurs in the proximal convoluted tubule, although all parts of the renal tubule participate.

9. Any wastes or excess essential ions that may have escaped glomerular filtration are removed from the blood of capillaries surrounding nephrons and added to the filtered fluid that will become urine. This process, known as secretion, occurs along the proximal convoluted tubule, loop of Henle, and collecting ducts. In addition to helping in waste removal, secretion also plays a role in regulating blood pH by removing excess hydrogen ions from the blood.

10. Kidney function is influenced by antidiuretic hormone, aldosterone, atrial natriuretic factor, and the renin-angiotensin system. Antidiuretic hormone (ADH), produced by the hypothalamus and released by the posterior pituitary, increases water reabsorption in the final portions of the distal convoluted tubule and collecting ducts of nephrons. Aldosterone, a steroid hormone secreted by the adrenal cortex, increases sodium reabsorption by the distal convoluted tubules and collecting ducts. Because water follows sodium, aldosterone indirectly increases water reabsorption by nephrons. Atrial natriuretic factor (ANP), a hormone released by cells of the right atrium of the heart, causes decreased water and solute reabsorption by the kidneys. The renin-angiotensin system is a series of steps by which the protein renin is converted to an active form, an-giotensin II, that stimulates (1) the adrenal cortex to release aldosterone, (2) the posterior pituitary to release ADH, (3) the thirst center in the hypothalamus to increase water intake, and (4) increases in glomerular filtration rate.

11. Juxtamedullary nephrons enable our kidneys to conserve water. Such nephrons maintain gradients in solute concentration of the interstitial fluid within the kidneys. From the cortex to the medulla, the interstitial fluid increases in solute concentration. This interstitial gradient is maintained by the loop of Henle that deposits NaCl in the renal medulla and the collecting duct that allows urea to diffuse out to the inner renal medulla. Maintenance of the interstitial gradient enables large amounts of water to move out of collecting ducts by osmosis and leads to the production of concentrated urine.

12. Renal failure is a decrease or complete cessation of glomerular filtration. Such failure may be acute, occurring over a period of a few hours or days, or chronic, developing over a period of months or years. Failure of the kidneys severely disrupts homeostasis, leading to conditions such as acidosis, anemia, edema, hypertension, and a toxic build-up of nitrogenous wastes in the blood. Treatments for renal failure include hemodialysis (using artificial devices to cleanse the blood) and kidney transplantation.

13. There are two ureters, one for each kidney. The ureters are tube-like organs that transport urine from the kidneys to the urinary bladder.

14. The urinary bladder is a muscular sac that receives urine from the ureters and stores it until the urine is expelled to the outside of the body by way of the urethra.

15. The urethra is a muscular tube that transports urine from the floor of the urinary bladder to outside the body. The urethras of males and females differ. The male urethra is much longer than the female urethra and transports reproductive fluid in addition to urine (the female urethra only transports urine).

16. Urination, the process by which the urinary bladder is emptied, entails involuntary and voluntary actions. Moderate filling of the bladder stimulates stretch receptors in the wall of the bladder, and impulses are sent along sensory nerves to the lower spinal cord. From the spinal cord, impulses move along motor nerves back to the bladder where they cause it to contract and the internal urethral sphincter, a thickening of smooth muscle at the junction of the bladder and urethra, to relax. As a result of these actions, stored urine is pushed into the urethra. If the time for urination is appropriate, the brain then permits voluntary relaxation of the external urethral sphincter, a band of skeletal muscle further down the urethra, and urine exits the body.

REVIEW QUESTIONS

1. List the components of the urinary system and their functions.
2. Describe the external and internal anatomy of the kidney.
3. List the ways in which kidneys maintain homeostasis.
4. Describe the structure of a nephron.
5. Explain filtration, reabsorption, and secretion by nephrons.
6. How do nephrons contribute to the regulation of blood pH?
7. Explain how each of the following influence kidney function: antidiuretic hormone, aldosterone, atrial natriuretic factor, and the renin-angiotensin system.
8. How is water conservation accomplished by the kidneys?
9. Describe acute and chronic renal failure. What treatments are available for those with severely impaired kidneys?
10. Describe the structure of the ureters and urinary bladder.
11. How do the urethras of males and females differ? What are the clinical implications of these differences?
12. Describe the process of urination.
13. What is urinary incontinence? What is urinary retention?

Reproduction and Development

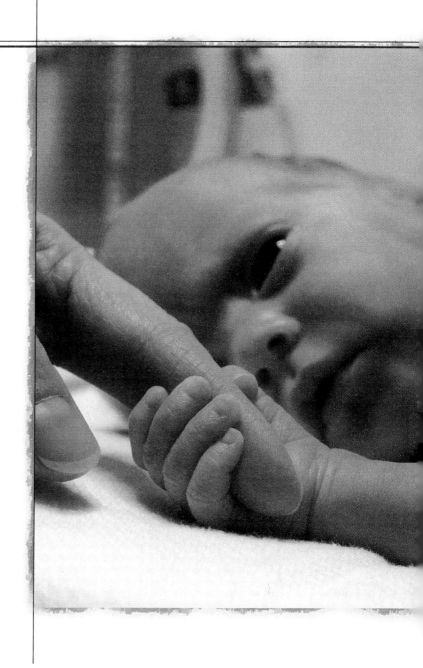

(© Henley and Savage/Tony Stone Worldwide)

Chapter 16

Reproductive Systems

(© Bachmann/Photo Researchers, Inc.)

In both the male and the female, the gonads serve two important functions: They produce the gametes (eggs and sperm—the cells that will form a new individual when they fuse) and they produce the sex hormones. In the male, the gonads are the **testes** and the gametes they produce are the spermatozoa (sperm). The male sex hormone produced in the testes is **testosterone**. The **ovaries** are the female gonads. The gametes they produce are the ova (singular, ovum). The female sex hormones they produce are **estrogen** and **progesterone**.

The Male Reproductive System

The male reproductive system consists of the testes, a system of ducts through which the sperm travel, the penis, and various accessory glands that produce secretions that help protect and nourish the sperm as well as providing a transport medium that aids the delivery of sperm to the outside of a male's body (Figure 16–1; Table 16–1).

The Testes

There are two testes, each about 5 cm (2 in.) by 2.5 cm (1 in.). The testes are located externally in a sac of skin called the **scrotum** where the temperature is several degrees cooler than it is within the abdominal cavity. The lower temperature is important in the production of healthy sperm.

Reflexes in the scrotum help to keep the temperature within the testes fairly stable. Cooling of the scrotum, such as might occur when a male jumps into frigid waters, for instance, triggers contraction of a muscle that pulls the testes closer to the warmth of the body. However, in a hot shower, the muscle relaxes and the testes hang low, away from the heat of the body. The skin of the scrotum is also amply supplied with sweat glands that help to cool the testes.

The testes begin development within the abdominal cavity on the back body wall, near the kidneys. Later, usually during the seventh or eighth month of fetal development, the testes descend, guided by strands of connective tissue. Each passes through a canal in the abdominal wall called the **inguinal canal** and enters the scrotum. Soon afterward, the inguinal canals are usually sealed with connective tissue.

In some male infants, about 3% of those born at full term and 30% of those born prematurely, the testes fail to descend before birth. This is not a cause for alarm because,

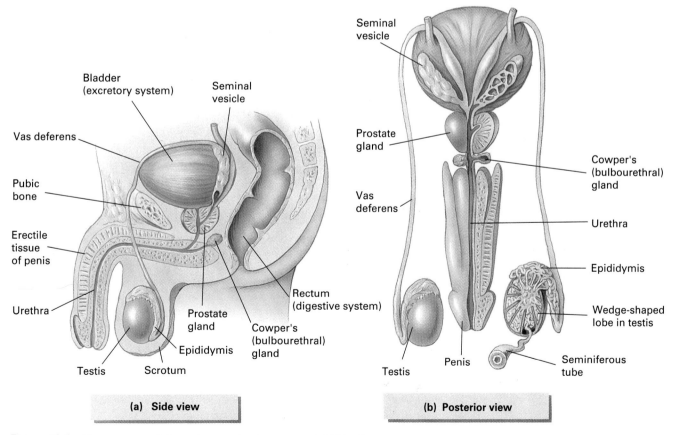

Figure 16–1   The anatomy of the male reproductive system. (a) Side view. (b) Posterior view.

Table 16–1 The Male Reproductive System

Structure	Function
Testes	Produce sperm and testosterone
Epididymis	Sperm storage and maturation
Vas deferens	Conducts sperm from epididymis to urethra
Urethra	Tube through which sperm leaves the body
Prostate gland	Produces secretions that make sperm mobile and that counteract the acidity of the female reproductive tract
Seminal vesicles	Produce secretions that make up most of the volume of semen
Cowper's glands	Produce secretions just before ejaculation; may lubricate; may rinse urine from urethra
Penis	Delivers sperm to female reproductive tract

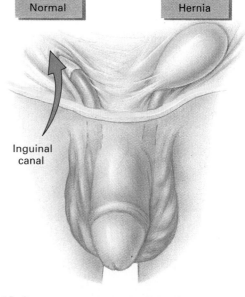

Figure 16–2 An inguinal hernia forms when a section of the intestines pushes into a weak area in the abdominal wall created when an inguinal canal, a narrow passageway through which a testis moved in its journey to the scrotum, fails to close properly.

even if they have not descended by birth, they usually descend on their own before the first birthday. However, in those cases when testes have not descended by the time the boy is 5- or 6-years-old, he faces two serious risks: sterility and an increased likelihood of developing testicular cancer. To reduce these risks, the testes can be made to descend with surgery or by administering certain hormones (for instance, human chorionic gonadotropic hormone, HCG).

Occasionally the inguinal canal doesn't close completely and a weak spot is left in the abdominal wall. Then, with an increase in pressure, such as might occur when lifting a heavy object, a section of the intestines may be pushed into the weak area of the abdominal wall, forming an inguinal hernia (Figure 16–2). (This is what doctors are looking for when they place a hand on a man's groin and say, "Turn your head and cough." Coughing increases abdominal pressure so that an inguinal hernia, if one exists, can be felt more easily.)

Sperm Production

Beginning at puberty, which usually occurs during the teenage years, a healthy male produces over 100 million sperm each day for the rest of his life. The sites of sperm production are the **seminiferous tubules**, which make up 80% of the mass of each testis. Their length is enormous. In fact, if they were placed end-to-end, they would stretch about a half mile. The seminiferous tubules lie within the testis, with one to four tubules in each of several hundred wedge-shaped compartments, or lobules (Figure 16–3a).

Figure 16–3 (a) The internal structure of the testis and epididymis. Resembling tangled masses ▶ of yarn, one to four highly coiled seminiferous tubules are packed into each of the several hundred compartments within each testis. The seminiferous tubules function as sperm factories, producing over a hundred million sperm each day. The sperm are then stored and mature within the epididymis. (b) A cross section of a seminiferous tubule showing the sequence of events, called spermatogenesis, that gives rise to the spermatozoa, or sperm. (c) As the cells that will become sperm develop, they are pushed from the outer wall of the tubule to the central canal, or lumen. The process of spermatogenesis involves meiosis, a type of cell division that begins with a cell containing two copies of every chromosome and ends with four cells each containing one copy of every chromosome, and spermiogenesis, structural changes in the cells. (d) The structure of a mature spermatozoon (sperm). The head contains the father's chromosomes, his genetic contribution to the next generation. Over the top of the head is a sac called the acrosome, which contains enzymes that will help digest a pathway through the layers surrounding the egg, assisting fertilization. The midpiece contains spirally arranged mitochondria that will provide metabolic energy to fuel the trip to the egg. The whip-like movements of the tail will propel the sperm. (b, © Professors P.M. Motta, K.R. Porter, and P.M. Andrews/Science Photo Library/Photo Researchers, Inc.; d, photo, © David M. Phillips/Visuals Unlimited)

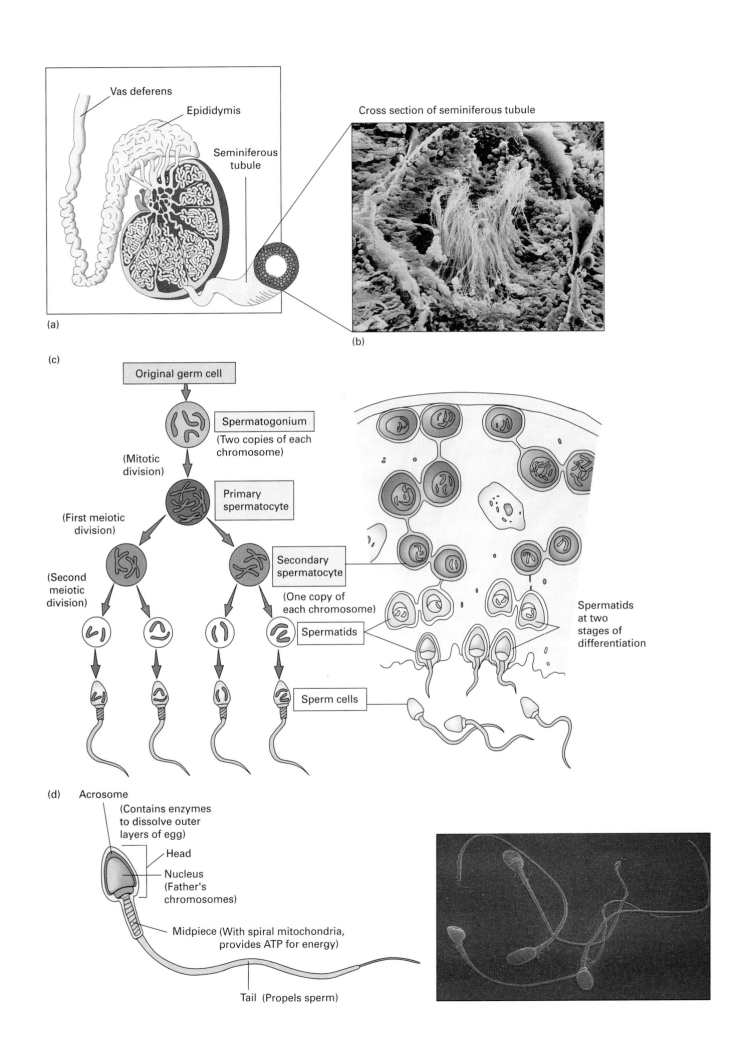

(a)

Vas deferens

Epididymis

Seminiferous tubule

Cross section of seminiferous tubule

(b)

(c)

Original germ cell

Spermatogonium (Two copies of each chromosome)

(Mitotic division)

Primary spermatocyte

(First meiotic division)

(Second meiotic division)

Secondary spermatocyte

(One copy of each chromosome)

Spermatids

Sperm cells

Spermatids at two stages of differentiation

(d)

Acrosome (Contains enzymes to dissolve outer layers of egg)

Head

Nucleus (Father's chromosomes)

Midpiece (With spiral mitochondria, provides ATP for energy)

Tail (Propels sperm)

The sequence of events within the seminiferous tubules that gives rise to sperm, called **spermatogenesis**, involves a change in the amount of genetic information within the cell as well as changes in the shape and functioning of the cells (Figure 16–3b and c). The alterations in genetic content occur during a special type of cell division, called meiosis, which reduces the number of chromosomes in a cell from two copies of each to only one copy of each. (The details of meiosis are described in Chapter 18.) Each of your body cells contains 23 pairs of chromosomes. One member of each pair came from your father and the other member from your mother. Each chromosome contains a portion of the instructions for making and maintaining your body. It is important, therefore, that one member of each pair (one complete set of instructions) be present in a gamete. However, it is equally important that only one member of each pair of chromosomes be in a gamete so that, when the egg and sperm nuclei fuse during fertilization, the newly created zygote (fertilized egg) does not have more than two copies of each chromosome.

The process of spermatogenesis begins in the outermost layer of each seminiferous tubule. It is here that undifferentiated germ cells called **spermatogonia** (singular, *spermatogonium*) develop. Each spermatogonium divides mitotically to produce two new spermatogonia. One of these remains at the periphery of the tubule and divides again to give rise to new spermatogonia. The other pushes deeper into the wall of the tubule, where it will enlarge and form a **primary spermatocyte**. The chromosomes within each primary spermatocyte replicate (make copies of themselves) and the copies remain attached to one another. Then the members of each pair of chromosomes separate. The result of the first meiotic division is two cells, called **secondary spermatocytes**, each containing one complete set of chromosomes (with a replicate of each chromosome still attached). The second meiotic division follows very quickly, forming two **spermatids** from each secondary spermatocyte. During this division the copies of each chromosome separate and each spermatid receives a single copy of each of the 23 chromosomes.

Although the spermatids have the chromosomal complement necessary for fertilization, numerous structural changes must still occur to create a cell capable of swimming to the egg and fertilizing it. During this process, called **spermiogenesis**, a spermatid is converted to a streamlined **spermatozoon**, or sperm cell, equipped to deliver the father's genetic contribution to the next generation.

The mature sperm cell has three distinct regions: the head, the midpiece, and the tail (Figure 16–3d). The head of the sperm is a flattened oval that contains little else besides the densely packed chromosomes. Positioned like a ski cap on the head of the sperm is the **acrosome**, a membranous sac containing enzymes. A few hours after the sperm have been deposited in the female reproductive system, the membranes of their acrosomes break down. The enzymes then spill out and digest the outer layers of the egg, assisting fertilization. Within the midpiece, mitochondria are arranged in a spiral. The mitochondria are the powerhouses of the cell that will provide energy in the form of ATP that is needed to fuel the movements of the tail. The tail contains contractile filaments. The whip-like movements of the tail propel the sperm during its long journey through the female reproductive system.

One cause of male infertility is a sperm count of less than 50 million sperm in each milliliter of semen. It may seem surprising that so many sperm are needed, when only one of them actually fertilizes the egg. However, it has been estimated that as few as 500 of the 150 million to 250 million sperm released into the female's body actually reach the egg. Why so few? Some of the sperm released are defective. For instance, some of these malformed cells may have two tails or a bent neck. In addition, some sperm are killed by the acidity of the female reproductive system and others are killed by roving defense cells that identify the sperm as intruders. Furthermore, the trip from the vagina, where sperm are deposited, to the fallopian tube, where fertilization usually occurs, is a very long one. To better appreciate the length of a sperm's journey, we can mentally enlarge all the structures to more familiar dimensions. If a sperm were enlarged to the height of an average man and the female reproductive tract were enlarged proportionally, the sperm's journey would be 8 miles long, a daunting adventure for even a human swimmer. Some sperm, seemingly normal in appearance, are simply not hardy enough to survive the trip. Of those that make it to the top of the uterus, roughly half enter the fallopian tube without the egg. During the last lap of the journey, the sperm must swim against a current created by the cilia that line the oviduct. Then, when they finally reach the egg, the enzymes of hundreds of acrosomes are needed to digest a pathway through the outer layers of the egg, enabling one sperm cell to penetrate and enter the egg.

Sperm counts are dropping in men throughout the world. In 1940, sperm counts were typically about 113 million per milliliter of semen (the fluid released when a man ejaculates). The average sperm count in 1990, however, was 66 million per milliliter of semen. Commenting on declining sperm numbers at a congressional hearing, Louis Guillette, a reproductive physiologist from the University of Florida, declared, "Every man in this room is half the man his grandfather was." There are many factors that could be playing a role in the sperm crisis. Pollutants such as polychlorinated biphcyls (PCBs) and DDT might be harmful to reproduction (see Environmental Concerns: *Environmental Estrogens*). Other researchers wonder whether the increasing incidence of certain sexually transmitted diseases, such as chlamydia, is to blame.

Hormone Production

The **interstitial cells**, which are located between the seminiferous tubules of the testis, produce the male steroid sex

Environmental Estrogens

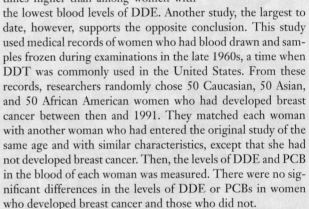

The world around us has become contaminated with dozens of synthetic chemicals that affect the activity of the sex hormones, estrogen and progesterone. Indeed, there are at least 45 environmental contaminants reported to cause changes in reproductive systems. These include pesticides, herbicides, fungicides, and industrial by-products. Some, including the pesticide DDT and bisphenol A (a breakdown product of plastics), mimic the effects of estrogen. Others, such as certain PCBs (polychlorinated biphenyls), block the effects of natural estrogens. Some pollutants bind to receptors for androgens and can feminize a male by blocking the action of natural androgens.

These hormone-modulating chemicals enter the environment when people put them there—the chemicals spew from incinerators or people spray them on plants to keep insects away or to kill undesirable plants. Before 1979, when their manufacture in the United States was banned, PCBs were used as insulators and coolants, as well as in plastics, adhesives, and paints. Like organochlorine pesticides, PCBs are very stable. When articles made from them were discarded, the PCBs soaked into the ground and were carried by groundwater to rivers, lakes, and oceans. The chemicals then passed along the food chain, from grazers to carnivore to carnivore, accumulating in body tissues at each level. Then, when people eat contaminated food, drink contaminated water, or breathe urban air, the chemicals enter their bodies. Once in the body, the estrogenic pollutants may threaten human reproductive health.

There is already evidence that chemicals with estrogenic properties, consumed as contaminants in food, can accumulate in humans. For example, seal and beluga whale blubber, which are loaded with DDT and PCBs, are a common component of the diet of certain Inuits living in northern Quebec. The breast milk of Inuit women who eat blubber is four to seven times higher in estrogenic pesticides than the breast milk of women in Quebec who don't eat blubber.

Sex hormones choreograph so many biological activities—development, anatomy, physiology, and behavior—that there is a myriad of ways that environmental estrogens could possibly wreak havoc with human reproduction. Some researchers argue that synthetic estrogenic chemicals cause cancers of the reproductive system, such as breast cancer and testicular cancer, as well as a variety of noncancer reproductive effects, including the development of female characteristics in males, decreased sperm counts, higher probability of endometriosis in women, and weakened immune systems. Although evidence of effects on other animal species is mounting, studies of effects on humans remain contradictory.

Because estrogen can fuel the growth of certain breast tumors, it seems reasonable to suspect that estrogenic pesticides that accumulate in the breast tissue might cause cancer. Studies on nonhuman animals produce clear results. Rats exposed to DDT have increased mammary tumors or testicular tumors, depending on their sex. Studies on humans, however, yield conflicting results. One study reported that the breast cancer rate among women with the highest blood levels of DDE (a breakdown product of the pesticide DDT) was four times higher than among women with the lowest blood levels of DDE. Another study, the largest to date, however, supports the opposite conclusion. This study used medical records of women who had blood drawn and samples frozen during examinations in the late 1960s, a time when DDT was commonly used in the United States. From these records, researchers randomly chose 50 Caucasian, 50 Asian, and 50 African American women who had developed breast cancer between then and 1991. They matched each woman with another woman who had entered the original study of the same age and with similar characteristics, except that she had not developed breast cancer. Then, the levels of DDE and PCB in the blood of each woman was measured. There were no significant differences in the levels of DDE or PCBs in women who developed breast cancer and those who did not.

The developing fetus is probably the most sensitive of all biological systems and, as organ systems take shape, the proper balance of sex hormones is critical. Too much estrogen at the wrong time can cause a genetically male fetus to look outwardly like a female, to have feminized sex organs, or to exhibit female behaviors later in life.

There are numerous examples of environmental estrogens feminizing males of many animal species. Many fish-eating birds, including bald eagles, terns, and gulls, with high levels of pesticides or PCBs have shown impaired reproductive functions. Consider, for example, the gulls in the Great Lakes region that managed to hatch from eggs with shells that had been thinned by DDT. Both males and females had grossly feminized reproductive systems. The behavior of the gulls was also odd. The male hormone testosterone is normally responsible for sex drive as well as aggressive behavior. These male gulls, however, showed little interest in mating and allowed the females to become the dominant members of the colony. Then, some female gulls began sharing a nest with other females. Researchers demonstrated that these effects were due to estrogenic pesticides by injecting pollution-free gull eggs with DDT. The young that hatched from them had the same array

(continued)

of reproductive abnormalities as those in the natural population that had been exposed to high DDT levels. These observations are consistent with the idea that the estrogenic effects of DDT chemically castrated the male gulls.

Alligators in Florida's fourth largest lake, Apopka, also have reproductive problems thought to be caused by estrogenic pesticides (Figure A). Their hatching rate is extremely low, 15% to 20%, whereas the hatching rate of alligator eggs in other Florida lakes is 70% to 80%. About half of the Apopka alligators that do hatch, die within 2 weeks. The males in this population have a ratio of estrogen to testosterone that is typical of a normal female and their penises are one-half to one-third the normal size. Apopka females have a ratio of estrogen to testosterone that is twice as high as normal. The problems of these alligators appear to be caused by an estrogenic pesticide, dicofol, that was produced by a lakeside chemical company. When dicofol was painted on alligator eggs from an unconta-

minated lake, the hormone levels of the hatchlings were almost identical to those in hatchlings from Lake Apopka.

The critics of the idea that human reproductive disturbances are caused by estrogenic pollutants argue that environmental estrogens are very weak, only one-hundredth to one-thousandth as strong as natural estrogen and the levels of environmental estrogens are very low compared with that of natural estrogen. Thus, the effects of environmental estrogens would be swamped by those of natural estrogen. Furthermore, the environment contains substances that act as estrogens and others that act as anti-estrogens. They may cancel one another out.

It is clear that we have polluted our environment with chemicals that have the potential of altering reproductive anatomy, physiology, and behavior. It is also clear that this pollution is already affecting a variety of animal species. What is not clear is how great a threat these chemicals are to humans or how we humans will decide to cope with a self-induced problem.

Figure A Alligators hatching near Lake Apopka, Florida, have reproductive problems caused by high levels of a pollutant that mimics the effects of estrogen. *(© Howard K. Suzuki, 1994)*

hormones, collectively called **androgens**. The most important androgen is **testosterone**, a hormone needed for sperm production and the maintenance of male reproductive structures.

The Duct System

Sperm produced in the seminiferous tubules next enter the **epididymis**, where they are stored and mature. The epididymis is a highly coiled tubule that spans the top and posterior surface of a testis. If it were uncoiled, the epididymis

would be 6 m (about 20 ft) long. Sperm that enter the epididymis look as if they are mature, but they cannot yet function as mature sperm. During their 2- to 3-week journey along the epididymis, the sperm become capable of fertilizing an egg and of moving on their own, although they do not yet do so. Sperm can be stored in the epididymis for about 1 month. Then they are either expelled through the penis during ejaculation of semen or they are reabsorbed.

Each **vas deferens** is a tube about 45 cm (18 in.) long that conducts sperm from the epididymis to the urethra. Some sperm may be stored in the part of the vas deferens closest to the epididymis. During ejaculation, sperm are propelled along the vas deferens by peristaltic waves of muscle contraction.

The **urethra** conducts urine from the urinary bladder or sperm from the vas deferens out of the body through the penis. Sperm and urine do not pass through the urethra at the same time.

The Accessory Glands

Interestingly, very little of the volume of semen is made up of sperm. Most comes from secretions of the accessory glands: the prostate gland, the paired seminal vesicles, and paired Cowper's (bulbourethral) glands.

About the size of a walnut, the **prostate gland** surrounds the urethra, just beneath the urinary bladder. Its secretions make up between 13% and 33% of the volume of semen. Prostate secretions are slightly alkaline and serve both to activate the sperm, making them fully motile, and to counteract the acidity of the female reproductive tract.

The prostate gland grows at different rates throughout a man's life. Growth is slow from birth until puberty. Then, spurred by increasing levels of testosterone, prostate growth is rapid during puberty. Growth continues, at a slower rate, until the man reaches 30 years of age. But then, between 30 and 45 years of age, there is little growth. Beginning in middle age, the prostate begins to enlarge. In fact, half of all men older than 50 have some degree of prostate enlargement, but this growth rarely causes problems before age 60. As the prostate enlarges, it often pushes against the urinary bladder, decreasing the amount of urine the bladder can hold and making it necessary to urinate more frequently. At the same time, however, the enlarged prostate may squeeze the urethra and restrict urine flow. Thus, urination for many older men is frequent and difficult. The standard treatment for an enlarged prostate is to insert an instrument through the urethra into the prostate and then to scrape out excess prostate tissue through the urethra. The enlargement of the prostate gland that accompanies aging is not related to prostate cancer, which is a growing concern for men.

The secretions of the **seminal vesicles** make up roughly 60% of the volume of semen. The secretion contains citric acid, fructose, amino acids, and prostaglandins.[1] Fructose is a sugar that provides energy for the sperms' long journey to the egg. Some of the amino acids coagulate the semen, which helps to keep the sperm within the vagina and also to protect the sperm within the clump from the acidic environment of the vagina. The prostaglandins serve to cut the viscosity of cervical mucus and to cause uterine contractions that assist the movement of sperm.

Cowper's glands (the bulbourethral glands) release a clear, slippery liquid immediately before ejaculation. The function of this secretion is not quite certain, but it may rinse the slightly acidic urine remnants from the urethra before the sperm pass through. Another idea is that it acts as a lubricant, easing the entrance of the penis into the vagina.

The Penis

The **penis** is a cylindrical organ whose role in reproduction is to deliver the sperm to the female tract. Its tip is enlarged, forming a smooth rounded head known as the glans penis, which has many sensory nerve endings and is important in sexual arousal. When a male is born, the glans penis is covered by a cuff of skin called the foreskin. The foreskin can be pulled back to expose the glans penis.

The surgical removal of the foreskin is called **circumcision**. Although 60% to 90% of American men are circumcised, only 20% of men in the rest of the world are. Circumcision is usually performed soon after birth. Although few people argue with the practice for religious reasons or because of cultural preference, some physicians question the medical value of circumcision.

The transfer of sperm to the female reproductive system is much easier when the penis is erect. An **erection**, which involves increases in the length, width, and firmness of the penis, is due to changes in the blood supply to the organ. Within the penis are three columns of spongy erectile tissue, which is a loose network of connective tissue with many empty spaces (Figure 16–4). During sexual arousal, the

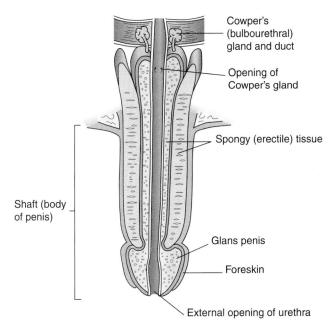

Figure 16–4   The structure of the penis. Three columns of spongy tissue fill with blood during sexual arousal, increasing the size of the penis. The engorged columns of spongy tissue push against their connective tissue casings, causing the penis to become erect.

[1]Prostaglandins are chemicals secreted by one cell that alter the activity of other cells.

Testicular and Prostate Cancers

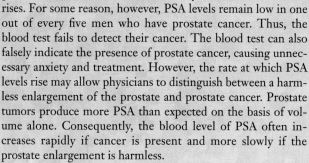

Testicular cancer affects only 1% of all men. It is, however, the most common form of cancer among men between the ages of 15 and 35 years and three-quarters of all the men who develop testicular cancer are younger than 50 years. It affects nearly 7000 American men each year and the incidence has been increasing worldwide over the past few decades. It is 30 to 50 times more likely in men whose testes did not descend into the scrotum or if they descended after 6 years of age. Because this cancer does not usually cause pain, it is important for every man to examine his testes each month to feel for a lump or a change in consistency. If a lump is felt, it is not necessarily cancerous because swellings in the epididymis are common. However, a physician should be consulted with the detection of any lump because the cure rate for tumors caught in the early stages is nearly 100%. Treatment usually involves removal of the diseased testis followed with radiation and chemotherapy. The healthy testis is usually left in place, and so the man is still potent and fertile.

Prostate cancer affects 1 of every 11 American males and kills 35,000 of them every year. The incidence of prostate cancer is rising and it is now the second highest cause of cancer deaths among men older than age 55. Part of the explanation for the increase is that men are living longer and prostate cancer usually affects older men.

There are two ways of detecting prostate cancer—a rectal exam and a blood test—and it is recommended that they be used together in the diagnosis. During a rectal exam for prostate cancer, a physician inserts a gloved finger into the rectum and, through the wall of the rectum, feels the prostate, which is located nearby (see Figure 16–1). A prostate gland that has merely enlarged with age feels diffuse and soft, but a cancerous prostate is firm or may contain a hard lump. One-third of the tumors detected by a rectal exam are already too large to be removed surgically. The blood test for prostate cancer measures the amount of a protein called prostate specific antigen (PSA). Because this protein is produced only by the prostate gland, the amount of PSA in the blood reflects the size of the prostate. As a tumor within the prostate grows, the blood level of PSA usually rises. For some reason, however, PSA levels remain low in one out of every five men who have prostate cancer. Thus, the blood test fails to detect their cancer. The blood test can also falsely indicate the presence of prostate cancer, causing unnecessary anxiety and treatment. However, the rate at which PSA levels rise may allow physicians to distinguish between a harmless enlargement of the prostate and prostate cancer. Prostate tumors produce more PSA than expected on the basis of volume alone. Consequently, the blood level of PSA often increases rapidly if cancer is present and more slowly if the prostate enlargement is harmless.

When prostate cancer is detected while it is still contained in the prostate, the chances of survival for at least 5 years are 95%. If the cancer has spread, the odds drop to a 50% chance of living another 2 or 3 years.

Surgical removal of the prostate is usually the first step in treating the cancer. In the past, this procedure often damaged nerves in the area, causing impotence and incontinence. Recently, however, the technique has improved greatly and those

arteries that pipe blood into the spongy tissue dilate (widen) and the spongy tissue fills with blood. This not only causes the penis to become larger it also causes the spongy tissue to squeeze shut the veins that drain the blood from the penis. As a result, the blood flows into the penis faster than it can leave, causing the penis to become larger and erect. Because the spongy tissue is enclosed in connective tissue sheaths, as the spongy tissue fills with blood, it pushes against the connective tissue casing, making the penis firm, like a water balloon.

Impotence, the inability to achieve or maintain an erection, generally leaves a man unable to have sexual intercourse. Affecting approximately 10 million men in the United States alone, it is hardly an uncommon problem. Indeed, it is not unusual for a man to experience impotence at some point in his life. There are a number of psychological causes for impotence, including worry, stress, a quarrel with the partner, and depression. However, there are also many physical causes. Nerve damage, such as often accompanies chronic alcoholism and sometimes diabetes, might be responsible. Since an erection depends on adequate blood supply, fatty deposits in the arteries (atherosclerosis) serving the penis can also cause impotence. Medications, especially certain drugs used to treat high blood pressure, antihistamines,

side effects can be prevented in most patients. Radiation therapy often follows surgery.

A more recent treatment is the implantation of radioactive pellets in the prostate. The implants may be as effective as surgery and radiation in treating early prostate cancer and they allow patients to resume activity the day after implantation.

Several experimental treatments for prostate cancer are on the horizon. Certain drugs, for instance, are now being tested that hold promise as nonsurgical treatments for an enlarged prostate. Because the male hormone testosterone stimulates the growth of prostate cancer, implants that inhibit the hormones that stimulate testosterone secretion are also being tested.

Although the frequency of surgery for prostate cancer that remains contained within the gland has increased dramatically in the past few years, it is not clear that treating prostate cancer results in a longer life. Most cases of prostate cancer grow very slowly. Thus, if a man is older than 70 at the time of diagnosis, he could easily die of some other cause before his prostate cancer kills him. Furthermore, the operation is certainly not risk-free, especially for older men. Indeed, nearly 2% of men of age 74 or older died within a month after surgery and almost 8% had major complications. Even men in their sixties with cancer contained in the prostate may add no more than a year to their life expectancy (often with a reduced quality of life) by aggressively treating their cancer. However, the picture is somewhat different for middle-aged men with prostate cancer. If a man in his forties or fifties is diagnosed with prostate cancer, it is likely that even a slow-growing tumor would have time to spread before the man died of some other cause and there is evidence that prostate cancer grows more rapidly in young men than in older ones. Consequently, it is often difficult to determine when the benefits of aggressive treatment will outweigh the risks.

Social Issue

The blood test for PSA often detects cancerous tumors in the prostate that are still microscopic. Nonetheless, routine screening for prostate cancer remains controversial. Because most prostate cancers grow slowly, some physicians worry that increased detection of small tumors will lead to risky treatment of men who would have died of some other cause before their prostate cancer killed them. On the other hand, early detection may save men from a painful death. How should the decision be made?

With prostate cancer on the rise, men might wonder how they can reduce their risk. One way might be to reduce the amount of red meat in the diet. Alpha-linolenic acid (a fat found in meat and butter) is associated with prostate cancer. Dietary fat increases the rate at which prostate cancer grows. Thus, dietary fat could cause a tumor that might otherwise have grown too slowly to cause problems to become life-threatening. Another way might be to move to a hot climate. It is thought that a vitamin produced as a result of exposure to sunshine may be protective since men who live in hot climates have a lower risk of prostate cancer. Also, a man considering having a vasectomy (a means of contraception in which each vas deferens is cut) might want to postpone his decision until the link between vasectomies and prostate cancer has been explored more thoroughly. Although several studies have shown that men who have had vasectomies have a higher rate of prostate cancer, it is not clear whether the vasectomy somehow *causes* the cancer or whether men who have had vasectomies are more likely to see a urologist and, therefore, are more likely to have their prostate cancer diagnosed.

antinausea and antiseizure drugs, antidepressants, sedatives, and tranquilizers, may cause the problem. Excessive alcohol consumption or marijuana use can also cause impotency.

The first step in treating impotency is to eliminate the cause of the problem, whenever this is possible. When it is not, as would be true if nerve damage were responsible, the most common treatment is a penile implant. Some implants are semirigid devices and other types are inflatable. A short-term solution is to inject drugs that cause blood vessels to dilate directly into the spongy tissue of the penis. Researchers have identified the chemical that triggers an erection (nitric oxide), bringing hope that someday many cases of impotence can be cured by noninvasive procedures, such as applying a patch or an ointment that contains a drug directly to the penis.

Hormones Involved with Male Reproductive Processes

Testosterone secretion by the interstitial cells of the testes is stimulated by **LH** (**luteinizing hormone**, which is sometimes called **ICSH, interstitial cell-stimulating hormone**). LH is secreted by the anterior pituitary gland, a pea-sized endocrine gland that lies at the underside of the brain.

With the onset of LH secretion at puberty, testosterone production begins and initiates many changes in the male's body. The male reproductive organs, including the testes and the penis, become larger. Testosterone also stimulates the growth of the long bones found in the arms and legs. It is, therefore, responsible for the growth spurt that accompanies puberty. Eventually, testosterone causes the closure of the growth regions of these bones and the young man stops growing taller. In addition, testosterone is responsible for the development and maintenance of the male secondary sex characteristics, which are features associated with "masculinity" but are not directly related to reproductive functioning. For example, the growth of muscles and the skeleton tends to result in wide shoulders and narrow hips. Patterns of hair growth begin to change. Pubic hair develops, as does hair under the arms. A beard appears, perhaps accompanied by hair on the chest. Meanwhile, the voice box enlarges and the vocal cords thicken, causing a deepening of the male voice. Testosterone also stimulates the activity of oil and sweat glands. Bacteria living on the skin are nourished by the secretions and the result can be acne and body odor.

Another hormone from the anterior pituitary gland, **FSH (follicle-stimulating hormone)**, stimulates spermatogenesis by making the cells that will become sperm more sensitive to the stimulatory effects of testosterone. FSH works by causing certain cells within the seminiferous tubules to secrete a protein that binds and concentrates testosterone.

The levels of male reproductive hormones are regulated by negative feedback cycles involving hormones from the hypothalamus, the pituitary gland, and the testis (Figure 16–5). The hypothalamus releases **GnRH (gonadotropin-releasing hormone)**, which stimulates the anterior pituitary to secrete FSH and LH. In turn, LH stimulates the production of testosterone by the interstitial cells of the testis. As testosterone levels rise, the release of both GnRH from the hypothalamus and of LH from the anterior pituitary are inhibited. The seminiferous tubules produce a hormone called **inhibin** in addition to the sperm. Inhibin production increases with sperm count. As its name implies, inhibin serves to inhibit the production of FSH from the anterior pituitary and it may also inhibit the hypothalamic se-

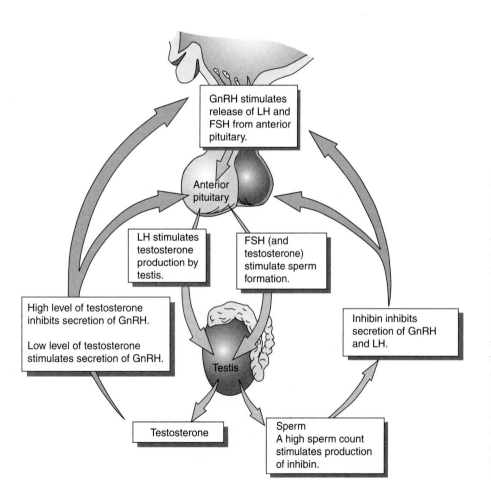

Figure 16–5 The feedback relationships among the hypothalamus, anterior pituitary, and the testes control the production of both sperm and testosterone. A regulating factor called gonadotropin releasing hormone (GnRH) from the hypothalamus stimulates the anterior pituitary to release follicle-stimulating hormone (FSH) and luteinizing hormone (LH). In turn, FSH stimulates production of sperm by the seminiferous tubules in the testes and LH stimulates testosterone secretion from the interstitial cells of the testes. As more sperm are produced, the seminiferous tubules also produce a hormone, inhibin, that exerts negative feedback control on the hypothalamus and pituitary, reducing FSH secretion. Testosterone also inhibits the hypothalamus and pituitary, causing LH levels to drop.

cretion of GnRH. As a result, the testosterone level and sperm production decline.

Critical Thinking

Anabolic steroids are taken by some male athletes to build muscles. A side effect of steroid abuse is a reduction in testis size. Considering that the anabolic steroids mimic the effect of testosterone, how can the shrinking testis size be explained?

The Female Reproductive System

The female reproductive system consists of the ovaries, the fallopian tubes, the uterus, and the vagina (Figure 16–6; Table 16–2).

The Ovaries

Each of a woman's two ovaries is almond-shaped and measures 5 cm by 2.5 cm (2 in. by 1 in.). The ovaries have two important functions: (1) to produce eggs in a process called **oogenesis** and (2) to produce the female hormones, **estrogen** and **progesterone**. The production of eggs and hormones are intimately related and occur in cycles, as we will see.

Before a female is born, the preparations for egg production begin. Early in fetal development, cells that may eventually develop into eggs migrate from the yolk sac (a structure that forms from embryonic membranes) to the ovaries, where they later develop into cells called **oogonia**. The oogonia divide mitotically and their numbers swell to several million oogonia. In the third month of fetal development, the oogonia enlarge, begin to store nutrients, and develop into **primary oocytes** (immature eggs). All the pri-

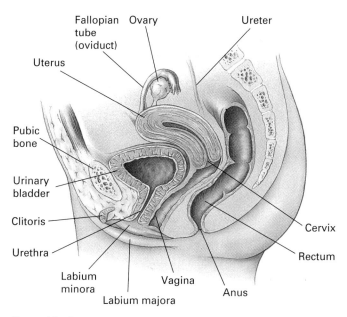

Figure 16–6 A side view of the female reproductive system.

mary oocytes then begin the meiotic process by making a copy of each of their chromosomes and the copies of each chromosome remain attached to one another. Each **primary oocyte** is surrounded by a single layer of flattened cells, called **follicle cells**, and the entire structure is called a **primary follicle**.

We see, then, that long before birth, all a woman's potential eggs have formed. Although she may have formed as many as 2 million primary follicles, only about 700,000 remain when she is born. The immature eggs remain in this state until she reaches puberty, usually at 10 to 14 years of age when her reproductive years begin. By the time of puberty, a female's lifetime supply of potential eggs has dwindled to between 300,000 and 400,000. The potential eggs age as the female does. Indeed, this may be one reason for

Table 16–2   The Female Reproductive System

Structure	Function
Ovary	Produce eggs and the hormones estrogen and progesterone
Fallopian tubes	Transport ovulated oocyte to uterus
Uterus	Receives and nourishes embryo
Vagina	Receives penis during intercourse; serves as birth canal
Clitoris	Contributes to sexual arousal
Breasts	Produce milk

the increased risk of genetic defects in children born to women who are older than 35.

Beginning at puberty, a group of primary follicles will continue development (Figure 16–7) at approximately monthly intervals. The follicle cells begin dividing, forming layers of cells, and secreting a fluid that contains estrogen. Soon, one of these primary follicles will become dominant. This one will continue development and the others will degenerate. The follicle cells continue dividing and fluid begins to accumulate between them. As the fluid accumulates, the wall of follicle cells splits. The inner layer of follicle cells directly surrounds the primary oocyte. The outer layer forms a balloon-like sphere enclosing the fluid and the oocyte. This structure grows rapidly. Within about 8 to 10 days after its development began, the follicle assumes its mature form, called a **Graafian follicle**. The mature follicle causes a spectacular bulge about 2 cm (nearly an inch) in diameter on the surface of the ovary. The primary oocyte then completes the first meiotic division, which it prepared for

years earlier. The pairs of chromosomes separate (with the copies still attached). The primary oocyte divides, forming two cells of unequal size—a large cell that contains most of the cytoplasm, called a **secondary oocyte**, and a tiny cell, called the **first polar body**. The polar body is essentially a garbage bag for one set of chromosomes. It has very few cellular constituents and plays no further role in reproduction. About 12 hours after the secondary oocyte has formed, the mature follicle pops, like a blister, releasing 5 to 10 ml of fluid, along with the oocyte mass, which is about the size of the head of a pin. The release of the oocyte from the ovary is called **ovulation** (Figure 16–8).

If the secondary oocyte is penetrated by a sperm, the second meiotic division takes place. The replicate chromosomes are separated. Cell division is again unequal. One set of chromosomes goes into a small cell, called the **second polar body**, and the other set into a large cell, the mature **ovum**, or egg. If fertilization occurs, the nucleus of the sperm, which contains one copy of each chromosome, and

(a) Stages of follicle development

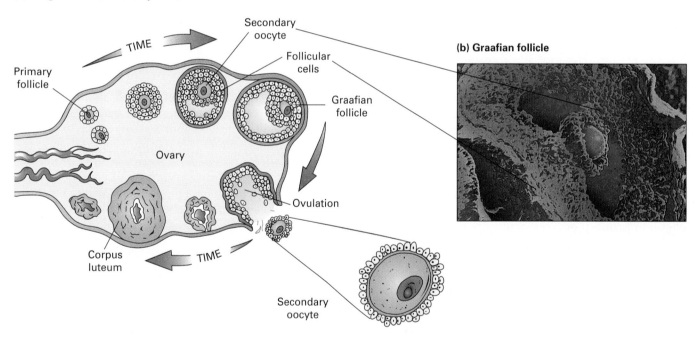

(b) Graafian follicle

Figure 16–7　The ovarian cycle. Each month several primary follicles begin to develop. A primary follicle consists of a primary oocyte surrounded by a single layer of follicle cells. The follicle cells divide and secrete a fluid that contains estrogen. One of the primary follicles becomes dominant and continues development and the rest degenerate. As the wall of follicle cells thickens, fluid accumulates between the follicle cells, eventually causing them to split into an inner layer surrounding the oocyte and an outer layer. The follicle continues to grow rapidly and forms the mature follicle, called a Graafian follicle, which bulges like a huge blister from the surface of the ovary. The primary oocyte completes the first meiotic division, forming a secondary oocyte and a small, nonfunctional cell called a polar body. The mature follicle ruptures, releasing the fluid and the secondary oocyte. The release of the oocyte from the ovary is called ovulation.　(b, © Bagavandoss/Photo Researchers, Inc.)

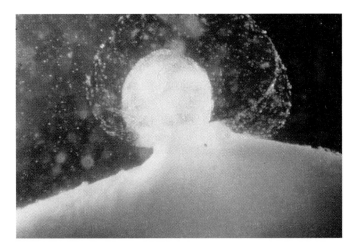

Figure 16–8   Ovulation.   *(© C. Edelman/La Villette/Photo Research-ers, Inc.)*

the nucleus of the ovum, which also contains one complete set of chromosomes, then fuse. The resulting cell, called a **zygote**, will divide repeatedly by mitosis and develop into the new individual.

The cells that formed the outer sphere of the mature Graafian follicle remain in the ovary and LH (luteinizing hormone) transforms these cells into an endocrine structure called the **corpus luteum** (which means yellow body). The corpus luteum secretes both estrogen and progesterone. Unless pregnancy occurs, the corpus luteum degenerates. If pregnancy occurs, the corpus luteum will be maintained by a hormone from the embryo called human chorionic go-nadotropin (HCG), as we will see later in this chapter.

The Fallopian Tubes

Two **fallopian tubes**, also known as the oviducts or uterine tubes, extend from the uterus and transport the oocyte from the ovary to the uterus. The end of each fallopian tube near-est the ovary is open and funnel-shaped. This end has many ciliated, finger-like projections that drape over the ovary, but they rarely directly contact it. When released from the ovary at ovulation, the oocyte is first cast into the abdominal cavity. If it does not enter the fallopian tube, the oocyte will be lost there. However, about the time of ovulation, the projections from the tubes begin to wave. The currents they create and those caused by the cilia lining the tubes help to draw the oocyte into the fallopian tube. Inside, the fallopian tube is a tunnel about twice the thickness of a human hair and the cilia there beat, creating a current toward the uterus. Fertilization usually takes place in the fallopian tube, near the ovary. The fertilized egg is swept along the fallopian tube toward the uterus, a distance of 10 cm (4 in.), by the beating cilia and the rhythmic muscular contractions of the fallopian tube.

The Uterus

The **uterus** is a hollow organ shaped like an inverted pear that receives and nourishes the embryo. In a woman who has never been pregnant, the uterus measures about 7.5 cm (3 in.) from top to bottom, about 5 cm (2 in.) wide, and about 2.5 cm (1 in.) thick. During pregnancy, the uterus expands to about 60 times this size as the fetus grows. After childbirth, the uterus never quite returns to its pre-pregnancy size.

The narrow neck of the uterus, which projects into the vagina, is called the **cervix**. The **vagina** is a muscular tube, about 8 to 10 cm (3 to 4 in.) long that opens to the outside of the body. The vagina receives the penis during sexual inter-course. Sperm that are deposited in the vagina can enter the uterus through an opening in the cervix and then swim to the fallopian tube to meet the egg. At birth, the infant is pushed through the cervix and then the vagina on its way to greet the world.

The wall of the uterus has two layers—a muscular layer called the **myometrium** (*myo-*, muscle; *metr-*, uterus; *-ium*, region) and a lining called the **endometrium** (*endo-*, within; *metr-*, uterus; *-ium*, region). The smooth muscle of the my-ometrium contracts rhythmically in waves during childbirth and forces the infant out. The endometrium becomes thicker and more vascular (more heavily penetrated by blood vessels) during each monthly cycle. If an embryo is formed, it implants (embeds) in the endometrium, where it will remain for the duration of the pregnancy. If the egg is not fertilized, the endometrium that built up during that cy-cle will be lost as menstrual flow.

Ectopic Pregnancy

If an embryo implants in an area other than the uterus, the condition is described as an ectopic (*ect-*, outside) pregnancy. The most common type of ectopic pregnancy is a tubal pregnancy, in which the embryo implants in a fallopian tube. The reported statistics on the frequency of ectopic pregnan-cies range from 1 in 40 to 1 in 300. The actual frequency is difficult to determine because some embryos that implant outside the uterus are reabsorbed by the woman's body with-out her even realizing that she was pregnant.

Symptoms of an ectopic pregnancy usually begin 7 or 8 weeks into the pregnancy, when the embryo is about an inch long and is severely stretching the narrow fallopian tube. At this point, a pregnancy test may or may not be positive. Some bleeding occurs in 50% to 80% of ectopic pregnan-cies. There may also be lower abdominal pain, which may be one-sided or diffuse. Some telltale symptoms may be de-tected during a routine examination during early pregnancy. The physician may notice that the uterus is smaller than it should be or be able to feel the enlarged fallopian tube.

A tubal pregnancy must be surgically terminated be-cause it places the life of the mother in danger. If the embryo is permitted to continue growing, it will eventually rupture the fallopian tube, which can cause the mother to bleed to death internally.

External Genitalia

The female reproductive structures that lie outside the vagina are collectively known as the **external genitalia** or the **vulva** (Figure 16–9). Overlying the pubic arch is a fatty mound called the **mons veneris** ("love mound"), which is covered with hair after puberty. Extending posteriorly from the mons veneris are two hair-covered folds of skin, the **labia majora** ("big lips"), which enclose two thinner skin folds, the **labia minora** ("little lips"). The anterior portions of the labia minora form a hood over the **clitoris**, a small, sensitive structure, composed largely of erectile tissue. The clitoris develops from the same embryological structure that, in a male, would form the tip of the penis and, like the penis, the clitoris has many nerve endings sensitive to touch. During tactile stimulation, the clitoris becomes swollen with blood and contributes to a woman's sexual arousal.

The Breasts

The **breasts**, or mammary glands, are present in both sexes, but only in females do they produce milk to nourish a newborn.

Inside the breast are 15 to 25 groups of milk-secreting glands. A milk duct drains each group through the nipple. Interspersed around the glands and ducts is fibrous connective tissue that supports the breast (Figure 16–10).

In a nonpregnant woman, most of the breast consists of the fibrous connective tissue and fat. However, the breast has no muscle. Consequently, exercise cannot develop the breasts. Exercise can only enlarge the pectoral muscles that underlie the breasts, creating the illusion that the breasts are larger.

The monthly changes in estrogen and progesterone levels prepare the breast for the possibility of pregnancy. Early in each cycle, estrogen stimulates the growth of the glands, ducts, and fibrous tissue of the breast. Progesterone then triggers the initial steps of the milk-secreting process. (Milk production does not begin for 2 to 3 days after the newborn begins suckling; see Chapter 17.) At the same time, blood flow to the breasts increases. Together, these changes may cause feelings of fullness and tenderness that peak just before a woman's period begins. If pregnancy does not occur, the secretions and cells are usually reabsorbed. The swelling and tenderness then subside. In women with fibrocystic breast disease (which, incidentally is not really a *disease*) the monthly build-up of the breast is too extensive or else the reabsorption process is insufficient. As a result, pockets of cell debris and trapped secretions form lumps in the breast that may be painful. Although these lumps are not cancerous, women with fibrocystic disease are four times more likely to develop cancer of the breast.

Breast Cancer

In industrialized countries, breast cancer is the most common form of cancer among women and the lifetime risk is

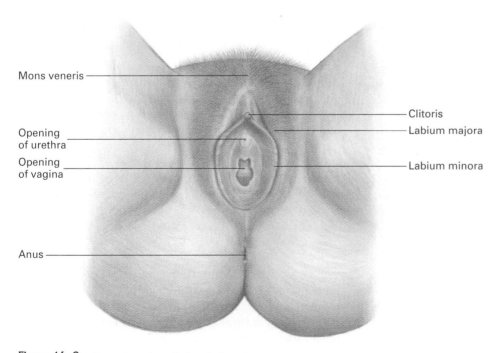

Mons veneris

Opening of urethra

Opening of vagina

Anus

Clitoris

Labium majora

Labium minora

Figure 16–9   The external genitalia of a female.

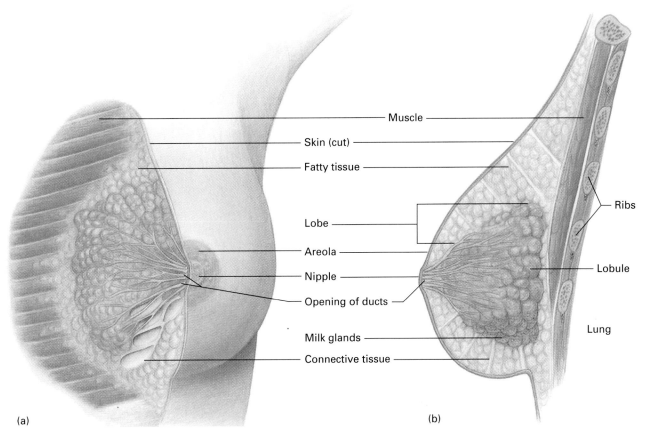

Figure 16-10   Breast structure. (a) front view (b) side view.

creeping steadily upward. One in nine American women will develop breast cancer at some point in her life. It kills approximately 46,000 women each year, making it the second biggest cancer killer in women after lung cancer. In fact, one American woman dies of breast cancer every 12 minutes. (Breast cancer also kills 300 American men each year.)

Breast cancer usually begins with the abnormal growth of the cells lining the milk ducts of the breast, but it sometimes begins in the milk glands themselves. Some breast tumors grow quite large without spreading. Other, more life-threatening types of breast cancer, aggressively invade surrounding tissues. Typically, cancerous cells begin to spread when the tumor is about 20 mm (about 3/4 in.) in diameter. At this point, they break through the membranes of the ducts or glands where they initially formed and move into the connective tissue of the breast. They may then move into the lymphatic vessels and/or blood vessels permeating the breast, either of which can transport the cells throughout the body.

Detecting Breast Cancer

Early detection is a woman's best defense against breast cancer. Monthly breast self-exam (BSE) is helpful in detecting a lump early (Figure 16-11). If a woman begins breast self-

exam in early adulthood, she becomes familiar with the consistency of her breast tissue. Later in life, it is easier to notice changes that might be telltale signs of breast cancer.

Mammograms, x-rays of breast tissue, are also helpful in early detection of breast cancer, because they can detect a tumor too small to be felt as a lump. In most cases, a breast tumor cannot be felt until it is more than 1 cm (0.4 in.) in diameter, but a mammogram can detect a lump less than 0.5 cm (0.2 in.) in diameter. New x-ray techniques have been developed that will soon provide images as clear as today's mammograms with less radiation exposure. In addition, computers will be used to scan the mammogram. These will make it easier to detect small tumors that previously might have been hidden in dense breast tissue.

Early detection can make the difference between life and death. A tumor large enough to be felt contains a billion or more cells and a few of them may have already spread from the tumor to other tissues of the body. After cancer cells spread, the woman's chance of survival decreases dramatically. An added benefit of mammograms is that they can detect tumors that are small enough to be removed by a type of surgery, called lumpectomy, that removes the lump but spares the breast. At later stages of breast cancer, the entire breast may have to be removed, a type of surgery called mastectomy.

How to Examine Your Breasts

① Stand in front of the mirror and look at each breast to see if there is a lump, a depression, a difference in texture, or any other change in appearance.

② Get to know how your breasts look and be especially alert for any changes in the nipples' appearance.

③ Raise both arms and check for any swelling or dimpling in the skin of your breasts.

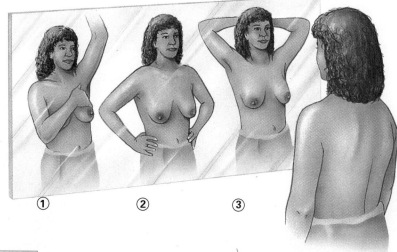

① ② ③

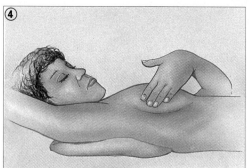

④

④ Lie down with a pillow under your right shoulder and put your right arm behind your head. Perform a manual breast examination. With the nipple as the center, divide your breasts into imaginary quadrants.

⑤ With the pads of the fingers of your left hand, make firm circular movements over each quadrant, feeling for any unusual lumps or areas of tenderness. When you reach the upper, outer quadrant of your breast, continue toward your armpit. Press down in all directions.

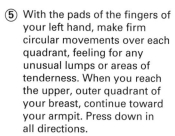

⑤

⑥ Feel your nipple for any change in size and shape. Squeeze your nipple to see if there is any discharge. Repeat from step 4 on the other breast.

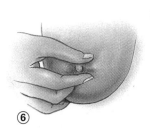

⑥

Figure 16-11   Breast self-exam can help a woman find lumps in her breast before they have spread to surrounding tissue. It also makes a woman familiar with the consistency of her breast tissue, making it easier for her to detect changes in consistency that might indicate cancer.

Social Issue

The value of mammography for women in their forties has become controversial. Central to the controversy is the fact that while mammography detects 9 of 10 lumps in the breast, 9 of 10 of the lumps detected in women in their forties will *not* be cancerous. As a result, many women in their forties who have mammograms will suffer a great deal of unnecessary emotional stress and undergo further tests, such as biopsies, that are costly. Furthermore, some studies indicate that the survival rate of women in their forties is not increased by early detection, perhaps because such relatively young women have a faster growing, more aggressive form of cancer. When a woman in her forties decides to have a mammogram, she must weigh the relative importance of cost, emotional well-being, and physical benefit. How do you think national policy decisions on recommendations for mammograms should be made?

Risk Factors

There are many things that increase a woman's risk of developing breast cancer that she can't do anything about. One of these is who her parents are. The genes she inherits from her parents can influence her risk of developing breast cancer. A woman whose mother had breast cancer is nearly twice as likely to develop breast cancer than one with no maternal history of breast cancer. If both her sister and her mother had breast cancer, a woman's risk is 2.5 times greater than someone with no family history of breast cancer.

There are at least two genes that increase a woman's risk of breast cancer. Although only 5% to 10% of all breast cancers are related to these genes, the unlucky women who inherit either of them have an 85% chance of developing breast cancer at some point in their lives. The normal forms of these genes are thought to produce proteins that put the brakes on tumor growth. The mutant forms, however, produce proteins that are not effective in stopping tumor growth.

Social Issue

The identification of specific genes that dramatically increase a woman's risk of developing breast cancer opens the possibility of developing tests that will detect these genes and warn a woman of her increased risk. This knowledge may be a burden to some women and their families. Because the odds of cancer are increased dramatically among carriers of the genes, a woman who learns that she has either gene may opt for a mastectomy before there are any signs of cancer. The emotional trauma, disfigurement, and expense might be unnecessary. If a woman with a family history of breast cancer is tested and learns that she does not carry the genes that increase her risk, but her sisters do, she may suffer "survivor's guilt." Should such tests be developed?

There is a common thread—exposure to estrogen—that runs through the tapestry of risk factors associated with breast cancer. You may recall that during each menstrual cycle, estrogen stimulates breast cells to begin dividing in preparation for milk production, in case the egg is fertilized. Excessive estrogen exposure, therefore, may push cell division to a rate characteristic of cancer.

One factor that influences estrogen levels is the number of times a woman ovulates during her lifetime, because estrogen is produced by both the maturing ovarian follicle and the corpus luteum. The number of times a woman ovulates is, in turn, affected by such factors as:

1. Age when menstruation begins. Ovulation usually occurs in each menstrual cycle. Thus, the younger a woman is when menstruation begins, the more opportunities there are for ovulation.
2. Menopause after age 55. The later menopause occurs in life, the more menstrual cycles a woman is likely to experience. Estrogen levels are low and ovulation ceases after menopause.
3. Childlessness and late age at first pregnancy. Ovulation does not occur during pregnancy. Thus, pregnancy gives the ovaries a rest. Furthermore, the hormonal patterns of pregnancy appear to transform breast tissue in a way that protects against cancer. As a result, delaying pregnancy until after age 30 or remaining childless increases a woman's risk of developing breast cancer later in life.
4. Breast feeding. Women who breast feed their infants have a 20% lower risk of developing breast cancer before they reach menopause. Nursing may guard against breast cancer by blocking ovulation or by causing physiological changes that leave the breast tissue more resistant to cancer-producing chemicals in the environment.
5. Exercise. Even moderate exercise can suppress ovulation in adolescents and women in their twenties (but, unfortunately, this is not as likely in older women). Suppression of ovulation may be the reason that women in their teens or twenties who exercise 3 or more hours a week reduce their risk of developing breast cancer by 30% compared with their "couch potato" peers. Continued moderate activity into the forties cuts the risk of breast cancer by 60%.

Other factors may also influence estrogen levels. For instance, obese women have higher estrogen levels than do thin ones. This is because estrogen is produced in fat cells, as well as by the ovaries. Indeed, obesity, especially if the fat is carried above the waist, translates to a threefold elevated risk of breast cancer. A weight gain as small as 10 pounds when a woman is in her twenties also increases her risk of breast cancer later in life.

Fatty foods directly affect estrogen levels, an observation that led researchers to believe that dietary fat might be a *controllable* factor that elevated breast cancer risk. Over the years, many studies on dietary fat and the development of breast cancer yielded conflicting results. However, the largest study to date found no link between the two. A low-fat diet may, however, reduce the likelihood of a recurrence of breast cancer.

Furthermore, some women have the cards stacked against them (or in their favor), because of effects that occurred before they were even born. While a fetus is growing within the mother's uterus, it is exposed to the mother's estrogen. Because estrogen is growth promoting, infants who are heavier at birth are thought to have been exposed to higher estrogen levels during development. Female infants who weighed more than 8 pounds at birth are, in fact, more likely than their lighter-weight peers to develop breast cancer later in life. In contrast, daughters born to women who experienced toxemia during pregnancy are much less likely to develop breast cancer than those whose mothers did not. Toxemia is a form of high blood pressure that occurs during pregnancy and is linked to the mother's low estrogen levels.

Certain organochlorine pesticides and ingredients in plastics mimic the effects of natural estrogen on the body. And, electromagnetic fields (EMFs) may elevate the body's own estrogen levels.

Ovarian and Menstrual Cycles

A woman's fertility is cyclic. At approximately monthly intervals, an egg matures and is released from an ovary. Simultaneously, the uterus is readied to receive and nurture the young embryo. If fertilization does not occur, the uterine provisions are discarded as menstrual flow. The ovaries and uterus will prepare again during the next cycle. The events in the ovary, known as the **ovarian cycle**, must be closely coordinated with those in the uterus, known as the **menstrual cycle** (or the uterine cycle). The menstrual cycle is known to some as the monthly miracle and to others as "the curse."

As we will see, the symphony of a female's fertility is orchestrated by the interplay of hormones. Events in both the uterus and the ovary are coordinated by interactions between hormones from the anterior pituitary gland, which is located in the brain, and from the ovary. The hormones from the pituitary, FSH and LH, cause the ovary to release its hormones, estrogen and progesterone. As in the male, the release of FSH and LH is regulated by a regulating hormone from the hypothalamus called GnRH (Figure 16–12).

Traditionally, the first day of menstrual flow is considered day 1 of the cycle because it is the point that is most easily noted. The bleeding usually lasts from 2 to 8 days, or an average of 5 days. Between 25 and 65 ml (4 to 6 tablespoons) of blood are lost. At this time, the ovarian hormones, estrogen and progesterone, are at their lowest levels, allowing the pituitary to produce its hormones, especially FSH. In turn, FSH causes a number of egg follicles in the ovary to develop and produce estrogen. Thus, even as the uterus loses the endometrial lining it prepared during the previous cycle in which no egg was fertilized, the egg that will be released in the next cycle is developing.

As the egg follicle develops and the number of follicle cells increases, estrogen levels rise. Estrogen causes the cells in the endometrial lining of the uterus to divide. These cells store glycogen, which could help nourish a future embryo during its early stages of development. In addition, estrogen inhibits the release of FSH through a negative feedback cycle. In spite of declining FSH levels, the follicle continues to grow because the follicle cells become increasingly sensitive to FSH. When the egg and follicle are nearly mature, the estrogen level rises rapidly and causes a sudden and spectacular release of LH (and FSH) from the pituitary.

The LH surge causes several important events. It causes the egg to undergo its first meiotic division. Next, LH trig-

gers ovulation and the egg bursts out of the ovary to begin its journey along the fallopian tube. Continued LH secretion then transforms the remaining follicle cells into the corpus luteum. The corpus luteum continues the estrogen secretion begun by the follicle cells and, importantly, also secretes progesterone.

Together estrogen and progesterone make the endometrial lining of the uterus a hospitable place for the embryo. The blood supply to the endometrium increases. Uterine glands develop that secrete a mucous material that can nourish the young embryo when it arrives in the uterus.

The rising estrogen and progesterone levels inhibit pituitary secretion of FSH and LH. As FSH levels decline, the development of new follicles is inhibited.

If fertilization does not occur, the corpus luteum will degenerate within about 2 weeks (14 plus or minus 2 days, regardless of the length of the menstrual cycle). Whether regression of the corpus luteum is caused by the declining levels of LH or a programmed destruction is not known. But, whatever its cause, the degeneration of the corpus luteum results in falling levels of estrogen and progesterone.

Progesterone is essential to the maintenance of the endometrium. As progesterone levels drop because of the degeneration of the corpus luteum, the blood vessels nourishing the endometrial cells collapse. The cells of the endometrium then die and are sloughed off along with mucus and blood, as menstrual flow. No longer inhibited by estrogen and progesterone, FSH and LH levels begin to climb. Thus, the cycle begins again.

If the egg is fertilized, however, the corpus luteum is maintained by a hormone from the embryo, called **human chorionic gonadotropin (HCG)**. Thus, if pregnancy occurs, HCG will prevent the degeneration of the corpus luteum, keeping estrogen and progesterone levels high enough to prevent endometrial shedding. HCG is detectable in the mother's blood within 7 to 9 days after fertilization and in her urine less than 2 weeks after fertilization. Pregnancy tests are designed to detect HCG. Between the second and third month of development the placenta has developed sufficiently to take over the production of estrogen and progesterone during the rest of the pregnancy. The corpus luteum then degenerates.

Menopause

A woman's fertility usually peaks when she is in her twenties and then gradually declines. By the time she reaches 45 to 55

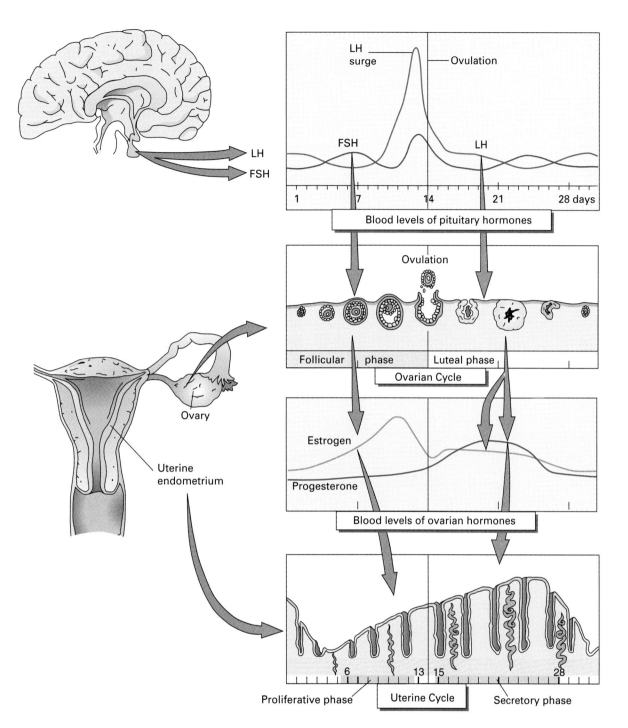

Figure 16–12 The ovarian and uterine cycles are orchestrated by the interplay of hormones from the hypothalamus, anterior pituitary, and the ovary. The hypothalamus produces gonadotropin-releasing hormone (GnRH), which stimulates the anterior pituitary to produce follicle-stimulating hormone (FSH) and luteinizing hormone (LH). FSH stimulates the development of a follicle with its maturing egg. The follicle cells, in turn, secrete estrogen, which begins the thickening of the uterine endometrium in preparation for embryo implantation. The rising level of estrogen triggers the release of a surge of LH from the pituitary gland. LH is the immediate stimulus for ovulation, after which it transforms the follicle cells remaining in the ovary into a temporary endocrine structure called the corpus luteum. The corpus luteum secretes both estrogen and progesterone. Progesterone further prepares the endometrium for implantation and maintains it. Progesterone and estrogen inhibit GnRH release from the hypothalamus and LH and FSH release from the pituitary. The corpus luteum degenerates in about 2 weeks (14 plus or minus 2 days) after ovulation. Consequently, estrogen and progesterone levels drop, allowing FSH and LH secretion to increase. Without progesterone, the endometrium breaks down and is lost as menstrual flow, marking the start of a new cycle.

years of age, few of the potential eggs prepared before birth remain in the ovary and those that do become increasingly less responsive to FSH and LH, the hormones that cause egg development and ovulation. The ovaries, therefore, gradually stop producing eggs and the levels of estrogen and progesterone fall. During this time, menstrual cycles become increasingly irregular. Eventually, ovulation and menstruation stop completely, an event called **menopause**.

The drop in estrogen levels has a number of physiological effects that can range in severity from annoying to life-threatening. At one end of the scale is the loss of a layer of fat that was formerly promoted by estrogen. This loss results in a reduction in breast size and the appearance of wrinkles. Estrogen also plays an important role in regulating a woman's body thermostat. So, as estrogen levels fall, many women experience hot flashes—waves of warmth spreading upward from the trunk to face. During a hot flash, the skin may redden and sweating may occur. Although a single hot flash lasts only a few minutes, hot flashes may occur periodically throughout the day and night. Indeed, "night sweats" caused by hot flashes may awaken a woman several times during the night. The absence of estrogen can cause vaginal dryness that can make sexual intercourse painful. Without estrogen, the male hormones produced by the adrenal gland predominate and can cause facial hair to grow.

More serious problems associated with the estrogen deficit that accompanies menopause include an increased risk of diseases of the heart and blood vessels and weak bones. Estrogen offers some protection against atherosclerosis, a condition in which fatty deposits clog the arteries. Thus, after menopause, this protective effect on the heart is lost. Estrogen is also important in the body's ability to absorb calcium from the digestive system and to deposit it in bone. Without calcium, bone becomes weak and porous—a condition known as osteoporosis (discussed in Chapter 5).

Possible Problems with the Female Reproductive System

Premenstrual syndrome (PMS) is a collection of symptoms that appear 7 to 10 days before a woman's period begins. These symptoms include depression, irritability, fatigue, and headaches. Indeed, 20% to 40% of all menstruating women feel that PMS significantly interferes with their daily living. Technically, a woman will get PMS only if she ovulates. Women on the pill and pregnant women, therefore, can't suffer from PMS. Symptoms must be cyclic. Thus, if a woman is always grouchy, the irritability is not caused by PMS.

Some researchers suggest that a progesterone deficiency is to blame for the symptoms of PMS. Progesterone, for example, has a calming effect on the brain. It also decreases fluid retention. It is reasoned, therefore, that in the days just before a woman's period begins, as progesterone levels plummet, the nervous system may be stimulated and fluids may be retained, causing an uncomfortable, bloated feeling.

Treatments for PMS are varied. For women whose symptoms are severe, drugs that elevate the levels of deficient neurotransmitters are effective. Some women with milder symptoms find relief from changes in diet. Caffeine, alcohol, fat, and sodium should be avoided. On the other hand, foods high in calcium, potassium, manganese, and magnesium should be increased. Aerobic exercise for at least half an hour stimulates the release of enkephalins and endorphins and may provide some relief.

Prostaglandins, chemicals used in communication between cells in many parts of the body, are the primary cause of **menstrual cramps**. Endometrial cells produce prostaglandins that, among other things, cause the smooth muscle cells of the uterus to contract, causing cramps. High levels of prostaglandins can cause sustained contractions, called muscle spasms, of the uterus. These muscle spasms may cut down blood flow and, therefore, the oxygen supply to the uterine muscles, resulting in pain.

Endometriosis is a condition in which tissue from the lining of the uterus is found outside the uterine cavity—commonly in the fallopian tubes, on the ovaries, or on the outside surface of the uterus, the bladder, or the rectum. We see, then, that endometrial tissue can move out the open ends of the fallopian tubes to the abdominal cavity. Endometrial tissue, wherever it is, grows and breaks down with the hormonal changes that occur over each menstrual cycle. These cyclic changes can cause extreme pain associated with the menstrual cycle.

Endometriosis is also associated with infertility. Researchers now suspect that infertility may result because women with endometriosis lack a particular protein (beta-3) on her endometrial cells. The protein is either directly involved in the implantation of the embryo, is necessary for continued development of the embryo, or is a marker for a chemical that is critical. In addition, bleeding from endometriosis can cause scar tissue to form in fallopian tubes or some endometrial tissue can cover the ovary and interfere with release of the egg; either can impair fertility.

Vaginitis is an inflammation of the vagina that is most commonly caused by one of three types of organisms: yeast (*Candida albicans*), a protozoan (*Trichomonas vaginalis*), or a bacterium (*Gardnerella vaginalis*). Whereas a yeast infection is generally not sexually transmitted, infections with *Gardnerella* or *Trichomonas* are usually transmitted through sexual intimacy with an infected person. Although vaginal discharge is always the primary sign of vaginitis, the nature of the discharge varies with the causative organism. The discharge accompanying a yeast infection, for instance, is white and curdy, like cottage cheese. It causes intense itching and sometimes burning and itching of the outer lips of the vagina. An infection with *Trichomonas* causes a watery, foamy greenish or yellowish discharge, vaginal itching, and pain. In contrast, the discharge caused by *Gardnerella* is gray or white in color and has a "fishy" odor. It does not usually cause itching. It is important to identify the cause of a vaginal infection because each has a different treatment.

Yeast infections are very common and occur when the normal, slightly acidic vaginal environment is disrupted, which allows the yeast cells normally present in the vagina to overgrow. Many factors can alter the pH of the vagina and lead to a yeast infection. One such factor is the levels of reproductive hormones, which fluctuate over a woman's menstrual cycle and also over her reproductive lifetime. Estrogen and progesterone increase the alkalinity of the vagina. This is why yeast infections are more likely to occur when hormone levels are high—at the time of ovulation, during use of the birth control pill, or during pregnancy. And, because the acidity of the vagina is normally maintained by bacteria known as lactobacilli that live in the vagina, anything that kills these bacteria, such as antibiotic treatment for an infection elsewhere in the body, may lead to a yeast infection.

Pelvic inflammatory disease (PID), which affects more than a million women a year in the United States alone, is a catch-all term used to refer to an infection of the pelvic organs. The symptoms of PID—abdominal pain or tenderness, low-back pain, pain during intercourse, abnormal vaginal bleeding or discharge, fever, or chills—can come on gradually or suddenly and vary in intensity from mild to severe.

PID can be caused by a number of organisms, but the most common culprits are the sexually transmitted bacteria that cause chlamydia and gonorrhea (discussed in Chapter 16A). The bacteria ascend from the vagina through the cervix to the uterus. There, they may infect the endometrium and then spread to the wall of the uterus and to the fallopian tubes. Near the ovary, the fallopian tube opens to the abdominal cavity, allowing the infection to spread into the abdominal cavity. The infection can be curbed by treatment with antibiotics, which kill the bacteria.

If PID is not treated quickly, its consequences can be long-lasting and severe: reduced fertility and an increased risk of ectopic pregnancy. As the infection spreads through the fallopian tubes, the body attempts to defend itself by secreting fibrous material to wall off the area. The resulting scar tissue can block the tubes. If both tubes are completely blocked, the woman is sterile, because the sperm cannot reach the egg. If a tube is partially blocked, the woman may have difficulty conceiving a child. Indeed, more than 11% of women who have had only one episode of PID are left infertile and 23% of those who have had two bouts are infertile. If pregnancy does occur, the scarring also increases the odds that it will be ectopic. An egg is many times larger than sperm. Thus, sperm may be able to swim past the scar tissue and fertilize the egg. The resulting embryo, however, being too big to pass the scar tissue and reach the uterus, may implant in the fallopian tube.

Birth Control

Sexual activity is actually a double-edged sword—one edge is the possibility of pregnancy and the other is the possibility of getting a sexually transmitted disease (STD). The way in which couples dodge one of the edges of the sword influences the chances of being cut by the other one. For this reason, as we discuss various means of contraception we will also consider whether they reduce the risk of spreading sexually transmitted infections.

There is a variety of means of birth control available today. Different methods interrupt the reproductive process in different places, have varying degrees of effectiveness, and have varying degrees of protection against the spread of STDs (Table 16–3).

Abstinence

Abstinence, that is, not to have intercourse at all, is the most reliable way to avoid both pregnancy and the spread of STDs. However, 60% of American teenagers have become sexually active by the time they are 19-years-old. Thus, abstinence does not always seem to be the most popular means of contraception.

Sterilization

Well before the end of their reproductive life spans, most people have had all the children they want. Other than abstinence, sterilization is the most effective way to ensure that pregnancy does not occur. However, unlike abstinence, sterilization offers no protection against STDs.

Sterilization in men involves an operation called a **vasectomy** that blocks the vas deferens on each side and prevents sperm from leaving the man's body (Figure 16–13a). The procedure, which usually lasts about 20 minutes, can be performed in a physician's office under local anesthesia. The physician simply makes small openings in the scrotum through which each vas deferens can be pulled, cut, a small segment removed, and at least one end sealed shut. Because sperm make up only about 1% of the semen, the volume of semen the man ejaculates is not noticeably reduced. His interest in sex is not lessened because testosterone, which is responsible for the sex drive, is released from the interstitial cells of the testis and carried around the body in the blood.

The risks of vasectomy are minimal. Less than 1% of the time, the ends of the vas deferens rejoin and provide an outlet for sperm. However, some recent studies suggest that a vasectomy may increase a man's chance of prostate cancer

Table 16–3 Methods of Birth Control

Method	Procedure	How It Works	% Failure Typical Use	% Failure Perfect Use	Risks	Protection From STDs
Abstinence	Abstain from sexual activity	Sperm never contacts egg	0	0	None	Yes
Sterilization						
Vasectomy	Cut and seal each vas deferens	No sperm in semen	<1	<1	Possible link to prostate cancer	None
Tubal ligation	Fallopian tubes blocked	Sperm can't reach egg	<1	<1	Infection from surgery	None
Hormonal methods						
Combination birth control pill	Hormone pill taken daily	Prevents egg development and release	3	0.1	Heart disease; stroke	None
Minipill	Hormone pill taken daily	Thickens cervical mucus; endometrium not properly prepared; sperm movement impaired; egg doesn't mature	13.2	1.1	Ovarian cysts	None
Depo-Provera	Progesterone injection every 3 months	Same as minipill	0.3	0.3	Possible linkage to osteoporosis	None
Norplant	Progesterone containing implants	Same as minipill	0.09	0.09	Abnormal vaginal bleeding, infection at implantation site	None
Intrauterine device (IUD)	Small plastic device inserted into uterus by physician	Interferes with both fertilization and implantation	<2	<2	Pelvic inflammatory disease	None
Barrier Methods						
Diaphragm	Inserted into vagina before intercourse	Covers cervix and prevents sperm from entering	18	6	No risks presently known	Some
Cervical cap	Inserted into vagina before intercourse	Same as diaphragm (partly by suction)	18	9	No risks presently known	Some
Male condom	Fits over penis before intercourse	Prevents sperm/penis from contacting vagina	12	3	No risks presently known	Latex–excellent; skin–not good
Female condom	Held in vagina by flexible rim rings	Prevents sperm/penis from contacting vagina	21	5	No risks presently known	Very good
Spermicides	Inserted into vagina before intercourse	Kills sperm for one hour after application	21	6	No risks presently known	Some
Fertility awareness	Abstain from sex on days that eggs and sperm may meet	Sperm never come in contact with egg	20		None	None
Calendar				9		
Ovulation method				3		
Symptothermal (body temp.; cervical mucus)				2		

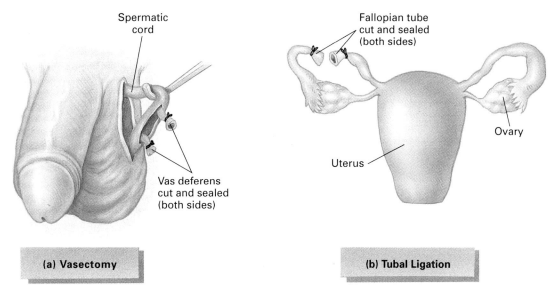

Spermatic cord

Vas deferens cut and sealed (both sides)

(a) Vasectomy

Fallopian tube cut and sealed (both sides)

Ovary

Uterus

(b) Tubal Ligation

Figure 16–13 Sterilization procedures prevent fertilization by cutting and sealing the tubes through which the gametes would travel and meet. (a) In a male, the vas deferens on each side is cut, a segment removed, and at least one end is sealed in a procedure called a vasectomy. (b) In a female, the fallopian tubes are blocked in a procedure called a tubal ligation.

later in life (see the Personal Concerns essay, *Testicular and Prostate Cancer*).

Female sterilization, called **tubal ligation**, involves blocking the fallopian tubes to prevent the egg and sperm from meeting (Figure 16–13b). Commonly the tubes are cut and the ends seared shut or mechanically blocked with clips or rings. Tubal ligation is frequently done using a procedure called a laparoscopy. In laparoscopy, two small incisions are made in the abdominal cavity. One is just below the navel and is used to insert a laparoscope, a light that illuminates the abdominal cavity. The other incision, near the pubic hairline, provides an entrance for the tubal sterilization instrument. Laparoscopy is generally performed in a hospital under general anesthesia. Because the abdominal cavity is opened, the risk of infection is greater following tubal ligation than it is after a vasectomy.

Sterilization should be considered permanent even though it is sometimes possible to reverse the procedure. At best, the reversal procedure is expensive and requires that the surgeon have special training in microsurgical techniques. Even then, success is not guaranteed. The pregnancy rate after a vasectomy has been reversed ranges from 16% to 79%, although well over 80% of the men have sperm in their semen. (The reduction in fertility after a successful vasectomy reversal is thought to be caused by antisperm antibodies that form while the vas deferens are blocked and sperm are absorbed into the bloodstream. The antisperm antibodies coat the sperm and immobilize them, even after vasectomy reversal.) After undergoing surgery to reverse tubal sterilization, 43% to 88% of the women become preg-

nant. Even when tubal ligation is successfully reversed, however, there remains an increased risk of ectopic pregnancy if fertilization is accomplished.

Hormonal Contraception

Hormonal contraception is currently available only to females. There are two basic types: the combined birth control pill and the progesterone-only means of contraception.

The Combination Birth Control Pill

When most people speak about "the pill" they are referring to the **combination birth control pill**, so named because it contains synthetic forms of both estrogen and progesterone, which mimic the effects of natural hormones that would ordinarily be produced by the ovaries (Figure 16–14). Among these effects is the suppression of FSH and LH release from the pituitary gland. Without these pituitary hormones, the egg does not mature and is not released from the ovary.

The pill has other effects that reduce the likelihood of pregnancy. For instance, it causes cervical mucus to become thick, making it difficult for sperm to enter the uterus. In addition, the endometrium is not properly prepared for successful implantation, because estrogen and progesterone are present throughout the cycle instead of sequentially, as happens normally.

The pill is highly effective in preventing pregnancy. The first-year failure rates among pill users vary from 0.1% among the most highly motivated and consistent users, to

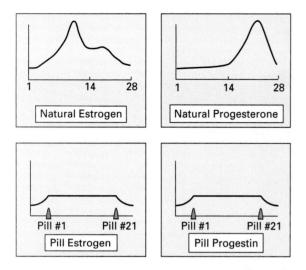

Figure 16–14 The combined birth control pill contains synthetic estrogen and progesterone, which are present then in steady, low levels throughout the cycle. In a menstrual cycle that is not regulated by the pill, the level of estrogen and progesterone is higher and the hormones are produced sequentially. The pill prevents pregnancy because the estrogen and progesterone inhibit the pituitary hormones FSH and LH. Therefore, the pill prevents the development of an egg, and if one should develop, the pill prevents ovulation.

3% among all typical users (a less motivated or less consistent group), to 4.7% among typical users younger than 22 years of age.

Almost all (7 of 8) pregnancies that occur in women who are on the pill result from forgetfulness. The pill only works if its hormones are present over the entire menstrual cycle. To be most effective, it should be taken at the same time every day. If a woman forgets to take it one day, the inhibition of pituitary hormones is lessened and the egg may develop and be ovulated. However, even if the pill is used correctly, it can fail when the hormones are expelled from the digestive system during vomiting or diarrhea. In addition, some drugs (for example, certain antibiotics and antihistamines) may reduce the pill's effectiveness.

For most healthy, nonsmoking women younger than 35 years of age, the pill is an extremely safe means of contraception. In fact, for these women pill use is safer than pregnancy. Consider a line 215 m long (700 ft), which is as tall as a 70-story building, as representing 100,000 young, nonsmoking pill users. A line representing the number of these women who die as a result of pill use would be less than a quarter of an inch long, whereas a line representing women who died from complications of pregnancy would be just less than an inch.

Nonetheless, a small number, 200 to 500 of the 10 million or so pill users in the United States, do die each year from a pill complication, so it should be noted as a possibility. The risk increases with age and with cigarette smoking

(Figure 16–15). Indeed, most deaths of pill users occur in women older than age 35 who smoke more than 15 cigarettes a day.

Pill use has side effects of varying degrees of severity. Many of these side effects are not usually of medical concern, but they may be uncomfortable enough for a woman to decide to stop using the birth control pill. They include headaches, breast tenderness, weight gain, and an increased frequency of vaginal infections.

Other side effects can be more serious. Problems with the circulatory system are the most important of the serious (sometimes fatal) complications of pill use. Most pill-associated deaths are caused by heart attack or stroke, which occur when the blood supply to the heart or a region of the brain is blocked. The pill increases the risk of heart attack or stroke in several ways. First, estrogen alters a protein responsible for blood clotting, sometimes causing clots to form. These clots can form in any blood vessel, and when they do, they can block blood flow, shutting off the blood supply to the cells served by that vessel. Second, by altering the types of lipids in the blood, progesterone can promote the deposit of fatty material in the arteries (atherosclerosis). These deposits are another way that blood flow through a vessel can be reduced. Third, the pill causes high blood pressure in about 1% to 5% of pill users. Although pill-related high blood pressure is not usually a direct threat to life, it is a contributing cause of heart disease and stroke.

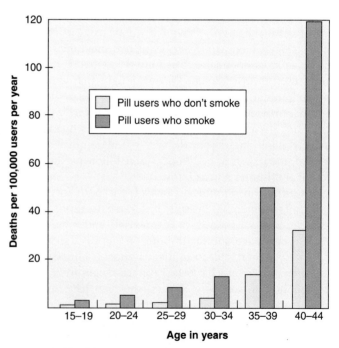

Figure 16–15 The birth control pill is an extremely effective way to prevent pregnancy that has few health risks for young, healthy women who do not smoke. The risks increase with the woman's age and the habit of cigarette smoking.

Pill users have a greater risk of getting certain STDs from an infected partner. One way the pill increases the likelihood of transmission of these diseases is by making the vaginal environment more alkaline, which favors the growth of bacteria that cause gonorrhea and chlamydia. Another reason that women on the pill have a greater risk of getting STDs is that it makes the user's cervix more vulnerable to disease-causing organisms. However, because the pill causes cervical mucus to become thicker, it decreases the chances that infections can spread to the uterus and cause pelvic inflammatory disease (PID).

Some, but not all, studies indicate that women on the pill have an increased risk of developing cervical cancer. Cervical cancer is often caused by the virus that causes genital warts, called the human papilloma virus (HPV). As was mentioned, the pill causes changes in the cervix that make it more vulnerable to disease-causing organisms. Thus HPV can gain easier entry to the cells. In addition, some studies suggest that progesterone can cause slightly abnormal cells infected with HPV to turn cancerous.

Not all the news about cancer and the pill is bad, however. A pill user's risk of developing two other forms of cancer—ovarian and endometrial—is reduced.

Progesterone-Only Contraception

There are currently three types of progesterone-only contraception: the **minipill**, **injections**, and **subcutaneous implants** (Norplant). As with the combined birth control pill, a woman must take the minipill faithfully every day. However, the minipill is generally less effective than the combined pill. Typical first year failure rates range between 1.1% and 13.2%. Progesterone injections (Depo-Provera), given every 3 months, on the other hand, are an extremely effective means of contraception. The first–year failure rate is only 0.3%. Progesterone-containing implants (Norplant) are slender, flexible silicone capsules about the size of a matchstick (Figure 16–16). Six implants inserted under the skin in a woman's upper arm are barely visible and provide extremely effective contraception for 5 years as the progesterone slowly diffuses out of the capsules. The first–year failure rate is a mere 0.09%.

There are several mechanisms through which progesterone-only contraceptives may prevent pregnancy. All types may prevent ovulation, but they vary in their ability to do this. They all cause thickening of cervical mucus, which makes it more difficult for the sperm to swim through it to reach the egg. In addition, without estrogen, the endometrium is not prepared properly for the implantation of an embryo should fertilization occur.

There are a few disadvantages to the progesterone-only contraceptives. They disturb the menstrual cycle, for instance. Consequently, a woman's periods may occur at irregular intervals or be skipped completely. In addition, weight gain of a pound or more a year can be a problem. Breast tenderness is another common complaint. Another important

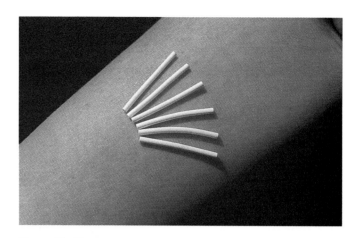

Figure 16–16 Progesterone-containing silicone rods (Norplant) are inserted just under the skin in the upper arm. The rods are shown here arrayed on the arm in the way that they would be inserted. This system provides highly effective birth control for 5 years. Fertility is restored as soon as the rods are removed. *(© Scott Camazine/Sue Trainor/Photo Researchers, Inc.)*

drawback is that, like all hormonal contraceptives, these provide no protection against sexually transmitted diseases.

Social Issue

Norplant is an extremely effective means of contraception. If widely used, how will it affect society as a whole? Will it prevent additional teenage pregnancies or will it be a blank check for uncontrolled sexual behavior? Is the fear of sexually transmitted diseases enough to promote a sense of responsibility?

Intrauterine Devices

The idea of an **intrauterine device** (IUD), a small device that is inserted into the uterus to prevent pregnancy, originated with camel drivers who placed stones in the uteri of female camels. During long treks across the desert a pregnant camel could, indeed, be an inconvenience. Fortunately, contraceptive technology has progressed since those days. Today's IUD is a small, plastic device inserted into the uterus by a physician.

There are currently two types of IUDs available in the United States. (Figure 16–17). One type has a fine copper wire wrapped around its vertical stem and the other type contains progesterone. They are both very effective in preventing pregnancy—better than 98%, for a period of 1 to 8 years, depending on the model. However, IUDs offer no protection against the spread of sexually transmitted diseases.

No one is certain how an IUD functions to prevent pregnancy, but studies suggest that it can interfere with both fertilization and implantation. Fertilization is inhibited

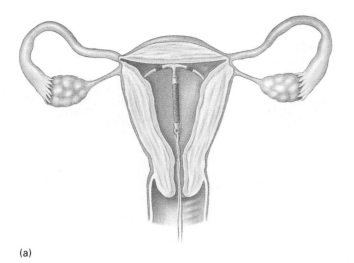
(a)

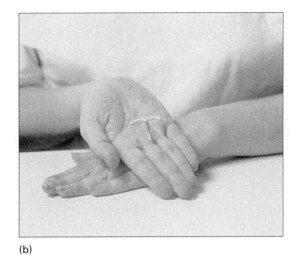
(b)

Figure 16–17 An intrauterine device is a small device that is inserted into the uterus by a physician. It prevents pregnancy by altering the uterine environment in ways that reduce the chances of fertilization and implantation (if fertilization does occur). *(b, © Gary Parker/Science Photo Library/Photo Researchers, Inc.)*

because the IUD reduces sperm motility and increases the rate at which the egg moves along the oviduct. If fertilization does occur, implantation generally does not, because the IUD also prevents the proper preparation of the uterine lining. But, if the embryo does implant, the IUD usually physically dislodges it.

The use of an IUD has some drawbacks. Insertion is often painful if the woman has not had children. An IUD often increases menstrual flow and cramping. It is occasionally expelled from the body (2% to 10% of the time), often without the woman's knowledge, leaving her without protection against pregnancy. Although an IUD can perforate the uterus, this happens very rarely. In fact, the incidence of perforation is about 1 in 2500. The most serious risk is that of pelvic inflammatory disease, which occurs when disease-causing organisms, usually sexually transmitted organisms, enter the uterus from the vagina during IUD insertion process. The risk of pelvic infection due to an IUD is low, less than 1%, but if it does occur, it can lead to sterility or ectopic pregnancy. Thus, an IUD is generally not considered a wise choice of birth control for a woman who may want children in the future or who has a high risk of exposure to STDs.

Barrier Methods of Contraception

Barrier methods of contraception include the diaphragm, cervical cap, contraceptive sponge, and male and female condom. They work, as their name suggests, by creating a barrier between the egg and sperm.

A **diaphragm** is a dome-shaped soft rubber cup on a flexible ring. It is inserted into the vagina before intercourse so that it covers the cervix, preventing sperm from entering (Figure 16–18). Before insertion, spermicidal cream or jelly should be added to the inner surface of the diaphragm. The

spermicide adds chemical protection, fills in gaps between the diaphragm and the wall of the vagina, and helps hold the diaphragm in place. Because a diaphragm covers the cervix it offers some protection to the woman against important STDs, including chlamydia and gonorrhea, and it lessens the risk of cervical cancer.

A diaphragm is at-the-time birth control. It must be inserted before intercourse (up to 2 hours before is acceptable) and *must* be left in place for at least 6 hours afterward, because sperm may remain alive within the vagina this long. If intercourse is repeated before the 6 hours are up, additional spermicide should be added to the vagina without removing the diaphragm.

Diaphragms come in a variety of sizes and it is important that a woman be properly fitted by a health-care professional. Fit is important because the diaphragm stays in place by pressing outward against the vaginal walls. If the device is too small, it can slip and leave the cervix unprotected. If it is too large, it will be uncomfortable.

A **cervical cap** is smaller than a diaphragm and fits snugly over the cervix, held in place partly by suction. The cap is one-third filled with spermicidal cream or jelly and inserted before intercourse. Like the diaphragm, it should be left in place for at least 6 hours afterward. The cap provides effective birth control for up to 48 hours.

Unlike a diaphragm or cervical cap, a contraceptive sponge or condom can be purchased without a prescription. A **contraceptive sponge** is about 2 inches in diameter and contains a spermicide. One side of the sponge has an indentation into which the cervix fits. The other side has a strap that makes removal easier. The sponge must be thoroughly moistened before use to activate the spermicide. It is then inserted and offers protection against pregnancy for the next 24 hours. The sponge is not currently available in the United States, but is available in Canada.

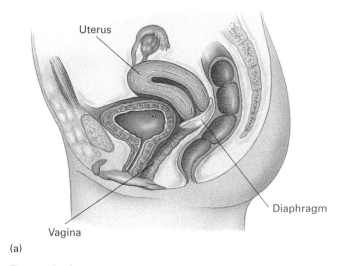

(a)

(b)

Figure 16-18 A diaphragm functions as a barrier to the cervix. It is used with spermicidal cream or jelly, which kills sperm. *(b, © SIU/Visuals Unlimited)*

The **male condom** is a thin sheath of latex or animal intestines that is rolled onto an erect penis, where it fits like a glove (Figure 16–19a). The sperm are trapped within the condom and cannot enter the vagina. The effectiveness of condoms in preventing pregnancy depends largely on how consistently they are used. The typical user failure rate is 12%, but when used properly on every occasion of intercourse, the failure rate is only 3%. Proper use of a condom includes withdrawing the penis and the condom from the vagina as soon as propriety permits so that the erection is not lost, allowing semen to escape. Condom rupture is relatively uncommon, and when it occurs, it is usually due to improper use. Some condoms come already lubricated with a spermicide to help prevent tearing and to kill sperm, in case some accidentally spill into the vagina.

Other than complete abstinence from sexual activity, the latex condom is the best means of preventing the spread of STDs available today. The animal skin membranes may offer greater sensitivity, but they do allow some microorganisms, particularly viruses, to pass. Latex has no pores and so microorganisms cannot pass from the infected secretions or surfaces of one partner to another. It is highly recommended that people who might be exposed to STDs use a condom for disease protection, even if they are already using another means of birth control to prevent pregnancy. Keep in mind, however, that a condom can only prevent disease

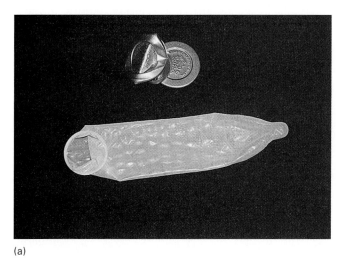

(a)

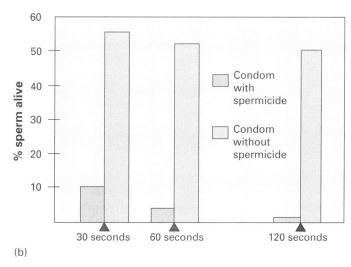

(b)

Figure 16-19 (a) The male latex condom, if used consistently and properly, is a reliable means of contraception. It is also the best means of preventing the spread of STDs available today for sexually active couples. Millions of people use latex condoms exclusively for protection from sexually transmitted infections. (b) Spermicide-containing condoms kill sperm quickly. *(b, © M. Long/Visuals Unlimited)*

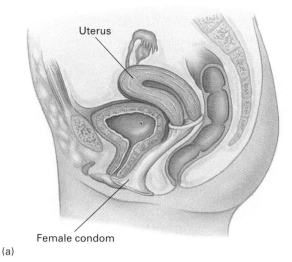

Uterus

Female condom

(a)

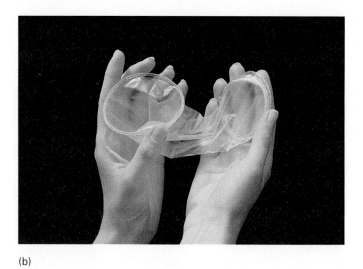

(b)

Figure 16–20 The female condom is a polyurethane sac with a flexible ring at each end to hold it in place. Although it is not as effective as its male counterpart in preventing pregnancy, results of preliminary studies suggest that it is as effective as the male condom in preventing the spread of sexually transmitted infections. *(b, © Scott Camazine/Sue Trainor/Photo Researchers, Inc.)*

transmission between the body surfaces it separates. Diseases can still be spread if contact occurs between other, unprotected body surfaces, such as the external genitalia or thighs.

The **female condom** is a loose sac of polyurethane, a type of clear plastic that resembles that of a food storage bag. At each end of the sac is a flexible ring (Figure 16–20). These help hold the device in place. The ring at the closed end of the sheath is inserted in much the same way as a diaphragm. The other one remains outside the vagina, where it helps keep the device from slipping inside and covers some of the female's external genitalia. Early studies on the effectiveness of the female condom suggest that the typical failure rate is approximately the same as for other barrier means of contraception, albeit somewhat higher than that of a male condom. It does, however, appear to be an effective means of preventing the spread of sexually transmitted infections. Female condoms offer women who cannot count on their male partners using condoms a means to protect themselves.

Critical Thinking

A latex condom provides an impenetrable barrier to disease-causing organisms. Yet, studies indicate that women who use a diaphragm or vaginal sponge routinely have lower rates of STDs than do women who rely on their male partners to use a condom. How can you explain this difference?

Spermicidal Preparations

Spermicidal preparations consist of a sperm-killing chemical in some carrier, such as foam, cream, jelly, film, or tablet

(Figure 16–21). When used without any other means of contraception, foams are the most effective form to use, because they are effective immediately and disperse more evenly. The sperm-killing effect lasts for about 1 hour after the spermicide has been activated. Spermicides have a typical user failure rate of 21%. They are particularly useful in boosting the contraceptive effects of other means of birth control, such as a condom or diaphragm.

Laboratory tests have shown that nonoxynol-9, the sperm-killing chemical in almost all spermicides in the United States, has been shown to kill the organisms responsible for many STDs, including gonorrhea, chlamydia, syphilis, trichomoniasis, genital herpes, and AIDS. Conditions in real life are not the same as in a laboratory, however.

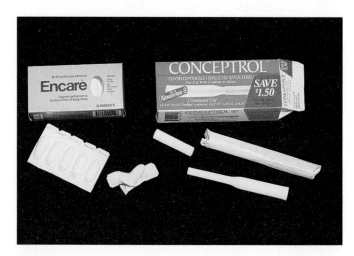

Figure 16–21 Spermicides contain a chemical that kills sperm, usually nonoxynol-9. They come in several forms, including foam, cream, and jelly. *(© M. Long/Visuals Unlimited)*

Consider, for instance, HIV, the AIDS-causing virus. Nonoxynol-9 kills the virus in laboratory tests, but it also damages the cells lining the vagina and this could increase susceptibility to the virus. Although clinical evidence confirms that spermicide protects against chlamydia and gonorrhea, the evidence on protection against HIV is contradictory. Still, spermicides provide women whose male partners are not cooperative in using latex condoms some control in STD prevention.

Fertility Awareness

Fertility awareness, which also goes by the names of **natural family planning** or the **rhythm method**, is a way to reduce the risk of pregnancy by avoiding intercourse on all days that might result in sperm and egg meeting. This sounds easier than it is. Sperm can live in the female reproductive tract for up to 3 days. An egg lives only 12 to 24 hours after ovulation. That means that there are only 4 days in each cycle during which fertilization might occur. But, which 4 days? Therein lies the problem. It is difficult enough to pinpoint when ovulation occurs, much less predicting it 3 days in advance. The situation is not entirely hopeless, however. In almost all women, regardless of their cycle length, ovulation occurs 14 (plus or minus 2) days *before* the onset of flow in the next cycle. Thus, if the woman has a stable cycle length, she can count back 14 days from the date she expects her next period and add two days on either side to determine the most likely days of ovulation. Intercourse should be avoided for 3 days before the earliest date on which ovulation might occur, because sperm may remain alive that long. It should also be avoided for 1 day after the latest date on which ovulation is expected, because the egg survives for 24 hours (Figure 16–22). Irregularities in cycle length further complicate the problem, of course.

The highest degree of success with the rhythm method is achieved if the time of ovulation is known and the couple avoids intercourse for at least 1 day after ovulation is *known* to have occurred. Body temperature changes can be a tip-off that ovulation has occurred. It rises under the influence of progesterone from the corpus luteum, which forms after ovulation. Thus, when the body temperature rises and remains elevated, it is likely that ovulation has passed and the infertile period has begun (Figure 16–23). Body temperature cannot, however, *predict* when ovulation will occur. The change in body temperature is small, usually no more than one-half to one degree, so a special thermometer is needed to detect it. In addition, because body temperature fluctuates during the course of the day and is easily influenced by many factors, it should be measured upon awakening, at the same time each day. Because the amount and consistency of cervical mucus change at the time of ovulation, these too can be used as indicators of ovulation.

Future Means of Birth Control

Advances in each type of birth control—IUDs, barrier methods, chemical methods, and hormonal methods—are

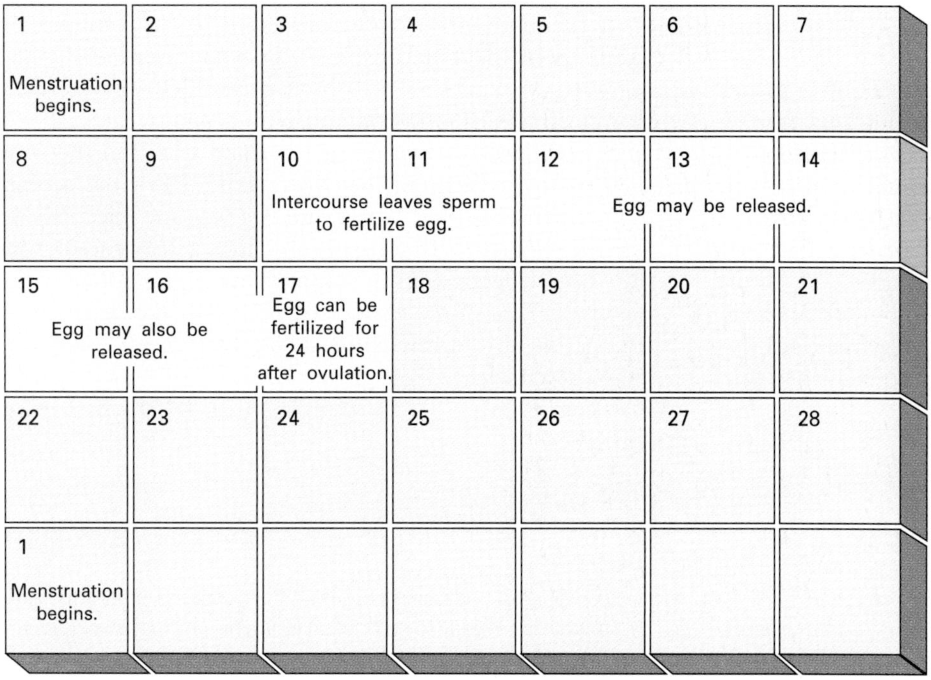

1 Menstruation begins.	2	3	4	5	6	7
8	9	10 Intercourse leaves sperm to fertilize egg.	11	12 Egg may be released.	13	14
15 Egg may also be released.	16	17 Egg can be fertilized for 24 hours after ovulation.	18	19	20	21
22	23	24	25	26	27	28
1 Menstruation begins.						

Figure 16–22 Couples using the rhythm method reduce the risk of pregnancy by avoiding intercourse at times that might result in viable sperm and egg meeting.

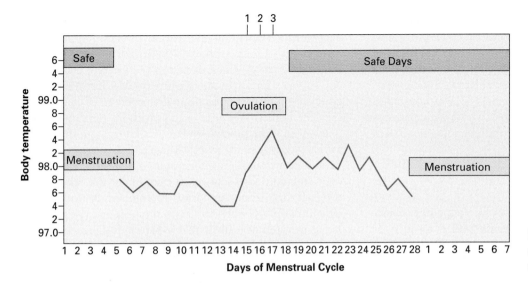

Figure 16-23 Body temperature changes during a typical menstrual cycle.

on the horizon. A new IUD may soon be available that contains a different synthetic progesterone (levonorgestrel). It can remain in place longer and causes less cramping and bleeding. A disposable diaphragm containing a spermicide that will be activated by the change in pH associated with the presence of semen is also under development, as are vaccines that might provide safe, reversible contraception. Among the vaccines being considered are those that would cause a woman's body to produce antibodies against HCG (a hormone produced by an embryo); against sperm, causing them to clump together, be immobile, or unable to fuse with an egg; or against an egg, making it impenetrable to sperm. Other vaccines, which would be used by males, would stimulate antibody production against hormones that play a role in sperm production, namely GnRH or FSH.

Many new hormonal means of contraception are being tested. For instance, new progestin skin implants that require fewer rods for the same effectiveness may soon be available. Biodegradable implants that would eliminate the need for surgical removal are also being developed. Two new injectables are being developed. One is longer-acting than Depo-Provera and the other contains estrogen, in addition to progestin, to help minimize disruption of normal bleeding patterns. Also under development are vaginal rings that would be inserted into the vagina like a diaphragm and provide continuous delivery of progestins.

Male birth control pills are also being tested. At first, researchers approached the problem of developing male hormonal contraception by trying to shut off testosterone production, because this is the hormone most directly responsible for sperm production. While testosterone inhibition does lower sperm count, it also tends to dampen a man's sex drive and can make it difficult to achieve an erection. In Italy, researchers have been successful in eliminating sperm in semen by administering a chemical that inhibits testosterone along with a dose of testosterone. FSH plays a role in sperm maturation and, therefore, provides another potential control mechanism. Gossypol, a chemical derived from cottonseed oil, blocks FSH action and, therefore, sperm maturation. This, too, proved to be less than ideal; it caused permanent sterility in 20% of the users. Other researchers working on a male birth control pill have targeted the releasing hormone from the hypothalamus (GnRH) that stimulates pituitary release of FSH and LH, and they are now testing a chemical that prevents GnRH from acting. This chemical is given along with testosterone injections so that the men are still able to get erections. In early studies on this drug's effectiveness, it stopped sperm production completely in 75% of the men tested, but only *decreased* sperm production in the remaining 25%. The effects, however, were reversible.

SUMMARY

1. The male reproductive system is composed of the testes, a series of ducts (the epididymis, vas deferens, and urethra), accessory glands (the prostate, seminal vesicles, and Cowper's glands), and the penis. The functions of these structures are given in Table 16–1.
2. The testes are located outside the body cavity in a sac called the scrotum. The temperature in the scrotum is several degrees cooler than body temperature. These cooler temperatures are important to proper sperm development. During development, the testes form within the abdominal cavity. They pass through the inguinal canal into the scrotum later in fetal development.
3. Within each testis, the production of gametes, called spermatogenesis, is a continuous process occurring within the seminiferous tubules. Sperm are produced from undifferentiated

germ cells by mitotic division followed by two meiotic divisions, which result in haploid gametes.

4. Each sperm has three distinct regions. The head of the sperm contains the male's genetic contribution to the next generation. The neck is packed with mitochondria, which produce ATP to power movement. The whip-like flagellum propels the sperm to the egg.

5. The male hormone testosterone is produced by the interstitial cells, located within the testes between the seminiferous tubules.

6. Semen is the fluid released when a man ejaculates. Most of the semen consists of the secretions of the accessory glands.

7. Impotence is the inability to achieve an erection. It is not uncommon for a man to experience impotence at some point in his life. There are psychological as well as physiological causes for impotence and there are several options for treatment.

8. Male reproductive processes are regulated by an interplay of hormones from the anterior pituitary gland in the brain (LH and FSH), from the hypothalamus (GnRH), and from the testis (testosterone and inhibin).

9. The female reproductive system consists of the ovaries, fallopian tubes, uterus, vagina, and external genitalia. See Table 16–2 for the functions of these structures.

10. A female is born with a finite number of primary follicles. A primary follicle is a primary oocyte (an immature egg) surrounded by a single layer of follicle cells. The primary oocytes will remain in this state until puberty, when (usually) one each month will continue development and be ovulated as a secondary oocyte.

11. After a secondary oocyte is ovulated, it will travel down the fallopian tubes into the uterus. If fertilization occurs, it will most likely occur in the portion of the fallopian tube nearest to the ovary. The fertilized egg (zygote) will travel along the fallopian tube into the uterus where it will implant in the endometrium of the uterus. If fertilization does not occur, the endometrial lining of the uterus will degenerate and be lost during menstruation.

12. An ectopic pregnancy is one in which the embryo implants in an area other than the uterus, most commonly in the fallopian tubes. This condition is life-threatening to the mother.

13. Hormones regulate the ovarian cycle (which prepares an egg for fertilization) and menstrual cycle (which prepares the endometrium of the uterus for implantation of the embryo). Day 1 of the menstrual cycle is the first day of menstrual bleeding. At this time the endometrium is being shed because the levels of estrogen and, in particular, progesterone are low. At the same time, FSH, secreted by the anterior pituitary, is stimulating the development of primary follicles. Follicle cells secrete estrogen. Thus, as the follicle develops, estrogen levels climb. Estrogen causes the endometrium to become thicker and more vascular. Ovulation is the release of the egg from the ovary. LH triggers ovulation. After ovulation, LH transforms the follicle cells remaining in the ovary after ovulation into a temporary endocrine structure called the corpus luteum. The corpus luteum secretes estrogen and progesterone. These hormones ready the endometrium for implantation. If the egg is not fertilized, the corpus luteum degenerates, causing the levels of estrogen and progesterone to fall, which leads to the onset of menstruation. However, if the egg is fertilized, the young embryo produces HCG, which maintains the corpus luteum. HCG is the hormone that pregnancy tests detect.

14. At menopause, which usually occurs between the ages of 45 and 55, menstruation and ovulation stop. As a result, the levels of estrogen and progesterone drop. Menopause has many psychological and physiological effects on a woman's body.

15. There are many birth control options for sexually active women and men. Refer to Table 16–3 for the type of contraceptive device, how it is used, how it works, its failure rate, risks, and the protection it offers against STDs. These methods are being improved upon and new methods are being tested for future means of birth control.

REVIEW QUESTIONS

1. Name the male and female gonads. What are the functions of these organs?
2. How is the temperature maintained in the testes? Why is temperature control important?
3. Name and describe the functions of the three regions of a sperm cell.
4. List reasons why so few sperm cells reach the egg in a female.
5. Trace the path of sperm from their site of production to their release from the body, naming each tube the sperm pass through.
6. Name the accessory glands and give their functions.
7. What is the function of the penis? Describe the process by which the penis becomes erect.
8. List the hormones from the hypothalamus, the anterior pituitary gland, and the testis that are important in the control of sperm production. Explain the interactions among these hormones.
9. List the major structures of the female reproductive system and give their functions.
10. Describe the ovarian cycle. Include in your description primary oocytes, primary follicles, Graafian follicles, and the corpus luteum.
11. What are the two layers to the wall of the uterus?
12. Describe the interplay among hormones from the anterior pituitary and from the ovaries that is responsible for the menstrual cycle.
13. What is an ectopic pregnancy? Why is it dangerous to the mother's health?
14. Describe the structure of breasts. What is their function?
15. What is menopause? Why does menopause lower estrogen levels? What are some effects of lowered estrogen levels?
16. Describe a vasectomy and a tubal ligation.
17. How does the combination birth control pill reduce the chances of pregnancy?
18. What health risks are associated with pill use?
19. What are three types of progesterone-only contraception? Which of these three has the lowest failure rate?
20. What is an IUD? How does it work to prevent pregnancy?
21. What are some barrier means of contraception? How do these work?

C h a p t e r

16A

Sexually Transmitted Diseases and AIDS

(© George Ranalli/Photo Researchers)

exuberantly transmitted diseases (STDs) have an impressive impact on humans, both in terms of their direct effects on the victims and in terms of their cost to society in general. Thirteen million infections occur each year and two-thirds of them affect people younger than age 25. The cost of treating these infections is in the millions of dollars.

We should also mention another trait of STDs. In an age when it seems that political correctness is all important, STDs are sexist. They affect women more severely than men, causing sterility, ectopic pregnancy, and cervical cancer.

One of the reasons that STDs are rampant is that people who have them are often unaware they are infected. Also, the symptoms of some STDs, such as syphilis, disappear without treatment, leading the person to believe that he or she is cured. However, this is not so, as we will see. It is important, therefore, that we become familiar with the symptoms, diagnosis, and treatment of STDs so that we may behave responsibly.

STDs Caused by Bacteria

We will consider three STDs that are caused by bacteria: chlamydia, gonorrhea, and syphilis.

Chlamydia

There are 3 to 10 million new cases of chlamydia each year, making it 3 times more common than gonorrhea and 30 times more common than syphilis. Chlamydial infections are rapidly becoming epidemic not only because they are highly contagious but also because they may not cause noticeable symptoms that would prompt the infected person to seek treatment. In fact, up to 75% of infected women and 20% of infected men have no symptoms at all. Indeed, many people learn that they have chlamydia only because they have been informed by a responsible partner diagnosed with this infection. If symptoms do develop, they may take weeks or months to appear. However, even without outward signs of infection, the disease can be passed to others.

Chlamydia is caused by a very small bacterium (*Chlamydia trachomatis*) that cannot grow outside a human cell. It is completely dependent on another cell for energy, because it cannot produce its own ATP molecules. Its name comes from the Greek word *chlamys*, which means "cloak." Indeed, while this bacterium is cloaked inside the host cell and mul-

tiplying profusely, the host cell provides camouflage. The infection begins when a healthy cell engulfs (phagocytizes) a bacterium. Once inside, the bacterium multiplies, eventually causing the cell to burst and release its load of new bacteria (Figure 16A–1). These can then infect any new, healthy cells they contact.

Symptoms of chlamydia depend largely on the site of the infection. *Chlamydia* can infect the cells of a variety of mucous membranes, including those of the urethra, vagina, cervix, fallopian tubes, throat, anus, and eyes. Worldwide, the major cause of blindness is infections by *Chlamydia trachomatis*.

The most common symptoms of chlamydia are those of urinary tract infection. When the urethra becomes inflamed, it causes a burning sensation during urination, itching or burning around the opening of the urethra, and a clear or white discharge from the urethra. The burning pain experienced while urinating generally spurs prompt medical attention. In males, the most common site of chlamydial infection is the urethra, since this is the mucous membrane most likely to be exposed to the bacteria during sexual intimacy. Chlamydial infections in women, however, are more likely to affect the pelvic organs than the urethra, because the vagina and cervix are their primary sites of contact during sexual intimacy. Unfortunately, pelvic infections are less likely to produce symptoms than are urinary tract infections, and the untreated infection may permanently damage a female's reproductive tract. Therefore, it is important that a man who is diagnosed with chlamydia inform everyone with whom he has been sexually intimate.

In women, a chlamydial infection generally begins in the vagina and cervix, but it often spreads to the fallopian

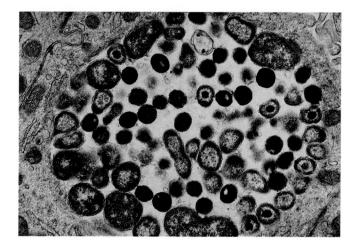

Figure 16A–1 *Chlamydia trachomatis* bacteria can be seen inside this human cell. *Chlamydia* cannot live outside a cell because they cannot produce their own ATP molecules. Once inside a cell, they multiply. Eventually the cell bursts and liberates many new bacteria that can then be taken into healthy cells. (*© David M. Phillips/Visuals Unlimited*)

tubes. It may even spread from the open ends of the fallopian tubes to the abdominal cavity. The general term for an infection of the pelvic organs is **pelvic inflammatory disease (PID)**. Chlamydia is, in fact, responsible for 50% to 90% of all cases of PID. Symptoms of PID include vaginal discharge, bleeding between periods (which is actually from the cervix), and pain during intercourse.

Chlamydia can have long-term reproductive consequences because it often causes scar tissue to form in the tubes that gametes must travel through. If the vas deferens or fallopian tubes are completely blocked, the man or woman is sterile. In a woman, partial blockage of the fallopian tube may allow the tiny sperm to reach the egg and fertilize it, but the blockage will not allow the much larger embryo to move past the scar tissue to reach the uterus. The embryo may then implant in the fallopian tube, resulting in an ectopic pregnancy that places the woman's life in danger. Sterility is more likely to occur in women than in men, because women are less likely to have symptoms that prompt treatment. Indeed, 20,000 women in the United States become sterile each year as a result of chlamydial infections. Most of these do not know they've had the disease until they try to become pregnant. Seventy-five percent of the women who are unable to get pregnant due to blocked fallopian tubes test positive for antibodies to chlamydia, indicating that they have or have had the disease.

Chlamydial infections during pregnancy place the fetus at risk. Specifically, chlamydia can cause the protective membranes around the fetus to rupture too soon and it can kill the fetus. In addition, newborns can be infected as they pass through an infected cervix or vagina at birth. If untreated, the child may remain infected for 2 or more years. Chlamydial infections picked up during birth can affect the mucous membrane of the eye (the conjunctiva), the throat, vagina, rectum, or lungs and can be fatal.

Diagnosis of chlamydia can now be done in a physician's office with quick, accurate tests. One of these is a urine test that detects the DNA of *Chlamydia*. In women, a Pap test performed to find precancerous cells on the cervix may also help detect chlamydia. In the past, diagnosis required swabbing the urethra of men or the cervix of a woman to wipe off cells and then culturing the cells for several weeks. In most cases, this is no longer necessary. Nonetheless, although the need to culture cells can delay treatment, it may be a more accurate way to detect chlamydia in symptomless women.

Considering the suffering chlamydia can cause, it is almost ironic that it can be so easily cured with antibiotics. The standard treatment is doxycycline taken each day for 7 days. A single dose of a newer drug, azithromycin, may prove to be a more effective treatment, because patients are more likely to complete treatment and be cured before resuming sexual relations.

Gonorrhea

One of the oldest known sexually transmitted diseases, gonorrhea is caused by the bacterium, *Neisseria gonorrhoeae*.

Gonorrhea, like chlamydia, primarily infects the mucous membranes of the genital or urinary tract, the throat, or the anus. The bacteria that cause gonorrhea can't survive long if exposed to air and so gonorrhea bacteria are generally transferred directly from an infected mucous membrane to an uninfected one during sexual intimacy. All that is required for transfer is contact between the membranes. Thus, a male need not ejaculate to infect his partner. A woman who has unprotected intercourse with a man with gonorrhea has a 50% chance of becoming infected. If she's on the pill, however, it is almost certain she will get it because the pill creates a hospitably alkaline environment for the bacteria. On the other hand, a man who has intercourse with a woman with gonorrhea has only about a 22% chance of becoming infected himself. The difference in transfer rates is most likely because of differences in the surface area of uninfected membrane that is exposed during intercourse—the small opening of the urethra in a male versus the entire vaginal lining and cervix in a female. Gonorrhea can also be spread to the lining of the anus during anal intercourse or to the throat during oral sex. Furthermore, it can infect the eyes, if the eye is touched after the infected area is touched. In the case of infants, eye infection can occur as the baby passes through an infected vagina or cervix during the birth process.

The symptoms of gonorrhea are similar to those described for chlamydia. In males, symptoms of urethritis (inflammation of the urethra) usually begin within 2 to 20 days of exposure. Typically, there is a discharge from the urethra that is thin and watery at first, but several days to a week later it becomes thicker and turns to yellow or white (Figure 16A–2). (This discharge is actually responsible for the name of the disease. An early Greek physician mistook the discharge for semen. As a result, the disease was named gonorrhea, from the Greek for "flow of seed.") Urination can produce a painful, burning sensation. As with chlamydia,

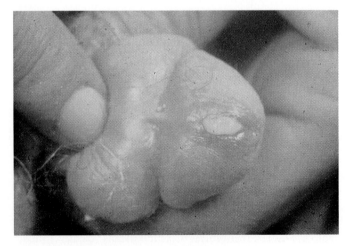

Figure 16A–2　A yellowish-white discharge from the urethra caused by gonorrhea. *(Science VU/Visuals Unlimited)*

gonorrhea can spread to the vas deferens, epididymis, prostate, or bladder. In women, the urethra may become infected, causing a discharge and burning sensation. However, it is more likely to infect the vagina and cervix and eventually result in PID. In the long run, gonorrhea can result in sterility in either sex and increase the risk of ectopic pregnancy in women. Rarely, gonorrhea can cause arthritis or fatal heart and liver infections.

Keep in mind that gonorrhea does not always produce symptoms. About 30% to 40% of infected men and 60% to 90% of infected women have no symptoms and are, therefore, unaware they have the disease. Nonetheless, they are contagious.

Gonorrhea can be diagnosed by examining a drop of pus from a man's urethra or the cervical secretions of a woman for the bacterium, by culturing cells obtained by swabbing the urethra or cervix, or by a urine test that looks for the DNA in *Neisseria*. In the first method, the so-called smear technique, the material is stained in a special way and studied under the microscope. The bacteria can usually be seen in the smear of infected cells. The culture technique has been the preferred method of diagnosis of gonorrhea in women. Some cervical secretions are placed in a growth medium that favors the growth of gonococcal bacteria. Initial studies of the new urine test suggest that it may be as accurate as the culture technique.

At one time gonorrhea could be easily cured with antibiotics, such as penicillin or ceftriaxone. Unfortunately, because of the misuse of antibiotics, some strains of the gonococcal bacteria are now drug-resistant. Therefore, it is very important to be retested for gonorrhea when the antibiotic regimen is completed to be sure the infection is cured. Because many people with gonorrhea also have chlamydia, it is not uncommon to automatically treat both infections at the same time. Immunity to gonorrhea does not develop, so you can become infected with each new exposure. It also means that if all partners are not treated and cured, the disease can "ping-pong" among them.

Critical Thinking

If a person with a bacterial infection stops taking the antibiotic when the symptoms go away, but before the entire prescribed treatment has been completed, the symptoms may recur. How can this misuse of antibiotics lead to drug-resistant strains of bacteria?

Syphilis

It may seem remarkable, but syphilis is still an important health concern for sexually active people, because it can be fatal if it is not treated. In 1995; nearly 69,000 people in the United States had syphilis.

Syphilis is caused by a corkscrew-shaped bacterium (a spirochete), called *Treponema pallidum*. As terrible as they

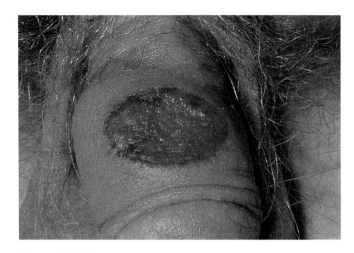

Figure 16A-3 The first stage of syphilis is characterized by a chancre, a hard, painless, crater-shaped bump at the place in the body where the bacteria entered, usually the genitals. *(© Biophoto Associates/Photo Researchers, Inc.)*

are, they are extremely delicate. These bacteria cannot survive drying or even minor temperature changes, so it is not possible to get syphilis from toilet seats, wet towels, or drinking glasses. You become infected by direct contact with an infected sexual partner. The bacteria can invade any mucous membrane or enter through a break in the skin. Unfortunately, they can also cross the placenta and infect the growing fetus if a pregnant woman is infected. Once inside the body the bacteria enter the lymphatic system and the bloodstream. Within hours the infection spreads throughout the body, but the person is totally unaware; there are no symptoms yet.

If untreated, syphilis progresses through three stages. The first stage of syphilis is characterized by a painless bump, called a **chancre**, that forms at the site of contact, usually within 2 to 8 weeks of the initial contact (Figure 16A-3). Because syphilis is sexually transmitted, the chancre usually appears on the genitals, but it can form anywhere on the body, since minute scratches on the skin will allow the bacteria to enter. The chancre appears as a hard, reddish-brown bump with raised edges that make it resemble a crater. It can be small or as large as a dime. The chancre lasts for one to a few weeks. During this time, it ulcerates, becomes crusty, and disappears. The disease, however, has not disappeared.

Unfortunately, a chancre is not always noticed. One reason is that it may form in places that can be difficult to see. In a woman, for instance, it commonly develops within the vagina or on the cervix, and since it is painless, it may go unnoticed. Another reason is that it is easy to mistake a visible chancre for something else—a pimple, an insect bite, or some other kind of local infection.

At about the same time as the chancre appears, the lymph nodes in the groin begin to swell as the immune system is activated. Again, though, there is no pain, so even this reaction may go unnoticed.

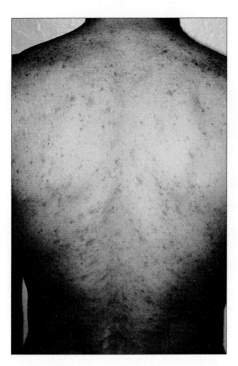

Figure 16A–4 During the second stage of syphilis a reddish-brown rash covers the entire body, including the palms of the hands and the soles of the feet. *(© Ken Greer/Visuals Unlimited)*

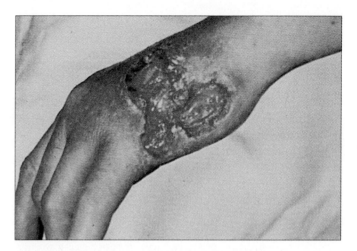

Figure 16A–5 Lesions, called gummas, shown here on the hand, are characteristic of the third stage of syphilis. These lesions can also form on the blood vessels and the bones. *(Science VU/ Visuals Unlimited)*

So how is the disease identified at this stage? During the primary stage of syphilis, diagnosis is made by identifying the bacterium in the discharge from a chancre. A special microscope and staining techniques are needed. Treatment at this stage is simple. It is done with an antibiotic, usually penicillin.

If syphilis is not detected and treated in the first stage, it will progress to the second stage, which is characterized by a reddish-brown rash covering the entire body (Figure 16A–4). The rash usually appears within a few weeks to a few months after the disappearance of the chancre (or chancres). Typically, this rash covers even the palms of the hands and the soles of the feet. The person with the rash may confuse it with an ordinary skin rash, such as might be caused by an allergy or German measles. The rash does not hurt or itch. Each bump eventually breaks open, oozes fluid, and becomes crusty. The ooze contains millions of bacteria and this is, therefore, the most contagious stage of syphilis.

Secondary syphilis usually has indications besides the rash and these may produce the first pain associated with the disease. At first there may be flu-like symptoms, including a slight temperature, chills, a sore throat, and gastrointestinal upset. Often the person aches all over and feels tired all the time. About this time, ulcers may appear on the mucous membranes of the genitals or the eyes. In addition, warty growths may appear in the genital area. Patches of hair may fall out from the head, eyebrows, and eyelashes.

Diagnosis at this stage of syphilis is done by a blood test. The most commonly used blood test is the VDRL (Venereal Disease Research Laboratory test), which measures the levels of antibodies in the blood. Antibodies are proteins produced by body cells to help defend against disease-causing organisms. The problem is the antibodies measured by the VDRL blood test are not specific to syphilis, so sometimes a person might have a positive VDRL but not have syphilis.[1] Another blood test looks for the antibodies specific for *Treponema* bacteria. Syphilis *can* be cured in the second stage with antibiotics such as penicillin, but as the disease progresses it becomes increasingly difficult to treat.

Again, the symptoms go away whether or not they are treated. However, in some people the symptoms recur periodically. In others, all outward signs of syphilis are gone for years. For the first 4 years of this symptomless period, however, the person is still contagious. During this period, the blood tests are positive for antibodies. Thus, the disease may be uncovered accidentally during a routine blood test.

The third stage brings the drama to a grisly conclusion. At this stage, lesions called **gummas** may appear on the skin or certain internal organs (Figure 16A–5). Gummas often form on the aorta, the major artery that delivers blood from the heart to the rest of the body. As a result, the artery wall may be weakened and may burst, causing the person to bleed to death internally. The *Treponema* bacteria can also infect the nervous system, damaging the brain and spinal cord. As a result, the person may lose sensations in the legs, which causes difficulty in walking, become paralyzed, or become

[1]Diseases that will result in a positive VDRL include malaria, mononucleosis, hepatitis, and leprosy.

insane. The disease may now also cause blindness by affecting the optic nerve, the iris, and other parts of the eye.

In its third stage, syphilis becomes more difficult to diagnose and very difficult to treat. General blood tests such as the VDRL may be negative, but specialized blood tests are usually positive. Treatment requires massive doses of antibiotics over a prolonged period. The damage that is done to the body cannot be repaired.

STDs Caused by Viruses

Genital Herpes

Genital herpes is caused by herpes simplex viruses (HSV). There are actually two types of HSV that cause similar sores, but they tend to be active in different parts of the body. Type 1 (HSV-1) is most commonly found above the waist, where it causes fever blisters or cold sores. Type 2 (HSV-2), on the other hand, is more likely to be found be-low the waist, where it causes genital herpes on the genitals, buttocks, or thighs. As a result of oral-genital sex with an infected person, however, HSV-1 can cause sores on the genitals and HSV-2 can cause cold sores (Figure 16A–6).

The herpesvirus is quite contagious and can be spread by direct contact with another person's infection (Figure 16A–7). Mucous membranes are most susceptible. The risk of an uninfected female getting herpes from unprotected sex with an infected male is greater than vice versa (33% vs. 5%), because it is easier for the virus to penetrate the delicate tissues of the female genital tract. Skin is a good barrier, unless there is a cut, abrasion, burn, acne, eczema, or other break in the skin. The contact, by the way, need not be sexual. A mother with a cold sore can transfer the virus when she kisses her baby. A person with herpes should avoid touching the infected area when possible. If a sore is touched, hands should be washed carefully to avoid infecting another part of the body or another person.

The first hints of genital herpes infection begin about 2 to 20 days (an average of 6 days) after the initial contact. The initial bout may be severe. There is often fever, aching mus-

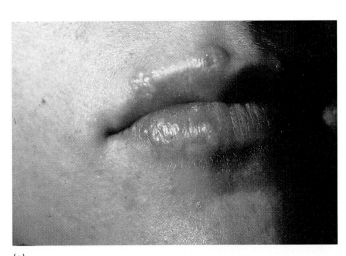

(a)

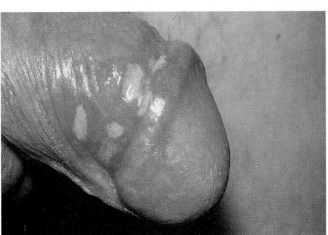

(c)

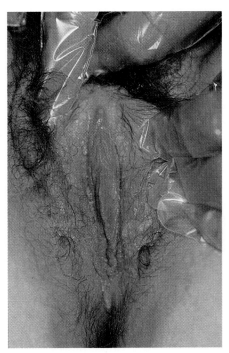

(b)

Figure 16A–6 Blisters formed by the herpes simplex virus. (a) Cold sores (also called fever blisters) are most often caused by HSV-1. Genital herpes, shown here on (b) the labia of a female and (c) the penis of a male, is usually caused by HSV-2. However, both HSV-1 and HSV-2 can cause oral herpes as well as genital herpes. *(a, © Ken Greer/Visuals Unlimited; b, © Biophoto Associates/Photo Researchers, Inc.; c, Biophoto Associates/Science Source/Photo Researchers, Inc.)*

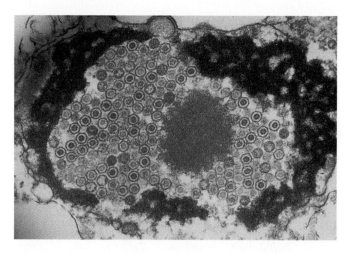

Figure 16A–7 An infection with the herpes simplex virus begins when the virus enters the body through a mucous membrane or any break in the skin. Inside the host cell, the virus produces many new copies of itself. Herpesviruses can be seen here inside a cell. The new copies can leave the cell slowly, providing no clues of shedding or the cell may rupture, releasing hundreds of new viruses. The new viruses can then infect an uninfected surface. *(© G. Musil/Visuals Unlimited)*

cles, and swollen glands in the groin. Soon blisters appear, accompanied by local swelling, itching, and possibly burning, especially if the blisters get wet during urination. The blisters form at the site of contact and last about 2 days before they ulcerate, leaving small painful sores.

The initial attack may last as long as 3 weeks, because the body has never been in contact with this virus before, and so there are no antibodies to fight off infection quickly. After this, antibodies will be present in case the virus becomes active again.

When the symptoms subside, it does not mean the virus is gone from the body. During dormant periods, the virus retreats to ganglia (clusters of nerve cells) near the spinal cord (Figure 16A–8). Then, at times of stress (emotional or physical), the virus may be reactivated and blisters may reform. About 30% of the people infected with herpes never have a recurrence. Those who do have recurrences usually have symptoms before the outbreak. The symptoms, called the prodrome, may be experienced in different ways—itching, tingling, throbbing, or pins and needles where the sores will appear. A recurrence is usually less severe than the initial outbreak. However, if the immune system is weakened, such as by HIV, the virus that causes AIDS, or by leukemia, the recurrence may be severe.

Many people apparently never develop symptoms of herpes or else the symptoms were so mild that they went unnoticed. Only about one in four of those people whose blood contains antibodies to HSV-2 (indicating that the virus had been in their bodies) can recall any previous signs of genital herpes. Nonetheless, these asymptomatic persons can un-

wittingly transmit the virus. In fact, *most* cases of genital herpes are transmitted by partners who never had symptoms or whose symptoms had been atypical and undiagnosed!

Genital herpes can be transmitted any time the viruses are being shed. The most contagious time is when active sores are present, because the number of viruses released from a sore is hundreds of times greater than is released at other times. Viral shedding begins to increase during the prodrome, so unprotected contact should certainly be avoided as soon as there is a hint of a recurrence, and shedding remains high for about 10 days after sores have healed. However, viral shedding also occurs intermittently during the intervals between outbreaks when no sores are apparent.

A latex condom provides protection against the spread of HSV *only* if it keeps the virus from reaching the uninfected surface. Viruses shed from uncovered areas, such as the scrotum, vulva, thighs, or buttocks, could still be transferred to an uninfected partner.

A herpes infection can cause problems during pregnancy. Most women with herpes have successful pregnancies and normal deliveries. In some cases, however, the infection spreads to the fetus as it is growing in the uterus and can cause miscarriage or stillbirth. HSV can also be transmitted to the newborn during vaginal delivery, especially if active sores are present. First-time HSV infection is a greater threat to the fetus than is a recurrence of a past infection. In the newborn, a herpes infection can cause local infections in the eyes, skin, or mucous membranes or full-blown infections in-

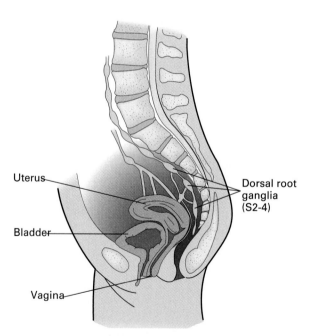

Uterus

Bladder

Vagina

Dorsal root ganglia (S2-4)

Figure 16A–8 Once a person is infected with the herpes simplex virus, the virus remains in his or her body for life. When the symptoms disappear, the virus follows nerve sheaths to ganglia (clusters of nerve cells) near the spinal cord.

volving the central nervous system. Full-blown infections are often fatal. Even survivors may have neurological problems. To avoid these problems, physicians generally recommend that the baby be delivered by cesarean section (surgical removal) if the mother has active sores at the time of delivery.

Clinicians diagnose herpes by examining the sores or by culturing and then testing the fluid from sores. In women with active sores, a stained smear of cervical secretions (as might be obtained from a Pap test) almost always reveals certain telltale cells (multinucleated giant cells with inclusions inside the nucleus). Also, blood tests can detect antibodies to the herpesviruses. However, blood tests may not be useful during an initial attack because the antibodies may not show up for several months. The presence of antibodies reveals previous HSV infections and can be useful in determining whether a pregnant woman has ever been exposed to the herpesvirus.

Although there is no cure for HSV, the antiviral drug acyclovir (Zovirax) is helpful in easing symptoms during the initial outbreak and in reducing the chances of recurrences. Acyclovir interferes with the ability of HSV to synthesize DNA and, therefore, to replicate. The result is that fewer viruses are shed. With acyclovir, sores are often less painful and heal more quickly. Unfortunately, acyclovir-resistant strains of HSV are beginning to crop up. Another antiviral drug, called Famvir, is now available for the treatment of recurrent herpes. Famvir speeds healing of herpes sores and reduces viral shedding.

A vaccine against HSV-2 is now being tested. The vaccine is a recombinant form of one of the glycoproteins (proteins with carbohydrates attached) in the coat of HSV-2. In the early trials, the vaccine did reduce the number of outbreaks of herpes, as compared with a control group. However, the vaccine does not seem to be as effective in reducing outbreaks as acyclovir. Nonetheless, it offers hope that vaccines will, someday, be useful in changing the course of chronic viral infections. It is hoped that vaccines will also help prevent the spread of HSV infection from an infected to a noninfected partner of a person with genital herpes. If a noninfected partner is vaccinated, the risk of becoming infected will be reduced.

Genital Warts

Genital warts may be caused by any of several different human papilloma viruses (HPV). These are not the same viruses that cause warts on the hands and feet. Although chlamydia holds the title of being the most common of all STDs in the United States, genital warts is the most common of the viral STDs. It is estimated that 3 million cases are diagnosed each year.

Several factors contribute to the prevalence of genital warts. HPVs are slow-growing viruses, and so there may be a long delay, an average of 2 to 3 months (although longer time periods are possible), between exposure to the virus and the formation of a wart. From the time of infection, months before warts actually appear, the virus can be spread, even though the person has no idea that he or she has been infected. Warts often form in locations where they are not likely to be discovered—the vagina, cervix, or anus. Furthermore, warts are highly contagious. About 80% of those people who have regular contact with a partner with genital warts will also develop warts.

Warts may look slightly different, depending on where they form. They look similar to warts that grow on other parts of the body when they grow on dry skin, such as the shaft of the penis or the outer vulva (Figure 16A–9). On

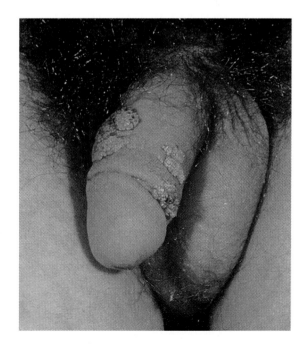

Figure 16A–9 Genital warts on the penis. *(© 1991 Steven J. Nussenblatt/Custom Medical Stock)*

Table 16A–1   Sexually Transmitted Diseases

	Symptoms	Diagnosis and Treatment	Effects
Chlamydia	First symptoms occur 7–21 days after contact. Up to 75% of women and 20% of men show no symptoms. Women: Discharge from the vagina Vaginal bleeding between periods Pain during urination/intercourse Abdominal pain accompanied by fever and nausea Men: Urethral discharge Pain during urination	Diagnosis: Urine test for Chlamydial DNA In women, Pap test Treatment: Antibiotics such as doxycycline and azithromycin	Long-term reproductive consequences such as sterility Infection can pass to infant during childbirth Can cause a rupture of the protective membrane surrounding the fetus
Gonorrhea	First symptoms occur 2–21 days after contact. About 30–40% of men and women show no symptoms Women: Vaginal discharge Pain during urination and bowel movement Cramps and pain in lower abdomen More pain than usual during menstruation Men: Thick yellow or white discharge from penis Inflammation of the urethra Pain during urination and bowel movements	Diagnosis: Examine penile discharge or cervical secretions Culturing cells Treatment: Some strains can be cured with penicillin or ceftriaxone	Can cause long-term reproductive consequences such as sterility Infection can pass to infant during birth Can cause heart trouble, skin disease, arthritis, and blindness
Syphilis	Stage 1: Occurs 2–8 weeks after contact. Chancre forms at site of contact. Swelling of lymph nodes in groin area Stage 2: Occurs 6 weeks to 6 months after contact Reddish-brown rash anywhere on the body Flu-like symptoms Ulcers or warty growths may appear Loss of patches of hair Stage 3: Lesions on skin and internal organs May affect nervous system Blindness Brain damage	Diagnosis: Identifying the bacteria from a chancre Blood test such as VDRL Treatment: Large doses of antibiotics over a prolonged period of time	Infection can pass to fetus during pregnancy Can cause heart disease, brain damage, blindness, and death
Genital Herpes	First symptoms appear 2–20 days after contact. Some people have no symptoms. Flu-like symptoms Small painful blisters that can leave painful ulcers Blisters go away but the virus remains Recurrences	Diagnosis: Examining and culturing fluid from sores Blood test for antibodies Treatment: Zovirax or Famvir can ease symptoms	Cannot be cured Infection can pass to fetus, causing miscarriage or stillbirth Can cause brain damage in newborns
Genital Warts	First symptoms appear 1–6 months after exposure. Small warts on sex organs May cause itching, burning, irritation, discharge, bleeding	Diagnosis: Appearance of growth In women, Pap test may help Treatment: For removal—freezing burning, laser, surgery	Formation of additional warts Closely associated with cervical and penile cancer Infection can pass to infant during childbirth

moist areas, however, they may first look like small pink bumps, roughly the size of a rice grain, but later they may grow together, forming clumps that look like a cauliflower. On the cervix, warts are often flat. Warts may cause itching, irritation, a foul-smelling discharge, and bleeding.

Diagnosis of warts is usually on the basis of their appearance. In women, a Pap smear is also helpful.

Treatments for genital warts are intended to kill the cells that contain the virus. These may destroy visible warts, but HPV may remain in nearby normal-looking tissue, which can cause new warts to form weeks or months after old ones have been destroyed. Methods for removing genital warts include: (1) freezing (cold cautery), (2) burning with electrical instrument (hot cautery), (3) laser (high-intensity light), (4) surgery, and (5) podophyllin (a chemical that is painted onto the warts and washed off after the prescribed time period, before it burns the skin).

The HPV viruses are closely linked to both cervical cancer in women and penile cancer in men. Indeed, HPV can be isolated in 90% of women with cervical cancer. For this reason, any woman who has ever had genital warts should have Pap tests at least once a year. HPV is also thought to be responsible for one-third of all cases of penile cancers in the United States (Table 16A–1).

AIDS

AIDS, or *a*cquired *i*mmune *d*eficiency *s*yndrome, is destined to leave its mark on many aspects of society, including medicine, science, law, economics, and education. The name of the condition is apt because, first, it is not inherited, but rather it is acquired. Second, the symptoms are those associated with a damaged immune system. Finally, a syndrome is a set of symptoms that tend to occur together and AIDS is certainly associated with a devastating set of symptoms.

We now know that AIDS is caused by a virus, the human immunodeficiency virus (HIV). As we will see, one of the primary targets of the HIV virus is helper T cells, which serve as the main switch for the entire immune response. The helper T cells activate both B cells (B lymphocytes), leading to the production of antibodies, and T cells (T lymphocytes), which are the basis of cellular defense mechanisms. Once HIV enters a helper T cell, the T cell stops functioning well, although this is not immediately apparent. HIV will, in the end, kill the infected helper T cell. The infection and eventual death of helper T cells cripple the immune system, giving disease-causing organisms that surround us all the time the opportunity to cause infection. These are the "opportunistic infections" that characterize AIDS and, in time, cause death.

A Brief History and Global Perspective

By the end of 1996, the cumulative number of AIDS cases in North America was 553,000 and approximately 62% of them had died. Also, considering the World Health Organization (WHO) 1996 estimate of 750,000 persons infected with HIV in North America, the incidence of AIDS is likely to continue to rise in the coming years (Figure 16A–10). Worldwide the problem is of even greater magnitude. There are 22.6 million HIV-infected persons, according to WHO estimates (Figure 16A–11). With approximately 14 million persons infected with HIV, Africa is currently the area hardest hit by HIV. The rate of new infections is probably greatest in areas of South and Southeast Asia.

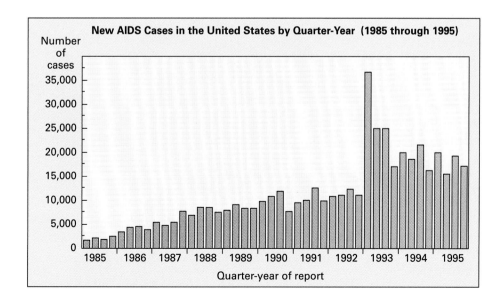

Figure 16A–10 The number of new AIDS cases in the United States for each quarter-year from 1985 through 1995, according to data from the Center for Disease Control (CDC).

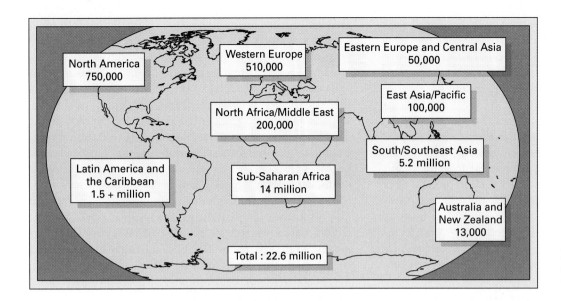

North America
750,000

Western Europe
510,000

Eastern Europe and Central Asia
50,000

East Asia/Pacific
100,000

North Africa/Middle East
200,000

South/Southeast Asia
5.2 million

Latin America and
the Caribbean
1.5 + million

Sub-Saharan Africa
14 million

Australia and
New Zealand
13,000

Total : 22.6 million

Figure 16A–11 The world-wide distribution of HIV infections.

One might wonder, then, where did HIV come from and how did it reach epidemic proportions in less than two decades? Currently, the leading idea regarding the origin of the strain of HIV responsible for most AIDS cases worldwide is that it evolved in Central Africa hundreds or thousands of years ago. If HIV is that ancient, why did it take until the 1970s and early 1980s before the first AIDS cases in the world were reported? Several factors may have contributed. HIV is known to mutate rapidly and may have genetically changed from a relatively benign to a more aggressive form within the last 20 to 30 years. Alternatively, HIV may have existed at low, stable levels in remote regions of Central Africa for a long time before being introduced into densely populated regions of Africa and the western world. In addition, social factors may have hastened the spread and perhaps even the aggressiveness of HIV. During the 1960s, while war, tourism, and trade brought the outside world into Africa's once isolated villages, the villagers were driven toward populated cities by drought and industrialization. Prostitution flourished, as did intravenous drug use. These trends *could* have provided an opportunity for the rapid evolution and spread of HIV. Once an aggressive form of HIV was introduced into the general population, it could spread rapidly because the virus can be passed during the many years before there is any indication of a problem.

HIV Structure

There are actually two classes of HIV viruses, and within each class there are many genetic strains. The first class, HIV-1, is responsible for the great majority of AIDS cases worldwide. A second type, HIV-2, is so far isolated in Africa. Here, we will simply refer to HIV, and in most cases, this will mean HIV-1.

The genetic material in HIV is RNA, rather than DNA. The RNA, along with several virus-specified enzymes, is encased in a protein coat. The RNA and protein coat together comprise the core of the virus (Figure 16A–12).

Surrounding the virus core is an outer covering or envelope consisting of glycoprotein units embedded in a lipid membrane. The lipid membrane is actually a piece of plasma membrane stolen from the previous host cell and altered for use by the virus. Each glycoprotein unit has two parts—a

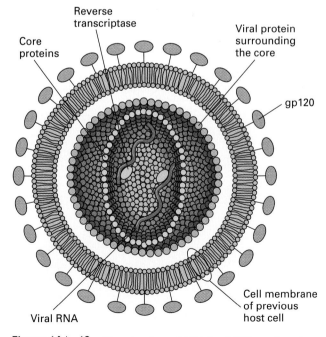

Reverse transcriptase

Core proteins

Viral protein surrounding the core

gp120

Viral RNA

Cell membrane of previous host cell

Figure 16A–12 The structure of HIV, the AIDS-causing virus.

spherical region, called gp120, and a stick-like region, called gp41, which projects into the lipid layer. The proteins of the viral envelope are responsible for binding the virus to the host cell.

Life Cycle of HIV

HIV binds to an uninfected cell when the spherical glycoprotein in its outer envelope (gp120) fits into a receptor on the host cell surface. The host cell receptor is a surface protein called CD4. Helper T cells and macrophages are the predominant cell types with these CD4 receptors. Thus, these cells are the most common targets of HIV. The usual function of CD4 receptors is in exchanging information between cells of the immune system. Nonetheless, as bad luck would have it, the viral surface protein fits into the CD4 receptors like a key in a lock. After the two proteins are properly docked, the contents of the virus enter the host cell (Figure 16A–13).

Once within the host cell, the RNA of HIV undergoes a process called reverse transcription. During this process the viral RNA is rewritten as dual strands of DNA. Since going from RNA to DNA is the reverse of the usual genetic information transfer, viruses that work this way are known as retroviruses (*retro-*, backwards).

The backwards writing of genetic information from RNA to DNA is performed by an enzyme called reverse transcriptase, which is inserted into the host cell along with viral RNA. It is interesting to note that reverse transcriptase tends to make errors as it copies RNA into DNA. Indeed, mistakes are made so frequently that variant forms of HIV arise rapidly and with serious consequences. These variants are responsible for drug-resistant strains, as we will see shortly. In addition, the diversity of forms may play an important role in the devastation of the immune system that characterizes an HIV infection. The immune system must form an army of cells specialized to attack each different variant of HIV. Although the immune system can fight several variants simultaneously, as the number of different HIV variants increases, there may be a point where there are simply too many to be fought effectively.

The newly formed DNA, which contains all the instructions necessary for producing thousands of new viruses, is then spliced into the host DNA. After it has been incorporated into the host cell DNA, that cell treats the HIV genes

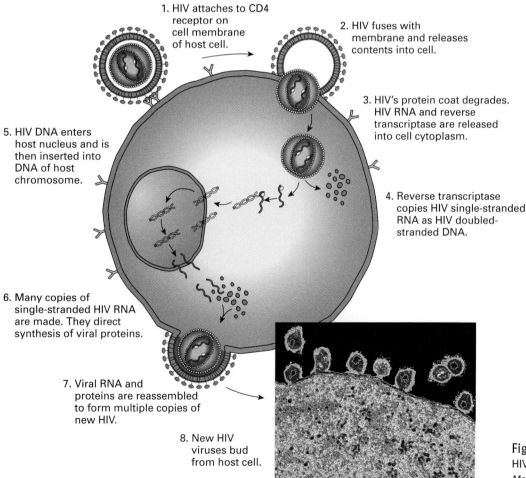

1. HIV attaches to CD4 receptor on cell membrane of host cell.

2. HIV fuses with membrane and releases contents into cell.

3. HIV's protein coat degrades. HIV RNA and reverse transcriptase are released into cell cytoplasm.

5. HIV DNA enters host nucleus and is then inserted into DNA of host chromosome.

4. Reverse transcriptase copies HIV single-stranded RNA as HIV doubled-stranded DNA.

6. Many copies of single-stranded HIV RNA are made. They direct synthesis of viral proteins.

7. Viral RNA and proteins are reassembled to form multiple copies of new HIV.

8. New HIV viruses bud from host cell.

Figure 16A–13 The life cycle of HIV. *(Photo, J. L. Carson/Custom Medical Stock Photo)*

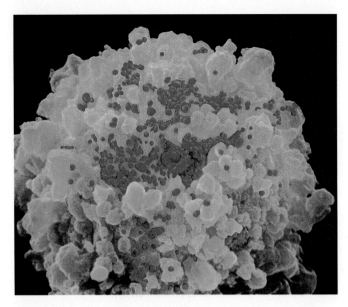

Figure 16A–14 HIV particles (red) on the surface of a lympho-cyte (green). *(© NIBSC/Science Photo Library/Photo Researchers, Inc.)*

as it would its own. Each time the cell reproduces, the viral DNA is copied along with its own.

The HIV genes can reside in a helper T cell chromosome for years, until the cell is activated to respond to some foreign antigen, such as another kind of virus, a fungus, or a parasite. At this point, the virus begins making copies of itself instead of allowing the cell to fight the invader. The viral genome is activated and turns the cell into a virtual virus factory. The newly formed viruses bud off from the host cell and can move through the bloodstream to infect new cells (Figure 16A–14).

Transmission

HIV is found in many bodily secretions—blood, semen, vaginal secretions, breast milk, saliva, tears, urine, cerebrospinal fluid, and amniotic fluid. However, HIV is not easily transmitted and only the first four of those listed contain HIV concentrations high enough to cause infection in another person.

Today, the major modes of transmission are sexual contact without a latex condom and intravenous drug use (Figure 16A–15). HIV cannot be transmitted by casual contact. It is obvious then that *most* HIV infections occur because a person engaged in a behavior that has a high risk of transmitting the virus. The good news that follows is that a person can drastically reduce his or her risk of getting HIV by never engaging in those behaviors.

HIV can enter the body through any break in the skin or a mucous membrane. Therefore, the virus could be transmit-

ted through unprotected sexual activity of any kind (that is, activity in which skin or mucous membranes of the vagina, vulva, penis, mouth, or anus of the partners is not separated by a barrier that the virus cannot penetrate). Sores or lesions caused by other sexual infections, including herpes, syphilis, and chancroid, provide an easy route of entry for HIV and, therefore, increase the risk of transmission. Anal intercourse is particularly risky for the receptive person, because HIV in the infected semen can directly infect cells in the rectum and anus and because macrophages roaming the mucous membranes of the anus may be directly infected.

Contact with infected blood is another means of HIV transmission. One way this transmission may occur is by receiving a transfusion of HIV-infected blood. Contaminated blood was the major means of transmission to hemophiliacs, who require the clotting factor from many different donors because they suffer from an inherited blood-clotting disorder. Since 1985, however, blood donors have been screened and the donated blood has been tested for HIV antibodies. Just before testing, in 1984, the risk of receiving contaminated blood was roughly 40 in 100,000. Today, the risk is only 2.25 in 100,000. Some risk remains because a recently infected donor may not have developed antibodies by the time the blood was donated. Although the risk of getting HIV from a transfusion with contaminated blood is remote, when the need for blood can be anticipated, such as for elective surgery, it may be wise for a person to donate his or her own blood ahead of time.

Intravenous drug users who share needles also have an increased risk of HIV infection. When injecting a drug, the user first draws blood into the syringe to be sure the needle

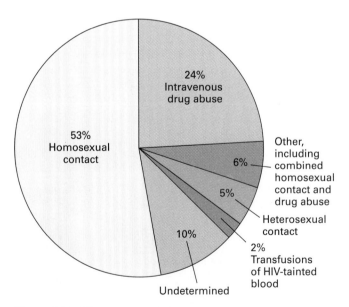

Figure 16A–15 Reported AIDS cases among males in the United States, by modes of transmission. *(Data from a 1996 CDC report.)*

is in a vein. The blood then mixes with the drug. If the syringe is then used by another person, that blood, which may be infected, could be injected directly into the bloodstream along with the drug.

A particular tragedy lies in the fact that pregnant women can infect their fetuses. In fact, an HIV-infected pregnant woman has about a 30% chance of spreading the infection to her fetus. Maternal and fetal blood supplies are brought close together, allowing nutrients and oxygen to be exchanged, but they are kept apart by membranes in the placenta. During the last months of fetal development, however, small tears may occur in the placenta and may allow the infection to spread to the bloodstream of the fetus. Bleeding during the birth process may also cause infection in the newborn. Because antibodies can cross the placenta, virtually all newborns of HIV-infected mothers test positive for HIV antibodies and are considered HIV positive (HIV+), even if the newborn is uninfected. The antibodies remain in the blood for about 15 months. There are now tests that can be done to determine whether a newborn actually harbors HIV or tests positive due to its mother's HIV antibodies.

Testing for HIV

Soon after a person is infected with HIV, his or her body usually begins to make antibodies against the virus. Antibodies can usually be detected within 8 weeks or so, but it may take 6 months or perhaps even years. An enzyme-linked immunosorbent assay (ELISA) is the most common test for antibodies to HIV. Currently, the ELISA test is better than 99.9% accurate in detecting HIV antibodies. By the end of 1994, there were at least nine different blood tests for HIV antibodies. Other tests look for HIV antibodies in urine or saliva. Home tests for HIV antibodies are now available. A sample of blood, urine, or saliva is collected in the privacy of one's home and then mailed to a testing laboratory. The results can be obtained by calling a counseling service with the reference number of the sample. If HIV antibodies are detected in a sample, another, more sensitive test, commonly the Western blot, is used to confirm the diagnosis. When antibodies are detected, the person is considered to be HIV+.

There are now other types of tests to detect HIV infection. One of these is used to look for one of the proteins in the HIV core. This test is usually done after an HIV+ woman gives birth to determine whether the virus had spread to the fetus. A newer test looks for strands of HIV RNA in blood. Thus, it reveals the number of viral particles in the bloodstream, referred to as the viral load. The viral load tells how many viruses the body is actually fighting, which, it turns out, is extremely valuable information. It allows physicians to predict the rate at which an HIV infection will progress. Disease symptoms appear sooner when the viral load is high. In addition, monitoring viral load allows researchers to judge the effectiveness of new antiviral drugs.

HIV Infection versus AIDS

Clinically, being HIV+ is not the same as having AIDS. The median time between HIV infection and the diagnosis of AIDS is 10 years, although some symptoms usually begin about 8 years after infection. By this time, the body has fought long and hard but is beginning to lose the battle. A diagnosis of AIDS is made when an HIV+ person develops one of the following conditions: (1) a helper T cell count below 200/µl of blood; (2) 1 of 26 opportunistic infections, the most common of which are *Pneumocystis carinii* pneumonia and Kaposi's sarcoma, a cancer of connective tissue that affects primarily the skin; (3) a loss of more than 10% of body weight (wasting syndrome); or (4) dementia (mental incapacity such as forgetfulness or inability to concentrate).

Sites of HIV Infection

The Immune System

HIV can infect any cell that has a CD4 receptor. The most important of these is the helper T cell. The hallmark of the disease is the progressive decline in helper T cell numbers. We know that any of several mechanisms might be responsible for the death of helper T cells. It may be that large numbers of viruses bursting out of a T cell kill the cell. Alternatively, the T cells may become so glutted with the DNA and RNA of the virus that they are unable to function and die. Another idea is that HIV causes the infected T cells to link together. The resulting clump of T cells would then be unable to mount an effective immune response. Then again, HIV infection may trigger an autoimmune response that causes other cells of the immune system to kill the helper T cells. There is also the possibility that HIV may cause the helper T cells to commit suicide. It seems that healthy T cells from HIV-positive people may self-destruct when they are activated by an antigen (much as a computer crashes when a virus it contains is activated). For some reason, when the antigen binds to the T cell, it sets into motion a program within the T cell that leads to death of the T cell.

HIV may even be able to suppress immune function and kill uninfected T cells (Figure 16A–16). You may recall that gp120 is the glycoprotein on the viral surface that binds to CD4 receptors of T cells and macrophages. It is thought that HIV may be able to release some of the gp120 molecules from its surface and these free molecules can then bind to CD4 receptors of healthy T cells. Such binding would have two possible effects. First, with gp120 filling their CD4 receptors, the uninfected T cells would be unable to communicate and mount an immune response. Second, the presence of the gp120 molecule in the receptors would make the T cells targets for attack by other immune responses.

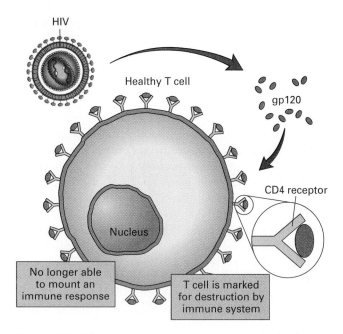

Figure 16A–16 HIV can suppress immune function and kill uninfected T cells. The protein from the HIV outer surface that fits into the CD4 receptors on helper T cells, gp120, may be freed from the virus's surface and bind to CD4 receptors on uninfected T cells. As a result of this binding, the uninfected T cell may be unable to mount an immune response and it may be marked for destruction by immune reactions.

Macrophages are the other major cell type that becomes infected with HIV. HIV does not kill a macrophage, but it does impair its ability to secrete the chemicals that activate the helper T cells. An infected macrophage may also act as a kind of Trojan horse by carrying the viruses within it to different organs, especially the brain, lungs, and bone marrow.

Suppressor T cells may also be infected. Such an infection is important because these cells normally shut down the immune response when the job is done. But as HIV kills suppressor T cells, their numbers drop and there may be unrestrained immune responses in which the immune system attacks not only invaders but its own cells as well, in autoimmune responses.

HIV may infect B cells, but only if the B cells are first infected with Epstein-Barr virus, the virus that causes mononucleosis. Although HIV infection does not seem to kill B cells or impair their ability to produce antibodies when properly stimulated, it does turn B cells into reservoirs of HIV virus.

Locations of lymphocyte production, such as the bone marrow and thymus gland, may also become infected. In this way, both types of cells (B cells and T cells) that form the essential lines of body defense become infected from the moment they form. The infected cells are ineffective in fighting disease from then on.

The lymph nodes are another important site of HIV infection. Clustered along lymphatic vessels, lymph nodes filter the lymph, helping to rid the body of infectious agents. Within the lymph nodes are tremendous numbers of macrophages and lymphocytes. Massive amounts of HIV can be found in the lymph nodes, even before symptoms of infection develop. HIV can replicate in cells with the lymph nodes and infect lymphocytes as they pass through.

The Brain

One of the symptoms of AIDS, dementia or mental incapacity, is due to the effects on the brain. When the brain becomes infected the symptoms can include forgetfulness, impaired speech, inability to concentrate, depression, seizures, and personality changes. Roughly 60% of people with AIDS have signs of dementia.

Why isn't the brain protected from HIV by the blood-brain barrier that allows only selected materials to travel from the blood into the brain itself? It is thought that the infection reaches the brain because an infected macrophage can become an infected monocyte and slip past the blood-brain barrier. Once incorporated into the brain, the monocytes can become microglia, which are cells that have important regulatory roles within the brain. Although nerve cells themselves are rarely infected, the connective tissue surrounding them is sometimes infected, and this often results in inflammation. However, inflammation or not, HIV infection seems to kill nerve cells. The neuron density in people with AIDS is almost 40% lower than in non-AIDS patients.

The Progression of an HIV Infection

An HIV infection usually progresses through a series of stages: the initial infection, an asymptomatic stage, initial disease symptoms, early immune failure, frank AIDS, and death (Figure 16A–17).

Initial Infection

During the initial stages of infection, the virus actively replicates and the circulating level of HIV rises. The body's immune system produces antibodies against the virus, in an attempt to eliminate it. Thus, after the initial stage of infection, generally between 6 weeks to 6 months later, antibodies to HIV can be detected in the blood. The person is then considered to be HIV+.

Many people have no symptoms when they first become infected. Others do experience some mild disease symptoms during the initial infection. The typical symptoms at the time of the initial infection fall into two categories. The first, and most common, type of symptoms are flu-like—swollen lymph glands throughout the body (lymphadenopathy), sore throat, fever, and perhaps a skin rash. These are symptoms typical of many viral infections and are, therefore, not helpful in diagnosing an HIV infection. The second type of

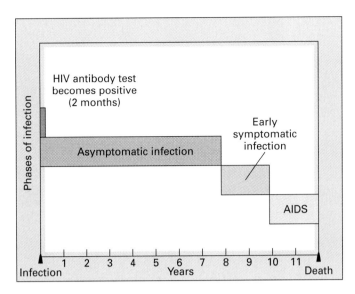

Figure 16A–17 An average time line between HIV infection, AIDS, and death.

symptoms includes those of brain infection (encephalopathy), in which there may be brain swelling or inflammation, especially of the outer protective membranes (the meninges). The brain inflammation may result from an immune response to infection or from effects of molecules from infected cells that affect brain cells. Whatever the cause, the results include headaches, fever, and problems concentrating, remembering, or solving problems.

Asymptomatic Stage

Weeks to months after the initial stage of infection, the person often feels well again, often for several years. This is the period of asymptomatic infection. During this stage of HIV infection, the immune system is able to mount a strong enough defense to control, but not conquer, the infection.

During the asymptomatic stage, HIV is far from idle. It is "hiding out" in certain lymphatic organs, including the spleen, tonsils, and adenoids, but most importantly in the lymph nodes, where it replicates. Consequently, the spaces between cells in the lymph nodes become packed with virus particles. Most helper T cells reside in the lymph nodes and many of them have the DNA form of HIV genes incorporated in their own chromosomes. Because helper T cells frequently circulate through the lymph nodes, this is an excellent place to encounter healthy T cells. Thus, millions of T cells become infected as they move through the lymph nodes on their way to other parts of the body.

Initial Disease Symptoms

Eventually, the immune system begins to falter as the virus gets the upper hand. Helper T cell numbers continue to

drop, and symptoms set in again, signaling the beginning of the end (Figure 16A–18). The initial disease symptoms fall into three classes.

The first class of symptoms includes those of wasting syndrome, which involves an otherwise unexplained loss in body weight of more than 10% of the total body weight, often accompanied by diarrhea. The weight loss is similar to that of a cancer patient. (This wasting is the source of the common name for AIDS in parts of Africa—slim disease.) Whereas healthy people who eat fewer calories than they use burn fat to make up the difference, it seems that people with AIDS have an energy deficit caused by an overactive metabolism, but their bodies target protein as the fuel to make up the shortage of calories. Thus, proteins in vital organs, the heart and lungs for instance, are sacrificed for calories. Also part of the wasting syndrome are fevers, which usually occur at night and cause periods of excessive perspiration, called night sweats.

A second class of symptoms is that of lymphadenopathy. In simpler words this means that the lymph nodes in the neck, armpits, and groin may swell and remain swollen.

The third class of symptoms is neurological. These may appear either because HIV has spread to the brain or because other organisms have caused a brain infection. Among the neurological symptoms are dementia, weakness, and paralysis caused by spinal cord damage, or pain, burning, or a tingling sensation, usually in hands or feet, caused by peripheral nerve damage.

Early Immune Failure

As T cell numbers continue their gradual decline, the body becomes increasingly vulnerable to infection (Figure 16A–19). Early signs of immune failure include the following:

1. Thrush is a fungal infection in the mouth caused by *Candida*, which is normally held in check by other microbes and the immune system. Thrush causes white, sore areas to form in the mouth that feel furry. Although antifungal drugs can be used to control thrush, it is difficult to eliminate completely. The fungus can spread down the esophagus, causing a burning sensation when the person eats.

2. Shingles is a painful rash caused by the virus that causes chickenpox (*Varicella zoster*). The virus can remain dormant in nerve cells long after a person recovers from chickenpox, and it may become active again when the immune system is compromised.

3. Hairy leukoplakia is an overgrowth of the papillae (the bumps that give the tongue a rough texture) that causes white plaques to form on the surface of the tongue. The overgrowth, which resembles cancer cells, seems to result from infection with Epstein-Barr virus, the same virus that causes mononucleosis.

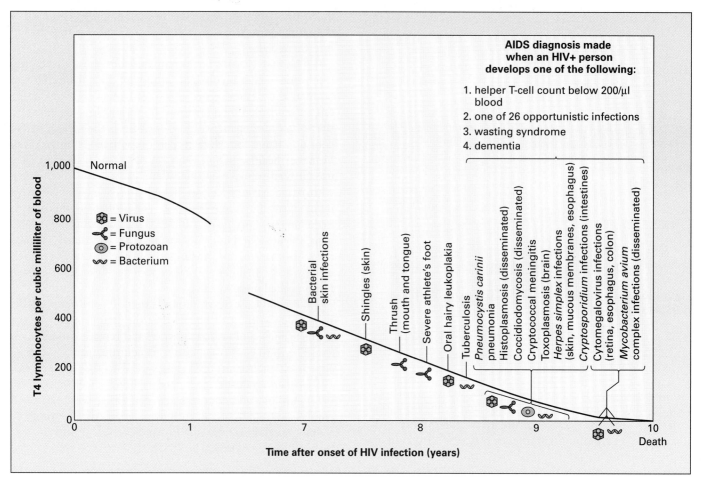

Figure 16A–18 As the numbers of helper T cells decline during an HIV infection, there is a typical progression of opportunistic infections. *(Adapted from "AIDS-Related Infections," by John Mills and Henry Masur, August, 1990, pages 52–53. ©1990 by Scientific American, Inc. All rights reserved.)*

Frank AIDS

Usually 8 or more years after the initial infection, when the helper T cell count has dropped to only 200/μl, AIDS (sometimes called frank, or full-blown, AIDS) begins. The rate of progression varies with the number of other organisms the person is exposed to and the health of the immune system at the start. When the immune system becomes this crippled, the body becomes vulnerable to certain characteristic fungal, protozoan, bacterial, and viral infections, as well as cancers.

One of the most common of the fungal infections is *Pneumocystis carinii*, which infects the lungs and is a leading cause of death in people with AIDS. It causes a dry cough and progressive shortness of breath. Three types of common soil fungi can also cause generalized infections in people with AIDS —histoplasmosis, coccidioidomycosis, and cryptococcus. In people with AIDS, these fungi can infect brain, skin, bone, liver, and lymphatic tissue.

Protozoa, single-celled organisms, can also cause serious illness when the immune system is weakened by HIV. For instance, one of these protozoans (*Cryptosporidium*) can infect the lining of the intestinal tract and cause long and severe bouts of diarrhea. Another protozoan (*Toxoplasma gondii*) is an intracellular parasite that can infect many different organs. When the brain becomes infected, symptoms similar to brain tumors develop, such as convulsions, disorientation, and dementia.

Important among the bacterial infections that commonly plague AIDS patients is tuberculosis. People with AIDS usually get an atypical form of tuberculosis that may infect both the lungs and the bone marrow.

Cytomegalovirus (CMV), a member of the herpes family, is a common virus that infects many people in childhood, but it can be reactivated in AIDS patients. For healthy people, infection with CMV does not cause major problems. Healthy children may have no symptoms at all and healthy

Wasting syndrome
- unintentional loss of more than 10% of body weight
- accompanied by severe diarrhea

Oral hairy leukoplakia
- white patches on tongue due to overgrowth of papillae
- caused by Epstein-Barr virus

Recurrent *Herpes simplex* infections
- recurring herpes blisters
- may last more than a month

Thrush
- overgrowth of yeast (*Candida*) in mouth or throat causing sore areas

PCP (*Pneumocystis carinii* pneumonia)
- dry cough, shortness of breath, fever, fatigue
- caused by a fungus

Cytomegalovirus (CMV)
- can cause blindness in persons with AIDS
- hepatitis
- inflammation of brain (encephalitis)

Toxoplasmosis
- inflammation of brain (encephalitis)
- due to a protozoan parasite

Kaposi sarcoma
- cancer of connective tissue that commonly affects skin
- appears as red or purple spots
- may be due to a virus of the herpes family

In women, gynecological problems:
- vaginal yeast infections
- pelvic inflammatory disease
- genital warts

Shingles
- painful rash consisting of small blisters
- caused by reactivation of chicken-pox virus

Lymphoma
- cancer of B cells in immune system
- affects brain, lymph nodes, GI tract, bone marrow, and liver
- may be due to reactivation of Epstein-Barr virus

Figure 16A–19   There are many HIV-related opportunistic infections.

adults typically get mononucleosis-like symptoms. When CMV infection recurs in people with AIDS, however, it can cause blindness, an adrenal hormone imbalance, or pneumonia.

Kaposi's sarcoma and lymphomas are cancers more common in AIDS patients than in the general population. In Kaposi's sarcoma, tumors form in connective tissue, especially in blood vessels and appear as pink, purple, or brown spots on the skin (Figure 16A–20). The spots are caused by the cancer's means of obtaining nourishment. The tumor

cells cause a network of blood vessels to form that tap into surrounding blood vessels. In addition, they cause neighboring blood vessels to leak and bathe the tumor cells in nourishing blood. The tumors spread and eventually may affect all linings of the body. Lymphomas are cancers of the B cells. It is thought that the lymphomas develop when genetic material of HIV is inserted into the DNA of a human chromosome near a certain cancer-causing gene and activates the gene. In AIDS patients, there may also be an unusual lymphoma that spreads to the brain.

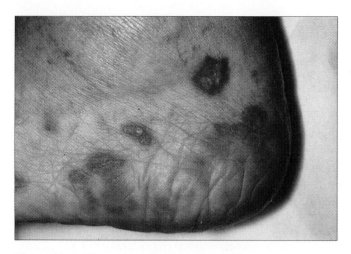

Figure 16A–20 Kaposi's sarcoma is a cancer common among persons living with AIDS in which the tumor cells obtain nourishment by forming a network of blood vessels to tap into nearby blood vessels. The blood vessels then leak blood that bathes the tumor cells. This results in the formation of purple, pink, or brown spots on the skin. *(Science VU/Visuals Unlimited)*

The frequency of Kaposi's sarcoma is 20,000 times higher among people with AIDS than among others. Why? There is no simple answer. There may, in fact, be several reasons that Kaposi's sarcoma afflicts so many AIDS patients. The suppression of the immune system is a likely contributing factor. One of the functions of healthy T cells is to roam the body looking for cells that have become cancerous and destroy them. The reduced immune surveillance caused by AIDS could allow cancer cells to proliferate. This is consistent with observations that Kaposi's sarcoma seems to plague people whose immune systems are suppressed for reasons other than AIDS. However, immunosuppression can't be the whole story. The incidence of Kaposi's sarcoma among people with AIDS is still about a 100 times higher than among people whose immune systems are suppressed for reasons other than AIDS. It turns out that HIV-infected lymphocytes produce chemicals that specifically stimulate the growth of Kaposi's sarcoma cells. There may be even more linking AIDS and Kaposi's sarcoma than just immunosuppression and growth factors. An infectious agent may also play a part. When researchers compared a sample of Kaposi's sarcoma tissue with a sample of normal healthy skin from AIDS patients with Kaposi's sarcoma, they found a piece of foreign DNA in about 93% of the Kaposi's sarcoma samples and only 6% of the healthy skin samples. The DNA fragment is thought to be a new virus in the herpes family, HHV8. More recently, researchers looked for antibodies to HHV8 in blood samples. HHV8 antibodies were found in 83% of 46 men with Kaposi's sarcoma, but on only 1% of 141 HIV-negative blood donors.

There are at least three other viruses that lead to some form of cancer in humans. Could this be another? If so, it would explain why, among HIV-positive people, gay and bisexual men (those who are likely to pick up a sexually transmitted virus along with HIV) are about 20 times more likely to develop Kaposi's sarcoma than are hemophiliacs and drug users.

Treatments for HIV Infection

The current strategy is to treat an HIV infection with drugs that slow the rate at which the HIV virus can make new copies of itself, as well as by treating opportunistic diseases and, in some cases, by helping to prevent their onset.

The original class of antiviral drugs used to slow the progression of an HIV infection includes those that act on reverse transcriptase. Recall that reverse transcriptase is the enzyme necessary for rewriting the RNA of HIV as DNA that can then be inserted into the DNA of the host cell chromosome. These drugs are chemically similar to certain building blocks of DNA. For example, azidothymidine, which is also called AZT, zidovudine, or Retrovir, is chemically similar to the DNA building block, thymidine. When reverse transcriptase rewrites the RNA that contains HIV genes as double-stranded DNA, it incorporates AZT into the DNA where thymidine should be. Unlike thymidine, however, AZT cannot link to the next building block in the chain. Thus, synthesis of the viral DNA stops and the virus cannot replicate. Fortunately, the enzymes responsible for synthesizing host cell DNA do not efficiently incorporate AZT. So, the host cell can continue to grow and divide, but viral replication is slowed. As a result, the patients feel better, have fewer bouts of infection, and may regain some of the mental and motor abilities that were lost due to HIV infection of the central nervous system.

AZT is not a wonder drug—it doesn't kill the virus, it remains effective for only a limited time, and it has side effects. Although AZT can slow down viral replication, it cannot destroy viruses that are not in the process of replication. Those latent viruses can begin active replication as soon as the drug's effectiveness wanes. Because HIV genes mutate rapidly, new forms of reverse transcriptase arise that are able to select the proper building block for DNA and ignore AZT. The AZT-resistant variants can replicate in spite of AZT, and so the number of these AZT-resistant variants steadily increases. Thus, AZT generally loses effectiveness after one to several years. Furthermore, AZT can cause serious side effects, including headaches, nausea, and a drop in red and white blood cells.

Two other drugs, DDI (dideoxyinosine) and DDC (dideoxycytidine), are chemically similar to other building blocks of DNA and, like AZT, are incorporated into the growing DNA strand by reverse transcriptase. DDC is often used in combination with AZT, posing a double threat to viral replication, because resistance to both drugs will require two separate mutations. DDI is often used to treat the people who cannot tolerate AZT.

A second class of antiviral drugs is the protease inhibitors. Recall that after the HIV genetic information has been inserted into the DNA of the host cell, it can begin producing proteins needed to form new copies of the HIV virus. The proteins initially produced are too big to be used, however. These proteins are cut to the proper size by an enzyme called a protease. Protease inhibitors block the action of this enzyme, preventing the preparation of the proteins for the assembly of new viruses. The first two protease inhibitors approved by the Food and Drug Administration, ritonavir and indinavir, dramatically reduced the number of viruses in the blood.

Combinations of drugs, sometimes referred to as drug cocktails, show great promise. Typically these cocktails consist of two inhibitors of reverse transcriptase and a protease inhibitor. The idea behind this drug combination is that protease inhibitor will slow the rate of formation of new viruses so that the development of resistance to the other drugs will be minimized. In the early studies, in more than 85% of the patients treated with these drug cocktails, the level of HIV in the blood dropped below the level that can be detected. Early studies suggest that these drug cocktails may help HIV-infected people to live longer.

A variety of other approaches to combating an HIV infection are under active investigation. Clearly, though, the best way to fight the devastation caused by HIV is to prevent it from establishing the initial infection. Theoretically, this could be accomplished by the development of a vaccine similar to those for other viral infections such as small pox or measles. Vaccines work by causing a person's body to produce antibodies against a disease without actually causing the disease. Unfortunately, there are several characteristics of HIV that have thwarted efforts to develop a vaccine:

1. The mutation rate in HIV is very high, especially for the proteins in its outer coat, the proteins that an antibody would recognize. Indeed, it has been estimated that each time reverse transcriptase rewrites the RNA of HIV as DNA there is at least one mutation. Furthermore, a person infected with HIV probably produces about a billion new virus particles a day. Thus, the vaccine could stimulate the production of antibodies against one protein, but it would fail to recognize the mutant protein.
2. There are many strains of both HIV-1 and HIV-2. Antibodies would have to recognize the exact strain entering the body.
3. HIV establishes a latent state of infection when its genetic material has become incorporated into the host DNA. Antibodies cannot attack the virus while it is hidden within the host cell.
4. HIV can pass directly from an infected cell to an uninfected one, going from one cell interior to another. Since antibodies can only attack viruses outside the cell, they cannot prevent the spread of the virus by this means of transmission.

Nonetheless, there have been several approaches to developing a vaccine to HIV. One approach—stimulating antibody production to the HIV surface protein gp120, the protein that allows HIV to enter host cells—has so far proved unsuccessful. Another approach is to use a weakened or killed HIV preparation to stimulate immune responses. Initial trials show promise. A third approach is to create a vaccine from the DNA that codes for one of the proteins on the surface coat of HIV. It turns out that injecting a gene will trigger an immune response for the protein whose synthesis is directed by the gene. The protein would then stimulate an immune response against HIV. The safety of a DNA vaccine is now being tested. Early studies on persons already HIV+ showed no serious side effects, and so the vaccine is now being tested on uninfected volunteers.

As we have seen, an infection with a retrovirus occurs in many steps. Each step presents a different Achilles heel that can be exploited to stave off HIV infection. One class of anti-HIV drugs under investigation may prevent the virus from entering cells. Recall, that HIV must bind to CD4 receptors on a host cell to enter the cell. Recombinant DNA techniques (discussed in Chapter 19) have been used to make large quantities of CD4 proteins. When these proteins are incubated with T cells or macrophages in test tubes they bind to all available CD4 receptors. In theory, this would prevent binding of HIV and subsequent infection. Although it is not clear how it worked, in test tube experiments, another drug, dextran sulfate, also seems to block HIV entry into the host cell. Unfortunately, events in real life do not always play out the way they do in test tubes; clinical trials did not show that dextran sulfate reduced HIV infection. An enzyme has been isolated that changes certain chemical bonds in the outer coat of HIV, rendering the virus unable to bind to CD4 receptors.

Another family of antiviral drugs interferes with particular proteins important in HIV replication. One protein, tat, determines the potency of HIV by influencing the rate of replication. Another, nef, slows down viral gene expression and helps keep HIV in a latent state. Viral proteases cut viral proteins into smaller, active pieces. Drugs are being developed to impede the course of HIV infection by interfering with each of these proteins.

Certain so-called antisense molecules are also being developed to combat HIV infection. These molecules would combine with the HIV single-stranded RNA, forming a double-stranded complex that would destroy the HIV RNA.

Another approach to treating AIDS is to repair the damaged immune system. For instance, it is hoped that someday there will be growth factors that can stimulate the division of helper T cells, thus restoring their numbers. You may recall from Chapters 10 and 12 that the cells that give rise to those of the immune system can be found in bone marrow and in blood from the placenta of a newborn. In the one test, a person whose immune system was already ravaged by HIV was given a bone marrow transplant from a baboon. Baboon immune cells are not susceptible to HIV, so it was hoped that the baboon cells would supply new, infection-resistant cells to the immune system. Unfortunately, the baboon cells were killed by the recipient's immune system. The chance of fetal cells being rejected is much lower than those from bone marrow. Thus, there is a glimmer of hope on the horizon that cells derived from placentas could someday be used to replace the damage immune system of a person living with AIDS.

We have covered a range of human problems here, ranging from annoyances to those that are life-threatening. To avoid these problems we must begin to take a measure of control of our own lives. We have reviewed the precautions necessary to lead a healthy life in terms of the conditions we've discussed. You know what to do and what not to do. The rest is a matter of personal responsibility.

REVIEW QUESTIONS

1. Why do some people infected with STDs fail to seek medical attention?
2. What are the STDs that are caused by bacteria? viruses?
3. Why is chlamydia rapidly becoming an epidemic?
4. What are the most common symptoms of chlamydia?
5. How can chlamydia be detected in women? In men?
6. What are the symptoms of gonorrhea? How is it diagnosed?
7. What symptoms characterize each of the three stages of syphilis?
8. What are the symptoms of genital herpes? How is genital herpes spread?
9. What are genital warts? How are they treated?
10. What are some potential long-term consequences of genital warts?
11. What is the difference between being HIV+ and having AIDS?
12. Describe the life cycle of HIV.
13. How does HIV "take over" the immune system?
14. Why isn't the brain protected from HIV by the blood-brain barrier?
15. Why are the diseases that typically occur in AIDS patients frequently called opportunistic infections?
16. What characteristics of HIV make it so difficult to develop a vaccine against it?

Development and Aging

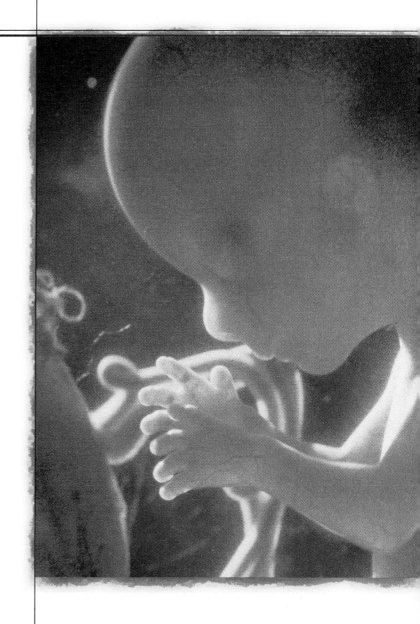

(© Petit Format/Nestle/Science Photo Library/Photo Researchers, Inc.)

A newborn enters the world outside the mother's womb and rests (Figure 17–1). How did this tiny human develop? How will this child's body change and grow while passing through infancy, childhood, adolescence, and adulthood? And what will the later years be like? In this chapter we describe the major milestones of human development.

The miracle of human development begins with fertilization, the union of a sperm and an egg, and culminates in the formation of an adult. Early development occurs within the female reproductive tract. Then, after about 266 days, birth occurs, marking the transition to development outside the mother's body. The period of development before birth is called the **prenatal period.** This period can be further subdivided into the pre-embryonic period (from fertilization through the second week), the embryonic period (from the third through the eighth weeks), and the fetal period (from the ninth week until birth). The period of development after birth is called the **postnatal period,** and this includes the stages of infancy, childhood, puberty, adolescence, and adulthood. Relative to prenatal development, postnatal development is a lengthy process, taking from 20 to 25 years to complete. Bodily changes do not cease in adulthood. Indeed, our bodies continue to change as part of the aging process. We begin our discussion of human development at the moment it all begins—when sperm meets egg.

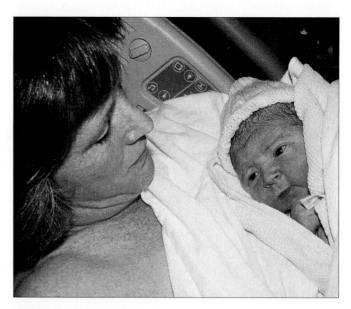

Figure 17–1   A newborn and his mother. The period of development before birth is the prenatal period, and the period after birth is the postnatal period.   *(© William Bemis)*

Prenatal Development

How do egg and sperm unite to form a single cell, and how does that cell become an embryo with all the organs of an adult human? What changes occur as the embryo becomes a fetus, and as a fetus becomes a newborn? In this section we consider the major milestones of prenatal development, what normally happens, and what can happen when things go wrong.

The Pre-Embryonic Period

The pre-embryonic period begins with **fertilization,** the union between an egg (technically a secondary oocyte) and a sperm. Fertilization takes about 24 hours from start to finish and usually occurs in a widened portion of the fallopian tube, not far from the ovary (Figure 17–2). The process of fertilization, as we will see, is a beautifully orchestrated per-

formance by egg and sperm, aided by the actions of the fallopian tubes and uterus.

Early Events and Fertilization

We begin with a consideration of how egg and sperm come to meet. The secondary oocyte, released by the ovary at ovulation, is swept into the fallopian tube by finger-like projections

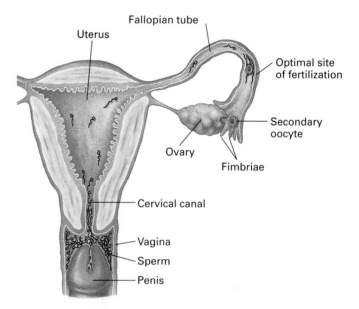

Figure 17–2   Fertilization. The secondary oocyte, released from the ovary at ovulation is swept into the fallopian tube where, following sexual intercourse, it may encounter sperm. The optimal site of fertilization is the widened portion of the fallopian tube.

called fimbriae (singular, *fimbria*; see Figure 17–2). Once inside, cilia and waves of peristalsis move the oocyte along the fallopian tube, away from the ovary, toward the uterus. While slowly drifting along the fallopian tube, the egg sends forth a chemical signal that attracts sperm. The trip made by sperm is a considerably more competitive venture than the movements of the typically lone egg. During sexual intercourse, from 200 million to 600 million sperm are deposited in the vagina and on the cervix. Using their whip-like tails, the sperm swim through the cervical canal and into the uterus where muscular contractions, perhaps stimulated by chemicals in the semen, help sperm move into the fallopian tube. Of the vast number originally deposited in the female reproductive tract, only about 200 sperm reach the site of fertilization, some after only a few minutes and others after almost an hour. Timing is everything when it comes to fertilization. Typically, oocytes cannot be fertilized more than 24 hours after their release from the ovary and most are fertilized within 12 hours. In addition, the vast majority of sperm do not survive more than 48 hours in the female reproductive tract.

Two critical processes—capacitation and the acrosome reaction—must precede fertilization. **Capacitation** is the process by which secretions from the uterus or fallopian tubes alter the surface of the acrosome, the enzyme-containing cap on the head of sperm. (The basic structures of a human sperm and secondary oocyte are shown, for review, in Figure 17–3). Usually requiring about 6 or 7 hours to complete, capacitation may involve removal of cholesterol from the plasma membrane of the sperm and a glycoprotein coat from the surface of the acrosome. Although details of capacitation remain somewhat of a mystery, it is clear that this process destabilizes the plasma membrane of sperm, a necessary prerequisite for the acrosome reaction and fertilization. Indeed, upon ejaculation, sperm are incapable of fertilizing eggs—they must first undergo capacitation.

The **acrosome reaction** occurs when capacitated sperm contact the corona radiata, a layer of cells surrounding the secondary oocyte (Figure 17–4). Once such contact occurs, perforations develop in the plasma membrane and outer acrosomal membrane of the sperm, and enzymes spill out of the acrosome. These enzymes disrupt attachments between cells of the corona radiata, allowing the sperm to pass through this layer to the layer below called the zona pellucida. Enzymes released from the acrosome also break apart the zona pellucida, creating a pathway by which the sperm can reach the oocyte.

Although several sperm may enter the zona pellucida, only one sperm usually crosses the layer, enters the oocyte, and fertilizes it. Two mechanisms prevent **polyspermy,** the abnormal condition in which more than one sperm enters an oocyte. First, when the sperm contacts the plasma membrane of the oocyte, it triggers a rapid depolarization of the membrane, making it impermeable to other sperm. Next, enzymes released by granules near the plasma membrane of the oocyte make the zona pellucida impermeable to other sperm. These blocks to polyspermy ensure equal genetic contributions from each parent and prevent abnormal numbers of chromosomes in the resulting embryo. Too many chromosomes, as would result if two or more sperm fertilized an oocyte, impair cell division and cause death of the embryo.

After penetration by one sperm of the zona pellucida, the plasma membranes of the sperm and oocyte fuse. The sperm's head and tail enter the cytoplasm of the oocyte, leaving behind the plasma membrane (see Figure 17–4). Following entry by the sperm, the oocyte undergoes the second meiotic division and is now technically considered an ovum. (Meiosis, discussed in Chapter 18, involves two divisions and results in the formation of gametes—eggs and sperm—with 23 chromosomes.) The nucleus of the ovum is called the female pronucleus, and the nucleus within the head of the sperm swells and is called the male pronucleus. Male and female pronuclei move toward each other and upon contact lose their nuclear membranes and fuse. The fertilized ovum is now called a **zygote.** Just visible to the unaided eye, this tiny speck of a cell contains genetic material from the mother (23 chromosomes) and father

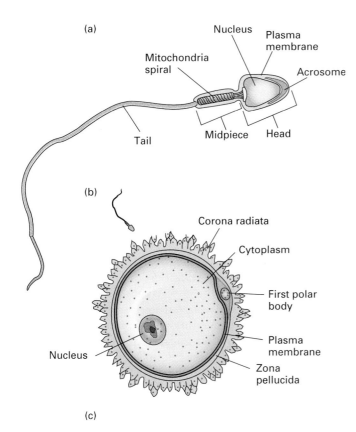

(a)

Nucleus

Plasma membrane

Mitochondria spiral

Acrosome

Tail

Midpiece   Head

(b)

Corona radiata

Cytoplasm

First polar body

Nucleus

Plasma membrane

Zona pellucida

(c)

Figure 17–3   Human gametes. (a) The main parts of a sperm. (b) A sperm drawn to the same scale as the secondary oocyte shown in c. (c) A secondary oocyte surrounded by the zona pellucida and the corona radiata.

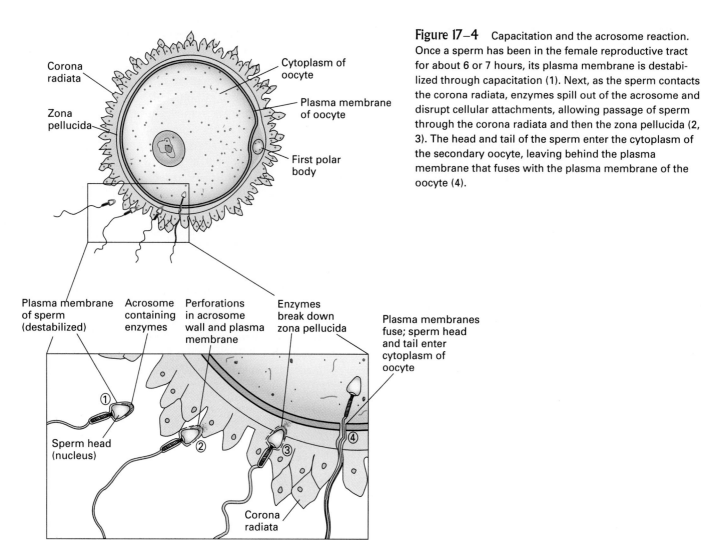

Figure 17-4 Capacitation and the acrosome reaction. Once a sperm has been in the female reproductive tract for about 6 or 7 hours, its plasma membrane is destabilized through capacitation (1). Next, as the sperm contacts the corona radiata, enzymes spill out of the acrosome and disrupt cellular attachments, allowing passage of sperm through the corona radiata and then the zona pellucida (2, 3). The head and tail of the sperm enter the cytoplasm of the secondary oocyte, leaving behind the plasma membrane that fuses with the plasma membrane of the oocyte (4).

(23 chromosomes). The events of fertilization are summarized in Figure 17–5.

Cleavage of the Zygote

The zygote is not a single cell for long. Indeed, about 1 day after fertilization, the zygote undergoes **cleavage**, a rapid series of mitotic cell divisions in which the zygote first divides into two cells, then four cells, then eight cells, and so on (Figure 17–6). Cleavage usually occurs as the zygote moves along the fallopian tube toward the uterus. The cleaving zygote becomes a **morula**, a solid ball of 12 or more cells whose name reflects its resemblance to the fruit of the mulberry tree. These early cell divisions do not result in an overall increase in size (the cells within the ball become progressively smaller as divisions occur), and thus the morula is about the size of the fertilized ovum. Increases in overall size are prevented by the tight-fitting zona pellucida that still covers the morula.

Identical twins, also called monozygotic twins ("from one zygote"), develop from a single fertilized ovum that splits in two at an early stage of cleavage. Such twins have identical genetic material and thus are always the same gender. **Fraternal twins**, on the other hand, occur when two oocytes are released from the ovaries and fertilized by different sperm. Such twins, also called dizygotic twins ("from two zygotes"), may or may not be the same gender and are as similar genetically as any siblings.

About 4 days after fertilization the morula enters the uterus and spaces begin to appear between the cells at the center of the ball. Fluid from the uterine cavity accumulates within the spaces, converting the morula into a hollow sphere of cells called a **blastocyst** (see Figure 17–6). The blastocyst contains the **inner cell mass**, a group of cells that will become the embryo, and the **trophoblast**, a thin layer of cells that will give rise to part of the placenta. The **placenta** is the organ that delivers oxygen and nutrients to the embryo and carries carbon dioxide and wastes away from it. Before implantation and development of the placenta, the blastocyst floats freely in the uterus where it is nourished by secretions from uterine glands. Increases in overall size of the blastocyst become possible because of degeneration of the zona pellucida. The ovarian cycle, fertilization, and the

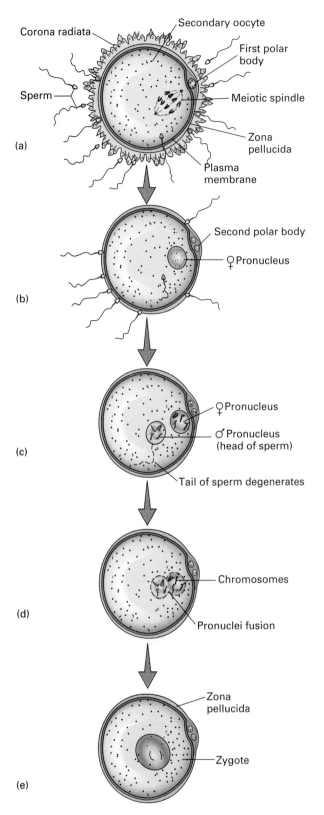

Figure 17–5 A summary of the events of fertilization. (a) Several sperm surround a secondary oocyte (only 4 of the 23 pairs of chromosomes within the oocyte are shown). A single sperm has entered the oocyte and the oocyte undergoes the second meiotic division, becoming an ovum. (b) The nucleus of the ovum becomes the female pronucleus. (c) The nucleus within the head of the sperm swells, forming the male pronucleus. (d) Male and female pronuclei fuse. (e) The resulting zygote.

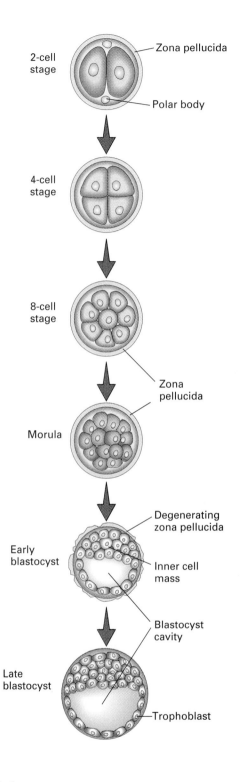

Figure 17–6 Cleavage of the zygote and formation of the blastocyst. A day or so after fertilization, the zygote undergoes a rapid series of mitotic divisions called cleavage. The zygote divides into two cells, then four cells, then eight cells, and it eventually forms a solid ball of cells known as a morula. Next, a fluid-filled cavity forms, changing the morula into a blastocyst.

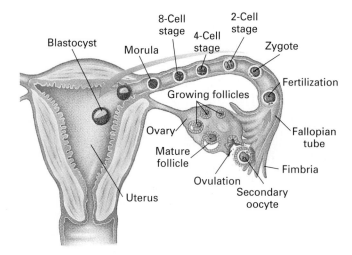

Figure 17–7 Summary of the ovarian cycle, fertilization, cleavage, and formation of the blastocyst.

Successful early development and implantation require precise conditions. Indeed, it is estimated that from one-third to one-half of all zygotes fail to become blastocysts and to implant, and they are either reabsorbed by cells of the endometrium or expelled from the uterus in an early spontaneous abortion. Causes of early spontaneous abortion include chromosomal abnormalities in the zygote and an inhospitable uterine environment for implantation. The latter might result from the presence of an intrauterine device (IUD) or inadequate production of estrogen and progesterone by the corpus luteum. Many women who spontaneously abort in the first few weeks following fertilization are unaware of their pregnancy, and they simply notice a delay in their menstrual period and an unusually heavy flow when the period arrives. **Infertility** is the inability to conceive (become pregnant) or to cause conception (in the case of males). Im-

events of the first 6 or 7 days of development are summarized in Figure 17–7.

Critical Thinking

Given what you know about the events necessary to produce identical and fraternal twins, how might triplets—two of which are identical and one fraternal—form?

Implantation of the Blastocyst

About 6 days after fertilization, the blastocyst attaches to the lining of the uterus (also called the endometrium) and begins to digest its way inward, surviving on nutrients released from the cells it destroys along the way. **Implantation** is the process by which an embryo becomes imbedded in the lining of the uterus (Figure 17–8). Normally, implantation occurs high up on the back wall of the uterus. Sometimes, however, a blastocyst implants outside the uterus and an **ectopic pregnancy** results. Although some extrauterine implantations occur on the cervix, ovary, or membrane surrounding the reproductive organs, the vast majority occur in the fallopian tubes. Most ectopic pregnancies occur as a result of impaired passage of the dividing zygote along the fallopian tube to the uterus. Factors that hinder passage of the zygote through the fallopian tube include pelvic tumors, structural abnormalities in the fallopian tubes, and scar tissue from surgery, pelvic inflammatory disease, or previous ectopic pregnancy. Women with tubal pregnancies often experience abnormal bleeding and intense abdominal pain caused by distension and irritation of the fallopian tube and surrounding membranes. If the tubal pregnancy is not diagnosed and the embryo and affected fallopian tube removed, tubal rupture and hemorrhage can be fatal to the mother.

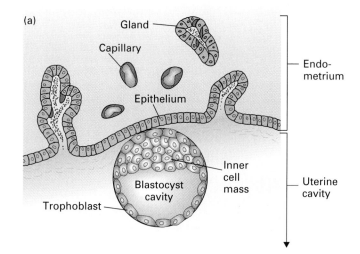

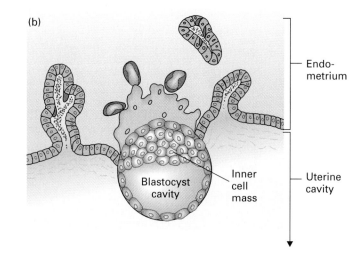

Figure 17–8 Implantation. (a) About 6 days after fertilization, the blastocyst attaches to the endometrium of the uterus, normally high up on the back wall. (b) One day later, the blastocyst begins to digest its way into the endometrium.

plantation is a major hurdle in the series of steps leading to a successful pregnancy. In the Personal Concerns essay, we describe the options available to infertile couples.

When conditions permit, the blastocyst completes implantation by the end of the second week of the pre-embryonic period. During this time, cells within the blastocyst produce human chorionic gonadotropin (HCG), a hormone that enters the bloodstream of the mother and is excreted in her urine. This hormone forms the basis for many pregnancy tests. Enough HCG is produced by the end of the second week to yield a positive pregnancy test.

Development of Extraembryonic Membranes

The amnion, yolk sac, chorion, and allantois are membranes that lie outside the embryo. Formed during the second or third week after fertilization, these membranes protect and nourish the embryo and later the fetus (Figure 17–9). The amnion surrounds the entire embryo, enclosing it in a fluid-filled space called the amniotic cavity. Amniotic fluid forms a protective cushion around the embryo that later can be examined as part of prenatal testing in a procedure known as amniocentesis (Chapter 18). Although the **yolk sac** is the primary source of nourishment for embryos in many species of vertebrates, in embryonic humans it does not provide nourishment and remains quite small (human embryos receive nutrients from the placenta). In humans the yolk sac is a site of blood cell formation and contains cells, called **primordial germ cells**, that migrate to the gonads where they differentiate into spermatogonia or oogonia, immature cells that will eventually become sperm and oocytes. The **chorion** forms the wall of the sac within which the yolk sac and the embryo within its amniotic cavity are suspended by a connecting stalk. The chorion becomes the embryo's major contribution to the placenta. Finally, the **allantois** is a small membrane that serves as a site of blood cell formation and later becomes part of the **umbilical cord**, the rope-like connection between the embryo and the placenta. The umbilical cord consists of blood vessels (two umbilical arteries and one umbilical vein) and supporting connective tissue.

Formation of the Placenta

The placenta forms from the chorion of the embryo and a portion of the endometrium of the mother. A major function of the placenta is to allow oxygen and nutrients to diffuse from maternal blood into embryonic blood and permit wastes such as carbon dioxide and urea to diffuse from embryonic blood into maternal blood. The placenta also produces hormones such as HCG, estrogen, and progesterone. Although placental hormones are essential for the continued maintenance of pregnancy, their high levels may be responsible for **morning sickness**, the nausea and vomiting experienced by some women early in pregnancy. Morning sickness is not restricted to the morning and may be caused, in part, by high levels of HCG from the placenta.

Formation of the placenta begins shortly after implantation and is completed by about the third month of pregnancy. During implantation, cells of the outer layer of the trophoblast rapidly divide and invade the endometrium. As these cells digest their way into the uterine lining, cavities form that fill with blood from maternal capillaries severed by the invading cells. Soon the inner layer of the trophoblast invades the endometrium and finger-like processes of the chorion, called **chorionic villi**, grow into the endometrium (refer again to Figure 17–9). The villi grow, divide, and continue their invasion of maternal tissue, leading to the formation of ever larger cavities and pools of maternal blood.

Chorionic villi contain blood vessels connected to the developing embryo. Thus, oxygen and nutrients within the pools of maternal blood can diffuse through the capillaries of the villi into the umbilical vein and travel on to the embryo. Wastes leave the embryo by the umbilical arteries, move into capillaries within the villi, and diffuse into maternal blood. In addition to providing exchange surfaces for diffusion of nutrients, oxygen, and wastes, some villi anchor the embryonic portion of the placenta to maternal tissue. Although maternal and fetal blood vessels are in close proximity within the placenta, under normal circumstances there is

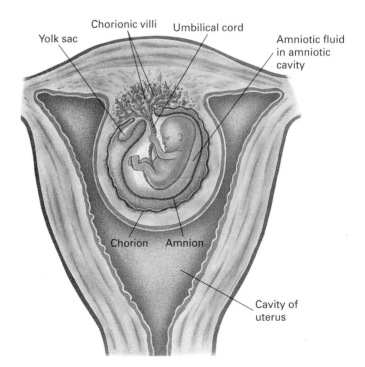

Figure 17–9 Extraembryonic membranes. The amnion, yolk sac, chorion, and allantois form during the second or third week after fertilization. Here the amnion, yolk sac, and chorion are shown about 10 weeks after fertilization; the allantois has been incorporated into the umbilical cord. Chorionic villi are finger-like extensions of the chorion that play a critical role in the formation and anchoring of the placenta.

Making Babies, But Not the Old-Fashioned Way

Some couples desperately want a child, but they cannot conceive one. The inability to become pregnant or to cause a pregnancy is called infertility. Infertility is not uncommon, striking about 1 in 6 American couples. In some of these couples, the woman is infertile. In such women, hormonal imbalances may disrupt ovulation or implantation, or scarring from disease may block the fallopian tubes and prevent passage of gametes. In women older than 40, aging eggs may be the problem. In other couples, it is the man who is infertile. Male infertility is often caused by production of few or sluggish sperm, making fertilization unlikely, if not impossible. Also, in a strange twist of fate, otherwise compatible couples may be incompatible at the cellular level. Women, for example, are sometimes allergic to their partner's sperm and produce antibodies that kill the sperm before their mission has been accomplished. Finally, in many of the most frustrating cases the cause of infertility is unknown. Can anything be done to help those who are infertile? For roughly half of couples who seek help the answer is "yes," although the road to reproduction may be long, expensive, and filled with ups and downs. In this essay we consider treatments for infertility and the ethical questions they raise.

Some cases of female infertility can be treated by administering hormones. For example, hormones that trigger ovulation may be given to women whose ovaries fail to properly release eggs. Women who tend to miscarry (spontaneously abort) may be given progesterone to enhance the receptivity of their uterine lining. However, many conditions—including scarred fallopian tubes and a low sperm count—do not respond to hormone therapy. What can be done in these cases?

Artificial insemination is one option available to couples whose infertility is caused by a low sperm count. In this procedure, sperm that have been donated and stored at a sperm bank are deposited (usually with a syringe) in the woman's cervix or vagina at about the time of ovulation. Sperm from the male member of the couple may be concentrated and then used. Alternatively, couples may use semen from an anonymous donor. (Artificial insemination with donor sperm is also frequently used by lesbian couples seeking to have children).

Some cases of infertility can be treated with in vitro fertilization (IVF), a technique made successful in England in 1978 with the birth of Louise Joy Brown, the first "test-tube" baby. Literally meaning "fertilization in glass," IVF involves placing eggs and sperm together in a glass dish in the laboratory. First, the woman is treated with hormones, such as follicle-stimulating hormone (FSH), to trigger superovulation. Superovulation is the ovulation of several oocytes, rather than the typical lone oocyte. An abundance of eggs is the key to many treatments of infertility. Large numbers of eggs can be fertilized to produce many embryos. Having several embryos available allows selection of the healthiest ones for immediate use and freezing of other healthy embryos for future use. Next, the oocytes are removed from the woman's ovaries with a suction device and placed into a dish that contains a sample of the man's sperm. If fertilization occurs, then the zygotes are transferred to a solution that will support further development. A few days later, when the embryos are at the 8-cell or 16-cell stage, one or more of them is transferred to the woman's uterus where it is hoped they will implant and complete development. Women with blocked fallopian tubes may conceive with IVF because the technique bypasses the fallopian tubes altogether. Costing about $8000 for each attempt, IVF is successful (that is, leads to a normal pregnancy) about 18% of the time. Blocked fallopian tubes can also be reopened surgically using either a laser beam or a tiny balloon.

Failure of embryos to implant is, in part, responsible for the relatively low success rate of IVF. (Implantation is also a major hurdle for fertile couples who lose about 1 in 3 pregnancies at this time.) The problem with IVF seems to be the rather violent squirt of the embryo into a reproductive tract that is already traumatized from hormone treatments and retrieval of eggs. Two new procedures avoid the embryo's problematic arrival in the uterus. In gamete intrafallopian transfer (GIFT), eggs and sperm are collected from a couple and then inserted into the woman's fallopian tube where fertilization may occur. Then, any resulting embryos drift naturally (and gently) into the uterus. The second procedure is zygote intrafallopian transfer (ZIFT). Here, eggs and sperm are collected and brought together in a dish in the laboratory. If fertilization occurs, then the resulting zygotes are inserted into the woman's fallopian tubes and from there they can travel on their own to the uterus. GIFT and ZIFT each cost about $10,000 and have success rates of 28% and 24%, respectively, although success rates of up to 50% have been reported from some clinics. Another possibility for couples in which the man has few sperm,

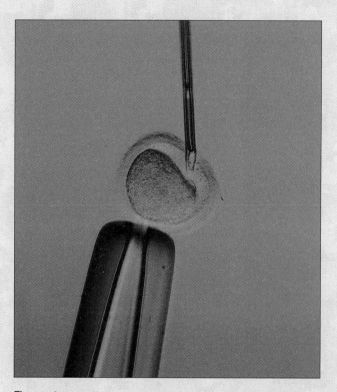

Figure A Intracytoplasmic sperm injection. An egg, held in place with a blunt-nosed pipette(bottom), is injected with a single sperm. *(© 1993 Richard G. Rawlins, Ph.D./Custom Medical Stock Photo)*

or sperm that lack the strength or enzymes necessary to penetrate an egg, is intracytoplasmic sperm injection (ICSI). In this procedure, a tiny needle is used to inject a single sperm cell into an egg (Figure A). First successful in humans in 1992, ICSI costs about $12,000 and has a success rate of 24%.

At the beginning of this essay we stated that about half of couples that seek treatment for infertility are eventually rewarded with the birth of a child. What about the remaining couples? Is there hope that some day they might also conceive? The answer is a decided "yes." Advances in reproductive research and technology are rapidly providing potential treatments for infertility. For example, scientists are currently working on techniques with mice in which spermatogonia, the

stem cells deep within the testes that produce sperm, are transferred from a fertile animal to a sterile one. The once sterile mouse can then sire offspring. Some day it might be possible to remove stem cells from a man with a low sperm count, encourage the cells to multiply under laboratory conditions, and then return them—now in much greater numbers—to their rightful owner who might then father children.

What about research with eggs? Although sperm and embryos are routinely frozen, eggs quickly become inviable when manipulated outside the body and are thus difficult to store. However, new techniques for freezing eggs are being developed. Such techniques might one day permit a young woman intent on pursuing a career to freeze her relatively young eggs. Years later, when her career is well-established, she might remove her eggs from the deep freeze and have them fertilized. Thus, the ability to successfully freeze eggs would allow women who postponed childbearing to avoid the problem of trying to conceive with old eggs. Other research focuses on using eggs from fetuses. Fetal ovaries are a rich source of eggs, containing about 2 million as compared with the estimated 400 eggs that may effectively be used by most women. Scientists are investigating the possibility of transplanting into infertile women tissue from the ovaries of fetuses made available from elective or spontaneous abortions. Fetal eggs might one day be stored at an egg bank where they could be used by infertile women. The babies produced from such eggs would be in the truly unusual position of having a biological mother who had never been born.

Obviously, current and future reproductive technology raises many ethical questions. Among them, do people who donate sperm or eggs have a claim to their biological offspring? What happens to viable embryos that are created in the laboratory but turn out to be "extra"? Do such embryos have a right to life? And what about when couples divorce or die—who gets the embryos then? And finally, is it in our long-term interest to help obviously defective sperm fertilize eggs or to have women old enough to be grandmothers bearing children? We need to think carefully and soon about these questions because many of the issues are already upon us.

no direct mixing of maternal and fetal blood. Indeed, all exchanges between mother and fetus occur across capillary walls. Nevertheless, this intimate connection between mother and fetus makes the fetus susceptible to maternal nutrition, habits, and lifestyle. Substances that cross the placenta include the AIDS virus, drugs, alcohol, caffeine, and toxins in cigarette smoke. The fully formed placenta and its internal structure are shown in Figure 17–10.

Usually the placenta is implanted in upper portions of the uterus. Sometimes, however, the placenta forms in the lower half of the uterus, entirely or partially covering the cervix. This condition, known as **placenta previa**, may cause premature birth or maternal hemorrhage and usually makes vaginal delivery impossible. In fact, about 75% of women with diagnosed placenta previa have their baby delivered by cesarean section before labor begins. Cesarean section, often shortened to C-section, is the procedure by which the fetus and placenta are removed from the uterus through an incision in the abdominal wall and uterus. The signs and symptoms of placenta previa include painless, bright–red bleeding late in pregnancy as the placenta pulls away from the expanding lower uterus. Suspected cases of placenta previa are confirmed with ultrasound, and bed rest either at home or in a hospital is usually recommended. It is

hoped that reduced maternal activity will aid in prolonging the pregnancy until the baby can survive outside the mother's body. In the past, placenta previa posed a serious health risk. Today, however, the vast majority of mothers with diagnosed placenta previa survive delivery without problems, as do their babies.

The Embryonic Period

The embryonic period extends from the third to the eighth week. It is a time of great change characterized by the formation of three distinct germ layers from which all tissues and organs develop. Two major milestones during this period include development of the central nervous system and the gonads. By the end of the embryonic period, all organs have formed and the embryo has a distinctly human appearance (Figure 17–11).

Gastrulation

"It is not birth, marriage, or death, but gastrulation which is truly the most important time in your life" (L. Wolpert, 1983, as quoted in J. M. W. Slack, 1983, *From Egg to Embryo: Determinative Events in Early Development.* Cambridge University Press, Cambridge, MA, p. 1). What is gastrulation

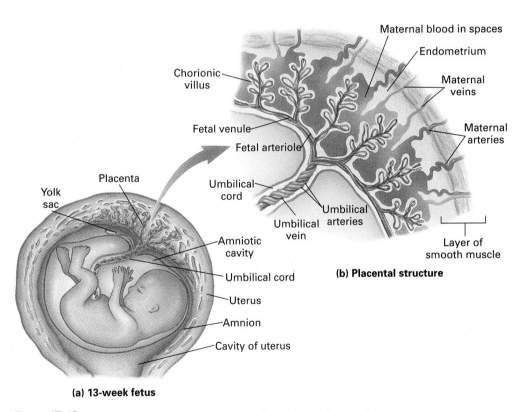

(a) 13-week fetus

(b) **Placental structure**

Figure 17–10   The placenta. Formed from the chorion of the embryo and the endometrium of the mother, the placenta permits oxygen and nutrients to diffuse from maternal blood into embryonic blood and allows wastes and carbon dioxide to diffuse from embryonic blood into maternal blood. (a) The fully formed placenta in a 13-week fetus. (b) The internal structure of the placenta.

and why is it considered, at least by one developmental biologist, to be a critical stage in development?

Morphogenesis, the development of body form, begins during the third week after fertilization. At this time, cells within the inner cell mass differentiate and migrate, forming three **primary germ layers** known as ectoderm, mesoderm, and endoderm (Figure 17–12a). The cell movements that establish the primary germ layers are called **gastrulation**, and the embryo during this period is called a **gastrula**.

All tissues and organs develop from the primary germ layers. More specifically, **ectoderm** forms the nervous system, epidermis, and epidermal derivatives such as hair, nails, and mammary glands. **Mesoderm** gives rise to muscle, bone, connective tissue, and organs such as the heart, kidneys, ovaries, and testes. **Endoderm** forms some organs and glands (for example, the pancreas, liver, thyroid gland, and parathyroid glands) and the epithelial lining of the bladder, respiratory tract, and gastrointestinal tract.

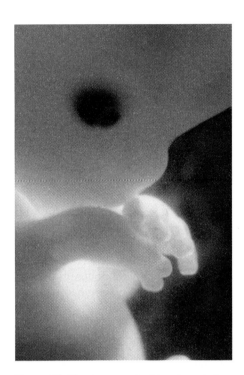

Figure 17–11 An 8-week-old embryo in its amniotic sac. By the end of the embryonic period, primordial tissues of all internal and external structures have developed and the embryo has a distinctly human appearance. *(© Petit Format/Nestle/Science Source/ Photo Researchers, Inc.)*

Figure 17–12 Formation of the central nervous system from ▶ ectoderm. Figures on the left are dorsal views (views of the embryo's back) and those on the right are cross sections. (a) The notochord induces overlying ectoderm to thicken and form the neural plate. Note also the three primary germ layers and their position with respect to the yolk sac and amniotic cavity. (b) The neural plate folds inward, forming the neural groove. (c) The neural folds grow upward. (d) The neural folds fuse to form the neural tube that will eventually develop into the brain and spinal cord. Neural crest cells, also of ectodermal origin, lie near the neural tube and form structures such as spinal and cranial nerves and components of the head. Somites are blocks of mesoderm that lie to the side of the neural tube and develop into vertebrae, connective tissue, and skeletal muscles of the neck and trunk.

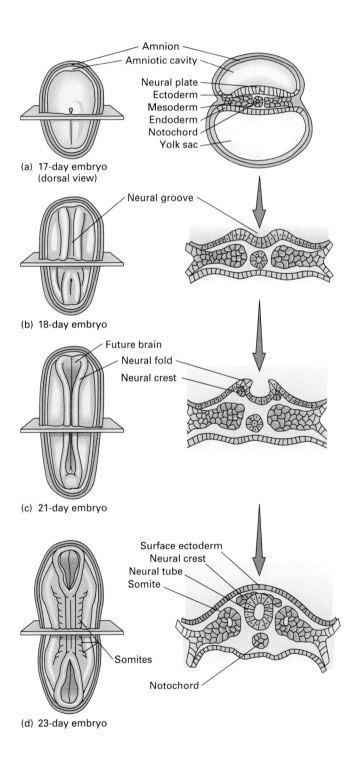

As gastrulation proceeds, a flexible rod of tissue called the **notochord** develops where the vertebral column will form. The notochord defines the axis of the embryo, gives the embryo some rigidity, and prompts overlying ectoderm to begin formation of the central nervous system (see the following discussion). During development, vertebrae form around the notochord that eventually degenerates, existing only as the pulpy, elastic material in the center of intervertebral disks.

Neurulation

A major milestone in embryonic development is formation of the central nervous system from ectoderm (see Figure 17–12). The developing notochord induces overlying ectoderm to thicken and form the **neural plate**. The neural plate then folds inward, forming a groove called the neural groove that extends the length of the embryo, along its back surface. The raised sides of the groove, known as neural folds, grow upward and eventually meet and fuse to form the **neural tube**. This series of events is called **neurulation**, and the embryo during this period is called a **neurula**. The anterior portion of the neural tube develops into the brain and the posterior portion forms the spinal cord. Other ectodermal cells, known as **neural crest cells**, occur near the neural tube. Neural crest cells migrate and differentiate, eventually forming spinal and cranial nerves, the meninges of the brain and spinal cord, and several skeletal and muscular components of the head. Alongside the neural tube, mesoderm cells organize into blocks called somites. **Somites** eventually form skeletal muscles of the neck and trunk, connective tissues, and vertebrae.

Failure of the neural tube to develop and close properly results in a birth defect called **spina bifida** ("split spine"). Infants born with spina bifida often have a cyst on their back within which lies an exposed portion of the spinal cord. Such infants face the constant threat of infection of the nervous system and may also experience areas of skin without sensation and paralysis of some muscles. The severity of spina bifida varies greatly. Although some neural tube defects are quite minor and can be improved through surgery after birth, others such as anencephaly (incomplete development of the brain) result in stillbirth or death shortly after birth. Maternal nutrition has an important impact on the incidence of defects of the neural tube. Specifically, taking vitamins and folic acid before conception appears to reduce the incidence of spina bifida that now strikes about 2 of every 1000 newborns in the United States.

Development of the Gonads

Among the 23 chromosomes provided by the mother's egg is an X chromosome, and among the 23 chromosomes provided by the father's sperm is an X or a Y chromosome. The X and Y chromosomes are called sex chromosomes and they are involved in determining the gender of human embryos. We will talk more about chromosomes and patterns of human inheritance in the next chapter. Here our focus is on how sex chromosomes influence whether embryos develop as males or females.

The gender of an embryo is determined at fertilization by the type of sperm (X-bearing or Y-bearing) that fertilizes the egg. If an X-bearing sperm fertilizes the egg, then the zygote is XX and will normally develop as a female. If, on the other hand, a Y-bearing sperm fertilizes the egg, then the zygote is XY and will normally develop as a male. Male and female embryos do not differ anatomically for the first few weeks. This situation changes, however, about 6 weeks after fertilization when a region of the Y chromosome— called Sry for sex-determining region of the Y chromosome—initiates development along the male track in XY embryos. Testes develop and soon begin producing testosterone, the sex hormone that directs development of male reproductive organs. Female embryos lack the Y chromosome, and in its absence, ovaries develop. Development of female reproductive structures is not influenced by hormones and will occur even if ovaries are absent. Thus, it is the absence of the Y chromosome and testosterone that leads to female development.

Today human sex determination by chromosomal means is well established, but such has not always been the case. Aristotle, for example, believed that the sex of a human embryo was determined by the temperature of the male partner during sexual intercourse. According to Aristotle, episodes of heated passion would produce sons, and more frigid encounters would yield daughters. So convinced was he of the role of temperature in human sex determination, that he advised elderly men (whom he assumed couldn't generate much heat) to have intercourse in the summer if they wished a male heir. Although we now know that temperature is unimportant in sex determination in humans, it does affect sex determination in other vertebrates such as turtles and alligators. Even in these animals, however, it is not the temperature of the male partner, but rather the temperature at which the fertilized eggs develop that determines the sex of offspring.

Social Issue

Methods are currently being developed to separate X-bearing sperm from Y-bearing sperm. Such "sperm sorting" would make possible the selection of a child's sex. Couples could decide whether to use X-bearing sperm to produce a female baby or Y-bearing sperm to produce a male baby. Once the choice had been made, the man would donate a semen sample and the sperm would be sorted according to whether they contained an X or Y chromosome. Then, the chosen sperm would be deposited within the woman's vagina or cervix at a time when pregnancy is likely to occur. (The process of depositing semen within the female reproductive tract is called artificial insemination.) Should parents be allowed to choose the sex of their children? What implications would sex selection have for families and for society?

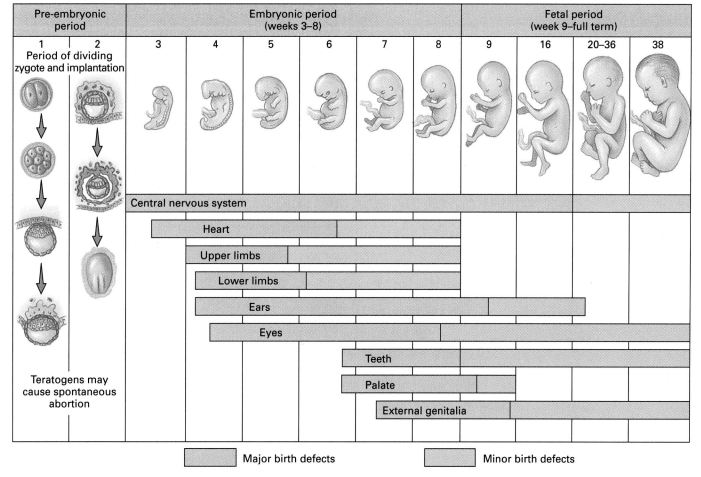

Pre-embryonic period		Embryonic period (weeks 3–8)						Fetal period (week 9–full term)			
1	2	3	4	5	6	7	8	9	16	20–36	38

Period of dividing zygote and implantation

Teratogens may cause spontaneous abortion

Central nervous system

Heart

Upper limbs

Lower limbs

Ears

Eyes

Teeth

Palate

External genitalia

Major birth defects   Minor birth defects

Figure 17–13 Critical periods in development. Red indicates stages when teratogens produce severe birth defects and yellow indicates stages when minor birth defects may be produced. Note that most severe defects occur as a result of exposure to a teratogen during the embryonic period when tissues and organs are forming.

Birth Defects

Developmental defects present at birth are called **birth defects**. Such defects involve structure, function, behavior, or metabolism and may or may not be hereditary. The study of birth defects is called **teratology**. Workers in the field of teratology often classify the causes of birth defects as genetic or environmental. Genetic causes include mutant genes or changes in the number or structure of chromosomes. Environmental causes include drugs, chemicals, radiation, and certain viruses such as herpes simplex and rubella (the cause of German measles). Environmental agents that disrupt development are called **teratogens**. Birth defects resulting from genetic causes are discussed in Chapter 18 and those associated with drugs are considered in Chapter 8A. Here we focus on how the embryo's age, level of exposure (dose), and genetic background influence the ability of a teratogen to disrupt development.

Teratogens have their greatest effects on the developing embryo during periods of rapid differentiation when cells and tissues acquire specific functions. Because organs and or-

gan systems develop at different times and rates, each has a critical period when it is most susceptible to disruption by teratogens (Figure 17–13). The central nervous system, for example, has a critical period extending from week 3 through week 16 after fertilization. During this time, the developing nervous system is highly sensitive and thus teratogens may cause major birth defects. After week 16, however, exposure to teratogens tends to induce relatively minor defects.

Another look at Figure 17–13 reveals that exposure to harmful agents during the first 2 weeks after fertilization is not known to cause birth defects. This is because most development during the pre-embryonic period centers on formation of extraembryonic structures, such as the amnion and chorion, rather than rapid differentiation of structures within the embryo. The developing zygote, blastocyst, and later embryo are not, however, immune to environmental influences. Indeed, rather than producing birth defects during the pre-embryonic period (a time when most women are unaware they are pregnant), teratogens may interfere with cleavage or implantation and cause a spontaneous abortion.

As mentioned previously, the level of exposure to a drug or chemical determines the impact of a teratogen on the developing embryo. Generally speaking, the greater the exposure to a teratogen during pregnancy, the more severe the resulting defect. The genetic makeup of the embryo also determines whether a teratogen will disrupt development. For example, some embryos exposed to a medication to prevent convulsions in their mothers show major defects, others display minor defects, and still others are unaffected. These differences in response are thought to result from genetic differences among the embryos.

Social Issue

In the late 1950s and early 1960s, the drug thalidomide was widely used in Europe as a sedative and as relief for morning sickness. Thalidomide turned out to be a potent teratogen, causing defects of the limbs (complete absence, abnormally small, or seal-like arms and legs) and abnormalities of the heart, ears, and urinary and digestive systems. About 12,000 infants whose mothers had taken thalidomide were affected. Today drug companies in the United States want to market thalidomide for treatment of leprosy and HIV. Should a drug known to cause severe birth defects be made available for treatment of seriously ill adults?

The Fetal Period: Growth and Circulation

The fetal period extends from the ninth week after fertilization until birth. During this period, tissues and organs continue to differentiate. Growth during the fetal period is extremely rapid, reflected in substantial increases in length and phenomenal increases in weight (Table 17–1). One striking change is that the rate of growth of the head slows relative to other regions of the body, and this alters the way the body is proportioned. At the start of the fetal period the head is about half the crown-rump length of a fetus. (Also known as "sitting height", crown-rump length is a measurement of the distance from the head to the rump.) In contrast, shortly before birth, the head makes up about one third of the crown-rump length (Figure 17–14). Change in the relative rates of growth of various parts of the body is called **allometric growth**. Such growth helps to shape humans and other organisms.

The fetal circulatory system differs from circulation after birth because several organs—lungs, kidneys, and liver, to name a few—are nonfunctional in the fetus and thus most blood is shunted past them through temporary vessels or openings. Recall that the fetus receives its oxygen and nutrients from maternal blood and gives up its carbon dioxide and wastes to maternal blood and that the site of exchange is the placenta. Next we will trace the path of blood flow from the placenta to the fetus and then back to the placenta, not-

Table 17–1 Changes in Length, Weight, and Physical Appearance During the Fetal Period[a]

Age (weeks)	Crown-Rump Length (mm)	Weight (gm)	External Characteristics
9	50	8	Eyes closed; genitalia not distinguishable as male or female; ears low-set
10	61	14	Fingernails begin to develop
12	87	45	Gender distinguishable; neck well-defined
14	120	110	Head erect; eyes face anteriorly; ears close to final position; legs well-developed; toenails begin to develop
16	140	200	Ears stand out from head
18	160	320	White greasy substance covers skin; fetal movement felt by mother
20	190	460	Head and body hair (lanugo) visible
22[b]	210	630	Skin wrinkled, translucent, and pink
24	230	820	Fingernails obvious; body is lean
26	250	1000	Eyes partially open; eyelashes present
28	270	1300	Eyes wide open; good head of hair often present; skin wrinkled
30	280	1700	Toenails obvious; body filling out
32	300	2100	Fingernails reach finger tips; skin is pink and smooth
36	340	2900	Body plump; lanugo hairs almost absent; toenails reach toe tips; limbs flexed
38	360	3400	Prominent chest; fingernails extend beyond finger tips

[a]Modified from Table 6–2 in K. L. Moore and T. V. N. Persaud, *The Developing Human.* 5th ed. Philadelphia: Saunders, 1993.

[b]Approximate minimum age at which fetuses may survive (with medical care) if born.

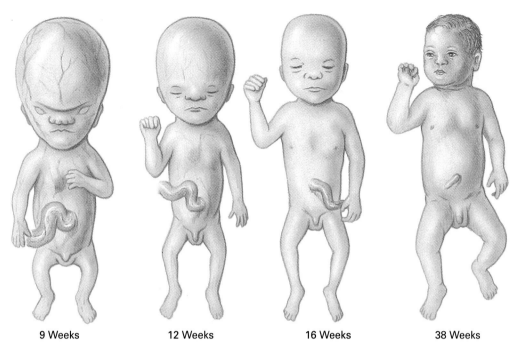

| 9 Weeks | 12 Weeks | 16 Weeks | 38 Weeks |

Figure 17–14   Allometric growth. Changes in body proportions occur during the fetal period when the rate of growth of the head slows relative to the rest of the body. Note that at 9 weeks the head makes up about half the crown-rump length of the fetus, but by 38 weeks it makes up only about one-third of this measurement. For comparison, all stages are drawn to the same total height.

ing the circulatory bypasses unique to life before birth. We end the discussion with a description of the changes in circulation that occur at birth when the organs of the newborn begin to function.

The umbilical vein (within the umbilical cord) carries blood rich in oxygen and nutrients from the placenta to the fetus (Figure 17–15a). Some of this blood flows to the fetal liver that produces red blood cells but does not yet function in digestion. Most of the blood from the placenta, however, bypasses the liver by means of a shunt called the **ductus venosus** and enters the inferior vena cava where it mixes with blood low in oxygen. This mixed blood then enters the right atrium of the heart. Some of the blood in the right atrium passes to the right ventricle and on to the lungs, as it does in circulation after birth, but most moves into the left atrium through a small hole in the wall between the atria called the **foramen ovale**. Like the liver, fetal lungs are not yet functional and need only a small amount of blood for growth and removal of wastes from their cells. Another shunt, called the **ductus arteriosus**, connects the pulmonary artery to the aorta and also functions to divert blood away from the lungs. Thus, blood flowing out of the right ventricle into the pulmonary artery is shunted away from the lungs and into the aorta where it travels to all areas of the fetus. Fetal blood must return to the placenta to rid itself of carbon dioxide and wastes and to pick up oxygen and nutrients. It does so by flowing through umbilical arteries that branch off large arteries in the legs. The two umbilical arteries run through the umbilical cord to the placenta. Another glance at Figure 17–15a reveals that much of the blood traveling through the fetus is moderate to low in oxygen. Indeed, the umbilical vein is the only fetal vessel that carries fully oxygenated blood.

At birth, when the lungs, liver, and other organs begin to function, several changes occur, converting fetal circulation to the postnatal pattern (Figure 17–15b). Blood flow to the placenta ceases when the umbilical cord is tied and cut off, the scar left by the cord becoming the baby's navel or "belly button." Within the infant, the umbilical arteries, umbilical vein, ductus venosus, and ductus arteriosus constrict, shrivel, and form ligaments, some of which disappear with age. The foramen ovale normally closes shortly after birth, leaving a small depression called the **fossa ovalis**. With the closure of the foramen ovale, blood from the right atrium passes to the right ventricle and on to the lungs that are now functional. In some newborns, the foramen ovale fails to close. Known as **blue babies**, these infants have a bluish appearance because much of their blood still bypasses the lungs and is low in oxygen. Fortunately, this defect can be corrected with surgery.

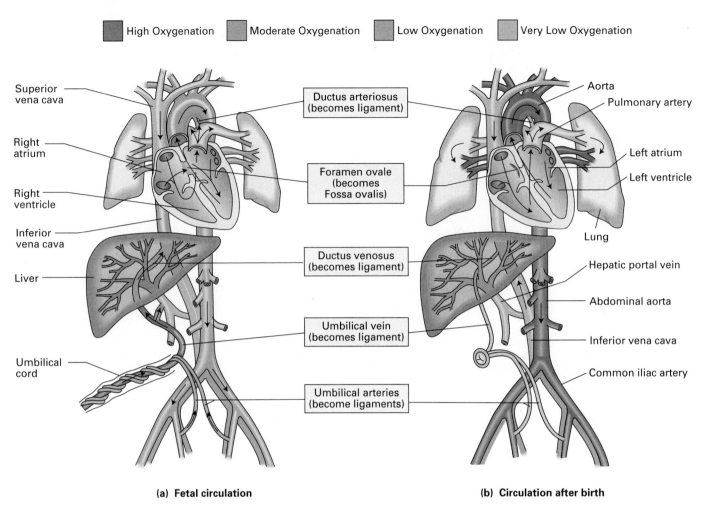

Superior vena cava

Right atrium

Right ventricle

Inferior vena cava

Liver

Umbilical cord

Ductus arteriosus (becomes ligament)

Foramen ovale (becomes Fossa ovalis)

Ductus venosus (becomes ligament)

Umbilical vein (becomes ligament)

Umbilical arteries (become ligaments)

Aorta

Pulmonary artery

Left atrium

Left ventricle

Lung

Hepatic portal vein

Abdominal aorta

Inferior vena cava

Common iliac artery

(a) Fetal circulation

(b) Circulation after birth

Figure 17–15 Fetal circulation and changes at birth. (a) Fetal circulation is characterized by a connection to the placenta (the umbilical cord) and several bypasses around organs that are not yet functional. (b) At birth, the umbilical cord is tied off and cut, and the bypasses close, allowing more blood to reach the now functional organs.

Childbirth

Birth, also called **parturition**, usually occurs about 38 weeks after fertilization. The process by which the fetus is expelled from the uterus and moved through the vagina and into the outside world is called **labor**, an appropriate term given the phenomenal effort often required.

Labor begins with contractions of the smooth muscle of the uterus. These contractions appear to be triggered by a complex interaction of several factors, among them (1) declining levels of maternal progesterone, (2) sufficient levels of maternal estrogen, (3) release of prostaglandins and the hormone relaxin from the placenta (relaxin, also released by the ovaries, facilitates delivery by dilating the cervix and relaxing the ligaments and cartilage of the pubic bones), and (4) secretion of oxytocin by the fetus and mother. Uterine contractions during **true labor** occur at regular intervals,

are usually painful, and intensify with walking. As true labor proceeds, contractions become more intense and the interval between them decreases. **False labor**, on the other hand, is characterized by irregular contractions that fail to intensify or change with walking. Expectant mothers who arrive at the hospital in false labor are sent home to await the arrival of true labor.

The Stages of Labor

True labor can be divided into three stages. The first stage, known as the **dilation stage**, begins with the onset of contractions and ends when the cervix has fully dilated to 10 centimeters (4 inches). During this stage, the amniotic sac usually ruptures, releasing amniotic fluid. Known commonly as "breaking the water," rupture of the amniotic sac is done deliberately by a doctor or midwife if it does not happen spontaneously. The second stage, the **expulsion stage**, begins with full dilation of the cervix and ends with delivery of the baby. Some physicians make an incision to enlarge the vaginal opening, just before passage of the baby's head. This procedure, known as an **episiotomy**, facilitates delivery and avoids ragged tearing (some people, however, believe that episiotomies are unnatural and often unnecessary). Once delivery is complete, the umbilical cord is tied off and cut. The newborn takes its first breath and the conversion of fetal circulation to the postnatal pattern begins. Labor, however, is not over. Indeed, the third (and final) stage of true labor is the **placental stage**. This stage begins with delivery of the newborn and ends when the placenta (now often called the afterbirth) detaches from the wall of the uterus and is expelled from the mother's body. Continuing and powerful uterine contractions aid in expulsion of the placenta and constriction of maternal blood vessels torn during delivery. Once labor has ended, any vaginal incision or tearing is sutured. The stages of labor are shown in Figure 17–16.

The vast majority of babies are born head first, facing the vertebral column of their mother. Some babies, however, are born "buttocks first" and such deliveries are called **breech births**. Because breech births are associated with difficult labors and umbilical cord accidents (compression or looping of the cord around the baby's neck), many physicians attempt to turn such babies into the head first position before delivery. **Turning** is done by a trained physician applying his or her hands to the expectant mother's abdomen and gently guiding the fetus into the head-down position. If turning is unsuccessful, then the baby may be delivered by cesarean section.

On average, newborns born 38 weeks after fertilization weigh 3400 grams (7.5 lb) and are called **full-term infants**. Amazingly, fetuses of at least 32 weeks of age typically survive if born. Such babies, called **premature infants**, require medical care because their organ systems have not matured sufficiently to take over the functions normally performed by the mother's body. Still other babies are born one or more weeks late. Because the placenta begins to deteriorate with time, labor is typically induced if pregnancies continue for 41 or more weeks after fertilization.

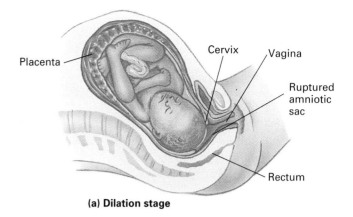

(a) Dilation stage

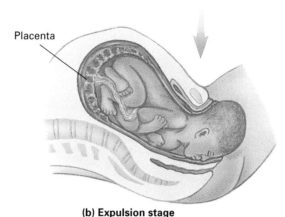

(b) Expulsion stage

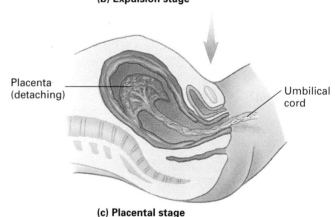

(c) Placental stage

Figure 17–16 The stages of labor. (a) The dilation stage begins with the onset of contractions and ends when the cervix is dilated 10 centimeters. (b) The expulsion stage begins with full dilation of the cervix and ends with delivery of the baby. (c) The placental stage begins with delivery of the baby and ends with expulsion of the placenta.

Critical Thinking

Compared with humans, our closest relatives the apes have a fairly easy time giving birth. Female apes, for example, have relatively large pelves (the plural of pelvis) and the small heads of their infants pass easily through roomy birth canals. Infant apes are born face up, usually with little fuss. Birth in humans is decidedly more difficult and risky. With its large head, the typical human fetus may spend hours making its way through a narrow birth canal. Furthermore, because the birth canal changes dimensions along its length, the fetus enters the canal facing sideways and then must turn to enter the world face down. What evolutionary trends among humans might have produced the tight squeeze of human birth?

Pain Relief During Childbirth

Most women experience pain during childbirth, although the level varies greatly and seems to depend, at least in part, on their level of fear and tension, the setting, and the support they receive. Typically, the more prepared and relaxed a woman is, the less pain she feels. This is why most health-care workers encourage expectant mothers and their partners to enroll in childbirth classes. Such classes provide information on the stages of labor and medical procedures associated with delivery and teach breathing and relaxation techniques.

Drugs may be administered to reduce the pain of childbirth. Analgesics—substances that relieve pain—may be administered intramuscularly (typically a shot in the buttocks) or intravenously. Demerol is a commonly used analgesic during childbirth. Also used are regional anesthetics, substances that produce loss of sensation in specific areas of the body. For example, in **epidural anesthesia**, an anesthetic is injected in the lower back into the space between the dura mater (the outer membrane covering the spinal cord) and the vertebral canal, blocking pain and motor impulses to the abdomen. Because motor impulses are blocked, epidural anesthesia makes pushing out the baby impossible. As a result, epidural procedures are performed fairly early in labor, rather than toward the end when pushing by the mother is critical for successful vaginal delivery. In addition, epidural anesthesia is associated with slowing the fetal heartbeat, necessitating continuous fetal monitoring. **General anesthesia** removes sensation of pain and produces unconsciousness. Although once a popular method for relieving pain during childbirth, general anesthesia today is used almost exclusively for emergency cesareans when it is preferred over regional anesthetics because of the speed with which it takes effect.

Many women rely on breathing or relaxation techniques rather than medications to relieve the pain of childbirth. Childbirth without the use of pain-killing drugs is often called **natural childbirth**. Whether to use medication during childbirth is a decision that is best made by weighing the costs and benefits to both mother and baby. Also, because it is impossible to predict the precise circumstances of labor and delivery, it is beneficial to keep an open mind.

Lactation

Lactation is the production and ejection of milk from the mammary glands. Recall from Chapter 6 that prolactin from the anterior pituitary gland promotes milk production, and oxytocin released from the posterior pituitary gland makes milk available to the suckling infant by stimulating milk ejection or let-down. The structure of the breast was described in Chapter 16.

Prolactin levels increase during pregnancy and reach their peak at birth. However, milk production does not begin during pregnancy because high levels of estrogen and progesterone inhibit the actions of prolactin. It is only when levels of estrogen and progesterone decline at birth that prolactin can initiate milk production, making milk available about 3 days later. In the interval after birth when milk is not yet available, the breasts produce **colostrum**, a cloudy fluid that contains protein and the sugar lactose, but no fat. Milk production is maintained after birth when suckling by the infant stimulates release of prolactin. Once suckling ends, so too does milk production.

Which is best—bottle or breast? Most people today believe that breast-feeding is best for infants. Advocates of breast-feeding point out that compared with commercial formulas based on cow's milk, breast milk is more digestible and much less likely to cause an allergic reaction, constipation, or overweight in infants. In addition, maternal milk and colostrum contain antibodies and special proteins that boost the immune system, making breast-fed babies less susceptible than bottle-fed babies to illness in the first few months of life. Finally, obtaining milk from a breast requires more effort than sucking on a bottle, and the greater effort is thought to promote optimum development of the infant's jaws, teeth, and facial muscles.

Breast-feeding is also beneficial to the mother. For example, nursing helps the mother's uterus return more rapidly to its prepregnant size because oxytocin stimulates uterine contractions, as well as milk let-down. Thus, when oxytocin is released in response to an infant's sucking, it helps the uterus shrink. Breast-feeding is also convenient and economical—bottles and formula do not have to be purchased or carried along while traveling. Perhaps most importantly, breast-feeding establishes prolonged periods of skin-to-skin contact between mother and infant and such contact is wonderful for both.

While we might think of breast-feeding as "natural" and "easy," it can be difficult during the first few weeks when mothers experience sore nipples and the feeding relationship is just being established. Indeed, many health-care professionals urge new mothers to give breast-feeding a fair trial, allowing 6 or more weeks to establish a successful breast-feeding relationship. Despite its benefits, breast feeding is not for everyone. Some women simply prefer bottle-feeding while others, such as those with a medical condition requiring medication, are warned against breast-feeding because many drugs pass into milk.

Postnatal Development

The period of growth and development after birth is known as the postnatal period. Because changes during the postnatal period are usually familiar to us, we include only a brief description of them.

Stages in postnatal development include infancy, childhood, puberty, adolescence, and adulthood. **Infancy** roughly corresponds to the first year of life. It is a time of rapid growth when total body length usually increases by one-half and weight may triple. Most 1-year-olds have several teeth and are testing their skills at walking. **Childhood** is the period from about 13 months to 12 or 13 years. Growth during this period is initially rapid and then slows. Toward the end of childhood there is a dramatic increase in growth known as a "growth spurt." Primary teeth (also called deciduous teeth) continue to come in, only later to be replaced with secondary (or permanent) teeth. **Puberty** is when secondary sexual characteristics such as pubic and underarm hair develop. This stage usually occurs slightly earlier in girls (from 12 to 15 years of age) than in boys (from 13 to 16 years of age). **Adolescence** is a period of rapid physical and sexual maturation that begins with puberty and ends by about age 17. The ability to reproduce is achieved during this stage. By the end of adolescence, physical, mental, and emotional maturity have typically been attained. **Adulthood** is generally reached somewhere between 18 and 21 years of age. Growth and formation of bone are usually completed by age 25, and thereafter developmental changes occur quite slowly. Our next focus is aging, a process that begins only a few years after growth has ceased.

Aging

Aging is the normal and progressive alteration in the structure and function of our bodies. Observable characteristics of aging—graying and sparse hair, wrinkles and sagging skin, reduced muscle mass, stooped posture, and loss of teeth—are familiar to us. However, we also age on the inside. For example, bones weaken due to loss of calcium and other minerals, blood vessels stiffen as collagen replaces elastic fibers in arterial walls, and nephrons decline in number and efficiency ultimately challenging efforts by the kidneys to maintain balance in the body's fluids. Aging also strikes our nervous system and sense organs as evidenced by our deteriorating ability to remember, see, hear, taste, and smell. Figure 17–17 summarizes some of the changes that typically occur with old age. What causes aging? Also, can we do anything to stall or reverse the process?

Possible Causes of Aging

There is no consensus among scientists as to what causes our bodies to age. Some, with a "whole body" view, suggest that aging is produced by changes in critical body systems. For example, aging might be prompted by a decline in function of the immune system or by changing levels of certain hormones. Other scientists seek explanations at the cellular level. One possibility is that there are genes for aging that when turned on slow or stop processes such as cell division. Continued cell division is necessary to replace cells that die and thus is essential to efficient functioning of tissues and organs. Supporting the contention that cessation of cell division is genetically programmed is the observation that cells grown in the laboratory do not divide indefinitely. In fact, such cells divide only a certain number of times and the number is correlated with the age of the individual that donated the cells (be the donor a roundworm, mouse, or human) and the life span of the species.

Other researchers postulate that highly reactive molecules known as free radicals disrupt cell processes and lead to aging. By-products of normal cellular activities, free radicals have an unpaired electron and thus readily combine with and thereby damage DNA, proteins, and lipids. Mitochondrial DNA seems particularly susceptible to damage from free radicals. Recall that mitochondria play a critical role in cell energetics, and thus damage to these organelles threatens the supply of energy needed to sustain cellular activities. It has also been suggested that aging is associated with a decline in the ability of cells to repair damaged DNA. An inability to "fix" DNA might lead to an accumulation of gene mutations and a deterioration in cell functions. Finally, some scientists have suggested that glucose—our main source of fuel—is the culprit. Glucose changes proteins such as collagen by causing cross-linkages to form between their molecules. This shackling together of protein molecules may cause the stiffening of connective tissue and heart muscle associated with aging.

Is there a primary cause of aging? Most scientists believe that aging is not caused by a single factor but rather several processes that interact to ensure our eventual deterioration and demise.

Achieving a Healthy Old Age

Aging is a normal biological process that, at present, cannot be stopped. However, we can stave off the disease and disability often associated with aging. The keys to a healthy old age lie in medical advances and lifestyle.

It is now possible for physicians to treat many conditions of old age. For example, worn out hip or knee joints can be replaced with artificial joints, and the clouding of vision caused by cataracts is now readily treated by replacing the old lens of the eye with an artificial lens. In addition, research is underway to determine if antioxidants (substances that inhibit the formation of free radicals) are effective in delaying certain aspects of aging. Potential antioxidants include the hormone melatonin and vitamins E and C. Some day, low doses of melatonin or high doses of vitamins E or C may be used to delay aging.

A healthy lifestyle can also lead to a more enjoyable and illness-free old age. Some components of a healthy lifestyle include proper nutrition, plenty of exercise and sleep, no smoking, and routine medical checkups. Although it is never too late to change lifestyle habits, the earlier healthy habits are followed the better.

Figure 17–17 Changes in the human body with age. The characteristics shown are typical of a 75-year-old person. (a) External changes. (b) Internal changes. The values in parentheses are the percentage of function left at 75 years of age relative to 100% function at 20 years of age.

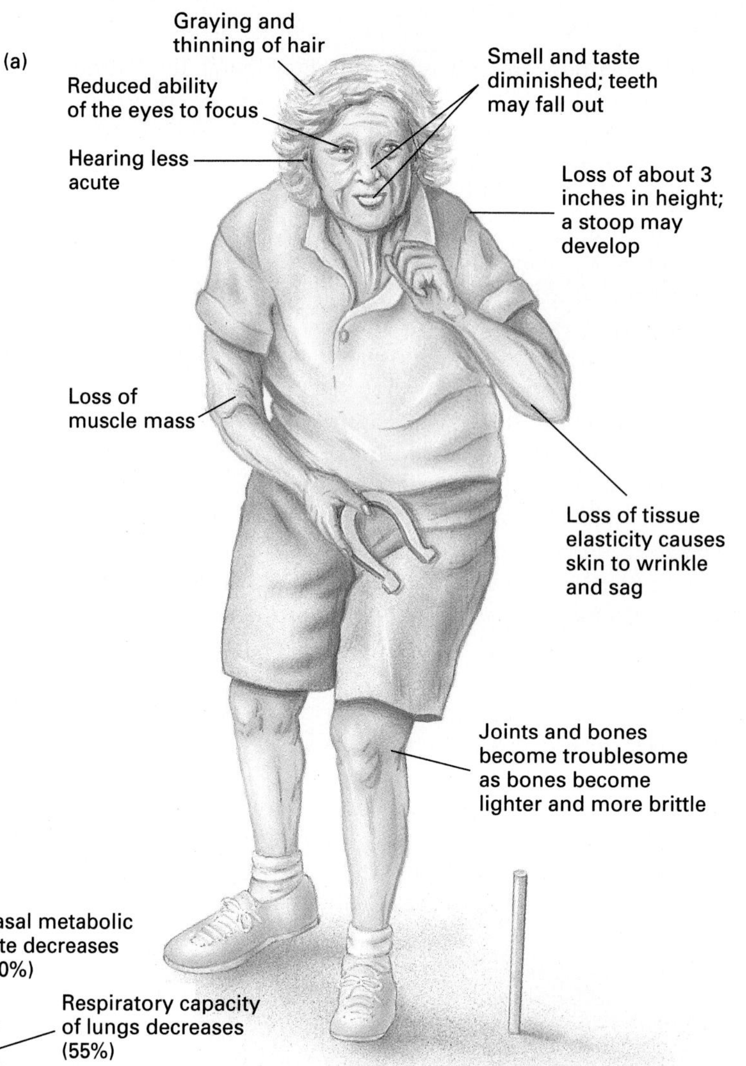

(a)

Graying and thinning of hair

Reduced ability of the eyes to focus

Hearing less acute

Smell and taste diminished; teeth may fall out

Loss of about 3 inches in height; a stoop may develop

Loss of muscle mass

Loss of tissue elasticity causes skin to wrinkle and sag

Joints and bones become troublesome as bones become lighter and more brittle

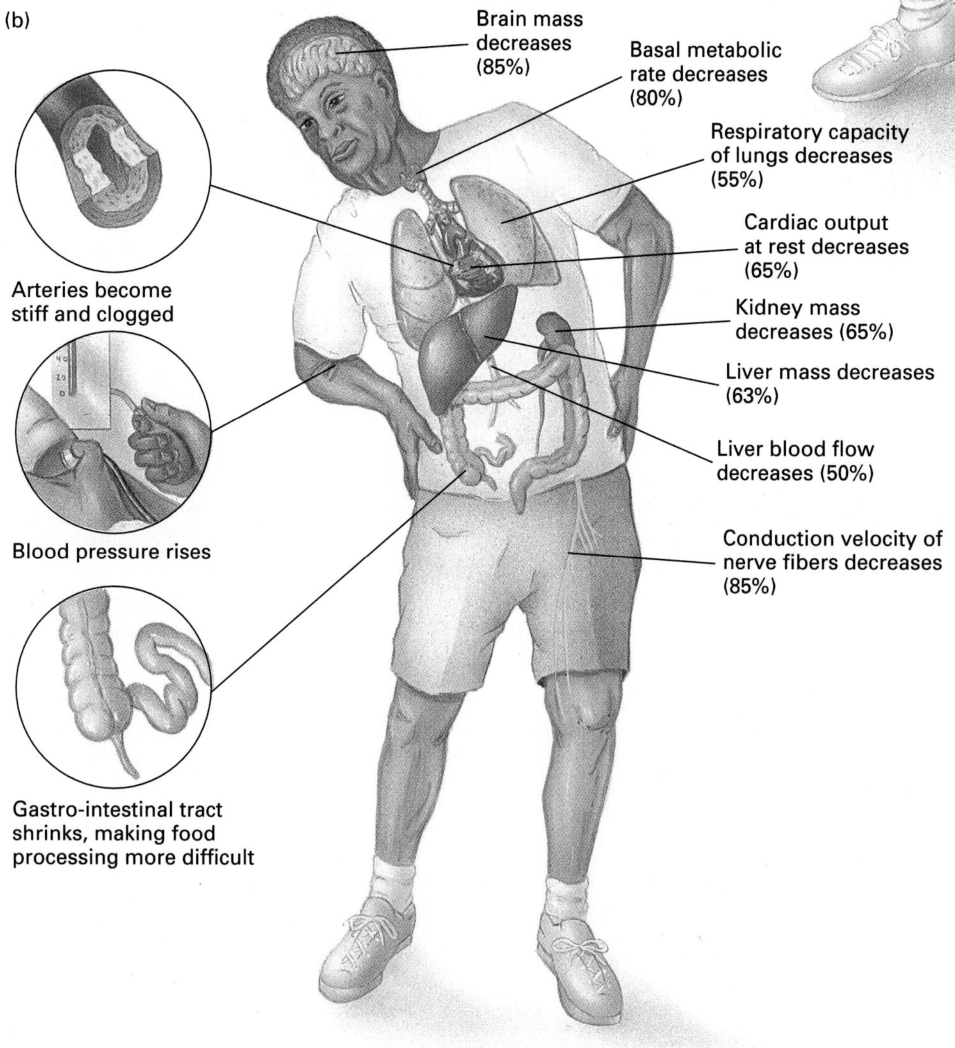

(b)

Brain mass decreases (85%)

Basal metabolic rate decreases (80%)

Respiratory capacity of lungs decreases (55%)

Cardiac output at rest decreases (65%)

Kidney mass decreases (65%)

Liver mass decreases (63%)

Liver blood flow decreases (50%)

Conduction velocity of nerve fibers decreases (85%)

Arteries become stiff and clogged

Blood pressure rises

Gastro-intestinal tract shrinks, making food processing more difficult

SUMMARY

1. The period of development before birth is the prenatal period. The prenatal period is divided into the pre-embryonic period (from fertilization through the second week), the embryonic period (from the third through the eighth weeks), and the fetal period (from the ninth week until birth). The period of development after birth is the postnatal period. Within the postnatal period are the following stages: infancy (from birth to 12 months), childhood (from 13 months to 12 or 13 years), puberty (from 12 to 15 years in girls and 13 to 16 years in boys), adolescence (from puberty to 17 years), and adulthood (generally reached by at least 21 years of age).

2. Fertilization is the union of an egg and a sperm (typically in the fallopian tubes) to form a single cell called the zygote. Before fertilization can occur, sperm must first undergo capacitation and the acrosome reaction. During capacitation, secretions from the female reproductive tract destabilize the plasma membrane of the sperm. During the acrosome reaction, enzymes spill out of the cap on the head of the sperm and digest a pathway through the layers surrounding the oocyte. A rapid depolarization of the oocyte's plasma membrane and changes in the zona pellucida ensure that only one sperm enters and fertilizes it. These mechanisms block polyspermy, the abnormal condition in which more than one sperm enters an oocyte.

3. Cleavage is a rapid series of mitotic cell divisions that transforms the zygote into a morula, a solid ball of cells. Next, a fluid-filled space appears within the ball of cells, changing the morula into a blastocyst. Within the blastocyst is the inner cell mass that will become the embryo and the trophoblast that will form part of the placenta. The placenta is the organ that delivers nutrients and oxygen to the embryo (and later the fetus) and carries wastes away. The placenta also produces hormones such as estrogen, progesterone, and human chorionic gonadotropin. Toward the end of the first week, the blastocyst begins implantation, the process by which the embryo becomes embedded in the endometrium of the uterus.

4. Extraembryonic membranes—the amnion, yolk sac, chorion, and allantois—lie outside the embryo and begin to form during the second or third week after fertilization. The amnion surrounds a fluid-filled cavity (the amniotic cavity) that protects the embryo from physical trauma. The yolk sac is a site of blood cell formation and contains primordial germ cells that migrate to the gonads where they form cells that mature into gametes. The chorion is the embryo's contribution to the placenta. The allantois is a site of blood cell formation that becomes part of the umbilical cord, the rope-like connection between the embryo and placenta.

5. The placenta forms from the chorion of the embryo and the endometrium of the mother. Cells of the embryonic trophoblast invade the endometrium, creating pools of maternal blood. Then, chorionic villi, finger-like processes of the embryo's chorion, grow into the endometrium and contact the pools of blood. Oxygen and nutrients within the pools of maternal blood can then diffuse into the capillaries within the villi and move into the umbilical vein and on to the embryo. Wastes leave the embryo by the umbilical arteries, move into capillaries within the villi, and diffuse into maternal blood. Although maternal and fetal blood vessels are in close proximity within the placenta, there is no direct mixing of blood.

6. Gastrulation includes the cell movements by which the primary germ layers—ectoderm, mesoderm, and endoderm—are established. Ectoderm forms the nervous system and the epidermis and its derivatives (hair, nails, oil glands, sweat glands, mammary glands). Mesoderm forms muscle, bone, connective tissue, and organs such as the heart, ovaries, and testes. Endoderm forms organs such as the liver, some endocrine glands, and the epithelial lining of the respiratory and gastrointestinal tracts.

7. Neurulation is the series of events leading to formation of the neural tube from ectoderm. The anterior portion of the neural tube forms the brain and the posterior portion forms the spinal cord. Other ectodermal cells called neural crest cells migrate and differentiate into spinal and cranial nerves, meninges, and skeletal and muscular components of the head. Somites, blocks of mesoderm organized alongside the neural tube, eventually form vertebrae, connective tissue, and skeletal muscles of the neck and trunk.

8. The gender of a human embryo is determined at fertilization by the type of sperm (X-bearing or Y-bearing) that fertilizes the egg (which is X-bearing). An XX zygote will develop into a female and an XY zygote into a male. In XY embryos, a region of the Y chromosome initiates development along the male track, leading to testes and production of testosterone. Female embryos lack a Y chromosome, and in its absence, ovaries develop.

9. Birth defects are developmental defects present at birth. Such defects may be caused by genetic factors (mutant genes or changes in the number or structure of chromosomes) or teratogens (environmental agents such as drugs or radiation). Teratogens have their greatest effects on the developing embryo during periods of rapid differentiation that typically occur during the embryonic period.

10. Growth during the fetal period is extremely rapid, leading to substantial increases in height and weight. However, rate of growth of the head slows relative to other parts of the body, producing change in the way the body is proportioned. Such change in the relative rates of growth of different parts of the body is called allometric growth.

11. Fetal circulation differs from circulation after birth in that most blood is shunted through temporary vessels or openings past organs such as the lungs and liver that are not yet functional. The shunts of fetal circulation include the ductus venosus (a vessel that diverts most blood away from the liver and into the inferior vena cava), the foramen ovale (an opening between the two atria that diverts blood away from the lungs), and the ductus arteriosus (a vessel that diverts blood away from the lungs by connecting the pulmonary artery to the aorta). Once organs begin to function at birth, the bypasses of fetal circulation begin to close.

12. Parturition occurs about 38 weeks after fertilization. Labor is the process by which the fetus is expelled from the uterus and moved through the vagina into the outside world. Labor occurs in three stages: (1) the dilation stage—from the onset of uterine contractions to full dilation of the cervix, (2) the expulsion stage—from full dilation of the cervix to delivery of the baby, and (3) the placental stage—from delivery of the baby to expulsion of the placenta.

13. The pain of childbirth can be relieved through analgesics (painkillers), anesthetics (substances that produce loss of sensation), and breathing and relaxation techniques.

14. Lactation is the production and ejection of milk from the mammary glands. At birth, when levels of estrogen and progesterone drop, prolactin initiates milk production. For the first few days after birth, the breasts produce colostrum, a cloudy fluid that contains protein and lactose, but no fat. Milk becomes available about 3 days after birth. Suckling stimulates the release of prolactin, and thus milk production is maintained as long as the infant suckles. Most people believe that breast-feeding is better than bottle-feeding for babies and mothers.

15. Aging is the progressive decline in the structure and function of the body. Possible causes of aging include (1) genetically programmed cessation of cell division, (2) damage to DNA, protein, and lipids caused by free radicals, (3) inability of cells to repair damaged DNA, and (4) alteration of proteins by glucose such that cross-linkages form and produce a stiffening in tissues. Although aging is a normal process, it is often associated with disease and disability. We can stave off disease and disability through medical advances and a healthy lifestyle.

REVIEW QUESTIONS

1. List the stages of prenatal and postnatal development. What ages and major developmental milestones are associated with each stage?
2. Describe capacitation and the acrosome reaction.
3. What is polyspermy and how is it prevented?
4. Describe implantation.
5. What are the functions of the four extraembryonic membranes?
6. What is the placenta and how does it form?
7. What is gastrulation?
8. What tissues and organs are formed from each of the three primary germ layers?
9. What is the neural tube and how is it formed?
10. How is the gender of a human embryo determined?
11. What are teratogens? Explain the relationship between the effects of teratogens and embryonic age.
12. Provide an example of allometric growth during human development.
13. How does fetal circulation differ from circulation after birth?
14. What are the three stages of labor?
15. Why is breast-feeding considered better than bottle-feeding for most infants and mothers?
16. What is aging? What are some possible causes of aging?

PART | 5

Genetics

Our understanding of the inheritance of human traits has bene-
fited from studies of large extended families. (© *Rafael Macia/
Photo Researchers, Inc.*)

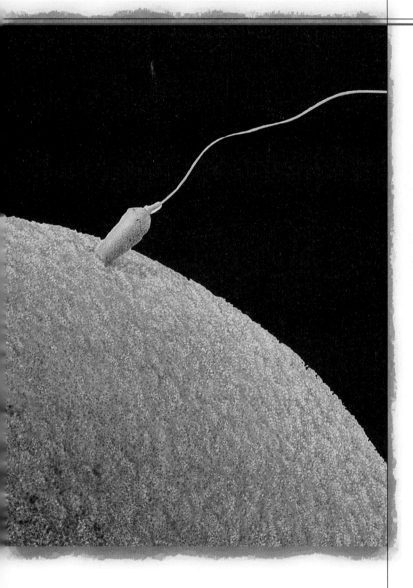

Chapter

18

Chromosomes, Meiosis, and the Principles of Inheritance

None of us remember the moment that most likely has had more influence over the rest of our lives than any other event—conception. One sperm from the swarm around your mother's egg pierced the egg membrane and penetrated the egg. That moment determined the genetic baggage that you have carried through life. There is no denying that the assortment of genes in your mother's egg and in the sperm that fertilized it is largely responsible for making you who you are today. As you flip through the pages of the family album, you may notice distinctive traits scattered through the generations. Physical characteristics may be the most obvious way in which heredity has shaped you, but they are a very small part of your genetic legacy. The biochemical reactions taking place in your cells, susceptibility to disease, certain behavior patterns, and even your life span are all influenced by the genes you received at the moment of conception. Although environment influences the expression of genes, your genes provide the basic outline for your possibilities and limitations.

In this chapter we will consider the genetic foundation that has been so important in shaping who you are. We will begin with the physical basis of heredity, the chromosomes. Next, we will see how the chromosomes are parceled out during the formation of an egg or sperm, so that we can understand how children of the same parents can be so similar and yet so different. Finally, we will learn how traits are passed through generations and how to predict the appearance of traits in future children.

Chromosomes

A **chromosome** is a combination of a DNA molecule, which contains the genetic information of a cell, and specialized proteins, primarily a group of proteins called histones. The information in DNA directs the development and maintenance of the body; the proteins are for support and control of gene expression. A **gene** is a segment of the chromosome's DNA that directs the synthesis of a protein, which in turn plays a structural or functional role within the cell. In this way, a gene determines the expression of a particular characteristic or trait. Each different kind of chromosome in a human cell contains a specific assortment of genes (Figure 18–1). Like beads on a string, genes are arranged in a fixed sequence along the length of specific chromosomes. The point on a chromosome where a particular gene is found is called its **locus** (plural, *loci*).

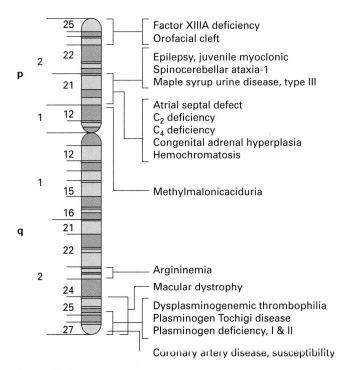

Figure 18–1 A map of a human chromosome. Each chromosome contains a specific assortment of genes in a specific order.

In the human body, **somatic cells**, that is, all cells except for eggs or sperm, have 46 chromosomes. Those chromosomes actually consist of two sets of 23 chromosomes—one set of hereditary information from each parent. Thus, each cell contains two chromosomes with genes for the same traits, called **homologous pairs** of chromosomes or homologues for short. Genes, then, also occur in pairs and the members of each pair are located at the same locus on homologous chromosomes. (Keep in mind, however, that the forms of the genes may differ. For example, one form may direct the formation of blue eyes and another, brown eyes. As we will see later in this chapter, the alternate forms of a gene are called alleles. Because homologous chromosomes may possess different alleles for the same traits, and there are many genes on one chromosome, the homologues are not usually identical.)

One pair of the 23 pairs, the **sex chromosomes**, is involved in determining whether the person is male or female. There are two types of sex chromosomes, X and Y. A person who is XX is genetically female and one who is XY is genetically male. The other 22 pairs of chromosomes are called the **autosomes**. The autosomes determine the expression of most of the inherited characteristics of a person.

Meiosis

As our bodies produce gametes (eggs or sperm), the chromosomes undergo a ritual dance called meiosis. **Meiosis** is a special type of cell division that occurs only in the gonads (ovaries or testes) and gives rise to gametes. Any cell with two sets of chromosomes is described as being **diploid**. As a result of meiosis, the eggs or sperm differ from somatic cells in that they are **haploid**, meaning they have a single set of chromosomes. When a sperm fertilizes an egg, a new cell, called a **zygote**, is created. Since the egg and sperm both contribute a set of chromosomes to the zygote, it is diploid. After trillions of mitotic cell divisions, the zygote eventually develops into a new individual (Figure 18–2).

Why Meiosis?

Meiosis serves two important functions. The first is to keep the chromosome number constant through generations. If gametes were instead produced by mitosis, they would be diploid. Then, when a sperm containing 46 chromosomes fertilized an egg with the same number of chromosomes, the zygote would have 92 chromosomes. There would be 184 chromosomes in the next generation zygote, formed by an egg and sperm each containing 92 chromosomes. The next generation, would have 368 chromosomes in each cell and the next one, 736. You can see how quickly the chromosome number would become unwieldy and, more importantly, al-ter the amount of genetic information in each cell. As we will see later in this chapter, even one extra copy of a single chromosome usually causes a fetus to die. Exceptions to this include Down syndrome, which results from an extra copy of chromosome 21, and Klinefelter syndrome, which results when a male is XXY.

A second function of meiosis is to increase genetic variability in the population. Later in this chapter we will consider the mechanisms through which this occurs. Genetic variability is important because it provides the raw material on which natural selection can act, leading to the changes in populations that are called evolution. The relationship between genetic variability and evolution will be discussed further in Chapter 20.

The Process of Meiosis

Meiosis and mitosis are preceded with the same event—the replication of chromosomes. Unlike mitosis, however, meiosis involves *two* divisions. In the first division, the chromosome number is reduced, because the two sets of chromosomes are separated into two cells, each bearing one complete set. In the second division, the replicated copies

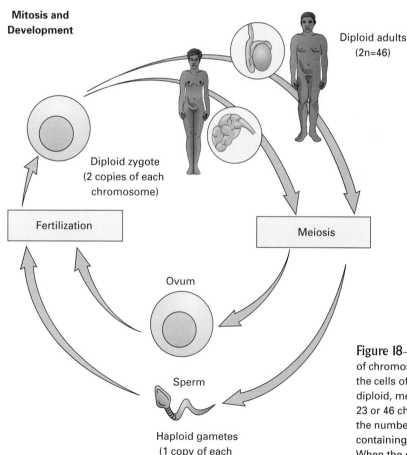

Mitosis and Development

Diploid adults
(2n=46)

Diploid zygote
(2 copies of each
chromosome)

Fertilization

Meiosis

Ovum

Sperm

Haploid gametes
(1 copy of each
chromosome)

Figure 18–2 Meiosis is important because it keeps the number of chromosomes constant from generation to generation. Most of the cells of the human body—all but the eggs and sperm—are diploid, meaning that they contain two sets of chromosomes (2 × 23 or 46 chromosomes). Meiosis is a type of cell division in which the number of chromosomes is halved, producing haploid cells containing only one set of chromosomes (23 chromosomes). When the sperm fertilizes an egg, their chromosome sets are joined, creating a diploid zygote.

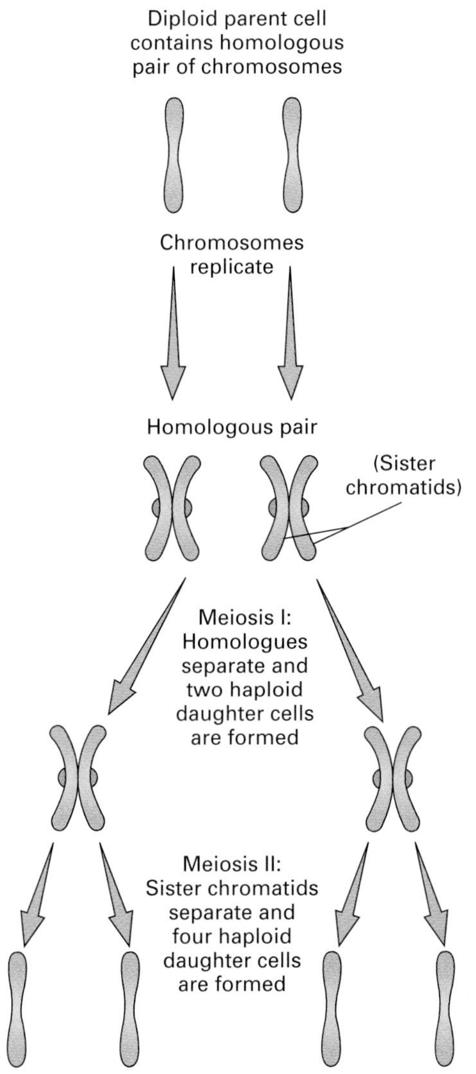

Diploid parent cell
contains homologous
pair of chromosomes

Chromosomes
replicate

Homologous pair

(Sister
chromatids)

Meiosis I:
Homologues
separate and
two haploid
daughter cells
are formed

Meiosis II:
Sister chromatids
separate and
four haploid
daughter cells
are formed

Figure 18–3 Meiosis reduces the chromosome number from the diploid number (2 copies of each chromosome) to the haploid number (1 copy of each chromosome). There are 2 meiotic divisions.

are separated. We see, then, that meiosis begins with one diploid cell and, two divisions later, produces four haploid cells. The orderly movements of chromosomes during meiosis ensure that each haploid gamete produced contains one complete set of chromosomes (Figure 18–3).

Changes in the shape and functioning of the four haploid cells result in functional gametes. In males, one diploid cell results in four functional sperm. In contrast, in females meiotic divisions of one diploid cell result in only one functional egg and three nonfunctional polar bodies. As a result, the egg contains most of the nutrients found in the original diploid cell, and these nutrients will nourish the early embryo.

As we fill in some details of meiosis, we will compare the chromosomes to volumes in a set of instruction manuals for running the body. To draw this analogy, consider that you inherit one complete 23-volume set of manuals from your mother and another complete 23-volume set from your father. In each cell, then, there is a total of 46 volumes.

In preparation for meiosis, each chromosome replicates and the two copies remain attached to one another by a centromere. As long as they remain attached, each strand of the chromosome is called a **chromatid**, just as in mitosis. Replication of chromosomes would be somewhat analogous to copying[1] each of the 46 volumes of manuals and holding the copies together with a rubber band. Each of the two copies would be considered as a chromatid as long as they were banded together.

The first meiotic division (meiosis I) is called **reduction division** because it produces two cells, each with 23 chromosomes (Figure 18–4). However, the daughter cells do not contain a random assortment of any 23 chromosomes. Instead, each daughter cell contains *one complete set of chromosomes*, with each chromosome consisting of two chromatids. In our analogy, each cell would have one 23-volume set of instruction manuals, with the copies of each volume still held together with a rubber band.

It is important that each cell receives a complete set of chromosomes or manuals. It would not do if one of the daughter cells had two copies of volume 3 and no copy of volume 6. Although there would still be 23 volumes present, part of the instructions (volume 6) would be missing.

The separation of chromosomes into complete sets that occurs during meiosis I occurs reliably because during prophase I (prophase of the first meiotic division) members of homologous pairs line up next to one another, a phenomenon called **synapsis** ("bringing together"). Continuing our analogy, volume 1 of the set of instruction manuals from your father would line up with volume 1 from your mother. Paternal volume 2 would pair with maternal volume 2, and so on. The matched homologous pairs then become positioned at the midline of the cell, the so-called **equatorial plate**, and attach to spindle fibers (metaphase I). Next, in anaphase I, the microtubules composing the spindle fibers shorten, causing the members of each pair to separate and pulling them to opposite ends of the cell. Cytokinesis now occurs, resulting in two daughter cells, each with one member of each chromosome pair (telophase I). Each chromosome still consists of two chromatids. Interkinesis, a brief interphase-like period, follows. Interkinesis differs from mitotic interphase in that there is no replication of DNA during interkinesis.

---

[1]As we will see in Chapter 19, replication is not strictly analogous to photocopying because when a chromosome is replicated (copied) each of the resulting chromosomes consists of half the original information and half a copy. If an instruction manual were "photocopied" in the manner a chromosome is replicated, the left half of a page might be from the original and the right half would then be a photocopy.

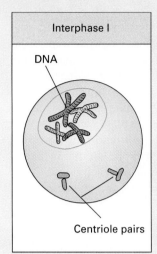

Interphase I

DNA

Centriole pairs

Pre-Meiotic Interphase: DNA replicates; copies remain attached to one another by centromere; each copy is called a chromatid.

Figure 18–4 Stages of meiosis.

**Meiosis I:
Separates homologues**

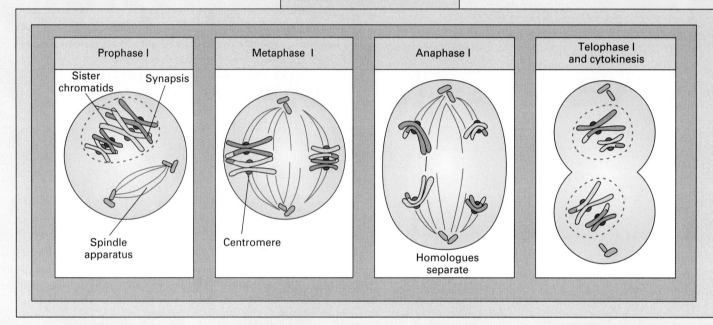

Prophase I

Sister chromatids   Synapsis

Spindle apparatus

Metaphase I

Centromere

Anaphase I

Homologues separate

Telophase I and cytokinesis

Prophase I:
Chromosomes condense; synapsis occurs (homologous chromosomes pair and become perfectly aligned with one another); crossing over takes place; spindle apparatus forms; nuclear membrane breaks down.

Metaphase I:
Homologous pairs of chromosomes are aligned at the equatorial plate; a spindle fiber from one pole attaches to one member of each pair while a spindle fiber from opposite pole attaches to the homologue.

Anaphase I:
Homologous pairs of chromosomes separate and move to opposite ends of the cell; each homologue still consists of two chromatids.

Telophase I:
One member of each homologous pair is at each pole; cytokinesis occurs (cell pinches in two) forming two daughter cells; they now have haploid number of chromosomes, but each chromosome still consists of two chromatids; nuclear membrane reforms; chromatin reforms.

During the second division (meiosis II), each chromosome lines up in the center of the cell independently (as occurs in mitosis) and the sister chromatids (attached replicates) comprising each chromosome separate. This would be analogous to separating the copies of each volume in the set of instruction manuals. Separation of replicates occurs in both daughter cells produced in meiosis I, resulting in four cells, each containing one copy of each chromosome. The events of meiosis II are similar to those of mitosis, except that there are only 23 chromosomes lining up independently in meiosis II and there are 46 chromosomes independently aligning in mitosis (Figure 18–5).

Interkinesis:
A brief interphase-like period; there is no replication of DNA during interkinesis.

Meiosis II:
Separates sister chromatids

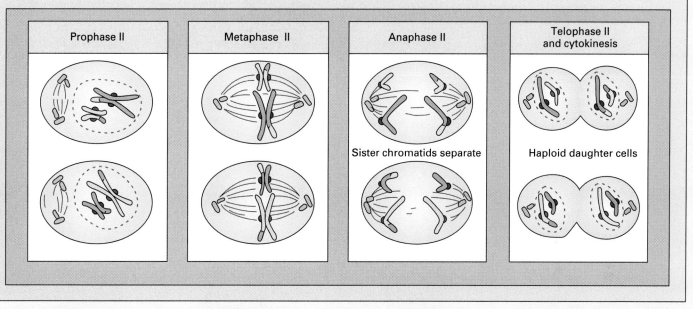

Prophase II	Metaphase II	Anaphase II	Telophase II and cytokinesis
		Sister chromatids separate	Haploid daughter cells

Prophase II:
Chromosomes condense again; nuclear membrane breaks down; spindle apparatus forms.

Metaphase II:
Chromosomes align on equatorial plate as in mitosis.

Anaphase II:
Centromeres of sister chromatids separate; chromatids of each pair are now called chromosomes; chromosomes move toward opposite poles of the cell.

Telophase II:
Nuclei form at opposite poles; cytokinesis occurs.

Critical Thinking

If you were examining dividing cells under a microscope, how could you determine whether a particular cell was in metaphase of mitosis or metaphase I of meiosis?

Genetic Recombination During Meiosis

At the moment of fertilization, when the nuclei of an egg and a sperm fuse, a new, truly *unique* individual is formed. Although certain family characteristics may be passed along, each child bears its own assortment of genetic characteristics

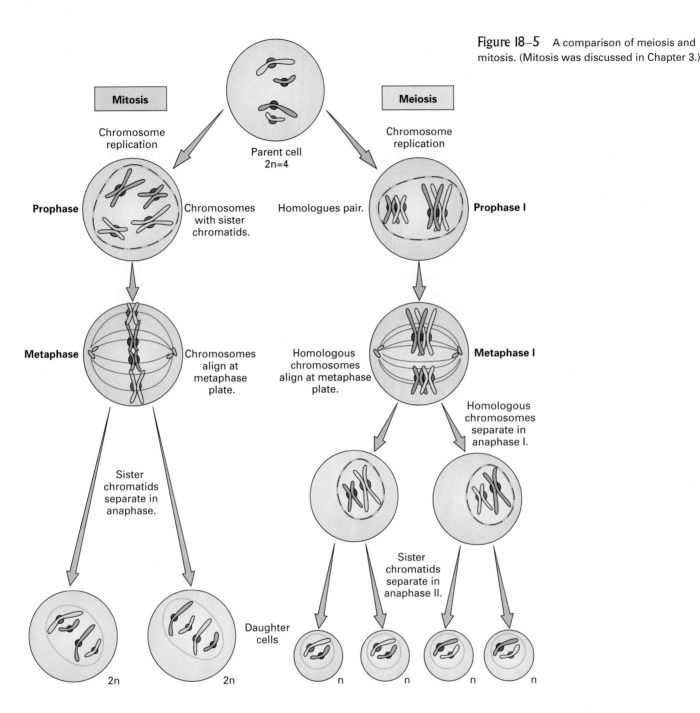

Figure 18–5 A comparison of meiosis and mitosis. (Mitosis was discussed in Chapter 3.)

Mitosis

Chromosome replication

Parent cell
2n=4

Prophase — Chromosomes with sister chromatids.

Metaphase — Chromosomes align at metaphase plate.

Sister chromatids separate in anaphase.

2n   2n

Meiosis

Chromosome replication

Homologues pair. — **Prophase I**

Homologous chromosomes align at metaphase plate. — **Metaphase I**

Homologous chromosomes separate in anaphase I.

Daughter cells

Sister chromatids separate in anaphase II.

n   n   n   n

(Figure 18–6). Only identical twins are exactly alike genetically, because they form from a single zygote.

Genetic diversity arises largely because of the shuffling of maternal and paternal forms of genes during meiosis. One way this mixing occurs is through a process called **crossing over**, in which corresponding pieces of chromatids of maternal and paternal homologues (nonsister chromatids) are exchanged during synapsis (Figure 18–7). After crossing over, the resulting chromatids have a mixture of DNA from the two parents. Since the homologues align gene by gene in synapsis, the exchanged segments contain genetic informa-

tion for the same traits. However, since the genes of the mother and those of the father may direct different expressions of the trait, attached or unattached earlobes for instance, the chromatids that result from crossing over have a new, novel combination of genes.

Independent assortment is a second way that meiosis provides for the shuffling of genes between generations (Figure 18–8). Recall that the homologous pairs of chromosomes line up at the equatorial plate during metaphase I. However, the orientation of the members of the pair relative to the poles of the cell is random. Thus, like the odds of a

Figure 18–6 Each child inherits a unique combination of maternal and paternal genetic characteristics because of the shuffling of chromosomes that occurs during meiosis. Thus, although family characteristics may be passed from generation to generation, no siblings are genetically identical (with the exception of identical twins). This photograph shows Mary and Eric Goodenough with their four sons: Derick, Stephen, David, and John. *(Judith Goodenough)*

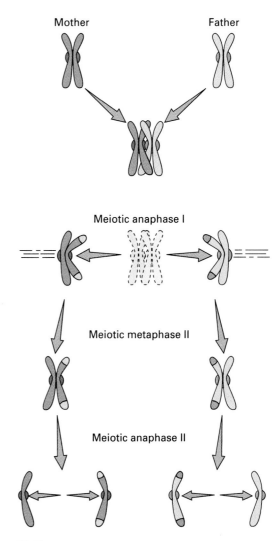

Figure 18–7 Crossing over. During synapsis, when the homologous chromosomes of the mother and the father are closely aligned, corresponding segments of nonsister chromatids are exchanged. Each of the resulting chromatids then has a mixture of maternal and paternal genetic information.

flipped coin coming up heads, there is a fifty-fifty chance that a given daughter cell will receive the maternal chromosome. Each of the 23 pairs of chromosomes orients independently during metaphase I. The alignments of all 23 pairs will determine the assortments of maternal and paternal chromosomes in the daughter cells. The general formula for determining the possible combinations of maternal and paternal chromosomes in gametes due to independent assortment is 2^n, where n is the haploid number. In humans, the haploid number is 23, so each person can produce 2^{23}, approximately 8 million, different gametes solely due to independent assortment.

Mistakes in Meiosis

Most of the time, meiosis is a precise process that results in the even distribution of chromosomes to gametes. Nonetheless, meiosis is not foolproof and occasionally a mistake is made. A pair of chromosomes or sister chromatids may adhere so tightly to one another that they do not separate during anaphase. As a result, both go to the same daughter cell and the other daughter cell has none of this type of chromosome (Figure 18–9). The failure of homologous chromosomes to separate during meiosis I or sister chromatids to separate during meiosis II is called **nondisjunction**.

What happens if nondisjunction creates a gamete with an extra or a missing chromosome and that gamete is then united with a normal gamete during fertilization? The excess or deficit of chromosomes also occurs in the resulting zygote. For instance, if the abnormal gamete has an extra chromosome, the resulting zygote will have three copies of one type of chromosome and two copies of the rest. This condition, in which there are three representatives of one chromosome, is called **trisomy**. If, on the other hand, a gamete that is missing a representative of one type of chromosome joins with a normal gamete during fertilization, the resulting zygote will have only one copy, rather than the normal two copies, of one type of chromosome. The condition in which there is only one representative of a particular chromosome in a cell is called **monosomy**. **Aneuploidy** is a general term that describes gametes or cells with too many or too few chromosomes compared with the normal

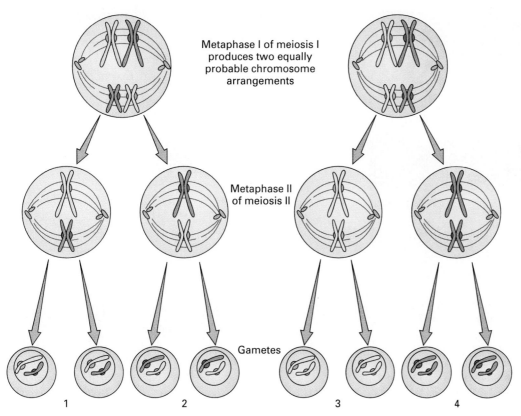

Metaphase I of meiosis I produces two equally probable chromosome arrangements

Metaphase II of meiosis II

Gametes

1 2 3 4

Figure 18–8 The alignment of homologous maternal and paternal chromosomes relative to the poles of the cell is random. The orientation of the homologues determines the mixture of maternal and paternal chromosomes in the resulting gametes. The members of each homologous pair orient independently of the other pairs. In this figure, one color symbolizes the chromosome from one parent and the other color indicates the chromosomes from the other parent. Notice that with only two homologous pairs, there are four possible combinations of chromosomes in the resulting gametes. Your cells have 23 pairs of chromosomes and so you could produce approximately 8 million genetically different gametes!

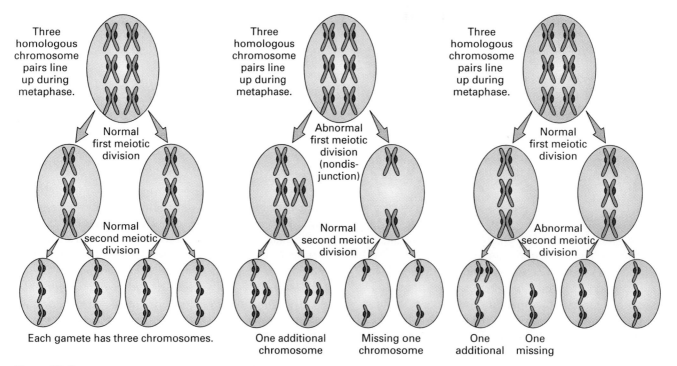

Three homologous chromosome pairs line up during metaphase.

Normal first meiotic division

Normal second meiotic division

Each gamete has three chromosomes.

Three homologous chromosome pairs line up during metaphase.

Abnormal first meiotic division (nondisjunction)

Normal second meiotic division

One additional chromosome

Missing one chromosome

Three homologous chromosome pairs line up during metaphase.

Normal first meiotic division

Abnormal second meiotic division

One additional

One missing

Figure 18–9 Nondisjunction is a mistake that occurs during cell division in which homologous chromosomes or sister chromatids fail to separate during anaphase. One of the resulting daughter cells will have an extra copy of one chromosome and the other will be missing that type of chromosome.

number. The imbalance of chromosomes usually causes abnormalities in development. Most of the time, the resulting malformations are severe enough to cause the death of the fetus, which will result in a miscarriage. Indeed, in about 70% of miscarriages, the fetus has an abnormal number of chromosomes.

When a fetus inherits an abnormal number of certain chromosomes, for instance chromosome 21 or the sex chromosomes, the resulting condition is not usually fatal. The upset in chromosome balance does, however, cause a specific syndrome. (A syndrome is a group of symptoms that generally occurs together.)

Down Syndrome

One in every 800 to 1000 infants born will have three copies of chromosome 21 (trisomy 21), a condition known as **Down syndrome**. The symptoms of Down syndrome include moderate to severe mental retardation, short stature or shortened body parts due to poor skeletal growth, and characteristic facial features (Figure 18–10). Individuals with Down syndrome typically have a flattened nose, a forward-protruding tongue that forces the mouth open, upward slanting eyes, and a fold of skin at the inner corner of each eye, called an epicanthal fold. Approximately 50% of all infants with Down syndrome have heart defects and many of them die as a result of this defect. Blockage in the digestive system, especially in the esophagus or small intestine, is also common and may require surgery shortly after birth.

The risk of having a baby with Down syndrome increases with the age of the mother. Indeed, a 30-year-old woman is twice as likely to give birth to a child with Down syndrome than is a 20-year-old woman. After age 30, the

(a)

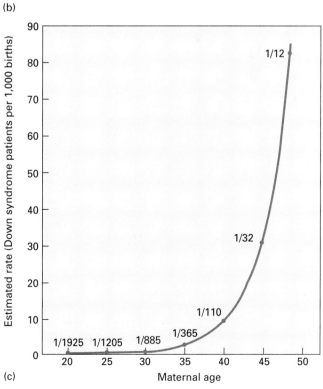

(b)

(c)

Figure 18–10 Three copies of chromosome 21 (trisomy 21), as seen here (a), causes Down syndrome. (b) A person with Down syndrome is moderately to severely mentally retarded and has a characteristic appearance, including a flattened nose, a tongue that protrudes forward, upward slanting eyes, and a fold of skin at the inner corner of each eye. (c) The incidence of Down syndrome increases with the age of the mother, especially after age 35.
(a, courtesy of Dr. Leonard Sciorra; b, © Bernd Wittich/Visuals Unlimited)

risk rises dramatically, and so at age 45 a mother is 45 times as likely to give birth to a Down syndrome infant as is a 20-year-old woman.

There are several possible explanations for the increased incidence of Down syndrome babies among older mothers. Potential eggs form and meiosis begins before a woman is even born. Although homologous chromosomes in potential eggs pair in prophase I before a woman is born, they do not separate until shortly before that egg is ovulated. It might be that the probability that chromosomes will remain stuck together during anaphase increases with the length of time they have been paired. Exposure to environmental factors, including ionizing radiation and viruses, that are known to increase the rate of nondisjunction may also play a role. The older a woman is, the longer her potential eggs have been exposed to these environmental factors. However, other studies suggest that the chances of a zygote forming with trisomy 21 may be the same regardless of the mother's age. The difference in the incidence of Down syndrome babies with maternal age may be because an older mother is more likely to carry an affected fetus to term than is a younger mother.

Nondisjunction of Sex Chromosomes

Like autosomes, sex chromosomes may fail to separate during anaphase. This error can occur during either egg or sperm formation. A male is chromosomally XY, and so when the X and Y separate during anaphase, equal numbers of X-bearing and Y-bearing sperm are produced. However, if nondisjunction occurs during sperm formation, half of the resulting sperm will carry both X and Y chromosomes, whereas the other resulting sperm will not contain any sex chromosome. Since a female is chromosomally XX, each of the eggs she produces should contain a single X chromosome. When nondisjunction occurs, however, an egg may contain two X chromosomes or none at all. When a gamete with an abnormal number of sex chromosomes is joined with a normal gamete during fertilization, the resulting zygote has an abnormal number of sex chromosomes (Figure 18–11).

Turner syndrome occurs in individuals who have only a single X chromosome (XO). Approximately 1 in 2000 female infants is born with Turner syndrome, but this represents only a small percentage of the XO zygotes that are formed, because most are lost as miscarriages. Such an individual has the external appearance of a female (Figure

(a) Nondisjunction during egg development

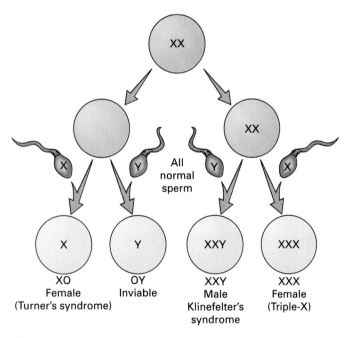

(b) Nondisjunction during sperm development

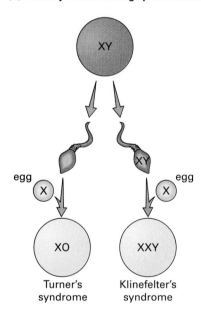

Figure 18–11 The sex chromosomes may fail to separate during formation of either eggs or sperm. During egg formation, nondisjunction will result in an egg that contains two X chromosomes or one with no sex chromosome. If the X and Y chromosomes fail to separate during sperm formation, half the resulting sperm will be XY and the other will have no sex chromosome. When a gamete with an abnormal number of sex chromosomes joins a normal gamete in fertilization, the resulting zygote has an abnormal number of sex chromosomes. Imbalances among sex chromosomes upset normal development of reproductive structures.

18–12). Before puberty, she looks like a normal, XX female. The only hint of Turner syndrome may be a thick fold of skin on the neck. As she ages, however, she generally becomes noticeably shorter than her peers. Her chest is wide and her breasts underdeveloped. In 90% of the women with Turner syndrome, the ovaries are also poorly developed, leading to infertility.

Injections of growth hormone will allow girls with Turner syndrome to reach a normal height and orally administered sex hormones at an appropriate age will promote breast development. Although the ovaries of a woman with Turner syndrome cannot produce eggs, pregnancy may be possible through in vitro fertilization, in which a fertilized egg from a donor is implanted in her uterus.

Klinefelter syndrome is observed in males who are XXY. Although the extra X chromosome can be inherited as a result of nondisjunction during either egg or sperm formation, it is twice as likely to come from the egg. As with Down syndrome, increased maternal age may increase the risk slightly.

Klinefelter syndrome is fairly common. Approximately 1 in 500 to 1 in 1000 of all newborn males is XXY. However, not all XXY males display the symptoms of having an extra X chromosome. In fact, many of them live their lives without ever suspecting that they are XXY. Signs that a male has Klinefelter syndrome do not usually show up until puberty. During the teenage years, the testes of an XY male gradually increase in size. In contrast, the testes of an XXY male remain small and do not produce an adequate amount of the male sex hormone, testosterone. As a result of the testosterone insufficiency, he may grow 2 to 4.5 inches taller than average and secondary sex characteristics, such as facial and

body hair, fail to develop fully. His arms and legs may be exceptionally long and his breasts may develop slightly. The penis is usually of normal size, but the testes don't produce sperm, so men with Klinefelter syndrome are usually sterile.

Testosterone injections, especially if they begin in adolescence, bring about changes that make a male with Klinefelter syndrome more similar to his XY peers—more developed muscles, deeper voice, more facial and body hair, and testicular development. In doing so, these treatments may also ease the emotional and psychological problems experienced by many males with Klinefelter syndrome. Sterility, however, remains irreversible.

Principles of Inheritance

How can an understanding of meiosis help us answer important questions about inheritance: Why your brother is the only sibling with Mom's widow's peak (a hairline that comes to a point on the forehead) and freckles? How you can have blue eyes when both your parents are brown-eyed? Will you be bald at 40, like Dad? Let's take a second look, then, to see how chromosomes, meiosis, and heredity are related.

As you may recall, chromosomes are made of DNA and protein. Certain segments of the DNA of each chromosome function as genes. A **gene** directs the synthesis of a specific protein that can play either a structural or functional role in the cell. In this way, the gene-determined protein can influence whether a certain **trait**, or characteristic, develops. For instance, the formation of your brother's widow's peak was directed by a protein coded for by a gene that he inherited from Mom.

There are different forms of genes, called **alleles**, and these produce different versions of the trait they determine. One of the genes that determines eye color, for instance, dictates whether the brown pigment melanin will be deposited in the iris. When this allele is present, eye color will range from green to dark brown, depending on the amount of melanin present. The other allele does not lead to melanin deposit. If neither homologue bears an allele for melanin deposition, eye color will be blue.

Individuals with two copies of the same allele for a gene are said to be **homozygous** for that trait. Those with different alleles for a given gene are said to be **heterozygous**. When the effects of a certain allele can be detected, regardless of whether an alternative allele is also present, it is described as **dominant**. Dimples and freckles are human traits dictated by dominant alleles. The allele whose effects are masked in the heterozygous condition is described as **recessive**. Because of this masking, only homozygous recessive alleles are expressed. Conditions such as cystic fibrosis, in which excessive mucus production impairs lung and pancreatic function, and albinism, in which the pigment melanin is

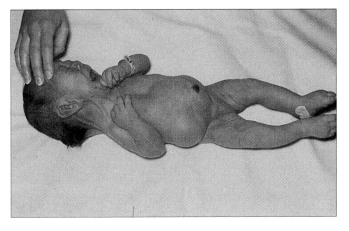

Figure 18–12 A female with Turner syndrome has one rather than the normal two X chromosomes. The signs of Turner syndrome are a skin fold at the neck, short stature, a wide chest, and underdeveloped breasts. Because the ovaries are underdeveloped, she is usually infertile. (© Dr. Irene Uchida, McMaster University, Ontario)

missing in the hair, skin, and eyes, result from recessive alleles. By convention, the dominant allele is designated with a capital letter and the recessive allele with a lowercase letter (*A* and *a*, for example).

We observe the dominant form of the trait whether the individual is homozygous dominant (*AA*) or heterozygous (*Aa*) for that trait. Thus, we cannot always tell which alleles are present. However, it is often useful to distinguish between an individual's genetic make-up and its appearance. **Genotype** refers to the precise alleles that are present, that is, whether the individual is homozygous or heterozygous for different gene pairs. **Phenotype**, on the other hand, refers to the observable traits of an individual.

Segregation of Alleles and Independent Assortment

We have seen that the members of each homologous pair of chromosomes segregate during meiosis I, each homologue going to a different daughter cell. Consider what this means for inheritance—the alleles for each gene segregate during gamete formation and so half the gametes bear one allele and half bear the other. This principle is known as the **law of segregation**.

Furthermore, each pair of homologous chromosomes lines up at the equatorial plate during meiosis I independently of the other pairs. The orientation of the paternal and maternal homologues relative to the poles of the cell is entirely random. What this means for inheritance is that each pair of alleles (on different chromosomes) segregates into gametes independently. This principle is the **law of independent assortment**.

Mendelian Crosses

During the nineteenth century, an Austrian monk by the name of Gregor Mendel worked out much of what we know today about the laws of heredity by performing specific crosses of pea plants. Although Mendel knew nothing about chromosomes, his ideas about the inheritance of traits are consistent with what we now know about the movement of chromosomes during meiosis and can be used to predict the outcome of crosses today.

Monohybrid Crosses

Mendel began by studying **monohybrid crosses**, those that consider the inheritance of a single trait from individuals differing in the expression of that trait. Consider, for example, the inheritance of freckles, a characteristic determined by a dominant allele. The individuals in the initial cross, in this case, a freckled-female who is homozygous dominant (*FF*) and a homozygous recessive male with no freckles (*ff*) are referred to as the **Parental (P) generation** (Figure 18–13). We know that alleles segregate during meiosis, but

since both parents are homozygous, each can produce only one type of gamete (as far as freckles are concerned). The female produces gametes with the dominant allele (*F*) and the male produces gametes with the recessive allele (*f*).

A **Punnett square** is a useful tool for determining the probable outcome of genetic crosses. To use a Punnett square, columns are set up and labeled to represent each of the possible gametes of one parent. Remember that the alleles segregate during meiosis. In this case, then, there would be two columns to represent the gametes of the male without freckles, each labeled with recessive alleles, *f*. In a similar manner, rows are established and labeled to represent all the possible gametes formed by the other parent, in this case, the woman with freckles. There would be two rows in this Punnett square, each labeled *F*. Each square is then filled in by combining the labels on the corresponding rows and columns. These squares represent the possible offspring of these matings, the so-called **first filial (F₁) generation**. All the F₁ children would have freckles and be heterozygous for the trait (*Ff*).

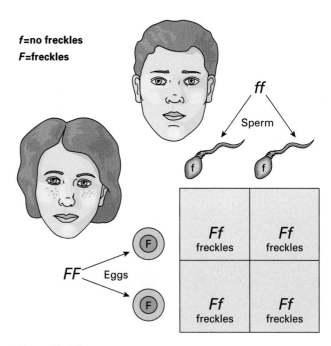

Figure 18–13 Gametes produced by a homozygous dominant female with freckles and a homozygous recessive male without freckles. This Punnett square illustrates the probable offspring from a cross between a homozygous dominant female with freckles and a homozygous recessive male without freckles. Each column is labeled with the possible gametes produced by the male. Each row is labeled with possible gametes produced by the female. Combining the labels on the corresponding rows and columns indicates the genotype of possible offspring. All the offspring from this cross will be heterozygous and, therefore, freckled.

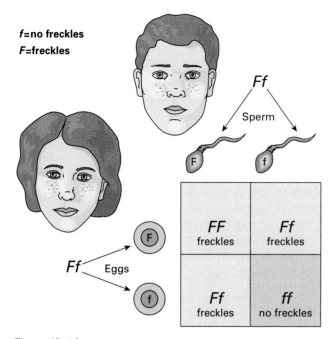

f=no freckles
F=freckles

Ff Sperm

Ff Eggs

	F	f
F	*FF* freckles	*Ff* freckles
f	*Ff* freckles	*ff* no freckles

Figure 18–14 If a heterozygous man with freckles mates with a heterozygous woman with freckles, the probable outcome of phenotypes will be 3 freckled children: 1 child with no freckles. The probable ratio of genotypes will be 1 homozygous dominant: 2 heterozygous: 1 homozygous recessive.

The offspring of the F₁ generation are the **second filial**, or **F₂, generation.** Suppose one of the F₁ offspring marries another person who is heterozygous for the freckle trait, and they have children. There would be two rows and two columns in the Punnett square, labeled with *F* and *f* (Figure 18–14). Combining the labels on rows and columns, we see that the probable ratio of genotypes in the F₂ generation would be 1 homozygous dominant: 2 heterozygous: 1 ho-

mozygous recessive. The ratio of phenotypes would be 3 freckled children: 1 without freckles.

Critical Thinking

At the beginning of the section on inheritance, a question was raised about why only one sibling (your hypothetical brother) inherited your mother's freckles. Implicit in the way the question was phrased was that there is at least one other sibling (you), who like your father, does not have freckles. Explain why the mother in this example must be heterozygous for the freckle trait.

Dihybrid Crosses

A **dihybrid cross**, one that considers the inheritance of two traits from individuals who are heterozygous for both traits, illustrates the law of independent assortment. Like freckles, a widow's peak is controlled by a dominant allele. What would the children look like if the parents were a woman who is homozygous for both freckles and a widow's peak (*FFWW*) and a man with no freckles and a straight hairline (*ffww*)? To answer that question, we begin by determining the possible gametes each mate can produce. Since they are homozygous for both traits, the woman can produce only gametes bearing dominant alleles (*FW*) and the man can produce only gametes with the recessive alleles (*fw*). All the children (F₁ generation) will have freckles and a widow's peak. They will, however, be heterozygous for each trait (*FfWw*).

Suppose one of these children marries someone who is also heterozygous for freckles and a widow's peak. Would it be possible for this couple to have a child that had neither freckles nor a widow's peak? Yes, but the chances are only 1 in 16. Let's see why. First determine the possible gametes that each parent could produce (Figure 18–15). Keep in

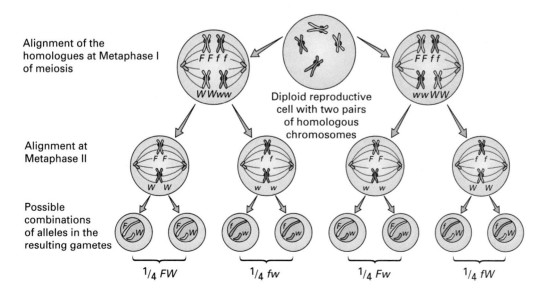

Alignment of the homologues at Metaphase I of meiosis

F F f f
W W w w

Diploid reproductive cell with two pairs of homologous chromosomes

F F f f
w w W W

Alignment at Metaphase II

F F / *w w*   *f f* / *w w*   *F F* / *w w*   *f f* / *W W*

Possible combinations of alleles in the resulting gametes

¼ *FW*   ¼ *fw*   ¼ *Fw*   ¼ *fW*

Figure 18–15 A person who is heterozygous for two genes that are located on different chromosomes can produce four different types of gametes, because the genes assort independently of one another during meiosis I.

mind that alleles for a gene segregate and that genes for different traits that are located on different chromosomes will segregate independently of one another. There are, therefore, four possible allele combinations in gametes produced by a person who is heterozygous for two different genes. In this case, the gametes are *FW*, *Fw*, *fW*, and *fw*. We construct a Punnett square with four columns labeled with the possible male gametes and four rows labeled with the possible female gametes. Each type of gamete has an equal chance of joining with any other type at fertilization. The labels on corresponding columns and rows are combined to indicate the possible gamete combinations. Notice in Figure 18–16 that the phenotypes of children in the next generation (F_2) will be 9 freckled, widow's peak: 3 freckled, straight hairline: 3 no freckles, widow's peak: 1 no freckles, straight hairline. We see, then, that the expected phenotypic ratio resulting from a dihybrid cross is 9:3:3:1.

F=freckles
f=no freckles
W=widow's peak
w=straight hair line

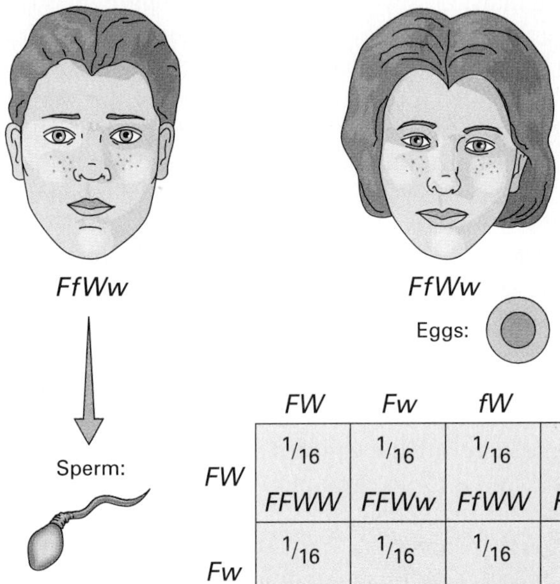

	FW	Fw	fW	fw
FW	1/16 FFWW	1/16 FFWw	1/16 FfWW	1/16 FfWw
Fw	1/16 FFWw	1/16 FFww	1/16 FfWw	1/16 Ffww
fW	1/16 FfWW	1/16 FfWw	1/16 ffWW	1/16 ffWw
fw	1/16 FfWw	1/16 Ffww	1/16 ffWw	1/16 ffww

- 9/16 freckled, widow's peak
- 3/16 freckled, straight hair line
- 3/16 no freckles, widow's peak
- 1/16 no freckles, straight hair line

Figure 18–16 A dihybrid cross is one that considers the inheritance of two traits. It illustrates the law of independent assortment, that is, that each allele pair is inherited independently of others found on different chromosomes. The phenotypes of the offspring of a dihybrid cross would be expected to occur in a 9:3:3:1 ratio.

Testcrosses and Pedigrees

When an individual shows a dominant phenotype, it is impossible to determine whether the genotype is homozygous or heterozygous without further information. When studying non-human organisms, geneticists use a **testcross** to determine the underlying genotype for a dominant phenotype. In a testcross, an individual with a dominant phenotype is mated with a homozygous recessive individual and the phenotypes of the offspring are considered (Figure 18–17). A homozygous recessive individual can produce only gametes carrying the recessive allele. If the individual with a dominant phenotype is homozygous, all its gametes will carry the dominant allele. Consequently, all offspring from such a mating will be heterozygous and display the dominant form of the trait. However, if the individual with the dominant phenotype is heterozygous for the trait, two types of gametes can be formed, one carrying the dominant allele and the other carrying the recessive allele. Thus, if any of the offspring of a testcross show the recessive phenotype, we know that the parent with the dominant phenotype is heterozygous for that trait.

It is sometimes important to determine the genotype of a human with a dominant phenotype for a particular trait. Although it would not be ethical to arrange a testcross, we can often deduce the unknown genotype by looking at the expression of the trait in the ancestors or descendants of the person in question. A chart showing the genetic connections among individuals in a family is called a **pedigree**. Family or medical records are used to fill in the pattern of expression of the trait in question for as many family members as possible. Pedigrees are useful, not just in determining an unknown genotype, but in predicting the chances that one's offspring will display the trait.

Genetic disorders are often determined by recessive alleles, and so knowledge of whether the parents carry the allele will help predict whether their child could be born with the condition. For example, **cystic fibrosis** (CF), a disorder in which abnormally thick mucus is produced, is controlled by a recessive allele. Approximately 1 in every 2000 infants in the United States is born with CF. It is a leading cause of childhood death and the average age of survival is 24 years. The three primary signs of CF are salty sweat, digestive problems, and respiratory problems. Many children with CF suffer from malnutrition because mucus clogs the pancreatic ducts, preventing pancreatic digestive enzymes from reaching the small intestine where they function. Thick mucus also plugs the respiratory passageways, making breathing

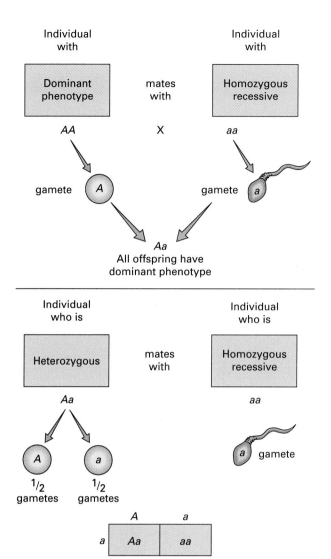

Figure 18–17 In a testcross an individual who displays the dominant phenotype is mated with an individual who is homozygous recessive. If the "dominant" individual is homozygous, all the offspring will show the dominant phenotype. However, if the "dominant" individual is heterozygous, half the offspring are expected to show the recessive form of the trait.

Figure 18–18 Cystic fibrosis is a condition in which abnormally thick mucus is produced causing serious digestive and respiratory problems. Percussion therapy, which involves forceful pounding on the back, breaks up mucus blocking the air tubules in the lungs, making breathing easier. People with cystic fibrosis generally die in their early twenties. This child with cystic fibrosis is using a therapeutic toy to enhance airflow. If she exhales with enough force, the ribbons wave. *(©1992 SPL/Custom Medical Stock Photo)*

difficult and increasing susceptibility to lung infections such as pneumonia (Figure 18–18).

Because CF is inherited as a recessive allele, children with CF are usually born to normal, healthy parents who had no idea they carried the trait. A **carrier** is someone who displays the dominant phenotype but is heterozygous for a trait and can, therefore, pass the recessive allele to descendants. Approximately 1 in 22 Caucasians in the United States is a carrier for CF. (CF is less common among Asians and rare among Black Americans.)

Consider the pedigree showing the inheritance of CF in eight generations of descendants of five ancestral couples (Figure 18–19). All the parents in the seventh generation (1–12) must have been heterozygous carriers of the CF allele. We know this because at least one of the children from each marriage was homozygous for the trait and was, therefore, born with CF. Of the children born to two carriers, we would predict that three-fourths would be unaffected and that one-fourth would have CF. In this pedigree, we see some children with CF in eight families. Of the 33 children in these eight families, 14 had CF, a greater number than would be expected. Based on the expected ratios from a monohybrid cross, we would predict that only 8 children, 25% of all the children of these carriers, would be affected. The actual number is biased toward affected offspring because it is so difficult to identify any heterozygous parents who, by chance, did not have an affected child.

What are the odds that a second child of the parents of a child with CF will be a carrier? Fifty-fifty. Why? We know that an unaffected sibling must have inherited a dominant allele from one of its parents. The question, therefore, is reduced to the chance of inheriting a recessive allele from the other heterozygous parent, which is 50%.

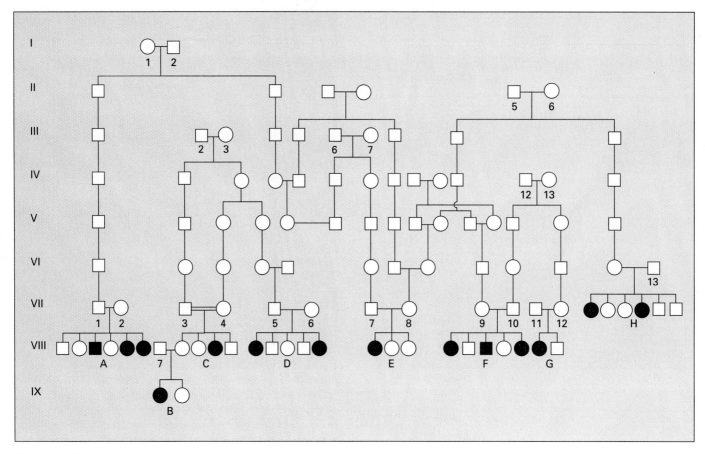

Figure 18-19 The inheritance of cystic fibrosis is shown in this pedigree. A pedigree shows the genetic relationships among individuals. A pedigree is constructed so that each generation occupies a different horizontal line, numbered from top to bottom with the most ancestral at the top. Males are indicated as squares and females as circles. A horizontal line connects the partners in a marriage. An affected individual is indicated with a solid symbol.

Critical Thinking

Is your hypothetical mother (who was described at the beginning of the section on inheritance) with freckles and a widow's peak homozygous or heterozygous for widow's peak? How do you know? (*Hint:* Construct a pedigree showing the appearance of a widow's peak in members of this hypothetical family. Fill in the known genotypes for as many members as possible.)

Dominance Relationships

What makes an allele dominant or recessive? In many cases, the dominant allele produces a normal, functional protein, but the recessive allele does not produce a protein or it produces it in an altered form that doesn't function properly. Consider the inheritance of **albinism**, the inability to produce the brown pigment melanin that normally gives color to the eyes, hair, and skin (Figure 18–20). Because of the

lack of melanin, an albino has pale skin and white hair. As a child, an albino has pink eyes, but the eye color darkens to blue in an adult. Because there is no melanin in the skin to protect against sunlight's ultraviolet rays, an albino is very vulnerable to sunburn and skin cancer.

The ability to produce melanin depends on an enzyme called tyrosinase. The dominant allele, which results in normal skin pigmentation, produces a functional form of tyrosinase. A single copy of the dominant allele can produce all of this enzyme that is needed. The recessive allele that causes albinism produces a nonfunctional form of tyrosinase and so melanin cannot be formed.

Codominance

The situations we have described so far involve **complete dominance**: in a heterozygote, the dominant allele produces a functional protein and its effects are apparent, but the recessive allele produces a less functional protein or none at all and its effects do not manifest. This is not always the case, however. Both alleles for certain traits produce

Figure 18–20 An albino lacks the brown pigment melanin in the skin, hair, and irises of the eyes. Albinism is controlled by a recessive allele that directs the production of a nonfunctional form of an enzyme that is necessary for the production of melanin. A single copy of the dominant allele produces enough of the functional form of this enzyme to generate sufficient melanin for normal pigmentation. (© Richard Dranitzke/Science Source/Photo Researchers, Inc.)

functional proteins. In this case, which is described as **codominance**, the effects of *both* alleles are separately apparent in a heterozygote.

Consider the alleles for normal and sickle-cell hemoglobin as an example of codominance. Hemoglobin is the pigment in red blood cells that carries oxygen. A red blood cell filled with normal hemoglobin (Hb^A) is a biconcave disk. The allele for sickling hemoglobin (Hb^S) produces an abnormal form of hemoglobin that is less efficient in binding oxygen. In the homozygous sickling condition ($Hb^S Hb^S$), called sickle-cell anemia, the red blood cells contain only the abnormal form of hemoglobin. When the oxygen content of the blood drops below a certain level, as might occur during excessive exercise or respiratory difficulty, the red blood cells become sickle-shaped and tend to clump together. The clumped cells can clog capillaries and break open, causing great pain. Vital organs may be damaged by lack of oxygen. In the heterozygote ($Hb^A Hb^S$), both alleles are expressed

Table 18–1   The Relationship Between Genotype and ABO Blood Groups

Genotype	Blood Type
$I^A I^A, I^A I^O$	A
$I^B I^B, I^B I^O$	B
$I^A I^B$	AB
$I^O I^O$	O

and so the red blood cells contain both types of hemoglobin. Such heterozygotes are said to have the sickle-cell trait. They are generally healthy, but sickling and clumping of red cells may occur if there is a prolonged drop in the oxygen content of the blood, such as might occur when traveling at high elevations. (Sickle-cell anemia is discussed in more detail in Chapter 10.)

Multiple Alleles

Many genes have more than two alleles. When three or more forms of a given gene exist, they are referred to as **multiple alleles**. Keep in mind, however, that one individual has only two alleles for a given gene, even if multiple alleles exist in the population. Just like alleles for any other gene, these segregate independently during meiosis.

The ABO blood groups provide an example of multiple alleles. Blood type is determined by the presence of certain polysaccharides (sugars) on the surface of red blood cells. Type A blood has the A polysaccharide, type B has the B polysaccharide, type AB has both polysaccharides, and type O has neither. The synthesis of each of these polysaccharides is directed by a specific enzyme. The enzyme, in turn, is specified by an allele of the gene.

The gene controlling ABO blood groups has three alleles, I^A, I^B, and I^O. Alleles I^A and I^B specify the A and B polysaccharides, respectively. When both these alleles are present, both polysaccharides are produced. I^A and I^B are, therefore, codominant. I^O is recessive to both I^A and I^B. The relationship between these alleles and the resulting blood type is shown in Table 18–1. (Blood types are discussed in more detail in Chapter 10.)

Critical Thinking

A man who has blood type AB is accused of fathering a child with blood type O. The mother has blood type B. Is it possible for the accused man be the father of this child? (*Hint:* What are the possible genotypes of the three people involved? Given each possible genotype of the parents, what gametes could each produce? Use Punnett squares to determine whether any combination of possible parental genotypes could produce a child with the same genotype as this child.)

Polygenic Inheritance

So far we have discussed traits that occur with distinct classes of possible phenotypes. In most cases, you have the trait or you don't, although its expression may be modified by the environment. In the case of multiple alleles, there may be several distinct classes, as in A, B, AB, and O blood types.

The expression of most traits is much more variable than this, however. Indeed, many traits, including height, skin color, and eye color, vary almost continuously from one extreme to the other. Environment can play a role in smoothing out variations. For instance, diet and disease influence adult height and exposure to sunlight darkens skin color. But, all environmental factors being equal, there is still considerable variation in the expression of certain traits. Such variation results from **polygenic inheritance**, that is, the involvement of two or more genes in the determination of the trait. The more genes involved, the smoother the gradations and the greater the extremes of expression.

Although human height is probably controlled by more than three separate genes, we will simplify things to see the variation in expression possible when as few as three genes, A, B, and C, are involved in determining a trait. Assume that the dominant alleles (A, B, C) of each gene add height and the recessive alleles (a, b, c) do not. How tall would one expect the children to be if both of their parents were of medium height and heterozygous for all three genes? As you can see in Figure 18–21, there would be six genetic height classes, ranging from very short to very tall. The probability of the children having either extreme of stature is slim, 1/64, but it is possible. It is most likely (a 20/64 chance), however, that the children will be of medium height, like their parents.

Skin color is also determined by several genes. The allele for albinism prevents melanin production, so if a person is homozygous recessive for this allele no melanin can be deposited in the skin. In addition, there are probably at least four other genes involved in determining the amount of melanin deposited in the skin. Two alleles of four genes would create nine classes of skin color, ranging from pale to dark.

Linkage

It is estimated that there are between 50,000 and 100,000 human genes packaged onto the 23 different kinds of chromosomes. Thus, each chromosome bears a great number of genes. Genes on the same chromosome tend to be inherited together, because an entire chromosome moves into a gamete as a unit. Genes that tend to be inherited together are described as being **linked**. We see, then, that linked genes do not *usually* assort independently.

Usually is emphasized here because there is a mechanism that can unlink genes on the same chromosome—crossing over. You may recall that crossing over occurs during prophase I of meiosis when the homologous chromosomes pair, aligning themselves so that the genes for the same traits are next to one another. Bridges form be-

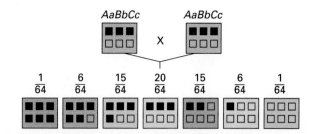

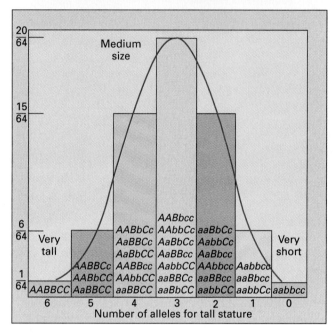

Figure 18–21 Human height varies in a continuous manner. One reason is that height is determined by more than one gene (polygenic inheritance). This figure shows the distribution of alleles for tallness in children of two parents of medium height, assuming that three genes are involved in the determination of height. The top line shows the parental genotypes and the second line indicates the possible genotypes of the offspring. Alleles for tallness are indicated with dark squares.

tween nonsister chromatids, and segments of DNA are exchanged. The result is an exact exchange of alleles. After crossing over occurs, the resulting chromatid can have a different assortment of alleles than its sister chromatid. For instance, consider the possible gametes that could be formed in a person who is heterozygous for linked genes A, B, C, and D. If no crossing over occurred, two types of gametes would be formed, ABCD and abcd. If crossing over occurred between genes B and C, however, four types of gametes would be formed—ABCD, abcd, ABcd, and abCD. Notice that the alleles have been recombined, creating two new kinds of gametes. We know that crossing over has occurred when we see a recombination of alleles in the offspring.

We can use the frequencies of recombination between genes to **map** the chromosome, that is, to indicate the order and relative positions of genes on a chromosome. Since genes are arranged in a linear series along a chromosome,

Cross	Offspring	
AD × ad	AD + ad (parental)	90%
	aD + Ad (recombinant)	10%
AC × ac	AC + ac (parental)	92%
	aC + Ac (recombinant)	8%
CD × cd	CD + cd (parental)	98%
	cD + Cd (recombinant)	2%

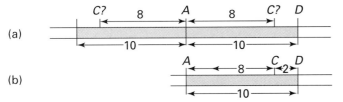

Figure 18–22 Like beads on a string, genes are arranged on a chromosome in linear sequence. Thus, the greater the distance between two genes, the greater the chance of a cross-over between them. The frequencies of recombination between genes on a chromosome can be used to map a chromosome, that is, to determine the relative positions of genes along a chromosome. The example illustrated assumes that genes A and D recombined in 10% of the offspring. In other words, 10% of the offspring were Ad or aD. We say, then, that A and D are 10 units apart on the chromosome. The example further assumes that A and C recombined in 8% of the offspring and C and D recombined in 2% of the offspring. The data tell us that A and C are 8 units apart. (a) However, so far we cannot tell whether C lies to the left or right of A. (b) The amount of recombination between C and D allows us to determine the correct order of genes. If the order were CAD, we would expect genes C and D to be recombined in 18% of the offspring. We find instead that 2% of the offspring show recombination between C and D. Thus, the order of genes must be ACD.

the greater the distance between two genes, the greater the probability they will be separated in crossing over. Thus, the relative frequencies of cross-overs and recombination between various genes can reveal the sequence of genes along the chromosome and the relative distance between them (Figure 18–22).

Sex-Linked Inheritance

You may recall that one pair of chromosomes is called the sex chromosomes. There are two kinds of sex chromosomes, X and Y. They are not truly homologous because the Y chromosome is much smaller than the X chromosome and they do not carry all of the same genes.[2] The Y chromosome

[2]X and Y are considered to be a homologous pair because they each have a small region at one end that carries some of the same genes. During meiosis, the tiny homologous region on X and Y will pair in synapsis. As a result, they segregate into gametes in the same manner that autosomes do.

carries very few genes, but it is most important in determining gender. If a particular gene on the Y chromosome is present, an embryo will develop as a male. In the absence of that gene, an embryo will develop as a female. In contrast, most of the genes on an X chromosome have nothing to do with sex determination. For instance, genes for certain blood clotting factors and for the pigments in cones (the photoreceptors responsible for color vision) are found on the X chromosome, but not on the Y chromosome. Furthermore, the X chromosome has about as many genes as an autosome, but the Y chromosome has relatively few. Thus, most genes on the X chromosome have no corresponding alleles on the Y chromosome and are known as **X-linked genes**.

Because most X-linked genes have no homologous allele on the Y chromosome, they have a different pattern of inheritance than do autosomes. A male is XY and therefore will express virtually all of the alleles on his single X chromosome, even those that are recessive. A female, on the other hand, is XX, and so she does not always express recessive alleles. As a result, the recessive phenotype is much more common in males than in females (Figure 18–23). Furthermore, a son cannot inherit an X-linked recessive allele from his father. To be male, a child must have inherited his father's Y chromosome, not his X chromosome. Consequently, a son can inherit an X-linked recessive allele only from his mother. A daughter, however, can inherit an X-linked recessive allele from either parent. If she is heterozygous for the trait, she will have a normal phenotype, but be a carrier for that trait.

Among the disorders caused by X-linked recessive alleles are red-green color blindness, two forms of hemophilia,

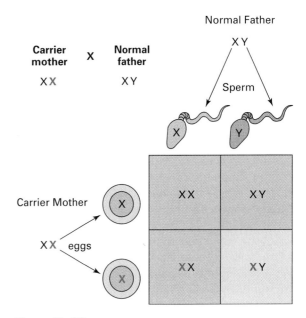

Figure 18–23 Genes that are X-linked have a different pattern of inheritance than do genes on autosomes, as seen in this cross between a carrier mother and a father who is normal for the trait. The recessive allele is indicated in red.

and Duchenne muscular dystrophy. Red-green color blindness, the inability to distinguish red and green was discussed in detail in Chapter 9. Hemophilia is a bleeding disorder caused by a lack of certain blood clotting factors. Hemophilia A is a lack of blood clotting factor VIII and hemophilia B (Christmas disease) is a lack of clotting factor IX. Hemophilia was discussed in Chapter 10.

Duchenne muscular dystrophy is an X-linked recessive condition in which there is progressive muscle weakness because the muscle cells break down and are gradually lost. The responsible allele produces a defect in an important protein (dystrophin) that gives support to the plasma membrane of muscle cells. It affects only boys, with rare exceptions. The first muscles affected are those of the shoulders, hips, thighs, and calves. Therefore, early symptoms include difficulty climbing stairs or rising to an upright position when bending over, falling easily, and a waddling gait. Difficulty in walking, which starts when the boy is between 1- and 3-years-old, gradually worsens until the boy is between 8- and 11-years-old, when walking becomes impossible. With time, the disease spreads to all muscles. Death usually occurs by age 20, because of heart or respiratory failure.

Critical Thinking

Explain why Duchenne muscular dystrophy is inherited from one's mother but is usually only expressed in sons. (*Hint:* Consider its mode of inheritance.)

Sex-Influenced Inheritance

The expression of certain *autosomal* genes is powerfully influenced by the presence of sex hormones, and so their expression differs in males and females. We describe these traits as **sex-influenced**.

Male pattern baldness, premature hair loss on the top of the head but not the sides, is an example of a sex-influenced trait.[3] Pattern baldness is much more common in men than in women because its expression depends on both the presence of the allele for baldness and the presence of testosterone, the male sex hormone. The allele for baldness, then, acts as a dominant allele in males, because of their high level of testosterone, and a recessive allele in females, because they have much lower testosterone levels. A male will develop pattern baldness whether he is homozygous or heterozygous for the trait. However, only women who are homozygous for the trait will develop pattern baldness. When a woman does develop pattern baldness, it is usually appears later in life than it does in a man. The allele is expressed in women because the adrenal glands produce a small amount of male hormone. After menopause, when the supply of es-

[3]Male pattern baldness and its treatments are discussed in more detail in Chapter 4.

trogen declines, adrenal male hormones may cause the expression of the baldness gene. However, balding in women may be merely thinning of hair.

Critical Thinking

Male pattern baldness was passed from father to son through at least four generations of the Adams family. [John Adams (1735–1826), the second U.S. President, passed it to his son, John Quincy Adams (1767–1848), the sixth U.S. President. He, in turn, passed the gene to his son, Charles Frances Adams (1807–1886), a diplomat, who passed it to his son, Henry Adams (1838–1918), a historian.] Explain why father-to-son transmission of the trait rules out X-linked inheritance.

Changes in Chromosome Structure

Chromosomes can break, allowing alterations in structure. Breakage can be caused by certain chemicals, radiation, or viruses. It also occurs as an essential part of crossing over. Although it doesn't happen often, chromosomes can be misaligned when crossing over occurs. Then, when the pieces reattach, one chromatid will have lost a segment and the other will have gained a segment.

The loss of a piece of chromosome is called a **deletion**. The most common type of deletion occurs when the tip of a chromosome breaks off and is not included in a daughter cell following cell division. Deletion of more than a few genes on an autosome is usually lethal, and the loss of even small regions causes disorders.

In humans, the most common deletion, the loss of a small region near the tip of chromosome 5, causes cri-du-chat syndrome (meaning "cry of the cat"). An infant with this syndrome has a high-pitched cry that sounds like a kitten meowing. The unusual sound of the cry is caused by an improperly developed larynx (voice box). Infants have a round face, wide-set, downward sloping eyes with epicanthal folds, and misshapen ears (Figure 18–24). Although the condition is not usually fatal, it does cause severe mental retardation.

An added piece of chromosome is called a **duplication**. The effects of a duplication depend on its size and position. In general, however, a small duplication is less harmful than a deletion of comparable size. There is a small region of chromosome 9 that can be duplicated, resulting in cells containing three copies of this segment. The result is mental retardation, accompanied by facial characteristics that may include a bulbous nose, wide-set squinting eyes, and a lopsided grin.

Genetic disorders also occur when certain sequences of three subunits of DNA (nucleotides) are duplicated multiple times. The fragile X syndrome provides an example. The

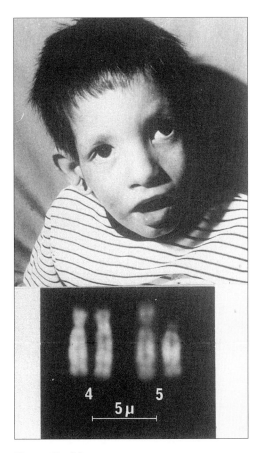

Figure 18-24 Occurring in one out of every 50,000 live births, cri-du-chat syndrome is the most common deletion found in humans. It is caused by the loss of a small region near the tip of chromosome 5. The name of the syndrome, which means "cry of the cat," is given because of the sound of the cry of affected babies. Infants have a characteristically round face and wide-set eyes. Some of the facial features persist into childhood. *(© Dr. Irene Uchida, McMaster University, Ontario)*

syndrome is so named because the long sequence of repeats makes the X chromosome fragile and easily broken. Besides making the chromosome fragile, the repeated subunits can shut down the activity of the entire chromosome. Fragile X syndrome is the most common form of inherited mental retardation, affecting roughly 1 in 1250 males and 1 in 2500 females. It is not known, however, exactly how fragile X syndrome causes the retardation. Other characteristics may include attention deficit, hyperactivity, large ears, long face, and flat feet.

Social Issue

The degree of mental retardation accompanying fragile X syndrome is variable. It affects nearly all males who inherit a fragile X chromosome. About a quarter of the girls with fragile X syndrome are retarded, although others may have only learning problems. Some cities are beginning to screen not just retarded children but also those with learning difficulties for fragile X chromosomes. Some specialists argue that identifying children with fragile X syndrome benefits the children because specialists such as speech therapists, counselors, and physicians can be called on to help. Genetic counseling can help families decide whether to have children. Others argue that there is no special treatment for children with fragile X syndrome, other than that available to all children with learning problems. After diagnosis, insurance companies may discriminate against the children and drop their health coverage. In your opinion, do the benefits of screening for fragile X syndrome outweigh the disadvantages?

Genomic Imprinting

Most traits are inherited in the manner we have so far described—if a particular allele or alleles are inherited, a particular phenotype results. As with so many things, however, there are exceptions to this rule. One exception is **genomic imprinting**, a phenomenon in which the expression of an allele depends on which parent contributed the allele. For example, two disorders—Prader-Willi syndrome and Angelman syndrome—can result from a deletion of a particular region of chromosome 15. Prader-Willi syndrome results if the deletion was inherited from the father and Angelman syndrome results if it was from the mother. Keep in mind that the affected person can be of either sex. Some degree of mental retardation occurs in both conditions. Infants with Prader-Willi syndrome are called "floppy babies," because they have weak muscles. Muscle tone gradually improves and with it comes a voracious appetite. When the child is old enough to move about on the floor, food must be raised to a higher place, where it cannot be reached. As the child grows older, the parents must learn patterns of food storage that are not typical of those in the average family—locking food away in cupboards and refrigerators. The extreme appetite leads to obesity, which becomes the most significant health problem during adulthood. Indeed, obesity is not the only life-threatening aspect of Prader-Willi syndrome. It can lead to high blood pressure, diabetes, high cholesterol, heart attacks, and strokes. Individuals with Angelman syndrome, which was once called the "happy puppet" syndrome because of the appearance and behavior of affected children, have severe mental retardation. Nonetheless, they have a happy disposition and laugh a lot. Their movements are jerky and repetitive, like those of a puppet. Even the face of a child with Angelman syndrome is puppet-like, with red cheeks, a large jaw, and a large mouth.

Genomic imprinting can be thought of as a form of temporary gene inactivation. It apparently determines whether the allele from the mother or the father will be active or silenced. Although it is not completely clear how genes become imprinted, one idea is that during gamete formation genes may be "tagged" by the attachment of a small molecular group (a methyl group) to the DNA. The methyl group may then promote the binding of an "imprinting factor," which could in turn prevent the allele from producing its protein. Whatever the mechanism of gene imprinting, the maternal and paternal imprints must be "erased" in the cells that will produce gametes, so that an individual can imprint its own genes in gametes according to *its* sex.

Detecting Genetic Disorders

There are more than 4000 disorders that have their roots in our genes. Tests are now available to look for predisposition to many of these genetic disorders and some tests can even confirm the presence of a suspected disease-related allele in a particular person. Knowledge about the presence or absence of a faulty gene can be very helpful to couples who are planning a family. Normal, healthy parents can carry recessive alleles for disorders such as cystic fibrosis and Tay-Sachs disease (a disorder of lipid metabolism that causes death, usually between the ages of 1 and 5). Although pedigree analysis can help prospective parents determine whether they *might* be carriers of recessive alleles, it can't answer the question with certainty. Yet this information would allow the parents to weigh the risks of passing a lethal allele to their children or to know that their child will not be affected.

Prenatal testing is usually recommended when a defective gene runs in the family or when the mother is older than 35, which increases the risk of problems due to nondisjunction. There are two available procedures for diagnosing genetic problems in the fetus—**amniocentesis** and **chorionic villi sampling (CVS)** (Figure 18–25). Although it is possible to look for more than 100 disorders with these procedures, tests are only run for those that are common and those that are of particular concern in that pregnancy. This is why reassuring results from either form of prenatal testing cannot guarantee a healthy baby, in spite of the high degree of accuracy with which they can detect genetic disorders.

In amniocentesis, a needle is inserted through the lower abdomen into the uterus and a small amount of amniotic fluid, 10 to 20 ml, is withdrawn. Because it is important that the needle not injure the fetus, umbilical cord, or the placenta, ultrasound is used to find the safest spot for insertion. Floating in the amniotic fluid are living cells that have sloughed off the fetus. These cells are grown in the laboratory for a week or two and then examined for abnormalities in the number of chromosomes and the presence of certain alleles that are likely to cause specific diseases. Biochemical tests are also done on the fluid to look for certain chemicals that are indicative of problems. For example, a high level of alpha-fetoprotein, a substance produced by the fetus, suggests that there is a problem with the development of the central nervous system (a neural tube defect). Amniocentesis

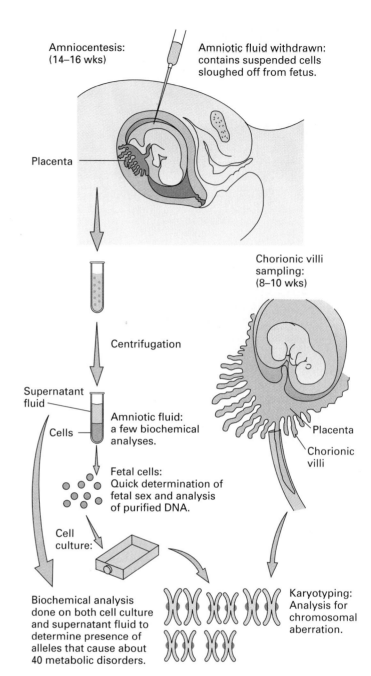

Figure 18–25 Amniocentesis and chorionic villi sampling (CVS) are procedures available for prenatal genetic testing. In amniocentesis, a needle is inserted through the lower abdomen into the uterus. A small sample of amniotic fluid is withdrawn. The fetal cells floating in the fluid are grown in tissue culture and then analyzed for certain genetic abnormalities. Biochemical tests performed directly on the fluid can reveal other problems, such as neural tube defects. In CVS, the physician inserts a small tube through the vagina and cervix to the area of the uterus where the chorionic villi are located and some tissue is gently sucked off the placenta. The cells are then analyzed for genetic abnormalities.

Gene Testing

Genetic screening involves the testing of people who have no symptoms to determine whether they carry genes that will influence their chances of developing certain genetic diseases. It is a technology whose time has come. Like a snowball rolling downhill, the practice is gaining momentum. However, genetic screening raises many ethical issues.

Among the advantages of gene testing is that, armed with information that one has a treatable or preventable condition, steps can be taken to reduce the risk of developing the disorder. It can also save suffering in future generations when the disorder is caused by a recessive allele that can remain hidden for generations or by a dominant allele that is not expressed until late in life. Consider, for instance, Tay-Sachs disease. This disorder causes the death of children, usually by the age of 5. The infant appears healthy at birth, but at about 6-months-old it gradually stops smiling, crawling, or turning over. Eventually the child becomes blind, paralyzed, and unaware of its surroundings. Tay-Sachs disease is especially prevalent in descendants from Jewish communities from eastern Europe. As a result of voluntary screening programs, the number of children born with Tay-Sachs disease has decreased by tenfold in many communities.

There is also a dark side to genetic testing. The psychological consequences of test results can be devastating. Many genetic disorders cannot be prevented or treated. One must wonder, then, do you really want to know *now* what will cause your death? Huntington disease is caused by a dominant allele that provides no hints of is existence until relatively late in life, usually past child-bearing years. About 60% of the people with Huntington disease are diagnosed between the ages of 35 and 50. The gene causes degeneration of the brain, leading to muscle spasms, personality disorders, and death, usually within 10 to 15 years. Because Huntington disease is caused by a dominant allele, a bearer has a 50% chance of passing it to children. Thus, a person whose parent died of Huntington disease could be relieved if a gene test does not detect the allele. But, it is equally likely that the allele *will* show up. Many of those at-risk for Huntington disease prefer to live without the knowledge of their possible fate.

A gene test confirming the risk of a serious disease can cause anxiety and depression in carriers and feelings of guilt in siblings who do not carry the allele. Much of the carrier's worry may be unnecessary, because predictive gene tests deal in probabilities, not certainties. Mutations in the *BRCA-1* gene on chromosome 17 increase a woman's risk of developing breast

cancer and ovarian cancer. Carriers have an 85% chance of developing breast cancer by age 65. A woman who learns she is a carrier must make medical decisions that could influence the length of her life. One option would be to have frequent mammograms in the hope of discovering breast cancer, if it should develop, early enough for treatment. Of course, there is a chance that cancer would not be detected before it spread, making effective treatment difficult. A second option would be to have a mastectomy quickly. However, the cancer may have already spread. Also, there is a 15% chance that she would not develop breast cancer.

There is also the threat that the results of gene tests will not remain private information, but instead be used by employers, as well as life and health insurers. As an employer, if you had information about the genetic make-up of prospective employees, would you choose to invest time and money in training a person who carried an allele that increased the risk of cancer, heart disease, Alzheimer's disease, or alcoholism? As an insurer, would you knowingly cover a carrier?

Since the results of gene testing can have both positive and negative consequences for those being tested and their families, who should decide whether screening should be done, for which genes, on whom, and in which communities? A flippant answer might be, "There oughtta be a law!" Should we leave ethical issues to judges and legislators? Should moral matters be decided by society or clergy or should they be personal decisions?

If gene testing is done, who should be told the results? If the affected person is an infant, should the parents *always* be told the results, even if the condition is poorly understood? How do we balance helping such children with the possibility of stigmatizing them?

We live in a world of limited resources. Thus, as soon as it is decided who should be tested, we must decide who pays the bill. Both testing and treatment are expensive. Should testing be done only when treatment or preventative measures are available? How much say should the agent that pays for the procedure have in who is tested and who receives medical treatment?

There are no easy answers. It is time for each of us to think about the issues raised by these and similar questions.

is generally safe for both the mother and the fetus, but there is a small risk of triggering a miscarriage or of the needle injuring the mother, which could cause infection or bleeding.

Amniocentesis is usually done between 14 and 18 weeks after the woman's last menstrual period, when there is enough amniotic fluid, about 250 ml (a cup), to minimize the risk of injuring the fetus. (A few medical centers are now able to perform an amniocentesis at 12 weeks of pregnancy.)

CVS involves taking a small piece of chorionic villi, which are small, finger-like projections of the part of the placenta called chorion, part of the placenta. Cells of the chorion have the same genetic composition as those of the fetus. Guided by ultrasound, a small tube is inserted through the vagina and the cervix to where the villi are located. Gentle suction is then used to remove a small tissue sample, which can be analyzed for genetic abnormalities.

There are pros and cons to CVS. Advantages are that it can be performed 6 to 8 weeks earlier in the pregnancy than amniocentesis and the results are available within a few days. Since more than 95% of the high-risk women who opt for prenatal tests receive good news, early diagnosis saves weeks of worry over the health of the fetus. Also, if there is a genetic problem in the fetus and the couple wishes to terminate the pregnancy, the procedure can be performed earlier in the pregnancy, when it is safer for the mother. When abortion is not chosen, early diagnosis allows more time to plan the safest time, location, and method of delivery. A disadvantage of CVS is that it has a slightly greater risk of triggering miscarriage than does amniocentesis.

If detected early enough, certain birth defects can be prevented. Congenital adrenal hyperplasia (an overgrowth of the adrenal glands), for instance, will cause a female fetus to develop abnormal genitalia, unless it is treated with hormones from week 10 to 16 of gestation. Early diagnosis with CVS can tell a physician whether hormone treatment is needed.

Newborns are routinely screened for phenylketonuria (PKU), an inherited metabolic disorder. This simple blood test can prevent mental retardation. People with PKU have inherited an allele responsible for a defective enzyme that prevents them from converting phenylalanine (an amino acid in food) to tyrosine. As a result, they have too much phenylalanine in their bodies and too little tyrosine, an imbalance that somehow causes brain damage. Although nothing can be done to correct the enzyme, brain damage can be prevented with a strict diet that excludes most phenylalanine-containing foods during early childhood.

Many predictive genetic tests are now available or are being developed. These tests identify people who are at risk of getting a disease, *before* symptoms appear. The procedure is simple and can usually be done with a small blood sample. When steps can be taken to prevent the disease, predictive gene tests can be lifesaving. For instance, colon cancer will develop in nearly everyone who has the alleles for familial adenomatous polyposis, a condition in which thousands of benign polyps grow in the intestine. If the colon is routinely inspected for polyps and they are removed, cancer can be prevented.

Other predictive gene tests look for alleles that might predispose one to a disorder. A protein that transports cholesterol in the blood, called ApoE, comes in three forms, each specified by a different allele. Having two alleles for one of these, ApoE-2, causes catastrophically high blood cholesterol levels, which can lead to heart attack and stroke. Knowing that a person had this genetic make-up, a physician could prescribe medication to lower blood cholesterol.

Social Issue

Two copies of another *ApoE* allele, *ApoE-4*, increase a person's risk of heart disease by 30% to 50%. It also nearly guarantees that Alzheimer's disease will develop by 80 years of age. Alzheimer's disease is an untreatable condition in which brain tissue degenerates. A person is gradually robbed of memories, of the ability to function normally in society, and eventually of life itself. Suppose that a gene test that is performed out of concern for a person's risk of heart disease reveals that two copies of *ApoE-4* are carried. In your opinion, is a physician with this knowledge morally obligated to tell the patient about the inevitability of Alzheimer's disease or to keep the secret? What, if any, factors in the patient's life should be considered in making this decision?

SUMMARY

1. A chromosome contains DNA and proteins called histones. A gene is a segment of DNA that codes for a protein that plays a structural or functional role in the cell. Genes are arranged along a chromosome in a specific order. Each of the 23 different kinds of chromosomes in human cells contains a specific sequence of genes. The particular location of a gene on a chromosome is called its locus.

2. Somatic cells (all cells except for eggs and sperm) are diploid—that is, they contain two sets of chromosomes, one from each parent. Homologous chromosomes carry genes for the same traits. In humans, the diploid number of chromosomes is 46, two sets of 23 homologous pairs. One pair of chromosomes, the sex chromosomes, determines gender. Males are XY and females are XX. The other 22 pairs of chromosomes are called autosomes. Eggs and sperm are haploid and contain only one set of chromosomes.

3. Meiosis, a special type of cell division that occurs in the ovaries or testes, begins with a diploid cell and produces four haploid cells that will become gametes (egg or sperm). Meiosis is important because it halves the number of chromosomes in gametes, thereby keeping the chromosome number constant between generations. When a sperm fertilizes an egg, a diploid

cell, called a zygote, is created. After many mitotic divisions, the zygote will develop into a new individual.

4. Before meiosis begins, the chromosomes are replicated and the copies remain attached to one other by centromeres. The attached replicated copies are called sister chromatids. There are two cell divisions in meiosis. During the first meiotic division (meiosis I), members of homologous pairs are separated. Thus, the daughter cells contain only one member of each homologous pair (although each chromosome still consists of two chromatids). During the second meiotic division (meiosis II), the sister chromatids are separated.

5. Nondisjunction is the failure of homologous chromosomes or sister chromatids to separate during cell division. It results in an abnormal number of chromosomes in the resulting cells. Nondisjunction of chromosome 21 can result in Down syndrome. Nondisjunction of the sex chromosomes can cause Turner syndrome (XO) or Klinefelter syndrome (XXY).

6. Genetic recombination during meiosis results in variation among offspring from the same two parents. One cause of genetic recombination is crossing over, in which corresponding segments of DNA are exchanged between maternal and paternal homologues, creating new combinations of alleles in the resulting chromatids. Crossing over occurs during synapsis, when maternal and paternal homologues pair with one another, aligning gene by gene. A second cause of genetic recombination is the independent assortment of maternal and paternal homologues into daughter cells during meiosis I. The orientation of the members of the pair relative to the poles of the cell determines which member a daughter cell will receive. Each pair aligns independently of the others. As a result, humans can create about 8 million genetically different gametes.

7. Different forms of a gene are called alleles. An individual who has two of the same alleles is said to be homozygous. An individual with two different alleles for a gene is said to be heterozygous. The allele that is expressed in the heterozygous condition is described as being dominant. The allele that is masked in a heterozygote is described as recessive. The precise alleles present in an individual is its genotype. The phenotype refers to the observable traits.

8. The law of segregation states that alleles for each gene separate during gamete formation so that half the gametes receive one allele and the other half receive the other allele.

9. A monohybrid cross is one in which only one trait is considered. If the Parental (P) generation consists of a homozygous dominant individual mated with a homozygous recessive, the genotypes of the F_1 offspring will be heterozygous and the phenotypes will be dominant. If the F_1 offspring are crossed with one another the genotypes of the F_2 generation will be 1 homozygous dominant: 2 heterozygous dominant: 1 homozygous recessive. Their phenotypes will be 3 dominant: 1 recessive.

10. The law of independent assortment states that each pair of alleles that are located on separate chromosomes separate independently of other pairs of alleles.

11. A dihybrid cross considers the inheritance of two traits simultaneously. If an individual that is homozygous dominant for two traits is mated with an individual that is homozygous recessive, the F_1 progeny will all have dominant phenotypes and be heterozygous. Matings among these F_1 progeny will produce F_2 offspring in a 9:3:3:1 ratio of phenotypes.

12. In a testcross, a homozygous recessive individual is mated with one who shows the dominant phenotype to determine its genotype. If any offspring show the recessive phenotype, the individual showing the dominant phenotype must be heterozygous for the trait. Pedigrees, which are constructed to show the genetic relationships among individuals in an extended family, are often useful in determining the unknown genotypes of humans showing dominant phenotypes.

13. When two alleles are codominant both are separately apparent in the phenotype. For example, people who have the sickle-cell trait are heterozygous for the gene. Their red blood cells contain both normal hemoglobin and sickling hemoglobin because both alleles are expressed.

14. When there are three or more alleles for a particular gene, they are called multiple alleles. ABO blood groups are determined by three alleles, I^A, I^B, and I^O. Blood types are determined by the presence of certain polysaccharides on the surface of red blood cells. I^A and I^B produce enzymes that result in the presence of polysaccharides, A and B respectively. I^A and I^B are codominant. I^O is recessive to both.

15. Many traits, including height, skin pigmentation, and eye color, are determined by more than one gene (polygenic inheritance). Such traits are variable in their manner of expression. When many genes are involved in determining a trait, the variation in expression can be continuous.

16. Linked genes are those that are inherited together, unless crossing over occurs between them. Genes are arranged in a linear order along the chromosome. The greater the distance between two genes, the greater the probability of cross-over between them. We can use the relative frequencies of crossovers between genes to determine their position along a chromosome, thereby creating a chromosome map.

17. An X-linked gene is one located on the X chromosome that has no corresponding allele on the Y chromosome. A recessive X-linked allele will always be expressed in a male, but it will only be expressed in the homozygous condition in a female. Red-green color blindness, hemophilia, and Duchenne muscular dystrophy are disorders caused by X-linked recessive alleles.

18. The expression of a sex-influenced trait depends on both the presence of the allele and the presence of sex hormones. Therefore, the expression of the allele depends on one's sex. For example, the expression of the allele for male pattern baldness requires the presence of testosterone.

19. The loss of a piece of a chromosome is called a deletion and the gain of a piece is called a duplication. Either chromosome abnormality can cause a genetic disorder.

20. Genomic imprinting causes genes to be expressed differently depending on which parent contributed the gene. It is a biochemical means of marking genes to be either active or silent. The same chromosomal deletion will cause Prader-Willi syndrome if it is inherited from the father's sperm or Angelman syndrome if inherited from the mother's egg.

21. Tests are available to determine whether someone is likely to develop a genetic disease. There are two prenatal gene tests: amniocentesis and chorionic villi sampling (CVS). In amniocentesis, a sample of amniotic fluid is taken. The fetal cells in the fluid are grown in the laboratory and then analyzed to look for genetic problems. Tests on amniotic fluid can reveal certain other developmental problems, such as neural tube defects. CVS involves sampling cells from the chorion of the placenta and analyzing them for genetic disorders.

1. Explain the relationship between genes and a chromosome.
2. What is the difference between a cell that is diploid and one that is haploid?
3. Which human cells are diploid? Which are haploid?
4. Why is meiosis important?
5. Describe the alignment of chromosomes at the equatorial plate during meiosis I and meiosis II. Explain the importance of these alignments in creating haploid gametes from diploid cells.
6. Explain how crossing over and independent assortment result in genetic recombination that causes variability among offspring from the same two parents.
7. Define nondisjunction. Explain how nondisjunction can result in abnormal numbers of chromosomes in a person.
8. What causes Down syndrome and what are the usual characteristics of the condition?
9. Differentiate between Turner syndrome and Klinefelter syndrome by describing the cause of the conditions and the characteristics of each.
10. State the law of segregation. Explain how this law relates to events that take place during meiosis I.
11. State the law of independent assortment. Explain how this law relates to events that take place during meiosis I.
12. Define a monohybrid cross and a dihybrid cross.
13. What is the ratio of genotypes and phenotypes in the F_2 generation resulting from a monohybrid cross between a homozygous dominant individual and a homozygous recessive one?
14. What is the ratio of phenotypes in the F_2 generation resulting from a dihybrid cross between two individuals who are both heterozygous for the same two traits.
15. What is a testcross? What could the results of a testcross reveal about the genotype of the parent showing the dominant trait?
16. What is a pedigree? What can a family pedigree reveal about the inheritance of a trait?
17. Using an example, explain what is meant by codominance.
18. Differentiate between multiple alleles and polygenic inheritance.
19. What are linked genes? Why are they usually inherited together? What can cause such genes to become unlinked?
20. Explain why the pattern of inheritance for recessive X-linked genes is different from the pattern for recessive autosomal alleles.
21. What two procedures are used for prenatal testing? How do they differ?

DNA and Biotechnology

Genetic engineering has brought us improved varieties of crops. *(© Science VU/Visuals Unlimited)*

"There is no substance so important as DNA ... it is the prime molecule of life.... The key to our optimism that all secrets of life are within the grasp of future generations of perceptive biologists is the ever accelerating speed at which we have been able to probe the secrets of DNA."[1]

Is this declaration boastful optimism or is it true that a thorough understanding of a mere molecule—DNA (deoxyribonucleic acid)—will put all secrets of life within our grasp? If it is true, can we humans deal with the information responsibly? Clearly, the search for an understanding of DNA is an endeavor that brings together science, society, politics, and ethics.

In this chapter, we will become more familiar with the structure and function of DNA to see how this substance can be the basis not just of our genetic inheritance and legacy but also of the evolutionary process that has provided the diversity of life we see around us. We will begin to understand that the importance of DNA on a personal level is that it directs the synthesis of specific proteins that play structural or functional roles in our bodies. Then, we will consider the technology that our understanding of DNA has already made available and some of the promises it holds for the future.

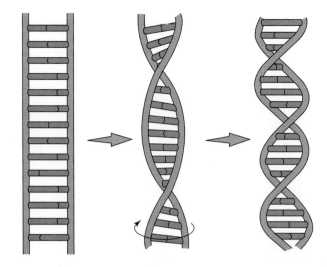

Figure 19–1 DNA is a double-stranded molecule that is twisted to form a spiral structure called a double helix.

Structure of DNA

DNA is sometimes called the thread of life—and a very slender thread it is. If the DNA from a single cell were unwound, it would be a mere 50-trillionths of an inch in diameter. With all the DNA strands fastened together end-to-end, the thread would stretch over 5 feet in length. DNA might even be considered the thread that ties all life together, because the DNA of organisms ranging from bacteria to humans is built from the same kinds of subunits. The order of the subunits encodes the information needed to make the proteins that build and maintain life.

DNA is a double-stranded molecule resembling a ladder that is gently twisted to form a spiral, called a double helix (Figure 19–1). Each side of the ladder and half of each rung are made from a string of repeating subunits called nu-

cleotides. A **nucleotide**, in turn, is composed of one sugar (deoxyribose), one phosphate, and one nitrogenous base. There are four different nitrogenous bases in DNA: two purines—adenine (A) and guanine (G)—and two pyrimidines—thymine (T) and cytosine (C). The sides of the ladder are composed of alternating sugars and phosphates and the rungs consist of paired bases. Adenine will pair only with thymine (an A-T pair) and cytosine only with guanine (a C-G pair). Each base pair is held together by weak hydrogen bonds (Figure 19–2). The pairing is specific because of the shapes of the bases and the number of hydrogen bonds that can form between them.

Because of the specificity of base pairing, the bases on one strand of DNA are always **complementary** to the bases on the other strand. We see then that the order of bases on one strand determines the sequence on the other strand. For instance, if the sequence of bases on one strand were CATATGAG, the complementary sequence on the opposite strand would be GTATACTC.

In the DNA of each human cell there are an astounding 3 billion base pairs. Although the pairing of adenine with thymine and cytosine with guanine is specific, the sequence of bases along the length of the DNA molecule can vary in a myriad of ways. As we will see, genetic information is encoded in the exact sequence of bases.

Replication

For DNA to be the basis of inheritance, its genetic instructions must be passed from one generation to the next. More-

[1]James D. Watson, Michael Gilman, Jan Witkowski, and Mark Zoller. 1991. *Recombinant DNA*, second edition. Scientific American Books, W. H. Freeman and Company. New York, N.Y.

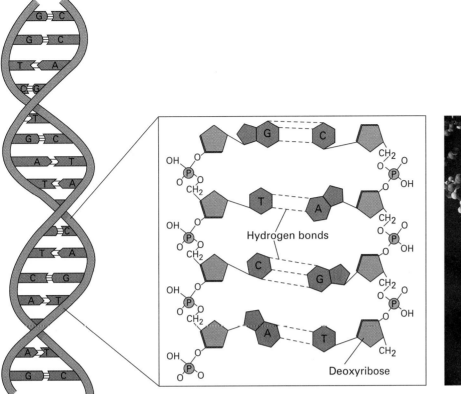

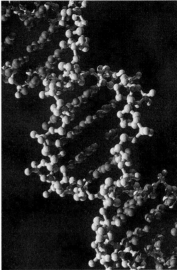

Figure 19–2 Three representations of DNA. The structure of DNA resembles a ladder twisted around itself to form a double helix. The sides of the ladder are composed of alternating molecules of phosphate and the sugar deoxyribose. Each rung is composed of a pair of nitrogenous bases, either adenine with thymine or cytosine with guanine. *(photo, © Kenneth Eward/BioGrafx/Science Source/Photo Researchers, Inc.)*

over, for DNA to direct the activities of each cell, its instructions must be present in every cell. These requirements dictate that DNA be copied before either meiotic or mitotic cell division. It is important that the copies be exact. The key to the precision of the copying process, or **replication**, is in the complementarity of the bases.

Replication begins when an enzyme breaks the hydrogen bonds that hold together the two nucleotide strands of the double helix, thereby "unzipping" and unwinding the two strands. As a result, the nitrogenous bases on the separated regions of each strand are exposed. Bases of free nucleotides then attach to complementary bases on the open DNA strands. Enzymes called **DNA polymerases** link the new nucleotides together to form a new strand. In this way, two DNA strands form, each identical to the original one. As each of the new double-stranded DNA molecules forms, it twists forming a double helix (Figure 19–3).

Each strand of the original DNA molecule serves as a template for the formation of a new strand. This is called **semiconservative replication** because in each of the new double-stranded DNA molecules one original strand is saved (conserved) and the other strand is new. Complementary base pairing creates two new DNA molecules that are identical to the parent molecule.

How DNA Works

We've seen how genetic information is passed accurately from cell to daughter cells and from generation to generation. The next obvious question is, "How does DNA issue commands that direct cellular activities?" The answer is that DNA directs the synthesis of **RNA (ribonucleic acid)**, which in turn, directs the synthesis of a protein (Figure

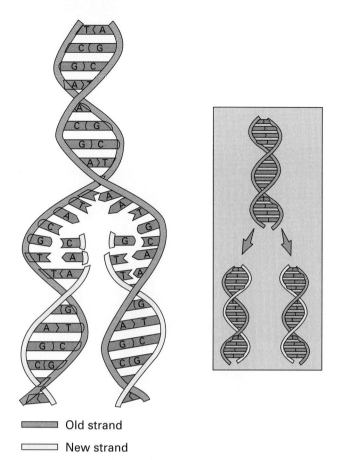

Old strand

New strand

Figure 19-3 DNA replication begins when an enzyme "unzips" the two strands of the parent molecule. Each of the parental strands serves as a template for the formation of a new strand. As complementary bases pair, free nucleotides become linked, forming a new strand. Two new double-stranded molecules are formed, each identical to the parent molecule. This method of replication is called semiconservative because the new DNA consists of one "old" strand and one "new" strand.

Like DNA, RNA is composed of nucleotides linked together, but there are some important differences between DNA and RNA. First, the nucleotides of RNA contain the sugar ribose, instead of deoxyribose, as in DNA. Second, in RNA the nucleotide uracil (U) pairs with adenine instead of thymine, as occurs in DNA. Third, most RNA is single-stranded.

The first step in converting the DNA message to a protein is copying the message as RNA, a process called **transcription**. The process begins with the unwinding of the region of DNA to be copied, which is orchestrated by an enzyme. The DNA message is determined by the order of bases. RNA nucleotides pair with their complementary bases—cytosine with guanine and uracil with adenine—with one strand of DNA serving as the template (Figure 19–5). The signal to start transcription is given by a specific sequence of bases on DNA, called the **promoter**. An enzyme called **RNA polymerase** binds with the promoter on DNA and then moves along the DNA strand, aligning the appropriate RNA nucleotides and linking them together. Another sequence of bases serves as a stop signal. When RNA polymerase reaches this signal, transcription ceases and the RNA molecule is released.

Three types of RNA are produced and each plays a different role in protein synthesis. **Messenger RNA (mRNA)**

19–4). The protein can be a structural part of the cell or an enzyme that then speeds up certain chemical reactions within the cell. The protein whose synthesis is directed by a gene is the molecular basis of the inherited trait; it determines the phenotype. The central dogma of biology is that DNA sequence specifies mRNA sequence and mRNA sequence specifies amino acid sequence of a protein. To gain a fuller appreciation for how DNA works, we will consider each step in slightly more detail.

Transcription: RNA Synthesis

Just as the CEO of a major company issues commands from headquarters instead of from the factory floor, DNA issues instructions from the cell nucleus and not from the cytoplasm, where the cell's work is done. RNA is the intermediary that carries the information encoded in DNA to the cytoplasm and directs the synthesis of the specified protein.

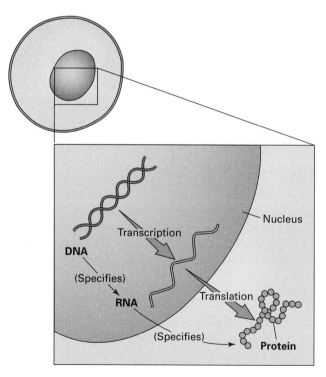

Figure 19-4 The central dogma of biology is that DNA specifies mRNA, which specifies a protein. The genetic instructions of DNA are rewritten (transcribed) in the form of RNA, which is then translated into a particular protein. The protein is often an enzyme that directs specific cellular activities.

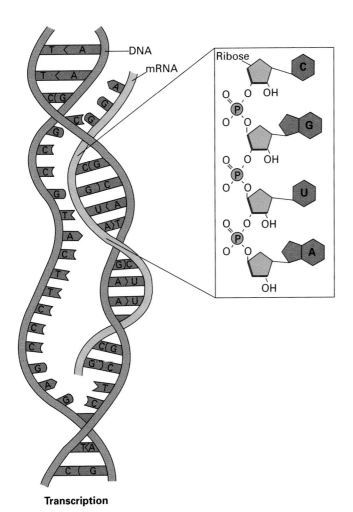

Transcription

Figure 19-5 During transcription, DNA serves as a template for the synthesis of RNA. A segment of DNA unwinds. RNA nucleotides pair with the complementary DNA bases and are linked together to form an RNA transcript, which is then released from the DNA. RNA contains the sugar ribose instead of deoxyribose as is found in DNA. In RNA, uracil replaces thymine as the complementary base to adenine. Unlike DNA, most RNA is single-stranded.

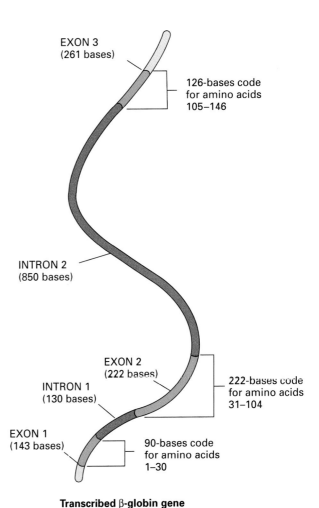

Transcribed β-globin gene

Figure 19-6 Certain regions of the mRNA transcript are not part of the instructions for producing a protein. These segments, called introns, are edited out of the RNA strand before it leaves the nucleus. The remaining regions, called exons, direct the synthesis of a protein. Shown here are the exons and introns of the gene that codes for the β protein chain in hemoglobin, the protein in red blood cells that carries oxygen in the body.

carries the DNA instructions for synthesizing a particular protein. The order of bases in mRNA specifies the sequence of amino acids in the resulting protein, as we will see. **Transfer RNA (tRNA)** binds to a specific amino acid and transports it to the appropriate region of mRNA. **Ribosomal RNA (rRNA)** combines with proteins to form the ribosomes, which are structures on which protein synthesis occurs.

Messenger RNA is usually processed before it leaves the nucleus. Most of the DNA between a promoter and stop signal includes regions that do not contain codes that will be translated into protein and, therefore, are not part of the gene itself. These unexpressed regions of DNA are called **introns,** for intervening sequences (Figure 19–6). Introns

are snipped out of the newly formed mRNA strand by enzymes before it leaves the nucleus. The remaining segments, called **exons** for expressed sequences, actually direct the synthesis of a protein.

Translation: Protein Synthesis

Messenger RNA is so named because it carries the DNA genetic message from the nucleus to the cytoplasm where it is translated into protein. Just as we might translate a message written in Spanish into English, **translation** converts the nucleotide language of mRNA into the amino acid language of a protein. Before examining the process of translation, we should become more familiar with the language of mRNA.

The Genetic Code

The **genetic code** is used to convert the linear sequence of bases in DNA to the sequence of amino acids in proteins. The "words" in the code, called **codons**, are sequences of three bases on mRNA that specify 1 of the 20 common amino acids or the beginning or end of the protein chain (Table 19–1). For instance, the codon UUC on mRNA specifies the amino acid phenylalanine. (The complementary sequence on DNA would be AAG.)

The four different bases in RNA (A, U, C, and G) could form 64 different combinations of three-base sequences. The number of possible codons, therefore, exceeds the number of amino acids. So, we find that several codons may code for the same amino acid. The codon AUG can either serve as a start signal to initiate translation or it can specify the amino acid methionine, depending on where it occurs in the mRNA molecule. In addition, three codons (UAA, UAG, and UGA) are stop codons that specify the end of a protein. If we think of the codons as genetic words, then a stop codon functions as the period at the end of the sentence.

Transfer RNA

A language interpreter translates a message from one language to another. Transfer RNA serves as an interpreter that converts the genetic message carried by mRNA into the language of protein, which is a particular sequence of amino acids. To accomplish this, a tRNA molecule must be able to recognize both the codon on mRNA and the amino acid that the codon specifies.

The structure of the tRNA molecule is important to its ability to serve as an interpreter of genetic messages (Figure 19–7). Each tRNA molecule is a relatively short strand of RNA consisting of about 90 nucleotides. The strand folds and twists upon itself, forming a hairpin-like structure that contains some double-stranded and some single-stranded regions. At the main bend of the "hairpin" is a sequence of three nucleotides called the **anticodon**, which binds to a codon on the mRNA molecule following base-pairing rules. Importantly, each type of tRNA molecule also binds to a particular type of amino acid. The specificity of this binding is ensured by enzymes. For example, a tRNA molecule with the anticodon AAA binds to the amino acid phenylalanine and ferries it to the mRNA molecule where the codon UUU is presented for translation. Phenylalanine will then be added to the growing amino acid chain.

Ribosomes

Ribosomes function as the workbenches on which proteins are assembled. A ribosome consists of two subunits, each

Table 19–1 The Genetic Code

FIRST BASE		SECOND BASE				THIRD BASE
		U	C	A	G	
U		UUU UUC Phenylalanine UUA UUG Leucine	UCU UCC UCA UCG Serine	UAU UAC Tyrosine **UAA Stop** **UAG Stop**	UGU UGC Cysteine **UGA Stop** UGG Tryptophan	U C A G
C		CUU CUC CUA CUG Leucine	CCU CCC CCA CCG Proline	CAU CAC Histidine CAA CAG Glutamine	CGU CGC CGA CGG Arginine	U C A G
A		AUU AUC Isoleucine AUA AUG Methionine Start	ACU ACC ACA ACG Threonine	AAU AAC Asparagine AAA AAG Lysine	AGU AGC Serine AGA AGG Arginine	U C A G
G		GUU GUC GUA GUG Valine	GCU GCC GCA GCG Alanine	GAU GAC Asparagine GAA GAG Glutamic Acid	GGU GGC GGA GGG Glycine	U C A G

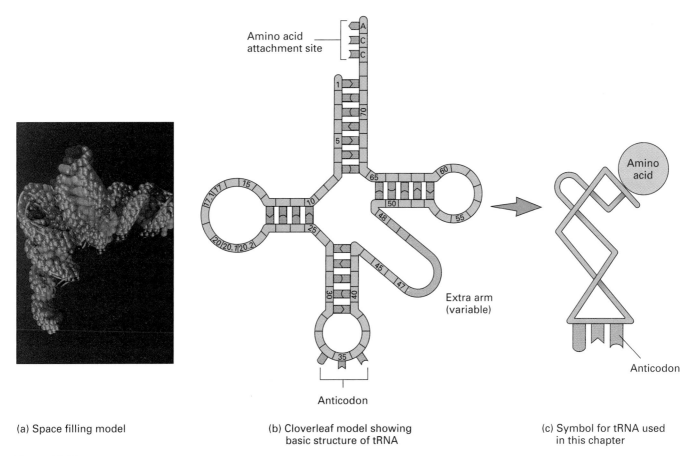

Amino acid
attachment site

A
C
C

Anticodon

Extra arm
(variable)

Amino
acid

Anticodon

Anticodon

(a) Space filling model

(b) Cloverleaf model showing
basic structure of tRNA

(c) Symbol for tRNA used
in this chapter

Figure 19–7 A tRNA molecule is a short strand of RNA that twists and folds on itself. The job of tRNA is to ferry a specific amino acid to the ribosome and insert it in the appropriate position in the growing peptide chain. The appropriate position is determined by a sequence of three, unpaired nucleotides on the tRNA molecule, called the anticodon. The anticodon pairs with a codon on mRNA, following complementary base pairing rules. *(a, © 1995 T.J. O'Donnell/Custom Medical Stock Photo)*

composed of rRNA and protein (Figure 19–8). The subunits form in the nucleus and are shipped to the cytoplasm. They remain separate except during protein synthesis.

The role of the ribosome in protein synthesis is to bring the tRNA and mRNA molecules close enough together to interact. The smaller subunit binds to mRNA. The ribosome has two binding sites for tRNA, called the P site and the A site. The tRNA holding the growing peptide chain binds to the P site, while the tRNA carrying the next amino acid to be added to the peptide chain binds to the A site. The tRNA that enters the A site is one that has the anticodon complementary to the mRNA codon exposed at the base of the A site.

Stages of Translation

Newly formed mRNA and ribosomal subunits move from the nucleus to the cytoplasm where protein synthesis (translation) occurs. Translation can be divided into three stages:

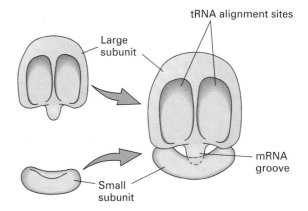

tRNA alignment sites

Large
subunit

Small
subunit

mRNA
groove

Figure 19–8 A ribosome consists of two subunits of different sizes. When the two subunits fit together to form a functional ribosome, a groove for mRNA is formed. The ribosome has two alignment sites for tRNA molecules. It also contains an enzyme that will promote the formation of a peptide bond between the amino acids that are attached to the tRNAs in the alignment sites.

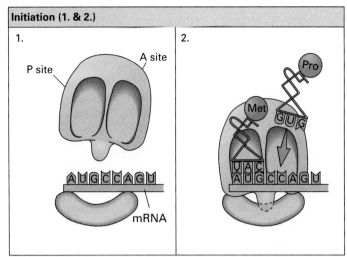

Initiation (1. & 2.)

1.

P site

A site

mRNA

AUGCCAGU

2.

Pro

Met

GUG

UAC

AUGCCAGU

Initiation: Translation begins. The small ribosomal subunit binds to start codon (AUG) on mRNA. The ribosomal subunits then assemble to form a functional ribosome and the tRNA with the appropriate anticodon (UAC) slides into the P site, allowing the codon and anticodon to pair. The next codon on mRNA is positioned in the A site.

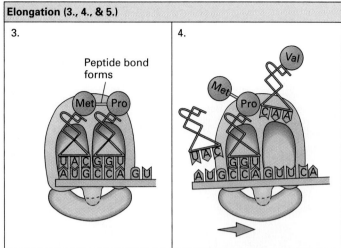

Elongation (3., 4., & 5.)

3.

Peptide bond forms

Met — Pro

UAC GGU

AUGCCA GU

4.

Val

Met

Pro CAA

UAC GGU

AUGCCA GUUCA

Elongation: Amino acids are added to the growing chain. A tRNA with an anticodon complementary to the next codon on mRNA enters the A site. A peptide bond forms between the adjacent amino acids, with the assistance of a ribosomal enzyme. The tRNA in the P site separates from its amino acid and leaves the ribosome. The remaining tRNA moves to the P site and the ribosome moves along the mRNA, and the next codon then enters the A site. A tRNA with the anticodon complementary to this codon then enters the A site. A peptide bond forms between the adjacent amino acids, the tRNA in the P site is discharged from the ribosome, and the ribosome moves along the mRNA. Amino acids are added until the ribosome reaches a stop codon.

Figure 19–9 Translation occurs in three steps: initiation, elongation, and termination.

initiation, elongation, and termination (Figure 19–9). During initiation the major players in protein synthesis come together. First the small ribosomal subunit attaches to the mRNA strand at the start codon, which is usually AUG. The start codon is aligned in the P site of the ribosome. The tRNA with the complementary anticodon, UAC, bearing the amino acid methionine quickly pairs with the AUG codon. The larger ribosomal subunit then joins the smaller one to form a functional, intact ribosome. When the ribosomal subunits come together, there is a groove between them, in which the mRNA is positioned. This initiation step is important because it determines the point from which all codons will be read during protein synthesis and, therefore, the order and identity of amino acids in the protein.

Elongation of the protein occurs as additional amino acids are added to the chain. With the start codon positioned in the P site, the next codon is aligned in the A site. The tRNA bearing an anticodon that will pair with the codon in the A site slips into place and the amino acid it bears binds to the previous amino acid with the assistance of enzymes. The tRNA in the P site then leaves the ribosome. Simultaneously, the tRNA in the A site, which is attached to the growing protein chain, moves to the P site. The anticodon on tRNA remains paired with the mRNA codon as it shifts from the A site to the P site. Thus, the mRNA strand is

pulled through the groove between ribosomal subunits, positioning the next codon in the A site. The appropriate tRNA slips into the A site and its amino acid binds to the previous one. Then the mRNA and ribosome move relative to one another and the next codon is aligned in the A slot. The addition of each amino acid takes about 60 milliseconds.

In this way, amino acids are added to the growing protein chain until a stop codon—UAA, UAG, or UGA—moves into the A site of the ribosome. There are no tRNA anticodons that pair with the stop codons. Instead, certain proteins, called release factors, slide into the A site and cause the protein, the mRNA strand, and the ribosomal subunits to separate from one another.

Numerous ribosomes may glide along an mRNA strand at the same time, each producing its own copy of the protein directed by that mRNA. As soon as one ribosome moves past the start codon another ribosome can attach. A cluster of ribosomes simultaneously translating the same mRNA strand is called a **polysome**.

Mutations

DNA is remarkably stable and the processes of replication, transcription, and translation generally occur with incredi-

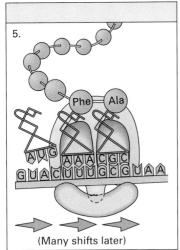

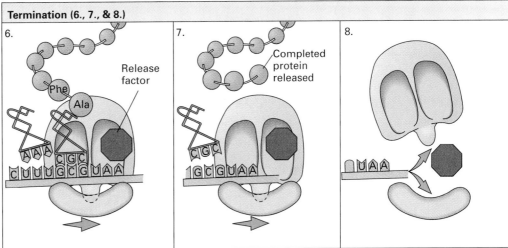

Termination (6., 7., & 8.)

5.

(Many shifts later)

6.

Release factor

Phe
Ala

7.

Completed protein released

8.

Termination: The newly synthesized protein is released. A stop codon (UAA, UAG, or UGA) is aligned in the A site. There are no tRNAs with anticodons complementary to stop codons. A release factor enters the A site. The tRNA in the P site separates from the new protein and both are released from the ribosome. The subunits of the ribosome then detach.

ble precision. However, sometimes DNA is altered and this can change its message. Such changes in DNA are called **mutations**. In the previous chapter, we discussed one type of mutation, called **chromosomal mutation**, in which whole sections of chromosomes become rearranged, duplicated, or deleted. Now that we are familiar with the chemical structure of DNA and how it directs the synthesis of proteins, we can consider another type of mutation—point mutation. A **point mutation** involves changes in one or a few nucleotides in DNA. Although a point mutation can occur in any cell, it can only be passed to one's children when it is present in a cell that will become an egg or a sperm. Mutations that occur in body cells can affect the functioning of that cell, sometimes with disastrous effects, but they cannot be transmitted to offspring.

One way in which a point mutation can occur is through the replacement of one nucleotide pair by another pair of nucleotides. Because of the redundancy of the genetic code, these substitutions do not always change the genetic message. For example, the third base pair in a DNA sequence that would be transcribed to the mRNA codon UCU could be altered so that the new codon read UCC without changing the genetic message since the codons UCU and UCC both specify the amino acid serine. This proccess is demonstrated in the table that follows.

DNA	Transcribed to RNA	Amino Acid Specified
A-T	U	
G-C	C	Serine
A-T	U	

Point Mutation in DNA	New Codon	Amino Acid Specified
A-T	U	
G-C	C	Serine
G-C	C	

Other substitutions may change the genetic message so that the wrong amino acid is added to the protein chain, but the change may not affect the functioning of that protein. Nonetheless, many substitutions do have drastic effects on the functioning of the resulting protein. A single base pair substitution in the DNA coding for hemoglobin causes sickle-cell anemia, a condition in which the red blood cells become distorted when the oxygen content of the blood is low. Whereas the normal hemoglobin DNA reads CTC, the DNA that results in sickle-cell hemoglobin reads CAC. The normal mRNA codon, GAG, specifies that the amino acid glutamic acid be incorporated into the protein at that point. However, the mutant mRNA codon, GUG, directs the incorporation of valine. This small change makes a big difference in the way that the resulting hemoglobin functions.

Another type of point mutation, one that is caused by the addition or deletion of a nucleotide, generally has more serious effects than do those caused by substitutions. This is because additions and deletions cause changes in the reading frame of the genetic message. *All* the triplet codons that follow the addition or deletion are likely to change. These **frameshift mutations** change the resulting protein. Consider the following example of a frameshift mutation:

Normal

Strand of DNA	
Transcribed	ACA CCT CTT TTT TAA ATT
mRNA	UGU GGA GAA AAA AUU UAA
Protein	Cys – Gly – Glu – Lys – Ile – stop

Frameshift Mutation

Original Strand	
of DNA	ACA CCT CTT TTT TAA ATT
	↑
	Deletion of A
New Transcribed	
Strand of DNA	ACC CTC TTT TTT AAA TT
mRNA	UGG GAG AAA AAA UUU AA
Protein	Trp – Glu – Lys – Lys – Phe –?

Regulation of Gene Expression

At the moment of your conception you received one set of chromosomes from your father and one set from your mother. The cell formed by the union of sperm and egg then began a remarkable series of cell divisions, divisions that continue in many types of body cells to this day. With each cell division, the genetic information was faithfully replicated and exact copies were parceled into the daughter cells. Thus, every cell in your body, except for eggs or sperm, contains a complete set of instructions for making every structure and performing every function in your body.

How, then, can liver, bone, blood, muscle, and nerve cells look and act differently from one another? The answer is deceptively simple: Only certain genes are active in a certain type of cell, and most are turned off. The active genes produce certain types of proteins and these determine the structure and function of the cell.

As is often the case in biology, the answer to one question prompts other questions. What controls gene activity? How are genes turned on or off? The answers to these questions are a bit more complex because gene activity is controlled at several levels.

Gene Regulation at the Chromosomal Level

At the chromosome level, gene activity is affected by **packing**, which is an intricate system of folding the DNA with small proteins called histones. First, the DNA strand is wrapped twice around a core of eight histone proteins and another histone holds the loops in place, creating a structure known as a **nucleosome** (Figure 19–10). The array of nucleosomes along a DNA molecule looks like beads on a string. Nucleosomes may play a role in controlling gene activity by preventing the enzymes needed for transcription from reaching the DNA.

In the next level of DNA packing, the string of nucleosomes becomes tightly coiled. This strand then becomes looped and twisted, forming a compact structure.

An example of how DNA packing can influence gene activity is provided by the inactivation of almost all the genes on one of the X chromosomes in each of a female's cells. You may recall that a female's cells have two X chromosomes. However, one of the X chromosomes in each body cell becomes highly condensed, forming a structure called a **Barr body**. Either one of the two X chromosomes can be inactivated in this way. Thus, a female who is heterozygous for a trait located on the X chromosome will express different alleles in different cells.

Gene Regulation at the Transcriptional Level

The most important way that gene activity is regulated occurs at the level of transcription, that is, by mechanisms that turn genes on or off, thus determining whether a gene will be actively transcribed into RNA. For example, in the next chapter we will consider how the loss of control over the activity of certain genes can lead to cancer.

One way that genes can be regulated at the transcriptional level is by a "master gene" that turns on a battery of other genes, which, in turn, leads to the production of proteins that cause the development of a particular cell type. For instance, a master gene in an immature muscle cell produces a protein called myoD1. MyoD1 then turns on genes on other chromosomes that produce the muscle-specific proteins needed to transform the immature cell into a muscle cell.

If master genes are likened to the ignition switch in a car, then other regions of DNA, called enhancers, are similar to the accelerator. Enhancers are segments of DNA that increase the *rate* of transcription of certain genes and, therefore, the amount of a specific protein that is produced.

Regulatory Chemicals

Chemical signals can also regulate gene activity. You may recall from Chapter 6 that one of the ways that certain hormones bring about their effects is by turning on specific genes. Steroid hormones, for instance, bind to receptors in the cytoplasm of a target cell. The hormone-receptor com-

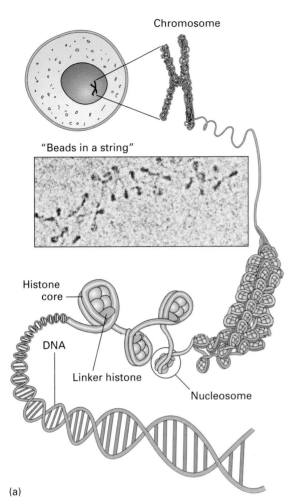

Chromosome

"Beads in a string"

Histone core

DNA

Linker histone

Nucleosome

(a)

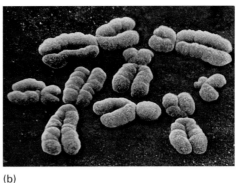

(b)

Figure 19-10 At the chromosome level, the regulation of gene activity involves an intricate system of folding the DNA with proteins called histones. Two loops of DNA are wrapped around a core of histones and held in place by another histone to form a nucleosome. Nucleosomes on a DNA strand look like beads on a string. Nucleosomes may regulate gene activity by preventing the enzymes needed to form RNA from reaching the necessary regions of DNA. The nucleosome string then coils on itself. This cord then becomes looped on itself and condenses further. (a, photo) The electron micrograph of nucleosomes shows a segment of chromatin released from the nuclei of chicken red blood cells, an image unstained in ice. *(a, photo, Courtesy of Jan Bednar and Christopher Woodcock, Biology Dept., University of Mass., Amherst; b, © Biophoto Associates/Science Source/Photo Researchers, Inc.)*

plex then diffuses into the nucleus and turns on specific genes. This is the way that the genes in cells that produce facial hair are turned on. It is the reason that your father may have a beard, but your mother probably doesn't even though she has the necessary genes to grow one. In this case, the male hormone testosterone binds to a receptor and turns on hair-producing genes. Facial cells of both men and women have the necessary testosterone receptors. However, women don't usually produce enough testosterone to activate the hair-producing genes, and so bearded women are rare.

Critical Thinking

Considering the mechanism by which testosterone turns on hair-producing genes in certain cells on the face, explain why female athletes who inject themselves with testosterone to stimulate muscle development can develop facial hair.

Besides hormones that turn on specific genes, there are regulatory proteins that can either activate or repress a gene.

An activator protein, for instance, can bind to DNA somewhere near the promoter for a gene. It then helps position and activate RNA polymerase so that transcription begins.

Gene Regulation at the RNA Processing Level

After it has been transcribed, the mRNA can be modified in various ways that alter the protein produced by a gene. Earlier in this chapter we discussed one way in which this occurs: The introns, regions of mRNA that do *not* code for protein are edited out of the mRNA transcript, leaving only the useful segments, which are called exons. During editing, the exons of a single gene can be linked together in different ways, resulting in different proteins.

Gene Regulation at the Level of Protein Synthesis

Gene activity can also be influenced at the level of translation (protein synthesis) so that large quantities of specific proteins can be produced quickly when they are needed. (An analogy might be preparing your automatic coffee maker before going to bed so that your morning "eye opener" will be ready soon after flipping the "on" switch.) Cells get ready for a future frenzy of protein synthesis by producing the necessary mRNA, often in large quantities. However, the mRNAs are "masked" by proteins that prevent their translation until the switch is thrown. For example, during the development of an ovum, vast quantities of the mRNAs coding for the proteins required for early embryonic development are produced in a masked state. Then, when the egg is fertilized by a sperm, the masking proteins are removed, and in a burst of activity, the needed proteins are synthesized.

Genetic Engineering

The manipulation of genetic material for our own purposes, the practice of **genetic engineering**, began almost as soon as we began to understand the language of DNA and how traits are inherited. Genetic engineering has already provided numerous benefits, including the production of hundreds of useful products. Benefits include antibiotics and certain hormones, improved diagnosis and treatment of human diseases, increased food production from plants and an-

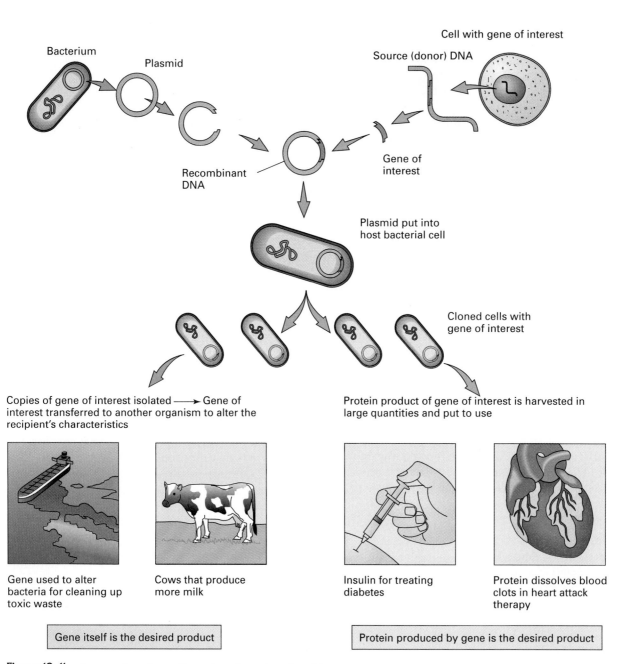

Figure 19-11   An overview of genetic engineering using plasmids.

imals, and insight into the growth processes of cells. Genetic engineering is a subset of a broader endeavor, **biotechnology**, that involves making a living cell perform a useful task in a controllable way.

The benefits of genetic engineering fall into two categories. First, it provides a way to produce large quantities of a particular gene product. The useful gene is simply transferred to another cell, usually a bacterium or yeast cell, that can be grown easily in large quantities. For example, large quantities of human insulin for treating diabetes have been generated by genetically engineered bacteria. Second, genetic engineering allows a gene for a desirable trait to be taken from one plant, animal, or microorganism and incorporated into an unrelated species. The recipient species then shows the desirable trait. For instance, flounder, a fish that lives in very cold water, has a gene that produces an antifreeze that protects its blood and tissues from freezing. This gene has been transferred to tomatoes so that they can withstand freezing temperatures without damage.

The implication of the ability to manipulate genes is immediately apparent. It will allow us to change our fates. If we consider life to be a poker game and genes the cards, we would no longer have to play the hand we are dealt.

Genetic engineers may be interested in producing vast quantities of a particular gene product, generating a new type of cell or organism with a particular ability that it did not previously have, or identifying or sequencing a particular gene. Whatever the goal, the means of achieving the ends are often very similar (Figure 19–11).

The basic idea behind genetic engineering is to put the "gene of interest," the one that produces the protein or trait that is desired, into another piece of DNA to create **recombinant DNA**, which is DNA from two sources. The recombinant DNA then carries the gene of interest into another cell, one that multiplies rapidly, producing many copies of the gene of interest. Then the harvest—large amounts of the gene product or many copies of the gene of interest—is reaped. Let's consider the procedure in slightly more detail, one step at a time.

Slicing and Splicing DNA

The **source DNA**, which contains the gene of interest, and the **host (vector) DNA**, which receives the transferred genes, are both cut at specific places by a **restriction enzyme**. (Restriction enzymes are produced by bacteria as a defense against invasion by viruses. Viruses cause infection by inserting their own DNA into the host cell and taking over its genetic machinery. Restriction enzymes evolved to chop up viral DNA in bacteria and render it harmless.) There are many different restriction enzymes. Each makes a staggered cut between specific base pairs, leaving several unpaired bases on the cut end of each DNA strand. The region of unpaired

bases is called a "sticky end" because of its tendency to pair with single-stranded regions of complementary base sequences on the ends of other DNA molecules that were cut with the same restriction enzyme (Figure 19–12).

The sticky ends are the secret to splicing together the gene of interest and the host DNA. The sticky ends of DNA from different sources will be complementary and stick together as long as they have been cut with the same restriction enzyme. The initial attachment between sticky ends is temporary, but they can be permanently pasted together by another enzyme, **DNA ligase**, which forms bonds between the sugars and phosphates that form the sides of the DNA

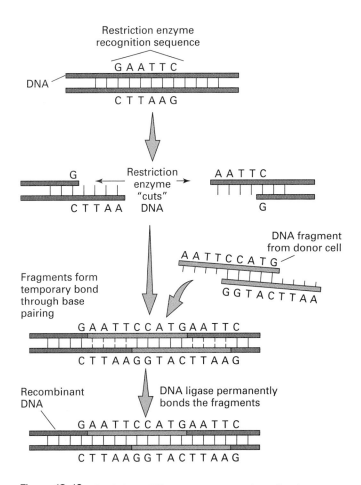

Figure 19–12 DNA from different sources can be spliced together using a restriction enzyme to make cuts in the DNA and then DNA ligase to hold the fragments together. A restriction enzyme makes a staggered cut at specific sequences of DNA, leaving a region of unpaired bases on each cut end. The region of single-stranded DNA at the cut end is called a sticky end, because it tends to pair with the complementary sticky end of any other piece of DNA that has been cut with the same restriction enzyme, even if the pieces of DNA came from different sources.

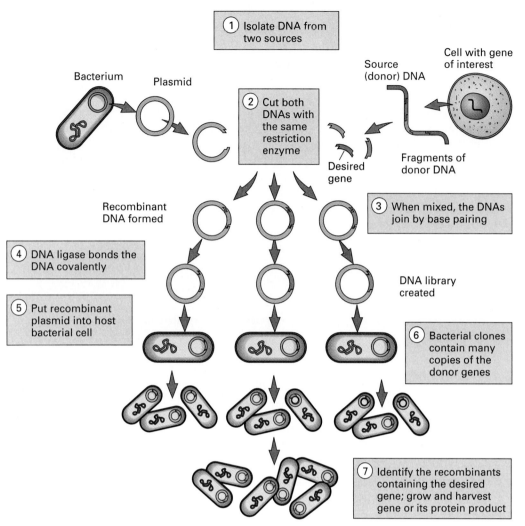

Figure 19-13 A restriction enzyme is used to chop donor (source) DNA into a mixture of fragments. Only some of those fragments contain the desired gene. The same restriction enzyme is used to open the ring of plasmid DNA. The sticky ends of donor DNA fragments join with those of the plasmid DNA. The result is a mixture of recombinant plasmids, each containing different genes from the donor DNA. The recombinant plasmids are then cloned in host cells.

ladder. The resulting recombinant DNA contains DNA from two sources (Figure 19–13).

Transferring the Recombinant DNA to a New Host Cell

Biological carriers that ferry the recombinant DNA to the host cell are called **vectors**. Bacterial **plasmids**[2] are small, circular pieces of self-replicating DNA separate from the bacterial chromosome and are common vectors. The source DNA and plasmid DNA are subjected to the same restriction enzyme. Fragments of source DNA, some of which will contain the gene of interest, will be incorporated into plasmids when their sticky ends join. The recombined plasmids can then be inserted into bacterial cells. A plasmid can replicate within the bacterium, creating multiple copies. Also,

each time a bacterium containing a plasmid divides, copies of the plasmid go to each daughter cell. Thus a **clone**, that is, a group of identical organisms all descended from a single ancestor, is created. In this case, all members of a clone carry the same recombinant plasmid. Bacteria reproduce rapidly, and so cloning can produce a billion copies of a particular desirable gene in only 10 hours.

Although the basic idea is usually the same, there may be variations on this theme. For instance, the gene of interest is sometimes combined with viral DNA. The viruses are then used as vectors to insert the recombinant DNA into a host cell. Hosts other than bacteria, including yeast or animal cells, can also be used.

Identifying the Recombinants with the Gene of Interest

A restriction enzyme cuts the source DNA into a mixture of many fragments, only some of which contain the gene of in-

---

[2]Plasmids seem to have evolved as a means of moving genes among bacteria. A plasmid can replicate itself and pass, with its genes, into another bacterium.

terest. Thousands of recombinant plasmids are then created, each containing a segment of the source DNA, and each recombinant is introduced into a bacterial cell. Because each cell bearing a recombinant plasmid contains different information, the resulting mixture of cloned recombinant DNA fragments is often called a **DNA library**.

We see, then, that only some of the recombinant DNA plasmids will contain the gene of interest. How can the correct gene be located? One way to find the gene of interest would be to look for the protein it produces. Another way would be to look for the gene itself by using a probe. In this case, a **probe** would be a single-stranded molecule of DNA or RNA with a sequence of bases complementary to the sequence in the gene of interest. During its synthesis, radioactive molecules are incorporated into the probe. Then the cells containing recombinant plasmids are treated so that the DNA splits into single strands. The radioactive probe is added and reveals the location of the gene of interest when it binds to the DNA by complementary base pairing.

Social Issue

Genetic engineering involves altering an organism's genes—adding new genes and traits to microbes, plants, or even animals. Do you think we have the right to "play God" and alter life forms in this way? In the 1980s the U.S. Supreme Court approved patenting of genetically engineered organisms, first microbes and later mammals. Do you think it is ethical to patent a new life form?

Applications of Genetic Engineering

Genetic engineering has already yielded many benefits that improve the quality of human life and even greater promises are made for the future. In some cases, a gene has been put into bacteria, yeast, or other cells so that large quantities of a particular gene product can be obtained. In other cases, a gene or a desirable trait in one species has been transferred to the cells of another species. These **transgenic organisms**, organisms that contain genes from another species, then show the desired trait.

Plants

Genetic engineering has produced plants that are resistant to certain diseases, such as the tobacco mosaic virus, as well as to insects and herbicides. For instance, a soil bacterium, *Bacillus thuringiensis*, produces a protein that kills an early stage in the life cycle of moth pests and the protein is often sprayed on plants as an insecticide. A natural insecticide has been produced by inserting a gene from this bacterium into the chromosomes of certain plants, including cotton and walnut trees. In the engineered plants, the insecticide is produced *within* the plants and so it kills only its intended targets, the insects that feed on those plants. An added bonus is

that the insecticide can't harm the environment because it is contained within the plant and not sprayed over a large area of plants. Besides insects, weeds can be a problem in agriculture. Many of us know how bothersome and time-consuming it is to weed the garden. When farming is a business, however, the equipment, fuel, and people needed to weed crop fields can also be expensive. Glyphosate is an herbicide that kills a wide variety of plants. Advantages of glyphosate include that it is biodegradable and that it doesn't accumulate in the food chain, harming top predators, such as humans. A gene for resistance to glyphosate has been introduced into certain crops. When crops with this gene are sprayed with glyphosate they survive without problems, but the weeds among them are killed.

Social Issue

Glyphosate-resistant crops have revolutionized agriculture because the herbicide lowers the cost of producing the crop with little harm to the environment because it breaks down quickly. However, critics worry that the development of herbicide-resistant crops will lead to increased and prolonged use of herbicides. This, in turn, could lead to more herbicide residues on the foods we eat and the long-term effects are unknown. What do you think? Should researchers continue to develop new herbicide-resistant crops? If there are long-term consequences, who should be held responsible?

We experience some of the benefits of genetic engineering at our dinner tables. Crops have been created that grow faster, produce greater yields, and have longer shelf lives. For instance, tomatoes have been engineered so that vine-ripened tomatoes won't get mushy during transportation to the market.

Animals

Genetic engineering has already made strides in bringing us healthier, more productive farm animals at lower cost. Vaccines have been created to protect piglets against a form of dysentery called scours, sheep against foot rot and measles, and chickens against bursal disease (a viral disease affecting the bursa and liver that is often fatal). Genetically engineered bacteria produce bovine somatotropin (BST), a hormone naturally produced by a cow's pituitary gland that enhances milk production. Injections of BST can boost milk production by nearly 25%.

Transgenic animals have been created by injecting a fertilized egg with the gene of interest. The embryo is then implanted in a female's reproductive system, where it continues development. In about 5% of the transgenic embryos, the gene becomes incorporated into the host cell DNA and begins producing its protein. In this way, extra copies of the

Figure 19-14 These pigs contain foreign genes that produce growth hormone. The growth hormone gene was introduced into the fertilized egg, became incorporated into the animal's own DNA, and now produces growth hormone. As a result, the animals grow faster and produce leaner meat. *(© Courtesy of the United States Department of Agriculture)*

gene for growth hormone have been introduced into cows and pigs, creating fast-growing animals that yield leaner meat (Figure 19–14).

Environment

Genetic engineering has also been used to improve or preserve the quality of the environment. Sewage treatment has been improved to lessen the amount of phosphate and nitrate discharged into waterways. Phosphate and nitrate can cause excessive growth of aquatic plants, which could choke waterways and dams, and algae, which can produce chemicals that are poisonous to fish and livestock. Microorganisms are also being genetically engineered to modify or destroy chemical wastes or contaminants so that they are no longer harmful to the environment. For instance, oil-eating microbes that can withstand the high salt concentrations and low temperatures of the oceans have proven useful in cleaning up after marine oil spills.

Pharmaceuticals

Genes have been put into a variety of cells, ranging from microbes to mammals, to produce proteins for treating allergies, cancer, heart attacks, blood disorders, autoimmune disease, and infections (Table 19–2). Among these products are

Table 19–2 Applications of Genetic Engineering

Gene Product	Function
Medical Use	
Human growth hormone	Promotes growth in children with underactive pituitary glands
Human insulin	Treatment for diabetes mellitus
Interleukin-2	Possible cancer treatment
Tissue plasminogen activator (tPA)	Used in heart attack therapy to dissolve blood clots
Interferons	Used in reducing effects of viral infections and possible cancer treatment
Tumor necrosis factor (TNF)	Kills certain cancer cells
Prourokinase	Treatment for heart attacks
Hepatitis B vaccine	Prevention of hepatitis B
Erythropoeitin	Treatment for anemia
Colony-stimulating factor	Treatment for leukemia
Factor VIII	Treatment for hemophilia
Serum albumin	Used to replace body fluids lost due to burns or illness
Agricultural and Animal Uses	
Porcine growth hormone (PGH)	Improves weight gain in hogs
Bovine growth hormone	Improves weight gain in cattle
Cellulase	Breaking down cellulose for animal feeds
Recreational Use	
Snomax®	Making snow for ski resorts

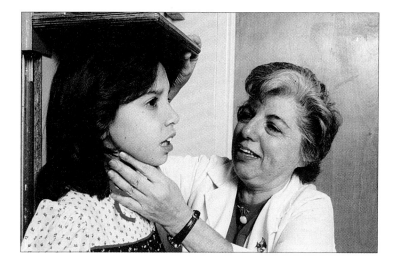

Figure 19–15 This girl has an underactive pituitary gland. The undersecretion of growth hormone would have caused her to be very short, even as an adult. However, growth hormone from genetically engineered bacteria has helped her grow to almost a normal height. *(© Genentech/Visuals Unlimited)*

hormones, including insulin for treating diabetes, and human growth hormone. Treatment with human growth hormone now allows children with underactive pituitary glands to reach nearly normal adult height (Figure 19–15).

Genetic engineering has also been used to produce vaccines. You may recall from Chapter 12 that a vaccine typically uses an inactivated bacterium or virus to stimulate the body's immune response to that type of organism. The idea is that the body will learn to recognize proteins on the surface of the infectious organism and mount defenses against any organism bearing those proteins. Because the organism used in the vaccine was rendered harmless, the vaccine can't trigger an infection. Vaccines work well, but the production of vaccines is time-consuming and expensive. The idea behind a genetically engineered vaccine is to put the gene for the surface protein of the infectious organism into bacteria. The bacteria would then produce large quantities of that protein, which could be used as a vaccine. There is no chance that the vaccine could cause infection because only the surface protein is used.

Gene Therapy

The problems associated with many genetic diseases result because a faulty gene fails to produce its normal protein product. The idea behind **gene therapy** techniques is that these genetic diseases could be controlled if healthy, functional genes were put into the body cells that were affected by the faulty gene, the so-called target cells.

How could a healthy gene be transferred to the affected cells? One way would be to use viruses. Viruses generally attack only one type of cell. For instance, an adenovirus, which causes the common cold, typically attacks cells of the respiratory system. You may recall that a virus consists of little more than genetic material, commonly DNA, surrounded by a protein coat. Once inside the cell, the viral DNA takes over the cell's metabolic machinery to produce its own pro-

teins. So, if the healthy gene were spliced into the DNA of a virus that had been rendered harmless, the virus would theoretically deliver it to the host cell and ensure that the desired gene product was produced. Indeed, this is just how gene therapy for cystic fibrosis was accomplished. Cystic fibrosis is a genetic disease that causes thick mucus to clog the air passages of the lungs. Once the gene for cystic fibrosis had been identified and isolated, the problem that remained was how to get a healthy form of the gene into the cells that need it. What was needed was a vector that would deliver the gene directly to the cells of the lungs. We are all familiar with an appropriate vector—the cold virus. In some of the first gene therapy trials, the healthy cystic fibrosis gene was delivered to the cells of the lungs by genetically engineered cold viruses. The results of the early trials were disappointing. The cold viruses delivered the gene to less than 1% of the targeted cells. Other methods of delivering the healthy gene are now being tried.

Another method of delivering a healthy gene has also been developed. The gene can be inserted into a plasmid and then the plasmid is enclosed in a capsule of fatty material called a liposome. Liposomes can fuse with cell membranes and deliver the genetic material inside the cell. Liposomes are also being tested as vectors in cystic fibrosis gene therapy trials.

Whatever the method of delivery, the effectiveness of gene therapy depends on the life span of the target cells. When the target host cell dies, the healthy gene is lost with it. The first condition to be treated with gene therapy was severe combined immune deficiency (SCID). The immune system of children with SCID is devastated, leaving them vulnerable to every passing germ. The cause of the problem is a mutant gene that prevents the production of an enzyme called adenosine deaminase (ADA). Without ADA, white blood cells never mature and die while still developing in the bone marrow. The first gene therapy trial began in 1990

(Text continued on page 550)

Cloning: Will There Ever Be Another Ewe?

What do the Dalai Lama and a lamb named Dolly have in common? They prompt us to look at the way we live our lives with a critical eye. Born on July 5, 1996, Dolly was the first sheep—indeed, the first animal—to be cloned *from an adult* (Figure A). (A clone is a genetically identical copy or copies of a cell or individual.) The goal of Dr. Ian Wilmut, the Scotsman whose work produced Dolly, was not to produce a copy of his favorite ewe but to develop techniques that would eventually lead to the production of animals that could be used as factories to pump out proteins that would be beneficial to humans, such as drugs or hormones. In other words, cloning could make genetic engineering more efficient. With the ability to clone, a scientist would need to engineer an animal with a particular desired trait only once. Then, when the animal was old enough for scientists to be sure it had the desired traits, the animal could be duplicated exactly, producing a multitude of copies. Cloning from an adult instead of from an embryo is advantageous because you can be certain about the traits it possesses. In other words, "What you see is what you get."

Besides its promised practical benefits, Wilmut's work has increased our understanding of gene regulation and cell differentiation. You may recall that each animal life begins with a single cell called a zygote. The zygote contains the genetic information needed to produce all the cell types of an adult. During development, cells become differentiated: that is, certain genes are turned on and others are turned off, producing adult cells that are specialized to perform specific functions in the body, including heart, liver, muscle, bone, and blood cells. If a cell divides after it has become differentiated, its descendent cells are also differentiated. For instance, skin cells only form new skin cells, not muscle cells. Therefore, many scientists previously thought that the changes in gene activity that occurred during differentiation were irreversible. Dolly changed those ideas, however, because she was cloned from a fully differentiated adult cell. The genetic material that directed Dolly's development came from a cell in the mammary gland of a 6-year-old ewe—a cell already specialized to secrete milk.

Dr. Wilmut created Dolly by taking mammary cells from the udder of a pregnant ewe (Figure B). (The cloned ewe is named after Dolly Parton, a country singer who is perhaps known as much for her mammary cells as she is for her voice.) The cells were kept in a culture medium with low concentrations of nutrients. It turns out that starving the cells was the key to creating a clone from an adult cell; it caused the mammary cells to stop dividing and turned off certain active genes. Next, the nucleus was sucked out of an egg cell from another ewe. The starved, shut-down mammary cell was then placed next to the enucleated egg cell and an electric pulse was used to cause the cell membranes to fuse, allowing the nucleus to move into the egg and activating the new cell to begin cell division. Because it had been starved, the state of the transplanted nucleus was the same as that of the recipient egg cytoplasm, and so a cell was created in which all components were synchronized and ready for cell division. After about 6 days, at the approximate age when a naturally created embryo would implant in the uterus, the cloned embryos were implanted into the uterus of a host ewe. In the experiment that created Dolly, there were 277 artificially created "zygotes," 29 of which began to develop. Several of these early embryos were were then implanted into each of 13 different host ewes. Only one lamb, Dolly, was born.

Figure A Dolly is the first sheep to be cloned from an adult. Although the techniques that produced Dolly were developed to produce an animal that could produce quantities of products useful to humans, such as certain drugs, the work has far-reaching ramifications. It has taught us that genes previously turned off as a cell differentiates to its adult form can be turned on again and orchestrate the development of a new individual. Dolly's birth has also prompted heated debate regarding the feasibility and morality of cloning humans.
(AP Wide World Photos)

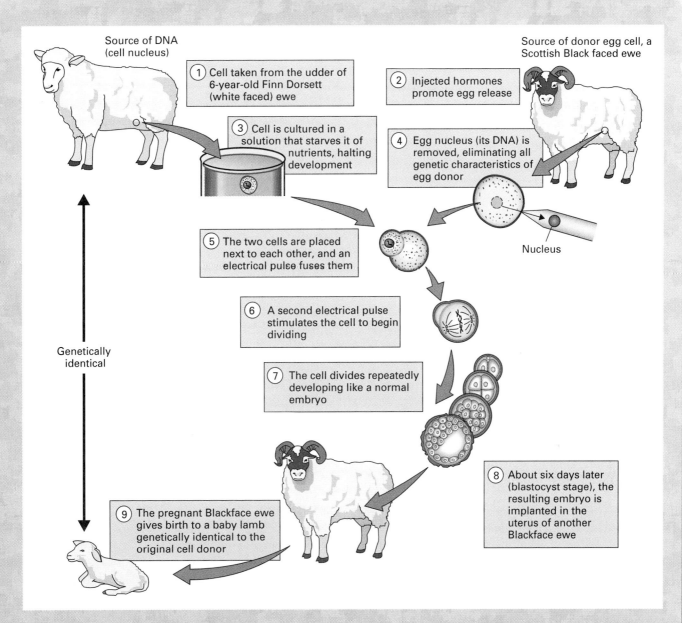

Figure B   The procedure used to clone an animal from an adult cell.

We still don't know for sure whether Dolly is one in 277 or one in a million, but her existence proves that it is *possible* to create a clone from an adult cell and this possibility raises many others. Will it be possible to clone humans? If so, what are the moral implications?

The central issue doesn't seem to be whether it is moral to create life in a laboratory. After all, in vitro fertilization and the freezing of embryos for later use have been practiced for years. Society has already had to confront issues such as what to do

with human embryos left in a petri dish at the end of the day or with unclaimed frozen embryos whose time in storage has expired. So, why does the idea of cloning humans stimulate such heated debated?

Cloning would provide an opportunity not just to create a new individual but to create an individual with a *known* genetic makeup, indeed specifically *for* that genetic makeup, in the hopes that it will lead to certain desirable traits. What is

(continued)

different about cloning is the ability to choose the traits considered valuable and produce multiple individuals with those traits. It promises the tantalizing possibility of controlling the destiny of the human race.

But which traits should be valued? Artistic talent? Scientific logic? Characteristics such as these would be difficult to reproduce with certainty because environmental influences, that is, life experiences, have as much, if not more, to do with their development as genes. After all, we already know that identical twins, who are in fact naturally produced clones, have different personalities and talents. Instead, cloning individuals with certain physical traits deemed desirable, such as height or strength, would have a more certain outcome. Could cloning then lead to the development of working classes, each specifically designed to excel at certain tasks? It would be up to society to prevent such abuses.

If it does become possible to clone humans someday, what might this technology mean to you and me? How would it affect *our* lives? It would not give us the power to create an exact copy of a specific person. Personality characteristics are determined by environment as well as genes. Although cloning could exactly replicate the genetic makeup of a specific person, there would be no way to replicate all the past environmental and social interactions in the life of the person being cloned, and so the clone would be a different "person" than the origi-

nal one. Thus, we could not replace a loved one who has died or decide that we should live forever. We could, however, produce an identical twin of a person.

Cloning could offer a solution to those medical problems that could be cured by a transplant—say bone marrow or a kidney—by providing an organ that has a matching tissue type. A family with a child in desperate need of a bone marrow transplant to cure leukemia might be glad to raise a second, identical child to save the life of the first.

In June, 1997, a commission established by President Clinton recommended that cloning a human should be a criminal offense, regardless of who pays for it or the reasons for the procedure. However, the commission did not make a decisive statement about cloning *research* involving human embryos. Instead, they continued the existing moratorium on federally-funded human embryo research and requested that the private sector honor it as well. This leaves open the question of whether research can be done that creates a human embryo clone as long as that embryo is destroyed. Furthermore, they ensured that the debate will continue by recommending that the ban be re-evaluated in 3 to 5 years.

We see, then, that science does affect society. If human cloning does become feasible, should it be done? For what reasons? Who should decide? These issues will be debated into the next century. What do you think?

---

when white blood cells of 4-year-old Ashanthi DeSilva, a child with SCID, were genetically engineered to contain the ADA gene and then returned to her tiny body (Figure 19–16). Her own gene-altered white blood cells began producing ADA, and her body defense mechanisms were strengthened. Ashanthi's life began to change. She wasn't ill as often as she had been before. She could play with other children. However, the life span of white blood cells is measured in weeks, and when the number of gene-altered cells declined, new gene-altered cells had to be infused. Then a technique was developed for isolating stem cells, the cells that give rise to all types of blood cells. The ADA gene was introduced into stem cells and these were transfused with the hope that the gene-altered stem cells would migrate to the bone marrow and begin producing normal white blood cells. In 1993, the first genetically engineered stem cells were transfused. In 1996, about half of Ashanthi's white blood cells carried the normal gene.

Although still in its infancy and considered highly experimental, gene therapy is being tested as treatment for a variety of forms of cancer, including cancers of the brain, lung, ovary, kidney, and skin (melanoma, an especially deadly form). As of 1996, there were 79 approved clinical

trials of cancer gene therapy in the United States. Many forms of cancer involve defects in genes that control cell division. Gene therapy for cancer involves several strategies, including (1) inserting functional genes to replace the faulty ones that cause unrestrained cell division, (2) inserting genes to stimulate the immune response against cancer cells, and (3) inserting genes into tumor cells that will kill only the cancerous cells while leaving the cells of surrounding tissue unharmed. We will consider these techniques in more detail in Chapter 19A, after we are more familiar with the genetic basis of cancer.

Repairing Faulty Information in mRNA

The kind of gene therapy that uses a functional gene to replace a defective one can be helpful if adding a functional protein is enough to alleviate the problem. A newer approach, which has not yet been tested on humans, is to leave the faulty gene in the DNA but to correct the mRNA message before it is translated to protein.

Earlier in this chapter we saw that DNA often contains regions called introns that are not translated into protein. The introns are transcribed to mRNA but are snipped out of

Figure 19-16 In 1990, 4-year-old Ashanthi DeSilva entered the medical record books as the first person to receive gene therapy, changing the way the world viewed genetic disease. Ashanthi was born with severe combined immune deficiency (SCID), which is caused by a mutant gene that fails to produce an enzyme needed for the maturation of white blood cells for body defense. Some of Ashanthi's own white blood cells were removed, genetically engineered to contain a healthy form of the gene, and then the gene-altered white blood cells were returned to her tiny body. Her immune system was strengthened and for the first time in her life she could go out into public places without fearing contact with people who might carry germs. She could play with other children and go to school. *(Ted Thai/*Time *Magazine/© Time, Inc.)*

the RNA by enzymes called **ribozymes** before the mRNA is translated to protein. A ribozyme recognizes a specific sequence of bases on mRNA, snips the RNA wherever those sequences are found, and then splices the cut ends back together.

It is hoped that ribozymes can be used to treat genetic diseases such as sickle-cell anemia. In sickle-cell anemia, a mutation causes an error in the production of one of the protein chains (beta globin) in hemoglobin. Traditional gene therapy, in which a normal beta globin gene would be added to the cell, wouldn't be effective for sickle-cell anemia, because the added gene would not fall into the cell's system of gene regulation. A ribozyme has been designed that would correct the defective RNA message before it was translated to beta globin, while leaving the mutant gene in its normal location in the chromosome so that globin production would still be precisely controlled. This ribozyme is being tested on mouse cells.

It is thought that before ribozymes are used to edit faulty genes, they will prove useful as antiviral drugs. Ribozymes will be especially helpful in combating infections caused by viruses that use RNA as their genetic information, HIV for instance. It is hoped that ribozymes will be able to cut the HIV RNA when it first enters a host cell and to cut any new viral RNA that may be generated with the host cell.

DNA Fingerprinting

DNA fingerprints, like the more conventional ones on fingers, can help identify individuals in a large population. They depend on the fact that there are many stretches of DNA that don't code for known traits. Some of these are composed of small, specific sequences of DNA that are repeated many times. The number of times these sequences are repeated varies considerably from person to person, from as few as four to several hundred repeats. Because the length of segments containing repeating sequences varies throughout the population, they can be used to help identify a specific person.

The first step in preparing a DNA fingerprint is to extract DNA from a tissue sample. The type of tissue doesn't matter, so sources that are readily available or that aren't too painful to remove, including blood, semen, skin, and hair follicles, are commonly used.

If only a minute tissue sample is available, a spot of blood from a crime scene for instance, it is helpful to amplify (increase) the amount of DNA by using a **polymerase chain reaction** (**PCR**). In the PCR, the DNA of interest is unzipped and mixed with nucleotides, primers (special short pieces of nucleic acid), and the enzyme DNA polymerase, which promotes DNA replication. Each time this is done the number of copies of the DNA of interest is doubled. In this way, billions of copies of the desired DNA can be produced in a short time.

The DNA can then be cut into pieces by using a restriction enzyme. Recall that a restriction enzyme recognizes a specific sequence of bases and cuts the DNA molecule apart wherever that sequence occurs. The result is a mixture of millions of tiny DNA fragments of many different sizes.

If we were to sort and arrange these DNA fragments according to size, the pattern that results from the DNA from one person would always be identical. *But,* if we were to compare the cut-up DNA from different people, the pattern would always be different. This is because the number and size of the fragments are determined by the sequence of bases along the DNA strand. The pattern of DNA fragments that have been cut by a restriction enzyme and sorted by size is called a **DNA fingerprint**.

We have seen how the DNA is cut into segments of varying length, but how can those fragments be sorted by size? The answer is a technique called gel electrophoresis, which takes advantage of the fact that DNA fragments have a slight negative charge. A drop of the suspension of DNA fragments is placed at one end of a thin layer of gel and immersed in a special solution; then, an electric current is applied. Because of the attraction between opposite charges, the DNA pieces slowly move away from the negative pole and toward the positive pole. The rate at which each fragment moves depends on its size. Smaller pieces move faster than larger ones. After a few hours, the pieces of cut up DNA in the sample will have separated into bands along a narrow path.

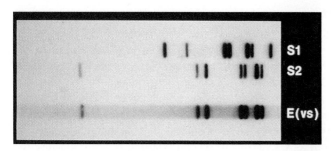

Figure 19-17 A DNA fingerprint can serve as a molecular mug shot left at the scene of a crime. Each strip shows a DNA fingerprint from a different source—blood from the crime scene and from seven suspects. The pattern of banding in a DNA fingerprint is determined by the sequence of bases in a person's DNA and is, therefore, unique to each person. A match of DNA fingerprints between the crime scene and a suspect can reveal the culprit with a high degree of certainty. Notice the match of DNA fingerprints between blood from the crime scene (the bottom line) and suspect number 2.   *(© 1989 ICI Americas, Inc./Cellmark Diagnostics)*

The pattern of bands on the gel is the DNA fingerprint. Each band represents a "pile" of DNA fragments of a certain length. The width of each band reflects the number of pieces of that specific length. There are a number of ways to make the pattern of bands visible. For instance, the gel can be soaked in a fluorescent dye that binds to DNA and glows in ultraviolet light.

DNA fingerprinting has many applications, but the most familiar is probably in crime investigations. Since 1987, when DNA fingerprints were first admitted as evidence in a trial, they have been admitted in over 2000 other court cases! Generally, the DNA fingerprint is created from a sample of tissue such as blood or hair follicles that was collected at the crime scene. A fingerprint can be produced from tissue left at the scene years before. That DNA fingerprint is then compared with the DNA fingerprints of various suspects. A match of DNA fingerprints reveals the culprit with a high degree of certainty (Figure 19–17).

Of course, the degree of certainty of a match between DNA fingerprints depends on how carefully the analysis was done. DNA fingerprints should look like a bar code, but sometimes the bars run together, lessening the certainty of a match. Because DNA fingerprints are being used as evidence in an increasing number of court cases each year, it is important that national standards be set to ensure the reliability of these molecular witnesses.

Human Genome Project

The **Human Genome Project** is a worldwide research effort to reach a common goal: determining the location of the estimated 100,000 genes along the 23 pairs of human chromosomes and sequencing (determining the order of) the estimated 3 billion base pairs that make up those chromosomes. By mid-June 1996, approximately half of the genes had been located. The DNA of model organisms, including the mouse, fruit fly, a nematode, and yeast, are also being mapped with the hope of gaining some insight into basic biology—the organization of genes within the genome, gene regulation, and molecular evolution. When this project was initiated in 1990, it was expected to take 15 years to complete at a cost of an estimated $200 million per year.

What is expected to be gained from this monumental effort? It is hoped that the information will help us diagnose, understand, and eventually treat many of the 4000 human genetic diseases. We have already identified and cloned the genes responsible for many genetic diseases, including Duchenne muscular dystrophy, retinoblastoma, cystic fibrosis, and neurofibromatosis. These isolated genes can now be used to test for their presence in people. As we saw in Chapter 18, gene tests can be used to identify people who are carriers for certain genetic diseases such as cystic fibrosis, allowing families to make choices based on known probabilities of bearing an affected child, for prenatal diagnosis, and for diagnosis before symptoms of disease begin. It is further hoped that after a disease-related gene has been identified, scientists can learn more about the protein it codes for and this can suggest ways to correct the problem. Also, finding a disease-related gene may open the door for gene therapy. We have seen how gene therapy is already being used to treat some human disorders, including SCID and cystic fibrosis. Future targets include hemophilia, diabetes, and Parkinson's disease.

Social Issue

Some people worry that once we know the location of every gene and have perfected the techniques of gene therapy, we will no longer limit gene manipulation to repairing faulty genes and begin to modify genes to *enhance* human qualities. We could begin to design our babies by choosing "good" genes. In doing so, we could improve the quality of life for all humanity. What do you think? Where should the line be drawn? Who should draw that line? Who should decide which genes are "good?"

In this chapter we have seen some of the ways that DNA may indeed hold the "secrets of life." We have also seen some of the ways that the unraveling of the secrets has alleviated some human suffering. At an ever accelerating rate, we are gaining knowledge about our genes and how they relate to the structure and function of our bodies and our behavior. Regardless of how that knowledge is used, it is bound to change the way we view ourselves.

SUMMARY

1. DNA consists of two strands of linked nucleotides, twisted together to form a double helix. Each nucleotide consists of a phosphate, a sugar, called deoxyribose, and one of four nitrogenous bases (adenine, thymine, cytosine, or guanine). Alternating sugar and phosphate molecules are linked, forming the two sides of the molecule. Two bases are paired in the interior of the double helix, like rungs on a ladder. Pairing is specific: adenine with thymine and cytosine with guanine.

2. DNA replication is semiconservative, that is, each new double-stranded DNA molecule consists of one old and one new strand. An enzyme "unzips" the two strands of the parent molecule, allowing each strand to serve as a template for the formation of a new strand. Complementary base pairing ensures the accuracy of replication.

3. Genetic information is transcribed from DNA to RNA and then translated into a protein.

4. RNA is assembled on DNA following base-pairing rules during a process called transcription. RNA differs from DNA in that the sugar ribose replaces deoxyribose and the base uracil replaces thymine. Most RNA is single-stranded.

5. Messenger RNA (mRNA) carries the DNA genetic message to the cytoplasm where it is translated into protein. The words in the genetic code are sequences of three RNA nucleotides, each called a codon. Each of the 64 codons specifies a particular amino acid or indicates the point where translation should start or stop.

6. Transfer RNA (tRNA) interprets the genetic code. At one end of the tRNA molecule is a sequence of three nucleotides called the anticodon that pairs with a codon on mRNA following base pairing rules. The tail of the tRNA molecule binds to a specific amino acid.

7. Each of the two subunits of a ribosome consists of ribosomal RNA (rRNA) and protein. A ribosome brings tRNA and mRNA together, allowing protein synthesis.

8. Translation begins when the two ribosomal subunits and an mRNA assemble with the mRNA sitting in a groove between the two subunits of a ribosome, attached to the ribosome at the start codon. During translation, the ribosome slides along the mRNA molecule, reading one codon at a time, and tRNA molecules ferry amino acids to the mRNA and insert them into a growing protein chain. Translation stops when a stop codon is encountered. The protein chain then separates from the ribosome.

9. Gene activity is regulated at several levels. At the chromosomal level, gene activity is regulated by chromosomal packing. DNA is wrapped around a core of histone proteins, forming a nucleosome. The nucleosome string becomes further coiled and looped. At the transcriptional level, gene activity can be affected by other segments of DNA. Genes can be turned on by master genes and the amount of RNA produced can be increased by other regions of DNA, called enhancers. Chemical signals such as regulatory proteins or hormones can also affect gene activity. After transcription, the introns in mRNA transcripts can be edited out and the exons combined in different ways to produce different proteins. Gene activity can also be regulated at the level of protein synthesis by the masking of mRNA that will be needed for a future burst of protein synthesis.

10. Genetic engineering is the manipulation of genetic material for practical purposes. It can be used to produce large quantities of a particular gene product or to transfer a desirable genetic trait from one species to another species or to another member of the same species.

11. Genetic engineering requires a restriction enzyme to cut the source DNA, which contains the gene of interest, and the host DNA at specific places, creating sticky ends composed of unpaired complementary bases. A vector is then needed to transfer the recombinant DNA to a host cell. Common vectors include bacterial plasmids and viruses. The host cell is a type of cell that reproduces rapidly, often a bacterium or yeast cell. With each host cell division, copies of the gene of interest enter each daughter cell. By introducing a gene from one species into a different one, transgenic plants and animals have been created.

12. Genetic engineering has had many useful applications in plant and animal agriculture, the environment, and in medicine. Gene therapy involves introducing a healthy form of a gene into body cells to correct problems caused by a defective one.

REVIEW QUESTIONS

1. Describe the structure of DNA.
2. Explain why complementary base pairing is crucial to exact replication of DNA.
3. Why is DNA replication described as semiconservative?
4. Explain the roles of transcription and translation in converting the DNA message to a protein.
5. In what ways does RNA differ from DNA?
6. What roles do mRNA, tRNA, and rRNA play in the synthesis of protein?
7. Define codon. What role do codons play in protein synthesis?
8. What type of molecule has an anticodon? What is the role of an anticodon?
9. Describe the events that occur during the initiation of protein synthesis, the elongation of the protein chain, and the termination of synthesis.
10. Define genetic engineering. Explain the roles of restriction enzymes and a vector in genetic engineering.
11. Describe some of the ways in which genetic engineering has been used to improve the quality of human life.
12. What is gene therapy? Name two of the genetic diseases that have been treated in this way.
13. What are ribozymes? What role do they play in correcting the mRNA message before it is translated into protein? How might they be used to correct faulty genes?
14. What is a DNA fingerprint?
15. What is the Human Genome Project? What are its intended benefits?

C h a p t e r

19A

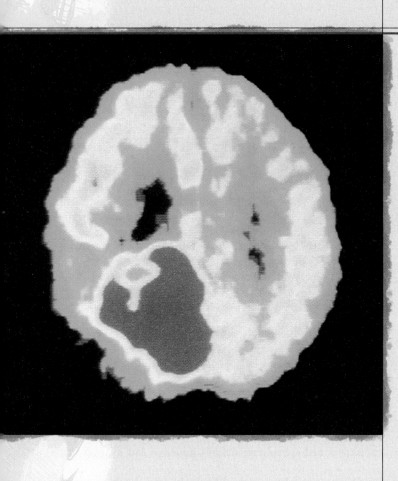

The red area in this PET scan of a brain is a tumor.
(© Science VU/Visuals Unlimited)

Cancer

Cancer, the "Big C," is perhaps the disease we humans dread the most—and with good reason. It touches the lives of nearly everyone. One of every three people in the United States will develop cancer at some point in his or her life and the other two are likely to have a friend or relative with cancer. The American Cancer Society estimates approximately 1,200,000 people will be diagnosed with cancer in the United States during the next 12 months.

We have learned a great deal about the basic nature of cancer in recent years, and the knowledge is beginning to pay off because we have a better idea how to prevent and treat it. As a result, the overall cancer death rate for many types of cancer considered as a group dropped by 3% between 1990 and 1995.

The Nature of Cancer

Although we tend to group them together, cancer, in fact, is a family of more than 100 different diseases, usually named for the organ in which they arise (Table 19A–1). However, all forms share one characteristic—uncontrolled cell division. Cancer cells behave like aliens taking over our bodies, but they are actually traitorous body cells.

An abnormal growth of cells can form a **tumor**, which is also called a **neoplasm**, meaning new growth. However, not all tumors are cancerous. In fact, tumors can be benign or malignant. Surrounded by a capsule of connective tissue, a **benign tumor** is an abnormal mass of tissue that usually remains at the site where it forms. Its cells do not invade surrounding tissue or spread to distant locations. In most cases, a benign tumor does not threaten health, because it can be completely removed by surgery. Benign tumors can be

harmful when they press on nearby tissues enough to interfere with their functioning. In some locations, however, such as in the brain, a benign tumor may be inoperable and can be life-threatening. Only **malignant tumors**, those that can invade surrounding tissue and spread to multiple locations throughout the body, are properly called cancerous.

Cells on their way to becoming cancerous typically look different from normal cells. The changes in shape, nuclei, and organization within tissues of precancerous cells are described as **dysplasia**. Their ragged edges give precancerous cells an abnormal shape. Their nuclei become unusually large, atypically shaped, and may contain increased amounts of DNA. As a group, precancerous cells form a disorganized clump and, significantly, have an unusually high percentage of cells in the process of dividing.

Eventually, the tumor will reach a critical mass consisting of about a million cells. Although still only a millimeter or two in diameter (smaller than a BB), the cells in the interior can't get a sufficient supply of nutrients and their own waste is poisoning them. This tiny mass is now called **carcinoma in situ**, which literally means "cancer in place."

Unless the mass is removed, at some point, perhaps years later, some of its cells will start to secrete chemicals that cause blood vessels to invade the tumor. This marks an ominous point of transition because the tumor cells now have supply lines bringing in nutrients to support continued growth and carrying away waste. Of equal importance, the tumor cells have an escape route—they can enter blood (or lymphatic) vessels and travel to numerous, distant locations throughout the body. Like cancerous seeds, the cancer cells that spread, or **metastasize**, can begin to grow into tumors in their new locations (Figure 19A–1).

As long as a tumor stays in place, it can grow quite large and still be easily removed by a surgeon. However, once cancer cells leave the original tumor they usually spread to so many locations that a surgeon's scalpel is no longer an effective weapon. At this point, chemotherapy or radiation is generally used to kill the cancer cells wherever they are hiding. The original tumor is rarely a cause of death. Instead,

Table 19A–1 Cancer Nomenclature by Tissue Type

Type of Cancer	Tissue Type
Carcinomas	Cancers of the epithelial tissues that infiltrate and metastasize. They include various skin cancers, such as squamous cell, basal cell, and melanoma.
Leukemias	Cancers of the bone marrow stem cells that produce the white blood cells.
Sarcomas	Cancers of muscle, bone, cartilage, and connective tissues. They may involve connective and muscle tissues of the bladder, kidneys, liver, lungs, parotid gland, and spleen.
Lymphomas	Cancers of the lymph tissues such as lymph nodes. These include Hodgkin's disease, non-Hodgkin's lymphoma, and lymphosarcoma.
Adenocarcinomas	Cancers of glandular epithelia, including liver, salivary glands, and breast.

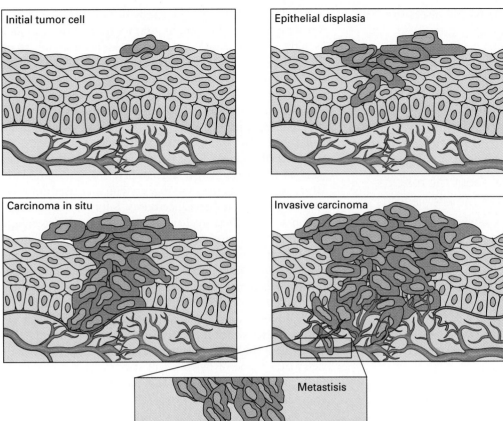

Initial tumor cell

Epithelial displasia

Carcinoma in situ

Invasive carcinoma

Metastisis

Figure 19A–1 A tumor forms when cells escape the normal controls over cell division. An ominous change occurs when the cancer cells successfully cause blood vessels to invade the tumor, delivering needed nutrients and growth factors to support continued growth and providing an "escape route" for cancerous cells. Cancer cells can leave the original tumor, spread (metastasize), and form multiple tumors at new locations throughout the body.

metastases, the tumors that form in distant sites in the body, are responsible for 90% of the deaths of people with cancer.

Once the cancer cells have separated from the original tumor, they usually enter the circulatory or lymphatic systems, which carry the renegade cells to distant sites. Thus, circulatory pathways in the body often explain the patterns of metastasis. Cancer cells escaping from tumors in most parts of the body, including the skin, encounter the next capillary bed in the lungs. Thus, many cancers spread to the lungs. However, blood leaving the intestines travels directly to the liver and so colon cancer typically spreads to the liver.

What does it matter if cancer spreads, you might ask. The important question is how it kills. The simplest answer is that cancer kills by interfering with the ability of body cells to function normally. For instance, cancer cells are greedy. They deprive normal cells of nutrients, thereby weakening them, sometimes to the point of death. Cancer

cells can also prevent other cells from performing their usual functions. In addition, tumors can block blood vessels or air passageways in the respiratory system or press on vital nerve pathways in the brain.

Cancer: The Renegade Cell

The 30 to 50 trillion cells in the human body generally work together, much as any organized society does. There are "rules" or controls that tell a cell when and how often to divide, to self-destruct, and to stay in place. However, cancer cells are outlaws. They evade the many controls that would keep order in the body. As a result, cancer cells can become hazardous to health. We'll consider the normal systems of checks and balances that regulate healthy cells and see how cancer cells are able to get around the safeguards to grow and spread out of control.

Loss of Restraints on Cell Division

The first step in the development of cancer is the loss of control over cell division. What goes wrong? The answer, we will see, lies in the genes. Cancer is fundamentally a genetic disease.

Genes Regulating Cell Division

Cell division is usually regulated by two types of genes: proto-oncogenes and tumor-suppressor genes. Proto-oncogenes stimulate cell division, usually by producing or affecting the function of growth factors (Figure 19A–2). In contrast, normal tumor-suppressor genes turn off cell division. Thus, healthy tumor-suppressor genes function like brakes on cell division. The combined activities of these two types of genes allow the body to control cell division so that

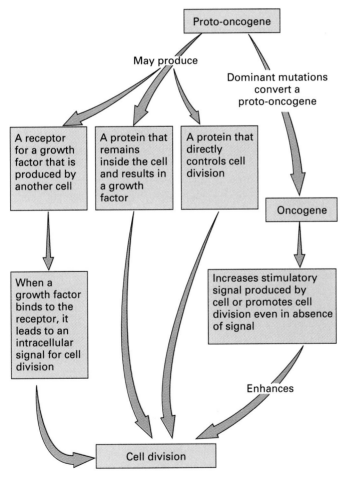

Figure 19A–2   Proto-oncogenes are healthy genes that stimulate cell division under the appropriate conditions, such as during growth and development or for replacing dead cells. An oncogene (a mutant proto-oncogene) accelerates cell division.

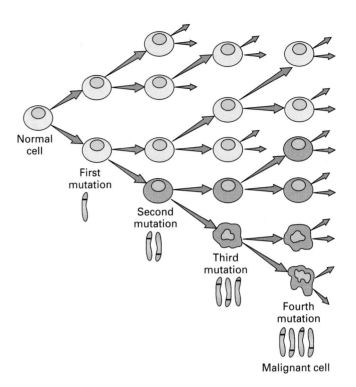

Figure 19A–3   Multiple mutations must occur before a cell becomes cancerous.

the body can grow and develop normally, defective cells can be repaired, and dead cells can be replaced.

When mutations change the instructions of these genes, the normal system of checks and balances that regulates cell division goes awry, which can result in the unbridled cell division that characterizes cancer. A mutation in a proto-oncogene can destroy regulation. The mutant gene is called an **oncogene** and it speeds up the rate of cell division. An oncogene does to the rate of cell division what a stuck accelerator would do to the speed of a car. The *ras* oncogenes are thought to be important in most pancreatic and colon cancers, as well as some lung cancers. Other oncogenes play a role in leukemia and many of the most deadly forms of breast and ovarian cancer.

A mutation in a tumor-suppressor gene can promote cancer by taking the brakes off cell division. The tumor-suppressor gene *p53* is especially important. It regulates another gene whose job is the production of a protein that keeps cells in a nondividing state. When *p53* mutates, cell division is no longer curbed. Mutant *p53* seems to be an important culprit in colon, breast, lung, brain, and bladder cancers. Mutation in another tumor-suppressor gene, called *RB*, causes retinoblastoma, which is a rare form of childhood eye cancer.

Damage must occur in *at least* two genes before cancer occurs (Figure 19A–3). For instance, colon cells must accumulate damage in at least one oncogene and three tumor-suppressor genes before they become cancerous. This explains why it is possible to inherit a predisposition to a certain form of cancer. A person who inherits only one mutant gene may be predisposed to cancer, but a second event, a mutation in at least one other gene, is required before uncontrolled cell division is unleashed.

Limited Life Span

Healthy cells have yet another safeguard against unrestrained cell division—a mechanism that limits the number of times a cell can divide during its lifetime. When grown in the laboratory, for instance, human cells divide only about 50 or 60 times before entering a nondividing state called senescence. Like sand running through an hourglass, no matter how quickly or slowly the events occur, there is a predetermined end.

How does a cell count the number of times it has divided? We aren't sure yet, but the answer might lie in **telomeres**, pieces of DNA at the tips of chromosomes that protect the ends of the chromosomes like the plastic pieces on the

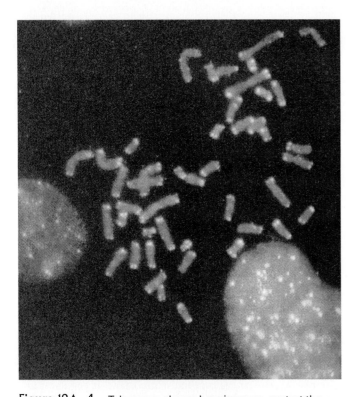

Figure 19A–4 Telomeres, shown here in green, protect the tips of chromosomes, like the small plastic pieces at the ends of shoelaces. Each time a cell replicates its DNA before it divides, a tiny slice of each telomere is removed. Eventually, the entire telomere has been shaved off. Then, with each round of cell division, a small amount of DNA in genes that may be important for the proper functioning of the cell will be shaved off the exposed tips of chromosomes. Thus, telomeres may serve as molecular counting mechanisms that limit the number of times a cell can divide. Cancer cells seem to retain the ability to construct new telomeres to replace the bits that are shaved off.   *(© Dr. Jerry Shay, Southwestern Medical Center, Dallas, TX and Rich Allsopp, Geron Corp., Menlo Park, CA)*

ends of shoelaces, or in **telomerase**, the enzyme that constructs the telomeres (Figure 19A–4). Soon after an embryo is fully developed, most cells stop producing telomerase, and so the construction of additional telomeres stops as well. Each time DNA is copied in preparation for cell division, a tiny piece of every telomere is shaved off, shortening the chromosome slightly. When the telomeres are completely gone, the chromosome tip is no longer protected and parts of important genes will be sliced off each time DNA is copied. Telomeres, then, may be the cell's way of limiting the number of times division can occur. When the telomeres are gone, time is up for that cell. Thus, telomere length may serve both as a gauge of a cell's age and as an indicator of how long that line of cells will continue to divide.

It is currently suspected that the "fountain of youth" that bestows immortality on cancer cells, allowing them to divide continuously, may be their unceasing production of telomerase. This enzyme reconstructs the telomeres after each cell division, stabilizing telomere length and protecting the important genes at chromosome tips. Although most types of mature human cells can no longer produce telomerase, the genes for telomerase apparently become turned on in cancer cells. In one study, the enzyme was present in 90 of 101 biopsies of human tumors.

Escape from Programmed Cell Death

When the genes that regulate cell division become faulty, back-up systems normally swing into play to protect the body from the renegade cell. One such system, **programmed cell death**, is a process by which cells activate a genetic suicide program in response to a biochemical or physiological signal. Activation of the so-called "death genes" prompts cells to manufacture proteins that they then use to kill themselves. Often, the condemned cells go through a predictable series of physical changes called **apoptosis**. During apoptosis, the outer membrane of the condemned cell boils up and pinches off to produce bulges called "blebs" (Figure 19A–5).

The defective DNA in cancer cells does not trigger genetic self-destruction. Although it isn't the only way that cancer cells evade this safeguard, a faulty tumor-suppressor gene *p53* is often at least partly responsible. Besides producing a protein that inhibits cell division, *p53* normally prevents the replication of damaged DNA and, if damage is detected, *p53* halts cell division until the DNA can be mended. If the damage is beyond repair, *p53* triggers the events that will lead to the programmed cell death. Because *p53* determines whether a cell repairs its DNA or self-destructs, this gene is sometimes called "the guardian of the genome."

In a cancer cell, however, a faulty *p53* gene fails to initiate the events leading to cellular self-destruction, and so the cells are free to proliferate out of control. Tumors containing cells with damaged *p53* grow aggressively and spread easily and quickly to new locations in the body. Mutations in

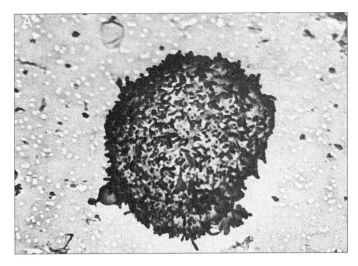

Figure 19A–5 Programmed cell death is a back-up system that protects the body from a cell in which the genes regulating cell division have been damaged. Defective DNA, for instance an oncogene or a faulty tumor-suppressor gene, normally triggers a genetic suicide program that causes the cell to self-destruct, as this cell is doing. As the cell goes through programmed cell death, its plasma membrane forms bulges called "blebs." Cancer cells are able to evade the body's protective mechanism of cellular self destruction. *(© Dr. Lawrence Schwartz, University of Massachusetts, Amherst)*

p53 also make it difficult to kill the cancer cells by radiation or chemotherapy, because these techniques are intended to damage the DNA of the cancer cell and trigger programmed self-destruction. However, the cancer cell is simply unable to self-destruct in response to DNA damage.

Another tumor-suppressor gene, *MTS1* (multiple tumor-suppressor gene), may be an even more important cancer gene than *p53*. Defects in *MTS1* are found in many types of cancer, including melanoma and cancers of the lung, breast, and brain. *MTS1* is a key player in the development of cancer because it regulates another gene, the so-called survival gene c-*myc*, which, in turn, decides whether the cell will divide or kill itself. The *MTS1* protein causes the survival gene c-*myc* to be switched off, allowing the cell to kill itself. Without the *MTS1* protein, the survival gene is switched on and the cell is instructed to divide. Researchers hope that the *MTS1* protein can be used as a drug that will cause tumor cells to self-destruct.

The Ability to Attract a Blood Supply

We have seen that cancer cells have escaped the normal cellular controls on division and are unable to issue the orders that would lead to their own death. Instead, they multiply and form a tumor.

When a tumor reaches a critical size, about a million cells, its growth will stop unless it can attract a blood supply that will deliver the nutrients needed to support its growth and remove waste. Cancer cells release special growth factors that cause capillaries to invade the tumor (Figure 19A–6). These tiny blood vessels will now be the lifeline of the tumor, removing wastes and delivering fresh nutrients and additional growth factors that will spur tumor growth. They also serve as pathways for the cancer cells to spread to other sites in the body.

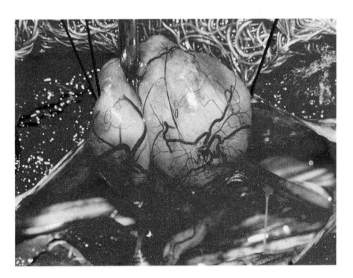

Figure 19A–6 Tumor cells release growth factors that cause blood vessels to penetrate the tumor mass. These vessels bring nutrients and additional growth factors and remove wastes. They also provide a pathway for cancer cells to leave the tumor and spread to other locations in the body. Here, blood vessels have formed in a tumor on the spinal cord. *(SIU/Visuals Unlimited)*

In a healthy person, blood vessel formation is uncommon; it is an event usually reserved for repairing cuts or other wounds. Unwanted blood vessels that invade tissues can cause damage. For instance, when blood vessels invade the eye's light-sensitive retina, blindness can result, and when they invade joints, they can cause arthritis. To prevent such damage, cells produce a protein that prevents blood vessels from spreading into tissues. The gene that normally produces this protein is by now familiar—*p53*. Mutations in *p53* can block the production of the protein that prevents

the attraction of blood vessels, and so blood vessels invade the tumor.

The Breakdown of Cell Adhesion

With blood vessels in place, the cancer cells can then begin to spread. Unlike most cells, cancer cells have the ability to travel about the body. This provides another example of their ability to escape cellular control mechanisms. You may recall from Chapter 3, that normal cells are "glued" in place

Table 19A–2 Control Mechanisms and Methods and Prevention of Cancer Cell Evasion

Mechanisms That Protect Cells from Cancer	Methods of Evasion Used by Cancer Cells	Treatments Aimed at Preventing Evasion Tactics
Genetic Controls on Cell Division Proto-oncogenes—stimulate cell division through effects on growth factors.	Oncogenes (mutant forms of proto-oncogenes) accelerate the rate of cell division.	Gene therapy
Tumor-suppressor genes—inhibit cell division.	Mutations in tumor-suppressor genes take the "brakes" off cell division.	Gene therapy
Limitations on the Number of Times a Cell Can Divide Telomeres protect the ends of chromosomes, but a fraction of each is shaved off each time the DNA is copied; when the telomeres are gone, DNA containing important genes is damaged when the DNA is copied, and the cell can no longer divide.	Genes to produce telomerase, the enzyme that reconstructs telomeres, are turned on in cancer cells, and so telomere length is stabilized.	Block telomerase activity or turn off the genes that produce it
Programmed Cell Death A genetic program that initiates events that lead to the death of the cell when damaged DNA is detected or another signal is received.	Mutations in tumor-suppressor genes: • Mutant gene *p53* no longer triggers cell death when damaged DNA is detected. • Mutant gene *MTS1* does not produce the protein that prevents the survival gene c-*myc* from initiating programmed cell death.	Gene therapy
Controls That Prevent the Formation of New Blood Vessels These controls are normally in effect except in a few instances such as wound healing.	Cancer cells produce growth factors that attract new blood vessels and proteins that counter the normal proteins that inhibit blood vessel formation.	Gene therapy
Controls that Keep Normal Cells in Place Cellular adhesion molecules (CAMs) hold cells in place; unanchored cells stop dividing and self-destruct.	Cancer cell oncogenes send a false message to the nucleus that cell is properly anchored.	Gene therapy to insert the gene that produces CAMs into tumor cells; using drugs to block certain CAMs on cancer cells so that they can't grip blood vessel walls during metastasis

by special molecules on their surfaces called cellular adhesion molecules (CAMs). Most cells must be anchored to a surface or another cell in order to proliferate. In fact, when many types of cells are unanchored they not only stop dividing, their genetic program for self-destruction is activated.

Cancer cells must become "unglued" from other cells in order to travel through the body. How do they do this? One way is by secreting enzymes that break down the glue that holds them and their neighbors in place. In this way, their anchors are broken and mechanical barriers, such as basement membranes, that would prevent metastasis are breached. But how can unanchored cancer cells continue dividing and evade self-destruction? Their oncogenes send a false message to the nucleus saying that the cell is properly attached (Table 19A–2).

Causes of Cancer

We've seen that many of the tactics a cancer cell uses to evade normal cellular safeguards are possible because of changes in genes. Those genetic changes are usually brought about by viruses or by mutations caused by chemical carcinogens or radiation.

Viruses

Some human cancers are caused by viruses (Table 19A–3). It is estimated that viruses cause about 5% of the cancers in the United States. You may wonder how viruses can alter the genetic makeup of a cell, leading to cancer. Recall that viruses usually cause infection by inserting copies of their genes into the DNA of the host cell chromosome. It turns out that oncogenes are found among the genes of a few viruses. When the viral oncogene is inserted into the host cell DNA, it behaves as a host oncogene would and the cell is one step closer to becoming cancerous. In other cases, the viral DNA may be inserted into the host DNA in a location that disrupts the functioning of one of the host genes. It could, for example, be inserted into a regulatory gene that controls a proto-oncogene, breaking the switch that turns the gene on or off. Still other viruses may interfere with the function of the immune system, lessening its ability to find and destroy cancer cells as they arise.

Chemical Carcinogens

A **carcinogen** is an environmental agent that fosters the development of cancer. Some chemicals, especially certain organic chemicals, cause cancer by causing mutations. As we saw in Chapter 19, a change in as little as one nucleotide in a DNA sequence can alter a gene's message. Thus, even a small alteration in DNA can wreak havoc with a cell's regulatory mechanisms and lead to cancer development.

Chemical carcinogens are around us most of the time—in our air, food, water, and physical surroundings. We can choose to avoid contact with some of these. For instance, tobacco smoke contains a host of chemical carcinogens[1] and the mixture is the most lethal carcinogen in the United States. Tobacco smoke is responsible for 30% and may contribute to as much as 60% of all cancer deaths in the United States. Excessive alcohol consumption is another cancer risk factor that can be easily avoided.[2] Other chemical carcinogens are more difficult to avoid. These include benzene, formaldehyde, hydrocarbons, certain pesticides, and chemicals in certain dyes and preservatives.

Other chemicals contribute to the development of cancer by stimulating cell division rather than by causing mutations. If a cancerous cell has already formed, this stimulation will speed up the rate at which the cancer progresses. Certain hormones can promote cancer in this way. For example, the female hormone estrogen stimulates cell division in the tissues of the breast and in the lining of the uterus (the endometrium). Sustained, high levels of estrogen are linked with breast cancer.[3] Also, the incidence of endometrial cancer was elevated in postmenopausal women whose hormone replacement therapy included estrogen alone. (Estrogen does not promote endometrial cancer when taken in combination with progesterone.)

Table 19A–3 Some Viruses Linked to Human Cancers

Virus	Type of Cancer
Human papilloma viruses (HPVs)	Cervical, penile, and other anogenital cancers in men and women
Hepatitis B and C	Liver cancer
Epstein-Barr virus	B cell lymphomas, especially Burkitt's lymphoma, nasopharyngeal carcinoma
Human T cell leukemia virus (HTLV-1)	Adult T cell leukemia
Cytomegalovirus (CMV)	Lymphomas and leukemias
Herpesviruses (HHV8)	Kaposi's sarcoma

[1]Tobacco smoke and its link leads to various cancers are discussed in greater detail in Chapter 14A.

[2]The link between alcohol consumption and cancer is discussed in greater detail in Chapter 8A.

[3]The link between estrogen and breast cancer is discussed in Chapter 16.

Radiation

Radiation can also cause cancer by causing mutations in DNA. It is impossible to avoid exposure to radiation from natural sources, such as the ultraviolet light from the sun, cosmic rays, radon, and uranium. However, we can take reasonable precautions to minimize our risks. For example, sunlight's ultraviolet rays cause skin cancer.[4] Although we probably wouldn't choose to spend our lives entirely indoors to reduce the risk of skin cancer, it would be wise to avoid sunbathing, as well as tanning lamps and tanning parlors. It would also be a good idea to use a sun screen whenever exposure to the sun is unavoidable.

Another natural source of radiation, radon gas, that forms as a decay product of uranium can increase the risk of developing lung cancer. Radon gas can seep into homes from underground. Radon itself isn't a major hazard. The problem is that radon decays into solid particles, which then stick on dust particles. Then, these might be inhaled and settle on lung surfaces where they decay and emit radiation. Over time, this can lead to lung cancer. The risk of developing lung cancer in the United States as a result of radon is estimated to be between 0.1% and 0.9%. Nonetheless, it is estimated that radon is responsible for 2% of total cancer deaths. (Smokers should be warned that the combination of radon and cigarette smoke can be especially harmful.)

The amount of uranium in rocks and, therefore, the amount of radon formed vary from place to place. Generally, the higher the uranium level in an area, the greater chances that houses there will have high levels of indoor radon. However, other factors, such as the porosity of soil and amount of underground water, also affect indoor radon levels. Indoor radon levels vary over a 1000-fold throughout the United States, so we are not all at equal risk. Test kits are available that will measure the level of radon in your home.

[4]Skin cancer and its relationship to ultraviolet light in sunlight are discussed in the Personal Concerns essay "Fun in the Sun?" in Chapter 4.

If the level is high, consult your health department about ways to reduce your household exposure to radon.

Prevention of Cancer

The good news is that there are some lifestyle changes that can greatly decrease your risk of developing cancer (Table 19A–4). Tobacco use and unhealthy diet are responsible for two-thirds of all cancer deaths in the United States. Tobacco smoke is the leading carcinogen and it is obvious how to modify that risk, so here we will focus on diet as a way to reduce your cancer risk.

A few simple food choices may reduce your risk of developing cancer. The best rules are to eat a well-balanced diet with all foods consumed in moderation. For instance, a high-fat diet is linked to several types of cancer, including colon and prostate cancers. Most people in the United States consume far too much fat. Thus, it is wise to reduce fat intake, especially saturated fat, which comes from animal sources such as red meat. Consuming large quantities of smoked, salt-cured, and nitrite-cured foods, including ham, bologna, and salami, increases the risk of cancers of the esophagus and stomach.

A diet rich in fruits and vegetables can reduce your cancer risk. These foods reduce colon cancer because they are high in fiber, which can dilute the contents of the intestine, bind to carcinogens, and reduce the amount of time the carcinogens spend in the intestines. A current "hot topic" is the role that antioxidant vitamins might play in protecting against cancer. As their name implies, antioxidants interfere with oxidation, a process that can result in the formation of molecules called free radicals that can damage DNA and lead to cancer. The three major antioxidants are beta-carotene, vitamin E, and vitamin C. The first two are common in red, yellow, and orange fruits and vegetables and the last abounds in citrus fruits among other sources.

Table 19A–4   Tips for Reducing Your Cancer Risk

1. Don't use tobacco. If you do, quit. Avoid exposure to secondhand smoke.
2. Reduce the amount of saturated fat in your diet, especially the fat from red meat.
3. Minimize your consumption of salt-cured, pickled, or smoked foods.
4. Eat at least five servings of fruits and vegetables every day.
5. Avoid excessive alcohol intake. One or two drinks a day should be the maximum.
6. Watch your caloric intake and keep your body weight proper for your height.
8. Avoid excessive exposure to sunlight: wear protective clothing and use sun screen.
9. Avoid unnecessary medical x-rays.
10. Have the appropriate screening exams on a regular basis. Women should have Pap tests and mammograms. Men should have prostate tests. All adults should have tests for colorectal cancer.

Critical Thinking

Studies have shown that a diet high in beta-carotene, which is found in many vegetables, is associated with a twofold decrease in lung cancer. However, a diet high in natural sources of beta-carotene is also likely to be low in fat and high in fiber. Why is it difficult to say for certain that beta-carotene is the factor in the vegetables that protects against cancer?

Change in bowel or bladder habit or function
A sore that does not heal
Unusual bleeding or bloody discharge
Thickening or lump in breast or elsewhere
Indigestion or difficulty swallowing
Obvious change in wart or mole
Nagging cough or hoarseness

Source: American Cancer Society.

Diagnosis of Cancer

Early detection is critical to cancer survival because successful treatment is much more likely before the cancer has spread. You know your own body better than anyone else does. For this reason, it is wise for everyone to be aware of cancer's seven warning signs, the first letters of which spell the word CAUTION (Table 19A–5). There are also many routine tests that can detect cancer in people without symptoms, some of which you can perform on yourself but others require a visit to a medical professional (Table 19A–6).

When cancer is suspected, there are many noninvasive techniques to confirm (or refute) the suspicion by examining blood, urine, feces, or other bodily secretions. For example, DNA analysis can pick up gene mutations associated with lung cancer in sputum, bladder cancer in urine, and colon cancer in feces. There are other telltale signs of cancer that can be detected in other tests. For instance, the enzyme telomerase is produced by cancer cells but rarely by normal ones. Researchers are hoping to develop a simple, noninvasive test, such as a urine test, that would detect telomerase and, therefore, be helpful in diagnosing certain cancers. Other proteins can also hint of cancer. Prostate cells produce prostate specific antigen (PSA). Thus, elevated blood PSA levels suggest the presence of prostate cancer.

Many new imaging techniques allow physicians to look inside the body and identify tumors. These include x-rays, computerized tomography (CT) scan, magnetic resonance imaging (MRI),[5] and ultrasound. A **biopsy** involves removing a small piece of tissue suspected to be cancerous. A biopsy is often done using a needle instead of surgery. In either case, cells are then examined under a microscope to see whether they have the characteristic appearance of cancer cells.

Treatment of Cancer

The conventional cancer treatments—surgery, chemotherapy, and radiation therapy—are still the mainstays of cancer treatment, but there are many new treatments that hold the promise of a brighter future.

[5]CT scans and MRIs are described in more detail in Chapter 8.

Surgery

When the tumor is accessible and can be removed without damaging vital surrounding tissue, surgery is usually performed to try to eradicate the cancer or remove as much as possible. If every cancer cell is removed, as can be done with early tumors (carcinoma in situ), a complete cure is possible. However, if cancer cells have begun to invade surrounding tissue or have spread to distant locations, surgery cannot "cure" the cancer. In cases of invasive cancer, surrounding tissue and perhaps even nearby lymph nodes may also be removed. Unfortunately, more than half of all tumors have already metastasized by the time of diagnosis, and so further treatment is necessary.

Radiation Therapy

If the cancer has spread from the initial site but is still localized, surgery is usually followed by radiation therapy. In some cases, localized tumors that are difficult to remove surgically without damaging surrounding tissue, such as cancer of the larynx (voice box), may be treated with radiation alone.

As we have seen, radiation damages DNA and extensive DNA damage can kill cells. The greatest damage is done to cells that are rapidly dividing. The intended targets of radiation, cancer cells, are actively dividing but so are several types of body cells, called renewal tissues, that normally continue dividing throughout life. These include cells of the reproductive system, cells that replace layers of skin, and cells that give rise to blood cells or hair. Unfortunately, radiation cannot distinguish between cancer cells and renewal tissue, and so good cells are sacrificed to kill the harmful ones.

Critical Thinking

Radiation kills all rapidly dividing cells, whether they are cancerous or part of renewal tissues. How might the death of renewal tissue cells be responsible for some of the side effects of radiation, including anemia (low red blood cell count), nausea, vomiting, diarrhea, hair loss, and sterility?

Table 19A–6   Recommended Cancer Screening Tests

Cancer-related checkup every 3 years

Should include the procedures listed below plus health counseling (such as tips on quitting cigarettes) and examinations for cancers of the thyroid, testes, prostate, mouth, ovaries, skin, and lymph nodes. Some people are at higher risk for certain cancers and may need to have tests more frequently.

Breast	• Exam by doctor every 3 years
	• Self-exam every month
	• One baseline breast x-ray between ages 35–40
	Higher Risk for breast cancer: Personal or family history of breast cancer, never had children, first child after 30
Uterus	• Pelvic exam every 3 years
Cervix	• Pap test—after 2 initial negative tests 1 year apart—at least every 3 years, includes women under 20 if sexually active
	Higher Risk for cervical cancer: Early age at first intercourse, multiple sex partners

Cancer-related checkup every year

Should include the procedures listed below plus health counseling (such as tips on quitting cigarettes) and examinations for cancers of the thyroid, testes, prostate, mouth, ovaries, skin and lymph nodes. Some people are at higher risk for certain cancers and may need to have tests more frequently.

Breast	• Exam by doctor every year
	• Self-exam every month
	• Breast x-ray every year after 50 (between ages 40–50, ask your doctor)
	Higher Risk for breast cancer: Personal or family history of breast cancer, never had children, first child after 30
Uterus	• Pelvic exam every year
Cervix	• Pap test—after 2 initial negative tests 1 year apart—*at least* every 3 years
	Higher Risk for cervical cancer: Early age at first intercourse, multiple sex partners
Endometrium	• Endometrial tissue sample at menopause if at risk
	Higher Risk for endometrial cancer: Infertility, obesity, failure of ovulation, abnormal uterine bleeding, estrogen therapy
Colon & rectum	• Digital rectal exam every year
	• Guaiac slide test every year after 50
	• Proctological exam—after 2 initial negative tests 1 year apart—every 3 to 5 years after 50
	Higher Risk for colorectal cancer: Personal or family history of colon or rectal cancer, personal or family history of polyps in the colon or rectum, ulcerative colitis

Chemotherapy

When it is thought that the cancer may have spread by the time of diagnosis, chemotherapy is often used. The drugs used generally reach all parts of the body and kill all rapidly dividing cells, just as radiation does. Some of the drugs used in chemotherapy block DNA synthesis, others damage DNA, and a few others prevent cell division by interfering with other cellular processes. The side effects are similar to those that accompany radiation therapy.

Immunotherapy

The body defense mechanisms, most importantly cytotoxic T cells of the immune system, continually search the body for abnormal cells, such as cancer cells, and kills those it finds. The goal of immunotherapy, then, is to boost the patient's immune system so that it becomes more effective in eliminating cancer cells. One form of immunotherapy involves the administration of factors normally secreted by lymphocytes, including interleukin-2 (that stimulates lymphocytes that attack cancer cells), interferons (that stimulate the immune system and also have direct effects on the tumor cells), and tumor necrosis factor (that has direct effects on cancer cells).

A newer, more experimental form of immunotherapy is to develop a vaccine that would stimulate T cells to attack and kill the cancer cells. Unlike most vaccines, the ones used here cannot *prevent* disease (cancer); they can only treat it. The vaccine strategy is now being tested against melanoma (a deadly form of skin cancer). The trials, now in their final phases, show promise. In one test, 47 people with melanoma were inoculated with a vaccine prepared from their own melanoma cells that had been previously inactivated by radiation. After 3 years, 60% of those who were vaccinated but only 20% of those who were unvaccinated are free of tumors.

Inhibition of Blood Vessel Formation

In recent years, there has been a major change in the way we think about cancer. Gone are the days when the only aim was killing the cancer cells. As we have learned more about the molecular biology of cancer, new ways of slowing its progression or dealing with it as a genetic disease have been developed.

The formation of blood vessels is a critical step in the life of a tumor, because they bring nourishment and provide a pathway for cell migration. Researchers are working on ways to cut these life lines and starve the tumor. A drug has been developed and is being tested in laboratory animals that turns off the biochemical switch that triggers the formation of blood vessels in tumors. With this treatment, some tumors shrank and others disappeared altogether.

Gene Therapy

Several strategies of gene therapy against cancer cells have been tried. One idea has been to insert normal tumor-suppressor genes into the cancerous cells. For instance, you may recall that gene *p53* normally triggers programmed cell death when DNA damage is detected, but this gene is often faulty in cancer cells. It is hoped that inserting a healthy form of *p53* into cancer cells will lead to their death, causing the tumor to shrink.

Another idea is to insert a piece of DNA into a cancer cell that will prevent an oncogene from exerting its effects. This so-called antisense DNA is a piece of DNA that is complementary to the mRNA produced by the oncogene. The antisense DNA would be expected to bind to the mRNA produced by the oncogene and prevent it from being translated into protein, which would prevent the effects of the oncogene. Currently, this is being tried as a means of treating leukemia.

One of the most promising gene therapy approaches, inserting a gene that makes tumor cells sensitive to a drug that will kill them, is now being evaluated as treatment for brain, ovarian, and prostate cancers. In this case, a viral gene for the enzyme thymidine kinase is inserted into the tumor cells. When the gene is expressed, the resulting enzyme makes the cell sensitive to a drug called ganciclovir. The drug is activated in only those cells that produce the enzyme coded for by the inserted gene. Thus, only the tumor cells are killed.

Another gene therapy approach is to stimulate the body's own defense mechanisms to increase their attack on the cancerous cells. You may recall from Chapter 12 that interleukin-2 is a chemical that activates body defenses, including those that are directed against cancerous cells. The idea is that inserting the gene for interleukin-2 into tumor cells would stimulate the immune system to attack these genetically altered cells. This approach is being tried as treatment for cancers of the skin, kidney, brain, lung, and colon.

In other trials, a gene for tumor necrosis factor (TNF) is inserted into a certain type of white blood cell called tumor-infiltrating lymphocytes. TNF is a protein that kills cells, healthy ones as well as those that are cancerous, by shutting off their blood supply. The idea is that the tumor-infiltrating lymphocytes will deliver the TNF gene to only the cancer cells so that the cells of the surrounding tissue will not be harmed.

One of the problems associated with chemotherapy as a means of treating cancer is that the chemicals that are administered also kill the cells in bone marrow that give rise to all blood cells, including those of the immune system, along with the cancerous cells. There are currently several clinical trials underway designed to determine whether a gene for resistance to chemotherapy drugs can be inserted into bone marrow cells so that they will be protected from the damaging effects of chemotherapy.

1. Explain why oncogenes and mutant tumor-suppressor genes can lead to the unrestrained cell division characteristic of cancer.
2. What are telomeres and how do they limit the times a normal cell can divide? How do cancer cells escape this limitation?
3. How does programmed cell death normally protect the body from cells with damaged DNA? How can cancer cells evade this fate?
4. Why is the formation of blood vessels a critical step in the development of a tumor?
5. What is metastasis and why does this influence the odds of cancer survival?

6. How can viruses, chemical carcinogens, and radiation cause cancer?
7. Name some dietary recommendations that will reduce one's risk of developing cancer.
8. How is cancer diagnosed?
9. Name the three conventional means of treating cancer. Explain how immunotherapy and gene therapy are being used to treat cancer.

Evolution and Ecology

(© David Muench/Tony Stone Images)

20

A Cro-Magnon wall painting.   *(© Science VU/Visuals Unlimited)*

Evolution: Basic Principles and Our Heritage

Imagine revolutionizing the pervasive view of the Earth and its inhabitants and forever changing the course of scientific thinking with insights gathered from a trip taken when 22-years-old. This is exactly what happened when Charles Darwin, a somewhat disenchanted student of medicine and theology, left Great Britain in 1831 to spend 5 years aboard the HMS *Beagle* (Figure 20–1). The goal of the voyage was to chart the coastline of South America. This goal allowed Darwin, an avid naturalist, to examine, describe, and collect plants and animals from the South American coast and nearby islands. In 1836 Darwin returned to Great Britain where he began to mull over all that he had seen and read. In declining health and limited to a few hours of work a day, 23 years passed before his book *On the Origin of Species* (1859) was published. In this book he proposed that the incredible diversity of life was the product of evolution. What is evolution and how does it work? How has evolution shaped species, including our own? These are some of the questions that we address in this chapter.

Basic Principles of Evolution

Evolution occurs on two levels, one small and one grand. **Microevolution** involves changes in the frequency (abundance) of certain alleles relative to others within a gene pool. The term **gene pool** refers to all of the alleles of all of the genes of all individuals in a population. Thus, microevolution is genetic change occurring within a population of organisms over time. (A **population** is a group of individuals of the same species living in a particular area.) **Macroevolution**, on the other hand, focuses on larger scale phenomena such as the evolution of mammals, mass extinctions, and patterns of change among groups of species. Whereas microevolutionary change is brought about by alleles increasing or decreasing in abundance within a population, macroevolutionary change might result from long-term changes in the climate or position of the continents. We begin our coverage of evolution at the level of populations and then expand to larger scale phenomena.

Sources of Variation in Populations

Look around and you will see variation in almost any population. For example, the pigeons that gather in a local park or campus display a bewildering array of plumage colors and patterns—some are light and dark gray with iridescent

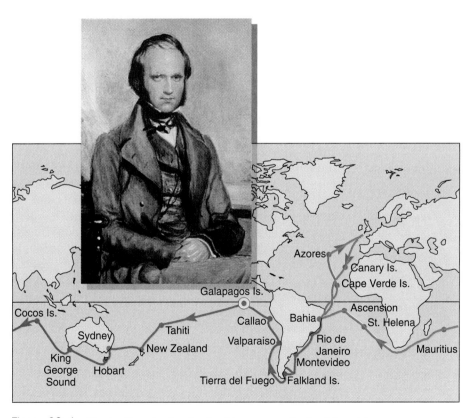

Figure 20–1   Charles Darwin shortly after his return to England from his voyage on the H.M.S. *Beagle.*   *(Charles Darwin illustration © William E. Ferguson)*

necks, others are all white or all black, and still others sport patches of brown and white (Figure 20–2). Consider also your classmates. They do not all look alike, and you probably do not precisely resemble your brothers or sisters (unless you are an identical twin, but even then, the resemblance is not necessarily precise at fine levels of detail). Let's take a closer look at what makes you and your sibling different—or any individuals in a population, for that matter. We begin by describing some of the ways that variation can appear in populations and then consider how the special traits of individuals can, under the right conditions, serve them well and send their genes into the next generation.

Mutations, changes in the DNA of genes, are one source of variation in populations. Such changes occur at a rather low rate in any set of genes. For example, the average mutation rate for genes in humans that lead to visible phenotypic effects has been estimated at one mutation per million gametes per generation. Because mutations occur at such a low rate, their contribution to genetic diversity in large populations is quite small. Mutations can appear spontaneously, as an inherent property of DNA, or they can be caused by outside sources, such as radiation or chemical agents.

Two processes that occur during meiosis—crossing over and independent assortment—generate considerable genetic variation. **Crossing over** is the reciprocal exchange of genetic material between nonsister chromatids during meiosis (specifically at prophase I when homologous chromosomes pair up side by side). Crossing over breaks up old combinations of alleles and produces new combinations from the two parents. **Independent assortment**, when homologous chromosomes and the alleles they carry segregate randomly into gametes during meiosis, creates mixes of maternal and paternal chromosomes in gametes. As a result of crossing over and independent assortment, the gametes of any one individual show substantial variation in their genetic makeup.

Finally, the gametes produced by meiosis are haploid, necessitating their union with another gamete to produce a diploid individual. Thus, **fertilization**, the union of a haploid egg and a haploid sperm to produce a diploid zygote, also contributes to genetic variation. This union of parental gametes produces a new individual with new combinations of alleles.

Causes of Microevolution

Recall that microevolution involves changes in the frequency of certain alleles relative to others within a gene pool. Some of the processes that cause microevolutionary change include genetic drift, gene flow, natural selection, and mutation.

Genetic Drift

When allele frequencies within a population change randomly because of chance alone, the evolutionary agent is called **genetic drift**. In large populations, genetic drift is usually negligible, but in small populations (anything less than about 100 individuals), chance events can cause allele frequencies to drift randomly from one generation to the next. Populations small enough for genetic drift to occur are often associated with two phenomena, the bottleneck effect and the founder effect.

Sometimes dramatic reductions in population size occur as a result of disasters that kill unselectively. For example, if a population experiences a flood or drought and most members die, then the genetic makeup of the survivors may not be representative of the original population—this is the **bottleneck effect**, named because the population experiences a dramatic decrease in size much like the size of a bottle decreases at the neck. Under these circumstances, by chance alone, certain alleles may be more or less common in the survivors than in the original population, and some alleles may have been eliminated all together resulting in reduced overall genetic variability among survivors. For example, remarkable uniformity in genetic makeup has been found in the cheetahs of South Africa. Their low genetic variability is thought to result from two drastic reductions in population size. The first reduction occurred during the last ice age, about 10,000 years ago, and the second occurred at the start of this century. Both reductions have been linked to over hunting by humans.

Genetic drift also occurs when a few individuals leave their population and establish themselves in a new, somewhat isolated place. By chance alone, the genetic makeup of the colonizing individuals is probably not representative of the population they left. Genetic drift in new, small colonies is called the **founder effect**. This effect has been demonstrated in humans where it typically results in a relatively high frequency of inherited disorders. For example, the

Figure 20–2 There is variation among individuals of populations. Here, such variation can be seen in the colors and patterns of plumage in pigeons. *(© Bruce Gaylord/Visuals Unlimited)*

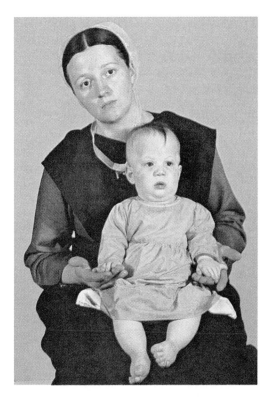

Figure 20–3 An Amish woman holding her child with Ellis-van Creveld syndrome, a disorder characterized by dwarfism and extra digits. The relatively high incidence of this genetic disorder among the Amish in Lancaster County, Pennsylvania, is thought to result, in part, from the founder effect. *(Courtesy of Victor A. McKusick, John Hopkins University)*

Afrikaners of South Africa are all descended from some 30 seventeenth-century Dutch families. As a result of genetic drift among the founding families, present-day Afrikaners suffer from an unusual set of recessive genetic diseases that are quite rare in other populations. Similarly, the relatively high incidence among the Amish in Lancaster County, Pennsylvania, of Ellis-van Creveld syndrome, an inherited condition associated with dwarfism and extra digits, is also thought to be due, in part, to the founder effect (Figure 20–3). Between 1720 and 1770, about 200 Amish people moved to Pennsylvania from Switzerland. Today the population stands at about 14,000. The recessive allele for Ellis-van Creveld syndrome is thought to have been carried by either a Mr. Samuel King or his wife who immigrated to Pennsylvania in 1744. The couple and their descendants were apparently quite prolific, causing the frequency of the deleterious allele to drift higher in subsequent generations.

Gene Flow and Speciation

Another cause of microevolution is **gene flow**, which occurs as a result of the movement of individuals into and out of populations. As these individuals come and go, they carry with them their unique sets of genes, and if they successfully interbreed with the resident population, then gene flow occurs.

The question of gene flow is important to the consideration of how species are formed. A **species** may be defined as a population or group of populations whose members are structurally and functionally similar and are capable of successful interbreeding under natural conditions. Such interbreeding must produce fertile offspring. Pockets, or localized populations of a species, exhibit varying levels of interbreeding with one another. If one population of a species should become geographically isolated from other populations, perhaps by being separated by a newly formed river, it may begin to take a separate evolutionary route and to accumulate a distinctly different set of allele frequencies and mutations from the other groups. Finally the two populations may become so different that we say **speciation**, the formation of a new species, has occurred. In any case, as long as there is even a small amount of gene flow between two populations, any tendency toward speciation is likely to be swamped by the arrival of genes from other populations. With the genes continually mixed by gene flow, there is no opportunity for a special, or distinct, set of genes to arise.

Natural Selection

Charles Darwin, in his book *On the Origin of Species* (1859), argued that species were not specially created, unchanging forms but rather were descendants of ancestral species. Present day species had thus evolved from past species. Darwin also proposed that such evolution could occur by natural selection. His ideas can be briefly summarized as follows:

1. There is individual variation within species and some of this variation is inherited.
2. Some individuals live longer and have more offspring than do others because of their particular inherited characteristics.
3. **Natural selection** is the differential survival and reproduction of individuals that results from genetically based variation in structure and function.
4. Evolutionary change occurs as the traits of individuals that survive and reproduce become more common in the population whereas those of less successful individuals become less common.

Given Darwin's ideas, how do we measure an individual's evolutionary success? Typically, success is measured by **fitness** (sometimes called Darwinian fitness), the average number of offspring left by an individual. Note that to succeed in evolution, survival is not enough—one must reproduce. Indeed, you could live to be over 100 years old, but if you did not reproduce in that time, your individual fitness would be zero. Some of the diseases or conditions discussed in earlier chapters are associated with zero fitness because they either cause premature death (premature, here, meaning death before reproduction) or sterility (inability to reproduce). For example, Tay-Sachs disease, an untreatable condition characterized by the accumulation of lipids in nerve cells, leads to death in early childhood. In the case of Tay-Sachs disease, then, fitness is zero because individuals die before reproduction is possible. Consider also Turner's syndrome, a condition that results when females have a single X chromosome. Here, fitness is zero because of sterility. In summary, according to Darwin's notion of evolution by natural selection, individuals with greater fitness have more of their genes represented in future generations.

A result of natural selection is that populations become better attuned to the particular environments in which they exist, a process called **adaptation**. We say that natural selection leads to populations becoming better adapted to their environment. But what if the environment changes? Those individuals that had become so finely attuned to it would lose their advantage. This is where variation comes in. Remember, we said that there will be variation in any population, whether we're discussing you and your sibling or pigeons in a park. So if the environment should change, then other kinds of individuals, bearing different alleles, may be selected by nature to leave more descendants. If the environment again stabilizes, then the population re-establishes itself around a new set of allele frequencies that meet the requirements of the new conditions.

Although natural selection often proceeds slowly and is difficult to observe and to measure, there are some cases in which we have been able to see it in progress. Perhaps the best-known case is that of the peppered moth, *Biston betularia*. The moths were well known to naturalists in nineteenth-century England as light-colored insects that often rested camouflaged against the lichens that covered the trees at that time. In 1845, however, a black peppered moth was found, an interesting oddity.

As England became increasingly industrialized, a cloak of soot began to fall over the countryside, killing the lichens and darkening the trees. Naturalists began finding more and more black peppered moths and fewer of the original, light-colored ones (Figure 20–4). By the 1950s, the light-colored moths were rare indeed and found only on trees in remote, unpolluted areas. An amateur naturalist and physician H. B. D. Kettlewell argued that the coloration of the moths protected them against predatory birds. He devised a series of experiments in which marked moths of both types were released in polluted and unpolluted areas and found that he recovered far more light moths from the unspoiled areas and more dark moths from the polluted woodlands. Then he clinched his argument by actually filming birds taking more of the moths that contrasted with their backgrounds.

The black form became predominant as conditions changed, just as the lighter form became more rare. In time, as the English countryside returned to a less polluted state, the trees began to lighten, and the moth population shifted back to the lighter form. Similar trends have been noted in a number of moths and butterflies in other places, including areas in the northeastern United States.

Another example of evolution by natural selection concerns the rise of antibiotic-resistant bacteria. Consider what happens when you take an antibiotic to treat a bacterial infection. Although many bacteria are killed by the treatment, some survive because they have genes that confer resistance to the antibiotic. The survivors reproduce and pass on the trait for resistance to future generations of bacteria. Over time we find that antibiotic resistance is more common in bacterial populations than before—in other words, natural selection has caused evolutionary change. Failure to take the full recommended dose of antibiotic and thereby allowing more resistant bacteria to survive may also set the stage for the evolution of bacterial resistance. Strains of antibiotic-resistant bacteria have been identified for many diseases including tuberculosis, bacterial dysentery, and urinary tract infections. Indeed, the evolution of drug-resistant strains of

(a)

(b)

Figure 20–4 Black and white forms of the peppered moth on (a) a light background and (b) a soot-blackened background. *(© Michael Tweedie/Photo Researchers, Inc.)*

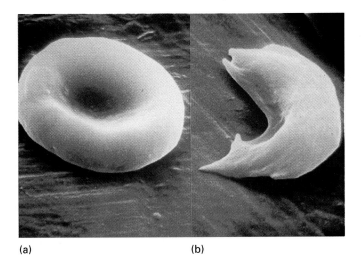

(a)   (b)

Figure 20–5 Sickle-cell anemia is caused by defective hemoglobin that causes the blood cells to assume a sickle-shape. (a) Normal red blood cells are shaped like biconcave disks. (b) Sickled red blood cells are long with irregular ends and, as a result, often clog small vessels. Sickle-cell anemia provides an example of heterozygote advantage. People heterozygous for sickle-cell anemia show resistance to malaria. Thus, in regions of the world where malaria is a problem, heterozygotes are favored over homozygous recessives (who suffer from sickle-cell anemia) and over homozygous dominants (who are more susceptible to malaria). Such an advantage maintains the allele for sickled cells in the population. *(© Stanley Flegler/Visuals Unlimited)*

microorganisms may prove to be one of the greatest threats to humankind.

How is genetic variation within a population maintained if natural selection continually removes unsuccessful individuals (and their alleles)? One mechanism by which variation is maintained within a population is **balanced polymorphism**, when natural selection maintains two or more alleles for a trait from one generation to the next. In some cases, a balanced polymorphism can exist when the environment changes frequently so that at certain times one allele is favored and at other times another allele for the same trait is favored. In other circumstances, the heterozygous condition may be favored over either homozygous condition. The latter mechanism is called **heterozygote advantage**.

An example of heterozygote advantage concerns sickle-cell anemia, a genetic disorder in humans in which hemoglobin (the iron-containing protein in red blood cells that transports oxygen) is defective, causing red blood cells to be sickle-shaped rather than the normal disk-shaped (Figure 20–5). The sickle-shaped cells clog small blood vessels, leading to pain and tissue damage from insufficient oxygen. The condition is caused by a recessive allele, and homozygous recessive individuals are often at such a disadvantage that they may not survive their teens and twenties. We might expect from such poor survival (and presumably poor repro-

duction) of homozygous recessive individuals that the allele for sickle-cell anemia would become increasingly uncommon and eventually disappear. Our expectation, however, would be incorrect for regions of the world where malaria is a problem. Heterozygous individuals are more resistant to malaria, and if they should contract the disease they have a much milder case, an important advantage in tropical and subtropical regions of Africa. (Apparently, the malarial parasite spends a portion of its life cycle in red blood cells and fairs less well when some sickle-cell hemoglobin is present.) In regions where malaria is a problem, heterozygotes are favored over homozygous recessive individuals who suffer from sickle-cell anemia and homozygous dominant individuals who are more susceptible to malaria. Thus, because heterozygotes have an advantage over both homozygotes, the allele for sickled cells continues to be maintained in these populations. In populations where malaria is not a problem—for example African Americans in the United States—the frequency of the allele for sickle-cell anemia is slowly declining.

Critical Thinking

Some people have suggested that natural selection no longer operates within the human species because modern medicine levels the playing field. Medical technology, they argue, helps "fit" and "unfit" individuals alike to survive and reproduce. Others believe that natural selection continues to operate in modern society, suggesting that some present-day conditions act as selective agents. Air pollution, for example, might select for individuals with genetic resistance to respiratory disease. Which argument do you favor? Can you think of a time in human development when natural selection is clearly acting?

Mutation

Mutations, you will recall, are rare changes in the DNA of genes. Mutations produce new alleles that when transmitted in gametes cause an immediate change in the collection of alleles in a population. Essentially the new (mutant) allele is substituted for another allele. If the frequency of the mutant allele increases in a population, then it is not because mutations are suddenly occurring with great frequency, but rather it might be that possession of the mutant allele confers some advantage and thus individuals with the mutant allele produce more offspring than those without it. In other words, the increased frequency of the mutant allele relative to others results from natural selection. Differences between populations brought about by mutation, natural selection, or genetic drift tend to be reduced by gene flow.

Now that we are familiar with microevolution and some of its causes, let's consider some macroevolutionary phenomena.

Macroevolution and Systematic Biology

Evolution on a grand scale is macroevolution. We now move from discussing evolution in terms of changes in the frequencies of alleles within populations, to major patterns of change involving groups of species. First we examine how species are named and then how their evolutionary histories can be interpreted and depicted.

Systematic Biology

Systematic biology or systematics is the discipline that deals with the naming, classification, and evolutionary relationships of organisms. A stable system of names is essential for communicating information about organisms. Our system for naming organisms has been used for more than 200 years, since its development by the Swedish naturalist, Carl Linnaeus. Linnaeus set out to classify the world's known organisms, and the hierarchical nomenclature that he developed (kingdom, phylum, class, order, family, genus, species) has proven extremely useful to succeeding generations.

The center of Linnaeus' naming scheme is the Latin binomial (two-part name), made up of the genus name followed by the species name. For example, humans are genus *Homo*, species *sapiens*. Because these names are Latin or are latinized forms of words from other languages, they are italicized and the first letter of the genus name is always capitalized (as with most proper names) but the species name is traditionally given in all lower case letters. Whereas microevolutionary research focuses on variation and change at or within the species level, macroevolutionary research focuses on differences between species, as well as genera, families, and the other so-called "higher taxa" (higher taxa being those above the species level).

Phylogenetic Trees

Phylogenetic trees are generalized descriptions of the history of life. They depict hypotheses about genealogical relationships among species or higher taxa. Trees can illustrate in simple graphic form concepts that are difficult to express in words, and so they have achieved great importance in evolutionary biology. We commonly encounter phylogenetic trees in textbooks or newspaper or magazine articles, and most of us give little thought to the data that underlie the construction of a tree or to the methods used to evaluate and select among many possible trees. Here we consider some of the basics of constructing phylogenetic trees.

A major goal of contemporary research in systematic biology is the development of data matrices to be used for construction of phylogenetic trees. Matrices for phylogenetic analysis can consist of columns of taxa and rows of "characters" that score the presence or absence of particular features in the taxa of interest. For example, consider a simple three-taxon matrix with a salmon (fish), frog (amphibian), and human (mammal). As character data, we might score for the presence or absence of two paired appendages (fins or limbs), the presence or absence of individual fingers or digits, and the presence or absence of hair (Figure 20–6a). We could also define countless other characters and score them as well, but the ones chosen for this example have a certain predictive value and familiarity. We can array these features on a phylogenetic tree as shown in Figure 20–6b. It is easy to see from this example how trees can be generated from character data and how character data can be well illustrated on trees. Now imagine how this might work were we to build much larger matrices—the same principles apply, but the work is so complex that sophisticated math and high-speed computers quickly become essential. The science of generating phylogenetic trees, particularly the approach

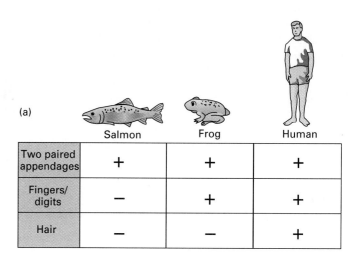

	Salmon	Frog	Human
Two paired appendages	+	+	+
Fingers/ digits	−	+	+
Hair	−	−	+

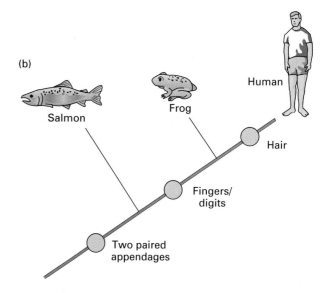

Figure 20–6 A phylogenetic tree depicts hypotheses about evolutionary relationships among organisms. Such a tree is constructed from a data matrix (a). A sample tree developed from the data matrix (b).

known as cladistics briefly explained here, has become one of the "hottest" areas in all of biology.

Rates of Evolutionary Change

The elegant Darwinian account of how microevolution proceeds explains much about what caused changes in the Earth's life over time, and its principles are easy to understand: Natural selection acts on variation to produce change over time. However, how fast does evolution occur?

When invoking Darwin's theory of evolution, scientists have generally assumed that such change occurred slowly, a process called **gradualism**. This assumption has led to some problems, such as the expectation that there should be abundant evidence of transitional forms in the fossil record. Forms like *Archaeopteryx*, the link between the first birds and their reptilian ancestors, should be quite commonplace (Figure 20–7). In reality, however, intermediate forms are uncommon and the fossil record documents that changes often occur rather rapidly after long periods of stability. In fact, most of the physical changes, such as those indicated by dramatically different fossils, appear to occur just after a species diverges from its immediate ancestor. These great changes, possibly brought on by only a few genetic alterations, are then followed by long periods of stability. When relatively

brief periods of evolutionary change ("brief" in geological time can be thousands of years) are interspersed with long periods of relatively little change, the process is known as **punctuated equilibrium**. Most scientists today recognize that sometimes evolution occurs in small steps and other times it occurs in great leaps. Gradualism and punctuated equilibrium are contrasted in Figure 20–8.

Divergence and Convergence

As populations (or species) are subjected over time to the various influences of the environment, they may change in dramatically different directions. Two populations subjected to different environments may become increasingly different, a process called **divergent evolution**. Divergent evolution commonly occurs when a population is divided by such geologic changes as mountain formation or continental drift. (Continental drift refers to the process by which continents slowly move around on the Earth's surface. The continents sit on large plates of the Earth's crust that float on the layer of molten rock beneath them. The theory that the outermost layer of the Earth consists of slowly moving crustal plates is called plate tectonics.) Such barriers limit gene flow, and the diverging populations may be exposed to increasingly different environments due to these geologic changes. For instance, when Darwin visited the Galapagos Islands off the western coast of South America 160 years ago, he observed marine iguanas (Figure 20–9). These lizards, although closely related to iguanas from the mainland of South America, have diverged greatly from their mainland cousins in appearance, habits, and diet: For example, they are the only lizards in the world to eat marine algae as their principal food item, diving many meters below the ocean's surface to graze. Marine iguanas have many adaptations that allow this unusual way of life. We infer that such a dramatic divergence occurred as these lizards evolved in isolation from mainland iguanas.

In other cases, two populations that are separated geographically may become more alike if they are subjected to similar environments. This process is called **convergent evolution**. Nature provides many striking examples of convergent evolution at many different levels of analysis. For example, pterosaurs (an extinct group of flying reptiles) and birds convergently evolved flight based on flapping the forelimbs. The flight surfaces used by the two groups, however, are different: a flap of skin suspended by the bones of the arm and a single extremely elongated finger in the case of pterosaurs and feathers in the case of birds.

Consider also the similarities between the kangaroo rats of North American deserts and similarly adapted—although evolutionarily distant—jerboas from the deserts of the Old World. These two lineages of rodents display, among other similarities, long hind limbs used for hopping (in some species, the hind limbs are four times as long as the front limbs) and a long naked tail with a tuft of hair at the tip. These specializations have presumably evolved indepen-

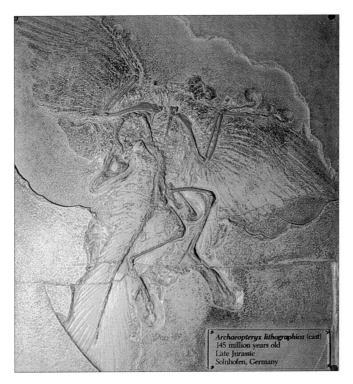

Figure 20-7 *Archaeopteryx.* These fossilized remains show an animal with a mix of bird-like traits (for example, feathers) and reptilian traits (for example, teeth and a long tail with many vertebrae). Such transitional forms are quite rare in the fossil record. *(© Dennis Drenner)*

Archaeopteryx lithographica (cast)
145 million years old
Late Jurassic
Solnhofen, Germany

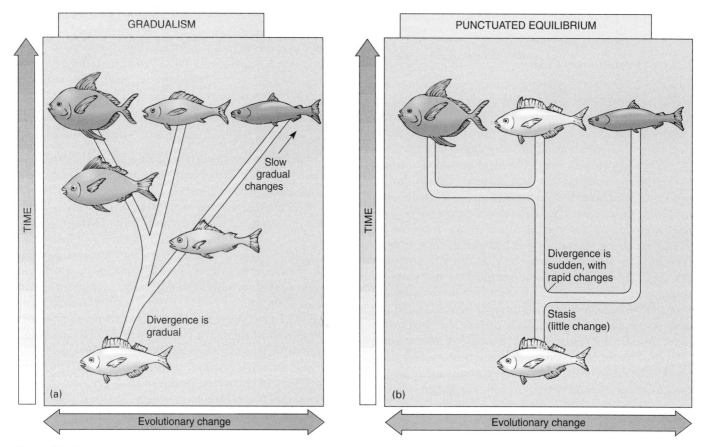

Figure 20–8 Rates of evolutionary change. (a) In some instances, evolution seems to occur as small, steady changes in species over time, a process known as gradualism. (b) In other cases, long periods of no evolutionary change are followed by bursts of rapid change and speciation, a process called punctuated equilibrium.

Figure 20–9 A marine iguana from the Galapagos Islands. These iguanas evolved in isolation and have diverged greatly from their relatives on mainland South America. (*© Joe McDonald/Visuals Unlimited*)

dently in the two groups in response to similar selection pressures associated with desert environments. Jumping may be essential to escaping predators in open country, hence the long, spring-like hind legs and the long tail for balance.

Evidence of Evolution

How do we know that evolution has occurred? The physical evidence of evolution surrounds us, and even a little training will help you learn to detect it. In general, physical evidence of evolution—such as that provided by fossils—contributes more to our understanding of macroevolutionary phenomena. However, some methodologies, such as comparative molecular biology, contribute to our understanding of both microevolution and macroevolution. Here are some examples of the basic data that document evolution.

The Origin and Early Evolution of Life

The Earth is estimated to be 4.5 billion-years-old. Evidence from physical and chemical changes in the Earth's crust and atmosphere suggests that life has existed on Earth for at least 3.5 billion years. The environment of the early Earth was very different from that of today and would have been an extremely hostile place for most organisms. The Earth's crust was hot and volcanic, with intense lightning and ultraviolet radiation and almost no oxygen in the atmosphere. Many changes that led to Earth's present environment were caused by living organisms and in this way the Earth and its life are inextricably linked. However, where did life come from in the first place?

In the 1920s, two evolutionists—A. I. Oparin of Russia and J. B. S. Haldane of Great Britain—independently suggested that conditions of the early Earth favored the synthesis of organic compounds from inorganic molecules. Specifically, they hypothesized that the low oxygen atmosphere of the primitive Earth encouraged the joining together of simple molecules to form complex molecules (oxygen tends to attack chemical bonds rather than to promote their formation). Oparin and Haldane reasoned further that the energy required for such synthetic reactions could have come from lightning and intense ultraviolet radiation striking the primitive Earth. (UV radiation is presumed to have been more intense during primeval times because young suns emit more UV radiation than do old suns and the early Earth lacked an ozone layer produced from oxygen, which today shields the Earth from much UV radiation.)

The Oparin-Haldane hypothesis was tested in 1953 by Stanley Miller and Harold Urey who re-created in their laboratory conditions presumed similar to those of the early Earth (Figure 20–10). By discharging electric sparks

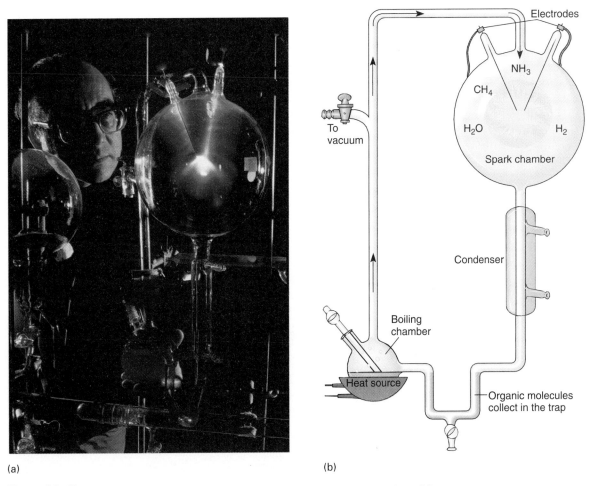

(a) (b)

Figure 20–10 Testing the Oparin-Haldane hypothesis. (a) Laboratory re-creation of the conditions of the early Earth used by Stanley Miller (shown) and Harold Urey. Using this apparatus, Miller and Urey demonstrated that simple organic compounds could be synthesized from inorganic molecules in the low oxygen environment of the primitive Earth. (b) Diagram of the apparatus. The primeval sea was simulated by water in the boiling chamber, the spark chamber contained gases similar to those in the primitive atmosphere, and the electrodes discharged sparks to simulate lightning. Newly formed organic molecules were collected in the trap at the bottom of the apparatus. *(a, © 1988 Roger Ressmeyer/Starlight)*

through an atmosphere containing methane, water, hydrogen gas, and ammonia, the scientists generated a variety of organic compounds, including amino acids. These results lent support to the Oparin-Haldane hypothesis and subsequent re-creations of the primeval Earth have yielded other amino acids as well as macromolecules such as sugars, lipids, and, under certain conditions, ATP.

Over long periods of time, these molecules accumulated, and with the growth of the oceans as the Earth's crust cooled, they formed a complex mixture in the earliest oceans. We do not know exactly how these molecules became organized into cells, but we theorize that the molecules first aggregated into droplet-like structures that displayed some properties of life, such as the ability to maintain an internal environment different from surrounding conditions. Some time later, genetic material may have been incorporated into compartments within the droplets, thus allowing the transfer of information from one generation to the next. Indeed, one constant feature throughout the evolution of life on Earth has been the presence of self-replicating molecules of nucleic acids—RNA and DNA. The stability of the genetic code across billions of years of Earth history remains one of the most powerful insights of our time. It is humbling to know that a remarkably constant and simple set of genetic codes applies to all life. This unity suggests a single common origin for life.

How did more complex cells arise from the earliest cells? The physical evidence for this comes from the study of cells living today. Lynn Margulis, an imaginative scientist at the University of Massachusetts at Amherst, incorporated information from a variety of fields and concluded that many of the organelles within eukaryotic cells appeared as other, smaller organisms became incorporated into the cells. For example, she proposed that mitochondria are the descendants of once free-living bacteria that invaded the ancient cells and formed a mutually beneficial relationship with them. She also suggested that chloroplasts were formed the same way. As evidence of her theory she noted that the genetic makeups of these cellular organelles often are closer to free-living bacteria than they are to other parts of the cells in which they are found. Today her hypothesis—called the endosymbiont hypothesis—is generally accepted as fact by the biological community.

In the Margulis hypothesis, the evolving cell is **autotrophic** (self-feeding). That is, thanks to the incorporation of the photosynthetic bacteria, the cell can make its own food, using the energy of sunlight. This is a major evolutionary step because natural selection would obviously favor any life forms that could make their own food. Before that, cells probably had to use the simple anaerobic processes for fermentation as they broke down certain kinds of molecules found in the surrounding environment, extracting their energy. However, with the autotrophs who could make their own food, evolution took off in a powerful new direction.

There are two kinds of autotrophs: **phototrophs**, which use the energy of light, and **chemotrophs**, which use the

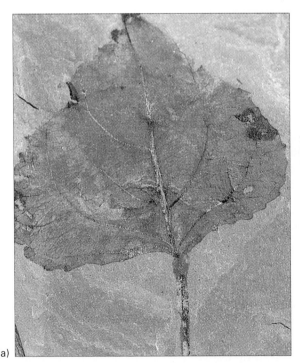

(a)

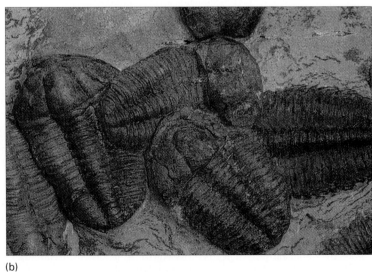

(b)

Figure 20–11 A sampling of past life in fossils. (a) The imprint of a leaf. (b) Fossil trilobites. Trilobites were marine arthropods that lived from 550 to 250 million years ago. (c) An insect in amber. *(a, Carlyn Iverson; b, © John Cancalosi/OKAPIA, 1991/Photo Researchers.)*

energy in chemicals. Both use carbon dioxide as a source of carbon and simple inorganic molecules as a source of hydrogen to add to the carbon. The most primitive living phototrophs get hydrogen from hydrogen sulfide. Other phototrophs get their hydrogen from water.

The problem with using water as a source of hydrogen for those early cells was that it released oxygen, a corrosive gas that tended to destroy whatever it came in contact with. So, as the early cells flourished, they created a sea of the poisonous gas. Many life forms undoubtedly fell before it. Others were forced to retreat to seabeds and other places the oxygen couldn't reach. In time, though, selective pressures favored cells that not only were immune to oxygen but that also came to use it in their metabolic processes.

By now, the Earth was alive with these autotrophic life forms. In their bodies they stored an enormous supply of nutrients. In time, another type of life evolved. These were the **heterotrophs** ("other feeders"), organisms that, instead of making their own food, captured and ingested the autotrophs, availing themselves of the autotrophs' stored food. Then, of course, yet other heterotrophs would have appeared that specialized in eating heterotrophs. With the advent of these groups, the trophic (feeding) cycles on Earth would have become much more diverse, setting the stage for the evolution of the many nutritive pathways we see today.

The Fossil Record

The Earth is littered with silent monuments to life that passed before (Figure 20–11). We find, for example, exquisitely veined leaves imprinted in ancient mud and tiny insects preserved in resin that dripped as sticky sap from some ancient tree. Also, we find mineralized bones and teeth, hardened remains that tell us much about the ancestry of today's vertebrates. These preserved remnants and impressions of past organisms are **fossils**. Most fossils occur in sedimentary rocks—such as limestone, sandstone, shale, and chalk—that form when accumulated minerals and organic particles are cemented together in quiet water.

Fossilization is the process by which fossils form (Figure 20–12). In a typical case, an organism dies and settles to the bottom of a body of water. If not destroyed by scavengers or decay, the organism is buried under accumulating layers of sediment. Whereas hard parts such as bones, teeth, and shells may be preserved if they become impregnated with minerals from surrounding water and sediments, soft parts usually decay or are destroyed or carried away. As new layers of sediment are added, the older (lower) layers solidify under the pressure generated by overlying sediments. Eventually, when the sediments are uplifted and the water disappears, wind may erode the surface of the rock formation and expose the fossil.

How do fossils provide evidence of evolution? Fossils of extinct organisms are interesting to the evolutionist because they show similarities to and differences from species living today. Similarities to other fossil and modern species can be

(a)

(b)

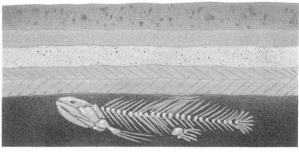

(c)

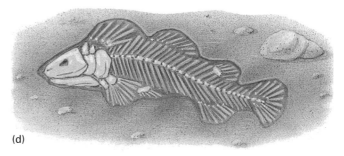

(d)

Figure 20–12 A typical sequence for fossilization. (a and b) An organism dies and is buried under water. (c) Providing the remains are not destroyed by predators, scavengers, or microorganisms, layers of sediment accumulate on top of the organism, whose hard parts become impregnated with minerals. (d) Eventually, uplift and erosion may expose the fossil.

used to assess the degree of evolutionary relationships. Often fossils reveal combinations of features not seen in any living forms that help us understand how major new adaptations arose. Sometimes we are lucky enough to find transitional forms that closely link ancient organisms to modern species. Consider a recently discovered fossil whale that had hind limbs and an aquatic lifestyle. This transitional form links older terrestrial forms with hind limbs to modern whales that are aquatic and lack hind limbs. Also, because fossils found in deeper layers of rock are typically older than those found in more superficial layers, we can obtain relative ages for fossils (that is, this organism is older than that organism) and study the chronological appearance of lineages. For example, the first fossil vertebrates to appear in deep (old) layers of rock were fishes, and then as we move closer to the surface we begin to recover fossil amphibians, then reptiles, then mammals, and finally birds. This chronological sequence for the appearance of the major groups of vertebrates has been supported by other lines of evidence, some of which we consider later in this section. Other large-scale evolutionary events revealed by the fossil record include changes in diversity over time, changes in faunas and floras as continents moved and climates changed, and changes in patterns of species origination and extinction.

Although the fossil record tells us much about past life, it has limitations. Our brief description of how fossils form suggests at least two drawbacks. First, fossils are rare. When most animals or plants die, their remains are eaten by predators or scavengers or are broken down by microorganisms. Even if a fossil should form, the chances are small that it will be exposed by erosion or other forces and not be destroyed by these forces before it can be discovered. Second, the fossil record represents a biased sampling of past life. For example, aquatic plants and animals have a much higher probability of burial under water than do terrestrial organisms, and thus preservation of aquatic organisms is more likely. Large animals with hard skeletons are far more likely to be preserved than are small animals without hard parts, and organisms from large, enduring populations are more likely to be represented in the fossil record than are those from small transient populations. Despite these limitations, fossils document that life on Earth has not always been the same as it is today, and the fact of these changes is potent evidence of evolution.

Biogeography

Biogeography is the study of the geographic distribution of organisms. Geographic distributions can reflect evolutionary history and relationship for the simple reason that related species are more likely to be found in the same geographic area than are unrelated species. In Australia, for example, we find many species of marsupials—mammals, such as opossums and kangaroos, with pouches—that today are minor components of the fauna in North and South America. Australia is an island, remote from other major continental land masses, and the presence of so many species of marsupials there suggests that they arose from distant ancestors whose descendants were not replaced by animals immigrating from other places. A careful comparison of the animals in a given place with those occurring elsewhere can

yield clues as to the relationship of the groups and if they have been separated, for how long.

New distributions of organisms occur by two basic mechanisms: **dispersal** and **vicariance**. Either the organisms move to new areas (dispersal) or the areas occupied by the organisms move or are subdivided (vicariance). Consider the case of the Australian marsupials—their evolutionary history involves both dispersal and vicariance. Fossil evidence suggests that marsupials evolved in North America and that some dispersed southward into South America, then to Antarctica, and later to Australia, to which it was attached at the time (Figure 20–13). Then, as Australia and other land masses slowly shifted to form the modern continental arrangement, the ancestors of today's marsupials were carried away from their place of origin to evolve in isolation in Australia. As the continental land masses separated and shifted, they carried populations of other living things with them as well. It should come as no surprise, then, that other types of organisms including trees, freshwater fishes, and even birds are quite different in Australia than in other parts of the world, reflecting its long history of isolation and continental movement.

Dispersal is also important in human evolution. For example, at the end of the last ice age, humans from Africa and southern Europe followed the retreating ice northward and colonized increasingly northern latitudes as the environment permitted.

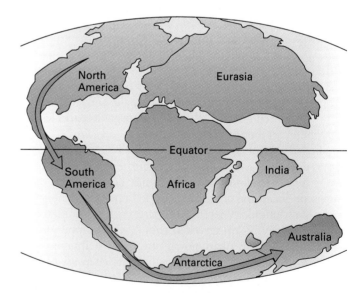

Figure 20–13 About 65 million years ago, marsupials appear to have left their place of origin in North America and dispersed southward into South America, Antarctica, and Australia, continents that were still close together. The subsequent drifting apart of the southern continents set Australia off on its own, allowing its marsupials to evolve in isolation. The story of marsupials and Australia thus involves both dispersal (movement of organisms) and vicariance (movement of continents).

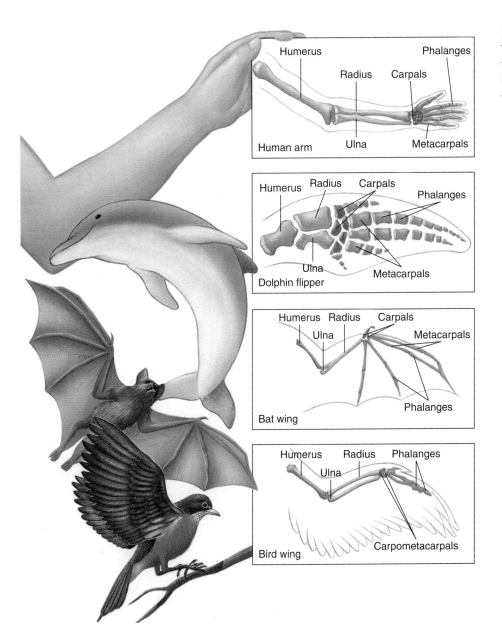

Figure 20–14 Homologous structures. The similarity of the bones of the forelimbs of birds, bats, dolphins, and humans suggests these organisms share a common ancestry.

Comparative Anatomy and Embryology

Comparative anatomy is exactly what it sounds like—the anatomies of different species are compared. This is one of the oldest disciplines in the natural sciences, and it retains great interest and vitality today. Comparative anatomists often look for traits shared in common, long considered a measure of relatedness. Put simply, those species with more shared traits are considered more likely to be related. For example, the vast majority of vertebrates—animals with backbones—alive today are united by the presence of jaws and paired fins or limbs. Features such as the jaws of vertebrates are often called shared derived features by comparative anatomists ("shared" because almost all vertebrates have them and "derived" because they are not present in the ancestors of vertebrates). As another example, a wide range of very different vertebrates share similar structures, such as the bones of forelimbs, suggesting that they share a common ancestry (Figure 20–14). Structures that have arisen from a common ancestry are called **homologous structures**.

Homologous structures usually arise from the same kind of embryonic tissue. Hence, comparative embryology, the comparative study of development, can be a useful evolutionary tool because common embryological origins can be considered as evidence of common descent. For example, 4-week-old human embryos closely resemble embryos of other vertebrates, including fish, complete with a tail and gill pouches (Figure 20–15). As development proceeds, the gill pouches of fish develop into gills whereas those of humans develop into other structures such as the eustachian tubes connecting the middle ear and throat. Nevertheless,

Fish   Salamander   Tortoise   Chicken   Pig   Cow   Rabbit   Human

Development

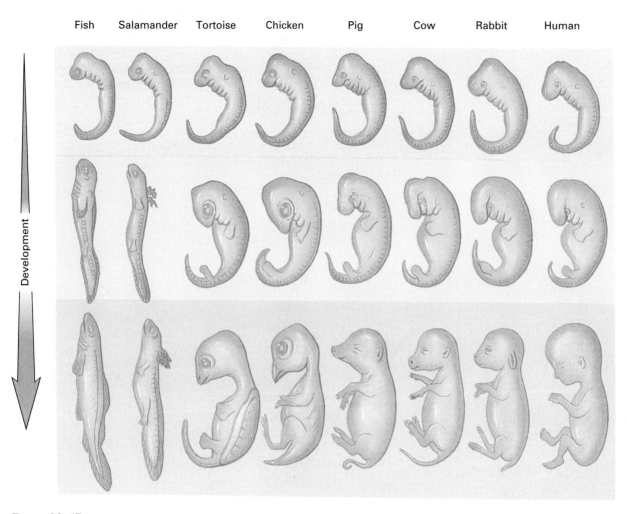

Figure 20–15   Resemblance early in development indicates common descent. A series of embryos of vertebrates shown at three comparable stages of development. Note that humans, like the other vertebrates shown, have gill pouches and a tail early in development. The embryos in the different groups have been drawn to the same approximate size to permit comparison.   *(Adapted from Romanes, G.T., 1910,* Darwin, and After Darwin. *Open Court, Chicago.)*

the fact that human and fish embryos, and all vertebrate embryos for that matter, look very similar at early stages of development indicates a common descent from an ancient ancestor.

Comparative Molecular Biology

Just as visible traits, such as forelimbs, can be compared, so can the molecules that are the basic building blocks of life. We have already mentioned that a common genetic code is potent evidence that all life is related. What can differences tell us? For example, scientists can compare the sequences of amino acids in proteins or the nucleotide sequences in DNA (Table 20–1). Such comparisons of DNA sequences tell us that only about 2.5% of our nucleotide sequences differ from those of chimpanzees. Expressed another way, we share about 98% of our genetic material with chimpanzees,

indicating that humans and chimpanzees diverged fairly recently from our most recent common ancestor. As with anatomical characters, molecular characters can be used to develop phylogenetic trees, and there is hope that the synthesis of molecular characters with traditional anatomical characters will provide new insights into the evolution of life.

The study of comparative molecular biology also suggests the existence of a **molecular clock**. The molecular clock hypothesis is based on the notion that single nucleotide changes in DNA, called point mutations, and the amino acid changes in proteins that can be produced by some point mutations occur with steady, clock-like regularity caused by background radiation and errors inherent in DNA copying. If this is true, then we can compare molecular sequences to estimate the ages of separation between

Table 20-1 Differences in DNA Sequences Indicating Phylogenetic Distances Between Pairs of Primate Species[a]

Species Pairs	Percentage Differences in Nucleotide Sequences
Human-chimpanzee	2.5
Human-gibbon	5.1
Human-Old World monkey	9.0
Human-New World monkey	15.8
Human-lemur (prosimian)	42.0

[a]Data from Stebbins, G. L. 1982. *Darwin to DNA, Molecules to Humanity.* W. H. Freeman, San Francisco, CA.

species: the more differences in sequence, the more time that has elapsed since the common ancestor. A comparison of such sequences tells us that humans and chimpanzees, for example, diverged from a common ancestor about 6 to 8 million years ago.

There is debate about the reliability of molecular clocks. Such debate stems from evidence that not only do different molecules have different mutation rates but different species may also have different rates. For example, some evidence suggests that humans have inherently slower mutation rates than do other mammals. Apparently, most mutations occur when DNA is being copied to make sperm that will then enter into fertilization, but sperm production is less frequent in longer-lived species, like humans. Finally, it is worth noting, that for the molecular clock to be useful, we need accurately dated fossil taxa and a good phylogeny.

We have now learned something about microevolution, macroevolution, and the evidence for evolution. Now let's look at our own past and see how humans have evolved.

Our Heritage

Where do we begin to discuss human evolution? With some ancestral thing lying in the mud at the bottom of an ancient seabed? The first lungfish that rose to the water's surface to gulp air? The first amphibians to crawl ashore? A dog-like reptile that gave rise to the earliest mammals? We can't just begin with humans because that would skip a host of prehistoric ancestors. So, let's begin with the primates, an order of mammals that includes monkeys, apes, and humans.

Critical Thinking

As we look through the stages of our ancestry, we will be reviewing the species that led to us. At some point the line *became* us. Before getting into the details of that ancestry, what qualities would you now say are necessary for a species to qualify as human?

Primate Evolution

The first primates probably arose about 65 million years ago from an insect-eating tree-living mammal similar to today's tree shrews (Figure 20–16). Indeed, primates share many characteristics thought to reflect an arboreal lifestyle specialized for the visual hunting and manual capture of insects. For example, characteristics such as flexible shoulder joints, retention of five functional digits on the front limbs and hind limbs, digits with exceptional mobility and sensitive pads on their ends, and toes and thumbs usually opposable to other digits would help in the pursuit and capture of insects along branches. A complex visual system and a large brain relative to body size would make possible the well-developed depth perception, hand-eye coordination, and neuromuscular control needed by arboreal insectivores. Also, most primates give birth to a single infant at a time and provide extensive parental care, perhaps reflecting the difficulty of carrying and rearing infants in trees, infants who must eventually learn to hunt insects. Humans are the most terrestrial of primates, although we retain many of the basic characteristics that evolved away from the ground—up in the trees.

Figure 20–16 A tree shrew. These modern animals resemble the arboreal, insect-eating mammals from which primates evolved about 65 million years ago. *(© Warren & Genny Garst/Tom Stack & Associates)*

Figure 20–17 Prosimians. Modern prosimians such as lemurs retain ancestral primate features. *(© Frans Lanting/Minden Pictures)*

There are two groups, or suborders, of modern primates—the prosimians and the anthropoids. The **prosimians** ("premonkeys") include the lemurs, lorises, pottos, and tarsiers (Figure 20–17). These modern species have been grouped together because they retain ancestral primate features. The **anthropoids** include monkeys, apes, and humans. There are two groups of monkeys. All New World monkeys are arboreal, have nostrils that open to the side, and have grasping (prehensile) tails. As the name suggests,

these animals occur in South and Central America. Examples of New World monkeys are spider monkeys, capuchins, and squirrel monkeys. The Old World monkeys lack prehensile tails, often having short tails or none at all, and have nostrils that open downward. Some Old World monkeys are arboreal and some are terrestrial. Examples of Old World monkeys are baboons and rhesus monkeys. New World and Old World monkeys are shown in Figure 20–18.

Also within the anthropoid suborder are four genera of modern apes: *Hylobates* (gibbons), *Pongo* (orangutans), *Gorilla* (gorillas), and *Pan* (chimpanzees). An example from each genus is shown in Figure 20–19. The final modern member of the anthropoid suborder is us, *Homo sapiens*. Apes and humans are termed **hominoids**. The term **hominid** is used to refer to members of two genera: *Australopithecus* and *Homo* (see the section Hominid Species that follows).

The earliest fossil primates of 65 million years ago resembled modern tree shrews. In time the descendants of these animals spread over the Old and New Worlds and differentiated into many forms as they encountered new environments—some ate insects whereas others ate plants, some were active during the day, and others were active at night. About 50 million years ago, anthropoids evolved from these early primates, although their ancestry is the subject of much debate. Early anthropoids invaded Africa. Later, some crossed the South Atlantic Ocean to South America and it is here that we find the first fossils of New World monkeys. The division between the lineages of Old World monkeys and hominoids (apes and humans) occurred about 40 million years ago. Early hominoids spread widely over the Old World and diversified.

The common ancestor of humans and chimpanzees was one of many anthropoids living in Africa. According to evidence from molecular biology, the lines leading to modern

(a) (b)

Figure 20–18 New World and Old World monkeys. (a) New World monkeys, such as the spider monkey shown here, are arboreal, have prehensile tails, and have nostrils that open to the side. (b) Old World monkeys, such as baboons, lack prehensile tails and have nostrils that open downward.
(a, Frans Lanting/Minden Pictures; b, © William E. Bemis)

humans and chimpanzees diverged from this ancestor only about 6 to 8 million years ago. Molecular data further indicate that beyond chimpanzees, gorillas are our next closest living relatives, then orangutans, and finally gibbons (Figure 20–20). Interestingly, these relationships among modern apes and humans were first suggested in 1863 by T. H. Huxley when he published his evidence based on comparative anatomical study of the four groups.

Let's now consider the origin and evolution of humans. We begin by describing the problems associated with studying human evolution and dispel some common misconceptions about our evolutionary history. Then we discuss some of the hominid species.

Human Origins and Controversy

Consideration of human ancestry can almost always arouse argument. Some arguments hinge on scientific disagreements, others on philosophical or religious perspectives. Many believe that the scientific questions are more interesting because they rely less on personal opinion. So, let's consider the scientific questions at this point.

Paleoanthropology is the scientific study of human origins and evolution. Many debates about human evolution stem from past practices of paleoanthropologists. For exam-
ple, in the past paleoanthropologists tended to give each new fossil discovery a new name and to consider it an example of a separate line. More recent studies indicate that many of these "finds" are probably of the same lineage. The confusion was compounded when some paleoanthropologists extrapolated from sketchy information, such as a partial skull or a few teeth. Another problem stems from variations in a population. Two specimens that differ slightly in some traits may be interpreted as different lines even when not warranted. Then, there is the problem of the paucity of fossils. There just haven't been enough hominid remains found to reconstruct a solid, unequivocal history of our species. Finally, many arguments undoubtedly arise because of certain misconceptions, as we see in the next section.

Misconceptions About Our Heritage

Not long after Charles Darwin published *On the Origin of Species* in 1859, arguments about human evolution began to fly. Darwin, in ill-health and not given to public controversy, retired from the fray, leaving his defense in the hands of such powerful intellectual forces as T. H. Huxley, one of the greatest zoologists of his time. In one heavily publicized debate with Archbishop H. Wilburforce, the cleric sought to enlist ridicule by asking Huxley whether he was related to

Figure 20–19 Apes. Modern apes are organized into four genera: (a) *Hylobates* (gibbons), (b) *Pongo* (orangutans), (c) *Gorilla* (gorillas), and (d) *Pan* (chimpanzees). A representative of each genus is shown. *(a, Visuals Unlimited/Joe McDonald; b and c, David J. Cross/Peter Arnold, Inc.; d, Frans Lanting/Minden Pictures)*

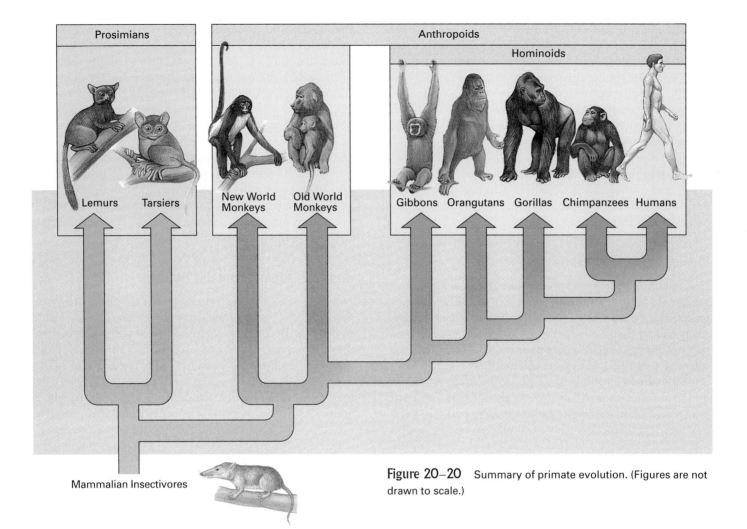

Prosimians | Anthropoids

Hominoids

Lemurs Tarsiers

New World Monkeys Old World Monkeys

Gibbons Orangutans Gorillas Chimpanzees Humans

Mammalian Insectivores

Figure 20–20 Summary of primate evolution. (Figures are not drawn to scale.)

the apes on his mother's or his father's side. The audience chuckled, but Huxley is said to have muttered, "The Lord hath delivered him into my hands." He then went on to say that he would rather be related to the apes than to a man who would not use his God-given ability to reason. Huxley carried the day.

Unfortunately, the idea that we have descended from chimpanzees or any of the other modern apes is a convenient fiction that has become embedded in the minds of many people. People visit the zoo, look at a chimpanzee and think, "This can't be true." (The chimp might hold the same opinion.) The problem is that no one ever said that to begin with. Humans and chimpanzees represent separate branches of the anthropoid tree, branches that diverged 6 to 8 million years ago. Thus, the common ancestor of humans and chimpanzees was different from any modern species of ape.

Another misconception is that modern humans evolved in an orderly, stepwise progression. We often see such a stepwise progression illustrated sometimes humorously, and its appeal lies in its simplicity. However, as is so often the case, simplicity is deceiving. The pathway to modern humans has been fraught with "hopeful experiments" leading to dead end after dead end and producing what looks like a family "bush" rather than an orderly progression to modern humans (see the next section).

Another misconception is that, as humans evolved, the various bones and organ systems evolved together at the same rate. This simply is not the case. There is no reason to believe that the human brain evolved at the same rate as, say, the appendix or the foot. Instead, various traits evolved at their own rates, a phenomenon known as **mosaic evolution**.

Hominid Species

Several evolutionary trends are apparent within hominids. **Bipedalism** (walking on two feet) is a characteristic that evolved early on and probably set the stage for the evolution of other characteristics, such as increases in brain size. Once the hands were freed from the requirements of locomotion, they could be used for tasks such as making tools. Other changes associated with upright posture include the S-shaped curvature of the vertebral column ("the lumbar curve"), modifications to the bones and muscles of the pelvis, legs, and feet, and positioning of the skull on top of the vertebral column (in the ancestral condition, the articulations between the skull and vertebral column were at the rear of the brain case). The faces of hominids also changed. For example, the forehead changed from sloping to vertical, the brow ridges and crests on the skull for muscle attachments were reduced, the jaws became shorter, and the nose more prominent.

The first hominids are placed in the genus *Australopithecus*, meaning "southern ape." The six species of *Australopithecus* that we discuss here are sometimes collectively termed australopithecines. The oldest known hominid remains (17 pieces, mostly teeth and a few bones) are from Ethiopia. This species, termed *Australopithecus ramidus*, has been dated at about 4.4 million-years-old, fairly close in time to the 6 to 8 million-year-old split between chimpanzees and humans estimated from molecular biology. Although *A. ramidus* is the most ape-like of the known hominid ancestors, it appears to display several characteristics of the human lineage, including bipedalism, reduced sexual dimorphism (difference between the sexes) of the canine teeth, and a human-like shoulder joint and elbow.

The most spectacular australopithecine fossil found to date is that of a young adult female assigned the name *Australopithecus afarensis* (because she was found in the Afar region of Ethiopia). When paleoanthropologists arranged her more than 60 pieces of bone, they estimated that she was only about 1 meter tall and perhaps weighed 30 kilograms (about 66 pounds) when she died (Figure 20–21). Her bones were dated to be 3.2 million-years-old and she was named "Lucy" because the Beatles' song, "Lucy in the Sky with Diamonds" was playing the night the paleoanthropologists celebrated their find. As more remains were found, it became apparent that the males of Lucy's species were somewhat taller (about 1.5 meters) and heavier (about 45 kilograms or 99 pounds) and that the brains of *A. afarensis* were similar in size to those of modern chimpanzees or gorillas (380–450 cc). Other remains of *A. afarensis* were found in Tanzania and dated at between 3.6 and 3.8 million-years-old. Although many aspects of the anatomy of *A. afarensis* suggest adaptations for tree-living, the remains also indicate bipedalism, leading some anthropologists to suggest that Lucy and other members of her species slept in trees and harvested fruits as they walked through forests.

About 3 million years ago, when *A. afarensis* had been in existence for nearly 1 million years, four new hominid species appeared in the fossil record. *Australopithecus africanus* is a species believed to have been a hunting and gathering omnivore. Three more robust hominids (*A. robustus, A. aethiopicus,* and *A. boisei*) are thought to have been savannah-dwelling vegetarians. *Australopithecus africanus* appeared delicate compared with the three robust species that had massive skulls, heavy facial bones, pronounced brows,

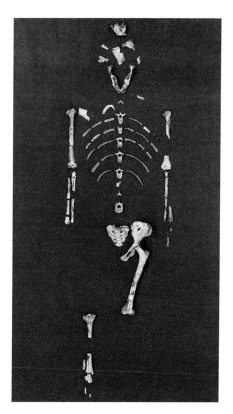

Figure 20–21 "Lucy." Lucy is a remarkably complete skeleton of a young female *Australopithecus afarensis,* a hominid in existence over 3 million years ago. *(Institute for Human Origins)*

and huge teeth. The large, flattened jaw teeth suggest that the robust forms were primarily plant eaters—plant material must be crushed to destroy the plant's protective cell walls and plants often contain abrasive materials that wear down the teeth. It is unclear whether Lucy's species, *A. afarensis,* gave rise to these other australopithecines or simply lived at the same time as they did.

Homo habilis ("skillful man"), the first member of our genus, appeared in the fossil record about 2.5 million years ago. These remains are highly variable, causing some researchers to question the taxonomic status of *H. habilis.* This hominid differed from *A. afarensis* not so much in body size or morphology but in brain size; the cranial capacities of *H. habilis* have been estimated at between 500 and 650 cc. Was *H. habilis* the first hominid to use stone tools? Simple stone tools, dating from 2.5 to 2.7 million years ago have been found in Africa, although whether they were used by *H. habilis* or *A. robustus* is unclear.

A new hominid, *Homo erectus* ("upright man"), appeared in the fossil record about 2 million years ago. Whereas the bones of the australopithecines and *H. habilis* are found only in Africa, *H. erectus* was a wanderer: Bones have been found in Africa, India, China, Indonesia, and Europe. It is believed that *H. erectus* originated in East Africa and coexisted there

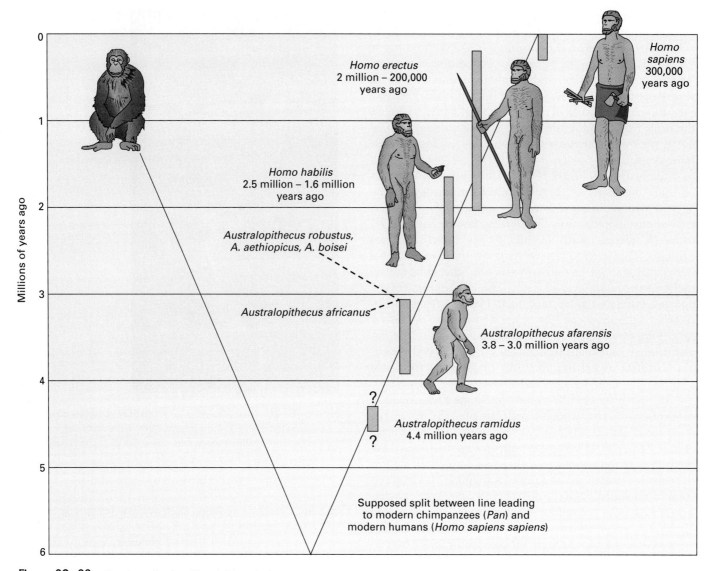

Figure 20–22 One hypothesis of hominid evolution.

Labels within figure:

Homo erectus
2 million – 200,000
years ago

Homo habilis
2.5 million – 1.6 million
years ago

Australopithecus robustus,
A. aethiopicus, A. boisei

Australopithecus africanus

Homo sapiens
300,000
years ago

Australopithecus afarensis
3.8 – 3.0 million years ago

Australopithecus ramidus
4.4 million years ago

Supposed split between line leading
to modern chimpanzees (*Pan*) and
modern humans (*Homo sapiens sapiens*)

Millions of years ago

for several thousand years with some of the robust australopithecines. *Homo erectus* differed from earlier hominids in being larger (up to 1.85 meters tall and weighing at least 65 kilograms or 143 pounds), displaying reduced sexual dimorphism because of substantial increases in female size, and having a brain volume overlapping that of modern humans. *Homo erectus* used more sophisticated tools and weapons than had other hominids, and there is evidence that this species used fire. Most researchers believe that the species *H. erectus* is the closest species to ourselves. *Homo erectus* disappeared between 200,000 and 300,000 years ago.

Now we come to our own species, *Homo sapiens* ("thinking man"), which evolved around 200,000 to 300,000 years ago. *Homo sapiens* differs from *H. erectus* in having a slightly larger brain and more robust skull and in possessing larger teeth and a less projecting jaw. However, *H. erectus* and

H. sapiens overlap in all of these characteristics. In other words, we differ from *H. erectus* only in a matter of measurements, not in the presence or absence of specific structures.

Did modern humans evolve several times in different regions? This idea, known as the **multiregional model**, suggests that modern humans evolved independently in several different areas from distinctive local populations of *H. erectus*. The alternate idea, known as the **monogenesis model** ("one origin"), suggests a single origin for all *H. sapiens*. According to the monogenesis model, modern humans evolved from *H. erectus* in Africa and only later migrated to other continents where they replaced the diverse descendants of *H. erectus*. Although debate continues, the monogenesis model has wide support.

Let's assume that *H. sapiens* arose from *H. erectus* in Africa about 300,000 years ago. These early forms of

H. sapiens share many features with their ancestor and are known as "archaic *H. sapiens*." About 200,000 years ago a population of *H. sapiens* developed a set of distinct features, features apparently adapted for life in cold climates. These humans with cold-adapted features were the so-called Neanderthals, discovered in Germany's Neander Valley. The Neanderthals were a widely dispersed group that ranged across Europe and Asia. Some lived in caves ("the cave men"), but others made temporary camps. Interestingly, Neanderthals had a larger brain case than we do although this may not have been correlated with intelligence because those brains had to service more massive bodies. They had larger bones, suggesting heavier musculature, and rather short legs. They also had thick brow ridges, large noses, broad faces, and well-developed incisors and canines. Neanderthals were so distinctive in their characteristics that most anthropologists assign them subspecies status, calling them *H. sapiens neanderthalensis*. Modern humans are known as *H. sapiens sapiens*.

Neanderthals inhabited Europe from about 200,000 years ago until they vanished about 30,000 years ago for still mysterious reasons. Some scientists suggest they were "outcompeted" or simply killed outright by a form of *H. sapiens sapiens* called Cro-Magnons. Others suggest that interbreeding between *H. sapiens sapiens* and *H. sapiens neanderthalensis* might have resulted in the loss of the Neanderthal phenotype. Although the extent and type of interactions between Neanderthals and Cro-Magnons is still debated, it is clear that the Cro-Magnons had superior tools and weapons, including bows and arrows, knives, and stone-tipped spears. It is generally believed they lived in moderate-sized groups of 50 or more. A possible scenario for human evolution is shown in Figure 20–22.

SUMMARY

1. Microevolution occurs within populations (groups of individuals of the same species that live in a particular area) and involves change in the frequency of certain alleles relative to others in the gene pool. A gene pool is a collection of all of the alleles of all of the genes of all of the individuals in a population. Macroevolution is large-scale evolution involving change among groups of species. Examples of macroevolutionary phenomena include changes in diversity or patterns of distribution of species brought about by long-term changes in climate or position of the continents.

2. Mutations, rare changes in the DNA of genes, are one source of genetic variation within populations. Other sources of variation include crossing over and independent assortment during meiosis. Crossing over is the reciprocal exchange of genetic material between nonsister chromatids. Independent assortment is the random segregation into gametes of homologous chromosomes. Finally, fertilization, the union of an egg and a sperm to produce a zygote with new combinations of alleles, also contributes to variation.

3. Causes of microevolution include genetic drift (random changes in allele frequencies due to chance within a population), gene flow (loss or gain of alleles from a population because of movement of individuals who successfully interbreed upon settling), natural selection (differential survival and reproduction of individuals caused by their particular inherited characteristics), and mutation.

4. Genetic drift is negligible in large populations. Populations small enough for genetic drift to occur are often associated with two circumstances—the bottleneck effect and the founder effect. The bottleneck effect occurs when a natural disaster kills many members of a population, so that by chance alone, the genetic makeup of survivors is different from that of the original population. The founder effect occurs when a few individuals leave a population and settle in a new location. Simply by chance, the genetic makeup of colonizing individuals may not be representative of the parental population.

5. A species is a population or group of populations whose members are capable of interbreeding to produce fertile offspring under natural conditions. Sometimes, populations of a species become geographically isolated and accumulate distinct sets of allele frequencies and mutations. Such populations may become so different that successful interbreeding is no longer possible, and we say that speciation, the formation of new species, has occurred. Even a small amount of gene flow makes speciation unlikely.

6. Charles Darwin argued that all life forms—past and present—are connected; in short, species living today are descendants of ancestral species. He also proposed that evolution could occur by natural selection. According to Darwin, evolutionary change occurs as the traits of successful individuals (those that survive and reproduce) become more common in a population whereas traits of less successful individuals become less common. An individual's evolutionary success can be measured by fitness, the average number of offspring left. It follows, then, that individuals with greater fitness have more of their genes represented in future generations.

7. Natural selection constantly adjusts any population to its prevailing environment, a process called adaptation. Such adjustments involve the removal of unsuccessful individuals and their alleles. One mechanism by which genetic variation within populations is maintained in the face of natural selection is balanced polymorphism. Balanced polymorphism is the maintenance by natural selection of two or more alleles for a trait from one generation to the next. Heterozygote advantage is an example of balanced polymorphism and occurs when the heterozygous condition is favored over either homozygous condition.

8. Systematic biology deals with the naming, classification, and evolutionary relationships of organisms. Organisms are named according to a scheme developed by Carl Linnaeus more than 200 years ago. In this scheme, each organism is given a two-part name consisting of the genus name followed by the species name. In addition, organisms may be classified using a hierarchy of increasingly more general categories. Thus, similar species are grouped in the same genus, similar genera (the plural of genus) in the same family, similar families in the same

order, similar orders in the same class, similar classes in the same phylum, and similar phyla (the plural of phylum) in the same kingdom.

9. A phylogenetic tree is a schematic diagram depicting genealogical relationships among species or higher taxa. Such trees represent hypotheses and are constructed from data matrices in which the taxa can be arrayed in columns and the characters (features) to be scored for their presence or absence in each taxon are arrayed in rows.

10. Evolutionary change may occur slowly and gradually (a process known as gradualism) or it may occur rapidly after long periods of no change (a process known as punctuated equilibrium).

11. Divergent evolution occurs when populations exposed to different environments become increasingly different. Convergent evolution refers to the independent development of similarity in species living in similar environments.

12. Examples of the basic data that document evolution include evidence on the origin and early evolution of life, the fossil record, and evidence from the fields of biogeography, comparative anatomy, embryology, and molecular biology.

13. The Earth is estimated to be 4.5 billion years old. Physical and chemical changes in the Earth's crust suggest life originated about 3.5 billion years ago. A. I. Oparin and J. B. S. Haldane suggested that conditions of the early Earth, particularly the low oxygen atmosphere, favored the synthesis of organic compounds from inorganic molecules. They further suggested that lightning and intense ultraviolet radiation from the sun provided the energy for such reactions. Laboratory re-creations of the conditions of the primitive Earth have yielded organic compounds, including amino acids. Such re-creations lend support to the Oparin-Haldane hypothesis.

14. Over time, it is believed, organic molecules accumulated in the early oceans and aggregated into droplets that displayed some of the properties of life. Then perhaps genetic material was incorporated into compartments within the droplets, permitting the transfer of information from one generation to the next. This simple set of genetic codes applies to all life, past and present, and thus suggests a single common origin for life. According to the endosymbiont hypothesis, more complex cells formed as smaller organisms were incorporated into the cells, forming organelles such as mitochondria, chloroplasts, and cilia. The early cells were autotrophic, capable of making their own food. Later, heterotrophs evolved, organisms that captured and ingested the autotrophs rather than making their own food. Eventually other heterotrophs evolved that specialized in eating heterotrophs. And so began the evolution of the many feeding relationships we see today.

15. Fossils are the preserved remnants and impressions of past organisms. The fossil record documents that life on Earth has not always been the same as it is today and thus provides evidence of evolution.

16. Biogeography is the study of the geographic distribution of organisms. New distributions of organisms occur either by dispersal (movement of organisms to new areas) or vicariance (when areas occupied by organisms move or are subdivided). In general, related species are more likely to be found in the same geographic area than are unrelated species. Thus, by comparing the animals in a given place with those occurring elsewhere, we may gain insights about the relationships of the groups and if they have been separated, for how long.

17. Comparative anatomy also provides evidence of evolution. Generally, species with more shared traits are considered more likely to be related. For example, the bones of the forelimbs are common to many different vertebrates, and this similarity in structure suggests the vertebrates share a common ancestry. Structures that have arisen from a common ancestry are called homologous structures and usually arise from the same embryonic tissue. Thus, common embryological origins can also be viewed as evidence of common descent.

18. Molecules that are the basic building blocks of life may also be compared for evidence of evolution. For example, scientists compare the sequences of amino acids in proteins or the nucleotide sequences in DNA of different species to gauge relatedness and time of divergence from a most recent common ancestor. The latter idea is based on the presumed existence of a molecular clock, the notion that single nucleotide changes in DNA and amino acid changes in proteins occur with steady, clock-like regularity.

19. Humans are primates. There are two suborders of modern primates—the prosimians (lemurs, lorises, pottos, and tarsiers) and the anthropoids (monkeys, apes, and humans). Monkeys include New World and Old World groups, and apes include gibbons, orangutans, gorillas, and chimpanzees. Modern humans, *Homo sapiens*, are the final member of the anthropoid suborder. The term hominoid refers to apes and humans and the term hominid refers to members of the genera *Australopithecus* and *Homo*.

20. Primates evolved from an insect-eating arboreal mammal about 65 million years ago. These early primates are grouped with modern lemurs, lorises, pottos, and tarsiers as prosimians. Anthropoids evolved from the early prosimians and invaded Africa. Some anthropoids crossed from Africa to South America and hence the New World monkeys. Old World monkeys and hominoids diverged about 40 million years ago. In Africa, humans and chimps diverged from a common ancestor about 6 to 8 million years ago. Some data suggest that chimps are our closest living relatives, followed by gorillas, orangutans, and finally gibbons.

21. Paleoanthropology is the study of human origins and evolution. Intense argument about human evolution stems from intraspecific variation, the paucity of fossils, and the tendency of some anthropologists to give each fossil a new name and to consider it a new line.

22. The following are three common misconceptions about human evolution: (1) Humans have descended from chimpanzees (when, in fact, humans and chimps represent separate branches that diverged from a common ancestor); (2) human evolution has occurred in an orderly progression from ancient to modern forms (in actuality, there have been several times when two or more species of hominids lived at the same time, and some of these lines were evolutionary dead ends); and (3) traits of humans evolved at the same rate. In the latter case, evidence indicates that traits of humans evolved at different rates, a phenomenon known as mosaic evolution.

23. The early hominids were placed in the genus *Australopithecus*. The oldest known hominid remains are of *A. ramidus*, about 4.4 million-years-old. Excellent specimens have been found of

A. afarensis ("Lucy," for example), a species about 3.8 million-years-old. About 3 million years ago, several new hominid species appeared (*A. africanus, A. robustus, A. aethiopicus,* and *A. boisei*). It is unclear whether *A. afarensis* gave rise to these new forms or simply coexisted with them.

24. About 2.5 million years ago, *Homo habilis* remains appear in the fossil record in Africa. The taxonomic status of these remains is questionable. About 2 million years ago, *H. erectus* remains appear in the fossil record. Fossils of *H. erectus* are not restricted to Africa. About 200,000 years ago, *H. erectus* disappeared and *H. sapiens* emerged. The multiregional model suggests that *H. sapiens* evolved independently in different regions from distinctive populations of *H. erectus*. The monogenesis model suggests that modern humans evolved from *H. erectus* in Africa and then migrated to other regions where they replaced the diverse descendants of *H. erectus*. Neanderthals are generally considered to have been a cold-adapted subspecies of *H. sapiens*.

REVIEW QUESTIONS

1. Distinguish between microevolution and macroevolution.
2. What are four sources of variation within populations?
3. How does genetic drift lead to microevolution?
4. Define speciation and relate it to gene flow.
5. Define natural selection. How can variation within populations be maintained in the face of natural selection?
6. Describe the binomial system by which organisms are named and the hierarchical system by which they are classified.
7. What is a phylogenetic tree?
8. Compare gradualism and punctuated equilibrium.
9. Define and provide examples of divergent and convergent evolution.
10. How might life have evolved from inorganic molecules to complex cells?
11. What is a fossil? Describe the process of fossilization and relate it to limitations of the fossil record.
12. How do new distributions of organisms arise?
13. Describe how comparative anatomy and embryology provide evidence of evolution.
14. What is a molecular clock?
15. Describe the two suborders of modern primates.
16. Why do discussions of human evolution inspire controversy?
17. What are three popular misconceptions about human evolution?
18. Describe the major species of hominids and trends in hominid evolution.

Chapter | **21**

(© Bob Winsett Photography)

Ecology, the Environment, and Us

We are but one of a host of species that share this small planet. Not only do we have many of the same opportunities as do the other species but we also face many of the same threats. If we are ever to understand the environment and our place in it, we must have some knowledge of its physical characteristics and of the other species that share its components as well. Here, then, let's take a look at this place we call home, keeping in mind that a lot of other creatures call it home. Let's pay particular attention to just how we and they, together, can influence the well-being of each other and see what lessons nature might have for our kind.

We can make two important observations as we view Earth from a distance (Figure 21–1). First, we see the Earth as isolated. With respect to materials, this observation is indeed accurate. Aside from an occasional meteorite, perhaps a bit of debris from space, and, according to a controversial hypothesis, sprinklings of cosmic rain from celestial "snowballs" that disintegrate high above the earth's surface, there is no source of new materials. In essence, "What you see is what you get." Many of the materials that came together to form our planet some 4.5 billion years ago have cycled from organism to organism and between the living and nonliving components of our world. A carbon atom in your body may once have been a part of a dinosaur or of Aristotle. In this chapter, then, we will consider some of the ways that materials are cycled. Second, the light reflected from Earth's surface reminds us that Earth receives one very important contribution—energy in the form of sunlight. As we will see, that energy is captured by green plants and transferred from organism to organism, sustaining nearly all life on Earth.

Keeping these observations in mind, we will begin with an overview of the part of the Earth where life exists, and then break that down into smaller, more manageable units. We will look not only at how the planet's physical traits influence life but how living things influence each other as well. Then we will consider how energy passes from one level to another in living systems. The great lesson we should learn from all this is the interdependence of living things and, in turn, their dependence on the nonliving world.

Our focus here, then, is ecology, which comes from the Greek words *oikos*, meaning home, and *logos*, mean-

Figure 21–1 A view of Earth from space shows us that Earth is isolated. Since there is no regular input of materials, important elements must be cycled from organism to organism and between the living and nonliving components of Earth. The only input to the system is energy from the sun, which sustains nearly all life on Earth. *(NASA)*

ing to study. Thus, ecology is the study of our home, the Earthly environment, including both its living (*biotic*) components and nonliving (*abiotic*) components. More precisely, **ecology** is the study of the interactions among organisms and between them and the environment. The key word is *interaction*. Ecologists are the scientists who study these interactions.

Levels of Ecological Organization

The **biosphere** is that part of the Earth in which life is found and it encompasses all of the Earth's living organisms. In essence, it is where light, minerals, water, and gases interplay in a thin veil over the Earth to produce environments that permit life to exist. The biosphere extends only about 7 miles above sea level and the same distance below, to the deepest trenches of the sea. If the Earth were the size of a basketball, the biosphere would have the depth of about one coat of paint. In this thin layer over one small planet, then, we find all of the *known* life in the entire universe.

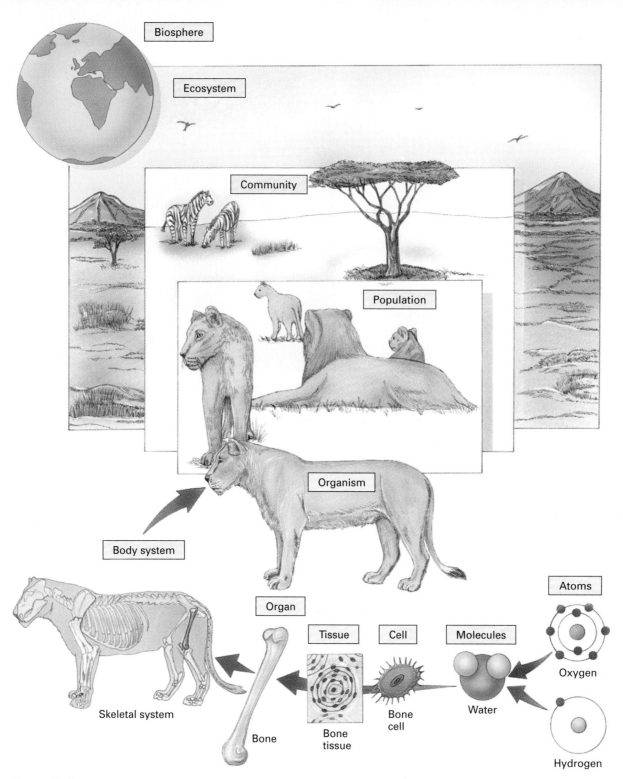

Figure 21–2 Levels of biological organization. All life exists within the biosphere, where the interplay of water, light, and gases permits its existence. The biosphere contains many ecosystems, which include all the organisms in a particular area along with their physical environment. Organisms of all species living in a particular area compose a community. A population is the members of one species that interact with one another in a particular area.

Because the biosphere is so encompassing, it is often divided into smaller, more easily managed units (Figure 21–2). **Ecosystems** are defined areas in which certain living organisms interact with their physical environment. The size of the defined area comprising an ecosystem varies with the interest of the ecologist. An ecosystem can be as large as the whole Earth, a particular forest, or even a single rotting log within a forest. Regardless of its size, an ecosystem is relatively self-contained; materials usually cycle within it.

A **community** is all living species in an ecosystem that can potentially interact. Thus, the organisms in a community must be in the same proximity. Communities can exist on land or in the water. For example, a community might be *all the species* that live and interact in an alpine meadow or in a coral reef. Whereas a community includes all living organisms in a geographic area, a **population** is all the individuals of the *same species* living in a distinct geographic area. A population might be the yellow-bellied marmots living in a meadow or the four-eyed butterfly fish living in a coral reef.

On a yet finer scale, we can consider an individual organism and describe it according to its niche (sometimes called *ecological niche*). The **niche** is defined as the organism's role in the ecosystem. The niche includes all physical, chemical, and biological factors that the organism requires to live, remain healthy, and reproduce. Such factors include the nature of the organism's food and how it obtains food, its predators, its specific needs for shelter, and its interactions with other species. The niche also includes the physical conditions, such as temperature, light, water, and oxygen, required for survival. The organism's **habitat**, that is, the place (physical area) where it lives, is also a part of its niche. It has been suggested that the habitat is an organism's address, whereas the niche is its profession.

Divisions of the Biosphere

The biosphere varies from place to place, and so, for convenience, it has been divided into units called biomes and aquatic realms. A **biome** is a relatively distinct terrestrial ecosystem that covers a large geographic area and is characterized by particular assemblages of plants and animals, climate, and soil, wherever it occurs on Earth. Tropical rain forests and deserts are examples of biomes.

Climate, particularly temperature and the availability of water, affects the distribution of organisms. Each species is adapted to a certain climate. Thus, we often find particular biomes in regions with similar climates. Because climate tends to vary somewhat predictably with latitude, we find that biomes form broad bands on the Earth's surface.

We should note that there is some disagreement among ecologists regarding how many biomes there are and what their characteristics are. Also, the picture can become confused because each biome generally changes gradually to the adjoining one with no clear lines of distinction. In Figure 21–3 we note several of the major biomes.

(a)

(c)

Figure 21–3 Selected biomes of the Earth. (a) The desert biome. Deserts receive less than 25 cm (10 in.) of rain each year, and this falls in very brief periods. Both plants and animals must be able to conserve water. Animals may tend to avoid the sun by foraging at night. (b) The taiga biome. The taiga is primarily composed of evergreen forests with highly variable rainfall, between 50 and 100 cm (20–40 in.). The needles help save water by providing little surface through which water can leave. (c) The tropical rain forest biome. Tropical rain forests may receive 200 to 1000 cm (80–400 in.) of rain each year. The largest is the Amazon rain forest. *(a, Stan Osolinski/Dembinsky Photo Associates; b, Charlie Ott/Photo Researchers, Inc.; c, Frans Lanting/Minden Pictures)*

Ecological Succession

Like everything else in this world, ecosystems change over long periods of time. The sequence of changes in the species making up a community is called **ecological succession**.

Primary succession occurs where no community previously existed. Such places may be on rocky outcroppings, where lava has covered everything, or where an island has appeared. When primary succession begins, no soil exists.

The first living things to invade such an area are called pioneer species. Among the most prominent of these are lichens, which are actually two species—a fungus and a photosynthetic organism, usually an alga. The fungus provides the attachment to the barren surface and retains water; the alga provides the food. The lichens secrete acid, which helps to break down the rock, beginning soil formation. In time, the lichens die and their remains mix with the rock particles, furthering soil formation in cracks and crevices. Wind-blown debris is snagged by the lichens, and water collects near their bases and is retained in the new soil layer. Insects come and some leave their corpses. Soil building continues as yet more visiting animals appear, enriching the new soil with their droppings (Figure 21–4a).

Soil building is a slow process, but once the "dirt" is in place, things happen quickly. Plants begin to appear (Figure 21–4b). Their roots force themselves into every crevice, and their leaves fall and decompose, as the pace accelerates. In some areas, trees eventually become the dominant plant form (Figure 21–4c). Each species that enters the area changes the environmental conditions slightly and changes the resources available for others, thus favoring some species and not others. The community that eventually forms and remains if no disturbances occur is called the **climax community**. A sand dune on the shores of Lake Michigan, for example, made it to a climax forest in only 1000 years.

When an existing community becomes cleared, either by natural means or by human activity, it undergoes a sequence in species composition known as **secondary succession**. It occurs in places with soil in place, usually on old deserted fields and farms that had been cleared and planted by

(a)

(b)

(c)

Figure 21–4 Primary succession is the sequence of changes in the species composition of a community that begins where no life previously existed. (a) This thin mat of lichen is helping to break down the bare rock, thus beginning soil formation. (b) After soil has begun to accumulate, other plant species appear that often include shrubs and dwarf trees. (c) Trees later become the dominant plant form. *(a, Courtesy of James D. Mauseth; b, Visuals Unlimited/Glenn N. Oliver; c, Wolfgang Kaehler)*

(a)   (b)

Figure 21–5 Secondary succession following the 1988 fires in Yellowstone National Park.
(a) Immediately after the fires, there are no visible signs of life. Dead trees stand as ghostly reminders
of the forest that had existed and gray ash covers the forest floor. (b) The following spring young
plants, such as these trout lilies, appear and begin the stages of secondary succession.   *(a, Ted and
Jean Reuter/Dembinsky Photo Associates; b, Stan Osolinski/Dembinsky Photo Associates)*

humans, then abandoned, or after fire, flood, wind, or grazing damage (Figure 21–5). The initial invaders, such as grasses, weeds, and shrubs, are gradually pushed out as other plants edge in from the surrounding community. In some areas, pines move in next, but these may eventually resign their dominance to hardwoods, such as oak. Thus, the area finally "heals" and the community that forms often resembles the one that existed before the disturbance (Figure 21–6). This recovery may be brief or extended. Grasslands may recover in 20 years, but wagon tracks left on the tundra over 100 years ago are still visible, scarred monuments to our rough treatment of this delicate environment.

Energy and Trophic Levels

Virtually all the energy that propels life on this small planet comes from the sun. Less than 0.1% of the sun's energy that reaches the Earth's surface worldwide is captured by living organisms. Nonetheless, the life that abounds on Earth owes its existence to that captured energy, which is shifted, shuffled, channeled, and scattered through various systems from one level to the next.

Food Chains and Food Webs

The flow of energy through the living world begins when light from the sun is absorbed by photosynthetic organisms,

such as plants, algae, and cyanobacteria. Photosynthesis essentially captures light energy and transforms it into the chemical energy of the sugar glucose, which is manufactured from carbon dioxide and water. The amount of light energy that is converted to chemical energy in the bonds of organic molecules during a given period of time is called gross **primary productivity**. Some of the energy stored in these glucose molecules is used by the photosynthesizers themselves in cellular respiration to fuel their own metabolic activities. Any remaining energy, called the net primary productivity, can be used for growth and reproduction. Once the photosynthesizer uses the energy in glucose to make new organic molecules, these molecules may then become food for an animal.

We see, then, that the photosynthesizers, called the **producers**, form the first **trophic level** (*trophic*, feeding). The second trophic level contains the **primary consumers**, that is, the herbivores (plant eaters). On the next level are the **secondary consumers**, carnivores that feed on herbivores. Then there are the carnivores that eat other carnivores. These form still higher trophic levels—tertiary and quaternary consumers. The great recyclers in the system are the **decomposers**, such as bacteria and fungi, and **detritovores** (detritus eaters), which consume dead organic material for energy, releasing inorganic material that can then be used by producers.

At one time, the feeding relationships responsible for the flow of energy through ecological systems was described as a **food chain**—a linear sequence in which A eats B, which

Years after cultivation	Dominant vegetation	
1	Crabgrass	
2	Horseweed	
3	Broomsedge	
5–15	Pine seedlings	
25–50	Pine forest (with developing understory of deciduous hardwoods—not shown)	
150	Oak-hickory climax forest	

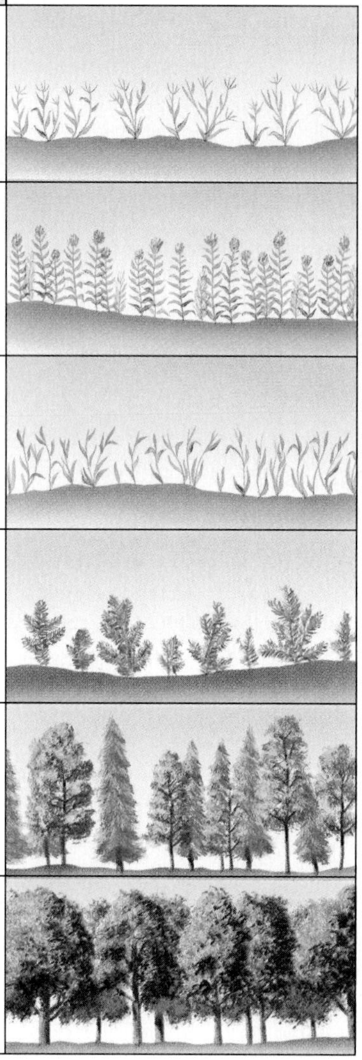

Figure 21–6 Representative stages in secondary succession on an abandoned field in North Carolina.

eats C, which eats D, and so on. However, now we know that such "chains" are too simplistic because many organisms eat at several trophic levels. To illustrate, consider the number of trophic levels on which humans feed. Whereas eating chicken makes us secondary consumers, eating tuna (a predatory fish) makes us tertiary consumers. Couldn't we put ourselves among the primary consumers as well? Don't we eat vegetables? (At least we're supposed to.) These more realistic patterns of interconnected food chains give rise to what are called **food webs**. Figure 21–7 describes part of a food web in a community where land meets water.

There are two kinds of food webs. The food webs we have described are examples of **grazing food webs**, in which the energy flow begins with the photosynthesizer and flows to herbivores and then to carnivores. In **detrital food webs**, however, energy flow begins with detritus (organic material

from the remains of dead organisms), which is eaten by a primary consumer. Detritus feeders are especially common in aquatic habitats. Crabs are an example of detritus feeders in a salt marsh. Terrestrial detritus feeders include earthworms, termites, and maggots (fly larvae). The two types of food chains are generally interconnected, as when a herring gull eats a crab or when a robin eats an earthworm.

Pyramids of Energy

It is important to note that energy is lost as it is transferred from one trophic level to the next. Although the efficiency of transfer can vary greatly among organisms, on average only 10% of the energy available at one trophic level is transferred to the next higher level. As a result, ecosystems rarely have more than four or five trophic levels.

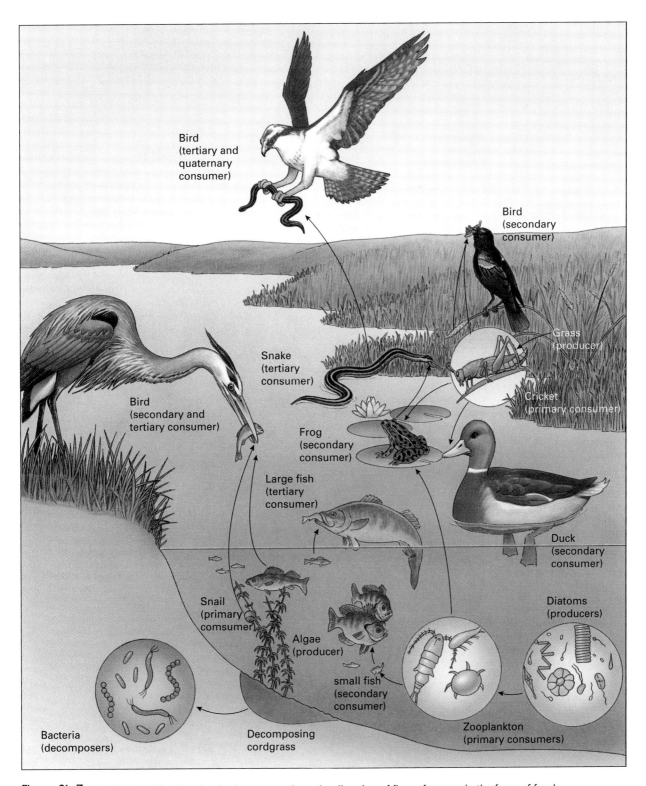

Figure 21-7 In this simplified food web, the arrows show the direction of flow of energy in the form of food.

What causes energy transfer between trophic levels to be so inefficient? As we sort through some of the answers to this question, keep in mind that only the energy that is converted to body mass is available to the next higher trophic level. Let's consider some of the energy losses that might oc-

cur between trophic levels (Figure 21-8). An animal must obtain its food—usually by grazing or hunting and these activities use energy. Furthermore, not all the food available at a given trophic level is captured and consumed. Some of the food eaten cannot be digested and is lost as feces. The

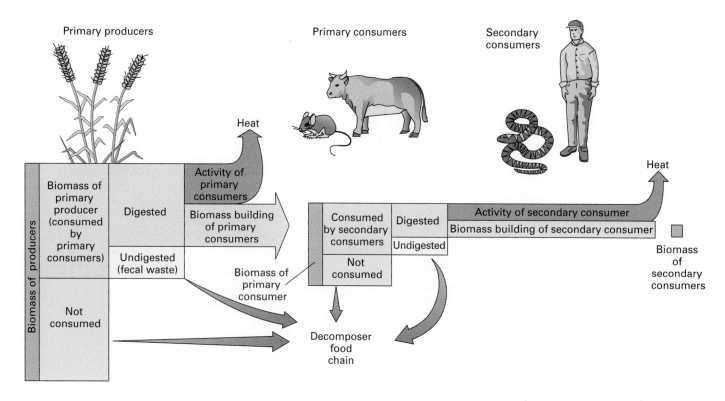

Figure 21–8 As energy flows through a food chain, only a small amount of it is stored as biomass and becomes available to the next higher trophic level. Some of the food energy is simply not captured, some is undigested and is lost in fecal waste, and some is used for cellular respiration. Only the remaining energy can be converted to biomass; it becomes available to the next higher trophic level.

energy in the indigestible material is unavailable to the next trophic level. Roughly one-third of the energy in the food that is digested is used by that animal as a source of energy for cellular respiration. The remaining energy can be converted to biomass and will be available to the next higher trophic level. As these losses become multiplied at successively higher trophic levels, a point is reached where there simply isn't enough food or energy to feed another level. The diminishing amount of energy available at each trophic level is sometimes depicted graphically, forming a **pyramid of energy** (Figure 21–9).

Pyramids of Numbers

Pyramids of numbers describe the counts of individuals at each trophic level. In many ecosystems, the number of individuals decreases at each higher trophic level (Figure 21–10). The declining numbers are caused by both the loss of energy between trophic levels and the tendency for predators to be larger than their prey. We see this in a grassland in which there are far more grass plants than there are herbivores, such as zebras and wildebeests, grazing on them and far fewer predators than herbivores. However, in some

ecosystems, certain forests for instance, pyramids of numbers can be partly inverted because the primary producers may be large trees that provide food for tremendous numbers of small insects.

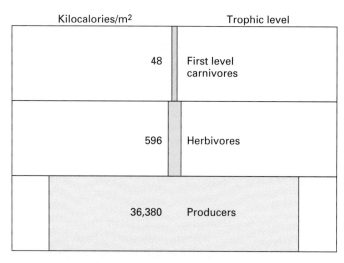

Kilocalories/m²		Trophic level
48		First level carnivores
596		Herbivores
36,380		Producers

Figure 21–9 A pyramid of energy. Note that each level contains less energy than the one below it.

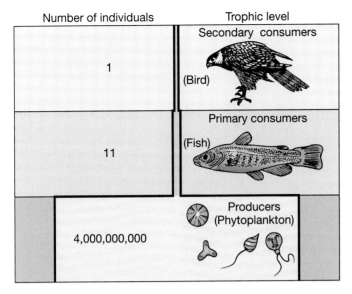

Number of individuals | Trophic level

Secondary consumers
(Bird)

1

Primary consumers
(Fish)

11

Producers
(Phytoplankton)

4,000,000,000

Figure 21-10 A pyramid of numbers. The number of organisms that can be supported at each successive trophic level is usually smaller than the number of organisms at the level below.

Ramifications of Ecological Pyramids

Biological Magnification

Chemicals that are essential to life—carbon, hydrogen, oxygen, nitrogen, and phosphorous—are passed from one trophic level to the next, being continuously recycled from one organism to the next. Organic molecules are broken down and then either metabolized for energy or put together to form the biomass of another individual.

However, certain molecules, including the chlorinated hydrocarbon pesticides such as DDT, heavy metals such as mercury, and radioactive isotopes, are broken down or excreted very slowly. What happens to molecules such as these? They tend to stay in the body, often stored in fatty tissues such as liver, kidneys, and the fat around the intestines. As more of these types of molecules are consumed, they accumulate and become concentrated in the animal's body.

The concentration of these substances in the bodies of consumers becomes magnified at each higher trophic level. This is because an important reason for eating is to obtain energy and, on average, only 10% of the energy available at one trophic level transfers to the next. To see the picture more clearly, let's consider a simplified example and assume that a nondegradable substance such as DDT was sprayed on plants, in some dilute concentration that we will represent as 1. An herbivore might eat many of these plants to stay alive. Once in the animal's body, the DDT stays there and accumulates. Thus, if we assume that the herbivore consumed 10 contaminated plants, the DDT concentration in its body would be 10 times greater than was present on one of the plants. When these DDT-containing herbivores are eaten by a predator, the DDT passes to its body and accu-

mulates. Since many herbivores must be consumed to keep the carnivore alive, the DDT concentration in the carnivore's body might be 100 times greater than that originally sprayed on the plants. Indeed, we might expect the substance to be 10 times more concentrated at each successive level on the trophic scale. This tendency of a nondegradable chemical to become more concentrated in organisms as it passes along the food chain is known as **biological magnification** (Figure 21-11).

Top carnivores are most likely to be hurt by nondegradable harmful substances in the environment. We have witnessed the decline in the numbers of certain birds that feed high on the food chain, including eagles, falcons, osprey, and pelicans. DDT causes the birds' livers to break down a hormone needed to deposit calcium in eggshells, leading to fragile, thin-shelled eggs that break when the parents try to incubate them. In the early 1970s, DDT use was banned in many countries. Since then, many populations of these birds have made dramatic recoveries.

If we are wise, humans will learn an important lesson from observations of the effects of biological magnification, because we too are top carnivores and we continue to pollute our environment with nondegradable, potentially harmful substances. In the 1950s in Minamata Japan, methyl mercury, a heavy metal, was released into Minamata Bay as industrial waste and the mercury accumulated in the food chain. Many people living in the region developed mercury poisoning because they ate fish contaminated with mercury. Today, DDT can still be found in human breast milk in areas throughout the world, and polychlorinated biphenyls (PCBs) are also found in human breast milk in some regions. The Great Lakes ecosystem contains low concentrations of many chemicals that tend to bioaccumulate, including PCBs and heavy metals. Studies are underway to determine whether biological magnification had harmful effects on humans living in the Great Lakes region.

Critical Thinking

Explain why the mercury levels in a lake may be low enough that the water is safe to drink, and yet fish from the same lake may be poisonous.

Feeding the Hungry and the Pyramid of Energy

The world's human population is growing at an alarming rate. The pyramid of energy suggests a way to feed the growing population more efficiently—eat lower on the food chain. Because only about 10% of the energy available on one trophic level becomes available to the next one, we see that:

10,000 calories of corn—can produce → 1000 calories of beef → can produce → 100 calories of human energy

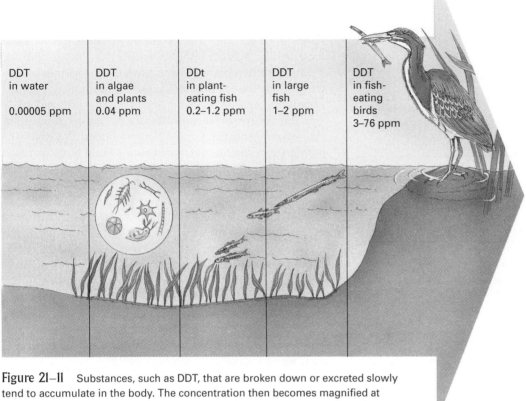

DDT
in water

0.00005 ppm

DDT
in algae
and plants
0.04 ppm

DDt
in plant-
eating fish
0.2–1.2 ppm

DDT
in large
fish
1–2 ppm

DDT
in fish-
eating
birds
3–76 ppm

Figure 2I–II Substances, such as DDT, that are broken down or excreted slowly tend to accumulate in the body. The concentration then becomes magnified at each successive trophic level, because the energy losses at each level require that a consumer eat many individuals at a lower trophic level to stay alive. The DDT concentration in the green heron at the end of this food chain is one million times greater than the DDT concentration in the water.

However, if humans began to eat one level lower on the food chain about ten times more energy would be available to them:

10,000 calories of grain (corn/wheat/rice etc.)—can produce →
1000 calories of human energy

We see, then, that more people could be fed and less land would have to be cultivated if we adopted a largely, or exclusively, vegetarian diet (Figure 21–12). This is one reason why densely populated regions of the world, including China and India, are primarily vegetarian. You may recall from Chapter 13A, that a vegetarian diet must contain the proper combinations of foods. Certain healthful combinations are common in different cultures of the world—for instance, in Central and South America, beans with rice or with corn or wheat tortillas; in Japan and China, rice and tofu; and in India, rice and lentils. It seems likely that, as the human population continues to expand, meat will become even more of a luxury throughout the world than it is today.

Critical Thinking

A traditional diet of Inuits involves a relatively long food chain:

diatoms → zooplankton → fish → seals → Inuits

Explain how the length of this food chain may be one factor contributing to the small population size of Inuit groups. Seal and beluga whale blubber contain high levels of DDT and PCBs. Explain why the breast milk of Inuit women who eat blubber is four to seven times higher in these chemicals than the breast milk of women in Quebec who eat lower in the food chain, on fish for example.

Cycling in Ecosystems

Resources on Earth are limited and the supply of many of them cannot be replenished. Life on Earth is demanding.

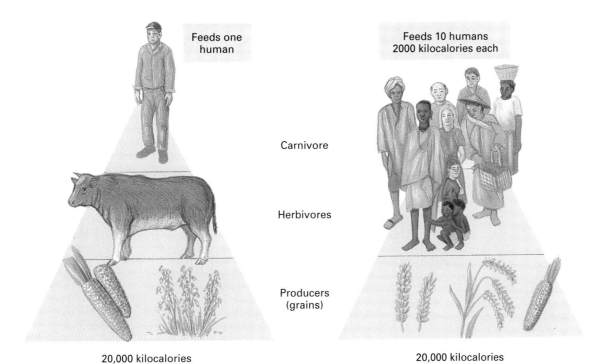

Feeds one
human

Feeds 10 humans
2000 kilocalories each

Carnivore

Herbivores

Producers
(grains)

20,000 kilocalories

20,000 kilocalories

Figure 21–12   Energy pyramids may hold an important lesson for humans. Because only approximately 10% of the energy available at one trophic level is available at the next higher level, a greater number of people could be fed if they ate a vegetarian diet rather than eating a diet containing meat. This is one reason that diets in densely populated regions of the world are largely vegetarian. As the world human population continues to grow, meat will become an even greater luxury than it is today throughout the world.

Many of Earth's reserves would be quickly depleted if it weren't for nature's cycling: Materials move through a series of transfers, from living to nonliving systems and back again (Figure 21–13).

Because of the constant cycling of matter, we can say that the molecules that make up your body at this very minute, that hold you together, and that enable you to go on are, in a sense, on loan. Upon your death they will be released into the biosphere, perhaps reappearing in some politician of the next century or in some delicate plant that graces your final resting place. Some find it reassuring to know that we are, in fact, a part of the great natural world and that, in a sense, the elements that compose our very bodies go on forever.

Let's look at some of the more important of these **biogeochemical cycles**, the recurring pathways of materials between living and nonliving systems.

Figure 21–13   Through biogeochemical cycles, matter cycles among living organisms and the physical environment. This is in contrast to energy, which flows through the ecosystem in one direction. ▶

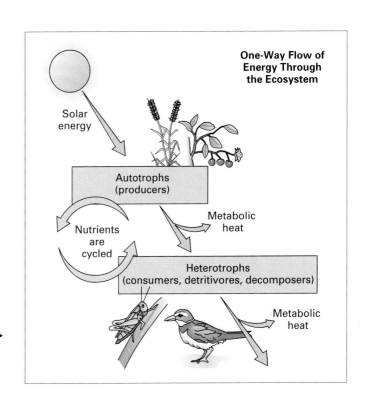

**One-Way Flow of
Energy Through
the Ecosystem**

Solar
energy

Autotrophs
(producers)

Metabolic
heat

Nutrients
are
cycled

Heterotrophs
(consumers, detritivores, decomposers)

Metabolic
heat

Cycling in Ecosystems   **603**

The Water Cycle

Each drop of rain teaches us something about water recycling. Indeed, water continuously cycles from the atmosphere in the form of rain, snow, sleet, or hail to the land, where it collects in ponds, lakes, or oceans, and back to the atmosphere as it evaporates. Most of the rain or snow that falls to Earth returns to the sea at some point. This cycle provides us with a renewable source of drinking water. Because water is so critical to life, large amounts temporarily pause in the bodies of living things. In living cells, water helps regulate temperature and acts as a solvent for many biological reactions. Oxygen is produced from water through the reactions of photosynthesis. Water also cycles back to the environment from living things, as plants return 99% of the water they absorb to the atmosphere in the process of transpiration. The water cycle is shown in Figure 21–14.

The Carbon Cycle

If you've ever seen a dead possum beside the road, you know something about carbon cycling. Put simply, in the carbon cycle, carbon moves from the environment, into the bodies of living things, and back to the environment (Figure 21–15). As we will see, life and the carbon cycle are intimately related. Carbon is essential to organisms because it is a part of molecules such as proteins, carbohydrates, fats, and nucleic acids. Also, certain processes of living organ-

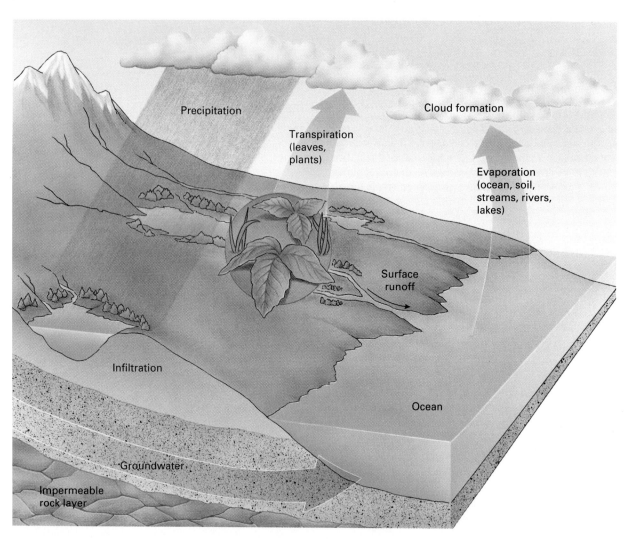

Figure 21–14 Diagram of the water cycle. Water, which makes up a large part of the mass of most organisms, is essential to life. Water cycles from the atmosphere to the land as precipitation and then back to the atmosphere as it evaporates from oceans and other bodies of water where it has collected. Water also returns to the atmosphere from plants in the process of transpiration. This cycling provides us with a renewable source of drinking water.

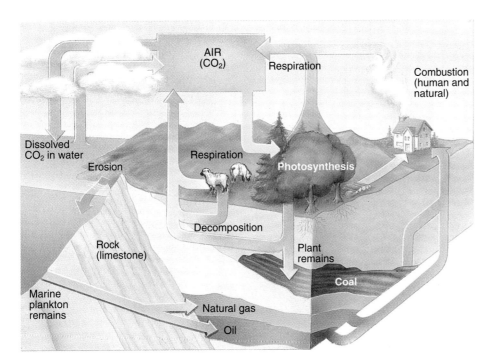

Figure 21-15 In the carbon cycle, carbon cycles between the environment and living organisms. Carbon dioxide is removed from the environment as producers use it to synthesize organic molecules in photosynthesis. The carbon in those organic molecules then moves through the food chain, serving as a carbon source for herbivores, carnivores, detritovores, and decomposers. Carbon is returned to the atmosphere in the form of carbon dioxide when organisms use the organic compounds in cellular respiration.

isms—photosynthesis and respiration—are critical to carbon cycling.

Carbon moves from the environment primarily during photosynthesis, as plants, algae, and cyanobacteria use carbon dioxide (CO_2) to produce sugars and other organic compounds. When the photosynthesizers are eaten by herbivores, the organic molecules then serve as a carbon source for the herbivores. They, in turn, use the carbon to produce their own organic molecules, which then serve as a carbon source for carnivores. When these organisms die, as did the possum beside the road, the organic molecules in their bodies will serve as a carbon source for decomposers. However, while alive, all organisms cycle carbon back to the environment through the reactions of cellular respiration, which breaks down organic molecules to CO_2.

In some cases, there may be a significant delay before carbon cycles back into the environment. For example, carbon may remain tied up in the wood of some trees for hundreds of years. Most of the carbon that has left the carbon cycle is thought to be in limestone, a type of sedimentary rock that formed from the shells of marine organisms that sank to the bottom of the ocean floor and became covered by sediments. Also, fossil fuels, so named because they formed from the remains of organisms that lived millions of years ago, are the vast stores of products of photosynthesis.

How does the carbon placed in long-term storage get returned to the environment? Three processes are involved—respiration, erosion, and combustion. The trees will eventually die and the natural process of decomposition will make the carbon available for new organisms, which will respire and release CO_2 to the atmosphere. The carbon in

limestone is recycled through erosion. Millions of years after it forms, sedimentary rock containing limestone can be lifted to the Earth's surface by the process of geological uplift. The limestone is then eroded by chemical and physical weathering, and carbon is returned to the environment and is available to cycle through the food chain once again. Combustion returns the carbon in fossil fuels to the environment. Today, fossil fuels such as coal, oil, and natural gas are being burned and the carbon they contain is being returned to the atmosphere as CO_2. However, during the eons in which these carbon molecules were stored as fossil fuels, the carbon cycle reached a new point of balance. The burning of fossil fuels is now upsetting the equilibrium by returning more CO_2 to the atmosphere than is currently being removed. We will consider some of the possible results of this imbalance later in this chapter when we discuss the greenhouse effect.

The Nitrogen Cycle

Nitrogen is a principal constituent of a number of critical molecules including proteins and nucleic acids. Unfortunately, it is often in short supply to living systems, so its cycling can be critical.

The nitrogen cycle involves five important steps (Figure 21–16):

1. **Nitrogen gas (N_2) in the atmosphere is converted to ammonia (NH_3) by nitrogen-fixing bacteria.** The largest reservoir is the atmosphere, composed of about 79% nitrogen gas. However, nitrogen gas cannot directly

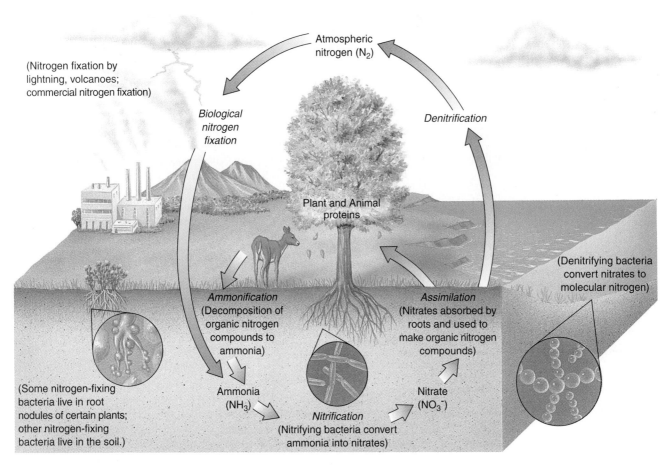

Atmospheric
nitrogen (N₂)

(Nitrogen fixation by
lightning, volcanoes;
commercial nitrogen fixation)

Biological
nitrogen
fixation

Denitrification

Plant and Animal
proteins

(Denitrifying bacteria
convert nitrates to
molecular nitrogen)

Ammonification
(Decomposition of
organic nitrogen
compounds to
ammonia)

Assimilation
(Nitrates absorbed by
roots and used to
make organic nitrogen
compounds)

(Some nitrogen-fixing
bacteria live in root
nodules of certain plants;
other nitrogen-fixing
bacteria live in the soil.)

Ammonia
(NH₃)

Nitrification
(Nitrifying bacteria convert
ammonia into nitrates)

Nitrate
(NO₃⁻)

Figure 21–16 Diagram of the nitrogen cycle. Atmospheric nitrogen can be converted to ammonia by nitrogen-fixing bacteria, including cyanobacteria. Nitrifying bacteria then convert the ammonia to nitrate, the main form of nitrogen absorbed by plants. Plants use nitrate to produce proteins and nucleic acids. Animals eat the plants and use the plant's nitrogen-containing chemicals to produce their own proteins and nucleic acids. When plants and animals die, their nitrogen-containing molecules are converted to ammonia by ammonifying bacteria. Denitrifying bacteria return nitrogen to the atmosphere.

interact with life. So, as you sit there, you are bathed in nitrogen gas without interacting with it. Before it can be used, nitrogen gas must be fixed, that is, converted to a form that living organisms can use—ammonia. This is done by nitrogen-fixing bacteria, such as those of the genus *Rhizobium*, that live in nodules on the roots of leguminous plants such as peas and alfalfa and, in aquatic environments, by certain cyanobacteria.

Farmers have known about the advantage of planting nitrogen-fixing crops for centuries. In fact, Greek writings from the second century extol the virtues of planting leguminous crops. In natural systems, nitrogen-fixing bacteria may be found on the roots of wild legumes, buckthorns, alder, and locust.

2. **Ammonia (NH₃) is converted to nitrate (NO₃⁻) by nitrifying bacteria.**

3. **The nitrogen in ammonia or nitrates is assimilated in** the proteins or nucleic acids of living organisms. The ammonia or nitrates are first absorbed by plants, and the nitrogen is used to form plant proteins and nucleic acids. The nitrogen then passes through the food chain and is incorporated into the nitrogen-containing compounds of animals.

4. **The nitrogen-containing compounds of living organisms are converted to ammonia in the process of ammonification.** One way in which ammonification takes place is by the formation of nitrogen-containing waste products urea or uric acid, which are excreted. A second way is by decomposition. Decomposers such as bacteria break down the waste products and dead bodies of plants and animals, producing ammonia, or ammonium (NH₄⁺), much of which they use themselves and the rest is released into the environment where it is available for plants.

5. **The nitrates that are not assimilated into living organisms are converted to nitrogen gas by denitrifying bacteria.** Denitrification, then, returns nitrogen to the environment.

Human Disturbances of Biogeochemical Cycles

Throughout most of history, biogeochemical cycles have worked just as we have described. However, human activity, particularly since the time of the Industrial Revolution, has disrupted the normal pattern of cycling. Some of these disturbances result from new technologies, but most are caused by the demands of an ever-increasing human population.

Disruptions to the Water Cycle

Water Shortages

The Earth has been called the "water planet" because we have so much of it. Still, relatively little is available for human use. You might think that at least the fresh water would be available, but much of it is tied up in ice or clouds, flowing in underground rivers, or otherwise inaccessible for immediate human use.

There are several problems associated with water. One is that in some areas there simply isn't enough. Indeed, the World Health Organization estimates that the lack of clean water for drinking, washing, and sewage disposal can be blamed for 80% of the disease in developing countries. Although water shortages in other parts of the world are more severe, there are regions in the United States, particularly in the agricultural regions of the south and west, where water shortages are being experienced. The water tables beneath the highly populated northeast United States are also dwindling dangerously.

In some places, water shortages are caused by drought, but in most regions the underlying cause is too many people drawing from a limited water supply. North Americans use an exceptionally large quantity of water. In the United States, each person uses about 7500 liters of water a day, whereas a person in a less developed country uses less than 1% of this amount.

You might respond to these numbers defensively and claim that *you* certainly don't use that much water. In a sense you would be right. Personal use accounts for only a small percentage of total water use. Most of the water is used in agriculture. In an effort to feed a growing human population, irrigation has become increasingly important as a way to make arid areas fruitful. The rest of the water is used primarily in industry for steam generation or the cooling of power plants.

Irrigation has pros and cons. On one hand, irrigation makes it possible to grow crops in areas that would otherwise be barren, allowing us to feed many people. On the other hand, in the long run, irrigation can make land unfit for agriculture. Irrigation water contains dissolved minerals. Whereas runoff from natural rainfall would carry salts away, irrigation water soaks into the soil. Then, when the water evaporates from the soil, the salts are left behind. The resulting accumulation of salts in the soil is called **salinization**. Worldwide, salinization destroys the fertility of 5000 square kilometers of irrigated land each year (Figure 21–17).

Figure 21–17 Although irrigation increases the productivity of land for a period of time, it can cause salinization, the buildup of salts in the soil, that eventually can make the land unproductive. Irrigation water contains salts that are left behind when the water evaporates. *(USDA/Soil Conservation Service)*

There are two main sources of water: surface water, such as rivers and lakes, and groundwater, which is water found under the Earth's surface in porous layers of rock, such as sand or gravel. When too much surface water is used in an area, ecosystems are affected. As much as 30% of a river's flow can usually be removed without affecting the natural ecosystem. However, in some parts of the southwestern United States, as much as 70% of the surface water has been removed. Such overdrawing of surface water can disrupt ecosystems. For example, much of the surface water that feeds Mono Lake in eastern California is now diverted to supply water to Los Angeles. Not only has Mono Lake's water level dropped 12.2 m (40 ft), but the lake's saltiness has nearly doubled. These changes have already affected the plants and animals living in or around the lake. If it becomes saltier, it will be unable to support any wildlife. Wetlands are often disastrously affected when surface water is diverted for human use. In Nevada, much of the water that once fed the wetlands in Stillwater National Refuge has been diverted for agriculture, causing the death of thousands of birds.

Critical Thinking

Conservationists sometimes buy wetlands, hoping to preserve habitat of endangered wildlife. In spite of the protection of conservationists, these wetlands sometimes dry up. Explain why it would also be important to buy the rights to the water that feeds the wetland.

Human use of groundwater is depleting the aquifers (porous layers of underground rock where groundwater is found) (Figure 21–18). Water is added to aquifers by rainfall or melting snow. In some areas, however, water is being removed from aquifers faster than it can be replenished. The Ogallala Aquifer underlies eight states in the Great Plains region of the United States. In many states, this aquifer is being emptied eight times faster than natural processes can refill it. In particularly dry regions, such as parts of Texas, New Mexico, Oklahoma, and Colorado, it is being drained 100 times faster than it can be replenished. The result is that the aquifer is much lower than it was only a few years ago and scientists expect that by the year 2000 Texas will lose the ability to irrigate half its farmland.

What steps can be taken to reduce water shortages? First, we should waste less water. Each of us could help by individual efforts. However, since most water is used in agriculture, the biggest benefit would come from improved irrigation methods. Between 1950 and 1980, for example, Israel reduced its water wastage from 83% to a mere 5%, primarily by changing from spray irrigation to drip irrigation. Second, the price of water in the United States could be raised. The cost of water in the United States is substantially below its cost in European countries. History has shown us that the consumer's interest in conserving water is directly related to its cost. For example, when an average person living in Tucson is compared with one living in Las Vegas, the one in Tucson pays twice as much for water and uses half as much. Clearly, economic and legal incentives for conserving water are needed.

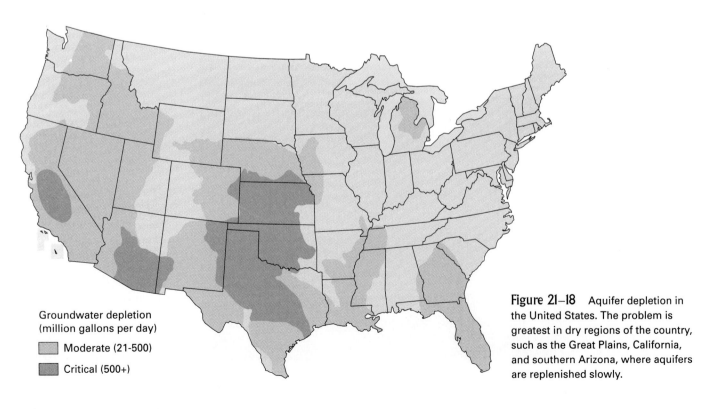

Groundwater depletion
(million gallons per day)

☐ Moderate (21-500)

■ Critical (500+)

Figure 21–18 Aquifer depletion in the United States. The problem is greatest in dry regions of the country, such as the Great Plains, California, and southern Arizona, where aquifers are replenished slowly.

Social Issue

Water Pollution

Another problem is that much of the water we have is polluted, often by our own activities. In a recent study of 80 American cities, 250 synthetic chemicals were found in the water supply. We have no idea what effect many of these chemicals have on life. Relatively few have been tested for their effects on cancer formation or birth defects.

Disruptions to the Carbon Cycle

For most of our history, the global production of carbon dioxide has roughly equaled its absorption by plants and algae. The Industrial Revolution, however, marked the end of that period as we began to burn fossil fuels and to cut the Earth's great forests. Burning fossil fuels returns carbon to the environment in the form of CO_2, after it was out of circulation for millions of years. Deforestation, the removal of a forest without adequate replanting, such as is occurring in areas of the Amazon rain forest, increases atmospheric CO_2 in two ways. First, it reduces the *removal* of CO_2 from the atmosphere through the trees' photosynthesis. Second, the trees are often burned after cutting, which *adds* CO_2 to the atmosphere. Currently, about 6 billion tons of CO_2 are added to the atmosphere each year, with about three-quarters of it coming from cars and factories and the remaining quarter largely from deforestation. In the past 1200 years, in fact, global atmospheric CO_2 levels have increased 25% (Figure 21–19).

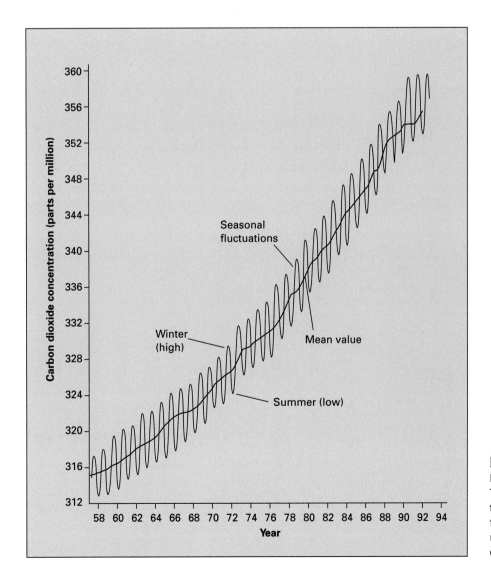

Figure 21-19 The concentration of CO_2 has been slowly increasing for many years. The increase is caused by CO_2 being added to the atmosphere through the burning of fossil fuels and by a reduction in the removal of CO_2 from the atmosphere by deforestation.

Human activities could cause an even more dramatic increase in atmospheric CO_2 than is observed. In fact, only about half of the extra CO_2 produced by human activity shows up in the atmosphere. It is thought that much of this CO_2 dissolves in the oceans, which serve as a sink, or reservoir, for CO_2. Nonetheless, 1 to 2 billion tons of carbon a year are unaccounted for. Scientists assume that some of this is taken up by trees in European forests that are now regrowing after being cut down in the last century. In addition, increased CO_2 usually stimulates plant growth and this would, in turn, cause increased removal of CO_2 from the atmosphere.

It is clear that atmospheric CO_2 is rising, and this raises concerns because CO_2 is one of the greenhouse gases that play a role in warming our atmosphere. Other greenhouse gases include methane, nitrous oxide, chlorofluorocarbons (CFCs), and ozone in the lower atmosphere. These gases are also accumulating in the atmosphere, largely because of human activity. Like the glass on a greenhouse, these gases allow the sunlight to pass through to the Earth's surface, where it is absorbed and radiated back to the atmosphere as longer-waved infrared radiation, or heat. Both greenhouse glass and greenhouse gases reflect the infrared radiation, thus trapping the heat (Figure 21–20). This is why there is a concern that the increase in atmospheric CO_2 could lead to a rise in temperatures throughout the world—global warming.

Is the Earth getting warmer? The answer isn't clear and seems to depend on where you measure the temperature. Measurements taken on a mountaintop in Hawaii suggest that global temperatures may be increasing, but those taken by satellite indicate that the lower atmosphere has actually cooled slightly in the past two decades. More specifically, according to ground-based measurements the Earth's atmosphere has warmed by 0.13°C per decade, but satellite measurements indicate that it has dropped by 0.05°C per decade. How would it be possible for the atmospheric temperature to drop in spite of increases in levels of atmospheric CO_2? It could be that factors such as the haze created by sulfur dioxide air pollution or dust from volcanoes are blocking some sunlight from entering the atmosphere, thus counteracting the effect of greenhouse gases.

Why is there such hoopla about global warming? Why should anyone really care whether the atmosphere becomes 1°C to 2°C warmer? Keeping in mind that some scientists don't agree that the Earth's atmosphere is getting warmer, let's consider some of the possible effects of global warming.

1. **Rising sea level.** An increase of just a few degrees in the overall temperature of the Earth could cause glaciers and polar ice caps to begin melting. At the same time, the ocean water would expand as it warmed. Although scientists disagree about the extent to which the sea level would rise, the estimates range between 0.2 meter and 2.2 meters by the year 2050. Indeed, the sea level along the Atlantic Coast of the United States has already risen about 30 cm. The rising sea level would have the greatest effect on coastal countries such as Bangladesh, Egypt, Vietnam, and Mozambique. Sea water could then cover a significant amount of land in these densely populated areas. For example, if the sea level rose even 1 meter, 17% of Bangladesh would be covered with water. In the United States, major cities such as New York, Miami, Jacksonville, Boston, San Francisco, and Los Angeles could be largely underwater.

Social Issue

If global warming occurs, it will affect some parts of the world and some cities in the United States more than oth-

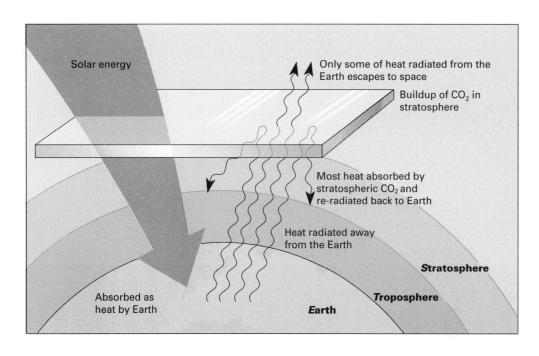

Figure 21–20 Carbon dioxide and other greenhouse gases promote global warming by trapping infrared radiation (heat) in the atmosphere, in much the same way the glass on a greenhouse traps heat. Thus, the global warming that may be occurring because these gases are accumulating in the atmosphere is known as the greenhouse effect.

ers. Coastal areas are often densely populated. Where will the people go when their homes and businesses are flooded? How should national policies be developed? How should international policies be developed? Who should pay for help for those in need?

---

2. **Changing weather patterns.** Global warming would cause changes in both temperature and rainfall patterns. You may recall that the distribution of biomes, each of which has characteristic plant and animal life, is primarily determined by climates, especially temperature and rainfall. If global warming occurs, some species will thrive and others may become extinct. There will be a shift in the agricultural regions of the world. Food production in certain areas, including the central plains of the United States and Canada, could drop. Areas that now produce enough food for export may not be able to produce enough for their own citizens. On the other hand, other regions, parts of India and Russia, for instance, might receive increased rainfall that would benefit agriculture.

Social Issue

The greenhouse gases are produced primarily by the more developed countries. However, the effects are likely to be felt throughout the world. The less developed countries are poorer and have less technical knowledge to deal with the problems caused by global warming. As the food-producing regions of the world shift with changing weather patterns, so will the distribution of "haves" and "have nots." How will we as a world community deal with the shifts in economic and political power brought on by global warming?

Although the possible consequences of global warming are dire, we aren't certain that the Earth is getting warmer, so what do we do? An important step, which would have other benefits besides slowing global warming, would be to reduce our production of greenhouse gases. We will soon run out of fossil fuels anyway, probably within your lifetime.

Therefore, it would be wise to improve energy efficiency of automobiles and appliances and to replace fossil fuels with other forms of energy, such as solar, wind, or wave power.

The good news is that *you* can make a difference. Whenever possible walk, don't drive or ride. Drive a fuel-efficient vehicle. Wear clothing that will minimize the use of heating or cooling systems in your home. In general, conserve energy. The environment will benefit, the exercise will improve your health, and you will save money at the same time.

Disruptions to the Nitrogen Cycle

Human activity is also disturbing the nitrogen cycle. As we have seen, in the natural cycle, nitrogen-fixing bacteria living in nodules on the roots of leguminous plants and certain cyanobacteria convert atmospheric nitrogen to ammonia, a form of nitrogen that organisms can use. The large-scale use of chemical fertilizer is only slightly more than 150 years old. However, during the past century, industry has doubled the amount of nitrogen gas it fixes during the production of fertilizer. Indeed, today industry fixes 80 million metric tons of nitrogen each year in the production of fertilizer. The demand for fertilizer has increased with the demand for food to feed a growing human population. A problem arises, however, because the natural dinitrification processes cannot keep pace with the rate of industrial nitrogen fixation and the amount of nitrogen that is added to the land in commercial fertilizers. The result is an excess of fixed nitrogen.

Some of that excess fixed nitrogen washes into nearby streams, rivers, ponds, or lakes as fertilizer runoff. Fertilizer runoff is one cause of **eutrophication**, the enrichment of water in a lake or pond by nutrients. The nutrient boost can lead to the explosive growth of photosynthetic organisms, including algae and cyanobacteria, and in shallow areas, weeds. The problem is that when these organisms die, organic material accumulates at the bottom of the lake. Decomposers then deplete the water of dissolved oxygen. Gradually, fish, such as pike, sturgeon, and whitefish, that require higher oxygen concentration in the water, are replaced by others, such as catfish and carp, that can tolerate lower levels of dissolved oxygen.

SUMMARY

1. Ecology is the study of the interactions among organisms and between organisms and their environment.
2. The biosphere, the part of the Earth in which life exists, includes all living organisms. Living organisms (biotic components) and the physical environment (abiotic components) within a defined area form an ecosystem. Whereas a community is all the organisms in an ecosystem that can potentially interact, a population is all the individuals of the same species within a community. An organism's niche is its role in the ecosystem.

3. A biome is a relatively distinct terrestrial ecosystem with a characteristic composition of organisms with a specific climate.
4. Ecological succession is the sequence of changes in communities over long periods of time. Primary succession occurs where no community previously existed. The first invaders are called pioneer species. A climax community eventually forms and remains as long as no disturbances occur. Secondary succession describes the sequences of changes that occur when an existing community is disturbed by human or natural means.

5. Most of the energy in living systems comes from the sun when that energy is absorbed and stored in the molecules of photosynthetic organisms, called producers. The energy that producers have stored in their molecules enters the animal world through plant-eating herbivores (primary consumers), which may be eaten by carnivores (secondary consumers), or by animals that may eat other carnivores (tertiary consumers). Detritovores and decomposers consume dead organic material, releasing inorganic material that can be used by producers. The position of an organism in these feeding relationships is referred to as a trophic level.

6. The feeding relationships that allow for the one-way flow of energy through the ecosystem and for the cycling of materials among organisms are called food chains or food webs. A food chain describes the simple linear sequence of who eats whom. Many food chains in an ecosystem usually interconnect to form a complicated pattern called a food web.

7. Pyramids of energy show the loss of energy from one trophic level to another. On average, only 10% of the energy available at one trophic level is available at the next higher level. Ecosystems generally have only four or five trophic levels.

8. Pyramids of numbers describe the number of individuals at each trophic level.

9. Biological magnification—the tendency for a nondegradable substance to become more concentrated in organisms as it passes along the food chain—is a consequence of the energy loss between trophic levels. Thus, top carnivores are most likely to be hurt by nondegradable toxic substances in the environment.

10. Because energy is lost with each transfer in the trophic scale, one way to feed more people would be for humans to adopt a largely vegetarian diet.

11. The water cycle is the pathway of water as it falls as precipitation; collects in ponds, lakes, and seas; and returns to the atmosphere through evaporation or transpiration.

12. The carbon cycle is the worldwide circulation of carbon from the abiotic environment (carbon dioxide in air) to the biotic environment (carbon in organic molecules of living organisms) and back to the air. Carbon enters living systems when photosynthetic orgainsms incorporate carbon dioxide into organic materials. Carbon dioxide is formed again when the organic molecules are used by the living organism for cellular respiration.

13. The nitrogen cycle is the worldwide circulation of nitrogen from nonliving to living systems and back again. Atmospheric nitrogen (N_2) cannot enter living systems. Nitrogen-fixing bacteria living in nodules on the roots of leguminous plants convert N_2 to ammonia (NH_3) that is then converted to nitrates by nitrifying bacteria. The ammonia and nitrates are then available to plants to use in their proteins and nucleic acids. The nitrogen is then transferred to organisms that consume the plants. Nitrogen-containing molecules can be converted to ammonia in the process of ammonification. Nitrates that are not assimilated into living organisms can be converted to nitrogen gas by denitrifying bacteria.

14. The water cycle can be disturbed as humans cause shortages through overuse of water supplies. Most water use is in agriculture, primarily for irrigation. Irrigation can cause salts to accumulate in the soil (salinization), which can make the land unusable for agriculture.

15. Humans have disturbed the carbon cycle through activities that have increased carbon dioxide levels in the atmosphere—burning fossil fuels, which directly adds carbon dioxide, and deforestation, which decreases the removal of carbon dioxide from the atmosphere. The rise of atmospheric carbon dioxide is a concern because it is a greenhouse gas (along with methane, nitrous oxide, CFCs, and ozone). These gases trap heat in the Earth's atmosphere and could cause global warming.

16. Humans have disrupted the nitrogen cycle through the industrial fixation of nitrogen to produce fertilizer. Natural dinitrifying processes cannot keep up with industrial nitrogen fixation, and so there is an excess of fixed nitrogen. Some of this fixed nitrogen washes into nearby waterways. Fertilizer runoff is one cause of eutrophication.

REVIEW QUESTIONS

1. Distinguish among biosphere, ecosystem, community, and population.
2. Explain the difference between an organism's niche and its habitat.
3. Define biome and list some examples of biomes.
4. What is ecological succession? How does primary succession differ from secondary succession?
5. Explain how energy flows through an ecosystem.
6. Define: producer, primary consumer, secondary consumer, and decomposer. Explain the role each plays in cycling nutrients through an ecosystem.
7. Explain why the feeding relationships in a community are more realistically portrayed as a food web than as a food chain.
8. Explain the reasons for the inefficiency of energy transfer from one trophic level to the next higher one. Why does this loss of energy limit the number of possible trophic levels?
9. What is meant by a pyramid of energy? What is meant by a pyramid of numbers?
10. Define biological magnification. Explain how biological magnification is a consequence of the energy loss between trophic levels. Why should humans be concerned about biological magnification?
11. Explain why more people could be fed on a vegetarian diet than on a diet in which meat provides most of the protein calories.
12. Describe the water cycle, the carbon cycle, and the nitrogen cycle.
13. Explain some of the ways that humans are disturbing the water cycle.
14. Which human activities are primarily responsible for the rising level of atmospheric carbon dioxide?
15. What is meant by the greenhouse effect? Explain why there is concern that the rising level of atmospheric carbon dioxide could lead to global warming.
16. If global warming does occur, what are some possible consequences?
17. What steps can be taken to slow the rate of global warming?

Chapter **22**

Human Population Dynamics

(© David M. Grossman/Photo Researchers, Inc.)

There is no way to discuss human populations without arousing concern. The concern stems from two kinds of information. First, the increase in our numbers is startling, and we just don't know how long it can go on (Figure 22–1). Notice in Table 22–1 that it took until 1804 for the human population to reach 1 billion, but since then we have been adding a billion persons to our numbers at a much faster rate. In mid-1996, the world population was 5.77 billion persons. It is expected that by 1998 our numbers will reach 6 billion people. Second, we have learned a great deal about population changes from observing other species, and the information we've gleaned in this way is not encouraging for our future. So let's look, now, at our own numbers and the influences that may be driving them. You may find some of this information alarming. Still, we humans are, by nature, problem solvers, so let's look at the facts as realistically as we can and see if we can begin to form some ideas leading to solutions. We will start by reviewing some of the general principles of population change.

Principles of Population Dynamics

Population dynamics describe how populations change. The human population, we can say, consists of all individuals alive at any point in time. This enormous population can be broken down into smaller groups. Thus, we can speak of the population of the United States or Colombia or China and we may break it down further and consider, say, urban American populations versus rural ones.

The basic principle of population dynamics is simple— population size changes when individuals are added and removed at different rates. A population grows when more individuals are added than are subtracted (Figure 22–2). Individuals can be added to a population by births or by **immigration**, the arrival of individuals from other populations. Individuals are removed from a population through death or by **emigration**, the exodus of individuals from a population.

If we are to consider the overall system, that is, the human population of the world, the effects of immigration and emigration on population size are eliminated. (At least we think that's a true statement and that no one from other "populations" is intermingling with us.) However, immigration and emigration can have dramatic effects on the population size of countries, cities, and regions. In 1996, more than 24.5 million people, 9.3% of the population in the United States, were born in another country. Also, immigra-

Figure 22–1 The human population has grown steadily throughout history, but since the Industrial Revolution, it has skyrocketed. *(Mike Andrews © 1993 Earth Scenes)*

Table 22–1 World Population Milestones

Number of Billions	Year That Number Was Reached	Years Required to Add That Billion
1	1804	
2	1927	123
3	1960	33
4	1974	14
5	1987	13
Predictions for Future		
6	1998	11
7	2009	11
8	2021	12
9	2035	14
10	2054	19
11	2093	39

Data from Population Division, Department for Economic and Social Information and Policy Analysis, World Population Growth from Year 0 to Stabilization. United Nations, New York; 1996.

tion is constantly changing the face of America (Figure 22–3). In 1950, 75% of minorities in the United States were African Americans, but it is expected that they will be outnumbered by Hispanics by 2010, and it is expected that within the next few decades the descendants of white Europeans will slip into minority status. These shifts in populations can have social, political, and economic effects.

Social Issue

The United States is currently struggling with several issues related to immigration and the use of public funds. Some states, California for instance, are debating whether the children of illegal immigrants should be allowed to attend public schools. Other states, Massachusetts for example, are questioning whether legal immigrants should be entitled to the same governmental health care provisions that citizens receive. The entire nation is wrestling with questions about the acceptance of other languages, such as Spanish. What do you think? How should these issues be decided?

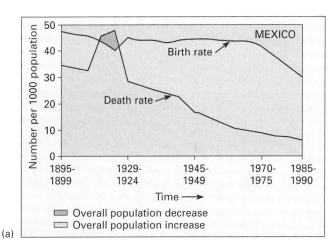

(a)

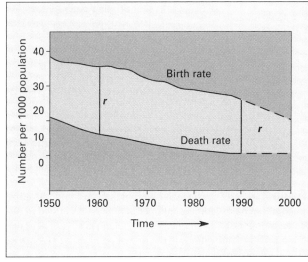

(b)

Figure 22–2 ▶ Population growth is determined by relative birth rates and death rates. When the birth rate exceeds the death rate, the population grows. (a) In Mexico, both the birth rate and the death rate have decreased. However, the death rate has declined more than the birth rate, so the population of Mexico has grown rapidly. (b) Worldwide, the birth rate has declined more than the death rate, and so the overall rate of growth has slowed.

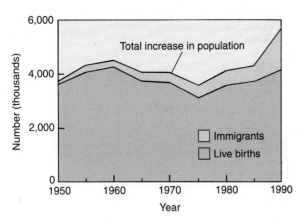

Figure 22-3 Births and immigration have both contributed to the population increase in the United States. The growth in population in the United States caused by immigration has been increasing since 1950.

Births

Each year, about 84 million persons are added to the human population. Although absolute numbers may be alarming and may help us understand the problems associated with population growth, a **rate** (the number of births or deaths per a specified number of individuals in the population during a specific time period) can be more informative. For example, we would consider the addition of 1000 individuals to a population differently if it occurred overnight than if it occurred during the period of a year. Furthermore, the addition of 1000 individuals would be more important if the initial population size were 100 than if it were 100,000. Expressing population growth as a *rate*, then, puts it into perspective. In 1996, the world's human population was growing at a rate of 1.48% a year.

However, rate is not enough to predict how quickly new individuals will be added to the population; we must also know the number of individuals in the starting population. Since added individuals also reproduce, population growth occurs in much the same way as compound interest accumulates in the bank—the more you start with, the more you end up with. At the same growth rate, then, the larger the size of the starting population, the more individuals are produced (Figure 22-4).

Besides growth rate and the size of the initial population, two other factors—the age structure of the population and the age when reproduction begins—are important in predicting future changes in population size. The age structure of a population is important because only individuals within a certain age range reproduce. Among humans, for instance, toddlers and octogenarians are counted as members of the population, but they do not reproduce. The **age**

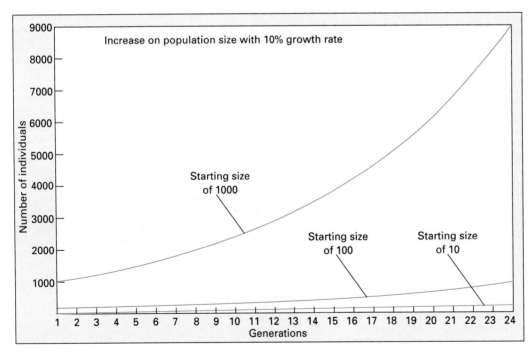

Figure 22-4 If the growth rate remains the same, the larger the size of the initial population the greater the number of individuals added to the population over the same amount of time. In all three populations shown here the growth rate is 10%, but the number of individuals in the population increases more rapidly as the population size increases from 10 to 100 to 1000.

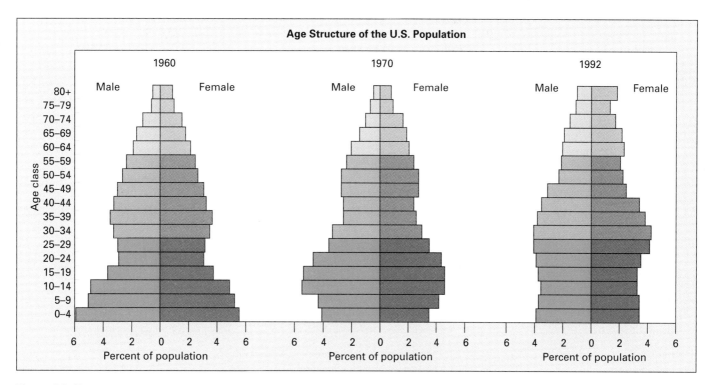

Figure 22–5 The age structure of the United States population has changed between 1960 and 1990. As members of the prereproductive class gradually enter their reproductive ages, they have an impact on the numbers added to the overall population.

structure of a population indicates the number of individuals of each age. The ages are often grouped into prereproductive, reproductive, and postreproductive categories. Generally, only those individuals of reproductive age add to the size of the population. However, the prereproductive group (those who are currently too young to reproduce) usually get older, and when they enter the reproductive class, they have children, adding additional members to the population. Thus, those born during a baby boom, such as occurred during the late 1940s and early 1950s in the United States, eventually reach reproductive age, which in humans is generally considered to be between the ages of 15 and 45, and have an effect on the rate at which numbers of individuals are added to the population (Figure 22–5). Even if the birth rate remains the same, the overall size of the population will grow more rapidly as the size of the reproductive class increases.

We see, then, that the relative size of the base of the age structure of a population determines how quickly numbers will be added to the population. Whereas a wide base to any age structure reflects a growing population, a narrow base is characteristic of a population that is getting smaller (Figure 22–6). A large base to any such pyramid spells trouble in terms of future demands on the country's commodities. Moderately stable nations, such as the United States, show a more consistent age distribution with some fluctuation, but with a small prereproductive population. A very stable country, such as Sweden, has an even smaller prereproductive base of those younger than 15 years of age.

Social Issue

The age structure of a population has many social and economic ramifications. It determines the so-called dependency ratio—the number of people in the working population compared with the number of people who must be supported (generally considered to be those younger than 15 and older than 65). In the United States today there are about two workers supporting each dependent. However, as the baby boomers approach retirement age, there will be fewer workers to support each dependent. If this support comes in the form of Social Security, won't the Social Security tax have to be raised? Within the next 30 years, the dependents older than 65 will outnumber those younger than 15. Should we invest in schools or nursing homes? Who decides?

The age at which a female has her first offspring has a dramatic impact on the rate at which a population grows. The younger she is when she begins to reproduce, the faster the population grows. When she has a child, the female moves from the prereproductive to the reproductive class. In fact, the age when reproduction begins is the most important factor influencing a female's overall reproductive potential. A woman

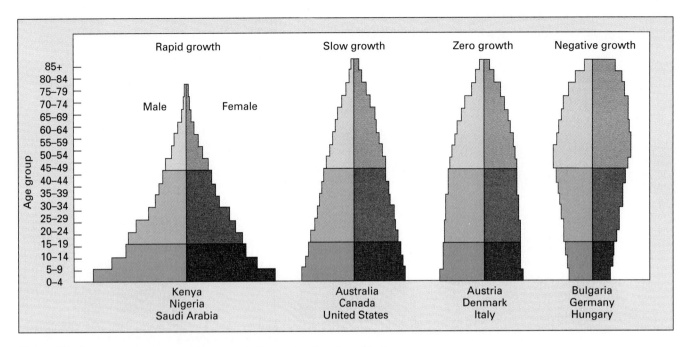

Figure 22–6 The age structure of a country indicates the direction of its future population growth: The broader the base of individuals who are still too young to reproduce, the more rapid the future growth of the population.

who has three children, one each year beginning when she is 13-years-old, and a woman who has five children, beginning when she is 30-years-old, have roughly the same effect on overall population growth. One factor that has been helping to slow the rate at which the population of the United States is growing is that an increasing proportion of women are postponing childbirth until their later reproductive years. Figure 22–7 shows the number of births per 1000 women in various age categories of their reproductive years in 1960, 1970, 1980, and 1990. Two important trends can be seen in this figure. First, the birth rate in the United States is declining for women of all ages. Second, the largest drop in birth rate occurred among women in their twenties. This trend is obvious when birth rates in each age group in 1960 are compared with those in 1990. One reason may be that increasing numbers of women want to establish professional careers before beginning motherhood. Regardless of the reasons, this trend has helped to slow the growth of the United States population.

Deaths

Worldwide, in spite of the steady decline in birth rate over the last 200 years, the human population has continued to grow because the drop in death rate has been greater than the drop in birth rate. The discovery of antibiotics was certainly instrumental in the decline of death rates. However, even more important were improvements in sanitation, hygiene, and nutrition.

Perhaps because of a human tendency to look on the positive side, the effects of death rate on population growth

are usually expressed as **life expectancy** (the average number of years a newborn is likely to live) or as **survivorship** (the number of individuals of a group born in the same year that is expected to be alive at any given age). Between 1900 and 1991, life expectancy of a person living in a developed country increased from 46 years to more than 70 years, primarily because of a drop in infant mortality from 40 per

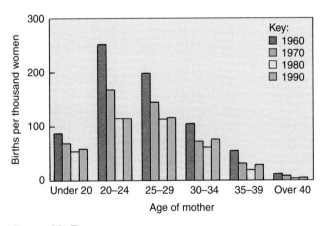

Figure 22–7 Fertility rates (births per 1000 women in one year) of American women in 1960, 1970, 1980, and 1990. The fertility rate of women in every age category has been steadily falling. In addition, there is a trend to delay childbirth until later in life. This is seen when the drop in the fertility rate of women 20 to 24 years of age between 1960 and 1990 is compared with the drop in fertility rate for women 30 to 34 years of age during the same time period.

1000 births to 10 per 1000 births. Indeed, it turns out that infant mortality is the most important factor in determining life expectancy.

Growth Rate

A population's growth rate is determined by the rates at which individuals are added to and subtracted from the population:

$$\text{Population growth rate} = \text{Birth rate} + \text{Immigration rate} - \text{Death rate} + \text{Emigration rate}$$

As we've seen, immigration and emigration can be important factors influencing the growth rate of some populations, but in our future discussions in this chapter we will pretend that their effects counterbalance one another.

In 1994, the birth rate (number of births per year per 1000 members of the population) in the world was 25. The death rate (number of deaths per year per 1000 members of the population) was 9. Thus, the growth rate of the world population was

$$\frac{\text{Birth rate}}{25/1000} - \frac{\text{Death rate}}{9/1000} = \frac{\text{Population growth rate}}{16/1000 = 1.6/100 = 1.6\%}$$

The growth rate in some parts of the world is greater than in other parts (Figure 22–8). The growth rate of the world population has been slowing steadily in the past decade. However, it is not *reversing*. (Another way of looking at it is that a bus heading toward a cliff at 60 miles an hour may slow to 50 miles an hour just before it goes over the edge. However, it doesn't change direction by slowing down a little. It will still go over the cliff.)

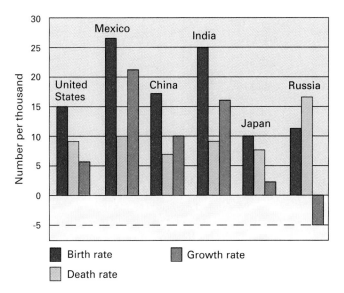

Figure 22–9 Gladys Davis is surrounded by her six great-grandchildren. She had four children, two of whom had children of their own, producing three children. They, in turn, produced these six great-grandchildren. By mid-1997, two of the children shown here had started their own families, producing five children. Although the size of each of these families was small, there are currently 15 living descendants of Jack and Gladys Davis. Thus, although they reproduced at close to replacement value, this family has grown sevenfold. How much has your family grown in the last few generations? *(Stephen Goodenough)*

Although a growth rate of a few percent a year may not seem startling, its importance becomes clearer when it is expressed as **doubling time**, that is, the number of years it will take for a population to double at that rate of growth. Think about your own family. How many living descendants do your grandparents have (Figure 22–9)?

To give you an idea of how important this measure is, if this page is 1/250 inch thick and were doubled (exponentially) only eight times it would be an inch thick. At 12 more doublings it would be as thick as a football field is long. At 42 times it would reach to the moon and at only 50 times would reach the 93 million miles to the sun. Obviously, knowing the doubling time of a population can yield some interesting predictions.

The simplest way to arrive at a doubling time is to divide the population's growth rate into 70 (a demographic constant value). Thus, if the world population continues to grow at the rate it did in 1996 (1.48%), the world population will double in just a little more than 47 years! This is how long it will be before we need two schools where there is now one and two roads, two dams, two telephone poles, two hospitals, two doctors, and two oil wells where there is now one. (One might ask, if we double the number of our oil wells, can we double our oil supply?) What happens to our

Figure 22–8 The growth rate of a population equals the birth rate minus the death rate (when the effects of immigration and emigration are ignored). The data for selected countries in 1996 are shown here.

quality of life in general? Where should the new roads go? Where do *you* live?

Doubling times vary widely in different countries, but in general, **less developed countries** (a euphemism for poorer countries) have faster doubling times than do **more developed countries** (richer countries). Do you see any irony in the fact that the countries less able to withstand more pressure are under the most pressure to care for rapidly increasing populations? Look at the doubling time for the world. How old will you be when the Earth's human population doubles? Your children?

Critical Thinking

A pair of cockroaches is placed on a small island. Assume that the doubling time for a population of cockroaches is 1 month. After 20 years the island contains half of the total number of roaches it could possibly hold. How much longer would it take until the island was completely filled with roaches? (*Hint:* What is the doubling time for the population?)

Patterns of Population Growth

The importance of population size can be seen in what is described as a J-shaped curve. Let's look at this curve and then compare it with what we will call the S-shaped curve.

The J-Shaped Curve

Let's return to an idea we introduced earlier in this chapter—the importance of population size on the rate at which new individuals are added to a population. Recall that *when the rate of growth remains constant*, the number of individuals in a population increases more rapidly the larger the size of that population. This unrestricted growth at a constant rate is called **exponential growth**, and when the size of a population growing in this fashion is shown graphically, we see a **J-shaped growth curve**.

Suppose you were to place a single bacterium on a petri dish containing nutrient agar. Given ideal conditions, the bacterium and each of its descendants could divide every 30 minutes. With each generation their numbers are effectively doubled. Two bacteria would produce 4, these would leave 8 and those 8 would leave 16, then 32 (Figure 22–10). Note that numbers are added to the population at an accelerating pace even though the growth rate remains constant, because the size of the population is increasing.

The S-Shaped Curve

Exponential growth occurs in ideal environments, where there is plenty of food and resources and adequate waste re-

moval. In the real world, however, these conditions rarely exist for long. What happens when the food supply begins to become depleted, when space runs out, or when wastes begin to accumulate? The answer to this question is of more than theoretical interest to humans. If you glance back at Figure 22–1, you will notice that the human growth curve is J-shaped. Let's see what we can learn from the changes in growth patterns of other organisms under more realistic conditions.

Although a population may be able to grow exponentially for a period of time, the environment simply can't support unlimited numbers of individuals of any species. The number of individuals of a given species that a particular environment can support for a prolonged period of time is called its **carrying capacity**. It is determined by such factors as availability of resources, including food, water, and space; ability to clean away wastes; and predation pressure.

Since there are environmental limits placed on population growth, we may wonder what might happen if a population overshoots the carrying capacity of the environment. One possibility is that environmental resources may become critically depleted and the population may crash. Population crashes are typical of populations of some types of insects and of most annual plants. Crashes can occur as part of natural cycles, as when numbers of grasshoppers swell each spring as they hatch from eggs, only to die each fall. These crashes are "programmed," part of a natural cycle, and one would certainly hope that our human demise will not be a part of some similar grand scheme. However, we should keep

Time (hours)	Number of individuals for curve (a)
10	1,048,576
9½	524,288
9	262,144
8½	131,072
8	65,536
7½	32,768
7	16,384
6½	8,192
6	4,096
5½	2,048
5	1,024
4½	512
4	256
3½	128
3	64
2½	32
2	16
1½	8
1	4
½	2
0	1

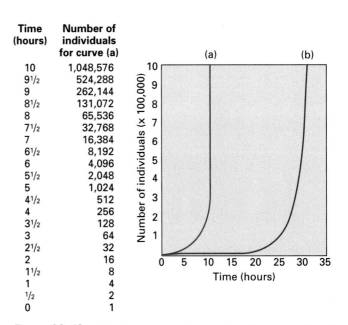

Figure 22–10 A J-shaped growth curve describes unrestricted growth at a constant rate. This curve shows the number of bacteria in a colony assuming that the bacteria divide every 30 minutes and all descendants survive.

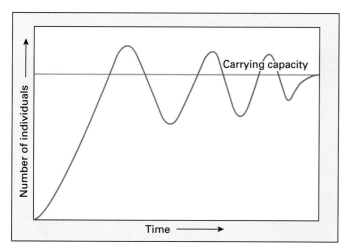

Figure 22–11 A logistic growth pattern produces an S-shaped growth curve. Population growth is often restrained by environmental factors, including availability of food, water, and space and the ability to prevent wastes from accumulating. Environmental factors such as these determine the environment's carrying capacity. Typically, population growth becomes slower the closer the population size comes to the carrying capacity. Eventually, growth levels off and fluctuates around the carrying capacity.

Figure 22–12 Food shortages may reflect density dependent population controls because they have a greater impact as population density increases. *(© Carl Purcell/Photo Researchers, Inc.)*

in mind that population crashes occur and there is no reason to believe that any species is immune, including our own.

More commonly, however, population growth begins rapidly, but it becomes increasingly slower as the population size approaches the carrying capacity of its environment because of environmental pressures, such as limited resources or accumulating wastes. Eventually, growth levels off and fluctuates slightly around the carrying capacity—a pattern that forms an **S-shaped growth curve** (Figure 22–11). This pattern is called **logistic growth**.

Regulation of Population Size

Now let's look at two ways populations can be reduced and place them in the context of what we have just learned.

Density-independent regulating factors include those events that bring about death that are not related to the density of individuals in a population. These are typically natural disasters, such as floods, mudslides, earthquakes, hurricanes, and fires. In Bhopal, India, not long ago, poisonous fumes escaped from a chemical plant and caused hundreds of deaths among the people in surrounding towns. Such events are all density-independent because adding or removing people from the population would not alter the event itself.

Density-dependent regulating factors include events that have a greater impact on the population as conditions become more crowded. For example, starvation may be a density-dependent effect. If only so much food is available,

an increasing population means less food to go around until, finally, some individuals begin to starve (Figure 22–12). The greater the density, the greater the effects of starvation on the population. Since disease-causing organisms and parasites spread more easily in crowded conditions, these too are density-dependent factors that regulate population sizes.

Human Influences on Our Population Growth Curve

Population Control

During the past three decades, human birth rates have declined throughout the world, especially in developing countries. Between 1965 and 1995, the average number of children per woman dropped by 65% in east Asia, about 33% in other parts of Asia, nearly 50% in Latin America, and 10% in Africa. In the United States and nearly all other developed countries, the average number of children per woman fell by about 40% and is now below the replacement level of 2.1 children per woman. (Replacement-level fertility is the number of children a couple must produce to replace

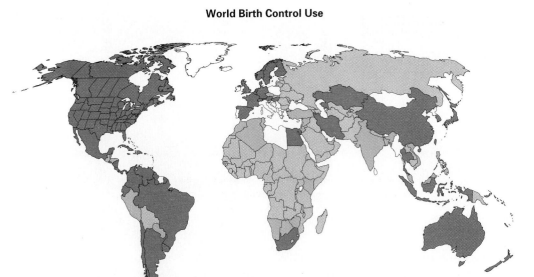

World Birth Control Use

Percent of women of reproductive age using modern birth control

☐ Less than 25 ☐ 25 to 44 ☐ 45 to 64 ☐ 65 or more ☐ No data

Figure 22–13 This map shows the use of modern methods of contraception during the late 1980s and early 1990s by both married women and women in nonmarital unions. The availability and use of modern contraceptive methods played a key role in reducing the birth rate throughout the world, especially in developing nations. *(Used with permission from Scientific American, Sept., 1996, p. 34)*

themselves. This is generally 2.1 children in developed countries, and it is 2.7 children in developing countries.)

Several factors have contributed to the drop in birth rate. An important factor is the increased use of modern contraceptive measures (Figure 22–13). Availability of reliable means of contraception has given many women an opportunity to decide for themselves when and if they want to have children. Education has also played a role. Not only has increased literacy brought reliable information about birth control methods to women but the demands of increased education have caused women to postpone childbearing until later in life. In some countries, government policies have also helped to lower birth rates. China, for instance, has dramatically lowered its birth rate, largely though universal access to family planning services (including abortion), economic incentives to have smaller families, and educational campaigns promoting one-child families.

Nonetheless, our numbers continue to climb. As we've seen in this chapter, our population will grow as long as the birth rate is greater than the death rate. Also, even if the birth rate is at replacement level, the population will grow in regions where the largest percentage of the population is in the prereproductive age group.

Alterations of Earth's Carrying Capacity

It is difficult to estimate the Earth's carrying capacity for humans. Indeed, the Earth's ability to support people is influenced both by natural constraints, such as limited resource availability, and by human choices on a variety of issues, including economics, politics, technology, and values. Thus, Earth's carrying capacity for humans is uncertain and constantly changing. For instance, technological advances, in agriculture and pollution control have raised the carrying capacity. At the same time, however, we have lowered the carrying capacity by consuming Earth's resources faster than they can be replaced.

Estimates of Earth's carrying capacity vary widely, ranging from 5 billion to 20 billion people. Those who accept the lower values point to resource depletion and pollution as evidence that we have already exceeded Earth's carrying capacity with a world population of more than 5.7 billion. Therefore, some people predict that our population will crash sometime during the twenty-first century because of starvation or war over the remaining resources. Other people are more optimistic. They point to the declining growth rate of the human population and estimate that the world population will stabilize at about 11.6 billion just after the year 2200.

Fallout from the Population Bomb

Whatever your personal predictions for the future of humankind, one thing should be clear—the world is becoming an increasingly crowded place. You may recall from the previous chapter that Earth is a closed ecosystem. Therefore, we run the risk of using up our supplies and making our world unlivable because of pollution. In fact, the size of the human population contributes to most of the problems we face today.

The People-Food Predicament

Most of us know what hungry means, but not hunger. Hungry is the feeling we get when lunch is late; hunger is a condition in which lack of food causes physiological changes within our body. Hunger can be brought on by two conditions: **undernourishment**, when not enough food is eaten, and **malnourishment**, when the diet is not balanced and the right foods are not eaten.

Hunger particularly affects children, pregnant women, and nursing mothers. Children are notably susceptible to two very dangerous nutritional diseases. Marasmus is caused by a diet low in proteins and calories. The result is wrinkled skin, a startling appearance of old age, and thin limbs caused by wasting muscles. A number of children get the disease after being fed baby formulas that were over diluted after they were received from international aid agencies. The misuse of the aid often causes the problem. Kwashiorkor, on the other hand, results from a protein deficiency. It often appears when a child is weaned from breast milk. One of the many symptoms of the disease is fluid retention (edema), which is typically seen as a pronounced swelling of the abdomen.

Chronic undernourishment (starvation) is associated with irreversible changes in growth and brain development. If children receive less than 70% of the required daily calories, their growth and activity fall below normal levels and the number of brain cells decreases and brain chemistry is altered. Thus, undernourishment can lower intelligence.

One reason for hunger is that food is unequally distributed throughout the world. North Africa, for example, produces very little food, because of a combination of drought, poor farming practices, and warfare. On the other hand, both food abundance and scarcity may exist in the same country. For example, southern Brazil produces a great deal of food, but the northern part of the country does not, and so the people living there may suffer from malnutrition. The United States is an important exporter (our population has not yet caught up with our food production), but even here we find hunger in certain areas.

Although factors such as droughts and wars contribute to the problems of feeding the world's hungry people, overpopulation is an important cause. Indeed, one of the greatest challenges facing us today is increasing food production at a rate that will feed the growing number of people. For instance, world grain production has been increasing fairly steadily, but it can't keep up with the growing population, and so the amount of grain per person is beginning to decrease (Figure 22–14). With the current rate of population increase, we must produce an *additional* 250 million metric

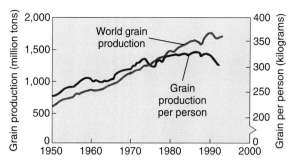

Figure 22–14 Although the world's food production has increased steadily, it has been outpaced by the growth of the world's population. The amount of food per person is beginning to decrease. *(Data from the U.S. Department of Agriculture)*

tons (280 million tons) each year, just to feed the new people that have been added to the population.

Agricultural production has increased largely because of the '**green revolution**'—the development of high-yield varieties of crops and the use of modern cultivation methods, including the use of farm machinery, fertilizers, pesticides, and irrigation. The green revolution has boosted crop production in many countries. For example, in the past Indonesia imported more rice than any other country, but now it has enough rice to feed its people, as well as some to export.

Nonetheless, the green revolution has a big price tag—high energy costs and environmental damage. Although crop yields may be four times greater than with more traditional methods, modern farming practices use as much as a hundred times more energy and mineral resources. It takes energy to produce fertilizers. In addition, fossil fuels are needed to provide power for tractors and combines and to install and operate irrigation systems. The environment has also been damaged by practices associated with the green revolution (Figure 22–15).

Fishing has long provided humans with an abundant food supply, but now many fish populations are being depleted to such levels that certain kinds of fish can no longer be caught in certain areas. Recent examples include Atlantic cod in New England and salmon in the northwest United States. The problem is that we haven't given the fish enough time to replace their numbers at their reduced population size. As numbers of fish have dwindled, the response of humans has been to fish harder, adding more boats and new techniques, including electronic searches. Part of the problem is that each fishing nation behaves as if it must get its share before its competitors make it impossible.

Biologist Garret Hardin has added to our understanding of the problem by describing "The Tragedy of the Commons." In Victorian England, commons were grazing areas, sometimes in town, where everyone could graze sheep. In this way, the sheep could be looked after and marketed efficiently. The system worked fine as long as everyone limited the amount of sheep they grazed. If someone added a few

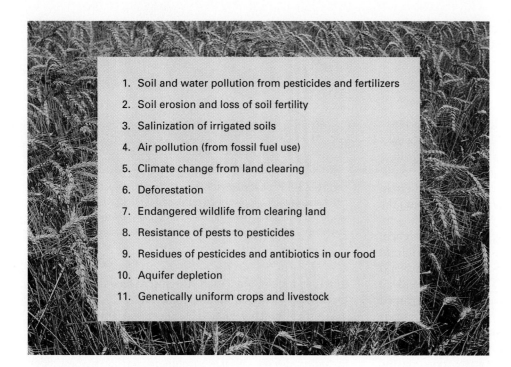

1. Soil and water pollution from pesticides and fertilizers

2. Soil erosion and loss of soil fertility

3. Salinization of irrigated soils

4. Air pollution (from fossil fuel use)

5. Climate change from land clearing

6. Deforestation

7. Endangered wildlife from clearing land

8. Resistance of pests to pesticides

9. Residues of pesticides and antibiotics in our food

10. Aquifer depletion

11. Genetically uniform crops and livestock

Figure 22–15 The green revolution has increased crop production, but it has also caused many environmental problems. *(Inga Spence/Tom Stack and Associates)*

extra sheep (as is bound to happen), that person would have a marketing advantage without doing too much harm. His neighbors couldn't allow him to have that advantage, though, so they would add a few sheep of their own to keep up and soon the area would be overgrazed and destroyed.

It has been suggested that the sea is a commons and that we are creating a modern tragedy by behaving selfishly. To illustrate, biologists have suggested that fish breeding stocks could be replenished if 29% of each catch were returned to the sea. However, as the number of fish has declined, fishermen have found it increasingly difficult to make a living. Thus, they would find it difficult to feed their own families if they returned a portion of their catch to the sea.

Scientists are presently working to introduce new foods that humans might be persuaded to eat. Genetic engineering may soon provide us with proteins formed from amino acids that closely fit our nutritional needs. Scientists have developed incaparina, a nutritious, vitamin-enriched mix of corn and cotton seed meal. Other researchers are culturing algae, developing rodent ranches and antelope ranches, finding ways to use "trash fish" (those usually thrown back), and finally, if you can believe it, working on ways to create edible slime grown on sewage sludge.

Pollution

We have discussed water pollution in Chapter 21, the health consequences of air pollution in Chapter 14, and acid rain as a result of air pollution in Chapter 2. Here we will focus on another consequence of air pollution: the destruction of the ozone layer.

Ozone, composed of three atoms of oxygen (O_3), can be a good thing or a bad thing. If it's close to the Earth, say any-

where in the first few miles above the surface, it's generally a bad thing. In fact, it is the primary component of photochemical smog, the noxious gas produced largely by sunlight interacting with air pollutants. Ozone irritates the eyes, skin, lungs, nose, and throat. (The effects of photochemical smog on the respiratory system are discussed in the Environmental Concerns essay in Chapter 14.) It can also damage forests and crops and dissolve rubber.

However, naturally produced ozone is an essential part of the stratosphere, a layer of the lower atmosphere that encircles the Earth about 10 to 45 kilometers (6–28 miles) above its surface. Without this layer of ozone, most forms of life could not exist on Earth. The ozone layer acts as a shield against ultraviolet (UV) radiation from the sun. UV radiation causes cataracts, aging of the skin, sunburn, and snow blindness, and it is the primary cause of skin cancer that kills 12,000 Americans each year. In addition, UV radiation inhibits the immune system and can interfere with plant growth.

The ozone in the atmosphere only reaches levels of about 1 molecule in 100,000. If it were all compressed into one layer, its depth would only be equal to the diameter of a pencil lead. Still, it is able to shield the Earth from excessive UV radiation.

In 1985, British scientists reported a sharp drop in the concentration of ozone over the Antarctic. This hole appears on a yearly cycle. It appears at the beginning of the Antarctic spring (September) and each year the ozone layer in this region has gotten thinner. Now ozone holes have been reported in the Arctic and over other parts of the world (Figure 22–16).

Chlorofluorocarbons (CFCs)—important in cooling systems, as an aerosol propellant, in the manufacture of plas-

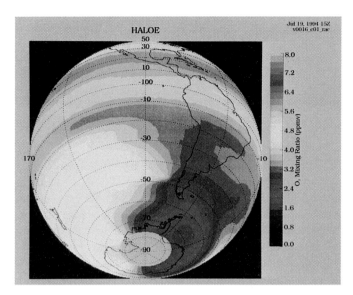

Figure 22–16 The ozone hole over Antarctica is shown in blue in this computer-generated image of part of the Southern Hemisphere. The green areas over South America and Central America indicate low ozone levels. Normal ozone levels are indicated in yellow. No ozone measurements were taken over the South Pole (indicated in white). *(NASA)*

tic foam such as Styrofoam, and as solvents in the electronic industry—are the primary culprit responsible for the destruction of the ozone layer. The CFCs drift up to the stratosphere. Here, UV radiation causes them to break down to chlorine, fluorine, and carbon. Then, under the conditions found in the stratosphere, chlorine can react with ozone, converting it to oxygen. Because chlorine is not altered in this reaction, a single chlorine molecule can destroy thousands of ozone molecules.

A number of laws are now in place to prevent the production or release of CFCs. Worldwide agreement to reduce the production of CFCs has been growing. In 1978 the United States banned the use of CFCs in certain propellants, and in 1987 several countries agreed to reduce CFC production by 50% by 1998. Then, in 1992, representatives of 93 countries agreed to ban CFCs by 1996. However, the problem is already ongoing. Tiny molecules of these chemicals are moving upward toward the stratosphere, where we can expect them to increase in concentration for decades to come. Although banned, their destructive effects will linger for years.

Resource Depletion

Soil Erosion and Desertification

As the population grows, land is cultivated, grazed, and deforested faster than it can recover. Wind and rain then carry away the topsoil and the productivity of farmland declines. In this way, over farming and overgrazing transform mar-

ginal farmlands to deserts, a process called **desertification** (Figure 22–17).

Deforestation

It is often the case that you don't know what you have until it's gone. This may be true for forests. Many of us live in areas where trees are so common that we take them for granted. However, trees, especially when they are grouped together to form forests, are an essential part of the global ecosystem. As we learned in the previous chapter, forests play an important role in water, carbon, and nitrogen cycles. Tree roots also reduce erosion by holding soil in place. If trees are cut down, rainwater runs off the land carrying away soil and causing floods. In addition, trees are part of local ecosystems that include an incredible variety of life. They also influence local and global climate, including temperature and rainfall. And, in hot weather, their shade is pleasant.

Why, then, are forests disappearing? The cause is simple—the human population is expanding. People need space in which to live. Wood is needed for homes and furniture. Livestock can't graze in forests. So, trees are cut and most are not replanted.

Deforestation is the removal of trees from an area without replacement. Deforestation is taking place in many regions of the world, including the United States. Before colonization, forests covered most of the eastern seacoast of the United States, but few would suggest that Boston should

Figure 22–17 Desertification is the process by which the overuse of farmland or, in this case rangeland, converts the region to desert. *(Visuals Unlimited/J. Alcock)*

be leveled and forests replanted. Many of the forests of the Pacific Northwest have been sacrificed; 90% to 95% of the forests have been cut.

Nonetheless, tropical forests are falling the fastest. Why? The first and most important reason that land is cleared is so native families can feed themselves. A second reason is commercial logging. The third reason is cattle ranching. Cattle can graze on the land for about six to ten years after trees have been cleared from a tropical forest and then shrubs called scrub savannah take over and make the area unsuitable for rangeland. Most of these cattle ranches are owned by foreign companies and the beef is often exported to fast food restaurant chains. Thus, the decisions we make about what to eat for lunch may have an indirect influence on the rate of tropical deforestation. Less important reasons include mining and the development of hydroelectric dams.

Whatever the reason for the tropical deforestation, soil fertility declines rapidly when these forests are cut. When it does, native people who depend on the soil for their living become even poorer. The entire region often experiences periods of drought alternating with floods.

Social Issue

Deforestation of tropical rain forests sometimes places the interests of native people at odds with those of developers. How can the interests of both parties, as well as the interests of the rest of us, be balanced? Who makes the decisions? Who guarantees fairness? Who looks at the "big picture"?

Loss of Biodiversity

Biodiversity can be described as species richness. In specific terms, **biodiversity** is the number and variety of all living things in a given area. Recently, an alarm has been raised by scientists worldwide because there are indications that globally, and especially in certain critical areas, biodiversity is decreasing. We are losing species, perhaps as many as 100 species each day, caused in large part by human activity.

Mass extinctions, the loss of many species from the Earth, have occurred previously, but those extinctions were driven by competition between similar species and by environmental change. This time, though, many other species are being forced to compete with us as we change the environment to suit ourselves, rendering it unsuitable for them.

We are not always doing this on purpose, but our ignorance is part of the problem. In earlier years, when we used DDT as an insecticide, we simply didn't know it was causing bird egg shells to thin so much that the nesting bird parents would break them. We also didn't know that releasing CFCs from our refrigerators and air conditioners would deplete the thin layer of ozone in the upper atmosphere that pro-

tects us from the sun's damaging ultraviolet rays. Still, our ignorance of what we are doing cannot undo the effects of our behavior.

How many species are there? The answer is that no one knows. Many scientists believe, however, that there are 8 million to 30 million species now in existence, with about 90% of these living on land. However, we are aware of only about 1.7 million of these species, and only 3% of these have been studied.

Where are these species? Most are in the tropics. Although tropical rain forests cover only 7% of the Earth's surface, they are home to between 50% and 80% of the Earth's species. It is estimated that 7 million to 8 million kinds of species live in the tropics. Nonetheless, most of the *known* species come from the temperate regions, because most scientists live in the temperate regions and they tend to investigate their own surroundings.

Habitat destruction is largely responsible for the loss of biodiversity, and the primary reason habitats are destroyed is to create living space and economic development of an expanding human population. It has been estimated that roughly nine-tenths of the original tropical rain forest will be destroyed in the next 30 years. This means that in the next 30 years, we could lose a quarter of all the species on Earth. (How old will you be?) By the end of the next century, perhaps two-thirds of the species living today will be gone. Evolutionary biologists tell us that, currently, we are losing species at the rate of 1000 to 10,000 times the average rate for the last 65 million years, since the extinction of the dinosaurs. However, the rain forests aren't the only type of habitat being destroyed. Lumbering and acid rain are destroying northern forests. Also, marine ecosystems are being destroyed by pollution, overfishing, and coastal development.

Social Issue

Most tropical rain forests, places where biodiversity is greatest, are located in developing nations. These regions are ecologically and economically important to all nations. The developing nations, however, are least able to afford the measures that will be needed to protect biodiversity. What steps would you suggest to promote the international cooperation needed to preserve biodiversity? Businesses in developed countries, which have the economic resources to invest in biodiversity and the search for medicines, are usually driven by known profit-making potential. The discovery of marketable medicines is uncertain. What measures can you suggest that would stimulate the interest of businesses in investing in the tropical rain forests? The interests of developing nations that own the tropical rain forests must also be protected. How can the interests of both parties be protected?

We might wonder, however, why we should care about the loss of species we didn't even know existed. Two reasons

Figure 22-18 The rosy periwinkle is a source of two anticancer drugs. The plant is found in Madagascar, a region where humans have destroyed 90% of the vegetation. *(Doug Wechsler)*

immediately come to mind. The first reason is that it preserves genetic diversity. That genetic diversity, encoded in the chromosomes of the vast array of species, is useful for crossbreeding. In this way, the traits of two different strains can be joined. For example, not long ago a new kind of corn was found on a Mexican hillside. (The farmer had to be persuaded not to plow it under to plant his usual corn crop.) The newly discovered corn was a perennial and proved to be resistant to a disease called corn smut, a fungal disease that causes the kernels to turn black and become oddly shaped. Plans were made to cross it with common forms of agricultural corn, and these sorts of experiments continue.

Interestingly, the usefulness of genetic diversity as a reservoir of genetic stores has increased dramatically with modern technology. With our recently acquired abilities in genetic engineering, specific genes can be isolated and moved into the genome of other kinds of organisms. Thus, the maintenance of genetic stores becomes even more important as we seek to maximize our crop yields or otherwise create new forms of organisms that might benefit society.

The second benefit to maintaining biodiversity lies in developing new kinds of medicines. About 25% of all known drugs come from plants. The rosy periwinkle found in Madagascar is a source of two anticancer drugs, vincristine and vinblastine, and the Mexican yam was once a source of oral contraceptives (Figure 22–18). A problem here is that most plants that have proven to be medicinally useful are found in the tropics, and this is an area of drastically increasing populations. Thus, we can expect greater rates of deforestation for the future. With 155,000 of the known 250,000 plants existing in tropical rain forests and less than 1% of the

known species having been tested for their medicinal value, we have no way of knowing what potential new medicines are being destroyed.

A Sustainable Future

In many ways, our grandmothers were right—older ideas *were* better, at least in some ways. Today's economy is built on waste. We seem to have an idea that the world is here for humans alone.

In the last two chapters we have seen that humans are part, a *small* part, of a large ecosystem. We can no longer think that we are the pinnacle of all life forms. We must realize that we are subject to the same principles that govern all other life, indeed, all of Earth. Instead of trying to conquer nature, we must work with natural laws. Above all, we must recognize that resources on Earth are limited and must be shared with all living organisms.

This dramatic turnabout in thinking won't be easy. It will require changes in the way governments and businesses operate and the ways in which people think. However, as was said earlier, humans are, by nature, problem solvers. We now have the opportunity to intelligently approach one of the greatest problems to ever confront humankind. We have to be optimistic that a solution is possible and to work together to achieve it. We can certainly expect an increasing sense of urgency as each year reveals some new facet to a problem that will increasingly demand our attention.

Maintaining Our Remaining Biodiversity

In the face of powerful evidence that we are undergoing a marked reduction in the Earth's biodiversity, there is some guarded optimism that we may be able to reverse the trend. Here, then, let's consider some of the measures that are being suggested.

First, both developed (richer) and developing (poorer) countries must carefully census their life forms. They need to increase their reference libraries and their taxonomic inventories of the species they harbor. The need is especially acute in tropical countries. Unfortunately, these are often the least likely to be able to afford such programs, so the effort must be worldwide, with developed countries subsidizing the research in less affluent areas.

Second, we must link economic development of impoverished regions, particularly the tropics, to conservation. Lending agencies must stipulate that certain areas must be set aside and left undeveloped. The decisions regarding which locations should be spared development should be made with the advice of ecologists so that the effects on the biodiversity of the region would be minimized. (One cannot help wondering what effect these rules would have had on the development of the United States.)

Third, education of those whose decisions would most immediately affect biodiversity should be a priority. For example, recent studies have shown that native people are able to make more money through sustainable use of the forest (as in gathering nuts and fruits) than would have been possible if the forests had been cleared for agriculture or ranching. In addition, it is imperative that local residents in threatened areas be provided information regarding the benefits of carefully controlled ecotourism. Many people in developing countries are

also learning that scientists worldwide are interested in the medicines and healing knowledge they have, so in a sense, the educational opportunities are reciprocal.

Fourth, biological research worldwide must begin to be done with heavy emphasis on multiple land-use planning. In other words, biologists must begin to work more closely with zoning and land-use personnel to maximize the use of areas being cleared or cultivated. It is suggested that crops be rotated to increase biodiversity and that single-species forests planted by paper companies, for example, might be better replaced by a more diverse forest that could harbor more species of animals. In this latter case, of course, the situation becomes more complicated than it seems because, as paper companies point out, why should they be asked to reduce their profits to increase biodiversity? This is one of the big problems with any ecological issue. We may know what's good for the environment, but it may not always be good for all concerned.

Social Issue

In the preceding example, in what way do you think paper companies could be enticed to plant mixed forests that are less profitable for them? Should we pay more for their products or give them tax breaks or outright subsidies to keep their profits high enough?

SUMMARY

1. Population dynamics is the description of how populations change. Population size changes when the number of individuals added through birth or immigration is different from the number of individuals removed through death or emigration.
2. The change in population is usually expressed as a rate: the number of births or deaths for a set number of individuals during a set time period. If we rule out immigration and emigration, we can say the growth rate of a population equals the birth rate minus the death rate. Population size is important in

determining how quickly new individuals will be added. Population size increases faster, the larger its initial size (at the same *rate* of increase).
3. The age structure of a population is important in determining future changes in population size. The age structure of a population is the number of individuals in each age category. Ages are commonly grouped into prereproductive, reproductive, and postreproductive categories. A growing population has a large base of prereproductive individuals. A stable population

has approximately equal numbers of individuals in each age category. A population that is getting smaller has a large proportion of individuals who are past reproductive age.

4. The age at which females have their first offspring is the most important factor in determining their reproductive potential and, therefore, has a great influence on the rate at which a population grows.

5. There has been a steady decline in death rate over the last 200 years, primarily because of a drop in infant mortality.

6. The doubling time for a population is the length of time it will take for the population to double in size at that rate of growth. In general, less developed countries have faster doubling times than do more developed countries.

7. A J-shaped growth curve results when a population grows exponentially and is not limited by environmental factors.

8. The carrying capacity of the environment is the number of individuals (of a particular species) it can support over a long period of time. It is determined by the amount of available resources. If a population overshoots its carrying capacity and depletes its resources, a population crash, in which the numbers drop rapidly, can follow.

9. Commonly, as a population grows and its size approaches the carrying capacity of the environment, the growth rate slows and the population size levels off, fluctuating around the carrying capacity. This so-called logistic growth produces an S-shaped growth curve.

10. Population size can be regulated by density-independent regulating factors, events such as natural disasters whose effects are not influenced by the size of the population, and by density-dependent regulating factors, which include factors that have a greater impact as the population becomes more crowded. Density-dependent factors include food availability and disease.

11. Humans have influenced the population growth curve by reducing the worldwide birth rate, largely through the increased use of modern contraceptive methods. Humans have raised the carrying capacity of the Earth through technological advances, but they have also decreased it through resource depletion and pollution.

12. The size of the human population contributes to many of the problems we face today.

13. Hunger can be caused by undernourishment (starvation) or by malnourishment (a diet that does not contain enough of the right kinds of nutrients). Hunger has its greatest effects on children, pregnant women, and nursing mothers.

14. It has proven difficult to increase food production at a rate that can keep up with the growing human population. The green revolution has increased agricultural production through the use of high-yield varieties of crops and modern cultivation methods. The green revolution also has high energy costs, depletes natural resources, and causes environmental damage. Many fishing areas are becoming depleted because of overfishing.

15. Air pollution in the form of chlorofluorocarbons (CFCs) is destroying the ozone layer in the lower atmosphere (the stratosphere). The ozone layer traps ultraviolet (UV) light, protecting life on Earth from the harmful effects of UV radiation.

16. Overgrazing and over farming are converting marginal farm lands to deserts, a process called desertification.

17. Deforestation—removing trees from an area without replacing them—is a growing problem throughout the world, especially in the tropical rain forests.

18. Biodiversity, the number and variety of living things, is being reduced dramatically, largely because of human activity. Most of the loss is occurring in the tropics and most is due to habitat destruction. Two practical reasons for concern over the loss of biodiversity are that these species could have genes that would someday prove useful or that they could be found to produce chemicals with medicinal qualities.

REVIEW QUESTIONS

1. List four factors that determine whether a population grows or declines.

2. Explain why population growth is usually expressed as a rate.

3. Explain why the size of the initial population influences how quickly it grows even when the growth rate remains constant.

4. What is meant by the age structure of a population? How does a population's age structure relate to predictions of its future growth?

5. What is meant by the doubling time of a population? Why is this an important measure?

6. Under what conditions is a population likely to have a J-shaped growth curve?

7. What is the carrying capacity of the environment? How does this relate to the S-shaped growth curve?

8. Differentiate between density-independent and density-dependent regulating factors. Give examples of each.

9. Describe some of the health consequences of hunger. Annual grain production has been increasing steadily, but the amount of grain available for each person has remained the same or decreased in recent years. Why?

10. What is the green revolution? What are its benefits and drawbacks?

11. What determines whether ozone is helpful or harmful? What is causing the destruction of the ozone layer?

12. What causes desertification?

13. Define deforestation. Where is it occurring most rapidly? What are the primary reasons for deforestation?

14. Define biodiversity. Where is it greatest? Why should we be concerned about the loss of biodiversity?

Glossary

Accommodation A change in the shape of the lens brought about by contraction of the smooth muscle of the ciliary body that changes the degree to which light rays are bent so that an image can be focused on the retina.

Acetylcholine A neurotransmitter found in both the central nervous system and the peripheral nervous system. It is the neurotransmitter released at neuromuscular junctions.

Acetylcholinesterase An enzyme that breaks the neurotransmitter acetylcholine into its inactive component parts, acetate and choline. Acetylcholinesterase stops the action of acetylcholine at a synapse.

Acid Any substance that increases the concentration of hydrogen ions in solution.

Acinar cells Exocrine cells of the pancreas that secrete digestive enzymes into ducts that empty into the small intestine.

Acromegaly A condition characterized by thickened bones and enlarged soft tissues caused by overproduction of growth hormone in adulthood.

Acrosome A membranous sac at the tip of a sperm cell that contains enzymes that facilitate sperm penetration into the egg during fertilization.

Acrosome reaction A process that must precede fertilization. It occurs when capacitated sperm contact the corona radiata, a layer of cells surrounding the secondary oocyte. Once such contact occurs, perforations develop in the plasma membrane and outer acrosomal membrane of the sperm, and enzymes spill out of the acrosome. These enzymes disrupt attachments between cells of the corona radiata and zona pellucida, allowing the sperm to reach the oocyte.

Actin The contractile protein that makes up the major portion of the thin filaments in muscle cells. An actin (thin) filament is composed of actin, troponin, and tropomyosin.

Action potential A nerve impulse. An electrochemical signal conducted along an axon. A wave of depolarization caused by the inward flow of sodium ions followed by repolarization caused by the outward flow of potassium ions.

Active immunity Immune resistance in which the body actively participates by producing memory B cells and T cells following exposure to an antigen, either naturally or through vaccination.

Active site A specific location on an enzyme where the substrate binds.

Active transport The movement of molecules across the plasma membrane from a region of lower concentration to that of higher concentration with the aid of a carrier protein and energy (usually in the form of ATP) supplied by the cell.

Adaptation The process by which populations become better attuned to their particular environments as a result of natural selection.

Adaptive trait A change in the structure or function of an organism that results from natural selection and that makes an individual better able to survive and reproduce in its natural environment.

Addison's disease An autoimmune disorder characterized by fatigue, loss of appetite, low blood pressure, and increased skin pigmentation resulting from hyposecretion of glucocorticoids and aldosterone.

Adenosine triphosphate (ATP) A nucleotide that consists of the sugar ribose, the base adenine, and three phosphate groups. ATP is the energy currency of all living cells.

Adipose capsule A protective cushion of fat beneath the renal fascia of each kidney.

Adolescence The stage in postnatal development that begins with puberty and ends by about age 17. It is a period of rapid physical and sexual maturation during which the ability to reproduce is achieved.

Adrenal cortex The outer region of the adrenal gland that secretes glucocorticoids, mineralocorticoids, and gonadocorticoids.

Adrenaline See epinephrine.

Adrenal medulla The inner region of the adrenal gland that secretes epinephrine and norepinephrine.

Adrenocorticotropic hormone (ACTH) Anterior pituitary hormone that controls the synthesis and secretion of glucocorticoid hormones from the cortex of the adrenal glands.

Adulthood The stage in postnatal development that is generally reached somewhere between 18 and 21 years of age. Growth and formation of bone are usually completed by age 25, and thereafter developmental changes occur quite slowly.

Afferent (sensory) neuron A nerve cell specialized to conduct nerve impulses from the sensory receptors *toward* the central nervous system.

Agglutinate To clump together.

Aging The normal and progressive alteration in the structure and function of the body. Aging is possibly caused by declines in critical body systems, disruption of cell processes by free radicals, slowing or cessation of cell division, and decline in the ability to repair damaged DNA.

Agonist The muscle that moves the bone. The action of the agonist is opposite that of the antagonist.

Agranulocytes The white blood cells without granules in their cytoplasm, including monocytes and lymphocytes.

AIDS   Acquired Immune Deficiency Syndrome. A diagnosis of AIDS is made when an HIV+ person develops one of the following conditions: (1) a helper T cell count below 200/mm³ of blood; (2) one of 26 opportunistic infections, the most common of which are *Pneumocystis carinii* pneumonia and Kaposi's sarcoma, a cancer of connective tissue that affects primarily the skin; (3) a loss of more than 10% of body weight (wasting syndrome); or (4) dementia (mental incapacity such as forgetfulness or inability to concentrate).

Albinism   A genetic inability to produce the brown pigment melanin that normally gives color to the eyes, hair, and skin.

Aldosterone   A hormone (the primary mineralocorticoid) released by the adrenal cortex that stimulates the reabsorption of sodium within kidney nephrons.

Allantois   The extraembryonic membrane that serves as a site of blood cell formation and later becomes part of the umbilical cord, the rope-like connection between the embryo and the placenta.

Allele   An alternative form of a gene. One of two or more slightly different versions of a gene that code for different forms of the same trait.

Allergen   An antigen that stimulates an allergic response.

Allergy   A strong immune response to an antigen (an allergen) that is not usually harmful to the body.

Allometric growth   The change in the relative rates of growth of various parts of the body. Such growth helps to shape humans and other organisms.

Allosteric site   A specific site on an enzyme that is away from the active site.

Alpha cells   The cells found in the islets of Langerhans of the pancreas that produce the hormone glucagon.

Alveolus   A thin-walled rounded chamber. In the lungs, the alveoli are the surfaces for gas exchange. They form clusters at the end of each bronchiole that are surrounded by a vast network of capillaries. The alveoli greatly increase the surface area for gas exchange.

Amino acids   The building blocks of proteins consisting of a central carbon atom bound to a hydrogen atom, an amino group (NH_2), a carboxyl group (COOH), and a side chain designated by the letter R. There are twenty different amino acids important to human life; some can be synthesized by our bodies (nonessential amino acids), whereas others cannot be synthesized and must be obtained from the foods we eat (essential amino acids).

Amniocentesis   A method of prenatal testing for genetic problems in a fetus in which amniotic fluid is withdrawn through a needle so that the fluid can be tested biochemically and the cells can be cultured and examined for genetic abnormalities.

Amnion   The extraembryonic membrane that encloses the embryo in a fluid-filled space called the amniotic cavity. Amniotic fluid forms a protective cushion around the embryo that later can be examined as part of prenatal testing in a procedure known as amniocentesis.

Anabolic steroids   Synthetic hormones that mimic testosterone and stimulate the body to build muscle and increase strength. Steroid abuse can have many dangerous side effects.

Anabolism   The building (synthetic) chemical reactions within living cells, as when cells build complicated molecules from simple ones. Compare with catabolism.

Analgesic   A substance, such as Demerol, that relieves pain.

Anaphase   In mitosis, the phase when the chromatids of each chromosome begin to separate, splitting at the centromere. Now separate entities, the chromatids are considered chromosomes and they move toward opposite poles of the cell.

Anaphylactic shock   An extreme allergic reaction that occurs within minutes after exposure to the substance that a person is allergic to and that can cause pooling of blood in capillaries, which causes dizziness, nausea, and sometimes unconsciousness as well as extreme difficulty in breathing. Anaphylactic shock can cause death.

Androgen   A steroid sex hormone secreted by the testes in males and produced in small quantities by the adrenal cortex in both sexes.

Anemia   A condition in which the blood's ability to carry oxygen is reduced. It can result from too little hemoglobin, too few red blood cells, or both.

Anesthesia   The drug-induced loss of the sensation of pain. It may be general or regional.

Aneuploidy   A general term describing the condition whereby gametes or cells contain too many or too few chromosomes compared with the normal number.

Aneurysm   A blood-filled sac in the wall of an artery caused by a weak area in the artery wall.

Angina pectoris   Choking or strangling chest pain, usually experienced in the center of the chest or slightly to the left, that is caused by a temporary insufficiency of blood flow to the heart. It begins during physical exertion or emotional stress, when the demands on the heart are increased and the blood flow to the heart muscle can no longer meet the needs.

Angioplasty   A procedure that widens the channel of an artery obstructed because of atherosclerosis. It involves inflating a tough, plastic balloon inside the artery.

Anorexia nervosa   An eating disorder characterized by deliberate self-starvation, a distorted body image, and low body weight.

Antagonist   The muscle that has an action opposite to the agonist and that relaxes to permit movement.

Antagonistic pairs   Muscles arranged in pairs so that the actions of the members of the pair are opposite to one another. This arrangement is characteristic of most skeletal muscles.

Anthropoids   The suborder of modern primates that includes monkeys, apes, and humans.

Antibody   An immunoglobin. A Y-shaped protein produced by plasma cells during an immune response that recognizes and binds to a specific antigen because of the shape of the molecule. Antibodies defend against invaders in a variety of ways including neutralization, agglutination and precipitation, or activation of the complement system.

Antibody-mediated immune responses   Immune system responses conducted by B cells that produce antibodies and that defend primarily against enemies that are free in body fluids, including toxins or extracellular pathogens, such as bacteria or free viruses.

Anticodon   A three-base sequence on tRNA that binds to the complementary base pairs of a codon on the mRNA.

Antidiuretic hormone (ADH)   A hormone manufactured by the hypothalamus but stored and released from the posterior pituitary. It regulates the amount of water reabsorbed by the distal convoluted tubules and collecting ducts of nephrons. ADH causes water retention at the kidneys and elevates blood pressure.

Antigen   A substance able to trigger an immune response, such as the production of antibodies.

Antigen-presenting cell A cell that presents the antigen to a helper T cell, initiating an immune response toward that antigen. The most important type of antigen-presenting cell is a macrophage.

Aorta The body's main artery that conducts blood from the left ventricle toward the cells of the body. The aorta arches over the top of the heart and gives rise to the smaller arteries that will feed the capillary beds of the body tissues.

Apical surface In epithelial tissue, the exposed part of the cell that faces the body cavity, the lining of an internal organ, or the outside of the body (as opposed to the anchored, basal surface).

Apoptosis A series of predictable physical changes in a cell that is undergoing programmed cell death. Apoptosis is sometimes used as a synonym for programmed cell death.

Appendicular skeleton The part of the skeleton comprising the pectoral girdle (shoulders), the pelvic girdle (pelvis), and the limbs (arms and legs).

Aqueous humor The fluid within the anterior chamber of the eye. It supplies nutrients and oxygen to the cornea and lens and carries away their metabolic wastes.

Arachnoid The middle layer of the meninges (the connective tissue layers that protect the central nervous system).

Arrector pili The tiny, smooth muscles attached to the hair follicles in the dermis.

Arteriole A small branch of an artery that is barely visible to the naked eye; a blood vessel located between an artery and a capillary. Arterioles serve to regulate blood flow through capillary beds to various regions of the body. They also regulate blood pressure.

Artery A muscular tube (blood vessel) that transports blood away from the heart toward the cells of body tissues.

Arthritis An inflammation of a joint.

Artificial insemination A treatment for infertility in which sperm that have been donated and stored at a sperm bank are deposited in the woman's cervix or vagina at about the time of ovulation.

Association neuron An interneuron. These neurons are located within the central nervous system between sensory and motor neurons and serve to integrate information.

Asthma A condition marked by spasms of the muscles of bronchioles making air flow difficult. It is often triggered by allergy.

Astigmatism Irregularities in the curvature of the cornea or lens that cause distortions of an image because the irregularities cause light rays to converge unevenly.

Atherectomy A technique for scraping lipid deposits associated with atherosclerosis away from an artery wall.

Atherosclerosis A narrowing of the arteries caused by thickening of the arterial walls and a build-up of lipid (primarily cholesterol) deposits. Atherosclerosis reduces blood flow through the vessel, choking off the vital supply of oxygen and nutrients to the tissues served by that vessel.

Atoms Units of matter that cannot be further broken down by chemical means; they are composed of subatomic particles, which include protons (positively charged particles), neutrons (with no charge), and electrons (with negative charges).

Atria (singular, *atrium*) The upper chambers of the heart that receive blood from the veins and pump it to the ventricles.

Atrial fibrillation Rapid, ineffective contractions of the atria of the heart.

Atrial natriuretic peptide (ANP) The hormone released by cells in the right atrium of the heart in response to stretching of the heart caused by increased blood volume and pressure. ANP decreases water and solute reabsorption by the kidneys, resulting in the production of large amounts of urine.

Atrioventricular bundle A tract of specialized cardiac muscle cells that runs along the wall between the ventricles of the heart and conducts an electrical impulse that originated in the SA node and was conducted to the AV node to the ventricles. The bundle forks into right and left branches and then divides into many other specialized cardiac muscle cells, called Purkinje fibers, that penetrate the walls of the ventricles.

Atrioventricular (AV) node A region of specialized cardiac muscle cells located in the partition between the two atria. It receives an electrical signal that spreads through the atrial walls from the sinoatrial node and relays the stimulus to the ventricles by means of a bundle of specialized muscle fibers, called the atrioventricular bundle, that runs along the wall between the ventricles.

Atrioventricular (AV) valves Heart valves located between the atria and the ventricles that keep blood flowing in only one direction, from the atria to the ventricles. The right AV valve consists of three flaps of tissue and is also called the tricuspid valve. The left AV valve consists of two flaps of tissue and is also called the bicuspid or the mitral valve.

Auditory tube See eustachian tube.

Autoimmune disorder An immune response misdirected against the body's own tissues.

Autonomic nervous system The part of the peripheral nervous system that governs the involuntary, unconscious activities that maintain a relatively stable internal environment. The autonomic nervous system has two branches: the sympathetic and the parasympathetic.

Autosomes The 22 pairs of chromosomes (excluding the pair of sex chromosomes) that determine the expression of most of the inherited characteristics of a person.

Autotroph An organism that makes its own food (organic compounds) from inorganic substances. The autotrophs include phototrophs, which use the energy of light, and chemotrophs, which use the energy in chemicals.

Axial skeleton The part of the skeleton comprised of the skull, the vertebral column, the breastbone (sternum), and the rib cage.

Axon A long extension from the cell body of a neuron that carries an electrochemical message away from the cell body toward another neuron or effector (muscle or gland). The tips of the axon release a chemical called a neurotransmitter that can affect the activity of the receiving cell. Typically, there is one long axon on a neuron.

Axon terminal The tip of a branch of an axon that releases a chemical (neurotransmitter) that alters the activity of the target cell.

B lymphocyte A type of white blood cell important in humoral immune responses that can transform into a plasma cell and produce antibodies.

Balanced polymorphism A phenomenon in which natural selection maintains two or more alleles for a trait in a population from one generation to the next. It occurs when the environment changes frequently or when the heterozygous condition is favored over either homozygous condition.

Barr body A structure formed by a condensed, inactivated X chromosome in the body cells of female mammals.

Basal cell carcinoma   The most common type of skin cancer occurring in the rapidly dividing cells of the basal layer of the epidermis.

Basal ganglia   A region of the cerebrum consisting of collections of cell bodies located deep within it called the white matter. They are important in coordinating movement and may also play a role in cognition and memory of learned skills.

Basal lamina   The layer of the basement membrane that is secreted by and lies closest to the epithelium; it helps anchor epithelial cells to connective tissue.

Basal metabolic rate (BMR)   A measure of the minimum energy required to keep an awake, resting body alive. It generally represents between 60% and 75% of the body's energy needs.

Basal surface   In epithelial tissue, the part of the cell that is anchored to the basement membrane (as opposed to the apical surface that is free and typically faces the body cavity, the lining of an internal organ, or the outside of the body).

Base   Any substance that reduces the concentration of hydrogen ions in solution.

Basophil   A white blood cell that releases histamine, a chemical that both attracts other white blood cells to the site and causes blood vessels to widen during an inflammatory response.

Benign tumor   An abnormal mass of tissue that usually remains at the site where it forms.

Beta cells   Cells found in the pancreatic islets of Langerhans that produce the hormone insulin.

Bicuspid valve   A heart valve located between the left atrium and ventricle. It is also called the mitral valve.

Bilirubin   A red pigment produced from the breakdown of the heme portion of hemoglobin by liver cells. It is excreted by the liver in bile.

Binary fission   A type of asexual reproduction in which the genetic information is replicated and then a cell, a bacterium for example, or organism divides into two equal parts.

Binomial nomenclature   The system for naming organisms developed by Carl Linnaeus. Each organism is given a genus name followed by a species name.

Biodiversity   The number and variety of all living things in a given area. It includes genetic diversity, species diversity, and ecological diversity.

Biofeedback   The use of artificial signals to provide feedback about unconscious visceral and motor activity particularly associated with stress.

Biogeochemical cycle   The recurring process by which materials (for example, carbon, water, nitrogen, and phosphorus) cycle between living and nonliving systems and back again.

Biogeography   The study of the geographic distribution of organisms. New distributions of organisms occur by dispersal (organisms move to new areas) and vicariance (areas occupied by the organisms move or are subdivided).

Biological magnification   The tendency of a nondegradable chemical to become more concentrated in organisms as it passes along the food chain.

Biome   A relatively distinct terrestrial ecosystem that covers a large geographic area and is characterized by particular assemblages of plants, animals, climate, and soil, wherever that ecosystem occurs on Earth.

Biopsy   The removal and examination, usually microscopically, of a piece of tissue to diagnose a disease, usually cancer.

Biosphere   The part of the Earth in which life is found. It encompasses all of the Earth's living organisms.

Biotechnology   The industrial manipulation of living cells that causes them to perform a useful task in a controllable way.

Bipedalism   Walking on two feet. The trait that evolved early in hominid evolution and set the stage for the evolution of other characteristics, such as increases in brain size.

Bipolar neuron   A neuron that has only two processes. The axon and the dendrite extend from opposite sides of the cell body. Bipolar neurons are receptor cells found only in some of the special sensory organs, such as in the retina of the eye and in the olfactory membrane of the nose.

Birth defects   Developmental defects present at birth. Such defects involve structure, function, behavior, or metabolism and may or may not be hereditary.

Bladder   Muscular sac-like organ that receives urine from the two ureters and temporarily stores it until release into the urethra.

Blastocyst   Stage of development consisting of a hollow sphere of cells. It contains the inner cell mass, a group of cells that will become the embryo, and the trophoblast, a thin layer of cells that will give rise to part of the placenta.

Blind spot   The region of the retina where the optic nerve leaves the eye and on which there are no photoreceptors. Objects focused on the blind spot cannot be seen.

Blood   Connective tissue that consists of cells and platelets suspended in plasma, a liquid matrix.

Blood pressure   The force exerted by the blood against the walls of the blood vessels. It is caused by the contraction of the ventricles and is influenced by vasoconstriction.

Blue babies   Newborns whose foramen ovale, the fetal opening between the right and left atria of the heart, fails to close. As a result, much of their blood still bypasses the lungs and is low in oxygen. The condition can be corrected with surgery.

BMR   See basal metabolic rate.

Bone   Strong connective tissue with specialized cells in a matrix comprised of collagen fibers and mineral salts.

Bone marrow   The soft material filling the cavities in bones. Yellow bone marrow serves as a fat storage site. Red bone marrow is the site where blood cells are produced.

Bone remodeling   The ongoing process of bone deposition and absorption in response to hormonal and mechanical factors.

Bottleneck effect   The genetic drift associated with dramatic, unselective reductions in population size so that by chance alone the genetic makeup of survivors is not representative of the original population.

Bowman's (glomerular) capsule   Cup-like structure surrounding the glomerulus of a nephron.

Brain   The organ composed of neurons and glial cells that receives sensory input and integrates, stores, and retrieves information.

Brain waves   The patterns recorded in an EEG (electroencephalogram) that reflect the electrical activity of the brain and are correlated with the person's state of alertness.

Breast   The front of the chest, especially either of the two protuberant milk-producing glandular organs (mammary glands) that in human females and other female mammals produce milk to nourish newborns.

Breech birth   Delivery in which the baby is born buttocks first rather than head first. It is associated with difficult labors and umbilical cord accidents.

Bronchi (singular, bronchus)   The respiratory passageways between the trachea and the bronchioles that conduct air into the lungs.

Bronchial tree   The term given to the air tubules in the respiratory system because their repeated branching resembles the branches of a tree.

Bronchioles   A series of small tubules branching from the smallest bronchi inside each lung.

Bronchitis   Inflammation of the mucous membranes of the bronchi, causing excess mucus and a deep cough.

Brush border   A fuzzy border of microvilli on the surface of absorptive epithelial cells of the small intestine.

Buffers   Substances that prevent dramatic changes in pH by removing excess hydrogen ions from solution when concentrations increase and adding hydrogen ions when concentrations decrease.

Bulbourethral glands (Cowper's glands)   Male accessory reproductive glands that release a clear slippery liquid immediately prior to ejaculation.

Bulimia   An eating disorder characterized by binge eating followed by purging by means of self-induced vomiting, enemas, laxatives, or diuretics.

Bursa (plural, bursae)   A flattened fibrous sac containing a thin film of synovial fluid that surrounds and cushions certain synovial joints. They are common in locations where ligaments, muscles, skin, or tendons rub against bone.

Bursitis   Inflammation of a bursa (a sac in a synovial joint that acts as a cushion). Bursitis causes fluid to build up within the bursa, resulting in intense pain that becomes worse when the joint is moved and cannot be relieved by resting in any position.

Calcitonin   A hormone secreted by the parafollicular cells of the thyroid gland when blood calcium levels are high that stimulates the removal of calcium from the blood and stimulates calcium deposition in bone.

Callus   A mass of repair tissue formed by collagen fibers secreted from fibroblasts or woven bone that forms around and links the ends of a broken bone.

Calyx   The cup-like extension of the renal pelvis. The urine produced by the kidneys drains to a calyx and then into the renal pelvis and out the ureter to the urinary bladder.

Canaliculi   Tiny canals that project outward from the lacunae in compact bone through which cytoplasmic projections of osteocytes extend, contacting other osteocytes.

Capacitation   The process by which secretions from the uterus or fallopian tubes alter the surface of the acrosome, the enzyme-containing cap on the head of sperm. Capacitation destabilizes the plasma membrane of sperm, a necessary prerequisite for the acrosome reaction and fertilization.

Capillary   A microscopic blood vessel between arterioles and venules with walls only one cell layer thick. It is the site where the exchange of materials between the blood and the tissues occurs.

Capillary bed   A network of true capillaries and a thoroughfare channel between an arteriole, which controls blood flow through the capillary bed, and a venule, which drains the capillary bed. When blood flows through the true capillaries, materials can be exchanged between the blood and the tissues. A thoroughfare channel acts as a shunt that allows blood to flow directly from an arteriole to a venule, bypassing the true capillaries.

Carbaminohemoglobin   The compound formed when carbon dioxide binds to hemoglobin.

Carbohydrates   Organic molecules that provide fuel for the human body. These molecules, which we know as sugars and starches, can be classified by size into the monosaccharides, disaccharides, and polysaccharides.

Carbonic anhydrase   An enzyme in the red blood cells that catalyzes the conversion of unbound carbon dioxide to carbonic acid.

Carcinogen   A substance that causes cancer.

Carcinoma in situ   A tumor that has not spread; "cancer in place."

Cardiac cycle   The events associated with the flow of blood through the heart during a single heartbeat. It consists of systole (contraction) and diastole (relaxation) of the atria and then of the ventricles of the heart.

Cardiac muscle   One of three types of muscle in the body, characterized by striated involuntary muscle that makes up the bulk of the walls of the heart.

Cardiovascular   Pertaining to the heart and blood vessels.

Carotene   An orange-yellow pigment found in foods such as carrots, apricots, and oranges. Carotene can accumulate in the dermis and in cells of the outermost layer of the epidermis.

Carpal tunnel syndrome   A syndrome whose symptoms may include numbness or tingling in the affected hand, along with pain in the wrist, hand, and fingers that is caused by repeated motion in the hand or wrist, causing the tendons to become inflamed and press against the nerve.

Carrier   An individual who displays the dominant phenotype but is heterozygous for a trait and can therefore pass the recessive allele to descendants.

Carrying capacity   The number of individuals of a given species that a particular environment can support for a prolonged period of time.

Cartilage   A type of connective tissue with a dense network of protein fibers in a gelatinous matrix.

Catabolism   Chemical reactions within living cells that break down complex molecules into simpler ones, releasing energy from chemical bonds. Compare with anabolism.

Cataract   A cloudy or opaque lens. Cataracts reduce visual acuity, and may be caused by glucose accumulation associated with Type I diabetes, excessive exposure to sunlight, and exposure to cigarette smoke.

CD4 cell   See helper T cell.

Cell adhesion molecules (CAMs)   Molecules that poke through the plasma membranes of most cells and help hold cells together to form tissues and organs.

Cell body   The part of a neuron that contains the organelles and nucleus needed to maintain the cells.

Cell-mediated immune responses   Immune system responses conducted by T cells that protect against cellular threats, including body cells that have become infected with viruses or other pathogens and cancer cells.

Cellular respiration   The oxygen-requiring pathway by which glucose is broken down by cells to yield carbon dioxide, water, and energy.

Cellulose   Structural polysaccharide found in the cell walls of plants. Humans lack the enzymes necessary to digest cellulose, and thus it passes unchanged through our digestive tract. Although cellulose has no value as a nutrient, it is an important form of dietary fiber known to facilitate the passage of feces through the large intestines.

Cementum   A calcified but sensitive part of a tooth that covers the root.

Central nervous system (CNS)   The brain and the spinal cord.

Centrioles   Paired structures within a centrosome. Each centriole is composed of nine sets of triplet microtubules arranged in a ring.

Centrosome   The region near the nucleus that contains centrioles. It forms the mitotic spindle during prophase.

Cerebellum   A region of the brain important in sensory-motor coordination. It is largely responsible for posture and smooth body movements.

Cerebral cortex   The extensive area of gray matter covering the surfaces of the cerebrum. It consists of billions of glial cells, nerve cell bodies, and unmyelinated axons and is often referred to as the "conscious" part of the brain.

Cerebral white matter   A region of the cerebrum beneath the cortex consisting primarily of myelinated axons that are grouped into tracts that allow various regions of the brain to communicate with one another.

Cerebrospinal fluid   The fluid bathing the internal and external surfaces of the central nervous system. It serves as a shock absorber, supports the brain, nourishes the brain, delivers chemical messengers, and removes waste products.

Cerebrovascular accident   See stroke.

Cerebrum   The largest and most prominent part of the brain composed of the cerebral hemispheres. It is responsible for thinking, sensory perception, originating most conscious motor activity, personality, and memory. It has three parts: the cerebral cortex, cerebral white matter, and the basal ganglia.

Cerumen   The waxy secretion of ceruminous glands that prevents foreign material from entering the eardrum.

Ceruminous glands   Modified apocrine sweat glands found in the tissue of the external ear canal that produce cerumen.

Cervical cap   A barrier means of contraception consisting of a small rubber dome that fits snugly over the cervix and is held in place partly by suction. It prevents sperm from reaching the egg.

Cervix   The narrow neck of the uterus that projects into the vagina whose opening provides a passageway for materials moving between the vagina and the body of the uterus.

Cesarean section   A procedure by which the baby and placenta are removed from the uterus through an incision in the abdominal wall and uterus. The term is often shortened to C-section.

Chancre   A painless bump that forms during the first stage of syphilis at the site of contact, usually within 2 to 8 weeks of the initial contact.

Chemical digestion   A part of the digestive process that involves breaking chemical bonds so that complex molecules are broken into their component subunits. Chemical digestion produces molecules that can be absorbed into the bloodstream and used by the cells.

Chemiosmosis   The process by which the inward flow of hydrogen ions across the inner membrane of the mitochondrion drives the synthesis of ATP from ADP and inorganic phosphate (Pi). Through chemiosmosis, the electron transport chain produces 32 molecules of ATP per molecule of glucose.

Chemoreceptor   A sensory receptor specialized to respond to chemicals. We describe the input from the chemoreceptors of the mouth as taste (gustation) and from those of the nose as smell (olfaction). Other chemoreceptors monitor levels of chemicals in body fluids, such as carbon dioxide, oxygen, and glucose.

Childhood   The stage in postnatal development that runs from about 13 months to 12 or 13 years of age. Growth during this period is initially rapid and then slows. Toward the end of childhood there is a dramatic increase in growth known as a "growth spurt."

Chitin   A structural polysaccharide found in the exoskeletons (hard outer coverings) of animals such as insects, spiders, and crustaceans.

Chlamydia   A genus of bacteria. In this text, it is an infection (usually sexually transmitted) caused by *Chlamydia trachomatus*, commonly causing urethritis and pelvic inflammatory disease. It may not cause symptoms, especially in women, and is a leading cause of sterility.

Cholecystokinin   A hormone secreted by the small intestine that stimulates the pancreas to release its digestive enzymes and the gallbladder to contract and release bile.

Chordae tendinae   Strings of connective tissue that anchor the atrioventricular valve to the wall of the heart, preventing the backflow of blood.

Chorion   The extraembryonic membrane that becomes the embryo's major contribution to the placenta.

Chorionic villi   Finger-like projections of the chorion that grow into the uterine lining of the mother during formation of the placenta and become part of the placenta.

Chorionic villi sampling (CVS)   A procedure for screening for genetic defects of a fetus by removing a piece of chorionic villi and examining the cells for genetic abnormalities.

Chromatid   One of the two identical replicates of a duplicated chromosome. The two chromatids that make up a chromosome are held together by a centromere and are referred to as sister chromatids. During cell division, the two strands separate and each becomes a chromosome in one of the two daughter cells.

Chromosomal mutation   A change in DNA in which a section of a chromosome becomes rearranged, duplicated, or deleted.

Chromosome   DNA (which contains the genetic information of a cell) and specialized proteins, primarily histones.

Chylomicron   A particle formed when proteins coat the surface of the products of lipid digestion, making lipids soluble in water and allowing them to be transported throughout the body.

Chyme   The semifluid, creamy mass created during digestion once the food has been churned and mixed with the gastric juices of the stomach.

Cilia   Extensions of the plasma membrane found on some cells, such as those lining the respiratory tract. They are usually shorter and much more numerous than flagella and move in a rowing motion.

Ciliary body   A portion of the middle coat of the eyeball near the lens that consists of smooth muscle and ligaments. Contractions of the smooth muscle of the ciliary body change the shape of the lens, which then focuses images on the retina.

Circumcision   The surgical removal of the foreskin of the penis, usually performed when the male is an infant.

Citric acid cycle   The cyclic series of chemical reactions that follows the transition reaction and yields two molecules of ATP (one from each acetyl CoA that enters the cycle) and several molecules of NADH and FADH$_2$, carriers of high-energy electrons that enter the electron transport chain. This phase of cellular respiration occurs in the matrix of the mitochondrion and is sometimes called the Kreb's cycle.

Cleavage A rapid series of mitotic cell divisions in which the zygote first divides into two cells, and then four cells, and then eight cells, and so on. Cleavage usually begins about one day after fertilization as the zygote moves along the fallopian tube toward the uterus.

Climax community The relatively stable community that eventually forms at the end of ecological succession and remains if no disturbances occur.

Clitoris A small, erectile body in the female that plays a role in sexual stimulation and is the developmental equivalent of the male penis.

Clonal selection The hypothesis that, by binding to a receptor on a lymphocyte surface, an antigen selectively activates only those lymphocytes able to recognize that antigen and programs that lymphocyte to divide, forming an army of cells specialized to attack the stimulating antigen.

Clone A population of identical cells descended from a single ancestor.

Cochlea The snail-shaped portion of the inner ear that contains the actual organ of hearing (the organ of Corti).

Codominance The condition in which the effects of both alleles are separately expressed in a heterozygote.

Codon A three-base sequence on mRNA that specifies one of the 20 common amino acids or the beginning or end of the protein chain.

Coenzymes Organic molecules such as vitamins that function as cofactors and help enzymes convert substrate to product.

Cofactors Nonprotein substances such as zinc, iron, and vitamins that help enzymes convert substrate to product. They may permanently reside at the active site of the enzyme or may bind to the active site at the same time as the substrate.

Colon The division of the large intestine composed of the ascending colon, the transverse colon, and the descending colon.

Coma An unconscious state caused by trauma to neurons in regions of the brain responsible for stimulating the cerebrum, particularly those in the reticular activating system or thalamus. Coma can be caused by mechanical shock, such as might be caused by a blow to the head, tumors, infections, drug overdose (from barbiturates, alcohol, opiates, or aspirin), or failure of the liver or kidney.

Combination birth control pill A means of hormonal contraception that consists of a series of pills that contains synthetic forms of estrogen and progesterone. The hormones in the pills mimic the effects of natural hormones ordinarily produced by the ovaries and inhibit FSH and LH secretion by the anterior pituitary gland and, therefore, prevent the development of an egg and ovulation.

Common cold An upper respiratory infection caused by one of the adenoviruses.

Community An assortment of organisms of various species living together in a defined habitat that can potentially interact.

Compact bone Very dense, hard bone, containing internal spaces of microscopic size and narrow channels that contain blood vessels and nerves. It makes up the shafts of long bones and the outer surfaces of all bones.

Complementary base pairing The process by which specific purine bases are matched with specific pyrimidine bases, adenine with thymine (in DNA) or with uracil (in RNA) and cytosine with guanine.

Complementary proteins A combination of foods each containing incomplete proteins that provide ample amounts of all essential amino acids when combined.

Complete proteins A protein that contains ample amounts of all the essential amino acids. Animal sources of protein are generally complete proteins.

Compound A molecule that contains atoms of different elements.

Computed tomography (CT scanning) A method of visualizing body structures, including the brain, using an x-ray source that moves in an arc around the body part to be imaged, thereby providing different views of the structure.

Concentration gradient A difference in the number of molecules or ions between two adjacent regions. Molecules or ions tend to move away from an area where they are more concentrated to an area where they are less concentrated. Each type of molecule or ion moves in response to its own concentration gradient.

Condom, female A barrier means of contraception used by a female that consists of a loose sac of polyurethane held in place by flexible rings (one at each end). It is used to prevent sperm from entering the female tract.

Condom, male A barrier means of contraception consisting of a thin sheath of latex or animal intestines that is rolled onto an erect penis, where it prevents sperm from entering the vagina. A latex condom also helps prevent the spread of sexually transmitted diseases.

Cone One of the three types of photoreceptors responsible for color vision.

Connective tissue Tissue that binds together and supports other tissues of the body.

Continuous ambulatory peritoneal dialysis (CAPD) A method of hemodialysis whereby the peritoneum, one of the body's own selectively permeable membranes, is used as the dialyzing membrane. It is an alternative to the artificial kidney machine during kidney failure.

Contraceptive sponge A barrier means of contraception consisting of a sponge containing spermicide.

Controlled experiment An experiment in which the subjects are divided into two groups, usually called the "control" group and the "experimental" group. Ideally, the groups differ in only the factor(s) of interest.

Convergence Turning toward a common point from different directions. In vision, the process by which the eyes are directed toward the midline of the body as an object moves closer.

Convergent evolution The process by which two populations that are separated geographically become more alike when subjected to similar environments.

Converging circuit A group of neurons arranged so that one neuron receives multiple inputs from one or many other neurons. Converging circuits have a concentrating effect.

Core temperature The temperature in body structures below the skin and subcutaneous layers.

Cornea A clear, transparent dome located in the front and center of the eye that both provides the window through which light enters the eye and helps bend light rays so that they focus on the retina.

Coronary artery bypass A technique for bypassing a blocked coronary blood vessel to restore blood flow to the heart muscle. Typically, a segment of a leg vein is removed and grafted so that it provides a shunt between the aorta and a coronary artery past the point of obstruction.

Coronary artery disease (CAD)   A condition in which fatty deposits associated with atherosclerosis form on the inside of coronary arteries, obstructing the flow of blood. It is the underlying cause of the vast majority of heart attacks.

Coronary circulation   The system of blood vessels that services the tissues of the heart itself.

Coronary sinus   A vessel that returns deoxygenated blood collected from the heart muscle to the right atrium of the heart.

Corpus callosum   A band of myelinated axons (white matter) that connects the two cerebral hemispheres so they can communicate with one another.

Corpus luteum   A structure in the ovary that forms from the follicles remaining in the ovary after ovulation and functions as an endocrine structure that secretes estrogen and progesterone.

Covalent bonds   A chemical bond formed when outer shell electrons are shared between atoms.

Covering and lining epithelium   The type of epithelial tissue that covers certain internal organs, constitutes the outer layer of the skin, and forms the inner lining of blood vessels and several body systems (digestive, urinary, reproductive, and respiratory).

Cowper's glands   See bulbourethral glands.

Cranial nerves   Twelve pairs of nerves that arise from the brain and service the structures of the head and certain body parts such as the heart and diaphragm.

Cranium   The skeletal portion of the skull that forms the cranial case. It is formed from eight (sometimes more) flattened bones including the frontal bone, two parietal bones, the occipital bone, two temporal bones, the sphenoid bone, and the ethmoid bone.

Creatine phosphate   A compound stored in muscle tissue that serves as an alternative energy source for muscle contraction.

Cretinism   A condition characterized by dwarfism, mental retardation, and slowed sexual development.

Cristae   Infoldings of the inner membrane of a mitochondrion.

Cross bridges   Myosin heads. Club-shaped ends of a myosin molecule that bind to actin filaments and can swivel, causing actin filaments to slide past the myosin filaments, which causes muscle contraction.

Cross tolerance   The development of tolerance for a drug that is not used, caused by the development of tolerance to another, usually similar, drug.

Crossing over   The breaking and rejoining of nonsister chromatids of homologous pairs of chromosomes during meiosis (specifically at prophase I when homologous chromosomes pair up side by side). Crossing over results in the exchange of corresponding segments of chromatids and increases genetic variability in populations.

Crown   The crown of a tooth. The part of a tooth that is visible above the gum line. It is covered with enamel, a nonliving material that is hardened with calcium salts.

CT scanning   See computed tomography.

Culture   Social influences that produce an integrated pattern of knowledge, belief, and behavior.

Cupula   A pliable gelatinous mass covering the hair cells within the ampulla of the semicircular canals of the inner ear and whose movement bends hair cells, triggering the generation of nerve impulses that are interpreted by the brain as movement of the head.

Cushing's syndrome   A condition characterized by accumulation of fluid in the face and redistribution of body fat caused by oversecretion of adrenal steroids.

Cyanosis   Bluish coloration of the skin that results from poorly oxygenated blood as a result of exposure to extremely cold temperatures or disorders of the circulatory or respiratory systems.

Cystitis   Inflammation of the urinary bladder caused by bacteria.

Cytokinesis   Division of the cytoplasm that occurs toward the end of mitosis.

Cytomegalovirus   Any of a group of highly host-specific herpesviruses that produce an infection characterized by unique large cells with nuclear inclusions. Cytomegalovirus infects many people in childhood but can be reactivated in people with AIDS.

Cytoplasm   The part of a cell that includes the cytosol (aqueous fluid within the cell) and all the organelles with the exception of the nucleus.

Cytoskeleton   A complex network of protein filaments within the cytosol that gives the cell its shape, anchors organelles in place, and functions in the movement of entire cells or certain organelles or vesicles within cells. The cytoskeleton includes microtubules, microfilaments, and intermediate filaments.

Cytosol   The aqueous solution inside the cell.

Cytotoxic T cells   A type of T lymphocyte that directly attacks infected body cells and tumor cells by releasing chemicals called perforins that cause the target cells to burst.

Dark adaptation   The increase in visual acuity that occurs in dim light after having been in bright light. It occurs as rhodopsin, the pigment in rods, is resynthesized.

Darwinian fitness   See fitness.

Deductive reasoning   A logical progression of thought proceeding from the general to the specific. It involves making specific deductions based on a larger generalization or premise. The statement is usually in the form of an "if ... then" premise.

Deforestation   Removing trees from an area without replacing them.

Dehydration synthesis   The process by which polymers are formed. Monomers are linked together through the removal of a water molecule.

Deletion   The loss of a nucleotide or segment of a chromosome.

Denaturation   The process by which changes in the environment of a protein, such as increased heat or changes in pH, cause it to unravel and lose its three-dimensional shape. Change in the shape of a protein results in loss of function.

Dendrite   A process of a neuron specialized to pick up messages and transmit them toward the cell body. There are typically many short branching dendrites on a neuron.

Dense connective tissues   Connective tissue that contains many tightly woven fibers and is found in ligaments, tendons, and the dermis.

Density-dependent contact inhibition   The phenomenon whereby cells placed in a dish in the presence of growth factors stop dividing once they have formed a monolayer. This may result from competition among the cells for growth factors and nutrients. Cancer cells do not display density-dependent contact inhibition and continue to divide, piling up on one another until nutrients run out.

Density-dependent regulating factor   One of many factors that have a greater impact on the population size as conditions become more crowded. It includes disease and starvation.

Density-independent regulating factor   One of many types of factors that regulate population size by causing deaths that are not related to the density of individuals in a population. This includes natural disasters.

Dentin A hard, bone-like substance that forms the main substance of teeth. It is covered by enamel on the crown and by cementum on the root.

Depo-provera A means of hormonal contraception that consists of an injection of progesterone every 3 months.

Depolarization A change in the difference in electrical charge across a membrane that moves it from a negative value toward 0 mv. During a nerve impulse (action potential), depolarization is caused by the inward flow of positively charged sodium ions.

Depth perception The ability to determine the relative positions of objects in the visual field and to see objects as three dimensional.

Dermal papillae The folded, upper surface of the papillary layer of the dermis that reaches up into the epidermis causing ridges on the skin surface.

Dermis The layer of the integument that lies just below the epidermis and is composed of connective tissue.

Desertification The process by which overfarming and overgrazing transform marginal farmlands and rangelands to deserts.

Desmosome A type of junctional complex that anchors adjacent cells together.

Detrital food webs Energy flow begins with detritus (organic material from the remains of dead organisms) that is eaten by a primary consumer.

Detrusor muscle A layer of smooth muscle within the walls of the urinary bladder. It plays a role in urination.

Diabetes insipidus A condition characterized by excessive urine production caused by inadequate ADH production.

Diabetes mellitus A condition characterized by excessive urine production in which large amounts of glucose are lost as a result of insulin deficiency.

Diaphragm A broad sheet of muscle that separates the abdominal and thoracic cavities. When the diaphragm contracts, inhalation occurs.

Diaphragm, contraceptive A barrier means of contraception that consists of a dome-shaped soft rubber cup on a flexible ring. A diaphragm is used in conjunction with a spermicide to prevent sperm from reaching the egg.

Diastolic pressure The lowest blood pressure in an artery during the relaxation of the heart. In a typical, healthy adult, the diastolic pressure is about 80 mm Hg.

Dihybrid cross A genetic cross that considers the inheritance of two traits from individuals who are heterozygous for both traits.

Dilation stage The first stage of true labor. It begins with the onset of contractions and ends when the cervix has fully dilated to 10 centimeters (4 inches).

Diploid The condition of having two sets of chromosomes per cell. Somatic (body) cells are diploid.

Disaccharides Molecules formed when two monosaccharides covalently bond to each other through dehydration synthesis. They are known as double sugars.

Diuretics Substances like alcohol that promote urine production.

Divergent evolution The process by which two populations subjected to different environments become increasingly different. It commonly occurs when a population is divided by such geologic changes as mountain formation or continental drift.

Diverging circuit A group of neurons arranged so that one neuron synapses with many other neurons. Diverging circuits usually have an amplifying effect.

DNA Deoxyribonucleic acid. DNA is the molecular basis of genetic inheritance in all cells and some viruses.

DNA fingerprint The pattern of DNA fragments that have been cut by a restriction enzyme and sorted by size. Each person has a characteristic, individual DNA fingerprint.

DNA library A large collection of cloned recombinant DNA fragments containing the entire genome of an organism.

DNA ligase An enzyme that catalyzes the formation of bonds between the sugar and the phosphate molecules that form the sides of the DNA ladder during replication, repair, or the creation of recombinant DNA.

DNA polymerase Any one of the enzymes that catalyze the synthesis of DNA from free nucleotides using one strand of DNA as a template.

Dominant allele The allele that is fully expressed in the phenotype of an individual who is heterozygous for that gene. The dominant allele usually produces a functional protein, whereas the recessive allele does not.

Dopamine A neurotransmitter in the central nervous system thought to be involved in regulating emotions and in the brain pathways that control complex movements.

Dorsal nerve root The portion of a spinal nerve that arises from the back (posterior) side of the spinal cord and contains axons of sensory neurons. It joins with the ventral nerve root to form a single spinal nerve, which passes through the opening between the vertebrae.

Doubling time The number of years required for a population to double in size at a given, constant growth rate.

Down syndrome A collection of characteristics that tend to occur when an individual has three copies of chromosome 21. It is also known as trisomy 21.

Drug cocktail A combination of drugs used to treat people who are HIV+. The combination usually includes a drug that blocks reverse transcription and a protease inhibitor.

Ductus arteriosus A small vessel in the fetus that connects the pulmonary artery to the aorta. It diverts blood away from the lungs.

Ductus venosus A small vessel in the fetus through which most blood from the placenta flows, bypassing the liver.

Duplication The duplication of a region of a chromosome that often results from fusion of a fragment from a homologous chromosome.

Dura mater The tough, leathery outer layer of the meninges that protects the central nervous system. Around the brain, the dura mater has two layers that are separated by a fluid-filled space containing blood vessels.

Dysplasia The changes in shape, nuclei, and organization of adult cells. It is typical of precancerous cells.

Ecological succession The sequence of changes in the species making up a community over time.

Ecology The study of the interactions among organisms and between organisms and their environment.

Ecosystem All the organisms living in a certain area, together with their physical environment.

Ectoderm The primary germ layer that forms the nervous system, epidermis, and epidermal derivatives such as hair, nails, and mammary glands.

Ectopic pregnancy A pregnancy in which the embryo (blastocyst) implants and begins development in a location other than the uterus, most commonly in the fallopian tubes (a tubal pregnancy).

Edema Swelling caused by the accumulation of interstitial fluid.

EEG See electroencephalogram.

Effector cells   Lymphocytes that are responsible for the attack on cells or substances not recognized as belonging in the body.

Efferent (motor) neuron   A neuron specialized to carry information *away from* the central nervous system to an effector, either a muscle or gland.

Elastic cartilage   The most flexible type of cartilage because of an abundance of wavy elastic fibers in its matrix.

Electrical gradient   A difference in electrical charge between two adjacent regions. Ions move away from regions with a similar charge and toward regions with opposite charges.

Electrical potential difference   A separation of positive and negative charges that can be used to do work. A potential difference can cause the ions to flow across a membrane, thereby generating an electric current.

Electrocardiogram (ECG or EKG)   A graphical recording of the electrical activities of the heart.

Electroencephalogram (EEG)   A graphical record of the electrical activity of the brain.

Electron transport chain   A series of carrier proteins embedded in the inner membrane of the mitochondrion that receives electrons from the molecules of NADH and FADH$_2$ produced by glycolysis and the citric acid cycle. During the transfer of electrons from one molecule to the next, energy is released and this energy is then used to make ATP.

Element   Any substance that cannot be broken down into simpler substances by ordinary chemical means.

Embolus   A blood clot that drifts through the circulatory system and can lodge in a small blood vessel and block blood flow.

Emigration   The departure of individuals from a population for some other area.

Emphysema   A condition in which the alveolar walls break down, thicken, and form larger air spaces, making gas exchange difficult. This results in less surface area for gas exchange and an increase in the volume of residual air in the lungs.

Enamel   A nonliving material that is hardened with calcium salts and covers the crown of a tooth.

Encephalitis   An inflammation of the meninges around the brain.

Endocardium   A thin layer of endothelial cells that lines the cavities of the heart and is continuous with the endothelial lining of all blood vessels.

Endochondral ossification   The development of bone tissue such that bone forms in replacement of cartilage. It occurs during development when the embryonic cartilage model of the skeleton is replaced by bone.

Endocrine glands   Glands that lack ducts and secrete their products (hormones) directly into the spaces or fluids just outside the cells.

Endocytosis   The process by which large molecules and single-celled organisms like bacteria enter cells. It occurs when a region of the plasma membrane surrounds the substance to be ingested, then pinches off from the rest of the membrane, enclosing the substance in a sac-like structure called a vesicle that is released into the cell. Two types of endocytosis are phagocytosis ("cell eating") and pinocytosis ("cell drinking").

Endoderm   The primary germ layer that forms some organs and glands (for example, the pancreas, liver, thyroid gland, and parathyroid glands) and the epithelial lining of the bladder, respiratory tract, and gastrointestinal tract.

Endometriosis   A painful condition in which tissue from the lining of the uterus (the endometrium) is found outside the uterine cavity.

Endometrium   The inner layer of the uterus consisting of connective tissue, glands and blood vessels. The endometrium thickens and develops with each menstrual cycle and is then lost as menstrual flow. It is the site of embryo implantation during pregnancy.

Endoplasmic reticulum   The network of internal membranes within eukaryotic cells. Whereas rough endoplasmic reticulum (RER) has ribosomes attached to its surface, smooth endoplasmic reticulum (SER) lacks ribosomes and functions in the production of phospholipids for incorporation into cell membranes.

Endorphin   A chemical released by nerve cells that binds to the so-called opiate receptors on the pain-transmitting neurons, reducing the release of substance P and quelling the pain. It is short for *endo*genous mor*phine*-like substance.

Endosymbiont hypothesis   The hypothesis that organelles such as mitochondria were once free-living prokaryotic organisms that were engulfed by primitive eukaryotic cells with which they established a symbiotic (mutually beneficial) relationship.

Endothelium   The lining of the heart, blood vessels, and lymphatic vessels. It is composed of flattened, tight-fitting cells. The endothelium forms a smooth surface that minimizes friction and allows the blood or lymph to flow over it easily.

Enkephalin   A chemical released by nerve cells that binds to the so-called opiate receptors on the pain-transmitting neurons, reducing the release of substance P and quelling the pain.

Enzymes   Substances (usually proteins, but sometimes RNA molecules) that speed up chemical reactions without being consumed in the process.

Eosinophil   The type of white blood cell important in the body's defense against parasitic worms. It releases chemicals that help counteract certain inflammatory chemicals released during an allergic response.

Epidermis   The outermost layer of the integument composed of epithelial cells.

Epididymis   A long tube coiled on the surface of each testis that serves as the site of sperm cell maturation and storage.

Epiglottis   A part of the larynx that forms a movable lid of hyaline cartilage covering the opening into the trachea (the glottis).

Epinephrine (adrenaline)   A hormone secreted by the adrenal medulla, along with norepinephrine, in response to stress. They initiate the physiological "fight-or-flight" reaction.

Epiphyseal plate   A plate of cartilage that separates the head of the bone from the shaft, permitting the bone to grow. In late adolescence the epiphyseal plate is replaced by bone, and growth stops.

Episiotomy   An incision made to enlarge the vaginal opening, just before passage of the baby's head at the end of the second stage of labor.

Epithelial tissue   One of the four primary tissue types. The tissue that covers body surfaces, lines body cavities and organs, and forms glands.

Equatorial plate   A plane at the midline of a cell where chromosomes line up during mitosis or meiosis.

Erythrocyte   A red blood cell. An enucleated biconcave cell in the blood that is specialized for transporting oxygen and assists in transporting carbon dioxide away from cells.

Erythropoietin   A hormone formed from a protein in blood plasma that stimulates red blood cell production and maturation. Formation is stimulated by a chemical from the kidney that is released when oxygen levels decline.

Esophagus A muscular tube that conducts food from the pharynx to the stomach using peristalsis.

Essential amino acid Any of the eight amino acids that the body cannot synthesize and, therefore, must be supplied in the diet.

Eukaryote A cell with a nucleus and extensive internal membranes that divide it into many compartments and enclose organelles. Eukaryotes include cells in plants, animals, and all other organisms except bacteria and cyanobacteria.

Eustachian tubes Small tubes that join the upper region of the pharynx (throat) with the middle ear. They help to equalize the air pressure between the middle ear and the atmosphere.

Eutrophication The enrichment of water in a lake or pond by nutrients.

Excitatory synapse A synapse in which the response of the receptors for that neurotransmitter on the postsynaptic membrane increases the likelihood that an action potential will be generated in the postsynaptic neuron. The postsynaptic cell is excited because it becomes less negative than usual (slightly depolarized), usually because of the inflow of sodium ions.

Exhalation Breathing out (expiration) involves the movement of air out of the respiratory system into the atmosphere.

Exhaustion phase The final phase of the general adaptation syndrome that occurs in response to extreme and prolonged stress. During the exhaustion phase, the heart, blood vessels, and adrenal glands begin to fail after the taxing metabolic and endocrine demands of the resistance phase.

Exocrine gland A gland that secretes its product through ducts onto body surfaces, into the spaces within organs, or into a body cavity. Examples include the salivary glands of the mouth and the oil and sweat glands of the skin.

Exocytosis The process by which large molecules leave cells. It occurs when products packaged by cells in membrane-bound vesicles move toward the plasma membrane. Upon reaching the plasma membrane, the membrane of the vesicle fuses with it, spilling its contents outside the cell.

Exophthalmos A condition characterized by protruding eyes that is caused by the accumulation of interstitial fluid.

Expiration The process by which air is moved out of the respiratory system into the atmosphere. It is also called exhalation.

Expiratory reserve volume The additional volume of air that can be forcefully expelled from the lungs after normal exhalation.

Exponential growth Unrestricted growth at a constant rate over a period of time. When exponential growth is plotted, it produces a characteristic J-shaped curve.

Expulsion stage The second stage of true labor. It begins with full dilation of the cervix and ends with delivery of the baby.

External auditory canal The canal leading from the pinna of the ear to the eardrum (tympanic membrane).

External intercostals The external layer of muscles of the rib cage. When the external intercostals contract, the rib cage is lifted and inhalation occurs.

External urethral sphincter A sphincter made of skeletal muscle that surrounds the urethra. This voluntary sphincter helps stop the flow of urine down the urethra when we wish to postpone urination.

Exteroceptor A sensory receptor that is located near the surface of the body and that responds to changes in the environment.

Extraembryonic membranes Membranes that lie outside the embryo where they protect and nourish the embryo and later the fetus. They include the amnion, yolk sac, chorion, and allantois.

F_1 generation The first filial generation. The progeny resulting from the first cross in a series.

F_2 generation The second filial generation. The offspring of the F_1 generation.

Facial bones The bones that form the face. They are composed of 14 bones that support several sensory structures and serve as attachments for most muscles of the face.

Facilitated diffusion The movement of a substance from a region of higher concentration to a region of lower concentration with the aid of a membrane protein that either transports the substance from one side of the membrane to the other or forms a channel through which it can move.

Fallopian tube One of two tubes that conduct the ova from the ovary to the uterus in the female reproductive system. It is also called an oviduct or a uterine tube.

Farsightedness A condition in which either the eyeball is too short or the lens is too thin, causing the image to be focused behind the retina.

Fascicle A bundle of skeletal muscle fibers (cells) that forms a part of a muscle. Each fascicle is wrapped in its own connective tissue sheath.

Fast-twitch fibers Muscle fibers that contract rapidly and powerfully, with little endurance. They have few mitochondria and large glycogen reserves. They depend on anaerobic pathways to generate ATP during muscle contraction.

Fat soluble vitamin A vitamin that does not dissolve in water and is stored in fat. This group includes vitamins A, D, E, and K.

Fatigue A state in which a muscle is physiologically unable to contract despite continued stimulation. Muscle fatigue results from a relative deficit of ATP.

Feces Waste material discharged from the large intestine during defecation.

Fermentation A pathway by which cells can harvest energy in the absence of oxygen. It nets only 2 molecules of ATP as compared with the approximately 36 molecules produced by cellular respiration.

Fertilization The union between an egg (technically a secondary oocyte) and a sperm. It takes about 24 hours from start to finish and usually occurs in a widened portion of the fallopian tube, not far from the ovary.

Fetal alcohol syndrome A group of characteristics in children born to mothers who consumed alcohol during pregnancy. It can include certain facial features including epicanthal fold, mental retardation, and slow growth.

Fever An elevated body temperature. Fever helps the body fight disease-causing invaders in a number of ways.

Fibrillation Rapid, ineffective contraction of the heart.

Fibrin A protein formed from fibrinogen by thrombin that forms a web that traps blood cells, forming a blood clot.

Fibrinogen A plasma protein produced by the liver that is important in blood clotting. It is converted to fibrin by thrombin.

Fibrocartilage Cartilage with a matrix dense in collagen fibers. Fibrocartilage is found around the edges of joints and the intervertebral disks.

Fight-or-flight response The body's reaction to stress or threatening situations by the stimulation of the sympathetic division of the autonomic nervous system.

Fitness   The average number of offspring left by an individual. It is sometimes called Darwinian fitness.

Flagellum   A tail-like motile structure composed of an extension of the plasma membrane containing microtubules in a 9 + 2 array found on some cells, such as sperm cells. It resembles a whip and performs an undulating motion.

Floating ribs   The last two ribs that do not attach directly to the sternum (breastbone).

Follicle stimulating hormone (FSH)   A hormone secreted by the anterior pituitary gland that in females stimulates the development of the follicles in the ovaries, resulting in the development of ova (eggs) and the production of estrogen, and in males stimulates sperm production.

Fontanels   The "soft spots" in the skull of a newborn. The membranous areas in the skull of a newborn that hold the skull bones together before and shortly after birth. The fontanels are gradually replaced by bone.

Food chain   The successive series of organisms through which energy (in the form of food) flows in an ecosystem. Each organism in the series eats or decomposes the preceding one. It begins with the photosynthesizer and flows to herbivores and then to carnivores.

Food web   The interconnection of all the feeding relationships (food chains) in an ecosystem.

Foramen magnum   The opening at the base of the skull (in the occipital bone) through which the spinal cord passes.

Foramen ovale   In the fetus, a small hole in the wall between the right atrium and left atrium of the heart that allows most blood to bypass the lungs.

Formed elements   Cells or cell fragments found in the blood. They include platelets, white blood cells, and red blood cells.

Fossa ovalis   The small depression left when the foramen ovale, the fetal opening between the right and left atria of the heart, closes shortly after birth.

Fossil   The preserved remnant or impression of a past organism. Most fossils occur in sedimentary rocks.

Founder effect   Genetic drift associated with the colonization of a new place by a few individuals so that by chance alone the genetic makeup of the colonizing individuals is not representative of the population they left.

Fovea centralis   A small region on the retina that contains a high density of cones but no rods. Objects are focused on the fovea for sharp vision.

Frameshift mutation   A mutation that occurs when the number of nucleotides inserted or deleted is not a multiple of three, causing a change in the codon sequence on the mRNA molecule as well as a change in the resulting protein.

Fraternal twins   Individuals that develop when two oocytes are released from the ovaries and fertilized by different sperm. Such twins may or may not be the same gender and are as similar genetically as any siblings. They are also called dizygotic twins.

Free radicals   Molecular fragments that contain an unpaired electron.

FSH   See follicle stimulating hormone.

Gallbladder   A muscular pea-sized sac that stores, modifies, and then concentrates bile. Bile is released from the gallbladder into the small intestine.

Gallstone   A concentration of bile salts and cholesterol that may form within the gallbladder or bile duct, causing a blockage of bile flow.

Gamete intrafallopian transfer (GIFT)   A procedure in which eggs and sperm are collected from a couple and then inserted into the woman's fallopian tube where fertilization may occur. If fertilization occurs, resulting embryos drift naturally into the uterus.

Ganglion (plural, ganglia)   A collection of nerve cell bodies outside the central nervous system.

Gap junction   A type of junctional complex that links the cytoplasm of adjacent cells through small holes, allowing physical and electrical continuity between cells.

Gastric glands   Any one of several glands in the stomach mucosa that contribute to the gastric juice.

Gastric juice   The mixture of hydrochloric acid (HCl) and pepsin released into the stomach.

Gastrointestinal (GI) tract   A long tubular system specialized for the processing and absorption of food that begins at the mouth and continues to the esophagus, stomach, intestines, and anus. Several accessory glands empty their secretions into the GI tract to assist digestion.

Gastrulation   Cell movements that establish the primary germ layers of the embryo. The embryo during this period is called a gastrula.

Gated ion channel   An ion channel that is opened to allow ions to pass through or closed to prevent passage by changes in the shape of a protein that functions as a gate.

Gene flow   Movement of alleles between populations as a result of the movement of individuals. It is a cause of microevolution.

Gene pool   All of the alleles of all of the genes of all individuals in a population.

Gene therapy   Treating a genetic disease by inserting healthy functional genes into the body cells that are affected by the faulty gene.

General adaptation syndrome (GAS)   A series of physiological adjustments made by our bodies in response to extreme stress.

General senses   The sensations that arise from receptors in the skin, muscles, joints, bones, and internal organs and include touch, pressure, vibration, temperature, a sense of body and limb position, and pain.

Genes   Segments of DNA that influence cell structure and function by specifying proteins to be synthesized.

Genetic code   The base triplets in DNA that specify the amino acids that go into proteins or that function as start or stop signals in protein synthesis. It is used to convert the linear sequence of bases in DNA to the sequence of amino acids in proteins.

Genetic drift   The random change in allele frequencies within a population. It is a cause of microevolution that is usually negligible in large populations.

Genetic engineering   The manipulation of genetic material for human practical purposes.

Genital herpes   A sexually transmitted disease caused by the herpes simplex virus (HSV) that is usually characterized by painful blisters on the genitals. It is usually caused by HSV-2 but can be caused by HSV-1.

Genital warts   Warts that form in the genital area caused by the human papilloma viruses (HPV). They also cause cervical cancer and penile cancer.

Genomic imprinting   A phenomenon in which the expression of an allele depends upon which parent contributed that allele.

Genotype   The genetic makeup of an individual. It refers to precise alleles that are present.

Giantism   A condition characterized by rapid growth and unusual height caused by abnormally high levels of growth hormone in childhood.

Glandular epithelium   A type of epithelial tissue that is a component of endocrine and exocrine glands and produces secretions.

Glaucoma   A condition in which the pressure within the anterior chamber of the eye increases as a result of the build-up of aqueous humor. It can cause blindness.

Glial cells   Nonexcitable cells in the nervous system that are specialized to support, protect, and insulate neurons.

Global warming   A long-term increase in atmospheric temperatures caused by a build-up of CO_2 and other greenhouse gases in the atmosphere. Greenhouse gases trap heat in the atmosphere.

Glomerulus   A tuft of capillaries within the renal corpuscle of a nephron.

Glottis   The opening to the airways of the respiratory system from the pharynx into the larynx.

Glucagon   The hormone secreted by alpha cells of the pancreas that elevates blood glucose levels.

Glucocorticoids   Hormones secreted by the adrenal cortex that affect glucose homeostasis, thereby influencing metabolism and resistance to stress.

Gluconeogenesis   The conversion of noncarbohydrate molecules to glucose.

Glycolysis   The splitting of glucose, a six-carbon sugar, into two three-carbon molecules called pyruvate. Glycolysis takes place in the cytosol of a cell and is the starting point for cellular respiration and fermentation.

Gnostic area   An association area of the cerebral cortex that blends the input from sensory association areas with stored sensory memories and assigns meaning to the experience. It is the general interpretation area.

GnRH   See gonadotropin-releasing hormone.

Goiter, simple   An enlarged thyroid gland caused by iodine deficiency.

Golgi complex   An organelle consisting of flattened membranous disks that functions in protein processing and packaging.

Golgi tendon organs   The highly branched nerve fibers located in the tendons that sense the degree of muscle tension.

Gonadocorticoids   The male and female sex hormones, androgens and estrogens.

Gonadotropin-releasing hormone (GnRH)   A hormone produced by the hypothalamus that causes the secretion of luteinizing hormone and follicle stimulating hormone from the anterior pituitary gland.

Gonorrhea   A sexually transmitted disease caused by the bacterium *Neisseria gonorrhoeae* that commonly causes urethritis and pelvic inflammatory disease. It may not cause symptoms, especially in women.

Graafian follicle   A mature follicle in the ovary.

Gradualism   The idea that evolutionary change occurs slowly over long periods of time.

Granulocytes   White blood cells with granules in their cytoplasm, including neutrophils, eosinophils, and basophils.

Graves' disease   An autoimmune disorder characterized by increased heart and metabolic rates, sweating, nervousness, and exophthalmos caused by hyposecretion of thyroid hormone.

Green revolution   The period of time during the twentieth century when crop yields were increased because of the development of high-yield varieties of crops and the use of modern cultivation methods, including the use of farm machinery, fertilizers, pesticides, and irrigation.

Greenhouse effect   An increase in atmospheric CO_2 that could lead to a rise in temperatures throughout the world. Greenhouse gases trap heat in the atmosphere.

Growth factors   A type of signaling molecule that stimulates growth by stimulating cell division in target cells.

Growth hormone (GH)   An anterior pituitary hormone with the primary function of stimulating growth through increases in protein synthesis, cell size, and rates of cell division. GH stimulates growth in general, especially bone growth.

Gumma   A large, unpleasant sore that forms during the third stage of syphilis.

Habitat   The natural environment or place where an organism, population, or species lives.

Hair cells   Sensory cells of the inner ear.

Hair root plexus   The nerve endings that surround the hair follicle and are sensitive to touch.

Haploid   The condition of having one set of chromosomes, as in eggs and sperm.

Haversion canal   The central canal in each osteon that runs longitudinally through the bone and contains blood vessels and nerves.

Haversion system   An osteon. The structural unit of compact bone that appears as a series of concentric circles of lacunae. The lacunae contain bone cells around a central canal containing blood vessels and nerves.

HCG   See human chorionic gonadotropin hormone.

Heart   A muscular pump that keeps blood flowing through an animal's body. The human heart has four chambers—two atria and two ventricles.

Heart attack   The death of heart muscle cells caused by an insufficient blood supply—a myocardial infarction.

Heart murmur   Heart sounds other than "lubb dup" that are created by turbulent blood flow. Heart murmurs can indicate a heart problem, such as a malfunctioning valve.

Heartburn   A burning sensation behind the breastbone that occurs when the acidic gastric juice backs up into the esophagus.

Heimlich maneuver   A procedure intended to force a large burst of air out of the lungs to dislodge material lodged in the trachea.

Helper T cell   (also known as a T4 cell or a CD4 cell, after the receptors on its surface) The kind of T lymphocyte that serves as the main switch for the entire immune response by presenting the antigen to B cells and by secreting chemicals that stimulate other cells of the immune system.

Hematocrit   The percentage of red blood cells in blood by volume. It is a measure of the oxygen-carrying ability of the blood.

Hemodialysis   Use of artificial devices, such as the artificial kidney machine, to cleanse the blood during kidney failure.

Hemoglobin   The oxygen-binding pigment in red blood cells. It consists of four subunits, each made up of an iron-containing heme group and a protein chain. Hemoglobin is responsible for the pinkish color of Caucasian skin.

Hemolytic disease of the newborn   A condition in which the red blood cells of an Rh+ fetus or newborn are destroyed by Rh+ antibodies previously produced in the bloodstream of an Rh− mother.

Heterotroph   An organism that cannot make its own food from inorganic substances and instead consumes other organisms or decaying material.

Heterozygote advantage   The phenomenon in which the heterozygous condition is favored over either homozygous condition. It maintains genetic variation within a population in the face of natural selection.

Heterozygous   The condition of having two different alleles for a particular gene.

Hilus   The notch in the concave border of each kidney where the ureter leaves the kidney and where nerves, lymphatic vessels, and blood vessels enter and exit the kidney.

Histology   The study of tissues.

Homeostasis   The ability of living things to maintain a relatively constant internal environment in all levels of body organization.

Hominids   Members of the genera *Australopithecus* and *Homo*.

Hominoids   Apes and humans.

Homologous chromosomes   A pair of chromosomes that bear genes for the same traits. One member of each pair came from each parent. Homologous chromosomes are the same size and shape and line up with one another during meiosis I.

Homologous structures   Structures that have arisen from a common ancestry.

Homozygous   The condition of having two identical alleles for a particular gene.

Hormonal implants (Norplant)   A means of hormonal contraception consisting of silicon rods containing progesterone that are implanted under the skin in the upper arm and prevent pregnancy for up to 5 years.

Hormone   A chemical messenger released by cells of the endocrine system that travels through the circulatory system to affect receptive target cells.

Host DNA   The DNA that is recombined with pieces of DNA from another source (that might contain a desirable gene) in the formation of recombinant DNA.

HSV-1   Herpes simplex virus 1. HSV-1 usually infects the upper half of the body and causes cold sores (fever blisters), but it can cause genital herpes if contact is made with the genital area.

HSV-2   Herpes simplex virus 2. HSV-2 causes genital herpes, but it can cause cold sores if contact is made with the mouth.

Human chorionic gonadotropin hormone (HCG)   A hormone produced by the cells of the early embryo (blastocyst) and the placenta that maintains the corpus luteum for approximately the first 3 months of pregnancy. HCG enters the bloodstream of the mother and is excreted in her urine. HCG forms the basis for many pregnancy tests.

Human Genome Project   A worldwide research effort to reach a common goal: determining the location of the estimated 100,000 genes along the 23 pairs of human chromosomes and sequencing the base pairs comprising these chromosomes.

Human papilloma virus (HPV)   One of the group of viruses that commonly cause genital warts.

Hyaline cartilage   A type of cartilage with a gel-like matrix that provides flexibility and support. It is found at the end of long bones as well as parts of the nose, ribs, larynx, and trachea.

Hybridoma   An antibody-producing cell produced by the fusion of a B cell and a cancer cell with the capacity to divide and reproduce virtually forever.

Hydrocephalus   A condition resulting from the excessive production or inadequate drainage of cerebrospinal fluid.

Hydrogen bond   A weak chemical bond formed between a partially positively charged hydrogen atom in a molecule and a partially negatively charged atom in another molecule or in another region of the same molecule.

Hydrolysis   The process by which polymers are broken apart by the addition of water.

Hyperglycemia   An elevated blood glucose level.

Hypertension   High blood pressure. A high upper (systolic) value usually suggests that the person's arteries have become hardened, a condition known as arteriosclerosis, and are no longer able to dampen the high pressure of each heart beat. The lower (diastolic) value is generally considered more important because it indicates the pressure when the heart is relaxing.

Hyperthermia   Abnormally elevated body temperature.

Hypodermis   The layer of loose connective tissue below the epidermis and dermis that anchors the skin to underlying tissues and organs.

Hypoglycemia   Depressed levels of blood glucose often resulting from excess insulin.

Hypothalamus   A small brain region located below the thalamus that is essential to maintaining a stable environment within the body. The hypothalamus influences blood pressure, heart rate, digestive activity, breathing rate, and many other vital physiological processes. It acts as the body's "thermostat"; regulates food intake, hunger, and thirst; coordinates the activities of the nervous system and the endocrine (hormonal) system; is part of the circuitry for emotions; and functions as a master biological clock.

Hypothermia   Abnormally low body temperature.

Hypothesis   A testable explanation for a specific set of observations that serves as the basis for experimentation.

Identical twins   Individuals that develop from a single fertilized ovum that splits in two at an early stage of cleavage. Such twins have identical genetic material and thus are always the same gender. They are also called monozygotic twins.

Immigration   Movement of new individuals from other populations into an area.

Immunoglobulin (Ig)   Any of the five classes of proteins that comprise the antibodies.

Implantation   The process by which a blastocyst becomes imbedded in the lining of the uterus. It normally occurs high up on the back wall of the uterus.

Impotence   The inability to achieve or maintain an erection long enough for sexual intercourse.

In vitro fertilization   A procedure in which eggs and sperm are placed together in a dish in the laboratory. If fertilization occurs, early-stage embryos are then transferred to the woman's uterus where it is hoped they will implant and complete development. It is a common treatment for infertility resulting from blocked fallopian tubes.

Incomplete proteins   Proteins that are deficient in one or more of the essential amino acids. Plant sources of protein are generally incomplete proteins.

Incontinence   A condition characterized by the escape of small amounts of urine when sudden increases in abdominal pressure, perhaps caused by laughing, sneezing, or coughing, force urine past the external sphincter. This condition is common in women, particularly following childbirth, an event that may stretch or damage the external sphincter making it less effective in controlling the flow of urine.

Incus   The middle of three small bones in the middle ear that transmit information about sound from the eardrum to the inner ear. The incus is also known as the anvil.

Independent assortment   The process by which homologous chromosomes and the alleles they carry segregate randomly into gametes during meiosis, creating mixes of maternal and paternal chromosomes in gametes. This is an important source of genetic variation in populations.

Independent assortment, principle of   A genetic principle that states that the alleles of unlinked genes (those that are located on different chromosomes) are randomly distributed to gametes.

Inductive reasoning   A logical progression of thought proceeding from the specific to the general. It involves the accumulation of facts through observation until the sheer weight of the evidence forces some general statement about the phenomenon. A conclusion is reached based on a number of observations.

Infancy   Stage in postnatal development that roughly corresponds to the first year of life. It is a time of rapid growth when total body length usually increases by one-half and weight may triple.

Infertility   The inability to conceive (become pregnant) or to cause conception (in the case of males).

Inflammation   A nonspecific body response to injury or invasion by foreign organisms. It is characterized by redness, swelling, heat, and pain.

Influenza   The flu. Influenza is a viral infection caused by any of the variants of influenza A or influenza B viruses.

Infundibulum   The short stalk that connects the pituitary gland to the hypothalamus.

Inguinal canal   A passage through the abdominal wall through which the testes pass in their descent to the scrotum and that contains the testicular artery and vein and vas deferens.

Inhalation   Breathing in. Inhalation (inspiration) involves the movement of air into the respiratory system.

Inhibin   A hormone produced in the testes that increases with sperm count and inhibits follicle stimulating hormone secretion and, therefore, inhibits sperm production.

Inhibitory synapse   A synapse in which the response of the receptors for that neurotransmitter on the postsynaptic membrane decreases the likelihood that an action potential will be generated in the postsynaptic neuron. The postsynaptic cell is inhibited because its resting potential becomes more negative than usual.

Inner ear   A series of passageways in the temporal bone that houses the organs for hearing and the sense of equilibrium.

Insertion   The end of the muscle that is attached to the bone that moves when the muscle contracts.

Insoluble fiber   Fiber that is mostly soluble in water. These fibers include cellulose, hemicellulose, and lignin.

Inspiration   Inhalation. The movement of air into the respiratory system.

Inspiratory reserve volume   The volume of air that can be forcefully brought into the lungs after normal inhalation.

Insulin   The hormone that reduces blood glucose levels and is secreted by beta cells of the pancreas.

Insulin shock   A condition that results from severely depressed glucose levels in which brain cells fail to function properly, causing convulsions and unconsciousness.

Integral proteins   Proteins embedded in the plasma membrane, either completely or incompletely spanning the bilayer.

Intercalated disks   Thickenings of the plasma membranes of cardiac muscle cells that strengthen cardiac tissue and promote rapid conduction of impulses throughout the heart.

Interferon   A type of defensive protein produced by T lymphocytes that slows the spread of viruses already in the body by interfering with viral replication. Interferons also attract macrophages and natural killer cells, which kill the viral-infected cell.

Interleukin 1   A chemical secreted by a macrophage that activates helper T cells in an immune response.

Interleukin 2   A chemical released by a helper T cell that activates both B cells and T cells.

Intermediate fibers   Components of the cytoskeleton made from fibrous proteins. They function in maintenance of cell shape and anchoring organelles such as the nucleus.

Internal intercostals   The internal layer of muscles between the ribs. Contraction of the internal intercostals lowers the rib cage and helps push air out during heavy breathing.

Internal urethral sphincter   Thickening of smooth muscle at the junction of the bladder and the urethra. This sphincter is involuntary and keeps urine from flowing into the urethra while the bladder is filling.

Interneuron   An association neuron. Neurons located within the central nervous system between sensory and motor neurons that serve to integrate information.

Interoceptor   A sensory receptor located inside the body where it monitors conditions. Interoceptors play a vital role in maintaining homeostasis. They are an important part of the feedback loops that regulate blood pressure, blood chemistry, and breathing rate. Interoceptors may also cause us to feel pain, hunger, or thirst, thereby prompting us to take appropriate action.

Interphase   The period between cell divisions when the DNA, cytosol, and organelles are duplicated.

Interstitial cells   Cells located between the seminiferous tubules in the testis that produce the male steroid sex hormones, collectively called androgens.

Intervertebral disks   Pads of cartilage that help cushion the bones of the vertebral column.

Intracytoplasmic sperm injection (ICSI)   A procedure in which a tiny needle is used to inject a single sperm cell into an egg. It is an option for treating infertility when the man has few sperm or sperm that lack the strength or enzymes necessary to penetrate the egg.

Intrauterine device (IUD)   A means of contraception consisting of a small plastic device that either is wrapped with copper wire or contains progesterone. It is inserted into the uterus to prevent pregnancy.

Intrinsic factor   A protein secreted by the stomach necessary for the absorption of vitamin B_{12} from the small intestine.

Ion   An atom or group of atoms that carries an electric charge resulting from the loss or gain of electrons.

Ion channel   A protein-lined pore or channel through a plasma membrane through which one type or a few types of ions can pass. Nerve cell ion channels are important in the generation and propagation of nerve impulses.

Ionic bond   A chemical bond that results from the mutual attraction of oppositely charged ions.

Ischemia   A temporary reduction in blood supply caused by obstructed blood flow. It causes reversible damage to heart muscle.

Islets of Langerhans   Small clusters of endocrine cells in the pancreas.

Isotopes   Atoms that have the same number of protons, but different numbers of neutrons.

J-shaped growth curve   A graphical representation of the growth of a population growing exponentially (unrestrained growth at a constant rate).

Jaundice   A condition in which the skin develops a yellow tone caused by the build-up of bilirubin in the blood and its deposition in certain tissues such as the skin. It is an indication that the liver is not handling bilirubin adequately.

Joint   A point of contact between two bones—an articulation.

Junctional complexes   Membrane specializations that attach adjacent cells to each other to form a contiguous sheet. There are three kinds of junctional complexes: tight junctions, desmosomes, and gap junctions.

Juxtaglomerular apparatus   The region of the kidney nephron where the distal convoluted tubule contacts the afferent arteriole bringing blood into the glomerulus. Cells in this area secrete renin, an enzyme that triggers events eventually leading to increased reabsorption of sodium and water by the distal convoluted tubules and collecting ducts of nephrons.

Kaposi's sarcoma   A cancer that forms tumors in connective tissue and manifests itself as pink or purple marks on the skin. It is common in people with suppressed immune systems, such as people living with AIDS, and is thought to be associated with a new virus in the herpes family, HHV8.

Karyotype   The arrangement of chromosomes based on physical characteristics such as length and location of the centromere. Karyotypes can be checked for defects in number or structure of chromosomes.

Keratinocytes   Cells of the epidermis that undergo keratinization, the process in which keratin gradually replaces the contents of maturing cells.

Ketoacidosis   A lowering of blood pH resulting from the accumulation of breakdown by-products of fats.

Kidney stones   Small, hard crystals formed in the urinary tract when substances like calcium (the most common constituent), uric acid, or magnesium ammonium phosphate precipitate from urine as a result of higher than normal concentrations. They are also called renal calculi.

Kidneys   Reddish, kidney bean-shaped organs responsible for filtering wastes and excess materials from the blood, assisting the respiratory system in the regulation of blood pH, and maintaining fluid balance by regulating the volume and composition of blood and urine.

Klinefelter syndrome   A genetic condition resulting from nondisjunction of the sex chromosomes in which a person inherits an extra X chromosome that results in an XXY genotype. The person is phenotypically male.

Labia majora   One of two elongated skin folds lateral to the labia minora. They are part of the female external genitalia.

Labia minora   One of two small skin folds on either side of the vagina and interior to the labia majora. They are part of the external genitalia of a female.

Labor   The process by which the fetus is expelled from the uterus and moved through the vagina and into the outside world. During true labor, uterine contractions occur at regular intervals, are often painful, and intensify with walking. Labor is usually divided into the dilation stage, expulsion stage, and placental stage. False labor is characterized by irregular contractions that fail to intensify or change with walking.

Lactation   The production and ejection of milk from the mammary glands. The hormone prolactin from the anterior pituitary gland promotes milk production, and the hormone oxytocin released from the posterior pituitary gland makes milk available to the suckling infant by stimulating milk ejection or let-down.

Lacuna (plural, lacunae)   A tiny cavity. It contains osteocytes (bone cells) in the matrix of bone and cartilage cells in the matrix of cartilage.

Langerhans cells   Macrophages that arise in bone marrow and migrate to the epidermis where they recognize and ingest foreign substances, eventually presenting them to white blood cells for final destruction.

Lanugo   Soft, fine hair that covers the fetus beginning about the third or fourth month after conception.

Large intestine   The final segment of the gastrointestinal tract, consisting of the cecum, colon, rectum, and anal canal. The large intestine helps absorb water, forms feces, and plays a role in defecation.

Laryngitis   An inflammation of the larynx in which the vocal cords become swollen and can no longer vibrate and produce sound.

Larynx   The voice box or Adam's apple. A box-like cartilaginous structure between the pharynx and the trachea held together by muscles and elastic tissue.

Latent period   The interval between the reception of the stimulus and the beginning of muscle contraction.

Leukemia   A cancer of the blood-forming organs that causes white blood cell numbers to increase. The white blood cells are abnormal and do not effectively defend the body against infectious agents.

Leukocytes   White blood cells. They are cells of the blood including neutrophils, eosinophils, basophils, monocytes, B lymphocytes, and T lymphocytes. Leukocytes are involved in body defense mechanisms and removal of wastes, toxins, or damaged, abnormal cells.

LH   See leuteinizing hormone.

Life expectancy   The average number of years a newborn is expected to live.

Ligament   A strong strap of connective tissue that holds the bones together, supports the joint, and directs the movement of the bones.

Limbic system   A collective term for several structures, including parts of the anterior thalamus, parts of the cerebral cortex, the hypothalamus, the olfactory bulb, the amygdala, and the hippocampus. It is important in governing emotions, including rage, pain, fear, sorrow, joy, and sexual pleasure.

Linkage   The tendency for a group of genes located on the same chromosome to be inherited together.

Lipids   Molecules including triglycerides, phospholipids, and steroids that do not dissolve in water.

Liver   A large organ that functions mainly in the production of plasma proteins, the excretion of bile, the storage of energy reserves, the detoxification of poisons, and the interconversion of nutrients.

Locus   The point on a chromosome where a particular gene is found.

Logistic growth   A pattern of population growth in which population growth begins rapidly but becomes increasingly slower as it approaches the carrying capacity of its environment until it levels off and fluctuates around the carrying capacity.

Longitudinal fissure   A deep groove that separates the cerebrum into two hemispheres.

Loose connective tissue   Connective tissue, such as areolar and adipose tissue, that contains many cells where the fibers of the matrix are fewer in number and more loosely woven than those found in dense connective tissue.

Lower esophageal sphincter   A ring of muscle at the juncture of the esophagus and the stomach that controls the flow of materials between the esophagus and the stomach. It relaxes to allow food into the stomach and contracts to prevent too much food from moving back into the esophagus.

Lumen   The hollow cavity or channel of a tubule through which the transported material flows.

Luteinizing hormone (LH)   A hormone secreted by the anterior pituitary gland that in females stimulates ovulation, stimulates the formation of the corpus luteum (which produces estrogen and progesterone), and prepares the mammary glands for milk production and, in males, stimulates testosterone production by the interstitial cells within the testes.

Lymph   Fluid within the vessels of the lymphatic system. It is derived from the fluid that bathes the cells of the body.

Lymph nodes   Small nodular organs found along lymph vessels that filter lymph. The lymph nodes contain macrophages and lymphocytes, cells that play an essential role in the body's defense system.

Lymphatic system   A body system consisting of lymph, lymphatic vessels, and lymphatic tissue and organs. The lymphatic system helps return interstitial fluid to the blood, transport the products of fat digestion from the digestive system to the blood, and assist in body defenses.

Lymphatic vessels   The vessels through which lymph flows. A network of vessels that drains interstitial fluid and returns it to the blood supply and transports the products of fat digestion from the digestive system to the blood supply.

Lymphocyte   A type of white blood cell important in nonspecific and specific (immune) body defenses. The lymphocytes include B lymphocytes (B cells) that transform into plasma cells and produce antibodies and T lymphocytes (T cells) that are important in defense against foreign or infected cells.

Lymphoid organs   Various organs that belong to the lymphatic system, including the tonsils, spleen, thymus, and Peyer's patches.

Lymphoma   Cancer of the lymphoid tissues. In people with AIDS, it commonly affects the B cells.

Lysosomal storage diseases   Disorders such as Tay-Sachs disease caused by the absence of lysosomal enzymes. In these disorders, molecules that would normally be degraded by the missing enzymes accumulate in the lysosomes and interfere with cell functioning.

Lysosomes   Organelles that serve as the principal sites of digestion within the cell.

Lysozyme   An enzyme present in tears, saliva, and certain other body fluids that kills bacteria by disrupting their cell walls.

Macroevolution   Large-scale evolutionary changes such as those that might result from long-term changes in the climate or position of the continents.

Macromolecules   Giant molecules of life such as nucleic acids, proteins, and polysaccharides. Macromolecules are formed by the joining together of smaller molecules.

Macrophage   A large phagocytic cell derived from a monocyte that lives in loose connective tissue and engulfs anything detected as foreign.

Magnetic resonance imaging (MRI)   A means of visualizing a region of the body, including the brain, in which the picture results from differences in the way the hydrogen nuclei in the water molecules within the tissues vibrate in response to a magnetic field created around the area to be pictured.

Malignant tumor   A cancerous tumor. An abnormal mass of tissue that can invade surrounding tissue and spread to multiple locations throughout the body.

Malleus   The first of three small bones in the middle ear that transmit information about sound from the eardrum to the inner ear. The malleus is also known as the hammer.

Malnourishment   A form of hunger that occurs when the diet is not balanced and the right foods are not eaten.

Mammary glands   The milk-producing glands in the breast.

Mast cells   Small, mobile connective tissue cells often found near blood vessels. In response to injury, mast cells release histamine, which dilates blood vessels and increases blood flow to an area, and heparin, which prevents blood clotting.

Mechanical digestion   A part of the digestive process that involves physically breaking food into smaller pieces.

Mechanoreceptor   A sensory receptor that is specialized to respond to distortions in the receptor itself or in nearby cells. Mechanoreceptors are responsible for the sensations we describe as touch, pressure, hearing, and equilibrium.

Medulla oblongata   Part of the brain stem containing reflex centers for some of life's most vital physiological functions: the pace of the basic breathing rhythm, the force and rate of heart contraction, and blood pressure. It connects the spinal cord to the rest of the brain.

Medullary cavity   The cavity in the shaft of long bones that is filled with yellow marrow.

Medullary rhythmicity center   The region of the brain stem controlling the basic rhythm of breathing.

Meiosis   A type of cell division that occurs in the gonads and gives rise to gametes. As a result of two divisions (meiosis I and meiosis II), haploid gametes are produced from diploid germ cells.

Meissner's corpuscles   Encapsulated nerve cell endings that are common on the hairless, sensitive areas of the skin, such as the lips, nipples, and fingertips, and tell us exactly where we have been touched.

Melanin   Pigment produced by the melanocytes of the skin.

Melanocyte-stimulating hormone (MSH)   The anterior pituitary hormone that stimulates melanin production.

Melanocytes   Spider-shaped cells located at the base of the epidermis that manufacture and store melanin, a pigment involved in skin color and absorption of ultraviolet radiation.

Melanoma   The least common and most dangerous form of skin cancer that arises in the melanocytes, the pigment-producing cells of the skin.

Melanosomes   Membrane-bound structures of melanocytes where the pigment melanin is formed.

Melatonin   A hormone secreted by the pineal gland.

Memory cell   A lymphocyte (B cell or T cell) of the immune system that forms in response to an antigen and that circulates for a long period of time; such cells are able to mount a quick immune response to a subsequent exposure to the same antigen.

Meninges   Three protective connective tissue membranes that surround the central nervous system: the dura mater, the pia mater, and the arachnoid.

Meningitis   An inflammation of the meninges.

Menopause   The end of a female's reproductive potential when ovulation and menstruation cease.

Menstrual cycle   The sequence of events that occurs on an approximately 28-day cycle in the uterine lining that involves the thickening and increased vascularization of the endometrium and the loss of the endometrium as menstrual flow.

Merkel cells   Cells of the epidermis found in association with sensory neurons where the epidermis meets the dermis.

Merkel disk   The Merkel cell-neuron combination that functions as a sensory receptor for light touch, providing information about objects contacting the skin. It is found on both the hairy and the hairless parts of the skin.

Mesoderm   The primary germ layer that gives rise to muscle, bone, connective tissue, and organs such as the heart, kidneys, ovaries, and testes.

Messenger RNA (mRNA)   A type of RNA synthesized from and complementary to a region of DNA that attaches to ribosomes in the cytoplasm and specifies the amino acid order in the protein. It carries the DNA instructions for synthesizing a particular protein.

Metabolism   The sum of all chemical reactions within the living cells.

Metaphase   In mitosis, the phase when the chromosomes, guided by the fibers of the mitotic spindle, form a line at the center of the cell. As a result of this alignment, when the chromosomes split at the centromere, each daughter cell receives one chromatid from each chromosome and thus a complete set of the parent cell's chromosomes.

Metastasize   To spread from one part of the body to another part not directly connected to the first part. Cancerous tumors metastasize and form new tumors in distant parts of the body.

Microevolution   Changes in the frequency (abundance) of certain alleles relative to others within a gene pool. The causes include genetic drift, gene flow, natural selection, and mutation.

Microfilaments   Components of the cytoskeleton made from the globular protein actin. They form contractile units in muscle cells.

Microtubules   Components of the cytoskeleton made from the globular protein tubulin. They are responsible for the movement of cilia and flagella.

Microvilli   Microscopic cytoplasm-filled extensions of the cell membrane that serve to increase the absorptive surface area of the cell.

Middle ear   An air-filled space in the temporal bone that includes the tympanic membrane (eardrum) and three small bones (the hammer, anvil, and stapes). It serves as an amplifier of sound pressure waves.

Mineralocorticoids   Hormones secreted by the adrenal cortex that affect mineral homeostasis and water balance.

Minipill   A birth control pill that contains only progesterone.

Mitochondrion   Organelle within which most of cellular respiration occurs in a eukaryotic cell. Cellular respiration is the process by which oxygen and an organic fuel such as glucose are consumed and energy is released and used to form ATP.

Mitosis   A type of cell division occurring in somatic cells in which two identical cells, called daughter cells, are generated from a single cell. The original cell first replicates its genetic material and then distributes a complete set of genetic information to each of its daughter cells. Mitosis is usually divided into prophase, metaphase, anaphase, and telophase.

Mitral valve   A heart valve located between the left atrium and ventricle that prevents the backflow of blood from the ventricle to the atrium. Also called the bicuspid valve.

Molecular clock hypothesis   The idea that single nucleotide changes in DNA (point mutations) and the amino acid changes in proteins that can be produced by some point mutations occur with steady, clock-like regularity caused by background radiation and errors inherent in DNA copying. It permits comparison of molecular sequences to estimate the ages of separation between species (the more differences in sequence, the more time that has elapsed since the common ancestor).

Molecule   A substance comprised of two or more atoms.

Monoclonal antibodies   Defensive proteins specific for a particular antigen secreted by a clone of genetically identical cells descended from a single cell.

Monocyte   The largest white blood cell. Monocytes are active in fighting chronic infections, viruses, and intracellular bacterial infections. A monocyte can transform into a phagocytic macrophage.

Monogenesis model   The idea that there was a single origin for all *Homo sapiens*; specifically, modern humans evolved from *Homo erectus* in Africa and only later migrated to other continents where they replaced the diverse descendants of *H. erectus*.

Monohybrid cross   A genetic cross that considers the inheritance of a single trait from individuals differing in the expression of that trait.

Monomers   Small identical molecules that join together to form polymers.

Mononucleosis   A viral disease caused by the Epstein-Barr virus that results in an elevated level of monocytes in the blood. It causes fatigue and swollen glands, and there is no available treatment.

Monosaccharide   The smallest molecular unit of a carbohydrate. Monosaccharides are known as simple sugars.

Monosomy   A condition in which there is only one representative of a chromosome instead of two representatives.

Mons veneris   A round fleshy prominence over the pubic bone in a female. Part of the female external genitalia.

Morning sickness   The nausea and vomiting experienced by some women early in pregnancy. It is not restricted to the morning and may be caused, in part, by high levels of the hormone human chorionic gonadotropin (HCG).

Morula   A solid ball of 12 or more cells produced by successive cleavages of the zygote. The name reflects its resemblance to the fruit of the mulberry tree.

Mosaic evolution   The phenomenon whereby various traits evolve at their own rates.

Motor (efferent) neuron   A neuron specialized to carry information *away from* the central nervous system to an effector, either a muscle or gland

Motor unit   The single axon from a motor neuron and all the muscle fibers (cells) it stimulates.

MRI   See magnetic resonance imaging.

mRNA   See messenger RNA.

Mucosa The innermost layer of the gastrointestinal tract. It secretes mucus that helps lubricate the tube, allowing food to slide through easily.

Mucus A sticky secretion that serves to lubricate body parts and trap particles of dirt and other secretions. It also helps protect the stomach from the action of gastric juice.

Multiple alleles Three or more alleles of a particular gene existing in a population. The alleles governing the ABO blood types provide an example.

Multiple sclerosis An autoimmune disease in which the body's own defense mechanisms attack myelin sheaths in the nervous system. As a patch of myelin is repaired, a hardened region called a sclerosis forms.

Multipolar neuron A neuron that has at least three processes, including an axon and a minimum of two dendrites. Most motor neurons and association neurons are multipolar.

Multiregional model The idea that modern humans evolved independently in several different areas from distinctive local populations of *Homo erectus*.

Muscle fiber A muscle cell.

Muscle spindles Specialized muscle fibers with sensory nerve cell endings wrapped around them that report to the brain whenever a muscle is stretched.

Muscularis The muscular layers of the gastrointestinal tract. These layers help move food through the gastrointestinal system and mix food with digestive secretions.

Mutations Changes in the DNA of genes that appear spontaneously, as an inherent property of DNA, or are caused by outside sources, such as radiation or chemical agents. They are rare and a source of genetic variation in populations.

Myelin sheath An insulating layer around axons that carry nerve impulses over relatively long distances that is composed of multiple wrappings of the plasma membrane of certain glial cells. In the central nervous system, the myelin sheath is formed by a type of glial cell called an oligodendrocyte. In the peripheral nervous system (the part outside the brain and spinal cord), the type of glial cell that forms the myelin sheath is the Schwann cell. The myelin sheath greatly increases the speed at which impulses travel and assists in the repair of damaged axons. Schwann cells forming the myelin sheath are separated from one another by short regions of exposed axon called nodes of Ranvier.

Myocardial infarction A heart attack. A condition in which a part of the heart muscle dies because of an insufficient blood supply.

Myocardium Cardiac muscle tissue that makes up the bulk of the heart. The contractility of the myocardium is responsible for the heart's pumping action.

Myofibril A rod-like bundle of contractile filaments (myofilaments) found in muscle cells.

Myofilament A contractile protein within muscle cells. There are two types: myosin (thick filaments) and actin (thin filaments).

Myoglobin An oxygen-binding pigment in muscle fibers.

Myometrium The smooth muscle layer in the wall of the uterus.

Myosin heads Club-shaped ends of a myosin molecule that bind to actin filaments and can swivel, causing actin filaments to slide past the myosin filaments, which causes muscle contraction. They are also called cross bridges.

Myxedema A condition characterized by swelling of the facial tissues because of the accumulation of interstitial fluids caused by hyposecretion of thyroid hormone.

Nasal cavity A chamber in the skull bounded by the internal and external nares.

Nasal conchae The three convoluted bones within each nasal cavity that increase surface area and direct airflow.

Nasal septum A thin partition of cartilage and bone that divides the inside of the nose into two nasal cavities.

Natural childbirth Childbirth without the use of pain-killing drugs.

Natural killer (NK) cells A type of cell in the immune system, probably lymphocytes, that roam the body in search of abnormal cells and quickly kill them.

Natural selection The differential survival and reproduction of individuals that results from genetically based variation in their anatomy, physiology, and behavior.

Nearsightedness Myopia. Nearsightedness is a visual condition in which the eyeball is elongated or the lens is too thick, causing the image to be focused in front of the retina.

Negative feedback The homeostatic mechanism by which the substance produced feeds back on the system, shutting down the production process.

Nephrons Functional units of the kidneys responsible for formation of urine. These microscopic tubules number 1 to 2 million per kidney and perform filtration (only certain substances are allowed to pass out of the blood and into the nephron), reabsorption (some useful substances are returned from the nephron to the blood), and secretion (the nephron directly removes wastes and excess materials in the blood and adds them to the filtered fluid that becomes urine).

Nephrotosis A painful condition that occurs when the renal fascia or adipose capsule is not substantial enough and a kidney slips from its normal position. It is also known as floating kidney. This condition is dangerous because the ureter of the displaced kidney may kink, preventing normal flow of urine from the kidney down the ureter to the bladder.

Nerve A bundle of parallel axons from many neurons. A nerve is usually covered with tough connective tissue.

Nervous tissue Tissue consisting of two types of cells, neurons and neuroglia, that comprise the brain, spinal cord, and nerves.

Neural crest cells Ectodermal cells near the developing central nervous system of an embryo that migrate and differentiate, eventually forming spinal and cranial nerves, the meninges of the brain and spinal cord, and several skeletal and muscular components of the head.

Neural tube The embryonic structure that gives rise to the brain and spinal cord.

Neuromuscular junction The area of contact between the terminal end of a motor neuron and the sarcolemma of a skeletal muscle fiber. When an action potential reaches the terminal end of the motor neuron, acetylcholine is released, triggering events that can lead to muscle contraction.

Neuron A nerve cell involved in intercellular communication. A neuron consists of a cell body, dendrites, and an axon. Neurons are excitable cells in the nervous system specialized to generate and transmit electrochemical signals called action potentials or nerve impulses.

Neuronal pool A group of neurons arranged in pathways called circuits and dedicated to a specific task.

Neurotransmitter A chemical released from the axon tip of a neuron that affects the activity of another cell (usually a nerve, muscle, or gland cell) by altering the electrical potential difference across the membrane of the receiving cell.

Neurulation A series of events during embryonic development when the central nervous system (brain and spinal cord) forms from ectoderm. During this period the embryo is called a neurula.

Neutrophil A phagocytic white blood cell important in defense against bacteria and removal of cellular debris. Most abundant of white blood cells.

New World monkeys A group of monkeys whose members are arboreal, have nostrils that open to the side, and have grasping (prehensile) tails. They occur in South and Central America and include spider monkeys, capuchins, and squirrel monkeys.

Niche The role of a species in an ecosystem. Includes the habitat, food, nest sites, etc. Describes how a member of a particular species uses materials in the environment and how it interacts with other organisms.

Node of Ranvier A region of exposed axon between Schwann cells forming a myelin sheath. In myelinated nerves the impulse jumps from one node of Ranvier to the next, greatly increasing the speed of conduction. This type of transmission is called saltatory conduction.

Nondisjunction Failure of the members of a pair of homologous chromosomes or the sister chromatids to separate during mitosis. Nondisjunction results in cells with abnormal chromosome numbers.

Norepinephrine (noradrenaline) A hormone secreted by the adrenal medulla, along with epinephrine, in response to stress. They initiate the physiological "fight-or-flight" reaction. Norepinephrine is also a neurotransmitter found in both the central and peripheral nervous systems. In the central nervous system, it is important in the regulation of mood, in the pleasure system of the brain, arousal, and dreaming sleep. Norepinephrine is thought to produce an energizing "good" feeling. It is also thought to be essential in hunger, thirst, and sex drive.

Norplant See hormonal implants.

Notochord The flexible rod of tissue that develops during gastrulation and signals where the vertebral column will form. The notochord defines the axis of the embryo, gives the embryo some rigidity, and prompts overlying ectoderm to begin formation of the central nervous system. During development, vertebrae form around the notochord. The notochord eventually degenerates, existing only as the pulpy, elastic material in the center of intervertebral disks.

Nuclear envelope The double membrane that surrounds the nucleus.

Nuclear pores Openings in the nuclear envelope that permit communication between the nucleus and the cytosol.

Nucleolus A specialized region within the nucleus that forms and disassembles during the course of the cell cycle. It plays a role in the generation of ribosomes, organelles involved in protein synthesis.

Nucleoplasm The chromatin and the aqueous environment within the nucleus.

Nucleosome The basic unit of DNA packing in eukaryotes in which DNA is wrapped two-and-a-half times around a core of eight histones and held in place by a linker histone.

Nucleotide A subunit of DNA composed of one five-carbon sugar (either ribose or deoxyribose), one phosphate group, and one of five nitrogenous bases (purines or pyrimidines). Nucleotides are the building blocks of nucleic acids (DNA and RNA).

Nucleus The command center of the cell containing almost all the genetic information.

Oil glands Glands associated with the hair follicles that produce sebum. They are also called sebaceous glands.

Old World monkeys A group of monkeys whose members lack prehensile tails, often having short tails or none at all, and have nostrils that open downward. Some Old World monkeys are arboreal and some are terrestrial. Examples include baboons and rhesus monkeys.

Oocyte A cell whose meiotic divisions will produce an ovum and three polar bodies.

Oogenesis The production of ova (eggs), including meiosis and maturation.

Oogonium (plural, oogonia) A germ cell in an ovary that divides, giving rise to oocytes.

Opsin One of several proteins that can be bound to retinal to form the visual pigments in rods and cones.

Optic nerve One of two nerves, one from each eye, responsible for bringing processed electrochemical messages from the retina to the brain for interpretation.

Organ A structure composed of two or more different tissues with a specific function.

Organ of Corti The portion of the cochlea that contains receptor cells that sense sound vibration. It is most directly responsible for the sense of hearing.

Origin The end of the muscle that is attached to the bone that remains relatively stationary during a movement.

Osmosis A special case of diffusion in which water moves across the plasma membrane or any other selectively permeable membrane from a region of lower concentration of solute to a region of higher concentration of solute.

Osteoarthritis An inflammation in a joint that is caused by degeneration of the surfaces of a joint caused by wear and tear.

Osteoblast A bone-forming cell.

Osteoclast A large cell responsible for the breakdown and absorption of bone.

Osteocytes Mature bone cells found in lacunae that are arranged in concentric rings around the Haversion canal. Cytoplasmic projections from osteocytes extend through canaliculi (tiny channels) that connect with other osteocytes.

Osteon An Haversion system. An osteon is the structural unit of compact bone that appears as a series of concentric circles of lacunae. The lacunae contain bone cells around a central canal containing blood vessels and nerves.

Osteoporosis A decrease in bone density that occurs when the destruction of bone outpaces the formation of new bone, causing bone to become thin, brittle, and susceptible to fracture.

Outer ear The external appendage on the outside of the head (pinna) and the canal (the external auditory canal) that extends to the eardrum. It functions as the receiver for sound vibrations.

Oval window A membrane-covered opening in the cochlea through which vibrations from the stirrup (stapes) are transmitted to fluid within the cochlea.

Ovarian cycle The sequence of events of the ovary that lead to ovulation. The cycle is approximately 28 days long and is closely coordinated with the menstrual cycle.

Ovaries The female gonads. The female reproductive organs that produce the ova (eggs) and the hormones estrogen and progesterone.

Ovulation The release of the secondary oocyte from the ovary.

Ovum A mature egg; a large haploid cell that is the female gamete.

Oxygen debt The amount of oxygen required after exercise to oxidize the lactic acid formed during exercise.

Oxyhemoglobin Hemoglobin bound to oxygen.

Oxytocin (OT) The hormone released at the posterior pituitary that stimulates uterine contractions and milk ejection.

Pacemaker The sinoatrial (SA) node. A region of specialized cardiac muscle cells located in the right atrium near the junction of the superior vena cava that sets the pace of the heart rate at about 70 to 80 beats a minute. The SA node sends out an electrical signal that spreads through the muscle cells of the atria, causing them to contract.

Pacinian corpuscle A large encapsulated nerve cell ending that is located deep within the skin and near body organs that responds when pressure is first applied. It is important in sensing vibration.

Packing An intricate system of folding the DNA with small proteins called histones. Packing plays a role in gene regulation.

Pain receptor A sensory receptor that is specialized to detect the physical or chemical damage to tissues that we sense as pain. Pain receptors are sometimes classified as chemoreceptors, because they often respond to chemicals liberated by damaged tissue, and occasionally as mechanoreceptors, because they are stimulated by physical changes, such as swelling, in the damaged tissue.

Paleoanthropology The scientific study of human origins and evolution.

Pancreas An accessory organ behind the stomach that secretes digestive enzymes, bicarbonate ions to neutralize the acid in chyme, and hormones that regulate blood sugar.

Papilla An indentation of connective tissue in the hair follicle that supplies nourishment to new cells.

Papillary layer The uppermost layer of the dermis consisting of loose connective tissue with elastic fibers.

Parasympathetic division The branch of the autonomic nervous system that is active during restful conditions. Its effects generally oppose those of the sympathetic nervous system.

Parathyroid hormone (PTH) A hormone released from the parathyroid glands that regulates blood calcium levels by stimulating osteoclasts to break down bone. PTH is secreted when blood calcium levels are too low.

Parental generation The parents—the individuals in the earliest generation under consideration.

Parkinson's disease A progressive disorder that results from the death of dopamine-producing neurons that lie in the heart of the brain's movement control center, the substantia nigra. Parkinson's disease is characterized by slowed movements, tremors, and muscle rigidity.

Parturition Birth, which usually occurs about 38 weeks after fertilization.

Passive immunity Temporary immune resistance that develops when a person receives antibodies that were produced by another person or animal.

Pathogen A disease-causing organism.

Pedigree A chart showing the genetic connections among individuals in an extended family that is often used to trace the expression of a particular trait in that family.

Pelvic inflammatory disease (PID) A general term for any bacterial infection of a woman's pelvic organs. PID is usually caused by sexually transmitted bacteria, especially those that cause chlamydia and gonorrhea.

Penis The cylindrical external reproductive organ of a male through which most of the urethra extends and that serves to deliver sperm into the female tract during sexual intercourse.

Pepsin A protein-splitting enzyme initially secreted in the stomach in the inactive form of pepsinogen that is activated into pepsin by hydrochloric acid.

Peptic ulcer A local defect in the surface of the stomach or small intestine characterized by dead tissue and inflammation.

Perforin A type of protein released by a natural killer cell that creates numerous pores (holes) in the target cell, making it leaky. Fluid is then drawn into the leaky cell because of the high salt concentration within, and the cell bursts.

Pericarditis An inflammation of the pericardial sac that encases the heart. It can cause a build-up of fluid within the pericardial sac that places a strain on the heart.

Pericardium The double-layered fibrous sac enclosing the heart. The outer layer provides a protective case that holds the heart in place. The inner layer is an integral part of the wall of the heart.

Perichondrium The layer of dense connective tissue surrounding cartilage that contains blood vessels that supply cartilage with nutrients.

Periodontitis Inflammation of the gums.

Periosteum The membranous covering that nourishes bone.

Peripheral nervous system The part of the nervous system outside the brain and spinal cord. It keeps the central nervous system in continuous contact with almost every part of the body. It is composed of nerves and ganglia. The two branches are the somatic and the autonomic nervous systems.

Peripheral proteins Proteins attached to the inner or outer surface of the plasma membrane.

Peristalsis Rhythmic waves of muscular contraction and relaxation in the walls of hollow tubular organs, such as the digestive organs, that serve to push contents through the tubes.

Peroxisome An enzyme-containing organelle within which chemical reactions occur that generate hydrogen peroxide. Such reactions have diverse functions ranging from the breakdown of fats to the detoxification of poisons and alcohol.

PET scan See positron emission tomography.

pH scale A scale for measuring the concentration of hydrogen ions. The scale ranges from 0 to 14, with a pH of 7 being neutral, a pH of less than 7 being acidic, and a pH of greater than 7, basic.

Phagocytes Scavenger cells specialized to engulf and destroy particulate matter, such as pathogens, damaged tissue, or dead cells.

Phagocytosis The process by which cells such as white blood cells ingest foreign cells or substances by surrounding the foreign material with cell membrane. It is a type of endocytosis.

Pharynx The space shared by the respiratory and digestive systems that is commonly called the throat. The pharynx is a passageway for air, food, and liquid.

Phenotype The observable physical and physiological traits of an individual. Phenotype results from the inherited alleles and their interactions with the environment.

Phospholipids   Important components of cell membranes. They have a nonpolar "water-hating" tail (comprised of fatty acids) and a polar "water-loving" head (comprised of an R group, glycerol, and phosphate).

Photoreceptor   A sensory receptor specialized to detect changes in light intensity. Photoreceptors are responsible for the sensation we describe as vision.

Phylogenetic trees   Generalized descriptions of the history of life. They depict hypotheses about genealogical relationships among species or higher taxa.

Physical dependence   A condition in which continued use of a drug is needed to maintain normal cell functioning.

Pia mater   The innermost layer of the meninges (the connective tissue layers that protect the central nervous system).

Piloerection   Contraction of the arrector pili muscles causing hairs to stand on end and form a layer of insulation.

Pineal gland   The gland that produces the hormone melatonin and is located at the center of the brain attached to the third ventricle.

Pinealocyte   Cell of the pineal gland that secretes melatonin.

Pinna   The visible part of the ear on each side of the head that gathers the sound and channels it to the external auditory canal.

Pinocytosis   A type of endocytosis in which droplets of fluid and the dissolved substances therein are engulfed by cells.

Pituitary dwarfism   A condition caused by insufficient growth hormone during childhood.

Pituitary gland (hypophysis)   The endocrine organ connected to the hypothalamus by the infundibulum. It consists of the anterior and posterior lobes.

Placenta   The organ that delivers oxygen and nutrients to the embryo and carries carbon dioxide and wastes away from it. The placenta is also called the afterbirth.

Placenta previa   The condition in which the placenta forms in the lower half of the uterus, entirely or partially covering the cervix. It may cause premature birth or maternal hemorrhage and usually makes vaginal delivery impossible.

Placental stage   The third (and final) stage of true labor. It begins with delivery of the baby and ends when the placenta detaches from the wall of the uterus and is expelled from the mother's body.

Plaque   A bumpy layer consisting of smooth muscle cells filled with lipid material, especially cholesterol, that bulges into the channel of an artery and reduces blood flow. Another type of plaque is a build-up of food material and bacteria on teeth that leads to tooth decay.

Plasma   A straw-colored liquid that makes up about 55% of whole blood. It serves as the medium for transporting materials within the blood. Plasma consists of water (91%–93%) with substances dissolved in it (7% or 8%).

Plasma cells   The effector cell produced from a B lymphocyte that secretes antibodies.

Plasma proteins   Proteins dissolved in plasma, including albumins, which are important in water balance between cells and the blood, globulins, which are important in transporting various substances in the blood, and antibodies, which are important in the immune response.

Plasmid   A small, circular piece of self-replicating DNA that is separate from the chromosome and found in bacteria. Plasmids are often used as vectors in recombinant DNA research.

Plasmin   An enzyme that breaks down fibrin and dissolves blood clots.

Plasminogen   A plasma protein. It is the inactive precursor of plasmin.

Platelet (thrombocyte)   A cell fragment of a megakaryocyte that releases substances necessary for blood clotting. It is formed in the red bone marrow from a precursor cell called a megakaryocyte.

Platelets   The cell fragments of the blood that function in clotting.

PMS   See premenstrual syndrome.

Podocytes   The cells in the inner membrane of Bowman's capsule with highly branched extensions that wrap around the glomerular capillaries. They are a part of the glomerular filter of a nephron.

Point mutation   A mutation that involves changes in one or a few nucleotides in DNA.

Polar body   Any of three small nonfunctional cells produced during the meiotic divisions of an oocyte. The divisions also produce a mature ovum (egg).

Polygenic inheritance   Inheritance in which several independently assorting or loosely linked genes determine the expression of a trait.

Polymer   A large molecule formed by the joining together of many smaller molecules of the same general type (monomers).

Polymerase chain reaction (PCR)   A technique used to amplify (increase) the quantity of DNA in vitro using primers, DNA polymerase, and nucleotides.

Polysaccharides   The complex carbohydrates formed when large numbers of monosaccharides (most commonly glucose) join together to form long chains through dehydration synthesis. Most polysaccharides store energy or provide structure.

Polysome   A cluster of ribosomes simultaneously translating the same mRNA strand.

Polyspermy   The abnormal condition in which more than one sperm enters an oocyte. It is usually prevented by a rapid depolarization of the plasma membrane of the oocyte as well as by the release of enzymes by the oocyte that make the zona pellucida impermeable to other sperm.

Population   A group of potentially interacting individuals of the same species living in a distinct geographic area.

Portal system   A system whereby a capillary bed drains to veins that drain to another capillary bed.

Positive feedback   The mechanism by which the substance produced feeds back on the system, further stimulating the production process.

Positron emission tomography (PET)   A method that can be used to measure the activity of various brain regions. The person being scanned is injected with a radioactively-labeled nutrient, usually glucose, that is tracked as it flows through the brain. The radioisotope emits positively charged particles, called positrons. When the positrons collide with electrons in the body, gamma rays are released. The gamma rays can be detected and recorded by PET receptors. Computers then use the information to construct a PET scan that shows where the radioisotope is being used in the brain.

Postnatal period   The period of development after birth. It includes the stages of infancy, childhood, puberty, adolescence, and adulthood.

Postsynaptic neuron   The neuron located after the synapse. The membrane of the postsynaptic neuron has receptors specific for certain neurotransmitters.

Precapillary sphincter   A ring-like muscle that acts as a valve that opens and closes a capillary bed. Contraction of the precapillary sphincter squeezes the capillary shut and directs blood through a thoroughfare channel to the venule. Relaxation of the precapillary sphincter allows blood to flow through the capillary bed.

Premenstrual syndrome (PMS)   A collection of uncomfortable symptoms, including irritability, stress, and bloating, that appears 7 to 10 days before a woman's menstrual period and is associated with hormonal cycling.

Prenatal period   The period of development before birth. It is further subdivided into the pre-embryonic period (from fertilization through the second week), the embryonic period (from the third through the eighth weeks), and the fetal period (from the ninth week until birth).

Presynaptic neuron   The neuron located before the synapse. The neuron that releases the neurotransmitter that affects the activity of the receiving cell.

Primary follicle   A spherical structure in the ovary that contains a primary oocyte and is surrounded by follicle cells.

Primary germ layers   Layers produced by gastrulation from which all tissues and organs form. They include ectoderm, mesoderm, and endoderm.

Primary motor area   A band of the frontal lobe of the cerebral cortex that initiates messages that direct voluntary movements.

Primary sensory area   A band of the parietal lobe of the cerebral cortex to which information is sent from receptors in the skin regarding touch, temperature, and pain and from receptors in the joints and skeletal muscles.

Primary spermatocyte   The original large cell that develops from a spermatogonium during sperm development in the seminiferous tubules. It undergoes meiosis and gives rise to secondary spermatocytes.

Primary structure   The precise sequence of amino acids of a protein. This sequence, determined by the genes, dictates a protein's structure and function.

Primary succession   The sequence of changes in the species making up a community over time that begins in an area where no community previously existed and ends with a climax community.

Probe   A single-stranded molecule of radioactive DNA or RNA with a sequence of bases complementary to the sequence in the gene of interest that is used to locate the desirable gene.

Prodrome   The symptoms that precede recurring outbreaks of a disease such as genital herpes.

Programmed cell death   A genetically programmed series of events that causes a cell to self-destruct.

Prokaryote   A cell that lacks a nucleus and other membrane-enclosed organelles. The prokaryotes include bacteria and cyanobacteria.

Prolactin (PRL)   Anterior pituitary hormone that stimulates mammary glands to produce milk.

Promoter   A specific region on DNA next to the "start" gene that controls the expression of the gene.

Prophase   In mitosis, the phase when the chromosomes begin to thicken and shorten, the nucleolus disappears, the nuclear envelope begins to break down, and the mitotic spindle forms in the cytoplasm.

Prosimians   The suborder of modern primates that includes the lemurs, lorises, pottos, and tarsiers.

Prostaglandins   The lipid molecules found in and released by the plasma membranes of most cells. They are often called the "local hormones."

Prostate gland   An accessory reproductive gland in males that surrounds the urethra as it passes from the bladder. Its secretions contribute to semen and serve to activate the sperm and to counteract the acidity of the female reproductive tract.

Proteins   The macromolecules composed of amino acids linked by peptide bonds. The functions of proteins include structural support, transport, movement, and regulation of chemical reactions.

Prothrombin   A plasma protein synthesized by the liver that is important in blood clotting. It is converted to an active form (thrombin) by thromboplastin that is released from platelets.

Protozoa   A group of single-celled organisms with a well-defined eukaryotic nucleus. Protozoa can cause disease by producing toxins or by releasing enzymes that prevent host cells from functioning normally.

Pseudopodium   The temporary extension of a cell used when feeding and moving. The name means "false foot."

Puberty   The stage in postnatal development when secondary sexual characteristics such as pubic and underarm hair develop. This stage usually occurs slightly earlier in girls (from 12 to 15 years of age) than in boys (from 13 to 16 years of age).

Pulmonary arteries   Blood vessels that carry blood low in oxygen from the right ventricle to the lungs where it is oxygenated.

Pulmonary circuit (or circulation)   The pathway that transports blood from the right ventricle of the heart to the lungs and back to the left atrium of the heart.

Pulmonary veins   Blood vessels that carry oxygenated blood from the lungs to the left atrium of the heart.

Pulp   The center of a tooth that contains the tooth's life-support systems.

Pulse   The rhythmic expansion of an artery created by the surge of blood pushed along the artery by each contraction of the ventricles of the heart. With each beat of the heart, the wave of expansion begins, moving along the artery at the rate of 6 to 9 meters per second.

Punctuated equilibrium   The idea that relatively brief periods of evolutionary change are interspersed with long periods of relatively little change.

Punnett square   A diagrammatic method used to determine the probable outcome of a genetic cross. The possible allele combinations in the gametes of one parent are used to label the columns and the possible allele combinations of the other parent are used to label the rows. The alleles of each column and each row are then paired to determine the possible genotypes of the offspring.

Pupil   The small hole through the center of the iris through which light passes to enter the eye. The size of the pupil is altered to regulate that amount of light entering the eye.

Purines   Nitrogenous bases (adenine and guanine) that contain a five-membered ring attached to a six-membered ring. They are the components of nucleotides.

Purkinje fibers   The specialized cardiac muscle cells that deliver an electrical signal from the atrioventricular bundle to the individual heart muscle cells in the ventricles.

Pyloric sphincter   A ring of muscle between the stomach and small intestine that regulates the emptying of the stomach into the small intestine.

Pyramid of energy   A graphical representation of the decreasing amount of energy available at each trophic level.

Pyramid of numbers   A graphical representation of the number of individuals at each trophic level.

Pyrimidines   Nitrogenous bases (cytosine, thymine, and uracil) that contain a six-membered ring comprised of carbons and nitrogens. They are the components of nucleotides.

Pyrogen   A fever-producing substance.

Quaternary structure   The shape of an aggregate protein. It is determined by the mutually attractive forces between the protein's subunits.

Radioisotopes   The isotopes that are unstable and spontaneously decay, emitting radiation in the form of gamma rays and alpha and beta particles.

Receptor potential   An electrochemical message (a change in the degree of polarization of the membrane) generated in a sensory receptor in response to a stimulus. Receptor potentials vary in magnitude with the strength of the stimulus.

Receptors   The protein molecules located in the cytosol and on the plasma membrane of cells that are sensitive to chemical messengers.

Recessive allele   The allele whose effects are masked in the heterozygous condition. The recessive allele usually produces a nonfunctional protein.

Recombinant DNA   Segments of DNA from two sources that have been combined in vitro and transferred to cells in which their information can be expressed.

Red blood cell (erythrocyte)   An enucleated biconcave cell in the blood that is specialized for transporting oxygen.

Red marrow   Blood-cell forming connective tissue found in the marrow cavity of certain bones.

Reduction division   The first meiotic division (meiosis I) that produces two cells, which each contain one member of each homologous pair (23 chromosomes with replicates attached in humans).

Reflex   A simple, stereotyped reaction to a stimulus.

Reflex arc   A neural pathway consisting of a sensory receptor, a sensory neuron, usually at least one interneuron, a motor neuron, and an effector.

Refractory period   The refractory period of a neuron is the interval following an action potential during which it cannot be stimulated to generate another action potential.

Relaxin   The hormone released from the placenta and the ovaries. It initiates labor and facilitates delivery by dilating the cervix and relaxing the ligaments and cartilage of the pubic bones.

Renal capsule   The innermost layer of connective tissue that protects the kidneys from trauma and infection.

Renal corpuscle   Portion of the nephron where fluid is filtered. It consists of a tuft of capillaries, the glomerulus, and a surrounding cup-like structure called Bowman's (glomerular) capsule.

Renal cortex   The outer region of the kidney, containing renal columns.

Renal fascia   The outermost layer of connective tissue on the kidneys. It anchors each kidney and adrenal gland to the abdominal wall and surrounding tissues.

Renal medulla   The inner region of the kidney. It contains cone-shaped structures called renal pyramids.

Renal pelvis   The innermost region of the kidney—the cavity within the renal sinus.

Renal sinus   The large space within a kidney, containing the renal pelvis, fat, and connective tissue.

Renal tubule   Site of reabsorption and secretion by the nephron. It consists of the proximal convoluted tubule, the loop of Henle, and the distal convoluted tubule.

Renin   An enzyme released by cells of the juxtaglomerular apparatus of nephrons. Renin converts angiotensinogen, a protein produced by the liver and found in the plasma, into another protein, angiotensin I. These actions of renin initiate a series of hormonal events that leads to increased reabsorption of sodium and water by the distal convoluted tubules and collecting ducts of nephrons.

Rennin   The gastric enzyme that breaks down milk proteins.

Replication   Copying from a template, as occurs during the synthesis of new DNA from preexisting DNA.

Repolarization   The return of the membrane potential to approximately its resting value. Repolarization of the nerve cell membrane during an action potential occurs because of the outflow of potassium ions.

Residual volume   The amount of air that remains in the lungs after a maximal exhalation.

Resistance phase   The phase of the general adaptation syndrome that follows the alarm reaction. It is characterized by anterior pituitary release of ACTH, TSH, and GH and the sustainment of metabolic demands by the body's fat reserves.

Respiratory distress syndrome (RDS)   A condition in newborns caused by an insufficient amount of surfactant, causing the alveoli to collapse and thereby making breathing difficult.

Resting potential   The separation of charge across the plasma membrane of a normal cell under homeostatic conditions. A neuron's resting potential is the electrical potential difference across the plasma membrane of a neuron that is not conducting an impulse. It is primarily caused by the unequal distributions of sodium ions, potassium ions, and large negatively charged proteins on either side of the plasma membrane. The resting potential of a neuron is about -70 mv.

Restriction enzyme   An enzyme that recognizes a specific sequence of bases in DNA and cuts the DNA into two strands at that sequence. Restriction enzymes are used to prepare DNA containing "sticky ends" during the creation of recombinant DNA. Their natural function in bacteria is to control the replication of viruses that infect the bacteria.

Reticular activating system (RAS)   An extensive network of neurons that runs through the medulla and projects to the cerebral cortex. It filters out unimportant sensory information before it reaches the brain and controls changing levels of consciousness.

Reticular lamina   The deeper of the two layers of basement membrane that anchor epithelial cells to connective tissue. The reticular lamina is secreted by cells of the connective tissue and gives the basement membrane its strength.

Reticular layer   The lower layer of the dermis that consists of dense connective tissue with collagen and elastic fibers.

Retina   The light-sensitive innermost layer of the eye containing numerous photoreceptors (rods and cones).

Retinal   A derivative of vitamin A that combines with an opsin (a protein) to form the light-absorbing pigments in rods and cones.

Retrovirus   Any one of the viruses that contains only RNA and carries out transcription from RNA to DNA (reverse transcription).

Reverberating circuit   A group of neurons arranged so that branches from a later neuron synapse with an earlier one, restimulating it and sending the impulse through the circuit again. This arrangement causes an impulse to cycle through the series continuously until one of the neurons in the circuit is inhibited or fails to fire because of fatigue. Reverberating circuits are thought to be important in maintaining wakefulness and in short-term memory.

Rheumatoid arthritis   An inflammation of a joint caused by an autoimmune response. It is marked by inflammation of the synovial membrane and excess synovial fluid accumulation in the joints, causing swelling, pain, and stiffness.

Rhythm method of birth control   A method of reducing the risk of pregnancy by avoiding intercourse on all days that might result in sperm and egg meeting.

Ribosomal RNA (rRNA)   A type of RNA that combines with proteins to form the ribosomes, structures on which protein synthesis occurs. The most abundant form of RNA.

Ribosome   A cell organelle involved in protein synthesis. Ribosomes consist of two subunits, each containing rRNA and proteins.

Ribozyme   An enzyme consisting of RNA that catalyzes reactions during RNA splicing. Ribozymes snip out introns before the protein is formed.

RNA   Ribonucleic acid. RNA is a single-strand nucleic acid that contains ribose (a five-carbon sugar), phosphate, adenine, uracil, cytosine, or guanine. RNA plays a variety of roles in protein synthesis.

RNA polymerase   One of the group of enzymes necessary for the synthesis of RNA from a DNA template. It binds with the promoter on DNA that aligns the appropriate RNA nucleotides and links them together.

Rod   One of the photoreceptors containing rhodopsin and responsible for black and white vision. Rods are extremely sensitive to light.

Root   The root of a tooth is the part below the gum line. It is covered with a calcified, yet living and sensitive connective tissue, called cementum.

Root canal   A channel through the root of a tooth that contains the blood vessels and nerves.

Round window   A membrane-covered opening in the cochlea that serves to relieve the pressures caused by the movements of the oval window.

rRNA   See ribosomal RNA.

Rugae   Folds in the mucosa of the lining of the empty stomach's walls that can unfold, allowing the stomach to expand as it fills.

S-shaped growth curve   A graphical representation of logistic growth (growth that levels off as the population size approaches the carrying capacity of the environment).

Salinization   An accumulation of salts in soil caused by irrigation over a long period of time that makes the land unusable.

Saliva   The secretion from the salivary glands that helps moisten and dissolve food particles in the mouth, facilitating taste and digestion.

Salivary amylase   An enzyme in saliva that begins the chemical digestion of starches, breaking them into shorter chains of sugars.

Salivary glands   Exocrine glands in the facial region that secrete saliva into the mouth to begin the digestion process.

Saltatory conduction   The type of nerve transmission along a myelinated axon in which the nerve impulse jumps from one node of Ranvier to the next. Saltatory conduction greatly increases the speed of nerve conduction.

Sarcomere   The smallest contractile unit of a striated or cardiac muscle cell.

Sarcoplasmic reticulum   An elaborate form of endoplasmic reticulum found in muscle fibers.

Schizophrenia   A mental illness characterized by hallucinations and disordered thoughts and emotions that is caused by excessive activity at dopamine synapses in one part of the brain (the midbrain). As a result, dopamine is no longer in the proper balance with glutamate, a neurotransmitter released by neurons in another brain region (the cerebral cortex).

Schwann cell   A type of glial cell in the peripheral nervous system that forms the myelin sheath by wrapping around the axon many times. The myelin sheath insulates axons, increases the speed at which impulses are conducted, and assists in the repair of damaged neurons.

Scientific method   A procedure underlying most scientific investigations that involves observation, formulating an hypothesis, making predictions, experimentation to test the predictions, and drawing conclusions. Experimentation usually involves a control group and an experimental group that differ in one or very few factors (variables). New hypotheses may be generated from the results of experimentation.

Sclera   The white part of the eye that protects and shapes the eyeball and serves as a site of attachment for muscles that move the eye.

Scrotum   A loose-fitting fleshy sac containing the testes.

Seasonal affective disorder (SAD)   A form of depression associated with winter months when overproduction of melatonin is triggered by short daylengths.

Sebum   An oily substance made of fats, cholesteral, proteins, and salts secreted by the oil glands.

Second messenger system   A system whereby water-soluble hormones that cannot pass through the lipid bilayer of the plasma membrane bind to a receptor on the surface of the target cell. Such bending activates a "second messenger" in the cytosol that relays the hormonal message.

Secondary spermatocyte   A haploid cell formed by meiotic division of a primary spermatocyte during sperm development in the seminiferous tubules.

Secondary structure   The bending and folding of the chain of amino acids of a protein to produce shapes such as coils, spirals, and pleated sheets. These shapes form as a result of hydrogen bonding between different parts of the polypeptide chain.

Secondary succession   The sequence of changes in the species making up a community over time that takes place after some disturbance destroys the existing life. Soil is already present.

Secretin   A hormone released by the small intestine that inhibits the secretion of gastric juice and stimulates the release of bicarbonate ions from the pancreas and the production of bile in the liver.

Segregation, law of   A genetic principle that states that the alleles for each gene separate (segregate) during meiosis and gamete formation, and so half of the gametes bear one allele and the other half bear the other allele.

Semicircular canals   Three canals in each ear that are oriented at right angles to one another and contain sensory receptors that precisely monitor any sudden movement of the head. They detect body position and movement.

Semiconservative replication   Replication of DNA in which the two strands of a DNA molecule become separated and each serve as the template for a new double-stranded DNA. Each new double-stranded molecule consists of one new strand and one old strand.

Semilunar valves Heart valves located between each ventricle and its connecting artery that prevent the backflow of blood from the artery to the ventricle. Whereas the cusps of the AV valves are flaps of connective tissue, those of the semilunar valves are small pockets of tissue attached to the inner wall of their respective arteries.

Seminal vesicles A pair of male accessory reproductive glands located posterior to the urinary bladder. Their secretions contribute to semen and serve to nourish the sperm cells, reduce the acidity in the vagina, and to coagulate sperm.

Seminiferous tubules Coiled tubules within the testes where sperm are produced.

Sensory (afferent) neuron A nerve cell specialized to conduct nerve impulses from the sensory receptors *toward* the central nervous system.

Sensory adaptation A gradual decline in the responsiveness of a sensory receptor that results in a decrease in awareness of the stimulus.

Sensory receptors The structures specialized to respond to changes in their environment (stimuli) by generating electrochemical messages that are eventually converted to nerve impulses if the stimulus is strong enough. The nerve impulses are then conducted to the brain, where they are interpreted to build our perception of the world.

Serosa A thin layer of connective tissue that forms the outer layer of the gastrointestinal tract. It secretes a fluid that reduces friction with contacting surfaces.

Serotonin A neurotransmitter in the central nervous system thought to promote a generalized feeling of well-being.

Sex chromosomes The X and Y chromosomes. The pair of chromosomes involved in determining gender.

Sex-influenced inheritance An autosomal genetic trait that is expressed differently in males and females, usually because of the presence of sex hormones.

Sickle-cell anemia A type of anemia caused by a mutation that results in a change in one amino acid in a globin chain of hemoglobin (the iron-containing protein in red blood cells that transports oxygen). Such a change causes the red blood cell to assume a crescent (sickle) shape when oxygen levels are low. The sickle-shaped cells clog small blood vessels, leading to pain and tissue damage from insufficient oxygen.

Simple diffusion The spontaneous movement of a substance from a region of higher concentration to a region of lower concentration.

Simple goiter An enlarged thyroid gland caused by iodine deficiency.

Sinoatrial (SA) node A region of specialized cardiac muscle cells located in the right atrium near the junction of the superior vena cava that sets the pace of the heart rate. It is also known as the pacemaker. The SA node sends out an electrical signal that spreads through the muscle cells of the atria, causing them to contract.

Sinusitis Inflammation of the mucous membranes of the sinuses making it difficult for the sinuses to drain their mucous fluid.

Skeletal (striated) muscle One of three types of muscle in the body, characterized by visible striations and conscious, voluntary control over contraction. It attaches to bones and forms the muscles of the body.

Skeleton A framework of bones and cartilage that functions to support and protect internal organs and to permit body movement.

Sliding filament model A model of the mechanism of muscle contraction in which the myofilaments actin and myosin slide across one another, causing the sarcomere to shorten. When enough sarcomeres shorten, the muscle contracts.

Slow-twitch fibers Muscle fibers that are specialized to contract slowly but with incredible endurance when stimulated. They contain an abundant supply of myoglobin and mitochondria and are richly supplied with capillaries. They depend on aerobic pathways to generate ATP during muscle contraction.

Small intestine The organ located between the stomach and large intestine responsible for the final digestion and absorption of nutrients.

Smooth muscle One of three types of muscle in the body, characterized by the lack of visible striations and by unconscious control over its contraction. It is found in the walls of blood vessels and airways and in organs such as the stomach, intestines, and bladder.

Sodium-potassium pump A molecular mechanism in a plasma membrane that uses cellular energy in the form of ATP to pump ions against their concentration gradients. Typically, each pump ejects three sodium ions from the cell while bringing in two potassium ions.

Soluble fiber A type of dietary fiber that either dissolves or swells in water. It includes the pectins, gums, mucilages, and some hemicelluloses. Soluble fiber has a gummy consistency.

Somatic cells All body cells except for gametes (egg and sperm). Somatic cells contain the diploid number of chromosomes, which in humans is 46.

Somatic nervous system The part of the peripheral nervous system that leads from the central nervous system to the skeletal muscles.

Somites Blocks formed from mesoderm cells of the developing embryo that eventually form skeletal muscles of the neck and trunk, connective tissues, and vertebrae.

Source (donor) DNA DNA containing the "gene of interest" that will be combined with host DNA to form recombinant DNA.

Special senses The sensations that include smell, taste, vision, hearing, and the sense of balance or equilibrium.

Speciation The formation of a new species.

Species A population or group of populations whose members are structurally and functionally similar and capable of successful interbreeding under natural conditions. Such interbreeding must produce fertile offspring.

Spermatid A haploid cell formed by mitotic division of a haploid secondary spermatocyte and that develops into a spermatozoon.

Spermatocyte A cell developed from a spermatogonium during sperm development in the seminiferous tubules.

Spermatogenesis The series of events within the seminiferous tubules that gives rise to physically mature sperm from germ cells. It involves meiosis and maturation.

Spermatogonium (plural, spermatogonia) The undifferentiated male germ cells in the sominiferous tubules that give rise to spermatocytes.

Spermicides   A means of contraception that consists of sperm-killing chemicals in some form of a carrier, such as foam, cream, jelly, film, or tablet.

Spermiogenesis   The changes that occur as a spermatid is converted to a spermatozoon (sperm cell).

Sphincter   A ring of muscle between regions of a system of tubes that controls the flow of materials from one region to another past the sphincter.

Sphygmomanometer   A device for measuring blood pressure. A sphygmomanometer consists of an inflatable cuff that wraps around the upper arm attached to a device that can measure the pressure within the cuff.

Spina bifida   Birth defect in which the neural tube fails to develop and close properly. Taking vitamins and folic acid before conception appears to reduce the incidence.

Spinal cord   A tube of neural tissue that is continuous with the medulla at the base of the brain and extends about 45 cm (17 in.) to just below the last rib. It conducts messages between the brain and the rest of the body and serves as a reflex center.

Spinal nerves   Thirty-one pairs of nerves that arise from the spinal cord. Each spinal nerve services a specific region of the body.

Spongy bone   The bone formed from a latticework of thin struts of bone with marrow-filled areas between the struts. It is found in the ends of long bones and within the breastbone, pelvis, and bones of the skull. Spongy bone is less dense than compact bone and is made of an irregular network of collagen fibers surrounded by a calcium matrix.

Sprain   A tear in a ligament (a strap of connective tissue that holds bones together).

Squamous cell carcinoma   The second most common form of skin cancer that arises in the keratinocytes as they flatten and move toward the skin surface.

Stem cell   A type of cell that can give rise to red blood cells, white blood cells, or platelets.

Steroid hormones   A group of closely related hormones chemically derived from cholesterol and secreted primarily by the ovaries, testes, and adrenal glands.

Steroids   Lipids, such as cholesterol, consisting of four carbon rings with functional groups attached.

Stomach   A muscular sac that is well designed for storage of food, liquefaction of food, and the initial chemical digestion of proteins.

Stratum basale   The deepest layer of the epidermis consisting of a single row of cuboidal to column-shaped cells characterized by rapid cell division.

Stratum corneum   The outermost layer of the epidermis that consists of 25 to 30 rows of flat, dead cells filled with keratin.

Stratum granulosum   The part of the epidermis located above the stratum spinosum that consists of three to five layers of flattened cells with granules of keratohyalin, a substance that contributes to the formation of keratin.

Stratum lucidum   The layer of the epidermis above the stratum granulosum found only in the palms and soles of the feet.

Stratum spinosum   The portion of the epidermis above the stratum basale that contains several layers of cube-shaped cells.

Strep throat   A sore throat that is caused by *Streptococcus* bacteria.

Stressors   Stimuli that produce stress.

Striated (skeletal) muscle   One of three types of muscle in the body, characterized by visible striations and conscious, voluntary control over contraction. It attaches to bones and forms the muscles of the body.

Stroke   A cerebrovascular accident. A condition in which nerve cells die because the blood supply to a region of the brain is shut off, usually because of hemorrhage or atherosclerosis. The extent and location of the mental or physical impairment caused by a stroke depend on the region of the brain involved.

Substance P   A neurotransmitter released by neurons within pathways involved in the sensation of pain.

Succession, ecological   The sequence of changes in the species making up a community over time.

Summation, wave   See wave summation.

Superovulation   The ovulation of several oocytes. It is usually prompted by administration of hormones.

Suppressor T cell   A type of T lymphocyte that turns off the immune response when the level of antigen falls by releasing chemicals that dampen the activity of both B cells and T cells.

Suprachiasmatic nucleus (SCN)   The region of the brain in the hypothalamus where the "master clock" controlling biological rhythms is believed to reside.

Surfactant   Phospholipid molecules coating the alveolar surfaces that prevent the alveoli from collapsing.

Survivorship   The number of individuals of a group born in the same year that is expected to be alive at any given age. It is one way to represent age-specific mortality.

Sweat glands   The eccrine and apocrine sweat glands of the skin. They are also called sudoriferous glands. Eccrine sweat glands aid in the elimination of wastes and regulation of body temperature, and apocrine sweat glands discharge an oily secretion onto hair follicles.

Sympathetic nervous system   The branch of the autonomic nervous system responsible for the "fight-or-flight" responses that occur during stressful or emergency situations. Its effects are generally opposite to those of the parasympathetic nervous system.

Synapse   The site of communication between a neuron and another cell, such as another neuron or a muscle cell.

Synapsis   The physical association of homologous pairs of chromosomes that occurs during prophase I of meiosis. The term literally means bringing together.

Synaptic cleft   The gap between two cells forming a synapse, for example, two communicating nerve cells.

Synaptic knob   A small bulb-like swelling of an axon terminal that releases neurotransmitter.

Synaptic vesicle   A tiny membranous sac containing between 10,000 and 100,000 molecules of a neurotransmitter. Synaptic vesicles are located in the synaptic knobs of axon terminals, and they release their contents when an action potential reaches the synaptic knob.

Synergistic muscles   Two or more muscles that work together to cause movement in the same direction.

Synovial cavity   A fluid-filled space surrounding a synovial joint formed by a double-layered capsule. The fluid within the cavity (synovial fluid) is a viscous, clear liquid that acts as both a shock absorber and as a lubricant between the bones.

Synovial fluid   A viscous, clear fluid within a synovial cavity that acts as both a shock absorber and as a lubricant between the bones.

Synovial joint A freely moveable joint. A synovial joint is surrounded by a fluid-filled space, called a synovial cavity. They are the most abundant types of joints in the body.

Syphilis A sexually transmitted disease caused by the bacterium *Treponema pallidum*. If untreated, it can progress through three stages and cause death. The first stage is characterized by a painless crater-like bump called a chancre that forms at the site where the bacterium entered the body.

Systematic biology The discipline that deals with the naming, classification, and evolutionary relationships of organisms. It is also called systematics.

Systemic circuit (or circulation) The pathway of blood from the left ventricle of the heart to the cells of the body and back to the right atrium.

Systole Contraction of the heart. Atrial systole is contraction of the atria. Ventricular systole is contraction of the ventricles.

Systolic pressure The highest pressure in an artery during each heartbeat. The higher of the two numbers in a blood pressure reading. In a typical, healthy adult, the systolic pressure is about 120 mm Hg.

T lymphocyte A type of white blood cell. Some T lymphocytes (T cells) attack and destroy cells that are not recognized as belonging in the body, such as an infected cell or a cancerous cell.

T4 cell A helper T cell. The kind of T lymphocyte that serves as the main switch for the entire immune response by presenting the antigen to B cells and by secreting chemicals that stimulate other cells of the immune system.

Tanning The build-up of melanin in the skin in response to UV exposure.

Taste bud A structure consisting of receptors responsible for sense of taste surrounded by supporting cells. Taste buds are located primarily on the surface epithelium and certain papillae of the tongue.

Taste hairs Microvilli that project into a pore at the tip of the taste bud that bear the receptors for certain chemicals found in food.

Tectorial membrane A membrane that forms the roof of the organ of Corti (the actual organ of hearing). It projects over and is in contact with the sensory hair cells.

Telomerase The enzyme that synthesizes telomeres.

Telomere A piece of DNA at the tips of chromosomes that protects the ends of the chromosomes.

Telophase In mitosis, the phase when nuclear envelopes form around the group of chromosomes at each pole, the mitotic spindle is disassembled, and nucleoli reappear. The chromosomes also become less condensed and more thread-like in appearance.

Temporomandibular joint syndrome (TMJ) A group of symptoms including headaches, toothaches, and earaches caused by physical stress on the mandibular joint.

Tendinitis Inflammation of a tendon caused by excessive stress on the tendon.

Tendon A strap of connective tissue that connects muscle to bone.

Teratogens Environmental agents, such as drugs, chemicals, radiation, and certain viruses, that disrupt development.

Teratology The study of birth defects.

Terminal hair Thick, strong hair found on the scalp, eyebrows, eyelashes, and after puberty, the underarms and pubic area.

Tertiary structure The three-dimensional shape of proteins formed by hydrogen, ionic, and covalent bonds between different side chains.

Testcross A genetic cross used to determine the underlying genotype for a dominant phenotype. The individual with the dominant phenotype is mated with a homozygous recessive individual. If any of the offspring show the recessive form of the trait, then the parent showing the dominant phenotype must have been heterozygous for the trait.

Testes The male gonads. The male reproductive organs that produce sperm and the hormone testosterone.

Testosterone A male sex hormone needed for sperm production and the maintenance of male reproductive structures. It is produced primarily by the interstitial cells of the testes.

Tetanus A smooth, sustained contraction of muscle caused when stimuli are delivered in such rapid succession that there is no time for muscle relaxation.

Thalamus A brain structure located below the cerebral hemispheres that is important in sensory experience, motor activity, stimulation of the cerebral cortex, and memory. It is composed of many nuclei, clusters of neurons, and nerve fibers, each one specializing in a different function.

Theory A broad-ranging explanation for some aspect of the universe that is consistent with many observations and experiments.

Thermoreceptor A sensory receptor specialized to detect changes in temperature.

Thoroughfare channel A blood vessel that serves as a shunt between an arteriole and a venule, allowing blood to bypass the true capillaries of a capillary network. Blood flow through the thoroughfare channel is regulated by precapillary sphincters.

Threshold The degree to which the voltage difference across the plasma membrane of a neuron or other excitable cell must change to trigger an action potential.

Thrombin A plasma protein formed from prothrombin by thromboplastin that is important in blood clotting. It converts fibrinogen to fibrin, which forms a web that traps blood cells and forms the clot.

Thrombus A stationary blood clot that forms in the blood vessels. A thrombus can block blood flow.

Thyroid gland The shield-shaped structure at the front of the neck that synthesizes and secretes thyroid hormone and calcitonin.

Thyroid-stimulating hormone (TSH) The anterior pituitary hormone that acts on the thyroid gland to stimulate synthesis and release of thyroid hormones.

Tidal volume The amount of air inhaled or exhaled during a normal breath.

Tight junction A type of junctional complex in which the membranes of neighboring cells are attached, forming a seal to prevent fluid from flowing across the epithelium through the minute spaces between adjacent cells.

Tissue A group of cells that work together to perform a common function.

Tolerance A progressive decrease in the effectiveness of a drug with continued use.

Tongue The large skeletal muscle studded with taste buds that aids in speech and eating.

Total lung capacity The total volume of air contained in the lungs after the deepest possible breath. It is calculated by adding the residual volume to the vital capacity.

Trabecula (plural, trabeculae) A supporting bar or strand of spongy bone that forms an internal strut that braces the bone from within.

Trachea   The tube that conducts air into the thoracic cavity toward the lungs. The trachea is reinforced with C-shaped rings of cartilage to prevent it from collapsing during inhalation and exhalation.

Trait   A phenotypically expressed characteristic.

Transcription   The process by which a complementary single-stranded messenger RNA (mRNA) molecule is formed from a single-stranded DNA template. As a result the information in DNA is transferred to RNA.

Transfer RNA (tRNA)   A type of RNA that binds to a specific amino acid and transports it to the appropriate region of mRNA. Transfer RNA acts as an interpreter between the nucleic acid language of mRNA and the amino acid language of proteins.

Transgenic organism   An organism that contains certain genes from another species that code for a desired trait. It can be created, for example, by injecting foreign DNA into an egg cell or early embryo.

Transition reaction   The phase of cellular respiration that follows glycolysis and involves pyruvate reacting with coenzyme A in the matrix of the mitochondrion to form acetyl CoA. The acetyl CoA then enters the citric acid cycle.

Translation   Protein synthesis. The process of converting the nucleotide language of mRNA into the amino acid language of a protein.

Transverse tubules   T tubules. The tiny, cylindrical inpocketings of the muscle fiber's plasma membrane that carry nerve impulses to almost every sarcomere.

Tricuspid valve   A heart valve located between the right atrium and ventricle that prevents the backflow of blood from the ventricle to the atrium.

Triglycerides   The lipids composed of one molecule of glycerol and three fatty acids. They are known as fats when solid and oils when liquid.

Trisomy   A condition in which there are three representatives of a chromosome instead of only two representatives.

tRNA   See transfer RNA.

Tubal ligation   A sterilization procedure in females in which each fallopian tube is cut and sealed to prevent sperm from reaching the eggs.

Tubal pregnancy   An ectopic pregnancy in which the embryo implants in a fallopian tube. This is the most common type of ectopic pregnancy.

Tuberculosis (TB)   A highly contagious disease caused by a rod-shaped bacterium, *Mycobacterium tuberculosis*. TB is spread by coughing.

Tumor (neoplasm)   An abnormal growth of cells. A tumor forms from the new growth of tissue in which cell division is uncontrolled and progressive.

Turner syndrome   A genetic condition resulting from nondisjunction of the sex chromosomes in which a person has 22 pairs of autosomes and a single, unmatched X chromosome (XO). The person is phenotypically female.

Tympanic membrane   The eardrum. A membrane that forms the outer boundary of the middle ear and that vibrates in response to sound waves.

Type I diabetes mellitus   An autoimmune disorder in which the beta cells of the pancreas are destroyed causing insulin deficiency.

Type II diabetes mellitus   A condition characterized by a decreased sensitivity to insulin caused by decreased numbers of insulin receptors on target cells.

Umbilical cord   The rope-like connection between the embryo and the placenta. It consists of blood vessels (two umbilical arteries and one umbilical vein) and supporting connective tissue.

Undernourishment   Starvation. A form of hunger that occurs when inadequate amounts of food are eaten.

Unipolar neuron   A neuron that has a single, continuous process (an axon) and the cell body lies off to one side. The dendrites are branches at one of the tips of axon. Most sensory neurons are unipolar.

Ureters   Tubular organs that carry urine from the kidneys to the urinary bladder.

Urethra   The muscular tube that transports urine from the floor of the urinary bladder to the outside of the body. In males, it also conducts sperm from the vas deferens out of the body through the penis.

Urethritis   Inflammation of the urethra caused by bacteria.

Urinalysis   An analysis of the volume, microorganism content, and physical and chemical properties of urine.

Urinary incontinence   Lack of voluntary control over urination. Incontinence is the norm for infants and children younger than 2- or 3-years-old, because nervous connections to the external urethral sphincter are incompletely developed. In adults, incontinence may result from damage to the external sphincter (often caused, in men, by surgery on the prostate gland), disease of the urinary bladder, and spinal cord injuries that disrupt the pathways along which travel impulses related to conscious control of urination. In any age-group, urinary tract infection can result in incontinence.

Urinary retention   The failure to completely or normally expel urine from the bladder. This condition may result from lack of sensation to urinate, such as might occur temporarily after general anesthesia, or from contraction or obstruction of the urethra, a condition caused, in men, by enlargement of the prostate gland. Immediate treatment for retention usually involves use of a catheter to drain urine from the bladder.

Urination   The process, involving both involuntary and voluntary actions, by which the urinary bladder is emptied. It is also called voiding or micturition.

Urine   The yellowish fluid produced by the kidneys. It contains wastes and excess materials removed from the blood. Urine produced by the kidneys travels down the ureters to the urinary bladder where it is stored until being excreted from the body through the urethra.

Uterus   A hollow muscular organ in the female reproductive system in which the embryo implants and develops during pregnancy.

Vaccination   A procedure that introduces a harmless form of the disease-causing organism into the body to stimulate immune responses against that antigen.

Vagina   A muscular tube in the female reproductive system that extends from the uterus to the vulva and serves to receive the penis during sexual intercourse and as the birth canal.

Vaginitis   An inflammation of the vagina.

Valence shell   The outermost energy shell of an atom. The number of electrons in the valence shell determines the type of chemical bond that forms between atoms.

Varicose veins   Veins that have become stretched and distended because blood is prevented from flowing freely and so it accumulates, or "pools," in the vein. A common cause of varicose veins is weak valves within the veins.

Vas deferens A tubule that conducts sperm from the epididymis to the urethra.

Vasectomy A sterilization procedure in men in which both the vas deferens are cut and sealed to prevent sperm from leaving the man's body.

Vasoactive intestinal peptide (VIP) A hormone released by the small intestine into the bloodstream that triggers the small intestine to release intestinal juices.

Vasoconstriction A decrease in the diameter of blood vessels, commonly of the arterioles. Blood flow through the vessel is reduced and blood pressure rises as a result of vasoconstriction.

Vasodilation An increase in the diameter of blood vessels, commonly of the arterioles. Blood flow through the vessels increases and blood pressure decreases as a result of vasodilation.

Vector A biological carrier that ferries the recombinant DNA to the host cell, usually a plasmid or virus.

Vein A blood vessel formed by the merger of venules that transports blood back toward the heart. Veins have distensible walls and so they serve as blood reservoirs, holding up to 65% of the body's total blood supply.

Vellus hair Soft, fine hair that persists throughout life covering most of the body surface.

Vena cava One of two large veins that empty oxygen-depleted blood from the body to the right atrium of the heart. The superior vena cava delivers blood from regions above the heart. The inferior vena cava delivers blood from regions below the heart.

Ventral nerve root The portion of a spinal nerve that arises from the front (anterior) side of the spinal cord and contains axons of motor neurons. It joins with the dorsal nerve root to form a single spinal nerve, which passes through the opening between the vertebrae.

Ventricles The two lower chambers of the heart that receive blood from the atria and pump it to the body. The ventricles function as the main pumps of the heart.

Ventricular fibrillation Rapid, ineffective contractions of the ventricles of the heart, which render the ventricles useless as pumps and stop circulation.

Venule A small blood vessel that receives blood from the capillaries. Venules merge into larger vessels called veins. The exchange of materials between the blood and tissues across the walls of a venule is minimal.

Vertebral column The backbone. It is composed of 26 vertebrae (7 cervical, 12 thoracic, 5 lumbar, 1 sacrum, and 1 coccyx) and associated tissues. The spinal cord passes through a central canal within the vertebrae.

Vestibular apparatus A closed fluid-filled maze of chambers and canals within the inner ear that monitors the movement and position of the head and functions in the sense of balance.

Vestibule A space or cavity at the entrance to a canal. In the inner ear, it refers to a structure consisting of the utricle and saccule.

Villi (sing. *villus*; tuft of hair) Small finger-like projections on the small intestine wall that increase surface area for absorption.

Virus A minute infectious agent that consists of a nucleic acid encased in protein. A virus cannot replicate outside a living host cell.

Vital capacity The maximal amount of air that can be moved in and out of the lungs during forceful breathing.

Vitiligo A condition in which melanocytes disappear from areas of the skin leaving white patches in their wake.

Vitreous humor The jelly-like fluid filling the posterior cavity of the eye between the lens and the retina that helps to keep the eyeball from collapsing and holds the thin retina against the wall of the eye.

Voltage-sensitive ion channels A gated ion channel that opens or closes in response to changes in voltage.

Water soluble vitamin A vitamin that is soluble in water. Water soluble vitamins include vitamin C and the various B vitamins.

Wave summation A phenomenon that results when a muscle is stimulated to contract before it has time to completely relax from a previous contraction.

White blood cells (leukocytes) Cells of the blood including neutrophils, eosinophils, basophils, monocytes, B lymphocytes, and T lymphocytes. They are involved in body defense mechanisms and removal of wastes, toxins, or damaged, abnormal cells.

X-linked genes Genes located on the X chromosome. Most X-linked genes have no corresponding allele on the Y chromosome and will be expressed in a male.

Yellow marrow A connective tissue found in the medullary cavities in the shafts of long bones that stores fat. It forms from red marrow and, if the need arises, it can convert back to red marrow and form blood cells.

Yolk sac The extraembryonic membrane that is the primary source of nourishment for embryos in many species of vertebrates. In embryonic humans, however, the yolk sac does not provide nourishment and remains quite small (human embryos receive nutrients from the placenta). In humans, the yolk sac is a site of blood cell formation and contains cells, called primordial germ cells, that migrate to the gonads where they differentiate into immature cells that will eventually become sperm and oocytes.

Zygote The diploid cell resulting from the joining of an egg nucleus and a sperm nucleus. The first cell of a new individual.

Zygote intrafallopian transfer (ZIFT) A procedure in which zygotes created by the union of egg and sperm in a dish in the laboratory are inserted into the woman's oviducts. Zygotes travel on their own from the oviducts to the uterus.

Index

Bold-face page numbers indicate pages on which the term is defined. Italic page numbers indicate an illustration.

Cancer *(continued)*
  types of,
    bladder, 557, 563
    brain, 550, 554, 555, 557, 559, 561, 565
    breast, 25, 152, 158, 345, 433, 442–446, 527, 557, 559, 561
    cervical, 405, 406, 453, 454, 461, 469
    colon, 25, 152, 528, 556, 557, 562, 563
    kidney, 550
    laryngeal, 563
    leukemia, 265, 266–268, 466, 550, 555(t), 557, 565
    liver, 561(t)
    lung, 25, 52, 395, 402–403, 404, 405, 406, 550, 557, 559, 562, 563
    lymphoma, 477, 555(t)
    melanoma, 90, 91, 550, 559, 565
    ovarian, 527, 550, 557
    pancreatic, 557
    penile, 469
    prostate, 158, 435, 436–437, 562, 563, 565
    retinoblastoma, 552, 557
    skin, 90–91, 520, 550, 556, 562, 624
    stomach, 345, 562
    testicular cancer, 430, 433, 435, 436–437
    uterine, 561
  radiation exposure and, 562
  treatments for, 555, 563, 565
  viruses and, 561
Candida albicans, 448, 475
Canines, 335, *337*
Capacitation, 483, *484*
CAPD, *see* continuous ambulatory peritoneal dialysis
Capillaries, 140, 379, 383, 487
Capillary, 140, *280*, **282**–284, 379, 383, 487
  bed, *282*, 283, 296
  cancer and, *559*
  exchange, 282, 283, *284*
  lymphatic, 302,
  structure of, *280*, *282*
Carbaminohemoglobin, 387, *388*
Carbohydrate, **26**(t)–27, 28, 31, 343, 346, 356(t), 359–362, 363, 371, 373, 604
  complex, 361, 363
  dietary recommendations, 361–362
Carbon, 593, 601, 604, 609
Carbon cycle, 604–*605*, 625
  human disruptions of, 609–611
Carbon dioxide, 268, 388, 389, 390, 408, 487, 597, 605, 609, 610
  blood transport, 387–*388*
Carbon monoxide, 399, 401, 404, 405, 406
  marijuana and, 223
Carbonic acid, 387, 388, 390, 412
Carbonic anhydrase, 387, *388*
Carboxypepsidase, 346(t)
Carcinogens, 158, 361, 477, **561**
  cigarette smoke and, 401, 402, 405, 406
Carcinoma, 555(t)
Carcinoma in situ, 555
Cardiac circulation, **290**–291
Cardiac cycle, **290**–291
Cardiac muscle cells, *59*, 81, *82*, *291*
Cardiopulmonary resuscitation (CPR), 298

Cardiovascular disease, 160, 295–301, 369, 404, 406, 448
  and blood cholesterol levels, 358
Cardiovascular system,
  risks to,
    alcohol and, 220
    cocaine and, *225*
Carnivores, 597–598, 601, 605
Carotene, skin color and, 89
Carotid arteries, 298
Carotid bodies, 390
Carpal tunnel syndrome, *118*
Carrier(s), genetic, 53, **519**, 523, 526
Carrier molecule, 42
Carrying capacity, **620**–621
Carrying capacity, Earth's, 622
Cartilage, 77-78, *79*, 382, 496
  model for skeleton, *110*
Catabolism, 7
Cataracts, 155, 237, 239, 404, 499, 624
Catheter, 421, 425
Cauda equina, 205
Cautery, 469
CD4 receptors, 471, 473, 479
Cecum, 350
Cefriaxone, 463
Cell, 38–70, 578
  eukaryotic, 578
  structure of, 39–59
Cell adhesion molecules (CAMs), **41**, 560–561
Cell body, *166*, 167
Cell cycle, 65–69(t), *66*
Cell death, 50
  programmed, 558-*559*
Cell division, 57, 343
  chemotherapy and, 565
  control of, 69
  estrogen and, 561
  limitations to, 558
  loss of control over, 555, 557–558
    regulation of, by genes, 557–558
Cell membrane, 357, 547; *see also* plasma membrane
Cell-cell recognition, 40–41
  organ transplants and, 40–41
Cell-mediated immune response, 315, *322*
Cellular respiration, **60**–63, *64*(t), 600
Cellulose, 27, *28*, 360
Cementum, 335, *337*
Central dogma, *534*
Central nervous system 192–208
Centrioles, *55*, 68
Centromere, 67, 507
Centrosome, *55*
Cerebellum, *194*, **200**
Cerebral cortex, 179–180, **193**, *194*, 195, 200, 201, 212, 257, 389
Cerebrospinal fluid, 205, 207–208, 390, 472
Cerebrovascular accident, 212; *see* stroke
Cerebrum, 193, *194*, 195–199
Cerumen, 95
Cervical cancer, 405, 406, 453, 454, 461, 469
Cervical cap, 450(t), 454
Cervical mucus, 451, 453, 457
Cervical secretions, 463, 467

Cervix, 145, 441, 449, 453, 461, 463, 467, 469, 483, 496, 528
Cesarean section, 467, 490
CFCs (chlorofluorocarbons), 91, 610, 624–625
Chanchroid, 472
Chancre, *463*, 464
Cheetahs, 570
Chemical bonds, 19–21(t)
Chemiosmosis, 61, *63*, *64*(t)
Chemoreceptors, 232, 233, 352, *390*
Chemotherapy, 267, 343, 436, 555, 559, 563, 565
  medical use of marijuana and, 226–227
Chemotrophs, 578–579
Chickenpox, 475
Childbirth, 139, 144–*145*, 446, 496–498
  age of mother and population growth, 617–*618*
  natural, 498
  pain relief during, 497–498
  spinal anesthesia and, 205
Childhood, 498–499
Chimpanzee, 585, 586, 587
China, 602
Chitin, 27, *28*
Chitosan, 27
Chlamydia, 432, 449, 453, 454, 456, 457, *461*–462, 468(t)
Chlamydia trachomatis, 461–462
Chloride ions, 388, 408
Chlorofluorocarbons (CFCs), 91, 610, 624–625
Chloroplasts, 578
Cholecystokinin (CCK), 353, 372
Cholesterol, 30–*31*, *84*, 137, 152, 220, 297, 346–348, 349, 357, 371, 373, 404, 483, 525, 528
  blood levels of, 528
Chondrocytes, 77–78, *79*
Chordae tendinae, **287**, *288*, 289
Chorion, 487, 493, 528
Chorionic villi, 487, 528
Chorionic villi sampling (CVS), *526*, 528
Choroid, *235*(t)–236
Chromatid, *66*–67, **507**, 508, 510, 511, 570
Chromatin, *45*–46, 68
Chromosomal mutation, 524
Chromosome, *45*–46, *66*, 478, 492, **505**, 511, 513, 515, 516, 552
  chromosome 5, 524
  chromosome 9, 524
  chromosome 15, 525
  chromosome 17, 527
  chromosome 21, 513
  human, map of, *505*
  mapping, 522–523
Chrymotrypsin, 346(t)
Chylomicron, 349
Chyme, 341, 343, 353
Cigarette smoke, 152, 295, 396, 499, 562
  contents of, 399–401, 400(t)
Cigarette smoking, 239, 499
  alcohol and, *220*
  birth control pill and, *452*
  carcinogens and, 561
  and cillia, *401*, 402, 403